Electronic Circuit Analysis

Electronic Circuit Analysis

B. Visvesvara Rao
Professor
Mahaveer Institute of Science and Technology
Hyderabad

K. Raja Rajeswari
Professor
Department of Electronics and Communication Engineering
Andhra University College of Engineering (Autonomous)
Andhra University
Vishakapatnam

P. Chalam Raju Pantulu
Divisional Engineer (Retired)
Control and Instrumentation
Andhra Pradesh Power Generation Corporation Limited

K. Bhaskara Rama Murty
Former Associate Professor
Department of Electronics and Communications Engineering
JNTU College of Engineering
Kakinada

Chennai • Delhi • Chandigarh

Editor— Acquisitions: Sojan Jose
Associate Editor—Production: M. R. Ramesh

ISBN 978-81-317-5428-3

First Impression

Published by Dorling Kindersley (India) Pvt. Ltd, licensees of Pearson Education in South Asia.

Head Office: 7th Floor, Knowledge Boulevard, A-8(A), Sector 62, Noida 201 309, UP, India.
Registered Office: 11 Community Centre, Panchsheel Park, New Delhi 110 017, India.

Compositor: MacroTex Solutions, Chennai
Printer: Batra Art Press

To my beloved wife, Late B. Laxmana Mani, whose unquestioned support, love and affection allowed me to dream big and deliver on a project of this scale

– B. Visvesvara Rao

Contents

Preface

Electronic Circuit Analysis is designed to serve as a text and reference book for two-semester courses of B.Tech. curriculum of various universities in India and abroad (B.S. with engineering major).

The working principles of electronic devices such as diodes and transistors (BJT, FET and MOSFET) are explained with simple text, diagrams, equations and worked-out examples to impart a deep understanding of the device functions.

From radio communication systems to cellular phones, all electronic equipments including computers, satellite communication systems, real-time embedded systems and Internet communication technologies use contraptions such as amplifiers, oscillators, power amplifiers, tuned amplifiers and power supplies as 'basic building blocks'. The analysis of the electronic circuits required to build such applications is presented in easily understandable methods to give students a firm foundation in the first principles and enable them to probe advanced and application topics such as very-large-scale integration (VLSI) and nano technologies.

The topics are disseminated in a clear step-by-step approach that provides teachers ample material to empower their students. They are also conceived for students to further their understanding through self-study.

Chapter 1 on Electron Dynamics presents the basic principles of movement of charge carriers, electrons, in electric and magnetic fields. Structural details of the cathode ray tube (CRT) and operation of the cathode ray oscilloscope are explained.

Chapter 2 on P-N Junction Diode Characteristics deals with the basic physics of the P–N junction diode and uses energy band diagrams to explain the flow of diode currents under different biasing conditions. Interpreting the behaviour of P–N junctions, diode characteristics and charge movement is basic to understand the function of transistors in different electronic applications.

Chapter 3 on Rectifiers, Filters and Voltage Regulators speaks about these fundamental components and their role in designing DC power supply circuits. Rectifier circuits are good examples for diode applications, and DC power supply is a core component supplying power to all electronic gadgets/products.

Chapter 4 on Characteristics of Transistor Devices (BJT, FET and MOSFET) expounds on the principal component of all electronic circuits—the Transistor—and the physics behind the different options of its assembly leading to the BJT, FET and MOSFET type of Transistors. The methods of obtaining Transistor V–I characteristics for each type are also spelt out. The Transistor's amplifier action is explained using its V–I characteristics and equations. Transistor characteristics are useful in the design and analysis of electronic circuits that are built with these core components.

Chapter 5 on Transistor Biasing and Stabilisation Circuits advances a qualitative discussion and analysis of the different types of biasing circuits of transistor amplifiers, including the establishment of a stable quiescent operating point (key design parameter) under DC/biasing/operating voltages. It also discusses compensation circuits to stabilise the operating point against variations in temperatures and biasing voltages.

Chapter 6 on Transistor (BJT) Amplifiers illustrates the basic configurations of BJT amplifiers and their h-parameter equivalent circuits. Continuing from the previous chapter, amplifier operation is analyzed under DC/biasing/operating and (input) signal voltages, leading to the development of different classes of amplifiers and their applications. Key parameters such as amplifier frequency response and bandwidth are calculated and analyzed.

Chapter 7 on Feedback Amplifiers elucidates the principles of different feedback techniques that modify the characteristic features of amplifiers. Various types of negative feedback topologies that lead to stabilisation of amplifiers for performance are discussed. Practical electronic circuits using these feedback techniques (such as voltage series and shunt, and current series and shunt) are examined.

Chapter 8 on Oscillators describes the basic operating principle of RC and LC oscillators for low- and high-frequency applications, respectively. It explains the different stable oscillator circuits using crystals (used as clocks in all electronic gadgets) in detail. It also explores the sweep circuit (used in major tube devices – radar displays, TV, CRO) built using UJT-based negative resistance oscillator.

Chapter 9 on FET and MOSFET Amplifiers delineates amplifier operations using FET and MOSFET devices, which are suitable for miniaturization and high-speed operations at low power and used as basic amplifier building blocks in VLSI and low-power electronic gadgets such as mobile phones, satellite and communication systems.

Chapter 10 on Multistage (Cascaded) Amplifiers takes a close look at building long multistage amplifiers, bridging basic amplifier blocks for signal gain and increased useful frequency bandwidth, cascading for optimum power transfer adjusted to different input and output impedance levels. High-input impedance circuits for high gain using Darlington pair transistors are detailed to demonstrate high-end applications. As in the other chapters, calculations and analysis support the discussion on designing and building of these multistage (cascaded) amplifiers.

Chapter 11 on Large Signal (Power) Amplifiers depicts the design of audio frequency (AF) amplifiers (used in smart and mobile phones, home theaters, radio and TV transmitters and receivers with speakers). Increased power is realised through push–pull amplifiers, the different variations (input and output powers and efficiency) of which are analyzed in detail in this chapter. Class C amplifiers (in radio and TV transmitters) with high efficiency and power are discussed. Modern versions such as Classes D and E amplifiers (home theaters), with lower noise levels and high speed operation, are also outlined.

Chapter 12 on High Frequency Transistor Circuits speaks about high-frequency equivalent circuits that use the hybrid-π model for BJT, FET and MOSFET, using key parameters such as junction capacitance. These are required to adapt to the present-day high-frequency applications (such as WiFi and 3G) that need specialised analysis and techniques.

Chapter 13 on Tuned Amplifiers spells out the basic principles of tuned amplifiers, single and double tuned amplifiers. These special circuits come into play when a TV or radio is tuned to a specific channel or station. The chapter also looks at synchronously tuned and stagger tuned amplifiers used for better and faster selectivity, while revealing the basic principles and techniques of circuit design to neutralise and stabilise high-frequency effects.

Chapter 14 on Switching and IC Voltage Regulators focuses on the analysis and design of different types of voltage regulators, three-terminal IC regulators and special types of voltage regulator circuits. With different voltages being used by different "circuit modules" inside electronic gadgets (laptops, computers) or special embedded systems (cable/SoHo LAN/WiFi gateway), these voltage regulators are used to "interface/translate" voltages across different circuit modules. The chapter also sketches the working of DC–DC regulator power supply circuits, switching regulators, SMPS and UPS.

Chapter 15 on Special Purpose Electronic Devices shows the structural details, principles of operation and applications of Tunnel Diode, Photo Diode (counter applications on belt conveyers and traffic counts), Varactor Diode (used in switching TV channels), Schottky Barrier Diode, Light Emitting Diode (in LED TV and lamps for cars, buildings and streets) and Silicon Control Rectifiers (used in power controls in industrial drives).

The objective of this book is to present uncomplicated procedures for electronic circuit analysis and to provide an insight into the chief principles governing them. To this end, its contents are reinforced by useful Web supplements in the form of PowerPoint slides that can be accessed at www.pearsoned.co.in/bvisvesvararao. While I have made every effort to provide a text that is error-free, it is possible that a few flaws might have crept in inadvertently. These, if detected, may be pointed out to the publisher, or directly to me at visrao@gmail.com. Comments and feedback on the topics discussed in this book are welcome.

ACKNOWLEDGEMENTS

I thank my beloved wife, Laxmana Mani, for her moral support and constant encouragement throughout my career as a teacher at JNTU College of Engineering, Kakinada. I thank her from the bottom of my heart for her infinite patience and strong support in bringing out this journey to reality – over a long period of 46 years.

I am indebted to my son, Satyam Bheemarasetti (technologist and entrepreneur), for providing suggestions on the technical language and flow of content in the book.

I am obliged to my grandson, Prithvi Bheemarasetti, for assisting me with research on technical material and preparation of equations and diagrams used in the book. My daughter-in-law, Lakshmi Lavanya, and my granddaughter, Lakshmi Jahnvi, provided moral support and encouragement when I was preparing the material for this book. I acknowledge their help.

I congratulate and thank my coauthors for their unstinted cooperation and support during the preparation of the book. I am grateful to S. V. S. Ganesh, son of Professor K. Raja Rajeswari, for his valuable suggestions on the preparation of the outlay of the book. I thank P. Nagavalli, wife of P. Chalam Raju Pantulu, and their children, P. N. V. Suresh and P. N. V. Satish, for their wholehearted encouragement and support for this project. I thank Professor K. Bhaskara Ramamurty's family members for their moral support.

Together, we have published three books in the field of Electronics and Communications Engineering:

1. Electronic Circuit Analysis (this book in 2011)
2. Signals and Systems (2009)
3. Electronic Devices and Circuits (Second Edition in 2007)

I owe my inspiration and encouragement to my first guru, Ganti Subrahamanyam (ME in Stanford University, under Professor F. E. Terman, Professor of Electrical Engineering and Dean, considered Father of Electronics), who was Professor and Principal at JNTU College of Engineering, Kakinada, and Professor N. Lakshminarayana, former Principal, JNTU Kakinada, who was my mentor during my career at Kakinada, and Professor D. Mallikarjuna Rao, Former Professor, JNTU Kakinada

My education and career intertwined with that of many wonderful leaders and educationalists. I acknowledge their influence and express my gratitude to those whose names appear foremost in my mind:

Dr. Y. Venkatrami Reddy, Former Vice-Chancellor, JNTU Hyderabad
Dr. Allam Apparao, Vice Chancellor, JNTUK Kakinada
Prof. M. Venkata Rao, George Mason University, USA
Prof. V. Ranga Rao, Former Rector, JNTU
Dr. R. Govinda Rajulu, Professor, IIIT Hyderabad
Prof. B. Satyam, Former Rector, Andhra University
Sri. G. Ramachandrayya, Retd. I. G., Police Wireless Communications
Sri. P. Satyanarayana, Retd. S. P

At JNTU Kakinada

Prof. V. V. S. Prasad, Former Principal
Prof. M. Rama Murty
Prof. M. Madhusudhana Rao
Prof. C. S. M. Sarma, Former Professor
Dr. K. Satyaprasad, Rector
Dr B. Prabhakara Rao, Director
Dr. E. V. Prasad, Former Principal
Dr. V. Ravindra, Registrar
Dr. Srinivasa Kumar, Director
Dr. I. Santhi Prabha, Director
Dr. K. Padma Raju, Director
Dr. M. Sailaja, Professor
Dr. A. Mallikarjuna Prasad
Dr. V. Kama Raju, Former Principal
Dr. A. Sree Ramarao, Former Principal
Dr. C. Penchalaiah, Former Principal
Dr. D. Anandamohan Rao, Former Principal
Dr. G. Raghuram, Professor
Prof. K. Anandamohan, Former Vice-Principal
Dr. Gandhi, Former Principal
Dr. P. Udaya Bhaskar, Principal
Dr. K. Murali Krishna, Professor
Dr. J. V. R. Murty, Professor
Dr. K. V. Ramana, Professor
Dr. B. Sarvesh, Professor

Dr. S. S. Tulasi Ram, Professor
Dr. P. Dakshina Murty, Professor

At JNTU Anantapur

Dr. K. Soundara Rajan, Former Rector, JNTUA
Dr. D. Rama Naidu, Professor
Dr. Ramana Reddy, Professor

At College of Engineering, Andhra University

Prof. G. S. N. Raju, Principal
Prof. G. Madhusudana Rao (Retired Principal)
Prof. K. V. V. S. Reddy
Prof. P. Mallikarkuna Rao
Prof. Y. Gopala Rao
Prof. G. Sasi Bhushana Rao
Smt. S. Santha Kumari
Dr. P. Rajesh Kumar
Dr. P. V. Sridevi
Smt. M. S. Anuradha
Smt. S. Aruna

At JNTU Hyderabad

Prof. M. R. K. Reddy, Former Director
Dr. L. V. A. R. Sarma, Director
Dr. P. Soma Sekhar , Director
Dr. L. Patap Reddy, Professor and Chairman, Board of Studies
Dr. Vinod Babu, Director
Dr. S. V. L. Narasimham, Professor
Dr. P. G. Krishna Mohan, Professor
Dr. Madhavi Latha, Professor
Dr. D. Sreenivasa Rao, Professor
Dr. Y. Yesu Ratnam Professor
Dr. A. Ramachandra Aryasri Director

There are many more leaders, teachers and students, who influenced and instilled in us a sense of responsibility to share what we learnt and taught, and taught and learnt, and we express our appreciation to each of them for making this grand project, a success.

I wholeheartedly thank Sojan Jose, Acquisitions Editor, Vijay Pritha R., Assistant Acquisitions Editor, Ramesh M. R., Associate Production Editor at Pearson Education and all others in their team who have helped in bringing out this book.

B. Visvesvara Rao

Chapter 1

ELECTRON DYNAMICS

Learning Objectives

- By the end of this chapter, you will be able to understand the field of *Electron Dynamics* – foundation for *Electronics Devices & Circuits.*
- *Electron dynamics* can be monitored and measured with the help of a 'Cathode Ray Oscilloscope' (CRO), using a Cathode Ray Tube (CRT). CRO is a versatile measuring instrument that is used to display and measure electrical signals and their waveforms.
- Different variations of CRT are used in many popular appliances/ applications, such as TV, radar, computer monitor, etc.

1.1 MOTION OF ELECTRONS IN ELECTRIC FIELDS

The subject of Electronics is all about *playing of electrons*. Since any play involves movement, it is necessary to study the *Dynamics of Electrons*.

1.1.1 Characteristics of 'Electron' Treated Conceptually as a Particle

Charge $q = 1.6 \times 10^{-19}$ C
Mass $m_e = 9.11 \times 10^{-31}$ kg

$$\frac{q}{m_e} = 1.759 \times 10^{11} \text{ C/kg}. \tag{1.1}$$

To have a feel for real numbers, note that 1 kg of electrons contain 1.1×10^{30} electrons and a coulomb of charge (−ve charge) rests with 6.25×10^{18} electrons approximately.

As all electrons, according to classical models, are charged particles, Electric and Magnetic Fields can induce motion in them. Once the solid nature of electrons is accepted, we can apply 'laws of dynamics' to electrons and study their movements (trajectories) in Electric and Magnetic Fields. Hence, this chapter is named as *Electron Dynamics*.

However, this simple model explains semiconductor devices (Diodes, Transistors, etc.) and vacuum tube-based devices (TV, Microwave, Magnetron, Cathode Ray Tube – CRT).

1.1.2 Force on Charged Particles in an Electric Field

By definition, force on a unit of *positive charge,* at any point in an Electric Field, is the *field intensity* ε at that point. For example, force on a unit of positive charge in a field of 1 V/m is 1 *N* (Newton).

The force f_q on a positive charge of q coulombs in a field of intensity ε V/m is given by the expression

$$f_q = q \cdot \varepsilon \text{ (N)}. \tag{1.2}$$

To determine the locus of a particle (electron), using Laws of Dynamics (Motion) in Physics, we can start with Eq. (1.2).

$$f_q = \varepsilon \cdot q = m \cdot a = m\frac{dv}{dt}. \tag{1.3}$$

According to Newton's Second Law of Motion, in Eq. (1.3), m is the mass, a is the acceleration, v is the velocity and t is the time. (In MKS system, m is in kg, a is in m/s^2, v is in m/s and t is in seconds.)

When the charged particle is an electron, force f_e acts in the opposite direction and results in:

$$f_e = -q \cdot \varepsilon \text{ (N)}. \tag{1.4}$$

1.1.3 Motion of Electrons in a Constant Electric Field (Initial Velocity along the Axis of the Field)

A simple case is – when the electron is situated between the plates of a set of parallel plate capacitors – shown in Fig. 1.1. We assume that the field is uniform, based on the fact that the distance (d) between the plates is small compared to the dimensions of the plates (considering any practical situation).

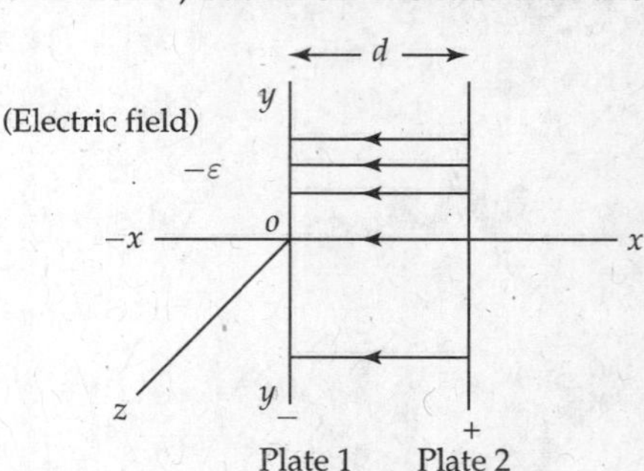

FIG. 1.1 *Parallel plates*

Key parameters are:

V_x = velocity of electrons in the x direction,

x = distance in the x direction,

t = time.

At $t = 0$, initial conditions of electrons are: initial velocity v_{0x} and initial position x_0. As shown in Fig. 1.1, there are no forces in the y and z directions, so there is no acceleration in the y and z directions. But zero acceleration can also mean constant velocity. In the one-dimensional model, since forces f_y and f_z are zero, and acceleration a_y and a_z are also zero, the electron moves in the x direction only.

Applying Newton's Second Law to the above case:

$$f = -q\varepsilon = m \cdot a_x = m\frac{dv_x}{dt} \tag{1.5}$$

$$\because \quad a_x = \frac{dv_x}{dt}; \quad dv_x = a_x\, dt$$

$$v_x = \int_{v_{0x}}^{v_x} dv_x = \int_0^t a_x\, dt$$

$$(v_x - v_{0x}) = a_x t$$

$$\text{or } v_x = v_{0x} + a_x t \quad (\text{similar to } v = u + at). \tag{1.6}$$

Again

$$dx = v_x\, dt$$

$$dx = (v_{0x} + a_x t)\, dt$$

$$\int_{x_0}^{x} dx = \int_0^t (v_{0x} + a_x t)\, dt$$

$$\therefore \quad x - x_0 = v_{0x} t + \frac{1}{2} a_x t^2$$

$$\therefore \quad x = x_0 + v_{0x} t + \frac{1}{2} a_x t^2 \quad \left(\text{Similar to } S = ut + \frac{1}{2}at^2\right). \tag{1.7}$$

Potential

Starting from Eq. (1.5):

$$a_x = -\frac{q}{m}\varepsilon = \frac{dv_x}{dt} \tag{1.8}$$

$$\because \quad dx = v_x\, dt. \tag{1.9}$$

Multiplying LHS of Eq. (1.8) by LHS of Eq. (1.9) and doing the same with RHS, the following result is derived:

$$-\frac{q}{m} \cdot \varepsilon\, dx = \frac{dv_x}{dt} \cdot v_x dt = v_x\, dv_x. \tag{1.10}$$

Integrating Eq. (1.10), subject to the initial conditions:

$$\frac{-q}{m} \int_{x_0}^{x} \varepsilon\, dx = \int_{V_{0x}}^{v_x} v_x\, dv_x. \tag{1.11}$$

The expression $-\int_{x_0}^{x} \varepsilon\, dx$ is nothing but the work done on the electron, in moving it against the field from x_0 to x, and by definition it is the voltage as expressed in Eq. (1.11):

$$\text{Voltage } V \equiv -\int_{x_0}^{x} \varepsilon\, dx. \tag{1.12}$$

Substituting Eq. (1.12) into Eq. (1.11), we obtain

$$\frac{q}{m}\cdot V = \int_{v_{0x}}^{v_x} v_x \cdot dv_x$$

$$\therefore \quad qV = \frac{1}{2}m(v_x^2 - v_{0x}^2). \tag{1.13}$$

Electron Volt (*eV*) is the (potential) energy associated with the electron, due to its presence in a field of potential difference (*V*) volts.

[*Energy* is expressed in *joules*, i.e., 10^7 ergs or 10^7 Dyne-cm. This is too big a unit in the context of Electron Dynamics. So a smaller and more practical unit is conceived. This smaller unit of energy is called the 'electron volt' expressed as 'eV'. If an electron falls through a potential difference of 1 V, its *potential energy* (*PE*) decreases by an electron volt. Thus, one electron volt eV = e × 1 V J = 1.6×10^{-19} J or [eV = 1.6×10^{-19} J.]

The RHS of Eq. (1.13) is clearly the kinetic energy (KE) associated with the electron and the LHS represents the PE (energy due to position). In other words, it is a statement reiterating the 'law of conservation of energy'. As the electron leaves the −ve plate, it has only PE. It acquires KE as it moves towards the +ve plate. Thus, it has some PE and some KE. At the +ve plate, it becomes fully KE. But always it is subject to the condition that W = PE + KE, where W is the total energy.

[An analogy can be drawn here to a stone on a wall at height h metres and freely falling due to a push. During the transit, it has some PE and some KE. The net energy is always the same. At any point x, the PE = mgx and the KE = $1/2\ mv_x^{\,2}$.

$$W = mgh(x) + \frac{1}{2}mv_x^2. \tag{1.14}$$

By definition, the PE at a point in an Electric Field is equal to the product of charge q and the potential at that point.

i.e., PE $\cong qV$.

From Eq. (1.13), if $v_{0x} = 0$ at $t = 0$, we get

$$qV = \frac{mv_x^2}{2} \quad \text{(or)} \quad v_x = \sqrt{2\left(\frac{q}{m}\right)V}\ \text{m/s}. \tag{1.15}$$

Lower case v is used for velocity and upper case V for potential or voltage.

1.1.4 Initial Velocity of the Electron Perpendicular to Electric Field

Look into Fig. 1.2, if an electron enters a decelerating field between two parallel plates with an initial velocity v_{0x}, the velocity decreases with time (as does the velocity of a stone thrown up against gravity). If conditions permit, the electron may reach the other plate (curve 1) or reverse its direction even earlier (curve 2).

In this case, the rules $v_x = [v_{0x} - a_x t]$ and $x = [v_{0x}t - (1/2)\,a_x t^2]$ apply. In case it cannot reach the upper plate, it returns back with a velocity of $|a_x t|$ after travelling a distance $d = (1/2)\,a_x t^2$.

It is neither necessary for the field to be uniform nor time invariant. The only difference it makes is that the proper expression has to be used. For instance, if the voltage varies with time t, $V(t)$ instead of v has to be used together with the relevant expression for $V(t)$ and the incremental expressions (d/dt...) have to be used.

Here, $V(t) = \frac{V}{T}t.$ (1.16).

If $V(t)$ is a linearly varying function as shown in Fig. 1.3:

$$V(t) = \frac{V}{T}t.$$

If it is sinusoidal and the angle θ is small ($\sin\theta$, where θ is small), the power series is utilised to evaluate the expression

$$\text{i.e. } \sin\theta = \theta - \frac{\theta^3}{3!} + \frac{\theta^5}{5!} - \cdots + \cdots$$

We approximate it to $\sin\theta = 2\pi ft$; Therefore, $V_m \sin\theta = V_m(2\pi ft)$.

The other rules remain the same. Since v is a function of time we have to use differentials instead of the simple formulae

$$a_y\,dt = dv_y \quad v_y\,dt = dy$$

$$\int_{v_0}^{v_y} dv_y = \int_0^t a_y\,dt \quad \& \quad y = \int_{y_0}^{y} dy = \int_0^t v_y\,dt. \quad (1.17)$$

If the charge carrier enters the field at an angle with an initial velocity, then it is resolved into the x- and y-components by using $v_x = v_\theta \cos\theta$ and $v_y = v_\theta \sin\theta$, where θ is the angle at which it enters the field (Fig. 1.4).

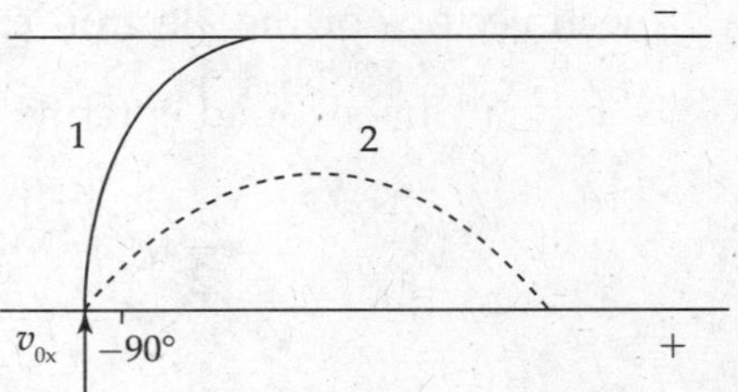

FIG. 1.2 *Electron trajectory*

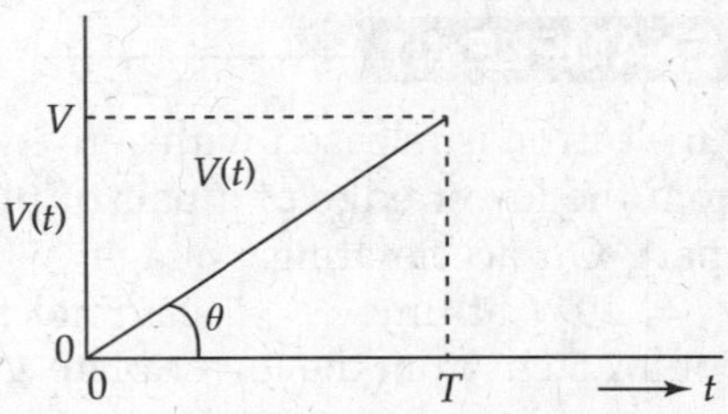

FIG. 1.3 *Voltage V(t) as a function of time*

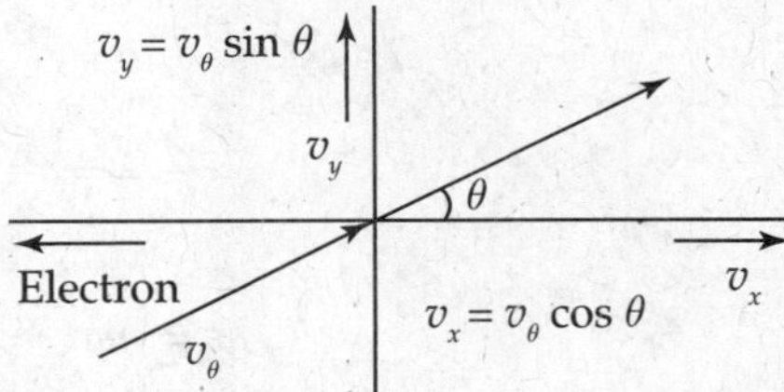

FIG. 1.4 *Charge carrier enters the field at an angle*

1.1.5 Two-dimensional Motion of Electrons

Motion of an electron is investigated, with an initial velocity in the x direction and a field in the y direction with the fields being uniform. No other fields exist in this region (Fig. 1.5). The initial conditions are:

$v_x = v_{0x}$; $x = 0$; $a_x = 0$; $\varepsilon_x = 0$

$v_y = 0$; $y = 0$; $a_y = a_y$; $\varepsilon_y = \varepsilon_y$

$v_z = 0$; $z = 0$; $a_z = 0$; $\varepsilon_z = 0$.

After it enters the field at $t = 0$. The velocity v_{0x} remains constant, since $a_x = \varepsilon_x = 0$ at $t = 0$.

So the distance in the x direction it travels is $x = v_{0x}t$:

$$\therefore\ t = \frac{x}{v_{0x}}.$$

But there is a constant acceleration in the y direction so that $v_y = a_y t$ and $y = \frac{1}{2}a_y t^2$.

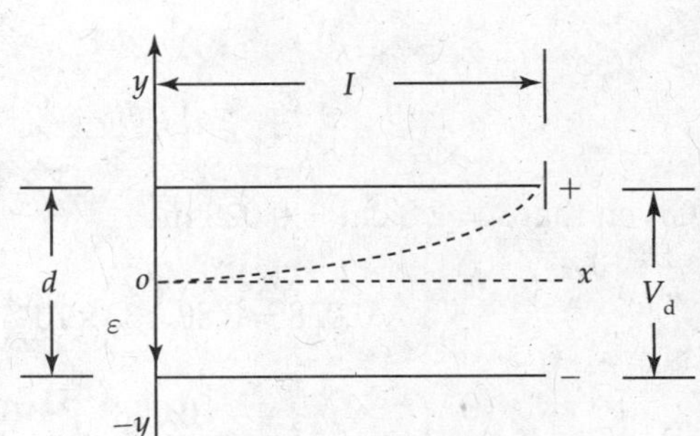

FIG. 1.5 *Two-dimensional electronic motion in a uniform electric field ε*

The trajectory of the electron can be known by finding the equation of motion. From $y = \frac{1}{2}a_y t^2$, y can be found in terms of v_{0x} from the equation, $x = v_{0x}\, t$:

$$y = \frac{1}{2}a_y t^2 = \frac{1}{2}a_y \frac{x^2}{v_{0x}^2} = \frac{1}{2}\left(\frac{a_y}{v_{0x}^2}\right)x^2 \quad \text{where } t = \frac{x}{v_{0x}}. \tag{1.18}$$

This equation is in the form of $y = \frac{dy}{dx}x^2 = x^2 \tan\theta$. So the path travelled by the electron is parabolic.

EXAMPLE 1.1

An electron is released with zero initial velocity ($V_0 = 0$) from the lower edge of a pair of plates, which are 3 cm apart. On accelerating voltage, $V_A = 0$ at $t = 0$ s and $V_A = 10$ V at time $T = 1$ μs. Find the *time to travel* to a point 2.8 cm from the lower plate.

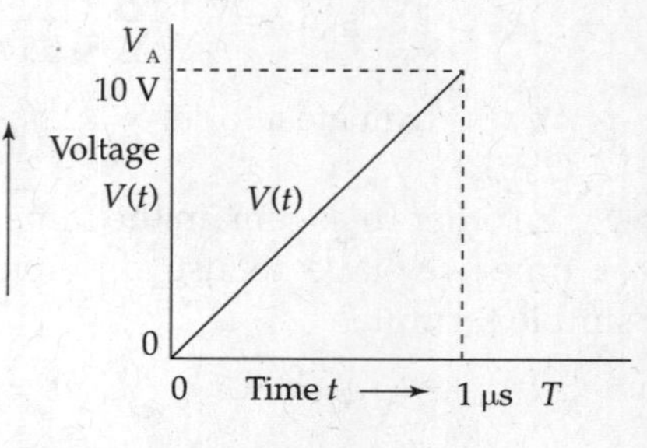

FIG. 1.6

Solution:

$$V(t) = \frac{V}{T}t = \frac{10\,t}{10^{-6}} = 10^7\, t$$

$$\varepsilon = \frac{V(t)}{d} = \frac{10^7 \times t}{0.03} = 33 \times 10^7 t$$

$$q\varepsilon = ma \;\Rightarrow\; \therefore\; a = \frac{q}{m}\cdot\varepsilon$$

$$\therefore\; a = 1.76 \times 10^{11} \times 33 \times 10^7 t$$

$$= 58.08 \times 10^{18} t \text{ m/sec}^2$$

In general and in this particular case

$$v = \int a\,dt = \int 1.76 \times 10^{18} \times 33 \cdot t \cdot dt$$

$$\therefore\; v = 1.76 \times 33 \times 10^{18} \times \frac{t^2}{2} = 0.88 \times 33 \times 10^{18} t^2 \text{ m/sec}$$

$$\because\; s = \int v\,dt \Rightarrow s = 0.88 \times 33 \times 10^{18} \times \frac{t^3}{3}. \text{ metres}$$

Given that $s = 2.8$ cm $= 0.028$ m

$$0.028 = 0.88 \times 33 \times 10^{18} \times \frac{t^3}{3} = 0.88 \times 11 \times 10^{18} \times t^3 = 9.68 \times 10^{18} \times t^3$$

$$\therefore\; t^3 = \frac{0.028}{9.68 \times 10^{18}} = \frac{0.028}{9.68} \times 10^{-18} \Rightarrow \quad t = 0.1428\,\mu\text{sec}$$

$\therefore$ Time t to reach 2.8 cm from the lower plate is 0.1428 μs.

EXAMPLE 1.2

Effective length of plates is 2 cm and accelerating voltage V_A for the electron beam is 1000 V. How much voltage is required between plates separated by 1 cm (0.01 m) to deflect an electron beam to 1°?

Solution: Given $V_A = 1000$ V, calculate deflecting voltage V_d.

Field intensity, $\varepsilon_d = \dfrac{V_d}{0.01}$ V/m.

Length of the plates $l = 0.02$ m and $\theta = 1°$

$$v_x = \sqrt{\frac{2eV_A}{m}}$$

$$\therefore \quad v_x^2 = \frac{2eV_A}{m}$$

$$a_y = \varepsilon.\frac{e}{m}$$

$$y = \frac{1}{2}a_y t^2$$

$$t = \frac{x}{v_0 x}$$

$$x = l = 0.02 \text{ m}$$

$$v_{0x}^2 = 2 \times 1000 \times \frac{e}{m}$$

$$y = \frac{1}{2} \times \frac{e}{m} \times \frac{V_d}{d} \times \frac{x^2}{v_{0x}^2}$$

$$\frac{dy}{dx} = \frac{e}{m} \times \frac{V_d}{d} \times \frac{x}{v_{0x}^2} \qquad (1)$$

$$\frac{dy}{dx} = \tan\theta = \tan 1° = 1.75 \times 10^{-2}.$$

Substituting values from the previous expressions in Eq. (1), we get

$$\frac{e}{m} \times \frac{V_d}{0.01} \times \frac{.02}{2 \times 1000 \times \frac{e}{m}} = \tan 1°$$

$$V_d = \frac{\tan 1° \times 0.01 \times 2000}{0.02} = 1000 \times \tan 1°$$

$$= 1000 \times 1.75 \times 10^{-2}.$$

Deflecting voltage $V_d = 17.5$ V.

EXAMPLE 1.3

An electron is emitted from a Thermionic Cathode with negligible initial velocity and is accelarated by a potential V_A of 1000 V between plates separated by a distance of 1 cm. Calculate the final velocity of the particle.

Solution: Given $V_A = 1000$ V and distance between plates $d = 1$ cm,

$$\text{Field intensity} = \frac{V}{d} = \varepsilon.$$

$$\varepsilon e = \frac{1}{2} m(v_f^2 - v_i^2)$$

$$\frac{e}{m} \times \varepsilon = \frac{1}{2} \times v_f^2$$

$$\because \quad v_i = 0 \quad {v_f}^2 = \frac{2eV}{m}$$

$$\because \quad \varepsilon = \frac{V}{d} = V \quad \text{When } d = 1 \text{ cm}$$

$$v_f^2 = \frac{2eV}{m};$$

$$\therefore \quad \text{Final velocity} \quad v_f = \sqrt{\frac{2eV}{m}};$$

$$\frac{e}{m} = 1.76 \times 10^{11}$$

$$v_f = \sqrt{2 \times 1.76 \times 10^{11} \times 1000}$$

$$= \sqrt{3.52 \times 10^{14}} = 1.876 \times 10^7 \text{ m/s}$$

$$\text{Final velocity } v_f = 1.876 \times 10^7 \text{ m/s}.$$

EXAMPLE 1.4

Distance between the plates of a parallel plate capacitor is 1 cm. If a direct voltage of 1000 V is applied between them, how long will it take for an electron to reach the positive plates?

Solution:

$$\varepsilon = \frac{V}{d} = \frac{1000}{0.01} = 10^5 \text{ V/m}$$

$$a = \frac{e}{m} \cdot \varepsilon = \frac{e}{m} \cdot \frac{1000}{0.01} = 1.76 \times 10^{11} \times 10^5$$

$$\therefore \quad a = 1.76 \times 10^{16} \text{ m/s}^2.$$

Distance between plates $d = 1$ cm $= 0.01$ m $= \frac{1}{2}at^2$; as initial velocity of electrons is zero.

$$\left[\begin{array}{l} \text{Given that } V = 1000 \text{ V} \\ e\varepsilon = ma \\ \therefore \quad a = \frac{e}{m}\cdot\varepsilon \\ y = \frac{1}{2}\cdot a_y t^2 \\ t = ? \end{array}\right]$$

From the equation, $S = ut + \frac{1}{2}\times at^2$

$$\text{When } u = 0 \Rightarrow S = d = \frac{1}{2}\times at^2$$

$$\therefore \quad d = 0.01 = \frac{at^2}{2}$$

$$0.02 = at^2$$

$$\therefore \quad t^2 = \frac{0.02}{a} \Rightarrow t = \sqrt{\frac{0.02}{a}}$$

$$= \sqrt{\frac{0.02}{1.76\times 10^{16}}} = 0.106\times 10^{-8}$$

$$= 1.06 \text{ ns}$$

The time taken to reach the positive plate = 1.06 ns.

EXAMPLE 1.5

Two large parallel metal plates are separated by a distance d of 1 cm with the upper plate being 200 V positive with respective to the lower plate. An electron with an initial velocity 10^6 m/sec is released upwards at the centre of the lower plate. Calculate the time of flight for the electron? What will be the velocity of the electron upon striking the upper plate? How much energy is conveyed to the upper plate?

Solution:

1. Time of flight for the electron:

Initial velocity of the electron in the upward direction $V_{0y} = 10^6$ m/s

Magnitude of Electric Field intensity, $\varepsilon = \frac{V}{d}$ V/m.

$$\varepsilon = \frac{200 \text{ V}}{1 \text{ cm}} = \frac{200}{1\times 10^{-2}} = 2\times 10^4 \text{ V/m}.$$

Acceleration, $$a=\frac{e}{m}\cdot\varepsilon=1.76\times10^{11}\times2\times10^{4}$$

$$a=3.52\times10^{15}\ \text{m/s}^2$$

$$S=ut+\frac{1}{2}\cdot at^2$$

$$\therefore\quad S=10^{-2}=10^{6}\cdot t+\frac{1}{2}\times3.52\times10^{15}\cdot t^2$$

$$\therefore\quad \frac{3.52\times10^{15}\cdot t^2}{2}+10^{6}\cdot t-10^{-2}=0$$

$$\text{Time } t=\frac{10^{-6}\pm\sqrt{(10^{6})^2+2\times3,52\times10^{15}\times10^{-2}}}{3.52\times10^{15}}$$

$$\therefore\quad t=\frac{-10^{6}\pm10^{6}\sqrt{1+10\times2\times3.52}}{3.52\times10^{15}}$$

$$=\frac{10^{6}(\sqrt{1+3.52\times20}-1)}{3.52\times10^{15}}=2.1\ \text{ns}.$$

2. Final velocity of electron upon striking the upper plate:

$$v_{\text{f}}=u+at=10^{6}+3.52\times10^{15}\times2.1\times10^{-9}$$

$$=10^{6}(1+3.52\times2.1)=8.4\times10^{6}\ \text{m/s}$$

3. Energy:

$$\text{KE}=\frac{1}{2}mv^2=\frac{1}{2}\times9.1\times10^{-31}\times(8.4\times10^{6})^2\ \text{J}.$$

$$\text{KE}=\frac{9.1\times10^{-31}\times10^{12}\times(8.4)^2}{2}$$

$$=\frac{70.56\times9.1\times10^{-19}}{2}=321\times10^{-19}\ \text{J}$$

$$\text{KE}=\frac{321\times10^{-19}}{1.6\times10^{-19}}=200.6\ \text{eV}$$

EXAMPLE 1.6

Two parallel plates of a capacitor are separated by 4 cm. An electron is initially at rest at the bottom plate. Voltage is applied between the plates, which increases linearly, from 0 to 8 V, in 0.1 ms. If the top plate is positive, determine (a) speed of the electron in 40 ns, (b) distance traversed by the electron in 40 ns.

Solution: Distance between the two plates of the capacitor = 4 cm = 4 × 10^{-2} m.

Electric field intensity, $\varepsilon = \frac{V}{d}\cdot\frac{t}{dt} = \frac{8\times t}{4\times10^{-2}\times0.1\times10^{-3}} = 2\times10^{6}\times t\ \text{V/m}$

We know that $a_x = \frac{e}{m}\times\varepsilon = 1.76\times10^{11}\times2\times10^{6}t\ \text{m/s}^2$

Then velocity $v = \int a\,dt = \int 3.52\times10^{17}t\,dt = 1.76\times10^{17}t^2\ \text{m/s}$

Velocity or speed attained by the electron in 40 ns

$$\text{Speed,}\quad s = 1.76\times10^{17}\times(40\times10^{-9})^2 = 281.6\ \text{m}$$

$$\text{Distance travelled,}\quad X = \int v\,dt = \int 1.76\times10^{17}\times t^2\,dt = \frac{1.76\times10^{17}\times t^3}{3}\ \text{m}$$

$$\text{Distance 'X' travelled in 40 ns} = \frac{1.76\times10^{17}\times(40\times10^{-9})^3}{3} = 37.5\times10^{-7}\ \text{m.}$$

1.2 ELECTROSTATIC DEFLECTION IN A CATHODE RAY TUBE

In a CRT, electrons from an Electron Gun are accelerated by a potential V_{ax} in the x direction and enter the transverse field of V_d known as V_a (accelerating voltage) with an initial velocity of

$$v_{0x} = \sqrt{\frac{2qV_a}{m}} = 5.93\times10^{5}\sqrt{V_a}\ \text{m/s.} \tag{1.19}$$

Due to deflecting voltage V_d, the electrons get deflected, reach the end of the plates at P and since there is no accelerating field beyond P, they continue in a linear path and touch the screen at P'' as shown in Fig. 1.7. The equation of motion up to P between the deflecting plates is as given by the equation $y = \frac{1}{2}\left(\frac{a_y}{v_{0x}^2}\right)x^2$. So the electrons will move in a parabolic path up to point P.

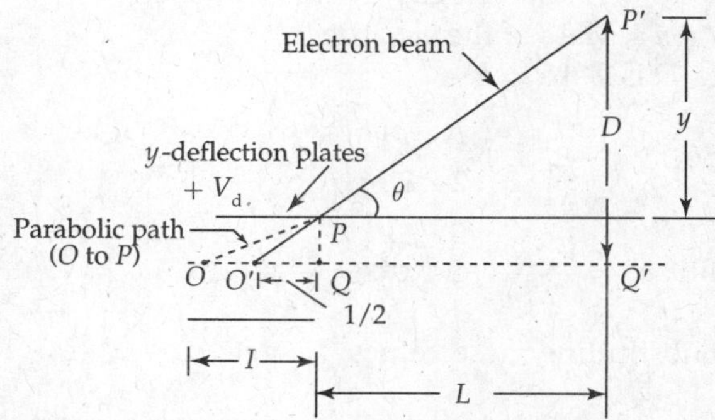

FIG. 1.7 *Electrostatic deflection in a cathode ray tube*

From there, the equations of the straight line PP' decides the motion of the electron and can be found from the equation of the straight line as per the geometry of Fig. 1.7. The straight line path from the edge of the deflecting plates to the screen is a tangent to the parabolic path between the plates at that point.

$$(y - y_1) = (\tan\theta)(x - x_1) = \frac{dy}{dx}(x - x_1) \tag{1.20}$$

$$y = \tan\theta\left(x - x_1 + \frac{y_1}{\tan\theta}\right) \tag{1.21}$$

At P, $x_1 = l$ and $y_1 = y$ at $l = \frac{1}{2}\left[\frac{a_y}{v_{0x}^2}\right] l^2$

But $\frac{dy}{dx} = \tan\theta = \frac{a_y \cdot l}{v_{0x}^2}$ at $y_1 = y_{\text{at } l}$

$$\therefore \quad y = \tan\theta\left(x - l + \frac{1}{2}\left[\frac{a_y}{v_{0x}^2}\right] l^2 \frac{v_{0x}^2}{a_y l}\right) = \tan\theta\left(x - \frac{l}{2}\right)$$

On simplification,

$$y = \tan\theta\left(x - \frac{l}{2}\right). \tag{1.22}$$

When $x = l/2$, $y = 0$ or the straight line PP' when projected backwards intersects the x-axis at O'. At $x = l/2$, that is O' bisects the line.

Deflection *D* *Deflection D* can be found from the geometry shown in Fig. 1.7. Key parameters are described as follows:

L = Distance from the centre of Deflection plates to screen, metres
l = Length of Deflection plates, metres
V_d = Deflection voltage, volts
d = Distance between plates, metres
V_a = Accelerating voltage, volts
a_y = Acceleration in the y direction
q = Charge of the electron
v_{0x} = Initial Velocity in the x direction
m = Mass of the electron.
D is y at $x = L + l/2$

So $D = \frac{a_y l}{v_{0x}^2}\left(L + \frac{l}{2} - \frac{l}{2}\right) = \frac{a_y l L}{v_{0x}^2}$

But $a_y = \frac{V_d}{d} \cdot \frac{q}{m}$ and $v_{0x}^2 = \frac{2qV_a}{m}$

Substituting these values, we get

$$D = \frac{V_d}{d} \cdot \frac{q}{m} \cdot \frac{m}{2q} \cdot \frac{lL}{V_a} = \frac{V_d \cdot lL}{2V_a \cdot d}. \tag{1.23}$$

Thus, Deflection D is proportional to the voltage from Eq. (1.23). This ratio of D to V_d is known as *Deflection Sensitivity S* (Eq. (1.24)):

$$S \cong \frac{D}{V_d} = \frac{l \cdot L}{2 \cdot V_a \cdot d}. \tag{1.24}$$

From Eq. (1.24) for Deflection Sensitivity S is independent of charge and mass of the electrons. Reducing d and V_a, increases the sensitivity S. In view of the construction of the equipment (TV tubes, CRT, etc.), L and l have their own design limitations.

EXAMPLE 1.7

Electrons emitted from a Thermionic Cathode, of a CRT gun, are accelerated by a potential of 400 V. Find Deflection Sensitivity *S*.

Solution: Given $L = 19.4$ cm, $l = 1.27$ cm, $d = 0.475$ cm and $V_A = 400$ V.

$$\text{Deflection sensitivity} \quad S = \frac{D}{V_d} = \frac{lL}{2V_a \times d}.$$

$L = 19.4 \times 10^{-2}$ m and $l = 1.27 \times 10^{-2}$ m; Accelerating Potential $V_A = 400$ V
Distance between plates $d = 0.475 \times 10^{-2}$ m

$$\therefore \quad S = \frac{19.4 \times 10^{-2} \times 1.27 \times 10^{-7}}{2 \times 400 \times 0.475 \times 10^{-2}} = 0.65 \text{ mm/V}.$$

1.3 MOTION OF ELECTRONS IN MAGNETIC FIELDS (MAGNETIC DEFLECTION)

We know that moving electrons constitute current. A current-carrying conductor produces a Magnetic Field, and so electrons are affected by Magnetic Fields. This property can be utilised to deflect electrons in the CRO, and this method proves better compared to electrostatic Deflection in some specific applications such as TV (Television) picture tubes.

It has been verified experimentally that if a current-carrying conductor of length *L* in metres is in a Magnetic Field of strength *B*, the conductor experiences a force f_m and is given by:

$$f_m = B\,I\,L \text{ Newtons,} \tag{1.25}$$

where Force f_m is in Newtons, Current *I* in Amperes and Magnetic Field *B* in Wb/m² or Tesla.

This is subject to the condition that the directions of *B* and *I* should be perpendicular to each other. Then the force f_m can be represented by the motion of a right-handed screw placed at the origin *O* and advanced into the plane containing *I* and *B* and moves through 90° from *I* to *B*

B and *I* need not necessarily be perpendicular to each other. In this case, the component of *I* resolved in a perpendicular direction to the direction of the Magnetic Field will be responsible for the force on the conductor. A word of caution is necessary in this context. Figure 1.8 represents the situation when the current is due to conventional positive charges. On the other hand, if the current is due to the electrons, the direction of motion is anti-parallel as shown by v^-. *It can be applied to moving electrons in any medium in the following manner:*

A Conductor of length *L* Contains *N* electrons.

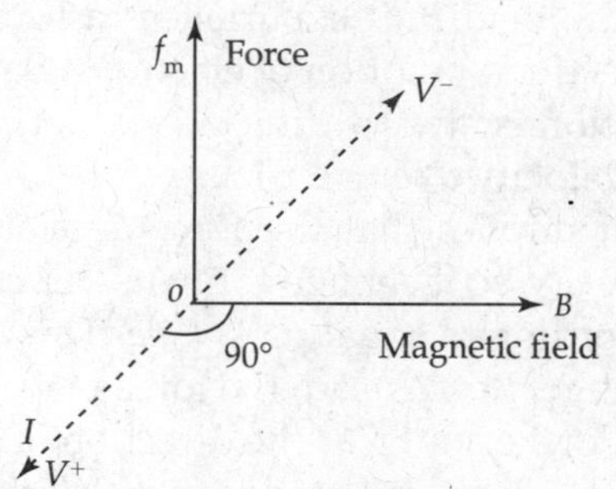

FIG. 1.8 *Direction of force f_m on a charged particle situated in a magnetic field*

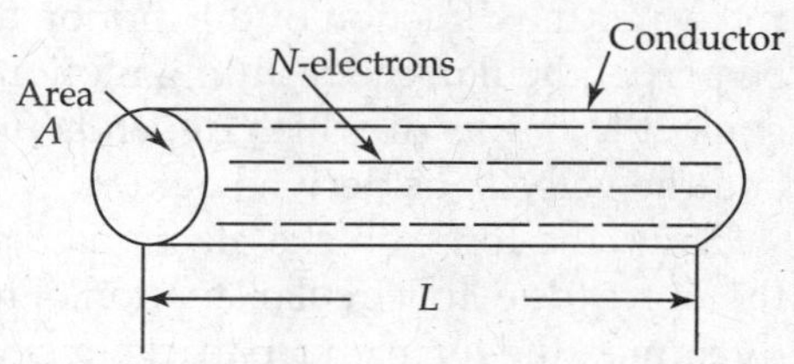

FIG. 1.9 *Conductor of length L*

If N electrons pass through a length of conductor L metres of cross-section A sq. m in time T seconds, as shown in Fig. 1.9, then the current I can be computed as

$$I = \frac{Nq}{T} \text{ Amps} \quad \text{(Rate of charge)} \tag{1.26}$$

$$I = \frac{Nev}{L} \text{ Amps,}$$

where T is in seconds, q is the charge of an electron and Nq is the total charge carried by electrons in the conductor in coulombs. Hence, 6.25×10^{18} electrons moving per second through a conductor contribute to a current of 1 A.

If the electrons move with a velocity v m/s, then T will be L/v seconds. Now substituting the value for current I, the force BIL will become

$$f_{\mathrm{m}} = BIL = B \times \frac{Nev}{L} \times L = BNev \quad \text{Newtons.}$$

Therefore, the force f_{m} per electron will be Bev:

$$f_{\mathrm{m}} = Bev \quad \text{Newtons.} \tag{1.27}$$

This Eq. (1.27) is a special case of the general equation:

$$f_{\mathrm{m}} = e(\bar{B}\bar{v}) = Bev \cdot \sin(\theta) \text{ with } \theta = 90°.$$

1.3.1 Process of Movement of an Electron in Magnetic Fields

$$f_{\mathrm{m}} = Bev \quad \text{Newtons.}$$

From Eq. (1.27), we understand that electrons experience motion when subjected to Magnetic Fields. This concept can be extended to a beam of electrons moving with a velocity v entering a Magnetic Field. If there is a velocity component v perpendicular to the direction of the field B, it experiences a force in a direction perpendicular to both the field and the direction of motion of electrons. So the electron movement depends on the condition that its motion is always directed perpendicular to the Magnetic Field B as well as to the direction of velocity v (current).

It implies that the force accelerates the electrons but does not affect the magnitude of velocity, so it changes only its direction. (If a force acts on an electron, it should be accelerated according to equation $f = ma$. This by no means implies that the magnitude of velocity has to change. But instead, the force may cause a change in direction.)

Furthermore, as the direction of force is perpendicular to the direction of motion, no work can be done on the electron. (Work done $W = f{\cdot}S \cos\theta = 0$; if $\cos\theta = 0$, then $\theta = 90°$. Since f and θ are not zero, only the distance S travelled in the direction of force is zero.)

Since it is necessary that the direction of the force and the direction of the motion are always to be perpendicular, the resulting motion should be a circle, the radius of which depends upon the field and the velocity as derived below:

Thus, the force on the electrons Bev is equal to the force due to circular motion. From classical dynamics, the force on a particle in circular motion is mv^2/R and these two should be equal.

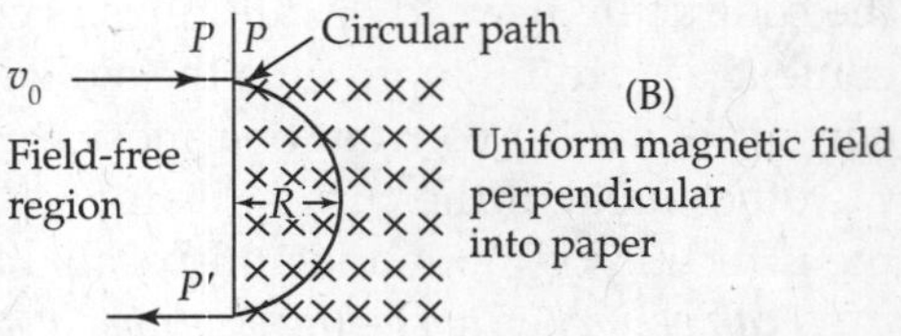

FIG. 1.10 *Circular motion of an electron in a transverse magnetic field*

$$f_{\mathrm{m}} = Bev = \left(\frac{mv^2}{R}\right) \text{ Newtons.}$$

$$\text{i.e.,} \quad R = \left(\frac{m \cdot v}{B \cdot e}\right) \quad (\therefore \quad R \propto v). \tag{1.28}$$

Therefore, the time period for one revolution T is as follows:

$$T = \frac{2\pi}{\omega} = \frac{2\pi m}{Be} \quad \left(\therefore \quad \omega = \frac{v}{R}\right) \tag{1.29}$$

Using data for ω from Eq. (1.29), we get

$$\omega = 2\pi f = \frac{2\pi}{T} \quad \text{and} \quad T = \frac{2\pi}{\omega} \left[\because \omega = \frac{v}{R} = \frac{veB}{mv} = \frac{eB}{m}\right].$$

where ω is the angular frequency.
From Eq. (1.29), the time period for one revolution T for one electron is

$$T = \frac{2\pi m}{eB} = \frac{6.28 \times 9.11 \times 10^{-31}}{1.6 \times 10^{-19} \times B}$$

$$= \frac{35.7 \times 10^{-12}}{B} = \frac{3.57 \times 10^{-11}}{B}.$$

It is found that the time period T is independent of the velocity, and so electrons travelling with different velocities can have the same time period of revolution. The radius R is dependent on both velocity and field, being directly proportional to the velocity and inversely proportional to field. Thus, electrons entering with a specific velocity travel circular paths of decreasing radius as the field increases.

The time period for one revolution T and ω are independent of speed and/or radius. As radius R is proportional to velocity ($R \propto v$), faster moving particles will traverse larger circles in the same time while a slower particle describes a smaller circle. This concept is used in focussing in instrumentation.

1.3.2 Motion of an Electron with a Velocity Component Each in Direction Parallel and Perpendicular to the Magnetic Fields (Motion of an Electron in Helical Paths)

Magnetic focussing:

- It is seen that if an electron enters the Magnetic Field perpendicular to the direction of field, the locus is a 'circle'. If it is parallel, no force acts and it continues to be at rest (If initial velocity of electron is zero) and maintains the same velocity v_0.
- If on the other hand, an electron enters the field at an angle θ to the field, *it will have a component each in directions parallel and perpendicular to the field*. The parallel component of the velocity 'v_θ sin θ' will

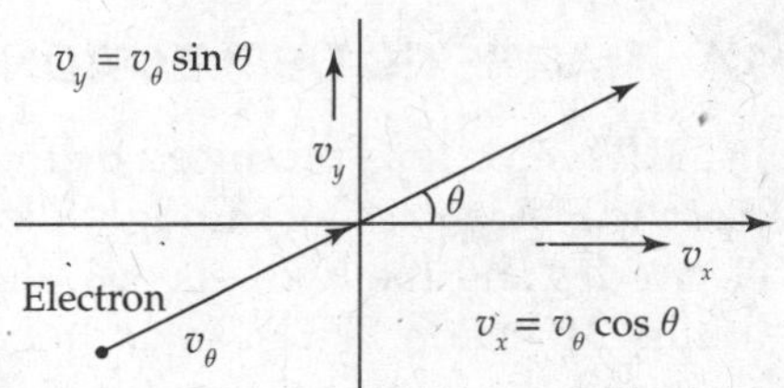

FIG. 1.11 *Electron entering the field at an angle* θ

not change, since it experiences no force in that direction. But the x-component, $v_\theta \cos(\theta)'$ experiences a force to cause it to move in a circle in the x–z plane. The resultant motion is helical path as shown in Fig. 1.11.

The electron describes a *helical path*, wherein the electron with *axial symmetry* (cylinder) moves along the *pitch of the helix*. *Pitch of the helix* is the 'displacement along the parallel component of the field', while the perpendicular component undergoes one revolution.

The pitch of the helix p is $(v_{0y} \times T)$, where T is the time period. Let the Pitch of the helix be P:

$$P = v_{0y} \cdot T = v_0 \cdot \sin\theta \cdot T = v_0 \cdot \sin\theta \cdot \frac{2\pi m}{Be}. \tag{1.30}$$

Electrons entering the field at different angles (perhaps due to mutual repulsion among them) will travel circular paths of varying radius, but may have the same pitch. This principle can be used to bring electrons of a diverging beam into focus at a point. As the field increases, as it is seen electrons travel in smaller and smaller circular paths with the same pitch and can be made to focus at a point on a screen, with the distance between the origin and the screen corresponding to integral number of pitches. This method of focussing the electrons to any point is called ***magnetic focussing***.

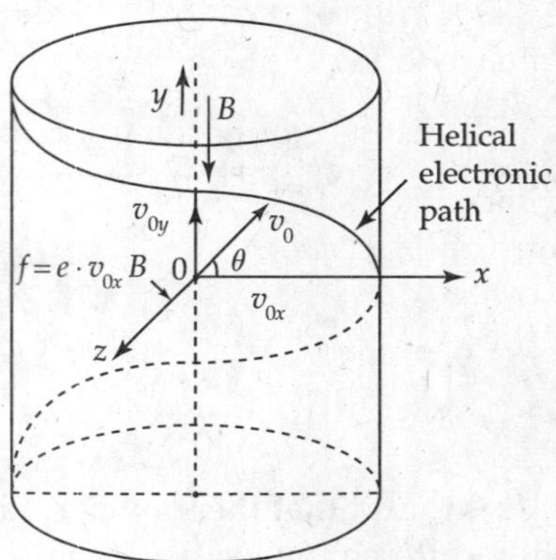

FIG. 1.12 *Helical path of an electron entering at an angle to the magnetic field B*

Advantages

- The previous discussion shows that electrons can be deflected and focussed. This ability of Deflection can be used for Deflections in CRTs, especially in TV tubes and computer monitors. This has additional advantages compared to electrostatic Deflection.
- For a given anode voltage, the Deflection Sensitivity is more for magnetic Deflection. Correcting factors can be introduced externally in the form of superimposed waveforms.

Disadvantage

- One disadvantage is that the Deflection for heavier unwanted negative ions is less and they cause sputtering (central dark spot on the screen due to heavy ion bombardment) of the screen. This can easily be avoided by using an ion trap in the form of aluminium coating on the inside of the screen.

1.4 MAGNETIC DEFLECTION IN A CATHODE RAY TUBE

In a CRT, Electrons produced by the Cathode K are accelerated to the fluorescent screen using Electric Fields passing through 'Electron Lens System' containing accelerating and focussing electrodes and the two sets of Deflection plates for display of the signal waveforms applied to the Deflection Plates. In addition, Magnetic Field may also be used for Deflection of the electron beam moving from the cathode to the CRT screen.

Figure 1.13 explains the magnetic Deflection in a CRT. A coil is wound on the constricted portion of the tube over a small length. Over this length, the field is considered to be uniform and the length of the field l is small compared to L, which is the distance between the centre of the Magnetic Field 0 and the screen of the CRT.

Electrons accelerated by an anode with voltage V_a enter the field with a velocity:

$$v = \sqrt{\frac{2eV_a}{m}} \quad (\therefore \; v = 5.93\times10^5\sqrt{V_a}\ \text{m/s}).$$

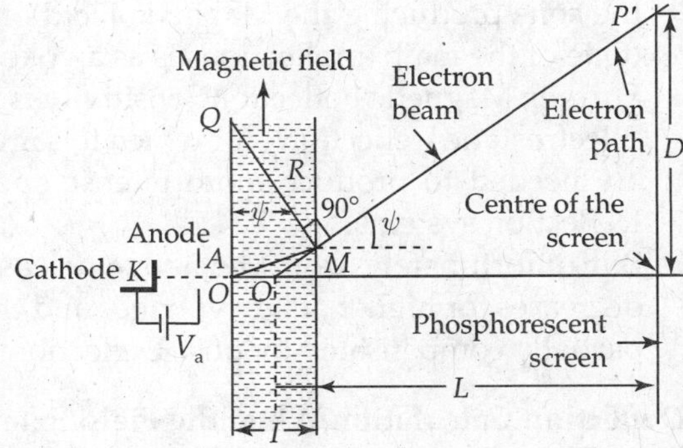

FIG. 1.13 *Magnetic deflection in a cathode ray tube*

Path of the electron beam is circular between the ends of the field that is 'O to M' and a straight line at an angle φ from the end of the field, which is 'M to P'' to the screen with reference to Fig. 1.13. l is the distance between the plates. L is the distance between the centre of the plates to the centre of the screen as shown in Fig. 1.13.

Since l is much less than L, $L \tan\varphi \cong D$, where D is the Deflection of the beam from the centre of the screen.

Tan $\varphi \cong \varphi$, since φ is very small.

$$\frac{l}{R} \quad (l = \text{arc length},\ R = \text{radius of circular path})$$

$$\varphi = \frac{l}{R} = \frac{leB}{mv} = \frac{D}{L} \tag{1.31}$$

and $R = \dfrac{mv}{eB}$ from Eq. (1.28).

$$\therefore \quad \text{Deflection } D = \frac{lLeB}{mv} \text{ metres.} \tag{1.32}$$

Substituting $v = \sqrt{\dfrac{2eV_a}{m}}$ into Eq. (1.32), we get

$$D = \frac{lLeB}{mv} = \frac{lLeB}{m\sqrt{\dfrac{2eV_a}{m}}} \text{ metres.}$$

i.e., D is proportional to B.

1.4.1 Magnetic Deflection Sensitivity

- Magnetic deflection sensitivity $S_m = \dfrac{D}{B}$ (By definition)

$$S_m = \frac{lL}{\sqrt{V_a}}\sqrt{\frac{e}{2m}} \text{ m/Wb/m}^2 \text{ or m/T.} \tag{1.33}$$

- Equation (1.33) for magnetic Deflection Sensitivity S_m of the CRT suggests that the quantity B, the Magnetic Field intensity, does not have any influence on S_m.
- The deflection sensitivity S_m increases with L, where L is the distance from the centre of the Magnetic Field and the CRT screen as shown in Fig. 1.13. This suggests for the location of

the coils producing the Magnetic Field on the CRT depends on the practical applications. (One of the main applications is as a yoke on TV tube.)

- Further, Magnetic Peflection Sensitivity is inversely proportional to $\sqrt{V_a}$. So it is better than 'Electrostatic Deflection', for which it is inversely proportional to V_a. Higher anode voltages are needed to produce more intense spots on the screen in both electric and magnetic Deflection systems.
- But unfortunately, from the above discussions, it is found that the Deflection Sensitivity decreases for higher anode voltages and thus they are conflicting requirements. This can be partially compensated by post-Deflection acceleration of the beam.

Deflection Calculation When the Field Extends Over the Entire Length of the Beam

From Fig. 1.14, it can be seen that the path of the electron OP' is the arc of a circle of radius R given by $R = \dfrac{mv}{eB}$ metres .

Velocity v can be known from the knowledge of the accelerating anode voltage V_a, where $v = 5.93 \times 10^5 \cdot \sqrt{V_a}$ m/s .

In this situation, Deflection (p to p') i.e. D can be known from the geometry in Fig. 1.14 by applying Pythagoras theorem, as

$$R^2 = (R-D)^2 + L^2$$

$$(R-D)^2 = R^2 - L^2$$

$$(R-D) = \sqrt{R^2 - L^2} \tag{1.34}$$

$$D = R - \sqrt{R^2 - L^2}.$$

An example for this type of phenomenon is found in nature in the form of the 'effect of Earth's Magnetic Field' on the Deflection sensitivities of CRTs. So shielding has to be provided to protect the CRT in a CRO from the effects of the Earth's Magnetic Field.

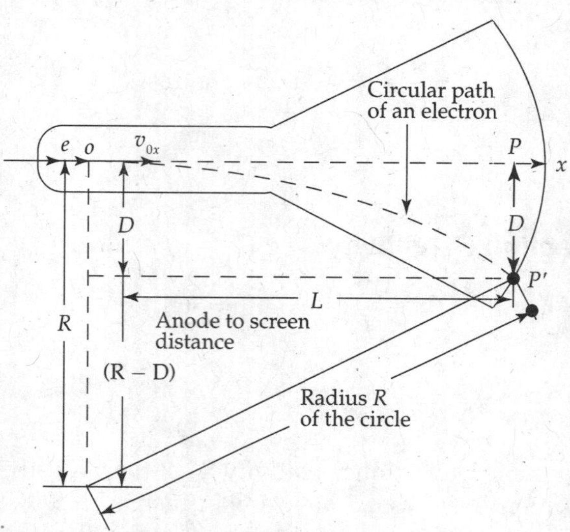

FIG. 1.14 *Circular path of an electron beam in a cathode ray tube*

The presence of the Earth's Magnetic Field adversely affects the performance of CRTs, since the actual Deflection is due to both Deflection field and Earth's Magnetic Field. By changing the orientation of the tube, so as to make the Earth's field parallel to the axis of the tube, the additional Deflection can be nullified.

EXAMPLE 1.8

Calculate the velocity and the KE acquired by an electron when accelerated by a voltage of 4900 V.

Solution: An electron is accelerated through a potential difference of 4.9 KV.

$$\text{Velocity of the electron} = 5.93\times10^{5}\sqrt{4900} = 5.93\times10^{6}\sqrt{49}$$

$$= 41.51\times10^{6}\ \text{m/s}.$$

$$\text{KE of the electron} = \frac{1}{2}mv^{2} = \frac{1}{2}\,9.109\times10^{-31}\times(41.51\times10^{6})^{2}$$

$$\text{KE of the electron} = 4.55\times10^{-31}\times1723\times10^{12} = 7840\times10^{-19}\ \text{eV}.$$

EXAMPLE 1.9

Electrons emitted from a Thermionic Cathode of a CRT are accelerated by a potential V_A of 400 V. Deflection Sensitivity S is 0.65 mm/V. What is the magnitude of transverse Magnetic Field over the entire length of the tube, in order to produce the same deflection as 30 V deflecting voltage on Y-plates? Distance L between the anode and the screen of CRT is 23.9 cm.

Solution: Deflection of the beam from the centre of the screen, $D = S \times V_d$
Data: Deflection Sensitivity of the tube $S = 0.65$ mm/V and deflecting voltage $V_d = 30$ V
Therefore, Deflection $D = 0.65 \times 10^{-2} \times 30 = 1.95$ cm.

$$(R-D)^{2} + L^{2} = R^{2}$$

$$(R-1.95)^{2} + (23.9)^{2} = R^{2}$$

$$R^{2} - 3.9R + 1.95^{2} + 23.9^{2} = R^{2}$$

$$\text{Radius } R = \frac{23.9^{2} + 1.95^{2}}{3.9}$$

$$= \frac{571.2 + 3.8}{3.9} = 147.4\ \text{cm}.$$

$$\text{Velocity } v = \sqrt{\frac{2eV_A}{m}} = \sqrt{2\times1.76\times10^{11}\times400}$$

$$= 5.93\times10^{5}\cdot\sqrt{400} = 11.86\times10^{6}\ \text{m/s}.$$

Due to magnetic deflection

$$\text{Radius } R = \frac{m\cdot v}{e\cdot B} \Rightarrow \quad \therefore \quad B = \frac{m\cdot v}{e\cdot R}.$$

Substituting the values for $\frac{m}{e}=\frac{1}{1.76\times10^{11}}$, $v=11.86\times10^{6}$ m/s, $R=147.4$ cm $=\frac{147.4}{10^{2}}$ m, we get

$$B=\frac{1}{1.76\times10^{11}}\times\frac{11.86\times10^{6}\times10^{2}}{147.4}=45.7\times10^{-6}\ \text{Wb}/\text{m}^{2}.$$

EXAMPLE 1.10

When an electron is placed in a Magnetic Field with a period of rotation $T=\frac{35.5}{B}\times10^{-12}$ secs so that the trajectory of an electron is a circle. (a) What is the radius described by an electron placed in a magnetic field, perpendicular to its motion when the accelerating potential is 900 V and $B = 0.01$ Wb/m². (b) What is the time period of rotation?

Solution:

a. Period of rotation: $T=\frac{35.5}{B}\times10^{-12}$ seconds.

Accelerating potential $V_A = 900$ V

Velocity of the electron $v=5.93\times10^{5}\times\sqrt{V_A}=5.93\times10^{5}\times\sqrt{900}=17.79\times10^{6}$ m/s

$$\text{Radius}=\frac{mv}{eB}\ \text{metres}$$

m = mass of the electron $= 9.109\times10^{-31}$ kg

v = velocity of electron $= 17.79\times10^{6}$ m/s

e = charge of an electron 1.602×10^{-19} C

B = magnetic flux density $= 0.01$ Wb/m²

$$\therefore\ R=\frac{9.109\times10^{-31}\times17.79\times10^{6}}{1.602\times10^{-19}\times0.01}=101.15\times10^{-4}\ \text{m}$$

$$=10115\times10^{-6}\cong1\ \text{cm}.$$

Therefore, radius of the circle = 1 cm.

b. Time period of rotation $T=\frac{35.5}{B}\times10^{-12}=\frac{35.5}{0.01}\times10^{-12}=35.5\times10^{-10}$ s.

EXAMPLE 1.11

An electron having an initial velocity v_0 of 5.93×10^{6} m/s enters a Magnetic Field of density B of 0.05 Wb/m², at an angle θ of 45° to the field. Predict the electron position after it has made one revolution in the field.

Solution: Initial velocity of electron $= v_0 = 5.93\times10^{6}$ m/s

The time T for one revolution around the circular path is

$$T=\frac{2\pi m}{Be}\ \text{sec}$$

$$T=\frac{2\pi m}{Be}=\frac{2\pi}{0.05\times1.76\times10^{11}}=\frac{2\pi\times10^{-9}}{8.8}=\frac{6.28\times10^{-9}}{8.8}=0.714\ \text{ns}.$$

Since l/T is the velocity of the electrons, the length l traversed by the electron during the time T is given as $l = v_{0x}T$.

$$v_{0x} = v_0 \cos\theta = v_0 \cos 45°$$

$$v_{0x} = 5.93\times10^6 \times \frac{1}{\sqrt{2}} = 0.707\times5.93\times10^6$$

$$l = v_{0x} \times T = 0.707\times5.93\times10^6 \times0.714\times10^{-9} = 0.5\times5.93\times10^{-3} \text{ m.}$$

$$l = 2.965\times10^{-3} \text{ m} = 0.2965 \text{ cm .}$$

EXAMPLE 1.12

A charged particle with three times the charge and mass two times that of an electron is accelerated through a potential difference V_A of 50 V, before it enters a uniform Magnetic Field of flux density B of magnitude 0.02 Wb/m², normally with the field. Find the velocity of the charged particle before entering the field, radius of the path and time for one revolution. Repeat the above calculations when an electron enters at an angle θ of 25°.

Solution: Given $\frac{e}{m} = 1.76\times10^{11}$, $B = 0.02 \text{ Wb/m}^2 = 2\times10^{-2} \text{ Wb/m}^2$ and $v = 5.138\times10^6$ m/s

a. Velocity $v = \sqrt{2\frac{eV_A}{m}}$ m/s

Velocity of charged particle before entering the field:

$$\therefore \quad v = \sqrt{2\frac{3e}{2m}\times50} = \sqrt{\left(\frac{3e}{m}\right)\times50} \quad \text{metres/sec.}$$

As per the data in the problem, $Q' = 3Q$ and $m' = 2$ m and $V_A = 50$ V

$$v = \sqrt{3\times1.76\times10^{11}\times50} = \sqrt{26.4\times10^{12}} = 5.138\times10^6 \text{ m/s .}$$

b. Radius R of the path, when charge particle enters normal to the Magnetic Field
If the particle enters at an angle θ of 25° with the field, $\theta = 25°$

$$\text{Using} \quad Bev = \frac{mv^2 \sin\theta}{R} \Rightarrow Be = \frac{mv\sin\theta}{R} \Rightarrow \quad \therefore \quad R = \left(\frac{m}{e}\right)\times\frac{v\sin\theta}{B}$$

$$R = \frac{5.138\times10^6}{1.76\times10^{11}\times2\times10^{-2}}\times\sin 25° = 1.46\times10^{-3}\times\sin 25°$$

$$R = 1.46\times10^{-3}\times\sin 25° = 0.617\times10^{-3} \text{ m.}$$

Time for one revolution:
Angular velocity in rad/s,

$$\omega = \frac{v}{R} \quad \text{where } R = \left(\frac{m}{e}\right)\frac{v}{B}$$

$$\therefore \quad \omega = \frac{eB}{m} \text{ rad/s.}$$

Time in seconds for one complete revolution is called period.

$$T = \frac{2\pi}{\omega} = \frac{2\pi m}{eB} = \left(\frac{2\pi}{B}\right)\cdot\frac{m}{e}.$$

For an electron,

$$T = \frac{2\pi}{B} \times \frac{1}{1.76\times10^{11}} = \frac{3.57\times10^{-11}}{2\times10^{-2}} = 1.785\times10^{-9}\ \text{s}.$$

Time for one revolution when $\theta = 25°$:

$$\omega = \frac{v\sin\theta}{R} \Rightarrow 2\pi f = \frac{v\sin\theta}{R}.$$

$$T = \frac{1}{f} = \frac{2\pi R}{v\sin\theta} = \frac{2\pi\times1.46\times10^{-3}\sin 25°}{5.13\times10^{6}\sin 25°} = 1.788\times10^{-9}\ \text{s}$$

1.5 COMPARISON BETWEEN ELECTROSTATIC AND MAGNETIC DEFLECTIONS

1.5.1 Electrostatic Deflection

$$\text{Electrostatic Deflection}\quad D = \frac{LlV_d}{2dV_a}, \tag{1.35}$$

where D is the Deflection, metres
L is the distance from centre of Deflection plates to screen, metres
l is the length of Deflection plates, metres
V_d is the deflection voltage, volts
d is the distance between plates, metres
V_a is the accelerating voltage.

1.5.2 Electrostatic Deflection Sensitivity

'Electrostatic Seflection Sensitivity' is defined as vertical Deflection of the beam per unit Deflection voltage:

$$S = \frac{D}{V_d} = \frac{(lL)}{(2dV_a)}\ \text{m/V}. \tag{1.36}$$

Deflection factor G of a CRO is the reciprocal of Deflection Sensitivity S:

$$G = \frac{1}{S} = \frac{2dV_a}{lL}\ \text{V/m}. \tag{1.37}$$

- Speed of Deflection is faster.
- For greater sensitivity, long plates with minimum distance between them are necessary. So the CRT will be long and beam potential will be less.
- Deflection plates limit the beam angle. To correct this, plates are bent or curved instead of being parallel.

- Segmented plates are used for large bandwidth.
- Maximum bandwidth is up to 350 MHz operations.
- Amplifiers with low current requirements are sufficient.
- CROs using CRTs are used in laboratories for display, measurement and analysis of signals.
- Electrostatic Deflection Sensitivity is independent of deflecting voltages V_d.
- Defection sensitivity is independent of mass and charge of the electrons.
- Reducing d and V_a increases the Electrostatic Deflection Sensitivity.
- L and l have limitations in view of the construction of the equipment.
- Deflection Sensitivity is inversely proportional to accelerating voltage V_a for electrostatic Deflection, and for magnetic Deflection Sensitivity is inversely proportional to $\sqrt{V_a}$. So for electrostatic Deflection larger voltages are needed to obtain the same sensitivity.
- Electrostatic Deflection suffers defocussing.

1.5.3 Magnetic Deflection

$$\text{Magnetic Deflection:} \quad D = \frac{lLB}{\sqrt{V_a}} \cdot \sqrt{\frac{e}{2m}} \text{ metres} \tag{1.38}$$

where l is the width of magnetic coil, metres
L is the length from centre of l to screen, metres
B is the magnetic flux density, Wb/m^2
V_a is the accelerating potential at the anode.

1.5.4 Magnetic Deflection Sensitivity

It is the ratio of magnetic Deflection D to the applied Magnetic Field B.

$$S_m = \frac{D}{B} = lL\sqrt{\frac{e}{2mV_a}} \text{ m/(Wb/m}^2\text{) or m/T.} \tag{1.39}$$

- Equation (1.39) for magnetic Deflection Sensitivity S_m of the CRT suggests that the quantity B, the Magnetic Field intensity, does not have any influence on it.
- The Deflection Sensitivity S_m increases with L, where L is the distance from the centre of the Magnetic Field to the CRT screen. This suggests the location of coils that produce Magnetic Field in a CRT, depending upon practical specifications.
- Magnetic Deflection Sensitivity depends on *e/m ratio*.
- Deflection D is directly proportional to B, and *so electrical parameters* such as voltage, frequency and current cannot be measured.
- Magnetic Deflection is associated with coils. So large currents are required for full screen display requiring more power dissipation in the system, making the unit bulky.
- As no Deflection plates are necessary, electron beam scan angle is wide and shorter tubes can be built.
- Magnetic Deflection Sensitivity is inversely proportional to $\sqrt{V_a}$. It is better than 'Electrostatic Deflection', which is inversely proportional to V_a.
- Magnetic Deflection is used in TV picture tubes and visual display units such as computer monitors.

1.6 ELECTROSTATIC FOCUSSING

Focussing of electron beam in a CRT, electron beams from accelerating anode system (travelling towards the CRT screen) tend to diverge due to mutually repulsive forces among electrons. In order to focus sharply on the CRT screen, methods used are:

- Electrostatic focussing,
- Magnetic focussing.

In order to understand the basic concept of *electron lens system* used for focussing of electrons, let us review the familiar concept of focussing a light beam. When a light beam arrives at the interface of two media, with different refractive indices μ_1 and μ_2, it *refracts*. This principle is used to focus optical beams through lenses (convex). Similarly, consider focussing electron beams through electrostatic lenses, formed by curving *equipotential surfaces* of different voltages.

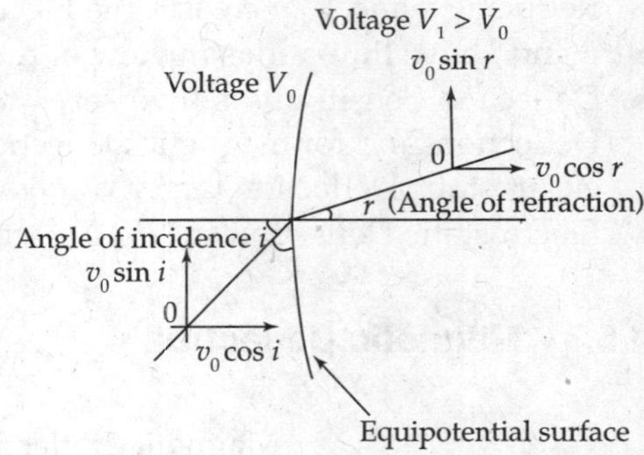

FIG. 1.15 *Concept of electron lens affect in a CRT*

Consider an electron approaching with a velocity v_0 at an equipotential surface, with voltages V_0 and V_1 on either side (Fig. 1.15). Energy E_0 of an electron is given by

$$E_0 = e \cdot V_0 = \frac{1}{2} \cdot \left[m \cdot v_0^2\right], \tag{1.40}$$

velocity v is proportional to $\sqrt{V}$.

$$\therefore \quad v_{ti} \cdot \sin(i) = v_{tr} \cdot \sin(r)$$

$$\text{However,} \quad v_{ni} \cdot \cos(i) \neq v_{nr} \cdot \cos(r)$$

$$\text{Thus,} \quad v_{ti} \cdot \sin(i) = v_0 \cdot \sin(i) \text{ and } v_{tr} \cdot \sin(r) = v_1 \cdot \sin(r)$$

$$\therefore \quad v_0 \cdot \sin(i) = v_1 \cdot \sin(r).$$

$$\therefore \quad \frac{\sin(i)}{\sin(r)} = \mu = \frac{v_1}{v_0} = \sqrt{\frac{V_1}{V_0}}. \tag{1.41}$$

Referring to Fig. 1.15, voltages on either side of the equipotential surface are V_0 and V_1, with $V_1 > V_0$. Since *electron velocity* v is proportional to $\sqrt{V}$, the electron enters on V_0 side with a velocity v_0 and crosses to the other side and proceeds with a velocity of v_1.

Since the particle enters at an angle $\angle i$ and velocity v_i, with velocity components (*tangential* and *normal*) of v_{ti} and v_{ni} (Fig. 1.15). It leaves at an angle $\angle r$ and velocity v_r with components v_{nr} and v_{tr} and without any change in the tangential component (no work).

Practical Focussing System (electrostatic focussing)

In designing lenses, equipotential surfaces have to be chosen to provide required electron paths. The equipotential surface shown in Fig. 1.16 is called a *double aperture lens*. This lens is a popular choice for CRO, Electron Microscope, etc. (Fig. 1.17).

Figure 1.18 shows an asymmetric lens system. The geometry is self-explanatory.

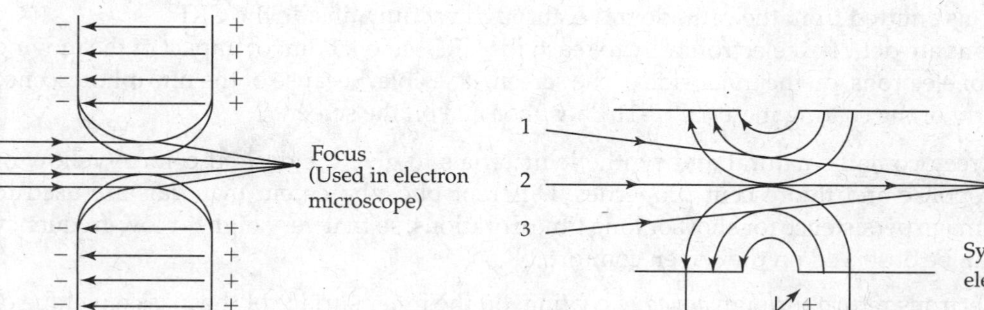

FIG. 1.16 *Practical focussing system (double aperture lens)*

FIG. 1.17 *Symmetrical electron lens*

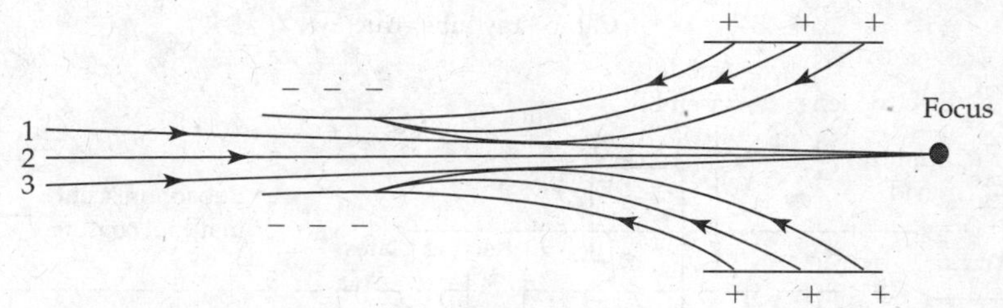

FIG. 1.18 *Assymetrical lens system*

1.7 CATHODE RAY OSCILLOSCOPE

1. CRT – core unit of a CRO
2. Power supply circuits – to provide (a) accelerating and focussing voltages and (b) voltages to other electronic circuits
3. Sweep circuits – to generate time Base voltages
4. Horizontal amplifier, Vertical amplifier for deflecting voltages and some more control circuits

In order to fully understand the various circuits in a CRO, detailed working knowledge of CRT is necessary.

Cathode Cathode is an indirectly heated type, where filament heater and the cathode are separate in their structure with an electrical insulation between them. 'Cathode' supplies electrons. Electrons are produced from the Cathode surface (source for electrons) due to thermionic emission process, governed by Eq. (1.42):

$$I = SAT^2 e^{-b\phi/T} \text{ Amps}, \tag{1.42}$$

I is the current due to emitted electrons from cathode, due to *Thermionic Emission*

S is the surface area of cathode (Cathode Assembly)

ϕ is the work function energy required for liberation or emission of electrons from the cathode A and b are constants of cathode material (Thorium oxide coated Tungsten)

T is the Absolute Temperature in °K.

- Electrons emitted from the cathode travel through vacuum, inside the CRT.
- Electrons are deflected electronically to reach the CRT screen. Point of impact of the moving beam of electrons, on the inner side of the screen, is visible, because of the phosphorescence property of the coating material (*P*-11 Phosphorous) on the screen.

The screen contains natural and synthetic materials to give the desired colour (yellow or green) response and fluorescent properties. Different phosphorescent materials are used to provide image persistence for short or long time durations, so that very high- or low-frequency signals can be observed on the screen comfortably.

- The electrons return through *aquadug* coating, on the inner surface of the glass envelope of the CRT (Fig. 1.19).
- Voltage waveforms can be observed on the CRT screen in association with additional electronic circuits in the CRO.

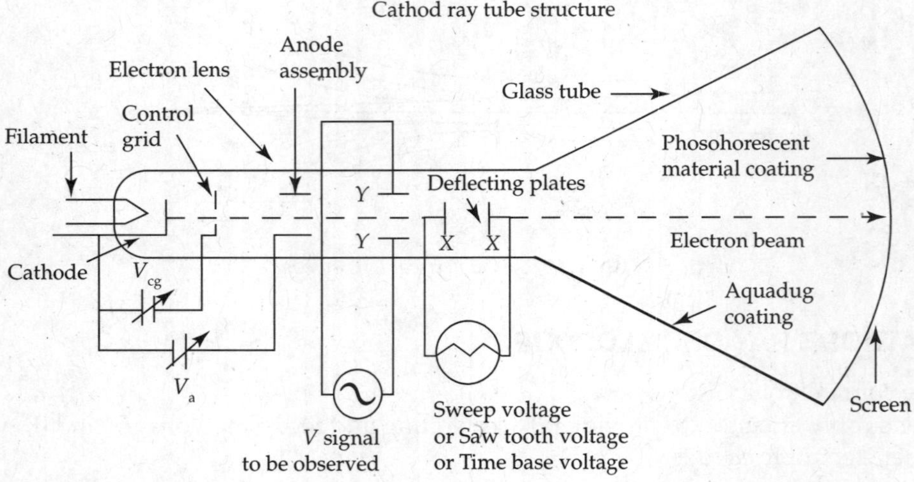

FIG. 1.19 *Structural details of a cathode ray tube*

Since electron's mass is quite small, so is its inertia. Thus, very high-frequency voltages can deflect a beam of electrons that are emitted from the cathode surface. Owing to the charge of electrons, *Electric or Magnetic Fields are varied in the region between the cathode and the screen to provide the required 'deflection and direction' to the electron beam.*

Electron Gun Electron Gun has systems for electron production and focussing the electron beam towards the CRT screen (coated with a phosphorescent material). Electron Gun consists of:

1. *Cathode* – Source for electrons using thermionic emission phenomenon,
2. *Control grid* – maintained at varying negative voltage to control the content of electron beam that is focussed on to the screen at later stages,
3. *Anode assembly* – for onward journey and focussing of electron beam onto the screen of the tube.

Control Grid Control grid *G* has a small hole in its centre and allows or transmits a pencil beam of electrons through it. The beam current and spot intensity on the CRO screen can be varied by the adjustment of control grid bias V_{cg}, where V_{cg} is maintained at a negative bias.

This function is attached to the front panel of CRO (CRO panel controls), providing *Intensity* or *brightness control*.

The electron beam, emerging from the hole of the control grid, is accelerated and focussed towards a small point on the screen, by anode assembly (that functions as an electron lens or Electric Field lens system). Adjusting the voltage on the first anode in the anode assembly is attached to CRO panel controls, providing *focussing control*. The complete beam forming assembly is known as the *Electron Gun*.

Deflecting Plates Electron beam, from the Electron Gun, passes between two pairs or sets of parallel plates and reaches the CRT screen, producing a spot of light, the colour of which depends on the type of phosphor-coating material on the inner surface of the screen.
The two sets of plates causing *deflection of the electron beam* are:

1. *Horizontal Deflection plates*, or *X-plates* or X-Deflection plates,
2. *Vertical Deflection plates* or *Y-plates* or Y-Deflection plates.

The *Sweep Voltage*, internally applied to the X-plates, controls the movement of electrons in the *horizontal* or *x direction*. The plates are named after the 'control of direction of movement of electrons'. Sweep Voltage is also known as *time base voltage*. Sweep Voltage is of *sawtooth wave shape*, shown in Fig. 1.26. The Sweep Voltage is internally connected to the X-plates. CRO has *INT/EXT sweep control* on the front panel. Keeping the control in *INT position*, includes the Sweep Voltage on to 'X-plates'. And keeping the control in *EXT position*, the Sweep Voltage is not to be connected to X-plates.

The *signal voltages to be observed* are applied to Y-plates. The applied voltages *control the movement* of electrons in the *vertical direction*. As the control or movement of electrons is in the *y* direction, the name for the plates is Y-plates. These two sets of deflecting plates are mutually perpendicular, as shown in Fig. 1.20.

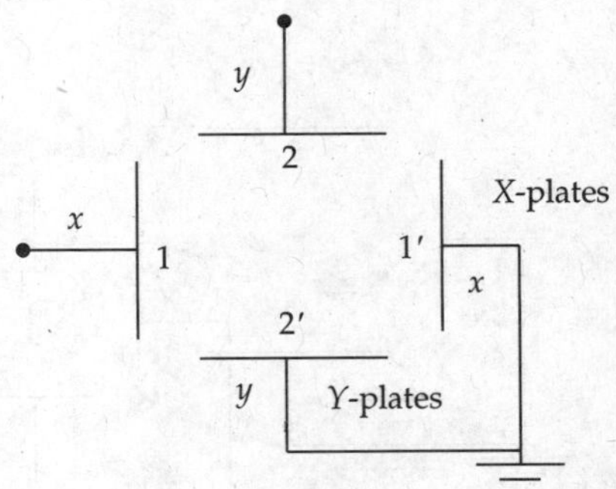

FIG. 1.20 *Mutually perpendicular deflecting plates (X-plates and Y-plates)*

X-Deflection Refer to Fig. 1.21. If no DC voltage is applied to X-plates, the electron beam spot on the screen is at point *A*. Applying a DC voltage to the X-plates, with right plate being positive with respect to the other, the spot shifts to point *B*. The amount of shift, of the electron spot, in the *x* direction is proportional to the applied voltage. The spot remains there as long as the applied voltage is present. The spot returns to point *A*, if the applied DC voltage is removed.

Y-Deflection: Observing AC Signals on a CRO Screen

The same reasoning applies to Y-plates. The only difference is that the electron beam moves (under the influence of fields) in the *vertical* or *y direction*. This means that Electric Field between Y-plates deflects the beam of electrons in a direction normal to the plane of the plates.

For example, if we want to see the waveform of a sinusoidal voltage applied to Y-plates, we have to provide the *time-axis* (*time base*) *voltage* or *Sweep Voltage* internally to *horizontal deflection plates* (Fig. 1.22). Since a sinusoid is a graph with the Y-coordinate representing the amplitude of the signal and X-coordinate representing the instance of time at which the instantaneous voltage V_{AC}, i.e. $V_{AC} = V_m \sin(\omega t)$.

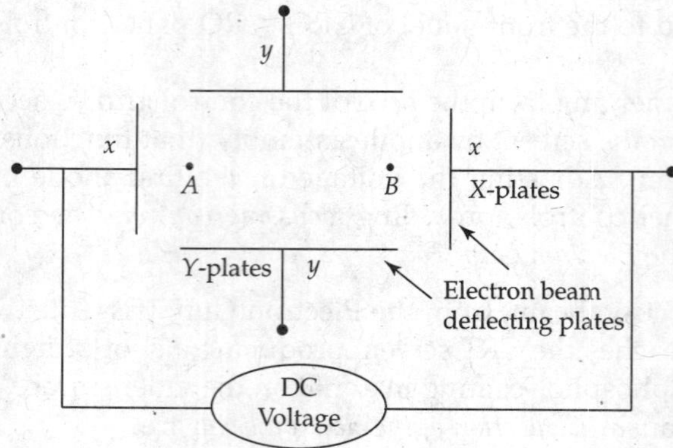

FIG. 1.21 *A spot movement from point A to point B due to applied DC voltage to X-plates of a CRT*

Electronic trace of a sinusoidal waveform is shown in Fig. 1.22. This means that a *pull* on electron beam in the x direction is proportional to *time base* or the *Sweep Voltage (ramp-type voltage)* applied to the X-plates, according to the following equation.

$$V(t) = \left[\frac{V}{T}\right]t. \tag{1.43}$$

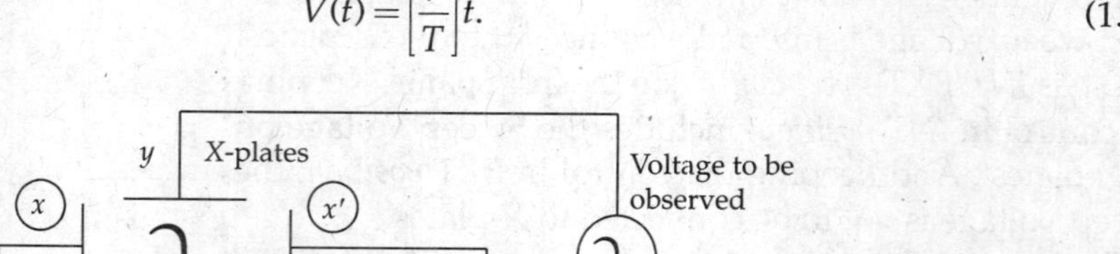

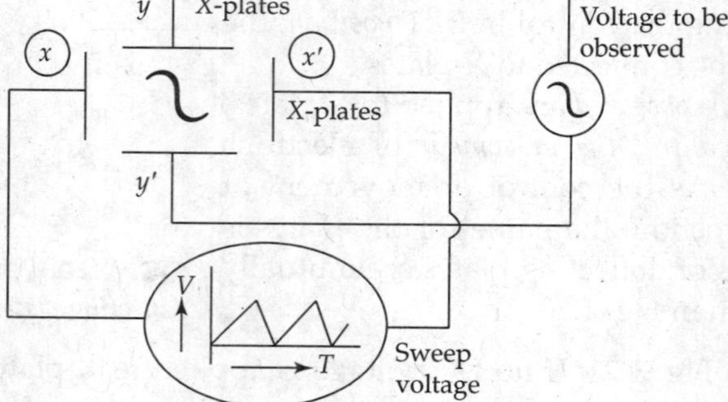

FIG. 1.22 *Vertical deflection due to voltage applied to Y-plates and sweep voltage applied to horizontal plates*

Features of Signal Waveforms on CRO Screen

Various features of signal waveforms on the CRO screen are described below:

1. *Unit time period* (T) of the signal represents passing of one cycle of signal on the CRO screen.
2. To get a continuous *trace of signal*, the cycle has to be repeated at a rate more than the persistence of human vision, at 24+ traces per second. In order to ensure continuity, the starting point of every new trace coincides with the *beginning* of the previous one. This is called *synchronisation*.
3. The repetitive trace can be obtained by using *Sweep Voltage* (in *triangular* or *sawtooth waveform*) from *Sweep Voltage Generator Circuit*. Sweep Voltage applied to X-plates controls the sweeping of electrons – movement of electrons in the x direction.
4. A simple method to generate Sweep Voltage, using UJT oscillator circuit follows.

Fluorescent Screen The front face of the CRT tube is the screen. Inside of the screen is coated with phosphorescent material that produces visible light on the impact of focussed electron beam on the screen. The luminescence of the total trace is produced due to the electron trace on the screen. The waveforms displayed on the screen are due to the signals applied to Y-plates.

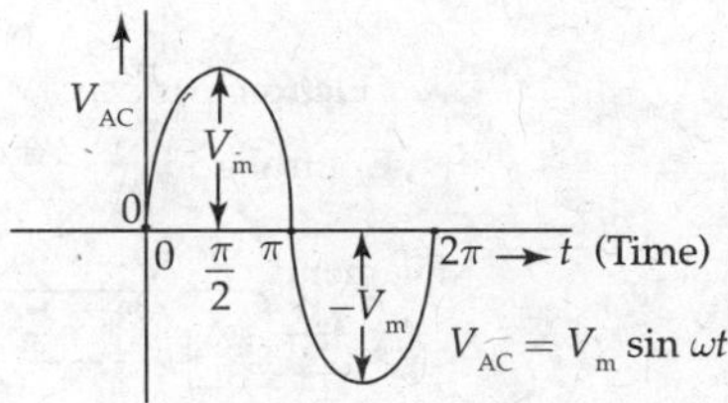

FIG. 1.23 *Trace of a sinusoidal waveform*

Aquadug Coating Aquadug coating is a conductive coating, as shown in Fig. 1.19. This conductive coating is connected to the grounded electrode in the CRT. This path provides return path for the electrons to complete the electric circuit (Fig. 1.23).

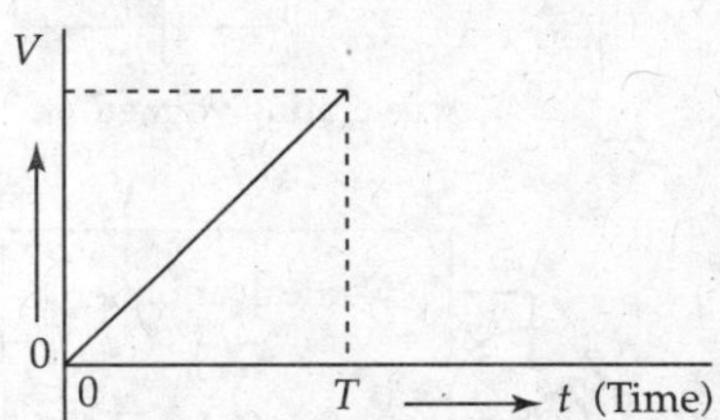

FIG. 1.24 *Time base (ramp) voltage*

Sweep Voltage Generator Circuit using UJT
Simple method of getting a repetitive triangular wave is shown in Fig. 1.25. The circuit is a UJT oscillator circuit. Working of the UJT oscillator circuit (Fig. 1.25) is explained as follows.

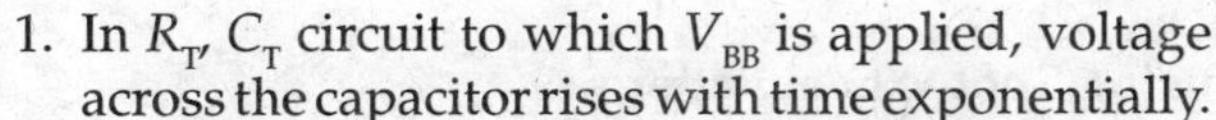

1. In R_T, C_T circuit to which V_{BB} is applied, voltage across the capacitor rises with time exponentially.
2. By the time, voltage across the capacitor reaches a value of V_P (peak voltage for the UJT to conduct), the UJT starts conducting and the capacitor starts discharging. The capacitor discharges through the input circuit path of the UJT from Emitter to 'Base-1' and to ground terminal in the conduction state.

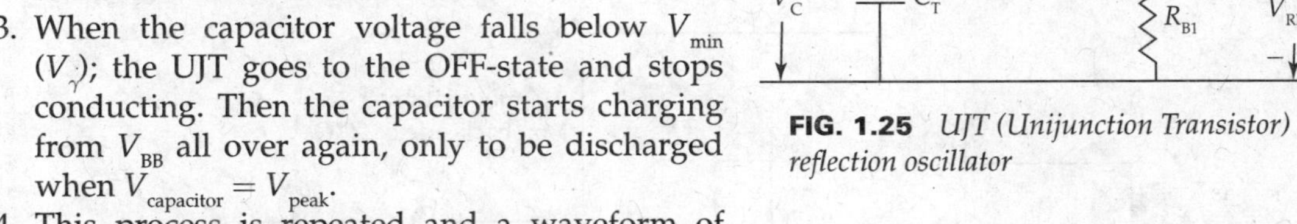

3. When the capacitor voltage falls below V_{min} (V_γ); the UJT goes to the OFF-state and stops conducting. Then the capacitor starts charging from V_{BB} all over again, only to be discharged when $V_{capacitor} = V_{peak}$.
4. This process is repeated and a waveform of voltage V_C is obtained across the capacitor C_T, in the UJT oscillator circuit. The waveform is shown in Figs. 1.26 and 1.27.

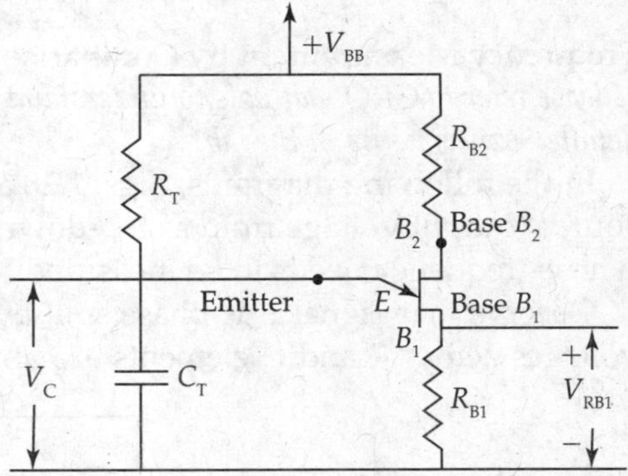

FIG. 1.25 *UJT (Unijunction Transistor) reflection oscillator*

Cathode Ray Oscilloscope Cathode ray oscilloscope consists of the following units:

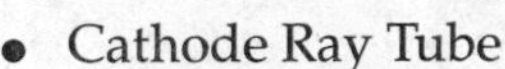

- Cathode Ray Tube
- Power Supply Circuit
- Sweep Circuit
- Vertical Amplifier
- Horizontal Amplifier
- Retrace Blanking Circuit
- Channel Control Circuits

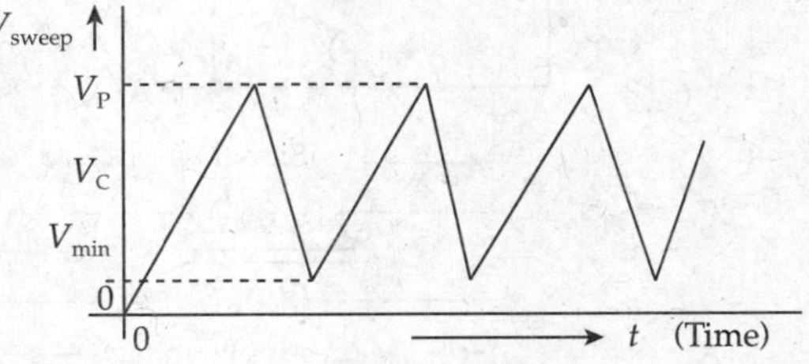

FIG. 1.26 *Sweep voltage across capacitor CT of reflection oscillator*

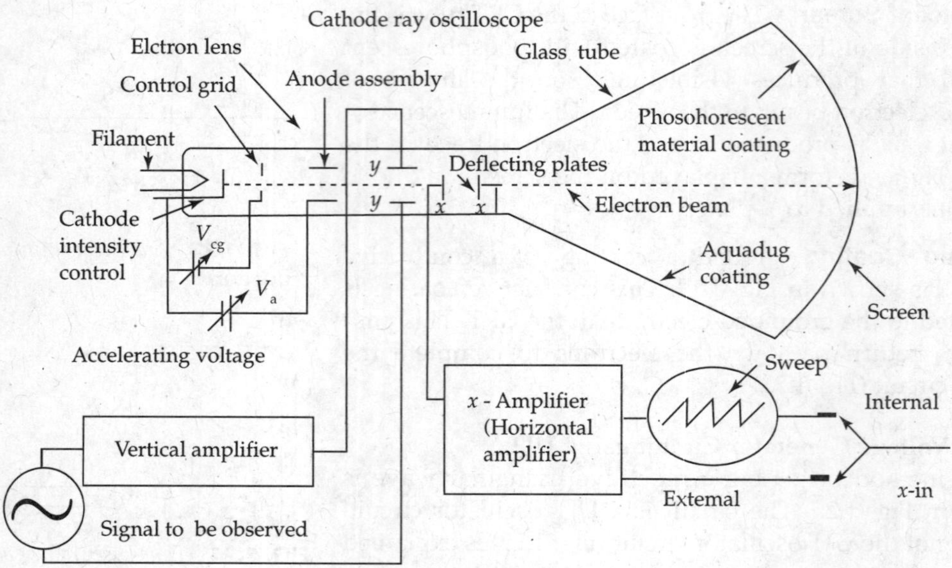

FIG. 1.27 *Main details of cathode ray oscilloscope*

Frequency Measurement by Comparison using Spot Wheel Pattern

**Please refer to CRO manuals to understand different components and functioning in order to conduct detailed experiments in the lab.*

In the following diagrams, Figs. 1.28 and 1.29 observe two signals, *Signal A* from a known source (output voltage from a step-down transformer) and *Signal B* from an unknown source whose frequency needs to be measured.

Observe that Signal *A* is phase split by an *RC* (resistor and capacitor) *network* and resultant voltages across *R* and *C* elements are applied to *X*- and *Y*-plates (of the CRO), to see a *circle*

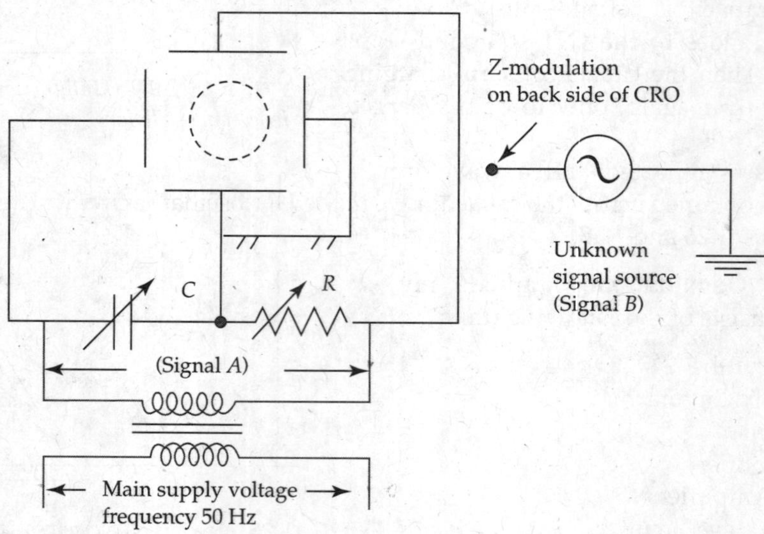

FIG. 1.28 *Frequency measurement of a signal using spot wheel pattern*

on the screen. When Signal B is applied to the *intensity control* (*Z-modulation* terminal on CRO rear side), if the signal amplitude is sufficiently large enough, this circle is split into a number of segments, with bright and dark spots on the screen. The number of segments on the circle gives the *multiple n*, by which frequency of Signal A is to be multiplied, to obtain the frequency of Signal B.

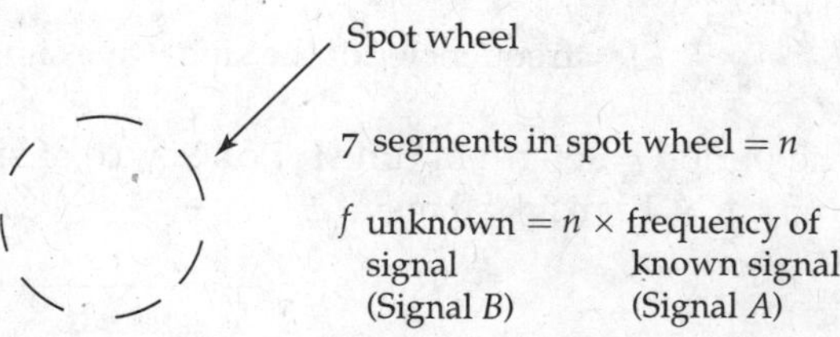

FIG. 1.29 *Frequency f_n determination of unknown signal using spot wheel pattern*

Z-modulation Z-modulation enables the signal trace on the screen to be switched from *brightness* to *darkness*, depending upon the frequency of the applied signal voltage. Any external signal cannot change the *intensity of the beam*, it can only be switched *ON* or *OFF*. In order to measure using the *Spot Wheel Pattern*, the unknown signal source should be applied to the input terminal marked Z-modulation on the backside of the CRO.

EXAMPLE 1.13

A sine wave is observed on a CRO screen as shown in Fig. 1.30. Time base setting is 10 ms/div and voltage setting is 0.5 V/div. Peak-to-peak signal height is 6 cm. Time period for one cycle of the signal is 5 cm. Calculate the peak voltage, *rms* voltage and the frequency of the observed sine waveforms.

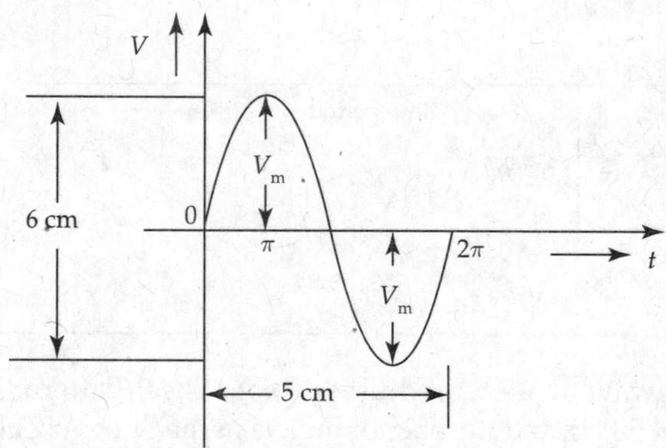

FIG. 1.30

Solution:

$$V_{\text{P-P}} = V_{\text{Peak-to-Peak}} = \text{Peak-to-Peak amplitude of sine wave}$$

$$= \text{Height of signal} \times \text{voltage setting.}$$

Amplitude of sine wave, $V_{\text{P-P}} = 6 \text{ cm} \times 0.5 \text{ V} = 3 \text{ V}$

$$V_{\text{peak}} \text{ or } V_{\text{max}} = \frac{V_{\text{P-P}}}{2} = \frac{3}{2} = 1.5 \text{ V}$$

$$V_{\text{rms}} = \frac{V_{\text{m}}}{\sqrt{2}} = 0.707 \times 1.5 = 1.06 \text{ V.}$$

$$f = \text{frequency } f \text{ of the sine wave signal} = \frac{1}{\text{Time period of the signal } (T)}$$

Time period = length of time for one cycle of signal × time base setting
$T = 5 \text{ cm} \times 10 \text{ ms} = 50 \text{ ms}$

$$f = \frac{1}{T} = \frac{1}{50 \times 10^{-3}} = \frac{1000}{50} = 20 \text{ Hz.}$$

EXAMPLE 1.14

A square wave from a pulse generator is observed on a CRO screen as shown in Fig. 1.31 with voltage control setting 0.5 V/div or 0.5 V/cm and time base setting of 1 μs/div. The time period $T = 0.5$ μs and height of the pulse = 4 cm. Calculate the amplitude and the frequency of the signal.

Solution Signal amplitude = 4 cm × 0.5 V = 2 V

$$\text{Frequency} = \frac{1}{T} = \frac{1}{0.5 \times 10^{-6}} = 2 \text{ MHz.}$$

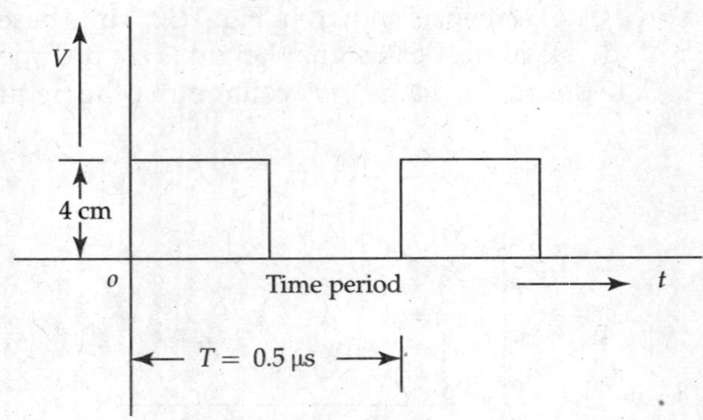

FIG. 1.31

EXAMPLE 1.15

The frequency of a sine wave is measured using Spot Wheel Pattern. If the standard signal source has a frequency of 50 Hz and number of breaks in the Spot Wheel Pattern is 6, calculate the frequency of the unknown signal.

Solution:

$$\frac{f_{\text{unknown}}}{f_{\text{known}}} = \text{Number of breaks in the spot wheel pattern}$$

$$f_{\text{known}} = 50 \text{ Hz}$$

$$\frac{f_{\text{unknown}}}{50} = 6$$

$$f_{\text{unknown}} = 6 \times 50 = 300 \text{ Hz.}$$

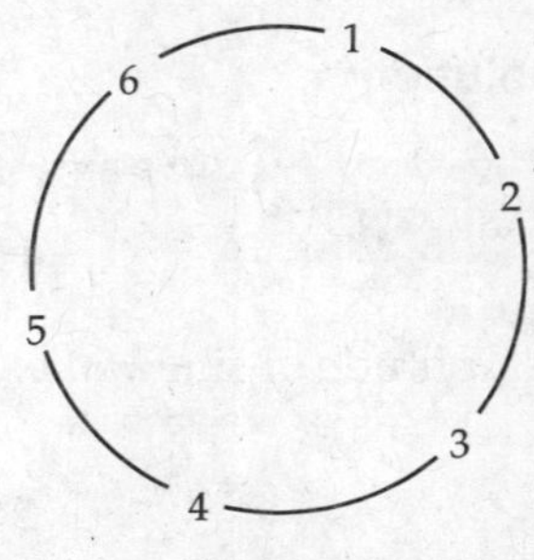

FIG. 1.32

SUMMARY

1. Movement of electrons in electric and magnetic fields in general is studied so that the velocity, KE and accelerations acquired by them can be calculated. The position of electrons and the time taken during movements are analysed with the help of laws of dynamics.
2. The trajectories of electron movements are studied for various conditions of electron velocities in different combinations of locations of electric and magnetic fields in the three-dimensional space. Electron dynamics is analysed for electrons moving in (1) simple electric fields, (2) Simple magnetic fields and (3) perpendicular electric and magnetic fields in varying situations.
3. Such study helps in understanding the investigations of electron movements through semiconductor devices such as Diodes, Transistors, FETs, SCRs, etc. and using them in various electronic circuits.
4. The study of electron dynamics in CRTs is analysed. This helps in the study of CRTs used in CROs. Knowledge of Electron dynamics is useful to underuse the other displays used in Radar, TV, Computer and Cell phone monitors, etc. which helped in civilisation of modern world.
5. CROs are used in Radar displays (Plan Position Indicators) (PPI).
6. In such applications, space constraints necessitates the use of short length tubes, and this is achievable only with magnetic deflection, due to its higher deflection sensitivity in comparison with Electrostatic deflection.
7. In the Laboratory applications for viewing waveforms at shorter distances, screens of sizes of about 10 cms are sufficient. They are realisable with tubes with Electrostatic deflection. For viewing TV images and in Radar displays larger screens are necessary designed using electrostatic deflection, such display tubes will become prohibitively large in dimensions and cumbersome to handle. Therefore the preferred deflection type is the higher deflection sensitivity that magnetic deflection affords and the short neck size that can be achieved. Their compact deflection yokes over short neck tubes in combination with large screens enabes smaller TVs and magnified display of enemy aircraft movement over a given target range in Radar applications that use PPI.

Questions for Practice

1. Derive an expression for electrostatic Deflection Sensitivity in a CRT.
2. Derive an expression for electromagnetic Deflection Sensitivity in a CRT.
3. Draw a neat sketch of CRO and explain the functions of each block.
4. Explain electrostatic focussing and practical focussing systems and explain the diagrams

5. Derive an expression for the trajectory of an electron moving in perpendicular electric and Magnetic Fields.
6. Derive the expression for the radius of the trajectory and period of rotation for an electron in a transverse Magnetic Field.
7. Write short notes on the following:
   (a) Sweep Voltage
   (b) Measurement of voltage using CRO
   (c) Motion of electron in a helical method
8. Mention the expression that is used to predict the parabolic trajectory of electron between two electrodes. Explain the significance of each term in it.
9. Mention the various controls on the front panel of a CRO and briefly explain them.

Multiple Choice Questions

1. The charge of an electron is ____________
   (a) 9.21×10^{-31} kg (b) 1.759×10^{11} C/kg
   (c) 1.6×10^{-19} C (d) 1.6×10^{-19} J
2. Velocity of an electron accelerated by an anode voltage of 100 V in a CRT is ____________
   (a) 59.5×10^5 m/s (b) 100×10^5 m/s
   (c) 60×10^5 m/s (d) 5.95×10^5 m/s
3. The 'time base' voltage ____________
   (a) is applied to the *X*-plates of CRT
   (b) is applied to the *Y*-plates of CRT
   (c) is applied to the control grid of CRT
   (d) provides intensity control
4. For an electron entering the Magnetic Field perpendicular to the direction of field, the locus is ____________
   (a) a circle (b) a helix
   (c) an ellipse (d) a parabola
5. The resultant path for an electron entering the Magnetic Field at an angle θ is ____________
   (a) a circle (b) a helix
   (c) an ellipse (d) a parabola
6. The 'Signal voltage' to be observed ____________
   (a) is applied to the *X*-plates of CRT
   (b) is applied to the *Y*-plates of CRT
   (c) is applied to the control grid of CRT
   (d) provides intensity control

7. To measure the amplitude of a signal wave form, using a CRO, the parameters to be considered are ______________
(a) height of the signal in centimetre and the voltage setting
(b) time base frequency and height of the signal
(c) frequency of the signal and voltage setting
(d) time period of the signal and voltage setting

8. During the measurement of Phase angle between two sine wave signals, if figure on the CRO screen is a circle, the phase difference between the signals is ______________
(a) 180° (b) 90°
(c) 270° (d) 360°

9. Electron emission from the cathode in the CRT in a CRO is due to the following phenomenon ______________
(a) field emission (b) secondary electron emission
(c) thermionic emission (d) radiation

10. The function of the Time base voltage in a CRO is to provide ______________
(a) horizontal movement of electrons between the deflection plates
(b) vertical movement of electrons between the deflection plates
(c) parabolic path
(d) brightness of signals on the screen

Answers to Multiple Choice Questions

1. (c)	2. (a)	3. (a)	4. (a)	5. (b)
6. (c)	7. (a)	8. (a)	9. (b)	10. (a)

Chapter 2

P–N JUNCTION DIODE CHARACTERISTICS

Learning Objectives

To get familiarity of structural details and fundamental concepts of

- Semiconductor materials and P–N junctions.
- The structural details and the working principles of Semiconductor Diode and its applications.

2.1 REVIEW OF SEMICONDUCTOR PHYSICS

- The electronics subject begins from the concepts of behaviour of charge carriers in electron devices and Integrated Circuits (ICs) under influence of electric fields.
- A model of an atom is shown in Fig. 2.1. The aspect of electron motion is analogous to the planetary motion in which the planets rotate round the sun. On similar lines, electrons move in closed stationary orbits around the positive nucleus in an atom.

2.1.1 Electron Configurations of Silicon and Germanium Atoms

- The shell structure and states occupied by electrons depend on the valence of material and its atomic number Z. Silicon and Germanium semiconductor materials are used for the manufacture of semiconductor devices.

Table 2.1

Element	Atomic number (Z)	Configuration
Silicon (Si)	14	$1s^2\ 2s^2\ 2p^6\ 3s^2\ 3p^2$
Germanium (Ge)	32	$1s^2\ 2s^2\ 2p^6\ 3s^2\ 3p^6\ 3d^{10}\ 4s^2\ 4p^2$

The distribution of electrons in the various orbits for Silicon and Germanium atoms is shown in Table 2.1 and in Figs. 2.1 and 2.3.

Electron configuration of Silicon atom (Fig. 2.1)

- The atomic number of Silicon atom is $Z = 14$. It contains 14 positive charges in the nucleus and 14 electrons that move about the nucleus in closed stationary orbits. The orbits are assumed to be concentric circles. Thus, each atom is electrically neutral (Zero charge for the atom as a whole). Hence, the Silicon material is an 'Electrically Neutral material'.
- The planetary model for the atom is considered only from the classical model. Each 'Silicon atom' has its electrons arranged in groups of energy levels or shells as the following:

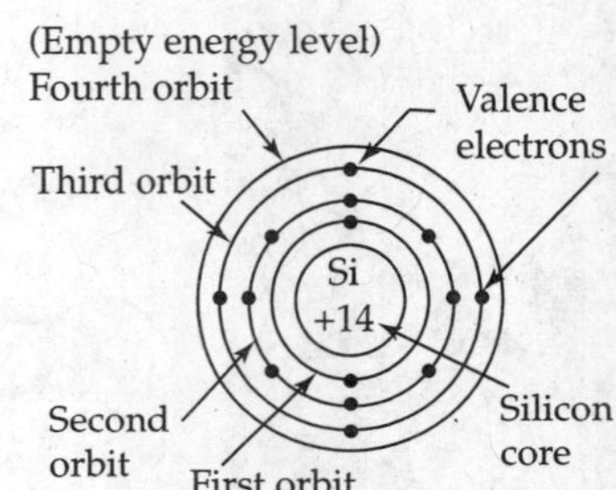

FIG. 2.1 *Electron configuration of silicon atom*

1. First orbit, the inner most energy level has 2 electrons (completely filled).
2. Second orbit has 8 electrons (completely filled).
3. Third orbit has the balance of 4 electrons (partially filled).
4. Energy levels starting from the fourth level are empty energy levels.
5. This last partially filled shell (third orbit) is called valence shell.
6. The 4 electrons in the third orbit (shell) are known as valence electrons.
7. Valence electrons are responsible for the chemical and electrical properties of the material.
8. Electrons extracted from valence shell and not subject to force of attraction of nucleus on them are called free electrons.

Silicon atom representation as a tetravalent material is shown in Fig. 2.2 as a basis to understand the concept of covalent bond formation etc. Silicon semiconductor, as a 'Tetravalent' material, has 'four valence electrons'. The force of attraction between the nucleus (core) and the electron inside the atom is given by

$$F = \frac{q^2}{4\pi\varepsilon_0 r^2} \text{ Newtons,} \tag{2.1}$$

where electronic charge q is in coulombs, r is the separation distance between electrons and nucleus (in an atom) in metres, the force F is in Newtons, ε_0 is the permittivity of free space in farads/metre, and permittivity of the free space $\varepsilon_0 = 8.849 \times 10^{-12}$ farads/metre.

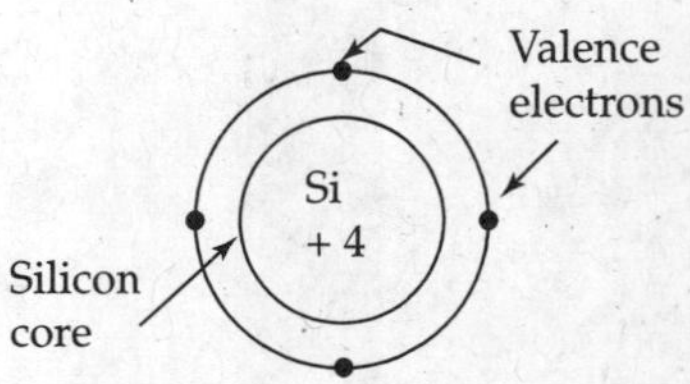

FIG. 2.2 *Representation of silicon atom with its valence electrons*

This force of attraction F between the nucleus and the electron is counter balanced by

$$\frac{mv^2}{r}\text{(centripetal force)},$$

where m is the electronic mass, $m = 9.109 \times 10^{-31}$ kg, v is the speed of the electron in the orbit, acceleleration $a = v^2/r$ and is directed towards the nucleus.

Then according to Newton's second law of motion

$$F = \frac{q^2}{4\pi\varepsilon_0 r^2} = \frac{mv^2}{r} \tag{2.2}$$

$$\therefore \quad \frac{1}{2}\cdot\frac{mv^2}{r} = \frac{q^2}{8\pi\varepsilon_0 r^2}$$

$$\text{Kinetic energy (KE)} = \frac{1}{2}mv^2 = \frac{q^2}{8\pi\varepsilon_0 r} \tag{2.2A}$$

The potential energy PE of the electron at a distance 'r' from the nucleus $= -\dfrac{q^2}{4\pi\varepsilon_0 r}$.

According to the conversation of energy, energy associated with the electrons

$$W = \text{Kinetic energy} + \text{Potential energy:}$$

$$\therefore \quad W = \frac{1}{2}mv^2 - \frac{q^2}{4\pi\varepsilon_0 r}, \tag{2.3}$$

where the energy W is in joules.

Substituting the value $\dfrac{1}{2}mv^2 = \dfrac{q^2}{8\pi\varepsilon_0 r}$ from Eq. (2.2A) into Eq. (2.3), we get

$$\text{Total energy of the electrons} \quad W = \frac{q^2}{8\pi\varepsilon_0 r} - \frac{q^2}{4\pi\varepsilon_0 r} = -\frac{q^2}{8\pi\varepsilon_0 r}. \tag{2.4}$$

Equation (2.4) shows the relation between the radius r (distance of electron in the circular orbit from the nucleus) and the energy W of the electrons. It also shows that the energy of the electron becomes less (i.e., more negative) as it approaches closer to the nucleus. The relation is given by Eq. (2.4A):

$$\text{Energy of an electron in the } n\text{th orbit} \quad W_n - \frac{13.6}{n^2}\ \text{eV}, \tag{2.4A}$$

where $n = 1, 2, 3$ and so on.

Electronic configuration of a Germanium atom (Fig. 2.3)

Germanium semiconductor atom has 'atomic number' Z = 32. It has 32 positive charges in the nucleus and 32 electrons in various shells containing 2, 8, 18 and 4 electrons. Germanium atom is electrically neutral. Germanium semiconductor as a whole is electrically neutral.

- First, second and third orbits are completely filled.
- Fourth orbit (shell) is partially filled.
- Energy levels from fifth orbit onwards are empty energy levels.

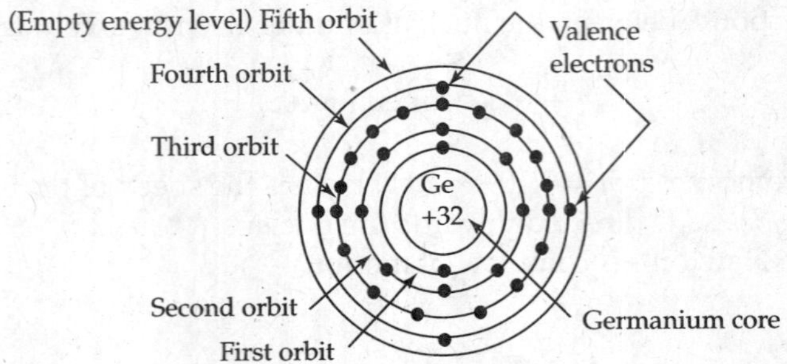

FIG. 2.3 *Electron configuration of germanium atom*

Germanium atom representation is shown in Fig. 2.4. It is a basis to know the formation of covalent bonds and so on. Germanium is also considered as a 'tetravalent' material, as it has 4 valence electrons in its outer incomplete shell. Thus, Silicon and Germanium materials are referred as tetravalent materials with similar electrical and chemical properties.

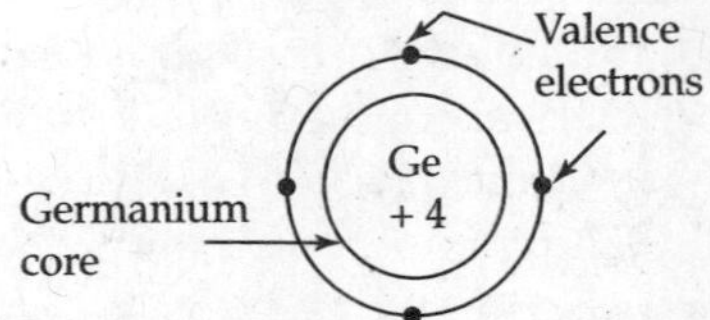

FIG. 2.4 *Representation of germanium atom with four valence electrons*

2.1.2 Energy-band Concepts of Materials

- The electron energy levels for a single free atom in a gaseous medium are discrete, since the atoms are sufficiently far apart. So the energy levels of individual atoms are not perturbed.
- The proximity of neighbouring atoms in solid media such as crystals does not appreciably affect the energy levels of inner shell electrons. But, groups of energy levels of outer shell electrons are changed due to the influence of electrons in the neighbouring atoms. They allow sharing of electrons among them to form covalent bonds between neighbouring atoms in the process of getting on to stable '8-electron configuration' in Silicon and Germanium semiconductors.

Sharing of outer shell electrons to form covalent bonds is shown in Fig. 2.5.

- *Valence band* The coupling between the outer shell electrons of the atoms results in a group or a band of closely spaced energy levels or states instead of the widely spaced energy levels of the isolated atoms. Because of the coupling between atoms in the crystals (As the inter-atomic distance is quite small in solid materials.) completely filled and partially filled energy levels are merged into an 'energy band' known as *Valence Band*.
- Top most energy level of Valence Band is E_V.
- Merging of empty energy levels in atoms form *Conduction Band* (top energy band).
- Lower most energy level of Conduction Band is E_C.
- Region between Conduction Band and Valence Band is known as *forbidden band gap* E_G, or *band gap* equal to $(E_C - E_V)$. It decreases with temperature.

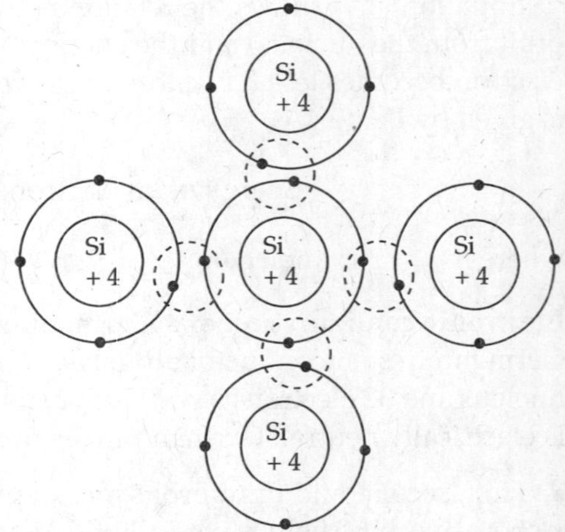

FIG. 2.5 *Covalent bonds about silicon atoms*

- The magnitude of the band-gap energy E_G predicts the type of the materials, such as conductors, semiconductors and insulators, which is discussed later.
- Energy-band diagrams (EBD) show the energy of electrons (in electron volts) associated with the energy levels on the y axis and the momentum (P) on the x axis. Energy of electrons is measured in eV (electron volts).
- The unit of *electron volt* (eV) is the energy acquired by an electron while falling through a potential difference of 1 V.

According to *quantum-mechanical theory,* when the energy band has all filled energy levels; electron there cannot contribute to electrical conduction. There is no open energy level to which they can move after absorbing any energy from the applied electric field. Therefore they do not absorb energy and do not become conduction electrons. Only the band containing the unfilled or empty energy levels is the Conduction band, to which electrons enter to contribute electrical conduction.

Conductivity of a pure semiconductor at 'Absolute-Zero temperature' is zero, since lower Valence Band is filled and there are no electrons in the upper Conduction Band.

At the ambient temperature, some electrons may acquire sufficient energy equal to or greater than the forbidden band-gap energy E_G and they will move to energy levels in the upper band. These electrons will be in an incompletely filled band and they can contribute to electrical conduction. While the *electrons* move to the *Conduction Band,* they leave *Holes in the Valence Band* (Holes were formed due to the formation of Hole–electron pairs during the process of breakage of covalent bands in Valence Band). Hole will have positive charge. Formation of *Hole–electron pairs* is shown in Fig. 2.6.

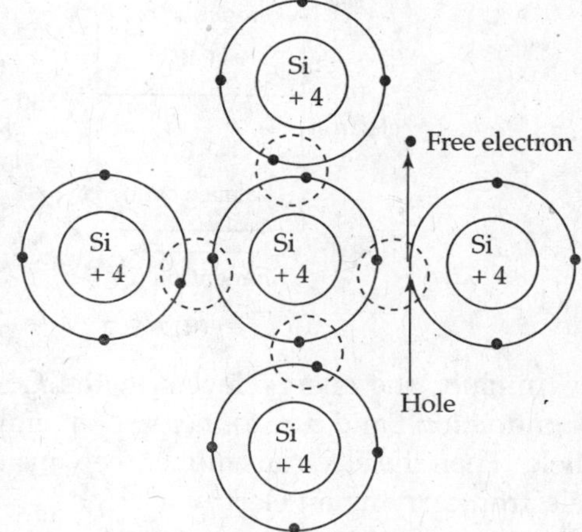

FIG. 2.6 *Formation of hole–electron pair*

Conductivity of 'intrinsic semiconductor' is due to the *Hole–electron pairs* formed during broken covalent bonds or due to supply of energy to free electrons to cross the forbidden band gap to enter the Conduction Band.

Resistivity of semiconductor material can be expressed as

$$\rho = Ae^{\frac{\Delta E_G}{kT}} \times 10^{-2}\ \Omega\text{-m}, \tag{2.5}$$

where A is a coefficient that varies slightly with temperature, and ρ is the resistivity of the semiconductor material. It is a function of temperature T and forbidden band-gap energy ΔE_G.

$$\bar{k} = \text{Boltzman constant} = 1.381\times10^{-23}\ \text{J/°K}$$

$$k = \frac{\bar{k}}{e}\ \text{Boltzman constant in electron volts/°K}$$

$$\therefore\ k = \frac{\bar{k}}{e} = \frac{1.381\times10^{-23}}{1.6\times10^{-19}} = 8.6\times10^{-5}\ \text{eV/°K}$$

Resistance across a standard mass and shape of a material at a given temperature is called the *resistivity* of the material. The reciprocal of *resistivity* is *conductivity* (σ).

2.2 ENERGY-BAND DIAGRAMS OF SEMICONDUCTOR MATERIALS

2.2.1 Classification of Materials

When voltages are applied, materials offer different values of electrical resistances to the passage of currents through them. On the basis of electrical resistances, materials are classified as *conductors, semiconductors and insulators.*

In solids, available energy states for the electrons form 'bands of energy levels' instead of discrete energy levels in atoms.

Conductors: Materials with adjacent or over-lapped conduction and Valence Bands with zero forbidden band-gap energy ($E_G = 0$) are known as *conductors.* EBD for a conductor material is shown in Fig. 2.7.

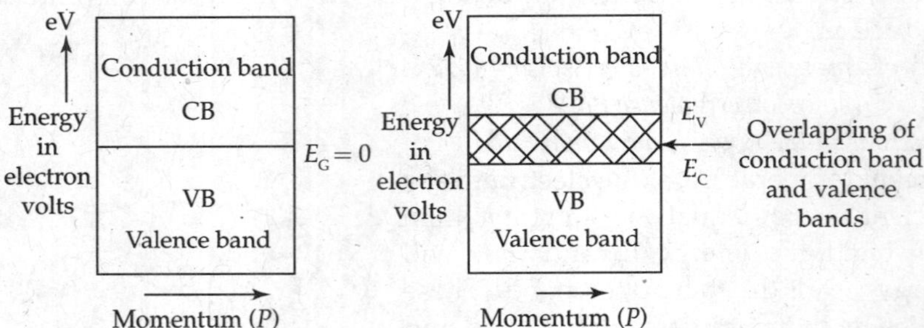

FIG. 2.7 *Energy-band diagrams for conductors*

Initially, the energy levels in the Conduction Band are empty. But, electrons enter the Conduction Band due to increase in temperature or energy acquired from an applied electric field. Then the electrons move freely inside the Conduction Band as charge carriers with each electron carrying an electron charge $q_n = 1.6 \times 10^{-19}$ C. So in a conductor, electric current can flow freely. Most familiar conductors are metals such as gold, silver and copper.

Semiconductors: Materials with small forbidden band-gap energy (E_G), around 1 eV, are called semiconductors. Silicon, Germanium and gallium arsenide are semiconductor materials. They are also known as *intrinsic or pure semiconductor* materials.

Semiconductor materials have some of the following features:

- Typical value of resistivity is of the order 0.6 Ω-m at the room temperature.
- The material has negative temperature coefficient of resistance. Resistance of the semiconductor material decreases with increasing values of temperatures.
- The addition or doping of trivalent or pentavalent materials to the intrinsic semiconductors (Silicon or Germanium) modulates the electrical conductivity σ of the semiconductor materials. This is the important feature for the fabrication of P- and N-type semiconductors, which are the backbone materials for semiconductor devices in electronic engineering technology.
- At 0°K, $E_{G0} = 1.12$ eV for a Silicon semiconductor material.
- For Germanium semiconductor, $E_{G0} = 0.785$ eV.
- $E_{G0} = 1.41$ eV for gallium arsenide.
- At room temperature (300°K), $E_G = 1.1$ eV for Silicon semiconductor.
- $E_G = 0.72$ eV for Germanium semiconductor.

- Forbidden band-gap energy

$$E_G = (E_C - E_V) \tag{2.6}$$

- At very low temperatures, the Conduction Band is practically empty. When the temperature is increased, the electrons in the top of Valence Band acquire sufficient thermal energy and move into the Conduction Band.

E_G is forbidden band-gap energy, E_C is the energy of the lower most energy level of Conduction Band and E_V is the energy of the top most energy level of Valence Band.

Silicon semiconductor has the forbidden band-gap energy $E_G = 1.12$ eV (Fig. 2.8). EBD for the Germanium semiconductor material is shown in Fig. 2.9. It has band-gap energy $E_G = 0.72$ eV.

- Silicon has wider forbidden band-gap energy compared with Germanium semiconductor material. This suggests that Silicon devices work up to higher temperatures with stable operation. Silicon devices are preferable for military and tropical country applications.
- Because of smaller forbidden band-gap energy, Germanium devices are limited to lower temperature applications.
- Typical band-gap energy in semiconductors is less than 2 eV.

Insulators: The materials with large forbidden band-gap energy $E_G > 6$ eV do not support conduction at all. Large forbidden band-gap energy between the Valence Band and the Conduction Band shown in Fig. 2.10 suggests that no electron can reach the Conduction Band. Such materials are known as insulators.

Insulators practically have no free electrons to act as charge carriers to support electrical conduction. Non-metallic solids such as glass, porcelain and mica behave as *insulators*. Their resistivity is very high, while conductivity is very low.

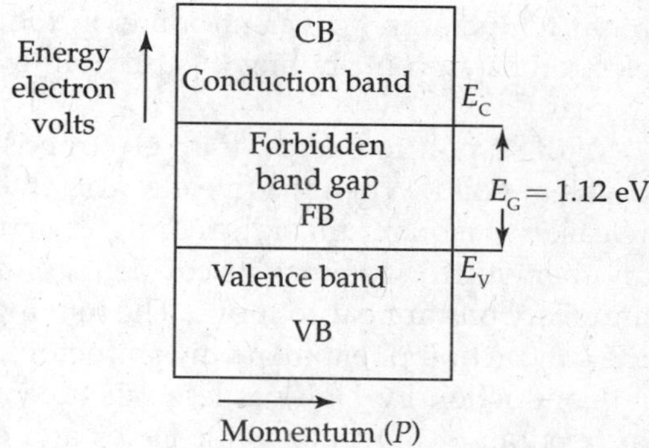

FIG. 2.8 *Energy-band diagram for silicon semiconductor*

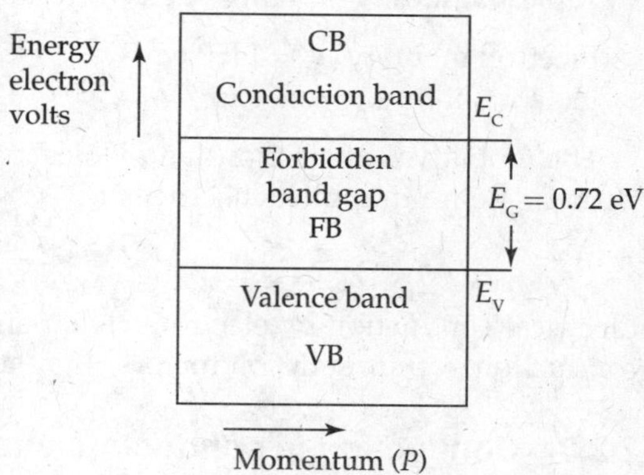

FIG. 2.9 *Energy-band diagram for germanium semiconductor*

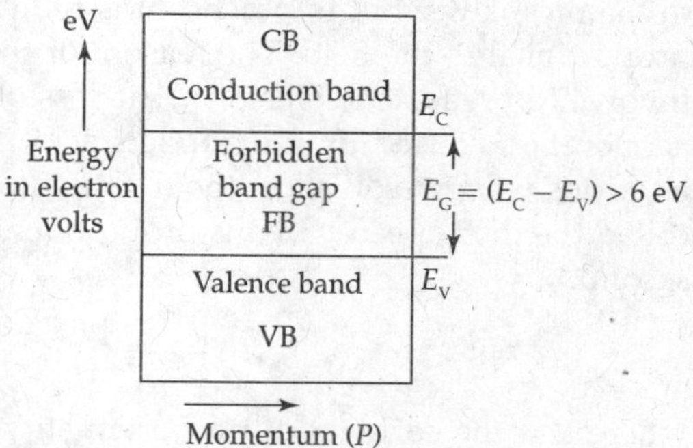

FIG. 2.10 *Energy-band diagram for insulator materials*

2.2.2 Conduction (Inverse of Resistance) in Intrinsic Semiconductors

Purest semiconductor is known as *intrinsic semiconductor*. At 0°K, semiconductor behaves like an insulator, because energies of the order of E_G cannot be acquired from an electric field. At room temperature, covalent bonds in the semiconductor may be broken into a few Hole–electron pairs, contributing to current flow through the material allowing the conductivity to increase.

With respect to energy, if an electron is given additional energy, it breaks away from its covalent bond. When the free electron enters a Hole in a Valence Band, this excess energy is released as a quantum of heat or light. In turn this quantum of energy may be reabsorbed by another electron to break its covalent bond and create a new Hole–electron pair. Thus Holes and electrons appear to move. The moving charge carries form current. Ohm's law governs the conduction phenomena in conductors and resistors.

Conduction by Holes is less when compared to that of electrons because of differences in freedom of movements for Holes and electrons, based on their mobility μ. The mobility of electrons μ_n is greater than the mobility of holes μ_p because of the differences in relative masses of electrons and Holes.

Typical values of mobility of electrons and Holes in semiconductors

- Electron mobility $\mu_n = 1450$ cm²/V-s ($1450 \times 10^{-4} = 0.145$ m²/V-s)
- Hole mobility $\mu_p = 550$ cm²/V-s ($550 \times 10^{-4} = 0.055$ m²/V-s) (2.7)

The mobility μ of electrons and Holes is defined as the velocity acquired by these charged particles per unit-applied electric field.

$$\mu = \frac{v}{E} \text{ m}^2/\text{V-sec.} \tag{2.8}$$

Electrical conduction by electron–Hole pairs generated by thermal energy is called intrinsic conduction in pure semiconductors, of either Silicon or Germanium.

2.2.3 Conduction in conductors and semiconductors

Mobility μ: In good conductors like metals, free electrons exist in abundance. They are supposed to be accelerated under the influence of electric or magnetic field as per ballistic (dynamics) laws. But in practice it is found that the electrons move with a constant velocity proportional to the field. The reason for this is the random nature of the electron movement involved in repeated collisions. The loss of energy during collisions is supplemented due to acceleration caused by the applied field E. Thus it is observed that the random motion of electrons when resolved in the direction of the field, the electrons acquire a constant speed called the drift speed v that is proportional to the field E (V /m) and velocity v is in metres/second.

$$\text{Thus,} \quad v \propto E$$

$$\therefore \quad v = \mu E \tag{2.9}$$

where μ is the constant of proportionality. μ is called as mobility. It is measured as m²/V-s. Mobility of electrons and Holes due to the influence of electric field is given in Eq. (2.10). Because of the lighter mass of electrons, electrons have large values of mobility μ_n compared to Hole mobility μ_p.

$$\mu = \frac{v}{E} = \frac{\left(\frac{\text{m}}{\text{s}}\right)}{\left(\frac{\text{V}}{\text{m}}\right)} = \frac{\text{m}^2}{\text{V-sec}}. \tag{2.10}$$

For a given excitation energy to electrons (due to applied field strength), electrons move faster in Germanium semiconductor when compared to Silicon semiconductor, because of small forbidden band-gap energy in Germanium semiconductors. So Germanium semiconductor devices find their use in high-frequency applications.

2.2.4 Current Density in a Conducting Medium

Currents in metals are due to the movement of charge carriers 'electrons'.

$$\text{Current density} \quad J = \frac{I}{A}\ \text{A/m}^2, \tag{2.11}$$

where I is the current in Amperes and A is the cross-sectional area of conducting medium in metre^2. Describing current density J as current per unit area has the advantage, since the dimensions of the conducting medium are not directly involved. Relation between current density and charge density ρ is described in the following:

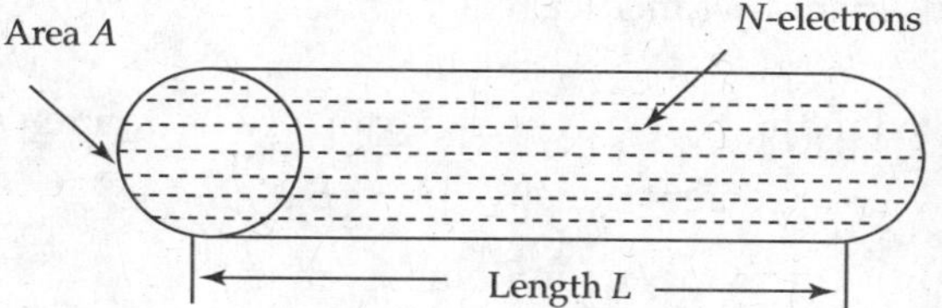

FIG. 2.11 *Electrons in a conducting medium*

Current density: Current I (Amperes) through a conductor by definition is *Charge* (in Coulombs)/*Time* (in seconds). Current is due to the movement of charges through a conducting medium in a given time. If, 1 C of charge moves through a conducting medium in 1 s, the resulting *current* is 1 A.

$\frac{1}{1.6\times10^{-19}} = 6.25\times10^{18}$ electrons carry 1 Coulomb of charge. So the movement of 6.25×10^{18} electrons for 1 s contributes to 1 A of current in a conductor.

$$\text{i.e., Current} \quad I = \frac{\text{charge}}{\text{time}} = \frac{qN}{T}, \tag{2.12}$$

where q is the charge of an electron and N is the number of electrons in a given volume. If the charge passes through a distance L (metres) in time T (seconds), through a conducting medium, then the velocity v with which the electrons move is L/T.

$$\text{i.e., Velocity} \quad v = \frac{L}{T} \quad \text{or} \quad \frac{1}{T} = \frac{v}{L}$$

$$\therefore \quad T = \frac{L}{v}. \tag{2.13}$$

Substituting the value of T from Eq. (2.13) in Eq. (2.12), we get

$$I = \frac{Nqv}{L}, \tag{2.14}$$

$$\therefore \text{ Current density} \quad J = \frac{I}{A} = \frac{Nqv}{LA}\ \text{A/m}^2 \tag{2.15}$$

$$\text{i.e.,}\quad J = nqv, \tag{2.16}$$

where $n = N/AL$ is the concentration of electrons that is the number of electrons per unit volume.

Using $v = \mu E$ in Eq. (2.16), we get

$$\text{Current density}\quad J = nq\mu E, \tag{2.17}$$

where μ is the mobility of charge carriers.

Current density J_p due to the movement of Holes $= pq\mu_p E$.

Drift current I through an area A due to movement of holes $= Apq\mu_p E$. (2.17A)

2.2.5 Conductivity and Resistivity of Semiconductor Materials

The value of conductivity of a material gives us an estimate of the extent to which a material supports the flow of current through it. Electrical conductivity depends upon the number of electrons available in the conduction process. The concept of conductivity is useful in many engineering applications including medical electronics.

$$J = nq\mu E$$

Equation (2.17) derived in the previous section can also be written as

$$\text{Current density}\quad J = \sigma E\ \text{A/m}^2, \tag{2.18}$$

$$\text{where } \sigma = nq\mu\ (\Omega)^{-1} - (\text{m})^{-1}\ \text{or mhos/m or Siemens/m} \tag{2.19}$$

is called as conductivity of the material.

$$\therefore\quad \sigma = \frac{J}{E}\ \text{Siemens/m.}$$

Thus, electrical conductivity of a material is defined as the ratio of current density J and electric field intensity E.

Conductivity of semiconductor materials increases with temperature, as an increase in temperature causes increase in conduction current. This is due to increase in broken covalent bonds that result in more charge carriers for current flow. So more electrons from Valence Band jump to Conduction Band with increase in temperature. The conductivity of semiconductors varies completely in the opposite way to that of metals.

Here it is found that current density (J) and field strength (E) are proportional to each other with σ as the constant of proportionality: $J \propto I$ and $E \propto v$.

So σ has the dimensions of Siemens/m as shown below:

$$\sigma = \frac{J}{E} = \frac{\left(\frac{I}{A}\right)}{\left(\frac{V}{d}\right)} = \frac{\left(\frac{\text{A}}{\text{m}^2}\right)}{\left(\frac{\text{V}}{\text{m}}\right)} = \frac{\left(\frac{\text{A}}{\text{V}}\right)}{\text{m}}\ \text{Siemens/m.}$$

As already explained, semiconductors contain two types of mobile charge carriers, electrons and Holes. In semiconductors, the conductivity depends upon the concentrations and mobility of both electrons and Holes (Fig. 2.11).

$$\text{Conductivity of semiconductors}\quad \sigma = q(n\mu_n + p\mu_p)\ \text{Siemens/m} \tag{2.20}$$

where n is the concentration (number) of electrons, p is the concentration (number) of Holes, μ_n is the mobility of electrons and μ_p = mobility of Holes.

In an intrinsic semiconductor $n = p = n_i$

$\therefore$ Conductivity of intrinsic semiconductor $= \sigma_i = qn_i[\mu_n + \mu_p]$ mhos/m (2.21)

$$\text{Resistivity} \quad \rho = \frac{1}{\sigma} = \frac{1}{q(n\mu_n + p\mu_p)}\ \Omega\text{-m} \tag{2.22}$$

$$\text{Resistivity of intrinsic semiconductor} = \rho_i = \frac{1}{qn_i[\mu_n + \mu_p]}\ \Omega\text{-m.} \tag{2.22A}$$

If the values for the mobility and concentrations of electrons and Holes are known, the conductivity of the materials can be estimated.

EXAMPLE 2.1

Calculate the values of conductivity and resistivity of intrinsic Silicon semiconductor with Hole mobility $\mu_p = 0.055$ m²/V-s and $\mu_n = 0.145$ m²/V-s. Assume that the number of electrons in the intrinsic semiconductor to be $1.5625 \times 10^{16}/\text{m}^3$.

Solution: Conductivity $\sigma_i = qn_i[\mu_n + \mu_p]$ Siemens/m

For an intrinsic semiconductor

$$n = p = n_i = 1.5625 \times 10^{16}\ /\text{m}^3$$

$$\therefore \quad \sigma_i = 1.6 \times 10^{-19} \times 1.5625 \times 10^{16} [0.145 + 0.055]$$

$$= 2.5 \times 10^{-3} \times 0.2 = 5.0 \times 10^{-3}\ \text{mhos/m}$$

$$\text{Resistivity} \quad \rho_i = \frac{1}{\sigma_i} = \frac{1}{5.0 \times 10^{-3}} = \frac{10^3}{5.0} = 200\ \Omega\text{-m.}$$

EXAMPLE 2.2

Calculate the conductivity of copper having density 8.9 g/cm³ and mobility 34.8 cm²/V-s. Atomic weight of copper is 63.57 while it has one valence electron per atom. Assume that the value of mass $M = 1.66 \times 10^{-27}$ kg [April/May 2007, set-I and May–June 2006].

Solution: Conductivity $\sigma = n_n q \cdot \mu_n$, where n_n is the number of electrons or charges per unit volume

$$\mu_n = \text{mobility of electrons} = 34.8\ \text{cm}^2/\text{V-sec}$$

$$q = \text{charge} = 1.6 \times 10^{-19}\ \text{C}$$

Number of copper atoms/unit volume

$$n = \frac{\text{density} \times \text{Avogadro's number}}{\text{atomic weight} \times \text{atomic mass unit}}$$

$$= \frac{8.9 \times 6.023 \times 10^{23}}{63.57 \times 1.66 \times 10^{-27} \times 10^3} = 0.5 \times 10^{47}.$$

(mass M is multiplied by 10^3 to convert its units into kg). Each atom of copper has one electron per atom. Number of electrons in copper $n_n = n \times$ number of valence electrons

$$\therefore \quad n_n = 0.5 \times 10^{47}\ \text{electrons/unit volume}$$

Conductivity $\sigma = n_n \cdot q \cdot \mu_n$ mhos/cm

$\sigma = 0.5 \times 10^{47} \times 1.6 \times 10^{-19} \times 34.8 = 27.84 \times 10^{28}$

Conductivity of copper material $\sigma = 2.78 \times 10^{29}$ mhos/cm.

2.2.6 Conduction in Semiconductors

At room temperature of 300°K, it requires an energy of $E_G = 1.12$ eV to break covalent bonds in Silicon material and $E_G = 0.7$ eV to break the covalent bonds in Germanium material and to produce some 'electron–Hole pairs'.

Even at room temperature, a few of the covalent bonds will be broken, leading to equal number of electrons and Holes in Conduction Band and Valence Band, respectively. Electrons in the Conduction Band and Holes in the Valence Band, in an intrinsic semiconductor, are shown in Fig. 2.12. Small dashes represent free or conduction electrons. Holes are represented by circles in valence band.

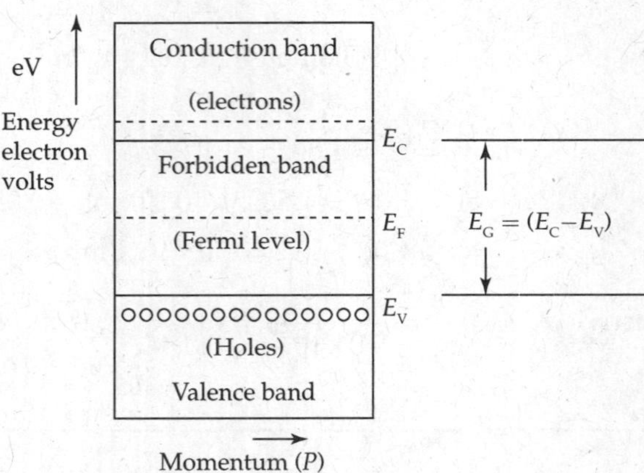

FIG. 2.12 *Energy-band diagram for an intrinsic semiconductor*

2.2.7 Fermi Level in Energy-Band Diagrams

Fermi level provides the information about electrical conductivity of materials. This level is *real* in conductors but *virtual* in semiconductors, where an energy gap exists between conduction and Valence Bands. Thus, Fermi level is a statistical quantity useful in determining the behaviour of materials in general.

In intrinsic semiconductors such as Germanium and Silicon, Fermi level exists approximately midway between conduction and Valence Bands. For Silicon semiconductor, Fermi level is 0.0073 eV below the middle of the band gap, at room temperature.

$$\text{Fermi level} \quad E_F \cong \frac{E_C + E_V}{2}. \tag{2.23}$$

Trivalent materials such as boron are added to intrinsic semiconductor (Silicon or Germanium) to form *P-type semiconductor*. The process of adding trivalent materials alters the conductivity of semiconductors. It is known as *Doping*. In P-type semiconductor, doping concentration introduces *Acceptor Energy Level* near the Valence Band and *Fermi Energy Level* shifts towards the Valence Band.

Pentavalent material such as phosphorus is added to intrinsic semiconductor (Silicon or Germanium) to produce *N-type semiconductor*. Adding pentavalent materials alters the conductivity of semiconductor materials. It is known as *Doping*.

In N-type semiconductor, doping concentrations introduce *Donor Energy Level* near the Conduction Band and *Fermi Energy Level* shifts towards the Conduction Band. Shifts in the Fermi energy levels, shown in various EBDs, clearly shows the movement of charge carriers in various semiconductor devices.

2.3 P- AND N-TYPE SEMICONDUCTORS

2.3.1 Extrinsic Semiconductors (Doped or Impure Semiconductors)

- If the symmetry of an intrinsic semiconductor is disturbed, by adding pentavalent (donor) or trivalent (acceptor) impurities to the pure semiconductor, *conduction* increases abnormally and the *conductivity* depends more on the doping atoms as explained below.
- Extra energy levels will be introduced in the forbidden band gap as a result of doping impure materials to the intrinsic semiconductors.
- Resulting N- or P-type semiconductors are known as *extrinsic semiconductor* materials.
- Extrinsic semiconductors can be manufactured with desirable characteristics with the addition of controlled magnitudes of doping levels (or impurities).

2.3.2 N-type Semiconductor (Donor-type Doping)

Adding pentavalent *phosphorus* material atoms into pure semiconductors (Silicon or Germanium) is known as *doping*. Added atoms replace some of the atoms of intrinsic material, which result in structures as shown in Fig. 2.13. Typical concentrations of doping are at the rate of one impure atom for every 10^5 or 10^6 atoms of pure semiconductor materials. But the actual doping concentrations depend upon conduction current requirements of materials to fabricate active devices.

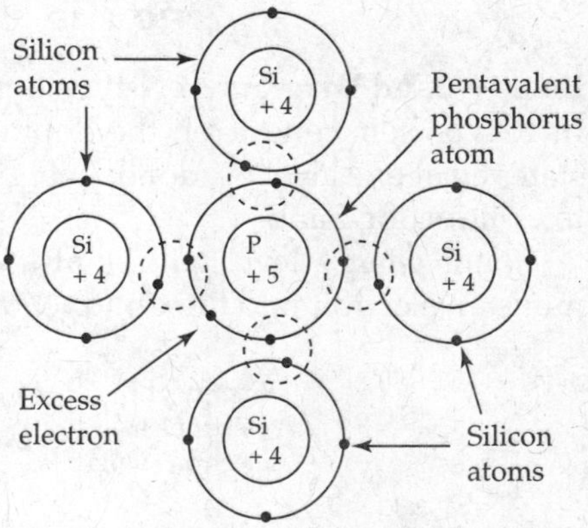

FIG. 2.13 *N-type semiconductor (donor type doping)*

- The conductivity of the impure semiconductor either can be increased or can be decreased by suitable doping levels.
- Excess electrons about phosphorus atoms contribute an 'energy level' nearer to the Conduction Band. Such energy level is known as donor energy level. The conduction in N-type semiconductor due to these excess electrons is known as 'excess conduction'.

Structure of N-type Semiconductor and Charge Profile

- In N-type semiconductors, majority carrier electrons cover the immobile positive ions and maintain charge neutrality of the material (Fig. 2.13).
- Electrons in N-type material are equal to the sum of the (1) electrons produced due to doping and (2) electrons that resulted in the process of broken covalent bonds at ambient temperature.

- Electrons will be larger in number.
- Electrons are the *majority carriers* in N-type semiconductors.
- Small number of *Holes* resulted during the process of broken covalent bonds become the *minority carriers* in the N-type of semiconductor materials.

Charge profile of N-type semiconductor (typical magnitudes of majority carrier electrons and minority carrier Holes) is shown in Fig. 2.14.

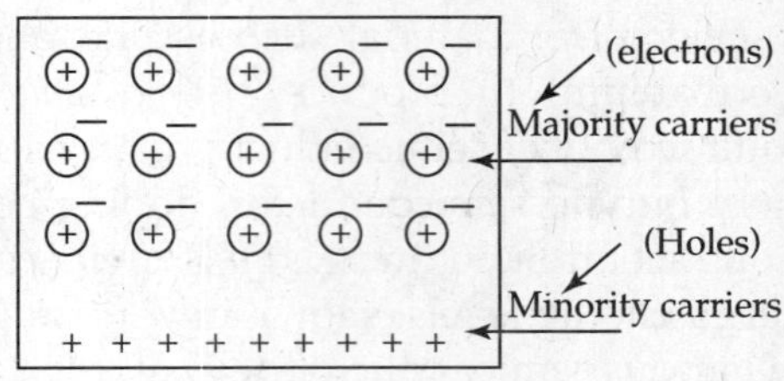

FIG. 2.14 *Mobile and immobile charges in N-type semiconductor*

Each electron carries a charge of 1.6×10^{-19} C. 6.25×10^{18} electrons contribute to 1 C of charge. So 6.25×10^{18} moving electrons contribute to 1 A of current in the material. Movement of 6.25×10^{17} electrons contributes to 100 mA of current in a conducting medium. These typical figures are shown just to provide the concept of the underlying mechanism of magnitudes of charges involved in the flow of currents.

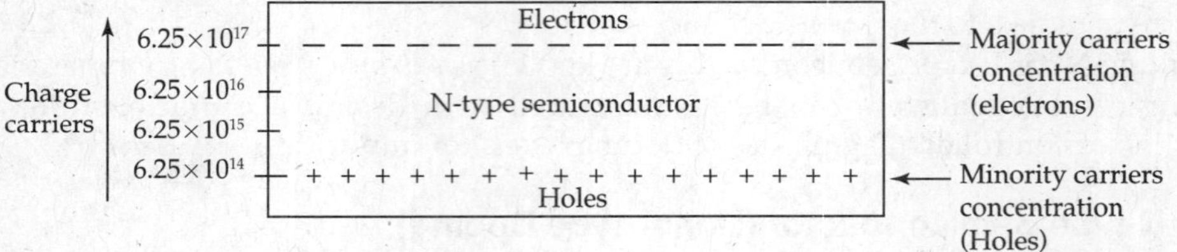

FIG. 2.15 *Charge profile of N-type semiconductor*

Energy-Band Diagram of N-type Semiconductor

In N-type semiconductor, the doping pentavalent atoms contribute a discrete filled energy states all at one level E_D (donor energy level) in the hitherto forbidden band-gap energy nearer to Conduction Band.

Donor energy level will be at a distance of 0.01–0.05 eV to Conduction Band. Thus to move an electron into the Conduction Band, a very small energy of the order of 0.01–0.05 eV

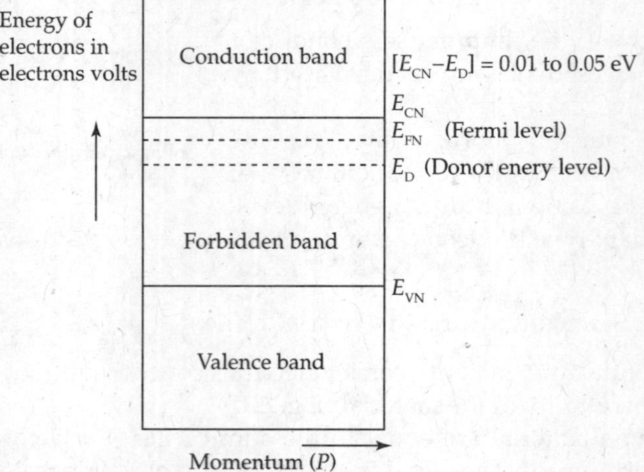

FIG. 2.16 *Energy-band diagram of N-type semiconductor*

(fraction of an electron volt) is sufficient to pull the electrons from the donor energy level into the Conduction Band, resulting in a highly increased conductivity. Hole production in the Valence Band decreases due to the lesser probability of electrons from the Valence Band to jump into the Conduction Band creating Holes.

As the probability of finding electrons at donor energy level nearer the Conduction Band increases, the Fermi energy level moves closer to the Conduction Band as shown in the EBD of Fig. 2.16.

The Fermi level moves up closer to E_{CN} and has the value of

$$E_{FN} = E_C - kT \ln\left(\frac{N_C}{N_D}\right) \tag{2.24}$$

$$\text{where } N_C = 4.82\times10^{21}\left[\frac{m_n}{m}\right]^{3/2} \times T^{3/2} \tag{2.25}$$

$$\text{and } N_D = N_C \cdot \left[e^{-\frac{(E_C - E_F)}{kT}}\right] \tag{2.26}$$

where m_n is the effective mass for electrons and m is the rest mass. N_C is a constant for the Conduction Band that depends on effective masses of Holes and electrons and temperature as well. Thus, electrons become the majority carriers and Holes are the minority carriers in N-type semiconductor.

2.3.3 P-type Semiconductor

Similarly, if trivalent impurity atoms such as *Boron* material atoms are introduced into the *Silicon* semiconductor (intrinsic material) the three peripheral electrons of boron atom form three covalent bonds with the neighbouring Silicon atoms. The fourth valence electron of neighbouring Silicon atom cannot form a covalent bond structure leaving a natural vacancy (Hole) as shown in Fig. 2.17. Compared to the environment it has one electron less to form a covalent bond thus leaving a relative unit positive charge called a Hole. These Holes can knock off electrons from the neighbouring atoms with as small energy as 0.05 eV and create Holes at a new position from where an electron is accepted to form a covalent bond structure. Thus the Hole is transferred from one atom to the other atom and thus virtually giving mobility to Holes in P-type semiconductors. It is to be cautioned that the Hole movement is indirect and conceptual. Typical doping atoms are boron, aluminum, gallium and iridium.

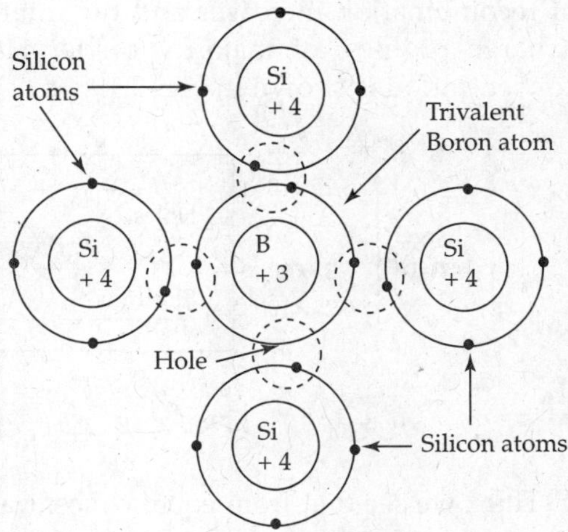

FIG. 2.17 *P-type semiconductor (acceptor type doping)*

Structure of P-material and Charge Profile

As has been observed in the previous discussion, a P-material consists of mobile positive Holes, one for each doping atom (acceptor atom). However, the *charge neutrality of P-type semiconductor* is retained, since for every Hole, there is an associated negative charge in the covalent bond structure and it is said that the mobile positive charge covers the immobile negative charge. Thus, a pictorial representation of the P-material can be arrived as shown in Fig. 2.18.

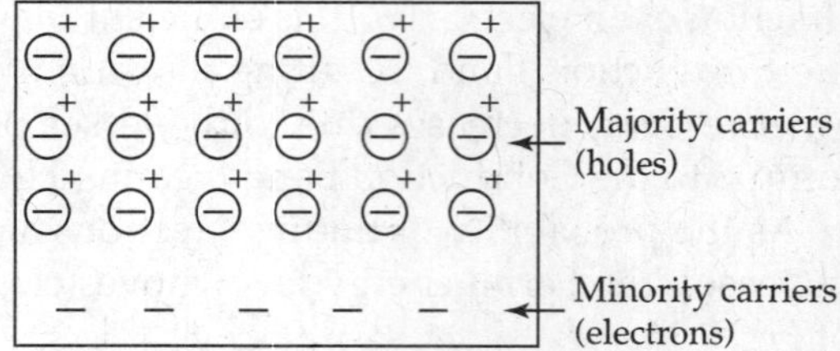

FIG. 2.18 *P-type material representation*

Mobile and Immobile Charges in a P-type Semiconductor

Un-encircled positive charges represent mobile Holes called *covering charges*, covering the immobile negative charges associated with covalent bond structure. This results in charge neutrality, since every mobile Hole covers the immobile negative ion.

But the structure is not complete in the sense that there will be a few mobile electrons due to the broken covalent bonds of the intrinsic material. Mobile carriers generated (Holes) due to doping will be many more compared with the intrinsic electrons. Mobile Holes are called the *majority carriers* and mobile electrons are called the *minority carriers*. The point of interest to note is that the *electrons in the P-type material will be less than what they would have been, had there not been doping.*

It can be justified as follows. Pure semiconductor has as many Holes as there are electrons and there is a certain probability of recombination between electrons and Holes. Yet in this process of generation and recombination, an average concentration of electron–Hole pairs exists. Doping increases the concentration of Holes in P-type material. Then the probability of recombination increases and the minority carrier concentration of electrons decreases. 'Charge profiles' of majority carrier Hole concentration and minority carrier electron concentration is shown in Fig. 2.19.

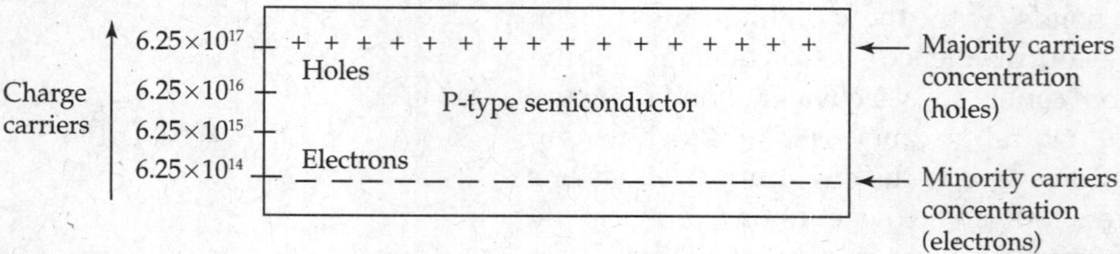

FIG. 2.19 *Charge profile of P-type semiconductor*

Thus, we see that from equal concentrations in a pure semiconductor, due to doping processes, the majority carrier concentration increases, while that of minority carrier concentration decreases. As shown in Figs. 2.17 and 2.19, the ratio of the new concentration for majority and minority carrier concentration is of the order of 1000. The majority carrier (Hole) contributed current (typically of the order of 100 mA) can be accounted by being contributed by 6.25×10^{17} Holes. Conspicuously, the minority carrier contributed current (typical value of 100 mA) can be accounted by being contributed by a carrier number of 3 orders less (i.e., 6.26×10^{14} electrons).

The current contribution due to Hole movement is known as deficit conduction, as Holes are formed due to deficiency of electrons due to covalent bond formations.

Energy-Band Diagram of P-type Semiconductor

The shift of the Fermi level (which is otherwise at the middle of the forbidden band gap in an intrinsic semiconductor) E_{FP} towards Valence Band in a P-type semiconductor and the shift of E_{FN} towards Conduction Band in an N-type semiconductor can be understood from the ongoing expressions for E_{FP} and E_{FN} in the following discussion.

Doping allows empty energy levels in the hitherto forbidden band gap at the same level that is E_A (acceptor energy level) (due to the doping atoms being disseminated into the material and being so far away from each other excluding the possibility of inter-atomic

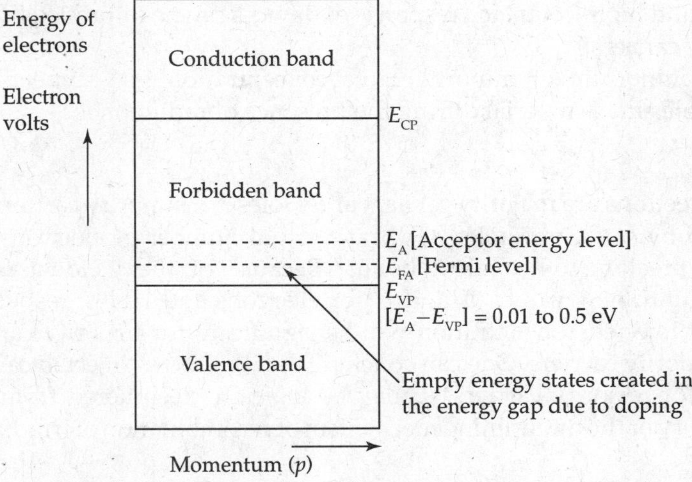

FIG. 2.20 *Energy-band diagram of P-type semiconductor*

influences). Now, for an electron to move from the Valence Band to an empty state, it need not travel all the way to the Conduction Band, but only up to acceptor energy level E_A, which is very close to boundary level E_V of Valence Band. In this process, with a very little energy (even at room temperatures) abundant numbers of Holes are created in the valance band, thus increasing the conductivity of P-type semiconductor. Incidentally, minority carrier concentration decreases due to the inability of electrons jumping the forbidden band gap and entering Conduction Band.

The Fermi level moves closer to the Valence Band and has a quantitative expression

$$E_{FP} = E_V + kT \cdot \ln\left[\frac{N_V}{N_A}\right] \tag{2.27}$$

$$\text{where } N_V = 4.82 \times 10^{21} \cdot \left[\frac{m_p}{m}\right]^{3/2} \cdot T^{3/2} \tag{2.28}$$

and m_p is the effective mass for Holes and m is the rest mass. N_V is a constant for Valence Band that depends on the effective masses of Holes and electrons and temperature as well.

Carrier Lifetime

Average concentrations of electrons and Holes in an intrinsic semiconductor remain more or less constant and equal. However, there is a continuous process of generation of Hole–electron pairs due to thermal agitation or irradiation and neutralisation due to opposite charges meeting each other (Hole–electron pairs disappear) in their random movements. Loss of electron–Hole pairs is due to the phenomenon called *recombination*. Generated electrons remain in the arena for a specific time called the *mean lifetime* τ_n before it disappears due to recombinations. So τ_n is called as carrier lifetime (for electron) (Fig. 2.20).

Similarly, τ_p is the lifetime for a Hole. τ_p and τ_n have another significance, when applied to concentrations, for instance τ_p corresponds to the time taken for the excess concentration of Holes to get reduced to $1/e$ of its enhanced value as shown in Fig. 2.21 (due to irradiation). Carrier lifetimes range from a few nanoseconds to hundreds of microseconds. Switching speed of semiconductors and high-frequency response of devices can be improved by decreasing the lifetime of minority carriers.

p = Thermal equilibrium of minority Hole concentration in N-type conductor. Total concentration of Holes in N-material during the presence of radiation is

$$\bar{p}_n = p_{n0}. \tag{2.29}$$

In an N-material, electrons are majority carriers and Holes are minority carriers. If this N-type material is irritated by say ultra violet (UV) rays, equal number of electrons and Holes are generated due to breaking of covalent bonds. Because of the incident UV energy, the concentrations go up to new values $\bar{n}$ and $\bar{p}$ for electrons and Holes, respectively. But this enhanced excess or injected concentration will be significant for majority carriers and quite insignificant for minority carriers. One can conclude that the excess injected carriers are of any consequence only for majority carriers. It may be logically deciphered from R–H–S (Reid–Hall–Shockley) theory or the underlining mechanism of recombination of irradiation generated excess carriers.

When UV source is switched off at time $t = 0$ s, then obviously the excess concentrations should come back to their thermal equilibrium values over time due to recombinations. As shown in Fig. 2.21, p_{n0} represents the thermal equilibrium concentration of Holes in N-material. Due to irradiation, p_{n0} becomes $\bar{p}_{n0}$.

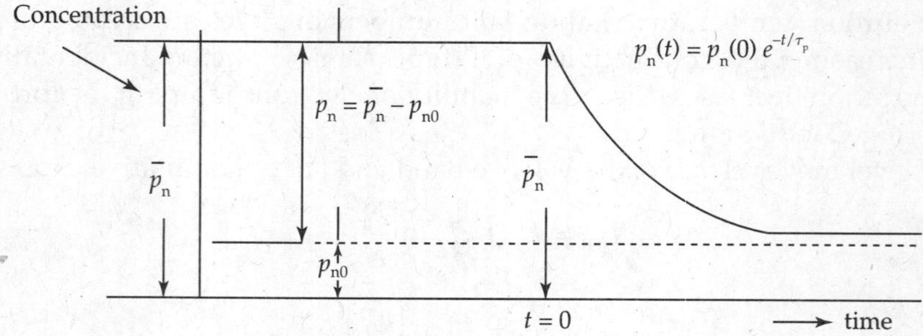

FIG. 2.21 *Variation of concentration of holes*

Excess concentration $p_{n0} = \bar{p}_{n0} - p_{n0}$. If at $t = 0$, the UV source is switched off and p_n is represented as $p_n(0)$. $p_{n0} = \bar{p}_{n0} - p_{n0}$ represents the excess minority carrier concentration, i.e. Holes in N-material.

When the radiation is switched off, that is the excess energy being removed; the excess concentrations disappear over time and return to their equilibrium values.

As shown in Fig. 2.21, $p_n(0) = p_n - p_{n0}$ will decrease exponentially with time given by the expression

$$p_n(t) = p_n(0) \cdot e^{-t/\tau_p} \tag{2.30}$$

$$n_p(t) = n_p(0) \cdot e^{-t/n}. \tag{2.31}$$

Similarly for P-material, as carriers move in a semiconductor material, they encounter opposite charges and recombine, thus decreasing the concentration with distance. Over a distance x, charge of excess minority carrier concentration can be represented by

$$p_n(x) = p_n(0) \cdot e^{-x/L_p}, \tag{2.32}$$

where $p_n(x)$ is representative of excess Hole concentration with distance, $p_n(0)$ represents the excess Hole concentration at $x = 0$ and falls off exponentially. As represented in the above equation, L_p is called the diffusion length for Holes. L_p represents the mean free path of a Hole before it recombines with an electron. This is otherwise the distance at which the excess Hole concentration falls off to a value $1/e$ of the concentration at $x = 0$.

$$\text{Similarly} \quad n_p(x) = n_p(0) e^{-x/L_n}, \tag{2.33}$$

where $n_p(x)$ is the concentration of excess minority electrons in the P-type semiconductors as a function of distance, $n_p(0)$, the concentration at $x = 0$ and L_n is the diffusion length for electrons as explained in Fig. 2.22.

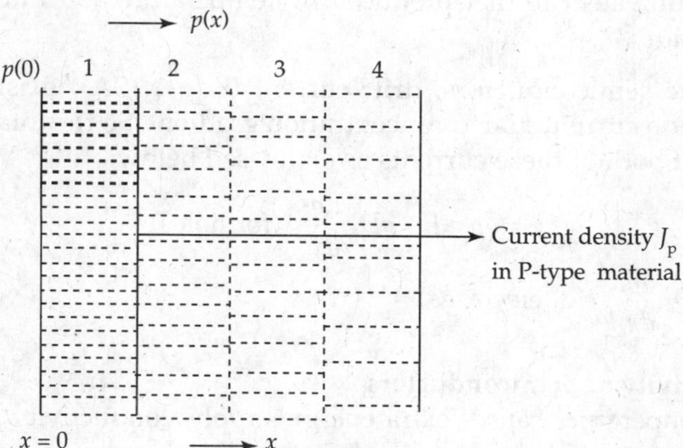

FIG. 2.22 *Diffusion of charges due to variation in charge density for the illustration of diffusion of holes and electrons through the P-type and N-type semiconductor materials of P–N diode*

Diffusion

Figure 2.22 shows a material in which concentration of charges is non-uniform and decreases with distance x. Variations in concentration of 'charges' with distance x is known as *concentration gradient*. It allows charge movement because of concentration variation. This process of movement of charges from regions of greater concentration to regions of smaller concentration is called *diffusion*. Due to this charge movement, a *diffusion Hole current density* J_p proportional to the concentration gradient $\frac{dp}{dx}$ appears. Such charge density is expressed as a relation.

$$\text{Diffusion hole current density} \quad J_p = -qD_p \frac{dp}{dx} \text{ Amps/m}^2 \tag{2.34}$$

$$\text{Diffusion current } I \text{ through an area} \quad A = -qAD_p \cdot \frac{dp}{dx} \text{ Amps,}$$

where D_p is the diffusion constant for Holes, measured as m²/s. The negative sign in Eq. (2.34) appears because of the fact that with distance x, the concentration of charges is decreasing.

Since electrons can be considered as duals of Holes, *diffusion electron current density* J_n for electrons is due to concentration gradient $\frac{dn}{dx}$.

$$J_n = qD_n \frac{dn}{dx} \text{ Amps/m}^2. \tag{2.35}$$

Einstein's Relation

Mobility μ and diffusion constant D are both thermodynamic phenomena and a correlation exists between them. The relation between them is *Einstein relationship.*

$$\frac{D_p}{\mu_p} = \frac{D_n}{\mu_n} = V_T \quad \left(V_T = \frac{kT}{e} = \frac{T}{11{,}600}; \; V_T = 26 \text{ mV at } 300°\text{K}\right), \tag{2.36}$$

where V_T is the voltage equivalent of temperature.

The significance of V_T can be explained as follows. Temperature causes thermal agitation resulting in movement of charges in a material V_T, can be expressed as a voltage, which also causes movement equivalent to that produced by temperature and thus it is called as *volt equivalent of temperature.*

Net current In the semiconductors, drift current (Eq. (2.17A)) exists due to potential gradient and diffusion current due to concentration gradient. So the total current should be the algebraic sum of both of these currents as expressed below:

$$J_p = q\mu_p pE - qD_p \frac{dp}{dx} \quad \text{(for holes)} \tag{2.37}$$

and $J_n = q\mu_n nE + qD_n \dfrac{dn}{dx}$ (for electrons). (2.38)

Temperature Sensitivity of Semiconductors

Since increase in temperature causes extra energy supply, conductivity σ of semiconductor increases with temperature as more and more covalent bonds are ruptured. *Energy gap* decreases as given by the following expressions:

1. E_G (T) = 0.785 – (2.33 × 10⁻⁴) T.
   At room temperature, E_G = 0.72 eV for Germanium semiconductor.
   At temperature of 600°K, E_G = 0.645 eV for Germanium semiconductor.
2. E_G (T) = 1.21 – (3.6 × 10⁻⁴) T.
   At room temperature 300 °K, E_G = 1.1 eV for Silicon semiconductor.
   At temperature of 600°K (300°K), E_G = 0.994 eV for Silicon semiconductor.

Though the velocity acquired by a charge carrier depends upon the electric field E through mobility ($v \propto E$ and $v_d/E = \mu$), its dependence on temperature is given by ($\mu \propto T^{-\eta}$), where η is the characteristic value of carriers.

2.4 MASS-ACTION LAW

Intrinsic semiconductors do not contain any impurities or doping atoms. At temperatures greater than absolute 0°K, increase in temperature provides enough energy for breaking covalent bonds into Hole–electron pairs. This results in the existence of equal number of free Holes in the Valence Band and free electrons in the Conduction Band (when the electrons reach the Conduction Band due to the acquired thermal energy). Hole and electron density or carrier concentration is represented by n_i.

Intrinsic concentration n_i of charges is a function of temperature in the semiconductor materials and is expressed by

$$n_i^2 = [A_0 T^2] e^{-\left[\frac{E_C - E_V}{kT}\right]} = \left[A_0 T^2\right] e^{-[E_{G0}/kT]} \tag{2.39}$$

where A_0 is the constant independent of temperature, $A_0 = 4.82 \times 10^{21}/\text{m}^3$; E_{G0} = energy band gap at 0°K, E_{G0} = forbidden band-gap energy at 0°K.

$(E_C - E_V) = E_{G0} = 0.785$ eV for Germanium and $E_{G0} = 1.21$ eV for Silicon materials.

k = Boltzman's constant = 1.38×10^{-23} J/°K, T = absolute temperature in °K.

$$\text{At room temperature} \quad T = 300°K,\ n_i^2 = 4.82 \times 10^{21} \times (300)^3 \cdot e^{-[1.21/kT]} \tag{2.40}$$

for Silicon semiconductor as already explained in Eq. (2.39), where n_i is the intrinsic concentration in a semiconductor material.

In a pure semiconductor, Hole and electron concentrations are equal. The product of electron concentration or density n and Hole concentration or density p is equal to the square of the intrinsic carrier concentration n_i as shown in Eq. (2.41):

$$n_i = n = p$$

$$n \times p = n_i^2 = 4.82 \times 10^{21} \cdot (T)^2 \cdot e^{-\left[\frac{E_C - E_V}{kT}\right]}. \tag{2.41}$$

This is called 'mass-action law'. This is also called 'law of mass-action'.

EXAMPLE 2.3

Find the concentration of Holes and electrons in a P-type Silicon semiconductor at 300°K, assuming resistivity as 0.02 Ω-cm. Assume the mobility of Holes $\mu_p = 475$ m²/V-s and intrinsic concentration $n_i = 1.45 \times 10^{10}$ per cm³ (May/June 2006, set-4).

Solution:

$$\text{Resistivity} \quad \rho = 0.02\ \Omega\text{-cm} = 0.02 \times 10^{-2} = 2 \times 10^{-4}\ \Omega\text{-m} \tag{1}$$

$$\text{Intrinsic concentration} \quad n_i = 1.45 \times 10^{10}\ /\text{cm}^3 \tag{2}$$

$$\therefore \quad n_i = 1.45 \times 10^{10} \times 10^6 = 1.45 \times 10^{16}\ /\text{m}^3 \tag{3}$$

$$\text{Hole mobility} \quad \mu_p = 475\ \text{m}^2/\text{Volt-sec} \tag{4}$$

$$\text{Resistivity} \quad \rho = \frac{1}{\sigma} = \frac{1}{pq\mu_p}$$

$$\text{Number of holes} \quad p_p = \frac{1}{\rho q \mu_p} = \frac{1}{2 \times 10^{-4} \times 1.6 \times 10^{-19} \times 475} = 1.52 \times 10^{20}\ /\text{m}^3$$

According to mass – action law,

$$p_p \cdot n_p = n_i^2.$$

$\therefore$ Number of electrons in P-type semiconductor $\quad n_p = \dfrac{n_i^2}{p}$ (5)

$\therefore$ Number of electrons in P-type semiconductor $\quad n_p = \dfrac{[1.45\times10^{16}]^2}{1.52\times10^{20}} = 1.38\times10^{12}\,/\,\text{m}^3$

Resistivity $\quad \rho = 0.02\ \Omega\text{-cm}$

Conductivity $\quad \sigma = \dfrac{1}{\rho} = \dfrac{1}{0.02} = 50\ \text{mhos/cm}$

Concentration of holes $\quad p = \dfrac{\sigma}{e\cdot\mu_p} = \dfrac{50}{1.6\times10^{-19}\times475}$

$$= 6.578\times10^{17}$$

Concentration of electrons $\quad n = \dfrac{n_i^2}{p} = \dfrac{(1.45\times10^{10})^2}{6.578\times10^{17}} = \dfrac{0.32\times10^{20}}{10^{17}}$

$$= 0.32\times10^{3}\,/\text{cm}^3.$$

EXAMPLE 2.4

Find the concentration of Holes and electrons in N-type Silicon semiconductor at 300°K assuming resistivity is 0.025 Ω-cm, μ_n = 1250 cm²/V-s. And n_i = 15 × 10¹³/cm³.

Solution: Resistivity $\quad \rho = 0.025\ \Omega\text{-cm}$

$\therefore$ Conductivity $\quad \sigma = \dfrac{1}{\rho} = \dfrac{1}{0.025} = 40\ \text{mhos/cm}$

Concentration of electrons $\quad n = \dfrac{\sigma}{e\cdot\mu_n} = \dfrac{40}{1.6\times10^{-19}\times1250} = 2.0\times10^{17}\,/\,\text{cm}^3$

Concentration of holes $\quad p = \dfrac{n_i^2}{n} = \dfrac{(15\times10^{13})^2}{2.0\times10^{17}} = \dfrac{225\times10^{26}}{2.0\times10^{17}} = 11.25\times10^{9}\,/\,\text{cm}^3$

2.5 CONTINUITY EQUATION (CONSERVATION OF CHARGE)

Assume a semiconductor material of volume $V = A \times dx$ (Fig. 2.23). Due to ambient temperature, electrons and Holes are continuously generated. These charges move randomly through the volume. During this movement, electrons and Holes encounter each other and recombine.

Since semiconductor has some finite resistance, which is not infinite; some free charges should be available. This implies that the generated Hole–electron pairs are more than those lost in recombinations. There can be a net charge flow in a specific direction through the conductor for example in the x direction (Fig. 2.23). Movement involves time and distance. So over a period of time there will be a change in concentration, which depends on distance also.

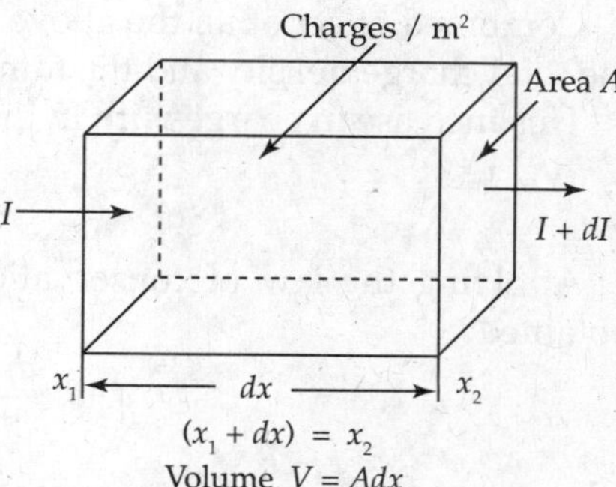

FIG. 2.23 *Representation of charge flow in the volume of a rectangular cube of a semiconductor material*

- Random movement of charges in a semiconductor at a given time may have a net flow of charge in one direction say positive x direction.
- This causes a change in concentration along the length of the conductor, creating a concentration gradient. In other words, whenever there is non-uniform doping concentration, it results in an inbuilt electric field gradient (potential difference) within the semiconductor to give rise to *diffusion current* (Eq. 2.34).
- Consequently, at two different points x_1 and x_2 along the x direction in a conductor, there will be different amounts of charge accumulation. At one point say x_1, charge may be more than those at the second point x_2, where $(x_1 + \Delta x) = x_2$, the charge may be less. This establishes a potential gradient (electric field). Resulting potential gradient or an equivalent battery opposes the further movement of the charges in that direction.
- Charge movement due to potential gradient or electric field is known as *drift*. Diffusion and drift processes are in constant opposition.
- Hence, the currents due to drift and diffusion processes will be equal and opposite, resulting in the conservation of charge over the volume.

This law of conservation of charge (charges can neither be created nor be destroyed) expressed as a relation is called the *continuity equation*. This shows that average current in the elemental volume is zero, when there is no external excitation.

'Continuity equation' can finally be understood that changes in current in a 'semiconductor medium' are due to the sum of the

- Difference between incoming and outgoing charges
- And the net charges that remain due to generation and recombination processes of Holes and electrons in the semiconductor over a period of time.

As shown in Fig. 2.23, if the concentration of Holes is p and τ_p *is the mean lifetime of the Hole*, the number of Holes lost due to recombination is p/τ_p per second. So the charge decreases within the same volume due to recombination.

1. The decrease in the charge within the volume V is equal to the product of charge of the Hole q, volume Adx and Holes per second per unit volume, which is

$$-qAdx\frac{p}{\tau_p}. \tag{2.42}$$

2. If due to thermal agitation, an amount of g electron–Hole pairs is generated per second, increase in charge within the volume is the product of charge of Hole q, volume Adx and the rate of regeneration g, which is

$$qAdx \cdot g. \tag{2.43}$$

3. In general, current should vary with distance in a semiconductor material. If I amperes of current enters the considered volume at x_1 and $(I + \Delta I)$ leaves the volume at x_2, decrease in charge in coulombs per second contributes to current of magnitude dI. (2.44)

Combined effect of all the above three processes shown by Eqs. (2.42)–(2.44) is to increase the total charge density and the number of coulombs per second.

This increase in charge with in the volume V is

$$qAdx\frac{dp}{dt} \tag{2.45}$$

Applying the law of conservation of charge to the above three processes, Eq. (2.46) is obtained:

$$qAdx\frac{dp}{dt} = -qAdx\cdot\frac{p}{\tau_p} + qAdx\cdot g - dI \tag{2.46}$$

where $-qAdx\ p/\tau_p$ is due to recombinations and $qAdxg - dI$ is due to thermal generation of charges.

The net Hole current should be the sum of the diffusion current (Eq. (2.34)) and the drift current (Eq. (2.17A)) as discussed earlier:

$$\therefore \quad I = -qAD_p\frac{dp}{dx} + qAp\mu_p E \tag{2.47}$$

where $-qAD_p\dfrac{dp}{dx}$ is the diffusion currect. (2.48)

and $qAp\mu_p E$ is the drift current. (2.49)

In Eq. (2.49), E is the field intensity within the volume.

Under equilibrium conditions, with no external excitation, Hole density should be a constant. Let this be p_0. So $I = 0$ and $\dfrac{dp}{dt} = 0$.

Substituting these values, we get

$$qAdx\cdot\frac{dp}{dt} = -qAdx\cdot\frac{p}{\tau_p} + qAdxg - dI \tag{2.50}$$

The result becomes

$$g = \frac{p_0}{\tau_p} \tag{2.51}$$

From Eq. (2.47), $I = -qAD_p \times \dfrac{dp}{dx} + qAp\mu_p E$

$$\frac{dI}{dx} = -qAD_p\cdot\frac{d^2p}{dx^2} + qAp\mu_p\cdot\frac{d(pE)}{dx} \tag{2.52}$$

$$\text{or } dI = -qAdx\times D_p\frac{d^2p}{dx^2} + qAdx\times p\mu_p\frac{d(pE)}{dx}. \tag{2.53}$$

Substituting these values in Eq. (2.46), we get

$$qAdx\cdot\frac{dp}{dt} = -qAdx\cdot\frac{p}{\tau_p} + qAdx\cdot\frac{p_0}{\tau_p} + qAdxD_p\frac{d^2p}{dx^2} + qAdx\cdot P\mu_p\frac{d(pE)}{dx}$$

$$\therefore \quad \frac{dp}{dt} = \frac{(p_0 - p)}{\tau_p} + D_p\frac{d^2p}{dx^2} + P\mu_p\frac{d(pE)}{dx}.$$

Since the concentration p is a function of time, distance and field, partial differentials should be used to indicate variation of p with time and distance.

$$\text{So } \frac{\partial p}{\partial t} = \frac{(p - p_0)}{\tau_\text{p}} + D_\text{p}\frac{\partial^2 p}{\partial x^2} + \mu_\text{p}\frac{\partial (pE)}{\partial x} \quad \text{(continuity equation)} \tag{2.54}$$

Equation (2.54) is called the continuity equation. Applying various boundary conditions, different results relating to behaviour of semiconductor devices can be obtained. This equation is used to determine the expressions for distribution of minority carriers in semiconductors. Three properties of minority carriers are mobility; diffusion constant and lifetime. They are determined by Haynes–Schockley experiment.

2.6 HALL EFFECT

When a slab of metal or semiconductor material carrying a current I (in x direction) is placed in a magnetic field B (in z direction) (magnetic field is perpendicular to the direction of current], an electric field E (ε) V/m (y direction) appears across the faces F_1 and F_2 of the semiconductor slab (in a direction perpendicular to both current and magnetic field) (Fig. 2.24).

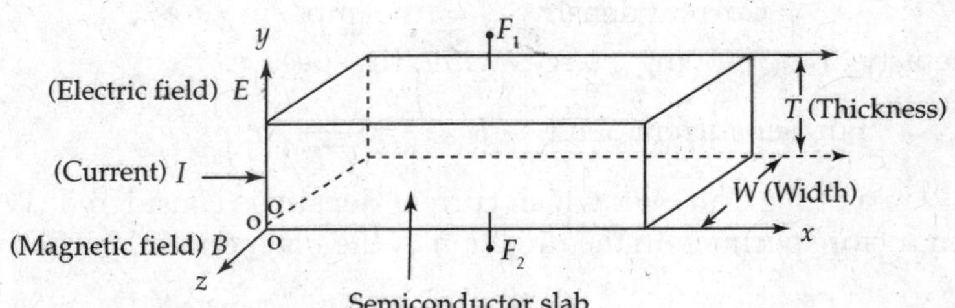

FIG. 2.24 *Illustration of hall effect phenomena*

Appearance of the voltage V_H across the two faces F_1 and F_2 of the semiconductor slab is known as Hall effect. Voltage V_H is simply called as *Hall voltage.*

$$\varepsilon = \frac{V_\text{H}}{T}\ \text{V/m}, \tag{2.55}$$

where T is the thickness of the semiconductor slab, V_H is the Hall voltage and ε is the electric field.

Appearance of the Hall voltage, V_H, can be explained as follows.

- Due to the current I flowing through the semiconductor slab situated in a transverse magnetic field B, a force F is exerted in a direction perpendicular to both B and I on the charges in the slab.
- Then the charges migrate towards the top face F_1 or bottom face F_2 depending upon the polarity of the charges.
- If the current carriers are electrons, they migrate towards the bottom face F_2 for the assumed direction of flow of current (I).
- Assumed current I may be due to Holes moving from left to right or due to free electrons moving from right to left through the semiconductor slab. Then the upper face F_1 will be positive with respect to the bottom face F_2.

- Charge carriers in semiconductor slab will be electrons if the material is N-type material and the polarity of V_H is such that F_1 will be positive with respect to F_2.
- If Hall voltage V_H between F_1 and F_2 is such that F_1 is positive, the inference is that the material is of N-type semiconductor.
- By extending the same logic, a P-material can be identified by the polarity of Hall voltage being in reverse direction with F_1 being negative and F_2 being positive as the Holes also move down towards face F_2.

Force f exerted on the electrons and Holes due to the induced electric field ε is

$$f_{VH} = Q \cdot \varepsilon \text{ (Newtons)} \tag{2.56}$$

Force f_B on the charge carriers due to the applied magnetic field B is

$$f_B = BQv \text{ (Newtons)} \tag{2.56A}$$

where v is the mean drift speed (m/s).

At an equilibrium position, the above two forces acting simultaneouly on the charges (electrons and Holes) will be equal.

$$\therefore \quad f_{VH} = Q \cdot \varepsilon = BQv \text{ (Newtons)} \tag{2.57}$$

$$\therefore \quad \varepsilon = Bv \text{ (V/m)} \tag{2.58}$$

$$\text{Current density} \quad J = \rho v \text{ Amps/m}^2, \tag{2.59}$$

where ρ is the density of the moving charges with drift speed v.

$$\text{Further current denity} \quad J = \frac{I}{A} = \frac{I}{WT} \text{ Amps/m}^2, \tag{2.60}$$

where $A = WT$ = area of the slab over which current density is considered (W = width of the metal or semiconductor specimen in the direction of the magnetic field B and in this case the z direction).

From Eqs. (2.55) and (2.58), we get

$$\text{Hall voltage} \quad V_H = \varepsilon \cdot T = BvT \text{ Volts.} \tag{2.61}$$

Further using $v = \dfrac{J}{\rho} = \dfrac{I}{\rho WT}$ in Eq. (2.61), we obtain

$$\text{Hall voltage} \qquad V_H = BvT = \frac{BIT}{\rho WT} = \frac{BI}{\rho W} \text{ Volts} \tag{2.62}$$

$$\therefore \quad \rho = \frac{BI}{V_H W} \tag{2.63}$$

$$\text{Hall coefficient} \quad R_H = \frac{1}{\rho} \tag{2.64}$$

$$\therefore \quad R_H = \frac{V_H W}{BI}. \tag{2.65}$$

(Note: Apart from the fact that Eq. (2.62) expresses V_H in terms of B, I, ρ and W, the practical way of measuring V_H involves directional notations also.)

- If the polarity of V_H is positive at the terminal F_1, then the charge carriers must be electrons and $\rho = ne$, where n is the electron concentration.
- If on the other hand, terminal F_2 becomes charged positively with respect to terminal F_1, the semiconductor must be P-type and $\rho = pe$, where p is Hole concentration.

- Using the concept of Hall effect, type of semiconductor materials can be identified as P- or N-type semiconductors.
  - *Concentration* and *nature of charge carriers* in materials can be determined using the Hall effect phenomenon.
- If the conduction in the material is primarily due to the charges of only one sign, the conductivity is related to mobility of the charges.

$$\sigma = \rho\mu \tag{2.66}$$

- Mobility μ of charge carriers can be calculated from Eq. (2.66A) from the measurements of conductivity and Hall coefficients.

$$\mu = \sigma R_{\text{H}} \tag{2.66A}$$

- Due to random movement of the charges, equation $R_{\text{H}} = \dfrac{V_{\text{H}}W}{BI}$ remains valid, when R_{H} is defined by

$$\frac{3\pi}{8\rho}. \tag{2.66B}$$

From Eqs. (2.65), (2.66A) and (2.66B), we get

$$\mu = \frac{8\rho}{3\pi}\cdot R_{\text{H}}. \tag{2.67}$$

Hall voltage V_{H} is directly proportional to magnetic field B (for defined value of current). Therefore, measuring the Hall voltage, magnetic field B can be determined.

Hall effect phenomenon can be used to determine the following parameters:

(1) P- or N-type semiconductor, (2) Conductivity of the material, (3) Mobility μ of the charge carriers and (4) Measurement of magnetic field.

2.6.1 Applications of Hall Effect

(1) To determine the type of semiconductor whether P-type or N-type. (2) To measure the mobility of charge carriers μ. (3) To measure the carrier concentration of semiconductor materials. (4) To measure the strength of magnetic flux density B. (5) Hall effect sensors can be used as proximity probes in instrumentation applications. (6) For measurement of speed of a turbo-generator in power plants. (7) To lock and unlock keyboard operations of computers through keyboard switches (Hall effect switch gives an output, if a magnet is taken near a special type of conductor).

2.7 QUALITATIVE THEORY OF P–N JUNCTION (OPEN CIRCUITED P–N JUNCTION)

The concept of P–N junction is essential to understand the working of P–N junction diode, Bipolar Junction Transistor (BJT) and Field Effect Transistor (FET) devices.

2.7.1 P–N Semiconductor Diode

P-type semiconductor is formed with trivalent material doping over one region of a semiconductor wafer and N-type semiconductor is formed with pentavalent material doping over the second region in proximity. The terminal connecting the P-type material is known anode and the terminal connecting to the N-material is cathode. The P–N material structure and symbol of P–N diode are shown in Fig. 2.25.

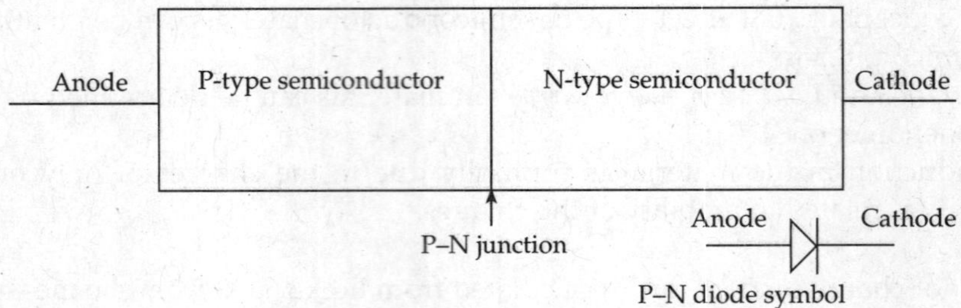

FIG. 2.25 *P–N semiconductor diode*

When the two blocks of P- and N-materials are in contact, charge profiles showing Hole and electron concentrations is shown in Fig. 2.26. Some processes start as following:

- *Concentration gradient among Holes* forms (about P–N junction) between large number of (majority carriers) Holes in P-type semiconductor and very few (minority carriers) Holes in N-type semiconductor. It causes for the diffusion of Holes from P- to N-material (Fig. 2.26).
- *Concentration gradient among electrons* forms (about P–N junction) between large number of (majority carriers) electrons in N-type semiconductor and very few (minority carriers) electrons in P-type semiconductor. It causes for the diffusion of electrons from N- to P-material (Fig. 2.26).
    - However, these processes cannot continue forever.
    - As the Holes and electrons migrate into the other regions, the hitherto covered immobile ions on either side of the junction get uncovered.
    - Mobile charges encountering opposite charges may recombine on either side of the P–N junction, where they initially meet (come across one another).
    - Such recombinations leave *immobile ions* on either side of P–N junction.
    - Positive immobile ions form nearer to the junction in N-material (because of donation of electrons from pentavalent atoms in N-material) (Fig. 2.28).
    - Negative immobile ions form nearer to junction in P-material (because of Holes (about trivalent atoms) accepting electrons during recombinations) (Fig. 2.28).
    - The region containing immobile ions is depleted of mobile covering charges. This area is known by various names: (1) depletion region, (2) transition region and (3) space charge region.
    - Depletion region width is of the order of 1 μ (Fig. 2.27).

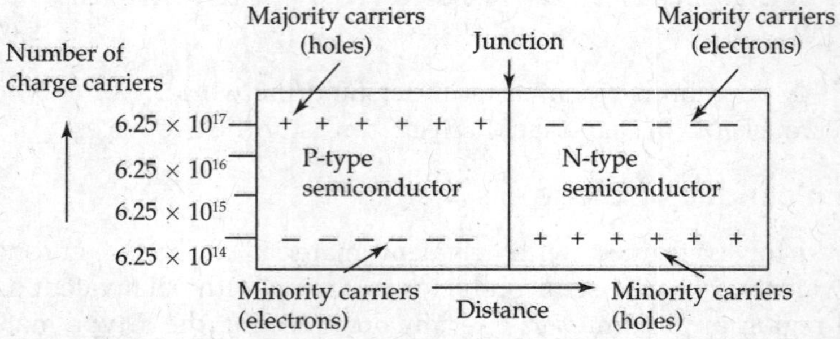

FIG. 2.26 *Concentration of holes and electrons in P-type and N-type materials charge profiles to illustrate the concept of diffusion process*

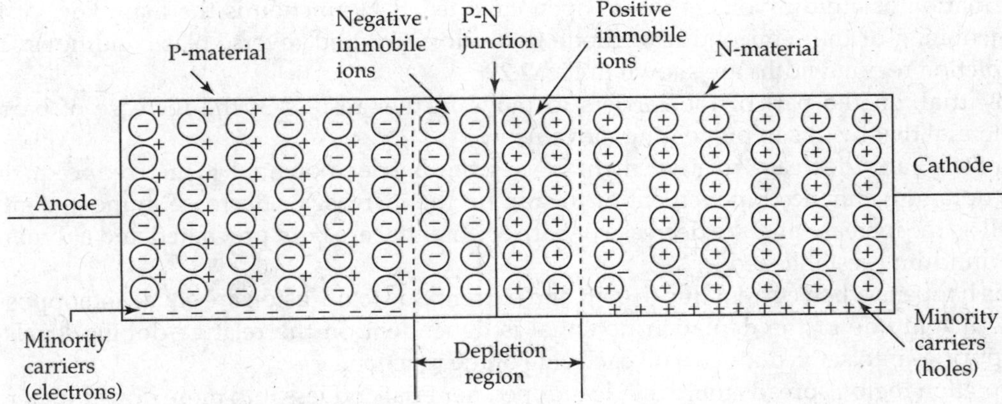

FIG. 2.27 *P–N junction with open circuit or no external excitation*

- The resulted uncovered immobile positive charges (immobile ions) in N-material side develop a potential gradient with fixed negative charges (immobile negative ions) on the P-material side around the junction in the depletion region.
- The built-in space charge opposes the movement of electrons from the N-region into the P-region (mobile negative electrons encountering immobile negative ions in the depletion region of P-type semiconductors).
- The same happens to Holes from P-material trying to move into N-type semiconductor. However, migration continues further till the immobile charges are sufficiently strong enough to prevent further progress of mobile charge movement.
- An equilibrium condition is established which creates a barrier for further movement of mobile charges. This barrier potential is known by various names: (1) space charge potential, (2) depletion potential, (3) transition potential, (4) contact potential and (5) built-in potential.

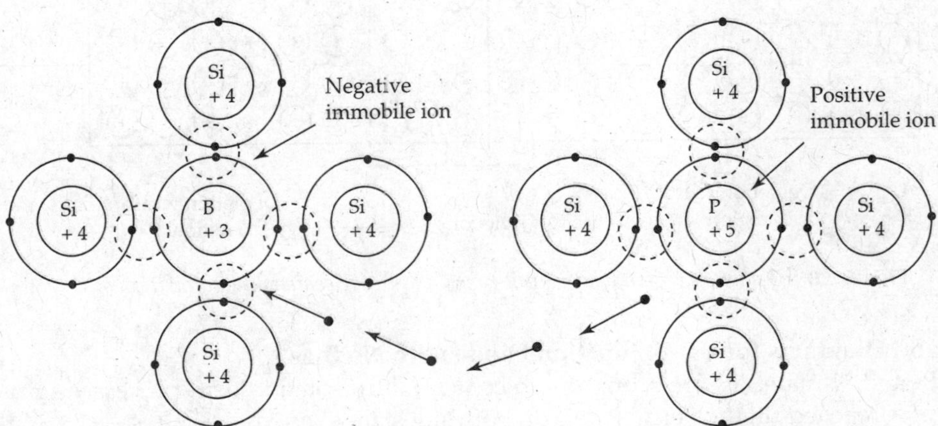

FIG: 2.28 *Formation of negative and positive immobile ions. Concept of the formation of an immobile negative ion about trivalent boron atom in P-type semiconductor and positive immobile ion about pentavalent phosphorus atom when an electron donated by pentavalent atom is accepted by a trivalent atom about the junction in depletion region of P–N diode*

- Formation of built-in voltage for an open circuited P–N junction is the main key for the functioning of the semiconductor diode under forward and reverse bias conditions. The depletion region widths are shown in Fig. 2.29.
- Any trial, on the part of the carriers to move further results in the increase of barrier potential that results in preventing movement.
- Similarly, any decrease in movement causes the barrier potential to get reduced, thus encouraging enhanced movement or to sum up the increase or decrease in movement as well as the increase and the decrease of barrier potential are both prevented and a dynamic equilibrium is established.
- This happens when current due to drift and current due to diffusion are equal and opposite. The magnitude of the depletion potential is dependent on the relative doping levels of impurities into semiconductor on each side of the junction.
- Depletion region spreads more into less doped material and less into more doped material. The reason is obvious since to get the same uncovering of immobile ions requires more penetration into the less doped region, whereas it requires less deep penetration into the more doped region.
- The immobile positive ions in depletion region in N-type semiconductor and the negative immobile ions in depletion region in P-type semiconductor form the fictitious potential E_0 around the P–N junction.
- This fictitious potential is known as contact or built-in potential. It is of the order of 0.2–0.3 V for Germanium semiconductor diode and 0.5–0.7 V for Silicon semiconductor diode (E_0 cannot be measured experimentally).

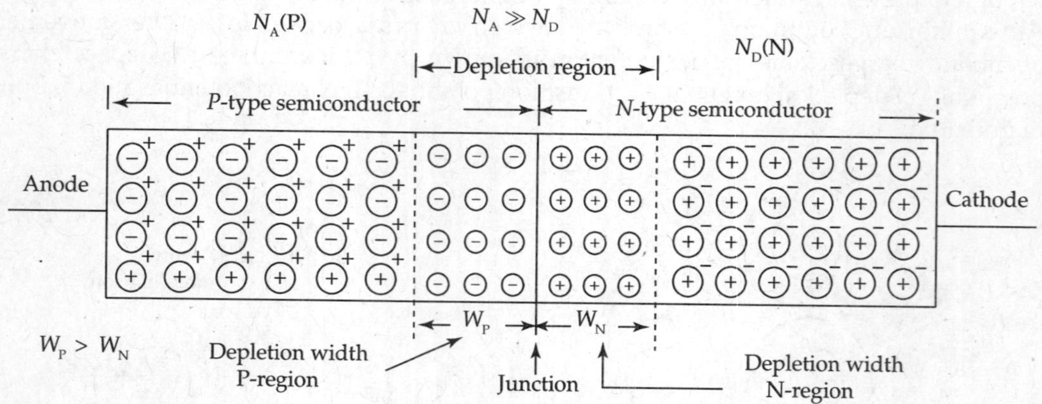

FIG. 2.29 *Depletion region widths in P-type and N-type semiconductor materials*

Energy-Band Diagrams for a P–N Junction Diode with No Bias

Once the P- and N-materials are brought into contact, diffusion and drift processes start and equilibrium is reached till the Fermi levels on both sides are aligned.

The above process pushes down both the conduction and the Valence Band energy levels producing a difference in energy levels equal to E_0.

This causes a potential difference V_0 to be developed across the space charge region and is called the contact difference of potential or built-in potential. The quantitative relation can

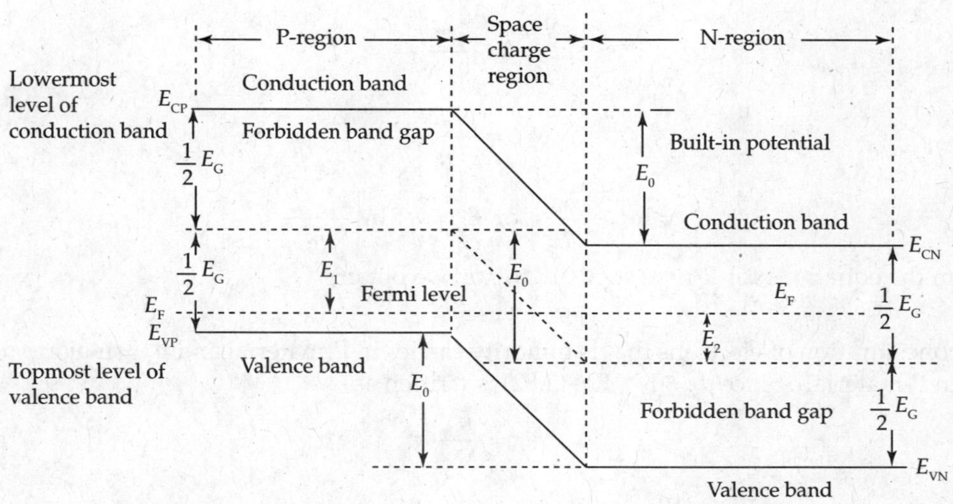

FIG. 2.30 *Energy-band diagram of P–N junction diode*

be obtained in terms of energy level as follows. According to the diagram, the shift in energy levels is given by

$$E_0 = E_{CP} - E_{CN} = E_{VP} - E_{VN} = E_1 + E_2 \tag{2.68}$$

E_0 represents the potential energy associated with the electrons at the junction. Contact Difference of Potential E_0 or V_0

$$E_F - E_{VP} = \frac{1}{2}E_G - E_1 \tag{2.69}$$

$$E_{CN} - E_F = \frac{1}{2}E_G - E_2 \tag{2.70}$$

Rearranging Eqs. (2.69) and (2.70), we get

$$E_1 = \frac{1}{2}E_G - (E_F - E_{VP}) \tag{2.71}$$

$$E_2 = \frac{1}{2}E_G - (E_{CN} - E_F) \tag{2.72}$$

From Fig. 2.30

$$E_1 + E_2 = E_0 \tag{2.73}$$

$$\therefore\ E_1 + E_2 = E_0 = E_G - E_F + E_{VP} - E_{CN} + E_F \tag{2.74}$$

$$\therefore\ E_0 = E_G + E_{VP} - E_{CN} \tag{2.75}$$

$$n \cdot p = N_C N_V e^{-(E_C - E_V)/kT} = N_C N_V e^{-(E_G)/kT} \tag{2.76}$$

And

$$np = n_i^2 \tag{2.77}$$

From Eqs. (2.76) and (2.77), we get

$$\therefore\ n_i^2 = N_C N_V e^{-(E_G)/kT} \tag{2.78}$$

$$e^{\frac{-(E_G)}{kT}} = \frac{n_i^2}{N_C N_V} \tag{2.79}$$

$$e^{\frac{E_G}{kT}} = \frac{N_C N_V}{n_i^2} \tag{2.80}$$

$$\frac{E_G}{kT} = \ln\left(\frac{N_C N_V}{n_i^2}\right) \text{ or } E_G = kT\ln\left(\frac{N_C N_V}{n_i^2}\right). \tag{2.81}$$

From the equations for P-type semiconductor, we obtain

$$n_p p_p = n_i^2 \tag{2.82}$$

(n_p = concentration of electrons that is minority carries in P-material and p_p = majority carrier holes in P-material). Since $p_p = N_A$, Eq. (2.82) is written as

$$n_p = \frac{n_i^2}{N_A} \tag{2.83}$$

$$n = N_C\, e^{-\left(\frac{E_C - E_F}{kT}\right)} \text{ and } \frac{n}{N_C} = e^{-\left(\frac{E_C - E_F}{kT}\right)}$$

$$\frac{N_C}{n} = e^{\left(\frac{E_C - E_F}{kT}\right)} \tag{2.84}$$

$$\therefore \quad \frac{(E_C - E_F)}{kT} = \ln\left(\frac{N_C}{n}\right)$$

$$\frac{(E_C - E_F)}{kT} = \ln\left(\frac{N_C}{n}\right) \tag{2.85}$$

For N-material

$$E_C = E_{CN} \text{ and } n = N_D \tag{2.86}$$

Substitute E_C and n from Eq. (2.86) into (2.85), we get

$$E_{CN} - E_F = kT\ln\left(\frac{N_C}{N_D}\right) \tag{2.86A}$$

Similarly the following equations can be obtained by suitable substitutions:

$$E_F - E_{VP} = kT\ln\left(\frac{N_V}{N_A}\right) \tag{2.87}$$

Substituting Eqs. (2.81), (2.86A), (2.87) into Eq. (2.75), we get
Built-in voltage V_0 or barrier potential = E_0 for diode

$$E_0 = kT\left(\ln\frac{N_C N_V}{n_i^2} - \ln\frac{N_C}{N_D} - \ln\frac{N_V}{N_A}\right)$$

$$E_0 = kT\left[\ln\frac{N_C N_V}{n_i^2} \cdot \frac{N_D}{N_C} \cdot \frac{N_A}{N_V}\right]$$

$$E_0 = kT\ln\left[\frac{N_D N_A}{n_i^2}\right] \tag{2.88}$$

$$E_0 = V_T \ln\left[\frac{N_D N_A}{n_i^2}\right] \text{V} \tag{2.89}$$

where voltage equivalent of temperature $V_T = \frac{\bar{k}T}{e} = kT$

Boltzman constant in eV/°K $= \frac{\bar{k}}{e} = 8.6\times10^{-5}$ eV/°K

and Boltzman's constant $\bar{k} = 1.38\times10^{-23}$ J/°K

Voltage equivalent of temp V_T at $T = 300$°K is equal to

$$V_T = \frac{\bar{k}T}{e} = \frac{1.38\times10^{-23}\times300}{1.6\times10^{-19}} = 2.58\text{ mV} \equiv 2.6\text{ mV}$$

$$\text{or}\quad V_T = kT = 8.6\times10^{-5}\times300 = 2.58\text{ mV} \equiv 26\text{ mV}.$$

Finally, E_0 depends on V_T, that is on temperature and also on the concentrations of P- and N-materials and the intrinsic concentration n_i. As doping level increases V_0 increases, and if doping is zero, as is the case with intrinsic semiconductor or any one type of extrinsic semiconductor, barrier potential V_0 is zero.

EXAMPLE 2.5

If $N_A = 10\ N_D$ and the doping level is such that there are 1.5 doping atoms per million of intrinsic atoms for Silicon; V_0 at 300°K will be 0.7 V as shown below.

Solution: Number of atoms/cm³ of Silicon material = 5×10^{22}

$$\therefore\ N_A = \frac{5\times10^{22}}{1.5\times10^{6}} = 3.33\times10^{16}$$

$$N_D = \frac{N_A}{10} = \frac{3.33\times10^{16}}{10} = 3.33\times10^{15}$$

$$V_0 = V_T\cdot\ln\left[\frac{N_D N_A}{n_i^2}\right]\text{ Volts}$$

We know that $V_T = 26$ mV at 300°K and $n_i = 1.5 \times 10^{10}$.

$$\therefore\ V_0 = 26\text{ mV}\cdot\ln\left[\frac{3.33\times10^{15}\times3.33\times10^{16}}{(1.5\times10^{10})^2}\right]$$

$$V_0 = 0.026\times\ln(4.9284\times10^{11}) = 0.026\times26.92 = 0.7\text{ V}.$$

EXAMPLE 2.6

In open circuited P–N junction, acceptor atom concentration $N_A = 2.5 \times 10^{16}/\text{m}^3$ and donor atom concentration $N_D = 2.5 \times 10^{22}/\text{m}^3$. Intrinsic carrier concentration $n_i = 2.5 \times 10^{19}/\text{m}^3$. Determine the value of contact difference of potential (Aug/Sep 2007, suppl exam).

Solution:

$$\text{Contact potential} \quad V_0 = V_T \cdot \ln \frac{N_D N_A}{n_i^2} \text{ V}$$

$$\therefore \quad V_0 = 26 \text{ mV} \cdot \ln \frac{2.5\times 10^{22} \times 2.5\times 10^{16}}{[2.5\times 10^{19}]^2} = 26 \text{ mV} \times \ln 1 = 0 \text{ V}.$$

EXAMPLE 2.7

The resistivity of the two sides of a step-graded Silicon junction is 5 Ω-cm (P-side) and 2.5 Ω-cm (N-side). Calculate the height of the potential barrier V_0. Consider mobility of holes μ_p = 475 cm²/V-s and μ_n = 1500 cm²/V-s. At the room temperature of 300°K and $n_i = 1.45\times 10^{10}$ atoms/cm³ (Aug/Sep 2007, suppl exam).

Solution:

$$\text{Boltzman constant} = \bar{k} = 1.381\times 10^{-23} \text{ J/°K}$$

$$\text{Boltzman constant } k = \frac{\bar{k}}{e} = \frac{1.381\times 10^{-23} \text{ J/°K}}{1.6\times 10^{-19} \text{ C}} = 8.62\times 10^{-5} \text{ eV/°K}$$

$$\text{Potential barrier} \quad V_0 = \frac{\bar{k}T}{e} \cdot \ln \frac{N_D N_A}{n_i^2} \text{ V} = kT \cdot \ln \frac{N_D N_A}{n_i^2} \tag{1}$$

$$\text{Resistivity of } P\text{-material} = \rho_p = \frac{1}{\sigma_p} = \frac{1}{pQ\mu_p} \ \Omega\text{-cm} \tag{2}$$

$$\therefore \quad \text{Number of holes in P-material} = p = \frac{1}{\rho_p Q \mu_p}$$

$$= \frac{1}{5\times 1.6\times 10\times 10^{-19} \times 475} = 2.63\times 10^{15} \tag{3}$$

$$\therefore \quad p = N_A = 2.63 \times 10^{15}$$

$$\text{Resistivity of } N\text{-material} = \rho_n = \frac{1}{\sigma_n} = \frac{1}{nQ\mu_n} \ \Omega\text{-cm} \tag{4}$$

$$\therefore \quad \text{Number of electrons in } N \text{ material} = n = \frac{1}{\rho_n Q \mu_n}$$

$$= \frac{1}{2.5\times 1.6\times 10\times 10^{-19} \times 1500} = 1.67\times 10^{15} \tag{5}$$

$$\therefore \quad n = N_D = 1.67 \times 10^{15}$$

$$\text{Boltzman constant in electron volts/°K} = \frac{\bar{k}}{e} = 8.62\times 10^{-5} \text{ eV/°K and } T = 300\text{°K}$$

$$\therefore \quad \frac{\bar{k}T}{e} = 8.62\times 10^{-5} \times 300 = 25.86 \text{ mV} \cong 26 \text{ mV}$$

Potential barrier $$V_0 = \frac{\bar{k}T}{e} \cdot \ln \frac{N_D N_A}{n_i^2} \text{ V}$$

$$\therefore \quad V_0 = 26 \times 10^{-3} \ln \frac{1.67 \times 10^{15} \times 2.63 \times 10^{15}}{[1.45 \times 10^{10}]^2} = 0.618 \text{ V}$$

Calculations can be simpler by using simple log for natural logarithms ln using the relation

$$[\ln(a)] = \frac{\log_{10}{}^{a}}{\log_{10}{}^{e}} = 2.303 \times \log_{10}{}^{a}$$

$$V_0 = \frac{\bar{k}T}{e} \times 2.303 \times \log \frac{N_D N_A}{n_i^2}$$

$$= 26 \times 10^{-3} \times 2.303 \times \log \frac{1.67 \times 10^{15} \times 2.63 \times 10^{15}}{[1.45 \times 10^{10}]^2}$$

$$= 0.618 \text{ V.}$$

2.7.2 Open Circuited Junction of P–N Diode

When an intrinsic semiconductor wafer (Silicon or Germanium) is doped with acceptor (P) and donor (N) impurities from either side, a P–N junction diode is formed. Due to very close contact between P- and N-type semiconductors, a junction is formed between them. Figure 2.31(1) shows the P–N diode with a junction, space charge region and P- and N-type materials with two external connecting terminals (anode and cathode).

- The mobile charges (majority carriers) from the two regions cross the junction due to concentration gradient between the charges shown in Fig. 2.31(2).
- This movement causes exposure of immobile charges, or ions on either side of the junction forming space charge region. It is also known as transition or depletion region. *The space charge region behaves as a dielectric or non-conductive layer*. It is of the order of 1 μ.
- Negative charges in P-material that were neutralised by the mobile Holes get uncovered due to the recombination of Holes with electrons (Fig. 2.31(3)).
- Similar situation arises with the immobile positive charges and the mobile electrons in the N-material (Fig. 2.31(2)).
- This leads to the formation of a potential gradient at the junction Fig. 2.31(3).
- Positive immobile ions with N-type semiconductor and the negative immobile ions with P-type semiconductor form a fictitious voltage known as contact voltage V_0. It is also known as diffusion or barrier or depletion or space charge or built-in potential designated as V_0.
- Contact potential V_0 is developed at the junction of the P–N diode when the diode is not provided with any external bias. It is of the order of 0.2–0.3 V for Germanium diodes and 0.5–0.7 V for Silicon diodes.
- Electrostatic field intensity and potential energy barriers for electrons and Holes at the junction contributing to the contact potential V_0 are as shown in Fig. 2.31(1)–(3).

(1) Diffusing charge profiles in P-type and N-type materials of P–N diode

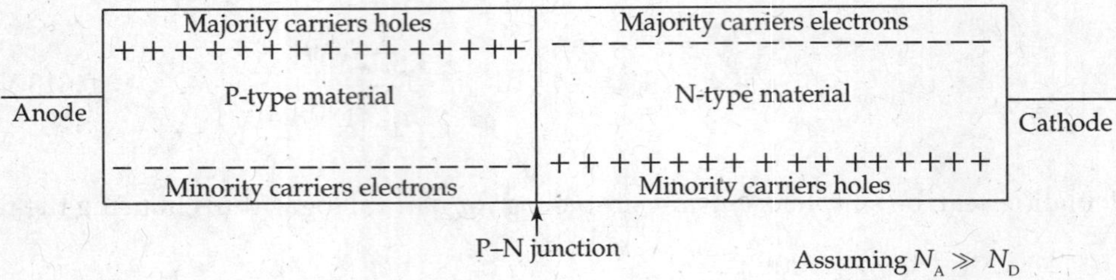

(2) P–N junction diode without external bias

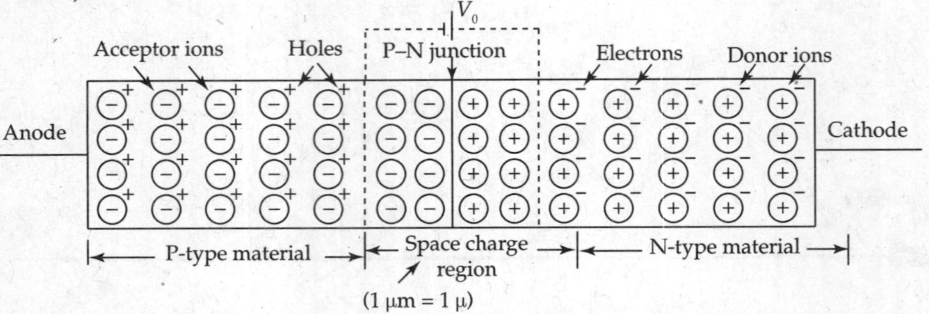

(3) Charge due to immoblie ions in space charge region

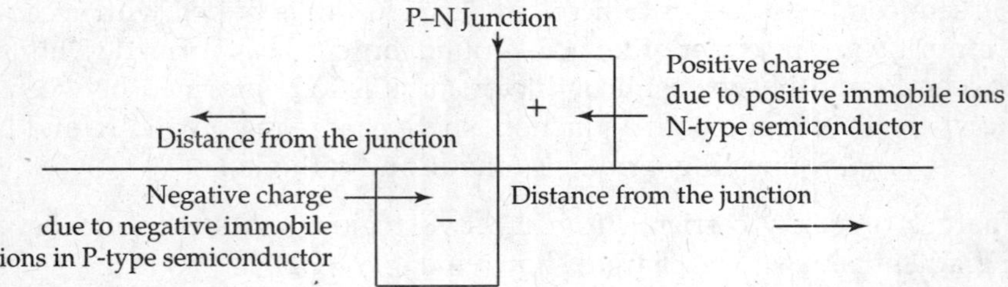

FIG. 2.31 *(1) (2) (3) concepts behind fictitious built-in voltage "V_0"*

- The space charge region will be of the order of 0.5-μm thickness on either side of P–N junction. Variation of charge density is shown in Fig. 2.31(3). This is a plot of charge density due to positive immobile ions in N-material and negative immobile ions in P-material.
- Figure 2.32(2) represents the variation of electric field with distance from the junction, which is proportional to $\int$ of charge density.

$$\frac{d^2V}{dx^2} = \frac{-\rho}{\varepsilon},$$

where ρ is the charge density, ε is the permittivity as obtained from Poisson's equation

$$\varepsilon = -\frac{dv}{dx} \quad \text{or} \quad \int \frac{\rho}{\varepsilon} dx \quad \text{[Fig. 2.31(b)].}$$

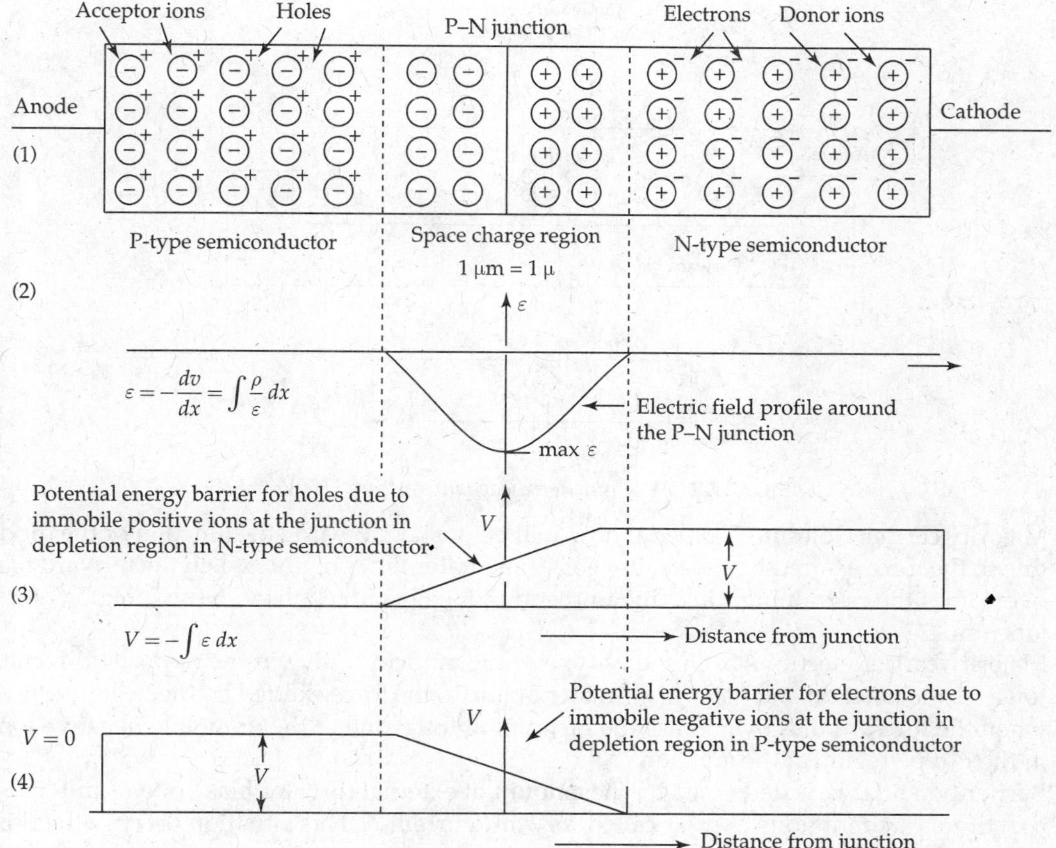

FIG. 2.32 *(1) Semiconductor diode without external bias (2) Sketch of electric field intensity (3) Potential energy barrier for holes in N-type semiconductor (4) Potential energy barrier for elections in P-type semiconductor*

In the N-material, the potential raises from left to right reaching a maximum value of V.

- Figure 2.32(3) represents the variation of potential energy barrier for Holes due to immobile positive ions at the junction in the depletion region of N-material.
- In the P-material, the potential raises from right to left reaching a maximun value of V. Figure 2.32(4) represents the variation of potential energy barrier for electrons due to immobile negative ions at the junction in depletion region of N-type material.

2.8 P–N JUNCTION DIODE (FORWARD BIAS AND REVERSE BIAS TO P–N JUNCTIONS)

2.8.1 Forward-biased P–N (Junction) Diode

When a diode is connected to a DC voltage V_f to make its P-material positive and N-material negative, the P–N junction diode is considered as *forward biased* (*DC source is known as bias*). *Forward bias opposes the built-in* (*fictitious*) *voltage* V_0 (Built-in voltage cannot be measured and it is only the concept arrived from experimental results.).

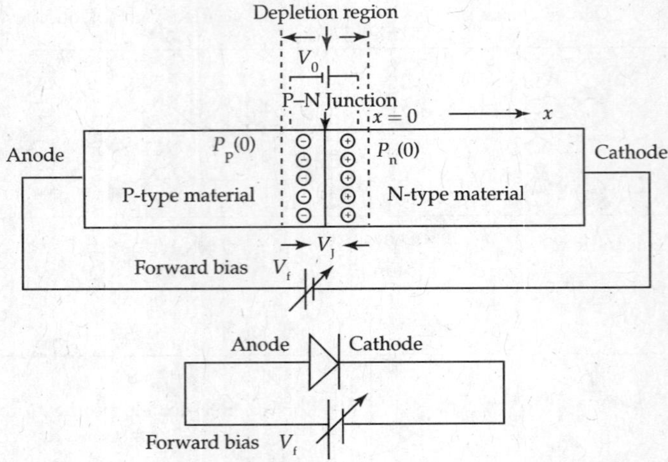

FIG. 2.33 *P–N junction diode with forward bias V_f*

- Majority carrier Holes in P-material now will be pushed towards the junction of the diode due to the force exerted by the positive of V_f and enter the N region, when the forward bias overcomes the restraining force due to the immobile positive ions in N-type region at the junction.
- Majority carrier electrons from the N-type semiconductor move in the opposite direction towards the junction and enter the P-material due to the force exerted by the negative of V_f, when the forward bias overcomes the restraining force due to the immobile negative ions in the P-type region at the junction.
- Barrier potential V_0 is decreased by the amount of external forward bias voltage and hence constitutes a continuous current called forward current I_f. This situation occurs when the forward bias voltage overcomes the restraining force by the contact or diffusion potential or voltage V_0.
- Reduction of 'space charge region width' of forward-biased diode (Fig. 2.34).

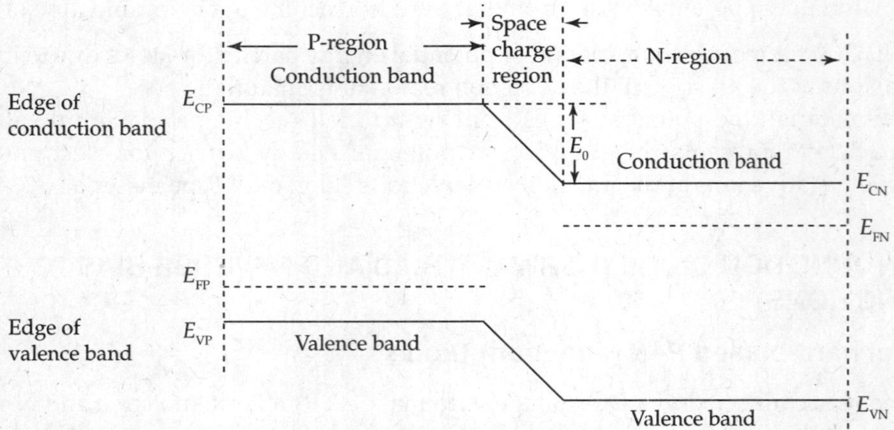

FIG. 2.34 *Energy-band diagram of a P–N junction diode win forward bias V_f, illustrating the reduction of space charge region*

Quantitative relationship between current and voltage is shown in the following manner.
From continuity equation, we can obtain expression for the current by considering the net current due to drift and diffusion phenomena. Diffusion current is due to concentration gradient among the charge carriers on either side of the P–N junction and drift current is due to potential gradient.

Diffusion current is due to Holes (being majority carriers) in P-material crossing the P–N junction and becoming the minority carriers (injected carriers) in the N-material region. The injected minority carrier current or the excess minority carrier current into the N-material can be defined as

$$I_{\text{pn}}(0) = Jp_{\text{n}}(0) \times A = -AqD_{\text{p}}\left(\frac{dp_{\text{n}}}{dx}\right) = \frac{\left[AqD_{\text{p}}p'_{\text{n}}(0)\right]}{L_{\text{p}}}. \tag{2.90}$$

$I_{\text{pn}}(0)$ represents the value of current at P–N junction for the Holes entering N-material region. This is actually $I_{\text{pn}}(x)$ when $x = 0$, where x is the distance from the junction.

On the P-material side, p_{p0} represents the equilibrium majority concentration of the Holes. As these enter the N-material region at the junction, they become $p_{\text{n}}(0)$.

This represents excess minority carriers (Holes) injected from P-material into the N-material and add to the local minority carriers in the N-material with distance. This falls off exponentially due to recombination with the majority carriers (electrons). This is represented as follows.

$$I_{\text{pn}}(0) = Jp_{\text{n}}(0) \cdot A = -Aq\left(D_{\text{p}}\frac{dp_{\text{n}}}{dx}\right) = \frac{(AqD_{\text{p}}p'_{\text{n}}(0))}{L_{\text{p}}}. \tag{2.91}$$

$$p_{\text{n}} = p_{\text{n}}(x) \cdot e^{-x/L_{\text{p}}} \tag{2.92}$$

[Equation (2.92) is obtained from continuity equation as shown below]. It is known that

$$\frac{dp_{\text{n}}}{dx} = -\frac{1}{L_{\text{p}}}p_{\text{n}}(x)\, e^{-x/L_{\text{p}}} \tag{2.93}$$

$$\text{at } x = 0, \quad \frac{dp_{\text{n}}}{dx} = -\left[\frac{1}{L_{\text{p}}}\right] \cdot p'_{\text{n}}(0), \tag{2.94}$$

where $p'_{\text{n}}(0)$ represents excess minority carriers entering into the N-material region. When Eq. (2.94) is substituted in Eq. (2.91), the resulting equation is

$$I_{\text{pn}}(0) = -AqD_{\text{p}}\left(-\frac{1}{L_{\text{p}}}\right)p'_{\text{n}}(0) = \frac{AqD_{\text{p}}p'_{\text{n}}(0)}{L_{\text{p}}}. \tag{2.95}$$

In Eq. (2.95), $p'_{\text{n}}(0)$, that is the magnitude of the injected minority carriers depends upon applied electrical potential. Quantitative relationship between $p'_{\text{n}}(0)$ and V is going to be derived in the succeeding section.

2.9 THE LAW OF JUNCTION

Across P–N junction of the semiconductor diode, electric field is very high (because of very small space charge region of order of 0.5 μ). The diffusion current is also very high due to very high concentration charge gradient. Since those oppose each other and large the net current density will be very small. So if the difference of these two large quantities is very small they should be very nearly equal:

J_p = Drift component of current – Diffusion component of current

$$J_p = q\mu_p p\varepsilon_p - qD_p \cdot \frac{dp}{dx} \equiv 0. \tag{2.96}$$

The next step is to equalise these two, the drift component of current and the diffusion component of current, as per the above reasoning:

$$q\mu_p p \cdot \varepsilon_p = qD_p \cdot \frac{dp}{dx} \tag{2.97}$$

$$\varepsilon_p = \frac{D_p}{\mu_p} \cdot \frac{1}{p} \cdot \frac{dp}{dx} = \frac{V_T}{p} \cdot \frac{dp}{dx} = -\frac{dV}{dx} \tag{2.98}$$

$$\because \quad \frac{D_p}{\mu_p} = V_T. \tag{2.99}$$

Therefore, in Eq. (2.98) p represents the equilibrium concentration of Holes on the P-material. The Holes enter at the P–N junction of the diode at $x = 0$ into the N-region and they become the injected minority carrier Hole concentration. Notionally, the concentration of Holes entering the N-region at the junction at $x = 0$ is termed as $p_n(0)$. The junction voltage is $(V_0 - V)$, where V_0 is the barrier or contact potential and V_f is the applied forward bias.

$$\varepsilon = -\frac{dv}{dx} = \left(\frac{V_T}{p}\right) \cdot \left(\frac{dp}{dx}\right) \tag{2.100}$$

From Eq. (2.100), we get

$$\frac{dp}{p} = -\frac{dv}{V_T} \tag{2.101}$$

Integrating Eq. (2.101), we get

$$\int_{p_p(0)}^{p_n(0)} \left(\frac{dp}{p}\right) = \int_0^{v_0} \frac{dv}{V_T} = \int_0^{(v_0 - v)} \frac{dv}{V_T} = [I_n \cdot p]_{p_p(0)}^{p_n(0)} = -\frac{(V_0 - V)}{V_T}$$

$$I_n \cdot p_n(0) - I_n \cdot p_p(0) = \frac{-(V_0 - V)}{V_T} = \frac{-V_0 + V}{V_T}$$

$$= [I_n p]_{p_p(0)}^{p_n(0)} = -\frac{(V_0 - V)}{V_T}$$

$$\frac{p_n(0)}{p_{p_0}} = e^{\frac{-(V_0 - V)}{V_T}} \tag{2.102);}$$

$$\therefore \quad p_n(0) = p_{p0} e^{-\frac{(V_0 - V)}{V_T}} \tag{2.103}$$

$$p_n(0) = p_{p_0} e^{\frac{-V_0 + V}{V_T}} = p_{p0} \cdot \left[e^{-V_0/V_T} \cdot e^{V/V_T}\right].$$

But it is known that $p_{p0} = p_{n0} \cdot e^{V_0/V_T}$.

From the derivation for V_0 in terms of concentration on P-material side and N-material side of the Holes, where p_{p0} represents holes in P-region, p_{n0} represents injected minority carriers in the N-region; under equilibrium conditions

$$p_{p0} = p_{n0} \cdot e^{V_0/V_T} \tag{2.104}$$

Or $p_{n0} = p_{p0} \cdot e^{-V_0/V_T}$; or combining Eqs., (2.103) and (2.104)
Equation for law of junction (2.105) follows:

$$\frac{p_n(0)}{p_{n0}} = e^{V/V_T} \tag{2.105}$$

$$p_n(0) = p_{n0} \cdot e^{V/V_T} \tag{2.106}$$

Equation (2.106) is called as *law of junction* at the boundary of the P–N junction formed by the P- and N-type semiconductor materials of the P–N diode. This law indicates that injected minority carrier Hole concentration $p_n(0)$ at the P–N junction into the N-material is obtained by multiplying the equilibrium minority concentration p_{n0} by e^{V/V_T}; that is the injected carrier current due to holes $p_n(0)$ exponentially increases with respect to the applied forward bias voltage V.

2.10 DIODE EQUATION (CURRENT COMPONENTS IN A P–N SEMICONDUCTOR DIODE)

For a forward-biased semiconductor (P–N) diode with forward bias V_f, Holes are pushed into the N-type semiconductor region and electrons are pushed into the P-type semiconductor region. The number of charge carries that move through the semiconductor materials on either side of the P–N junction contributes to various components of currents. At the same time, it is to be remembered that the magnitudes of currents, under the application of voltages to the P–N diode, increase or decrease along with charge distributions.

Consider the charge flow in the N-material region due to the injected excess minority carrier Hole concentration, $p'_n(0)$, equilibrium minority carrier Holes, p_{n0}, and the total minority carriers, $p_n(0)$. $p^1_n(0)$ is the excess minority carriers entering N-type region $= p_n(0) - p_{n0}$, where $p_n(0)$ is the total minority carriers and p_{n0} is the equilibrium minority carriers in the N-type region.

So the total should be the sum of the above.

$$p_{n0} + p'_n(0) = p_n(0) \tag{2.107}$$

$$I_{pn}(0) = -AqD_p \frac{dp}{dx} = AqD_p \cdot \frac{p'_n(0)}{L_p} \tag{2.108}$$

As already derived in Eq. (2.90)

$$\therefore \quad p'_n(0) = p_n(0) - p_{n0} \tag{2.109}$$

Using Eq. (2.106) for law of junction in Eq. (2.109), we get

$$p'_n(0) = p_{n0} \cdot e^{V/V_T} - p_{n0} \tag{2.110}$$

$$p'_n(0) = p_{n0}(e^{V/V_T} - 1) \tag{2.111}$$

$$I_{pn}(0) = \frac{AqD_p p_{n0}}{L_p}(e^{V/V_T} - 1). \tag{2.112}$$

Now substituting this value of $p'_n(0)$ in Eq. (2.108) for current $I_{pn}(0)$

Similarly, the majority carrier electrons crossing from N-region into P-region constitute a current I_{np} (0) as mentioned in Eq. (2.113):

$$I_{np}(0) = \frac{AqD_n n_{p0}}{L_n}(e^{V/V_T} - 1). \tag{2.113}$$

The total current I contribution from the semiconductor diode is the sum of the two quantities $I_{pn}(0)$ and $I_{np}(0)$ due to the holes moving from P- to N-material contributing to current $I_{pn}(0)$ from P- to N-side and the current $I_{np}(0)$ from P- to N-side due to the movement of electrons from N- to P-material through the semiconductor diode considered from Eqs. (2.112) and (2.113):

$$I = I_{pn}(0) + I_{np}(0) \tag{2.114}$$

$$\therefore \quad I = I_{pn}(0) + I_{np}(0) = Aq(e^{V/V_T} - 1)\left[\frac{D_p p_{n0}}{L_p} + \frac{D_p n_{p0}}{L_n}\right] \tag{2.115}$$

$$I = I_0(e^{V/V_T} - 1), \tag{2.116}$$

$$\text{where} \quad I_0 = Aq\left[\frac{D_p p_{n0}}{L_p} + \frac{D_n n_{p0}}{L_n}\right]. \tag{2.117}$$

I = Forward current of the diode I_0 is called reverse saturation current.

If the applied forward bias is much greater than V_T, $I = I_0 \cdot e^{V/V_T}$. Then the current I becomes I_f, for the forward-biased semiconductor diode.

This condition is basically obtained since $V_T = kT/e = 26$ mV at T = 300°. When voltage V $\gg V_T$ with V positive current I increases exponentially with voltage.

The junction potential or the built-in voltage becomes

$$V_J = (V_0 - V) \tag{2.118}$$

as already explained, where V_0 is the contact potential and V is the applied external potential. If V is positive, junction potential will be decreased by the amount V from V_0 and the barrier height is lessened (contact potential or barrier voltage V_0 is reduced with a consequence of reduction in the depletion region width W_f).

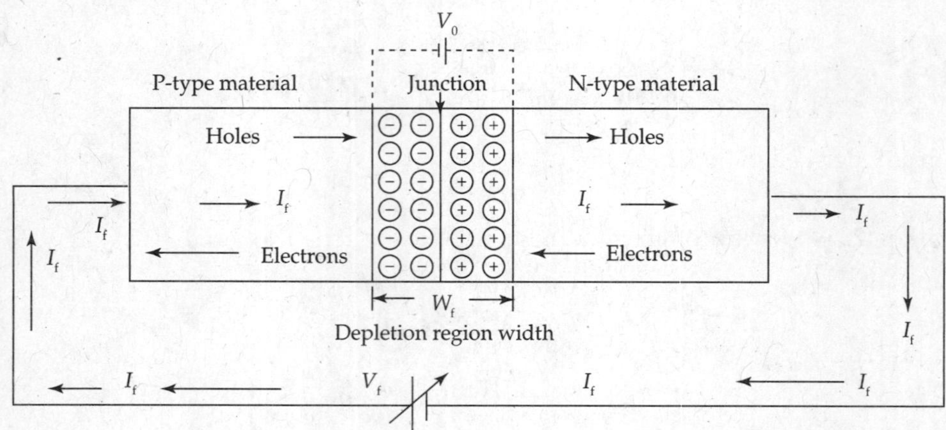

FIG. 2.35 *P–N semiconductor diode with forward bias V_f*

The discussion by now makes it clear that the current I known as the forward current I_f through the forward-biased semiconductor diode is mainly due to the *flow of majority carriers* through the device (Fig. 2.35).

In a forward-biased P–N diode, the forward current I_f enters the P-material side as hole current and leaves the N-material side as electron current of the same magnitude:

$$I_f = I_0(e^{V/V_T} - 1) \quad \text{(diode equation).} \tag{2.119}$$

Equation (2.119) *representing the forward current of the diode is known as diode equation.*

2.10.1 Current Components in a Reverse-Biased Diode (Reverse-Biased P–N Junction Diode)

Figure 2.36 shows the method of reverse biasing the P–N diode. Negative terminal of voltage V_r is connected to P-type material and positive terminal of the voltage V_r is connected to N-type material of P–N diode to reverse bias the P–N diode.

The reverse bias is with the same polarity as the built-in voltage. So the built-in voltage increases and results in an increase in depletion region width W_r shown in Fig. 2.37. Energy-band diagram (Fig. 2.37) shows this feature.

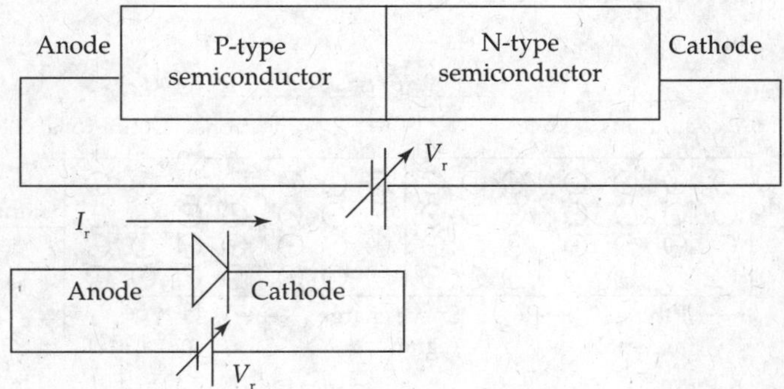

FIG. 2.36 *P–N diode with reverse bias V_r*

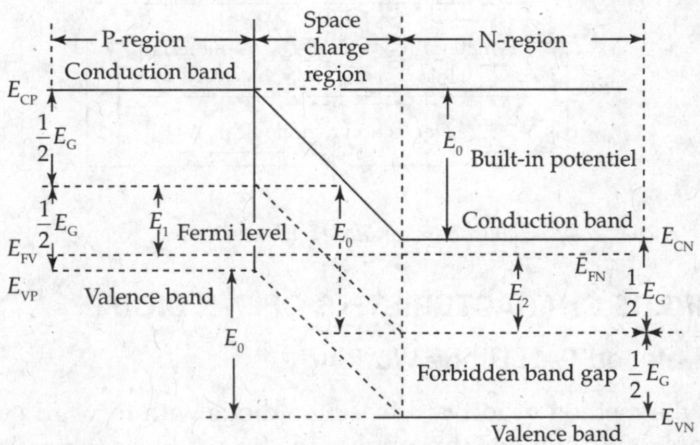

FIG. 2.37 *Energy-band diagram of P–N junction diode with reverse bias V_r with an increase in built-in potential E_0 and the width of the space charge region*

Energy-Band Diagram for P–N diode with Reverse Bias
If V is negative and $\gg V_T$, then the current through the device, $I = -I_0$, since e^{-V/V_T} substituted with $|V| \gg V_T$ is negligible. It means that under reverse-biased condition the current reaches a saturation value I_0, if $|V| \gg V_T$.
If V is negative

$$V_J = V_0 - (-V) = (V_0 + V). \tag{2.120}$$

So the barrier height is raised with a consequence of increase in the depletion region width W_r about the junction of the P–N diode.

The majority carriers, Holes from P-material and the majority carrier electrons from the N-material move away from the P–N junction due to the pulling forces applied by the reverse bias. This results in uncovering more immobile charges near the junction resulting in more immobile ions about the junction. Therefore, the barrier potential V_0 and the depletion region width (W_r) are increased as shown by the diode equation (Eq. (2.119)).

Reverse saturation current I_r or I_0 flows through the diode from N- to P-materials due to the movement of minority charge carriers through the device. Hence, the reverse saturation current I_0 or I_r or I_S is negligibly small. It is a few microamperes for a Germanium device and a few nanoamperes for a Silicon device.

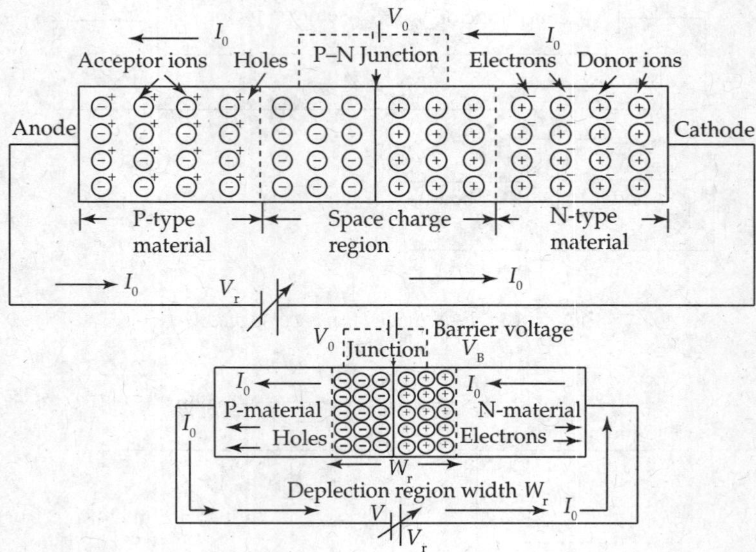

FIG. 2.38 *Semiconductor diode with reverse bias voltage V_r*

2.11 VOLT–AMPERE CHARACTERISTICS OF P–N DIODE

2.11.1 Forward-biased P–N Diode Working

Figure 2.39 shows the method of forward biasing a diode with forward bias V. But V_f is the actual voltage across the P–N diode that applies the forward bias to the P–N diode, because of the use of the current-limiting resistor R_L in the circuit. Forward bias reduces the depletion region width causing for movement of charge carries through the device.

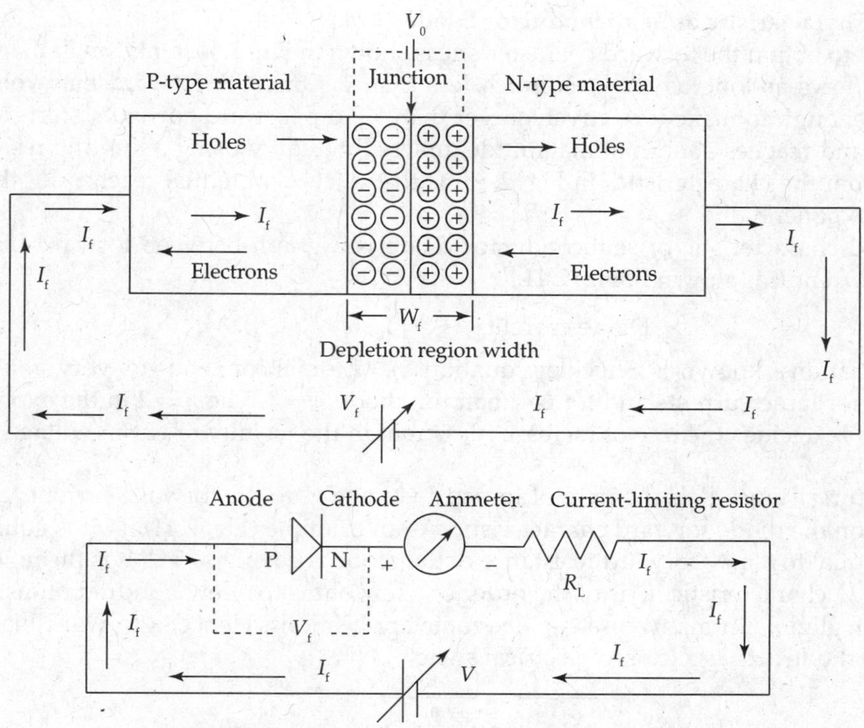

FIG. 2.39 *P–N semiconductor diode with forward bias V_f*

From the expression

$$I_f = I_0(e^{V_f/V_T} - 1) \quad \text{(Shockley equation)}, \tag{2.121}$$

it can be seen that the diode conduction does not start, till the forward bias voltage V_f reaches a particular value. This can also be seen from the forward characteristic shown in Fig. 2.41.

Semiconductor diode conduction can be explained qualitatively by considering the two types of forces on the majority charge carriers on either P- or N-type semiconductor materials.

One force is due to the applied forward bias $F_{V_f} = e \cdot V_f$. It pushes the majority carriers say Holes in P-material by positive of V_f into N-material. At the same time, the force $F_{V_0} = e \cdot V_0$ due to positive immobile ion charge in N-material (due to space charge about junction) restrains the entry of Holes from P- into N-materials as long as F_{V_f} is less than F_{V_0}.

Once F_{V_f} is larger than F_{V_0}, Holes from the P-material enter the N-material. They continue their journey to negative of forward bias V_f. These Holes will be supplemented by positive of V_f into the device to maintain continuous flow of charge carriers through the device and the circuit. Flow of majority carriers in this process contributes to the forward current I_f (The situation will be similar with majority carrier electrons from the N-material.).

Minimum forward bias voltage required for diode conduction is known as the cut-in voltage V_γ. It is also known as offset voltage or threshold voltage. It is of the order of 0.2–0.3 V for Germanium devices such as diode, transistors, etc. The cut-in voltage is of the order of 0.5–0.7 V for Silicon diodes, transistors, etc at 300°K.

Forward Characteristic of Semiconductor Diode

The circuit to obtain the forward characteristic is shown in Fig. 2.40. Only small magnitude of current I_f flows as long as e^{V_f/V_T} is much less than 1. Once the forward bias voltage V_f is sufficient enough to make it relatively larger than V_T, the forward current I_f starts increasing suddenly and reaches abnormal magnitude, unless limited by using a limiting resistance R_L. As seen from the characteristics in Fig. 2.41, as the junction potential reaches V_0, the current increases exponentially.

Forward characteristic of semiconductor diode is a graph between forward bias V_f and forward current I_f as shown in Fig. 2.41.

$$\text{Diode current} \quad I_D = I_0 \cdot (e^{V_f/\eta V_T} - 1). \tag{2.121A}$$

This equation is known as 'Shockley equation'. $\eta = 2$ for Silicon diode for very small currents and $\eta = 1$ for large currents and for Germanium diode $\eta = 1$. The $\eta = 2$ in the expression for Silicon diode decides the rate of increase in current in the vicinity of cut-in voltage as shown in Fig. 2.41.

Forward resistance R_f is the ratio of forward voltage V_f to the forward current I_f at a point as shown on the diode forward characteristic. As an example (Fig. 2.41) if V_{f1} is equal to 0.6 V and I_{f1} is equal to 8 mA, forward resistance of the diode is equal to 75 Ω. It is the inverse slope of the $I_f - V_f$ characteristic. R_f is of the order of a few ohms to a few hundred ohms. It will be virtually negligible in many cases in electronic applications. Hence, a forward-biased semiconductor diode acts as a 'closed electrical switch'.

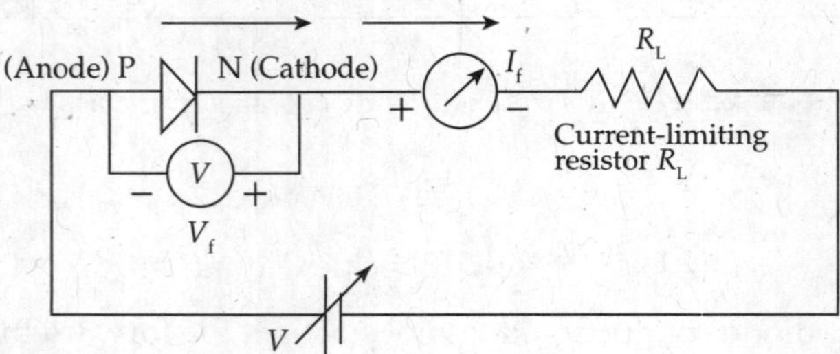

FIG. 2.40 *Forward biased P–N diode circuit to obtain forward characteristic*

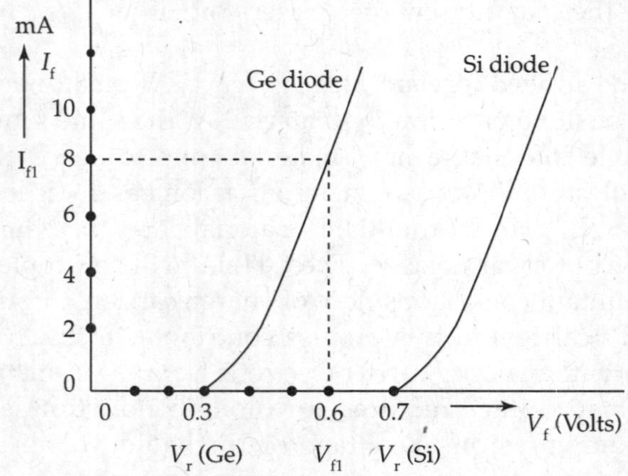

FIG. 2.41 *Forward characteristics of germanium and silicon diodes*

2.11.2 Reverse-biased Semiconductor Diode Working

When a diode is reverse biased by V_r as shown in Fig. 2.42, the current I_r will be of the order of a few nanoamperes for Silicon diode and a few microamperes for Germanium diode, because the reverse current I_r is due to the flow of minority carriers through the devices. The current I_0 or I_S known as 'reverse saturation current' is independent of the reverse bias voltage up to a certain value. It suddenly shoots high resulting in the breakdown of the diode. The voltage at which the diode breaks down is known as 'breakdown voltage', V_{br}. Breakdown voltage rating will be provided in manufacturer's data manuals of semiconductor diodes. For normal applications, this breakdown region is inoperable and destroys the diode.

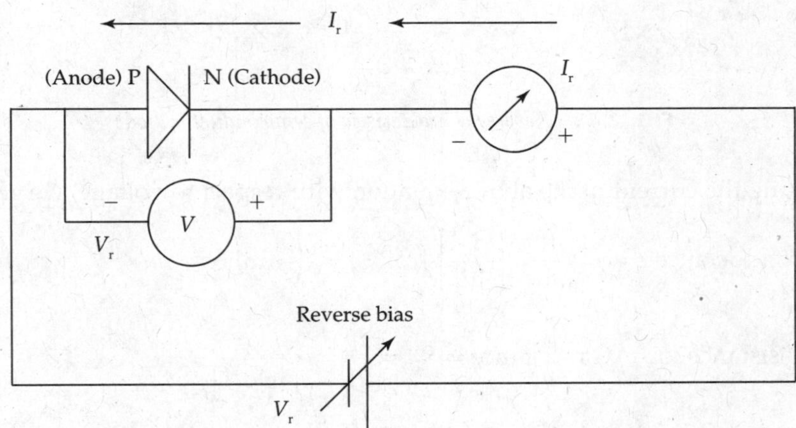

FIG. 2.42 *Reverse biased P–N diode circuit to obtain reverse characteristic*

Sudden increase in the reverse current is due to breaking of covalent bonds resulting in rushed current due to high field of the reverse voltage V_r. By adjusting the parameters of the device, this region can be made operable, as is the case with the so-called Zener or breakdown diode. Below this knee region reverse resistance R_r is of the order of mega-ohms for Germanium diode and hundreds of kilo-ohms for Silicon diodes.

Reverse characteristic of semiconductor diode

When the diode is operated with reverse bias V_r well below the diode rated breakdown voltage, the semiconductor diode works as open electrical switch with very high value of reverse resistance. The reverse current I_r will be of the order of a few hundred microamperes for a Germanium diode and a few hundred nanoamperes for a Silicon diode. The reverse characteristic is shown in Fig. 2.43.

Considering V_{r1} of magnitude 25 V and the corresponding reverse current I_{r1} of magnitude 50 μA on the diode reverse characteristic, the value of reverse resistance R_r, which is the ratio of V_{r1} and I_{r1} is equal to 500 kΩ.

Dynamic or AC resistance R_D can be calculated as follows:

From Eq. (2.116), $I = I_0(e^{V/\eta V_T})$ for a forward-biased diode.

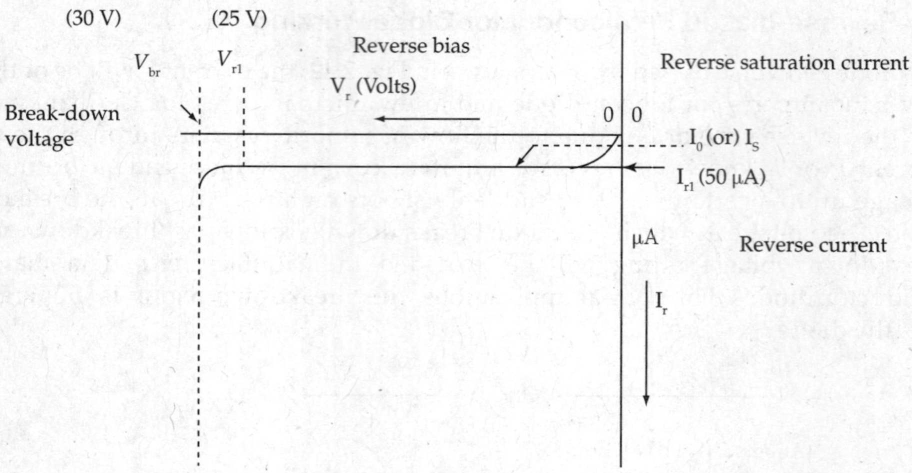

FIG. 2.43 *Reverse characteristic of semiconductor diode*

Differentiating the current in the above equation with respect to voltage, we get

$$\text{Dynamic conductance} \quad g_D = \frac{dI}{dv} = \left(\frac{I_0}{\eta V_T}\right) \cdot \left(e^{V/\eta V_T}\right) = \frac{I_f}{\eta V_T} \text{ mhos} \tag{2.122}$$

$$\text{Dynamic resistance or AC resistance} = R_D = \frac{dv}{dI} = \frac{\eta V_T}{I_0\left[e^{V/\eta V_T}\right]}$$

$$\therefore \quad R_D = \frac{\eta V_T}{[I_f + I_0]} \equiv \frac{\eta V_T}{I_f}\, \Omega \tag{2.122A}$$

When $\eta = 1$ for germanium device $R_D \cong \frac{V_T}{I_f}\, \Omega$

When $\eta = 2$ for silicon device $R_D \equiv \frac{V_T}{2I_f}\, \Omega$

where R_D is also known as 'differential resistance' from the mathematical operation.

- At room temperatures for a Germanium diode, for a forward current of 26 mA and V_T of 26 mV at room temperature of 300°K. Hence, dynamic conductance of Germanium diode under forward-biased condition $= g_f = dI/dv = 1$ mho.
- For a Silicon diode ($\eta = 2$ for small currents) for a forward current of 26 mA and V_T of 26 mV. Conductance $G_C = dI/dv$ is 0.5 mhos. Therefore, dynamic conductance of Silicon diode under forward-biased condition $= g_f = dI/dv =$ 0.5 mhos.

Dynamic or AC forward resistance $R_{Df} = \frac{dv}{dI} = \frac{\eta V_T}{I_f} = 1\,\Omega$ *for a Germanium diode ($\eta = 1$) and 2 Ω for a Silicon diode ($\eta = 2$)*

Dynamic resistance R_D of a semiconductor diode can be calculated from its characteristics as shown in Fig. 2.44.

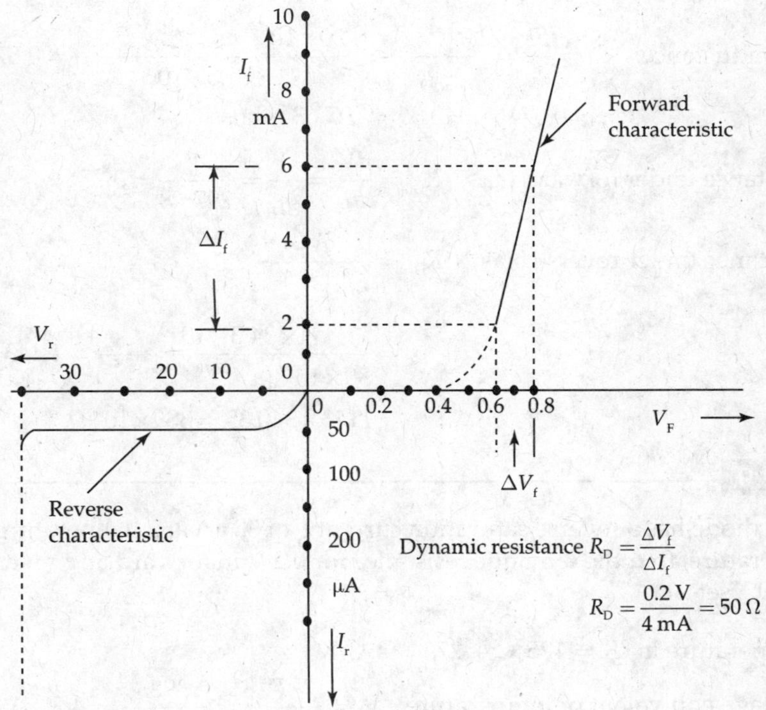

FIG. 2.44 *Illustration for dynamic resistance R_D from diode characteristics*

2.11.3 Diode Ratings or Specifications

(1) Peak forward current; (2) Max anode to cathode voltage during conduction; (3) Max reverse voltage (peak inverse voltage) when the diode is not conducting; (4) Peak current specification restricting the value of filter capacitance that can be used with diode as rectifier; (5) Breakdown voltage rating of diodes.

EXAMPLE 2.8

A P–N junction diode has a reverse saturation current of 30 μA at a temperature of 125°C. At the same temperature, find the dynamic resistance for 0.2 V bias in forward and reverse directions.

Solution: Forward current $I = I_0\left[e^{V/\eta V_T} - 1\right]$,

where I is the current under forward bias to the diode, I_0 is the reverse saturation current, V is the applied voltage to the diode.

$$V_T = \text{Voltage equivalent of temp } \frac{\bar{K}T}{e} = \frac{T}{11{,}600}$$

At temperature $T = (273 + 125) = 398^\circ\text{K}$,

$$V_T = \frac{T}{11{,}600} = \frac{398}{11{,}600} = 34.3 \text{ mV}$$

Dynamic conductance $g_D = \frac{dI}{dv} = \frac{I_0 \cdot [e^{V/\eta V_T}]}{\eta V_T} = \frac{30\times 10^{-6} \times e^{(0.2)/(1\times 34.3\times 10^{-3})}}{1\times 34.3\times 10^{-3}} = 0.8746\times e^{5.83}$

$\therefore \; g_D = 0.8746\times 340.35 = 297.68$ mhos

Dynamic resistance under forward bias $R_{Df} = \frac{dV}{dI} = \frac{1}{g_D} = \frac{1}{297.68} = 3.35\,\Omega$

Dynamic resistance under reverse bias $R_{Dr} = \frac{\eta V_T}{I_0 \left[e^{-V/\eta V_T}\right]}$

$$= \frac{1\times 34.3\times 10^{-3}}{30\times 10^{-6}[e^{-(0.2)/(1\times 34.3\times 10^{-3})}]} = 1.143\times 10^3 \times e^{5.83}$$

$$= 1143\times 340.35 = 389\times 10^3\ \Omega.$$

EXAMPLE 2.9

A P–N junction diode has a reverse saturation current I_r of 30 μA at a temperature of 125°C. At the same temperature, find the dynamic resistance for 0.2 V in forward and reverse directions (April/May 2007, set-2).

Solution: Temperature in °K = 125°C + 273 = 398°K

Voltage equivalent of temperature $V_T = \frac{T}{11,600} = \frac{398}{11,360} = 34.3$ mV

Forward current $I_f = I_r\left[e^{V_f/\eta V_T} - 1\right]$

Assumnig silicon diode $\eta = 2,$

Forward current $I_f = I_r\left[e^{V_f/\eta V_T} - 1\right]$

Differentiating the above equation w.r.t. V_f, we get

$$\frac{\delta I_f}{\delta V_f} = \frac{I_r}{2V_T} \cdot e^{V_f/2V_T}$$

Dynamic resistance of the diode $R_D = \frac{\delta V_f}{\delta I_f} = \frac{2V_T}{I_r} \cdot e^{-V_f/2V_T}\ \Omega$

$$= \frac{2\times 34.3\times 10^{-3}}{30\times 10^{-6}} \cdot e^{-(0.2)/(2\times 34.3\times 10^{-3})}$$

$$= 2286.66\times e^{-2.915} = 12.389\ \Omega$$

Dynamic resistance when reverse biased $R_{Dr} = \frac{2V_T}{I_r} \cdot e^{V_r/2V_T}$

$$= \frac{2\times 34.3\times 10^{-3}}{30\times 10^{-6}} \cdot e^{(0.2)/(2\times 34.3\times 10^{-3})} = 42.21\ \text{k}\Omega.$$

2.12 TEMPERATURE DEPENDENCE OF *V–I* CHARACTERISTICS (DIODE CURRENT)

Since I_0 and V_T are temperature dependent, both are taken into consideration for arriving at variation of I_0 with temperature. It is already seen that n_i^2 is temperature dependent and it is expressed as

$$I_0 = K \cdot T^m e^{-V_{G0}/\eta V_T},$$

where V_{G0} is the voltage numerically equivalent to band-gap energy and V_T is the volt equivalent of temperature. It is of the order of 26 mV at room temperature.

In the expression for I_0, the diffusion constants are involved and varying inversely with temperature. The temperature dependence of I_0 can be expressed as $I_0 = K_1 T^2 e^{-V_{G0}/\eta V_T}$. In the so far discussion, the generation and recombination of electron–hole pairs in the space charge region are neglected and are true for Germanium but not for Silicon. So the expression for the current I has to be modified as $I = I_0(e^{V/\eta V_T} - 1)$, where $\eta = 1$ for large currents and $\eta = 2$ for small currents. Also it is practically found that I_0 is directly proportional to n_i but not n_i^2. All these conditions can be expressed to get $I_0 = K_2 T^{1.5} \cdot e^{-V_{G0}/\eta V_T}$.

Generally, this relation can be expressed as

$$I_0 = KT^m \cdot e^{-V_{G0}/\eta V_T}. \tag{2.123}$$

For a Germanium device, $\eta = 1$, $m = 2$ and $V_{G0} = 0.785$ V. And for a Silicon device, $\eta = 2$, $m = 1.5$ and $V_{G0} = 1.21$ V.

Taking logarithms on both sides of Eq. (2.123), we get

$$\log I_0 = \log K + m \log T\left(-\frac{V_{G0}}{\eta V_T}\right). \tag{2.124}$$

We know that

$$V_T = \frac{T}{11{,}600}. \tag{2.125}$$

Using the value of V_T from Eq. (2.125) in Eq. (2.124), we get

$$\log I_0 = \log K + m \log T - \left(\frac{V_{G0}(11{,}600)}{\eta T}\right). \tag{2.126}$$

Differentiating Eq. (2.126) with respect to temperature T, we get

$$\frac{1}{I_0} \cdot \frac{dI_0}{dT} = 0 + \frac{m}{T} + \frac{V_{G0}(11{,}600)}{\eta T^2}. \tag{2.127}$$

Again using the value of V_T from Eq. (2.125) in Eq. (2.127), we get

$$\frac{1}{I_0} \cdot \frac{dI_0}{dT} = \frac{m}{T} + \frac{V_{G0}}{\eta T V_T}. \tag{2.128}$$

The reverse saturation current doubles for every 10°C rise in temperature both for Germanium and for Silicon devices approximately.

Using Eqs. (2.119) and (2.127), it can be derived that V is also dependent on temperature and approximately varies as shown in the following expressions:

$$\frac{dV}{dT} = -2.3 \text{ mV/°C} \quad \text{for silicon devices,} \tag{2.129}$$

$$\frac{dV}{dT} = -2.1\,\text{mV}/^\circ\text{C} \quad \text{for germanium devices.} \tag{2.130}$$

For practical design considerations, it is assumed as

$$\frac{dV}{dT} = -2.5\,\text{mV}/^\circ\text{C} \quad \text{for both silicon and germanium devices.} \tag{2.131}$$

Equation (2.131) suggests that dV/dT decreases with increasing temperatures for the semiconductor diodes. As temperature increases, the forward bias voltage of a diode decreases. As a result, the forward characteristic of a diode moves to left. For every 1°C increase in temperature, the forward voltage decreases approximately by 2 mV. An increase in temperature causes an increase in intrinsic carrier density, a decrease in band gap causes an increased value of diode current. It decreases the carrier mobility causing a decrease in current. All these three factors lead to a net result that is with increase in temperature, diode current increases. An increasing diode current leads to power dissipation in the diode, which ultimately leads to further increase in temperature.

EXAMPLE 2.10

A certain P–N junction diode has a leakage current of 10^{-13} A at room temperature of 27°C and 10^{-9} A at 125°C. The diode is forward biased with a constant current source of 1 mA at room temperature. If current is assumed to remain constant, calculate the barrier voltage at room temperature and at 125°C.

Solution:

$$\frac{1}{I_0}\cdot\frac{dI_0}{dT} = \frac{V_{G0}}{\eta V_T}$$

$$\therefore \text{ Barrier voltage } V_{G0} = \frac{\eta V_T}{I_0}\cdot\frac{dI_0}{dT} \tag{1}$$

Assume $\eta = 1$ for low values of device currents of 1 mA.

$$V_T = \frac{kT}{e} = \frac{T}{11,600} = \frac{300}{11,600} = 25.86\text{ mV} \equiv 26\text{ mV}$$

At room temperature

$$T = 27 + 273 = 300^\circ\text{K}$$

$$I_0 = 10^{-13}\text{ A}$$

$$dI_0 = (10^{-9} - 10^{-13})$$

$$= \left[\frac{1}{10^9} - \frac{1}{10^{13}}\right] = \frac{[10^4 - 1]}{10^{13}}$$

$$\equiv \frac{10^4}{10^{13}} = 10^{-9}\text{ A.}$$

$$dT = (125 - 27) + 273 = 98 + 273 = 371^\circ\text{K}$$

∴ At temperature of 300°K, from Eq. (1),

$$V_{G0} = \frac{1 \times 26 \times 10^{-3} \times 10^{-9}}{10^{-13} \times 371} = \frac{260}{371} = 0.7 \text{ V}$$

At temperature of 125°C

$$T \text{ in °K} = 125 + 273 = 398\text{°K}$$

At temperature of 398°K

$$V_T = \frac{T}{11{,}600} = \frac{398}{11{,}600} = 34.3 \text{ mV}$$

∴ At 398°K, barrier voltage

$$V_{G0} = \frac{1 \times 34.3 \times 10^{-3} \times 10^{-9}}{10^{-9} \times 398} = 0.086 \text{ mV.}$$

The calculations show that with the increase in temperature the barrier voltage decreases and the current through the device or the material increases.

EXAMPLE 2.11

For the reverse saturation current of I_0 or I_S of 5 μA, calculate the forward current I_f at room temperature of 300°K, for applied voltages V_f of 0.25 V and 0.35 V for both Germanium and Silicon diodes, respectively.

Solution:

Voltage equivalent of temperature $V_T = \frac{T}{11{,}600} = \frac{300}{11{,}600} = 0.026 \text{ V} = 26 \text{ mV}$

$\eta = 1$ for Germanium diode and $\eta = 2$ for Silicon diode. The reverse saturation current I_0 or $I_S = 5$ μA.
Forward current I_f for Germanium diode:

$$I_f = I_0\left[e^{V_f/\eta V_T} - 1\right]$$

When $V_f = 0.25$ V, $\eta = 1$ and $V_T = 26$ mV.

$$\therefore \quad I_f = 5 \times 10^{-6}\left[e^{0.25/26\times10^{-3}} - 1\right]$$

$$= 5 \times 10^{-6}(e^{9.6153} - 1) = 74.95 \text{ mA}$$

Forward current I_f when $V_f = 0.35$ V, $\eta = 1$ and $V_T = 26$ mV.

$$I_f = 5 \times 10^{-6} \cdot \left[e^{0.35/26\times10^{-3}} - 1\right] = 5 \times 10^{-6}\left[e^{13.46} - 1\right]$$

$$= 701.86 \text{ mA} = 0.7 \text{ A}$$

Forward current I_f for Silicon diode:

$$I_f = I_0\left[e^{V_f/\eta V_T} - 1\right].$$

When $V_f = 0.25$ V, $\eta = 2$ and $V_T = 26$ mV.

$$I_f = 5 \times 10^{-6}[e^{0.25/2\times26\times10^{-3}} - 1] = 5 \times 10^{-6}(e^{4.80} - 1) = 602.55 \text{ μA}$$

Forward current I_f When $V_f = 0.35$ V; $\eta = 2$ and $V_T = 26$ mV

$$I_f = 5\times10^{-6}[e^{0.35/0.052} - 1] = 5\times10^{-6}\times(e^{6.73} - 1) = 4.18\ \mu\text{A}.$$

EXAMPLE 2.12

Obtain the factor by which reverse saturation current of a Germanium diode is multiplied when the operating temperature increased from 20°C to 90°C.

Solution: The reverse saturation current of a diode doubles for every 10°C rise in temperature. If $I_0 = I_{01}$ at $T = T_1$

$$I_0(T) = I_{01}\times 2^{(T-T_1)/10}$$

$$I_0(90°\text{C}) = I_0(20°\text{C})\times 2^{(90-20)/10} = I_0(20°\text{C})\times 2^7$$

$$\therefore\ I_0(90°\text{C}) = 128 \times I_0(20°\text{C})$$

∴ The multiplication factor is 128.

EXAMPLE 2.13

Calculate the ratio of current for forward bias voltage of 0.04 V to the current for the same magnitude of reverse bias.

Solution: The ratio of the current I_f for forward bias $V = 0.04$ V to the reverse current I_r for the same magnitude of the reverse bias voltage $V = 0.04$ V

$$\frac{I_f}{I_r} = \frac{e^{V/\eta V_T} - 1}{e^{-V/\eta V_T} - 1}$$

$$\frac{I_f}{I_r} = \frac{e^{0.04/0.026} - 1}{e^{-0.04/0.026} - 1} = \frac{e^{1.538} - 1}{e^{-1.538} - 1} = \frac{3.655}{-0.785} = 0.4656.$$

EXAMPLE 2.14

The voltage across a Silicon diode at room temperature of 300°K is 0.7 V when 2 mA of current flows through it. If the voltage increases to 0.75 V, calculate the diode current when the voltage equivalent of temperature $V_T = 26$ mV $= 0.026$ V.

Solution: Voltage across Silicon diode, $V_D = 0.7$ V, current through the diode, $I = 2$ mA, $V_T = 26$ mV. To calculate I_0

$$\text{Diode current}\quad I = I_0\left[e^{V_D/V_T} - 1\right]$$

$$\therefore\ I_0 = \frac{I}{\left[e^{V_D/V_T} - 1\right]}$$

$$I_0 = \frac{2\times10^{-3}}{\left[e^{0.7/0,026} - 1\right]} = \frac{2\times10^{-3}}{e^{26.92} - 1}$$

$$= \frac{2\times10^{-3}}{(4.911\times10^{11} - 1)} = \frac{2\times10^{-3}}{4.911\times10^{11}} = 0.4\times10^{-14}$$

Reverse saturation current calculated, $I_0 = 0.4 \times 10^{-14}$.
If the diode voltage increases to 0.75 V

$$I = I_0\left[e^{V_D/V_T} - 1\right] = 0.4\times10^{-14}\left[e^{0.75/0.026} - 1\right]$$

$$\therefore\ I = 0.4\times10^{-14}\left[e^{28.846} - 1\right] = 0.4\times10^{-14}\left[3.37\times10^{12} - 1\right]$$

$$I = 0.4\times10^{-14}\times3.27\times10^{12} = 13.08 \text{ mA}$$

When the diode voltage has increased to 0.75 V, current $I = 13.08$ mA.

2.13 TRANSITION AND DIFFUSION CAPACITANCES (DIODE JUNCTION CAPACITANCES)

P–N semiconductor diode without any bias voltage under open circuit conditions has depletion region of width W as shown in Fig. 2.29. The semiconductor diode when forward biased has a reduced depletion region width of magnitude W_f as shown in Fig. 2.33. The semiconductor diode when reverse biased has an increased depletion region width of magnitude W_r as shown in Fig. 2.35. This clearly shows that depletion region widths under the three situations vary with the unbiased situation to forward bias and reverse bias operations of the semiconductor diodes.

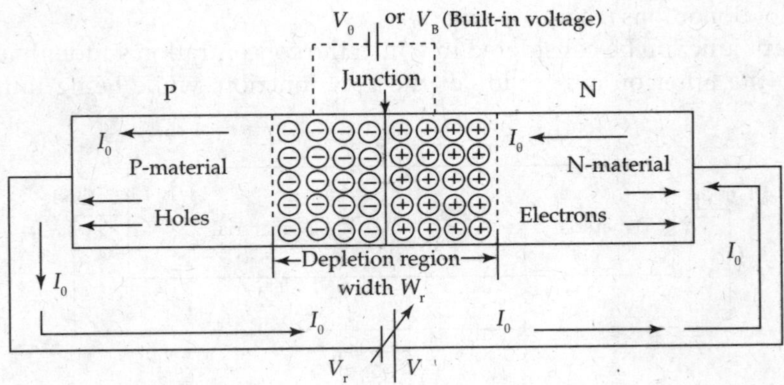

FIG. 2.45 *Semiconductor diode with reverse bias voltage V_r*

So the depletion regions about the P–N junctions are dependent on the voltage V applied to the P–N diode. Typical situation is shown in Fig. 2.45. Depletion region or space charge region width on either side of the P–N junction is formed by the acceptor ions (negative immobile ions) on P-material side and donor ions (positive immobile ions) on N-material side with junction voltage or contact voltage or barrier potential across the junction. The two semiconductor materials outside the depletion region containing charges act as parallel plates and the depleted region or the transition region or the space charge region behaves as the dielectric for the junction capacitances of the semiconductor devices. The junction capacitances of P–N diodes present problems at high frequencies since $x_c = 1/2\pi fc$ and can produce unwanted phenomena. There are two types of intrinsic capacitances associated with P–N junction diode.

1. C_T = Transition or depletion capacitance is voltage dependent under reverse-biased conditions. It is also known as junction capacitance. It has physical characteristics like a parallel plate capacitor.
2. C_D = Diffusion capacitance is current dependent under forward-biased conditions. It is also known as storage capacitance. It derives its name as both positive and negative charges are in the space charge region of a diode.

2.13.1 Space Charge Capacitance or Transition Capacitance C_T

When a P–N diode is reverse biased, the reverse saturation current I_0 is very much negligible and the reverse resistance is large. The contact potential or barrier voltage V_0 and the depletion region width W_r about the P–N junction increase.

The expression for C_T is given by

$$C_T = \frac{\varepsilon_0 \varepsilon_r A}{d} = \frac{\varepsilon_0 \varepsilon_r A}{W_r}, \tag{2.132}$$

where $\varepsilon_0 \varepsilon_r$ corresponds to dielectric constant and d corresponds to the depletion region width (W_r).

The general expression for capacitance C can be obtained as follows:

$$\text{Capacitance} \quad C = \frac{dq}{dv}, \tag{2.133}$$

where q is the charge in Coulombs, V is the voltage, N_A is the density of acceptor ions and N_D is the density of donor ions (charge densities).

Let an abrupt junction be considered in which the concentration suddenly changes from one region to the other on either side of the P–N junction, while being uniform in each

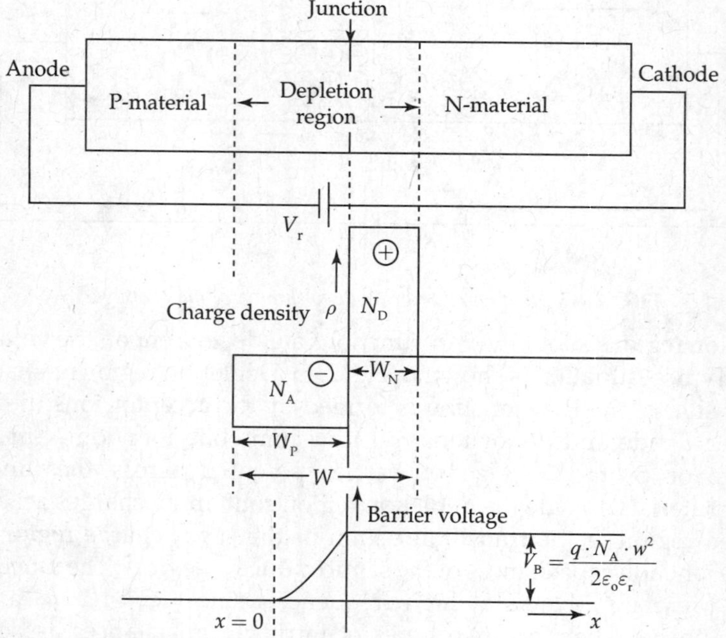

FIG. 2.46 *Charge density and barrier potential in reverse biased P–N junction diode*

region (step gradient). It is assumed that $N_D \gg N_A$. The charge density profile as a function of distance from the P–N junction is shown when N_A = concentration of acceptor ions and N_D = concentration of donor impurities. Also the potential gradients are as described in Fig. 2.46.

Since the net charge has to be conserved

$$qN_A W_p = qN_D W_n, \tag{2.134}$$

where W_p and W_n are the widths of space charge regions of P- and N-side materials. Since $N_A \ll N_D$, $W_p \gg W_n$ tends to 0.

From the Poisson equation, we have the relationship between potential and charge density ρ:

$$\frac{d^2V}{dx^2} = \frac{\rho}{\varepsilon_0 \varepsilon_r} = \frac{qN_A}{\varepsilon_0 \varepsilon_r}, \tag{2.135}$$

where ε_0 is the permittivity of the free space, ε_r is the relative permittivity of the materials, ρ is the charge density, N_A is the acceptor ion impurity concentration and q is the charge.

Integrating twice Eq. (2.135) w.r.t to x yields

$$V = \frac{\rho x^2}{2\varepsilon_0 \varepsilon_r} = \frac{qN_A x^2}{2\varepsilon_0 \varepsilon_r}. \tag{2.136}$$

At $x = W$, $V = V_B$, the barrier potential with reverse bias.

$$\therefore\ V_B = \frac{qN_A w^2}{2\varepsilon_0 \varepsilon_r} \tag{2.137}$$

Capacitance C of a capacitor can be expressed in terms of charge Q and voltage V, where Q is the charge stored in the volume AW, A is the area of the junction and W is the depletion region width.

$$C = \frac{dQ}{dV} \tag{2.138}$$

$$\text{Transition Capacitance } C_T = \frac{dQ}{dV_B} \tag{2.139}$$

$$Q = \text{volume} \times \text{charge density}.$$

$$Q = qN_A AW. \tag{2.140}$$

From Eq. (2.137), we get

$$C_T = \frac{dQ}{dV_B} = qN_A A \cdot \frac{dW}{dV_B}. \tag{2.141}$$

$$\text{Depletion region width } W = \sqrt{\frac{2\varepsilon_0 \varepsilon_r}{qN_A}} \cdot V_B^{1/2}. \quad \text{from Eq. (2.137)} \tag{2.142}$$

From Eq. (2.142), it is clear that the width of the depletion region W increases with the applied reverse bias voltage V_B. Substituting the value of W from Eq. (2.142) in Eq. (2.141)

$$C_T = \varepsilon_0 \varepsilon_r A \sqrt{\frac{qN_A}{2\varepsilon_0 \varepsilon_r}} \cdot V_B^{1/2} = \sqrt{\frac{\varepsilon_0 \varepsilon_r qN_A}{2}} \cdot AV_B^{1/2} = KV_B^{1/2} \tag{2.143}$$

$$\text{where } K = \sqrt{\frac{\varepsilon_0 \varepsilon_r qN_A}{2}} \cdot A. \tag{2.144}$$

It is also known from Eq. (2.141) that

$$C_T = \frac{\varepsilon_0 \varepsilon_r A}{W} = \frac{\varepsilon A}{W}, \tag{2.145}$$

where permittivity of the semiconductor $\varepsilon = \varepsilon_0 \cdot \varepsilon_r$.

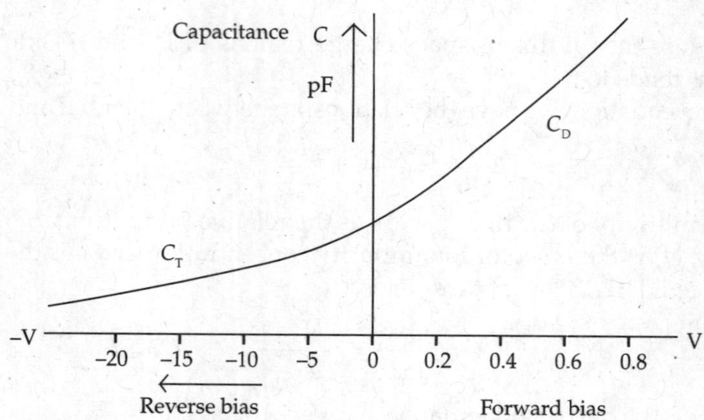

FIG. 2.47 *Capacitance variation profile under forward and reverse bias conditions*

Junction capacitance, the transition capacitance C_T of P–N diode is a function of the depletion region width W. It varies as square root of reverse bias voltages ($V_B = V_0 + V_R$), which is approximately equal to V_R to the diode. If the reverse bias voltage V_R applied to the diode is increased, there is a corresponding increase in the depletion region width W and the values of transition capacitance get decreased proportionately. Typical value of transition capacitance C_T is of order of 20 pF. The expression for C_T in Eq. (2.145) also shows that the expression for the junction capacitance is similar to the expression for capacitance of a normal capacitor. Transition capacitance is much larger, when the diode is forward biased as depletion region becomes narrower.

In parallel plate capacitor, charge resides on the plates of a capacitor. In case of transition capacitance, charges are located in between the two conducting P- and N-type layers.

2.13.2 Diffusion or Storage Capacitance C_D

The storage capacitance refers to the rate of charge of storage of minority carriers as the diode voltage is changed. Current-dependent capacitance under forward-biased constants is designated by C_D.

Under the assumption that $N_A \gg N_D$, concentration of minority carrier is a function of distance x from the P–N junction.

It becomes equal to $I_{pn}(0)$ at the junction, i.e., $x = 0$. Since N_A is very much greater than N_D, the entire current can be considered to be of holes (i.e., injected minority carriers into the N-side region). The charge Q is the product of Aq and dashed region of the curve as in Fig. 2.48 as per the notation. Then

$$Q = \int_0^\infty Aqp'_n(0) \cdot e^{-x/L_p} dx = Aqp'_n(0) \cdot (L_p). \tag{2.146}$$

The diffusion capacitance C_D as per the definition is

$$C_D = \frac{dQ}{dV},$$

$$\therefore \quad C_D = AqL_p \cdot \frac{dp'_n(0)}{dV}. \tag{2.147}$$

$$\therefore \quad \frac{dp'_n(0)}{dV} = g \cdot \frac{L_p}{D_p Aq}. \tag{2.148}$$

Substituting for $\frac{dp'_n(0)}{dV}$ from Eq. (2.152) in Eq. (2.147), diffusion capacitance C_D becomes

$$C_D = \frac{AqL_p g L_p}{D_p Aq} = \frac{L_p^2 g}{D_p}. \tag{2.149}$$

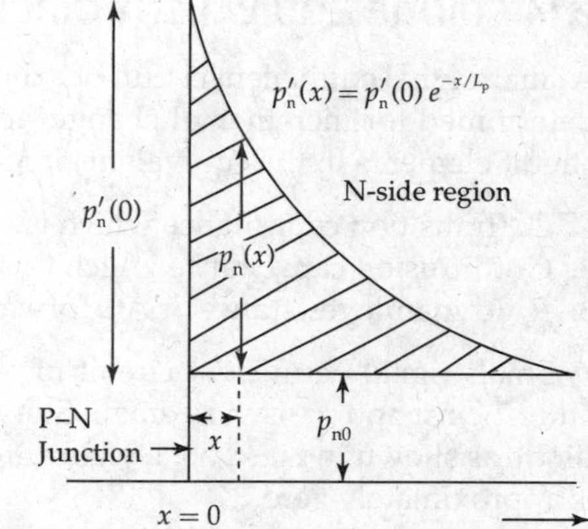

FIG. 2.48 *Current components contributing to diffusion capacitance C_D*

Since g is current dependent the diffusion capacitance C_D or storage capacitance C_S is a current-dependent capacitance.

$$\because L_p^2 = D_p \tau_p, \quad \text{where } \tau_p \text{ is the mean lifetime of the charge (holes).} \tag{2.150}$$

Substituting the value of $\tau_p = L_p^2 / D_p$ in Eq. (2.149),

$$C_D = \tau_p \cdot g \quad \Rightarrow g = \frac{C_D}{\tau_p}. \tag{2.151}$$

Dynamic resistance $R_D = \frac{\eta V_T}{I}$.

Dynamic conductance $g = \frac{I}{\eta V_T}$

$$\text{Hence} \quad C_D = \frac{\tau_p I}{\eta V_T} \tag{2.152}$$

Generally, C_D will be of the order of 0.1–1 μF. Diffusion capacitance will be larger than the value of the transition capacitance C_T that will be of the order 20 pF. This difference in the values of the two capacitances relates to the lowered depletion region width for forward-biased P–N junction (C_D is large) and the increased transition or depletion region width for the reverse-biased P–N junction (C_T is less than C_D). So the large value of diffusion capacitance can be a serious limitation for the use of forward-biased P–N junctions in high-frequency circuits. C_D is directly proportional to the forward current. For Germanium diode, $\eta = 1$ and if $\tau_p = 20$ μs and current I is 26 mA. Diffusion capacitance $C_D = 20$ μF.

This apparently large value of capacitance C_D is not as bad in effect, since diode forward resistance is of the order of a few ohms that results in the time constant of μ seconds. For a forward-biased diode, $C_D \gg C_T$. Therefore, C_T can be neglected. For a reverse-biased diode, $C_T \gg C_D$. Therefore, C_D can be neglected.

Transition capacitance exists for a diode for both forward- and reverse-biased conditions, whereas the diffusion capacitance exists only when the diode is forward biased.

2.14 DIODE EQUIVALENT CIRCUITS

A small signal equivalent circuit of a diode is obtained on the basis of values of circuit elements determined for incremental changes in the DC voltage applied to the diode. There are three circuit elements in the equivalent circuit of the diode, viz:

1. C_T (transition capacitance which is voltage dependent)
2. C_D (diffusion capacitance which is current dependent)
3. R_D (dynamic resistance or AC forward resistance of a diode).

Small signal equivalent circuit of a forward-biased diode is shown in Fig. 2.49. Both C_D and C_T exist and R_D is very small. Storage capacitance does not exist in case of reverse-biased diode as shown in Fig. 2.50. Dynamic resistance is very high, as the reverse saturation current is approximately zero.

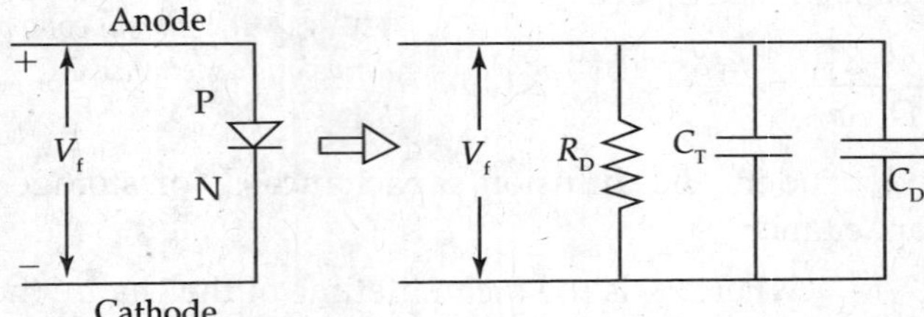

FIG. 2.49 *Small signal equivalent circuit of a diode when forward biased*

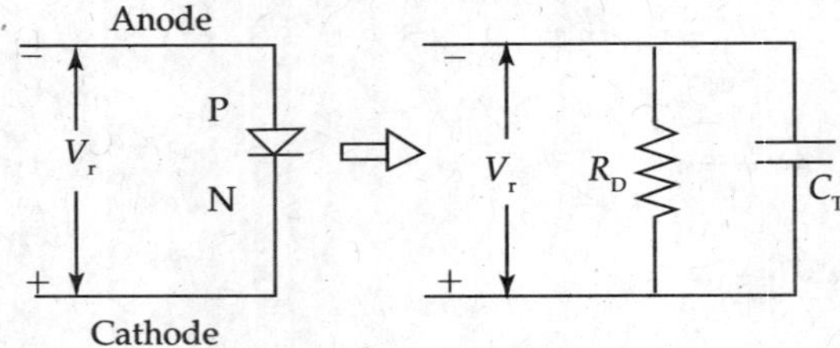

FIG. 2.50 *Small signal equivalent circuit of diode when reverse biased*

As the junction capacitances are very small, the effect of these capacitances can be ignored at low-frequency range up to several hundreds of kHZ. The diode can be modelled as a resistor of value R_D. At high frequencies, the junction capacitance and R_D are to be taken into consideration (Figs. 2.51 and 2.52).

2.15 BREAKDOWN MECHANISMS OF SEMICONDUCTOR DIODES

- The reverse-biased breakdown voltage of a P–N junction can be varied by suitable or necessary concentrations of doping for P- and N-type materials. By varying the doping concentrations, diodes with specific breakdown voltages ranging from less than 1 V to several hundred volts can be manufactured.
- If the reverse bias applied to a P–N junction is increased to a large value, the electric field intensity increases to a very high value.
- If the transition region or the depletion region width is of the order of a few nanometres, voltages in the range of fraction of a volt can create very high field intensities.

- If the doping is heavy, the depletion region width decreases further and the electric field intensity becomes higher.

If $v = 0.5$ V and depletion or transition width $W = 0.01\ \mu$, the field intensity is

$$\text{Electric field intensity} \quad \varepsilon = \frac{V}{W} = \frac{0.5\text{ V}}{0.01\times10^{-6}} = 5\times10^{7}\text{ V/m}.$$

A field of such high magnitudes can directly disrupt the covalent bonds and enhance the current to high proportions. In addition, the disrupted covalent bonds producing electron–Hole pairs will allow the electron–Hole pair to get accelerated by these high fields and a landslide can occur. This is called *Avalanche multiplication.* The line of demarcation between Zener and Avalanche breakdowns is very small, while a distinction can be made between the two phenomena.

Zener effect (tunnelling phenomenon) It is considered to be a narrow junction high field phenomenon similar to field emission. Because of very high field intensities of the order of 10^7 V/m some electrons are pulled across the forbidden band gap from valance band energy levels to Conduction Band in diodes having narrow depletion region widths. The electron tunnels through the barrier as a wave instead of acquiring the sufficient band-gap energy to cross the space charge region as a particle. This phenomenon is known as the Zener effect. Total scenario occurs when the filled energy levels on the P-type region get face to face with vacant energy levels on the N-type region of the diode and electrons tunnel through the transition region about the P–N junction under reverse bias operation in Zener Diodes.

Avalanche Breakdown On the other hand in a wider junction due to larger mean free paths the charge carriers acquire kinetic energy, sufficient enough to disrupt the covalent bonds and start a cumulative process. This process of charge carrier multiplication is similar to

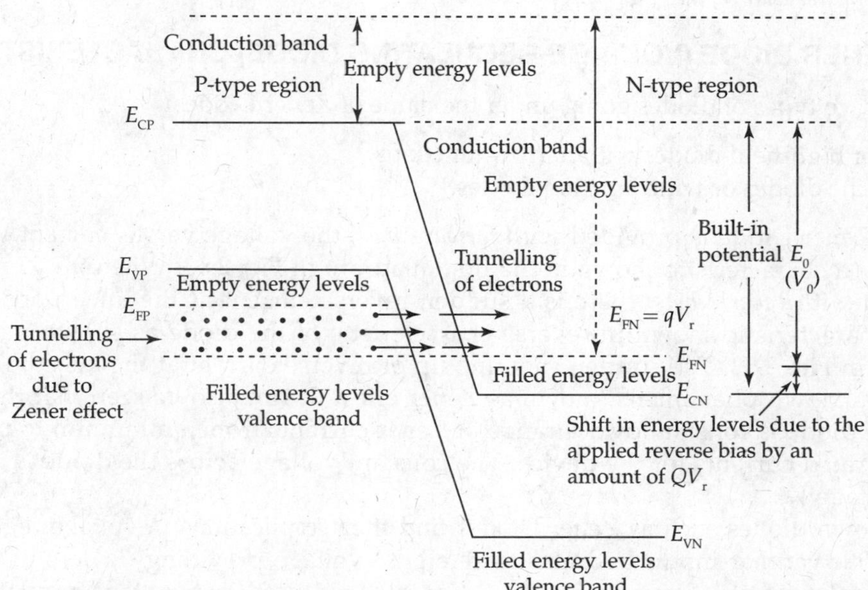

FIG. 2.51 *Energy-band diagram of reverse biased zener diode*

secondary electron emission as in X-ray production. This is known as Avalanche effect (impact ionisation). Paradoxically, the Zener breakdown occurs at lower voltages and the Avalanche breakdown occurs at higher voltages.

Breakdown diodes are designed for a required specific breakdown voltages. For well designed breakdown diodes, breakdown will be sharp and current after breakdown will essentially be independent of voltage (rated breakdown voltage). Forward and reverse characteristics of a Zener Diode are shown in Fig. 2.53.

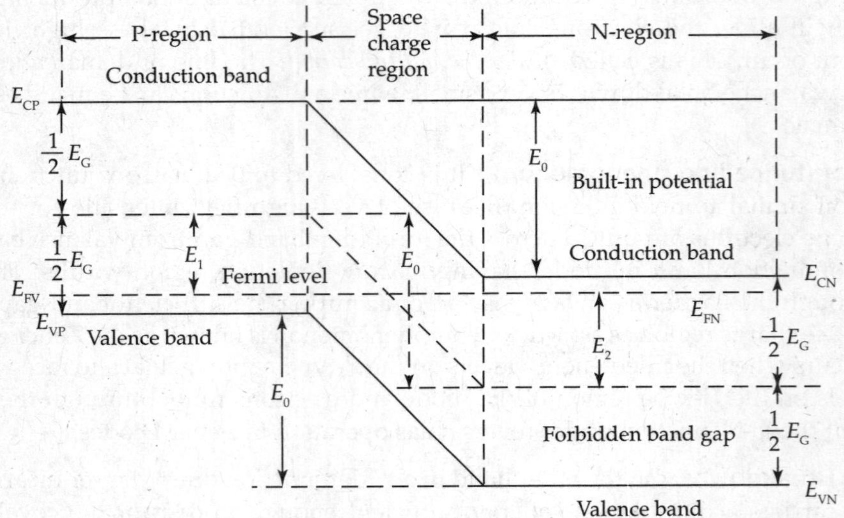

FIG. 2.52 *Energy-band diagram of P–N junction diode with reverse bias V_r with an increase in built-in potential E_0 and the width of the space charge region*

2.16 ZENER DIODE (VOLTAGE-REGULATING DIODE) CHARACTERISTICS

There are two types of diodes going under the name of Zener Diodes:

- Zener or high-field diodes with narrow junctions,
- Avalanche diodes or wide junction diodes.

When Zener Diode is provided with *forward bias*, the voltage versus current variations appears as a characteristic shown in the first quadrant in Fig. 2.53. When the Zener Diode is applied with *reverse bias*, there is a sudden reverse current at the knee portion of the reverse characteristic shown in reverse characteristic of the diode as shown in the third quadrant in Fig. 2.53. The sudden shooting up of reverse current at the knee point is due to the breakdown phenomena known as Zener effect. It can also be seen that the voltage across the diode is constant over a range of Zener currents from a minimum to maximum values of rated currents for the device. The constant voltage across the diode is known as Zener voltage V_Z.

Breakdown diodes such as Zener Diodes find their applications in voltage regulators in circuits with varying inputs and also as reference voltage providing devices in regulated power supply circuits. For example, 12 V Zener Diode used in a voltage regulator circuit holds the circuit's output voltage at 12 V while the input voltage to the regulator circuit varies

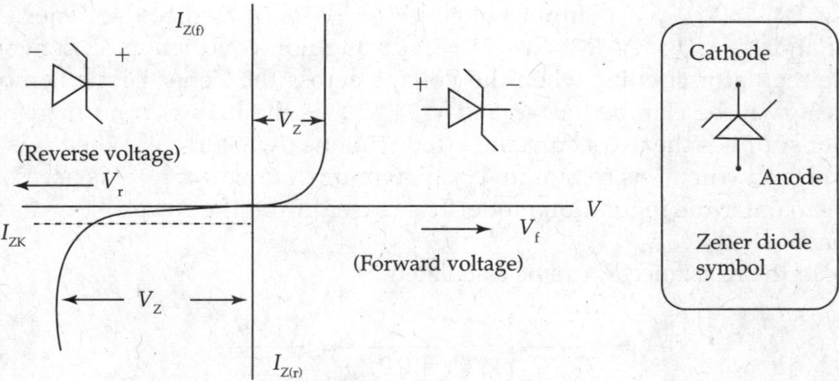

FIG. 2.53 *Forward and reverse characteristics of Zener diode*

around 12 V. V_Z and I_{ZK} are the minimum necessary current and voltage to enter the Zener region and below this current, it cannot act as a regulating device. Above $|I_{ZK}|$, the voltage drop across the Zener Diode is V_Z and is independent of current. The maximum allowable current depends on doping and is specified by the manufacturer. It acts as a voltage regulator above $I_{Z(min)}$ and below $I_{Z(max)}$ safely.

The Zener Diode draws minimum current when load current is maximum and maximum current when the load current is minimum. Zener Diode regulates only when the voltage across it is at least V_Z as obtained from the supply and this should always be kept in mind in designing the circuits with Zener Diodes.

Heavy doping causes the current to suddenly shoot up in close proximity of V_r as can be observed in Fig. 2.53. Regarding reverse characteristic below the breakdown region when I_{ZK} is crossed, current increases in the negative direction virtually parallel to *y* axis. This shows no change in voltage over wide current swings or changes as though the voltage drop across the Zener Diode is constant. Zener current and diode resistance adjust among themselves to keep *IR* drop constant across the device. If a Zener Diode were to be used in a circuit with varying voltages and currents, the voltage across a load can be maintained constant becoming immune to fluctuations in supply voltages over designed range of Zener currents.

In Fig. 2.54, V_i is the unregulated input voltage and V_o is the regulated output voltage. V_Z is the Zener Diode voltage drop. *I* is the current from the source. I_Z is the current through the Zener Diode. I_L is the load current.

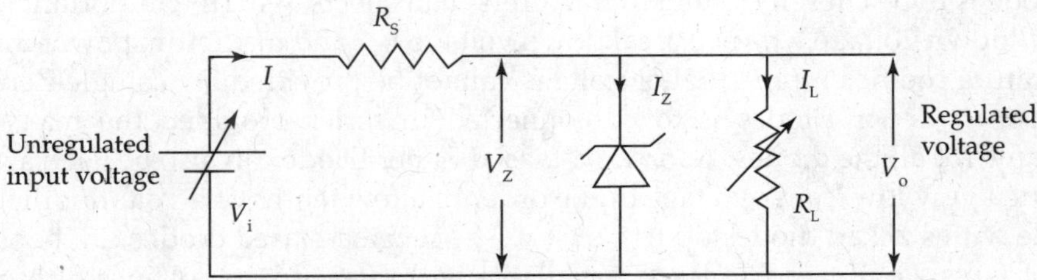

FIG. 2.54 *Voltage regulator circuit using Zener diode*

As long as $(V_i - IR_S) < V_Z$, output voltage $V_o = (V_i - IR_S)$, since the Zener Diode does not conduct and $V_o = (V_i - I_L R_S)$. Also $I_Z = 0$ during non-conduction of Zener Diode). In Zener Diode regulator circuits, when the voltage across the Zener Diode is more than or equal to V_Z, V_o will be clamped at V_Z and $V_o = V_Z$. As the load current increases, it is as though Zener supplies the extra current by reducing its own current. When I_L is maximum I_Z is minimum and when I_L is minimum I_Z is maximum. However, $I_{Z(min)}$ is necessary for the Zener Diode to enter the regulation mode. The current from the supply source I is the sum of the currents I_Z and I_L.

The value of the resistance R_S can be calculated as

$$R_S = \frac{(V_i - V_Z)}{(I_L + I_Z)} = \frac{(V_i - V_Z)}{(I_L + 0.2 I_L)} = \frac{(V_i - V_Z)}{1.2 I_L}. \qquad (2.153)$$

With normal operating currents of the Zener Diode, I_Z is chosen as equal to $0.2\, I_L$.

Power P_Z dissipated in the Zener Diode can be calculated as

$$P_Z = I_Z \cdot V_Z = \left[\left(\frac{(V_i - V_Z)}{R_S} \right) - I_L \right] V_Z \text{ W}. \qquad (2.154)$$

Zener Diode must be selected with larger power rating than the power dissipation rating P_Z that is calculated from Eq. (2.154), so that the maximum power dissipation and the temperature ratings of the device are not exceeded.

Applications of Zener Diode:

(1) Constant voltage regulation, (2) Surge protection, (3) Clamping and clipping voltages, (4) Voltage reference in precision equipment such as regulated power supply circuits, (5) Switching operation and (6) For meter protection.

Advantages:

1. Smaller size, cheaper and has long life and rugged.
2. Provides good regulation over a wider range of currents.

Drawbacks:

1. Power loss associated with A_S for light and heavy load current and poorer efficiency.
2. Output voltage depends upon the breakdown voltage of Zener Diode.

Specifications:

A Zener Diode is to be chosen from manufacturers data sheets, with the important parameter such as breakdown voltage V_Z, knee breakdown voltage V_{ZK} and maximum power dissipation, and temperature coefficient. If a higher voltage cannot be provided by a single Zener Diode, more number of Zener Diodes is to be connected in series provided the max allowable current is same for all the diodes. A forward-biased Zener Diode can also be used as a voltage regulator but a very low Zener voltage of the order of a few tenths of a volt. But the packages are available with stacked diodes up to 1 or 2 V. The reverse-biased diodes can be used up to hundreds of volts, i.e. Zener Diodes are available with breakdown voltages of hundreds of volts singly or in package.

SUMMARY

1. Formation of P-type and N-type semiconductor material formation in Silicon and Germanium Semiconductor materials are discussed.
2. Diodes, Transistors, FETs and MOSFETs (all devices) are made of P-type and N-type materials in different structures.
3. P–N Junction Diode structure and its characteristics are well discussed.
4. Zener diode device characteristics and its applications in voltage regulator circuits is introduced.
5. Diode applications in Rectifier circuits are analysed in next chapter.

Questions for Practice

1. (a) Derive an expression for total diode current starting from Boltzmann relationship in terms of the applied voltage (Nov 2010, JNTUH).

   (b) The reverse saturation current of a Silicon P–N junction diode is 50 nA at an operating temperature of 27°C. Compute the dynamic forward and reverse resistances of the diode for the applied voltages of 0.8 and 0.4 V respectively.
2. Distinguish between drift and diffusion currents in semiconductors.
3. State and derive the 'continuity equation'.
4. State Hall effect phenomenon and mention its applications.
5. Derive the expression for V_B (contact potential or barrier voltage) in terms of doping densities in P–N junction diode.
6. Derive the expression for *diffusion current density* in P-type semiconductor.
7. Derive the expression for *drift current* in P-type semiconductor diode.
8. (a) Explain the operation of Silicon P–N junction diode and obtain the forward bias and reverse bias volt–ampere characteristics (Nov 2010, JNTUH).

   (b) Obtain the transition capacitance C_T of a junction diode at a reverse bias voltage of 12 V if C_T of the diode is given as 15 pF at a reverse bias of 8 V. Distinguish between transition and diffusion capacitances.
9. Derive the expression for diode forward current equation.
10. Derive the expression for dynamic resistance of a diode.
11. The current through Silicon diode, $I_f = 60$ mA for a forward bias of $V_f = 0.6$ V. Calculate the static resistance of the diode.
12. The reverse saturation current I_S at $T = 300$°K of a P–N junction Germanium diode is 5 μA. Calculate the forward bias voltage V_f to be applied across the junction to obtain a forward current I_f of 50 mA.

13. Define transition capacitance of a semiconductor diode and derive the expression for transition capacitance.
14. Define diffusion capacitance of a semiconductor diode and derive the expression for diffusion capacitance.
15. If $\tau = 10$ μs, $I = 78$ mA, find C_D for a Silicon diode $\eta = 2$ and $V_T = 26$ mV.
16. Explain the significance of various parameters in the equivalent circuit of a diode under both forward and reverse bias operations.
17. Explain Zener effect and explain how it is similar to field emission?
18. Explain Avalanche effect and how it is similar to secondary electron emission?
19. Draw diagram of a simple voltage regulator circuit using Zener Diode and explain its working.
20. Draw and explain the Zener Diode characteristics?

Multiple Choice Questions

1. Majority carriers in N-type semiconductor are ______________
   (a) electrons (b) Holes (c) valence electrons
2. Majority carriers in P-type semiconductor are ______________
   (a) electrons (b) Holes (c) valence electrons
3. Cut-in voltage of a Germanium diode is ______________
   (a) 0.2 V (b) 0.5 V (c) 0.4 V (d) 0.7 V
4. Cut-in voltage of a Silicon diode at room temperature is ______________
   (a) 0.2 V (b) 0.5 V (c) 0.4 V (d) 0.6 V
5. Reverse-biased P–N junction diode has the following feature ______________
   (a) very narrow space charge region
   (b) large value of depletion region
   (c) low resistance element
   (d) large current flow
6. Forward-biased P–N junction diode has the following feature ______________
   (a) very narrow space charge region
   (b) large value of depletion region
   (c) low resistance element
   (d) large current flow
7. P–N junction diode has an application of ______________
   (a) bidirectional switch
   (b) unidirectional switch
   (c) controlled switches
   (d) none of the previous ones

8. Forbidden band energy gap in a Silicon semiconductor is of the order of ____________
   (a) zero (b) 1.1 eV
   (c) > 6 eV (d) 0.7 eV
9. Forbidden band energy gap in an insulator is of the order of ____________
   (a) zero (b) 1.1 eV
   (c) > 6 eV (d) 0.7 eV
10. Semiconductor diode can be used as ____________
    (a) rectifier (b) amplifier
    (c) oscillator (d) unilateral switch

Answers to Multiple-Choice Questions

1. (a)	2. (b)	3. (a)	4. (d)	5. (b)
6. (a)	7. (a)	8. (b)	9. (c)	10. (a)

Chapter 3

RECTIFIERS, FILTERS AND VOLTAGE REGULATORS

Learning Objectives

To get familiarity of structural details and fundamental concepts of

- DC sources as the supply voltages for amplifiers, radios, TVs, computers etc.
- DC sources may be battery for portable equipment but are very expensive for heavy use. An alternative is to convert cheaply available AC voltage from power lines to DC voltages.
- Principles of working of Rectifier circuits
- Use of Filter circuits to convert unidirectional voltage to DC voltage
- Working principles of 'regulator circuits' to produce stable DC output voltage

3.1 INTRODUCTION

Main blocks of typical power supply (Fig. 3.1)

Rectifier circuits:

- Rectification is the process to convert AC voltage (sinusoidal voltage) into unidirectional voltages using Diodes (polarity sensitive switches).
- Devices that convert alternating voltage into unidirectional voltage are known as *Rectifiers*.
- Rectifier circuits could be simple Half-Wave Rectifier or Full-Wave Rectifier (FWR) or Bridge Rectifier circuits.

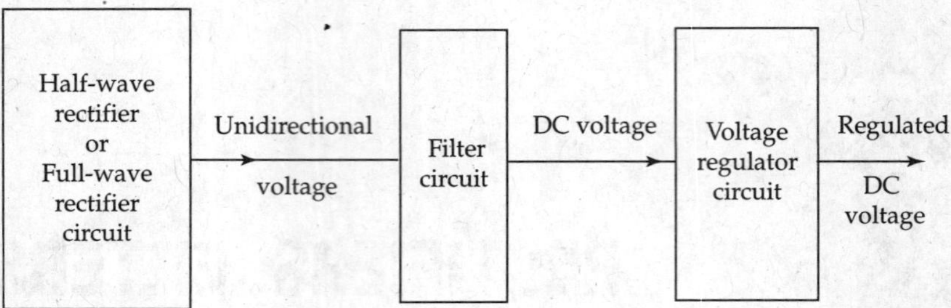

FIG. 3.1 *Block diagram of regulated power supply*

Filter circuits:

- Output of a Rectifier circuit is a pulsating DC (unidirectional voltage with half sinusoids) consisting of a DC component as well as superimposed ripple.
- In most of the applications, pure DC or a DC with tolerable ripple voltage is needed.
- Filter Circuits reduce the ripple (unwanted AC component) to the required levels.

Voltage regulator circuits:

- Voltage regulator is an electronic regulator designed to automatically maintain a constant DC output voltage.
- Basic building block of almost every power supply unit used in electronic circuits, whenever there is a stringent need of constant output DC voltage.

3.2 HALF-WAVE RECTIFIER CIRCUIT (HWR CIRCUIT WORKING PRINCIPLES)

Half-Wave Rectifier Circuit (Fig. 3.2) has

- Low-voltage AC power source using a step-down Transformer,
- Semiconductor Diode for rectification and
- Variable load resistor R_L.

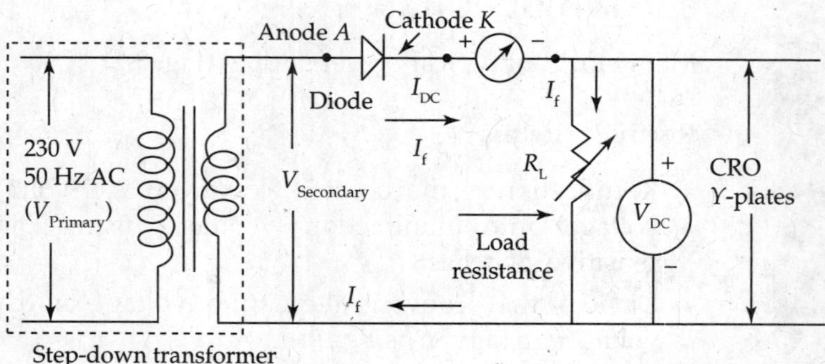

FIG. 3.2 *Half-wave rectifier circuit*

Low-voltage AC source

It is a step-down Transformer that reduces the mains power line AC sinusoidal voltage at 230 V 50 Hz to required level of voltage depending on the application, for example 6 V or

9 V and so on as the secondary voltage. This is necessary because supply voltages needed in electronic circuit applications are much smaller than the mains supply voltage.

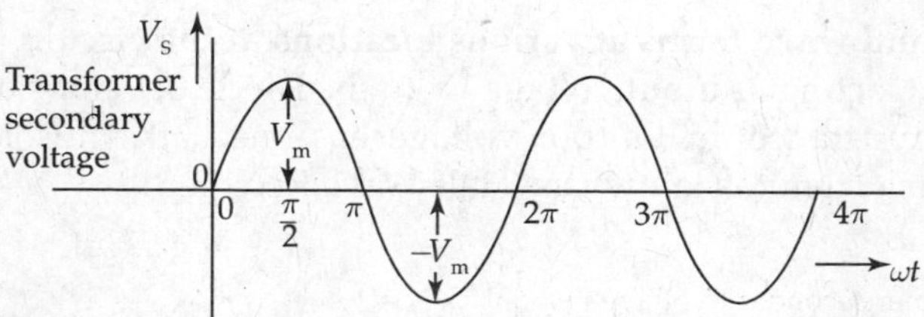

FIG. 3.3 *Transformer secondary voltage $V_S = V_m \sin \omega t$ where V_m may be 6 V or 9 V and so on depending upon the design requirement*

3.2.1 Semiconductor Diode Rectifier

- During positive half cycle (0–π interval) of Transformer secondary voltage, Diode *D* is forward biased. It conducts only when its anode is positive.
- Diode conducts during positive half cycles in each cycle of sinusoidal voltage. Forward current I_f (sinusoidal variation) flows through Diode and load resistance R_L.
- Output voltage is varying DC containing DC average value and AC component (unwanted signal component) called as ripple voltage.
- Output voltage contains only one half sinusoids for each cycle of AC voltage.
- Such rectifier circuit with the above features is known as 'Half-Wave Rectifier'.

Analysis:

Input voltage to Rectifier Diode is Transformer Secondary voltage: $V_S = V_m \cdot \sin \omega t$. During positive half cycle '0 to π' of the supply voltage, anode (P-material) of the Diode is positive, and hence, the Diode is forward biased.

Forward current I_f flows through the Rectifier circuit.

$$I_f = \frac{V_m \sin \omega t}{(R_L + r_f + r_s)} = \frac{V_m \sin \omega t}{R_L} \quad \text{if } r_f \text{ and } r_s \text{ are neglected.}$$

(where V_m is the maximum voltage of sine wave.)

Forward resistance r_f of the Diode and Transformer secondary winding resistance r_s are generally small in comparison with load resistance R_L.

$$\therefore \quad R_L + r_f + r_s \equiv R_L$$

$$I_f = \frac{V_m \sin \omega t}{R_L} = I_m \sin \omega t, \quad \text{where } I_m = \frac{V_m}{R_L}. \tag{3.1}$$

Current I_f flows through R_L. Then potential V_{out} develops across R_L.

$$V_{L\,(DC)} = V_{out} = I_f \times R_L \tag{3.2}$$

During negative half cycle (i.e., during π–2π interval) of secondary voltage V_S; Diode is reverse biased. Reverse current I_r flows through the circuit but it is of practically negligible value. $I_r \cong 0$.

Voltage developed across the Diode under reverse bias condition should be less than 'break down voltage' (Maximum reverse voltage across the Diode, it has to withstand) for safe operation. Maximum voltage V_m across the Diode under reverse bias (without breakdown of the Diode) is known as *peak inverse voltage* (*PIV*) of the Diode.

Voltage across $R_L = I_r \times R_L \cong 0$ during negative half cycle of input AC voltage.

Definition of HWR and wave forms at various locations of the circuit:
Alternating voltage V_S acting as input voltage V_{in} to the Diode develops unidirectional output voltage across load resistance R_L. Output voltage has one half sinusoids for each cycle of (input) AC signal. Such circuit is known as 'Half-Wave Rectifier'.

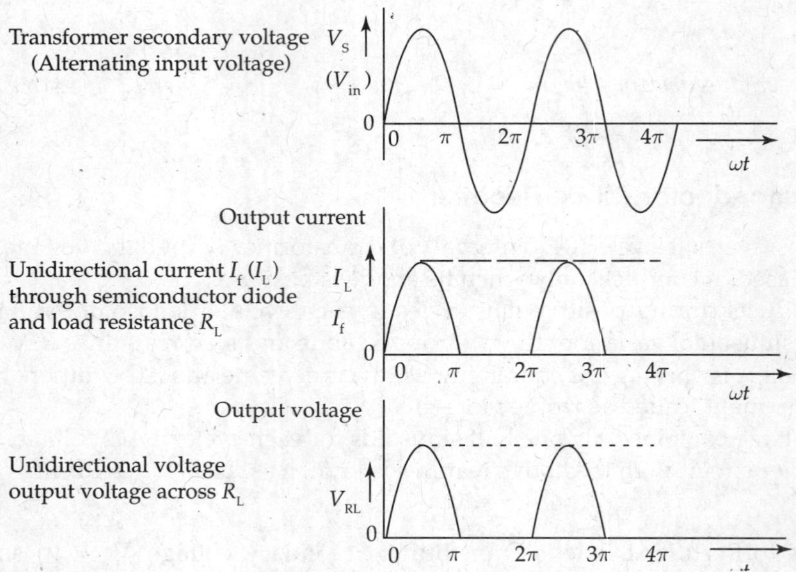

FIG. 3.4 *Signal waveforms at various points in the half-wave rectifier circuit*

Signal waveforms (Fig. 3.4) at various points of the HWR circuit (Fig. 3.2) can be observed using cathode ray oscilloscope (CRO).

Some assumptions made in the analysis:

- Semiconductor Diode is assumed to have negligibly small forward resistance r_f.
- During reverse bias interval, reverse saturation current I_r approximates to zero.
- Transformer secondary winding resistance r_s and the leakage reactance are neglected.
- Cut-in voltage $V_\gamma = 0.7$ V for Silicon Diode and it is 0.3 V for germanium Diode. Since the normal input voltages to Diodes are large in comparison with V_γ, cut-in voltage V_γ is taken to be zero.

From the signal waveform shown in Fig. 3.4.

Forward current I_f, the DC component or the average value of the current I_{DC}

$$I_{DC} = \frac{1}{2\pi}\int_0^{2\pi} I_f \cdot d\omega t = \frac{1}{2\pi}\left[\int_0^{\pi} I_f \cdot d\omega t + \int_{\pi}^{2\pi} I_r \cdot d\omega t\right]$$

$$I_{DC} = \frac{1}{2\pi}\left[\int_0^{\pi} I_m \cdot \sin(\omega t)\, d\omega t + \int_{\pi}^{2\pi} 0 \cdot d\omega t\right]$$

$$I_{DC} = \frac{I_m}{2\pi}[-\cos(\omega t)]_0^{\pi} = [-(-1)-(-1)] = \frac{I_m}{2\pi} \times 2$$

$$I_{DC} = \frac{I_m}{2\pi} \times 2 = \frac{I_m}{\pi} = 0.318 I_m \quad (3.3)$$

If $I_m = \frac{V_m}{R_L}$ is substituted in Eq. (3.3), we obtain

$$I_{DC} = \frac{V_m}{\pi R_L} \quad (3.4)$$

$$I_{DC} = \frac{V_m}{\pi(R_L + r_f)} \quad \text{if forward resistance } r_f \neq 0. \quad (3.5)$$

DC or average value of 'output voltage' across the load resistance R_L

DC component of current I_{DC} can be measured by DC ammeter in series with load R_L in the rectifier circuit. DC voltage across R_L is $V_{DC} = I_{DC} \times R_L$.

$$V_{DC} = \frac{I_m}{\pi} \cdot R_L = \frac{V_m}{\pi(r_f + R_L)} \cdot R_L$$

As r_f is considered to be approximately zero, $(r_f + R_L) \equiv R_L$.

$$\therefore \quad V_{DC} = \frac{V_m}{\pi} = 0.318\, V_m \quad (3.6)$$

where V_{DC} is the average value of the positive half cycle of V_S or V_{in}. The AC power P_{AC} supplied to the circuit is given by

$$P_{AC} = I_{rms}^2 \times R_L. \quad (3.7)$$

3.2.2 Effective or rms Value of Current (I_{rms})

Total value of rms voltage includes the AC and the DC components.
Effective or rms value of the current I_{rms} is

$$I_{rms} = \sqrt{\frac{1}{2\pi}\left[\int_0^{\pi} (I_m \cdot \sin \omega t)^2\, d\omega t + \int_{\pi}^{2\pi} (0\, d\omega t)\right]}$$

$$\therefore \quad I_{rms} = \sqrt{\frac{1}{2\pi}\left[\int_0^{\pi} I_m^2 (\sin^2 \omega t\, d\omega t\right]}$$

$$I_{rms} = \sqrt{\frac{1}{2\pi}\left[I_m^2 \int_0^{\pi}\left(\frac{1-\cos 2\omega t}{2}\right) d\omega t\right]} = \frac{I_m}{2} \quad (3.8)$$

$$V_{rms} = I_{rms} \times R_L = \frac{I_m}{2} \times R_L \quad (3.9)$$

$$\therefore \quad V_{rms} = \frac{V_m}{2} \quad (3.10)$$

$$\text{and } I_{rms} = \frac{I_m}{2}. \quad (3.11)$$

3.2.3 Efficiency of Rectification for Half-Wave Rectifier circuit

Efficiency of rectification η_r is defined as the ratio of DC output power (P_{DC} output), across the load R_L to the AC input power (P_{AC} input).

In this calculation, the effect of forward resistance r_f is ignored. If it is taken into consideration the rectification efficiency η will be less and varies with load.

$$P_{DC}(\text{output}) = I_{DC}^2 \times R_L = \left(\frac{I_m}{\pi}\right)^2 \times R_L \tag{3.12}$$

$$P_{AC}(\text{input}) = I_{rms}^2 \times R_L = \left(\frac{I_m}{2}\right)^2 \times R_L \tag{3.13}$$

$$\text{Rectification efficiency} \quad \eta_r = \frac{\left(\frac{I_m}{\pi}\right)^2 \times R_L}{\left(\frac{I_m}{2}\right)^2 \times R_L} = \frac{4}{\pi^2} = 0.406 \tag{3.14}$$

$$\%\ \text{Rectification efficiency} = 40.6. \tag{3.15}$$

% Rectification efficiency of HWR is approximately equal to 40.6 (Eq. (3.14)). This is of no consequence in normal low power circuits but it has to be considered in high power rectifier circuits.

Ripple frequency 'Ripple Frequency' of HWR is the number of half sinusoids per second in the output voltage waveform observed across load resistance R_L using a CRO. For one cycle of AC input signal waveform, one half sinusoids are observed across R_L. As the supply frequency f_S is 50 Hz, 50 half sinusoids can be observed per second. Therefore, ripple frequency f_s for HWR is 50.

Ripple factor (γ) By definition

$$\gamma = \frac{\text{rms value of rectified AC current}}{\text{average value of current of rectified signal}} \tag{3.16}$$

$$\gamma = \frac{I'_{rms}}{I_{av}} \quad \text{where } I_{rms}^2 = (I'_{rms})^2 + (I_{DC}^2) \tag{3.17}$$

I_{rms} = rms value of AC signal. I'_{rms} = rms value of rectified AC signal and represents the unwanted AC component relative to the desired DC under no load condition as measured using multimeters or CRO ripple content of HWR's output is large.

$$I_{rms}^2 = I'^2_{rms} + I_{DC}^2$$

$$I'^2_{rms} = I_{rms}^2 - I_{DC}^2$$

$$I'_{rms} = \sqrt{\left(I_{rms}^2 - I_{DC}^2\right)}$$

$$\text{Ripple factor} \quad \gamma = \frac{I'_{rms}}{I_{DC}} = \sqrt{\left(\frac{I_{rms}^2 - I_{DC}^2}{I_{DC}^2}\right)} \tag{3.18}$$

$$\gamma = \sqrt{\left(\frac{I_{rms}}{I_{DC}}\right)^2 - 1} = \sqrt{\left(\frac{I_m/2}{I_m/\pi}\right)^2 - 1} \tag{3.19}$$

$$\gamma_{\text{HWR}} = \sqrt{\frac{\pi^2}{4} - 1} = 1.21. \tag{3.20}$$

A good filter stage is clearly essential to reduce the value of ripple. This circuit is not useful to produce more uniform DC voltage, but is used in cheaper electronic circuits, for example a battery charging circuit uses a simple HWR circuit.

3.2.4 Peak Inverse Voltage: PIV for Diodes in HWR

- During reverse bias condition, diode has to withstand a maximum voltage V_m that exists across total secondary winding of transformer. This maximum voltage V_m across Diode under reverse bias condition (non-conducting Diode) is known as peak inverse voltage (PIV) rating of rectifier Diode. Therefore, PIV $= V_{m.}$
- Diode has to be selected with its breakdown voltage greater than V_m. Peak inverse voltage criteria for rectifier Diodes vary for different types of rectifier circuits.

3.2.5 Voltage Regulation

Voltage regulation is defied as the ratio of variation of DC output voltage ($V_{\text{No Load}} - V_{\text{Load}}$) and no load DC voltage ($V_{\text{No Load}}$) for load variations (variations in DC load current I_{Load}), where $V_{\text{No Load}}$ = No Load DC voltage (when load current is zero) and V_{Load} = load voltage at specific load currents:

$$\text{Regulation } \% = \frac{V_{\text{No Load}} - V_{\text{Load}}}{V_{\text{No Load}}} \times 100 \tag{3.21}$$

$$V_{\text{DC(No Load)}} = \frac{V_m}{\pi} \tag{3.22}$$

$$V_{\text{DC(Load)}} = V_{\text{DC(No Load)}} - I_L(r_f + r_T) = \frac{V_m}{\pi} - I_L(r_f + r_T), \tag{3.23}$$

where r_s or r_T is the secondary winding resistance of the transformer and r_f is the forward resistance of the Diode.

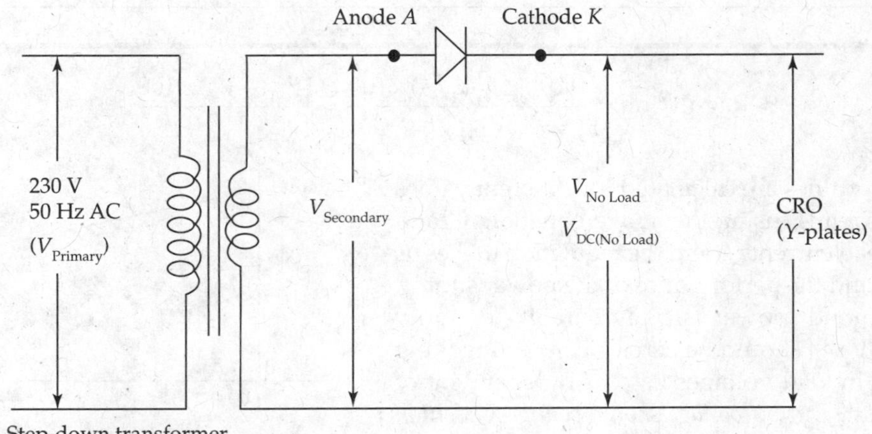

FIG. 3.5 *Half-wave rectifier circuit to measure no load voltage* $V_{\text{No Load}}$

Measurement of No-Load Voltage (Fig. 3.5)

To measure the no load DC voltage ($V_{No\ Load}$) of the HWR circuit, load resistance is disconnected from the previous circuit of Fig. 3.2. It means that output is open circuited or load resistance R_L can be considered as infinity. Then, no load voltage $V_{DC(No\ Load)}$ can be measured using a voltmeter or a CRO:

$$V_{DC(No\ Load)} = \frac{V_m}{\pi} \text{ (theoretical value).}$$

Measurement of Load Voltage

For different values of load resistance R_L that is for different values of load current I_L, the load voltages V_{Load} are measured using a CRO in the circuit shown in Fig. 3.6.

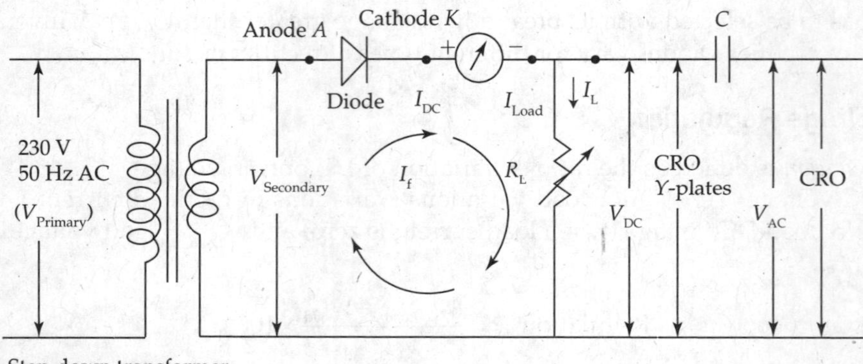

FIG. 3.6 *Half-wave rectifier circuit (to measure various voltages to determine 'regulation')*

Observations to determine regulation characteristic of HWR circuit				
No load voltage =				
S. No	**Load resistance R_L (Ω)**	**Load current I_L (mA)**	**Load voltage V_{Load}**	**% Regulation**

$$\% \text{ Regulation} = \frac{(V_{No\ Load} - V_{Load}) \times 100}{V_{No\ Load}} = \frac{(V_{NL} - V_L) \times 100}{V_{NL}}.$$

Regulation values are calculated for different values of load current I_L using the above equation. Graph between load currents I_L and the regulation makes us to understand the performance of the power supply.

For a good power supply circuit, quantity $|V_{No\ Load} - V_{Load}|$ should be zero or constant making variations in load voltages zero or independent of load current. *Output voltage should be independent of load current at least over a desired range of load currents for ideal voltage regulation.*

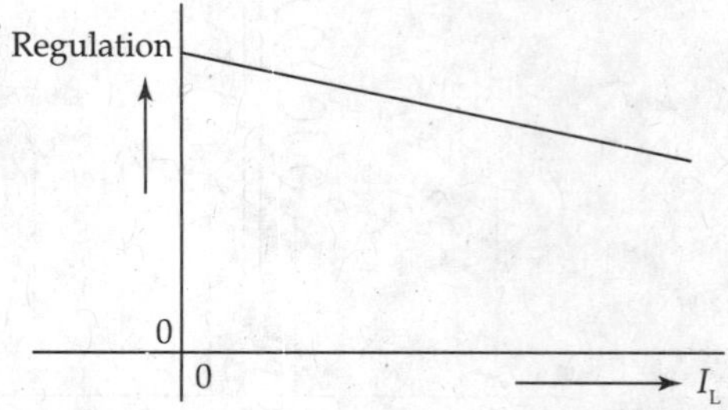

FIG. 3.7 *Voltage regulation curve (wrt load current)*

Measurement of ripple voltage (Fig. 3.6)

V_{DC} and V_{AC} are measured at nominal I_L for predicting the value of ripple factor. Ripple factor varies with I_L. Capacitor blocks DC and allows AC component or ripple voltage to be measured. Varying the load resistance R_L causes variations in load current I_L. These in turn causes variations in DC voltage V_{DC} and ripple AC voltages V_{AC}. So, varying R_L note down the corresponding variations in I_L, V_{DC} and V_{AC}. The observations are noted down in the tabular form. Values of ripple content shows the performance feature of the rectifier circuit in the process of conversion of AC into uniform DC voltage.

Observations to calculate ripple factor

S. No	Load current I_L (mA)	DC voltage V_{DC} (V)	Ripple AC voltage (V)	Ripple factor (γ)

HWR circuit is the basic block for building the other rectifier circuits.

Various features of HWR circuit

S. No.	Parameter	Value
1	DC load current I_{DC}	$\frac{I_m}{\pi} = 0.318\,I_m$
2	rms load current I_{rms}	$\frac{I_m}{2}$
3	DC output voltage V_{DC}	$\frac{V_m}{\pi}$
4	rms output voltage	$\frac{V_m}{2}$
5	% Rectification efficiency	40.6
6	% Transformer utilisation factor	28.6
7	Ripple factor	1.21
8	Peak inverse voltage	V_{max}
9	Ripple frequency	50 Hz

3.2.6 Transformer Utilisation Factor for Half-Wave Rectifier Circuit

In the design of DC power supply circuits, specifications of power transformer ratings are necessary for the manufacturing process of a transformer, in addition to the estimate of the ratings of the other components in the circuit. To decide the number of turns of the primary and secondary windings, gauze of the wire, insulation and so on, preliminary estimate of primary and secondary voltages, DC power to load and type of rectifier circuit are necessary.

Transformer utilisation factor (TUF) is defined as the ratio of 'DC power supplied to the load' to the 'rating of the AC voltage of the transformer secondary winding'.

$$\text{TUF} = \frac{\text{DC power output to the load}}{\text{AC rating of transformer secondary winding}} \tag{3.24}$$

$$\text{The rated voltage of secondary winding} = \frac{V_m}{\sqrt{2}} \tag{3.25}$$

$$\text{rms value of current through secondary winding of transformer} = \frac{I_m}{2} \tag{3.26}$$

$$\text{AC power rating of the transformer secondary winding} = \frac{V_m}{\sqrt{2}} \times \frac{I_m}{2} \tag{3.27}$$

$$\text{DC power supplied to the load for HWR} = \frac{V_m}{\pi} \times \frac{I_m}{\pi} = \frac{V_m \times I_m}{\pi^2}. \tag{3.28}$$

Substituting Eqs. (3.27) and (3.28) in Eq. (3.24), we get

$$\text{TUF} = \frac{\left(\frac{V_m}{\pi} \times \frac{I_m}{\pi}\right)}{\left(\frac{V_m}{\sqrt{2}} \times \frac{I_m}{2}\right)} = \frac{2\sqrt{2}}{(\pi)^2} = 0.286 \tag{3.29}$$

$$\%\ \text{TUF} = 28.6. \tag{3.30}$$

The rectifier efficiency and TUF are different. This is clear from their definitions.

3.2.7 Demerits of Half-Wave Rectifier Circuit

- Rectification efficiency is low only 40.6%.
- Ripple factor = 1.21. Therefore, ripple content is high.
- A good filter stage is clearly essential to reduce the value of ripple.
- TUF = 28.6%.
- Low value of utilisation of transformer ratings.

EXAMPLE 3.1

Transformer of a HWR has a secondary voltage of 30 V (V_{rms}) with winding resistance r_s of 10 Ω. Semiconductor Diode in the circuit has a forward resistance r_f of 10 Ω. Determine: (a) no load DC voltage, (b) DC output voltage when the load current $I_L = 25$ mA, (c) percentage regulation at this load current, (d) ripple voltage across the load, ripple frequency f_s and ripple factor, (e) DC power output, (f) AC power input, (g) power conversion efficiency and (h) PIV, Diode has to withstand.

Solution: For HWF circuit

a.
$$V_{rms} = \frac{V_m}{2}$$

$$\therefore\ V_m = 2\,V_{rms} = 2 \times 30 = 60\ \text{V}$$

No load DC voltage $V_{DC(\text{No Load})} = \frac{V_m}{\pi} = \frac{60}{\pi} = 19.1\ \text{V}.$

b. When the load current $I_L = 25$ mA, consider the voltage drops across r_s and r_f due to the flow of load current through them.

$$\text{Then } V_{\text{DC(Load)}} = \frac{V_m}{\pi} - I_{\text{DC}}(r_s + r_f)$$

$$V_{\text{DC(Load)}} = \frac{60}{\pi} - 25\times10^{-3}(10+10)$$

$$V_{\text{DC(Load)}} = 19.1 - 0.5 = 18.6 \text{ V}.$$

c. Percentage regulation at this load current

$$\%\text{ Regulation} = \frac{(V_{\text{No Load}} - V_{\text{Load}})}{V_{\text{Load}}}\times 100$$

$$= \frac{(19.1-18.6)}{18.6}\times100 = 2.688$$

$$\%\text{ Regulation} = 2.688$$

$$\text{Another formula to calculate \% regulation} = \frac{(r_s + r_f)}{R_L}\times 100$$

Load resistance R_L can be calculated using the formula

$$I_L = \frac{V_{\text{DC}}}{R_L}$$

$$\therefore\ R_L = \frac{V_{\text{DC}}}{I_L} = \frac{19.1}{25\times10^{-3}} = 764\ \Omega$$

$$\%\text{ Regulation} = \frac{(r_s + r_f)}{R_L}\times100$$

$$= \frac{(10+10)\times100}{764} = 2.62.$$

d. Ripple factor $\gamma = \sqrt{\left[\frac{V_{\text{rms}}}{V_{\text{DC}}}\right]^2 - 1} = \sqrt{\left[\frac{30}{19.1}\right]^2 - 1} = 1.211$

Also for resistance load form factor $F = \frac{V_L(\text{rms})}{V_L(\text{DC})} = \frac{30}{18.6} = 1.612.$

e. DC power output $P_{\text{DC}} = I_{\text{DC}}^2 \times R_L = (25\times10^{-3})^2\times764 = 0.477$ W.

f. AC input power $P_{\text{AC(input)}} = I^2_{\text{rms}}(r_s + r_f + R_L)$

$$\text{where } I_{\text{rms}} = \frac{I_m}{2} = \frac{V_m}{2R_L}$$

$$\therefore\ P_{\text{AC(input)}} = \frac{V_m^2}{4\times R_L^2}\times(r_s + r_f + R_L)$$

$$= \left[\frac{(60)^2\times(10+10+764)}{4\times(764)^2}\right] = 0.486 \text{ W}.$$

g. % Power conversion efficiency $\eta = \frac{P_{DC(output)}}{P_{AC(input)}} \times 100$

$$\% \eta = \frac{0.477}{0.486} \times 100 = 98.1\%.$$

h. PIV = 60 V (PIV is equal to the maximum voltage across secondary winding of the transformer which has to be with stood by the Diode under reverse bias (without breakdown of the device).

EXAMPLE 3.2

A Diode has an internal resistance of 20 Ω and 1000 Ω load from 110 V_{rms} source of supply voltage. Calculate: (May/June 2006, set-3, JNTU) (a) Efficiency of rectification; (b) Percentage regulation from no load to full load.

Solution: The provided data (*A* Diode) indicates that the circuit is a HWR

a. Efficiency of rectification

$$I_m = \frac{V_m}{(R_L + r_f)} = \frac{V_{rms} \cdot \sqrt{2}}{(R_L + r_f)}$$

$$\therefore \; I_m = \frac{110 \times \sqrt{2}}{(1000 + 20)} = \frac{155.56}{1020} = 152.5 \text{ mA}$$

$$I_{DC} = \frac{I_m}{\pi} = \frac{152.5 \text{ mA}}{\pi} = 48.52 \text{ mA}$$

$$\text{DC power output } P_{DC} = I_{DC}^2 \times R_L$$

$$= (48.52 \times 10^{-3})^2 \times 1000 = 2.354 \text{ W}$$

$$\text{AC power input } P_{AC} = I_{rms}^2 (R_L + r_f)$$

$$P_{AC} = \left[\frac{I_m}{2}\right]^2 \times (R_L + r_f)$$

$$= \left[\frac{152.5 \times 10^{-3}}{2}\right]^2 \times (1000 + 20) = 5.93 \text{ W}$$

$$\% \text{ Efficiency of rectification} = \eta_r \times 100 = \left[\frac{P_{DC}(\text{output})}{P_{AC}(\text{input})}\right] \times 100$$

$$\eta_r = \left[\frac{2.354}{5.93}\right] \times 100 = 39.7\%.$$

b. $$\text{Percentage regulation} = \left(\frac{(V_{\text{No Load}} - V_{\text{Full Load}})}{V_{\text{No Load}}}\right) \times 100$$

$$\text{No load voltage } V_{\text{No Load}} = \frac{V_m}{\pi}$$

$$\therefore \; V_{NL} = \frac{110 \times \sqrt{2}}{\pi} = 49.5$$

$$\text{Load voltage } V_{\text{DC(load)}} = \frac{V_{\text{m}}}{\pi} - I_{\text{DC}}(r_{\text{s}} + r_{\text{f}})$$

$$\therefore \quad V_{\text{DC(load)}} = \frac{110 \times \sqrt{2}}{\pi} - 48.52 \times 10^{-3} \times 20$$

$$= 49.5 - 0.97 = 48.53 \text{ V}$$

$$\% \text{ Regulation} = \frac{(V_{\text{No Load}} - V_{\text{Load}})}{V_{\text{No Load}}} \times 100$$

$$= \frac{(49.5 - 48.53)}{49.5} \times 100$$

$$= \frac{0.97}{49.5} \times 100 = 1.96\%.$$

3.3 FULL-WAVE RECTIFIER CIRCUIT

3.3.1 Various Components of Full-Wave Rectifier Circuit (Fig. 3.8)

- Mains transformer with centre tapped secondary winding.
- Two semiconductor Diodes D_1 and D_2 (rectifying Diodes).
- Load resistance R_L.

DC output power from rectifier circuit is increased by using the two Diodes so that the two half sinusoids of AC signal (in one cycle) provide the increase in energy. As the two half sinusoids of the complete or full waveform of one cycle of the AC voltage are used in this circuit function, the circuit is known as 'full-wave rectifier'.

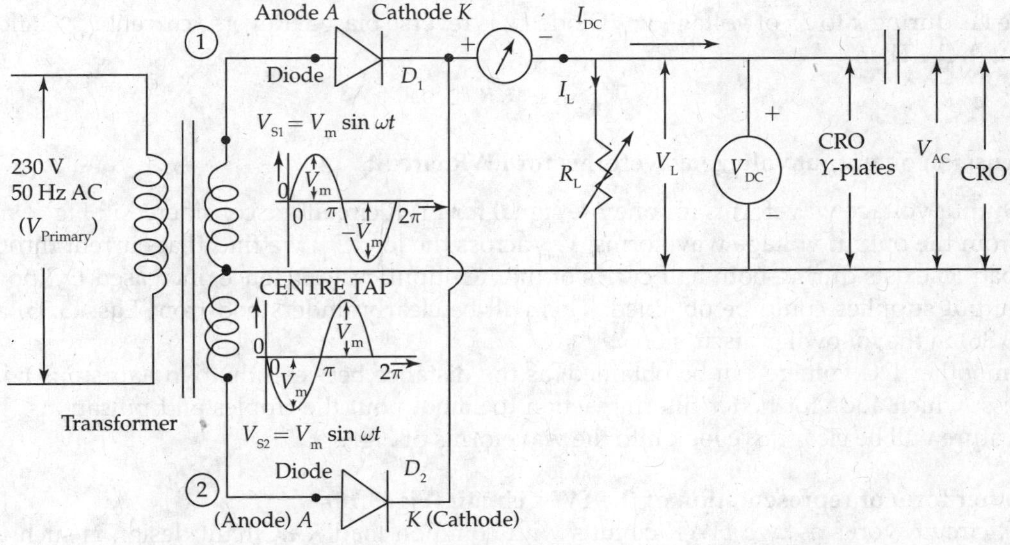

FIG. 3.8 *Full-wave rectifier circuit (to understand principle of working)*

3.3.2 Mains Transformer for Low-voltage Supply

- Primary winding of the transformer is connected to the mains AC voltage source. AC mains voltage is 230 V. Primary voltage $V_P = 230 \sin(2\pi \cdot ft)$, where the frequency ($f$) of the supply voltage is 50 Hz in India.
- Secondary winding of the transformer has a centre tap. The two secondary voltages V_{S1} and V_{S2} are equal in magnitude and 180° out of phase (Fig. 3.8).
- Upper half winding of the transformer secondary is in clockwise direction and the lower half of the winding is in anti-clockwise direction. So, the two induced voltages V_{S1} and V_{S2} are equal and 180° out of phase.

3.3.3 Working of Full-Wave Rectifier Circuit

During the positive half cycle i.e. the interval 0–π of the input voltage V_{S1}, Diode D_1 is forward biased. Forward current I_f flows through load resistance R_L and hence develops output voltage

$$V_{out} = I_f \times R_L. \tag{3.31}$$

At the same time, during the interval '0–π' of input voltage V_{S2}, Diode D_2 is reverse biased. Hence, reverse current $I_{r2} \cong 0$. Reverse current flows through R_L and

$$V_{out} = I_{r2} \times R_L \cong 0. \tag{3.32}$$

During negative half cycle i.e. from π to 2π time period of voltage V_{S2}, polarities of voltage across the secondary winding are such that Diode D_2 is forward biased. Forward current from Diode D_2 is I_{f2}. Current I_{f2} flows through R_L and develops DC output voltage V_{out} across load resistance R_L.

$$\text{DC output voltage} \quad V_{out} = I_{f2} \times R_L. \tag{3.33}$$

The two currents flow in the same direction through the load resistance. During negative half cycle i.e. during π to 2π of voltage V_{S1}, Diode D_1 is reverse biased. Reverse current $I_{r1} \cong 0$ flows through R_L. Then

$$V_{out} \cong I_{r.1} \times R_L \cong 0. \tag{3.34}$$

Discussion on output voltage waveforms for FWR circuit

- Output voltage waveforms for one AC signal for FWR circuit are considered in Fig. 3.9.
- From the output voltage waveforms, V_{out} across the load R_L, we find that current through load R_L exists during both half cycles of the AC-input cycles. Hence, increased DC power output supplies could be obtained. This will be clearly understood from Eqs. (3.35) and (3.36) in the following discussions.
- Smoother DC voltage can be obtained, as the distance between the two half sinusoids is less, which adds for better filtering action to smooth out the ripples and pulsations. This feature will be clear if we look into the waveforms of Fig. 3.9.

Another form of representation of the FWR circuit (Fig. 3.10)

FWR circuit works as two HWR circuits with common load R_L. Circuit design is such that conduction through two Diodes occurs in alternate half cycles of AC input signal. Two

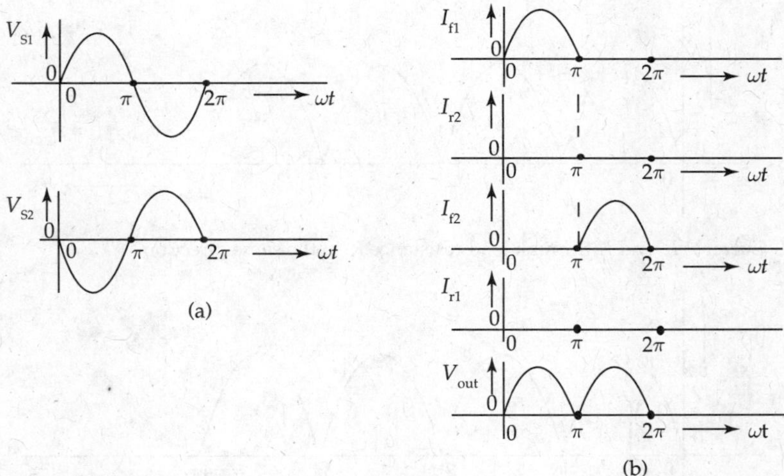

FIG. 3.9 *(a) Transformer secondary voltages; (b) signal waveforms of full-wave rectifier circuit*

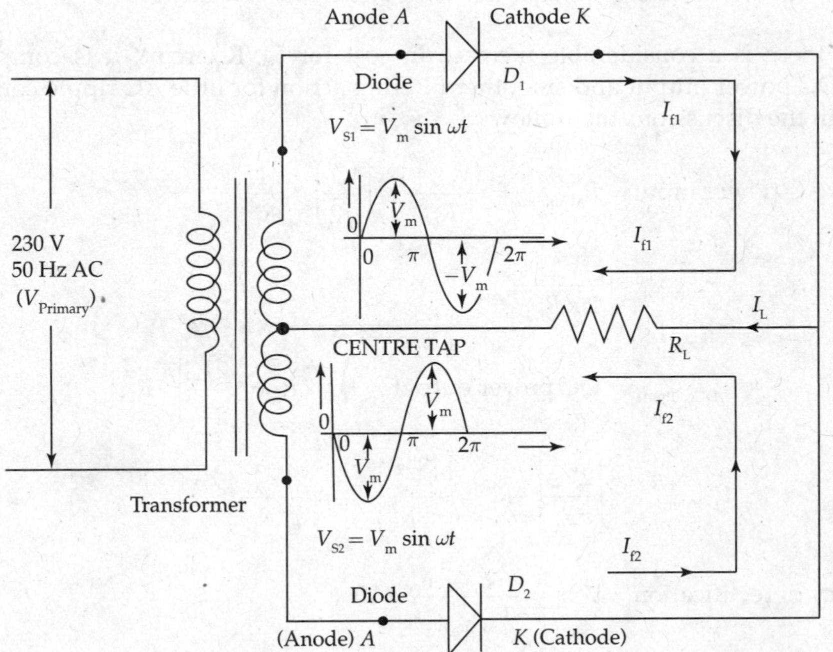

FIG. 3.10 *Full-wave rectifier circuit*

currents I_{f1} and I_{f2} flow in the same direction through load as seen by arrows on current flow lines. Output voltage waveform contains two half sinusoids in two half cycles of one cycle of AC input signal in the first quadrant only (Fig. 3.9). FWR produces more DC power (double to HWR circuit) (Eq. (3.35)) with increased value of rectification efficiency (derived in the following section).

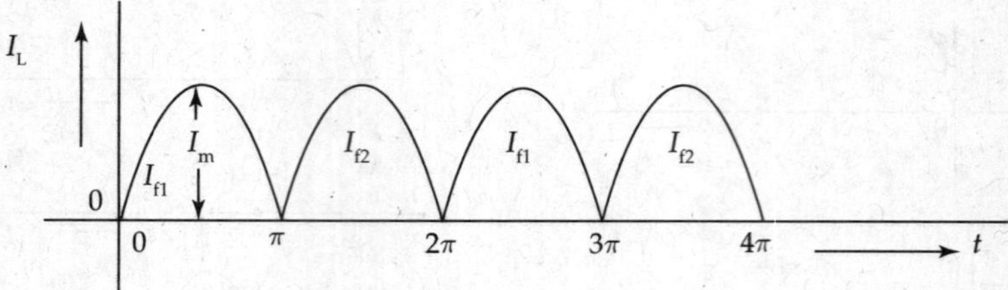

Currents I_{f1} and I_{f2} through load resistance R_L in the same direction

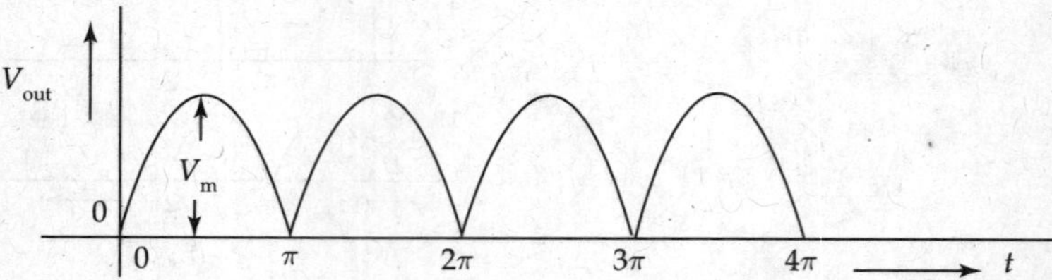

Output voltage V_{out} wave forms across R_L due to I_L through it

FIG. 3.11 *Output waveforms in full-wave rectifier circuits*

Even if there is a considerable increase in cost for FWR circuit, it is compensated by increased DC power output and smoother filtering action for little AC ripple content, which is clear from the discussions that follow:

$$\text{Total AC power input} = P_{AC(input)} = \frac{(V_{rms})^2}{R_L} = \left(\frac{V_m}{\sqrt{2}}\right)^2 \times \frac{1}{R_L}$$

$$P_{AC(input)} = \frac{V_m^2}{2 \times R_L} \tag{3.35}$$

$$P_{DC(output)} = \text{DC power output} = I_{DC}^2 \times R_L = \left[\frac{2I_m}{\pi}\right]^2 \cdot R_L \tag{3.36}$$

$$P_{DC(output)} = \left[\frac{2V_m}{\pi R_L}\right]^2 \cdot R_L = \left[\frac{4V_m^2}{\pi^2 R_L^2}\right] \cdot R_L = \frac{4V_m^2}{\pi^2 R_L} \tag{3.37}$$

$$\%\ \text{Efficiency of rectification} = \%\ \eta_r = \frac{P_{DC(output)}}{P_{AC(input)}} \times 100$$

$$= \frac{\left[4V_m^2/\pi^2 R_L\right]}{\left[V_m^2/2R_L\right]} = \frac{8 \times 100}{\pi^2} = 81\%. \tag{3.38}$$

Rectification efficiency of FWR circuit is twice that of HWR circuit.

Measurement of ripple factor Measuring DC voltage V_{DC} and AC ripple voltage V_{AC}, the ripple factor of FWR can be calculated as the ratio of output AC to DC voltages. Theoretical value will be 0.48. Please refer to Fig. 3.8.

Significance of ripple factor For a FWR, we get 0.48 V of unwanted AC for every 1 V of wanted DC, whereas for a HWR every 1 V of wanted DC we get 1.21 V of unwanted AC component. Filtering circuits reduces this ripple voltage better (because of the reduced distance between the sinusoids) and produce ripple-free output voltage.

Peak Inverse Voltage

Under reverse bias condition, each Diode (in FWR circuit) has to withstand a maximum voltage '$2V_m$' that exists across total secondary winding of transformer. This maximum voltage $2V_m$ across each Diode under reverse bias condition is known as 'peak inverse voltage' (PIV) rating. So, in FWR Rectifier circuit PIV = $2V_m$, each Diode has to be selected with its breakdown voltage greater than $2V_m$. So, the cost of the Diodes increases.

- Two voltages in the two half windings of the secondary winding of transformer add together. So, total secondary voltage between the points (1) and (2) is '$2V_m$'. (Fig. 3.10)
- From the circuit in Fig. 3.12, it will be clear that the maximum voltage that exists across a reverse-biased Diode D_2 is '$2V_m$'. So, PIV is '$2V_m$' for Diodes in the FWR circuit. Diodes with higher break-down voltages are to be selected
- Similarly, maximum voltage that exists across the reverse-biased Diode D_1 is '$2V_m$'.
- So, PIV is '$2V_m$' for the two Diodes in FWR circuit.

In high-power DC supply sources using an FWR circuit with centre-tapped secondary winding transformer, cost and size of the transformer increase abnormally. So, only low-power DC supplies use FWR configuration with centre-tapped secondary transformer

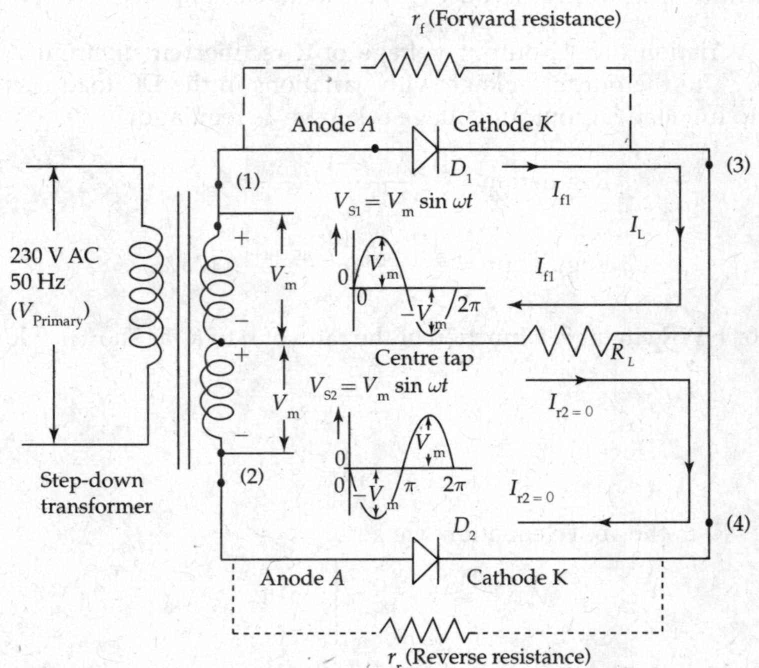

FIG. 3.12 *Peak inverse voltage (PIV) concept in a full-wave rectifier circuit*

Various features of FWR circuit

S. No.	Parameter	Value
1	DC load current I_{DC}	$\frac{2I_m}{\pi}$
2	rms load current I_{rmS}	$\frac{I_m}{\sqrt{2}}$
3	DC output voltage V_{DC}	$\frac{2V_m}{\pi}$
4	rms output voltage	$\frac{V_m}{\sqrt{2}}$
5	% Rectification efficiency	81.2
6	% Transformer utilisation factor	81.2
7	Ripple factor	0.483
8	Peak inverse voltage	$2V_{max}$
9	Ripple frequency	100 Hz

EXAMPLE 3.3

Define the concept of voltage regulation in rectifier circuits and also define percentage regulation. Verify with necessary equations that the regulation of FWR and HWR circuits depends on the ratio of forward resistance r_f to the load resistance R_L.

Solution: The variation of DC output voltage of a rectifier circuit from no load output voltage $V_{DC\ (No\ Load)}$ to the output voltage with variations in the DC load current expressed with reference to full load DC output voltage is known as regulation.

$$\text{Regulation} = \frac{[V_{\text{No Load}} - V_{\text{Full Load}}]}{V_{\text{Full Load}}}$$

$$\%\ \text{Regulation} = \frac{[V_{\text{No Load}} - V_{\text{Full Load}}]}{V_{\text{Full Load}}} \times 100\%.$$

% Regulation for HWR circuit is a function of the ratio of r_f to R_L as shown below.

$$I_{DC} = \frac{(V_m/\pi)}{(r_f + R_L)}$$

$$\therefore\ I_{DC}(r_f + R_L) = \frac{V_m}{\pi}$$

Using $V_{DC} = I_{DC} \cdot R_L$ in the above equation, we get

$$V_{DC} = \left[\frac{V_m}{\pi} - I_{DC} \cdot r_f\right] \text{ and } V_{DC\ (\text{No Load})} = \frac{V_m}{\pi}$$

$$\therefore\ \%\ \text{Regulation} = \frac{[V_{\text{No Load}} - V_{\text{Full Load}}]}{V_{\text{Full Load}}}$$

$$\% \text{ Regulation} = \frac{\left[\frac{V_m}{\pi} - \frac{V_m}{\pi} + I_{DC} \cdot r_f\right]}{I_{DC} \cdot R_L} \times 100\% = \frac{r_f}{R_L} \times 100\%.$$

The above derivation is also true for FWR circuit.

EXAMPLE 3.4

A power supply having output resistance of 2 Ω supplies a full-load current of 100 mA to a 50 Ω load. Find the percent voltage regulations and no load output voltage of the supply.

Solution: Given Output resistance $R_o = 2\,\Omega$, full-load current $I_{FL} = 100$ mA, full-load resistance $R_{FL} = 50\,\Omega$. We know that

$$V_{FL} = I_{FL} \cdot R_{FL} = 100 \times 10^{-3} \times 50 = 5.0 \text{ V}$$

Percentage voltage regulation:

$$V \cdot R\,\% = \frac{R_o \times I_{FL}}{V_{FL}} \times 100\% = \frac{2 \times 100 \times 10^{-3}}{5} \times 100\% = 4\%$$

No load output voltage:

$$V_{NL} = \frac{V_{FL}(R_{FL} + R_o)}{R_{FL}} = \frac{5(50+2)}{50} = 5.2 \text{ V}.$$

EXAMPLE 3.5

A FWR has a centre-tapped transformer 100-0-100 V. Each one of the Diode is rated at I_{max} of 400 mA and I_{av} of 150 mA. Neglecting voltage drop across the Diodes, find (a) the value of the load resistance that gives the largest DC power output, (b) DC output voltage, (c) DC load current and (d) PIV of each Diode. (June 2005, set-1)

Solution: V_m = maximum value of secondary voltage = $\sqrt{2} \times V_{rms} = \sqrt{2} \times 100 = 141.4$ V
Assuming the Diode is operated safely at 80% of the maximum value.

$$\text{Rated current} = \frac{80}{100} \times 400 \text{ mA} = 320 \text{ mA}$$

a. Value of load resistance R_L (that gives maximum DC Power output) $= \dfrac{V_m}{\text{rated current}}$

$$R_L = \frac{141.4}{320 \times 10^{-3}} = 441.88\,\Omega$$

∴ Load resistance $R_L = 441.88\,\Omega$.

b. DC load voltage $\quad V_{DC} = 2\dfrac{V_m}{\pi} = 2 \times \dfrac{141.4}{\pi} = 89.98$ V.

c. DC load current $\quad I_{DC} = \dfrac{V_{DC}}{R_L} = \dfrac{89.98}{441.8} = 0.2$ A.

d. PIV of each diode $\quad 2 \cdot V_m = 2 \times 141.4 = 282.8$ V.

Transformer utilisation factor for full-wave rectifier circuit
Specifications of the power transformer ratings are necessary for the manufacturing process of a transformer, in addition to the estimate of the ratings of the other components in the power supply circuit design. TUF is defined as the ratio of 'DC power supplied to the load' to the 'rating of the AC voltage of the transformer secondary winding'.

For FWR circuit

$$\text{TUF} = \frac{\text{DC power output to the load}}{\text{AC rating of transformer secondary winding}} \tag{3.39}$$

$$\text{The rated voltage of secondary winding} = \frac{V_m}{\sqrt{2}} \tag{3.40}$$

$$\text{The rms value of current through secondary winding of transformer} = \frac{I_m}{\sqrt{2}} \tag{3.41}$$

$$\text{AC power rating of the transformer secondary winding} = \frac{V_m}{\sqrt{2}} \times \frac{I_m}{\sqrt{2}} = \frac{V_m \cdot I_m}{2} \tag{3.42}$$

$$\text{Output DC voltage} \quad V_{DC} = \frac{2V_m}{\pi} \tag{3.43}$$

$$\text{DC current through load} \quad I_{DC} = \frac{2I_m}{\pi} \tag{3.44}$$

$$\text{DC power supplied to the load for FWR} = \frac{2V_m}{\pi} \times \frac{2I_m}{\pi} = \frac{4 \times V_m \cdot I_m}{\pi^2}. \tag{3.45}$$

Substituting Eqs. (3.41)–(3.45) in Eq. (3.46), we get

$$\text{Transformer utilisation factor (TUF)} = \frac{\left(\frac{2V_m}{\pi}\right) \times \left(\frac{2I_m}{\pi}\right)}{\frac{V_m}{\sqrt{2}} \times \frac{I_m}{\sqrt{2}}} \tag{3.46}$$

$$= \frac{8}{(\pi)^2} = 0.812$$

% TUF = 81.2.

There is an increase in utilisation of transformer secondary voltage, because two Diodes in FWR circuit conduct alternately in positive and negative half cycles of transformer secondary voltage.

Advantages:

1. Reduced ripple factor of 48% when compared to 121% in HWR circuit.
2. Increased value of TUF.
3. Decrease in the cost of filtering circuit.
4. More nearer to constant DC output voltage.

3.3.4 Half-wave and Full-wave Rectifier Circuits (Practical Circuit for Measurements)

Aim:

1. HWR circuit connections are made as per the circuit diagram
2. Use the oscilloscope to see the input and output waveforms
3. Determine the ripple voltage and ripple frequency
4. Determine the voltage regulation characteristic

Apparatus:
Circuit PCB, Multimeters, Ammeter 0–200 mA (DC), CRO.

Circuit:

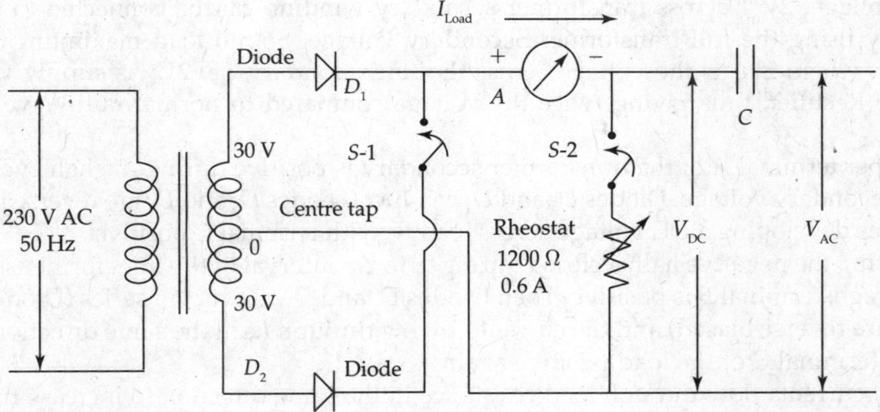

FIG. 3.13 *Half-wave and full-wave rectifier circuits using SPST switches*

Procedure:

1. Circuit connections are made as per the circuit diagram.
2. Open the switch *S*-1 for HWR operation.
3. Open the switch *S*-2 to disconnect R_L for open circuit condition and note the open circuit DC voltage, which is 'No-load DC voltage'.
4. Close the switch *S*-2 to include load resistance R_L (Rheostat 1200 Ω; 0.6 A).
5. By varying the load resistance R_L (Rheostat), note down the variations in the load current, (I_L) DC voltage (V_{DC}) and the AC ripple voltages (V_{AC}).
6. Observations are noted in the following table.

Observations:

S. No.	I_{Load} (mA)	V_{DC}	V_{AC}	Ripple factor	Regulation
1					

Calculations:

1. Ripple factor $= \gamma = \dfrac{V_{AC}}{V_{DC}}$ and
2. Regulation $= \dfrac{V_{No\ Load} - V_{Load}}{V_{No\ Load}}$.

Graph: Draw the regulation characteristic between I_{Load} and Regulation

* For Full-Wave Rectifier circuit experiment, repeat all the above steps closing switch *S*-1.

3.4 BRIDGE RECTIFIER CIRCUIT (FULL-WAVE RECTIFIER)

- Four Diodes are connected similar to the configuration of a Wheatstone Bridge. So, Rectifier circuit in Fig. 3.14 is known as a 'Bridge Rectifier'. Its performance is similar to 'Full-Wave Rectifier circuit' without the requirement of a centre tap on secondary winding of the transformer.
- Full voltage '$2V_m$' across transformer secondary winding can be connected to a Bridge, thereby using the full transformer Secondary voltage. So, no load maximum voltage is $4V_m/\pi$, where $2V_m$ is the voltage across the full secondary, i.e. $2V_m$ is supply voltage to Bridge Rectifier, thus giving twice the voltage compared to normal Full-Wave Rectifier circuit.
    - o Upper terminal *A* of the transformer secondary is positive during the half cycle '0 to π' of secondary voltage, Diodes D_2 and D_4 conduct (Diodes D_3 and D_1 are reverse biased) in series developing a DC voltage across load R_L with terminal *C* positive.
    - o During the negative half cycle i.e. during 'π to 2π' interval of the transformer secondary voltage, terminal *B* is positive. Then Diodes D_3 and D_1 conduct in series (Diodes D_4 and D_2 are reverse biased) and the current flowing through R_L in the same direction making the terminal C of the load positive again.
    - o Two currents flow through load resistance in the same direction to increase the output DC voltage. Secondary current is present during both halves of the secondary voltage. Thus the transformer is fully utilised.
- We observe that in both the situations discussed by now, there are two Diodes in series in each of the conduction paths and currents flow in the same direction through the load resistance R_L.

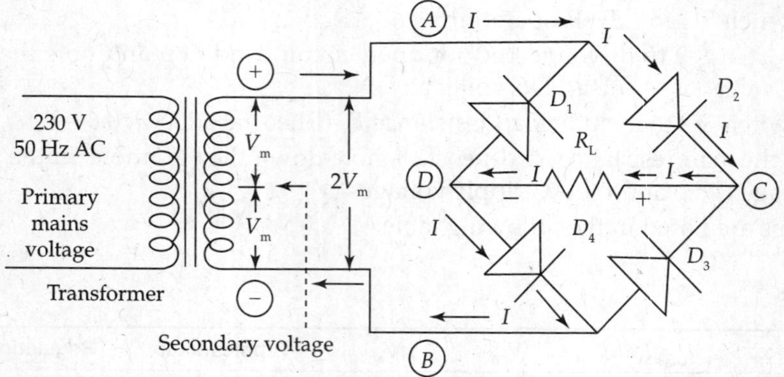

FIG. 3.14 *Bridge rectifier circuit*

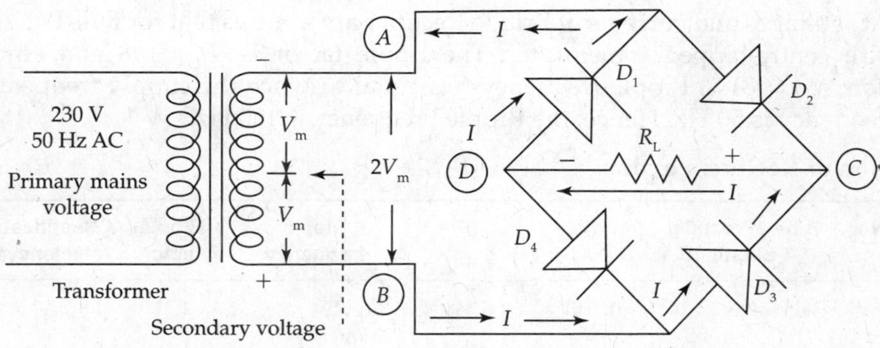

FIG. 3.15 *Bridge rectifier circuit*

Advantages

- The circuit develops twice as much DC as compared to FWR circuit with centre-tapped secondary winding for the same secondary of the transformer.
- With reduced size of Diodes (due to reduced PIV V_m), the use of four Diodes in the Bridge Rectifier circuit has become popular as a package.
- Ripple factor and voltage regulations are same as that of Full-Wave Rectifier.
- PIV across each Diode is V_m, similar to HWR circuit.

Disadvantages

- Since the two Diodes are in series during conduction, twice the voltage drop across them reduces the output voltage.
- Bridge Rectifier circuit needs four Diodes.

Signal waveforms across load resistance R_L for the applied input signal voltage and for different situations of Diode conductions in Bridge Rectifier circuits are shown in Fig. 3.16.

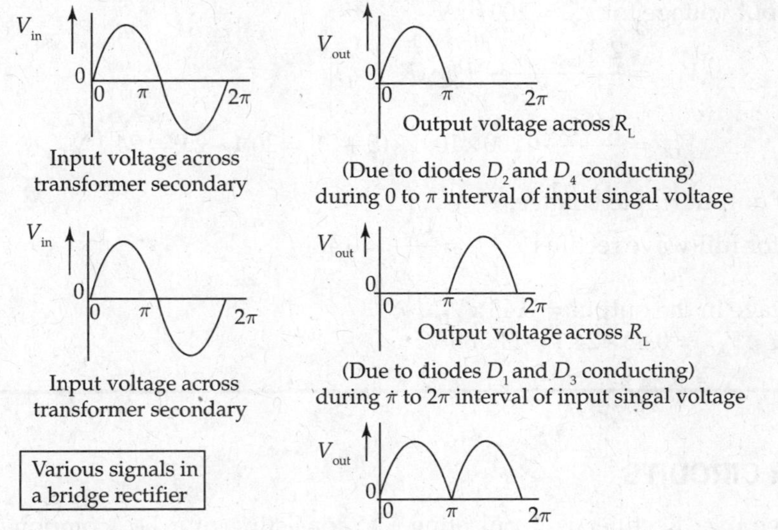

FIG. 3.16 *Output voltage across R_L for bridge rectifier acting as a F. W. Rectifier*

DC and AC voltages and currents of Bridge Rectifier are same as that for Full-Wave Rectifier circuit with centre-tapped transformer. The ripple factor is $\gamma = 0.48$ and efficiency of rectification is $\eta = 81\%$. Ripple frequency f_r is equal to twice the supply frequency f_S. The supply frequency is 50 Hz. Hence, the Ripple Frequency is 100 Hz.

Comparison of Rectifier circuits:

S. No.	Type of rectifier circuit	V_{DC}	PIV	Ripple frequency	Ripple factor	Rectification efficiency (%)
1	Half-wave	$0.318V_{max}$	V_{max}	Supply frequency f_S	1.21	40.6
2	Full-wave	$0.636V_{max}$	$2V_{max}$	$2f_S$	0.48	81.2
3	Bridge rectifier	$0.636V_{max}$	$2V_{max}$	$2f_S$	0.48	81.2

EXAMPLE 3.6

Bridge Rectifier circuit has Diodes with forward resistance, $r_f = 50\ \Omega$, transformer with secondary voltage $V_S = 30 \sin(\omega t)$, $I_{DC} = 200$ mA and transformer Secondary winding resistance $r_s = 5\ \Omega$. Calculate DC output voltage and ripple voltage.

Solution: Bridge Rectifier Circuit
Data: Secondary Voltage $V_{(sec)\ rms} = 30$ V. Secondary resistance $r_s = 5\ \Omega$
Forward resistance of each Diode, $r_f = 5\ \Omega$, $I_{DC} = 200$ mA.

$$V_{(sec)\ rms} = 30 \text{ V.}$$

$$\therefore\ V_{(sec)\ peak} = 1.414 \times V_{(sec)\ rms} = 1.414 \times 30 = 42.42 \text{ V.}$$

Peak Full-Wave Rectified Voltage $= V_{(sec)\ peak} - 2 \times V_{\gamma} = 42.42 - 2 \times 0.7 = 41$ V

V_{DC} = DC output voltage for $I_{DC} = 200$ mA

$$V_{DC} = \frac{2 \cdot V_{(sec)\ peak}}{\pi} - [I_{DC} \cdot (R_S + r_f)]$$

$$\therefore\ V_{DC} = \frac{2 \times 41}{\pi} - [200 \times 10^{-3} \times (5+5)] = 26.1 - 2.0 = 24.1 \text{ V.}$$

Calculation of output ripple voltage:

Ripple factor for full-wave rectifier $\quad \gamma = \dfrac{V_{AC}}{V_{DC}} = 0.48$

$\therefore$ Ripple voltage in the output $= 0.48 \times V_{DC}$
Ripple voltage $\quad V_{AC} = 0.48 \times 24.1 = 11.568$ V.

3.5 FILTER CIRCUITS

Output voltage of a Rectifier is a pulsating DC consisting of a DC component as well as superimposed ripple (AC content). In most cases, a pure DC or a DC with tolerable ripple is

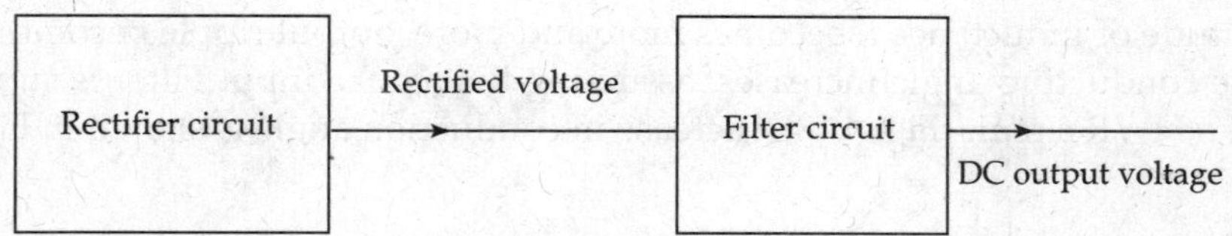

FIG. 3.17 *Combination of rectifier and filter circuits for DC source*

needed. To eliminate or reduce the ripple (unwanted AC component) content to the required level 'Filter circuits' similar to 'coffee Filters' are used. Thus, a Filter circuit is necessary to smoothen out the AC ripple.

Reactive elements Inductors and Capacitors respond to DC and AC differently. So, they can be used to reduce AC content relative to DC in rectified output voltage.

- Filtering process is done using Low Pass Filter circuits.
- Filter circuits contain Capacitors, Inductors and a combination of them such as Shunt Capacitor Filter, Choke Input Filter, *L*-Section Filter and π-section Filters depending on the purity of DC required or the level of ripple content that can be of no problem in a particular application.

Power supply Filter reduces magnitudes of all alternating components or ripple content in rectified output waveforms and passes the DC content. Ripple factor is used as a measure of effective functioning of a Filter circuit. Ultimately for good Filter circuits ripple factor should be as small as possible indicating the reduction of AC component in the Filtering process. Typical Filter circuits are illustrated below.

3.6 HALF-WAVE RECTIFIER WITH INDUCTOR FILTER (CHOKE INPUT FILTER)

Let us recall the formula

$$e = -\frac{d\phi}{dt}. \tag{3.47}$$

An Inductor opposes changes in current. It stores energy in its magnetic field, when the current is above the average value. It delivers energy to load when current falls below the average value. Thus, the pulsations (ripple content) in rectified output voltage fed to the Filter circuit can be smoothened out by keeping a series Inductor in the Filter circuit as shown in Fig. 3.18.

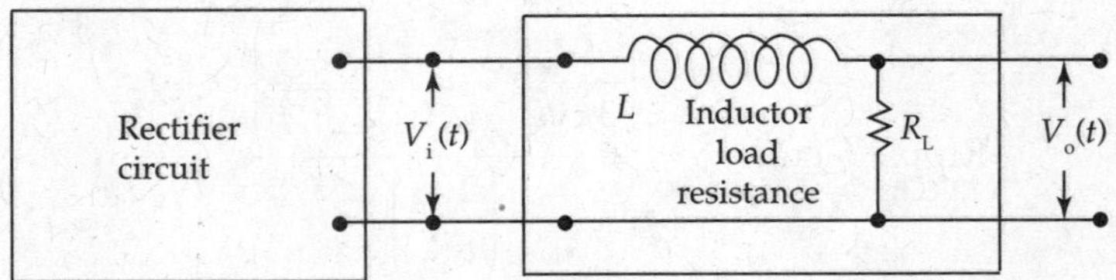

FIG. 3.18 *Inductor filter circuit*

$$V_0(t) = \frac{V_i \cdot R_L}{(R_L + j\omega L)} \angle\phi = \frac{V_i \cdot R_L}{\sqrt{(R_L^2 + \omega^2 L^2}} \cdot \tan^{-1}\left[\frac{\omega L}{R_L}\right]. \tag{3.48}$$

As the magnitude of Inductance L becomes more and more, output ripple becomes lesser and lesser and the conducting angle increases, assuming that choke input Filter is supplied with the output of a HWR circuit (Fig. 3.18). Increase in conduction angle is shown in Fig. 3.19.

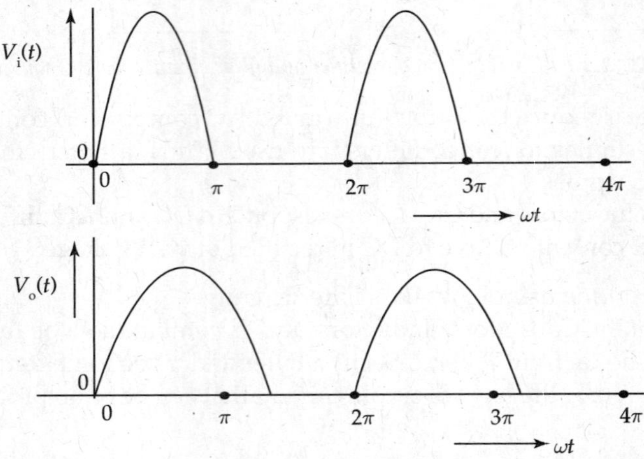

FIG. 3.19 *Effect of inductor on rectifier output waveform*

3.6.1 Function of the 'Inductor Filter'

Inductor offers very low impedance R for the DC component and offers a very high impedance Z which is $\sqrt{(R^2+(\omega L)^2}$ for AC component of the rectified output. As such AC components are blocked to a large extent and, at the same time, DC component is easily allowed to the load. External output contains fewer ripples, which means ripple content is attenuated much by filters.

For a Full-Wave Rectifier circuit, the ripple frequency is $2f_S$ and the pulsating output load current can be considered to contain

$$\text{AC component} = \frac{4V_m \cdot \cos(2\omega t - \phi)}{3\pi\sqrt{R_L^2 + 4\omega^2 L^2}} \tag{3.49}$$

$$\text{DC component} = \frac{2V_m}{\pi R_L} = \frac{0.638V_m}{R_L}. \tag{3.50}$$

If $(4\omega^2 \cdot L^2/R_L{}^2)$ is much greater than unity, then

$$\text{Ripple factor} \quad \gamma = \frac{\left(\dfrac{4V_m}{3\pi\sqrt{2}} \cdot \dfrac{1}{\sqrt{R_L^2 + 4\omega^2 L^2}}\right)}{\left(\dfrac{2V_m}{\pi R_L}\right)}$$

$$= \frac{\sqrt{2}R_L}{3\sqrt{R_L^2 + 4\omega^2 L^2}} = \frac{R_L}{3\sqrt{2} \cdot \omega L}$$

$\therefore$ Ripple factor will be small for low values of load or for large values of load currents and high values of inductances (L).

3.7 HALF-WAVE RECTIFIER CIRCUIT WITH CAPACITOR FILTER

Simplest and most popular Filter circuit is a Capacitor connected in parallel with Load Resistance R_L. Capacitor connected in shunt across load resistance of HWR circuit provides a bypass path for AC component or ripple content present in output voltage of Rectifier circuit and ripple gets attenuated. This Filter circuit is simple and cheap.

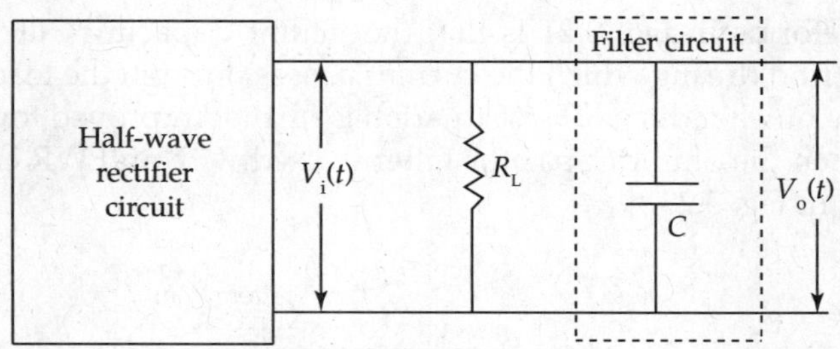

FIG. 3.20 *Half-wave rectifier circuit with capacitor filter*

- Capacitor Filter smoothes the load current by storing energy in one part of the cycle of signal and releases load current during next part of cycle. It is preferred over 'Choke Input Filters', because it occupies less space, light in weight and requires Diodes with lower values of PIV (Peak Inverse Voltage) rating.
  - 'Capacitor Input Filter' depends upon peak detection, whereas 'choke input Filter' depends on average detection.
  - Physically, Capacitor C gets charged to the peak value of the rectified voltage V_m during the period of conduction of the Diode during the interval '0 to $\pi/2$'. The Capacitor discharges through R_L from the interval '$\pi/2$ to 2π' (as the Diode is reverse biased at this juncture, it offers infinite reverse resistance. Then the Capacitor gets disconnected from the transformer secondary voltage and the Diode does not provide any path for Capacitor to discharge any charge through the Diode) after the rectified voltage decreases from its peak value in a time period '$\pi/2$ to 2π' with a time constant CR_L.

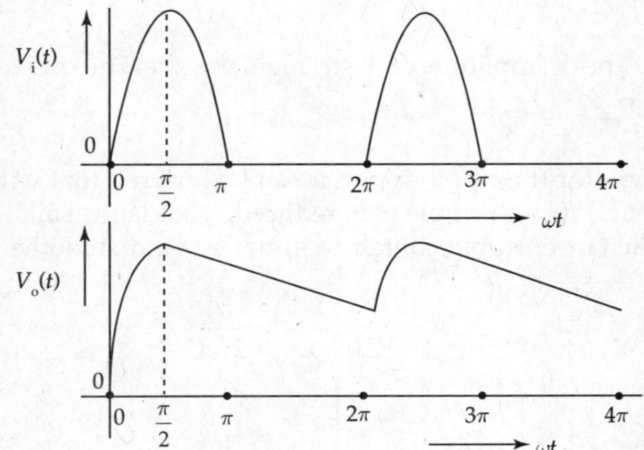

FIG. 3.21 *Effect of shunt capacitor filter on rectifier output waveform*

If CR_L is much larger than T (the time period of the input voltage), the ripple output voltage can be considered as a triangular wave superimposed on the DC. Larger the value of the capacitance C, greater is the reduction in ripple. *Performance of the Capacitor Filter circuit is clear from the waveforms shown in Fig. 3.21.*

3.8 FULL-WAVE RECTIFIER CIRCUIT WITH CAPACITOR FILTER

Inference from waveforms in Fig. 3.21 is that the 'Shunt Capacitor Filter' used with HWR prolongs the time period during which the current passes through the load resistance and the ripple content is very much reduced. This situation is further improved towards the reduction of ripple level V_r when the Shunt Capacitor Filter is used with an FWR circuit as seen in the waveform diagrams in Fig. 3.23.

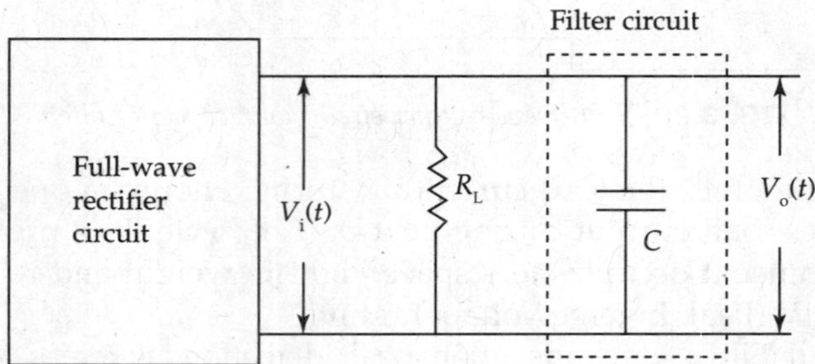

FIG. 3.22 *Full-wave rectifier circuit with capacitor filter*

Capacitor Filter (Fig. 3.23) uses the rectified output voltage $V_i(t)$ from an FWR. Circuit Typical waveforms across 'Shunt Capacitor Filter' with Full-Wave Rectifier circuit are shown in Fig. 3.23. 'Charging time' T_1 is much less than 'discharge time T_2'. Total time $(T = T_1 + T_2)$ is approximately equal to $T_2 = (1/2f)$. But V_r is the peak-to-peak amplitude of the AC voltage across the Capacitor obtained by the charging of the capacitance by the current I_{DC} in a time T.

$$V_{DC} = V_m - \frac{V_r}{2},$$

where V_r is the peak-to-peak amplitude of the triangular wave. The rms value of the triangular wave is known to be $V_{rms} = \frac{V_r}{2\sqrt{3}}$.

From this expression for the ripple factor, it can be inferred that with increasing values of the capacitance C and R_L, the ripple gets reduced. That is for smaller values of I_{DC} and larger values of capacitance, ripple content will be less and smoother will be the output waveform.

$$\because \quad V_r = \frac{Q}{C} = \frac{I \cdot T}{C} = \frac{I_{DC}}{2f \cdot C}$$

$$\text{Ripple factor } \gamma = \frac{V'_{rms}}{V_{DC}} = \frac{\left(\frac{V_r}{2\sqrt{3}}\right)}{V_{DC}} = \frac{\left(\frac{I_{DC}}{2\sqrt{3} \cdot 2f \cdot C}\right)}{I_{DC} \cdot R_L} = \frac{1}{4\sqrt{3}f \cdot C \cdot R_L}.$$

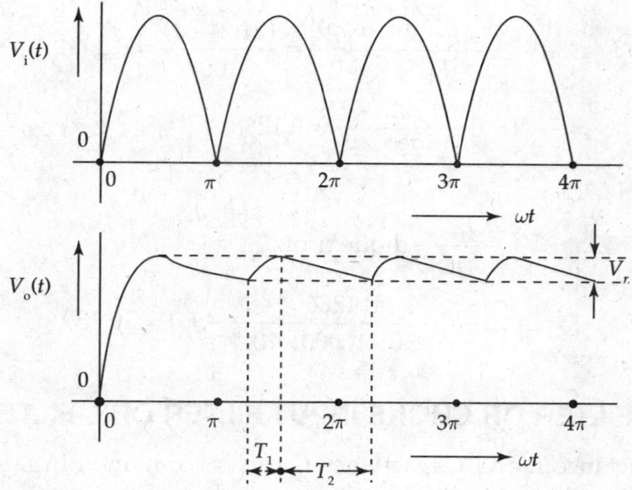

FIG. 3.23 *Full-wave rectifier with shunt capacitor filter output voltage waveforms*

EXAMPLE 3.7

A Full-Wave rectified voltage of 18 V peak is applied across 500 μF Filter Capacitor. Calculate ripple and DC voltages, if the load takes a current of 100 mA. (Aug/Sep 2006, suppl set-2; April/May 2007, set-4)

Solution: Data: $C = 500\ \mu\text{F}$, $I_{DC} = 100$ mA, $V_{max} = 18$ V. Assume that supply freq $f_S = 50$ Hz.
DC voltage

$$V_{DC} = V_m - \frac{I_{DC}}{4f \cdot C} = 18 - \frac{100 \times 10^{-3}}{4 \times 50 \times 500 \times 10^{-6}} = 18 - 1 = 17\ \text{V}$$

Load resistance

$$R_L = \frac{V_{DC}}{I_{DC}} = \frac{17}{100 \times 10^{-3}} = \frac{17 \times 10^3}{100} = \frac{17000}{100} = 170\ \Omega$$

Ripple factor

$$\gamma = \frac{1}{4\sqrt{3} f \cdot C \cdot R_L} = \frac{1}{4 \times 1.732 \times 50 \times 500 \times 10^{-6} \times 170} = 0.03396 \equiv 0.034$$

EXAMPLE 3.8

Ideal transformer with Secondary voltage (15-0-15) is used with FWR with Diodes having a forward drop of 1 V. Load resistance is 100 Ω. A Capacitor of 10,000 μF value is used as a Filter across the load resistance. Calculate the DC load current and voltage. (June 2005, set-4; Nov/Dec 2005, set-4; Aug /Sep 2006, set -3)

Solution: Transformer secondary voltage is 15-0-15 V_{rms}. This corresponds to 21.21-0-21.21 peak value, since $V_m = \sqrt{2} \times V_{rms} = 21.21$ V.

$$I_m = \frac{[V_m - V_d(\text{diode drop})]}{R_L} = \frac{(21.21-1)}{100} = 0.2021 \text{ A}$$

$$I_{DC} = \frac{2I_m}{\pi} = \frac{2\times0.2021\times7}{22} = 0.1286 \text{ A}$$

Assume that $f = 50$ Hz.

$$V_{DC} = V_m - \frac{I_{DC}}{4fC} - \text{diode drop}$$

$$= 21.21 - \frac{0.1286}{4\times50\times10000\times10^{-6}} - 1.0 = 20.15 \text{ V}.$$

3.9 *L*-SECTION FILTER OR CHOKE INPUT FILTER OR *L*-FILTER

Merits of both Inductance *L* and Capacitance *C* can be combined in *L*-Section Filter or the so-called *L–C* Filter or 'Choke Input Filter'. Inductor is added prior to and in series with the 'Capacitor Filter' as in the circuit shown in Fig. 3.24.

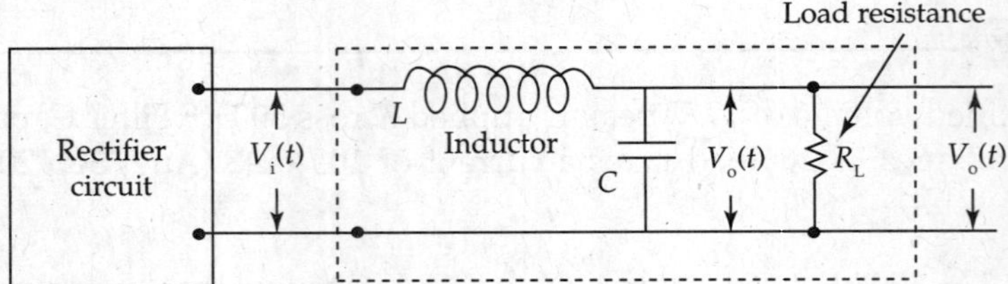

FIG. 3.24 *Rectifier with L-Section filler circuit*

Series Inductor *L* drops AC, which is further bypassed by capacitance C. This action reduces the ripple and contributes to the Filtering action. The final expression for ripple factor of *L–C* Filter is

$$\gamma = \left(\frac{v_{rms}}{V_{DC}}\right)_{\text{rectifier}} \cdot \left[\frac{\left(\frac{1}{\omega C}\right)}{\left(\omega L - \frac{1}{\omega C}\right)}\right] = \frac{\sqrt{2}}{3}\cdot\frac{1}{(\omega^2 LC - 1)} = \frac{0.471}{(\omega^2 LC - 1)} \cong \frac{0.471}{\omega^2 LC}. \tag{3.51}$$

For a Full-Wave Rectifier since the ripple frequency is 2*f*, the above expression for ripple factor gets modified to

$$\gamma = \frac{\sqrt{2}}{3}\frac{1}{2\omega L}\cdot\frac{1}{2\omega C} = \frac{0.118}{\omega^2 LC}. \tag{3.52}$$

Here, the ripple factor is independent of the load currents. Hence, *L*-section Filter is used in applications having wide variations of load currents.

EXAMPLE 3.9

A Full-Wave Rectifier supplies a load requiring 300 V at 200 mA. Calculate the transformer secondary voltage for (a) Capacitor input Filter using a Capacitor of 10 μF. (b) A choke input Filter using a choke of 10 Henries and a capacitance of 10 μF. Neglect the resistance of the choke. (May/June 2006, sets-1, 2 and 4; Aug/Sep 2006, suppl sets-1, 3 and 4; April/May 2007, set-3)

Solution: Given $V_{DC} = 300$ V and $I_{DC} = I_L = 200$ mA. Assume that $f = 50$ Hz.
a. For Capacitor input Filter, $C = 10$ μF

$$V_{DC} = V_m - \frac{I_{DC}}{4f \cdot C}$$

$$\therefore \; V_m = V_{DC} + \frac{I_{DC}}{4f \cdot C}$$

$$= 300 + \frac{200 \times 10^{-3}}{4 \times 50 \times 10 \times 10^{-6}} = 400 \text{ V}$$

The rms value of transformer secondary voltage is

$$\frac{V_m}{\sqrt{2}} = \frac{400}{\sqrt{2}} = 282.8 \text{ V.}$$

b. For 'Choke Input Filter', Inductor $L = 10$ Henries and Capacitor $C = 10$ μF. We know that

$$V_{DC} = 2\frac{V_m}{\pi}$$

$$\therefore \; V_m = \frac{V_{DC} \times \pi}{2} = \frac{300 \times 22}{7 \times 2} = 471.43 \text{ V}$$

Therefore, the rms value of secondary voltage is

$$\frac{V_m}{\sqrt{2}} = \frac{471.43}{\sqrt{2}} = 333.3 \text{ V.}$$

EXAMPLE 3.10

Explain the cause of 'surge' in Rectifier circuits using Capacitor Filter and how is the current limited? (May 2004, set-2)

Solution: When a sinusoidal alternating voltage is rectified, resulting output is a unidirectional voltage consisting of half sinusoids in only one direction above the time axis. It is not the desired DC voltage. Hence, Filter circuits are used to achieve the goal of conversion of AC to DC using the combination of Rectifier circuit and Filter circuits.

One of the basic and simple Filter circuits for general purposes is the simple 'Capacitor Filter' and a load resistor R_L across one of the Rectifier circuits whether a HWR or a FWR or a Bridge Rectifier depending on the demand of various features of the individual circuits.

Consider HWR with simple Capacitor Filter circuit. During the interval '0°–90°' or the first half sinusoid, the Filter Capacitor is almost charged to the peak value of the rectified half sinusoid when the Diode is forward biased. Voltage across the Capacitor $V_C = V_{P(sec)} - V_D$. During the interval 90°–180° of the positive half sinusoid and during the negative half cycle applied to the anode of the diode, the voltage across the Capacitor is of such polarity that the Diode is reverse biased. So, the Capacitor discharges through R_L till such time the Diode is forward biased again. The cycles of events repeat in the operation of the Filtering process to reduce the AC content and in turn produce DC voltage to the designed level.

Diode in HWR circuit with shunt Capacitor Filter does not conduct continuously as explained above, but repeatedly allows pulses of current to recharge the Capacitor each time Diode is conducting under forward-biased situations at different instances of time. *Current pulse is known as repetitive surge current. Highest surge current occurs when AC supply is first switched ON to Rectifier circuit. At the time of switching on Capacitor acts as a short circuit and surge current is at its maximum value.*

So as to limit this surge current in Rectifier circuits a small value of surge current limiting resistor R_S of the order of 250 Ω with high wattage power dissipation capability is connected in series path to the Diode in the Rectifier circuit. Value of the surge current limiting resistor R_S is calculated as the ratio of the peak secondary voltage to the peak value of the expected or rated surge current from the Diode specifications.

Regulation Characteristic

Graph between V_{DC} and I_{DC} is called the 'Regulation Characteristic'. Figure 3.25 shows the voltage regulation characteristics of *L*-section and Capacitor input Filter circuits.

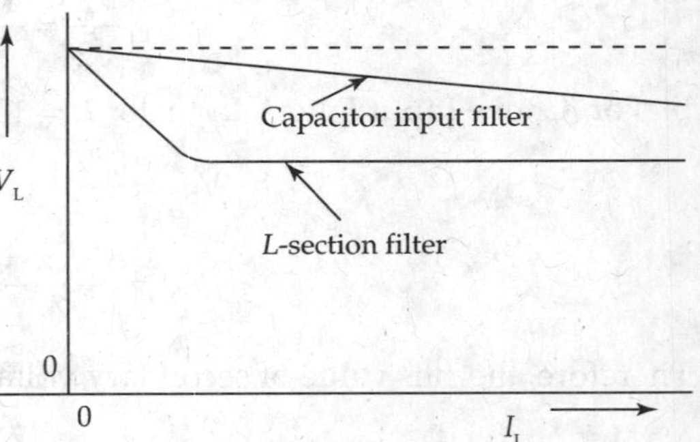

FIG. 3.25 *Regulation characteristic*

3.9.1 Analysis of an LC Filter

'Regulation', 'Ripple factor' and 'Critical Inductance' of an LC Filter

The so far discussed 'Series Inductor Filter' and 'shunt Capacitor Filter' are not capable of providing sufficiently low ripple factors. Low value of ripple factor can be achieved by using 'LC Filter' circuits as analysed below.

LC Filter (Choke Input Filter) is a combination of an Inductor Filter and a shunt Capacitor in parallel with a resistor as shown in Fig. 3.26. To achieve ripple independent of load current variations, an LC Filter is preferable. Inductor readily passes DC components of rectified output and at the same time offers high impedance to higher harmonics of AC as inductive reactance $X_L = \omega \cdot L$. The remaining harmonic components are bypassed by the shunt Capacitor C and at the same time Capacitor C offers infinite impedance for DC. Thus lower ripple factor is obtained.

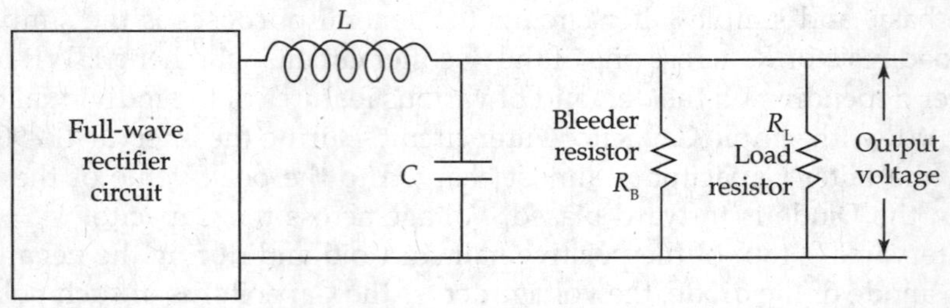

FIG. 3.26 *LC Filter circuit with bleeder resistor*

Regulation Ideal power supply provides constant output voltage irrespective of the variations of load current. In practice, output voltage from a DC power supply changes with variations in load current. Performance measure of a power supply is the regulation. Regulation characteristics shown in Fig. 3.22 show the variations in DC output voltage with load current variations for different Filter circuits.

Consider that the Filter is fed by rectified output from a Full-Wave Rectifier (Fig. 3.26).

∴ Input voltage to LC Filter = output voltage of Rectifier = V:

$$V = \frac{2V_n}{\pi} - \frac{4V_n \cos 2\omega t}{3\pi}. \tag{3.53}$$

Neglecting the series resistance with the Inductance, approximate value of DC voltage V_{DC}: DC input voltage to LC Filter = the DC component of Rectifier output:

$$V_{DC} = \frac{2V_m}{\pi}.$$

A more accurate value of DC output voltage V_{DC} is given below:

Assume that R = Sum of the resistances of the conducting Diodes, Inductor L and secondary winding of the power transformer.

$$V_{DC} = \frac{2V_m}{\pi} - I_{DC} \cdot R. \tag{3.54}$$

Ripple Factor

Assuming that capacitive reactance $X_C = 1/2\omega C$ is much smaller than the load resistance R_L and the reactance of the choke '$X_L = 2\omega \cdot L$' is large compared to parallel impedance of $X_C || R_L$, the entire alternating current due to second harmonic passes through the Capacitor and nothing through the other components. These conditions are necessary to restrict or block the passage of higher order AC components or the ripple content through the Filter circuit components to the load. Just like a coffee Filter, the electronic Filters also filters out the higher order harmonics, that means, it blocks the higher order harmonics, so that the output voltage is a pure DC voltage that is useful for practical appliances such as DC sources.

Therefore, the peak value of the second harmonic current through Capacitor C is $\frac{4V_m}{3\pi(X_L + X_C)}$, where the angular frequency of the second harmonic is 2ω.

Therefore, the peak value of the second harmonic current through Capacitor C is

$$\frac{4V_m}{3\pi\left[2\omega L - \frac{1}{2\omega C}\right]} = \frac{8V_m \omega \cdot C}{3\pi(4\omega^2 \cdot LC - 1)}. \tag{3.55}$$

The rms value of the second harmonic component of current through Capacitor C is

$$I_{rms} = \frac{1}{\sqrt{2}} \cdot \frac{8V_m \omega \cdot C}{3\pi\left[4\omega^2 \cdot LC - 1\right]}. \tag{3.56}$$

Effective voltage drop across R_L

V'_{rms} = effective voltage drop across the Capacitor C

$$V'_{\text{rms}} = I_{\text{rms}} \cdot X_C = \frac{I_{\text{rms}}}{2\omega \cdot C} = \frac{4V_m}{3\pi\sqrt{2} \cdot (4\omega^2 \cdot LC - 1)} \tag{3.57}$$

$$\text{Ripple factor} \quad \gamma = \frac{V'_{\text{rms}}}{V_{DC}} = \frac{\left(\dfrac{4V_m}{3\pi\sqrt{2}\left[4\omega^2 \cdot LC - 1\right]}\right)}{\left(\dfrac{2V_m}{\pi}\right)}$$

$$\therefore \quad \gamma = \frac{\sqrt{2}}{3(4\omega^2 \cdot LC - 1)} = \frac{0.471}{(4\omega^2 \cdot LC - 1)} \tag{3.58}$$

$$= \frac{\sqrt{2}}{3(4w^2 \cdot LC - 1)} \cong \frac{\sqrt{2}}{3(4\omega^2 \cdot LC)} = \frac{\sqrt{2}}{3 \times 2\omega L \times 2\omega C} = \frac{\sqrt{2}X_C}{3X_L}$$

$$\therefore \quad \gamma = \frac{\sqrt{2}X_C}{3X_L} = \frac{\sqrt{2}}{3 \times (2\pi f)^2 \cdot LC} = \frac{4.77}{LC} \tag{3.59}$$

when frequency $f = 50$ Hz, L in Henries and C in microfarads. The ripple factor is independent of load and the condition $4\omega^2 \cdot LC = 1$ is to be avoided.

Critical Value of Inductance L_C

One of the main properties of Inductor is that its reactance '$X_L = 2\pi \cdot fL$' increases for higher order frequency components and with increase in magnitudes of Inductance. Increasing the Inductance effectively Filters AC components and allows for continuous flow of DC current. At a particular value of Inductance known as critical Inductance L_C, there will be continuous flow of current to the load resistance. Calculation for the required magnitude of L_C is shown below.

Assume that the current flows through the Inductor over the complete cycle. Then the DC component of current $\geq$ peak AC current (Angular frequency of AC component is 2ω.):

$$I_{DC} = \frac{2V_m}{\pi \cdot R_L} \geq \frac{\dfrac{4V_m}{3\pi}}{Z},$$

where $Z = 2\omega L$ (second harmonic impedance of the filter).

$$\frac{2V_m}{\pi \cdot R_L} \geq \frac{\dfrac{4V_m}{3\pi}}{2\omega L}$$

$$\text{i.e.,} \quad \omega L = X_L \geq \frac{R_L}{3}$$

$$\therefore \quad L \geq \frac{R_L}{3\omega}. \tag{3.60}$$

Critical value of Inductance L_C

$$L_C = \frac{R_L}{3\omega} = \frac{R_L}{6\pi f}$$

At frequency $f = 50$ Hz

$$L_C = \frac{R_L}{3\times 2\times \pi\times 50} = \frac{R_L}{942.86}. \tag{3.61}$$

In practical circuits, the value of Inductance L should be kept more than this critical Inductance calculated above, so that current always passes over the complete AC *cycle.*

Design considerations We have obtained a condition that $X_L \geq R_L/3$ for DC current to pass through the entire cycle. But under no load condition, R_L is infinite. So, DC current I_{DC} becomes zero. Therefore, the required Inductance must be infinite. Then the LC Filter circuit functions as a simple shunt Capacitor Filter. Then under this no load situation, the output voltage is simply V_m. A small bleeder resistance R_B has to be connected in parallel with the Capacitor to meet the specified condition and maintain good regulation. The bleeder resistance draws a minimum current through the choke, in case of situations when the load resistance is disconnected. Bleeder resistance helps in maintaining continuous flow of minimum current to avoid sudden fluctuations in output voltage.

Then required magnitude of $R_B \leq 942.86\ L_C$.

Another way of taking care of the situations of maintaining constant DC output voltage for varying load currents is that a 'swinging choke' (swinging choke is an iron core Inductor, whose Inductance value is a function of DC current passing through it) is used instead of simple Inductor L. *Swinging chokes provide high value of Inductance at low values of load currents and small Inductance at high values of currents so that for varying load currents, maintenance of critical Inductance condition is satisfied.*

I_{DC} = Average current due to second harmonic

$$I_{DC} = \frac{2I_m}{\pi} = \frac{4}{3\pi}\cdot\frac{V_m}{X_L}$$

$$I_m = \frac{2}{3\pi}\cdot\frac{V_m}{X_L}, \quad \text{but } I_m = \frac{2}{3\pi}\cdot\frac{V_m}{(R_X + R_B)}$$

where R_X is the choke resistance $= 2\omega\cdot L$, R_B is the bleeder resistance and $R_X \ll R_B$.

$$\therefore\quad I_m = \frac{2}{3\pi}\cdot\frac{V_m}{R_X} = \frac{2}{3\pi}\cdot\frac{V_m}{2\omega L}$$

$$\therefore\quad \text{Bleeder resistance}\quad R_B = 3\omega L. \tag{3.62}$$

Bleeder resistance R_B maintains minimum current and improves voltage regulations.

3.10 MULTIPLE *L*-SECTION FILTER

Two or more LC Filters are cascaded to reduce ripple voltage for good filtering. Such cascaded Filter sections are known as 'multiple LC Filter' or 'multiple *L*-section Filters'.
Assume the reactance of Capacitors is smaller than reactance of Inductors and current passes throughout the entire cycle.

Each *L*-section Filter reduces the ripple by a factor $1/(4\omega^2\cdot LC - 1)$. This is already derived in the beginning of Section 3.9.1. The ripple factor of multiple *L*-section Filter shown in Fig. 3.27 is equal to

$$\gamma_{MLF} = \frac{\sqrt{2}}{3(4\omega^2\cdot L_1C_1 - 1)(4\omega^2\cdot L_2C_2 - 1)\cdots(4\omega^2\cdot L_nC_n - 1)}.$$

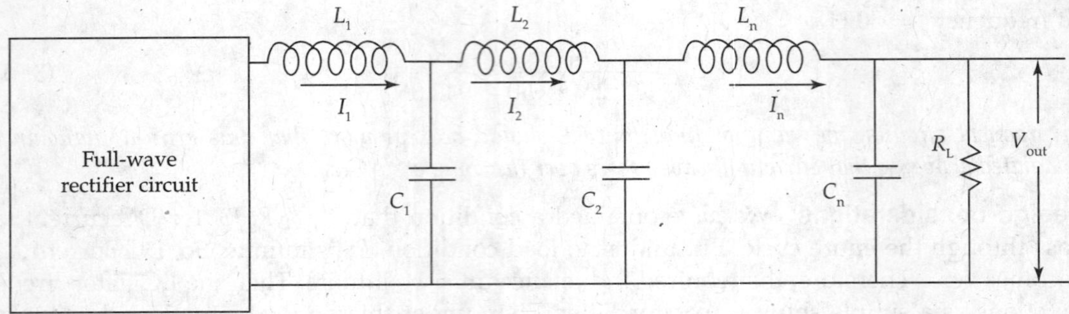

FIG. 3.27 *Full-wave rectifier with multiple section filter*

If the Inductors and Capacitors in the multiple *L*-section Filter circuit are identical for all *n* sections; then the overall ripple factor is given by

$$r = \frac{\sqrt{2}}{3} \cdot \frac{1}{(4\omega^2 \cdot LC - 1)^n} \cong \frac{\sqrt{2}}{3(4\omega^2 \cdot LC)^n}$$

$$(4\omega^2 \cdot LC)^n = \frac{\sqrt{2}}{3\gamma} = \frac{0.4714}{\gamma}$$

$$\therefore \quad 4\omega^2 \cdot LC = \left[\frac{0.4714}{\gamma}\right]^{1/n}$$

Thus the ripple factor can be made small. For the supply frequency $f = 50$ Hz, the value of $LC = 2.531 \times 10^{-6}\left(\frac{0.471}{\gamma}\right)^{1/n}$.

According to the practical requirements, Inductance *L* will be of the order of a few henries and the Capacitor is of the order of microfarads, then

$$LC = 2.531 \times 10^{-6}\left(\frac{0.471}{\gamma}\right)^{1/n}. \tag{3.63}$$

3.11 π-SECTION FILTER

π-section Filter (Fig. 3.28) is a combination of Capacitor input and *L*-section Filters. π-Section Filter circuit is most popular. The ripple is very much reduced by the double Filtering action. The expression for the ripple factor is

$$\gamma = \frac{\sqrt{2}X_C^2}{R_L \cdot X_L^2}.$$

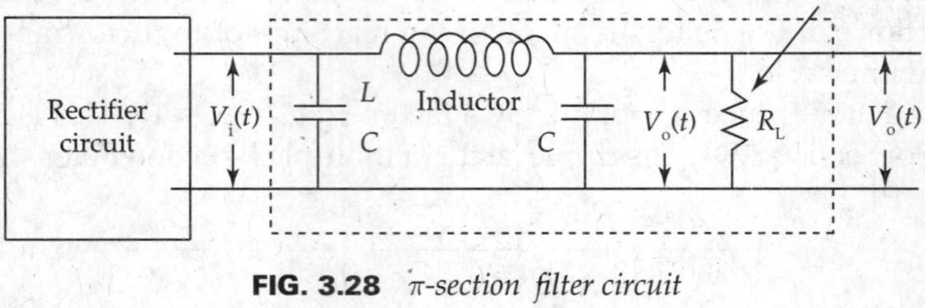

FIG. 3.28 *π-section filter circuit*

Figure 3.29 shows the regulation characteristics of various types of Filters.

$$V_{DC} = V_m.$$

- For *L*-section or π-section Filters No load DC voltage $V_{DC} = V_m$ (Fig. 3.29).
- As the current increases for an *L*-Section Filter, the voltage quickly drops and then changes more smoothly, since the Inductance *L* smoothes the varying current. Also, the ripple is less dependent of load current. This *L*-section Filter circuit is useful for large fluctuating loads.
- On the other hand, a π-section Filter has always a relatively high DC voltage up to I_{Lo}, the optimal load current. Thus, π-section Filter circuit is preferable for light (small) fixed load currents, since it provides high DC relatively.

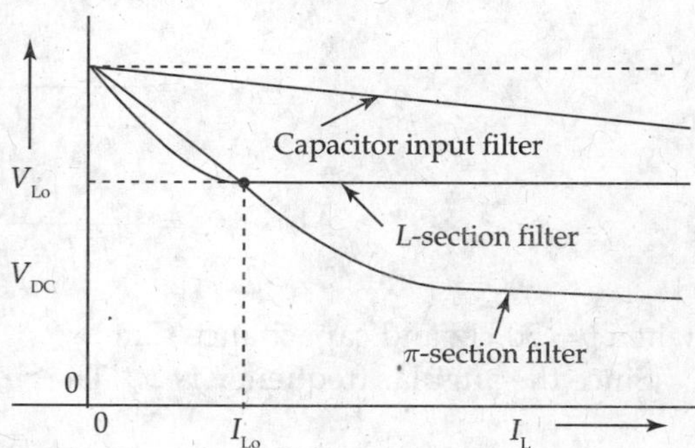

FIG. 3.29 *Regulation characteristics for various filters*

For low-current circuits, a Resistor replaces the Inductor element. This *R*–*C* Filter circuit is useful only if IR drop across the resistor is not much and also if regulation is not of much importance in the application especially for low-voltage supplies like the Transistor-regulated power supplies. The π-section Filter is preferable to single Capacitor input Filter circuit.

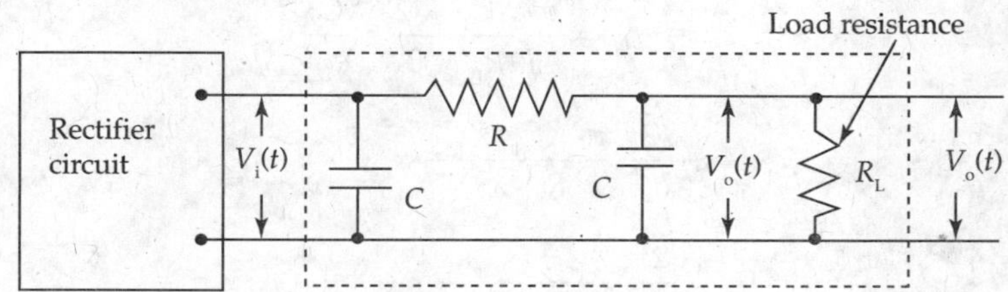

FIG. 3.30 *π-section filler circuit using R-C elements*

3.12 ANALYSIS OF π-SECTION FILTER (CLC FILTER, CAPACITOR INPUT FILTER)

The π-section Filter in Fig. 3.31 consists of the following two Filter components:

1. Shunt Capacitor input Filter formed by Capacitor C_1,
2. Followed by choke input Filter formed by *L* and C_2.

The π-section Filter provides higher output voltage (that approaches the peak value of AC potential of the source) at light loads and very smooth output. Capacitor C_1 offers a low impedance path for harmonics of Rectifier output. At the same time, it offers high impedance to DC and second harmonic content. Output across C_1 is a triangular with vertical sides. The remaining ripple content is reduced by LC Filter LC_2.

- The ripple factor with π Filter is γ_π.
- The ripple formed by C_1 is γ_C.
- The ripple factor formed by LC_2 Filter is γ_L.

From the previous analysis, ripple factor of 'Shunt Capacitor Filter' is given by

$$\gamma_C = \frac{1}{4\sqrt{3}\cdot f\cdot C_1R_L} = \frac{\pi}{2\sqrt{3}\cdot 2\pi\cdot f\cdot C_1R_L}$$

$$= \frac{\pi}{2\sqrt{3}\cdot\omega\cdot C_1R_L} = \frac{\pi\cdot X_{C1}}{\sqrt{3}R_L}$$

$$\text{Also}\quad \gamma_C = \frac{1}{4\sqrt{3}f\cdot C_1R_L} = \frac{2887}{C_1\cdot R_L} \tag{3.64}$$

when $f = 50$ Hz and capacitance C in μ, f

Since the angulár frequency is 2ω. The ripple factor formed by LC_2 is given as

$$\gamma_L = \frac{\sqrt{2}}{3[4\omega^2\cdot LC_2 - 1]} \cong \frac{\sqrt{2}}{3[4\omega^2\cdot LC_2]}$$

$$= \frac{\sqrt{2}}{3}\cdot\frac{X_{C_2}}{X_L}$$

Therefore, the ripple factor of π-section Filter is given by

$$\gamma_\pi = \gamma_C\cdot\gamma_L = \frac{\pi\cdot XC_1}{\sqrt{3}R_L}\cdot\frac{\sqrt{2}}{3}\cdot\frac{XC_2}{X_L}$$

$$= \frac{\sqrt{2}}{3\sqrt{3}}\cdot\frac{\pi\cdot XC_1\cdot XC_2}{R_L\cdot X_L}$$

$$= \frac{\sqrt{2}\pi}{3\sqrt{3}}\cdot\frac{1}{2\omega C_1}\cdot\frac{1}{2\omega C_2}\cdot\frac{1}{2\omega L}\cdot\frac{1}{R_L}$$

$$= \frac{\sqrt{2}}{24\sqrt{3}}\cdot\frac{\pi}{\omega^3\cdot C_1C_2\cdot LR_L}$$

$$\gamma_\pi = \frac{0.1068}{\omega^3\cdot C_1C_2\cdot LR_L}. \tag{3.65}$$

For frequency, $f = 50$ Hz, Inductance L in Henries and Capacitance in microfarads.

$$\gamma_\pi = \frac{3.45\times10^3}{C_1C_2\cdot LR_L} = \frac{3450}{C_1C_2\cdot LR_L}. \tag{3.66}$$

Ripple Factor of π-Section Filter used with Half-Wave Rectifier Circuit

If the rectified output to the Filter circuit is from a HWR, then the reactance of Capacitor and Inductor elements are calculated at the angular frequency ω only. Then the ripple factor of π-Section Filter fed from HWR output is as follows:

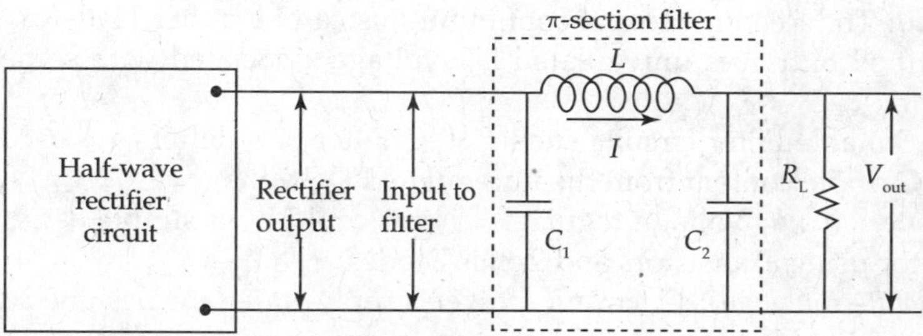

FIG. 3.31 *Half-wave rectifier with π-section filter*

$$\gamma_\pi = \gamma_c \cdot \gamma_L = \frac{\pi \cdot XC_1}{\sqrt{3}R_L} \cdot \frac{\sqrt{2}}{3} \cdot \frac{XC_2}{X_L}$$

$$= \frac{\sqrt{2}}{3\sqrt{3}} \cdot \frac{\pi \cdot XC_1 \cdot XC_2}{R_L \cdot X_L}$$

$$= \frac{\sqrt{2}\pi}{3\sqrt{3}} \cdot \frac{1}{\omega C_1} \cdot \frac{1}{\omega C_2} \cdot \frac{1}{\omega L} \cdot \frac{1}{R_L}$$

$$= \frac{\sqrt{2}}{3\sqrt{3}} \cdot \frac{\pi}{\omega^3 \cdot C_1 C_2 \cdot L R_L}$$

$$\gamma_\pi = \frac{0.854}{\omega^3 \cdot C_1 C_2 \cdot L R_L}.$$

For supply frequency $f = 50$ Hz, Inductance L in Henries and capacitances C_1 and C in farads.

$$\gamma_\pi = \frac{0.854}{(2\pi \times 50)^3 C_1 C_2 \cdot L R_L} = \frac{2.754 \times 10^3}{C_1 C_2 \cdot L R_L} = \frac{2754}{C_1 C_2 \cdot L R_L} \tag{3.67}$$

when the input to the filter is from the output of a HWR.

3.13 VOLTAGE REGULATORS

Stable source of DC voltages from power supply circuits is necessary for the operation of electronic gadgets and equipment. Design of power supplies is simplified with the latest technology devices using Integrated circuits (IC) accompanied by a few external discrete components in the total power supply circuit. They are reliable and stable in operation. They are inexpensive and working with them is easy and comfortable.

3.13.1 Building Blocks of a Voltage-regulated Power Supply (Fig. 3.32)

Working principles of various blocks required to obtain constant DC voltage

1. *AC mains supply voltage*: 220 V 50 Hz derived from the utility mains voltage.
2. *Step-down transformer*: Step-down transformer depends on the output DC voltage.

3. *Rectifier circuit*: The Rectifier circuit configurations can be either Half-wave or Full-Wave circuits. Rectified output is unregulated DC voltage associated with some unwanted AC ripple content.
4. *Filter circuit*: Suitable Filter circuits such as L-C Filter or C-Filters are used to remove the unwanted AC ripple content from the unregulated DC voltage.
5. *Voltage regulator circuit*: Voltage regulator circuit could be a simple series regulator or a shunt regulator using Transistors and Zener Diodes.
6. *Protection circuits (optional)*: Electronic power supply units are designed with protection circuits to provide safe operation for electronic circuits. For laboratory DC supplies and voltage stabiliser circuits, simple fuses provide short circuit protection. But normally current limiting and short circuit protection circuits are used for DC supplies used in industrial automation, instrumentation and mobile communication equipment.
7. *Efficiency*:
   - Performance of power supply unit depends upon the efficiency of providing DC output power for the supplied AC power input. Efficiency is the ratio of DC output power to AC input power.

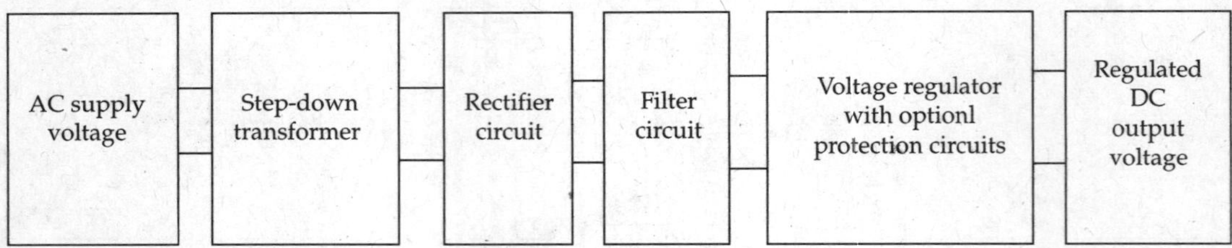

FIG. 3.32 *Building blocks of AC to DC voltage regulator circuit*

Usage of various blocks will be decided by precision requirement for DC output voltage. One example is current limiting and short circuit protection circuits are not used in simple battery eliminators used for portable instruments.

Necessity of voltage regulators in DC power supplies

A well-designed regulated power supply is used to achieve reliable and safe operation of electronic equipment overcoming the limitations in unregulated power supplies. Finally, main use of a voltage regulator circuit is to provide *constant stabilised DC output voltage in spite of the following drawbacks that may be present in the power supply system.*

1. Variations in mains supply voltage V_S.
2. Variations in load current I_L due to changes in load resistance of electronic circuit to which the DC source supplies the constant voltage.
3. Temperature (T) variations.

Advantages of voltage regulator circuits over unregulated power supply

1. Maintaining constant DC supply voltage to electronic appliances.
2. It minimises the damage to the electronic circuits.
3. Protective circuitry like constant current limiting and fold back current limiting against over loads and short circuit conditions can be easily incorporated.

The quality of a voltage regulator is determined by the following two quantities:

1. Line voltage regulation
2. Load voltage regulation

Line voltage regulation:

- Line voltage regulation to maintain constant output voltage whenever variations occur in mains input or supply voltage is known as 'Line regulation'.
- When there are variations in mains supply voltage, corresponding variations in DC power supplies cause damages to electronic appliances such as fridges, washing machines, TV receivers and so on.
- 'Line regulation' is defined as the ratio of change in the output voltage ΔV_{out} for certain change ΔV_{in} in the input line voltages from higher to lower values of line voltages. It is a measure of performance of the power supply. It is expressed in millivolts per volt.

Definition of normal line regulation

$$\% \text{ Normal Line regulation} = \left[\frac{\Delta V_{out}}{\Delta V_{in}}\right] \cdot 100\% \tag{3.68}$$

Calculation of line regulation for 1 V change in line supply voltage can be made using the following expressions:

$$\% \text{ Line regulation for 1 V increase in line voltage} = \left[\frac{\Delta V_{out} / V_{out}}{\Delta v_{in}}\right] \cdot 100\%$$

$$= \frac{(V_{oh} - V_{oN}) / V_{oN}}{V_{hS} - V_{NS}} \times 100\% \tag{3.69}$$

$$\text{(or) } \% \text{ Line regulation for 1 V decrease in line voltage} = \frac{(V_{oN} - V_{ol}) / V_{oN}}{V_{LS} - V_{NS}} \times 100\% \tag{3.70}$$

where V_{oh} is the output voltage at higher value of supply voltage, V_{ol} is the output voltage at lower value of supply voltage, $V_{out\ normal} = V_{oN}$ is the output voltage at normal value of supply voltage, V_{hS} is the high value of line supply voltage, V_{LS} is the low value of line supply voltage and V_{NS} is the normal value of line supply voltage.

EXAMPLE 3.11

A voltage regulator is designed to provide a constant DC output voltage of 12 V. If the line supply voltage increases to 240 V from the normal supply voltage of 230 V, the output voltage changes to 12.12 V. Determine the normal line regulation and the line regulation for 1 V change line supply voltage expressed as %/Volt.

Solution:

Change in output voltage $\Delta V_{out} = V_{oh} - V_{oN} = 12.12 - 12.0 = 0.12 \text{ V}$

Change in input voltage $\Delta V_{in} = V_{hS} - V_{NS} = 240 - 230 = 10 \text{ V}$

$$\text{Normal line regulation} = \frac{\Delta V_{out}}{\Delta V_{in}} \times 100 = \frac{0.12}{10} \times 100 = 0.1\%$$

$$\begin{aligned}\%\ \text{Line regulation for 1 V increase in supply voltage} &= \frac{\Delta V_{out} / V_{out}}{\Delta V_{in}} \times 100\% \\ &= \frac{V_{oh} - V_{oN} / V_{oN}}{(V_{hS} - V_{NS})} \times 100\% \\ &= \frac{(0.12/12)}{10} \times 100 = 0.1\% / \text{V}.\end{aligned}$$

Load voltage regulation:

- Load regulation to maintain constant output voltage in spite of the variations in the load current is known as 'Load regulation'.
- Varying load resistance causes variations in load current. The load resistance may vary during the operation of an electronic circuit. One example could be the variation in speech volume and brightness variations of its screen during day light and night times in a cell phone. So, a voltage regulator circuit maintains constant output voltage in spite of the variations in the load current.
- 'Load regulation' is defined as the ratio of variation in output voltage from No load voltage to full load voltage ($V_{NL} - V_{FL}$) to the output voltage at full load V_{FL}:

$$\%\ \text{Load regulation} = \left[\frac{V_{NL} - V_{FL}}{V_{FL}}\right] \cdot 100\%, \tag{3.71}$$

where V_{NL} is the output DC voltage when the load current is zero ($I_L = 0$ mA) and V_{FL} is the output DC voltage when the load current is maximum.

Both line and load regulation parameters should be zero for ideal voltage regulators. In practice, they should be as small as possible.

Voltage regulation characteristic Voltage regulation characteristic feature of a power supply can be predicted from a graph between variations in load voltage to variations in load current (Fig. 3.33).

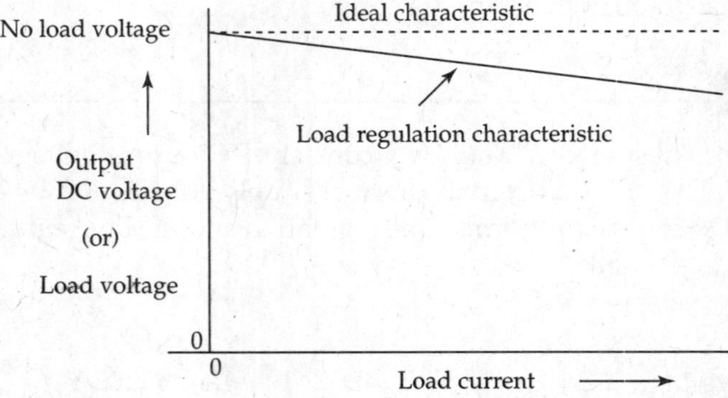

FIG. 3.33 *Voltage regulation characteristic with load variations*

Comparison between line regulation and load regulation

Line regulation

1. It maintains a constant output voltage at the regulator output terminals, when the input voltage changes.
2. It is defined as the % change in output voltage to the change in input voltage.
3. It is a percentage of output voltage.
4. It is expressed as millivolts per volt.

Load regulation

1. It maintains a constant output voltage at the regulator output terminals whenever the load current changes.
2. It is defined as the percentage change of the output voltage to the change in load current.
3. It is a percentage change in output voltage at no load to full load currents.
4. It is expressed in ohms.

EXAMPLE 3.12

A 50 V power supply has a line regulation of 0.2%, when the supply voltage is 75 V. Find the supply voltage for the output voltage to increase to 52 V?

Solution:

$$\text{Line regulation} = \frac{\Delta V_{out}}{\Delta V_{in}} \times 100 = 0.2\%$$

$$\therefore \quad \frac{\Delta V_{out}}{\Delta V_{in}} = \frac{[V_{out2} - V_{out1}]}{[V_{in2} - V_{in1}]} = \frac{[52-50]}{V_{in2} - 75} = 0.2$$

$$\therefore \quad [V_{in2} - 75] = \frac{(52-50)}{0.2} = \frac{2}{0.2} = 10$$

$$\therefore \quad V_{in2} = 10 + 75 = 85 \text{ V}.$$

The input voltage should increase from 75 V to 85 V to cause the output voltage to increase from 50 V to 52 V.

3.14 SIMPLE VOLTAGE REGULATOR CIRCUIT USING ZENER DIODE

Introduction A simple voltage stabiliser is an electronic device such as Gas Diode, Zener Diode or Avalanche Diode, which produce constant voltages across the devices for whatever changes that may occur in the line voltages, load currents and temperature, within the specified operating ranges for the intended applications of the power supplies. The voltage characteristic shown in Fig. 3.32 for Zener Diode is an example for the concept of how a Zener Diode works as a voltage stabiliser or regulator.

Zener Diode as voltage regulator and voltage source

- Zener Diode characteristic explains the concept of voltage regulation and use of Zener Diode as simple voltage regulator.
- Zener Diode operating in break-down region (when Zener Diode is reverse biased) develops a constant voltage V_Z across the device over specified conduction currents from minimum Zener current ($I_{Z(min)}$) to maximum Zener current ($I_{Z(max)}$).

Zener voltage V_Z is almost a straight line in the voltage–current characteristic of a Zener Diode in Fig. 3.34. Zener Diodes are available with stabilisation voltages V_Z ranging from a few volts to a few hundred volts. For a Zener Diode to operate as a voltage regulator, it should operate in the constant voltage region of the reverse characteristic of the Diode between $I_{Z(min)}$ and $I_{Z(max)}$.

The above feature suggests the use of Zener Diodes in voltage regulator circuits:

- To provide stabilisation to terminal voltage V_Z in simple shunt regulator circuits,
- To provide reference voltage V_R in linear voltage regulator circuits.

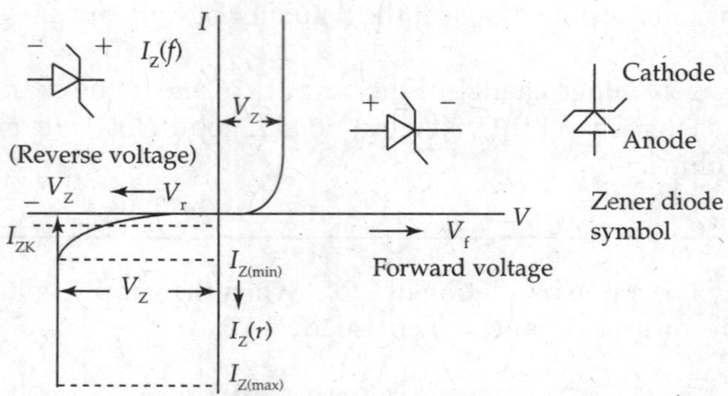

FIG. 3.34 *Forward and reverse characteristic of zener diode*

Simple Zener Diode shunt regulator circuit

The main purpose of the design of a 'voltage regulator' circuit is to supply a constant DC output voltage. Normal Diode used in Rectifier circuits are not suitable as regulator elements. Many simple DC power supplies regulate the output voltage using a simple voltage regulator circuit using a Zener Diodes.

Hence, the electronic devices such as Zener Diodes form as one of the main circuit elements for the stabilisation of the output DC voltages as shown in Fig. 3.35. There may be variations in supply voltage or load current I_L in the circuit, but the performance of the Zener Diode has to take care of them as a shunt voltage regulator and supply constant DC output voltage V_{out}.

Explanation of the working of simple Zener regulator circuit

In Fig. 3.35, V_{in} is the input voltage. It is an unregulated DC voltage from a Rectifier plus Filter circuit combination. This input source voltage is applied to a series combination of a resistor R_S and a Zener Diode. The cathode of the Zener Diode is connected to the positive terminal of the input voltage, so that the Zener Diode is reverse biased. A load resistance R_L is connected in parallel to the Zener Diode. As the Zener Diode is connected in shunt with load resistance R_L, this circuit is known as Zener shunt regulator. V_{out} is the output voltage across the load resistance R_L. I_S is the current from the source. I_Z is the current through the Zener Diode. I_L is the load current.

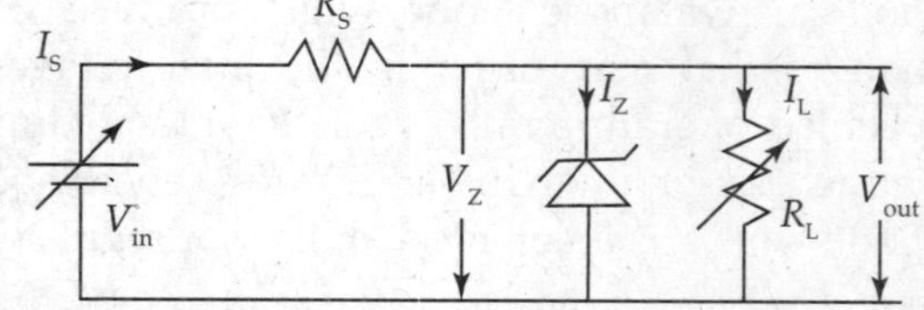

FIG. 3.35 *Voltage regulator circuit using Zener diode*

- As long as $(V_{in} - I \cdot R_S) <$, the output voltage V_o will be $V_{out} = (V_{in} - I \cdot R_S)$, since the Zener Diode does not conduct. Also, $I_Z = 0$ during the non-conduction time of the Zener Diode.
- In Zener Diode regulators, when the voltage across the Zener Diode is more than or equal to V_Z, V_{out} will be clamped at V_Z and $V_{out} = V_Z$.
- As the load current increases, it is as though Zener Diode supplies the extra current by reducing its own current.
- When the load current I_L is maximum, I_Z is minimum.
- When the load current I_L is minimum, the Zener current I_Z is maximum
- However, minimum value of Zener Diode current $I_{Z(min)}$ is necessary for the Diode to enter the regulation mode and maintain constant terminal voltage.
- The current from the supply source I_S is the sum of the currents I_Z and I_L.

Method of determination of component value of series resistance R_S

The value of the resistance R_S can be calculated as follows:

$$R_S = \frac{(V_{in} - V_Z)}{(I_L + I_Z)} = \frac{(V_{in} - V_Z)}{(I_L + 0.2 I_L)} = \frac{(V_{in} - V_Z)}{1.2 I_L}. \tag{3.72}$$

With normal operating currents of the Zener Diode, I_Z is chosen as equal to $0.2 \cdot I_{Load}$. Power P_Z dissipated in the Zener Diode can be calculated using the following equation:

$$P_Z = I_Z \cdot V_Z = \left[\left(\frac{(V_i - V_Z)}{R_S}\right) - I_L\right] \cdot V_Z \quad \text{Watts.} \tag{3.73}$$

A Zener Diode must be selected with larger power rating than the power dissipation rating P_Z that is calculated from Eq. (3.73), so that the maximum power dissipation and the temperature ratings of the device are not exceeded.

Applications of Zener Diode:

(1) Constant voltage regulation, (2) Surge protection, (3) Clamping and clipping voltages, (4) Voltage reference in precision equipment such as regulated power supply circuits, (5) Switching operation and 6. Meter protection.

Advantages:

(1) It is a simple circuit with few components. (2) It is a stable voltage source. (3) Smaller size, cheaper and has long life and rugged. (4) Provides good regulation over a wider range of currents. (5) The over riding ripple associated with input unregulated DC is reduced at the output terminals by a factor of r_Z / V_{in} , where r_Z is the incremental resistance of Zener Diode and V_{in} is the input supply voltage to Zener regulator.

Disadvantages:

1. Power loss associated with light and heavy load current in the current limiting resistor and Zener internal resistance.
2. Maximum load current is limited to $I_{Z(max)} - I_{Z(min)}$.
3. It is a simple low current regulator. For high current applications, a series pass power Transistor and error amplifier are necessary.
4. Poorer efficiency.
5. Output terminal voltage depends upon the 'break-down voltage' of Zener Diode.

Specifications Selection of a Zener Diode from manufacturers data sheets, with the following important parameter such as

1. Break-down voltage V_Z (terminal voltage of Zener Diode). The reverse-biased Diode can be used up to hundreds of volts, i.e. Zener Diodes are available with break-down voltages of hundreds of volts singly or in package.
2. Knee break-down voltage V_{ZK}.
3. Maximum power dissipation and temperature coefficient. If higher voltage cannot be provided by a single Zener Diode, more number of Zener Diodes is to be connected in series provided the max allowable current is same for all the Zener Diodes.
4. A forward-biased Zener Diode can also be used as a voltage regulator but a very low Zener voltage of the order of a few tenths of a volt. But the packages are available with stacked Diodes up to 1 or 2 V.

Zener Diode Shunt regulators are used for low power loads only. They cannot meet the higher load power demands. Power supplies for large amounts of powers are more sophisticated circuits using discrete components such as Zener Diodes, Transistors, silicon-controlled Rectifiers and operational amplifiers.

EXAMPLE 3.13

Calculate the value of the series resistance R_S in the given Zener Diode shunt regulator circuit. Data: $V_{in(min)} = 15.0$ V, $V_{in(max)} = 22$ V, $V_Z = 9.0$ V and the load resistance $R_L = 180\ \Omega$. Also, calculate the power dissipations across series resistance and the Zener Diode

Solution: Value of series resistance R_S is calculated using the condition that the voltage drop across the series resistance R_S caused by the load current must be less than the difference between the minimum supply voltage $V_{in(min)} = 15$ V and the Zener voltage $V_Z = 9.0$ V.

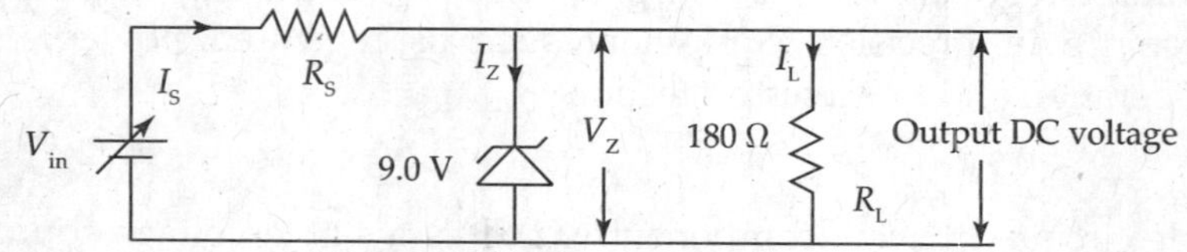

FIG. 3.36 *Zener diode circuit*

$$\text{Load current } I_L = \frac{V_Z}{R_L} = \frac{9.0}{180} = 50 \text{ mA}$$

$$R_S = \frac{[V_{in} - V_Z]}{(I_L + I_Z)} = \frac{(15-9)}{1.2 \times I_L}$$

$$= \frac{(15-9)}{1.2 \times 50 \times 10^{-3}} = \frac{6 \times 10^3}{60} = \frac{6000}{60} = 100\ \Omega$$

Voltage drop across R_S due to $I_L = I_L \times R_S = 50 \times 10^{-3} \times 100 = 5$ V

Difference between $V_{in(min)}$ and $V_Z = (V_{in(min)} - V_Z) = (15 - 9) = 6$ V

$\therefore$ The condition that $I_L \times R_S < (V_{in}\ (\text{min}) - V_Z)$ is satisfied.

In the regulator circuit shown in Fig. 3.36, the voltage across the series resistance R_S is the difference of the input voltage $V_{in} = 15$ V and the constant output $V_o = V_Z = 9$ V. Therefore, $V_{RS} = 15 - 9 = 6$ V.

$$\text{Power dissipation in series resistance} = \frac{V_{RS}^2}{R_S} = \frac{6^2}{100} = \frac{36}{100} = \frac{360}{1000} = 360 \text{ mW.}$$

This is the max power dissipation P_{RS} in the series resister = 360 mW
Power dissipation across the zener Diode = $V_Z \cdot I_{Z(max)} = 9 \times 50 \times 10^{-3} = 0.45$ W.
(assuming that $I_{Z(max)} = I_{L(max)} = 50$ mA and assuming that $I_{L(min)} = 0$ mA)

EXAMPLE 3.14

A partially Filtered voltage (unregulated DC voltage) with peak ripple voltage of 25 V with $V_{DC} = 100$ V from a Rectifier is to be applied to a 50 V Zener Diode with ratings $I_{Z(max)} = 40$ mA, $I_{Z(min)} = 5$ mA. Find the value of maximum and minimum currents through the Zener Diode if the load current I_L is set at 25 mA; $R_S = 3.75$ kΩ. Will the Zener Diode regulate? If it does not regulate, what value of R_S is needed for proper regulation to absorb fluctuation of voltage.

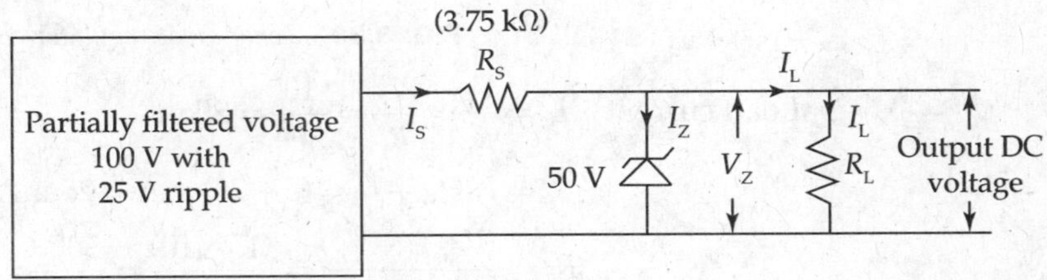

FIG. 3.37 *Zener diode circuit*

Solution:

$$\text{Load resistance } R_L = \frac{50 \text{ V}}{25 \text{ mA}} = 2 \text{ k}\Omega$$

when $I_{Z(max)}$ occurs at (100 + 25) = 125 V
$I_{Z(min)}$ occurs at (100 – 25) = 75 V.
Current through $R_S = I_{RS} = [(I_Z + I_{L(max)}] = (25 \text{ mA} + 40 \text{ mA}) = 65$ mA
But for $R_S = 3.75$ kΩ

$$\text{Soure current } I_S = \left[\frac{(125-50)}{(3.75\times10^3)}\right] = \left[\frac{75\times10^{-3}}{3.75}\right] = 20 \text{ mA}$$

Then the value of $I_S = 20$ mA and hence the Zener Diode does not regulate. So the value if R_S is to be changed or reduced, R_S can be calculated as follows:

I_Z is minimum, i.e. 5 mA, when V_L is maximum. Then, $I_L = 25$ mA. So the net source current through the resistor I_S is 25 mA + 5 mA = 30 mA.

$$R_S \times 30 \text{ mA} = (75 - 50) = 25 \text{ V.}$$

$$\therefore\ R_S = \frac{25}{30} \times 10^3 = 0.833\ \text{k}\Omega = 833\ \Omega$$

$$I_{S(max)} = \frac{125-50}{833} = \frac{75}{833} = 0.09\ \text{A} = 90\ \text{mA}.$$

But when V_L is maximum, i.e. 100 + 25 = 125 V, then I_S is maximum and it is calculated as

$\therefore\ I_{Z(max)} = 90 - 25 = 65$ mA.

The required $I_{Z(max)}$ is 65 mA, but this Zener Diode has $I_{Z(max)}$ of 40 mA only. So, this Zener Diode cannot be used.

$\therefore$ A Zener Diode with a max current ($I_{Z(max)}$) of 65 mA should be used.

EXAMPLE 3.15

For the following Zener Diode regulator circuit of Fig. 3.38, the supply voltage V_S varies from 15 V to 20 V. Zener diode with 10 V, 20 W capacity is used in the circuit. Calculate the value of the resistance R_S and power dissipation in it. Assume Zener breakdown occurs at a current of 5 mA.

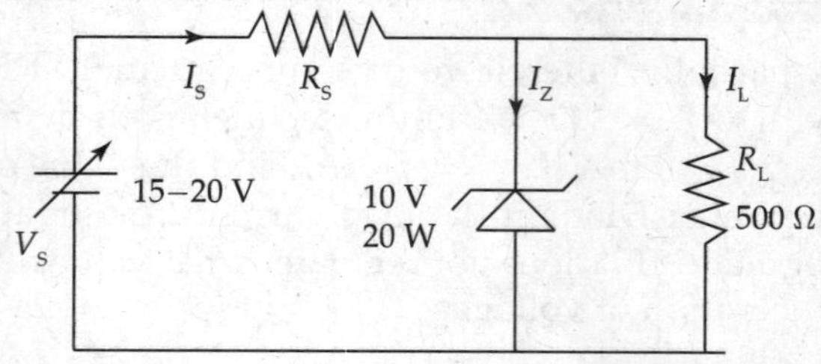

FIG. 3.38 *Zener diode regulator circuit*

Solution:

$$\text{Load current}\ I_L = \frac{V_L}{R_L} = \frac{V_Z}{R_L} = \frac{10}{500} = 20\ \text{mA}$$

$$I_S = I_{Z(min)} + I_L = (5+20)\ \text{mA} = 25\ \text{mA}$$

$$R_S = \frac{(V_{S(min)} - V_Z)}{I_S} = \frac{(15-10)}{25\times10^{-3}} = \frac{5000}{25} = 200\ \Omega$$

$$\text{When } V_S \text{ is minimum power dissipation in } R_S = \frac{(15-10)^2}{200} = 0.125\ \text{W}$$

$$\text{When } V_S \text{ is maximum power dissipation in } R_S = \frac{(20-10)^2}{200} = 0.5\ \text{W}.$$

EXAMPLE 3.16

Define the concept of Regulation in Rectifier circuits and also define percentage regulation. Verify with necessary equations that the regulation of FWR and HWR circuits depends on the ratio of forward resistance r_f to the load resistance R_L.

Solution: The variation of DC output voltage of a Rectifier circuit from no load output voltage $V_{DC(No\ Load)}$ to the output voltage with variations in the DC load current expressed with reference to Full load DC output voltage is known as Regulation.

$$\text{Regulation} = \frac{[V_{\text{No Load}} - V_{\text{Full Load}}]}{V_{\text{Full Load}}}$$

$$\%\ \text{Regulation} = \frac{[V_{\text{No Load}} - V_{\text{Full Load}}]}{V_{\text{Full Load}}} \times 100\%$$

For HWR, % Regulation is a function of the ratio of r_f to R_L as shown:

$$I_{DC} = \frac{(V_m/\pi)}{(r_f + R_L)} \qquad \therefore \quad I_{DC}(r_f + R_L) = \frac{V_m}{\pi}$$

Using $V_{DC} = I_{DC} \cdot R_L$ in the above equation, we get

$$V_{DC} = \left[\frac{V_m}{\pi} - I_{DC} \cdot r_f\right] \text{ and } V_{DC(No\ Load)} = \frac{V_m}{\pi}$$

$$\therefore \quad \%\ \text{Regulation} = \frac{[V_{No\ Load} - V_{Full\ Load}]}{V_{Full\ Load}}$$

$$\%\ \text{Regulation} = \frac{\left[\frac{V_m}{\pi} - \frac{V_m}{\pi} + I_{DC} \cdot r_f\right]}{I_{DC} \cdot R_L} \times 100\% = \frac{r_f}{R_L} \times 100\%.$$

Above derivation is also true for Full-Wave Rectifier circuit.

Classification of Voltage regulators

Voltage regulators are broadly classified into two types: (1) Linear regulators and (2) Switching-mode regulators.

Linear voltage regulators Linear regulators work in the linear region of electronic devices to deliver designed output voltages.

Linear voltage regulators are further classified into two types: (1) Feedback type and (2) Non-feedback type.

- Feedback regulators are again of two types. They are (1) Series regulator and (2) Shunt regulator.
- A ferro resonant regulator is Non-feedback type regulator.

Switching regulators Switched mode power supplies (SMPS) use switching regulators. SMPS uses the switching operations of a Transistor so that higher efficiencies of power conversion are obtained. Move details of these regulators are discussed in seperate chapter 14.

3.15 BLOCK DIAGRAM OF SERIES VOLTAGE REGULATOR

1. In a series voltage regulator, a linear device such as a Transistor/FET/SCR that is considered as a control element is connected in series between the unregulated input DC voltage and regulated output DC voltages.
2. A feedback circuit containing a sampling network and comparator circuit regulates the working of the control element so as to maintain constant DC output voltage despite the variations in load or input supply voltage.

Thus, a series voltage regulator circuit has four basic building blocks shown in Fig. 3.39. They are: (1) Sampling network, (2) Reference voltage, (3) Comparator circuit and (4) Control element.

Sampling network Simple voltage divider network with resistors 'R_1 and R_2' is generally used as a sampling network (Sampler). It is normally connected across the regulator output

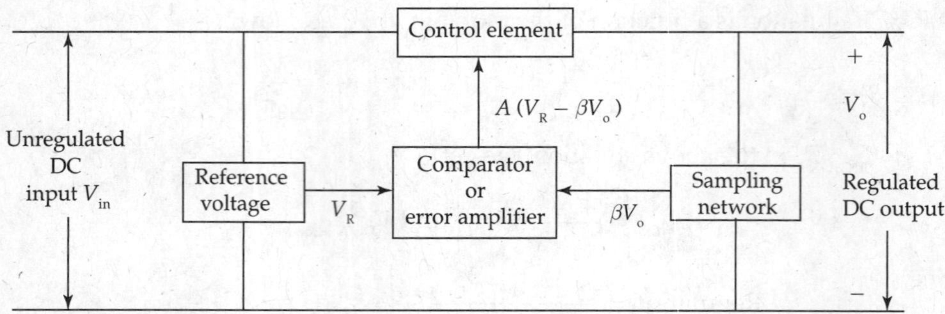

FIG. 3.39 *Various building blocks of 'series regulator' circuit*

terminals. Components of the sampling network should be of the same material as Diodes and should be operated at the same temperature.

Reference voltage source 'Zener Diode' with low-temperature coefficient is the usual reference voltage source. When the Zener Diode is operated at the break-down region, the device functions as a constant voltage source V_R. The voltage across the Diode remains constant for the rated current capabilities of the device. A battery connected internally can also be used as a voltage reference element.

Comparator circuit or error amplifier A Common Emitter Transistor amplifier or an operational amplifier or a differential amplifier is generally used as a comparator circuit. The sampled signal from the output voltage $\beta \cdot V_o$ is compared with the reference voltage V_R and the amplified error signal $[A(\beta \cdot V_o - V_R)]$ is applied to a control element for correcting the variations in output voltage.

Control element Control element is normally a power Transistor (BJT or FET) capable of handling power to be dissipated in the regulator circuit. The controlling voltage regulator Transistor is in 'series' path between the input and the output ports. Hence, this circuit is known as 'Series Voltage Regulator' circuit. Control element is a Transistor Emitter follower (Common Collector Transistor configuration). Depending upon the requirement of positive or negative output voltage V_o, either an NPN or PNP Transistor is used.

Currents through Transistor are controlled by the error signal from error amplifier circuit. When the output of the voltage regulator is short circuited, the series regulator Transistor consumes maximum power. So, the maximum Collector current $I_{C(max)}$ of the selected Transistor in the design process should be greater than the short circuit current that is estimated at the beginning. 'Super-Beta' Transistor configuration (Darlington Pair) is preferable to handle large currents

3.16 SERIES VOLTAGE REGULATOR CIRCUITS

3.16.1 Series Transistor Voltage Regulator Circuit (Emitter follower regulator)

- The input voltage to the regulator circuit is an unregulated DC voltage (from the Rectifier and Filter circuit combination).
- Unregulated DC voltage is connected to the series control Transistor T_1.
- Transistor works as 'Emitter follower'.

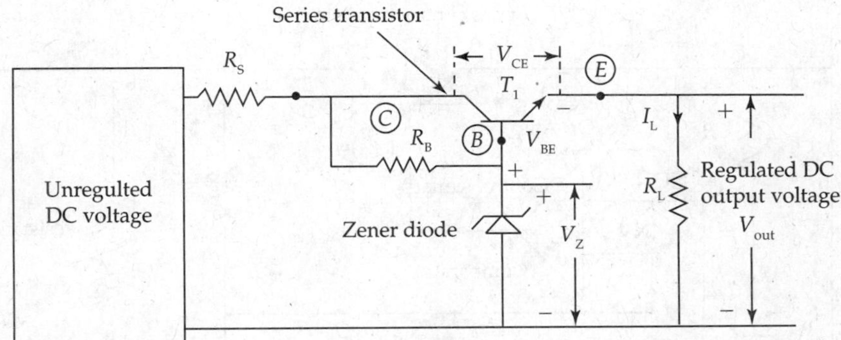

FIG: 3.40 *Series Transistor Voltage Regulator (Emitter Follower Regulator) circuit*

- Transistor Emitter is the output terminal. It is connected to the load terminal, where the regulated output voltage V_{out} is obtained.
- Zener Diode is connected at the Base terminal of the Transistor.
- The output voltage $V_{out} = (V_Z - V_{BE})$, where V_Z is the terminal reference voltage across the Zener Diode and V_{BE} is the forward bias to the Transistor T_1.

From the circuit, it is evident that output voltage V_{out} is compared with the internal reference voltage, V_Z. So, $V_{out} = V_Z - V_{BE}$. Therefore, $V_{BE} = (V_Z - V_{out})$.

The difference voltage V_{BE} controls the voltage V_{CE} across the series control Transistor, which automatically corrects the output voltage variations so that the output voltage attains to a stable and regulated voltage.

If the output voltage $V_{out} = V_Z - V_{BE}$ decreases, then the forward bias$V_{BE} = V_Z - V_{out}$ increases. This increase in the forward bias to the Transistor increases its Collector current I_C. So, the voltage V_{CE} across the series Transistor decreases. This decrease in the voltage V_{CE} automatically increases the outputvoltage according to the equation $V_{out} = V_{in} - V_{CE}$. Thus output DC voltage regulation to constant output voltage is obtained.

On similar lines, if the output voltage $V_{out} = V_Z - V_{BE}$ increases, then the forward bias $V_{BE} = V_Z - V_{out}$ decreases. This decrease in the forward bias to the Transistor decreases its Collector current I_C. So, the voltage V_{CE} across the series Transistor increases. This increase in the voltage V_{CE} automatically decreases the output voltage according to the equation $V_{out} = V_{in} - V_{CE}$. Thus output DC voltage regulation to constant output voltage is obtained.

Advantages:

1. It provides better regulation than a simple Zener Diode voltage regulator.
2. It has a lower output resistance and is capable of delivering larger output.
3. By employing super alpha configuration of Transistors (Darlington pair), the output resistance of the circuit can be made lower to deliver still higher powers.

3.16.2 Operational-amplifier as Comparator in 'Series Voltage Regulator Circuit')

Circuit operation:

- Changes in the load DC voltage (output V_o) cause changes in the voltage drops across the potential divider resistors R_1 and R_2..

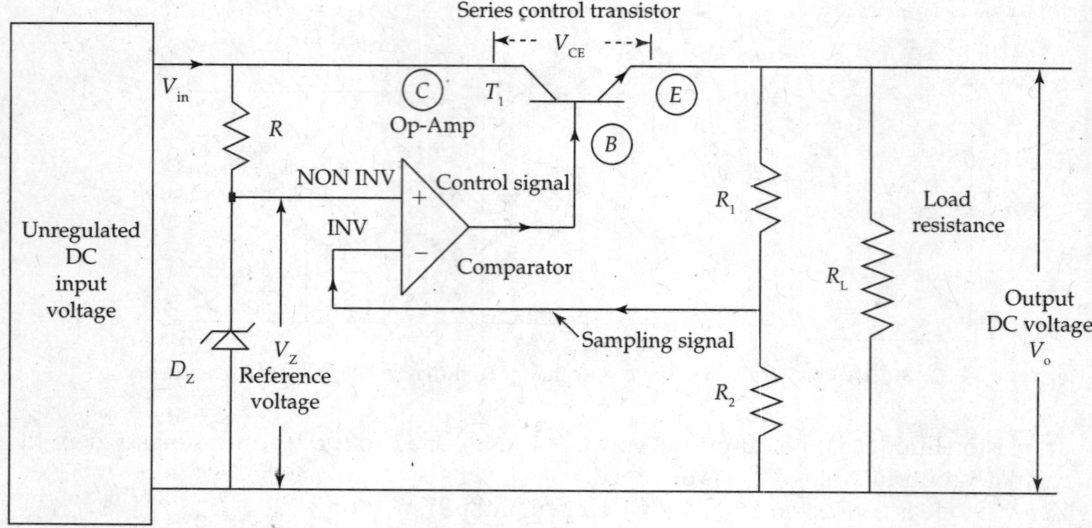

FIG. 3.41 *Op - amp as comparator in series voltage regulator circuit*

- Sampled signal (feedback signal V_f) voltage V_S is fed to inverting terminal (INV) of the operational amplifier (Op-Amp).
- Reference voltage V_Z across the Zener Diode (temperature compensated Zener) is connected to the non-inverting terminal (NON-INV) of the Op-Amp.
- Resistor R is designed to provide sufficient reverse bias to Zener Diode to keep it in breakdown and function as a reference voltage source V_Z.

$$\text{Feedback or sampled signal} \quad V_f = V_S = \frac{V_o \times R_2}{[R_1 + R_2]} = \beta \cdot V_o \tag{3.74}$$

$$\text{where feedback factor } \beta = \frac{R_2}{[R_1 + R_2]} \tag{3.75}$$

- The error signal $(V_Z - V_S)$ is amplified by the op-amp.
- Output voltage (Control signal) of the Op-Amp is fed to the Base of the series pass control Transistor T_1. The control signal changes the biasing conditions of T_1. It causes variations in Collector current. Series pass Transistor acts as a variable resistor for changes in its Collector currents. Voltage drop across Collector to Emitter V_{CE} occurs in such a way so as to maintain the load voltage constant.
- The output regulated voltage

$$V_o = \left[1 + \frac{R_1}{R_2}\right] \cdot V_Z \text{ V}. \tag{3.76}$$

This result is obtained by neglecting the Base to Emitter voltage drop of the control Transistor.

Merits of 'Series Regulator' Circuits

(1) They are used for high voltage medium current applications, with a nominal voltage drop across the series element. (2) They can be used to feed variable loads. (3) Circuit protection

features like fold-back current limiting and short circuit protection can be easily implemented. (4) Power dissipation is less and efficiency is more when compared to shunt regulators. (5) Simple design with good regulation. (6) Low output resistance. (7) Low temperature co-efficient.

Demerits of Series Voltage Regulator Circuits

- Complex circuit that needs more components. So, cost is high.
- Small value of difference between input and output voltages has to be maintained.
- Additional protection circuits are necessary to safeguard overload and short circuits

Basic series regulator circuit illustrating the above concepts:

3.16.3 Analysis of Series Voltage Regulator Circuit of Fig. 3.42:

- Assume that the output voltage V_o has increased.
- Then the sampled signal $\beta \cdot V_o$ increases.
- This sampled signal and the reference voltage V_R across the Zener Diode are applied to the error amplifier Transistor T_2.
- The increased sampled signal fed to the Base terminal of the Transistor T_2 causes an increase in V_{BE2}.
- The change in V_{BE2} is proportional to change in sampled signal. So, this increase in forward bias causes an increase in the Collector current I_{C2}.
- This causes a decrease in the Base current I_{B1} of the first Transistor so as to maintain current I_3 through R_3 constant, because $I_3 = I_{B1} + I_{C2}$.
- The decrease in the Base current I_{B1} of the Transistor T_1 causes a consequent reduction in its Collector current I_{C1}.
- The decrease in I_{C1} causes an increase in V_{CE1} the voltage between Collector and Emitter of the Transistor T_1.
- This causes reduction in output voltage so that output voltage is maintained constant.

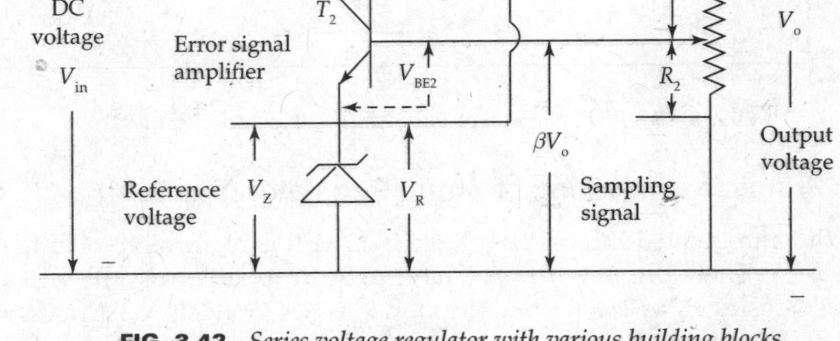

FIG. 3.42 *Series voltage regulator with various building blocks*

Finally, the important characteristic of any voltage regulator circuit is its *Transient response.* Transient response provides us the knowledge of the amount of *Time lag* required for the output voltage to reach steady-state conditions after sudden increase or decrease (changes) in the load currents for changes in the load resistance.

3.17 BLOCK DIAGRAM OF SHUNT VOLTAGE REGULATORS

Shunt voltage regulator circuit is another form of linear feedback type voltage regulator. The regulating (controlling) device is connected in the shunt path across the load resistance. Shunt voltage regulator does this function by shunting away the current from the load to regulate the output voltage to a constant value.

3.17.1 Block Diagram of Shunt Voltage Regulator Circuit (Fig. 3.43)

- Unregulated DC input voltage provides the source current I_S. It is clear from the circuit in Fig. 3.43 that the source supplies the shunt current I_{Sh} to the control element and the load current I_L. Therefore, $I_S = I_{Sh} + I_L$. The load current flowing through the load resistance develops and provides the output voltage according to the design of the shunt voltage regulator circuit.
- If the load voltage changes due to variations in load conditions, the sampling circuit provides a feedback signal to the comparator circuit. Comparator circuit compares the feedback signal with the reference voltage and sends a control signal to the control Transistor. The control signal draws increased shunt current and reduces the load current. Reduced load current brings back the load voltage to the designed constant output voltage. Thus providing a shunt path to the increased load current, the output is regulated to constant designed output voltage.

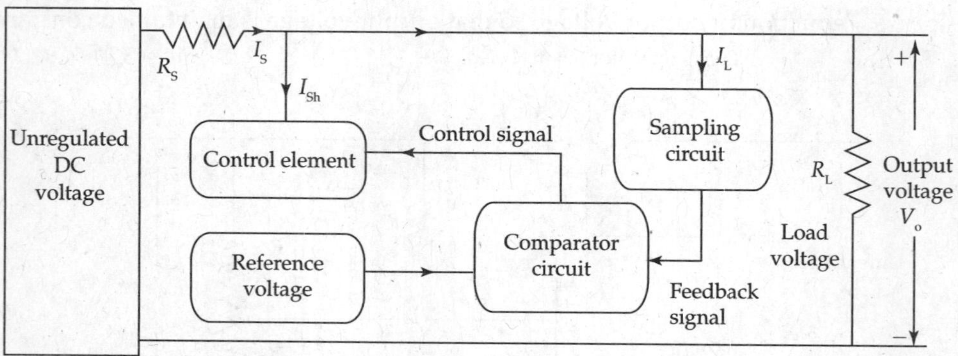

FIG. 3.43 *Block diagram of shunt voltage regulator circuit*

3.17.2 Basic Principle of Working of Shunt Regulator (Fig. 3.44)

Assuming that the unregulated DC input voltage is 15 V. It is required to obtain a regulated output voltage of 9 V. Therefore, 15 – 9 = 6 V have to be dropped across the series resistance R_S in the circuit. Choosing R_S as 2 Ω, current through R_S is 3 A $[6\text{ V}/2\,\Omega = 3\text{ A}]$. Assuming load resistance $R_L = R_P$. The current flowing through each of these resistances is $3\text{ A}/2 = 1.5\text{ A}$

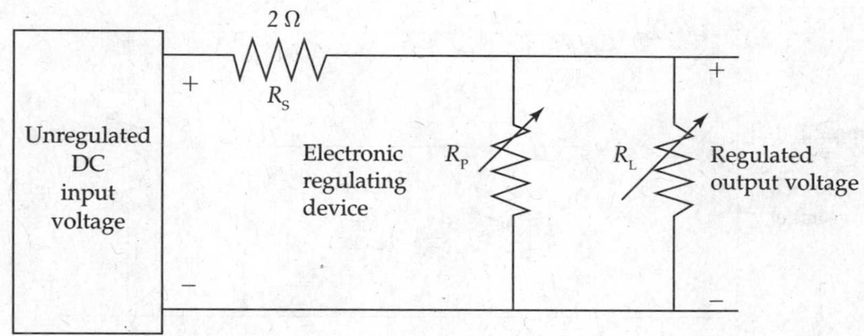

FIG. 3.44 *Voltage shunt regulator circuit*

If the load resistance is increased, the current through R_L decreases. Assuming that the current through $R_L = 1$ A, the current through R_S is $3 + 1 = 4$ A. Now, the voltage drop across $R_S = 4 \times 2 = 8$ V. As a result, the output voltage drops to $15 - 8 = 7$ V. The regulating device now comes into action by sensing the decrease in output voltage and it decreases its resistance allowing a current of 2 A. Then total current flow through $R_S = 2 + 1 = 3$ A, restoring back to its normal voltage drop across R_S to $3 \times 2 = 6$ V. Then the output voltage restores back to 9 V. The resistance of the regulating device decreases, when the load resistance increases. Then the output voltage restores to 9 V.

When there is a decrease in load resistance an opposite change takes place in the regulating device. When R_L is decreased and load current is 2 A, then the flow of current through R_S is $3 + 2 = 5$ A, causing a voltage drop across R_S of magnitude $5 \times 2 = 10$ V. The output voltage is now $15 - 10 = 5$ V. This change in output is sensed by the regulating device and its resistance increases the current through R_P to a value of $1.5 - 0.5 = 1.0$ A. The total current through R_S is $1 + 2 = 3$ A, with a voltage drop of $3 \times 2 = 6$ V. The DC output voltage is now restored to steady voltage of $15 - 6 = 9$ V.

The basic principle of shunt voltage regulator uses the sensing of regulating device to changes in load current and providing compensation by opposite changes through it, so as to maintain the regulating output voltage to a steady value. In an electronic voltage regulator, regulating Transistor controls automatically in a continuous manner, whenever there are changes in input voltages or changes in load conditions.

3.18 SHUNT VOLTAGE REGULATOR CIRCUITS

3.18.1 Shunt Transistor Voltage Regulator Circuit

This is another form of linear feedback regulator circuit using control element in the shunt path as shown in Fig. 3.45. The circuit contains (1) Control device Transistor T, (2) Fixed resistor R_S, (3) Current limiting resistor R, (4) Zener Diode and (5) Load resistance R_L.

Current limiting resistor R and Zener Diode provide a constant reference voltage for biasing the Collector–Base junction for the shunt Transistor T. Voltage across it is the important factor for total circuit operation. It is maintained at a constant value. The amount of forward bias to the Transistor affects its total resistance. A change in unregulated DC input voltage or a change in load current causes a variation in the current drawn by the control Transistor and regulates the output at a constant value.

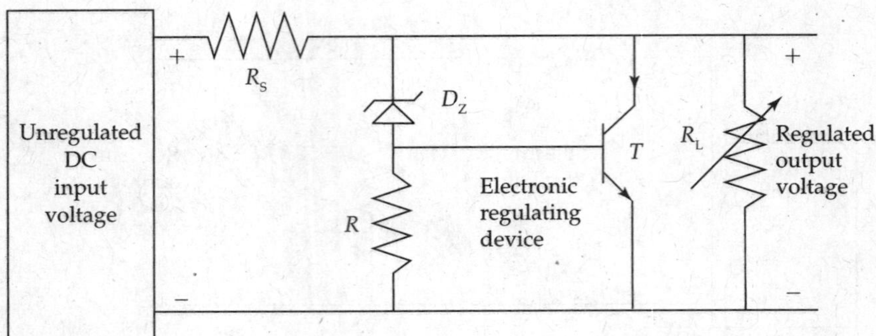

FIG. 3.45 *Voltage shunt regulator circuit using transistor*

- An increase in the input voltage causes an increase in the forward bias to the Transistor lowering the resistance of the Transistor. It results in more current flow through the shunt path. This increases the current through R_S causing more voltage drop across it, which results a drop in output voltage to maintain a constant output.
- An increase in the load current causes drop in output voltage. It reduces the forward bias to the Transistor that increases the resistance of the Transistor. The load current in the shunt path decreases. As a result, the current through R_S decreases so as to keep the output voltage constant.

A large value of R_S will have a large voltage drop across it, which is undesirable. This problem is overcome by using a constant current source in place of R_S to improve the performance of the regulator.

Advantages:

(1) It is a simple circuit with a few components resulting greater economy. (2) It is used when the load is relatively constant. (3) It has inherent short circuit protection. (4) Transient over voltage problems is quickly solved. (5) Under full load conditions, its efficiency is high, because the shunt regulator draws minimum current, maximum current is diverted through R_L. (6) The current gain of the regulator can be improved by using Darlington pair Transistors.

Disadvantages:

Under light load conditions, majority of the current is drawn by the shunt device, resulting in lower efficiency.

Applications:

1. They are implemented in many ways in spacecraft power systems and solar shunt arrays.
2. They are used to charge special batteries such as Lithium ion batteries, Silver-platted batteries mainly used in aerospace and defense applications

3.18.2 Operational Amplifier as Comarator in Shunt Voltage Regulator Circuit

The schematic diagram of Operational amplifier (Op-Amp) shunt regulator shown in Fig. 3.46 has R_1 and R_2 resistor combination as the sampling arrangement and senses a part

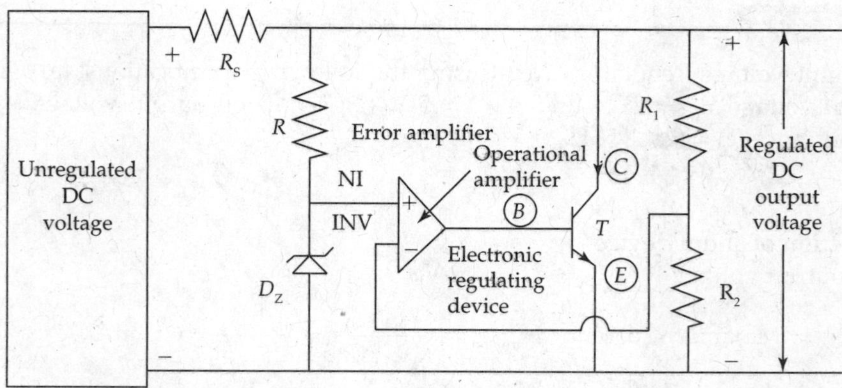

FIG. 3.46 *Voltage shunt regulator circuit using operational amplifier*

of output voltage $\beta \cdot V_o$. Sampled voltage is fed to inverting terminal of Operational amplifier. Reference voltage from the Zener Diode is fed to non-inverting terminal of *Operational amplifier*. Transistor T is the shunt-regulating device shunted across the load. Amplified error is connected to Base terminal of the shunt Transistor.

- A decrease in output voltage causes a decrease in error signal fed to shunt Transistor. This decreases the Collector current, which is equivalent to an increase in R_{CE} (Collector to Emitter resistance). R_{CE} acts as a voltage divider with R_S. Voltage drop across R_S decreases as less current is diverted from load. Thereby, output voltage is maintained constant at its designed value.
- Feedback voltage to the Op-amp increases, whenever the output voltage increases due to changes in load resistance. This is compared with reference voltage and amplified error causes conduction of control Transistor to increase. Then the load current I_L decreases and maintains constant output voltage.

Comparison between 'Series and Shunt voltage Regulator' circuits

S. No.	Series regulator	Shunt regulator
1	Control element is connected in series to load resistance R_L to stabilise the output voltage	Control element is connected in shunt or parallel load resistance R_L to stabilise the output voltage
2	Power dissipation is less and hence efficiency of the circuit is high	Power dissipation is more and hence efficiency of the circuit is less
3	It can be used to feed variable loads	Used with relatively constant loads (small changes in current and voltages)
4	Complex circuit requiring more components and so cost is large	Simple circuit with a few components. So the cost of the circuit is less
5	Additional protective circuit is necessary to safe guard overload and short circuit	In built short circuit protection, additional protective circuits are also used to enhance safety and reliable operation
6	High-voltage and low-current applications	Used at relatively constant loads

EXAMPLE 3.17

Design a Shunt Voltage Regulator Circuit using the following specifications: (a) Unregulated DC (1) input voltage $V_{in} = 25$ V and $\Delta V_{in} = 5$ V, (2) Regulated output voltage $V_{out} = 12$ V; (b) Load resistance $R_L = (25 \pm 5)\ \Omega$.

Solution:

Step 1: Selection of shunt device
Maximum output voltage $V_{out(max)} = V_{out} = 12$ V

$$\text{Maximum current} \quad I_{out(max)} = \frac{V_{out(max)}}{R_L - \Delta R_L} = \frac{12}{[25-5]} = 600 \text{ mA}$$

Power dissipation to be handled by the shunt device $= P_D$

$$P_D = V_{out} \cdot I_{out(max)} = (12 \times 600) \text{ mA} = 7.2 \text{ W}.$$

A Silicon transistor (NPN type) rating above maximum voltage.
Current and power dissipation have to be chosen from the manufacturer's datasheet.
Assume transistor $h_{fe} \rangle\ 100$.

Step 2: Selection of zener Diode.

$$\text{Zener voltage} \quad V_Z = [V_{out} - V_{BE}] = 12 - 0.7 = 11.3 \text{ V}$$

$$\because \quad h_{fe} = 100, \quad I_{B(max)} = \frac{600 \text{ mA}}{100} = 6 \text{ mA}$$

For reliable breakdown of Zener Diode, assuming minimum current of 2 mA power dissipation has to be chosen. With an excess magnitude of 11.3 V $\times$ 6 mA = 67.8 mW.
A Zener Diode of 400 mW dissipation rating could be selected:

Step 3: Value of resistance

$$R = \frac{V_{BE}}{\text{minimum Zener current}} = \frac{0.7 \text{ V}}{2 \text{ mA}} = 350\ \Omega$$

Commercially available 330 Ω resistor of 1/2 W rating is selected.

Step 4: Value of R_{SC} has to be calculated at maximum load current $I_{out(max)}$

$$V_{in} - \Delta V_{in} = V_{out} + R_{SC} \cdot I_{out(max)}$$

$$25 - 5 = 12 + R_{SC}(600 \text{ mA})$$

$$\therefore \quad R_{SC} = \frac{8}{600 \text{ mA}} = 13.33\ \Omega.$$

R_{SC} can be chosen of value of 15 Ω with 1 W power dissipation rating.

3.19 CURRENT LIMITING TECHNIQUES

'Power Supplies' using voltage regulators are subjected to accidental overloads if load resistance is reduced to a very low value or due to accidental short circuits. As, they possess

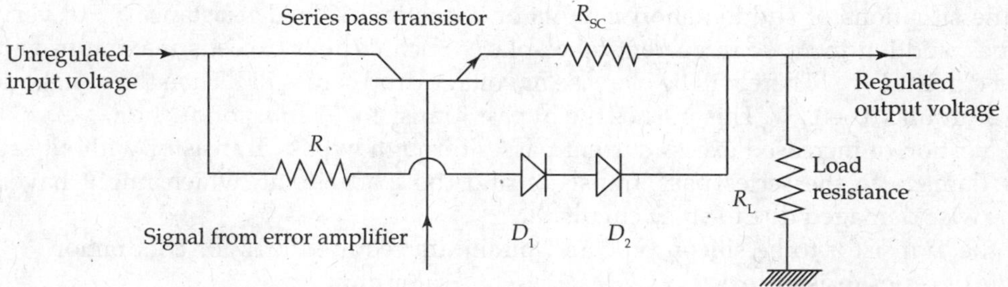

FIG. 3.47 *Constant current limiting protection circuit*

low output impedance, overloading causes damage to components of regulated power supplies. Usage of series fuse cans safe guard to some extent. But in some cases, such protection is of no use; as the thermal time lag of the power Transistor is much smaller than the fuses. As the series Transistor may be damaged due to thermal stress in a very small time interval, fuse will not be able to provide sufficient protection.

To enhance the safe and reliable operation of a power supply special electronic protection circuits such as (1) Constant current limiting and (2) Fold back current limiting are used.

Constant current limiting Resistor R_{SC} is added in between the load and regulated output as shown in Fig. 3.47.

Two Diodes are connected across the input and output of the regulator circuit. In the normal operation, the two Diodes will not conduct. In the case of overload/short circuit, the load current increases to beyond maximum of $I_{L(max)}$ causing more voltage drop across the resistance R_{SC}. Then the two Diodes are forward biased and start conducting. The conducting Diodes provide a bypass path for a part of the increased load current when the load current exceeds $I_{L(max)}$. Thus the series pass Transistor is protected against instantaneous damage to it and also safeguards some components in output circuit.

Current *i* through R_{SC} can be calculated using the following equation:

$$I_{R\,(SC)} = \left|\frac{[2V_D - V_{BE}]}{R_{SC}}\right| \text{mA.}$$

Another type of protection circuit is shown in the Fig. 3.48.

- Normally, the load current through R_L will be limited to maximum safe operating current. Then the voltage drop across the short circuit protection resistor R_{SC} is not sufficient enough to forward bias the additional by-pass path Transistor.

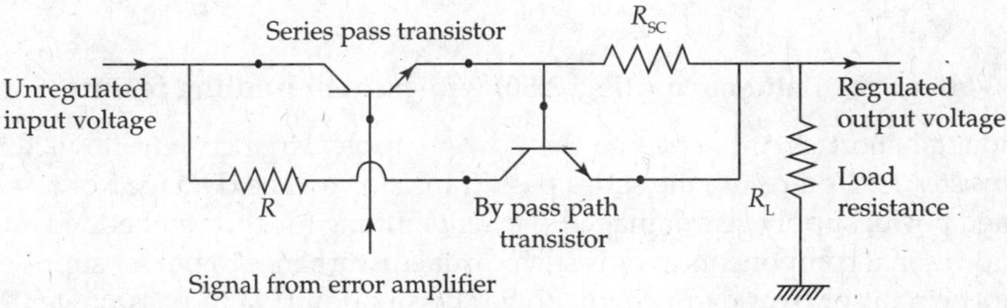

FIG. 3.48 *Constant current limiting protection circuit using another transistor for by-pass path*

- In the situations of sudden short circuits or reduction of load resistance R_L to very low values, sudden increase in current takes place. Such current exceeds maximum value of load current $I_{L(max)}$. There will be increase in voltage drop across R_{SC}. Then the voltage across R_{SC} is around 0.6–0.7 V. This pushes the bypass Transistor T_2 into conduction.
- The portion of increased excess current passes through bypass Transistor without causing any damage to the series pass Transistor and the load circuit, which might have been otherwise damaged due to short circuit.
- Let the Transistor to be silicon type and minimum required bias for conduction = 0.7 V. Value of series resistance $R_{SC} = 0.7\text{ V}/I_{L(max)}$ at a safer limit.

Protection circuits' in Figs. 3.47 and 3.48 assure the protection of series pass Transistor even under short circuit conditions. But these circuits suffer from two disadvantages of current limiting at low values of load currents and excessive power dissipation across the series pass Transistor. Hence, they are not suitable for high current regulation circuits.

It can be seen from Fig. 3.49 that output voltage remains constant till the load current increases to $I_{L(max)}$. Beyond $I_{L(max)}$ the output voltage drops to zero volts. Short circuit current is slightly greater than $I_{L(max)}$. Current limiting starts its function, when the output current exceeds the maximum rating by 10–20%.

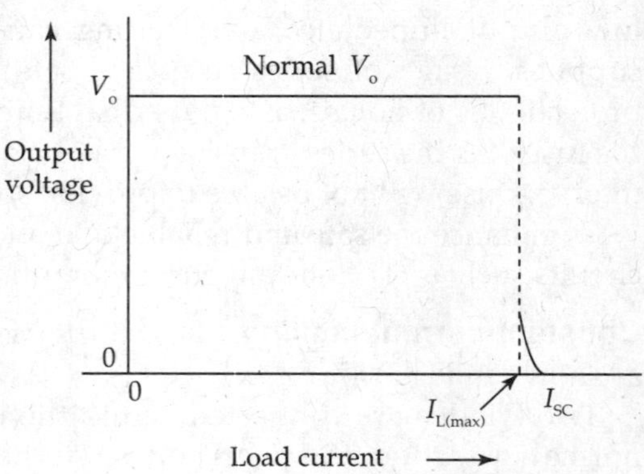

FIG. 3.49 *Variations of output voltage V_0 for different load currents*

Disadvantage of current limiting To obtain a sufficient safe operating area (SOA) to the pass Transistor, it must have a higher current capability than is needed for normal operation.

Current Limiting

Short circuit or overloading power supplies causes excess current flow through regulator Transistors. When overload currents exceed the maximum rated load current, controlling Transistor T_2 gets damaged. To avoid such situations, current flow through the control Transistor is limited deliberately by providing a bypass path for the excess currents to avoid damage to regulator supplies and provide safe operation. Such *current bypassing feature is considered as current limiting*. Current limiting protects the regulator from damage by holding the maximum output current at a constant level $I_{L(max)}$ that protects the regulator device from damage.

Series Pass Voltage regulator circuit (Fig. 3.50) with current limiting feature

- If an accidental short circuit occurs to the power supply, large current flows through *series power Transistor* T_1. As a result, the series pass Transistor will be damaged or a component in unregulated power supply gets damaged. Current limiting function embedded into the circuit through additional components avoids such casual disturbance to power supply circuit.
- When short circuit or overload occurs, power dissipation (P_D) in pass power Transistor is the product of input voltage V_{in} and maximum value of load current.

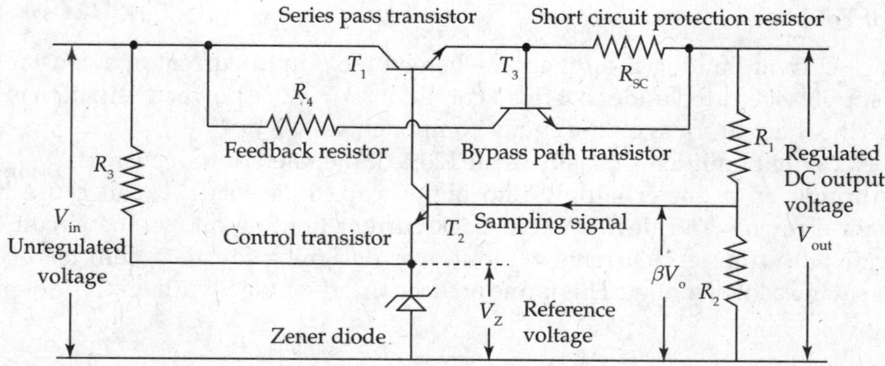

FIG. 3.50 *Series pass voltage regulator circuit with current limiting action*

$$P_D = V_{in} \times I_L(\text{max}) \text{ Watts.}$$

- When power supply is operating normally, voltage drop across *short circuit* (*SC*) *sensing or protection resistor* R_{SC} is small and Transistor T_3 will be in non-conduction or OFF-state. In case of over current flow at the output port, increased current flow through resistor R_{SC} produces a voltage drop larger than rated maximum voltage drop across it. Increased voltage drop across R_{SC} increases forward bias to Transistor T_3 and switches it into conduction or ON-state.
- Overload current now finds a bypass path through Transistor T_3 and passes through the *feedback resistor* R_4. Then the Base voltage of T_1 decreases resulting in reduction of output voltage. Designing a protection circuit to withstand overload stress needs circuit component selection with over maximum capacity design. Including fold back (reduce) current limiting feature to reduce or limit the current.
- When voltage across R_{SC} is around 0.6–0.7 V, current limiting activity starts due to turning ON feature of the *By-Pass Path Transistor* T_3 as explained above. This process further decreases the Base drive for the *Series pass Transistor* T_1. Op-amp can be used to replace *Control Transistor* T_2 to improve the performance of current limiting feature in the circuit.

Voltage regulator circuit using op-amp and current limiting feature (Fig. 3.51)

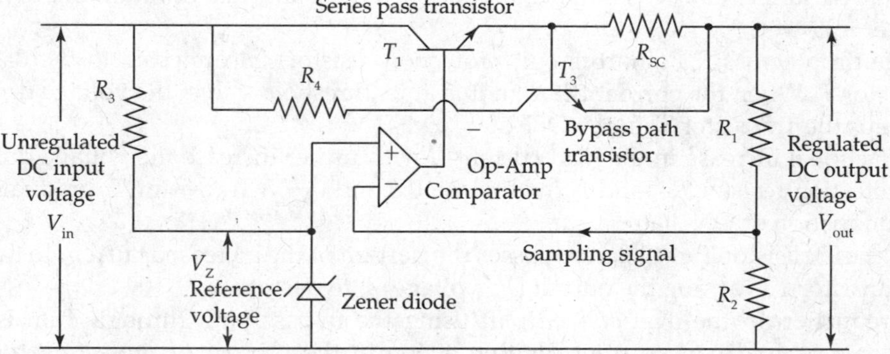

FIG. 3.51 *Op-amp voltage regulator circuit using current limiting features*

Features of Fold back current limiting

- Fold back current limiting is a function whereby the output current of a constant voltage power supply is reduced under overload conditions to reduce power dissipation in the load.
- Reduces the thermal stress on the circuit components.
- Fold back current limiting is usually set at 120% of the rated output current $I_{L(max)}$.
- While turning on a power supply into highly capacitive loads, it can cause havoc on the circuit elements and devices. Fold back current limiting allows the circuit elements to operate with transient currents without over designing for maximum load condition, thereby saving a lot of money. This is one of the main advantages with this type of protection technique.

Fold back Current limiting series voltage regulator circuit (Fig. 3.52)

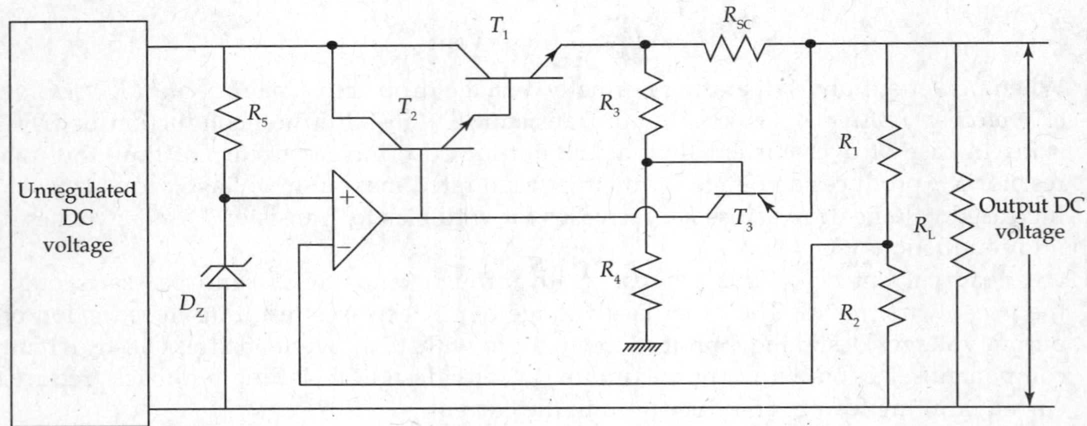

FIG. 3.52 *Fold back current limiting series voltage regulator circuit*

Circuit operation:

- When the load is less compared to the rated maximum allowed current, the Transistor T_3 (in fold back limiting circuit) in Fig. 7.10 is in the off-state.
- Voltage divider network of resistors R_3 and R_4 provides fold back current limiting action.
- Potential divider circuit senses the voltage at the output terminal (Emitter) of the Transistor (Emitter follower) T_1.
- Voltage drop across R_{SC} (short circuit protection resistor) is connected to the Base of the Transistor T_3 When the current through it is at its threshold value, the voltage drop across R_{SC} keeps the Transistor T_3 in the OFF-state.
- Any fractional increase in the load current I_L will further increase the voltage drop across the protection resistor R_{SC} and is in between 0.6 and 0.7 V. Transistor T_3 gradually comes into conduction or ON-state.
- When the Transistor T_3 is ON, it bypasses the portion of the increased current to the Base of the Transistor T_2, causing the output DC voltage V_o to decrease.
- Such reduction in the overload current using the bypass path through Transistor T_3 is considered as fold back current limiting action in the process of protecting the control (pass) Transistor from damage.

- A decrease in output voltage means, the voltage across R_{SC} also decreases. As a result, the current through R_{SC} decreases. More current is shared to the Base of the Transistor T_3. Now the reduction in output voltage reduces the load current to a safe operating value.

Fold back current versus output voltage response of a voltage regulator (Fig. 3.53)
From the Fold back response curve, the load voltage remains constant until a rated output current $I_{L(max)}$ is reached. In the event of over or excess current due to short circuit or overload condition, the extra current is bypassed through Transistor T_3 and feedback resistor R_4, protecting the series control power Transistor from damage. Now the power dissipation in the pass Transistor is minimised due to reduction or fold back current limiting action.

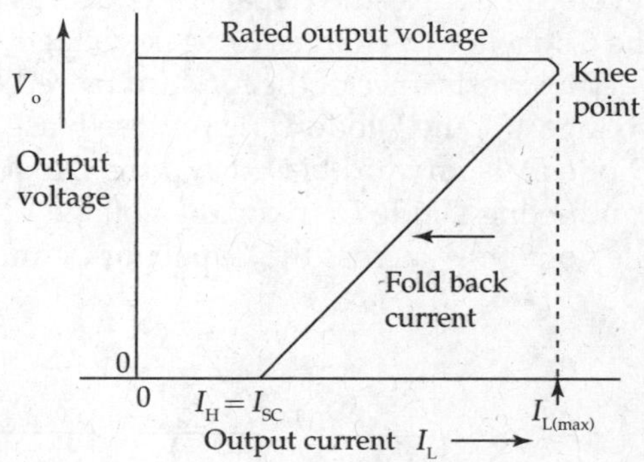

FIG. 3.53 *Fold back current response of voltage regulator*

Specifications of Voltage regulator circuits

(1) Mains input voltage, (2).. Maximum DC output voltage, (3) Maximum DC output current, (4) Low drop out voltage, (5) Type of protection arrangement and (6) Type of regulator circuit.

3.20 VOLTAGE MULTIPLIER CIRCUITS

Introduction Voltage multipliers are special class of AC/DC converters manly used to convert AC supply to higher DC voltages than obtained by conventional Rectifier circuits. The high voltage low current voltage multipliers are mainly used in CROs, TV picture tubes, video display units, flash gun used in photography, particle accelerators, X-ray equipment and so on.

It is not easy to boost up DC into high DC voltage. So, the high DC voltages are obtained from AC/DC conversion, which are a multiple of peak input voltage such as $2V_P$, $3V_P$, $4V_P$ and so on. They mainly employ cascaded arrangement of peak Rectifiers or slicer followed by peak Rectifiers employing Capacitors and Diodes.

3.20.1 Voltage Doublers

There are two types of voltage doublers.

- Half-Wave type or Cascaded type,
- Full-Wave type.

Half-Wave Voltage Doubler Circuit
A Half-Wave voltage doubler produces an output voltage, which is approximately double the peak voltage of the input signal waveform. All other higher order voltage multiplier circuits can be formed from the basic Half-Wave doubler multiplier circuits.

Half-Wave or cascaded type voltage doubler circuit is shown in Fig. 3.54. It consists of a transformer to provide the input voltage V_{in} and two sections with each section having a combination of a Capacitor and a diode.

Explanation of Half-wave doubler circuit operation (Fig. 3.54)

The input voltage to the network is a sinusoidal signal obtained from a transformer. During the positive half cycle of the input voltage, the Diode D_1 is forward biased and the Diode D_2 is reverse biased, as the Capacitor C_1 acts as a short circuit, when the circuit is switched on. Then the Capacitor C_1 charges to peak voltage V_m or V_P ($V_{in} = V_m \cdot \sin(\omega t)$ or $V_P \cdot \sin(\omega t)$) through the very low resistance of the forward biased Diode D_1. During the negative half cycle of the input voltage V_{in}, the Diode D_1 is reverse biased and it will not conduct. But, at the same time the Diode D_2 is forward biased and the Diode D_2 conducts. The Capacitor C_2 charges through the conducting Diode D_2. Now the voltage V_{C2} across the Capacitor C_2 is $2V_m$, which is the sum of the voltage V_m across the Capacitor C_1 and the input voltage V_m.

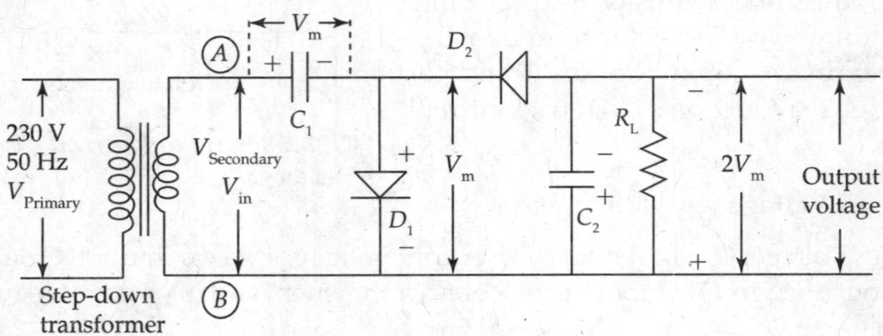

FIG. 3.54 *Half-wave voltage doubler circuit*

Thus, the voltage across the Capacitor C_2 is $2V_m$. This voltage across the Capacitor C_3 can be taken as the output voltage for the Half-Wave voltage doubler circuit. This output voltage V_{DC} is two times the maximum or the peak value of the input signal voltage, $V_{DC} = 2V_m$ or $2V_P$. Ripple frequency f_r of the output voltage is equal to the supply frequency f_S. This action continues cycle-by-cycle with C_1 being fully charged to V_P on each positive half cycle and charging C_2 to a voltage $2V_P$.

The voltage doubler circuit can be visualised as a combination of (1) Clamping circuit with (C_1 and D_1) and (2) Rectifier Filter circuit (C_2 and D_2).

Input signal waveform to voltage doubler circuit and Output voltage of a voltage doubler circuit are shown in Figs. 3.55 and 3.56.

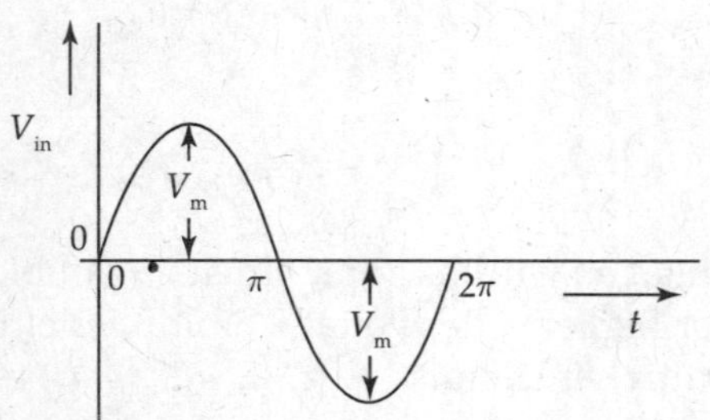

FIG. 3.55 *Input signal wave form to voltage doubler circuit*

FIG. 3.56 *Output signal of voltage doubler circuit*

Applications:

(1) Electronic voltmeters, (2) Cathode Ray Tubes, (3) They are useful when load current is relatively light and requires a voltage higher than available from a standard transformer, and (4) By cascading several half-wave voltage multiplier circuits, higher order voltage Multiplier circuits can be employed.

Full-Wave Voltage Doubler

In the Full-Wave voltage Doubler circuit (Fig. 3.56), the Diodes are connected to the same voltage source V_m but in the opposite direction.

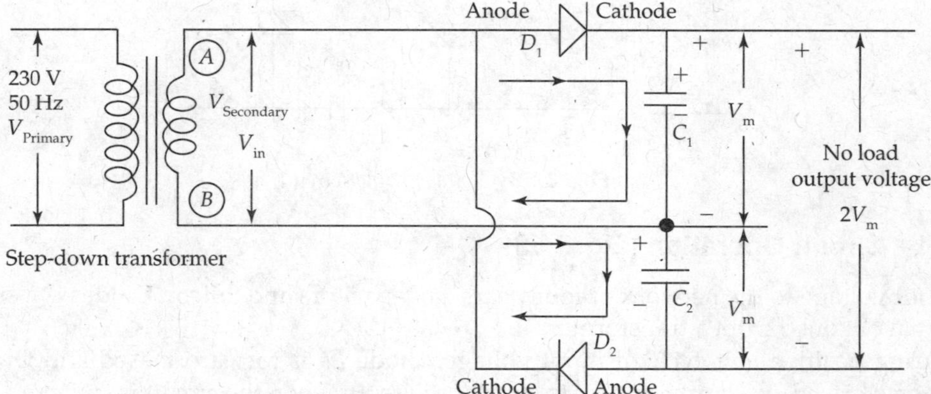

FIG. 3.57 *Full-wave voltage doubler*

Working of Full-Wave voltage doubler circuit (Fig. 3.57)

Full-Wave voltage doubler circuit is shown in Fig. 3.57. When the upper terminal *A* of the transformer secondary is positive, the semiconductor Diode D_1 conducts and charges the Capacitor C_1 to the peak voltage V_m of the input voltage. On the other hand, when the lower terminal *B* of the transformer secondary is positive the Diode D_2 conducts and charges the Capacitor C_2 to the full input voltage V_m as shown in Fig. 3.57. Meanwhile, the charge acquired by C_1 is retained, since the voltage across the Diode D_1 is due to the reverse bias. Hence the voltages across both the Capacitors are with such polarities that the DC output voltage, V_{DC} is equal to twice the peak input voltage. Therefore, $V_{DC} = 2V_m$. Thus during each half cycle, one of the two output Capacitors get charged. So, the circuit is known as Full-Wave voltage doubler. Ripple frequency f_r is equal to twice the supply frequency. This increase in ripple frequency decreases the value of the Filtering Capacitors and Filtering process becomes less costly. PIV rating of each Diode is only $V_m = V_P$.

Net output voltage is $2V_m$ under no load conditions. When loaded, the Capacitor discharges depending on R_L. Output voltage falls quickly as time constant decreases with decrease in R_L, i.e. increased load current. For fixed light loads, output voltage is almost twice as that can be obtained for a Full-Wave Rectifier with centre-tapped transformer and half winding. Capacitances need stand V_m and the Diodes have to withstand $2V_m$.

Drawbacks (1) AC supply transformer is always required, (2) This circuit cannot be used to produce higher order multiplier circuits other than two.

3.21 VOLTAGE TRIPLER

By cascading another section of the combination of a Capacitor and a Silicon Diode to the Half-Wave Doubler circuit, 'Voltage Tripler circuit' can be obtained as shown in Fig. 3.58.

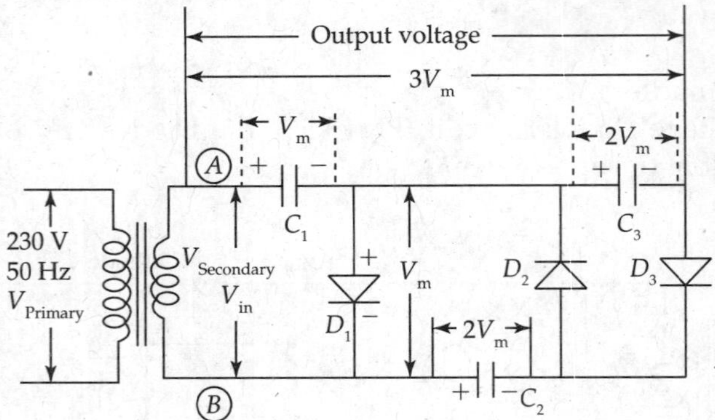

FIG. 3.58 *Voltage tripler circuit*

3.21.1 Circuit Operation (Fig. 3.58)

- Input voltage to the network of four pears of Capacitors and Silicon Diodes is a sinusoidal signal obtained from a transformer.
- During positive half cycle of input voltage, Diode D_1 is forward biased and Diode D_2 is reverse biased, as Capacitor C_1 acts as a short circuit, when the circuit is switched on. Then Capacitor C_1 charges to voltage V_m or V_P ($V_{in} = V_m \cdot \sin(\omega t)$ or $V_P \cdot \sin(\omega t)$) through very low resistance of the forward biased Diode D_1.
- During first negative half cycle of input voltage V_{in}, the Diode D_1 is reverse biased and it will not conduct. At the same time, Diode D_2 is forward biased and Diode D_2 conducts. Capacitor C_2 charges through conducting Diode D_2. Now the voltage V_{C2} across Capacitor C_2 is $2V_m$, which is the sum of the voltage V_m across Capacitor C_1 and input voltage V_m.
- During positive half period of next cycle of input voltage, Diode D_3 becomes forward biased. Now, Capacitor C_3 charges to voltage $2V_m$, which is the voltage already available with Capacitor C_2.
- From the schematic diagram of 'Voltage Tripler' circuit, we observe that voltage across the two Capacitors C_1 and C_3 is $3V_m$. So, voltage across the two Capacitors C_1 and C_3 is taken as the output voltage of the voltage Tripler circuit.
  Output voltage V_{DC} is three times the value of the maximum amplitude (V_m) or peak amplitude (V_P) of the input signal. So, the output voltage $V_{DC} = 3V_m$ or $3V_P$. Thus, we find that higher voltages can be obtained from voltage multiplier circuits.

Input signal waveform of voltage Tripler circuit (Fig. 3.59)

Output voltage from 'Voltage Tripler' circuit fluctuates on alternate half cycles. A good Filter circuit has to be added to it to produce smooth output DC voltage.

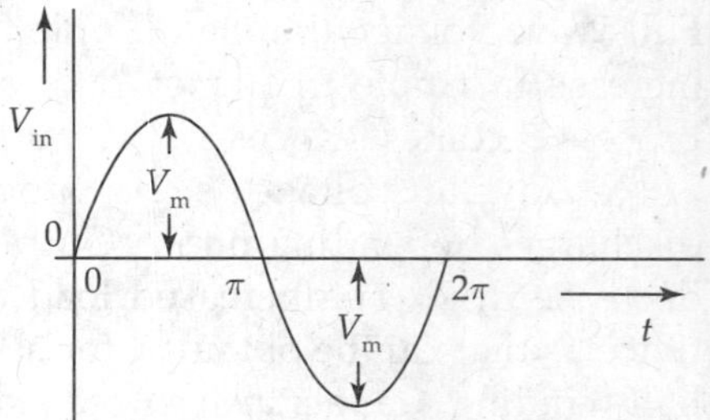

FIG. 3.59 *Input signal waveform to voltage trippler circuit*

3.22 VOLTAGE QUADRUPLER

Cascaded arrangements of two voltage doubler circuits form a quadrupler circuit as shown in Fig. 3.60. The process of cascading several voltage doubler circuits can obtain higher DC voltages from low AC supply. Main draw back of voltage multiplier circuits is that voltage regulation decreases with increase in additional stages of doubler circuits. So good regulation circuit has to be provided.

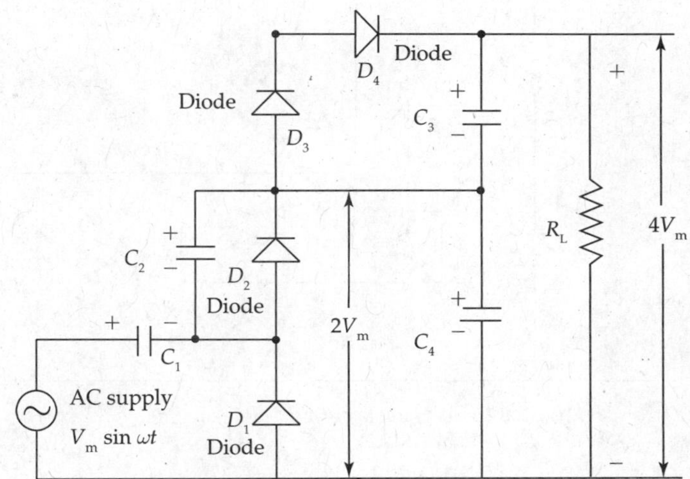

FIG. 3.60 *Voltage quadrupler circuit*

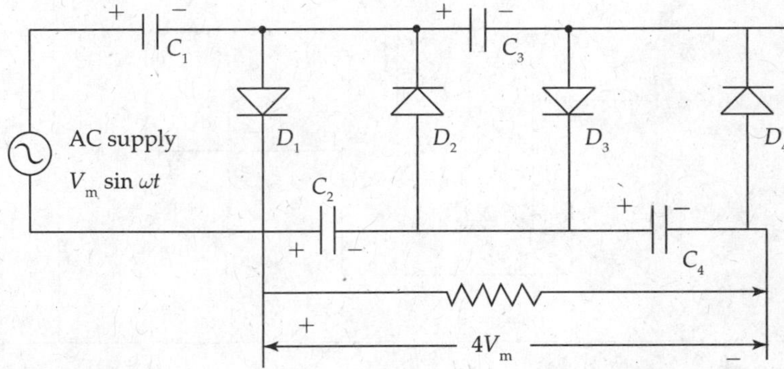

FIG. 3.61 *Voltage quadrupler circuit (re oriented circuit)*

3.23 ADJUSTABLE VOLTAGE REGULATORS

- Unregulated DC voltage V_{in} (rectified and filtered voltage) with ripple V_r is fed to an Operational amplifier. Reference voltage V_Z is connected to Non-Inverting input terminal of Op-amp.
- Sampling network has two resistors R_1 and R_2 at the out terminals of Op-Amp. Part of output voltage V_{out} is sampled and fed back to Inverting terminal of Operational amplifier. The Operational amplifier provides gain and adjustable output voltage to the load R_L.

Current boosting to work at higher currents above the rated current of op-amp (Fig. 3.62) Figure 3.63 is an adjustable voltage regulator using op-amp. In such circuit, the circuit operation is limited to the maximum current, which is the rated current of the operational amplifier. Certain applications demand the regulator to work at increased load currents. Such application is solved by using a power Transistor in the series path of the current boosting adjustable voltage regulator circuit as shown in Fig. 3.63.

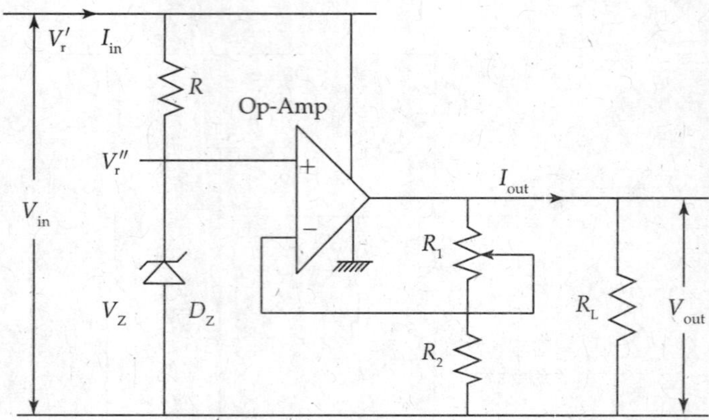

FIG. 3.62 *Adjustable output voltage regulator*

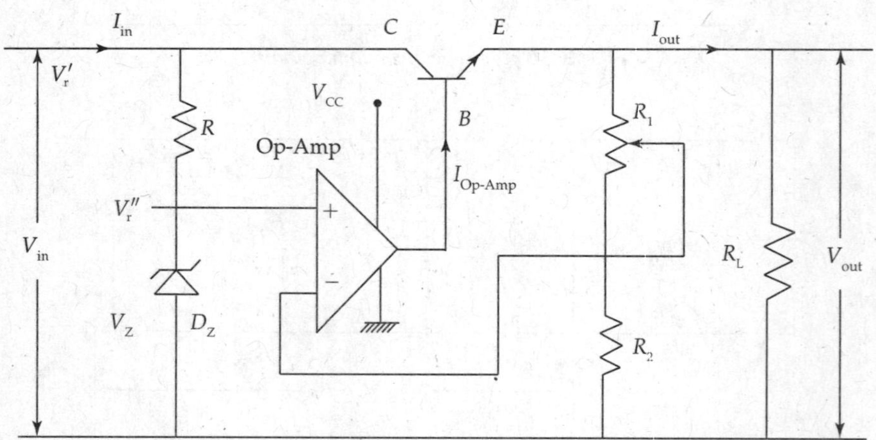

FIG. 3.63 *Linear voltage regulator with current boosting and adjustable voltage*

Input current ≈ Output current ≈ β. (maximum output current of the op-amp)

$$I_{in} \cong I_{out} \cong \beta \cdot I_{OP\text{-}AMP} = \frac{V_{out}}{R_L}$$

$$I_{out}(\max) = \beta \cdot I_{(OP\text{-}AMP)}.$$

Voltage regulator with current boost, adjustable output and short circuit protection Current sense resistor R_{SC} and a general-purpose Transistor provide protection against overload and short circuit as shown in Fig. 3.64.

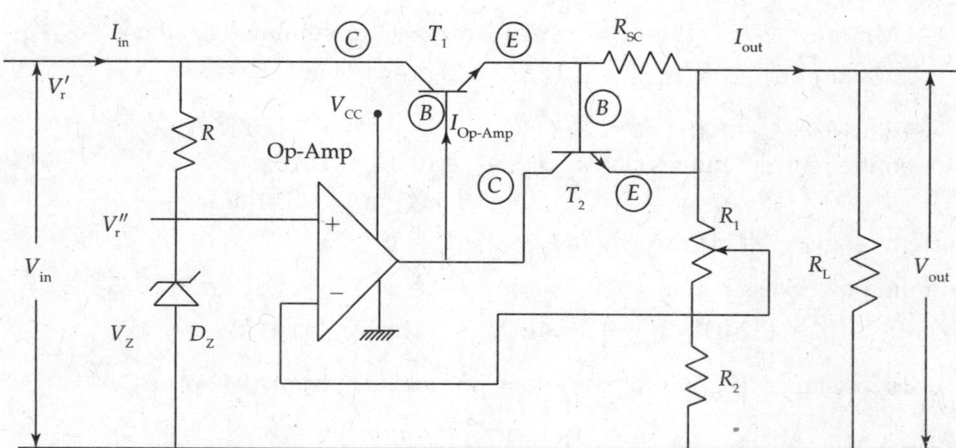

FIG. 3.64 *Linear voltage regulator with current boosting adjustable voltage and short circuit protection*

Voltage drop across $R_{SC} = V_{BE}(T2) < 0.7$ V.

R_{SC} is to be chosen when Iin. $= I_{out} \leq I_{out(max)}$. As a result, the Transistor T_2 is OFF and the current passes through the Transistor T_1, which will be in ON-state.

- In the event of overload or short circuit $I_{in} = I_{out(max)}$
- Voltage drop across $R_{SC} = V_{BE}$ (T2) > 0.7 V. As a result Transistor T_1 is ON, causing T_1 to turn OFF. There will be no current flow through the Transistor T_1 and load resistance R_L. Output voltage drops to zero protecting the Transistor T_1.
- When the overload or the short circuit condition is cleared, normal operation of the circuit is restored.

EXAMPLE 3.18

Design a voltage regulator with the following data: Unregulated DC input voltage $V_{in} = 20$ V, Zener Diode rating, $V_Z = 9.3$ V, $P_D = 400$ mW, $I_{Z(min)} = 2$ mA. Determine the following parameters for the regulator circuit: (a) Nominal output voltage, (b) Value of resistor R, (c) Load current range, (d) Maximum power dissipation of Transistor and (e) Value of R_{SC} and its power rating

Solution:

a. Nominal output voltage $V_{out} = V_Z + V_{BE} = 9.3 + 0.7 = 10$ V

Resistor R has to supply a current of 2 mA to the Zener Diode to conduct.

b. $R = \dfrac{V_{in} - V_Z}{I_{Z(min)}} = \dfrac{(20 - 9.3)}{2\text{ mA}} = \dfrac{11.3}{2 \times 10^{-3}} = 5.65\ \Omega.$

Commercially available resistor of value 510 Ω has to be selected.

c. Maximum allowable Zener current $\dfrac{P_D}{V_Z} = \dfrac{400\text{ mW}}{9.3} = 43$ mA

Load current range is the difference between maximum and minimum currents through the shunt path provided by the transistor.

Minimum $I_B = I_Z - I_{in} = 2\text{ mA} - 2\text{ mA} = 0$ (assuming $I_{in} = 0\text{ mA}$)

Maximum $I_B = (I_{Z(max)} - I_{in}) = (43 - 2)\text{ mA} = 41\text{ mA}$

Assuming transistor $h_{fe} = 100$

Transistor emitter current ranges $(1 + h_{fe}) I_B = (1 + 100) I_B = 101 I_B$

$= 101 \times 41\text{ mA} = 4141\text{ mA}.$

Load current range = 0–4141 mA where $I_{out(max)}$ is 4141 mA

d. Maximum power dissipation rating of transistor.

$$PD(tr) = V_{out} \times I_Z = 10\text{ V} \times 4141\text{ mA} = 41.41\text{ W}.$$

e. The series resistance RSC has to pass a maximum load current of 4141 mA.

$$\therefore \; R_{SC} = \frac{[V_{in} - V_{out}]}{I_{out(max)}} \frac{[20-10]}{4141\text{ mA}}$$

$$= \frac{10}{4141\text{ mA}} = \frac{10 \times 10^3}{4141} = 2.4\,\Omega$$

Power dissipation by $R_{SC} = I_{SC}^2 \times R_{SC}$

$$= (4141 \times 10^{-3})^2 \times 2.4 = 41.15\text{ W}$$

EXAMPLE 3.19

Determine minimum and maximum values for series resistor, required for a Zener Diode regulator with an output voltage of 5.6 V, if the supply voltage varies from 10 V to 50 V. Maximum load current is 20 mA and minimum Zener current is 3 mA.

Solution:

$V_{in(min)} = 10\text{ V}$, $V_{in(max)} = 50\text{ V}$, $I_{L(min)} = 0\text{ mA}$, $I_{L(max)} = 20\text{ mA}$, $I_{Z(min)} = 3\text{ mA}$, $I_{Z(max)} = 20\text{ mA}$ and $V_Z = 0.6\text{ V}$.

$$R_{S(max)} = \left[\frac{V_{in(min)} - V_Z}{I_{L(max)} + I_{L(min)}}\right]$$

$$= \left[\frac{(10-5.6)}{(20+3) \times 10^{-3}}\right] = \frac{4.4}{23} \times 10^3 = 191\,\Omega$$

$$R_{S(min)} = \left[\frac{V_{in(min)} - V_Z}{I_{L(min)} + I_{Z(max)}}\right]$$

$$= \frac{(10-5.6)}{(0+20 \times 10^{-3})} = \frac{4.4}{20 \times 10^{-3}} = 220\,\Omega$$

EXAMPLE 3.20

In Fig. 3.65 shown Input voltage, $V_i = 20\text{ V}$, $R_S = 200\,\Omega$ and $V_Z = 12\text{ V}$, $V_{BE} = 0.65\text{ V}$. Find output voltage, Collector to Emitter voltage of the Transistor and the current in the 200 Ω resistor.

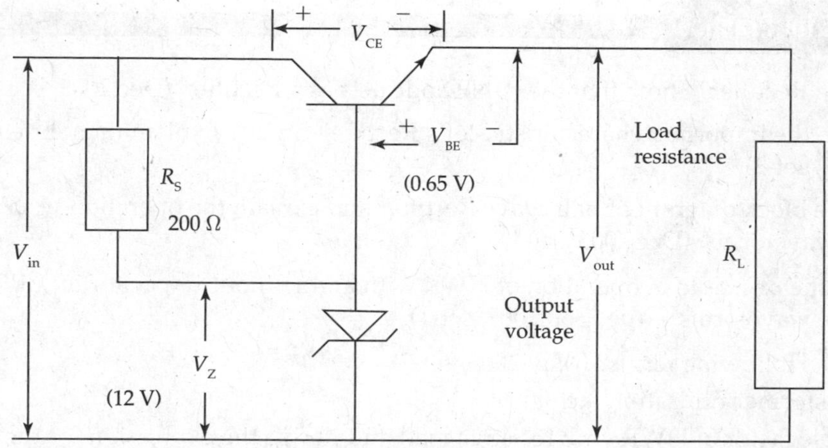

FIG. 3.65 *For example: 20*

Solution:

$$\text{Output voltage } V_{out} = V_o = V_Z - V_{BE}$$
$$= (12 - 0.65) = 11.35\text{ V}$$

$$\text{Using Kirchoff's law } V_{CE} = (V_i - V_o)$$
$$= (20 - 11.35) = 8.65\text{ V}$$

Voltage drop across 200 Ω resistor $R_S = V_{RS}$

$$V_{RS} = (V_i - V_Z) = (20 - 12) = 8\text{ V}$$

$$\text{Current through 200 Ω resistor } (R_S) = I_{RS} = \frac{V_{RS}}{R_S} = \frac{8\text{ V}}{200\ \Omega} = 0.04\text{ A}$$

SUMMARY

1. The concepts of HWR, Full-Wave Rectifier and Bridge Rectifier circuits that convert Alternating voltages from the power transformer are discussed. The output voltages of these rectifier circuits are unidirectional voltages and contain ripple content. Their output voltage would not represent pure DC voltages that required for Electronic circuits.
2. Filter circuits using simple capacitor filter to π-section filters and multiple section filters that use *R-L-C* elements for smoothening the rectified output voltages from rectifier circuits. Ripple content will be reduced.
3. Simple Zener Shunt regulator Circuit, Series Voltage Regulator circuit and Shunt Voltage Regulator circuits that produce stable and more uniform DC voltages (taking care of the variations in load currents and source voltages) required for practical circuits.

Questions for Practice

1. What is a Rectifier? Show that the P–N Diode acts as a Rectifier. (Dec 2003, set-1)
2. What are the important characteristics of a Rectifier circuit? Explain them briefly. (Aug/Sep 2007, set-2)
3. Draw the block diagram of half-wave Rectifier and explain the operation with the help of neat waveforms. (Dec 2003, set-1)
4. Explain the principle of operation of HWR with and without Capacitor input Filter and draw the waveforms. (Aug/Sep 2007, set-1)
5. Define the following terms? (May 2003, set-2)
   (a) Transformer utilisation factor
   (b) Ripple factor of HWR with resistive load and derive the expression for the same.
6. Define the following terms of a HWR with resistive load:
   (a) Ripple factor, (b) Peak Inverse voltage, (c) Efficiency. (May 2003, set-1; May 2004, set-4)
7. Derive the expression for ripple for π-section Filter used with HWR. (May 2004, set-3)
8. Show that for a HWR circuit $\eta = \dfrac{40.6\%}{\left[1+\dfrac{R_f}{R_L}\right]}$, where r_f is the forward resistance of the Diode and R_L is the load resistance. (Dec 2004, set-2)
9. Draw the circuit diagram for a Full-Wave Rectifier. Explain the operation of the circuit with relevant waveform. (May 2004, set-1)
10. Derive the expression for maximum efficiency of a Full-Wave Rectifier circuit? (Dec 2004, set-2)
11. Derive the expressions for ripple factor of HWR and Full-Wave Rectifier. (May 2004, set-2; Dec 2005 set-1)
12. Define percentage regulation and prove that the regulation of both Half-Wave and Full-Wave Rectifier is given by $\%\ \text{regulation} = \dfrac{R_F \times 100\%}{R_L}$. (May 2004, set-4)
13. Explain the circuit diagram of a Full-Wave Rectifier with Inductor Filter. (Aug/Sep 2006, set-2)
14. Derive the expression for the ripple factor for a Full-Wave Rectifier with Inductor Filter. (Dec 2004. set-1)
15. Discuss a Full-Wave Rectifier with π-Filter. (June 2005, set-3)
16. Draw and explain the circuit diagram of FWR with *L*-section Filter.
17. Draw circuit diagram of Full-Wave Rectifier with Inductor Filter. (May 2007, set-4)
18. What is the ripple factor if a power supply of 220 V, 50 Hz is to be Full-Wave Rectified and filtered with a 220 μF capacitor before delivering to a resistive load of 120 Ω?

Compute the value of the capacitor for the ripple factor to be less than 15%. (Nov 2010, EDC for ECE, EEE, CSE, EIE, IT, MCT, of JNTUH)

19. (a) Derive the expression for ripple factor of a Full-Wave Rectifier with and without a Capacitor filter? (Nov 2010, EDC for ECE, EEE, CSE, EIE, IT, of JNTUH)
    (b) Compute the average and RMS load currents, TUF of an unfiltered centre-tapped Full-Wave Rectifier specified below. Input voltage to transformer = 220 V 50 Hz. Step-down ratio of centre-tapped transformer = 4:1 (Primary to each section Secondary). Sum of transformer secondary winding in each secondary segment and Diode forward resistance = 100 Ω, Load resistance R_L = 220 Ω.

20. (a) Define Ripple factor and Form factor. Establish a relation between them?
    (b) Explain the necessity of Bleeder resistance in an *L*-Section filter and with a Full-wave Rectifier? (Nov 2010, EDC for ECE, EEE, CSE. EIE, IT, of JNTUH)
    (c) Compute ripple factor of an *L*-section Choke input filter used at the output of a Full-Wave Rectifier. Inductor and capacitor values of filter are given as 10 H and 8.2 μF respectively.

21. (a) List out the merits and demerits of Bridge type Full-Wave rectifiers over Centre-tapped secondary type Full-Wave Rectifier?
    (b) The Secondary voltages of a centre-tapped transformer are given as 60-0-60 V. The total resistance of secondary coil and resistance of each section of transformer secondary is 62 Ω. Compute the following for a load resistance of 1 kΩ. (a) Average Load current, (b) Percentage load regulation, (c) Rectification Efficiency, (d) Ripple factor for 240 V, 50 Hz supply to primary of transformer.
    (c) What is Bleeder resistance in *L*-Section filters (EDC-NOV 2010 JNTUH)

Multiple Choice Questions

1. The Silicon Diode used for rectification is ______________.
   (a) P–N Diode (b) tunnel Diode
   (c) Zener Diode (d) LED

2. In IC regulator circuits, the mostly used Filter is ______________.
   (a) *L* Filter (b) C-Filter
   (c) *LC* Filter (d) *RC* Filter

3. The Silicon diode popularly used for voltage regulator in power supplies is ______________.
   (a) P–N Diode (b) Zener Diode
   (c) tunnel Diode (d) Schottky Diode

4. The function of a series pass Transistor in a voltage regulator circuit is ______________.
   (a) to maintain the output voltage constant
   (b) to improve the voltage regulation
   (c) to enhance the power handling capability
   (d) to provide series feedback

5. As compared to a Full-Wave Rectifier using two Diodes, the four Diode Bridge Rectifier has the prominent advantage of ____________.
   (a) higher current capability
   (b) lower peak inverse voltage requirement
   (c) lower ripple factor
   (d) higher efficiency
6. Peak-to-Peak ripple voltage of a Full-Wave Rectifier with Capacitor Filter is ____________.
   (a) $\frac{I_{DC}}{2fC}$ (b) $\frac{V_m}{\pi}$ (c) $\frac{2V_m}{\pi}$ (d) None of these
7. VA rating of a transformer secondary winding of Full-Wave Bridge Rectifier is ____________.
   (a) 3.40 $V_L \cdot I_L$ (b) $V_I + V_R$
   (c) 1.23 $V_L \cdot I_L$ (d) none of these
8. Using Darlington pair as a series pass element ____________.
   (a) the output voltage can be kept constant
   (b) output current can be kept constant
   (c) temperature stability can be improved
   (d) voltage stability factor decreases
9. In a centre tap Full-Wave Rectifier, 30 V is the peak voltage between the centre tap and one of the ends of the secondary winding of the transformer. Peak Inverse voltage (PIV) across the Diode under reverse bias condition is ____________.
   (a) 60 V (b) 42 V (c) 50 V (d) 30 V
10. The simplest and most economical filter circuit is ____________.
    (a) *C*-input Filter (b) *L*-input Filter
    (c) LC-type *L*-input Filter (d) Cascaded LC Filter
11. In a Zener and Avalanche breakdown the current flows due to ____________.
    (a) majority carriers (b) minority carriers
    (c) both majority and minority carriers (d) none of these
12. The difference between series and shunt regulator is ____________.
    (a) position of control element
    (b) type of sampling network
    (c) type of error detector
    (d) amount of current to be handled

Answers to Multiple-Choice Questions

1. (a) 2. (b) 3. (b) 4. (a) 5. (b) 6. (a)
7. (c) 8. (c) 9. (a) 10. (a) 11. (c) 12. (a)

Chapter 4

CHARACTERISTICS OF TRANSISTOR DEVICES (BJT, FET AND MOSFET)

Learning Objectives

To get familiarity of structural details and fundamental concepts of

- NPN and PNP Transistors, its characteristics and *h*-parameters.
- The *h*-parameter analysis of Transistor amplifiers.
- Field Effect Transistors, its characteristics and amplifier concepts.
- Different types of MOSFET devices and their characteristics.
- Unijunction Transistor (UJT) characteristics.

4.1 INTRODUCTION

- The invention of Transistor (semiconductor electronic device) at (AT & T) Bell Telephone Laboratories at New Jersey, USA, in 1948, revolutionised the manufacture of electronic devices resulting in various applications of Electronics and Communication Engineering, computers, Internet, many ASIC (Application-Specific IC) applications, and satellite and wireless communications and systems – around the universe *20th century*.
- Transistors are the basic building blocks of all electronic circuits using discrete and nanotechnology. For example, a microprocessor installed in a laptop computer uses millions of Transistors using VLSI technology.
- Accelerated advances in electronics technologies, embedded and applied into every possible sector of Engineering and Science,

are helping humans to reach everyone, improve the lives and productivity on earth, and continue to extend our boundaries beyond every limit known to mankind.

4.1.1 Common Types of Transistors Used in Electronic Circuits

Transistors operate on very low power, weigh less and ensure long battery life within rugged mechanical assembly. Transistors are made of P- and N-type semiconductor materials. Family of Transistors is as follows:

1. **Bipolar Junction Transistor (BJT)**
   a. NPN Transistor
   b. PNP Transistor
2. **Junction Field Effect Transistor (JFET)**
   a. N-Channel JFET
   b. P-Channel JFET
3. **Metal Oxide Semiconductor Field Effect Transistor (MOSFET)**
   a. Enhancement MOSFET
   b. Depletion Enhancement MOSFET
4. **Unijunction Transistor (UJT)**

4.2 BIPOLAR JUNCTION TRANSISTOR (BJT): STRUCTURE OF MATERIALS

Transistor is a semiconductor device and has three semiconductor layers or regions, with connecting terminals to external circuits:

- *Emitter* (*E*) *layer*, connecting as an *Emitter*
- *Base* (*B*) *layer* (centre region of the Transistor), connecting as a *Base*
- *Collector* (*C*) *layer*, connecting as a *Collector*

4.2.1 Transistor Symbol and Terminology

A Transistor symbol is made of three terminals – vertical line is *Base*, and two angular lines are *Collector* and *Emitter.* Transistor type (PNP or NPN) is known by the flow direction of Emitter Current on the Emitter terminal on the Transistor symbol. If the arrowhead points 'in', it is a PNP Transistor (Fig. 4.1), and if it points 'out', it is an NPN Transistor (Fig. 4.2). Identification of Transistor terminals and its electrical data is available on Transistor manufacturers' data sheets.

4.2.2 NPN Transistor and Structure of Semiconductor Material

Transistor operates with *currents* caused as a result of the generation and control of movement of the 'two charge carriers' (*Holes* and *Electrons*) through its semiconductor material. Functionality of the Transistor depends further on biasing voltages at 'two P–N junctions J_1 and J_2'. So the Transistor is known as *Bipolar Junction Transistor* (BJT).

NPN Transistors are more popular because of their ease of manufacture, resulting in high availability. Furthermore, electron that flows through the devices favours high-frequency applications of NPN Transistors, suitable to present day technologies.

Figures 4.1 and 4.2 show the Transistors with three layers and two P–N junctions J_1 and J_2.

- *P–N junction J_1*, between Base and Emitter regions, is known as the *input junction* or the *Emitter junction.*
- *P–N junction J_2*, between Base and the Collector regions, is known as *output junction* or the *Collector junction.*
- In order for the Transistor to work as an 'amplifying device', DC Voltages at the input/Emitter junction has to be 'forward-biased' and the output/Collector junction 'reverse-biased'.

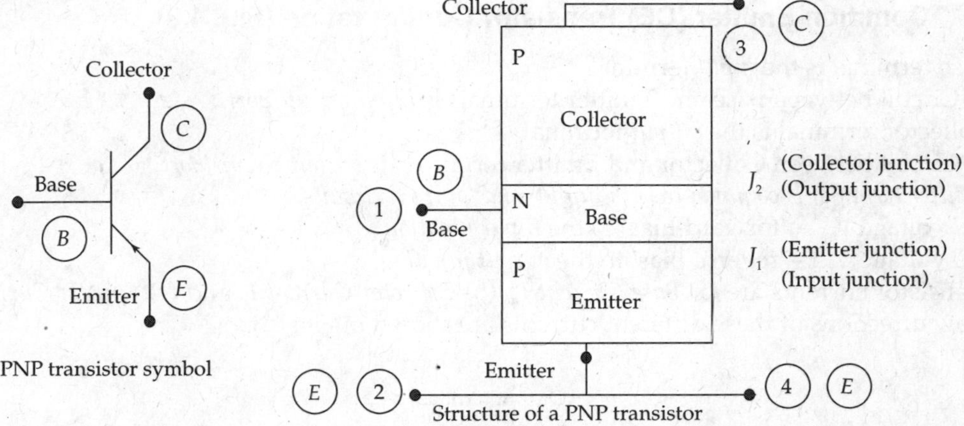

FIG. 4.1 *PNP transistor symbol and materials representation of PNP transistor*

Transistor is operated as a four-terminal circuit in various applications such as:

- Amplification of electrical signals – in audio/video amplifier, radio, television, radar, mobile phone, satellite communication, oscillator and so on.
- Electronic switching in digital circuits – microprocessor chips using billions of Transistors using VLSI technology to create computers (desktops, large servers, laptops), smart phones and so on.

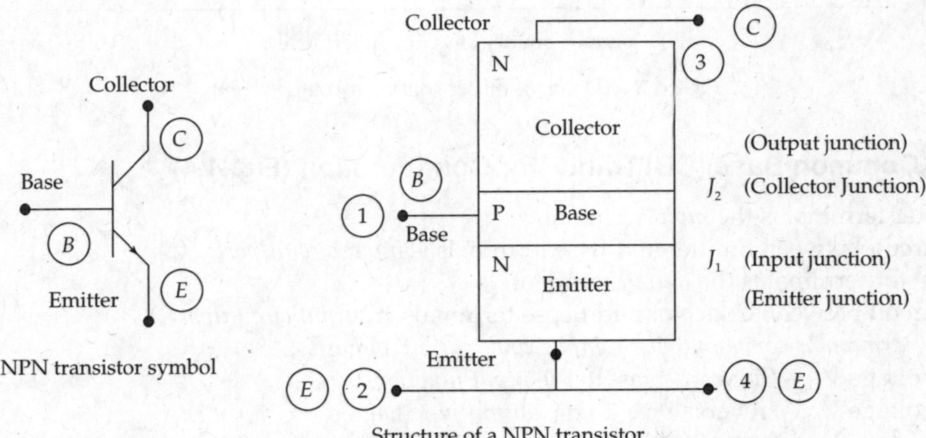

FIG. 4.2 *NPN transistor symbol and structure of semiconductor materials*

4.3 DIFFERENT CONFIGURATIONS OF BIPOLAR JUNCTION TRANSISTOR

Transistor functions depending on DC biasing voltages at the two PN junctions and one of the following three types of configuration:

1. Common Emitter Transistor (*CE* Transistor operation)
2. Common Base Transistor (*CB* Transistor operation)
3. Common Collector Transistor (*CC* Transistor operation)

4.3.1 Common Emitter (CE) Transistor Configuration (Fig. 4.3)

- Base terminal is the *input* terminal.
  - o Circuit between Base and Emitter terminals is *input circuit/port*.
- Collector terminal is the *output* terminal.
  - o Circuit between Collector and Emitter terminals is *output circuit/port*.
- *Emitter terminal is common to both 'input' and 'output' circuits.*
- DC voltage V_{BE} – 'forward-bias' to the *input junction*.
- DC voltage V_{CE} – 'reverse-bias' to the *output junction*.
- Transistor currents are (a) *Base Current* I_B, (b) *Collector Current* I_C and (c) *Emitter Current* I_E. Flow directions of these different currents are shown in Fig. 4.3.

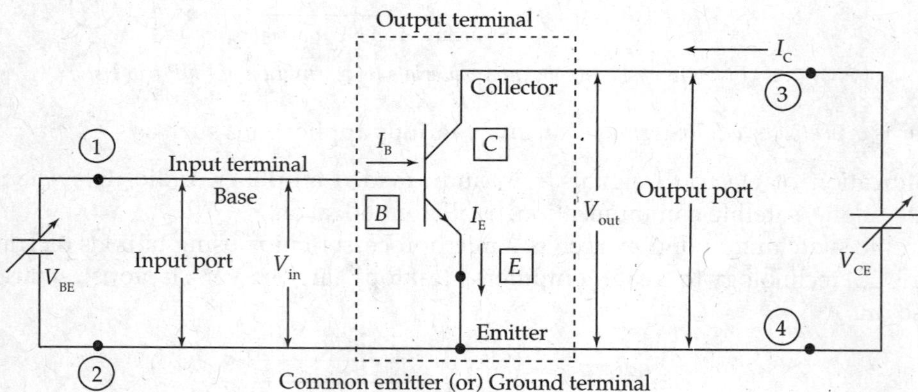

FIG. 4.3 *Common emitter transistor configuration*

4.3.2 Common Base (CB) Transistor Configuration (Fig. 4.4)

- Emitter terminal is the *input* terminal.
  - o Circuit between Emitter and Base terminals is *input circuit/port*.
- Collector terminal is the *output* terminal.
  - o Circuit between Collector and Bbase terminals is *output circuit/port*.
- *'Base' terminal is common to both 'input' and 'output' circuits.*
- DC voltage V_{EB} – 'forward-bias' to the *input junction*.
- DC voltage V_{CB} – 'reverse-bias' to the *output junction*.
- Transistor currents are (a) *Base Current* I_B, (b) *Collector Current* I_C and (c) *Emitter Current* I_E. Flow directions of these currents are shown in Fig. 4.4.

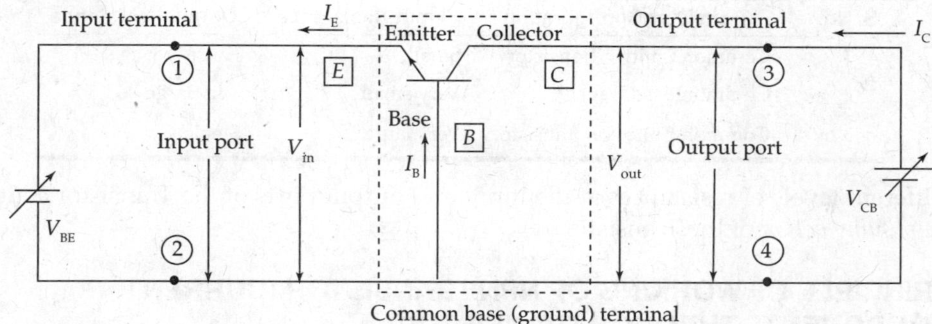

FIG. 4.4 *Common base connected (operated) transistor model*

4.3.3 Common Collector (CC) Transistor Configuration (Fig. 4.5)

- Base terminal is the *input* terminal.
  - Circuit between Base and Collector terminals is *input circuit/port.*
- Emitter terminal is the *output* terminal.
  - Circuit between Emitter and Base terminals is *output circuit/port.*
- *'Collector' terminal is common to both 'input' and 'output' circuits.*
- DC voltage V_{BC} – 'forward-bias' to the *input junction.*
- DC voltage V_{EC} – 'reverse-bias' to the *output junction.*
- Transistor currents are (a) *Base Current* I_B, (b) *Collector Current* I_C and (c) *Emitter Current* I_E. Flow directions of these currents are shown in Fig. 4.5.

Transistor is used as a *two-port* or *four-terminal network,* in the three types of configuration and its operation is described above. The name for the Transistor device is an *acronym* using the two words Transfer and resistor (Transfer + resistor = Transistor).

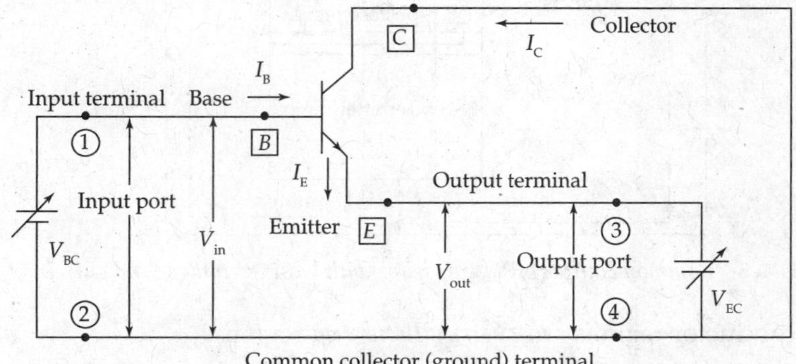

FIG. 4.5 *Common collector transistor configuration*

Different levels of resistances at the input and output ports of a Transistor are obtained for amplifier action with the following biasing voltages:

- *Input junction J_1 of the Transistor is 'forward-biased'.*
- *Output junction J_2 of the Transistor is 'reverse-biased'.*

S. No.	Transistor Configuration	Input Resistance	Output Resistance
1	Common Emitter Transistor	Small	Large
2	Common Base Transistor	Very small	Very large
3	Common Collector Transistor	Very large	Small

These different levels of resistances at the input and output ports of the Transistor contribute to the *amplifying action* of the Transistor.

4.4 PRINCIPLE OF WORKING OF NPN TRANSISTOR (CURRENT COMPONENTS THROUGH TRANSISTOR)

Structure and biasing voltages for a Common Emitter NPN Transistor (Fig. 4.6):

- *Emitter region* is made up of N-type semiconductor material (heavily doped) to provide sufficiently large Emitter Current.
- *Base region* is a very thin P-type semiconductor material and least doped with a smaller area of cross-section.
- *Collector region* is again an N-type semiconductor material (moderately doped). The Collector region is larger in area than Emitter and Base regions.

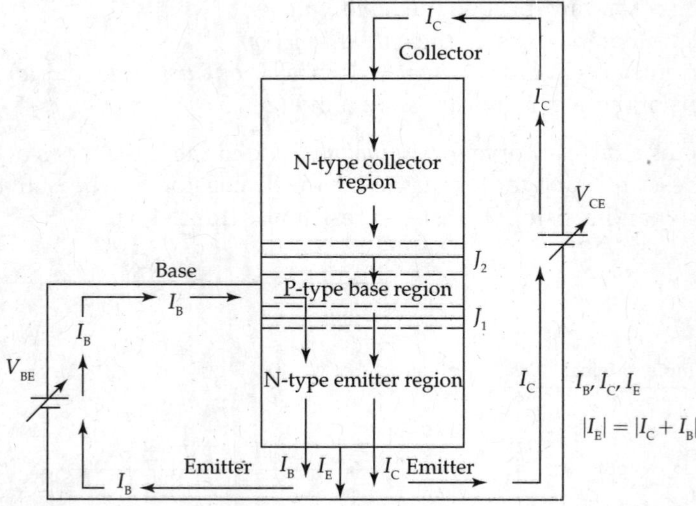

FIG. 4.6 *Common emitter NPN transistor with biasing voltages and currents I_B, I_C, I_E*

The three semiconductor regions are *electrically neutral* regions.

- V_{BE} – *forward-bias* voltage to the *input* (Emitter) junction
- V_{CE} – *reverse-bias* voltage to the *output* (Collector) junction
- *Depletion regions* – appear across the two junctions (*input* and *output*, *Emitter* and *Collector*) proportional to biasing voltages
- *Contact or built-in potential 'V_0'* – exists across the two junctions.
- Once the biasing voltages are applied, the conventional currents flow from positive terminal to the negative terminal of the voltage Sources, contributing to the three currents – (a) *Base*

Current 'I_B', (b) *Collector Current 'I_C'* and (c) *Emitter Current 'I_E'*. Flow of these currents is explained in detail in the following sections.

In order to understand the operation of NPN Transistor, consider two diodes (NP and PN) connected by a back-to-back virtual connection (Fig. 4.7).

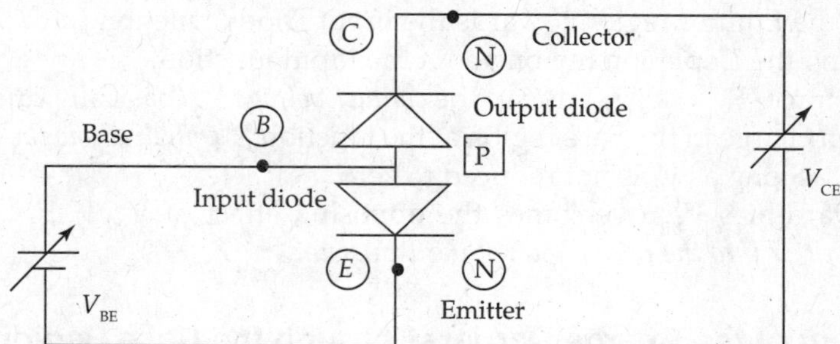

FIG. 4.7 *Representation of two back to back connected PN diodes for NPN transistor*

Input junction between Emitter and Base regions (NP diode) of the Transistor is forward-biased by DC voltage V_{BE}. *Output junction* between Base and Collector regions (PN diode) of the Transistor is reverse-biased by DC voltage V_{CE}.

4.4.1 Movement of Majority Carriers from the Emitter into the Base Regions in the Transistor

- Figure 4.8 shows an NPN Transistor in CE operation with bias voltages.

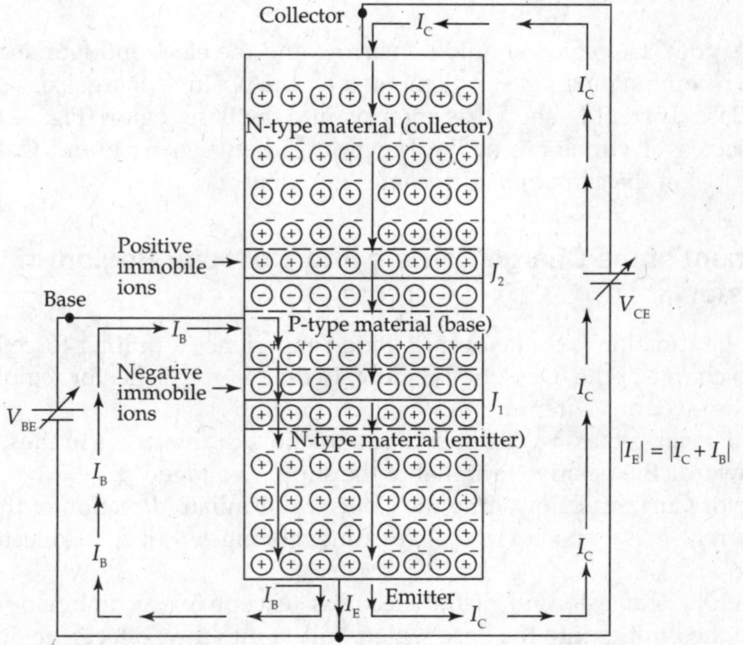

FIG. 4.8 *Common emitter NPN transistor with biasing voltages and currents I_B, I_C, I_E (movement of charges)*

- Depletion regions appear across the two junctions proportional to biasing voltages.
- *Contact Voltage* (or *built-in potential*) 'V_0' exists across the two junctions.
- When the forward-bias V_{BE} is increased, two forces (F_{VBE} and F_{V0}) act on the majority carrier electrons in the N-type Emitter of the NPN Transistor.
- Force '$F_{VBE} = e \cdot V_{BE}$' (*e* is the charge of the electron) due to the forward-bias V_{BE}, *pushes the electrons* in the Emitter region towards the input Diode junction J_1 (of the input diode), thereby reducing the Depletion region about the input junction.
    - Restraining force '$F_{V0} = e \cdot V_0$', due to the *contact voltage* V_0 (basically due to the negative immobile ion charge in the Base region at the junction J_1) *will disallow the electrons* into the Base region, as long as V_0 is not reduced to zero.
- Once the forward-bias V_{BE} overcomes the opposing effect of V_0, $V_{BE} > V_0$, *Electrons are injected from Emitter into the Base region of the Transistor.*

4.4.2 Movement of the Charge Carriers Through the Base Region in the Transistor

- *Injected electrons* from the Emitter into the Base region are known as *minority carriers*, based on the nomenclature of the P-type Base region.
- There will be *charge gradient* among the injected electrons and the small quantity of originally existing electrons in the P-type Base region.
- So the movement of the newly *injected electrons* from the Emitter junction to the other end of the Base region is by *diffusion process*.
- During this course of journey, some electrons *recombine* with the *Holes* in the Base material. These recombinations are less in number, as the available Holes for recombination are also less. This is because of the smaller area of the cross-section of the Base material and light *doping*.
- Hence, the P-type Base region should be narrow and the electron lifetime should be long. Holes lost in recombination are supplemented by the positive charges of V_{BE}, contributing to very low Base Current I_B, shown as entering into the Base region (Figs. 4.6 and 4.8).
- Once the injected electrons are at the Junction J_2, between Base and Collector regions, reverse-bias 'V_{CE}' at the *output junction* comes into play.

4.4.3 Movement of the Charge Carriers into 'Collector Region' of the Transistor

- Electrons at the Junction J_2 of the output Diode experience a pulling force by the positive immobile ion charge at the 'Depletion region', in the N-type Collector region.
- As a result, the electrons enter into the 'Collector region'.
- Because of the reverse-bias 'V_{CE}' at the output junction, the electrons in the Collector region are pulled towards the positive terminal of the supply voltage V_{CE}.
- So the Collector Current I_C flows into the Collector terminal (direction of the conventional Collector Current I_C is opposite to the electrons moving out of the Collector terminal) as shown in Fig. 4.6.
- For a good NPN Transistor, all of the electrons (except for recombination in the Base), injected from the Emitter into the Base region, collect into the Collector region contributing to Collector Current 'I_C' (nearly equal to 'I_E').

- As shown in Fig. 4.6, both Base Current 'I_B' and Collector Current 'I_C' flow out of the Emitter terminal, adding up to form the Emitter Current 'I_E'. The flow of the Emitter Current is outside the Transistor. Therefore,

$$\therefore \quad -I_E = I_B + I_C.$$

$$\text{Mathematically,} \quad |I_E| = |I_B + I_C|.$$

- Common Emitter Transistor provides *large current gain*, as evidenced by large Collector Current (in the output circuit) 'I_C' and small Base Current 'I_B'.
- The *Current Gain* is the ratio of output Collector Current 'I_C' to input Base Current 'I_B', defined as 'Beta' (β) for the CE Transistor (usually $\gg 1$):

$$\text{Forward current gain of common emitter transistor } \beta = \frac{I_C}{I_B}.$$

4.5 WORKING OF NPN TRANSISTOR AND TRANSISTOR CURRENTS

A sample CE Transistor circuit is shown in Fig. 4.9, with two DC biasing voltage Sources (two Transistor power supplies), three DC ammeters to measure variations of Collector Current, Base Current and Emitter Current.

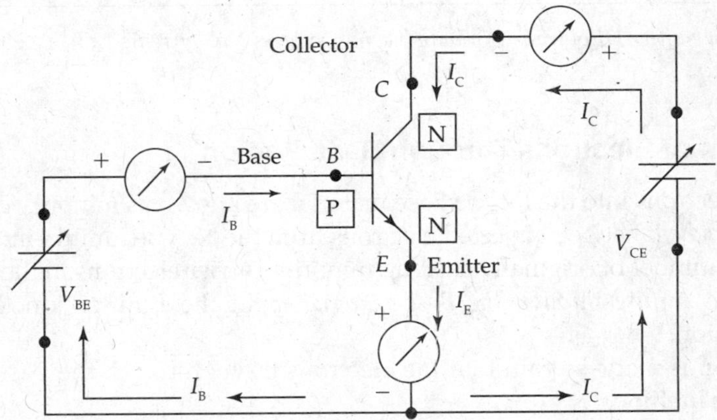

FIG. 4.9 *Common emitter transistor circuit*

4.5.1 Injection of Electrons from the Emitter Region into the Base Region

- Forward-bias V_{BE} to the junction J_1 reduces the *Depletion region width* W_{EB} about J_1. V_{BE} causes the movement of majority carrier electrons from the Emitter into the Base region, when it overcomes the junction (barrier) voltage.
- Emitter region is heavily doped and so *large numbers of electrons are injected* from the Emitter into the Base region. Hence, the Emitter is considered as the *Source of Electrons*'.
- At the same time, Holes move from the Base into the Emitter region. The Numbers of Holes that move from the Base into the Emitter are deliberately made less, using small area of cross-section of the Base with *light doping*. Hence, the Hole current into Emitter is practically zero.
- So the Emitter Current I_E is contributed mainly by the injected electrons from the Emitter into the Base region, which move further into the Collector region (Fig. 4.10).

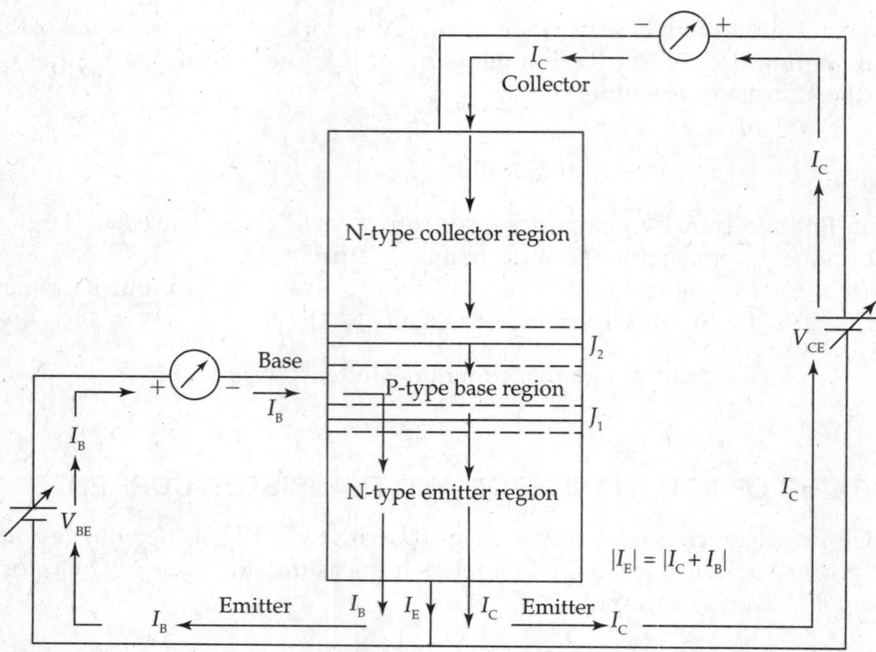

FIG. 4.10 *Common emitter NPN transistor with biasing voltages and currents I_B, I_C, I_E. For understanding the principles of working*

4.5.2 Diffusion of Electrons Through Base Region

- The injected electrons into the P-type Base material are known as minority *carriers*. There is a *concentration gradient* between injected electrons from the N-type Emitter into the Base region and the small number of originally existing minority carrier electrons in the P-type base.
- So the *electrons diffuse through the Base material* from the Emitter junction towards the Collector junction.
- The Base region is made so thin that the electrons travel through the Base material in less time than their lifetime.
- During this diffusion process, some electrons recombine with the Holes. The numbers of recombination are deliberately made less by reducing the width of the Base region and its light-doping concentrations so that the probability of recombination is less. The thickness of the Base region is made less than the diffusion length of the electrons.
- The lost Holes due to recombination are supplemented by the positive of V_{BE} and enter through the Base terminal contributing to the Base Current I_B due to the Holes entering the Base region.

4.5.3 Collection of Electrons into the Collector Region

- The remaining electrons enter the Collector region. Electrons are collected into the Collector region due to the positive nature of the reverse-bias voltage V_{CE}.
- The Collector region is much larger than the Emitter region to reduce the junction heating due to the power dissipation at the Collector junction.

- The electron that flows through the Collector circuit contributes to the Collector Current I_C.
- Power developed at the output junction is the product of V_{CE} and I_C.
- The Base Current will be of small magnitude when compared with Collector Current. Therefore, Emitter Current is approximately equal to Collector Current.

4.5.4 Emitter Current

- Emitter Current I_E is the sum of Base Current I_B and Collector Current I_C, because the two currents flow in the same direction through Emitter terminal.
- Reverse leakage current I_{CE0} or I_{C0}, which exists due to the reverse-bias to the output junction J_2, also adds to the Emitter Current. Therefore,

$$I_E = I_B + I_C + I_{CE0}.$$

4.5.5 Components of Current Through Common Base Transistor

Figure 4.11 shows various components of current through the Transistor device and external paths. Depletion region widths illustrate *Base width modulation.*

Transistor operates in the active region (due to fixation of quiescent operating point, discussed in Chapter 5, Methods of Biasing Transistor) when the Base–Emitter junction J_1 is forward-biased and the Base–Collector junction J_2 is reverse-biased, irrespective of the 'Transistor configuration'.

Emitter Current I_E is the sum of the majority carrier electron current and the minority carrier Hole current in the NPN Transistor (shown in Fig. 4.11).

$$\therefore \quad I_E = I_{NE} + I_{PE}.$$

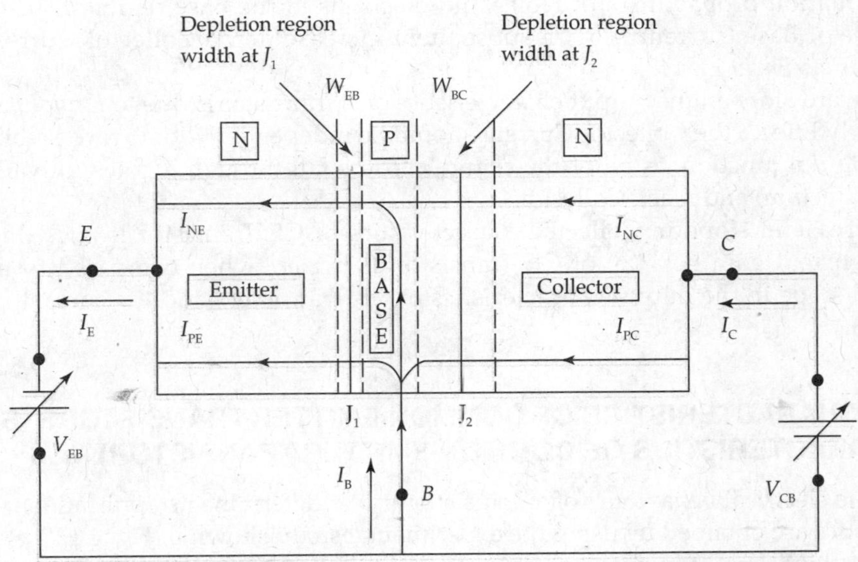

FIG. 4.11 *Various current components and depletion region widths to illustrate base width in a NPN transistor*

Except for few electrons that recombine with Holes in the Base region, all of the electrons that cross the Emitter–Base junction reach the Collector. So the Collector Current component is I_{NC}. There is a small minority carrier current I_{CO}, due to reverse-bias of the Base–Collector junction. Net Collector Current will be $I_C = I_{CO} - I_{NC}$), as long as Emitter Current I_E is not equal to zero.

When $I_E = 0$, the Collector Current $I_C = I_{CO}$, since I_{NC} (injected carrier current) is zero. Emitter injection efficiency as per definition

$$\gamma = \frac{\text{Current of injected carriers at the emitter base junction}}{\text{Total emitter current}} = \frac{I_{NE}}{I_{NE} + I_{PE}} \cong \frac{I_{NE}}{I_E}.$$

4.6 BASE WIDTH MODULATION AND EARLY EFFECT

- In Fig. 4.11, the (NPN) Transistor operates as an amplifying device, as the bias voltage V_{EB} forward biases the input junction J_1 (of the NPN Transistor) and the bias voltage V_{CB} reverse biases at the output junction. The quiescent operating point is in the active region (explained in Transistor Output Characteristics, Chapter 5).
- The reverse-biased 'Base to Collector junction' produces *Depletion regions* into the Base and Collector regions, with widths inversely proportional to the doping levels. Thus the spread of the *transition regions* (or *space charge regions*) into the Base region is much more into the Collector region, which becomes relatively negligible. In addition, almost all the Depletion regions are spread out in the Base region only. This reduces the effective width of the *Base region* proportional to the applied reverse potential or bias voltage across 'Base–Collector junction' J_2. This type of change in the effective Base width, based on variations of reverse-biasing voltages at the output junction of the Transistor, is known as *Base width modulation*.
- Effective Base width reduces with the increase in reverse-bias, reducing the recombination probabilities of Holes and electrons in the Base region. It results in the decrease of Base Current I_B, with subsequent increase in the Collector Current I_C. This is known as *Early Effect*.
- The upward slope in the output characteristics of a Transistor is due to *Early Effect*. But for this 'Early Effect', the Collector Current should be independent of the reverse-bias applied to the output junction, as all of the charge carriers fall through a potential valley (at the output junction) and reach Collector.
- The increase in slope in Collector characteristics of CE Transistor is *more* due to large (>>1) current gain beta (β) of CE Transistor. It is *more* when compared to the *smaller* upward slope in the output characteristics of CB Transistor, since its current gain alpha (α) (<1).

4.7 *V–I* CHARACTERISTICS OF COMMON EMITTER TRANSISTORS (STATIC CHARACTERISTICS OF COMMON EMITTER TRANSISTOR)

To know the electrical behaviours of a Transistor and to determine its applications, Transistor characteristics are obtained by using the experimental setup shown in Fig. 4.12.

V–I characteristics at input and output ports of the Transistor are known as 'static characteristics of the Transistor'.

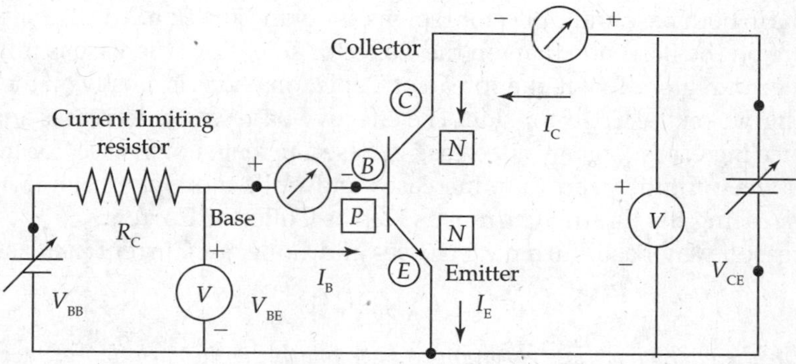

FIG. 4.12 *Common emitter transistor circuit to obtain input and output characteristics*

4.7.1 Input Characteristics of a Common Emitter Transistor

- *Input characteristics* of a CE Transistor are between V_{BE} and I_B (Voltage and currents of CE Transistor at its input port) for constant values of V_{CE}.
- Initially, maintain the Collector supply voltage V_{CE} at the output port of the Transistor at 5 V. Varying V_{BE} in increments, the magnitudes of input voltage V_{BE} and current I_B are measured for various values of V_{BE}.
- Now V_{CE} is changed to say 10 V. Varying V_{BE} in convenient increments, the magnitudes of V_{BE} and I_B are measured and tabulated.

V_{CE} = constant voltage (for example, V_{CE} = 5 V)		
S. No.	**Input voltage, V_{BE} (V)**	**Base Current, I_B (mA)**

- Using the observed data, the 'input characteristics' of the Transistor can be drawn. Typical characteristics are shown in Fig. 4.13
- Input characteristics of a Transistor are nothing but the forward characteristics of the input junction PN diode.
- Up to and below V_γ (cut in voltage) of the input junction (the Emitter junction), the Base Current I_B is virtually zero. From V_γ onwards, Emitter junction is progressively forward-biased and the input current starts increasing slowly first and rapidly due to the exponential relation between voltage and current (forward-biased semiconductor Diode current equation). Input characteristics are similar to the forward characteristics of a P–N diode.
- For the higher V_{CE} (10 V), the reverse-bias on the output junction (Collector junction) increases. The increase in reverse potential increases the Depletion

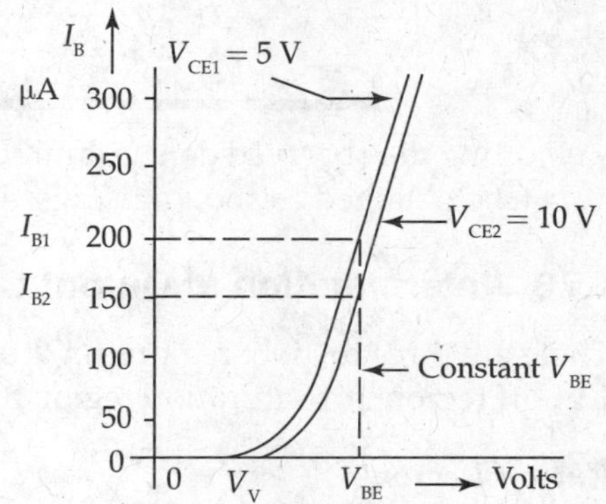

FIG. 4.13 *Input characteristics of common emitter transistor*

region widths in both Base and Collector regions about the junction. The increase in Depletion width is more on the Base side than on the Collector side, since the Base is much less doped than the Collector region. To wit, the spread of Depletion width is mostly on the Base side and literally negligible on the Collector side. The effective Base width decreases and the effective Base resistance increases causing a decrease in Base Current shown as I_{B2} for increased V_{CE2}.

The field between Emitter and Base increases and pulls more electrons from the Emitter region, thus increasing the Emitter Current as well as Collector Current.

That is the reason why Base Current decreases and Collector Current increases:

$$I_C(\uparrow) = I_E(\uparrow) - I_B(\downarrow).$$

As observed in the characteristics, variation of Base width due to variation of reverse-bias on the Base–Collector junction is called base-width modulation or Early Effect. The input characteristics show that for higher V_{CE}, the curve is pushed to the right or for a given V_{BE}; I_{B1} for V_{CE1} is larger than I_{B2} for V_{CE2}, where V_{CE2} is greater than V_{CE1}.

4.7.2 Output Characteristics of a Common Emitter Transistor

- Output characteristics of a CE Transistor are between V_{CE} and I_C (Voltage and currents of CE Transistor at its output port) for constant values of I_B.
- Using the experimental setup shown in Fig. 4.12, the Transistor output characteristics are obtained according to the following procedure:
- Initially, with $I_B = 0\ \mu$A (zero V_{BE}), varying V_{CE} in convenient increments, values of V_{CE} and the resulting Collector Current I_C magnitudes are observed. Now the Collector Current will be simply equal to I_{CEO} or I_{CO}, the reverse saturation current, because the forward-bias V_{BE} to Emitter junction is zero.
- V_{BE} is adjusted to choose a convenient value of I_B say I_{B1} and is kept constant. V_{CE} is varied in convenient steps, and the pairs of values of V_{CE} and I_C are observed and tabulated.
- For different sets of constant values of I_B, say I_{B2}, I_{B3} and I_{B4}, variations in I_C for varying magnitudes of V_{CE} are observed and tabulated.

I_B = Constant current (for example, 25 mA)		
S. No.	**Voltage, V_{CE} (V)**	**Collector Current, I_C (mA)**

- Plotting the observed data, a family of curves known as the Transistor output characteristics can be obtained as shown in Figs. 4.14 and 4.15.

4.7.3 Interpretation of Output Characteristics of Common Emitter Transistor

There are three distinct regions on the output characteristics (Figs. 4.14 and 4.15). They are (a) Cut-off region, (b) Saturation region and (c) Active region.

Cut-off Region

- The region between output characteristic with $I_B = 0\ \mu$A and V_{CE} *x*-axis is known as *cut-off region* on the graph of Output characteristics of Transistor.

- Collector Current I_C is almost zero in the cut-off region and equal to I_{CEO} (reverse saturation current between Emitter and Collector with Base open circuited). This reverse saturation current I_{CEO} or I_{CO} adds to Emitter Current.
- Transistor acts as an *open switch* when biasing voltages are fixed in the cut-off region.

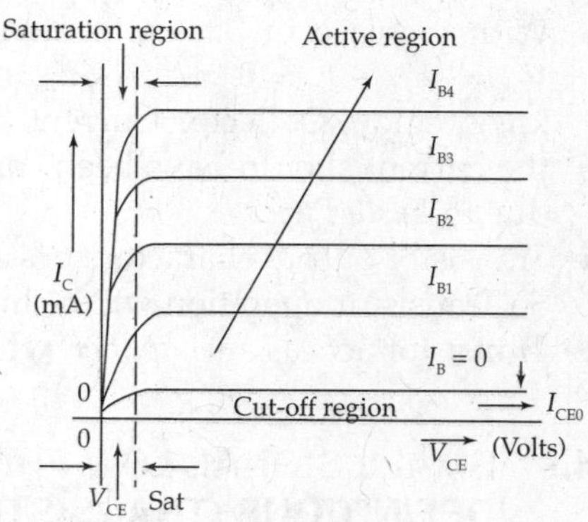

FIG. 4.14 *Common emitter transistor output characteristics*

Saturation Region

- The region between a line joining the entire knee points (on the graph of Output characteristics) and Collector Current 'I_C' *y*-axis is known as *saturation region.*
- Collector Current varies largely over a 'small region' wherein V_{CE} changes very less. V_{CE} is of the order of few tenths of a volt. It is called as $V_{CE(sat)}$. It is of about 0.3 V for a Germanium Transistor and 0.7 V for a Silicon Transistor.
- In this region, the Transistor acts as a *closed switch,* with negligible voltage drop across it. The resistance $r_{CE(sat)}$ in this region is only about a few ohms.
- Transistor acts as a *closed switch,* when biasing voltages are fixed in this region.

Active Region

- Other than cut off and saturation regions on the graph of Output characteristics, remaining region is known as *active region.*

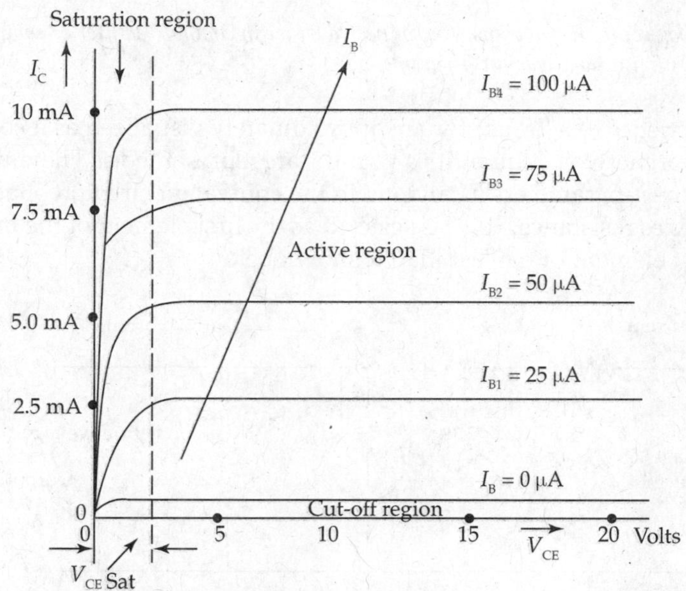

FIG. 4.15 *Typical output characteristics of a CE transistor*

- With increase in Base Current I_B, output characteristics shift up. For a given I_B with $V_{CE} = 0$ V – $I_C = 0$ mA. As V_{CE} increases, I_C increases up to knee points and beyond the knee region, Collector Current (I_C) increases with V_{CE} with a small upward rise. In fact, the current should have been constant but for the variation or modulation of Base width due to *Early Effect*.
- Transistor output characteristics are equally spaced for equal increments of Base Current I_B. So Transistor operation will be linear in the *active region*.
- Transistor acts as an *amplifier*, when biasing voltages are fixed in the active region

4.8 SMALL SIGNAL LOW-FREQUENCY TRANSISTOR PARAMETER DEFINITIONS (TRANSISTOR *h*-PARAMETERS)

An amplifier can be considered as a two-port (four-terminal) network, with conveniently defined parameters, which find practical use. For the active device, the Transistor, the hybrid parameters are useful. Transistor is basically a current amplifier, but natural Sources are voltage type in nature. So the following expressions are used to represent the input and output voltages and currents. Consider the Transistor as shown in Fig. 4.16.

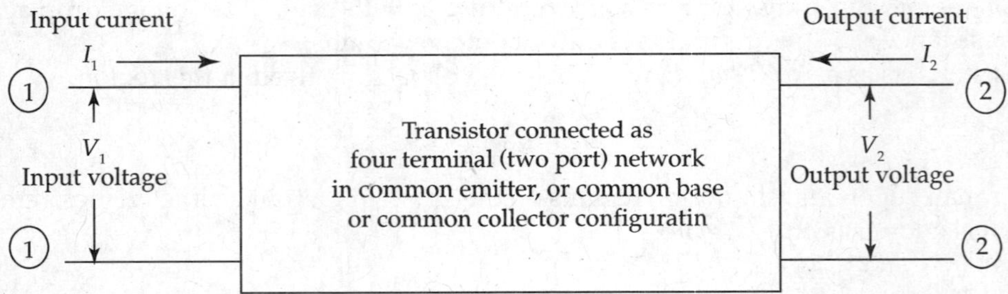

FIG. 4.16 *Block box containing a transistor connected either in common emitter or common base or common collector configuration to define h-parameters*

The input characteristics of a Transistor are approximately visualised as a constant voltage curves. So the input port network's quantities V_1 and I_1 are represented as Thevinin's equivalent network along with the *h*-parameters h_{11} and h_{12}. In the equivalent circuit, voltage Source V_1 is represented with a series resistance. So V_1 is selected as the first element of the first matrix, and I_1 is shown as the first element of the last matrix shown below.

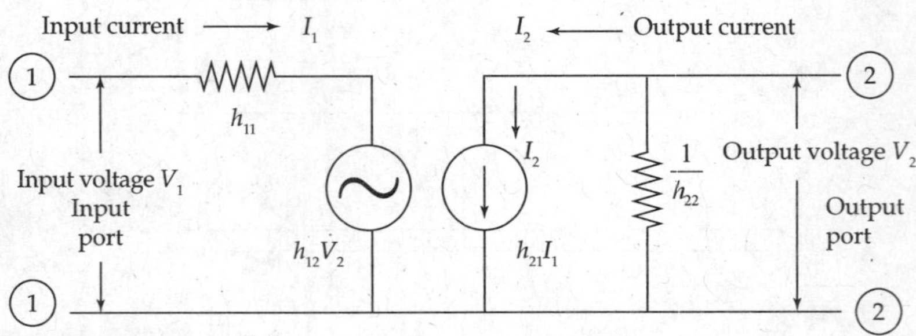

FIG. 4.17 *h-parameter network representation of a transistor*

Similarly, the output characteristics of a Transistor are approximately visualised as constant current lines. So the output port network's quantities I_2 and V_2 are represented as Norton's equivalent network along with the h-parameters h_{21} and h_{22}. In the equivalent circuit, the current Source I_2 is represented with a shunt resistance. So I_2 is selected as the second element of the first matrix, and V_2 is shown as the second element of the last matrix.

The network representation of the active device Transistor shown in Fig. 4.17 is represented by Eqs. (4.1) and (4.2):

$$V_1 = h_{11} \cdot I_1 + h_{12} \cdot V_2 \tag{4.1}$$

$$I_2 = h_{21} \cdot I_1 + h_{22} \cdot V_2. \tag{4.2}$$

Putting the equations in the matrix form, we get

$$\begin{bmatrix} V_1 \\ I_2 \end{bmatrix} = \begin{bmatrix} h_{11} & h_{12} \\ h_{21} & h_{22} \end{bmatrix} \begin{bmatrix} I_1 \\ V_2 \end{bmatrix},$$

where h_{11}, h_{12}, h_{21} and h_{22} are h-parameters that are known as 'hybrid parameters'.

Using some boundary conditions, h-parameters are obtained from Eqs. (4.1) and (4.2).

- In Eqs. (4.1) and (4.2), if V_2 is made zero, i.e. output port is short-circuited, two h-parameters 'h_{11}' and 'h_{21}' can be determined as explained below.

1. Input resistance h_{11}...(h_i)

$$h_{11} = \frac{V_1}{I_1}\bigg|_{V_2 = 0}, \tag{4.3}$$

where h_{11} is a ratio of voltage 'V_1' to current 'I_1' which is the ratio of voltage to current. So it has the dimensions of a resistance. Both the voltage 'V_1' and the current 'I_1' are the parameters of the input port. So h_{11} is termed *input resistance* 'h_i' represented by $h_i = V_1/I_1$. The unit of 'h_i' is ohms.

2. Forward current gain h_{21}...(h_f)

In the same way,

$$\text{Forward current gain} \quad h_{21} = \frac{I_2}{I_1}\bigg|_{V_2 = 0}, \tag{4.4}$$

where h_{21} is a dimensionless quantity representing the ratio of output current 'I_2' to the input current 'I_1' called as *forward current gain*. This is represented by $h_f = I_2/I_1$. 'h_f' is a ratio of two currents. So it is a dimension-less quantity.

- In Eqs. (4.1) and (4.2), if $I_1 = 0$ or open circuiting the input port, two more parameters 'h_{12}' and 'h_{22}' can be determined as explained below.

3. Reverse voltage transfer ratio h_{12}...(h_r)

$$\text{Reverse voltage transfer ratio} \quad h_{12} = \frac{V_1}{V_2}\bigg|_{I_1 = 0}, \tag{4.5}$$

where h_{12} is a *reverse voltage transfer ratio* and is denoted by 'h_r'. It is a ratio of two voltages. So it is a *dimensionless quantity* and gives an idea as to the extent of the unwanted voltage transfer from the output to the input port, since the signal amplifiers should be preferably unilateral in transfer of energy from the input port into the output ports, but not the other way round.

4. Output conductance $h_{22}\ldots(h_o)$

$$\text{Finally, output conductance } h_{22} = \frac{I_2}{V_2}\bigg| I_1 = 0, \tag{4.6}$$

where h_{22} represents the ratio of the output current 'I_2' to the output voltage 'V_2'. It is designated as *output conductance*, 'h_o'. h_o is measured in mhos or Siemens.

Thus, as the *h-parameters possess a mixture of units*, as defined by Eqs. (4.3)–(4.6), they are known as *hybrid parameters*.

- When applied to analysis of amplifiers with alternating signals, $V_2 = 0$ and $I_1 = 0$ represent constant DC values of voltage at the output port and current at the input port.
- As the Transistor characteristics are not entirely linear, the values of the *h*-parameters change from point to point and are defined over small linearised regions and hence are called *small signal parameters*.

In this context input resistance $$h_i = \left|\frac{\Delta V_i}{\Delta I_i}\right| V_2 = \text{constant } (\Omega) \tag{4.7}$$

Forward current gain $$h_f = \left|\frac{\Delta I_0}{\Delta I_i}\right| V_2 = \text{constant (no units)} \tag{4.8}$$

Reverse voltage transfer ratio $$h_r = \left|\frac{\Delta V_i}{\Delta V_0}\right| I_1 = \text{constant (no units)} \tag{4.9}$$

Output conductance $$h_o = \left|\frac{\Delta I_0}{\Delta V_0}\right| I_1 = \text{constant (mhos).} \tag{4.10}$$

Transistor amplifiers are operated either in CE or in CB or in CC Transistor operations as explained in Section 4.2.

Three sets of h-parameters are obtained with the second subscript to the h-parameters designating the grounded or the common terminal in Transistor amplifier operations.

Names of hybrid parameters for the three types of Transistor configurations

h-parameter	CE Transistor	CB Transistor	CC Transistor
Input resistance (ohms)	h_{ie}	h_{ib}	h_{ic}
Reverse voltage transfer ratio	h_{re}	h_{rb}	h_{rc}
Forward current gain	h_{fe}	h_{fb}	h_{fc}
Output conductance (mhos)	h_{oe}	h_{ob}	h_{oc}

The measurements of these *h*-parameters from the input and output characteristics of the Transistors are discussed in Section 4.13. The typical values of the *h*-parameters are discussed below. In the design of Transistor circuits, selection of the Transistor configuration and the type of Transistor are made depending upon the values of the *h*-parameters and the Transistor specifications discussed in Section 4.9.

4.9 h-PARAMETER DEFINITIONS FOR COMMON EMITTER TRANSISTOR

Common Emitter Transistor with voltages and currents is shown in Fig. 4.18. The *h*-parameter equivalent circuit of CE Transistor is shown in Fig. 4.19. The network representation of the active device Transistor shown in Fig. 4.19 is represented by Eqs. (4.11) and (4.12):

$$V_{BE} = h_{ie} \cdot I_B + h_{re} \cdot V_{CE} \tag{4.11}$$

$$I_C = h_{fe} \cdot I_B + h_{oe} \cdot V_{CE}. \tag{4.12}$$

Putting the equations in the matrix form, we get

$$\begin{bmatrix} V_{BE} \\ I_C \end{bmatrix} = \begin{bmatrix} h_{ie} & h_{re} \\ h_{fe} & h_{oe} \end{bmatrix} \begin{bmatrix} I_B \\ V_{CE} \end{bmatrix}, \tag{4.13}$$

where h_{ie}, h_{re}, h_{fe} and h_{oe} are *h*-parameters for CE Transistor.

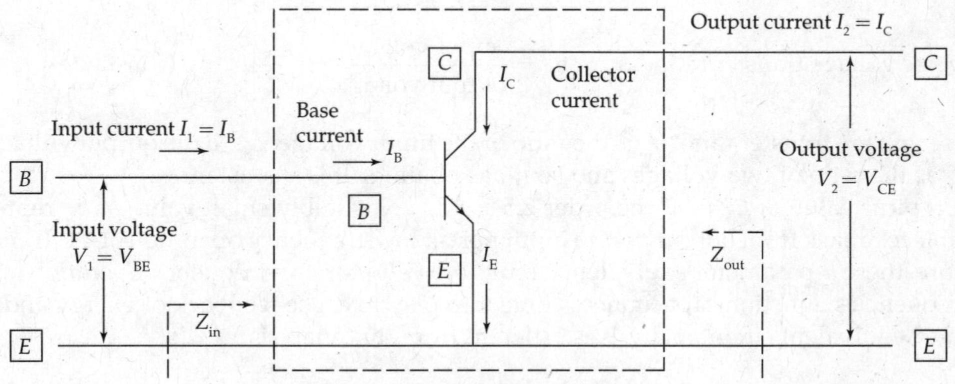

FIG. 4.18 *Common emitter transistor with voltages and currents as a four terminal network to define the hybrid parameters h_{ie}, h_{re}, h_{fe}, f_{oe}*

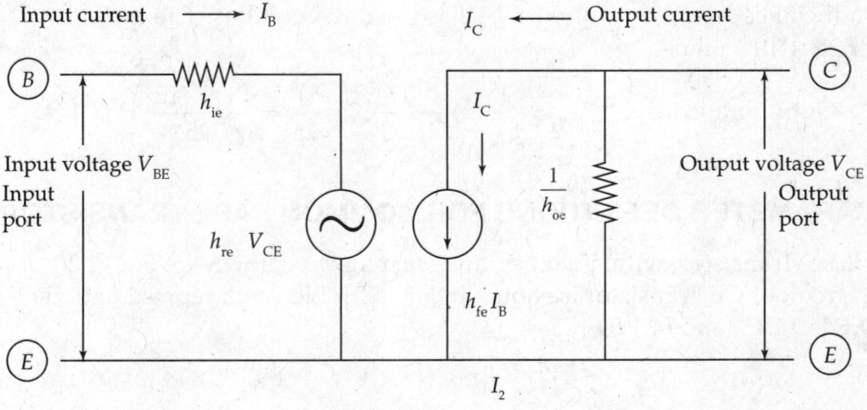

FIG. 4.19 *h-parameter network representation of common emitter transistor*

1. Input resistance $h_{11} = h_{ie} = \frac{\text{input voltage}}{\text{input current}} = \left|\frac{V_{BE}}{I_B}\right| V_{CE} = 0\ (\Omega)$, (4.14)

where h_{ie} is the ratio of voltage input voltage V_{BE} to input current I_B and so h_{ie} is the input resistance of CE Transistor. Input resistance h_{ie} has the units of ohms. The typical value for h_{ie} is of the order 1100 Ω.

2. Forward current gain $h_{21} = h_{fe} = \frac{\text{output current}}{\text{input current}} = \left|\frac{I_C}{I_B}\right| V_2 = 0\ (\text{constant})$, (4.15)

where $h_{21} = h_{fe}$ represents the ratio of two currents and so it has no units. It is a constant. h_{fe} represents the ratio of the output current I_C to the input current. It is known as forward current gain β for CE Transistor.

$$\therefore\ h_{fe} = \beta.$$

Since Collector Current is very much larger than Base Current, the values of β are large. The typical value of β is 50.

3. Reverse voltage transfer ratio $h_{12} = h_{re} = \frac{\text{input voltage}}{\text{output voltage}} = \left|\frac{V_{BE}}{V_{CE}}\right| I_B = 0\ (\text{constant})$ (4.16)

'Reverse voltage transfer ratio' h_{re} is the ratio of the input voltage V_{BE} to the output voltage V_{CE}. $h_{12} = h_{re}$ is the ratio of two voltages and so it has no units. It is a constant.

The typical value of h_{re} is of the order 2.5×10^{-4}. Negligibly small value of h_{re} represents negligible feedback from output port to input port at low-frequency operation of CE Transistor. Therefore, these *h*-parameter equivalent circuits are valid for low-frequency operation only. At high frequencies, junction capacitances come into play to cause feedback of energy and high-frequency equivalent circuit analysis is different from this analysis.

4. Output conductance $h_{22} = h_{oe} = \left|\frac{I_2}{V_2}\right| = \left|\frac{I_C}{V_{CE}}\right| I_B = 0\ (\text{mhos})$ (4.17)

where h_{oe} represents the ratio of the output current i_C to the output voltage V_{CE}. h_{oe} is the 'output conductance' at the output port. h_{oe} has the units in mhos. The typical value for h_{oe} is of the order 25×10^{-6} mhos.

$$\text{Output impedance } Z_{out} = \frac{1}{h_{oE}} = \frac{1}{25\times 10^{-6}} = \frac{10^6}{25} = \frac{1000\times 10^3}{25} = 40\ \text{k}\Omega.$$

4.10 *h*-PARAMETER DEFINITIONS FOR COMMON BASE TRANSISTOR

Common Base Transistor with voltages and currents is shown in Fig. 4.20. *h*-parameter equivalent circuit of CB Transistor is shown in Fig. 4.21. Network representation of Transistor is given by Eqs. (4.18) and (4.19):

$$V_{EB} = h_{ib} \cdot I_E + h_{rb} \cdot V_{CB} \tag{4.18}$$

$$I_C = h_{fb} \cdot I_E + h_{ob} \cdot V_{CB}. \tag{4.19}$$

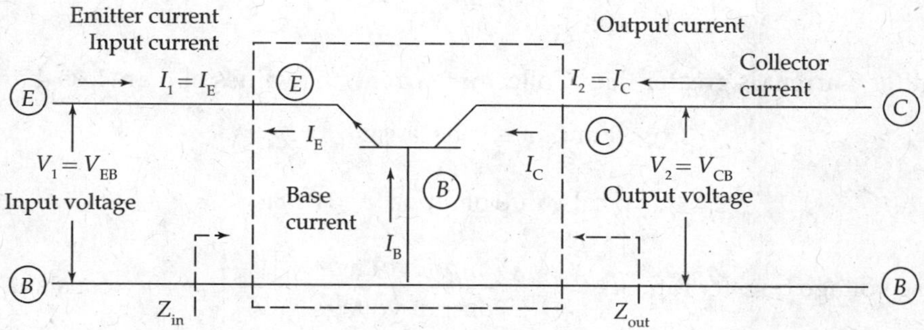

FIG. 4.20 *Common base transistor with voltages and currents as a four terminal network to define the hybrid parameters h_{ib}, h_{ob}, h_{fb}, f_{rb}*

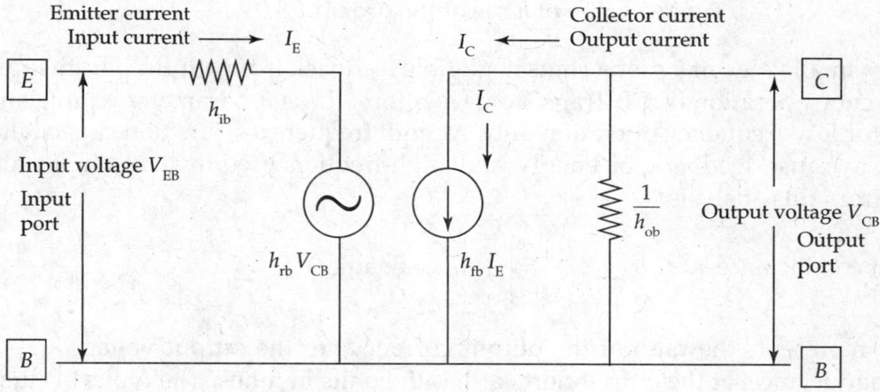

FIG. 4.21 *Small signal low frequency h-parameter equivalent circuit representation of common base transistor*

Putting the equations in the matrix form, we get

$$\begin{bmatrix} V_{EB} \\ I_C \end{bmatrix} = \begin{bmatrix} h_{ib} & h_{rb} \\ h_{fb} & h_{ob} \end{bmatrix} \begin{bmatrix} I_E \\ V_{CB} \end{bmatrix}, \tag{4.20}$$

where h_{ib}, h_{rb}, h_{fb} and h_{ob} are *h*-parameters for CB Transistor.

1. Input resistance $h_{11} = h_{ib} = \dfrac{\text{input voltage}}{\text{input current}} = \left|\dfrac{V_{EB}}{I_E}\right| V_{CB} = 0\ (\Omega)$, (4.21)

where h_{ib} is the ratio of the voltage input voltage V_{EB} to the input current I_E and so h_{ib} is the input resistance of CB Transistor. Input resistance h_{ib} has the units in ohms. The typical value for h_{ib} is of the order 20 Ω.

2. Forward current gain $h_{21} = h_{fb} = \dfrac{\text{output current}}{\text{input current}} = \left|\dfrac{I_C}{I_E}\right| V_{CB} = 0\ (\text{constant})$, (4.22)

where $h_{21} = h_{fb}$ represents the ratio of two currents and so it has no units. It is a constant. h_{fb} represents the ratio of the output current I_C to the input current I_E. It is known as forward current gain α (alpha) for CB Transistor.

$$\therefore \; h_{fb} = \alpha.$$

Since Emitter Current is greater than Collector Current, the values of α are less than 1

$$\alpha < 1 \text{ and } \alpha \equiv 1, \text{ because } I_E \cong I_C.$$

$$\text{Typical value of alpha } \alpha = 0.98. \tag{4.23}$$

3. Reverse voltage transfer ratio $h_{12} = h_{rb} = \dfrac{\text{input voltage}}{\text{output voltage}} = \left|\dfrac{V_{EB}}{V_{CB}}\right| I_E = 0 \;\; \text{(constant)}$ (4.24)

'Reverse voltage transfer ratio' h_{rb} is the ratio of the input voltage V_{EB} to the output voltage V_{CB}. $h_{12} = h_{rb}$ is the ratio of two voltages and so it has no units. It is a constant.

$$\text{Typical value of } h_{rb} \text{ is of the order of } 3.0 \times 10^{-4}. \tag{4.25}$$

Negligibly small value of h_{rb} represents negligible feedback from output port to input port at low-frequency operation of CB Transistor. Therefore, these h-parameter equivalent circuits are valid for low-frequency operation only. At high frequencies, junction capacitances come into play to cause feedback of energy and high-frequency equivalent circuit analysis is different from this analysis.

4. Output conductance $h_{22} = h_{ob} = \left|\dfrac{I_2}{V_2}\right| = \left|\dfrac{I_C}{V_{EB}}\right| I_E = 0 \text{ (mhos)},$ (4.26)

where h_{ob} represents the ratio of the output current I_C to the output voltage V_{CB}. h_{ob} is the 'output conductance' at the output port, h_{ob} has the units in mhos. The typical value for h_{ob} is of the order 0.5×10^{-6} mhos.

Therefore, the output impedance $Z_{out} = \dfrac{1}{h_{ob}} = \dfrac{1}{0.5 \times 10^{-6}} = \dfrac{10 \times 10^{-6}}{5} = 2.0 \text{ M}\Omega.$ (4.27)

From the relative magnitudes,

$$Z_{in} = 20\ \Omega \text{ and } Z_{out} = 2\text{ M}\Omega.$$

Common Base Transistor works as impedance transformer and negligible reverse voltage transfer ratio suggests its use as high-frequency amplifier.

4.11 *h*-PARAMETER DEFINITIONS FOR COMMON COLLECTOR TRANSISTOR

Common Collector Transistor with voltages and currents is shown in Fig. 4.22. h-parameter equivalent circuit of CC Transistor is shown in Fig. 4.23. The representation of the Transistor is represented by Eqs. (4.28) and (4.29):

$$V_{BC} = h_{ic} \cdot I_B + h_{rc} \cdot V_{EC} \tag{4.28}$$

$$I_E = h_{fc} \cdot I_B + h_{oc} \cdot V_{EC}. \tag{4.29}$$

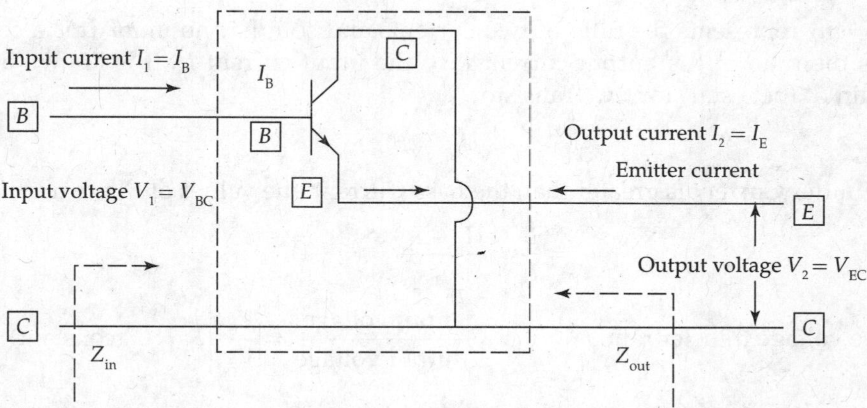

FIG. 4.22 *Common collector transistor with voltages and currents as a four terminal network to define the hybrid parameters h_{ic}, h_{rc}, h_{fc}, f_{oc}*

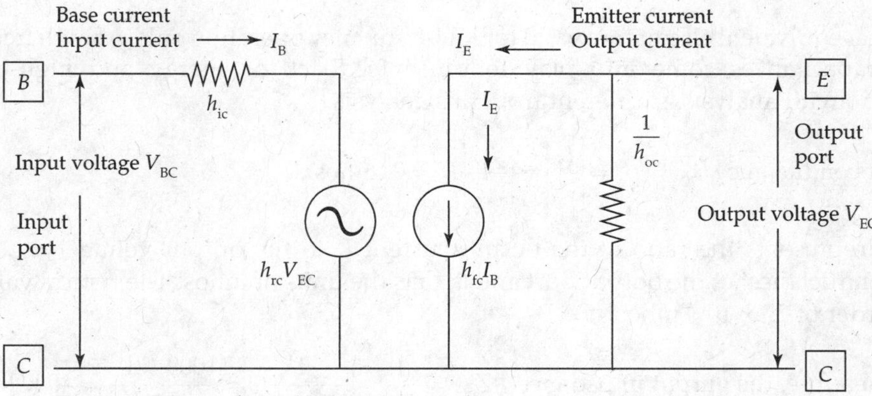

FIG. 4.23 *Small signal low frequency h-parameter equivalent circuit representation of common collector transistor*

Putting the equations in the matrix form, we get

$$\begin{bmatrix} V_{BC} \\ I_E \end{bmatrix} = \begin{bmatrix} h_{ic} & h_{rc} \\ h_{fc} & h_{oc} \end{bmatrix} \begin{bmatrix} I_B \\ V_{EC} \end{bmatrix}, \tag{4.30}$$

where h_{ic}, h_{rc}, h_{fc} and h_{oc} are *h*-parameters for CC Transistor.

1. Input resistance $h_{11} = h_{ic} = \dfrac{\text{input voltage}}{\text{input current}} = \left|\dfrac{V_{BC}}{I_B}\right| V_{EC} = 0\ (\Omega)$, (4.31)

where h_{ic} is the ratio of the input voltage V_{BC} to the input current I_B. Therefore, h_{ic} is the input resistance of CC Transistor. Input resistance h_{ic} has the units in ohms. The typical value for h_{ic} is of the order 1100 Ω.

2. Forward current gain $h_{21} = h_{fc} = \dfrac{\text{output current}}{\text{input current}} = \left|\dfrac{I_E}{I_B}\right| V_{EC} = 0$ (constant), (4.32)

where $h_{21} = h_{fc}$ represents the ratio of two currents and so it has no units. It is a constant. h_{fc} represents the ratio of the output current I_E to the input current I_B. It is known as forward current gain β^* (beta star) for CC Transistor.

$$\therefore\ h_{fc} = \beta^*.$$

Since Emitter Current is greater than the Base Current, the values of β^* are much larger:

$$\beta^* = \frac{I_E}{I_B} = \frac{(I_C + I_B)}{I_B} = (\beta + 1). \tag{4.33}$$

3. Reverse voltage transfer ratio $h_{12} = h_{rc} = \dfrac{\text{input voltage}}{\text{output voltage}} = \left|\dfrac{V_{BC}}{V_{EC}}\right| I_B = 0.$ (constant) (4.34)

'Reverse voltage transfer ratio' h_{rc} is ratio of the input voltage V_{BC} to the output voltage V_{EC}. $h_{1.2} = h_{rc}$ is the ratio of two voltages. So it has no units. It is a constant.

Typical value of h_{rc} is of the order of $\cong 1.0$. (4.35)

h-parameter equivalent circuits are valid for low-frequency operation only. At high frequencies, junction capacitances come into play to cause feedback of energy and high-frequency equivalent circuit analysis is different from this analysis.

4. Output conductance $h_{22} = h_{oc} = \left|\dfrac{I_2}{V_2}\right| = \left|\dfrac{I_E}{V_{EC}}\right| I_B = 0$ (mhos), (4.36)

where h_{oc} represents the ratio of the output current I_E to the output voltage V_{EC}, h_{oc} is the 'output conductance' at the output port and h_{oc} has the units of mhos. The typical value for h_{oc} is of the order of 25×10^{-6} mhos.

Therefore, the output impedance $Z_{out} = \dfrac{1}{h_{oc}} = \dfrac{1}{25\times 10^{-6}} = \dfrac{1000\times 10^3}{25} = 40\ \text{k}\Omega.$ (4.37)

The different sets of parameters are inter-related and inter-convertible.
For instance

$$h_{ie} = \frac{\Delta V_{BE}}{\Delta I_B} = \frac{\Delta V_{BE}}{\left(\dfrac{\Delta I_E}{(1+\beta)}\right)}$$

using $I_E = I_B + I_C$

$$\frac{I_E}{I_B} = (1+\beta)$$

$$\therefore\ I_B = \frac{I_E}{(1+\beta)}$$

$$\therefore\ h_{ie} = h_{ib}(1+\beta) = h_{ib}(1+h_{fe})$$

$$\therefore\ h_{ib} = \frac{h_{ie}}{(1+\beta)} = \frac{h_{ie}}{(1+h_{fe})}. \tag{4.38}$$

$$\text{Similarly, } h_{fb} = \frac{\Delta I_C}{\Delta I_E} = \frac{\Delta I_C}{\Delta I_B(\beta+1)} = \frac{\beta}{\beta+1}$$

$$\text{or } h_{fb} = \frac{h_{fe}}{[1+h_{fe}]}. \tag{4.39}$$

For CC parameters

$$h_{fc} = \frac{\Delta I_E}{\Delta I_B} = \frac{\Delta I_B + \Delta I_C}{\Delta I_B} = (1+h_{fe})$$

$$\therefore \quad h_{fc} = (1+h_{fe}) = (1+\beta) \tag{4.40}$$

$$h_{ic} = h_{ie}$$

$$h_{oc} = h_{oe}.$$

4.12 COMPARISONS OF CE, CB, CC TRANSISTOR CONFIGURATIONS

Salient features of CE, CB, CC Transistor operations
Comparisons of various parameters of CE, CB and CC Transistor configurations

Parameter	CE Transistor	CB Transistor	CC Transistor
Type of amplifier	Inverting voltage amplifier	Non-inverting voltage amplifier	Non-inverting voltage amplifier
Input resistance h_i	$h_{ie} = 1100\ \Omega$	$h_{ib} = 10\text{–}20\ \Omega$	$h_{ic} = 1100\ \Omega$
Output resistance $\frac{1}{h_o}$	$\frac{1}{h_{oe}} = 40\ \text{K}\Omega$	$\frac{1}{h_{ob}} = 2\ \text{M}\Omega$	$\frac{1}{h_{oc}} = 40\ \text{K}\Omega$
Forward current gain h_f	$h_{fe} = \beta > 25$	$h_{fe} = \alpha < 1$	$h_{fc} = \beta^* = (1+\beta)$
Reverse voltage Transfer ratio h_r	$h_{re} = 25 \times 10^{-3}$	$h_{rb} = 3.0 \times 10^{-6}$	$h_{rc} \approx 1$
Voltage gain	Very large	Reasonable gain	Less than 1

4.13 DETERMINATION OF *h*-PARAMETERS FROM TRANSISTOR CHARACTERISTICS

1. *Determination of input resistance R_{in} or h_{ie} from input characteristics (Fig. 4.24)*

Input resistance R_{in} or h_{ie} of the CE Transistor is defined from the forward resistance of the input Diode as $\Delta V_{BE}/\Delta I_B$ and can be determined from the input characteristics shown in Fig. 4.24.

For $V_{BE1} = 0.59$ V, there is a corresponding $I_{B1} = 100\ \mu$A.

For $V_{BE2} = 0.7$ V, $I_{B2} = 200\ \mu$A.

Therefore, increment in Base voltage $\Delta V_{BE} = (V_{BE2} - V_{BE1}) = (0.7 - 0.59) = 0.11$ V.

In addition, corresponding increment in Base Current $\Delta I_B = (I_{B2} - I_{B1}) = (200 - 100)\ \mu$A.

This is the slope of the input characteristic for a specific value of V_{CE} (Fig. 4.16).

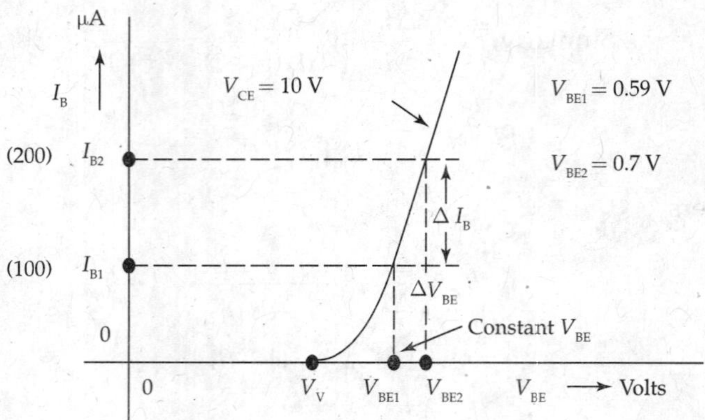

FIG. 4.24 *Input characteristic of common emitter transistor to determine h_{iE} or R_{in}*

By definition $$h_{ie} = R_{in} = \frac{\Delta V_{BE}}{\Delta I_B}\bigg| V_{CE} = \text{constant } \Omega. \tag{4.41}$$

$$h_{ie} = R_{in} = \frac{\Delta V_{BE}}{\Delta I_B} = \frac{(0.7-0.59)}{(200-100)\times 10^{-6}} = \frac{0.11}{100\times 10^{-6}} = 1.1 \text{ k}\Omega.$$

2. *Determination of reverse voltage transfer ratio h_{re} from input characteristics (Fig. 4.25)*
A constant I_B line is drawn on the input characteristics of CE Transistor in Fig. 4.25, to determine the reverse voltage transfer ratio h_{re} defined as below. h_{re} is the ratio of change in Base voltage V_{BE} to the change in Collector voltage V_{CE} for a constant I_B.

Consider some typical values from the input characteristic for a constant $I_B = 200$ μA, $V_{BE2} = 0.7$ V, $V_{BE1} = 0.575$ V, $V_{CE2} = 15$ V and $V_{CE1} = 10$ V.

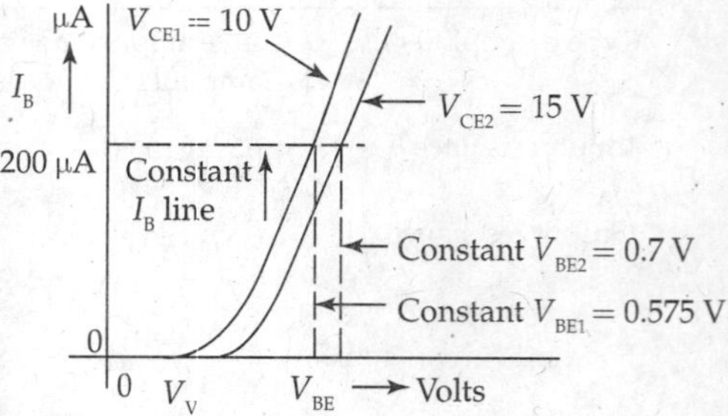

FIG. 4.25 *Input characteristics of common emitter transistor to determine h_{re}*

$$h_{re} = \left|\frac{\Delta V_{BE}}{\Delta V_{CE}}\right| I_B = \text{constant}. \tag{4.42}$$

$$\therefore\ h_{re} = \left[\frac{V_{BE2}-V_{BE1}}{V_{CE2}-V_{CE1}}\right] = \left[\frac{0.7-0.575}{15-10}\right] = \frac{0.125}{5} = 25\times 10^{-3}.$$

3. *Determination of β or h_{fe} of CE Transistor from its output characteristics (Fig. 4.26)*
Definition of current gain β of a CE Transistor: The incremental current gain β is also called h_{fe} and is obtained in the following way. For a CE Transistor, for a given V_{CE}, the ratio of incremental Collector Current ΔI_C and incremental Base Current ΔI_B is defined as the small signal forward current gain β or h_{fe}.

Calculation of current gain of a CE Transistor from its output characteristics:

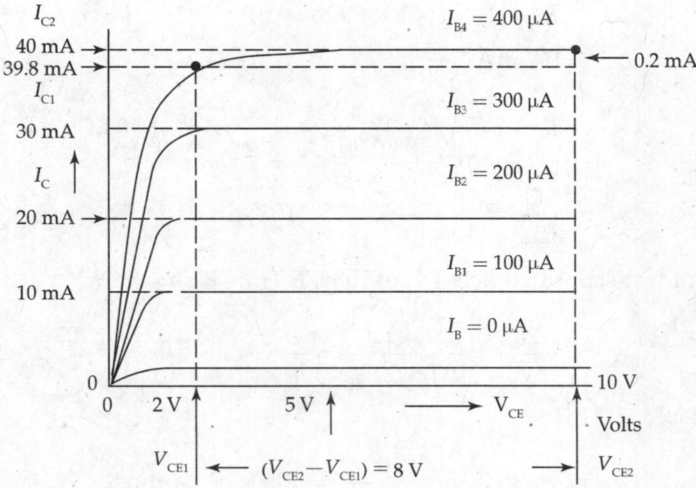

FIG. 4.26 *Transistor output characteristics to determine h_{fE} and h_{oE}*

Forward Current gain 'β' can be determined from the output characteristics of the CE Transistor as shown in Fig. 4.26.

On output curve for $I_{B2} = 200\ \mu A$; $I_{C2} = 20$ mA.

On the output curve for $I_{B1} = 100\ \mu A$; $I_{C1} = 10$ mA.

$$\Delta I_C = (I_{C2} - I_{C1}) = (20 - 10) \times 10^{-3} = 10 \text{ mA}.$$

$$\Delta I_B = (I_{B2} - I_{B1}) = (200 - 100) \times 10^{-6} = 100\ \mu A.$$

Current gain of CE Transistor h_{fe} or $\beta = \Delta I_C / \Delta I_B$

$$\beta \text{(beta)} = \frac{10 \times 10^{-3}}{100 \times 10^{-6}} = 100.$$

4. *Determination of $R_{out} = 1/h_{oe}$ of Transistor from its output characteristics (Fig. 4.26)* R_{out} is defined as the slope of V_{CE} versus I_C characteristic for a constant I_B.

$$\text{Thus,} \quad R_{out} = \left.\frac{\Delta V_{CE}}{\Delta I_C}\right|_{I_B = \text{constant}} \quad \text{ohms} \tag{4.43}$$

$$\therefore \quad h_{oE} = \frac{1}{R_{out}} = \left|\frac{\Delta I_C}{\Delta V_{CE}}\right| I_B = \text{constant} \quad \text{mhos or Siemens.} \tag{4.44}$$

Assuming the following data from CE Transistor output characteristics (Fig. 4.26), output conductance and output resistance are calculated as given below.

$$V_{CE2} = 10 \text{ V}; V_{CE1} = 2 \text{ V}$$

$$\Delta V_{CE} = V_{CE2} - V_{CE1} = (10 - 2) = 8 \text{ V}$$

$$I_{C2} = 40 \text{ mA}; I_{C1} = 39.8 \text{ mA}$$

$$\Delta I_C = I_{C2} - I_{C1} = (40 - 39.8) \times 10^{-3}$$

$$\Delta I_C = 0.2 \text{ mA}$$

$$\frac{1}{h_{oe}} = R_{out} = \frac{\Delta V_{CE}}{\Delta I_C} = \frac{8 \text{ V}}{0.2 \text{ mA}} = \frac{80}{2} \times 10^3 = 40 \text{ k}\Omega$$

$$h_{oe} = \frac{1}{R_{out}} = \frac{1}{40 \times 10^{-3}} = 25 \times 10^{-6} \text{ mhos.}$$

Determination of transconductance (g_m) of the CE Transistor:

By definition,
$$g_m \triangleq \frac{\Delta I_C}{\Delta V_{BE}} = \frac{\Delta I_C \cdot \Delta I_B}{\Delta I_B \cdot \Delta V_{BE}} = \frac{\beta}{h_{ie}} \tag{4.45}$$

As $\beta = 100$ and $h_{ie} = 1.1 \text{ k}\Omega$

$$g_m = \frac{h_{fe}}{h_{ie}} = \frac{\beta}{h_{ie}} = \frac{100}{1.1 \times 10^3} = 90.9 \times 10^{-3} \text{ mhos.}$$

4.14 COMMON BASE TRANSISTOR CHARACTERISTICS AND PARAMETERS

In CB Transistor configuration, Base terminal is grounded and forms the common terminal for both input and output ports. The mode of operation of the Transistor is called the CB mode of operation shown in Fig. 4.27.

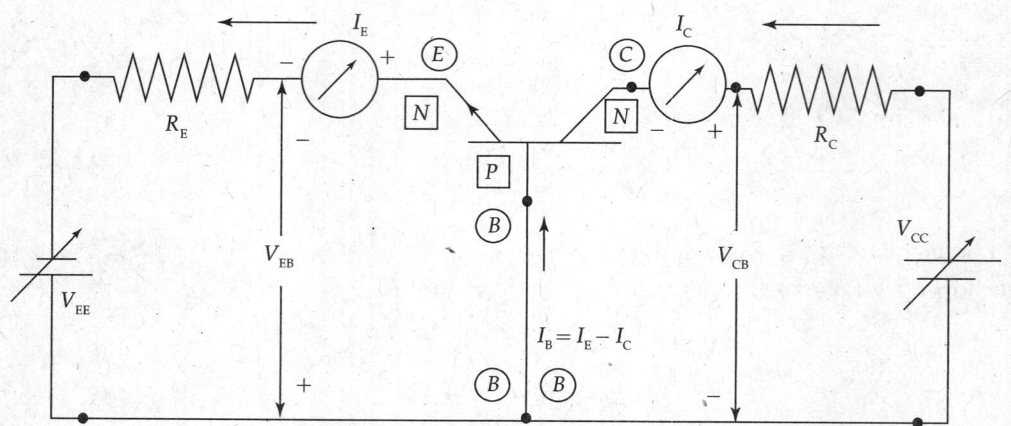

FIG. 4.27 *Common base transistor configuration*

For the Transistor to act in the active region of its output characteristics, it requires that the Emitter junction is forward-biased by V_{EB} and the output junction is reverse-biased by V_{CB} shown in Fig. 4.27.

Input current is the Emitter Current I_E and the output current is the Collector Current I_C. The current through the Base $|I_B| = |I_E - I_C|$. The current gain 'α' of the CB-operated Transistor is defined as the ratio of the change in Collector Current ΔI_C to the change in Emitter Current ΔI_E with constant Collector to Base voltage V_{CB}.

$$\text{Current gain } \alpha = \left.\frac{\Delta I_C}{\Delta I_E}\right| V_{CB} = \text{constant} \tag{4.46}$$

Current gain $\alpha = \left.\frac{\Delta I_C}{\Delta I_E}\right|$ is less than unity, since $I_C \langle I_E$.

Since Emitter Current is larger than the Collector Current, current gain 'α' of CB Transistor is always less than unity. The difference in magnitude between the Emitter and Collector currents ($I_E - I_C = I_B$) is very small. So the current gain 'α' (alpha) of the *CB Transistor is less than 1* and always close to unity.

4.14.1 Input Characteristics of Common Base Transistor

The experimental setup to obtain the data for input and output characteristics is shown in Fig. 4.27. The input characteristics of CB Transistor are between V_{EB} and I_E; the input voltage and current variations maintain constant V_{CB}, the reverse-bias to the Collector junction. The input characteristics are shown in Fig. 4.28.

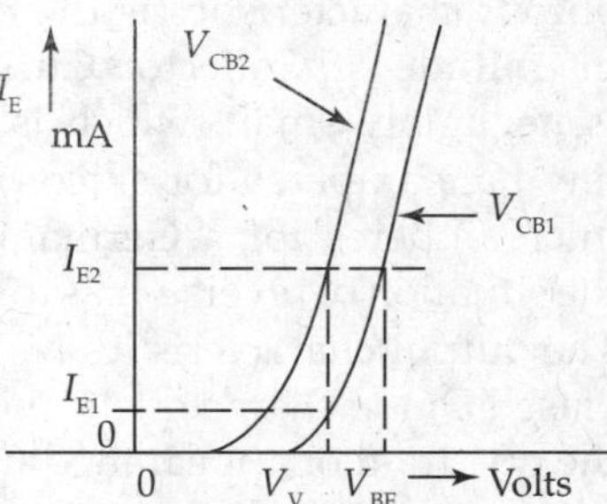

FIG. 4.28 *Input characteristics of common base transistor*

Input characteristic of CB Transistor is nothing but the forward characteristic of Emitter–Base diode. For a given V_{CB}, as V_{EB} increases, beyond the cut in voltage V_γ the Emitter Current I_E increases as in a forward-biased Diode (Fig. 4.28). After passing the cut-in voltage, 'V_γ', for small changes of V_{EB}, large changes of I_E occur.

For a constant V_{EB}, increasing the reverse-bias voltage V_{CB} can increase the Emitter Current I_E, when the 'Early Effect' comes into play. So the knee voltage (cut-in voltage or threshold voltage) for conduction of the Emitter diode lowers and large changes in I_E can be obtained at lower V_{EB}. This can be seen on the input characteristics that for a higher value of V_{CB2}, the characteristic moves to the left of V_{CB1} characteristic. Due to 'Base width modulation' or 'Early Effect', when V_{CB} is more I_E increases due to increased field at the Collector junction and it requires less V_{EB} to draw the same Emitter Current I_E and thus the curve for V_{CB2} moves to the left of that for V_{CB1}.

The slope of V_{EB} versus I_E characteristic (input characteristic) is the input resistance represented by 'h_{ib}' of the CB-operated Transistor. The input resistance of the CB Transistor is defined as

$$h_{ib} = \left.\frac{\Delta V_{EB}}{\Delta I_E}\right| V_{CB} = \text{constant } (\Omega). \tag{4.47}$$

The input resistance h_{ib} will be of the order of a few ohms that is about 10–20 Ω. This is very low when compared to the input resistance h_{ie} of CE Transistor, which is typically 1 kΩ. The reason for this difference in magnitudes of the input resistances is clear from the following equation:

$$h_{ib} = \frac{\Delta V_{EB}}{\Delta I_E} \quad \text{whereas} \quad h_{ie} = \frac{\Delta V_{BE}}{\Delta I_B} \ (\Omega). \tag{4.48}$$

Since I_B, the Base Current, is a few orders less than the Emitter Current I_E, the input resistance 'h_{ib}' of CB Transistor is so many orders less than 'h_{ie}'. In addition, the magnitudes of I_E and I_B

in the expressions for the input resistances of the CE and CB configurations of the Transistor confirm the same.

4.14.2 Output Characteristics of Common Base Transistor

For a constant magnitude of Emitter Current I_E, the graph between the Collector Current I_C and V_{CB} is called the output characteristic. A family of output characteristics is obtained for different values of I_E as shown in Fig. 4.29.

It is already familiar that for a given Emitter Current I_E the forward-biased input junction, a small current I_B, goes to the Base and the remaining current goes to the Collector as the Collector Current I_C, according to the equation $I_E = -(I_B + I_C)$.

The region between the output characteristic for $I_{E1} = 0$ mA and V_{CB} axis on the CB Transistor output characteristics is the cut-off region. The magnitude of Collector Current is I_{CB0}, which is negligibly small, which is of the order of a few nanoamperes for silicon device and a few microamperes for a Germanium device under the situation of reverse-bias to Collector junction. The output characteristics resemble the reverse-biased Diode characteristic with reorientation of the quadrant of appearance and the parameters.

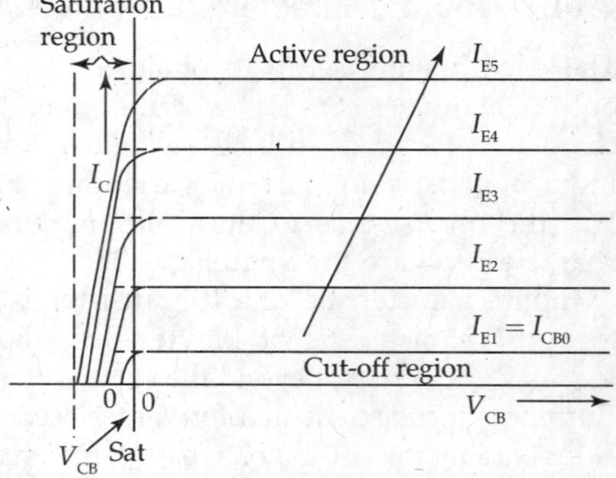

FIG. 4.29 *Output characteristics of CB transistor*

When the Collector terminal is open, the entire Emitter Current I_E goes to the Base and the Collector Current $I_C = 0$. When V_{CB} is zero, the electrons (for NPN Transistor) that cross the Base region face a field at Collector-to-Base junction J_2 that neither opposes nor aids allowing all the electrons to continue their journey into the Collector circuit, and the Collector Current I_C for a given Emitter Current I_E becomes maximum even at the starting point of the characteristic, that is for $V_{CB} = 0$ V.

Now if V_{CB} increases, the Base width modulation occurs and a very slight increase in the Collector Current I_C is observed. The reason for this virtually constant I_C with variations in V_{CB} is that the electrons with reverse-bias at the Collector face is an accelerating potential, and they move with constant velocities or experience acceleration, but the quantum of Collector Current 'I_C' changes by a very little magnitude.

The relation between the Collector Current and the Emitter Current is defined as

$$I_C = -\alpha \cdot I_E + I_{C0}, \tag{4.49}$$

where I_{C0} is the reverse saturation current, which is $\cong 0$ with $I_E = 0$ mA.

$$\therefore \quad \text{Alpha } \alpha = \frac{-(I_C - I_{C0})}{I_E} \text{ is less than unity.} \tag{4.50}$$

Current gain 'α' is known as forward current gain of 'common Base Transistor'.

As compared to the CE Transistor configuration, the phenomenon of 'Early Effect' is less prominent in CB Transistor operation, since the forward current gain alpha 'α' is less than unity and change in Collector Current I_C is negligibly small.

The slope of Collector Current I_C versus V_{CB} characteristic (output characteristic) is almost zero. So the output impedance of the CB mode Transistor runs into mega ohms.

Grounding of the Base terminal produces perfect isolation between the input and the output ports, since it acts as a shield between the 'Emitter and Collector'. There is no common path between the two ports. In other words, the Emitter to Collector inter-electrode capacitance virtually becomes zero.

The very low input impedance of the 'Common Base Transistor' configuration makes it unsuitable to be driven by non-ideal voltage Sources (whose Source impedance is not negligible). All practical voltage Sources are non-ideal. But this CB mode Transistor can be used with current Sources, but natural current Sources are rare and so the CB configuration finds less practical application.

Common Base Transistor is used in high-frequency applications because of its ability to provide perfect isolation between input and output ports of the Transistor.

Explanation of how the Transistor is considered as current-controlled device: The Transistor operates in the active region when the Base–Emitter junction J_1 is forward-biased and the Base–Collector junction J_2 is reverse-biased irrespective of the Transistor configuration. Emitter region is heavily doped, and so large numbers of electrons are injected from the Emitter into the Base region. That is how the Emitter is considered as the source of electrons. The Emitter Current I_E is the sum of the majority carrier electron current and the minority carrier hole current in the NPN Transistor.

All the electrons that cross the Emitter–Base junction do not reach the Collector, since there are some recombinations with holes in the Base region contributing to very small magnitude of the Base Current 'I_B' and the remainder enters the Collector and the Collector Current component is I_{NC}. There is a small minority carrier current due to reverse-bias of the Base–Collector junction designated as the current I_{CO}.

The net Collector Current will be $I_C = (I_{CO} - I_{NC})$, if I_E is not equal to zero. When $I_E = 0$, the Collector Current $I_C = I_{CO}$, since I_{NC} representing the injected carrier current is zero as $I_E = 0$.

Emitter injection efficiency as per definition:

$$\gamma = \frac{\text{Current of injected carriers at the emitter base junction}}{\text{Total emitter current}} = \frac{I_{NE}}{I_{NE} + I_{PE}} \cong \frac{I_{NE}}{I_E}.$$

Derivation of expression for Collector Current:

$$I_C = \beta I_B + (1 + \beta) I_{CO}$$

for a CE Transistor operated in the active region of output characteristics.

Collector Current I_C in a CE Transistor:

$$I_C = -\alpha \cdot I_E + I_{CO} \tag{4.50A}$$

Using the expression for Emitter Current $I_E = -(I_B + I_C)$ in Eq. (4.50 A), we get

$$I_C = \alpha \cdot (I_B + I_C) + I_{CO}$$

Rearranging the terms in the above equation, we get

$$I_C(1 - \alpha) = \alpha \cdot I_B + I_{CO}.$$

$$\therefore \quad I_C = \frac{\alpha}{(1-\alpha)} \cdot I_B + \frac{I_{C0}}{(1-\alpha)}. \tag{4.50B}$$

$$\text{We know that } \beta = \frac{I_C}{I_B} = \frac{I_C}{(I_E - I_C)} = \frac{\alpha}{(1-\alpha)} \tag{4.50C}$$

and also $\alpha = \dfrac{I_C}{I_E} = \dfrac{I_C}{(I_B + I_C)} = \dfrac{\beta}{(1+\beta)}$.

$$\therefore \quad (1-\alpha) = \left[1 - \frac{\beta}{(1+\beta)}\right] = \frac{1}{(1+\beta)} \tag{4.50D}$$

Substituting the expressions $\dfrac{\alpha}{(1-\alpha)} = \beta$ and $(1-\alpha) = \dfrac{1}{(1+\beta)}$ in Eq. (4.50B), we get

$$I_C = \beta \cdot I_B + (1+\beta) \cdot I_{C0}. \tag{4.50E}$$

This expression will be used in the derivations for stability factors '*S*' in Chapter 5, Biasing Circuits.

4.15 BIASING CIRCUIT FOR PNP TRANSISTOR IN COMMON EMITTER CONFIGURATION

Figure 4.30 shows a PNP Transistor in CE configuration with biasing voltages to obtain the Transistor characteristics. Emitter Current flowing into the device terminal on the Transistor symbol indicates that the Transistor is a PNP Transistor. Voltage V_{BE} is the forward-bias to the Emitter junction and voltage V_{CE} is the reverse-bias to the Collector junction of the Transistor. They are the necessary biasing voltages for the Transistor to act as an amplifying device. The design of biasing voltages depends upon the Transistor applications.

PNP Transistor with its structural materials and biasing voltages is shown in Fig. 4.31. When the input junction is forward-biased by V_{BE}, the majority carrier Holes are injected

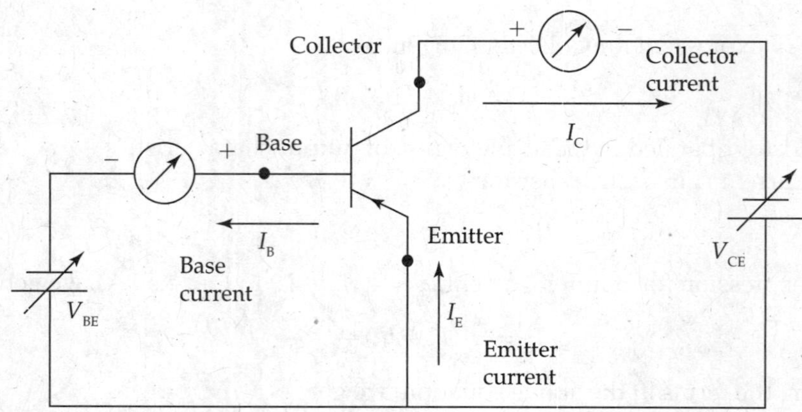

FIG. 4.30 *Common emitter (PNP) transistor with biasing voltages to obtain input and output characteristics*

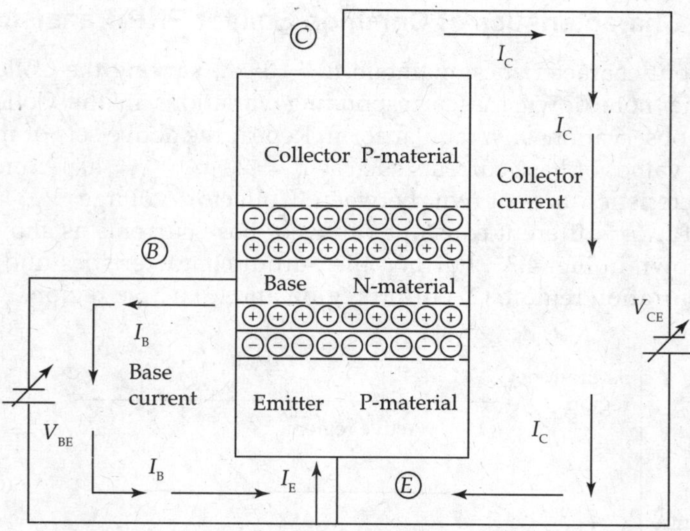

FIG. 4.31 *Structure of common emitter PNP transistor with biasing voltages V_{BE} and V_{CE}*

from the P-type (Emitter) into the N-type (Base), overcoming the cut-in voltage of the input junction (Cut-in voltage is seen on Transistor input characteristics in Fig. 4.32), as majority carriers contribute to the current through the Emitter junction. The injected minority carrier holes into the Base move by diffusion from the Emitter junction to the Collector junction. During this journey, some electrons of thin and lightly doped N-type Base recombine with the diffusing holes contributing to the Base Current, I_B. As most of the Emitter Current has to reach out to the Collector to function as a good Transistor, the recombination of holes and electrons in the Base material is made deliberately less by light doping and of smaller area of cross-section of the Base region.

As the Collector junction is reverse-biased by V_{CE}, the holes, the minority carriers in the Base move into the Collector region contributing to the Collector Current. Thus, the Emitter Current is equal to the sum of the Base Current and the Collector currents. On the output characteristics, shown in Fig. 4.33, a slight upward slope of the Collector Current is seen, which is due to the 'Early Effect' due to the reverse-bias to the output junction.

4.15.1 Input Characteristics of Common Emitter PNP Transistor

In the experimental setup (Fig. 4.30), maintain $V_{CE} = 5$ V, varying the forward-bias voltage V_{BE} in increments, and note down corresponding variations in Base Current I_B. These measurements are noted in a tabular form. Input characteristic can be drawn between the measurements V_{BE} and I_B from the tabular form. Another input characteristic can be drawn from another set of measurements for a second constant voltage V_{CE2}. Typical input characteristics are shown in Fig. 4.32.

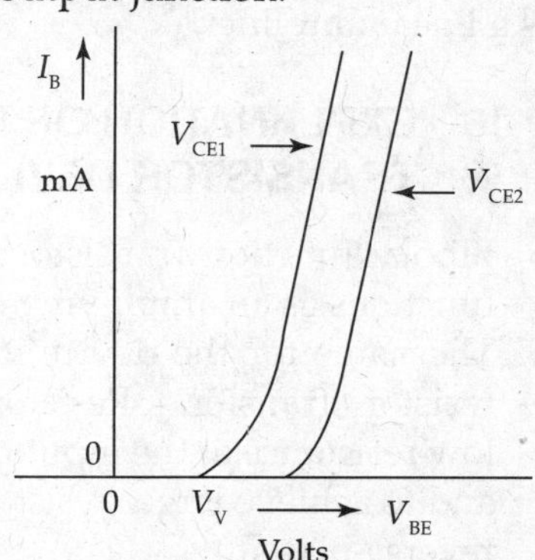

FIG. 4.32 *Input characteristics of common emitter transistor*

4.15.2 Output Characteristics of Common Emitter PNP Transistor

To obtain the output characteristics, maintain $I_{B1} = 0$ μA, varying the Collector voltage V_{CE} in increments, and note down the corresponding variations in the Collector Current I_C. Note down these observations in a tabular form. Repeat the above sets of measurements for different constant values of Base currents such as $I_{B1} = 20$ μA, $I_{B2} = 40$ μA and so on.

Output characteristics are drawn between Collector voltage V_{CE} and the output Collector Current I_C for different constant values of Base currents as shown in the output characteristics shown in Fig. 4.33. Further, the output characteristics find three regions of operations: (1) saturation region, (2) cut-off region and (3) active region.

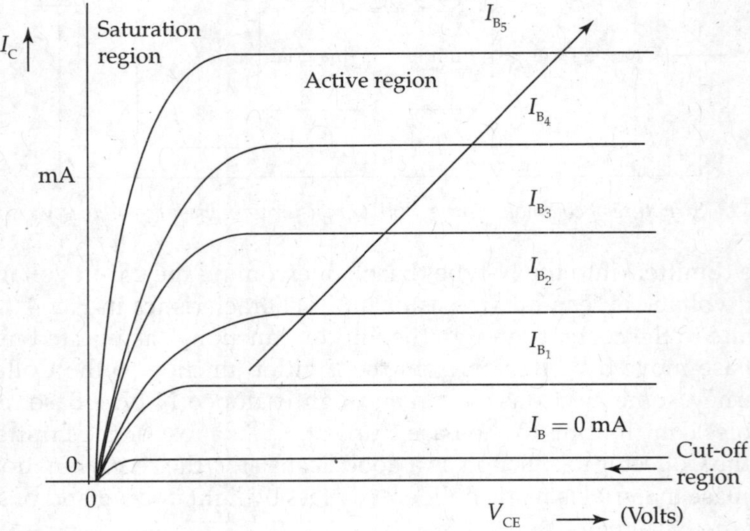

FIG. 4.33 *PNP common emitter transistor output characteristics*

For the Transistor to be operated as an amplifier, the biasing voltages known as the DC operating conditions are fixed in the active region of the characteristics by a quiescent operating point '*Q*'. If the operating conditions are fixed for Class-A amplifier, then amplifier is a linear amplifier.

4.16 EXPLANATION OF THE NEED OF BIASING VOLTAGES FOR THE TRANSISTOR DEVICES

- Bipolar Junction Transistor is used as a two-port network or as a four-terminal network. It functions as an amplifying device along with other applications.
- The name for the device itself is an acronym formed from the two words Transfer and resistor (Transfer + Resistor = Transistor). For achieving different levels of resistance say low resistance at the input port, the input junction of the Transistor is to be forward-biased and to achieve larger resistance at the output port of the Transistor output junction is to be reverse-biased.
- The different levels of resistances at the input and output ports of the Transistor contribute to the amplifying action of the Transistor. Thus for the Transistor to act as an amplifying

device, there is a need for biasing the input (Emitter junction) and output (Collector junction) junctions of the Transistor.

- An amplifier has two sets of voltages: (1) DC operating voltages for the Transistor to operate in the active region for the above reasons and (2) AC signal voltages for amplification.
- Further, the quiescent operating point or the DC operating point 'Q' of the Transistor is located on the DC load line drawn on the output characteristics of the Transistor, depending upon the class of operation of the Amplifier.
- For linear operation of the amplifier, for example in Class-A operation, the quiescent operating point is located at the middle of the DC load line for the device operation to be in the active region of the output characteristics and symmetrical input and output voltage swings.
- Fixation of the quiescent operating point 'Q' that is the quiescent Collector Current 'I_{CQ}', Base Current I_{BQ} and the voltage between Collector and Emitter 'V_{CEQ}' of the Transistor is fixed by the biasing voltages for the Emitter junction and the Collector junctions of the Transistor.
- AC signal voltages at the input port of the Transistor modulate the forward-bias and cause variations in the forward-bias (DC voltage). These varying DC voltages cause variations in the input Base Current.
- Collector Current is an amplified version of input current. This varying Collector Current produces increased output voltage at the output port of the amplifier.

EXAMPLE 4.1

NPN Transistor has $\alpha = 0.98$, $I_{C0} = 2$ μA and $I_{E0} = 1.6$ μA. The Transistor is used in CE connection with $V_{CC} = 12$ V and $R_L = 4$ kΩ. What is the minimum Base Current required in order that Transistor enters into saturation (Aug./Sep. 2007, Set. no. 2)?

Solution:

$V_{CC} = 12$ V, $R_L = 4$ kΩ, $V_{CC} - I_C \cdot R_L = V_{CE}$

Minimum value of Base Current $I_{B(sat)}$ required for the Transistor to enter into saturation region on the output characteristics occurs when $V_{CE(sat)} = 0$ V.

Therefore, assuming $V_{CE(sat)} = 0$ V:

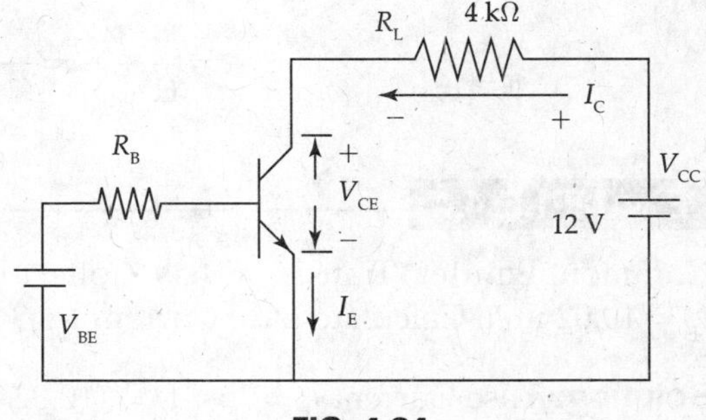

FIG. 4.34

Saturation collector current $= I_{C(sat)}$

$$= \frac{V_{CC}}{R_L} = \frac{12\text{ V}}{4\text{ k}\Omega} = 3\text{ mA}.$$

$$\alpha = 0.98 = \frac{I_C}{I_E}$$

$$\therefore\ I_E = \frac{I_C}{0.98} = \frac{3\times10^{-3}\times10^2}{98} = \frac{300}{98} = 3.06\text{ mA}.$$

$$I_B = (I_E - I_C) = (3.06 - 3.0)\times10^{-3} = 0.06\text{ mA}.$$

Minimum value of Base Current required for the Transistor to enter into the saturation region on the output characteristics = $I_{B(sat)}$

$$\therefore \quad I_{B(sat)} = 0.06 \text{ mA.}$$

EXAMPLE 4.2

A Transistor connected in CE configuration has Base Current 'I_B' of magnitude 0.25 mA and Collector Current $I_C = 50$ mA. Calculate the value of Emitter Current I_E through the Transistor. Also calculate the values of current gains of the Transistor in CE and CB modes of operation.

Solution: Emitter Current $I_E = I_C + I_B$

$$I_E = (50 + 0.25) \text{ mA} = 50.25 \text{ mA}$$

$$\therefore \text{ Current gain of common emitter transistor} = \beta = \frac{I_C}{I_B} = \frac{50\times 10^{-3}}{0.25\times 10^{-3}} = 200$$

$$\text{Current gain of common base transistor } \alpha = \frac{I_C}{I_E} = \frac{50\times 10^{-3}}{50.25\times 10^{-3}} = 0.995.$$

EXAMPLE 4.3

If a Transistor has a value of $\beta = 50$ and Collector Current of 10 mA, determine the value of Emitter Current and calculate the value of alpha of the Transistor.

Solution: Collector Current $I_C = 10$ mA
Current gain $\beta = 50$

$$\therefore \quad \text{Base current } I_B = \frac{I_C}{\beta} = \frac{10\times 10^{-3}}{50} = 0.2 \text{ mA.}$$

Emitter Current $I_E = I_C + I_B = (10 + 0.02)$ mA $= 10.02$ mA

$$\alpha = \frac{\beta}{(1+\beta)} = \frac{50}{(1+50)} = \frac{50}{51} = 0.98.$$

EXAMPLE 4.4

Common Emitter Transistor has Collector Current $I_C = 10$ mA and Emitter Current $I_E = 10.02$ mA. Calculate Base Current I_B, β and α values of the Transistor.

Solution: Base Current $I_B = I_E - I_C = (10.02 - 10) \times 10^{-3} = 0.02$ mA

$$\beta = \frac{I_C}{I_B} = \frac{10\times 10^{-3}}{0.02\times 10^{-3}} = \frac{10}{0.02} = 500.$$

$$\alpha = \frac{I_C}{I_E} = \frac{10\times 10^{-3}}{10.02\times 10^{-3}} = \frac{10}{10.02} = 0.998.$$

EXAMPLE 4.5

Find out the values of various currents through the Transistor with voltages as shown in Fig. 4.35. Collector supply voltage V_{CC} = 20 V and the Base voltage V_{BB} = 4 V. Transistor current gain $\beta = 500$. Assume $V_{BE} = 0.7$ V. Also calculate the Collector voltage V_C.

Solution: Various currents through the Transistor are Collector Current I_C, Base Current I_B and Emitter Current I_E. Emitter Current $I_E = (I_C + I_B)$

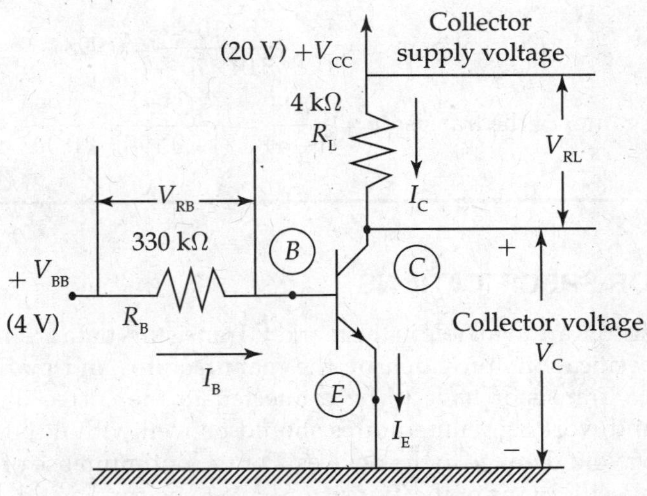

FIG. 4.35 *Transistor circuit*

In the Base Emitter circuit using *KVL* equation,

$$V_{BB} = I_B \cdot R_B + V_{BE}$$

$$V_{BB} - V_{BE} = I_B \cdot R_B. \quad (1)$$

From Eq. (1), we get

$$I_B = \frac{[V_{BB} - V_{BE}]}{R_B} = \frac{(4-0.7)}{330\times10^3} = \frac{3.3\times10^{-3}}{330}$$

$$= \frac{3300\times10^{-6}}{330} = 10\times10^{-6} = 10\,\mu\text{A}$$

$$\therefore \quad I_C = \beta\cdot I_B = 500\times10\times10^{-6} = 5\text{ mA}$$

$$I_E = I_C + I_B = (5+0.01)\times10^{-3} = 5.01\text{ mA.}$$

In the output circuit loop using *KVL* equation

$$V_{CC} = I_C \cdot R_L + V_C. \quad (2)$$

From Eq. (2), Collector voltage $V_C = V_{CC} - I_C \cdot R_L$

$$V_C = (20 - 5\times10^{-3}\times4\times10^3) = (20 - 20) = 0\text{ V.}$$

EXAMPLE 4.6

For a CB-operated Transistor, Emitter Current $I_E = 5.01$ mA and the Base Current is 0.01 mA. Calculate the values of Collector Current I_C, current gain 'α' and the current gain 'β'.

Solution:

$$\text{Emitter current } I_E = I_C + I_B \tag{1}$$

From Eq. (1), Collector Current $I_C = I_E - I_B = (5.01 - 0.01) \times 10^{-3} = 5$ mA

$$\text{Current gain } \alpha = \frac{I_C}{I_E} = \frac{5\times10^{-3}}{5.01\times10^{-3}} = 0.998.$$

$$\text{Current gain } \beta \text{ of the transistor} = \frac{\alpha}{(1-\alpha)} = \frac{0.998}{(1-0.998)} = \frac{0.998}{0.002} = 499.$$

4.17 TRANSISTOR SPECIFICATIONS

Many types of Transistors are available in the market. Transistors data are available with some of the following specifications provided by the manufactures and marketing companies. The three leads of a Transistor have to be connected in the correct way, and operating voltages, currents and working temperatures should be well within the maximum ratings of the Transistors to avoid damage to the devices. For the optimum use of the Transistor, the technical data and specifications of the Transistor have to be considered.

- *General data provide the details of the nomenclature of the Transistor number.*

1. The first letter of the Transistor number indicates whether it is a germanium or silicon Transistor. For example, the first letter 'B' of the Transistor BC107 indicates that it is a silicon Transistor. Similarly, the first letter 'A' of the Transistor AC126 indicates that it is a germanium Transistor. Once we know the Transistor material, we can determine its applications and operating temperatures. Silicon Transistors are more popular in use.
2. Second letter of the Transistor number C indicates that it is low-power audio frequency Transistor. Second letter D indicates that it is high-power audio frequency Transistor. Second letter F indicates that is a low-power high-frequency Transistor. Thus, some of the common applications of Transistors in audio frequency amplifiers, power amplifiers, radio frequency amplifier circuits and so on can be known.
3. If the Transistor number contains TIP as beginning letters, it indicates that it is a Texas Instruments power Transistor.
4. Identification of NPN- or PNP-type Transistor is made from the structural details of the Transistor. As the polarities of the operating voltages are different for the two types of Transistors, during replacement of a faulty Transistor with a good Transistor, replacement should be made with the exact type of Transistor.
5. Two or three digit numbers after the first two letters indicate the manufacturer's identification number.
6. Pin configuration details for identification of the Transistor lead 'Emitter, Base and Collector terminals' for making electrical connections in the circuit.

7. Once the design of an electronic circuit in a project is done with a particular Transistor, after some time same types of Transistors may not be available in the market. Then, Transistors with similar electrical properties will be available as possible substitutes for suitable replacements.

- Maximum ratings or limits of typical operating voltages and currents for safe utility of the device. Transistor fails if the operating voltages, currents and the temperatures exceed the maximum ratings specified by the vendors.
  1. Maximum Collector Current $I_{C(max)}$
  2. $V_{CE(max)}$ is the maximum voltage across the Collector and the Emitter. This rating is more important as the reverse-bias voltage to the output junction of the Transistor must be well below $V_{CE(max)}$ rating to avoid the break down.
  3. $P_{t(max)}$ suggests the total allowed power at the amplifier output and the power dissipation ($P_C = I_{CQ} \cdot V_{CE}$) at the Collector junction of the Transistor in the amplifier circuit. Heat sinks are used to ventilate the heat generated by the Transistors so that within the maximum limit $P_{t(max)}$, maximum power can be realised.
  4. No such considerations are needed for the voltages and currents at the Transistor input circuit, as the operating power levels are small at input circuit.
- Application of the device in amplifier, oscillator and switching circuits suggests the know how of the minimum and maximum values of the current gain β of the Transistor. For example for the design of an oscillator circuit, minimum value of h_{fe} or β of the Transistor has to be assured to satisfy the 'Barkhausen conditions' for oscillations in the circuit.
- Typical Transistor parameters such as h-parameters, hybrid-π parameters, operating voltages and alpha-cutoff frequency are also provided in Transistor specifications in the catalogues. Bandwidth specifications for the Transistor amplifiers with Transistor configurations will be of great guide in the electronics circuit designs and safe operations (Figs. 4.34 and 4.35).

4.18 HIGH-FREQUENCY LINEAR MODELS FOR THE COMMON EMITTER TRANSISTOR

Common Emitter Transistor circuit is mostly used in practical circuits. So CE Transistor model valid at high frequencies is discussed. The circuit is known as hybrid-π or Giacoletto model (Fig. 4.36). It is applicable for both low and high frequencies.

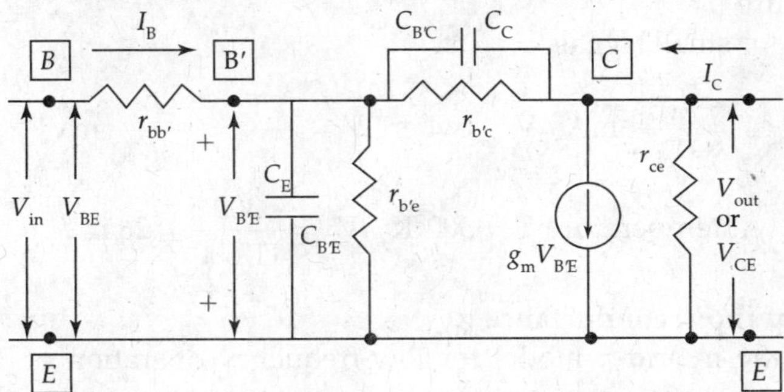

FIG. 4.36 *Hybrid-π equivalent circuit of common emitter transistor considered at high frequencies*

4.18.1 Hybrid-π or Giacoletto Model

Transistor circuit analysis using this hybrid-π model agrees with experimental results at all frequencies, where the Transistor gives reasonable amplification. All parameters (resistances and capacitances) in this model are assumed to be independent of frequency.

Elements of the Hybrid-π Model

1. $r_{bb'}$ is the Base spreading resistance between the external Base terminal B and the effective Base point B'. (Base spread resistance) $r_{bb'}$ is of the order of 100 Ω.
2. $r_{b'e}$ represents the resistance of the forward-biased Emitter junction. It is of the order of 1000 Ω.
3. $C_{B'E}$ or C_E is the capacitance of the *forward-biased input junction*. This diffusion capacitance accounts for the excess minority carrier storage in the Base region. C_E is of the order of 100 pF. It is placed in parallel with $r_{b'e}$.
4. $r_{b'c}$ represents the resistance of the reverse -biased Collector junction. It is of the order of 2–4 MΩ.
5. $C_{B'C}$ or C_C is the capacitance of the reverse-biased Collector junction. C_C is of the order of 2 pF.
6. g_m, $V_{B'E}$: Small changes in voltage $V_{B'E}$ across the Emitter junction result in small signal Collector Current $g_m \cdot V_{B'E}$. This is accounted as current generator $g_m \cdot V_{B'E}$ (when Collector is shorted to Emitter) (Fig. 4.36).
7. $r_{ce} = 1/g_{CE}$: The conductance between Collector and the Emitter is g_{CE}.

Typical values of hybrid-π parameters of CE Transistor at high frequencies

Parameter	$r_{bb'}$	$r_{b'e}$	$C_{B'E}$ or C_E	$C_{B'C}$ or C_C or C_0	$r_{b'c}$	r_{ce}	g_m
Value	100 Ω	1000 Ω	100 pF	2 pF	3–4 MΩ	100 kΩ	50 milli mhos

Determination of hybrid-π parameters from the published data by manufacturers:

Manufacturers supply data that characterise their Transistors. The determinations of hybrid-π parameters from commonly published data are considered.

1. Determination of transconductance g_m:

Transconductance 'g_m' can be calculated from the value of the Collector Current 'I_{CQ}' at the DC or quiescent operating current decided with the class of operation of amplifiers and the operating temperature 'T'.

In the linear region for small signals:

$$g_m = \frac{|I_{CQ}|}{V_T} \text{ mA/V} \quad \left[\text{where } V_T = \frac{KT}{e} = \frac{T}{11{,}600} \text{ mV}\right] \tag{4.51}$$

$$\text{At temperature } T = 300°\text{K} \quad V_T = \frac{300}{11{,}600} \cong 26 \text{ mV}. \tag{4.52}$$

2. Determination of input conductance $g_{b'e}$:

Figure 4.37 shows the hybrid-π model for low-frequency operation of Transistors. At low frequencies, capacitances have very high impedances. Their effects are negligible in the circuit performance.

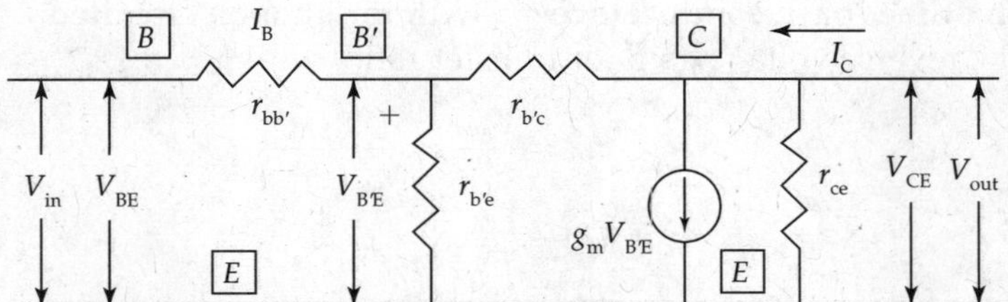

FIG. 4.37 *Hybrid-π equivalent circuit of a transistor considered at low frequencies*

From typical values of circuit components of CE Transistor, $r_{b'c} \gg r_{b'e}$. Hence, Base Current I_B flows through $r_{b'e}$. Therefore, $V_{B'E} = r_{b'e} \cdot I_B$.

The short circuit Collector Current $I_C = g_m \cdot V_{B'E}$, $\therefore \quad I_C = g_m \cdot V_{B'E} = g_m \cdot I_B \cdot r_{b'e}$.

The short circuit current gain h_{fe} is defined by

$$h_{fe} = \frac{I_C}{I_B}\bigg| V_{CE} = 0. \tag{4.53}$$

$$\text{i.e.,} \quad h_{fe} = \frac{g_m \cdot V_{B'E}}{I_B} = \frac{g_m \cdot I_B \cdot r_{b'e}}{I_B} = g_m \cdot r_{b'e}$$

$$\therefore \quad r_{b'e} = \frac{h_{fe}}{g_m} \text{ and also } g_{b'e} = \frac{g_m}{h_{fe}} \tag{4.54}$$

Transistor manufacturers usually specify $\beta_0 = h_{fE}$ (4.55)

$$r_{b'e} = \frac{h_{fe}}{g_m} = \frac{\beta_0}{g_m}. \tag{4.56}$$

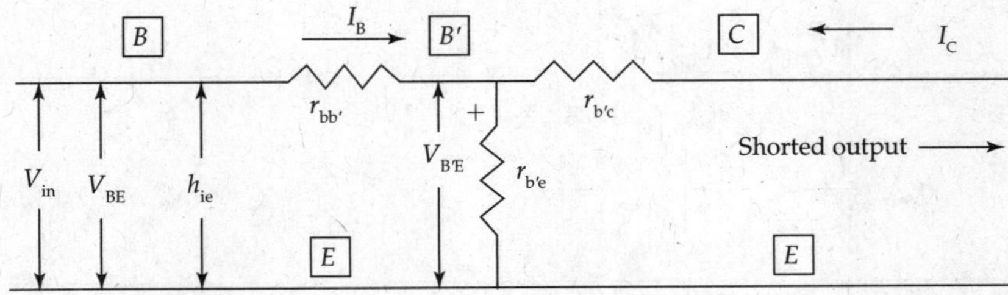

FIG. 4.38 *Hybrid-π equivalent circuit of a transistor with output shorted*

3. Determination of $r_{bb'}$:

From Fig. 4.38, the input resistance with output shorted is h_{ie}.

Under such condition, $r_{b'e}$ is parallel with $r_{b'c}$

$$\therefore \quad r_{b'e} \parallel r_{b'c} \cong r_{b'e} \text{ since } r_{b'c} = 4\ \text{M}\Omega \gg r_{b'e} = 1\ \text{k}\Omega \tag{4.57}$$

$$h_{ie} = (r_{bb'} + r_{b'e})$$

$$\therefore \quad r_{bb'} = (h_{ie} - r_{b'e}) = \left[h_{ie} - \frac{\beta_0}{g_m}\right] \tag{4.58}$$

From typical values of $h_{ie} = 1100\ \Omega$ and $r_{b'e} = 1\ \text{k}\Omega$, $r_{bb'} = (1100 - 1000) = 100\ \Omega$

4. Determination of feedback conductance $g_{b'c}$ with input open circuited

h_{re} is defined as the reverse voltage gain (or transfer ratio) with $I_B = 0$.

$$h_{re} = \frac{V_{B'E}}{V_{CE}} = \frac{r_{b'e}}{(r_{b'c} + r_{b'e})}$$

$$\therefore \; h_{re}(r_{b'c} + r_{b'e}) = r_{b'e}$$

$$h_{re} \cdot r_{b'c} = r_{b'e}(1 - h_{re})$$

$$\because \; h_{re} \cong 10^{-4}, \quad \text{i.e.,} \quad h_{re} \ll 1$$

$$\text{Then} \quad (1 - h_{re}) \cong 1.$$

Therefore, $h_{re} \cdot r_{b'c} = g_{b'e}$.

$$\therefore \; r_{b'c} = \frac{r_{b'e}}{h_{re}} \quad \text{or} \quad g_{b'c} = \frac{1}{r_{b'c}} = \frac{h_{re}}{r_{b'e}} = h_{re} \cdot g_{b'e} \tag{4.59}$$

When $h_{re} = 10^{-4}$, $r_{b'c} = \frac{r_{b'e}}{10^{-4}} = 10^4 \; r_{b'e}$ which means that $r_{b'c} \gg r_{b'e}$

that is the reverse resistance $r_{b'c}$ will be of the order of mega ohms.

5. Determination of output conductance $g_{ce} = \frac{1}{r_{ce}}$:

$$I_C = \left[\frac{V_{CE}}{r_{ce}}\right] + \left[\frac{V_{CE}}{r_{b'c} + r_{b'e}}\right] + g_m \cdot V_{B'E}$$

With input open circuited, output conductance is defined as h_{oe} for $I_B = 0$ with $I_B = 0$, $V_{BE} = h_{re} \cdot V_{CE}$

$$I_C = \left[\frac{V_{CE}}{r_{ce}}\right] + \left[\frac{V_{CE}}{r_{b'c} + r_{b'e}}\right] + g_m \cdot h_{re} \cdot V_{CE}$$

$$h_{oe} = \frac{I_C}{V_{CE}} = \left[\frac{1}{r_{ce}}\right] + \left[\frac{V_{CE}}{r_{b'c} + r_{b'e}}\right] + g_m \cdot h_{re} \tag{4.60}$$

$$h_{oe} = g_{ce} + g_{b'c} + g_m \cdot h_{re} \quad [\because \; (r_{b'c} + r_{b'e}) \cong r_{b'c}] \text{ and } g_{b'c} = \frac{1}{r_{b'c}} \tag{4.61}$$

using ($g_m = g_{b'e} \cdot h_{fe}$) in the above equation

$$h_{oe} = g_{ce} + g_{b'c} + g_{b'e} \cdot h_{fe} \cdot h_{re} \quad [\because (r_{b'c} + r_{b'e}) \cong r_{b'c}] \text{ and } g_{b'c} = \frac{1}{r_{b'c}}$$

$$\therefore \; \text{Output conductance} \quad g_{ce} = [h_{oe} - g_{b'c} \cdot h_{fe}] \tag{4.62}$$

From Eq. (4.61), we get

$$g_{ce} = [h_{oe} - g_m \cdot h_{re}] \tag{4.63}$$

neglecting $g_{b'c} = \frac{1}{r_{b'c}} \ll 1$.

6. Determination of $C_{B'C}$:

The value of the capacitance of the reverse-biased Collector-to-Base function C_{OB} is often specified in the manufacture's Transistor's data. $C_{B'C} = C_{OB}$ is the capacitance measured between the Collector and Base leads with the Emitter lead open circuited (for signal frequencies) $C_{B'C} = C_{OB}$.

7. Determination of $C_{B'E}$ or C_E:

Experimentally, $C_{B'E}$ is determined from a measurement of f_T, the frequency at which the CE short circuit current gain drops to unity.

$$C_{B'E} \text{ or } C_E = \frac{g_m}{2\pi f_T}. \tag{4.64}$$

Validity of Hybrid-π Model The hybrid-π model is valid under dynamic conditions, when the rate of change of V_{BE} is small enough so that the Base incremental current ΔI_B is also small. It was also proved that the hybrid-π model is valid for frequencies up to approximately $f_T/3$, where 'f_T' is the frequency at which current gain of an amplifier is unity. f_T is also known as unity gain frequency. f_T incidentally is the Gain-Bandwidth product of an amplifier, which will be explained more later.

EXAMPLE 4.7

Determine the hybrid-π parameters of a Transistor operating at Collector Current $I_C(Q) = 2$ mA, $V_{CEQ} = 20$ V and $I_{BQ} = 20$ μA. Transistor specifications are $\beta_0 = 100$; unity gain frequency $f_T = 50$ MHz; $C_{OB} = 3$ pF; $h_{iE} = 1.4$ kΩ; $h_{re} = 2.5 \times 10^{-4}$. $h_{oe} = 25$ μmhos. Assume that the operating temperature is 300°K.

Solution:

a. $g_m = \dfrac{|I_{CQ}|}{V_T} = \text{mA/V}$, where $V_T = \dfrac{kT}{e} = \dfrac{T}{11,600} \cong 26 \text{ mV}$ when $T = 300°\text{K}$

$\therefore$ Transconductance $g_m = \dfrac{2\text{ mA}}{26\text{ mV}} = \dfrac{2\times10^{-3}}{26\times10^{-3}} = \dfrac{1}{13}$ mhos when $I_{CQ} = 2$ mA.

b. $r_{b'e} = \dfrac{\beta_0}{g_m} = \dfrac{100\times13}{1} = 1300\ \Omega$.

c. $r_{bb'} = (h_{ie} - r_{b'e}) = (1400 - 1300) = 100\ \Omega$.

d. $r_{b'c} = \dfrac{r_{b'e}}{h_{re}} = \dfrac{1300}{2.5\times10^{-4}} = 5.2\times10^{6}\ \Omega$.

e. $g_{ce} = [h_{oe} - g_m \cdot h_{re}] = \left[25\times10^{-6} - \dfrac{1}{13}\times2.5\times10^{-4}\right]$

$\therefore\ g_{ce} = \left[\dfrac{13\times25\times10^{-6} - 250\times10^{-6}}{13}\right] = \left[\dfrac{(325-250)\times10^{-6}}{13}\right]$

$\therefore\ g_{ce} = \dfrac{75\times10^{-6}}{13} = 5.77\times10^{-6}$ mhos.

$$\therefore \quad r_{ce} = \frac{1}{g_{ce}} = \frac{13}{75} \times 10^6 = 173.3\ \text{k}\Omega.$$

f. Output junction capacitance $C_{B'C} = C_{OB} = 3$ pF.

g. Input junction capacitance $C_{B'E} = C_E = \dfrac{g_m}{2\pi f_T} = \dfrac{1}{13 \times 2\pi \times 50 \times 10^6} = 245$ pF.

4.19 APPLICATIONS OF BJT AS A SWITCH

A Transistor acts as a switch.

4.19.1 Transistor as an Open Switch

- It acts as an open switch, when the voltage V_{BE} between the Base and the Emitter of the Transistor is either zero or reverse-biased so that it operates in the 'cut-off region' of its output characteristics.
- In the output circuit, $V_{CE} = V_{CC} - I_C\ R_L$.
- When $V_{BE} = 0$ V, Base Current I_B is zero and Collector Current I_C is zero.
- Therefore, $V_{CE} = V_{CC}$.
- When the Transistor is not conducting, Collector voltage V_{CE} will be equal to the supply voltage V_{CC}.
- Then the Transistor is considered as a switch in the off condition.
- Power dissipation by the Transistor is zero as the Collector Current through the Transistor is zero, when the Transistor is in the open switch operation.

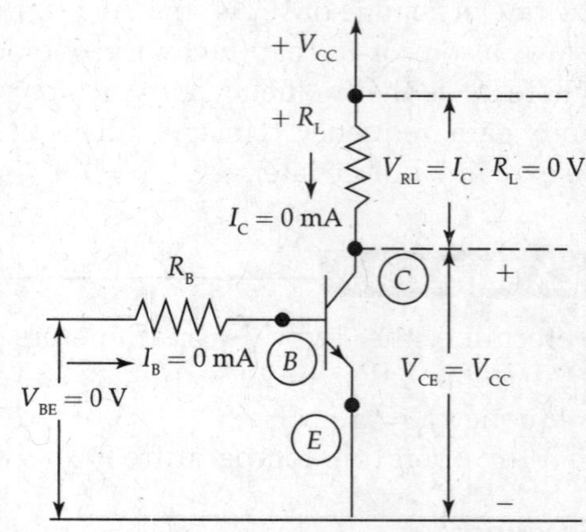

FIG. 4.39 *Common emitter transistor function as a 'open switch'*

4.19.2 Transistor as a Closed Switch

- It acts as a closed switch, when the voltage V_{BE} between the Base and the Emitter of the Transistor is positive that is forward-biased to the Emitter junction so that it operates in the 'saturation region' of its output characteristics.
- In the output circuit, $V_{CE} = V_{CC} - I_C \cdot R_L$.
- Forward-bias voltage V_{BE} is sufficiently large so that the resulting Collector Current I_C causes the voltage drop $I_C \cdot R_L$ equals the supply voltage V_{CC}.
- Then $V_{CE} = V_{CC} - I_C \cdot R_L = V_{CC} - V_{CC} = 0$ V.
- When the Transistor is conducting Collector to Emitter voltage V_{CE} will be zero.
- Then the Transistor is considered as a switch in the 'ON' condition.

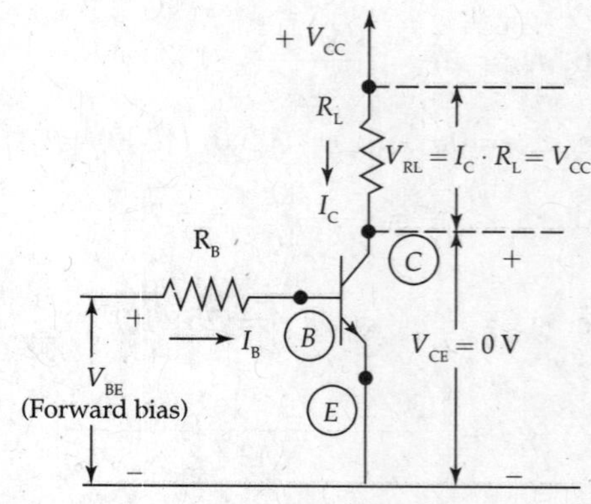

FIG. 4.40 *Common emitter transistor function as a 'closed switch'*

- Power dissipation by the Transistor $P_D = V_{CE} \cdot I_C = 0. I_C = 0$ W, when the Transistor acts as an 'ON' switch.
- Finally, the power dissipations in the Transistor take place only during the transitions between open and closed switch operations (Figs. 4.39 and 4.40).

4.19.3 Junction Field Effect Transistor

There are two types of field effect Transistors.

- JFET (J FET is also known as Junction Gate FET)
- MOSFET (Metal Oxide Semiconductor Field Effect Transistors) (MOSFETs are also called Insulated Gate FET (IGFET) from the Structure and Operation of MOSFET devices).
- Field Effect Transistor is a unipolar device. It has only one type of current carriers, the majority carriers for electrical conduction of the device.
- If the current carriers through FET are 'Electrons', it is known as N-Channel JFET.
- If the current carriers through FET are 'Holes', it is known as P-Channel JFET.

The circuit symbols of N- and P-channel FETs are shown in Fig. 4.41.

- Just like a Transistor, FET device also has three terminals namely: 'Source', 'Gate' and 'Drain'. But in a FET, the Source and Drain terminals are interchangeable. But the Emitter and Collector terminals cannot be interchanged in Transistor devices.

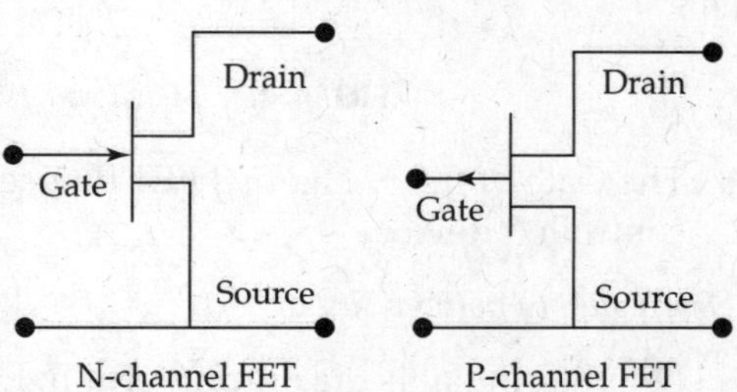

FIG. 4.41 *Device symbols for N-channel and P-channel FETs*

4.20 TYPICAL STRUCTURAL DETAILS OF JFET

FET structure (Fig. 4.42):

- FET device consists of N-type semiconductor bar with two small P-type material regions on either side of the bar.
- Upper end of the bar has an external terminal known as 'Drain'.
- Lower end of the bar has an outer terminal known as 'Source'.
- The two P-type regions on both sides of the bar are connected together to function as 'Gate' terminal.
- N-type region between the two P-type regions is known as N-channel. This type of JFET device is known as N-channel FET.
- The structural details for the FET are shown in Fig. 4.42.
- There are no separate input and output junctions as in the case of Transistors.

Gate supply voltage V_{GS} or V_{GG}

- Gate Terminal is the input terminal of the FET device.
- Gate Voltage V_{GS} (DC voltage) is connected between Gate and Source terminals of N-channel FET device. Negative terminal of the voltage Source is connected to the Gate terminal and positive terminal to the Source terminal. The two P–N Junctions between Gate and Source are thus reverse-biased by DC voltage V_{GS}.

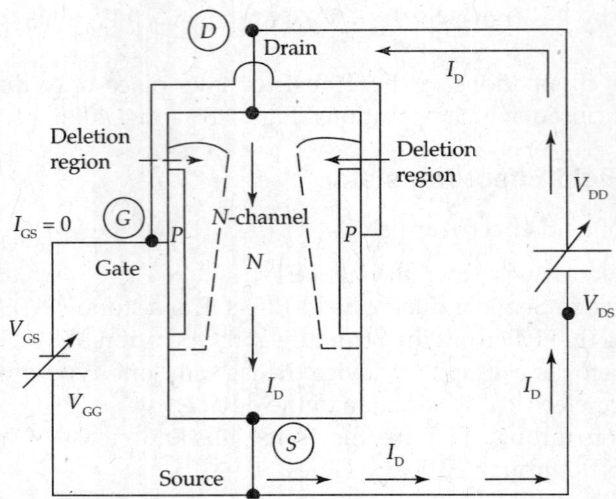

FIG. 4.42 *Structural details of N-channel field effect transistor*

- The Gate of the N-channel FET device draws negligible current $I_G = 0$. So FET is a voltage-controlled device.

Drain supply voltage V_{DS} or V_{DD}

- Drain terminal is the output terminal of the FET device.
- Source terminal is common to both input and output circuits of the FET for common Source FET operation.
- Drain supply voltage V_{DS} (DC voltage) is connected between Drain and Source terminals of N-channel FET device. Positive terminal of the voltage Source V_{DS} is connected to the Drain terminal and the negative terminal of the DC Source is connected to the Source terminal as shown in Fig. 4.42.
- Electrons are the majority carriers in the N-type channel of the FET.
- Electrons flow from Drain to Source in the output circuit between Drain and the Source for various values of 'V_{DS}'.
- The conventional Drain Current 'I_D' flows from the positive terminal of 'V_{DS}' through the device from the Drain to the Source and then completes its path to the negative terminal of 'V_{DS}' as shown in Fig. 4.42.

4.21 WORKING OF JFET

Initially, when $V_{GS} = 0$ V and $V_{DS} = 0$ V, the Depletion region widths (of unbiased junctions) about the junctions are uniform. With the increases in voltages, V_{DS} for different constant values of V_{GS}, the situation is that the reverse-bias 'V_{GS}' is more towards Drain end than the Source end at the two P–N junctions in the device. (Varying magnitudes of the reverse-biases are due to the flow of the Drain Current 'I_D' through the device channel across increased areas of cross-sections of the channel from the Source to the Drain, thus producing different voltage drops across the channel.) These varying reverse-biases cause varying magnitudes of Depletion region widths about the two junctions. Increased reverse-bias magnitudes from Source to Drain cause increased Depletion region widths towards the Drain as shown in Fig. 4.42.

With the Drain shorted to the Source and a negative voltage 'V_{GS}' applied between Gate and Source, two reverse-biased junctions appear on both sides of the Gate and Depletion regions spread into the N-type slab. As the reverse voltage between Drain and Source increases, conducting area of the channel is reduced. Drain Current decreases with increasing reverse-bias and when it is high enough the channel virtually closes and small reverse current flows. This occurs when I_D is almost zero. Then the channel is pinched off (cut off with $V_{DS} = 0$ V).

Using the experimental setup of Fig. 4.43, output (Drain) characteristics (Fig. 4.44) and the transfer characteristics (Fig. 4.45) of FET device can be obtained. Using these characteristics, the basic properties (Device constants) of a FET can be determined.

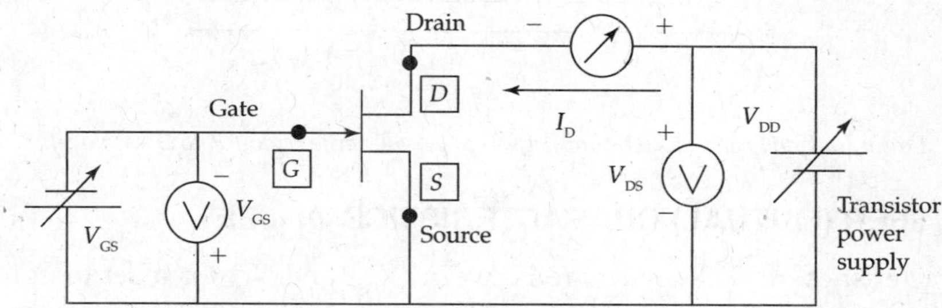

FIG. 4.43 *Experimental set up to obtain FET characteristics*

4.21.1 Output (Drain) Characteristics of JFET Device

Output characteristics of the FET device are between the Drain to Source Voltage 'V_{DS}' and Drain Current 'I_D' for a set of constant Gate to Source voltages V_{GS}. Initially, voltage V_{GS} is kept at 0 V. It is kept constant. Then varying the voltage V_{DS} in increments, corresponding incremental values of I_D are tabulated.

Repeat the above procedure for a set of constant values of V_{GS} such as ($V_{GS} = -1$ V), ($V_{GS} = -2$ V), ($V_{GS} = -3$ V) and ($V_{GS} = -4$ V). The measurements are recorded in the following table of observations.

Measurements to be made for obtaining the output characteristics of JFET

	Gate to source DC voltage, $V_{GS} = 0$ V	
S. No.	**Drain to source voltage, V_{DS} (V)**	**Drain current, I_D (mA)**

Using the observations made in the table, a family of output characteristics is drawn as shown in Fig. 4.44. *Drain Resistance* 'r_d' can be calculated from the output characteristics in the pinch-off region. *Amplification factor* 'μ' can also be determined from the output characteristics shown in Fig. 4.48. Method of calculations of 'μ' is shown in Fig. 4.48.

From the definitions of 'μ', 'g_m' and 'r_d', it can be determined that amplification factor is the product of g_m and r_d ($\mu = g_m \cdot r_d$). FET parameters are defined in Section 4.24. Maximum gain of FET amplifiers is decided by maximum value of the amplification factor of FETs (FET characteristics).

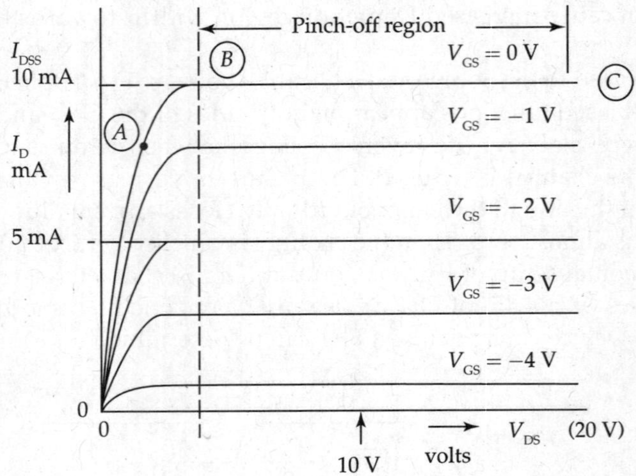

FIG. 4.44 *Drain (output) characteristics of JFET 0 to A → Ohmic region A to B → Non-linear region*

4.22 TRANSFER (MUTUAL) CHARACTERISTICS OF JFET

Transfer characteristics of FET device are between 'V_{GS}' and 'I_D' for a set of constant 'V_{DS}' (Fig. 4.45). Initially, V_{DS} is kept constant say at 10 V. Varying 'V_{GS}' from 0 to the pinch-off voltage 'V_P' of the device in increments (V_P is the Gate Voltage when 'I_D' becomes zero), corresponding increments of 'I_D' are noted and tabulated.

Measurements to be made for obtaining the transfer characteristics of JFET		
	Drain to Source Voltage, V_{DS} = 10 V	
S. No.	**Gate to Source Voltage, V_{GS} (V)**	**Drain Current, I_D (mA)**

$$\text{Drain current } I_D = I_{DSS}\left[1-\frac{V_{GS}}{V_P}\right]^2 \text{ mA.} \tag{4.65}$$

I_{DSS} = Drain Saturation Current.
Drain Saturation Current is the Drain Current when $V_{GS} = 0$ V [Eq. (4.65)].
V_{GS} = Gate to Source Voltage.
V_P = Pinch-off voltage (when $V_{GS} = V_P$, Drain Current is zero from Eq. (4.65)).
Figure 4.45 shows the transfer characteristic drawn for the measurements made above. It can be observed that when $V_{GS} = 0$ V, Drain Current I_D is equal to the Drain Saturation current. For increasing negative Gate Voltages, Drain Current goes on decreasing. At Gate Voltage V_{GS} (off) known as the *Pinch-off voltage*, Drain Current becomes zero.

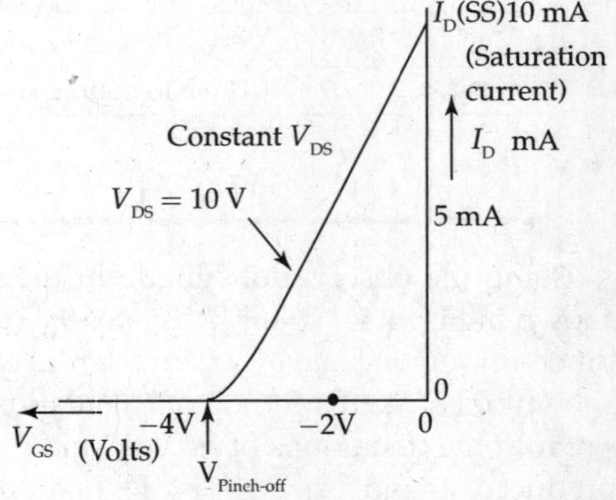

FIG. 4.45 *JFET transfer or mutual characteristic*

This procedure may be repeated for another constant voltage V_{DS} to obtain another transfer characteristic. The transfer characteristic is shown in Fig. 4.45. Transconductance 'g_m' can be calculated from the *transfer or mutual characteristic.*

4.23 DRAIN (OUTPUT) CHARACTERISTICS OF FIELD EFFECT TRANSISTOR

4.23.1 Ohmic Region 0 to *A* on the Output Characteristics

- V_{GS} is maintained constant at certain values (for example, $V_{GS} = 0$ V) as shown in Fig. 4.46. At each constant value of V_{GS}, as V_{DS} is increased up to a few volts, Drain Current I_D linearly increases from the origin '0'to '*A*'.
- The majority carrier electrons flow from the Source to the Drain through N-type channel and then complete their path through the Drain circuit. So the conventional Drain Current flows from the Drain to the Source through the device and then through the Drain circuit.
- The channel resistance is uniform and acts as a semiconductor resistor. The channel resistance limits the Drain Current. The resistance increases with V_{DS} as shown on the output characteristics of the FET device in Fig. 4.46. This region from '0' to '*A*' on the output characteristics is known as *Ohmic region.*

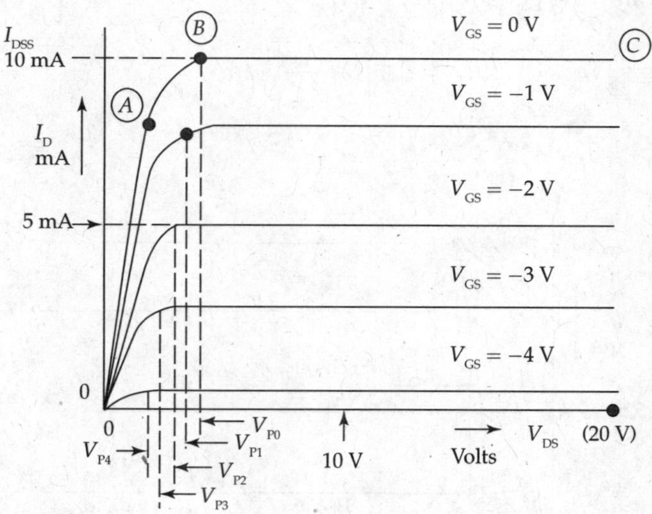

FIG. 4.46 *JFET characteristics with different values of pinch-off voltages depending upon the values of* V_{GS}

4.23.2 Nonlinear Region *A* to *B* on the Drain Characteristics

For each V_{GS}, $V_{DS}/I_D = r_d$, where r_d is the Drain Resistance. In this region of operation, the FET device behaves like a Voltage Variable Resister (VVR) or Voltage-Dependent Resistor (VDR).

For higher values of V_{DS}, the channel resistance becomes non-uniform, during the region '*A* to *B*' on the output characteristic, since the Depletion region width changes along the length of the channel from Source to Drain as shown in Fig. 4.42. This region is known as *nonlinear region.*

4.23.3 PINCH-OFF REGION *B* TO *C* ON THE DRAIN CHARACTERISTICS

At the Source end, the Depletion region width is minimum and at the Drain end it is maximum, because of varying reverse-bias voltages (due to varying voltage drops between the Source and the Drain along the channel region due to the flow of the Drain Current I_D) at the Gate–Source and Gate–Drain junctions, respectively, that can be seen from Fig. 4.42. Thus for each V_{GS} on the Drain characteristics, with increasing values of V_{DS}, the magnitudes of the Drain Current I_D increases up to a Drain to Source voltage V_{DSP} simply known as the pinch-off voltage V_P, *the pinch-off voltage* V_P.

The Drain Current starts increasing slowly and finally reaches a constant value or I_D reaches a saturation value (where the Depletion region widths come closer, but never overlap for the prevention of disappearance of reverse-bias to maintain the device operation) above the Knee portion from the pinch-off voltage V_P as shown in Fig. 4.46. The region 'B to C' on the output characteristic is the *pinch-off region*, the active region for *amplifier operation.*

4.23.4 Drain Saturation Current I_{DSS}

When $V_{GS} = 0$ V, Drain Current I_D will be maximum over that is reached, and it is called *Drain Saturation Current* I_{DSS}. This is also clear from Eq. (4.66). The relation between I_D, I_{DSS}, V_{GS} and V_P is given by the following equation (Fig. 4.47).

$$I_D = I_{DSS}\left[1 - \frac{V_{GS}}{V_P}\right]^2 \text{ mA.} \tag{4.66}$$

When $V_{GS} = 0$ V, $I_D = I_D$ (SS).

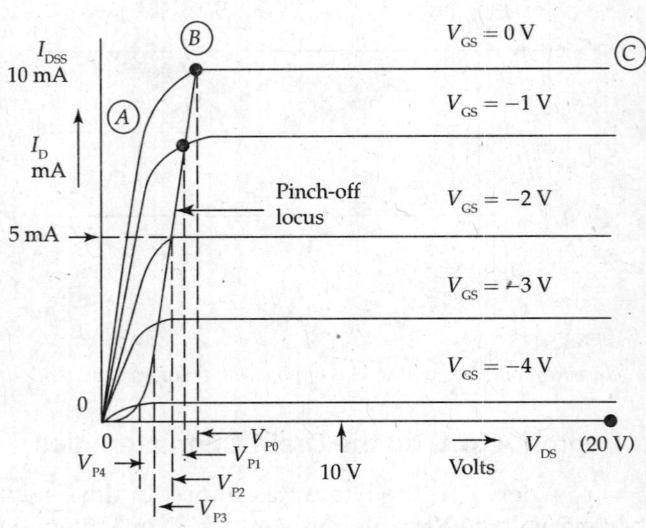

FIG. 4.47 *JFET characteristics with pinch-off locus*

For a given V_{GS}, V_{DS} versus I_D characteristics (output characteristics) have four regions:

1. Linear region from '0' to 'A' on the output characteristic of JFET is known as VDR region or simply *Ohmic region*.

2. The knee portion from 'A' to 'B' on the JFET output characteristic, in which the device resistance suddenly changes from a low value to a high value, exhibits a nonlinear behaviour. This is called as the *nonlinear region.*
3. From 'B' to 'C' portion of the JFET output characteristic, the curve is almost parallel to the x-axis (indicating the saturated or constant values of Drain Current I_D) and is known as the *Pinch-Off region* and the characteristic in this region exhibits a high resistance of the order of a few hundred kilo ohms and it is called as (one of the FET device parameters or constants) 'Drain Resistance r_d'. This device internal resistance is used in the representation of the small signal AC equivalent circuit for the analysis of the FET amplifier operation. The pinch-off region also suggests the voltage-controlled current Source (VCCS) behaviour of the device in practical circuit applications.
4. Beyond the point 'C', when the voltage V_{DS} becomes large enough to break down, the junction causing a sudden high rise of Drain Current, I_D. This *break-down region* is never used in practice, and so it is not shown on the output characteristics of JFET device.

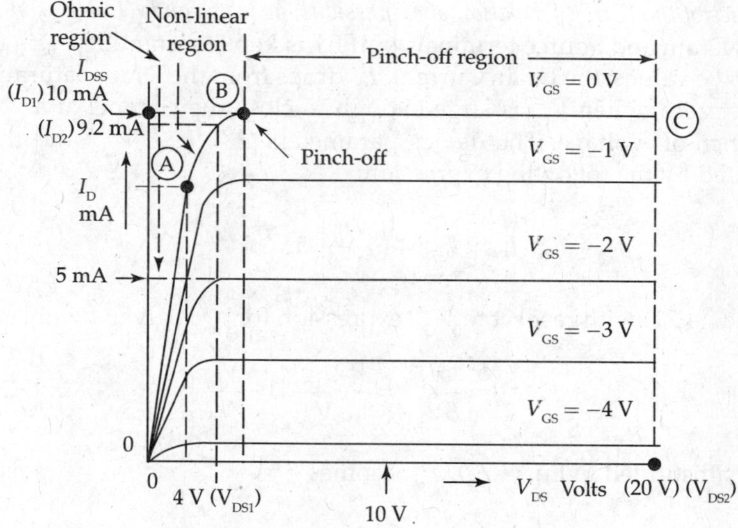

FIG. 4.48 *Drain characteristics or output characteristics JEFT characteristics with different regions of operation*

4.24 DEFINITIONS OF FET CONSTANTS

The three FET device constants are (1) Transconductance 'g_m', (2) Drain resistance 'r_d' and (3) Amplification factor 'μ':

$$\text{Transconductance } g_m = \frac{\Delta I_D}{\Delta V_{GS}} |V_{DS} = \text{constant} \rightarrow \text{(mhos)} \tag{4.67}$$

$$\text{Drain resistance } r_d = \frac{\Delta V_{DS}}{\Delta I_D} |V_{GS} = \text{constant} \rightarrow (\Omega) \tag{4.68}$$

$$\text{Amplification factor } \mu = \frac{\Delta V_{DS}}{\Delta V_{GS}} |I_D = \text{constant} \rightarrow \text{(constant) (no units)} \tag{4.69}$$

From these definitions, $\mu = \dfrac{\Delta V_{DS}}{\Delta V_{GS}} = \dfrac{\Delta V_{DS}}{\Delta I_D} \times \dfrac{\Delta I_D}{\Delta V_{GS}} = r_d \cdot g_m$

$$\therefore \quad \mu = g_m \cdot r_d \quad \text{(relation between } \mu, g_m \text{ and } r_d\text{).} \tag{4.70}$$

Calculation of drain resistance of FET from the drain characteristics

Drain Resistance 'r_d' can be calculated from the JFET output or Drain characteristics. Drain Resistance is the inverse of the slope of the output characteristic. Drain Resistance is the ratio of the increment in Drain to Source Voltage 'ΔV_{DS}' between 'V_{DS2}' and 'V_{DS1}' and corresponding increment in Drain Current 'ΔI_D' between 'I_{D2}' and 'I_{D1}' that are obtained from graphs. From the characteristics in Fig. 4.48, it is clear that large change in the Drain to Source Voltage produces a small change in the Drain Current.

Increment in Drain Voltage = $\Delta V_{DS} = [(V_{DS2} - V_{DS1})] = 20 - 4 = 16$ V

Increment in Drain Current $\Delta I_D = [I_{D2} - I_{D1}] = 10 - 9.2 = 0.8$ mA.

Drain Resistance 'r_d' is equal to 20 kΩ, which is the ratio of ΔV_{DS} to ΔI_D.

Transfer Characteristics of the JFET (derivation of expressions for transconductance g_m and g_{m0})

Voltage 'V_{DS}' across Drain and Source terminals of JFET is kept constant. If V_{GS} is varied from zero to higher negative values, the Drain Current 'I_D' drops from the Drain Saturation Current 'I_{DSS}' to reach almost zero, when V_{GS} is large enough to close the channel at a voltage V_P. It is known as the pinch-off voltage. The device parameters 'I_D', 'I_{DSS}', V_{GS}' and the pinch-off voltage 'V_P' are related by the following expression:

$$I_D = I_{DSS}\left[1 - \frac{V_{GS}}{V_P}\right]^2 \text{ mA.} \tag{4.71}$$

Differentiating Eq. (4.71) with respect to V_{GS}, expression for g_m is given by Eq. (4.75):

$$g_m = \frac{\partial I_D}{\partial V_{GS}} = -\frac{2I_{DSS}}{V_P}\left[1 - \frac{V_{GS}}{V_P}\right]. \tag{4.72}$$

When $V_{GS} = 0$ is substituted in Eq. (4.72), g_m becomes g_{m0}

$$\text{Then} \quad g_{m0} = -\frac{2I_{DSS}}{V_P} \tag{4.72A}$$

$$g_m = g_{m0}\left[1 - \frac{V_{GS}}{V_P}\right] \tag{4.73}$$

where $g_{m0} = -\dfrac{2I_{DSS}}{V_P}$.

From Eq. (4.71), we get

$$\left[1 - \frac{V_{GS}}{V_P}\right] = \sqrt{\frac{I_D}{I_{DSS}}}. \tag{4.74}$$

From Eqs. (4.72) and (4.74), we get

$$g_m = \frac{2\sqrt{I_D \cdot I_{DSS}}}{V_P}. \tag{4.75}$$

When $V_{GS} = V_P$, the pinch-off voltage, I_D becomes zero (This is another way of describing the pinch-off voltage on the Transfer characteristics of the JFET device), according to the above mathematical expression. This variation of Drain Current with variations in V_{GS} for constant value of V_{DS} is shown on the following Transfer or Mutual characteristics shown in Fig. 4.49. This characteristic is a nonlinear characteristic. Mutual or Transfer conductance 'g_m' can be calculated from the Transfer characteristic or Mutual characteristic of the JFET device (Fig. 4.49).

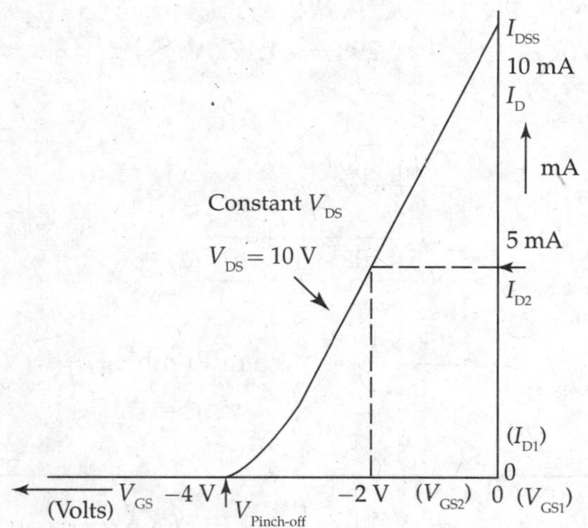

FIG. 4.49 *JFET transfer or mutual characteristic to define g_m*

From the transfer characteristics shown in Fig. 4.49, the Transconductance 'g_m' can be determined. It is the ratio of the increment 'ΔI_D' to the increment in the Gate Voltage 'ΔV_{GS}'. The change in the Drain Current 'ΔI_D' from 'I_{D1}' to 'I_{D2}' is 5 mA from the graph. The change in the Gate Voltage 'ΔV_{GS}' from 'V_{GS1}' to 'V_{GS2}' is 2 V. So 'g_m' is equal to 2.5 milli mhos. Transconductance is the slope of the Transfer characteristic of the FET device (Fig. 4.49).

Amplification factor 'μ' is the product of 'g_m' and 'r_d' from the definitions of the device constants. In these calculations, $r_d = 20 \times 10^3\ \Omega$ and $g_m = 2.5$ milli mhos. Therefore, the amplification factor $\mu = g_m \times r_d = 20 \times 10^3 \times 2.5 \times 10^{-3} = 50$. These values for the device constants are only for the sake of illustration of the method of calculations. Maximum gain of a JFET amplifier is limited to the value of amplification factor, μ.

Square law nature of the device Transfer characteristic suggests its use in mixer circuits of *radio receiver circuits*. Once we know the FET structure, device working and the device parameters, the device application as an Amplifier is considered for simple illustration of application of the FET device.

EXAMPLE 4.8

A FET has Drain Saturation Current I_{DSS} of 10 mA and Quiescent point Drain Current I_D is 5 mA with pinch-off voltage $V_P = -4$ V. Calculate the value of V_{GS}. Calculate the value of Transconductance g_m.

Solution:

a.
$$I_D = I_{DSS}\left[1-\frac{V_{GS}}{V_P}\right]^2 \text{ mA}$$

$$\left[1-\frac{V_{GS}}{V_P}\right]^2 = \frac{I_D}{I_{DSS}} = \frac{5\times10^{-3}}{10\times10^{-3}} = \frac{1}{2}$$

$$\left[1-\frac{V_{GS}}{-4}\right] = \left[1+\frac{V_{GS}}{4}\right] = \sqrt{\frac{1}{2}} = 0.707$$

$$\therefore \quad V_{GS} = 4(0.707-1) = -4\times0.293 = -1.172 \text{ V.}$$

b. Calculation of g_m

$$g_m = \frac{2\sqrt{I_D\cdot I_{DSS}}}{V_P}$$

$$= \frac{2\sqrt{5\times10^{-3}\times10\times10^{-3}}}{4}$$

$$= \frac{\sqrt{50}\times10^{-3}}{2} \cong 3.5 \text{ milli mhos.}$$

EXAMPLE 4.9

From the transfer curve of a FET device, the changes in Drain Current 'I_D' are from 10 to 5 mA, for a change in Gate Voltage V_{GS} from 0 to –2 V. What is the value of Drain Saturation Current and also calculate Transconductance g_m?

Solution: Drain current at V_{GS} = 0 V will be maximum and it is known as drain saturation current I_{DSS} through the FET.

$$\therefore \quad I_D = I_{DSS} \text{ when } V_{GS} = 0 \text{ V}$$

From the given data,
$$I_{DSS} = 10 \text{ mA}$$

$$\Delta I_D = (10-5)\times10^{-3} = 5 \text{ mA}$$

$$\Delta V_{GS} = (2-0) = 2 \text{ V}$$

$$\therefore \quad g_m = \frac{\Delta I_D}{\Delta V_{GS}} = \frac{5 \text{ mA}}{2 \text{ V}} = 2.5 \text{ milli mhos.}$$

EXAMPLE 4.10

From the output characteristic of a FET device for a constant V_{GS} = 0 V, changes in Drain Current 'I_D' are from 10 to 9.5 mA for a change in Drain Voltage V_{DS} from 2 to 12 V. Calculate the value of Drain Resistance 'r_d'.

Solution: Drain resistance $r_d = \dfrac{\Delta V_{DS}}{\Delta I_D}\ \Omega$

Incremental change in drain voltage $\Delta V_{DS} = (12 - 2) = 10$ V
Incremental change in drain current $\Delta I_D = (10 - 9.5) \times 10^{-3} = 0.5$ mA

$$\therefore \quad r_d = \frac{\Delta V_{DS}}{\Delta I_D} = \frac{10}{0.5 \times 10^{-3}} = \frac{10 \times 10 \times 10^3}{5} = 20 \text{ k}\Omega.$$

EXAMPLE 4.11

If transconductance of a FET device = 2.5 milli mhos and drain resistance 'r_d' is equal to 20 kΩ, calculate the value of amplification factor 'μ' of the FET.

Solution: Amplification factor $\mu = g_m \times r_d$

$$\mu = 2.5 \times 10^{-3} \times 20 \times 10^3 = 50.$$

4.25 COMPARISON BETWEEN FIELD EFFECT TRANSISTOR AND TRANSISTOR

Junction Field Effect Transistor (JFET)	Bipolar Junction Transistor (BJT)
FETs are unipolar devices. Currents through the FET are due to the movement of only one type of majority charge carriers. Electrons in N-channel FET device or holes in P-channel FET devices.	*BJTs are bipolar devices.* Transistor currents are due to the movements of both the majority and minority charge carriers in both NPN and PNP Transistors.
Availability of matched pairs is difficult.	*Availability of matched pairs is possible.* Availability of matched pair Transistors makes push–pull amplifier operation possible for increased powers and efficiency.
Gain-bandwidth product of FET amplifiers is less.	*Gain-bandwidth product of Transistor amplifiers is larger because of larger amplifier gains.*
No thermal runaway with FETs. FETs have negative temperature coefficient for increased drain currents causing less power dissipation in the device and avoids thermal run-away.	*Thermal runaway exists for BJTs.* Increased Collector currents cause increased power dissipation at the Transistor junctions leading to thermal runaway in BJTs.
Currents through the FET device are controlled by the electric field along the channel. From such behaviour, the name field effect Transistor is derived	Transistors are nonlinear devices as evident from the device characteristics. Transistor operation is considered as linear for small signal operations of amplifiers.
The input electrode current (gate current) of FET devices is negligibly small.	The control electrode current is high with the forward-biased input junction of BJT.
FET devices are smaller in size than Transistors. In a given volume, more FET devices can fit in than Transistors.	
High input impedance for FETs. The input junction between the gate and the source is reverse-biased and hence the input resistance of the FET device is very large. This high input resistance of the FET device is an advantage over Transistor devices, as they drain less power from the source.	*Low input impedance for BJTs.* Emitter junction of the Transistor is forward-biased, and hence the input resistance is very low. So power drain from the source may be larger than FETs.

Table *Cont'd*

Junction Field Effect Transistor (JFET)	Bipolar Junction Transistor (BJT)
When very high input resistance amplifier stage is to be designed, FET amplifier stage is preferred over bipolar Transistor amplifier stage.	
No feedback path in FET. There is no energy feedback from the output port to the input port of FET devices.	*Feedback paths exist in BJT.* *h*-parameter 'h_r' indicates the existence of the level of energy transfer from the output port to the input port of Transistors.
FET is a voltage-controlled current source. The input voltages between gate and source terminals control the current through the device. So it is a voltage-controlled current source (VCCS).	*Transistor is a current-controlled current source.* The input Base Current controls the output Collector Current for device operation. So it is a current-controlled current source (CCCS).
Wider bandwidths for FET amplifiers. Voltage gain is limited to the amplification factor 'μ' of the device, which has low values. So the amplifier circuits using FET devices have low orders of gain and hence the bandwidth of FET amplifiers is large.	*Small bandwidths for Transistor amplifiers.* The amplifier circuits using Transistors have large values of current and voltage gains. So the bandwidth of Transistor amplifiers is small.
Low noise level	Noise level is very high.
FETs act as fast switches because no voltage drop exists across the ON switch.	BJTs have finite saturation voltages across the ON switches.
The Transistor has another disadvantage that its input impedance is low and requires ideal voltage source for driving it. Other alternative is to use a buffer amplifier in the form of 'Emitter follower' as an impedance matching network. For example, at the front end of cathode ray oscilloscope (CRO).	

4.26 METAL OXIDE SEMICONDUCTOR FIELD EFFECT TRANSISTOR

Two types of MOSFET devices are available:

1. Enhancement mode MOSFET or E-MOSFET (MOSFET with induced channel)
2. Depletion Enhancement mode MOSFET (DE MOSFET or MOSFET with built-in channel)

Integrated Circuit (IC) technology has revolutionised the semiconductor devices and RLC components; fabrication or manufacture in batch processing with small size and at low cost and phenomenal reduction in size and increase in packaging densities are achieved. Reliability of working is increased, resulting in improved performance. MOSFET devices find their use in almost all latest devices using digital technology.

A silicon wafer of 1 sq. in size can accommodate up to 400 IC chips of surface area 50 by 50 mil. Each chip again can contain 50 separate components to make up 20,000 components in a single wafer. If 10 wafers are used in a single batch processing 2,00,000 components can be manufactured.

Like the JFET, the MOSFET is a lower-power semiconductor device that has high-input impedance as first generation electron devices or vacuum tubes and Low-power requirements of Transistors. Both MOSFET and JFET have Drain, Source and Gate electrodes in common. Both devices have conduction channel with its resistance controlled by Gate to Source Voltages. There are also P- and N-type channel MOSFETs; complementary MOSFET known as CMOS FET device uses both. The Gate–Source path in a JFET is a reverse-biased junction by V_{GS}. In a MOSFET, a thin layer of insulating material such as SiO_2 is placed over the channel region before the fabrication of the Gate electrode. The insulating layer is very thin, which is of the order less than 1 μm; the field produced by Gate Voltage still penetrates through insulating layer and modulates the conduction of the channel. Input resistance of MOSFET is much higher than that for JFET because of the SiO_2 insulation layer between the Gate and the Channel. Input resistance of MOSFET is of the order of $10^{14}\ \Omega$. Since there is an insulator in the input circuit of a MOSFET, gate potential is not restricted in polarity. There are therefore two possible modes of operation for a MOSFET. As mentioned earlier, non-restriction on polarities of Gate Voltage, Enhancement mode and Depletion mode of operations is possible.

4.26.1 Some Basic Steps Involved in the Manufacturing Process of MOSFET

Considering the fabrication details of the enhancement MOSFET (MOSFET with induced N-type channel) the following steps are involved. Basic MOSFET with N-type channel is formed on a P-type Silicon wafer substrate. Source and Drain of the MOSFET device with N-type-induced channel are N- type islands that are formed or diffused into a lightly doped P-type substrate.

Aluminium metal on the SiO_2 layer functions as the Gate electrode. Thus, the insulating SiO_2 layer separates the Gate and the conducting (induced) N-type channel. With these basic construction features of the MOSFET device, the device is also known as insulated Gate Field Effect Transistor (IGFET). Thus, MOSFET finds three layers of semiconductor materials for Gate, Source and Drain with insulating layers using SiO_2.

Basic MOSFET with N-type channel is formed on a P-type Silicon wafer substrate.

Step 1 The starting material is a P-type substance or wafer upon which an insulating layer of silicon dioxide SiO_2 is grown in the following second step.

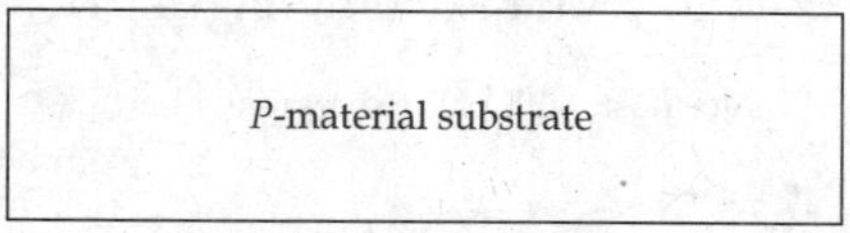

P-material substrate

FIG. 4.50 *(Base material) P-type substrate of MOSFET*

Step 2 Upon P-type substrate, SiO_2 layer is grown as shown in Fig. 4.51. SiO_2 layer is grown over the P-type substrate. SiO_2 (silicon dioxide) material has the fundamental property of preventing the diffusion of impurities through it. The thickness of the SiO_2 layer is of 0.5 μm (5000 Å).

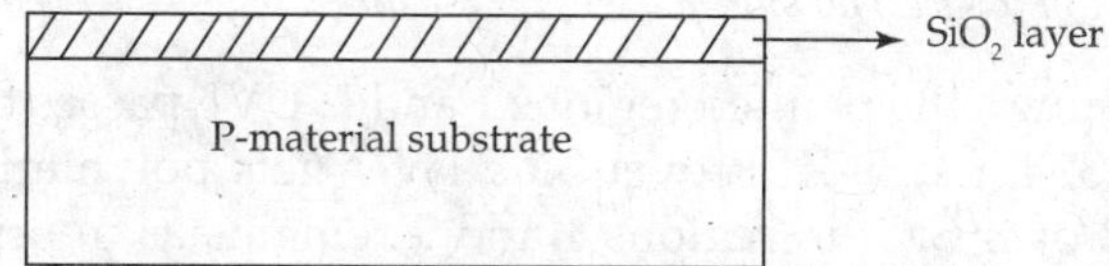

FIG. 4.51 *SiO_2 layer formation on P-material substrate*

Step 3 Kodak Photo Resist material (KPR) is coated on the SiO_2 layer (Fig. 4.52).

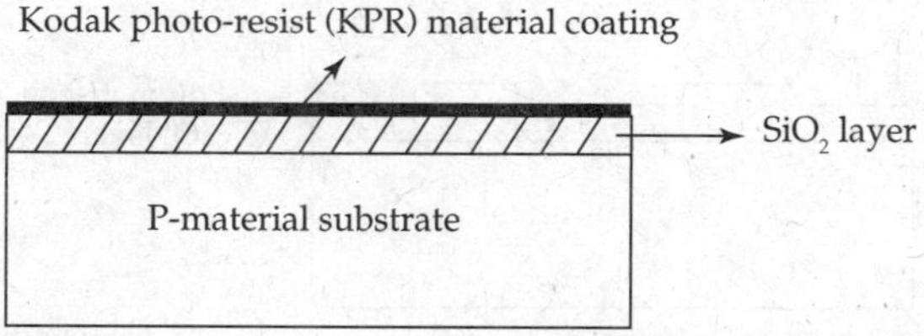

Kodak photo-resist (KPR) material coating

SiO2 layer

P-material substrate

FIG. 4.52 *Kodak Photo Resist (KPR) coating on SiO_2 layer*

To make photolithography-etching process over SiO_2 layer to provide two windows for the formation of two islands of N-type material (inside the substrate material), Kodak Photo Resist material (KPR) is coated over SiO_2 layer. (KPR is also known as a photosensitive material.)

Step 4 A high-resolution photo mask with painted regions at locations 1 and 2 is placed on the SiO_2 layer as shown in Fig. 4.53.

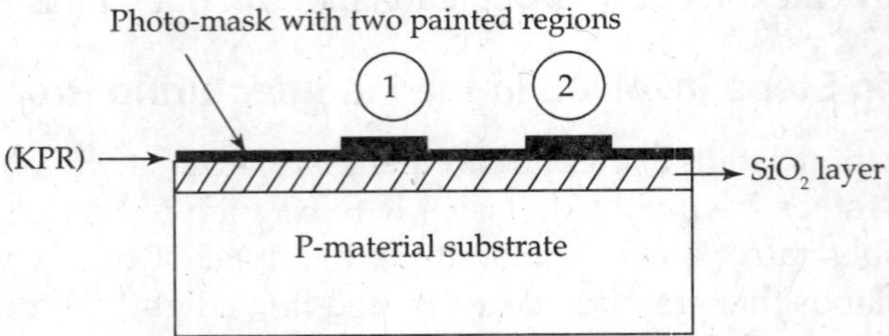

FIG. 4.53 *Photo-mask with two painted regions for forming windows for diffusing N-type materials to form source and drain*

A very high-resolution photo mask, with two painted regions at 1 and 2 locations, is placed on the silicon wafers (Fig. 4.53). The painted regions will not allow ultra violet light (UVL) to pass through them. (The number of painted areas corresponds to the number of windows and islands to be formed for the device or the circuit component.) In this MOSFET device, two windows are needed for *source and drain* island regions, so there are two painted areas on the photo mask.

Step 5 The photo mask is exposed to UVL radiation (Fig. 4.54).

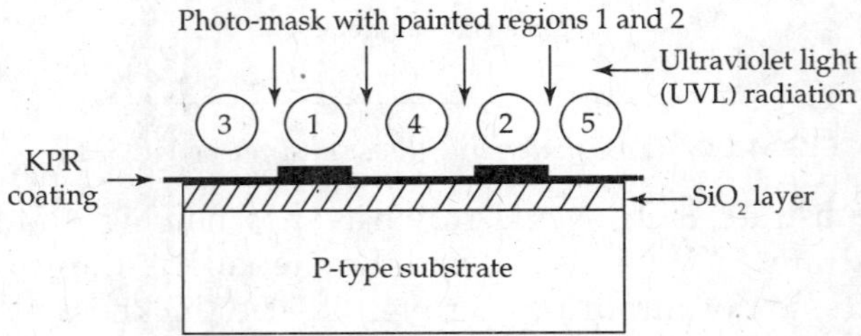

FIG. 4.54 *Photo mask (Rolex sheet) on silicon wafer with painted regions is exposed to ultra violet light*

UVL does not pass through the painted regions 1 and 2. UVL passes through the mask over the unpainted areas on 3, 4 and 5. KPR over SiO_2 layer gets polymerised over the exposed areas for UVL. The KPR over SiO_2 in regions 1 and 2 remains as in original situation and is dissolved by using etching solutions easily.

Step 6 Etching process to form two windows at regions 1 and 2 is shown in Fig. 4.55.

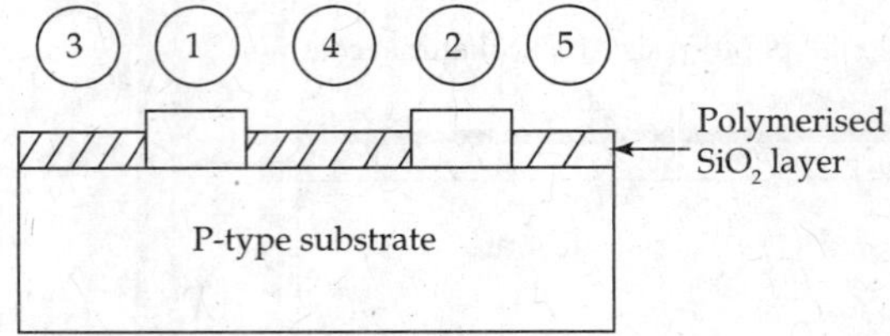

FIG. 4.55 *Etching process for the formation of two windows at regions 1 and 2*

Regions 3, 4 and 5 on SiO_2 layer get polymerised and cannot be dissolved by etching solution. Regions 1 and 2 on SiO_2 layer are not polymerised. So using etching solution, regions 1 and 2 of SiO_2 are etched away and two windows are formed (Fig. 4.56).

P-type substrate regions are exposed at the two window areas to facilitate pentavalent material diffusion through them for the formation of N-type of islands.

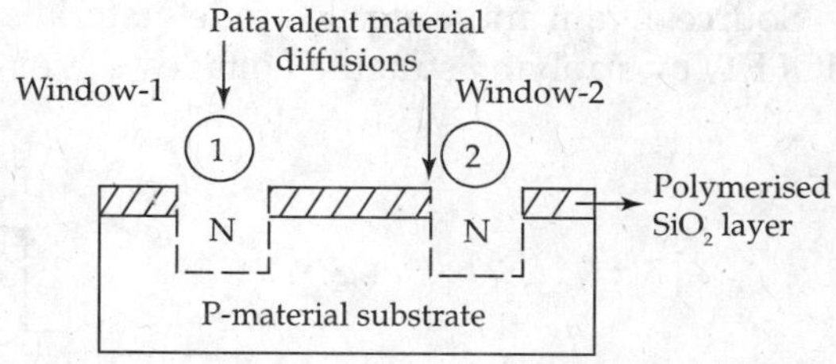

FIG. 4.56 *N-material diffusions through the two windows 1 and 2 to form source and drain islands*

Step 7 N-type diffusions through the two windows to form source and drain.
Through the two windows, pentavalent impurity atoms of phosphorus material are diffused into the dotted window areas and two N-type islands are formed. The insulating material below the Gate area does not allow diffusion through it and so the region (channel) between Drain and Source is unaffected during diffusion process.

As long as no electron channel is formed between Source and Drain, there is no conduction path between the Drain and the Source for the device function. When there is no voltage on the Gate for MOSFET without built-in channel, it acts as an open switch.

Step 8 Formations of N-type source and drain islands (Fig. 4.57).
N-type Island (1) acts as *source* and N-type Island (2) functions as *Drain*.

Aluminium metallisation forms *Gate, Source and Drain electrodes* of *MOSFET* device. Source and the Drain terminals are symmetrically placed to Gate electrode. There are other processing steps that involve re-oxidation and further etching to fabricate the total device. A silicon dioxide (SiO_2) layer between the *induced channel* connecting Drain, Source and Gate functions as an *insulating layer*. That is the reason why the MOSFET is also called as Insulated Gate Field Effect Transistor (IGFET).

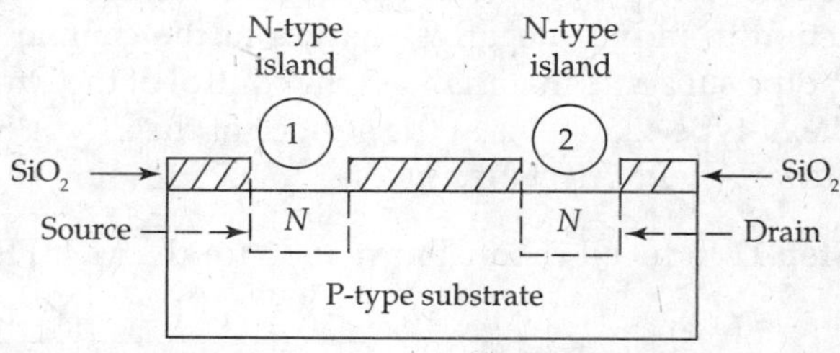

FIG. 4.57 *Formation of source and drain regions (islands)*

Step 9 Structural details of the MOSFET (IGFET) device are shown in Fig. 4.58.

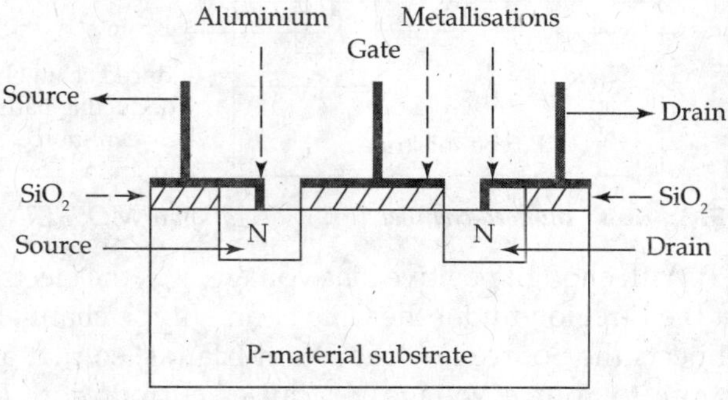

FIG. 4.58 *Structure of MOSFET*

Source, Drain and Gate electrode materials function as enhancement MOSFET (EMOSFET) (IGFET) by applying suitable voltages V_{DD} or V_{DS} and V_{GS} (Fig. 4.59).

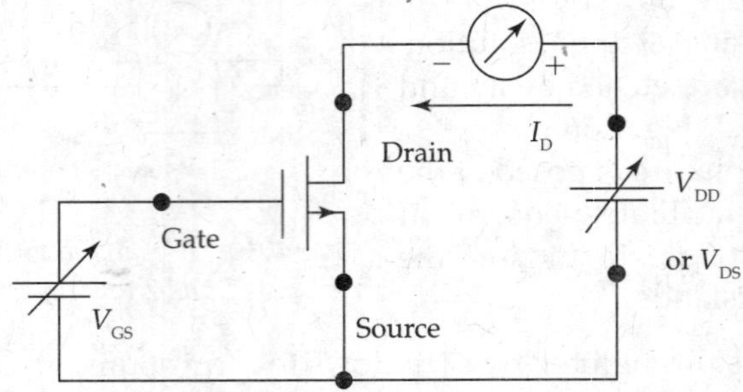

FIG. 4.59 *MOSFET with biasing voltages [N-channel MOSFET (IGFET)]*

Step 10 MOSFET with the Biasing voltages is shown in Fig. 4.59.
Voltages V_{DD} (V_{DS}) and V_{GS} are the DC voltages for the operation of enhancement MOSFET device with both biasing voltages positive with respect to Source. (Enhancement-mode MOSFET device is also known as *off* MOSFET.) The positive charges ($+V_{GS}$) on the metallic Gate induce corresponding negative charges in the Channel in the P-type material between Source and Drain on the other side of the SiO_2 layer, just as in a capacitor (Aluminium metal of Gate and P-type substrate with SiO_2 insulating layer as dielectric contribute to capacitor action to induce negative charges in the channel between the Source and the Drain in the P-type substrate material). As the control of the Drain Current due to mobile electrons through the N-type channel is at the surface of the P-type Semiconductor, the device is also known as Surface Field Effect Transistor.

Step 11 Induced channel formation for the MOSFET (Fig. 4.60).

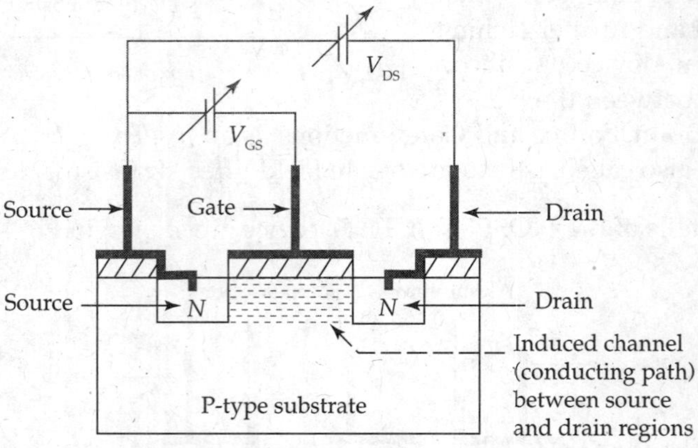

FIG. 4.60 *Induced channel during operation of MOSFET*

Working of MOSFET: With enough positive bias voltage, V_{GS}, the electron enhancement will be sufficient to convert the P-region under the Gate to an N-type channel. So the formation of the *induced channel* connects the Source and the Drain Islands, so that a conduction channel operates the device. Drain to Source Voltage V_{DS} causes current flow I_D due to the flow of induced electrons along the channel from the Drain to the Source.

From the Transfer characteristic between the parameters V_{GS} and I_D of the MOSFET shown in Fig. 4.61, it can be observed that a minimum Gate to Source Voltage V_{GST} or V_{TH} is required to induce mobile electrons and form a channel between the Source and the Drain to establish current flow between the two N-type of islands. This minimum voltage 'V_{TH}' is known as the Threshold voltage for the MOSFET. The current I_{DSS} is very small and is of the order of a few microamperes that can be known from the device specifications at the specified Threshold voltage V_{TH} or V_{GST}. Varying Gate to Source Voltages V_{GS} controls the induced mobile electrons into the channel and thus vary the conductance of the induced channel.

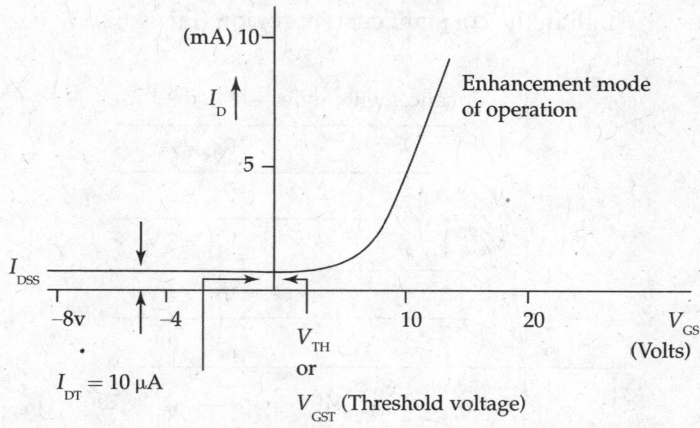

FIG. 4.61 *Transfer characteristic of enhancement mode MOSFET*

The increasing positive voltages on the Gate electrode enhance the conduction electrons in the N-type channel and in turn enhance the magnitudes of the Drain Current I_D. So the device is called as Enhancement-Mode MOSFET or simply E-MOSFET. For this Enhancement MOSFET, as long as the Gate-to-Source Voltage is less than the specified threshold voltage V_{TH}, the MOSFET is normally in the off condition or Off state. Once the positive voltage sufficiently exceeds the threshold voltage V_{TH}, N-type channel is induced and the device switches into the conduction or the ON-state. These binary states of switching of the normally off MOSFET device have good number of applications in digital electronic circuits. The additional feature of isolation or insulation of Gate from the Source and the Drain regions provides infinite input impedance for the device, as the Gate current is negligibly small.

The device parameters, the Drain Current I_D, V_{GS} and V_{TH}, are connected by the following mathematical expression:

$$I_D = K(V_{GS} - V_{TH})^2 \tag{4.76}$$

where the value of the constant 'K' depends on the device geometry details. 'K' has the units of mA/(V)^2 and typical value of 0.3 is common.

From this equation for the Drain Current relating the Threshold voltage, it is also clear that the device geometry parameter 'K' has a control on the Threshold voltage. In Practical digital circuits, the magnitudes of ON voltages depend upon the applications. So the ON-state voltage of the MOSFET device that is dependent upon the threshold voltage V_{TH} is in turn decided by the design constraints on the MOSFET device to provide a suitable threshold voltage for the desired practical applications.

4.27 OUTPUT CHARACTERISTICS FOR AN N-CHANNEL ENHANCEMENT-MODE MOSFET

Output characteristics of MOSFET device are graphs of the Drain Current I_D that will flow as V_{DS} is varied for a number of values of V_{GS}.

For an Enhancement-mode MOSFET, $I_D = 0$ for $V_{GS} = 0$ V. Hence, it almost touches V_{DS}-axis. The device characteristics are similar to JFET device output characteristics. As V_{DS} is increased; initially I_D increases till knee points on output characteristics and beyond knee points drain current I_D remains almost constant, even if 'V_{DS}' is increased. The output impedance of the MOSFET device is very high in the constant current region (Fig. 4.62).

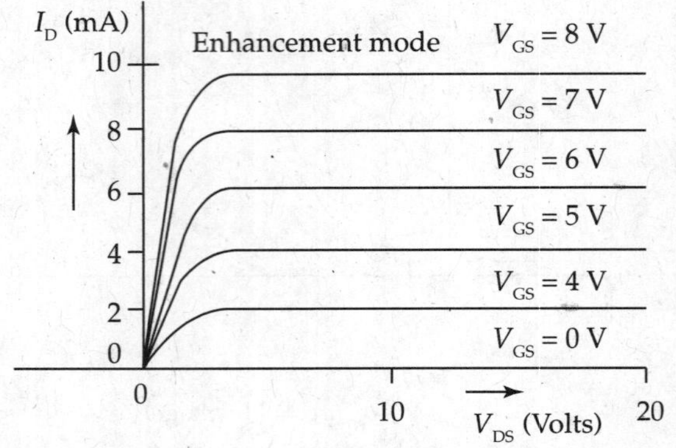

FIG. 4.62 *Output characteristics of N-channel E-MOSFET*

4.28 DEPLETION ENHANCEMENT MOSFET (DE MOSFET) (MOSFET WITH BUILT-IN CHANNEL)

A channel with charges corresponding to majority carriers of *Source and Drain* semiconductor materials is diffused between Source and Drain to form the Built-in channel for establishing an initial conducting path between them as shown in Fig. 4.63.

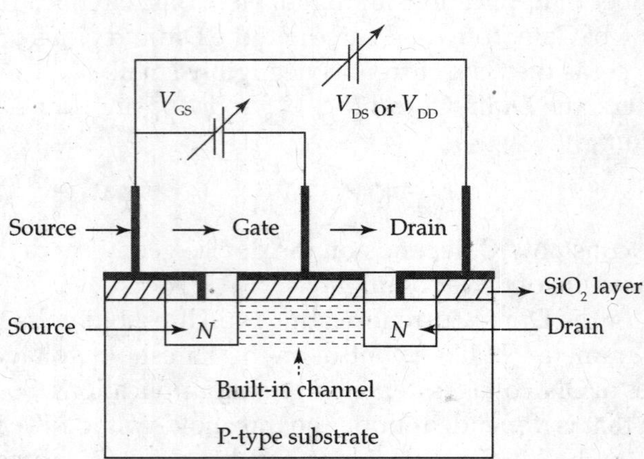

FIG. 4.63 *Structure of an N-channel (built-in channel) depletion enhancement mode MOSFET (DE MOSFET)*

DE MOSFET or Depletion- mode MOSFET is also known as a normally ON MOSFET, because with zero Gate to Source Voltage V_{GS}, the two N-type islands, the *Drain and the Source* are connected by N-Channel. Negative charges or voltage '$-V_{GS}$' on the metallised *Gate* induce corresponding positive charges in the N-Channel on the other side of the SiO_2 layer. Induced positive charges may also be thought of as repelling negative charges in the N-channel, with a consequent reduction in the conductivity of the N-channel for Depletion mode of operation of the device.

In addition to being operated as a Depletion-mode device, DE MOSFET can also be operated as an Enhancement device. Positive Gate Voltages 'V_{GS}' induce negative charges into the already Built-in N-type channel. This enhances the channel conductivity, resulting in more Drain Current, 'I_D' by applying V_{DS}.

Output characteristics for DE MOSFET are shown in Fig. 4.64 both for the enhancement mode of operation and for the Depletion mode of operation of the device. Transfer characteristics of the N-channel DE MOSFET both for the 'enhancement mode' and for the 'Depletion mode' are shown in Fig. 4.65.

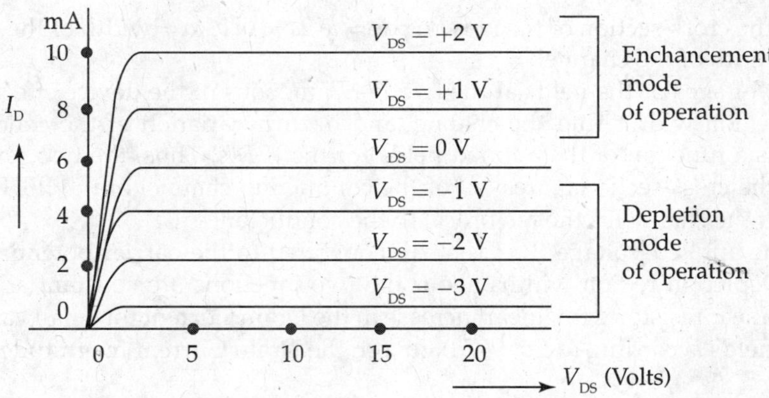

FIG. 4.64 *Output characteristics of DE MOSFET*

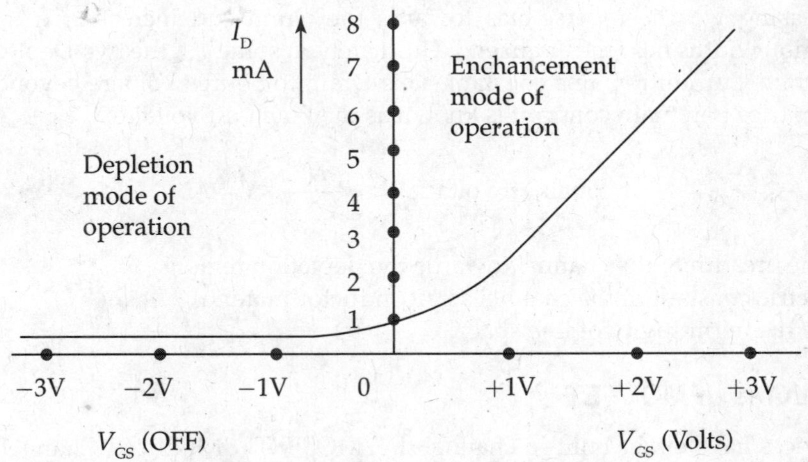

FIG. 4.65 *Transfer characteristic of N-channel DE MOSFET*

4.29 COMPARISONS OF JFET AND MOSFET WITH RESPECT TO VARIOUS FEATURES

4.29.1 Features of N-Channel JFET

Consider an N-Channel JFET device. N-type semiconductor bar of cross-sectional area '*A*' and length '*L*' has the Drain terminal at the upper end and Source terminal at the lower end. The N-type semiconductor bar is a sandwich between the two small areas of P-types at left and right sides of the bar to have combined function as Gate electrode. The N-type semiconductor bar has conductance 'G_C' of magnitude given by Eq. (4.77) and functions as a channel for device currents:

$$\text{Conductance } G_C = \frac{en\mu_n A}{L} \text{ Siemens} \tag{4.77}$$

where n is the electron concentration in the channel of the semiconductor,
μ_n is the mobility of the electrons,
e is the charge of the electrons,
A is the area of the cross-section of the N-type bar $= w\,2a = w\,b$, $w =$ width of the channel and $2a$ or b is the breadth of the channel.

The applied voltages or the fields at the two P–N junctions of the device effect by varying the Depletion region widths into the channel and in turn variation of the conductance '*G*' of the channel is a function of the applied fields for the JEFTs. Thus, the two reverse-biased junctions vary the cross-sectional area '*A*' of the conducting channel of the JFET. Hence, the resistance 'R_C' of the channel is the reciprocal to the conductance 'G_C'.

As long as the applied voltage 'V_{DS}' is small compared to the barrier potential of the two junctions, the Depletion region widths about the two junctions are constant and hence the channel acts as linear resistor with linear increase in the Drain Current for small values of 'V_{DS}'.

The applied fields according to Eq. (4.78) govern the Drain Current magnitudes:

$$I_D = I_{DSS}\left[1 - \frac{V_{GS}}{V_P}\right]^2. \tag{4.78}$$

With increasing V_{DS} the reverse-bias towards the Drain end increases, with increasing Depletion region widths near the Drain end. Ultimately, at some V_{DS} the two Depletion regions merge and Drain Current remains constant. This Drain to Source Voltage beyond which the Drain Current tries to remain constant is known as the Pinch-off voltage, 'V_P':

$$\text{Pinch-off voltage } V_P = \frac{eN_D a^2}{2\varepsilon_S} \text{ V}$$

where $2a$ is the breadth of the channel towards the depletion regions,
ε_S is the dielectric constant of the channel semiconductor material,
$\varepsilon_S = 12\varepsilon_0$, ε_0 is the permittivity of free space.

4.29.2 Features of MOSFET

MOSFET devices have either built-in channel (DE MOSFET) or Induced Channel (MOSFET) for the device functioning. Voltages between the Gate and the Source terminals 'V_{GS}' through capacitor action vary the carrier concentrations (electrons for N-type channel MOSFET or holes for P-type channel MOSFET) in the channel. This turn varies the conductivity of the channel and the device currents or channel currents.

The Gate electrode is isolated or insulated from the other electrode structures. So the input and the output ports are isolated with MOSFET amplifiers. The device is also known as IGFET. The MOSFET device consists of metal, Silicon dioxide layer as Insulator, N- or P-type semiconductor substrate to form the Source and the Drain Islands and a channel containing either electrons or holes as current charge carriers to modulate the conductivity of the channel:

$$I_D = K[V_{GS} - V_{TH}]^2,$$

where K depends on the device structure and V_{TH} is the threshold voltage for device conduction.

MOSFET circuits occupy less space on the silicon wafer and consume less power than BJTs. They are suitable for VLSI circuits, such as CPU, RAM and a variety of ICs.

4.29.3 Some MOSFET Family Device Symbols

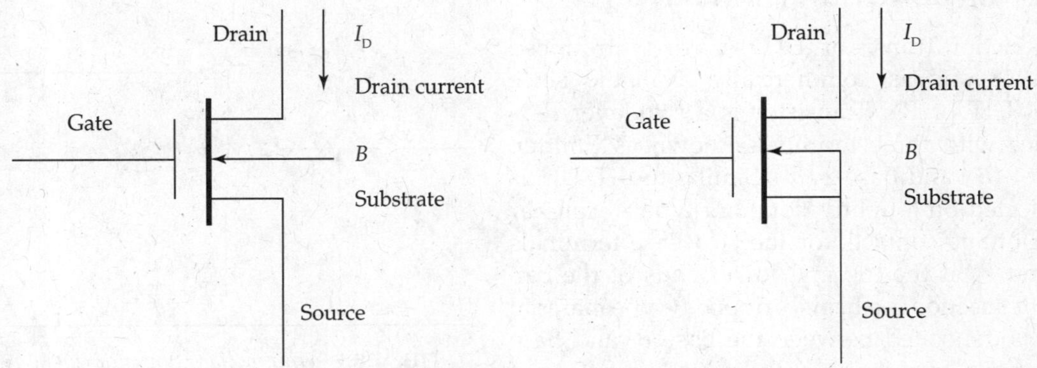

N-channel depletion mode MOSFET symbols

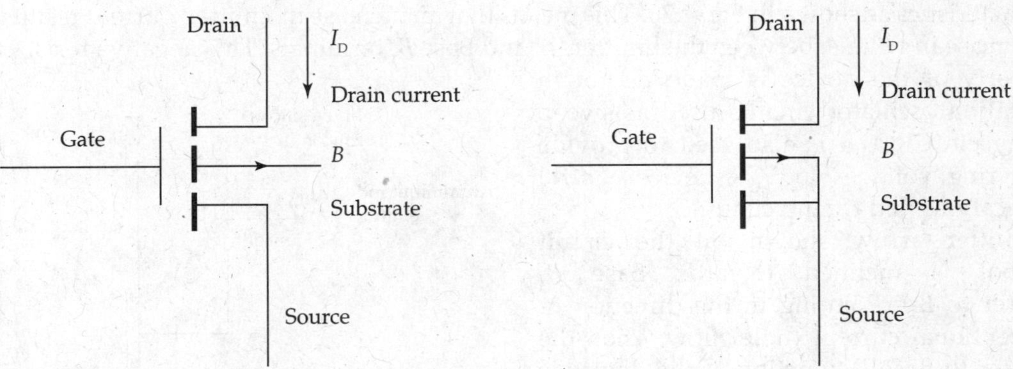

P-channel enhancement mode MOSFET symbols

Comparison between MOSFET and BJT devices

1. MOSFET Transistor consumes less power than BJTs.
2. A binary threshold Gate Voltage operates MOSFETs. Below the threshold voltage, MOSFETs are non-conducting. In a BJT, non-conduction is a gradual mechanism.
3. MOSFET operation follows square law, whereas for BJTs the voltage–current relationship is logarithmic.
4. For analogue applications, BJTs have higher gain factors.
5. In microcircuits, BJTs require more chip area and it requires tubs to isolate the devices. MOSFETs are smaller and less costly for fabrication, because of heavy doping and thick oxide in the region between adjacent devices. They require few fabrication steps. So MOSFETs are less costly for fabrication.
6. BJTs find most of their applications in analogue electronic circuits.
7. MOSFETs find their applications in digital electronics.
8. For the same channel length and Bandwidth MOSFETs are superior to FETs for high-frequency operation.
9. MOSFETs suffer from Gate break down, whereas in BJTs such break down is absent.

4.30 UNIJUNCTION TRANSISTOR

Unijunction Transistor (UJT) was first introduced in 1948 and is commercially available since 1952. UJT or PN Transistor is a two-layer P–N device with three terminals known as Emitter, Base-1 (B_1) and Base-2 (B_2). Similar to BJT, UJT is fabricated on a lightly doped N-type silicon bar with ohmic contacts for the two Base terminals B_1 and B_2 at the top and lower ends of the bar. Emitter section is a heavily doped P-type material that is deposited between the Base B_1 and Base B_2 regions, possessing only one P–N junction of small area (Fig. 4.67). As the device has only a single junction, it is known as a UJT.

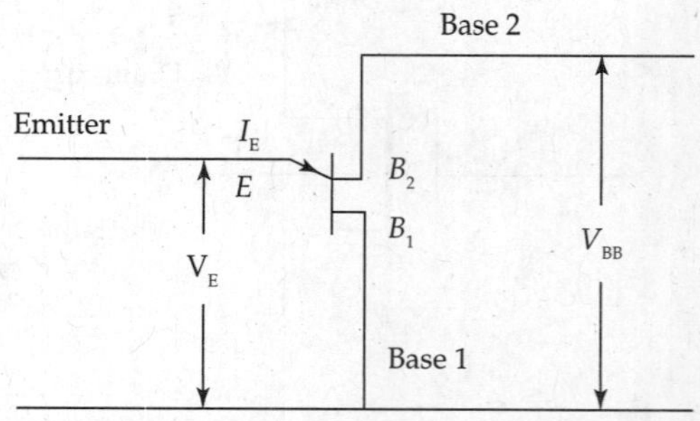

FIG. 4.66 *Unijunction transistor (UJT)*

The UJT is a versatile semiconductor device, which exhibits negative resistance characteristics as shown in Fig. 4.70. This means that an increase in Emitter Current results in a decrease in voltage between the Emitter '*E*' and Base B_1 terminals. This negative resistance property of the device is made use of in relaxation oscillator circuits used as sweep voltage in CRO. It is also used to provide triggering voltages to SCR device control applications and timing circuits.

Emitter arrow shown on the circuit symbol is inclined towards Base B_1 (nearer to B_2). Pointing in the direction of conventional current (hole) flow when the Emitter to Base B_1 junction of the device is forward-biased by the *Emitter voltage* 'V_E'.

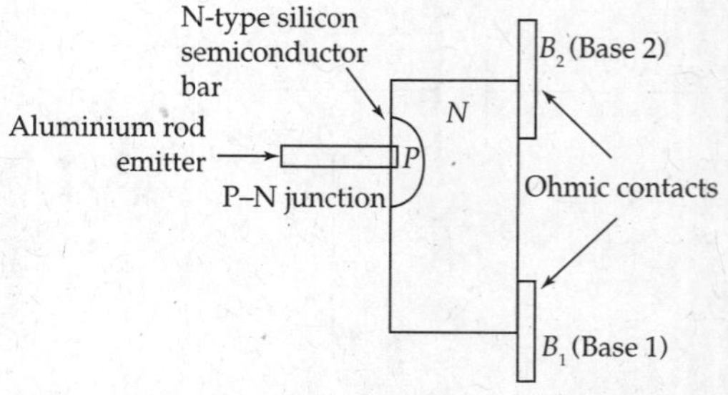

FIG. 4.67 *Structural details of unijunction transistor*

The main operational difference between FET and UJT is that the Gate surface of FET is much larger than the Emitter junction of UJT. The structural details of UJT are shown in Fig. 4.67. High resistively N-type silicon (lightly doped/increased resistance characteristic) of 8 × 10 × 35 mils called Base '*B*'.

UJT was originally known as double-Base diode. Aluminium rod is alloyed to silicon semiconductor slab closer to Base B_2 contact as shown in Fig. 4.68.

UJT circuit with biasing voltages to obtain the device characteristics:
To obtain UJT characteristics and understand the working of the device, Voltage V_{BB} is applied between Base B_1 and Base B_2. Voltage V_{EE} is used to provide a voltage 'V_E' that is applied between the Emitter '*E*' and the Base 'B_1' (Fig. 4.68).

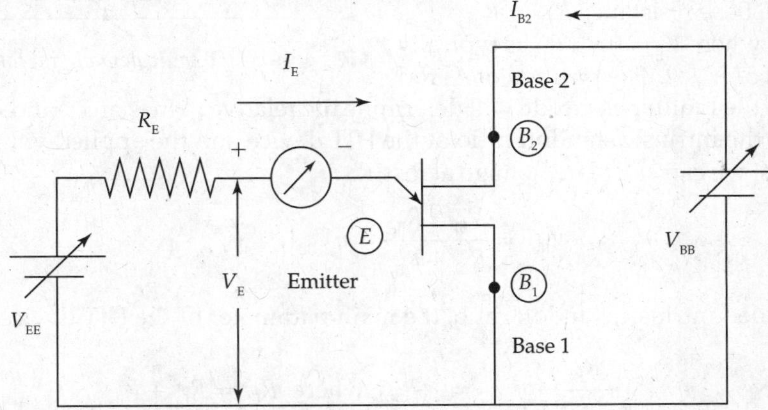

FIG. 4.68 *UJT biasing voltages to obtain UJT characteristics*

Maintain a constant supply voltage V_{BB} of say '0 V'. Varying the Emitter voltages in increments, the corresponding value of Emitter Current 'I_E' is noted down. The measurements are tabulated below.

Measurements to obtain UJT characteristics		
Supply voltage between Base B_2 and Base $B_1 = V_{BB} = 0$ V		
S. No.	**Emitter voltage, V_E (V)**	**Emitter Current, I_E (mA)**

Above procedure is repeated with measurements of Emitter voltage and Emitter Current variations for different constant values of $V_{BB} = 5$ V, 10 V, 15 V and 20 V and so on. The observations are tabulated.

UJT characteristics can be drawn from the tabulated data. During the observations, the salient features such as peak voltage V_P, peak current I_P, valley voltage V_V, valley currents I_V, saturation voltage, etc. are noted down, keeping in view of the expected characteristic of the UJT device.

Equivalent circuit of UJT device:
Equivalent circuit of UJT (Fig. 4.69) has two resistors R_{B1} (variable resistance) and R_{B2} (fixed resistance) that form a voltage divider during the operation of UJT and a single Diode

representing the P–N junction between the Emitter and Base B_1. That is the reason why the device is called a UJT.

R_{B1} may vary from 5 to 50 Ω for a corresponding change of I_E from 0 to 50 mA during the device operation. The resistance R_{B2} is a fixed internal resistance between the Emitter and Base B_2. The resistance R_{B1} is the internal variable resistance of the silicon bar between the Emitter and Base B_1 that depends on the operation or ON–OFF states of the UJT device. Inter-electrode Base resistance $R_{BB} = R_{B1} + R_{B2}$ ($I_E = 0$), that is when '$V_E = 0$ V'. R_{BB} is typically in the range of 5–10 kΩ. The position of Al rod functioning as the Emitter electrode will determine the relative values of R_{B1} and R_{B2} with $I_E = 0$.

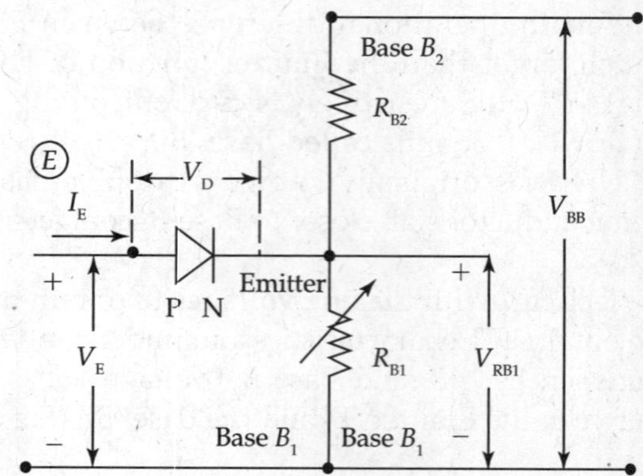

FIG. 4.69 *UJT equivalent circuit for device operation*

Knowing η the intrinsic standoff ratio of the UJT device and the applied voltage V_{BB} across Base terminals, we can calculate voltage across R_{B1}:

$$V_{R_{B_1}} = \frac{V_{BB} \times R_{B_1}}{R_{B_1} + R_{B_2}} = \eta V_{BB}\big|_{I_E=0}, \tag{4.79}$$

where 'η' is called intrinsic standoff ratio, the main parameter of the UJT device.

$$\eta = \frac{R_{B_1}}{(R_{B_1} + R_{B_2})}\bigg|_{I_E=0} = \frac{R_{B_1}}{R_{BB}}, \quad \text{where } R_{BB} = R_{B_1} + R_{B_2}. \tag{4.80}$$

Typical values of 'η' have a range from about 0.5 to 0.8 for most UJTs.
Power dissipation = 300 mW; RMS I_E = 50 mA; peak I_E = 2 A.
Emitter reverse voltage = 30 V. Inter-Base voltage = 35 V.

Principle of working of the UJT device, when the Emitter voltage $V_E \leq V_{Peak}$

Initially, when V_{BB} is switched on, and the Emitter voltage $V_E = 0$ V, voltage across R_{B1} will be of such a polarity that will reverse-bias the P–N junction between the Emitter and Base B_1. Only reverse current 'I_{E0}' of the order of a few microamperes will be present through the device in the Emitter circuit and the UJT device is in the off-state. The device is in the cut-off region of the characteristic shown in Fig. 4.70.

Principle of working of the UJT device, when the Emitter voltage $V_E \geq V_{Peak}$

For applied potential, V_{EE}, with increased voltages, when V_E is greater than $V_{RB1} = \eta V_{BB}$ by the forward voltage drop across the Diode V_D (V_D = 0.35–0.7 V, as UJT has silicon semiconductor bar), as shown in Figs. 4.67 and 4.68, Diode (P–N junction between the Emitter and Base B_1) is forward-biased and the UJT will fire. Emitter Current I_E will begin to flow through R_{B1} in the Emitter circuit due to the heavy injection of majority carrier holes from the Emitter into the Base B_1. The Emitter firing potential V_P is given by $V_P = \eta V_{BB} + V_D$.

Calculation of the peak voltage V_{Peak}:

If $\eta = 0.8$, $V_{BB} = 15$ V and $V_D = 0.7$ V.
Peak voltage $V_P = \eta V_{BB} + V_D = 0.8 \times 15 + 0.7 = 12.7$ V.

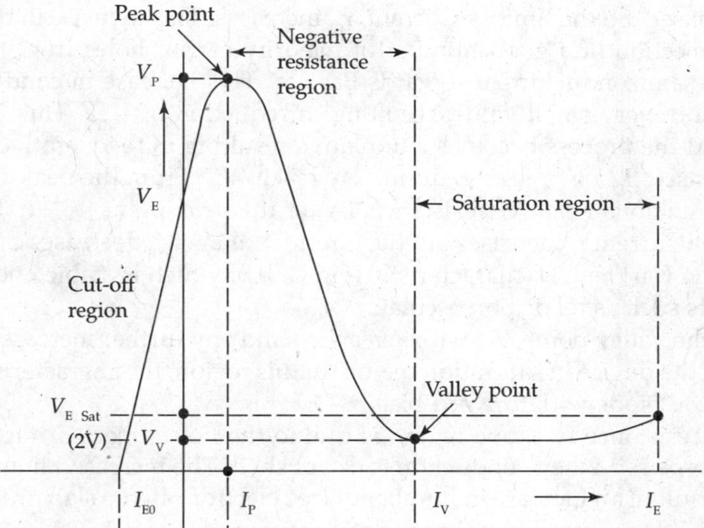

FIG. 4.70 *UJT characteristic*

UJT Emitter characteristics and various regions marked on it:

UJT Characteristics for different values of V_{BB} are shown in Fig. 4.71. UJT characteristics are shown in Fig. 4.71 for $V_{BB} = 5$ V, 10 V, 20 V and 30 V.

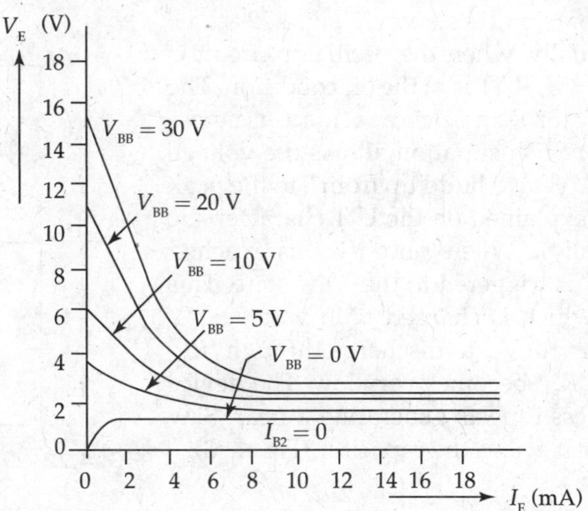

FIG. 4.71 *Unijunction transistor characteristics*

For Emitter potential to the left of the peak point, the magnitude of I_E is never greater than I_{E0} (μA). The current I_{E0} corresponds very closely to the reverse leakage current I_{CO} of the conventional bipolar Transistor. UJT will be in the off-state during this operation and this region is called the cut-off region (Fig. 4.71).

Once the voltage 'V_E' is greater than or equal to the peak voltage 'V_P', the Emitter potential V_E will drop with an increase in I_E. At this stage, the P–N junction between the Emitter and Base

B_1 is forward-biased. So the Emitter Current 'I_E' increases. This increase in the Emitter Current is due to the injection of a good number of majority carrier holes from the P-type Emitter region into the semiconductor bar towards Base B_1. This increase in conduction reduces the resistance R_{B1} to a very small value, resulting in reduction of V_{RB1}. This forward biases the Diode more and the process becomes cumulative resulting in regenerative action. The result is that for increased 'I_E', 'V_{EB}' falls to ultimately reach 'V_V'. From the peak voltage 'V_P' to 'V_V', the UJT device exhibits negative resistance. (After the peak point 'V_P' up to the valley point 'V_V', the Emitter Current I_E increases but the Emitter voltage V_E decreases causing the negative resistance region (on the UJT characteristic (Fig. 4.71)), which is stable enough to be used in practical circuits such as relaxation oscillator.

Eventually, the valley point 'V_V' will be reached and any further increase in I_E with increase in V_E will place the device in saturation region. In this region, the characteristics approach that of semiconductor Diode with forward-bias.

Decrease in resistance in active region is due to the holes injected into N-type slab from Aluminium P-type rod when conduction is established. The increased hole current in N-type material will result in an increase in number of free electrons in the slab producing an increase in conductivity G and a corresponding drop in resistance: $(R \downarrow = 1/G)$.

The three other important parameters of the UJT are I_P, V_V and I_V.

$V_P = \eta V_{BB} + V_D$. Increase in V_{BB} causes an increase in V_P.

4.31 APPLICATION OF UJT DEVICE AS AN OSCILLATOR

UJT relaxation oscillator circuit is shown in Fig. 4.72 as one of its applications. Initially, when the oscillator circuit is switched on (at time $t = 0^+$), UJT is in the off condition. The combination of the resistor R and the capacitor C is across the supply voltage V_{BB}. Such situation allows the voltage across the capacitor 'V_C' (V_E) to build up from 0 to the peak voltage 'V_P' ('V_P' is as explained on the UJT characteristic (Fig. 4.54) exponentially). When once 'V_P' is reached ($V_P \geq \eta \cdot V_{BB} + V\gamma$), UJT is triggered to the 'ON' state. This allows the capacitor, which is charged with voltage 'V_P' that is 'V_C', has gone up to 'V_P' to discharge through 'R_{B1}' almost instantly since 'R_{B1}' becomes very low. The device turns off and the process repeats generating a near 'Saw Tooth' voltage waveform across the capacitor (Fig. 4.73).

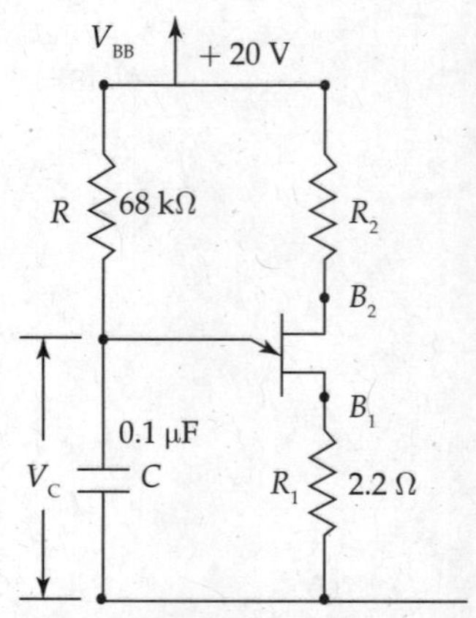

FIG. 4.72 *Oscillator circuit using UJT*

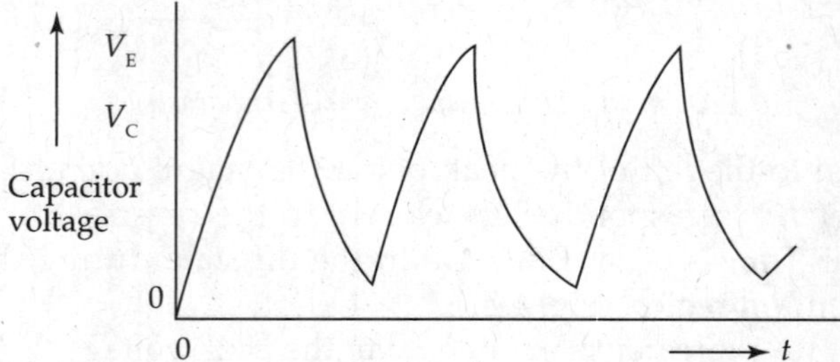

FIG. 4.73 *UJT oscillator output voltage*

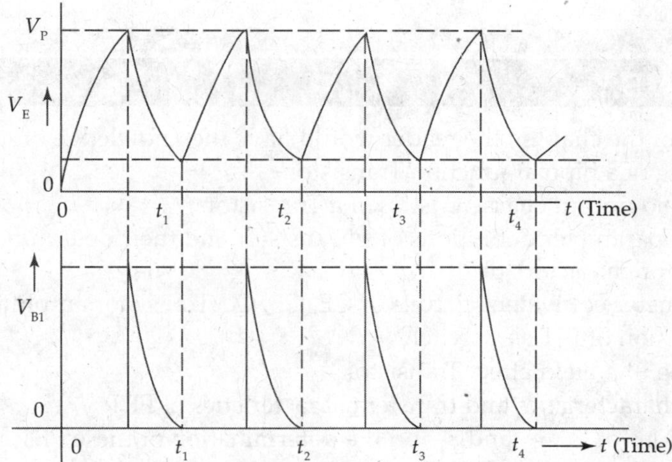

FIG. 4.74 *Waveforms of UJT oscillator*

The voltage across the Capacitor 'C' due to charging through 'R' to the supply voltage V_{BB} while the UJT is in the off-state and discharging through the conducting UJT through R_{B1} and R_1 is according to the following equation:

$$V_C(t) = V_{BB}\,[1 - e^{-t/R_E C}]. \tag{4.81}$$

In this UJT relaxation oscillator circuit of Fig. 4.56, the resistances 'R_{B1}' and 'R_{B2}' are predicted by the intrinsic standoff ratio 'η' of the device. The resistor 'R' and the capacitor 'C' determine the frequency of oscillations of UJT.

The output voltages of UJT can be used as Trigger voltages to SCR and its other family of devices. Typical values are $\eta = 0.55$, $R_{BB} = 10\ \text{k}\Omega$, $V_V = 1.2$ V, $I_V = 5$ mA, $I_P = 50\ \mu$A and $R_{B1} = 200\ \Omega$ during discharge. Frequency of oscillations is calculated as given below:

$$f = \frac{1}{T} = \frac{1}{RC \cdot \ln \dfrac{1}{[1-\eta]}}$$

$$= \frac{1}{68\times 10^3 \times 0.1\times 10^{-6}\ \ln \dfrac{1}{[1-0.55]}}$$

$$f = \frac{10^4}{68\ \ln(1.818)} = \frac{147}{\ln(1.818)}\ \text{Hz}.$$

Calculation of range of values of R_1 to ensure working of UJT as an oscillator:

$$\left[\frac{V_{BB} - V_V}{I_V}\right] < R_1 < \left[\frac{V_{BB} - V_P}{I_P}\right] = \frac{(10-2)}{5\times 10^{-3}} < R_1 < \frac{(10-6)}{50\times 10^{-6}}.$$

$\therefore$ Value of R_1 lies between 1.6 and 80 kΩ in this range of values.

SUMMARY

At the end of the chapter, the reader would have the knowledge of

1. Working of a bipolar junction Transistor
2. Input and output characteristics of a Transistor
3. Hybrid parameter definitions of a Transistor and their determinations from the Transistor characteristics
4. *h*-parameter equivalent circuits of CE, CB, CC Transistor amplifier circuits
5. Application of BJT as a switch
6. Working of a Field effect Transistor
7. Drain characteristics and transfer characteristics of FET
8. Definitions of 'μ', g_m and r_d and the determination of these FET constants from the FET characteristics
9. FET application as an amplifier
10. Typical fabrication steps for a MOSFET device
11. Silicon-controlled rectifier operation
11. UJT operation and its application as a sweep circuit

Questions for Practice

1. Draw the symbol and structural diagram of a Transistor showing the different current components of a Collector Current I_C of BJT and substantiate the name BJT given to a Transistor.
2. Draw the Transistor output characteristics showing the influence of Early Effect for a Common Emitter operated Transistor and explain the phenomenon of Early Effect.
3. Show the various operating regions on the Transistor output characteristics and predict the region of operation on the output characteristics of a Common Emitter operated Transistor so that the Transistor has to be used as a linear amplifier.
4. Draw the small signal low frequency equivalent circuit of a Transistor.
5. Define the *h*-parameters of the Transistor.
6. Mention the main advantages of JFET over BJT.
7. Justify the reasons for JFET to function as voltage-controlled voltage source.
8. Draw the equivalent circuit of JFET device and discuss the reasons for such representation?
9. Draw the output characteristics of FET device and explain various operating regions on the device characteristics?
10. Define the JFET device parameters and establish the relation between them. Explain the methods of their determination from the FET characteristics?
11. Mention the main advantage of DEMOSFET over E-MOSFET.

12. Explain the concept of Threshold voltage for the working of an E-MOSFET device and mention the expression relating the threshold voltage V_{TH} and the other parameters of the device E-MOSFET.
13. Draw the Structure diagram of N-Channel JFET device and explain the various current flow mechanisms for the device to work as an amplifier.
14. Draw the Structure diagram of P-Channel JFET device and explain the various current flow mechanisms for the device to work as an amplifier.
15. Draw a circuit showing the biasing voltages on a circuit diagram to obtain the JFET Characteristics and explain the method of obtaining the device characteristics.
16. Draw a typical Transfer characteristic of JFET device with $I_{DSS} = 10$ mA and $V_P = -4$ V and show the method of determination of g_m. Show the various salient features on the device characteristics.
17. Define Drain Resistance 'r_d' and of a JFET and determine the value of 'r_d' if $\Delta V_{DS} = 10$ V and $\Delta I_D = 0.2$ mA when $V_{GS} = -2$ V (constant). Draw FET output characteristics. Show the determination of Drain Resistance.
18. What is meant by Pinch-off? Draw the necessary diagram showing the shape of Depletion region at pinch-off in JFET device?
19. Explain (a) Threshold voltage, (b) Gate break down with reference to MOSFETs.
20. Compare the performance features of MOSFET, JFET and BJT giving the advantages and disadvantages of each device?

Multiple Choice Questions

1. Operating point represents ______________
   (a) values of I_C and V_{CE} when signal is applied
   (b) the magnitude of signal
   (c) zero signal values of I_C and V_{CE}
   (d) none of the above
2. Analogue Electronic circuits in amplifiers use the Transistor normally in ______________
   (a) active region (b) saturation region
   (c) cut-off mode (d) inverse active mode
3. When the Transistor is used as an electronic switch, the mode of operation for the Transistor to be in the ON condition is ______________
   (a) active mode (b) saturation mode
   (c) cut-off mode (d) inverse active mode
4. When a BJT is used in switching operation, the modes of operation are ______________
   (a) active mode (b) saturation mode and cut-off mode
   (c) cut-off mode (d) inverse active mode
5. Current gain α of a common Base Transistor is defined as the ratio of the currents ______________
   (a) collector to Emitter currents (b) collector to Base currents
   (c) emitter to Base currents (d) emitter to Collector currents

6. Value of current gain α of a common Base Transistor is ______________
   (a) less than 1 (b) greater than 1
   (c) 500 (d) 1

7. Beta (β) of a Transistor has current gain ______________
   (a) 1 (b) greater than 1
   (c) less than 1 (d) infinity

8. Current gain β (beta) of a Transistor is the ratio of ______________
   (a) collector to Emitter Current (b) collector to Base Current
   (c) emitter to Base Current (d) emitter to Collector Current

9. Slight increase in Collector Current with increase in reverse-bias V_{CE} to a Transistor is known as ______________
   (a) early Effect (b) hall effect
   (c) kirk effect (d) punch through

10. In a CE Transistor amplifier, maximum supply voltage is limited to ______________
    (a) avalanche break down of Base Emitter (input) junction
    (b) C–B break down with Emitter open
    (c) C–E break down with Base open
    (d) zener break down of E–B junction

11. Pinch-off voltage in a field effect Transistor is ______________
    (a) drain voltage that makes drain current zero
    (b) gate to source voltage that makes drain current I_D zero
    (c) gate to source voltage that makes the source current zero
    (d) none of these

12. When the gate terminal of N-channel FET is applied with negative voltage ______________
    (a) drain current increases (b) drain current decreases
    (c) no change in drain current (d) drain current becomes zero

13. JFET differs from BJT in the following aspect.
    (a) high input impedance (b) negative resistance device
    (c) higher output resistance (d) lower input resistance

14. Threshold voltage of N-channel MOSFET can be increased by ______________
    (a) increasing the channel dopant concentration (GATE)
    (b) decreasing the channel dopant concentration
    (c) reducing the Gate oxide thickness
    (d) reducing the channel length

15. If a Transistor is operating with both of its junctions forward-biased, but with the Collector–Base junction forward-bias greater than the Emitter–Base forward-bias, its operating point is in the (GATE) ______________
    (a) forward active mode (b) reverse saturation mode
    (c) reverse active mode (d) forward saturation mode

16. A BJT is said to be operating in the saturation region if ____________
 (a) both the junctions revere biased
 (b) base–Emitter junction is reverse-biased and B–C junction is forward-biased
 (c) B–E junction is forward-biased and B–C junction is revere biased
 (d) both the junctions are forward-biased (GATE)

17. UJT characteristics are ____________
 (a) single-valued function of current
 (b) multi-valued function of current
 (c) multi-valued function of voltage
 (d) none of these

18. UJT is also known as ____________
 (a) current-controlled device
 (b) voltage-controlled device
 (c) relaxation oscillator
 (d) none of these

19. Silicon dioxide is used in the fabrication of MOSFETs as ____________
 (a) contact material
 (b) insulating layer between the gate and the channel
 (c) diffusing element
 (d) none of the above

20. The volt–ampere (V–I) characteristic of UJT is ____________
 (a) similar to tunnel diode in some aspects
 (b) linear between the peak point and the valley point
 (c) similar to Common Emitter Transistor characteristics
 (d) similar to FET characteristics

Answers to Multiple Choice Questions

1. (b)	2. (a)	3. (b)	4. (c)	5. (a)
6. (a)	7. (b)	8. (b)	9. (a)	10. (c)
11. (b)	12. (b)	13. (a)	14. (b)	15. (c)
16. (d)	17. (b)	18. (c)	19. (b)	20. (a)

Chapter 5

TRANSISTOR BIASING AND STABILISATION CIRCUITS

Learning Objectives

To study

- Basic concepts of an Amplifier
- Biasing a Transistor to act as an Amplifier
- Various methods of Biasing BJT, FET and MOSFET devices
- Stability of Amplifier operation

5.1 BASIC CONCEPTS OF AN AMPLIFIER

One of the important applications of BJT/FET is to function as an *Amplifier*. Each Amplifier stage consists of an active device BJT/FET, '*R*, *L*, *C*' components, signal source and DC voltages. An Amplifier is considered as a four-terminal network (Fig. 5.1) with two ports (input and output ports).

1. A signal to be amplified is applied to 'input port' of Amplifier and resulting output signal appears at its 'output port'.
2. When output signal is larger than input signal, the circuit works as an *Amplifier*.
3. Ratio of output voltage to input voltage of an Amplifier is known as *voltage amplification* (A) or *voltage gain*.

$$\text{Voltage gain or Amplification} \quad A = \frac{\text{Output voltage}}{\text{Input voltage}} = \frac{V_{\text{out}}}{V_{\text{in}}}$$

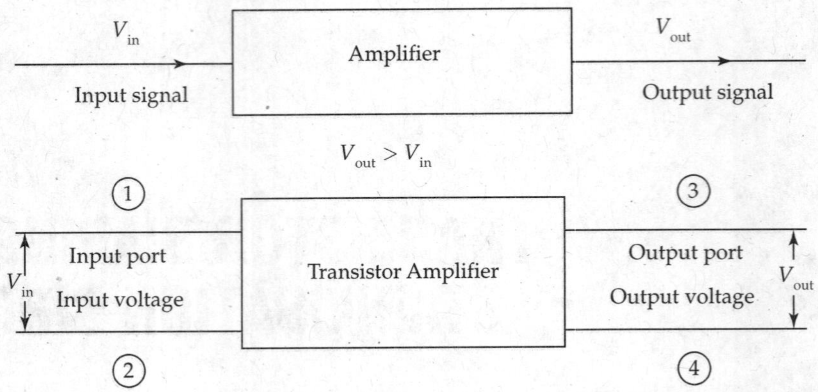

FIG. 5.1 *Concept of amplifier*

One simple application of audio Amplifier is *public address system* (Fig. 5.2).

Frequency components of a speech signal must be amplified by the same strength – uniformly – so that the *fidelity* (faithful reproduction) or quality of speech is maintained at the loudspeaker. Class-A operation of the Amplifier helps with this need, by limiting signal variations, to linear portion of active device characteristics.

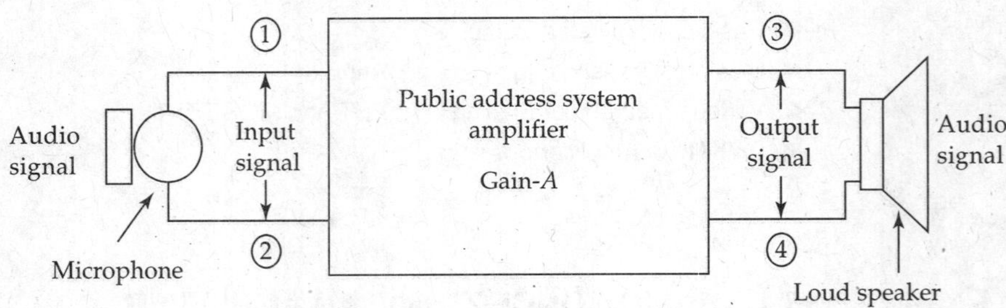

FIG. 5.2 *Audio amplifier system*

5.2 NEED FOR BIASING TRANSISTOR

For a Transistor (BJT) to work as an Amplifier, input junction of the Transistor has to be forward biased by V_{BE} and its output junction has to be reverse biased so that the Transistor operates in the **'Active region'** of its *output characteristics.*

CE Transistor Amplifier circuit is shown in Fig. 5.3. Amplifier has two sets of operating voltages – DC and AC voltages. Collector supply voltage and the two resistors *fix* DC operating conditions in this *Fixed-Bias Circuit.*

- 'Input Signal' (V_{in}) is applied between the input pair of terminals (Base- and Emitter-input port) through Input Coupling Capacitor (C_{in}) which also filters any DC component to enter input circuit.
- 'Output Voltage' (V_{out}) is obtained at the output pair of terminals (Collector and Emitter-output port) through the Output Coupling Capacitor (C_{out}) which filters any DC component into the output circuit.
- The two coupling capacitors (C_{in} and C_{out}) are used to separate the DC and AC quantities in the Amplifier circuit, hence also known as 'Blocking Capacitors'.

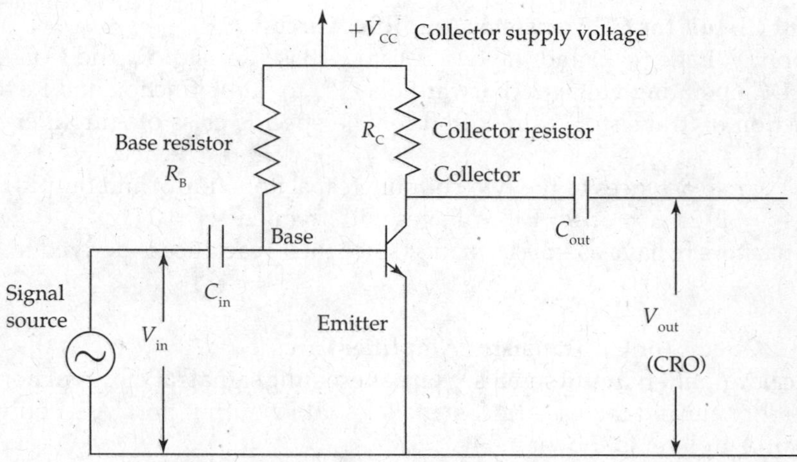

FIG. 5.3 *Common emitter transistor amplifier circuit*

Transistor Amplifier has two types of operating conditions, with DC and AC voltages in this case study. So the Amplifier analysis can be done using DC- and AC-equivalent circuits as shown in Figs. 5.4 and 5.5, respectively.

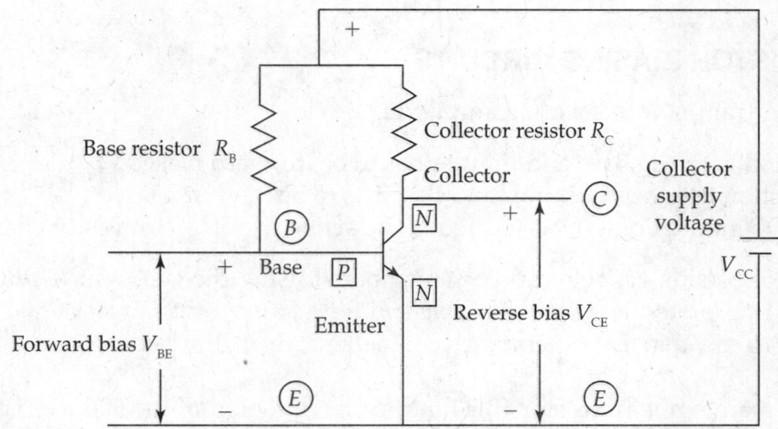

FIG. 5.4 *DC-equivalent circuit for common emitter transistor amplifier circuit*

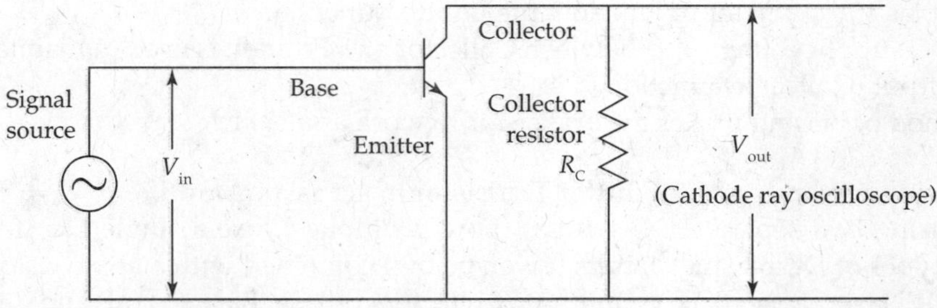

FIG. 5.5 *AC-equivalent circuit for common emitter transistor amplifier circuit*

DC-equivalent Circuit for CE Transistor Amplifier Circuit

Collector Supply Voltage (V_{CC}) and the two resistors (Base Resistor R_B and Collector Resistor R_C) fix up the DC operating voltages (Forward bias V_{BE} to input junction and Reverse bias V_{CE} to output junction of Transistor) as required for the specific class of Amplifier operation as shown in Fig. 5.4.

X_{Cin} and X_{Cout} are reactances of the two coupling capacitors (input and output).

X_{Cin} and $X_{Cout} = 1/2\pi fC = \infty$ for DC voltages with frequency $f = 0$ Hz.

The two capacitors behave as – *open circuits* (since their reactance is derived to be ∞ for DC voltages).

AC-Equivalent Circuit for CE Transistor Amplifier Circuit

In most practical Amplifiers, input signals V_{in} are alternating signals at the input port. Amplifiers develop proportionate and increased AC signals V_{out} at its output port. AC-Equivalent circuit of CE Transistor Amplifier is shown in Fig. 5.5.

In AC signal operations, capacitors function as 'short circuits' and couple input and output AC signals.

- Input Coupling Capacitor 'C_{in}' couples input signal source to input port – between Base and Emitter for 'amplification'.
- Output Coupling Capacitor 'C_{out}' couples amplified output signal (between Collector and Emitter) to external load.

5.3 TRANSISTOR BIASING CIRCUITS

In order for the Transistor to act as an amplifying device

1. Input junction between Base and Emitter has to be 'forward biased'.
2. Output junction between Collector and Base has to be 'reverse biased'.
3. Quiescent Operating Point '*Q*' is located in the active region of Transistor characteristics.

This mode of biasing scheme can be remembered using the acronym *IFOR active*. 'IFOR active' can be interpreted as *I for input junction, F for forward-bias, O for output junction* and *R for reverse-bias*, so that DC operating parameters are in the active region of Transistor characteristics.

In order to maintain a Transistor (BJT) in the active region of its characteristics, a simple method is to use *two separate DC sources* (Fig. 5.6).

- First DC supply V_{BB}, between Base and Emitter, with a series resistor R_B, is used to maintain forward-bias V_{BE} to the input junction, the Emitter junction.
- Second DC supply voltage V_{CC}, between Collector and Emitter, is used to maintain reverse-bias to output (Collector) junction.
- This method of biasing makes the Transistor to work as an *amplifying device.*

Various biasing methods of CE Emitter Transistor to act as an Amplifier

Instead of using two separate DC sources (power supplies), use a single DC source V_{CC} to build three types of DC biasing circuits (given below) combined with a few resistors.

(1) Fixed-Bias or Base Bias circuit, (2) Collector-to-Base Bias circuit and (3) Potential (voltage)-Bias circuit or self-biasing circuit.

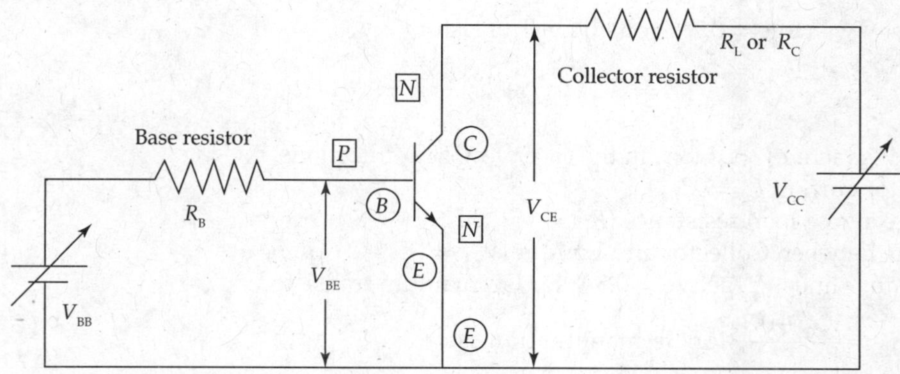

FIG. 5.6 *Two separate DC voltage sources to bias NPN transistor*

5.4 FIXED-BIAS CIRCUIT (BASE BIAS CIRCUIT) FOR COMMON EMITTER TRANSISTOR

Resistance R_B, in Fig. 5.6, is reoriented to provide required Transistor biasing with single DC source V_{CC}, as shown in Fig. 5.7.

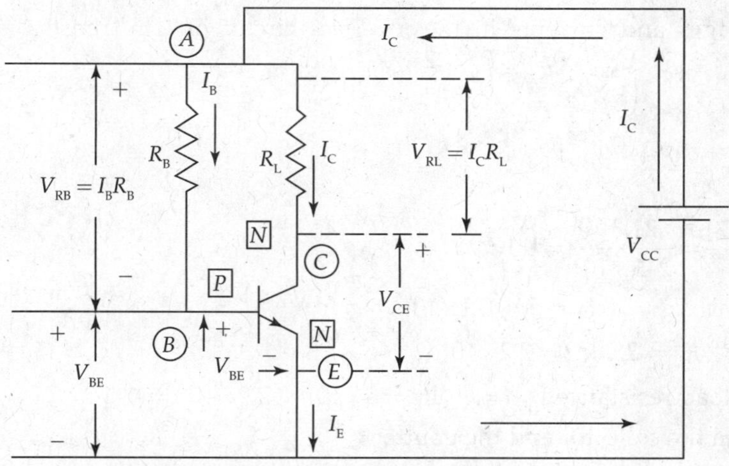

FIG. 5.7 *Fixed-bias circuit for common emitter transistor*

1. Choice of V_{CC}, R_L R_C and R_B are designed to provide forward-bias V_{BE} to Emitter junction and reverse-bias V_{CE} to Collector junction.
2. In addition, V_{CC} should be less than $V_{CE(max)}$ (maximum voltage rating that can be applied to Collector junction without causing breakdown to output junction, as per manufacturer specifications).

In Fig. 5.7, using KVL equations for input side loop ABEA:

$$V_{CC} = I_B \cdot R_B + V_{BE}$$

$$\therefore \quad I_B = \left(\frac{V_{CC} - V_{BE}}{R_B}\right) = \left(\frac{V_{CC} - 0.7}{R_B}\right),$$

because V_{BE} active = 0.7 V for silicon Transistor.

$$\therefore \quad I_B \cong \frac{V_{CC}}{R_B}$$

when V_{BE} (active) $\ll V_{CC}$ as in the case of most applications.
Collector current $I_C = \beta I_B$
Voltage across load resistance $(R_L) = V_{RL} = I_C R_L$
Voltage between Collector and Emitter $[V_{CE} = (V_{CC} - I_C R_L)]$.
On similar lines V_{BE} active = 0.3 V for Germanium transistor.

Then $I_B = \dfrac{V_{CC} - 0.3}{R_B}$ for Germanium transistor.

$\therefore \quad I_B \cong \dfrac{V_{CC}}{R_B}$ when V_{BE} active $\ll V_{CC}$ for Germanium transistor.

Base current I_B is constant. So this type of biasing network is called as 'Fixed-Bias Circuit'/ 'Constant Current Biasing Circuit'.

EXAMPLE 5.1

Determine voltages and currents in the *fixed-bias* circuit of CE Transistor (Fig. 5.7) with $V_{CC} = 22$ V and $\beta = 50$, $V_{BE} = 0.7$ V, $R_B = 213$ kΩ and $R_L = 2.2$ kΩ.

Solution:

Base current $\quad I_B = \dfrac{[V_{CC} - V_{BE}]}{R_B}$

$$\therefore \quad I_B = \frac{[22 - 0.7]}{213 \times 10^3} = \frac{21.3 \text{ V}}{213 \times 10^3} = 0.1 \text{ mA}$$

Collector current $\quad I_C = \beta \cdot I_B = 50 \times 0.1 \times 10^{-3} = 5$ mA

Load resistance $\quad R_L = 2.2\ \text{k}\Omega = 2.2 \times 10^3\ \Omega$

Voltage across load resistance $V_{RL} = I_C \cdot R_L = 5 \times 10^{-3} \times 2.2 \times 10^3 = 11$ V

Voltage between the collector and the emitter = $V_{CE} = V_{CC} - I_C \cdot R_L$

$\therefore \quad V_{CE} = 22 - 11 = 11$ V.

EXAMPLE 5.2

Calculate the DC bias voltages and currents in the Fixed-Bias Circuit with Emitter resistor of CE Transistor (Fig. 5.8) (neglect V_{BE} of transistor).

Solution: Assume that the voltage between Base and Emitter junctions $V_{BE} = 0$ V.
Applying KVL at the input of the circuit shown in Fig. 5.8,

$$V_{CC} = [I_B R_B + I_E R_E] = [I_B R_B + (I_C + I_B) R_E] \quad [\because \quad V_{BE} = 0\text{V (Data)}]$$

$$\therefore \quad V_{CC} = \frac{I_C}{\beta} R_B + \left[I_C + \frac{I_C}{\beta}\right] R_E = \left[\frac{I_C R_B}{\beta} + \left[I_C + \frac{I_C}{\beta}\right] \times R_E\right]$$

$$\Rightarrow V_{CC} = \left[\frac{R_B}{\beta} + \frac{R_E}{\beta} + R_E\right] \times I_C$$

$$\therefore \text{ Collector current } \quad I_C = \frac{V_{CC}}{\left[R_E + \dfrac{(R_B + R_E)}{\beta}\right]}$$

$$= \frac{20}{\left[1\times 10^3 + \dfrac{400\times 10^3 + 1\times 10^3}{100}\right]}$$

$$= \frac{20\times 100}{501\times 10^3}$$

$$= \frac{20}{5.01\times 10^3} = 3.992 \text{ mA}$$

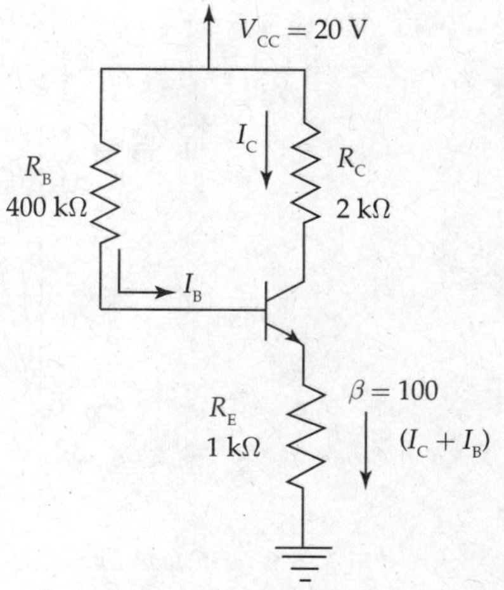

FIG. 5.8 *Fixed-bias circuit with emitter resistor*

Now, applying the KVL at the output of the circuit shown in Fig. 5.8,

$$V_{CC} = I_C R_C + V_{CE} + (I_C + I_B)\, R_E$$

$$\Rightarrow V_{CE} = \left(V_{CC} - I_C R_C - \left(I_C + \frac{I_C}{\beta}\right) \times R_E\right)$$

$$\therefore\ V_{CE} = \left[V_{CC} - I_C\left[R_C + R_E + \frac{R_E}{\beta}\right]\right]$$

$$= 20 - 3.992\times 10^{-3}\left[2\times 10^3 + 1\times 10^3 + \frac{1\times 10^3}{100}\right] = 7.984 \text{ V.}$$

To maintain DC operating point '*Q*' based on the previous equations, following design constraints are to be followed. The choice of R_L and R_B is not entirely arbitrary, but it is subjected to some design constraints based on Transistor ratings.

1. **Transistor Specification $V_{CE\ (max)}$:** Collector supply voltage (V_{CC}) should always be less than $V_{CE(max)}$, as prescribed by the manufacturer specifications.
2. **Power dissipation $P_{D\ (max)}$ rating for the Transistor**
   $V_{CC} \cdot I_C(Q)$ [where $I_C(Q)$ is the quiescent component of collector current]
   Power dissipation at Transistor Collector junction should always be less than the allowed maximum power dissipation specification. This design constraint is represented by a power dissipation curve (in the shape of a hyperbola) (Fig. 5.9).
3. Third constraint requires that DC Load Line XY, drawn on the Transistor output characteristics, should always be below the power dissipation curve as shown in Fig. 5.9.

Procedure to locate DC/*Q*-point

From DC-equivalent circuit of Fig. 5.7 at output circuit loop ACEA using Kirchoff's Voltage (KVL) law,

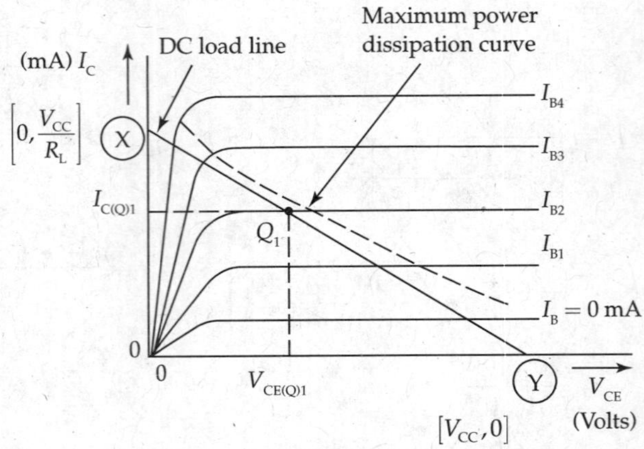

FIG. 5.9 *DC load line, quiescent operating point and maximum power dissipation curve*

$$V_{CC} = V_{RL} + V_{CE} \tag{5.1}$$

$$V_{CC} = I_C \cdot R_L + V_{CE} \tag{5.2}$$

$$\therefore \quad V_{CC} - V_{CE} = I_C \cdot R_L \quad \text{(DC Load line equation).} \tag{5.3}$$

Equation (5.3) represents the equation of a straight line known as DC Load Line equation. This line has a negative slope of $1/R_L$ as shown in Fig. 5.9.

Coordinates of X and Y of the load line are calculated using DC Load Line equation that can be drawn on Transistor output characteristics.

At point 'X', $V_{CE} = 0$ V. Substituting $V_{CE} = 0$ in the load line equation

$$\text{Collector current} \quad I_C = \frac{V_{CC}}{R_L}.$$

Then 'X' co-ordinates are

$$V_{CE} = 0\text{ V} \quad \text{and} \quad I_C = \frac{V_{CC}}{R_L}. \tag{5.4}$$

At point 'Y', Collector current $I_C = 0$ mA, then from Eq. (5.3), $V_{CE} = V_{CC}$.

Then the co-ordinates of the point 'Y' are $V_{CE} = V_{CC}$ and $I_C = 0$ mA.

Using the coordinates of X and Y, DC Load Line XY can be drawn on Transistor output characteristics. Power dissipation curve can be drawn by calculating V_{CE} and I_C, from maximum power dissipation rating for the selected Transistor (discussed in more detail in Chapter 6). Power dissipation curve is drawn on the output characteristics and it will have the shape of a hyperbola. DC Load Line should always be drawn below the maximum power dissipation curve, so that Transistor operating features will not exceed the maximum operating voltages and currents specified by the manufacturers.

- *DC operating point*, also known as *quiescent operating point 'Q' (bias point)*, is determined graphically knowing the Amplifier operation class (discussed in Chapter 11).
- For example, for Class-A operation of an Amplifier, quiescent operating point is fixed in the middle of the DC Load Line. The signal operations will be linear and distortion content is least.

5.5 STABILITY FACTOR

- In order to set the premise, it is necessary to discuss the stabilisation of DC operating conditions of the Transistor, in order to function as a stable Amplifier.
- DC operating point shifts due to temperature variations as well as variations due to device parameters in case of replacement of the active device.

$$\because \quad I_C = -\alpha \cdot I_E + I_{C0} \text{ and } I_C = \beta \cdot I_B + (\beta + 1)I_{C0}. \tag{5.5}$$

Even if Base current I_B is constant I_C can change due to changes in β, I_{C0} and V_{BE}. If a Transistor fails and is replaced by a new Transistor, for same I_B, Collector current $I_C = \beta_2 \cdot I_B$, where β_2 is β of the new transistor and may be more or less than the previous value β of the old Transistor. Similarly, at different temperatures T_1, T_2 and T_3, Collector current I_C and Quiescent Operating Point Q are shown as follows:

At T_1, I_C (1) • Q_1
At T_2, I_C (2) • Q_2
At T_3, I_C (3) • Q_3
as shown in Fig. 5.10.

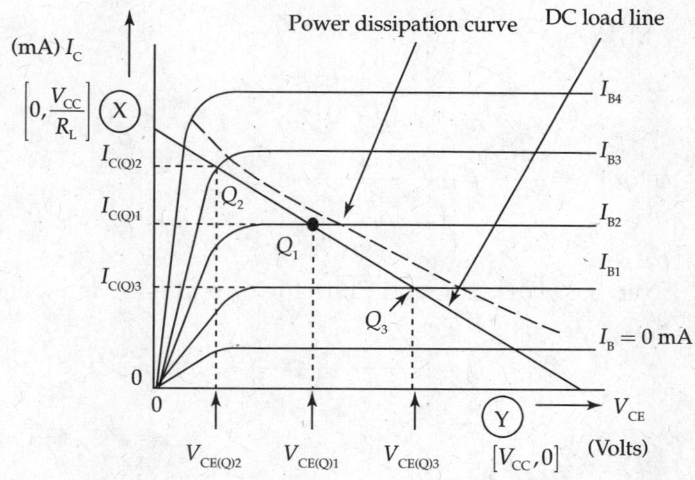

FIG. 5.10 *Locating Q-point on DC load line and maximum power dissipation curve*

These changes or shifts in quiescent operating point 'Q' are main causes for unstable operation in Transistor Amplifiers. This instability is measured by stability factor 'S'.

From the previous discussions, it is understood that Q_1 is the DC operating point at a particular temperature T_1 and beta of the Transistor $\beta = \beta_1$. If temperature increases to T_2 operating point shifts up to Q_2. Similarly if the temperature falls to T_3 or Transistor beta changes to β_3 for a replaced new Transistor, the operating point shifts to Q_3.

These shifts in operating points are stabilised or guarded with some additions in the circuit designs that are discussed in this chapter.

Before analysing the circuits, stability of operating point 'Q' is defined and criteria for stability need to be arrived at.

Definition of Stability Factor

Stability factor defines or accounts for incremental changes in the Collector current ∂I_C for incremental changes in the leakage currents ∂I_{C0} in the Transistor either due to changes in working temperatures T of the device or due to changes in β of the Transistor with their aging or replaced Transistor may have changes in values of β and changes in V_{BE}.

$$\text{Stability factor} \quad S = \frac{\partial I_C}{\partial I_{C0}} \tag{5.6}$$

$$I_C = \beta \cdot I_B + (\beta + 1) I_{C0} \tag{5.7}$$

Differentiating this equation with respect to I_C

$$\frac{\partial I_C}{\partial I_C} = \beta \frac{\partial I_B}{\partial I_C} + (\beta + 1) \frac{\partial I_{C0}}{\partial I_C} \tag{5.8}$$

$$1 = \beta \frac{\partial I_B}{\partial I_C} + (\beta + 1) \frac{\partial I_{C0}}{\partial I_C} \tag{5.9}$$

$$(\beta + 1) \frac{\partial I_{C0}}{\partial I_C} = 1 - \beta \frac{\partial I_B}{\partial I_C} \tag{5.10}$$

$$\frac{\partial I_{C0}}{\partial I_C} = \frac{\left[1 - \beta \left(\frac{\partial I_B}{\partial I_C}\right)\right]}{(\beta + 1)} \tag{5.11}$$

$$\text{Now, the stability factor} \quad S = \frac{\partial I_C}{\partial I_{C0}} = \frac{(\beta + 1)}{\left[1 - \beta \cdot \left(\frac{\partial I_B}{\partial I_C}\right)\right]}. \tag{5.12}$$

In the Fixed-Bias Circuit, Base current I_B and Collector current I_C are independent:

$$I_B = \left[\frac{V_{CC} - V_{BE}}{R_B}\right] \cong \frac{V_{CC}}{R_B} \quad \text{for an active transistor} \tag{5.13}$$

$$\therefore \quad \left[\frac{\partial I_B}{\partial I_C} = 0\right] \tag{5.14}$$

Substituting Eq. (5.14) into Eq. (5.12), Stability Factor (S) for fixed-bias circuit is given by

$$S = \left[\frac{(\beta + 1)}{\left[1 - \left(\frac{\partial I_B}{\partial I_C}\right)\right]}\right] = [\beta + 1]. \tag{5.15}$$

5.6 COLLECTOR-TO-BASE BIAS CIRCUIT TO CE TRANSISTOR

In Fig. 5.11, resistor R_B is connected between Collector and Base, improving the stability of quiescent operating point 'Q'. But this circuit is rarely used and obsolete, because of undesirable feedback through R_B connecting output and input ports. *Amplification has to be a unidirectional process* from the input port to the output port of the Amplifier. For Collector-to-Base bias circuit in Fig. 5.11, Loop ACBEA.

$$V_{CC} = V_{RL} + V_{RB} + V_{BE}\,(\text{active}) \qquad (5.16)$$

Considering $V_{BE} = 0.7$ V

$$V_{CC} = (I_C + I_B)\cdot R_L + I_B \cdot R_B + 0.7$$

$$V_{CC} = I_B(R_L + R_B) + I_C \cdot R_L + 0.7 \qquad (5.17)$$

Differentiating Eq. (5.17) with respect to Collector current I_C, we get

$$\frac{\partial V_{CC}}{\partial I_C} = 0. \qquad (5.18)$$

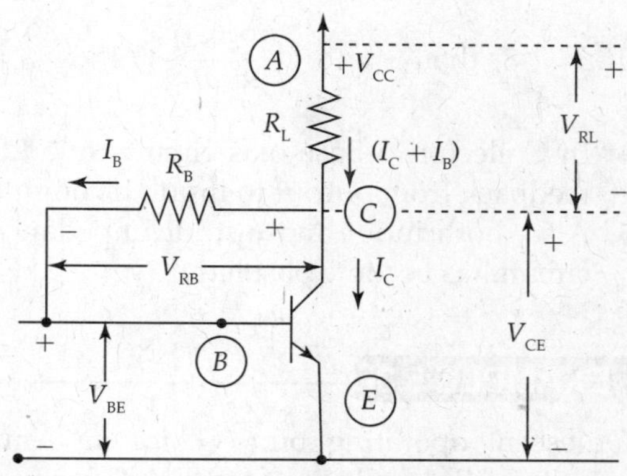

FIG. 5.11 *Collector-to-base bias circuit for CE transistor*

$$\therefore \quad \frac{\partial I_B}{\partial I_C}[R_L + R_B] + R_L + 0 = 0$$

$$\Rightarrow \quad \frac{\partial I_B}{\partial I_C} = -\frac{R_L}{[R_L + R_B]} \qquad (5.19)$$

Substituting $-R_L/[R_L + R_B]$ for $\partial I_B/\partial I_C$ in Eq. (5.15) for S, we get

$$\text{Stability factor} \quad S = \frac{\partial I_B}{\partial I_{CO}} = \left[\frac{(\beta+1)}{1-\beta\left(\dfrac{\partial I_B}{\partial I_C}\right)}\right] = \left[\frac{(\beta+1)}{1+\beta\left(\dfrac{R_L}{R_L + R_E}\right)}\right] \qquad (5.20)$$

Hence, stability factor S for Collector-to-Base bias circuit is much smaller than that for Fixed-Bias Circuit, making the operation more stable.

If R_B is much less than R_L, it can be approximated that $(R_L + R_B) \approx R_L$

$$\therefore \quad \text{Stability factor} \quad S = \frac{(\beta+1)}{(\beta+1)} \approx 1. \qquad (5.21)$$

If this condition is satisfied, it dramatically improves stability. But this is not a practical situation. When R_B is very small I_B becomes excessive, saturating the Transistor.

In practical circuits, $R_B = 15\,R_L$; substituting it in Eq. 5.20,

$$S = \frac{(\beta+1)}{\left[1+\beta\left(\dfrac{R_L}{(R_L + R_B)}\right)\right]}$$

$$= \frac{(\beta+1)}{\left[1+\beta\left(\dfrac{R_L}{R_L + 15R_L}\right)\right]}$$

$$= \frac{(\beta+1)}{\left[1+\beta\cdot\dfrac{1}{16}\right]} = 16\frac{(\beta+1)}{(\beta+16)}$$

If $\beta = 48$, then $S = 16 \times \frac{(48+1)}{(48+16)} = 16 \times \frac{49}{64} = 12.25.$

- In Collector-to-Base bias circuit, an additional disadvantage is the unavoidable negative feedback from output to input, through R_B, resulting in reduced gain.
- Also, both input and output impedances get reduced due to feedback through R_B. This circuit has become obsolete.

EXAMPLE 5.3

Quiescent operating pint 'Q' for an Amplifier circuit with Collector feedback bias circuit (Fig. 5.12) fixed at I_C = 4.9 mA and V_{CE} = 11 V, Transistor β = 49, Bias V_{BE} = 0.7 V. Calculate I_B and bias resistor R_B.

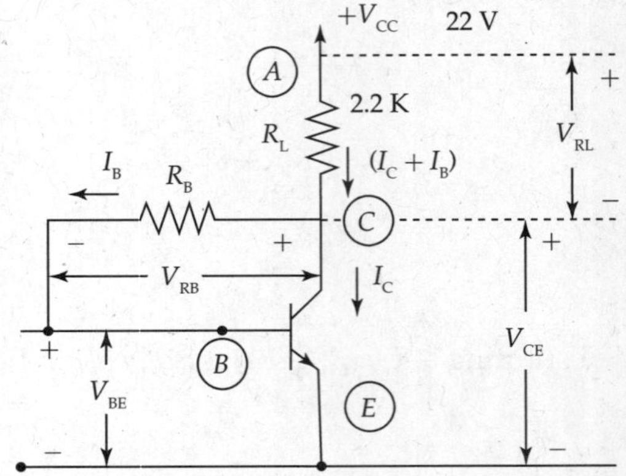

FIG. 5.12 *Collector-to-base bias circuit for NPN common emitter (CE) transistor*

Solution: Collector to emitter voltage

$$V_{CE} = V_{CC} - (I_C + I_B)R_L$$

Base current $I_B = \frac{I_C}{\beta} = \frac{4.9 \text{ mA}}{49} = 0.1 \text{ mA}$

$$\therefore \quad V_{CE} = V_{CC} - (I_C + I_B) \cdot R_L$$
$$= 22 - (4.9 + 0.1) \times 10^{-3} \times 2.2 \times 10^3$$
$$= 22 - 11 = 11 \text{ V}$$

Voltage across $R_B = V_{RB} = V_{CE} - V_{BE} = 11 - 0.7 = 10.3$ V

$\therefore$ Base resistor $R_B = \frac{V_{RB}}{I_B} = \frac{10.3 \text{ V}}{0.1 \times 10^{-3}}$

$$= 103 \times 10^3 \ \Omega = 103 \text{ k}\Omega.$$

Various voltages and currents in Collector-to-Base bias circuit to CE Transistor with R_E (Fig. 5.13)

Various voltages at Base, Emitter and Collector can be calculated as shown below. Measuring these voltages, DC or quiescent operating conditions can be designed for desired Amplifier performance.

V_{CC} = Collector supply voltage
V_{CE} = Voltage between Collector and Emitter
V_{RC} = Voltage across Collector resistor $R_C = (I_C + I_B)R_C = (\beta + 1)I_B \cdot R_C$
V_{RE} = Voltage across $R_E = V_E = I_E \cdot R_E = (I_C + I_B)R_E = (\beta + 1)I_B \times R_E$
$V_{RB} = I_B \times R_B$
Base voltage $V_B = V_{BE} + V_{RE}$
Collector voltage $V_C = (V_{CE} + V_{RE})$

$$V_{CC} = V_{RC} + V_{RB} + V_{BE} + V_E$$

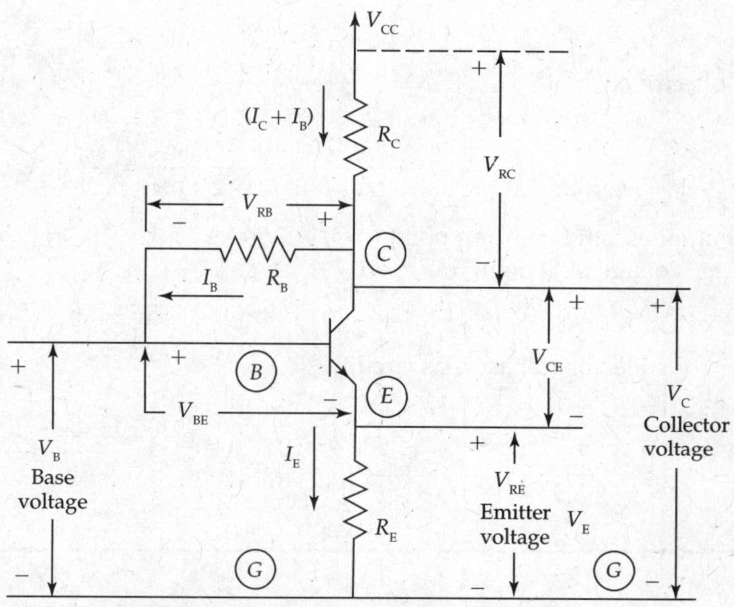

FIG. 5.13 *Collector-to-base bias transistor circuit with emitter resistor*

$$\therefore \quad V_{CC} = (\beta + 1)I_B \cdot R_C + I_B \times R_B + V_{BE} + (\beta + 1)I_B \cdot R_E$$

$$\text{Base current} \quad I_B = \frac{(V_{CC} - V_{BE})}{[(\beta+1)(R_C + R_E) + R_B]}$$

$$\text{Collector Current} \quad I_C = \beta \cdot I_B$$

$$\text{Emitter curent} \quad I_E = (I_C + I_B) = (\beta + 1) \cdot I_B$$

$$\text{At the output loop,} \quad V_{CC} = V_{RC} + V_{CE} + V_E$$

$$\Rightarrow \quad V_{CC} - V_{CE} = V_{RC} + V_E$$

$$\text{Collector current} \quad I_C = \frac{(V_{CC} - V_{CE})}{(R_C + R_E)} \quad \text{neglecting the base current}$$

If the collector resistance $R_C = R_L$ (Load resistance)

$$I_C = \frac{(V_{CC} - V_{CE})}{(R_L + R_E)}.$$

Even though the changes in bias conditions are much less than that of simple fixed-bias circuit, these changes are unacceptable. Biasing circuit with voltage-feedback-bias arrangement is inferior to potential- or voltage-divider-bias circuit.

EXAMPLE 5.4

For an NPN Transistor in CE configuration with Collector-to-Base bias circuit, Collector supply voltage V_{CC} = 10 V, Collector resistor R_C = 2 kΩ, and Base resistor R_B = 100 kΩ. Calculate parameters at quiescent operating point '*Q*' and stability factor *S*. Transistor Beta $\beta = 50$ and V_{BE} = 0.7 V.

Solution:

$$\text{Collector current} \quad I_C = \beta \cdot I_B = \left[\frac{\beta \cdot (V_{CC} - V_{BE})}{[R_B + (\beta+1) \cdot R_C]}\right]$$

$$\therefore \quad I_C = \left[\frac{50 \times (10 - 0.7)}{100 \times 10^3 + (50+1) \times 2 \times 10^3}\right] = 2.3 \text{ mA}$$

Collector current at quiescent operating point $I_C(Q) = 2.3$ mA
Collector-to-emitter voltage at Q point $= V_{CE}(Q)$

$$V_{CE}(Q) = [V_{CC} - I_C(Q) \times R_C] = [10 - 2.3 \times 10^{-3} \times 2 \times 10^3] = [10 - 4.6] = 5.4 \text{ V}$$

Stability factor 'S' for collector-to-base bias circuit

$$S = \left[\frac{(\beta+1)}{1 + \left(\frac{\beta \cdot R_C}{R_C + R_B}\right)}\right] = \left[\frac{(50+1)}{1 + \left(\frac{50 \times 2 \times 10^3}{(2+100) \times 10^3}\right)}\right] = \frac{51}{1.98} = 25.75 \cong 26.$$

5.7 POTENTIAL (VOLTAGE)-DIVIDER-BIAS TO CE TRANSISTOR

Potential-divider-bias circuit (Fig. 5.14) is used in most of the applications, as the circuit operation is more stable due to stable quiescent operation helped by Emitter resistor R_E:

$$V_{Th} = \frac{V_{CC} \times R_2}{(R_1 + R_2)} = V_{BG} \tag{5.22}$$

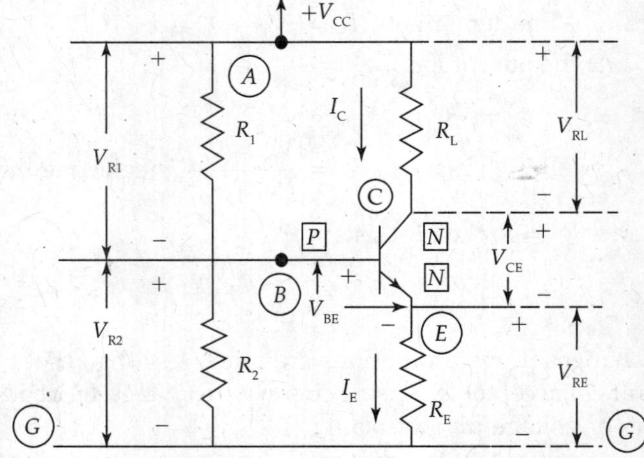

FIG. 5.14 *(Self-biasing circuit) Potential-divider-bias circuit*

Applying Thevinin's theorem in the circuit of Fig. 5.14,

$$V_{Th} = V_{BG} = \frac{V_{CC} \times R_2}{R_1 + R_2}. \tag{5.23}$$

V_{Th} is a potential division of V_{CC} between R_1 and R_2, i.e., open circuit voltage (Fig. 5.15).

In order to get equivalent Thevinin's resistance R_{Th}, short circuit Voltage Source V_{CC}.

Then the resulting circuit looks as shown in Fig. 5.16.

$$R_{Th} = R_B = R_1 \parallel R_2 = \frac{R_1 \cdot R_2}{[R_1 + R_2]} \tag{5.24}$$

R_{Th} then is the parallel combination of R_1 and R_2.

In the loop BDGEB of Fig. 5.17,

$$V_{Th} = R_B \cdot I_B + V_{BE} + (I_C + I_B)R_E$$

$$\therefore \quad V_{Th} = I_B(R_B + R_E) + I_C \cdot R_E + V_{BE} \tag{5.25}$$

Differentiating Eq. (5.25) with respect to I_C,

$$0 = \frac{\partial I_B}{\partial I_C}(R_B + R_E) + R_E + 0 \tag{5.26}$$

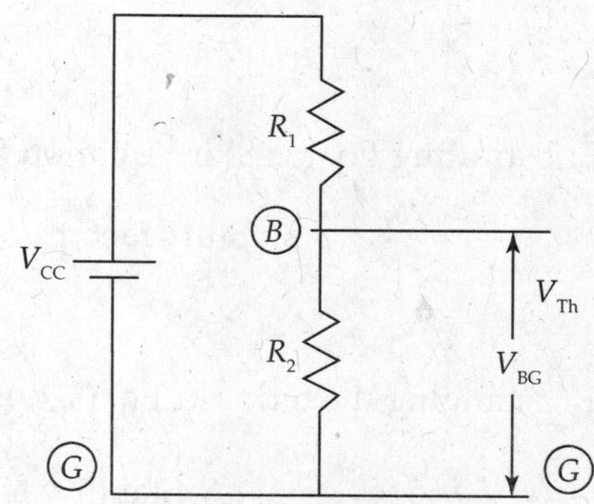

FIG. 5.15 *Thevinin's equivalent circuit at input port of the transistor to find the open circuit voltage V_{Th} or V_{BG}*

$$\text{which means } \frac{\partial I_B}{\partial I_C}(R_B + R_E) = -R_E \tag{5.27}$$

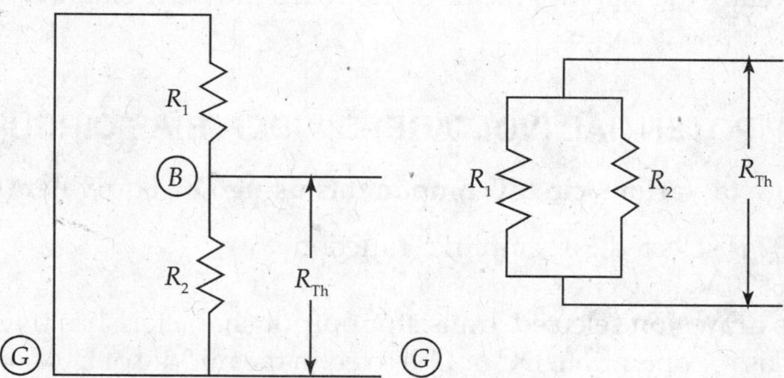

FIG. 5.16 *Thevinin's equivalent circuit at input port of the transistor to find R_{Th} $R_{Th} = R_B = R_1 \parallel R_2$*

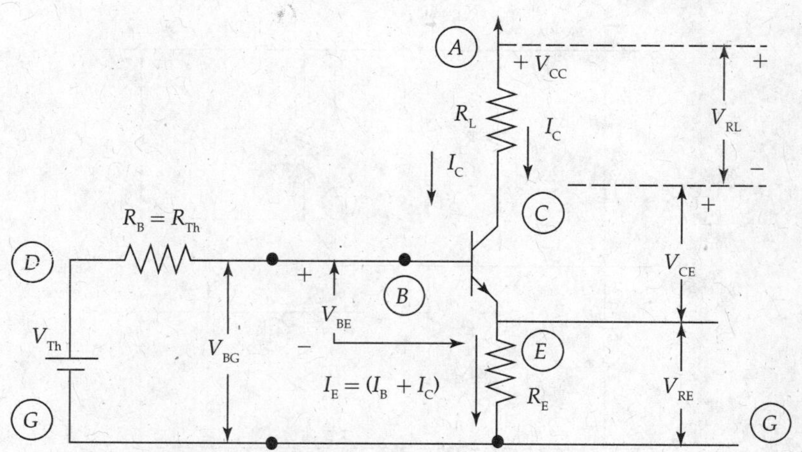

FIG. 5.17 *Thevinin's equivalent circuit suitable for analysis*

$$\therefore \quad \frac{\partial I_B}{\partial I_C} = -\frac{R_E}{(R_B + R_E)} \tag{5.28}$$

Substituting Eq. (5.28) in equation for stability factor '*S*',

$$\text{Stability factor} \quad S = \frac{\partial I_C}{\partial I_{C0}} = \frac{(\beta+1)}{\left(1 - \beta \cdot \left(\frac{\partial I_B}{\partial I_C}\right)\right)} = \frac{(\beta+1)}{\left(1 + \beta \cdot \left(\frac{R_E}{R_B + R_E}\right)\right)} \tag{5.29}$$

Rearranging the terms of Eq. (5.29),

$$\text{Stability factor} \quad S = \frac{(\beta+1)(R_B + R_E)}{R_E(\beta+1) + R_B} \tag{5.30}$$

$$\text{Stability factor } S \text{ can be rewritten as } S = \frac{(1+\beta)\left[1 + \frac{R_B}{R_E}\right]}{(1+\beta) + \left[\frac{R_B}{R_E}\right]} \tag{5.31}$$

So, inference is that – the stability factor S varies around 1 for small values of (R_B/R_E) and $(1 + \beta)$ when (R_B/R_E) is very large.

5.8 DESIGN OF POTENTIAL (VOLTAGE)-DIVIDER-BIAS CIRCUIT (FIG. 5.18)

Analysis and design of various circuit components using DC-equivalent circuit

- Transistor BC 107 is selected for Amplifier function
- V_{CC} is selected as 20 V
- DC Load Line is drawn on selected Transistor output characteristics (Fig. 5.19)
- For Amplifier Class-A operation, DC or *Q* is fixed in the middle of DC Load Line connecting points *A* and *B*.
- As Point *Q* is located in the middle of DC Load Line,

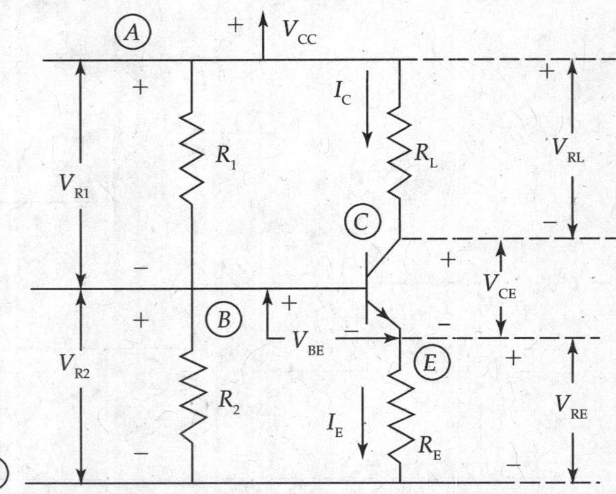

FIG. 5.18 *Analysis and design of self-biasing circuit or potential-divider-bias circuit*

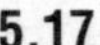

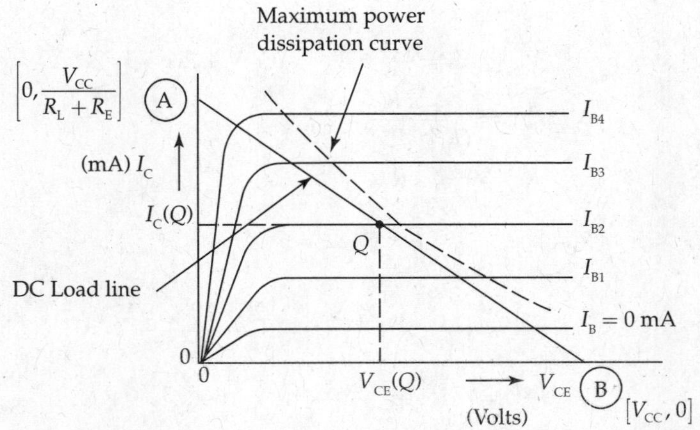

FIG. 5.19 *DC load line, quiescent operating point Q and maximum power dissipation curve*

$$V_{CE}(Q) = 0.5\ V_{CC} = 0.5 \times 20 = 10\ \text{V}.$$

- Emitter voltage is assumed as: $V_{RE} = 0.1\ V_{CC} = 0.1 \times 20 = 2$ V.
- At the output port of the DC-equivalent circuit, $V_{CC} = V_{RC} + V_{CE}(Q) + V_{RE}$

So, $V_{RC} = V_{CC} - V_{CE}(Q) - V_{RE} = V_{CC} - 0.5\ V_{CC} - 0.1\ V_{CC}$

$$\therefore\ V_{RC} = 0.4\ V_{CC} = 0.4 \times 20\ \text{V} = 8\ \text{V}.$$

- Selecting the quiescent value of Collector current $I_C(Q) = 2$ mA, Design R_1 and R_2, R_C and R_E

1. Design of R_C

$$V_{RC} = 0.4\ V_{CC} = 0.4 \times 20 = 8\ \text{V}$$

Assume $I_C(Q) = 2.0$ mA

Voltage across $R_C = V_{RC} = I_C(Q) \times R_C = 8.0$ V

$$\therefore\ R_C = \frac{V_{RC}}{I_C(Q)} = \frac{8}{2\ \text{mA}} = 4\ \text{k}\Omega.$$

2. Design of R_E

Emitter voltage $V_E = V_{RE} = 0.1\ V_{CC} = 0.1 \times V_{CC} = 0.1 \times 20 = 2$ V

$$\therefore\ V_{RE} = I_E \times R_E \cong I_C(Q) \times R_E = 2.0\ \text{V}$$

$$\therefore\ R_E = \frac{V_{RE}}{I_C(Q)} = \frac{2\ \text{V}}{2\ \text{mA}} = 1\ \text{k}\Omega.$$

3. Design of R_2

Assume the value of β of the Transistor as 100

Input resistance of the Transistor $\cong \beta R_E$

$$\text{Assume } R_2 \le \frac{\beta R_E}{10}$$

$$\text{Then } R_2 = \frac{\beta R_E}{10} = \frac{100 \times 1 \times 10^3}{10} = 10\ \text{k}\Omega.$$

4. Design of R_1

$$V_{R1} = V_B = V_{BE} + V_E = 0.7\text{ V} + 2.0\text{ V} = 2.7\text{ V}$$

$$V_B = \frac{V_{CC} \times R_2}{(R_1 + R_2)} = 2.7\text{ V}$$

$$V_B(R_1 + R_2) = V_{CC} \times R_2$$

$$\therefore \quad (R_1 + R_2) = \frac{V_{CC}}{V_B} \times R_2$$

$$\Rightarrow R_1 = \left[\frac{V_{CC}}{V_B} \times R_2 - R_2\right] = \left[\frac{20 \times 10 \times 10^3}{2.7} - 10 \times 10^3\right]$$

$$= \left[74 \times 10^3 - 10 \times 10^3\right] = 64\text{ k}\Omega.$$

Final design values of CE Transistor Amplifier for Class-A operation are as follows. $R_1 = 64$ kΩ, $R_2 = 10$ kΩ, $R_C = 4$ kΩ and $R_E = 1$ kΩ. CE Transistor Amplifier with the designed parameters is shown in the Circuit (Fig. 5.20).

The circuit in Fig. 5.20 shows design values of previously determined components.

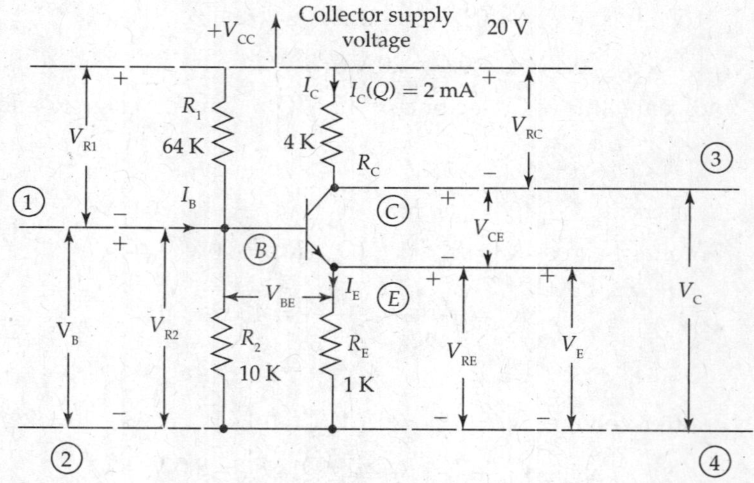

FIG. 5.20 *Potential-divider-biasing circuit with design values of the circuit components*

EXAMPLE 5.5

Draw circuit diagram of Self-Biasing circuit for a Germanium Transistor. Data: $V_{CC} = 20$ V, $R_C = 2$ K, $R_E = 100\ \Omega$, $R_1 = 100$ kΩ, $R_2 = 5$ kΩ, $\beta = 50$. Calculate $I_C(Q)$, $V_{CE}(Q)$ and Stability Factor S.

Solution: Potential-divider-bias (or) Self-Biasing Circuit

In the potential-divider-bias circuit (Fig. 5.21), Collector supply voltage V_{CC}, in association with R_1 and R_2, provides the bias and R_E provides bias stability.

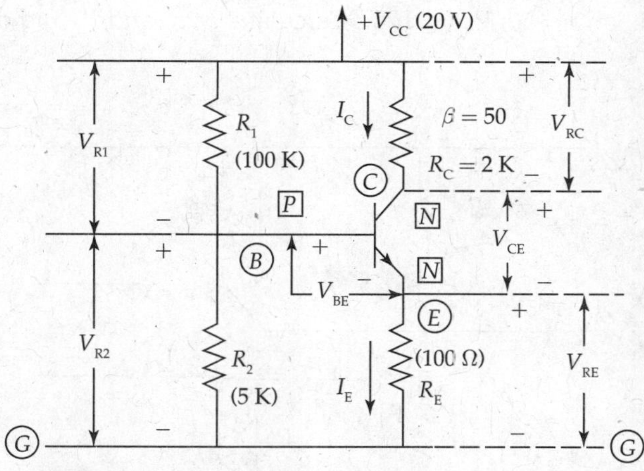

FIG. 5.21 *(Self-biasing circuit) Potential-divider-bias circuit*

$$\text{Stability factor } S = \frac{(1+\beta)\left[1+\dfrac{R_B}{R_E}\right]}{(1+\beta)+\dfrac{R_B}{R_E}} \quad (1)$$

$$\beta = 50;\ R_B = R_1 \parallel R_2 = 100\text{ K} \parallel 5\text{ K} = 4.76\text{ k}\Omega$$

$$\therefore\ S = (51)\frac{\left[1+\left(\dfrac{4.76\times 10^3}{10^2}\right)\right]}{51+\left(\dfrac{4.76\times 10^3}{10^2}\right)} = \frac{(51)\times 48.6}{98.6} = 25.13 \cong 25$$

$$\text{Voltage across } R_2 = V_{R2} = \frac{V_{CC}\times R_2}{[R_1+R_2]} = \frac{20\times 5\times 10^3}{[100+5]\times 10^3} \cong 0.95\text{ V.}$$

$$\therefore\ \text{Voltage } V_{RE} \text{ across } R_E = V_{R2} - V_{BE} = 0.95 - 0.3$$

$$V_{RE} = 0.65\text{ V (using } V_{BE} = 0.3\text{ V)}$$

$$\therefore\ I_E \cong I_C = \frac{V_{RE}}{R_E} = 6.5\text{ mA}$$

$$\therefore\ I_C(Q) = 6.5\text{ mA}$$

$$\text{Collector-to-Emitter Voltage } V_{CE}(Q) = [V_{CC} - I_C(Q)\cdot(R_C + R_E)]$$

$$V_{CE}(Q) = [20 - 6.5\times 10^{-3}(2\times 10^3 + 0.1\times 10^3)] = [20 - 13.65] = 6.35\text{ V.}$$

EXAMPLE 5.6

Design a CE Transistor Amplifier with voltage-divider-bias circuit for Class-A operation. Q-Point is chosen at $V_{CE} = 12$ V and $I_C = 2$ mA. Load resistance $R_L = 4.7$ kΩ; Stability factor S

should be ≤ 5; Transistor $\beta = 50$; $V_{BE} = 0.7$ V. Calculate R_1, R_2 and R_E and draw the circuit with designed components.

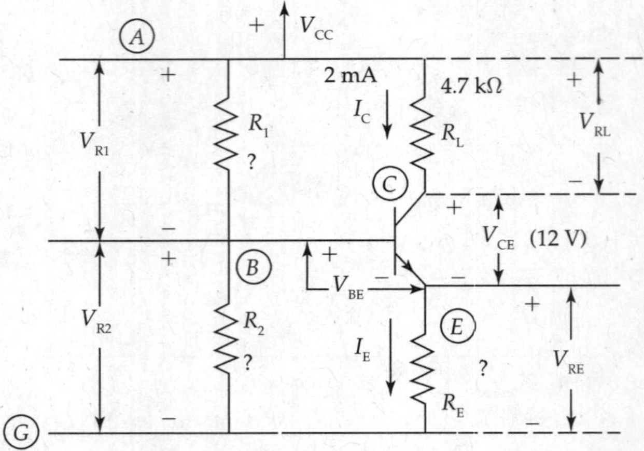

FIG. 5.22 *Design of potential-divider-bias for a class-A transistor amplifier*

Solution: For a Class-A Amplifier, quiescent or DC operating point is fixed in the middle of DC Load Line, so that $V_{CE} = 0.5\ V_{CC}$. $V_{CE} = 12$ V. Therefore, $V_{CC} = 24$ V, V_{BE} (active) $= 0.7$ V, $\beta = 50$ and $R_L = 4.7$ kΩ, Stability factor $S = 5$.

a. Calculation of Emitter resistor R_E:

$$V_{CC} - V_{CE} = I_C[R_L + R_E] \qquad (1)$$

Substituting the values of $V_{CC} = 24$ V, $V_{CE} = 12$ V, $I_C = 2$ mA and $R_L = 4.7$ kΩ,

$$(24 - 12) = 2\times10^{-3}(4.7\times10^3 + R_E)$$

$$\therefore\quad R_E = \frac{(12 - 9.4)}{2\times10^{-3}} = 1.3\times10^3\ \Omega$$

$[R_B/R_E]$ can be calculated using the expression for S from Eq. (5.29).

b. Calculation of R_B:

Substituting the values of $R_E = 1.3$ kΩ, $S = 5$, $\beta = 50$ in Eq. (5.29),

$$S = \frac{(\beta+1)}{\left[1+\beta\cdot\dfrac{R_E}{(R_B+R_E)}\right]} = \frac{51}{\left[1+50\dfrac{R_E}{(R_B+R_E)}\right]} = 5$$

$$\therefore\quad 1+\left[50\frac{R_E}{R_B+R_E}\right] = \frac{51}{5}$$

It becomes $\left[\dfrac{50\cdot R_E}{R_B+R_E}\right] = 9.2$ using $\left[\dfrac{51}{5} - 1\right] = 9.2$.

$$50\cdot R_E = [9.2\ R_E + 9.2\ R_B]$$

$$\Rightarrow\quad R_B = \frac{40.8\times1.3\times10^3}{9.2} = 5.76\ \text{k}\Omega \quad (\text{using } R_E = 1.3\times10^3\ \Omega)$$

c. Calculations for R_1 and R_2:

$$\text{Voltage across } R_E = V_{RE} = I_E \cdot R_E = 2\times10^{-3}\times1.3\times10^{3} = 2.6\text{ V} \quad (2)$$

$$V_{R2} = V_{BB} = (V_{BE} + V_{RE}) \text{ from the circuit in Fig. 5.23} \quad (3)$$

$$\therefore\ V_{BB} = 0.7 + 2.6 = 3.3\text{ V}$$

$$R_1 = \frac{V_{CC}}{V_{BB}}\times R_B = \frac{24\times5.76\times10^3}{3.3} = 41.9\text{ k}\Omega \quad (4)$$

$$R_2 = \frac{R_1\cdot R_B}{R_1 - R_B} = \frac{41.9\times10^3\times5.76\times10^3}{(41.9-5.76)\times10^3} = 6.68\text{ k}\Omega$$

with these calculated values, transistor circuit is shown in Fig. 5.23.
Biasing circuit with design components (solution to Example 5.6)

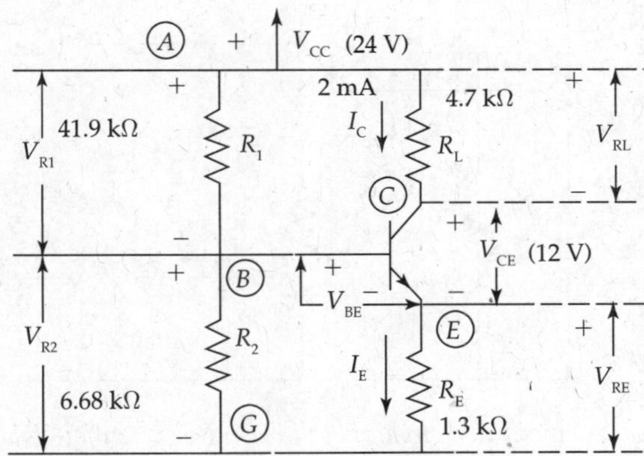

FIG. 5.23 *Design of potential-divider-bias circuit for Class-A transistor amplifier with designed values of R_1 R_2 R_E*

5.9 BIAS COMPENSATION CIRCUITS USING DIODES AND THERMISTORS

- For a Transistor to operate as an amplifying device, Emitter junction is forward biased and Collector junction is reverse biased, using one of the three biasing methods (a single DC source with few resistors). But only the potential-divider-biasing circuit or self-biasing circuit provides stable Q for better circuit response.
- But there is a loss of gain or amplification due to negative feedback through R_E in the process of stabilisation of DC operating conditions. In certain applications, this loss of gain may be a considerable disadvantage in the circuit operation.
- One of the simplest designs of electronic circuits is to counteract the effect of changes in V_{BE} (due to temperature variations or replacement of active device with another value of cut-in voltage) is to make the V_E much greater than the required forward-bias to the Emitter junction.
- But, some compensation methods or techniques are used to improve the stability of Q and thus resulting in extremely stable operating point meaning stable DC biasing voltages to the Emitter and Collector junctions of the Transistor.

A compensating semiconductor diode D_C applied with forward-bias V_{FD} is included in the Emitter path of the Transistor biasing circuit shown in Fig. 5.24. The diode to be used for compensation of V_{BE} should be of the same semiconductor material as the Transistor so that the voltage–temperature coefficient is the same. When V_{BE} changes by a magnitude of ΔV_{BE} with changes in temperature, the voltage across the diode D_C changes by ΔV_D since $\Delta V_D = \Delta V_{BE}$, the corresponding changes will cancel each other and compensation for changes in temperature takes place:

$$I_C \cong I_E = \frac{(V_{R2} + \Delta V_D - V_{BE})}{R_E} \cong \frac{V_{R2}}{R_E}.$$

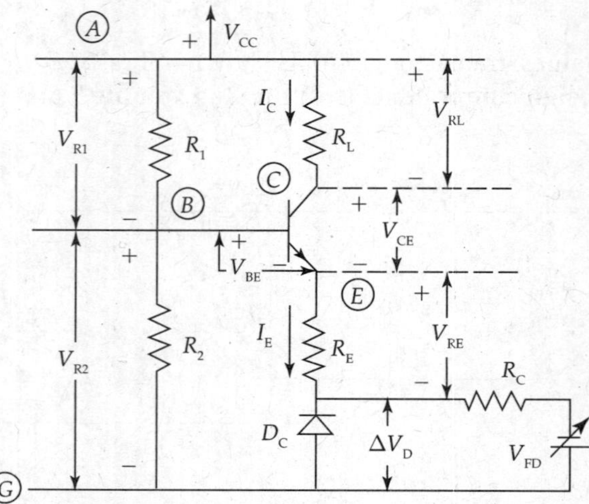

FIG. 5.24 *Diode compensation circuit in self-biasing or potential-divider-bias circuit*

Diode Compensation for Variations in I_{CO} – In case of germanium Transistors, changes in reverse saturation current I_{CO} with changes in temperature cause a corresponding significant changes in the Collector current I_C that cause instability of biasing voltages of the Transistor decided by the quiescent operating point. This instability reduced by introducing a germanium diode between the Base and Emitter path (the germanium diode is reverse biased by the voltage V_{BE} of the Transistor) for nullifying the increases in the reverse saturation currents with temperature changes as shown in Fig. 5.25.

Assume the current through the reverse-biased diode $I_{RD} = I_{CO}$.

Then the Base current $I_B = (I - I_{RD})$ (From Fig. 5.25) (5.32)

Using the expression for I_B in the equation for Collector current I_C,

$$I_C = \beta \cdot I_B + (\beta + 1)I_{CO} \tag{5.33}$$

$$I_C = \beta(1 - I_{RD}) + (\beta + 1)I_{CO}$$

$$I_C = \beta \times I - \beta \times I_{RD} + \beta \times I_{CO} + I_{CO} \tag{5.34}$$

Equation (5.34) proves that the 'Reverse saturation current' (I_{RD}) of compensating diode nullifies the variations in I_{CO} of the Transistor, which maintains constant Collector current (I_C) and provide stable operation of the device.

Simple Diode Compensation for fixed bias circuit

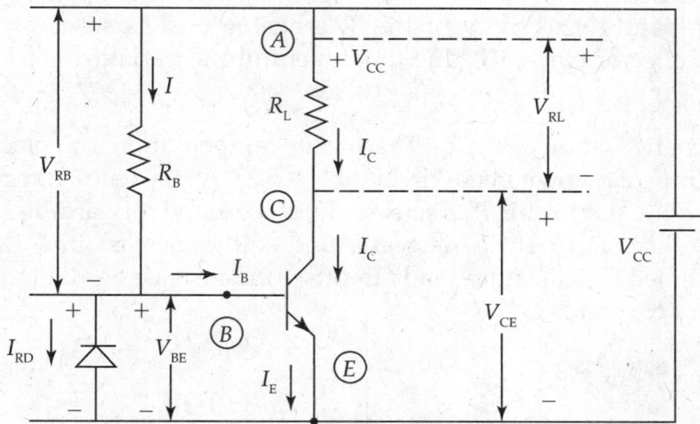

FIG. 5.25 *Fixed-bias circuit for CE silicon transistor with silicon diode for compensation of* I_{CO}

5.10 THERMISTOR COMPENSATION

- The Circuit in Fig. 5.26 shows one method of Transistor parameter variation compensation using temperature sensitive resistive elements such as *Thermistors* rather than diodes.
- The resistance of Thermistor devices changes with temperature. They use ceramic-like semiconductors with high thermal coefficients of resistance having high sensitivity to temperature variations.
- Thermistor has a negative temperature coefficient, where the resistance R_T of the device decreases exponentially with increase in temperature. Thermistor is connected in the CE potential-divider-bias circuit between positive V_{CC} and the Emitter point of the Transistor.
- As the temperature T rises, the resistance R_T of the Thermistor (due to the negative temperature coefficient property of the Thermistor) decreases and the current fed through R_T into the Emitter resistor R_E increases. Since the voltage drop across R_E is in the direction

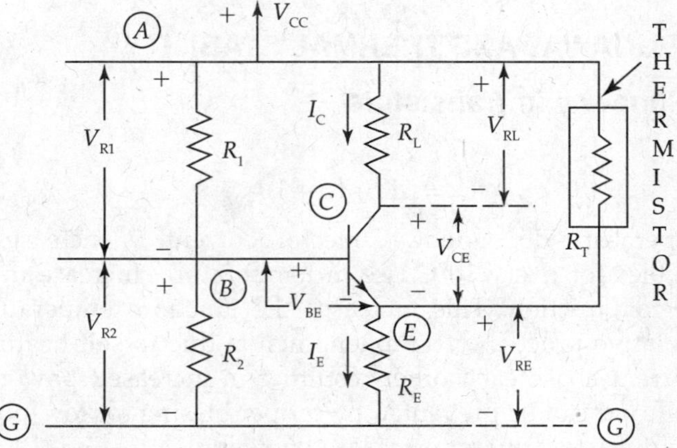

FIG. 5.26 *Thermistor for bias compensation in self-biasing circuit*

to reverse-bias the Transistor Emitter–base junction and reduces the Collector current to the previous designed value.

- Thus, the temperature sensitivity of the Thermistor R_T acts so as to compensate the change in the Collector current I_C due to temperature T, variations in I_{CO}, V_{BE} or Beta of the Transistor.

An alternative configuration using the Thermistor compensation is to place the Thermistor R_T across the resistor R_2 as shown in the circuit of Fig. 5.27. As the temperature T increases, the voltage drop across the thermistor R_T decreases and hence the forward-biasing Base voltage reduced. Hence, Collector current I_C decreases and so the increase in I_C due to rise in the temperature is nullified. This feature tends to offset the increase in the temperature due to increase in Collector current.

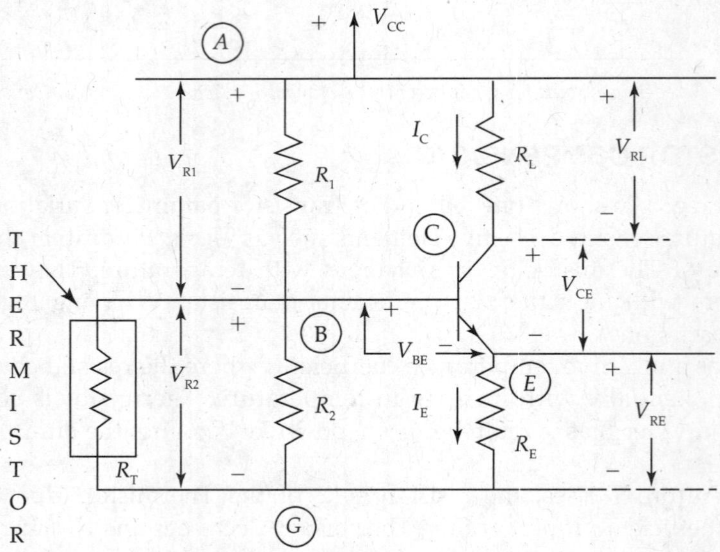

FIG. 5.27 *Thermistor at input port for bias compensation in self-biasing circuit*

5.11 THERMAL RUNAWAY AND THERMAL STABILITY

5.11.1 Thermal Runaway in Transistors

In a CE Transistor

$$I_C = \beta I_B + (\beta + 1) I_{CO}. \tag{5.35}$$

Since I_{CO} is temperature dependent, Collector current I_C increases with increase in temperature. I_{CO} doubles for every 10°C rise in temperature. Increase in I_C increases power dissipation at Collector junction. This increases the junction temperature causing further rise in I_C. This cumulative process or the phenomenon due to self-heating, in which rise in temperature and current chase each other resulting in increased power dissipation (P_C) is called *Thermal runway* and can be prevented by proper biasing for low-power circuits and by using heat sinks for Transistors operating at large powers.

In this context, a term called thermal resistance 'θ' is defined in the following section.

15.11.2 Thermal Runaway and Thermal Resistance θ

Power dissipation P_C in watts at the Collector junction is proportional to variations in temperature at the Collector junction with reference to ambient temperature, i.e., $(T_J - T_A)$:

$$P_C \propto (T_J - T_A), \tag{5.36}$$

where T_J is the junction temperature in °C and T_A is the Ambient temperature in °C.

The proportionality between P_C and $(T_J - T_A)$ can be converted into equality by introducing a constant 'θ' so that

$$\theta \times P_C = (T_J - T_A), \tag{5.37}$$

where 'θ' is called the *Thermal resistance* and has a dimension of temperature in °C per watt of power dissipation. The size of the Transistor and the device heat transfer methods to surroundings determine the magnitudes of Thermal resistance.

$$\theta = \frac{(T_J - T_A)}{P_C} \ \text{°C/W}$$

$$\therefore \quad P_C = \frac{(T_J - T_A)}{\theta} \tag{5.38}$$

Differentiating P_C with respect to T_J,

$$\frac{dP_C}{dT_j} = \frac{1}{\theta} \tag{5.39}$$

This gives the relation between the thermal conductivity and power dissipation change dP_C with respect to junction temperature change dT_J as long as

$$\frac{dP_C}{dT_J} < \frac{1}{\theta}. \tag{5.40}$$

This condition must be satisfied to prevent thermal runaway and to safeguard the device.

- If thermal conductivity is more than the rate of power dissipation with respect to temperature, thermal runaway is prevented. This can be justified as follows. More thermal conductivity means carrying away the heat from the Transistor junction into surroundings as quickly as generated.
- As long as heat is radiated away, thermal runaway is prevented. If it is not possible to directly radiate away heat by having proper ventilation, special heat sinks, which carry the heat, have to be designed.
- In one configuration, the device is enclosed (embedded) in the heat sink. Other configurations of heat sink designs are also possible for example in power Transistors. The Collector terminal of the Transistor is connected to the Transistor case and it can be fixed on a heat sink insulated from the ground.

In the low-power devices, thermal runaway can be prevented by careful selection of the quiescent or DC operating point. The power generated at the Collector junction $P_C(J)$ under no excitation signal condition is the product of $I_C(Q)$ and $V_{CE}(Q)$ where

$$V_{CE}(Q) = V_{CC} - I_C(Q)[R_L + R_E]$$
$$P_C(J) = V_{CE}(Q) \cdot I_C(Q)$$
$$P_C(J) = V_{CC} \cdot I_C(Q) - I_C^2(Q)[R_L + R_E],$$

where $V_{CC} \cdot I_C(Q)$ is the DC power supplied by the Collector supply voltage V_{CC}. Part of this power is consumed as power dissipation (P_{DC}) in two resistors R_L and R_E,

where $P_{DC} = I_C^2(Q) \cdot [R_L + R_E]$.

5.12 CONDITION FOR THERMAL STABILITY

The condition to prevent thermal runaway is

$$\frac{1}{\theta} > \frac{\partial P_C}{\partial T_J} = \frac{\partial P_C}{\partial I_C} \cdot \frac{\partial I_C}{\partial T_J} \tag{5.41}$$

The Eq. 5.41 can be written as

$$\frac{1}{\theta} > \frac{\partial P_C}{\partial I_C} \cdot \frac{\partial I_C}{\partial I_{C0}} \cdot \frac{\partial I_{C0}}{\partial T_J}$$

But Stability factor $S = \dfrac{\partial I_C}{\partial I_{C0}}$ (5.42)

$$\text{and } \frac{\partial I_{C0}}{\partial T_J} = 7\% \text{ per °C}$$

$$(I_{C0}) = 0.07\, I_{C0}. \tag{5.43}$$

This is because I_{C0} doubles for every 10°C rise in temperature. In other words, I_{C0} changes by 7% = 0.07 per °C for both silicon and Germanium transistors.

Using the above data, condition to prevent thermal runaway can be written as

$$\frac{1}{\theta} > \frac{\partial P_C(J)}{\partial I_C} (S)\ (0.07\, I_{C0}). \tag{5.44}$$

This condition can be applied to the equation for the power

$$P_C(J) = V_{CC} \cdot I_C(Q) - I_C^2(Q) \cdot (R_L + R_E). \tag{5.45}$$

Differentiating $P_C(J)$ with respect to $I_C(Q)$, we get

$$\frac{\partial P_C(J)}{\partial I_C(Q)} = V_{CC} - 2 \cdot I_C(Q)[R_L + R_E]. \tag{5.46}$$

For Class-A Amplifier with resistive load,

$$V_{CE}(Q) = V_{CC}/2 \text{ or } V_{CC} = 2\, V_{CE}(Q) \tag{5.47}$$
$$V_{CE}(Q) = 2 \cdot V_{CE}(Q) - I_C(Q)(R_L + R_E)$$

$$\therefore \quad V_{CE}(Q) = I_C(Q) \cdot (R_L + R_E)$$

$$\text{But } P_C(J) = V_{CC} \cdot I_C(Q) - I_C^2(Q)(R_L + R_E)$$

$$\frac{\partial P_C(J)}{\partial I_C(Q)} = [V_{CC} - 2\,V_{CE}(Q)]$$

$$\frac{1}{\theta} > [V_{CC} - 2\,V_{CE}(Q)]\,(S)\,[0.07\,I_{C0}] \tag{5.48}$$

- For this inequality to be satisfied, it requires $V_{CC} - 2\,V_{CE}(Q) > 0$, that is $V_{CE}(Q) < V_{CC}/2$. This is the requirement to avoid thermal runway. That is, the operating point is not chosen as $V_{CE}(Q) = V_{CC}/2$, but such that $V_{CE}(Q) < V_{CC}/2$.

EXAMPLE 5.7

CE Transistor with Collector-to-Base bias circuit has $V_{CC} = 10$ V, $R_C = 2$ kΩ and $R_B = 100$ kΩ. Calculate quiescent point and stability factor S.

Solution:

$$I_C = \beta \cdot I_B = \frac{\beta(V_{CC} - V_{BE})}{R_B + (\beta + 1) \cdot R_C}$$

$$\therefore \quad I_C = \frac{50 \times (10 - 0.7)}{100 \times 10^3 + (50 + 1) \times 2 \times 10^3}$$

$$= \frac{465}{202 \times 10^3} = 2.3 \text{ mA}$$

$$\therefore \quad I_C(Q) = 2.3 \text{ mA}$$

$$V_{CE}(Q) = [V_{CC} - I_C(Q) \times R_C]$$

$$= [10 - 2.3 \times 10^{-3} \times 2 \times 10^3]$$

$$= [10 - 4.6]$$

$$\therefore \quad V_{CE}(Q) = 5.4 \text{ V}$$

$$\text{Stability factor} \quad S = \frac{(\beta + 1)}{\left[1 + \left(\frac{\beta \cdot R_C}{R_C + R_B}\right)\right]}$$

$$= \frac{(50 + 1)}{\left[1 + \frac{50 \times 2 \times 10^3}{(2 \times 10^3 + 100 \times 10^3)}\right]}$$

$$= \frac{51}{\left[1 + \frac{100 \times 10^3}{102 \times 10^3}\right]} = \frac{51}{[1 + 0.98]}$$

$$= \frac{51}{1.98} = 25.75 \cong 26$$

5.13 BASIC FET AMPLIFIER CIRCUIT

Consider small signal N-Channel FET Amplifier circuit (Fig. 5.28).

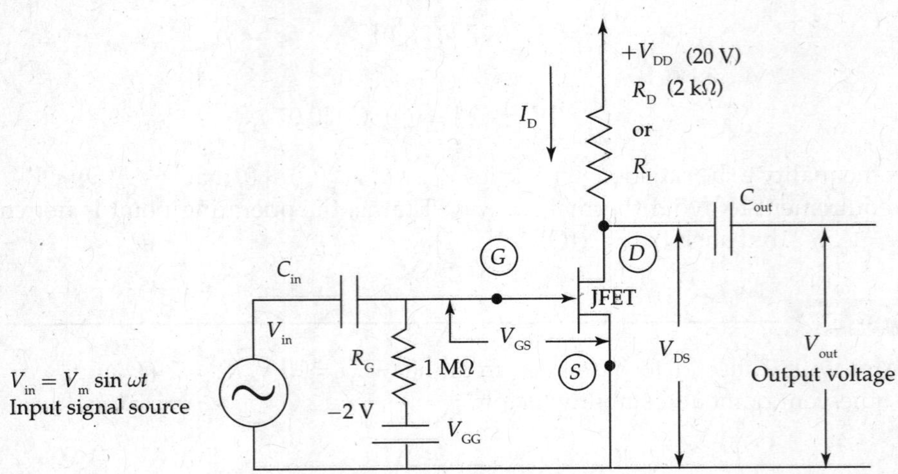

FIG. 5.28 *Basic JFET amplifier circuit*

In this Amplifier circuit, DC voltages are

1. Gate-to-source Bias Voltage V_{GG} used to set the DC voltage V_{GS} (based on the type of the Amplifier and its class of operation).
2. Drain Supply Voltage V_{DD} is set so that the Drain Current I_D flowing through FET device causes a voltage across load resistance $R_L = I_D \cdot R_L = V_{RL}$ which is about half the supply voltage V_{DD} to satisfy the Class-A operation of the Amplifier.
   Bias (DC voltage) V_{GG} is arranged such that the Gate Terminal of N-Channel FET is made negative relative to Source, because the requirement is that Gate-channel diode junction of FET is to be reverse biased so that input resistance R_{in} is very large (advantage over BJT devices) and facilitates non-ideal driving voltage Amplifiers.
3. Gate–source voltage V_{GS} is established by bias voltage V_{GG}. Because there may be a gate current of the order of few nanoamperes, voltage across gate resistor R_G $(1 \times 10^{-9} \times 1 \times 10^{6} = 10^{-3}\ \text{V})$ is negligibly small and the Gate–Source voltage V_{GS} is virtually unchanged at −2 V.

$$\text{For Amplifier Class-A operation, } V_{GS} = (1/2)V_P = 0.5(V_P). \tag{5.49}$$

Fixing up DC/*Q*-point to determine Amplifier class of operation

Assume the following output characteristics (static characteristics) for JFET device shown in Fig. 5.29. Output characteristics of the FET device can be obtained experimentally or by a curve tracer using a catode ray oscilloscope (CRO).

When JFET is biased with voltages for required Amplifier class of operation, steady-state values of Drain Current I_D and Voltage V_{DS} are related by DC/Static Load Line equation from the Amplifier analysis.

After obtaining the DC-equivalent circuit of Fig. 5.28, DC Load Line equation can be written as follows. (Discussed in detail in the Amplifier Chapter)

$$V_{DD} = V_{DS} + I_D R_L \text{ (DC Load Line equation)} \tag{5.50}$$

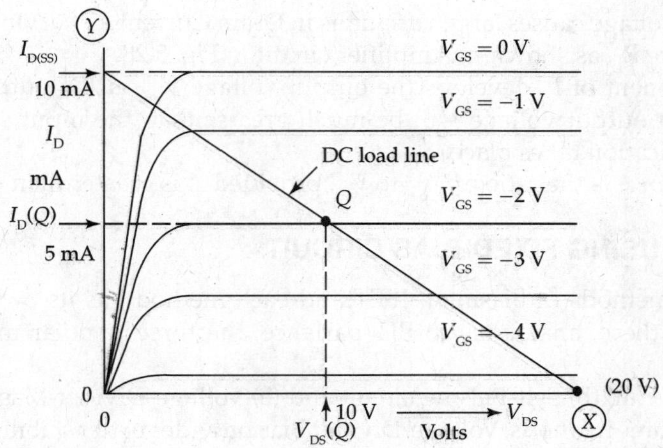

FIG. 5.29 *JFET output characteristics with DC load line and the quiescent operating point (Q) for Class-A operation of amplifier*

1. *Y*-coordinates of DC load line are $V_{DS} = 0$ V and Drain Current $I_D = (V_{DD}/R_L)$.
2. Coordinates of the point (*X*) of the line are $V_{DS} = V_{DD}$ Volts and $I_D = 0$ mA.
3. DC Load Line can be drawn by joining the two points *X* and *Y* on the output characteristics of the FET device.
4. For the required DC/Quiescent point (*Q*) operation of the Amplifier, Drain Current and the voltage V_{DS} are obtained by superimposing the DC Load Line on the output characteristics.
   - If an input signal, $V_{in} = V_m \cdot \sin \omega t$ is applied at Amplifier input port (Fig. 5.28), variations in the input signal voltage cause variations in negative voltage applied to the Gate Terminal (Fig. 5.30).

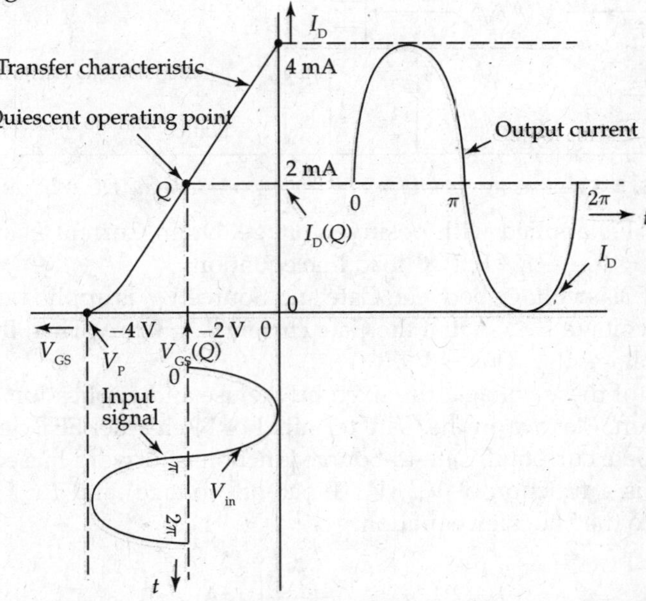

FIG. 5.30 *Class-A operation of JFET amplifier using 'mutual characteristic' of a JFET device*

- Varying DC voltage causes large variations in Drain Current I_D. Varying I_D flows through load resistance R_L, as shown in Amplifier circuit in Fig. 5.28.
- Signal component of I_D develops the output voltage V_{out} at the output port. It can be observed that output voltage will be much greater than the input signal voltage and signal amplification takes place.
- Amplifier Gain A is the ratio of V_{out} to V_{in}, provided A is greater than unity.

5.14 FET BIAS USING FIXED-BIAS CIRCUIT

There are several methods of biasing a JFET and each method has its own advantages and disadvantages. Of these, analogous to BJT parlance, the three popular methods of biasing JFET devices are:

(a) Fixed-Bias circuit, (b) Self-Biasing circuit and (c) Voltage-Divider-Bias circuit.

Amongst these three methods, Voltage-Divider Bias provides better stabilisation of quiescent operating point against variations in JFET parameters $I_{D(SS)}$ (Drain saturation current), V_P (pinch-off voltage for FET device), K (device structure constant) and V_{Th} (threshold voltage). The three types of biasing methods are discussed as follows.

Biasing FET using Fixed-Bias

Fixed-Bias Circuit for JFET is shown in Fig. 5.31. FET device is applied with Drain-to-source voltage V_{DS} using a resistor R_D (Drain resistor) and the DC voltage source V_{DD}.

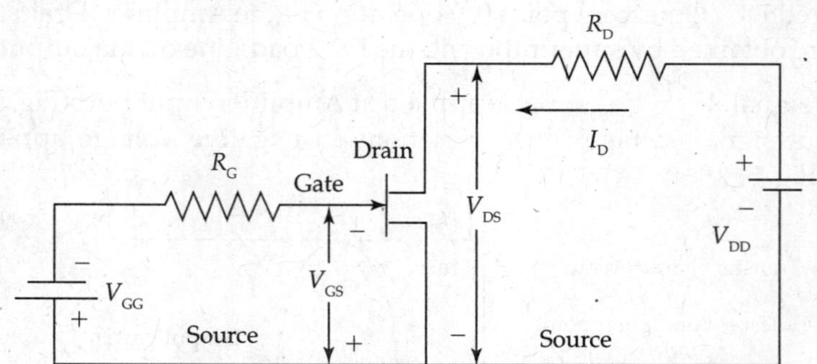

FIG. 5.31 *Biasing circuit for a FET using two separate DC voltages*

The Drain terminal is applied with positive voltage. Drain Current I_D and resistor R_D are chosen such that $V_{DS} = V_{DD} - I_D \cdot R_S$ (DC load line equation).

Similarly, Reverse-Bias voltage between Gate and Source V_{GS} is applied using Gate resistor R_G and Gate supply voltage V_{GG}, so that the Gate current $I_G = 0$ mA. Then Biasing voltage V_{GS} is equal to Supply voltage V_{GG}. That is $V_{GS} = V_{GG}$.

The biasing levels of these voltages are fixed by the use and application of Amplifiers and their class of operation. (Arrow on the Gate terminal of N-channel FET device indicates the direction of flow of Gate current if Gate-to-Source junction is forward biased.)

Drain Current I_D is a function of V_{GS}, V_P (Pinch-off voltage) and $I_{D(SS)}$ (Drain Saturation Current), according to the Shockley equation:

$$I_D(Q) = I_{D(SS)} \left[1 - \frac{V_{GS}}{V_P}\right]^2 \text{ mA.} \tag{5.51}$$

For an Amplifier, voltages corresponding to Q are fixed as follows:

$$V_{GS}(Q) = V_{GG} = V_G \text{ and } V_{DS} = [V_{DD} - I_D R_D]. \tag{5.52}$$

Method of fixing the magnitudes of DC voltage and currents for FET devices are already discussed in the previous section, when the FET works as an Amplifier.

EXAMPLE 5.8

Determine the voltages and currents at the Q-point from Fixed-Bias Circuit for FET device shown in Fig. 5.32. Data given: $I_{D(SS)} = 12$ mA and $V_{GS(OFF)} = -4$ V.

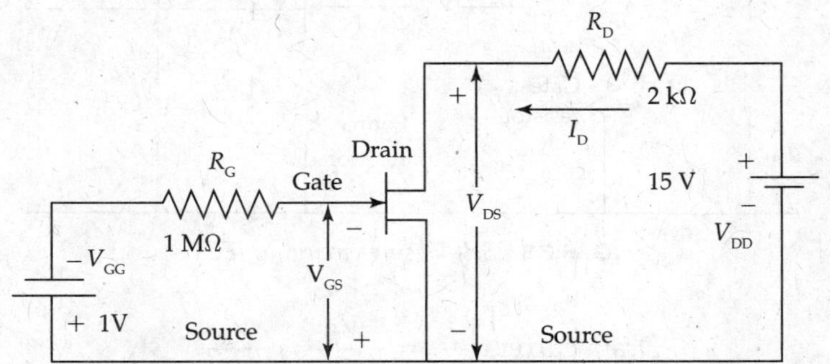

FIG. 5.32 *Biasing circuit for a FET using two separate DC voltages*

Solution: Since gate current $I_G = 0$ mA, $V_{GS} = V_{GG} = -1$ V.
From Shockley's equation,

$$\text{Quiscent component of drain current} \quad I_D(Q) = I_{D(SS)}\left[1 - \frac{V_{GS}}{V_P}\right]^2 \text{ mA}$$

$$I_D(Q) = 12 \text{ mA}\left[1 - \left[\frac{(-1)}{(-4)}\right]\right]^2 = 12 \text{ mA}\left(\frac{9}{16}\right) = 6.75 \text{ mA}$$

Drain to source voltage at Q-point $\quad V_{DS}(Q) = V_{DD} - I_D \cdot R_D$

$$\therefore\ V_{DS}(Q) = [15 - (6.75 \times 10^{-3})2 \times 10^3] = (15 - 13.5) = 1.5 \text{ V}.$$

5.15 SELF-BIASING CIRCUIT FOR FET

Self-Biasing circuit of FET is analogous to that of BJT. It differs from earlier Fixed-Bias circuit as shown in Fig. 5.33. Resistor R_S is connected between the Source and Common terminals. It eliminates the need for separate V_{GS}. When V_{DD} is applied to the circuit, Drain current I_D flows through R_S. Circuit is designed so that voltage drop $(I_D \cdot R_S)$, across R_S, provides required magnitude of reverse-bias voltage for Gate-to-Source junction. Hence, this type of biasing circuit is known as 'Self-Biasing circuit'.

Gate-to-Source junction is reverse biased. So the gate current $I_G = 0$. The Gate Terminal will be at the ground potential due to absence of Gate current and the potential drop across R_G, i.e., voltage $V_G = 0$ V. Hence, $V_{GS} = V_{RS} = I_D \cdot R_S$.

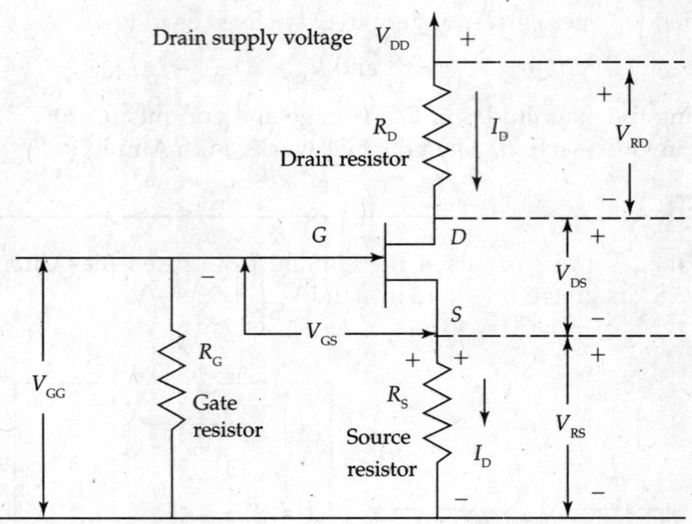

FIG. 5.33 *Self-biasing circuit for JFET*

$$\text{Drain current} \quad I_D = -\frac{V_{GS}}{R_S} = -\frac{V_{RS}}{R_S}. \tag{5.53}$$

The self-biasing line is a straight line (Fig. 5.34).
Drain Current for FET device is given by Shockley equation

$$I_D = I_{D(SS)}\left[1 - \frac{V_{GS}}{V_P}\right]^2 = \left[1 - \frac{V_{GS}}{V_{GS(OFF)}}\right]^2 \text{ mA.} \tag{5.54}$$

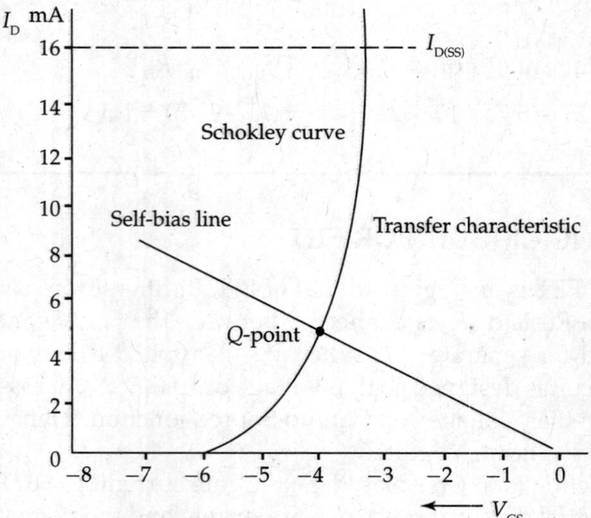

FIG. 5.34 *Locating Q-point*

The curve for Drain Current using Shockley equation is parabolic transfer curve relating the Drain Current in one port (output port) with Gate-to-source voltage V_{GS} at input port.

The two curves in Fig. 5.34 intersect at quiescent or DC operating point 'Q' [$I_D(Q)$, $V_{GS}(Q)$]. From the Eqs. (5.53) and (5.54)

$$I_{D(SS)}\left[1-\frac{V_{GS}}{V_{GS(OFF)}}\right]^2 = -\frac{V_{GS}}{R_S} \tag{5.55}$$

For N-Channel JFET $V_{GS(OFF)}$ is negative and $V_P = [V_{GS(OFF)}]$.

$$\therefore \quad I_{D(SS)}\left[1-\frac{V_{GS}}{V_P}\right]^2 = -\frac{V_{GS}}{R_S} \tag{5.56}$$

Treating Eq. (5.56) as a quadratic in V_{GS},

$$\left[\frac{I_{D(SS)}\times R_S}{V_P^2}\right]V_{GS}^2 + \left[\frac{2I_{D(SS)}}{V_P}+1\right]V_{GS} + I_{D(SS)}\cdot R_S = 0. \tag{5.57}$$

Equation (5.49) is of the form

$$[a\cdot V_{GS}^2 + b\cdot V_{GS} + c] = 0, \tag{5.58}$$

$$\text{where} \quad a = \frac{I_{D(SS)}\cdot R_S}{V_P^2}, \quad b = \left[\frac{2I_{D(SS)}\times R_S}{V_P}+1\right] \quad \text{and} \quad c = I_{D(SS)}\times R_S. \tag{5.59}$$

$$\text{For N-channel FET Gate voltage, } V_{GS} = \frac{-b+\sqrt{b^2-4ac}}{2a}. \tag{5.60}$$

$$\text{For P-channel FET Gate voltage, } V_{GS} = \frac{-b-\sqrt{b^2-4ac}}{2a}. \tag{5.61}$$

After determining V_{GS}, Drain current can be calculated using Eq. (5.53):

$$I_D = -\frac{V_{GS}}{R_S}. \tag{5.62}$$

$$\text{Drain-to-source voltage} \quad V_{DS} = [V_{DD} - I_D\cdot(R_D + R_S)]. \tag{5.63}$$

5.16 VOLTAGE-DIVIDER-BIAS CIRCUIT FOR FET

When using a BJT, Voltage-Divider Bias makes the circuit to behave independent of beta (β) changes and Transistor characteristics, so that the Amplifier with Voltage-Divider-Bias Circuit functions more stable (Fig. 5.35). In a similar way, FET Amplifier with Voltage-Divider-Bias Circuit works more stable. By using the Voltage-Divider rule,

$$\text{Gate voltage} \quad V_G = V_{DD}\cdot\frac{R_2}{R_1+R_2}.$$

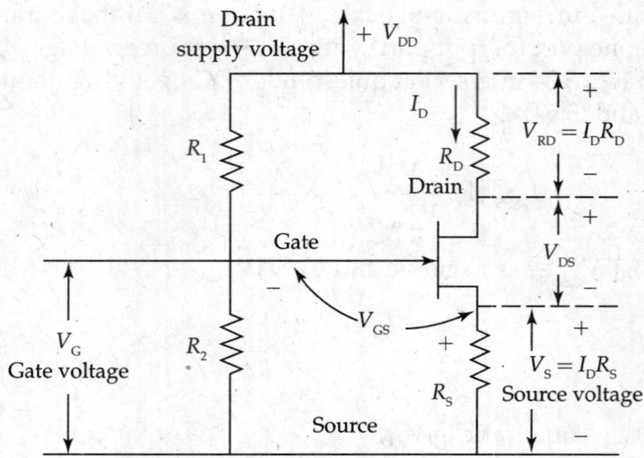

Constant voltage divider bias circuit for FET

FIG. 5.35 *Biasing circuit for N-channel FET*

$$\text{Gate-to-source voltage} \quad V_{GS} = V_G - V_S. \tag{5.64}$$

$$\therefore \quad V_{GS} = [V_G - I_D \cdot R_S] \tag{5.65}$$

$$\therefore \quad \text{Drain current } I_D = \frac{[V_G - V_{GS}]}{R_S} \tag{5.66}$$

Using Schokley equation,

$$I_D = I_{D(SS)}\left[1 - \frac{V_{GS}}{V_P}\right]^2. \tag{5.67}$$

Therefore, from Eqs. (5.66) and (5.67),

$$\frac{[V_G - V_{GS}]}{R_S} = I_{D(SS)} \times \left[1 - \frac{V_{GS}}{V_P}\right]^2 \tag{5.68}$$

$$\left[\frac{(V_G - V_{GS}}{R_S}\right] = I_{D(SS)}\left[1 + \frac{V_{GS}}{V_P}\right]^2$$

$$[V_G - V_{GS}] = I_{D(SS)} \cdot R_S \left[1 + \frac{2 \cdot V_{GS}}{V_P} + \frac{V_{GS}^2}{V_P^2}\right]$$

$$\therefore \quad \left[\frac{I_{D(SS)} \cdot R_S}{V_P^2}\right] \cdot V_{GS}^2 + \left[\frac{2 \cdot I_{D(SS)} \cdot R_S}{|V_P|} + 1\right] \cdot V_{GS} + I_{D(SS)} \cdot R_S - V_G = 0$$

The above equation is a quadratic equation similar to

$$a \cdot V_{GS}^2 + b \cdot V_{GS} + c = 0,$$

$$\text{where } a = \frac{I_{D(SS)} \cdot R_S}{V_P^2}, \tag{5.69}$$

$$b = \frac{2 \cdot I_{D(SS)} \cdot R_S}{|V_P|} + 1, \tag{5.70}$$

$$c = I_{D(SS)} \cdot R_S - |V_G|. \tag{5.71}$$

For N-channel FET,

$$V_{GS} = \frac{-b + \sqrt{b^2 - 4ac}}{2a}. \tag{5.72}$$

For a P-channel FET,

$$V_{GS} = \frac{b - \sqrt{b^2 - 4ac}}{2a}. \tag{5.73}$$

From the calculated values of V_{GS} at the Q-point,

$$\text{Drain current} \quad I_D = \left[\frac{V_G - V_{GS}}{R_S}\right] \text{mA} \tag{5.74}$$

$$\text{and Drain-to-source voltage} \quad V_{DS} = [V_{DD} - I_D(R_D + R_S)]. \tag{5.75}$$

EXAMPLE 5.9

Biasing circuit for N-Channel JFET has, $R_1 = 350\ \Omega$, $R_2 = 100\ \Omega$, drain resistance $R_D = 1.5\ \text{k}\Omega$, source resistance $R_S = 2.3\ \text{k}\Omega$, supply voltage $V_{DD} = 15$ V, Drain Saturation Current $I_{D(SS)} = 15$ mA, Pinch-off voltage $= V_P = -4.5$ V. Calculate I_D, V_{DS} and V_{GS} at Q. Also determine g_m.

Solution:

$$\text{Gate voltage} \quad V_G = \frac{V_{DD} \cdot R_2}{(R_1 + R_2)} = \frac{15 \times 100 \times 10^3}{(350 + 100) \times 10^3} = 3.333 \text{ V}$$

$$\text{Pinch-off voltage} \quad V_P = -4.5 \text{ V}$$

$$\therefore \quad |V_P| = 4.5 \text{ V}$$

$$\text{For the N-channel FET,} \quad V_{GS} = \frac{-b + \sqrt{b^2 - 4ac}}{2a}, \tag{1}$$

$$\text{where} \quad a = \frac{I_{D(SS)} \cdot R_S}{V_P^2} = \frac{15 \times 10^{-3} \times 2.3 \times 10^3}{(4.5)^2} = \frac{34.5}{20.25} = 1.7037$$

$$b = \left[\frac{2 \times I_{D(SS)} \cdot R_S}{V_P} + 1\right] = \left[\frac{2 \times 15 \times 10^{-3} \times 2.3 \times 10^3}{4.5} + 1\right]$$

$$= \left[\frac{69}{4.5} + 1\right] = 16.3333$$

$$c = [I_{D(SS)} \times R_S - V_G] = [(15 \times 10^{-3} \times 2.3 \times 10^3 - 3.3333]$$

$$= [34.5 - 3.3333] = 31.1667.$$

Substituting the value of a, b and c in Eq. (1), $V_{GS} = -2.63$ V.

$$\text{Drain current } I_D = \left[\frac{V_G - V_{GS}}{R_S}\right] = \left[\frac{3.33-(-2.63)}{2.3\times10^3}\right] = 2.59 \text{ mA}$$

$$\text{Drain-to-source voltage } V_{DS} = [V_{DD} - I_D \cdot (R_D + R_S)]$$

$$\therefore\ V_{DS} = [15 - 2.59\times10^{-3}(1.5\times10^3 + 2.3\times10^3)] = [15 - 9.842] = 5.158 = 5.16 \text{ V}$$

$$\text{Transconductance } g_m = \left[\frac{2\sqrt{I_D \cdot I_{D(SS)}}}{V_P}\right] = \frac{2\sqrt{2.59\times10^{-3}\times15\times10^{-3}}}{4.5} = 0.28 \text{ millimhos.}$$

CS FET Amplifier using Voltage-Divider Bias and its DC-equivalent circuit

Circuit diagram of a FET Amplifier using potential-divider bias and its DC-equivalent circuit are shown in Figs. 5.36 and 5.37.

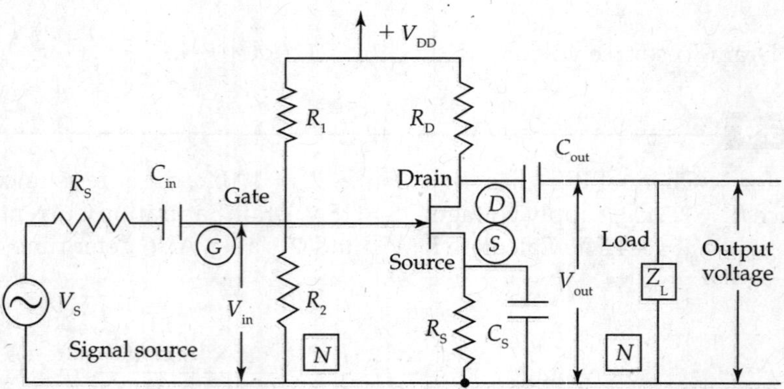

FIG. 5.36 *Common source FET amplifier circuit*

To provide necessary negative voltage V_{GS}, better biasing scheme is Voltage-Divider-Bias Circuit, where R_1 and R_2 form a voltage divider together with R_S and the supply voltage V_{DD}. Component design and arrangement are shown in Fig. 5.36. It provides required negative voltage V_{GS} (according to the design of Amplifier class of operation) at the Gate of the JFET device and Drain-to-Source voltage V_{DS} at Q.

$$V_{GN} = \frac{V_{DD} \cdot R_2}{(R_1 + R_2)} = \frac{V_{DD} \cdot R_1 \cdot R_2}{(R_1 + R_2) \cdot R_2} = \frac{V_{DD} \cdot R_G}{R_2},$$

where $R_G = \dfrac{R_1 R_2}{R_1 + R_2}$.

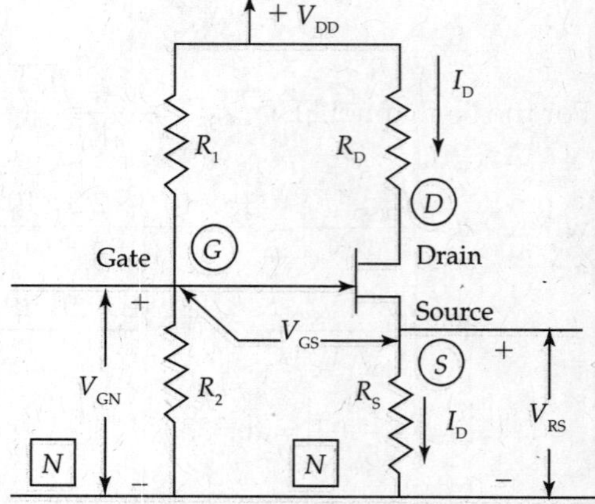

FIG. 5.37 *DC-equivalent circuit of common source JFET amplifier circuit*

For specified 'Q' of JFET Amplifier circuit (I_D, V_{GS}) and chosen values of V_{GN} and R_G, required values of R_2, R_1 and R_S are calculated from the DC-equivalent circuit:

$$R_S = \frac{V_{SN}}{I_D} = \frac{V_{GN} - V_{GS}}{I_D} \tag{5.76}$$

$$R_2 = \frac{R_G V_{DD}}{V_{GN}} \quad \text{and} \quad R_1 = \frac{R_G \cdot R_2}{R_2 - R_G}$$

The effect of any shift in V_{GS} is reduced by making $|V_{SN}|$ large compared to $|V_{GS}|$.

Same potential-divider-bias circuit can be used to EMOSFET device Amplifier circuit (Fig. 5.38). It's DC-equivalent circuit (Fig. 5.39) provides more clarification for the method of biasing. For DEMOSFET, it needs two types of polarity voltages and this circuit is not suitable.

Common Source MOSFET Amplifier using Voltage-Divider Bias and DC-equivalent circuit of CS MOSFET Amplifier

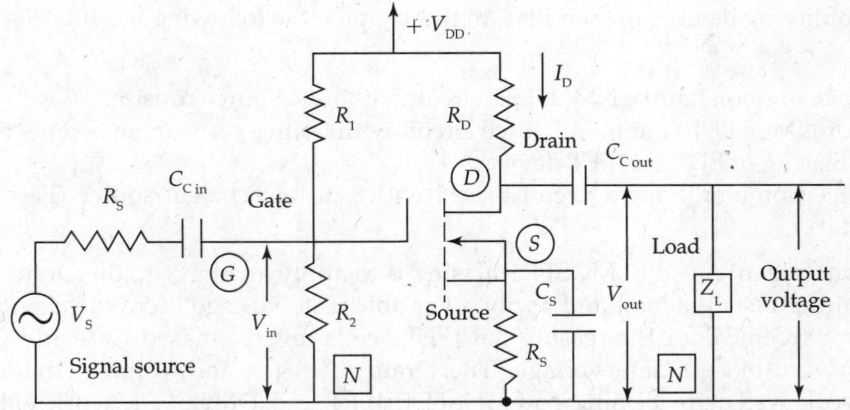

FIG. 5.38 *Common source enhancement MOSFET amplifier circuit*

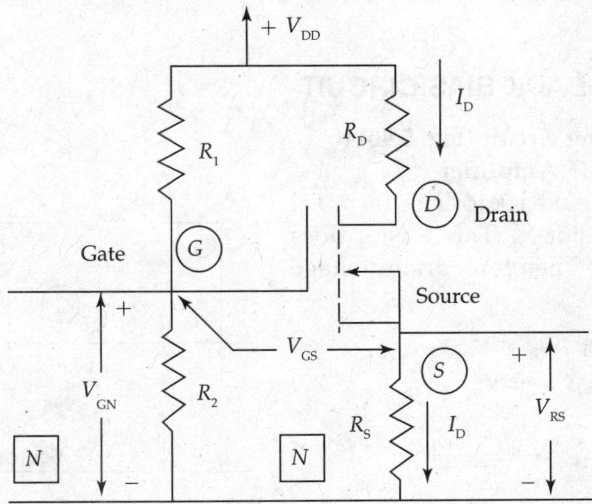

FIG. 5.39 *DC-equivalent circuit of CSMOSFET amplifier circuit*

5.17 BIASING OF ENHANCEMENT MOSFET

MOSFET device can be biased in two different types of environments for the device to work as an Amplifier: (1) Discrete type and (2) VLSI circuit environments. Methods of biasing are different in these environments. While designing discrete version of MOSFET circuits, biasing resistors are used. Whereas, in integrated MOSFET circuit design, other MOSFET devices are used for resistors, so that the circuit is totally built using MOS devices and VLSI scaling can easily be achieved.

If the MOSFET device is a DMOSFET, Gate-to-Source voltage V_{GS} can be positive or negative. Zero bias of DMOSFET is unique and it will not work with EMOSFET, FET or a bipolar device (BJT). Self-Biasing and Current Source-Biasing techniques can also be used with DMOSFET. The Depletion MOSFET (DMOSFET) devices can be operated as EMOSFET devices also. Enhancement mode of operation increases the conductivity of the channel, resulting in more Drain Current I_D for a given V_{GS}. Because of simpler construction and smaller size, EMOSFET devices are quite useful while designing VLSI circuitry. It has advantages over all other amplifying devices in their bias. In this chapter, the following biasing techniques are discussed.

1. Biasing of Common Source NMOSFET circuit without a Source resistor.
2. Biasing of NMOSFET Common Source circuit with Source resistor, analogous to Voltage-Divider Biasing of BJT and JFET devices.
3. NMOSFET Common Source circuit biased with constant current source (Feedback type bias circuit).

Basic principle involved in MOSFET biasing is to apply a voltage to the Drain through a current limiting resistive load and apply a variable Gate voltage. Zero bias is not suitable for JFET or EMOSFET for the reason that JFET needs reverse-biased Gate and EMOSFET needs a forward-biased Gate voltage. The Drain feedback and Voltage-Divider-Biasing methods work well with biasing of EMOSFET, JFET and DMOSFET, work with current sources but not with EMOSFET for the reason that a positive voltage V_{GS} causes bipolar unit into saturation.

5.18 DRAIN FEEDBACK BIAS CIRCUIT

Drain feedback biasing circuit (Fig. 5.40) to Enhancement MOSFET Amplifier

Drain voltage V_D is fed back to Gate Terminal through feedback resistor R_F. The resistor does not carry any current. Therefore, drain voltage $V_D = V_G$.(Gate voltage).

Gate-to-Source voltage $V_{GS} = V_D$

$$= V_{DD} - I_D \cdot R_D \quad (5.77)$$

$$\therefore \text{ Drain current } \quad I_D = \frac{V_{DD} - V_{GS}}{R_D} \quad (5.78)$$

$$\because \text{ Drain current } \quad I_D = K \cdot [V_{GS} - V_T]^2 \quad (5.79)$$

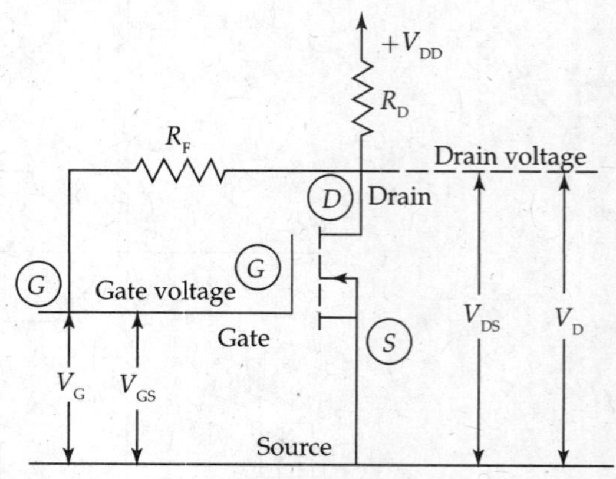

FIG. 5.40 *Feedback bias circuit using resistor R_F to enhancement MOSFET device amplifier*

From the above two equations on simplification, we get

$$K \cdot R_D \cdot V_{GS}^2 + (1 - 2 \cdot K \cdot R_D \cdot V_T) \cdot V_{GS} + (K \cdot R_D \cdot V_T^2 - V_{DD}) = 0 \quad (5.80)$$

This equation is of the form

$$a \cdot V_{GS}^2 + b \cdot V_{GS} + c = 0, \quad (5.81)$$

$$\text{where } a = K \cdot R_D \quad (5.82)$$

$$b = [1 - 2 \cdot K \cdot R_D \cdot V_T] \quad (5.83)$$

$$c = [K \cdot R_D \cdot V_T^2 - V_{DD}] \quad (5.84)$$

$$\therefore \; V_{GS} = \frac{[-b + \sqrt{b^2 - 4ac}]}{2a} \quad \text{for N-channel MOSFET} \quad (5.85)$$

$$V_{GS} = \frac{[b + \sqrt{b^2 - 4ac}]}{2a} \quad \text{for P-channel MOSFET} \quad (5.86)$$

where *a*, *b*, *c* parameters are denoted by Eqs. (5.81), (5.82) and (5.83).

Having determined V_{GS} drain current I_D can be calculated as follows:

$$I_D = \frac{[V_{DD} - V_{GS}]}{R_D} \quad (5.87)$$

$$\text{and } V_{DS} = V_{GS} \quad (5.88)$$

EXAMPLE 5.10

In Drain feedback-biased enhancement MOSFET Amplifier, determine the magnitudes of V_{GS}, V_D and V_{DS} with data given: $V_{DD} = 16$ V, $R_D = 2.7$ kΩ, $R_F = 1$ MΩ, $V_{Th} = 2$ V and $K = 20$ mA/V².

Solution:

$$I_D = \left[\frac{V_{DD} - V_{GS}}{R_D}\right] \quad \text{and} \quad I_D = K(V_{GS} - V_T)^2$$

$$V_{GS} = \frac{\left[-b + \sqrt{b^2 - 4ac}\right]}{2a},$$

where $a = K \cdot R_D = 20 \times 10^{-3} \times 2.7 \times 10^3 = 54$; $b = (1 - 2 \cdot KR_D |V_{Th}|) = (1 - 2 \times 54 \times 2) = 215$

$c = K \cdot R_D, V_{Th}^2 - V_{DD} = 54 \times 4 - 16 = 200$

$$\therefore \; V_{GS} = \frac{-b + \sqrt{b^2 - 4ac}}{2a} = \frac{-215 + \sqrt{(215)^2 - 4 \times 54 \times 200}}{2 \times 54} = 1.5 \text{ V}$$

$$\text{Drain current} \quad I_D = \frac{V_{DD} - V_{GS}}{R_D} = \frac{16 - 1.5}{2.7 \times 10^3} = 5.4 \text{ mA}$$

$V_{DS} = V_{GS} = 1.5$ V from Drain feedback bias circuit Fig. 5.40.

5.19 POTENTIAL-DIVIDER-BIASING CIRCUIT FOR EMOSFET

Potential-Divider-Bias for Enhancement MOSFET without Source resistor (Fig. 5.41)

$$\text{Gate voltage} \quad V_G = V_{GS} = V_{DD} \frac{R_2}{(R_1 + R_2)} \tag{5.89}$$

Assuming Gate-to-Source voltage V_{GS} is greater than the Threshold voltage V_{Th}, biasing of the enhancement type MOSFET is carried in the saturation region

$$\text{Drain current} \quad I_D = K[V_{GS} - V_{Th}]^2 \tag{5.90}$$

where V_{Th} = Threshold voltage is the minimum voltage required for device conduction and K is the Conduction parameter.

From the above equations, Drain Current I_D at Q can easily be calculated.

From the circuit (Fig. 5.41), $V_{DS}(Q)$ can be calculated from the following equation:

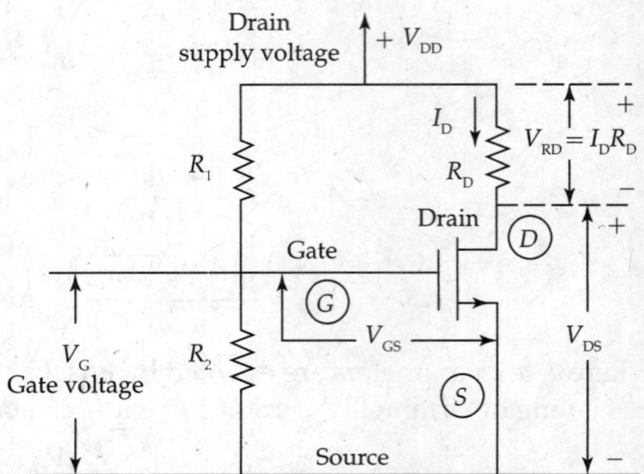

FIG. 5.41 *Biasing circuit for enhancement type common source NMOSFET amplifier without source resistor*

$$\text{Drain-to-source voltage} \quad V_{DS} = [V_{DD} - I_D \cdot R_D] \text{ Volts} \tag{5.91}$$

Potential-divider biasing for Enhancement type MOSFET with source resistor

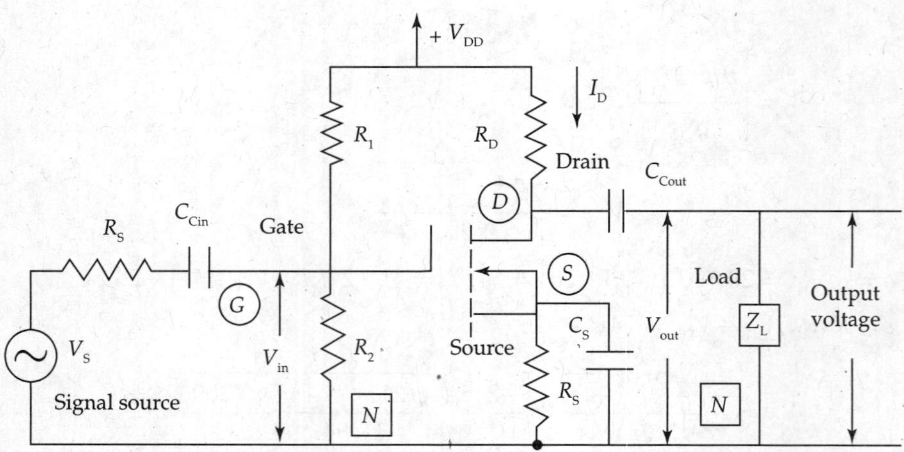

FIG. 5.42 *Common source enhancement MOSFET amplifier circuit*

Figure 5.43 shows DC-equivalent circuit of Common Source EMOSFET Amplifier, used for fixing up DC biasing voltages and currents at Q. After introducing R_S, circuit analysis is similar to the analysis of Voltage-Divider-Bias Circuit using FET (Fig. 5.37).

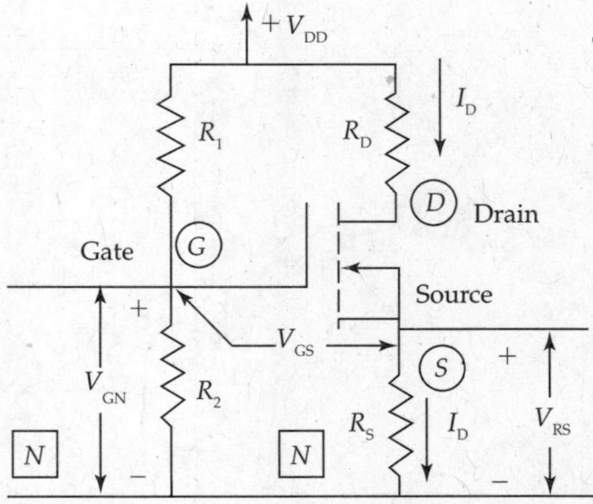

FIG. 5.43 *DC-equivalent circuit of CSMOSFET amplifier circuit with source resistance*

As, the Gate current I_G is zero,

Gate voltage $V_G = \left[V_{DD} \cdot \frac{R_2}{[R_1 + R_2]}\right]$ (5.92)

Source voltage $V_S = I_D \cdot R_S$ (5.93)

Gate-to-source voltage $V_{GS} = (V_G - V_{RS})$
$= (V_G - I_D \cdot R_S)$

$\therefore \quad I_D = \frac{V_G - V_{GS}}{R_S}$ (5.94)

For enhancement type MOSFET,

Drain current $I_D = K \cdot [V_{GS} - V_{Th}]^2$ (5.95)

From Eqs. (5.94) and (5.95),

$$K \cdot [V_{GS} - V_{Th}]^2 = \left[\frac{V_G - V_{GS}}{R_S}\right] \tag{5.96}$$

$$\therefore \quad K \cdot R_S \cdot V_{GS}^2 + [1 - 2 \cdot K \cdot R_S \cdot V_{Th}] \cdot V_{GS} + [K \cdot R_S \cdot V_{Th}^2 - V_G] = 0, \tag{5.97}$$

where K is the conductivity parameter of the device A/V^2, V_{Th} is the Threshold voltage. The above quadratic equation is of the form

$$a \cdot V_{GS}^2 + b \cdot V_{GS} + c = 0, \tag{5.98}$$

$$\text{where } a = K \cdot R_S, \quad b = [1 - 2 \cdot K \cdot R_S \cdot |V_{Th}|] \text{ and } c = [K \cdot R_S \cdot V_{Th}^2 - V_G] \tag{5.99}$$

The solution of the above quadratic equation gives the values of

$$V_{GS} = \left[\frac{-b + \sqrt{b^2 - 4ac}}{2a}\right] \quad \text{for N-channel MOSFET device} \tag{5.100}$$

$$V_{GS} = \left[\frac{b + \sqrt{b^2 - 4ac}}{2a}\right] \quad \text{for P-channel MOSFET device} \tag{5.101}$$

Values of a, b and c are calculated using Eq. (5.99). Having determined V_{GS}, from Eq. (5.101) Drain current I_D can be calculated from the following equation:

$$I_D = \frac{V_G - V_{GS}}{R_S} \tag{5.102}$$

Then Drain-to-source voltage $V_{DS} = [V_{DD} - I_D(R_D + R_S)]$ (5.103)

Alternatively, the quiescent operating point can be determined at the intersection of parabolic transfer characteristic curve and the bias line (Fig. 5.44).

Amplifier stability is a function of bias line slope (Fig. 5.44).

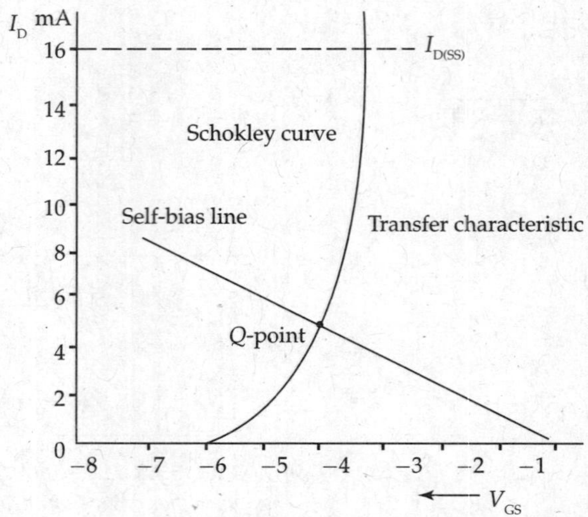

FIG. 5.44 *Locating Q-point*

EXAMPLE 5.11

Determine $V_{GS} \cdot I_D$ and V_{DS} for Voltage-Divider-Biased EMOSFET. Amplifier circuit, given that $R_1 = 2.2$ MΩ, $R_2 = 1.2$ MΩ, $R_D = 2.7$ kΩ, $R_S = 1.5$ kΩ, $K = 2$ mA/V², $V_{Th} = 2$ V, and $V_{DD} = 17$ V.

Solution:

$$V_G = V_{DD} \cdot \frac{R_2}{(R_1 + R_2)} = \frac{(17)(1.2\times10^6)}{(2.2+1.2)\times10^6} = 6\text{ V}$$

$$\text{Drain current} \quad I_D = \frac{(V_G - V_{GS})}{R_S} = K\cdot(V_{GS} - V_{Th})^2 \tag{1}$$

On simplification of the above equations,

$$K\cdot R_S \cdot V_{GS}^2 + (1 - 2\cdot K\cdot R_S \cdot V_{Th})V_{GS} + (K\cdot R_S \cdot V_{Th}^2 - V_G) = 0 \tag{2}$$

Equation (2) is of the form

$$a\cdot V_{GS}^2 + b\cdot V_{GS} + c = 0,$$

$$\text{where } a = K\cdot R_S = (2\times10^{-3})(1.5\times10^3) = 3$$

$$b = (1 - 2\cdot K\cdot R_S \cdot |V_{Th}|) = [1 - 2\cdot(2\times10^{-3})(1.5\times10^3)(2)] = -11$$

$$c = [K\cdot R_S \cdot V_{Th}^2 - V_G] = [(2\times10^{-3})\cdot(1.5\times10^3)(2)^2 - 6] = 6$$

$$\therefore\ V_{GS} = \frac{-b + \sqrt{b^2 - 4ac}}{2a} \quad \text{for N-channel MOSFET}$$

$$V_{GS} = \frac{-(-11) + \sqrt{(-11)^2 - 4\times3\times6}}{2\times3} = 3\text{ V}$$

$$I_D = \frac{(V_G - V_{GS})}{R_S} = \frac{(6-3)}{1.5\times10^{-3}} = 2\text{ mA}$$

$$V_{DS} = V_{DD} - I_D(R_D + R_S) = 17 - (2\times10^{-3})(2.7 + 1.5)\times10^3 = 8.6\text{ V.}$$

Biasing circuits discussed so far are not suitable for biasing MOSFET Amplifiers in IC environment, because resistors are extensively used in these circuits. Resistors require larger area on the chip, hence expensive and have large tolerances. Additional MOSFETs can be used in place of resistors. They are biased with current sources, which are less expensive and require small area. Large coupling and bypass capacitors are also not used due to the reasons of chip area considerations.

Some MOSFET devices are supplied with fourth electrode-substrate, not connected to the Source but brought out as a lead. It could be used as a control terminal. When the substrate terminal is not connected to the Source for some reasons, the substrate is applied with a potential. In such a case, the substrate to the channel diode should not be forward biased. It should be held – more *negative* than the Source terminal for N-channel MOSFET, whereas more *positive* than the Source terminal for P-channel MOSFET.

SUMMARY

1. Amplifier operation of Transistors is discussed in the beginning of the chapter. It is explained that DC and AC voltages co-exist in the Amplifiers. DC voltages are used in order to provide Biasing voltages to active devices (BJT, FET and MOSFET devices).
2. Three types of biasing schemes – (1) Fixed-Bias, (2) Feedback-Bias and (3) Voltage-Divider-Biasing circuits (for BJT, FET and MOSFET) are discussed, giving their relative merits and demerits in Amplifier operations, keeping in view of operational stability.
3. Dependency of Stability of Amplifier operations on biasing voltages, design using '*Q*' (Quiescent/DC operating point) and its location on DC load line are explained. Method of locating *Q* on DC load line on device output characteristics has important bearing on Amplifier class of operation.
4. Amplifier design is addressed with a top-down approach starting from Transistor specifications (power dissipation curve) and limiting factors for location of DC load line, *Q*-point and derived information about DC voltages to be fixed. These factors are in turn guidelines for the design of most of electronic circuits. Amplifier performance features ultimately depend upon biasing and stabilisation of *Q*-point and DC voltages.

Questions for Practice

1. Draw Fixed-Biasing circuit for NPN Transistor and explain its working using required equations. Also explain the need for biasing Transistors.
2. Draw Collector-to-Base biasing circuit for NPN Transistor and explain the working using required equations.
3. Draw the circuit of Voltage-Divider Biasing for NPN Transistor and explain the working using necessary equations.

4. A silicon Transistor is used in CE Amplifier with potential-divider-bias arrangement, with $V_{CC} = 18$ V, $R_C = 0.8$ kΩ, $R_S = 0.2$ kΩ and $\beta = 100$. Quiescent operating point 'Q' is chosen with $I_C(Q) = 4.5$ mA and $V_{CE}(Q) = 9$ V. Calculate R_1, R_2 and R_E, assuming a stability factor 'S' $= 10$.
5. DC-equivalent circuit of CE Amplifier with Collector-to-Base-bias configuration has $V_{CC} = 18$ V with component values of $R_C = 3.4$ kΩ and $R_L = 0.6$ kΩ. Collector-to-Base resistance $R_B = 300$ kΩ, $V_\gamma = 0.7$ V and $\beta = 100$. Calculate Collector, Emitter and Base voltages.
6. Voltage-Divider-Bias Circuit for an Amplifier has $V_{CC} = 14$ V, $R_1 = 21$ kΩ, $R_2 = 7$ kΩ, $V_{BE} = 0.5$ V, $R_C = 0.8$ kΩ and $R_E = 0.6$ kΩ. Calculate Base, Emitter, Collector voltages and V_{CE}.
7. Discuss design aspects of finding circuit components of a Voltage-Divider-biasing configuration for Transistor to be used in a linear Amplifier.
8. Draw various types of biasing circuits used for FET devices and explain them.
9. Draw various types of biasing circuits used for MOSFET and explain them.

Multiple Choice Questions

1. Stable Transistor biasing configuration.
   (a) fixed-bias circuit (b) collector-to-base bias circuit
   (c) self-biasing circuit (d) voltage-divider-bias circuit
2. Location of Q on Transistor output characteristics for an Amplifier.
   (a) cut-off region (b) saturation region
   (c) active region (d) none
3. Location of Q-point on the DC load line for Class-A Amplifier operation.
   (a) on collector current I_C axis (b) on voltage V_{CE} axis
   (c) middle point of DC load line (d) top 75% on DC load line
4. Value of Stability factor for Fixed-Bias Transistor circuit is ________.
   (a) 1 (b) $(\beta + 1)$ (c) β (d) $\left[\dfrac{(\beta+1)}{1+\dfrac{\beta R_C}{(R_C+R_B)}}\right]$
5. At pinch-off point V_P on FET transfer characteristic, Drain current I_D is ________.
   (a) maximum (b) zero (c) none (d) minimum
6. Type of bias for Gate-to-Channel junction of FET to offer very high input resistance ________.
   (a) zero bias (b) forward bias (c) reverse bias
7. For the Transistor to work as an amplifying device nature of biasing schemes is ________.
   (a) forward bias to emitter junction and reverse bias to collector junction
   (b) RB to emitter junction and RB to collector junction
   (c) FB to emitter junction and FB to collector junction
   (d) RB to emitter junction and FRB to collector junction

8. Value of Stability factor for Collector-to-Base-bias Transistor circuit is ________________.

   (a) 1 (b) $(\beta + 1)$ (c) β (d) $\left[\dfrac{(\beta+1)}{1+\dfrac{\beta R_C}{(R_C+R_B)}}\right]$

9. Value of Stability factor for Voltage-Divider-Bias Transistor circuit is ________________.

   (a) 1 to $(1 + \beta)$ (b) $(\beta + 1)$ (c) β (d) $\left[\dfrac{(\beta+1)}{1+\dfrac{\beta R_C}{(R_C+R_B)}}\right]$

10. Type of voltages for biasing Transistors

    (a) AC voltage (b) DC voltages (c) combination of both AC and DC voltages

Answers to Multiple-Choice Questions

1. (d)	2. (c)	3. (c)	4. (b)	5. (b)
6. (c)	7. (a)	8. (d)	9. (a)	10. (b)

Chapter 6

TRANSISTOR (BJT) AMPLIFIERS

Learning Objectives

After reading this chapter, you will be conversant with

- Amplification of AC signals showing signal waveforms
- Transistor Amplifier configurations and their performance parameters
- Design considerations for Amplifiers
- Transistor Amplifiers of Different configurations
- Analysis of Transistor Amplifiers
- Single-stage Amplifiers

6.1 INTRODUCTION

- Consider time-varying electrical signals from Cell phones, Home theatre system, Heart rate monitor, TV, Radio, Radar, Airplanes, Satellites and so on. Such signals in various applications are normally very weak. These signals are of the order of a few micro volts or milli volts with small energy, and they need reliable signal *conditioning and processing* for practical use.
- An Amplifier performs the task of simplest signal processing known as *signal amplification*. Linear amplifiers using BJTs and FETS produce an output signal faithfully without distortion, preserving the signal waveform. Working principles of *Transistor amplifiers* are discussed in this chapter.
- Amplifiers are four-terminal circuits using active devices, 'Transistors', 'R, L, C' components, signal source and DC source.

The *four terminals* are considered as two ports viz. *input port and output port.* A signal waveform to be amplified is connected to the input port of the Amplifier. Response to the applied signal appears at the output port.

- When the output signal is *larger than the input signal waveform*, the circuit is said to function as an *Amplifier*. Ratio of output voltage and input voltage of an Amplifier is known as *Amplification 'A' or voltage gain of the Amplifier* (Fig. 6.1).

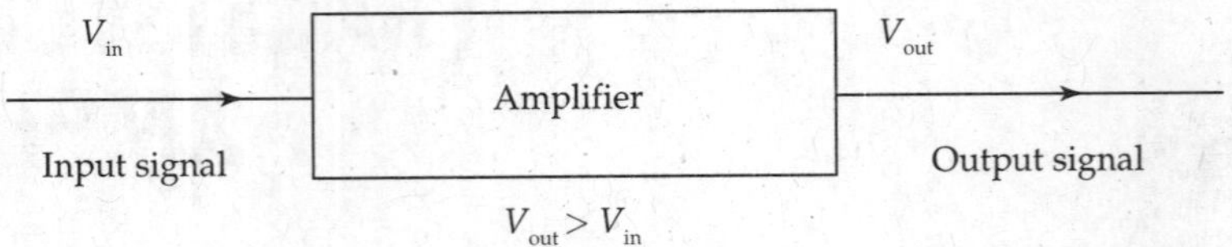

FIG. 6.1 *Concept of an amplifier*

$$\text{Voltage Gain or Amplication} = A = \frac{\text{Output Voltage}}{\text{Input Voltage}} = \frac{V_{out}}{V_{in}}$$

Single-stage linear Amplifiers were initially used to compensate for signal losses over long distance communication lines serving as repeaters or booster Amplifiers. Later on, the applications spread to audio and video Amplifiers in *Radio and Television signal transmission and reception, satellite communications systems* and so on.

Consider a common application of an audio Amplifier in *public address systems*. Electrical signal, for example, the output of a microphone (transducer at the input port), is connected to the Amplifier input port terminals. Amplified response (output) signal is connected to the load such as a loud speaker (transducer at the output port) (Fig. 6.2).

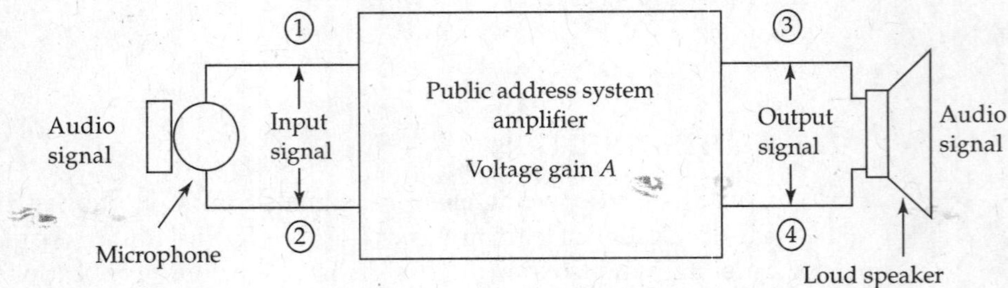

FIG. 6.2 *Audio amplifier system*

If the Amplifier output signal is without any changes, other than the increase in the signal amplitude in it, the Amplifier is considered to be an ideal or a linear Amplifier. This is normally not the situation, as the nonlinear characteristics of the active devices such as BJTs, FETs or Vacuum Tubes may introduce changes in some features of electrical signals during their passage through them. Such unwanted changes in signals are considered as *distortion*. It is discussed in later sections.

One practical situation is that all the frequency components of the speech signal in Public Address System (Fig. 6.2) must be amplified to the same strength to minimise changes in the output signal waveform so that the fidelity or the quality of the speech is maintained at the loudspeaker output.

6.2 CLASSIFICATION OF AMPLIFIER CIRCUITS

6.2.1 Applications of Amplifier Circuits

- Direct-coupled Amplifiers (DC Amplifiers): Amplify signals from 0 Hz, e.g., ECG instruments and other medical instruments.
- Audio frequency (AF) Amplifiers: Amplify signals from 20 Hz to 20 kHz, e.g., Audio Amplifier is used to amplify music or speech signals. Such as home theatre and radio.
- Video or pulse Amplifiers: Amplify TV and Radar signals from DC to 7.5 MHz, e.g., radio frequency (RF) Amplifiers are used to amplify radio signals from a few kilo hertz to hundreds of mega hertz in satellite systems and to increase cellular telephone range.
- RF Amplifiers are used to amplify radio signals ranging from a few KHz to hundreds of MHz, mostly used in radio Transmitters and Receivers.
- Microwave Amplifiers are used to amplify signals in radio transmitters.
- Very high-frequency Amplifiers (signals of 30–300 MHz).
- Ultra high-frequency Amplifiers (signals of 300–3000 MHz).
- Super high-frequency Amplifiers (signals of 3000–30,000 MHz).
- Optical Amplifiers: amplify signals used in optical communications.

Amplifier circuits are described in many ways based on:

- Method of Quiescent or DC- and AC-operating conditions of the Amplifiers
  - *Class A Amplifiers*: The biasing voltages and the signals have their operating voltage levels such that the output voltage exists for the entire duration of the signal. They are also known as small signal Amplifiers, as the signals operate on the linear range of the device characteristics. Linear circuit models are used for the analysis of BJT Amplifiers hybrid (h) parameter equivalent circuits.
  - *Class B Amplifiers*: Biasing voltages and the signals for amplification have their operating voltage levels such that output signals flow through the Transistors for half the time duration of the signal cycle. Standby signal power is zero. So they are used in satellite system Amplifiers. Input signal swings increase. So large powers are possible.
  - *Class AB Amplifiers*: Biasing voltages and signal voltage swings are adjusted such that Transistors (active devices) conduct for more than half cycle and less than full cycle of the applied AC signals. Mostly used in push–pull power Amplifiers to avoid crossover distortion.
  - *Class C Amplifiers*: To realise large powers from active devices in Class C Amplifiers, the Q-point biasing voltages and the AC signal swings are such as to make the conduction angles of the active devices between 120° and 160°. Output signal at the output terminal will be in the form of pulses and to realise continuous signal as output; tuned circuits are used in Class C Amplifier. In essence, the output signals are pulsed on for some portion of the half cycle, instead of existing continuously for the entire half cycle. As the conduction period of the active devices is much small, larger amounts of output power can be realised by using RF-tuned circuits (flywheel effect) that overcome the no conduction intervals by Class C pulsed operation.
  - Class D operation is used in switching *power Amplifiers* for pulsed input voltages. Here the active devices in the Amplifier are rapidly switched on and off at least twice for each cycle based on *sampling theorem*. Transistors in the Amplifiers are either completely on or completely off according to the pulse inputs. So DC power dissipation in the circuit is

almost negligible. Class D operation has theoretical efficiency of 100%. But using present day devices power conversion efficiencies in Class D Amplifiers of the order of 90% only are possible. Semiconductor devices have made the development of high fidelity, full audio range Class D Amplifiers.

- The type of Amplifier circuits
  - Voltage Amplifier (small signal Amplifier) increases the input voltage
  - Current Amplifier (large signal Amplifier) increases input signal current
  - Power Amplifier (large signal Amplifier) increases voltage and current
  - Video Amplifier
  - Audio Amplifier
  - Optical Amplifiers for optical communication and so on
  - Transconductance Amplifier
  - Transresistance Amplifier
- Type of load for the practical use of the circuits
  - *Tunable Amplifiers*: Normally referred as tuned Amplifiers with tuned or tunable LC circuits in the output circuits of the Amplifiers as load for practical application of the Amplifier circuits. They are narrow band Amplifiers used as RF Amplifiers and intermediate frequency (IF) Amplifiers used in radio receivers and communication receivers and in radio transmitter circuits
  - *Untuned Amplifiers*: Audio and video Amplifiers
- The type of interstage coupling of multistage Amplifiers
  - Resistance capacitance-coupled Amplifiers
  - Transformer-coupled Amplifiers
  - DC Amplifiers
- Number of stages of Amplifier circuits
  - Single-stage Amplifier
  - Cascaded or multistage Amplifiers
- Common terminal of active device used in amplifier circuits
  - Common Emitter (CE), common Base (CB) and common Collector (CC) Amplifiers, when the active device used in the Amplifier circuits is a Transistor.
  - Common source, common gate and common drain Amplifiers, when the active device used in the Amplifier circuits is a FET or MOSFET.
  - Common cathode, common grid and common plate Amplifiers, when the active device used in the Amplifiers is a vacuum tube.
- Phase relationship between output and input voltages
  - *Inverting Amplifier*: Output and the input signals voltages are 180° out of phase, as is the situation in CE Transistor and common source FET Amplifiers.
  - *Non-inverting Amplifier*: Output and input voltages are in phase as is the situation in Emitter follower and source follower circuits.
- Magnitudes of input signal amplitudes
  - Small signal Amplifiers
  - Large signal Amplifiers

6.3 SINGLE-STAGE COMMON EMITTER TRANSISTOR AMPLIFIER

Amplifier (Fig. 6.3) has one step of amplification process. CE *Transistor Amplifier has Emitter terminal common to both input and output circuits.*

In a Transistor Amplifier there will be two totally different sets of conditions:

- One set is the DC-biasing conditions to the Transistor.
- Second set is the AC signal conditions during amplification of signals.
- DC and AC signal conditions co-exist in the Transistor Amplifier circuits.

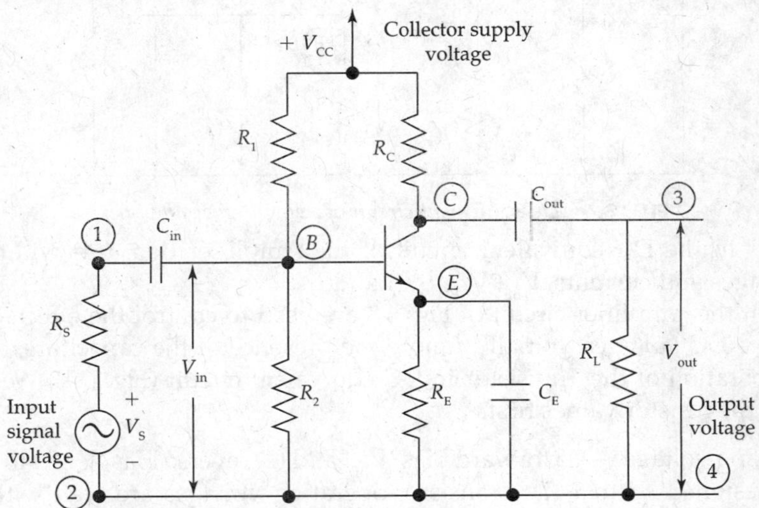

FIG. 6.3 *Single stage common emitter transistor amplifier*

AC input signal voltage superimposes on the DC or quiescent operating conditions (biasing) to achieve the varying DC conditions, which ultimately result in amplified signals from the Amplifier circuits.

- The DC-operating conditions for a Transistor to act as an amplifying device are
  - Voltage 'V_{BE}' to forward bias the Emitter (input) junction of a Transistor.
  - Voltage 'V_{CE}' to reverse bias the Collector (output) junction of a Transistor.
  - Transistor is made to operate in the active region of its output characteristics.

V_{CC}, R_1, R_2, R_C (or load resistance R_L) and R_E are designed to provide the required magnitudes of forward bias V_{BE} and reverse bias V_{CE} to the Transistor, based on the class of operation of Amplifiers. Emitter resistor 'R_E' also helps bias stability.

Various voltages in the *DC equivalent circuit of the Amplifier* (Fig. 6.4) are

- Collector supply voltage V_{CC},
- Collector voltage V_C,
- Base voltage V_B,
- Emitter voltage V_E.

Component value of R_C and the magnitude of V_{CC} should be such that V_{CE}, under no circumstances, should become larger than the parameter $V_{CE(max)}$ specified by the manufacturers in the data sheets of selected Transistor. This constraint is necessary to avoid the break down of the output junction, if the voltage V_{CE} exceeds the break down voltage $V_{CE(max)}$.

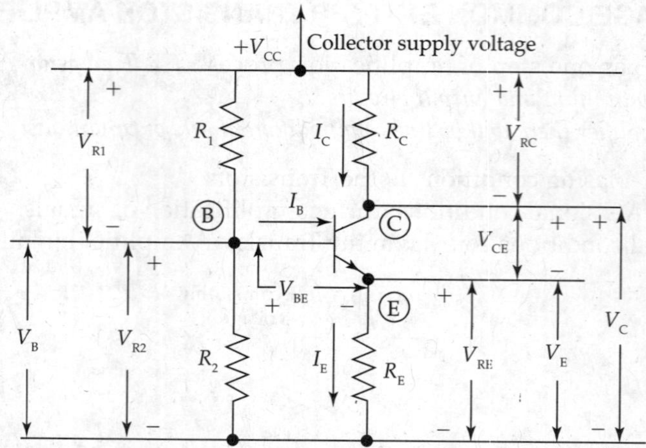

FIG. 6.4 *DC equivalent circuit of a single stage amplifier*

- The resistors in the DC equivalent circuit of the Amplifier (Fig. 6.4) will determine the biasing voltages and currents, V_{BE}, V_{CE}, I_B, I_C and I_E.
- Capacitors in the Amplifier circuit of Fig. 6.3 are used to control the effects of AC signals and block the DC voltages. Initially, ignore the presence of the capacitors, as they do not affect the operation of the Transistor in DC equivalent circuit (Fig. 6.4). We can therefore calculate all the resistor values first.

The two biasing voltages – (1) forward bias 'V_{BE}' and (2) reverse bias 'V_{CE}' – to the Transistor are properly designed so that the Transistor-operating voltages are well within the active region of the Transistor output characteristics for linear Amplifier operation.

Mostly Transistor voltage Amplifiers are CE type and use resistive loads and operate under 'class A' condition for linear operation. The input impedance for various values of load resistances and the output impedance for various values of source resistance for CE Transistor operation will be fairly constant. Therefore, CE Transistor Amplifier configuration is normally preferred.

Figure 6.5 shows one of the methods of obtaining the forward bias. Voltage across R_2 and voltage across R_E are such that $V_{BE} = (V_{R2} - V_{RE})$. For an NPN Transistor, $V_{R_2} > V_{RE}$ to satisfy the forward bias condition that is Base should be more positive with respect to the Emitter of the Transistor.

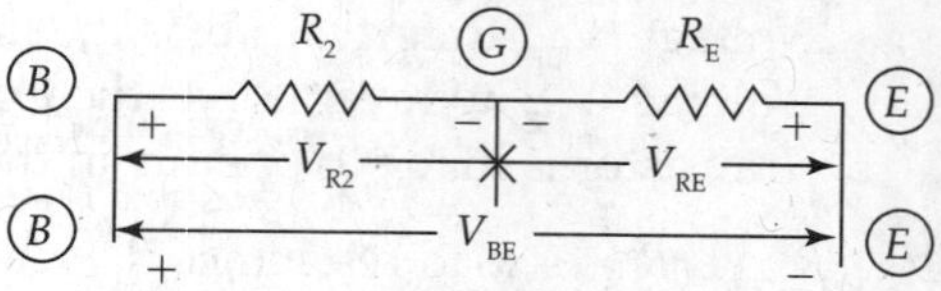

FIG. 6.5 *V_{BE} for emitter junction of CE transistor amplifier*

V_{BE} forward biases input junction of the Transistor. Input signal 'V_S' is an alternating signal source in nature that is connected to the input port of the Amplifier (Fig. 6.3). Input signal 'V_S' is coupled to the Base through a coupling capacitor C_{Cin} or C_{in} or C_C or C_B. Thus, the capacitor C_{in} blocks the DC bias V_{BE} from entering the signal source but allows AC signal into the input port. X_{CC} (input) should be as small as possible compared to Z_{in} and if this is not possible Z_{in} should be at least 10 times larger than X_{CC} reactance of the input coupling capacitor.

$$X_{CC}(\text{input}) = \frac{1}{10} \cdot Z_{in} \tag{6.1}$$

$$V_{be} = \frac{V_S \cdot Z_{in}}{Z_{in} + X_{CC}(\text{input})} \tag{6.2}$$

- Neglecting the source resistance R_S, the above equation requires that X_C (input) $\to$ zero. Then maximum voltage will be available between the Base and the Emitter. The series coupling capacitors are so selected as to act as effective short circuits to AC signals, while they act as open circuit for DC biasing voltages.
- Capacitor C_E across R_E should have a reactance less than 1/10 of the value of R_E at the lowest frequency of the signal Base band to be amplified. This is justified since the maximum reactance occurs at the lowest frequency and decreases with increasing frequency:

$$X_{CE} = \left(\frac{1}{10}\right) R_E. \tag{6.3}$$

Once it is an effective short circuit at lowest frequency of the signal to be amplified, it is more effective short circuit at all higher frequencies. C_E keeps the Emitter grounded (for CE Transistor configuration) for AC signals.

- Similarly output coupling capacitor C_{out} or C_C (out) or C_B or C_C should become perfect short circuits for AC signals and perfect blocks for DC so that AC and DC voltages are well programmed for operation.
- Now the AC input signal V_{in} is super imposed on the DC bias V_{BE} and the instantaneous voltage V_{be} will be

$$V_{be} = V_{BE}(\text{DC bias}) + V_{in}(\text{AC input signal}) \tag{6.4}$$

and see Figs. 6.6 and 6.7 for signal operation.

In Eq. (6.4), V_{be} is effective changing DC between the Base and the Emitter. To avoid notational ambiguity, the following quantities are defined below.

V_{be} is the varying voltage between the Base and the Emitter. It is the sum of DC bias 'V_{BE}' and instantaneous value of the super imposed AC input signal '$V_m \sin \omega t$'. Thus, V_{be} is instantaneous value of the superimposed signal, where V_{BE} is the operating quiescent (DC) bias between Base and Emitter, $V_{BE} = (V_{R2} - V_E)$.

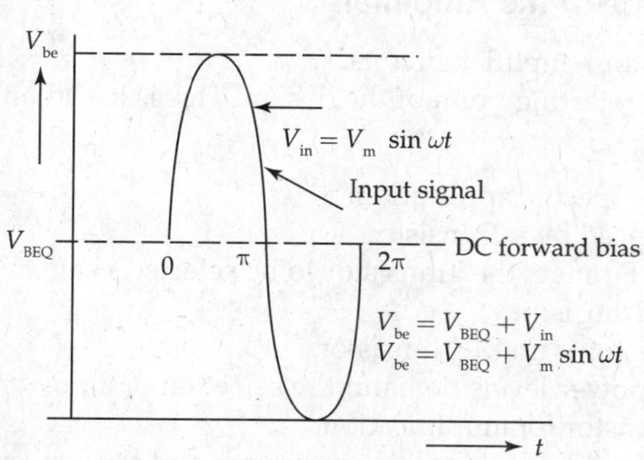

FIG. 6.6 *Effective voltage V_{be} between base and emitter of CE transistor amplifier*

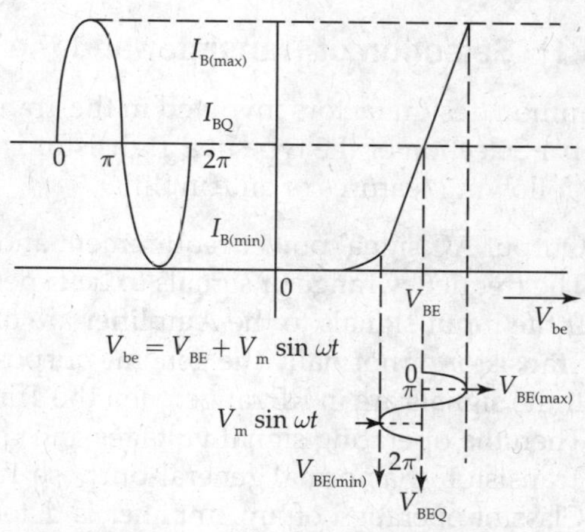

FIG. 6.7 *Input signal variations about the forward bias $V_{BE(Q)}$ in CE transistor amplifier*

Varying DC voltage between the Base and the Emitter causes the Base current to vary sinusoidal. So the Collector current ($I_C = \beta I_B$) varies from its quiescent value between ($I_{CQ} - I_{C(min)}$) and ($I_{CE(max)} - I_{CQ}$) shown in Fig. 6.11. The varying Collector current develops a voltage across the resistance R_L that again varies between $V_{C(max)}$ and $I_{C(min)}$. Voltage across R_L is the amplified output voltage. The waveforms are shown in Fig. 6.11.

The required output AC voltage is developed across 'R_C'. Due to this presence of R_C, the effective load resistance R_L becomes '$R_L \| R_C$' for the purposes of calculation of gain etc.

Though it is convenient for analysis to consider $C \rightarrow \propto$, not more than the value needed to make

$$X_{C_{in}} = \frac{1}{10} Z_{in}, \quad X_{CE} = \frac{1}{10} R_E \quad \text{and} \quad X_{C_{out}} = \left(\frac{1}{10}\right) R_L \tag{6.5}$$

are used. Otherwise, the initial charging current of the capacitor may exceed the maximum rated current of the device and the device blows off.

6.4 DESIGN PROCEDURE TO FIX UP THE DC-OPERATING CONDITIONS

Let us select a 'Transistor' that has the following specifications:

I D: BC107: NPN Transistor

- P_D: Power dissipation capability of the Transistor.
- $V_{CE(max)}$: Maximum Collector to Emitter voltage.
- h_{FE} or β of a Transistor at I_C and I_B from the output characteristics.
- f_T is the frequency at which the Amplifier gain is unity.
- f_T is the Gain-Bandwidth product of Amplifiers.
- Use the Transistor always well under the maximum values of specifications.
- First we identify a Transistor device that provides voltage gain, power gain and efficiency and distortion characteristics according to the desired specifications of an Amplifier.
- Class of operation of Amplifiers to determine the circuit efficiency before fixing up the actual power input and power output conditions.

6.4.1 Selection of Transistor and the Type of the Amplifier

Common design factors involved in the Transistor Amplifier circuits.

Step 1 Selection of the type of a Transistor is the starting point of the design. This is based on the following features of an Amplifier.

- Output AC signal power requirement and its specific application.
- The frequency range of signals to be experienced by a Transistor.
- If the input signals to the Amplifier are of AF range, the Transistor to be selected is an AF Transistor or normally the general-purpose Transistor.
- If the signals are in RF range, then the Transistor is an RF Transistor.
- Then the operating signal voltages and the power levels demand the selection of a power Transistor or a normal general-purpose Transistor for amplification.
- Class of operation of an Amplifier is determined by the specific application and the power level requirement. As a general example for a linear AF Amplifier (Preamplifier) or the voltage Amplifier in a Transistor or radio receiver, the DC conditions to be fixed are for Class A operation.

- If the output power requirement is of moderate level, normal general-purpose Transistor with required average current gain 'h_{FE}' or 'β' is selected, because of variations in their values shown in the data sheets provided by the manufacturers.
- In these empirical selection criteria, expression for current gain $A_I = -\frac{h_{FE}}{1 + h_{0E} \times R_L} \cong -h_{FE} = \beta$ is used. Current gain β of selected Transistor should be larger than the required current gain of Transistor Amplifier.

Once the Transistor selection criteria are done with, the specifications of β and $V_{CE(max)}$ are noted down along with the Transistor characteristics from the Transistor data sheets or Transistor characteristics are obtained practically

6.4.2 Criteria to Fix Up the DC Biasing Conditions for a Transistor Amplifier

Step 2

- For normal Amplifier operation to fix up the reverse bias to the output junction of the Transistor, Collector supply voltage V_{CC}, R_C and R_E are selected so that Quiescent operating voltage V_{CEQ} is much less than $V_{CE(max)}$ rating of the Transistor, and Transistor failure is avoided.
- An important point is that the DC conditions 'V_{CEQ}' and 'I_{CQ}' of the Transistor Amplifier circuit are fixed by the selection of class of operation of the Amplifier for a specific application.
- For Class A operation, the quiescent point is selected at the mid point of the total operating signal range so that equal or symmetric output signal swings occur during the Amplifier operation. At the same time, the other extremes of signal swings should not affect the output signal wave shape.
- DC load line is to be drawn below the power dissipation curve (V_{CE} versus I_C curve satisfying the value of their product to be equal to the maximum power dissipation rating of the device) drawn on Transistor output characteristics. This is required, because the power dissipation by the Transistor does not exceed its maximum power dissipation handling capacity.

6.5 POWER DISSIPATION CURVE AND DC LOAD LINE

- A DC load line is drawn below the parabolic power dissipation curve drawn on the output characteristics of the selected Transistor (Fig. 6.11). Coordinates of the two end points *A* and *B* of the DC load line are obtained by the DC load line (Eq. (6.1)) obtained from the DC equivalent circuit of the Amplifier of Fig. 6.4.
  DC load line equation:

$$V_{CE} = [V_{CC} - I_C(R_C + R_E)]. \tag{6.6}$$

Coordinates of the point *A* on the DC load line on current axis are 'V_{CE}' $= 0$ V and

$$I_C = \frac{V_{CC}}{(R_C + R_E)}. \tag{6.7}$$

The coordinates of the point *B* on the DC load line on the voltage axis are

$$V_{CE} = V_{CC} = 20 \text{ V and } I_C = 0 \text{ mA}. \tag{6.8}$$

(Assume $V_{CC} = 20$ V. Ref. Fig. 6.11.)
Joining the two end points '*A*' and '*B*', the DC load line is obtained.

- For Class A operation of the Amplifier, the quiescent operating point '*Q*' is selected at the middle of the DC load line drawn below the maximum allowable power dissipation curve that can be drawn from the maximum power dissipation rating of the Transistor provided in the data manual.
- Once the quiescent operating point is selected at the middle of the DC load line, the DC voltage 'V_{CE}' between the Collector and the Emitter is half the supply voltage that is $V_{CEQ} = 0.5, V_{CC} = 10$ V.
- Now, the magnitudes of the Collector current 'I_C' and the Base current 'I_B' are obtained from quiescent operating point '*Q*'.

Typical values of corresponding DC component of Collector current 'I_{CQ}' = 2 mA, and the Base current 'I_{BQ}' is 20 μA as shown in Figs. 6.10 and 6.11.

As an empirical rule, the Emitter voltage 'V_E' is taken as one-tenth the supply voltage $V_{CC} = 20$ V. That is, $V_E = 0.1 \times V_{CC} = 0.1 \times 20 \text{ V} = 2$ V.

As $I_{CQ} = 2$ mA and $V_E = I_E \times R_E$ are approximately $= I_{CQ} R_E$ (6.9)

$$V_E = I_E \cdot R_E \cong I_{CQ} \cdot R_E \tag{6.9}$$

$$\therefore\ R_E = \frac{V_E}{I_{CQ}} = \frac{2\text{ V}}{2\times 10^{-3}} = 1\text{ k}\Omega.$$

Similarly at point '*A*', $I_C = 2 \times I_{CQ} = 2 \times 2 \times 10^{-3} = 4$ mA.
At point '*A*', $I_C = 4$ mA

$$I_C = \frac{V_{CC}}{(R_C + R_E)} \tag{6.10}$$

$$\therefore\ (R_C + R_E) = \frac{V_{CC}}{I_C} = \frac{20\text{ V}}{4\text{ mA}} = 5\text{ k}\Omega.$$

Using the value of $R_E = 1$ kΩ

$$R_C = 4\text{ k}\Omega.$$

Then the Collector voltage $\quad V_C = V_{CE} + V_E = 10 + 2 = 12$ V.

6.6 DESIGN OF CIRCUIT COMPONENTS OF BIASING CIRCUIT

The pair of resistors R_1 and R_2 in association with the Collector supply voltage 'V_{CC}' provides voltage to the Base of the Transistor through the potential divider arrangement by resistor R_1 (resistor between the +ve of supply voltage and the Base terminal) and resistor R_2 (resistor between the Base terminal and the ground point or the CE point). Voltage at the Base terminal 'V_B' should be greater than the Emitter voltage 'V_E' by a magnitude of cut-in voltage or knee voltage required to forward bias the input junction by V_{BE} which is of the order of 0.5–0.7 V for Silicon Transistor (0.2–0.3 V for germanium Transistor) so that the 'Transistor is turned on' due to the set DC or quiescent operating conditions for the Transistor:

$$V_B = V_E + V_{BE} = 2.0 + 0.7 = 2.7\text{ V}$$

$$\text{(DC) Base voltage } V_B = \frac{V_{CC} \times R_2}{(R_1 + R_2)} \tag{6.11}$$

$$\therefore \quad V_B = \frac{20 \times R_2}{(R_1 + R_2)} = 2.7 \text{ V}.$$

Parallel combination of resistors 'R_1' and 'R_2' ($R_1 \| R_2$) should be much less than the input resistance βR_E at the input port of the Amplifier (R_E is the resistance connected between the Transistor Emitter terminal and AC ground point). It is required to maintain constant Base current irrespective of wider variations in current gain 'β' of the Transistor.

From the previous discussions, the Base voltage $V_B = [V_E + V_{BE}] = 2.7$ V with a standard assumption for $V_{BE} = 0.7$ V for a silicon Transistor which can work up to higher operating temperatures because of wider forbidden band gap energy of Silicon semiconductor, as we have initially selected a Silicon Transistor of NPN type.

Once we know the various levels of voltages at different points on the Transistor Amplifier, next process is to decide the operating currents and then calculate the values of resistors and capacitors in the Amplifier circuit configuration.

Various DC voltages are

$$V_{CC} = 20 \text{ V}, V_C = 12 \text{ V}, V_E = 2 \text{ V and } V_B = 2.7 \text{ V}.$$

Method of calculating R_1 and R_2 using stability factor 'S'

Assume stability factor $S = 5$, $\beta = 126$ and $R_E = 1$ kΩ (already calculated) as shown in Fig. 6.4 are the associated calculations. Using Eq. (6.12),

$$S = \frac{(\beta + 1)}{1 + \beta \dfrac{R_E}{R_E + R_B}} \tag{6.12}$$

$$5 = \frac{(126 + 1)}{1 + 126 \cdot \dfrac{1 \times 10^3}{1 \times 10^3 + R_B}} \tag{6.13}$$

$$5(1 \times 10^3 + R_B + 126 \times 10^3) = 127(1 \times 10^3 + R_B) \tag{6.14}$$

$$R_B = \frac{508 \times 10^3}{122} = 4.16 \times 10^3 \tag{6.15}$$

Using the values of $R_B = 4.16 \times 10^3$, $V_{CC} = 20$ V and $V_{BB} = 2.7$ V in Eq. (6.15), we get

$$\therefore \quad R_1 = \frac{V_{CC}}{V_B} \times R_B = \frac{20}{2.7} \times 4.16 \times 10^3 = 30.8 \times 10^3 \ \Omega \text{ (Using } R_{BB} = R_B) \tag{6.16}$$

$$R_2 = \frac{R_1 R_B}{R_1 - R_B} = \frac{30.8 \times 10^3 \times 4.16 \times 10^3}{(30.8 - 4.16) \times 10^3} = 4.81 \times 10^3 \ \Omega \tag{6.17}$$

Practical values of the resistors in the Amplifier circuit are $R_C = 4$ kΩ, $R_E = 1$ kΩ, $R_1 = 31$ kΩ and $R_2 = 5$ kΩ.

All the above factors are to be taken care during the design of the component values: *Typical Practical Circuit:* Once the selection of Transistor, design for Collector supply voltage V_{CC} and the resistor values are determined, introduction of an AC signal into the input circuit for amplification and the resulting signal variations are discussed below.

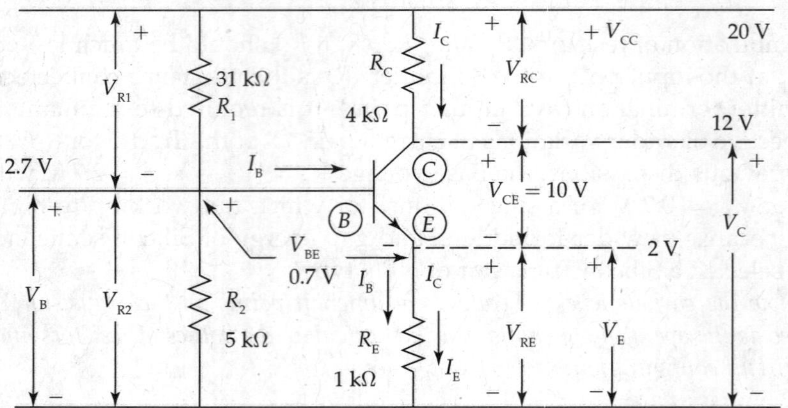

FIG. 6.8 *DC equivalent circuit of a common emitter transistor amplifier*

6.7 COMMON EMITTER TRANSISTOR AMPLIFIER WITH AC SIGNAL OPERATIONS

- Input signal V_{in} is applied at the input terminals 1 and 2 of the input port of the Amplifier (Fig. 6.9) through input coupling capacitor C_{in}.
- Amplified AC signal can be collected at the terminals 3 and 4 of the output port of the Amplifier after the output coupling capacitor C_{out}.
- To avoid signal loss, due to probable negative feedback through R_E, the Emitter resistor 'R_E' is bypassed by 'C_E' and keeps the Emitter terminal at AC ground.

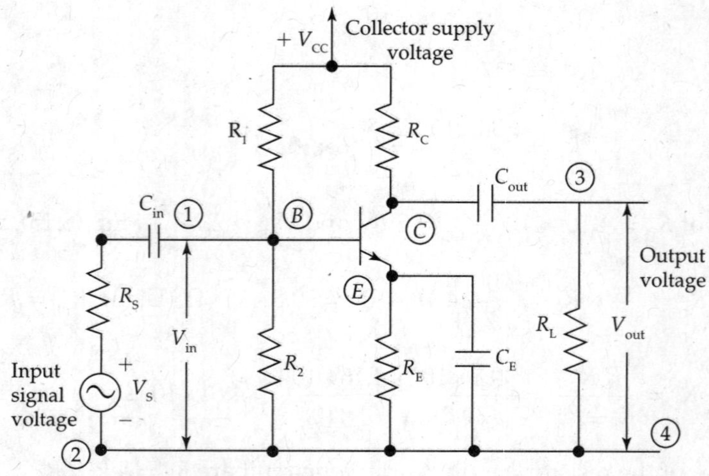

FIG. 6.9 *Single stage resistance capacitance coupled amplifier*

Effective input signal voltage 'V_{be}' between Base and Emitter of the Transistor is due to the superimposition of the input signal voltage V_{in} on the forward bias voltage 'V_{BE}'. This means that DC and AC signal conditions coexist in an electronic Amplifier system.

Thus, the small signal AC voltage 'V_{in}' is superimposed on the forward bias 'V_{BE}' which results in variations in the DC voltage levels between the 'Base and the Emitter'. This varying DC in the forward bias to the input junction of the Transistor results in varying Base current 'I_B' as shown in Fig. 6.7.

Component values of the three capacitors in the circuit could be selected so that the capacitive reactances are virtual short circuits (for the AC signals) in comparison with the resistances at the nodes of signal paths. Capacitive reactances are calculated at the lowest frequency of the band of frequencies of the signals to be amplified.

$$X_{Cin} = \frac{1}{2\pi f C_{in}} \langle\langle R_1 \parallel R_2 \parallel h_{ie} \tag{6.18}$$

$$X_{Cout} = \frac{1}{2\pi f C_{out}} \langle\langle R_L \tag{6.19}$$

$$X_{CE} = \frac{1}{2\pi f C_E} \langle\langle R_E \cong \frac{1}{10 \times R_E} \tag{6.20}$$

An amplifying device thus accepts a varying input signal and produces an output signal that varies in the same way as the input signal but has larger signal amplitude.

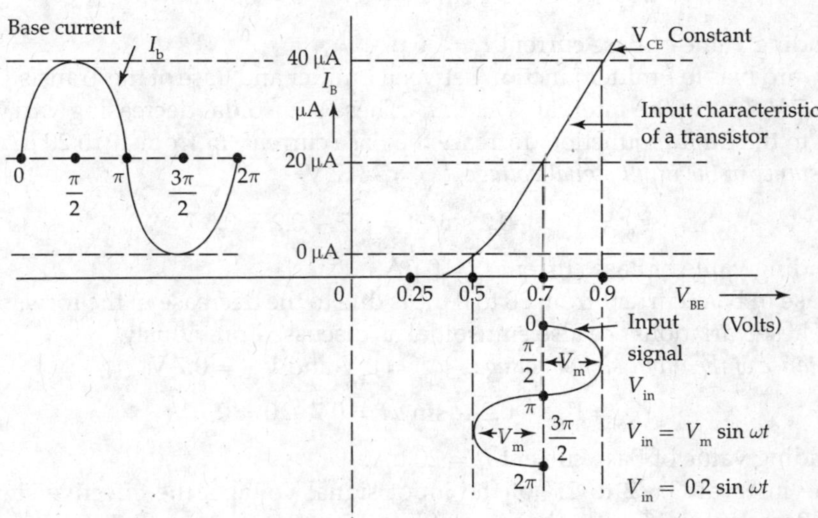

FIG. 6.10 *Variations in base current with variations in input signal voltage by causing variations in forward bias*

Let us consider the following typical magnitudes for V_{in} and V_{BE} shown in Fig. 6.10. The input signal is a sine wave of maximum amplitude 'V_m' of 200 mV.

$$\therefore \quad V_{in} = V_m \sin \omega t \tag{6.21}$$

$$V_{in} = 200 \times 10^{-3} \sin \omega t = 0.2 \sin \omega t$$

The DC forward bias $V_{BE} = 0.7$ V.

The affective signal 'V_{be}' between the Base and Emitter terminals of the Transistor is

$$V_{be} = V_{BE} + V_{in}$$

$$V_{be} = V_{BE} + V_m \cdot \sin(\omega t)$$

$$V_{be} = V_{BE} + 200 \cdot m \cdot v \cdot \sin(\omega t)$$

$$V_{be} = V_{BE} + 0.2 \cdot \sin(\omega t) \tag{6.22}$$

At '0' instance of the input signal voltage, $V_{in} = 0$ V and $V_{BE} = 0.7$ V.

$$\therefore \quad V_{be} = V_{BE} + 0.2 \times \sin \omega t = 0.7 + 0 = 0.7 \text{ V.}$$

Corresponding value of Base current $I_B = 20$ μA.

At π/2 instance of the input signal voltage, $V_{in} = 0.2$ V.

$$\therefore \quad V_{be} = V_{BE} + 0.2 \times \sin \omega t = 0.7 + 0.2 = 0.9 \text{ V.}$$

Corresponding value of Base current $I_B = 40$ μA.

The increase in Base current from 20 to 40 μA is due to the increase in the forward bias from 0.7 to 0.9 V. During the interval '0 to $\pi/2$' of the input signal voltage, the effective signal voltage between the Base and the Emitter increases from 0.7 to 0.9 V sinusoidal. This in turn causes a varying forward bias to the Emitter or the input junction of the Transistor. This varying DC forward bias voltage causes sinusoidal variations in the input Base current from 20 to 40 μA.

At 'π' instance of the input signal voltage, 'V_{in}' = 0 V and $V_{BE} = 0.7$ V.

$$\therefore \quad V_{be} = V_{BE} + 0.2 \sin \omega t = 0.7 + 0 = 0.7 \text{ V.}$$

Corresponding value of Base current $I_B = 20$ μA.

So the forward bias to Emitter junction between Emitter and Base of the Transistor decreases from 0.9 to 0.7 V during the interval '$\pi/2$ to π'. These sinusoidal decreasing variations in the forward bias to the Emitter junction decrease the Base current 'I_B' from 40 to 20 μA sinusoidal.

At 3π/2 instance of the input signal voltage, $V_{in} = -0.2$ V.

$$\therefore \quad V_{be} = V_{BE} + 0.2 \sin \omega t = 0.7 - 0.2 = 0.5 \text{ V.}$$

Corresponding value of Base current $I_B = 0$ μA.

The decrease in Base current from 20 to 0 μA is due to the decrease in the forward bias from 0.7 to 0.5 V. These variations are also sinusoidal as discussed previously.

At '2π' instance of the input signal voltage, 'V_{in}' = 0 V and $V_{BE} = 0.7$ V.

$$\therefore \quad V_{be} = V_{BE} + 0.2 \times \sin \omega t = 0.7 + 0 = 0.7 \text{ V.}$$

Corresponding value of Base current $I_B = 20$ μA.

During the interval '$3\pi/2$ to 2π' of the input signal voltage, the effective signal voltage between the Base and the Emitter increases from 0.5 to 0.7 V sinusoidal. This in turn causes varying forward bias to the Emitter or the input junction of the Transistor. This varying DC forward bias voltage causes sinusoidal variations in the input Base current 'I_B' from 0 to 20 μA.

The total situation is such that the introduced varying input signal voltage is causing variations in the affective voltage between the Base and the Emitter of the Transistor, which is in turn causing varying DC voltage as varying forward bias to input junction of the Transistor. These variations in the DC voltage produce variations in the input current, the Base current 'I_B', as shown in Fig. 6.10.

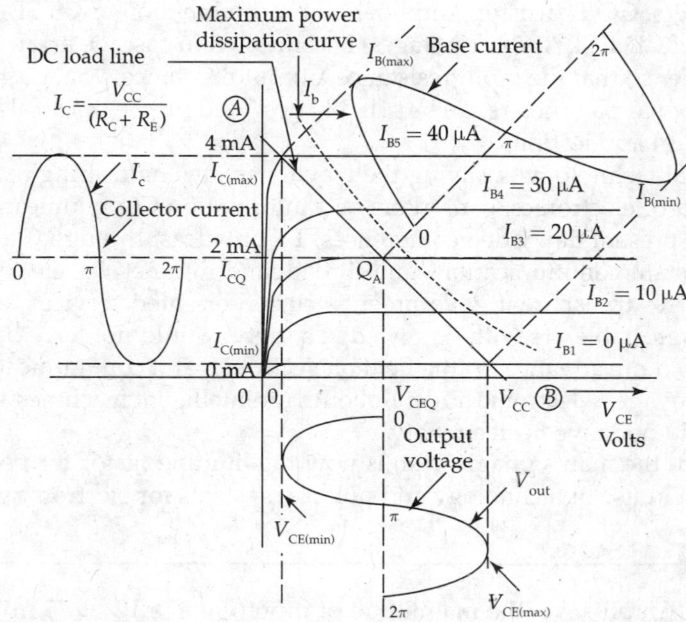

FIG. 6.11 *Signal variations of I_B, I_C and V_{out} in CE transistor amplifier*

Thus, the varying DC biasing voltage levels to the Transistor junctions result a signal output current I_C with increased magnitudes as output current $I_C = \beta I_B$. Variations in input Base current, output Collector current and the output voltage are shown in Fig. 6.11.

From the signal waveforms, it can be observed that the output voltage is having 180° phase shift with input signal voltage. Furthermore, the quiescent operating point Q_A is chosen at the middle of the DC load line for linear or Class A Amplifier operation.

AC equivalent circuit of the CE Transistor Amplifier (Fig. 6.12)

Output current I_{out} is the Collector current I_C. This current flowing through the load resistance R'_L produces the output voltage V_{out}. Therefore, output voltage between the Collector and the Emitter $V_{out} = -I_{out} \times R'_L$, where R'_L is the parallel combination of the resistors R_C and R_L.

- Output voltage is 180° out of phase to input signal voltage in CE Transistor Amplifiers. So the circuit is also known as Inverting Amplifier. Using a 'cathode ray oscilloscope', input voltage V_{IN} and output voltage V_{OUT} can be measured.
- Typical CE Transistor Amplifier (Inverter circuit) circuit has moderately large Voltage gain 'A_V', Current gain 'A_I' and has output and input resistances in the ranges of kilo-ohms depending upon Transistor biasing conditions and signal levels.

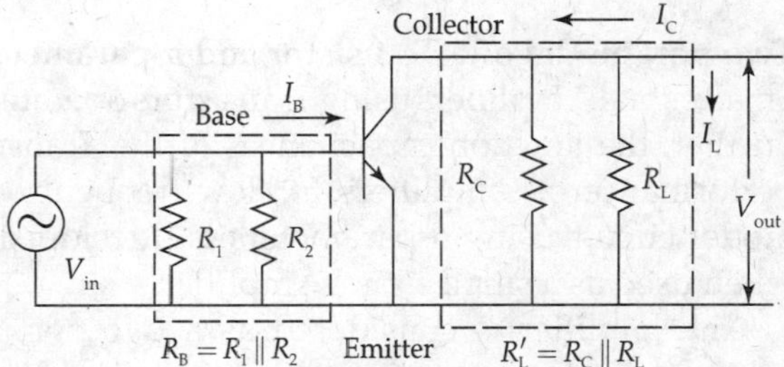

FIG. 6.12 *AC equivalent circuit for single stage transistor amplifier*

A combination of more than one Amplifier stage among CE Transistor Amplifiers, CB

Transistor Amplifiers, CC Transistor Amplifiers (or a combination of CS FET Amplifiers, CG FET Amplifiers and CD FET Amplifier stages) is connected in cascade arrangement to form a multistage Amplifier so that the input resistance, output resistance, voltage gain, current gain and frequency response parameters can be suitably designed for the practical specifications for the required practical applications.

Present day VLSI technology is paving the way for many engineering marvels of the 21st century in the field of electronics, computers and Information Communication Technologies without which the present day video conferences, TV broadcast through Cable and satellites, Cell phones for instant communications and 'Broadband Internet' for electronic governance all over the world for transparent governments' citizen-oriented services to common man/woman to lead comfortable life with 'knowledge society' would not have been possible. But for these and many more advances in the field of electronics and communication engineering, the Inter-galactic voyages, electronic trade, Robotics for intelligent machines and the electronic governments would not have been a reality.

As I understand, the man's imagination is now the limiting factor for peace on the globe because of incoherent use of technology and political systems for electronic governance.

EXAMPLE 6.1

In a CE Transistor Amplifier, if the magnitude of input voltage V_{in} = 25 mV and the output voltage is 2.5 V, calculate the voltage gain of the Amplifier.

Solution: For a CE Transistor Amplifier, output voltage V_{out} = 2.5 V.
Input signal voltage V_{in} = 25 mV.

$$\text{Voltage gain } A = \frac{V_{out}}{V_{in}}$$

$$\therefore \text{ Voltage gain } A = \frac{2.5\text{ V}}{25\text{ mV}} = \frac{2.5\times10^3}{25} = \frac{2500}{25} = 100.$$

6.8 THE *h*-PARAMETERS OF THE TRANSISTOR

6.8.1 Small Signal Amplifier Analysis of Bipolar Junction Transistor at Low Frequencies

Two port model of a Transistor and *h*-parameters

Small signal Amplifiers using Transistors operate in the linear region of its output characteristics. Further, the junction capacitances of the Transistor do not have considerable effects on the performance of Amplifiers for low-frequency signals. In such scenario, hybrid parameter model circuit using *h*-parameters is developed for Transistors, to estimate its performance when used as a small signal Amplifier.

An Amplifier is considered as a two-port (four-terminal) network, using a Transistor connected in any one of the three configurations Viz. CE/CB/CC Transistor. Initially, the *h*-parameters are defined for a Transistor in CE/CB/CC model in a block box and then applied to all the three configurations (CE/CB/CC) for their analysis. Consider the Transistor (BJT)

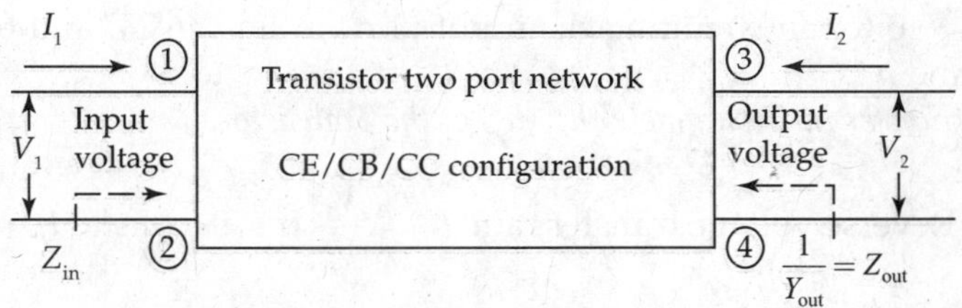

FIG. 6.13 *Transistor as a two port network*

as two-port network with variables V_1, I_1, V_2, I_2 shown in Fig. 6.13. Relations between input and output voltages and currents at each port in matrix form are arrived from the nature of input and output characteristics of the transistor. (Approximate consideration of the input characteristic as constant voltage curve V_1 forms the first element of the first matrix and current I_1 taken as the first element of the third matrix. Similarly, the output characteristics can be considered as constant current characteristics. Current I_2 becomes the second element of the first matrix and V_2 becomes the second element of the third matrix as shown in matrix forms (Eq. 6.23).)

$$\begin{bmatrix} V_1 \\ I_2 \end{bmatrix} = \begin{bmatrix} h_{11} & h_{12} \\ h_{21} & h_{22} \end{bmatrix} \begin{bmatrix} I_1 \\ V_2 \end{bmatrix} \tag{6.23}$$

Transistor is basically a current Amplifier but natural sources are voltage type in nature. So the following expressions are used to represent the input and output voltages and currents at each port in the following matrix form.

Expanding the matrices, Eqs. (6.24) and (6.25) are obtained:

$$V_1 = h_{11} \cdot I_1 + h_{12} \cdot V_2 \tag{6.24}$$

$$I_2 = h_{21} \cdot I_1 + h_{22} \cdot V_2 \tag{6.25}$$

Applying some boundary conditions, *h*-parameters can be defined and determined.

- V_2 is made zero (output port is short-circuited) in Eqs. (6.24) and (6.25) to define h_{11} and h_{21}.

(1) *Input resistance parameter h_{11} (h_i) of the Transistor:*

$$\text{Input resistance } h_{11} \triangleq \frac{V_1}{I_1}\bigg|_{V_2} = 0\ (\Omega) \tag{6.26}$$

h_{11} has the dimensions of resistance which pertains to the input port. So it can be termed *input resistance* h_i represented by $h_i = \frac{V_1}{I_1}$. The unit of input resistance 'h_i' is Ohms

(2) *Forward current gain parameter h_{21} (h_f) of the Transistor:*

$$\text{Forward current gain } h_{21} \triangleq \frac{I_2}{I_1}\bigg|_{V_2} = 0\ \text{(constant)} \tag{6.27}$$

h_{21} is a dimensionless quantity, which is the ratio of output current to input current. It is known as short circuit current gain. h_f is also called *forward current gain*. 'h_f' is a constant. It is represented by $h_f = \frac{I_2}{I_1}$.

- Now making $I_1 = 0$ (open circuiting the input port) in Eqs. (6.24) and (6.25) to define the parameters h_{12} and h_{22}.

(3) *Reverse voltage transfer ratio parameter* h_{12} (h_r) *of the Transistor:*

$$\text{Reverse voltage transfer ratio } h_{12} \triangleq \left.\frac{V_1}{V_2}\right| I_1 = 0 \text{ (constant)} \tag{6.28}$$

h_{12} is a *reverse voltage transfer ratio* which is named as 'h_r'. It is a constant. This in fact represents the unwanted voltage transfer from the output to the input, since Amplifiers should be preferably unilateral in transfer of energy from input port to the output ports, but not the other way round.

(4) *Output conductance parameter* h_{22} (h_o) *of the Transistor:*

$$\text{Output conductance } h_{22} \triangleq \left.\frac{I_2}{V_2}\right| I_1 = 0 \text{ (mhos)} \tag{6.29}$$

h_{22} represents the admittance of the output port. It is designated as *output conductance, '*h_o*'*, where h_o is measured in mhos or Siemens.

h-parameters possess a mixture of units, and hence are known as hybrid parameters.

When applied to analysis of Amplifiers with alternating signals $V_2 = 0$ and $I_1 = 0$, they represent constant DC values of voltage at the output port and current at the input port. Since the Transistor characteristics are not entirely linear, *h*-parameters change from point to point and are defined over small-linear regions. Hence, they are called small signal parameters.

6.8.2 Definitions of *h*-parameters of the Bipolar Junction Transistors

$$\text{Input resistance } h_i = \left|\frac{\Delta V_i}{\Delta I_i}\right| V_2 = \text{constant } (\Omega). \tag{6.30}$$

$$\text{Forward current gain } h_f = \left|\frac{\Delta I_0}{\Delta I_i}\right| V_2 = \text{constant} \tag{6.31}$$

$$\text{Reverse voltage transfer ratio } h_r = \left|\frac{\Delta V_i}{\Delta V_0}\right| I_1 = \text{constant} \tag{6.32}$$

$$\text{Output conductance } h_o = \left|\frac{\Delta I_0}{\Delta V_0}\right| I_1 = \text{constant (mhos)} \tag{6.33}$$

6.8.3 *h*-parameters for CE/CB/CC configurations of the Transistors

Transistor Amplifiers can have three configurations of the Transistor with its one of the terminals grounded to act as a common terminal to both input and output ports. The other two terminals forming the input and output terminals are subjected to the original definition. They are:

1. CE Transistor Amplifier configuration
2. CB Transistor Amplifier configuration
3. CC Transistor Amplifier configuration

Notation for the Transistor h-parameters for the three models is shown in tabular form:

The h-parameter designations of BJT configurations

h-parameter	CB	CE	CC
h_i	h_{ib}	h_{ie}	h_{ic}
h_r	h_{rb}	h_{re}	h_{rc}
h_f	h_{fb}	h_{fe}	h_{fc}
h_o	h_{ob}	h_{oe}	h_{oc}

1. CE Transistor configuration

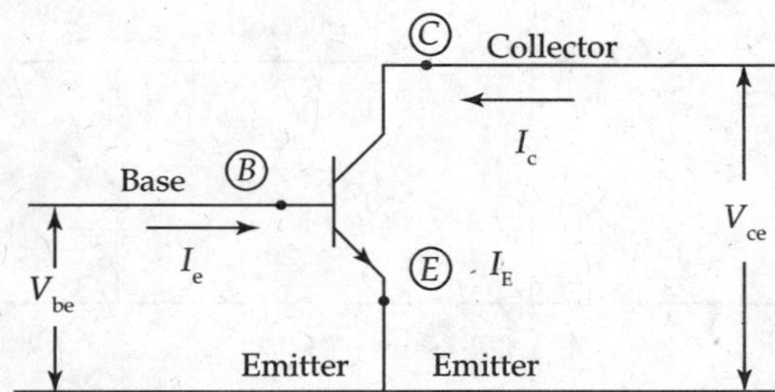

FIG. 6.14 *Common emitter transistor configuration*

- Input characteristic of the Transistor can be approximated by constant voltage curve, and the input port can be represented by a Thevinin voltage source $h_{re} \times V_{CE}$ with its source resistance h_{ie} in the Transistor equivalent circuit shown in Fig. 6.15.
- Similarly, the constant output current in output characteristics in active region (Transistor biasing in active region for the Transistor to act as amplifying device) suggests that output port of the Transistor can be represented by Norton equivalent circuit with current generator $h_{fe} \times I_B$ and parallel resistance $1/h_{oe}$ shown in Fig. 6.15.

h-parameter equivalent circuit of the CE Transistor

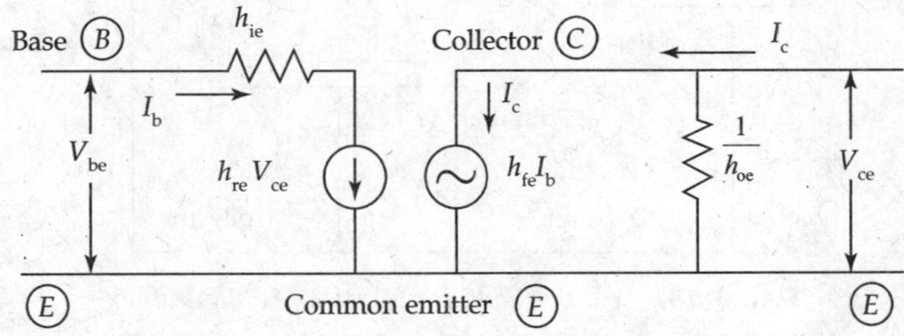

FIG. 6.15 *Common emitter transistor h-parameter equivalent circuit*

Equations for CE Emitter Transistor h-parameter equivalent circuit are as follows:

$$V_{be} = h_{ie} \cdot I_b + h_{re} \cdot V_{ce} \tag{6.34}$$

$$I_c = h_{fe} \cdot I_b + h_{oe} \cdot V_{ce} \tag{6.35}$$

for CE Transistor Amplifier.

2. CB Transistor configuration

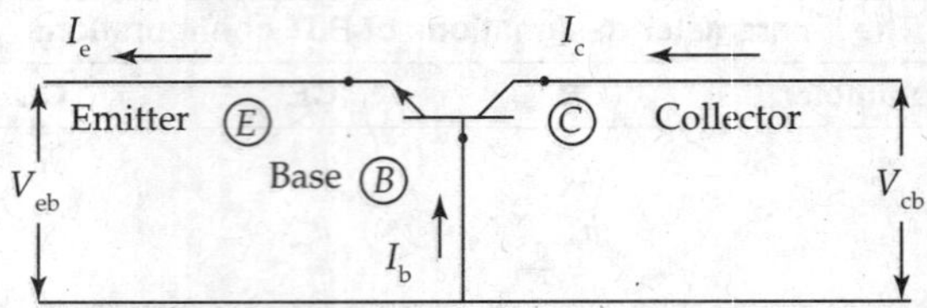

FIG. 6.16 *Common base transistor*

h-parameter equivalent circuit of CB Transistor

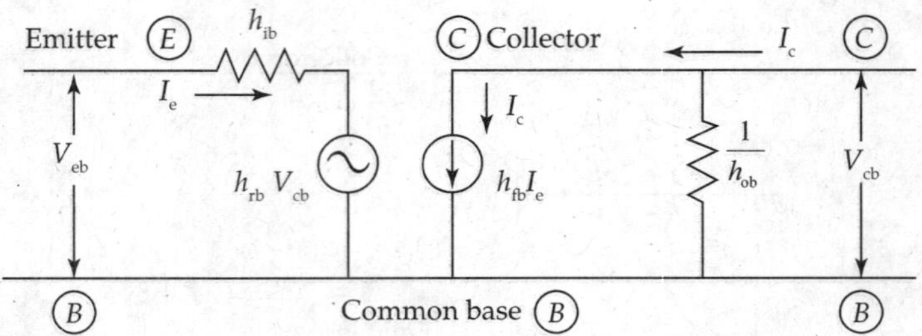

FIG. 6.17 *Common base transistor h-parameter equivalent circuit*

$$V_{eb} = h_{ib} \cdot I_e + h_{rb} \cdot V_{cb} \tag{6.36}$$

$$I_c = h_{fb} \cdot I_e + h_{ob} \cdot V_{cb} \tag{6.37}$$

h-parameter equations relate input and output voltages and currents, which are used to derive the expressions for current gain, A_I, input resistance h_{ib}, voltage gain A_V and output conductance h_{ob}.

3. CC Transistor configuration

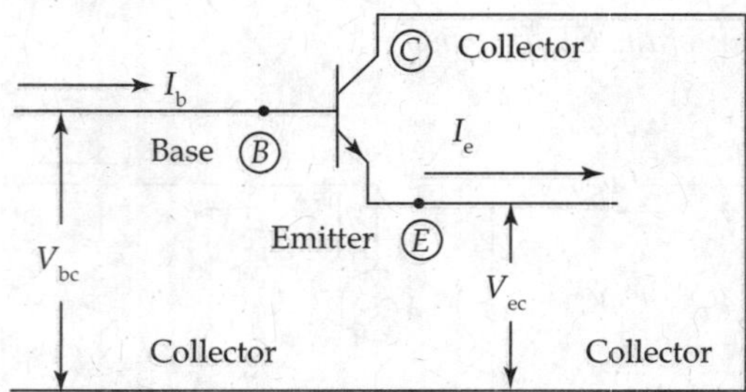

FIG. 6.18 *Common collector transistor configuration*

h-parameter equivalent circuit for CC Transistor configuration

$$V_{bc} = h_{ic} \cdot I_b + h_{rc} \cdot V_{ec} \tag{6.38}$$

$$I_e = h_{fc} \cdot I_b + h_{oc} \cdot V_{ec} \tag{6.39}$$

The above *h*-parameter equations relate input and output voltages and currents, which are used to derive the expressions for the current gain A_I, input resistance h_{ic}, voltage gain A_V and the output conductance h_{oc}.

Three sets of h-parameters are obtained with the second subscript to the hybrid parameters, designating the grounded terminal of the Transistor.

- For CE Transistor, the h-parameters are h_{ie}, h_{fe}, h_{re} and h_{oe}.
- For CB Transistor, h-parameters are h_{ib}, h_{fb}, h_{rb} and h_{ob}.
- For CC Transistor, h-parameters are h_{ic}, h_{fc}, h_{rc} and h_{oc}.

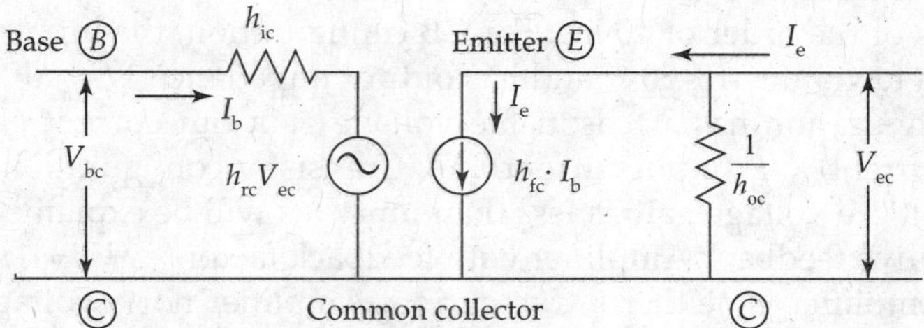

FIG. 6.19 *Common collector transistor h-parameter equivalent circuit*

The different sets of parameters are inter-related and inter-convertible.

$$\text{For instance,}\quad h_{ie} = \frac{\Delta V_{be}}{\Delta I_b} = \frac{\Delta V_{be}}{\left[\dfrac{\Delta I_e}{(1+\beta)}\right]}. \tag{6.40}$$

$$\therefore\quad h_{ie} = (1+\beta)\cdot h_{ib}$$

where $I_b = \dfrac{I_e}{(1+\beta)}$; $I_e = (I_b + I_c)$

Dividing the above equation on both sides by I_b, we get

$$\frac{I_e}{I_b} = \frac{(I_b + I_c)}{I_b} = (1+\beta)$$

$$\therefore\quad I_b = \frac{I_e}{(1+\beta)} \tag{6.41}$$

From Eq. (6.41), we get

$$h_{ib} = \frac{h_{ie}}{(1+\beta)} = \frac{h_{ie}}{(1+h_{fe})} \tag{6.42}$$

$$\text{Similarly}\quad \alpha = h_{fb} = \frac{\Delta I_c}{\Delta I_e} = \frac{\Delta I_c}{\Delta I_b(\beta+1)} = \frac{\beta}{\beta+1} \tag{6.43}$$

$$\text{(or)}\quad h_{fb} = \frac{h_{fe}}{(1+h_{fe})} \tag{6.44}$$

For CC parameters

$$h_{fc} = \frac{\Delta I_e}{\Delta I_b} = \frac{(\Delta I_c + \Delta I_b)}{\Delta I_b} = (h_{fe}+1)$$

$$\therefore\quad h_{fc} = (1+h_{fe}) = (1+\beta) \tag{6.45}$$

h_{ic} and h_{oc} are equal to h_{ie} and h_{oe}, respectively.

6.8.4 Comparison of CE, CB, CC Configurations of Transistors

Between CE, CC and CB Transistor configurations as per chosen directions of positive and negative polarities, CE Transistor configuration is considered to be an inverting voltage Amplifier, where as for the same chosen polarities, CB and CC configurations form non-inverting Amplifiers.

CE Transistor configuration has input impedance h_{ie} of the order of 1 kΩ and output impedance $1/h_{oe}$ is of the order of 40 kΩ. For CB configuration, the input impedance h_{ib} will be of the order of a few ohms 10–20 Ω and has output impedance $1/h_{ob}$ of the order of 2 MΩ. CB Transistor configuration has a reasonable voltage gain but current gain h_{fb} is less than unity (I_C (output current) $< I_E$ (input current)). CC Transistor configuration has a current gain of $h_{fc} = (1 + h_{fe})$, but the voltage gain is less than unity (as will be explained later it forms the voltage series negative feedback Amplifier with feedback factor $V_f/V_0 = 1$).

CE Transistor Amplifier is neither a true current Amplifier nor a voltage Amplifier, but a bit of both. CB Transistor Amplifier is an almost ideal current controlled current generator, since its input impedance is low and can be connected to a current source. Since its output impedance is high it can act as a current source, i.e., in effect a current controlled current source. CC configuration due to unity feedback factor has very high input impedance and very low output impedance and can act as a voltage controlled voltage source.

Although the active devices BJTs and FETs have different Bases of physical operation, once their circuit models replace these devices, their frequency response and other features can be analysed simultaneously. The following *h*-parameter model is for Bipolar Junction Transistors.

6.9 TRANSISTOR AMPLIFIER ANALYSIS USING *h*-PARAMETER EQUIVALENT CIRCUITS

There are several circuit parameters that are shared by all Amplifiers, whatever may be the type of an Amplifier. They are as follows:

1. Current gain: A_I
2. Input resistance: Z_{in}
3. Voltage gain: A_V
4. Output resistance: Z_{out}
5. Power gain: A_P

Performance characteristics of Amplifiers are analysed below:

Expressions for current gain A_I, input impedance Z_{in}, voltage gain A_V, output impedance Z_{out} and power gain A_P for the Transistor can be derived in a general way for all the three configurations of the Transistor. The parameters are suitably adopted for individual configurations, by changing the second subscript on the *h*-parameters as discussed here.

Just like equations can be framed for circuits, circuits can be formed from equations.

Considering the equations

$$\text{Input voltage} \quad V_{in} = h_i \cdot I_{in} + h_r \cdot V_{out} \tag{6.46}$$

$$\text{Output current} \quad I_{out} = h_f \cdot I_{in} + h_o \cdot V_{out} \tag{6.47}$$

Figure 6.20 shows the *h*-parameter equivalent circuit representing Eq. (6.46). This *h*-parameter model is useful for low frequencies only. For high-frequency operation, the

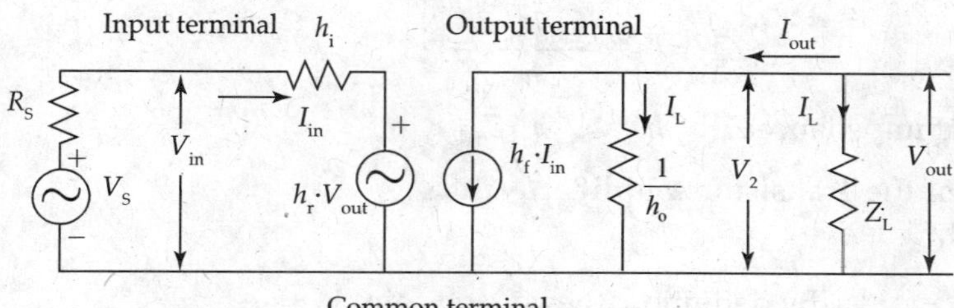

FIG. 6.20 *Small signal low frequency transistor equivalent circuit using h-parameters*

junction capacitançes of the Transistor and the external capacitors are to be added to the high-frequency Amplifier equivalent circuits using *h*-parameters. Once hybrid model is drawn for a Transistor, remaining circuit components in Amplifier circuit are added for complete analysis in predicting the Amplifier performance.

When Z_L is connected across the output port current i_L flows through Z_L.

$$\text{Therefore the output voltage } V_{out} = I_L \cdot Z_L = -I_{out} \cdot Z_L \tag{6.48}$$

Current gain A_i

$$\text{By definition, the current gain } A_I = \frac{I_L}{I_{in}} = \frac{-I_{out}}{I_{in}} \tag{6.49}$$

from the circuit shown in Fig. 6.20.

$$I_{out} = h_f \cdot I_{in} + h_0(-I_{out} \cdot Z_L) \tag{6.50}$$

$$\therefore \quad I_{out}(1 + h_o \cdot Z_L) = h_f \cdot I_{in} \tag{6.51}$$

$$\frac{I_{out}}{I_{in}} = \frac{h_f}{(1 + h_0 \cdot Z_L)} \tag{6.52}$$

$$\therefore \quad \text{Current gain } A_I = \frac{I_L}{I_{in}} = \frac{-I_{out}}{I_{in}} = \frac{-h_f}{(1 + h_o \cdot Z_L)} \tag{6.53}$$

Input impedance Z_{in} of the Transistor Amplifier

Input impedance (Z_i) Z_{in}

From Eq. 6.46,

$$V_{in} = h_i \cdot I_{in} + h_r \cdot V_{out} \tag{6.54}$$

$$\therefore \quad Z_{in} = \frac{V_{in}}{I_{in}} = h_i + \frac{h_r \cdot V_{out}}{I_{in}}$$

But $V_{out} = -I_{out} \cdot Z_L$

$$\therefore \quad \text{Input impedance } Z_{in} = h_i + h_r \frac{-I_{out}}{I_{in}} \cdot Z_L \tag{6.55}$$

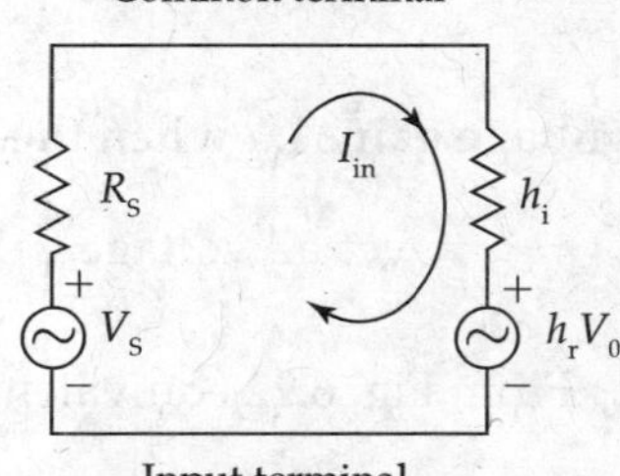

FIG. 6.21 *h-parameter equivalent circuit of the input port of CE transistor amplifier*

Using the value of $A_I = \frac{-I_{out}}{I_{in}} = -\frac{h_f}{[1 + h_0 \cdot Z_L]}$ in Eq. (6.55), we get

$$Z_{in} = h_i - \frac{h_f \cdot h_r \cdot Z_L}{[1 + h_0 \cdot Z_L]} \tag{6.56}$$

(OR) Input impedance $Z_{in} = h_i + A_i \cdot h_r \cdot Z_L$. (6.57)

Voltage gain A_V of the Transistor Amplifier

Voltage gain A_V

$$\text{By definition, the voltage gain } A_V = \frac{V_{out}}{V_{in}} \tag{6.58}$$

$$\text{But } V_{out} = -I_{out} \cdot Z_L \text{ and } V_{in} = I_{in} \cdot Z_{in} \tag{6.59}$$

Substituting the values of V_{out} and V_{in} from Eq. (6.59) into Eq. (6.58), we get

$$\text{Voltage gain } A_v = \frac{-I_{out} \cdot Z_L}{I_{in} \cdot Z_{in}} = A_i \frac{Z_L}{Z_{in}} \tag{6.60}$$

Output impedance Z_0 (Z_{OUT}) of the Transistor Amplifier

Output impedance Z_0. To find Z_0 which is defined as

$$Z_o = \left.\frac{V_0}{I_0}\right| V_S = 0. \quad \text{Also } Z_o = \frac{1}{Y_0} \tag{6.61}$$

$$\text{Output current } I_{out} = h_f \cdot I_{in} + h_o \cdot V_{out} \tag{6.62}$$

$$Y_o = \frac{I_{out}}{V_{out}} = h_f \cdot \frac{I_{in}}{V_{out}} + h_o \quad \text{with } V_S = 0. \tag{6.63}$$

From Fig. (6.21) with $V_S = 0$,

$$h_r \cdot V_{out} = -(R_S + h_i) \cdot I_{in} \tag{6.64}$$

$$\therefore \quad \frac{I_{in}}{V_{out}} = \frac{-h_r}{(R_S + h_i)} \tag{6.65}$$

$$Y_o = \frac{I_L}{V_{out}} = \frac{1}{Z_o} = h_o - \frac{h_f \cdot h_r}{(R_S + h_i)} \tag{6.66}$$

$$\text{Output impedance } Z_o = \frac{(R_S + h_i)}{(h_o \cdot R_S + \Delta h)} \tag{6.67}$$

where $\Delta h = h_{oe} \cdot h_{ie} - h_{fe} \cdot h_{re}$.

Voltage gain A_{VS} when the signal source has finite resistance R_S

$$\text{Over all voltage gain } A_{VS} = \frac{V_{out}}{V_S} = \frac{V_{out}}{V_{in}} \times \frac{V_{in}}{V_S} = A_V \cdot \frac{V_{in}}{V_S}$$

From Fig. 6.22 equivalent circuit,

$$V_{in} = \frac{V_S \times Z_{in}}{(R_S + Z_{in})}$$

$$\therefore \quad \frac{V_{in}}{V_S} = \frac{Z_{in}}{(R_S + Z_{in})}$$

$$\text{Hence,}\quad A_{VS} = A_V \cdot \frac{V_{in}}{V_S} = A_V \cdot \frac{Z_{in}}{(R_S + Z_{in})} \tag{6.68}$$

A_{VS} can also be written as

$$A_{VS} = A_V \cdot \frac{Z_L}{Z_L} \cdot \frac{Z_{in}}{(R_S + Z_{in})} = A_I \cdot \frac{Z_L}{(R_S + Z_{in})} \tag{6.69}$$

From Eq. (6.68), it is evident that $A_{VS} = A_V$ when source resistance $R_S = 0$. This means that A_{VS} is the overall voltage gain when voltage source has some finite resistance, whereas A_V is the voltage gain of the Transistor Amplifier with an ideal voltage source. Similarly, if R_S is equal to Z_{in} in Eq. (6.68), then $A_{VS} = 0.5\, A_V$.

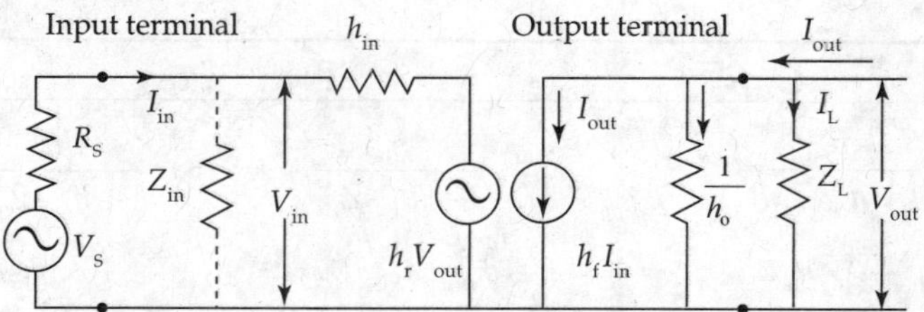

FIG. 6.22 *h-parameter equivalent circuit of transistor smplifier with voltage source having R_S*

Current gain A_{IS} when input current source I_S has finite resistance

In the equivalent circuit shown in Fig. 6.23, the current source I_S has finite source resistance, R_S. Overall current gain A_{IS} is calculated as follows:

$$A_{IS} = \frac{I_{out}}{I_S} = \frac{I_{out}}{I_{in}} \cdot \frac{I_{in}}{I_S} = A_I \cdot \frac{I_{in}}{I_S}$$

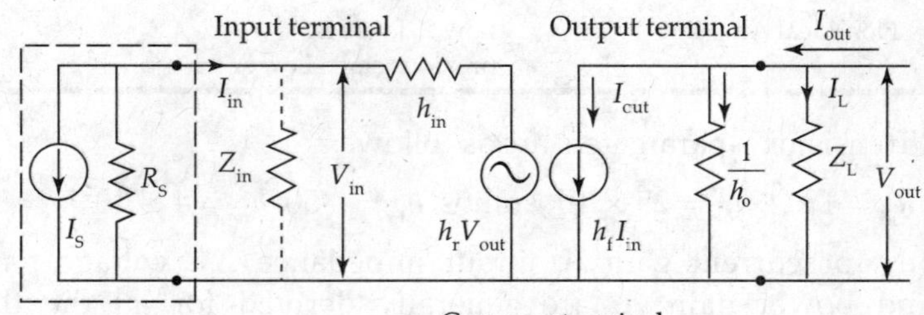

FIG. 6.23 *h-parameter equivalent circuit of transistor amplifier with current source having R_S*

From the equivalent circuit (Fig. 6.23),

$$I_{in} = \frac{I_S \cdot R_S}{(R_S + Z_{in})}$$

$$\therefore\quad A_{IS} = A_I \cdot \frac{R_S}{(R_S + Z_{in})}.$$

Using $\dfrac{I_{in}}{I_S} = \dfrac{R_S}{(R_S + Z_{in})}$ $A_{IS} = A_I$ when the current source has very large resistance.

Power gain A_P of Transistor Amplifiers

As long as the load resistance $Z_L = R_L$ a resistor

Output power $P_{out} = V_{out} \times I_{out}$ Watts

AC input power $P_{in} = V_{in} \times I_{in}$ Watts

$$\text{Power gain } A_P = \frac{P_{out}}{P_{in}} = \frac{V_{out} \cdot I_{out}}{V_{in} \cdot I_{in}} = A_V \cdot A_I \tag{6.70}$$

The Amplifier performance can be completely assessed by the four basic relations for A_I, A_V, Z_{in} and Z_{out} derived below.

Comparison of performance of CE, CB and CC Transistor Amplifiers

Parameter	CE	CB	CC
A_I	$\frac{-h_{fe}}{1+h_{oe} \cdot Z_L}$	$\frac{-h_{fb}}{1+h_{ob} \cdot Z_L}$	$\frac{-h_{fc}}{1+h_{oc} \cdot Z_L}$
Z_i	$h_{ie} + A_{Ice} h_{re} Z_L$	$h_{ie} + A_{Icb} h_{rb} Z_L$	$h_{ie} + A_{Icc} h_{rc} Z_L$
A_v	$A_{I\,ce} \cdot \frac{Z_L}{Z_{ie}}$	$A_{I\,cb} \cdot \frac{Z_L}{Z_{ib}}$	$A_{I\,cc} \cdot \frac{Z_L}{Z_{ic}}$
Y_0	$h_{oe} - \frac{h_{fe} \cdot h_{re}}{R_S + h_{ie}}$	$h_{ob} - \frac{h_{fb} \cdot h_{rb}}{R_S + h_{ib}}$	$h_{oc} - \frac{h_{fc} \cdot h_{rc}}{R_S + h_{ic}}$
Merits	Inverting Amplifier with reasonable voltage and current gain	Non-inverting Amplifier with high Z_0	Non-inverting Amplifier with large A_I. Excellent as an impedance transformer
Demerits	Non-ideal Amplifier	Low Z_i loads a previous stage Low A_I	$A_V < 1$

Typical values of Transistor h-parameters are as follows:

$$h_{ie} = 1\ \text{k}\Omega,\ h_{oe} = 25 \times 10^{-6}\ \text{mhos},\ h_{fe} = +50,\ h_{re} = 4 \times 10^{-4}.$$

The expressions for current gain A_I, input impedance Z_{in}, voltage gains A_V, output impedance Z_o and power gain A_P are generally derived for all the three Transistor configurations of the Amplifier. So for CE Transistor Amplifier the above expressions will be with the second subscript taken as '*e*':

$$\text{Current gain } A_I = \frac{-h_{fe}}{1+h_{oe} \cdot Z_L};$$

Input impedance $Z_{in} = h_{ie} + A_{Ice} h_{re} Z_L$

$$\text{Voltage gain} \quad A_V = A_{Ice} \cdot \frac{Z_L}{Z_{ie}}$$

$$\text{Output impedance} \quad Y_o = h_{oe} - \frac{h_{fe} \cdot h_{re}}{R_S + h_{ie}}$$

Comparison of the features of CE and CC Amplifier circuits with reference to voltage gain, Current gain, Input resistance and output resistance

Both voltage gain and current gains are much greater than unity for CE Amplifier, and so the CE Transistor Amplifier circuit is most popularly used in practice

$$\text{Current gain} \quad A_{\text{Ice}} = -\frac{h_{\text{fe}}}{1+h_{\text{oe}} \cdot Z_{\text{L}}} \quad \text{and} \quad A_{\text{Vce}} = A_{\text{I}} \cdot \frac{Z_{\text{L}}}{Z_{\text{in}}}.$$

Whereas, the voltage gain is approximately unity for CC Amplifier. From the above two equations for current gain for both CE and CC Amplifiers current gain variations with variations in load resistance are similar.

$$\text{Current gain} \quad A_{\text{Icca}} = -\frac{h_{\text{fC}}}{1+h_{\text{OC}} \cdot Z_{\text{L}}} \cong (1+h_{\text{fe}}) \text{ and } A_{\text{Vcca}} = A_{\text{I}} \cdot \frac{Z_{\text{L}}}{Z_{\text{IN}}}.$$

The expressions for input impedance Z_{ince} and Z_{incca} are as follows:

$$Z_{\text{ince}} = h_{\text{ie}} + A_{\text{I}} \cdot h_{\text{re}} \cdot Z_{\text{L}} \text{ whereas } Z_{\text{incca}} = h_{\text{ic}} + A_{\text{I}} \cdot h_{\text{rc}} \cdot Z_{\text{L}}.$$

As $h_{\text{re}} \cong 0$, and if substituted in the above equation for Z_{ince}, the input resistance of CE Transistor Amplifier is approximately equal to h_{ie} of the Transistor which is of the order of 1 kΩ and varies moderately with variations in the load resistance 'R_{L}', whereas the input resistance of CC Amplifier is very high and is of the order of megaohms, because $h_{\text{rc}} \cong 1$ and when substituted in the equation for Z_{incca}, the input resistance of CC Amplifier is very high and also due to voltage series negative feedback introduced into the CC Amplifier circuit.

The expressions for the output impedances of CE and CC Amplifiers are as follows:

$$\therefore \quad Z_{\text{outcca}} = \frac{V}{I} = \frac{h_{\text{ie}} \cdot I_{\text{B}}}{(1+h_{\text{fe}}) \cdot I_{\text{B}}}$$

$$= \frac{h_{\text{ie}}}{(1+h_{\text{fe}})} \cong \frac{h_{\text{ie}}}{h_{\text{fe}}} = \frac{1}{g_{\text{m}}} \quad \text{for common collector amplifier.}$$

From the above equation for CC Amplifier, the output impedance is very low. As the CC Amplifier has very low output impedance of the order of few tens of ohms and very large input resistance of the order of mega ohms, CC Amplifier is used to couple between high impedance source and low impedance loads and as impedance transformation circuits and unity gain Buffer Amplifier circuits:

$$Y_{\text{out}} = h_{\text{oe}} - \frac{h_{\text{fe}} \cdot h_{\text{re}}}{R_{\text{S}} + h_{\text{ie}}}.$$

The output impedance for CE Transistor Amplifier is the reciprocal of output admittance. Furthermore, the reverse resistance of the output junction of the CE Transistor has high resistance. So the output resistance of CE Transistor Amplifiers is reasonably large and is of the order of few thousands of ohms.

$$\text{Current gain} \quad A_{\text{I}} = \frac{-h_{\text{fc}}}{1+h_{\text{oc}} \cdot R_{\text{E}}} = \frac{1+h_{\text{fe}}}{1+h_{\text{oc}} \cdot R_{\text{E}}} \equiv (1+h_{\text{fe}}) \quad \text{provided } h_{\text{oc}} \cdot R_{\text{E}} \ll 1.$$

Above equation represents the current gain of CC *Transistor Amplifier.*

EXAMPLE 6.2

Find the values of A_I, A_V, A_{VS}, A_{IS}, Z_i and Z_o for the following circuit. Typical values of Transistor h-parameters are as follows: $h_{ie} = 1\ \text{k}\Omega$, $h_{oe} = 25 \times 10^{-6}$ mhos, $h_{fe} = +50$, $h_{re} = 4 \times 10^{-4}$.

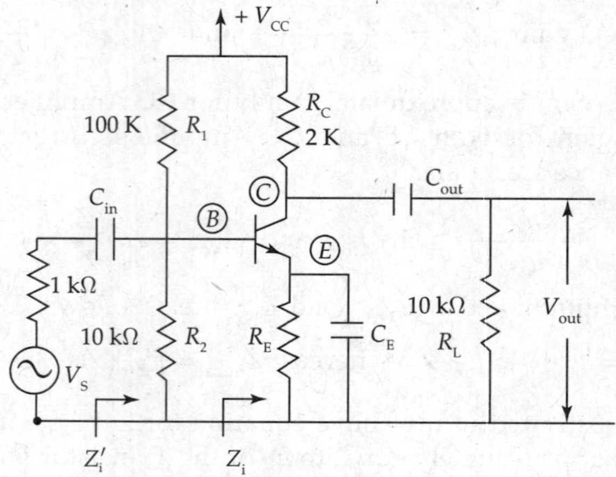

FIG. 6.24 *Common emitter transistor amplifier circuit*

Solution: $h_{fe} = +50$, $h_{ie} = 1\ \text{k}\Omega$, $h_{re} = 4 \times 10^{-4}$, $h_{oe} = 25 \times 10^{-6}$ mhos

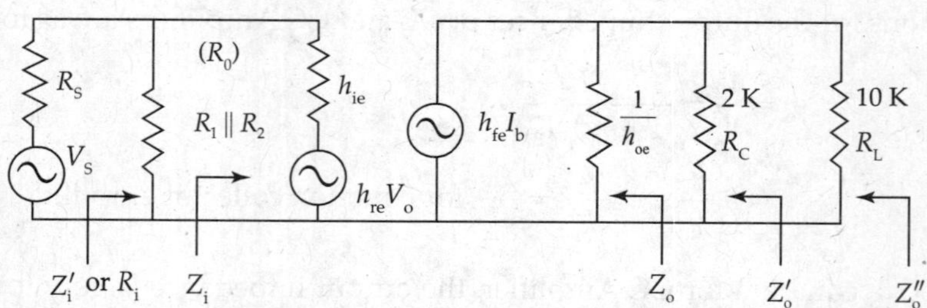

FIG. 6.25 *Small signal low frequency equivalent circuit using h-parameters*

From Fig. 6.25, Since $R_C = 2\ \text{k}\Omega$ and $R_L = 10\ \text{k}\Omega$

The effective load Z_L for Amplifier

$$= R_C \| R_L = \frac{2\text{ K} \times 10\text{ K}}{2\text{ K} + 10\text{ K}} = \frac{20\text{ K}}{12} = 1.67\ \text{k}\Omega$$

$$A_I = \frac{-h_{fe}}{1 + h_{oe} \cdot Z_L} = \frac{-50}{1 + 25 \times 10^{-6} \times 1.67 \times 10^3} = -48$$

$$Z_{in} = h_{ie} + A_I \cdot h_r \cdot Z_L$$

$$= 10^3 - 48 \times 4 \times 10^{-4} \times 1.67 \times 10^3$$

$$= 1000 - 32 = 968\ \Omega = 0.968\ \text{k}\Omega$$

$$A_V = \frac{A_I \cdot Z_L}{Z_{in}} = \frac{-48 \times 1.67 \times 10^3}{0.968 \times 10^3} = -82.8 \approx -83$$

$$Y_o = h_{oe} - \frac{h_{fe} \times h_{re}}{R_S + h_{ie}} = \frac{h_{oe} \cdot R_S + h_{oe} \cdot h_{ie} - h_{fe} \cdot h_{re}}{R_S + h_{ie}}$$

$$\text{if } \Delta h = h_{ie} \cdot h_{oe} - h_{fe} \cdot h_{re}$$

$$Y_o = \frac{h_{oe} \cdot R_S + \Delta h}{R_S + h_{ie}}$$

$$Z_o = \frac{R_S + h_{ie}}{h_{ie} \cdot R_S + \Delta h} = \frac{2 \times 10^3}{3 \times 10^{-2}}$$

$$= 67 \times 10^3 = 67\ \text{k}\Omega$$

$$Z_o'' = \frac{67 \times 10^3 \times 1.67 \times 10^3}{68.67 \times 10^3} = 1.63 \times 10^3$$

$$= 1.63\ \text{k}\Omega \quad (\because Z_o'' = Z_o \| Z_L)$$

$$Z_i' = \frac{Z_i \cdot R_B}{Z_i + R_B} = \frac{0.968 \times 10^3 \times R_B}{0.968 \times 10^3 + R_B} \rightarrow (R_B = R_1 \| R_2)$$

$$R_B = \frac{100 \times 10^3 \times 10 \times 10^3}{100 \times 10^3 + 10 \times 10^3} = \frac{10^9}{110 \times 10^3}$$

$$= \frac{1000 \times 10^3}{110} = 9.1\ \text{k}\Omega$$

$$Z_i' = \frac{0.968 \times 10^3 \times 9.1 \times 10^3}{9.1 \times 10^3 + 0.968 \times 10^3} = \frac{8808.8}{10.068} = 875\ \Omega$$

$$= R_i \rightarrow \left(\because Z_L' = \frac{Z_{in} \cdot R_B}{(Z_{in} \cdot R_B)} \right)$$

$$A_{VS} = \frac{V_o}{V_S} = \frac{V_o}{V_i} \times \frac{V_i}{V_S} = \frac{V_o}{V_i} \times \frac{R_i}{R_S + R_i}$$

$$= A_V \times \frac{R_i}{R_S + R_i} = \frac{-83 \times 0.875 \times 10^3}{1 \times 10^3 + 0.875 \times 10^3} \approx -39$$

A_{VS} refers to voltage gain taking the signal source resistance R_S into consideration

$$A_{IS} = A_I \times \frac{R_S}{R_S + R_i} = -48 \times \frac{10^3}{1.875\ \text{K}} = 25.6 \approx -26$$

A_{IS} refers to current gain taking R_S into consideration.

6.10 COMMON EMITTER TRANSISTOR AMPLIFIER ANALYSIS

Figure 6.26 shows CE Transistor Amplifier circuit. Figure 6.27 shows the h-parameter equivalent circuit of CE Transistor Amplifier.

Equations for the CE Transistor h-parameter equivalent circuit are as follows:

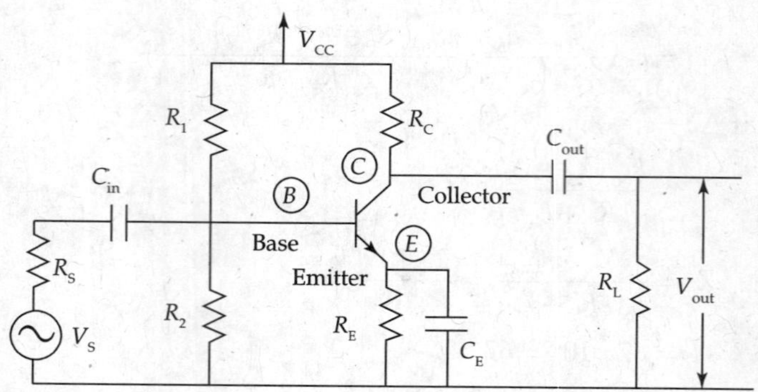

FIG. 6.26 *Common emitter transistor amplifier circuit*

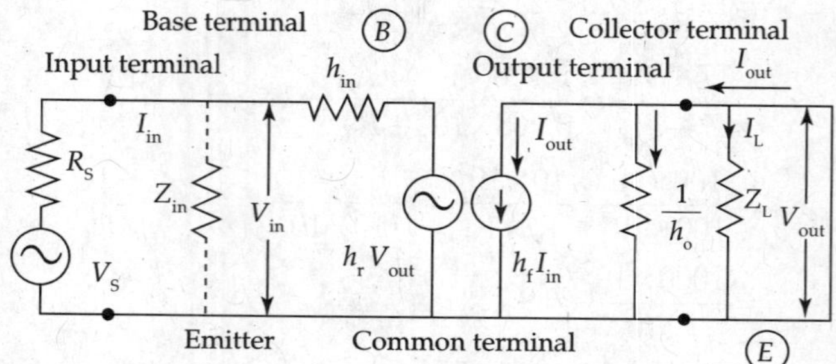

FIG. 6.27 *h-parameter equivalent circuit of common emitter transistor amplifier circuit*

$$V_{be} = h_{ie} \cdot I_b + h_{re} \cdot V_{ce} \tag{6.71}$$

$$I_c = h_{fe} \cdot I_b + h_{oe} \cdot V_{ce}. \tag{6.72}$$

From the generally derived expressions for A_I, Z_{in}, A_V, Z_o and power gain A_P for all the three Transistor Amplifier configurations, equations for CE Transistor Amplifier will be with the second subscript taken as '*e*',

$$\text{Current gain} \quad A_I = \frac{-h_{fe}}{1 + h_{oe} \cdot Z_L} \tag{6.73}$$

$$\text{Input impedance} \quad Z_{in} = h_{ie} + A_{Ice} \cdot h_{re} \cdot Z_L \tag{6.74}$$

$$\text{Voltage gain} \quad A_V = A_{I\,ce} \cdot \frac{Z_L}{Z_{ie}} \tag{6.75}$$

$$\text{Output impedance} \quad Y_o = h_{oe} - \frac{h_{fe} \cdot h_{re}}{R_S + h_{ie}} \tag{6.76}$$

EXAMPLE 6.3

Find the values of A_I, A_V, A_{VS}, A_{IS}, Z_i and Z_0 for the following circuit. Typical values of Transistor *h*-parameters are as follows: $h_{ie} = 1\ \text{k}\Omega$, $h_{0e} = 25 \times 10^{-6}$ mhos, $h_{fe} = +50$, $h_{re} = 4 \times 10^{-4}$.

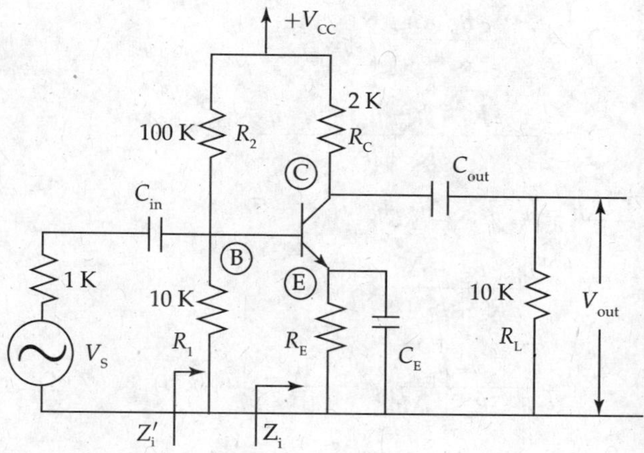

FIG. 6.28 *Common emitter transistor amplifier circuit*

Solution:

$h_{fe} = +50, h_{ie} = 1\ k\Omega, h_{re} = 4 \times 10^{-4}, h_{oe} = 25 \times 10^{-6}$ mhos

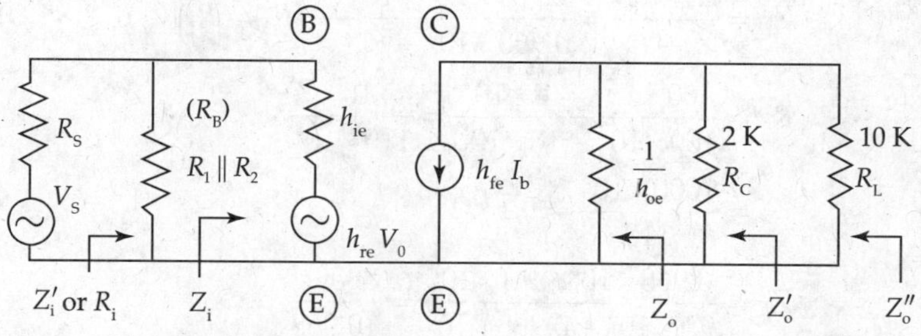

FIG. 6.29 *Small signal low frequency equivalent circuit using h-parameters*

From Fig. 6.28, since $R_C = 2\ k\Omega$ and $R_L = 10\ k\Omega$

The effective load Z_L for Amplifier

$$= R_C \| R_L = \frac{2\ K \times 10\ K}{2\ K + 10\ K} = \frac{20\ K}{12} = 1.67\ k\Omega$$

$$A_I = \frac{-h_{fe}}{1 + h_{oe} \cdot Z_L} = \frac{-50}{1 + 25 \times 10^{-6} \times 1.67 \times 10^3} = -48$$

$$Z_{in} = h_{ie} + A_I \cdot h_r \cdot Z_L = 10^3 - 48 \times 4 \times 10^{-4} \times 1.67 \times 10^3$$

$$= 1000 - 32 = 968\ \Omega = 0.968\ k\Omega$$

$$A_V = \frac{A_I \cdot Z_L}{Z_{in}}$$

$$= \frac{-48 \times 1.67 \times 10^3}{0.968 \times 10^3} = -82.8 \approx -83$$

$$Y_o = h_{oe} - \frac{h_{fe} \times h_{re}}{R_S + h_{ie}}$$

$$Y_o = \frac{h_{oe} \cdot R_S + h_{oe} \cdot h_{ie} - h_{fe} \cdot h_{re}}{R_S + h_{ie}}$$

$$\text{if } \Delta h = h_{ie} \cdot h_{oe} - h_{fe} \cdot h_{re}$$

$$Y_o = \frac{h_{oe} \cdot R_S + \Delta h}{R_S + h_{ie}}$$

$$Z_o = \frac{R_S + h_{ie}}{h_{ie} \cdot R_S + \Delta h} = \frac{2 \times 10^3}{3 \times 10^{-2}} = 67 \times 10^3 = 67\ \text{k}\Omega$$

$$Z_o'' = \frac{67 \times 10^3 \times 1.67 \times 10^3}{68.67 \times 10^3} = 1.63 \times 10^3$$

$$= 1.63\ \text{k}\Omega \quad (\because Z_o'' = Z_o \| Z_L)$$

$$Z_i' = \frac{Z_i \cdot R_B}{Z_i + R_B} = \frac{0.968 \times 10^3 \times R_B}{0.968 \times 10^3 + R_B} \rightarrow (R_B = R_1 \| R_2)$$

$$R_B = \frac{100 \times 10^3 \times 10 \times 10^3}{100 \times 10^3 + 10 \times 10^3} = \frac{10^9}{110 \times 10^3}$$

$$= \frac{1000 \times 10^3}{110} = 9.1\ \text{k}\Omega$$

$$Z_i' = \frac{0.968 \times 10^3 \times 9.1 \times 10^3}{9.1 \times 10^3 + 0.968 \times 10^3} = \frac{8808.8}{10.068}$$

$$= 875\ \Omega = R_i \rightarrow \left(\because Z_L' = \frac{Z_{in} \cdot R_B}{(Z_{in} \cdot R_B)}\right)$$

$$A_{VS} = \frac{V_o}{V_S} = \frac{V_o}{V_i} \times \frac{V_i}{V_S} = \frac{V_o}{V_i} \times \frac{R_i}{R_S + R_i}$$

$$= A_V \times \frac{R_i}{R_S + R_i} = \frac{-83 \times 0.875 \times 10^3}{1 \times 10^3 + 0.875 \times 10^3} \approx -39$$

A_{VS} refers to voltage gain taking the signal source resistance R_S in to consideration

$$A_{IS} = A_I \times \frac{R_S}{R_S + R_i} = -48 \times \frac{10^3}{1.875\ \text{K}} = 25.6 \approx -26$$

A_{IS} refers to current gain taking R_S into consideration.

EXAMPLE 6.4

For a single-stage Transistor Amplifier $R_S = 1\ \text{k}\Omega$ and $R_L = 10\ \text{k}\Omega$. The h-parameter values are $h_{fe} = 50$, $h_{ie} = 1.1\ \text{k}\Omega$, $h_{re} = 2.5 \times 10^{-4}$ and $h_{oe} = 25\ \mu\text{A/V}$. Find A_I, A_V, A_{VS}, A_{IS}, power gain A_P, R_i and R_o for CE Transistor configuration (JNTU, Nov. 2006).

Solution:

$$\text{Current gain } A_I = \frac{-h_{fe}}{1+h_{oe}\cdot Z_L} = \frac{-50}{1+25\times10^{-6}\times10\times10^3}$$

$$= \frac{-50}{1.25} = -40$$

$$\text{Input resistance } Z_{in} = R_{in} = h_{ie} + A_I \cdot h_{re} \cdot Z_L$$

$$= 1.1\times10^3 - 40\times2.5\times10^{-4}\times10\times10^3$$

$$= 1.1\times10^3 - 0.1\times10^3 = 1.0\times10^3\ \Omega$$

$$\text{Voltage gain } A_V = A_I \times \frac{Z_L}{Z_{in}} = -40\times\frac{10\times10^3}{1.0\times10^3} = -400$$

$$\text{Output conductance } Y_o = h_{oe} - \frac{h_{re}\cdot h_{fe}}{R_S + h_{ie}}$$

$$= 25\times10^{-6} - \frac{2.5\times10^{-4}\times50}{1\times10^3 + 1.1\times10^3}$$

$$= 25\times10^{-6} - \frac{125\times10^{-4}}{2.1\times10^3}$$

$$= 25\times10^{-6} - 5.95\times10^{-6}$$

$$= (25-5.95)\times10^{-6}$$

$$\cong 19.05\times10^{-6}\ \text{mhos}$$

$$\therefore \quad \text{Output impedance } Z_o = \frac{1}{Y_o} = \frac{1}{19.05\times10^{-6}}$$

$$= \frac{10^6}{19.05} = \frac{1000\times10^3}{19.05} = 52.5\times10^3\ \Omega$$

Voltage gain A_{VS} taking source resistance 'R_S' into account

$$A_{VS} = A_V \times \frac{R_{in}}{R_{in}+R_S}$$

$$= -400\times\frac{1.0\times10^3}{1.0\times10^3 + 1\times10^3}$$

$$= -400\times\frac{1.0}{2.0} = -200.$$

From the calculations for A_{VS}, we observe that the finite value of source resistance reduces the overall gain of an Amplifier. When the source resistance and the input resistances are equal, the gain of the overall Amplifier is reduced to half from voltage gain A_V.

$$\text{Current gain} \quad A_{IS} = \frac{A_I \cdot R_S}{R_{in}+R_S} = -\frac{40\times1\times10^3}{1\times10^3+1\times10^3} = -20$$

$$\text{Power gain} \quad A_P = A_V \times A_I = -400\times40 = 16000$$

EXAMPLE 6.5

A single-stage CE Transistor Amplifier circuit has $R_S = 1$ kΩ, $R_C = 1$ kΩ, $Z_L = R_L = 4$ kΩ, Transistor $h_{fe} = 50$, $h_{oe} = h_{re} \cong 0$, $h_{ie} = 1$ kΩ. Calculate the values of current gain A_I, input resistance Z_{in}, output resistance Z_{out}, voltage gain A_V and power gain for the Amplifier.

Solution:

(1) Current gain $A_I = -\dfrac{h_{fe}}{[1 + h_{oe} \cdot Z_L]}$

From the given data, $h_{fe} = 50$ and $h_{oe} \cong 0$

$$\therefore \text{ Current gain } A_I = -\frac{50}{1+0} = -50$$

(2) Input impedance $Z_{in} = R_{in} = h_{ie} + A_I \cdot h_{re} \cdot Z_L$

From the given data, $h_{ie} = 1$ kΩ and $h_{re} \cong 0$

$$\therefore \text{ Input resistance } Z_{in} = 1 \times 10^3 + 0 = 1 \times 10^3 = 1\text{ k}\Omega$$

(3) Output resistance $Z_{out} = R_o = R_C \| R_L$

Given data $R_C = 1$ kΩ and $R_L = 4$ kΩ

$\therefore$ Output resistance $R_o = 1\text{ k}\Omega \| 4\text{ k}\Omega$

$$R_o = \frac{1 \times 10^3 \times 4 \times 10^3}{[1+4] \times 10^3} = \frac{4}{5} \times 10^3 = 800\ \Omega$$

(4) Voltage gain $A_V = A_I \times \dfrac{Z_L}{Z_{in}}$

From calculations, $A_I = 50$, $Z_L = 4$ kΩ and $Z_{in} = 1 \times 10^3$

$$A_V = 50 \times \frac{4 \times 10^3}{1 \times 10^3} = 200$$

(5) Power gain $A_P - A_V \times A_I = 200 \times 50 = 10{,}000$.

(6) Voltage gain taking source resistance R_S into account known data $R_{in} = 1$ kΩ and $R_S = 1$ kΩ

$$A_{VS} = A_V \times \frac{R_{in}}{[R_{in} + R_S]}$$

$$= 200 \times \frac{1 \times 10^3}{[1 \times 10^3 + 1 \times 10^3]} = 200 \times \frac{1}{2} = 100$$

6.11 COMMON BASE TRANSISTOR AMPLIFIER ANALYSIS

The CB Transistor Amplifier is also known as the grounded Base Amplifier. It is called the *common-Base* configuration because, for AC signal source and the load, the Base of the Transistor is a common connection point, Emitter is the input terminal and Collector is the output terminal as shown in Fig. 6.30.

The capacitor 'C_2' between the Base and the ground acts as an effective short for AC signals and so the Transistor Base is at effective ground as shown in the CB Amplifier circuit. Input signal 'V_S' is applied between the Emitter (input terminal) and the Base (ground or the common terminal) through the input coupling capacitor C_E (C_{in}) at the Emitter terminal.

Output voltage is taken between the Collector (output terminal) and CB terminal through the output coupling capacitor 'C_C' (C_{out}) at output port of Amplifier. AC equivalent circuit clearly shows the input AC signal connected between the Emitter and the Base, while the amplified output voltage is available across the output port, which are the Collector and the Base terminals.

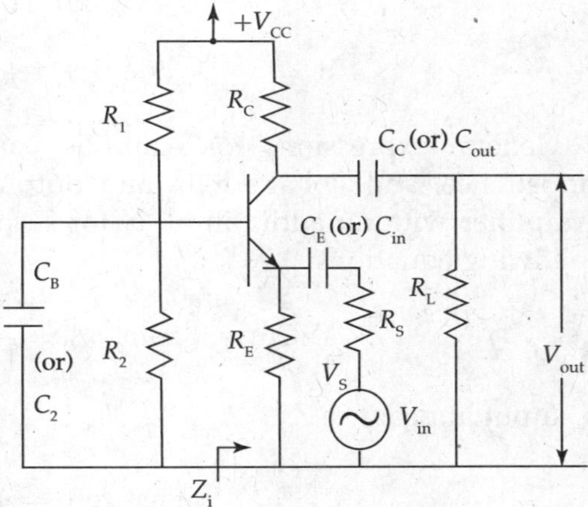

FIG. 6.30 *Common base transistor amplifier*

CB Transistor current gain alpha (α) is the ratio of output current (the Collector current I_C) and the input current (the Emitter current I_E). As the output Collector current is always less than the input Emitter current, the current gain alpha is less than unity. So this Amplifier can produce a voltage gain but no current gain between the input and the output signals.

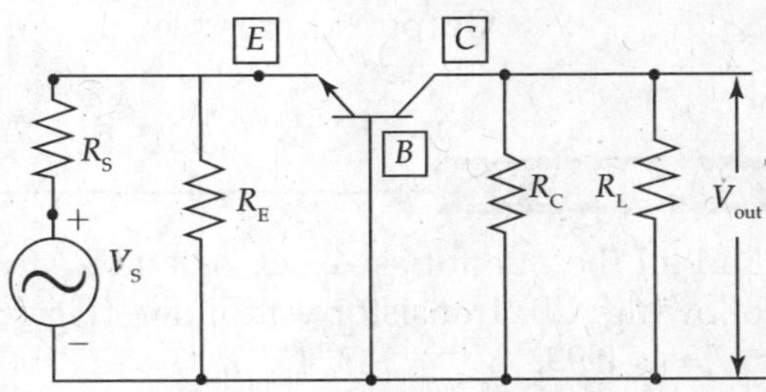

FIG. 6.31 *AC equivalent circuit of common base transistor amplifier*

The CB Amplifier has very small input impedance and output impedance is of the order of the load resistance, which is directly connected across the Collector and the Base. Because the input Emitter current and output Collector currents are approximately equal, the stray input capacitance of the Transistor is not effected or boosted by the 'Miller effect' while, it will be predominant in CE Transistor Amplifier.

The CB Amplifier is often used at high frequencies where it provides more voltage amplification and isolation between the input and output ports. Because of isolation between the input and output ports of the Amplifier, there will be negligible amount of feedback from the output port to the input port. CB Amplifier is highly stable at very high frequency signal amplification. It is also used as current buffer Amplifier, as the current gain 'α' is very close to unity.

h-parameter equivalent circuit of CB Amplifier is shown in Fig. 6.32. *h*-parameter equations for CB Transistor Amplifier.

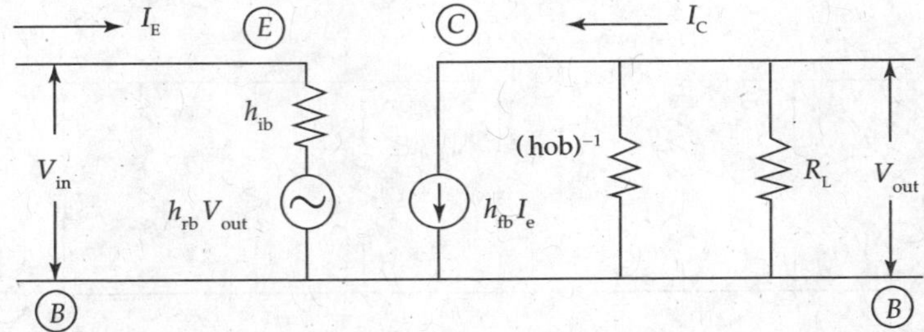

FIG. 6.32 *h-parameter equivalent circuit of common base transistor amplifier*

$$V_{eb} = h_{ib} \cdot I_e + h_{rb} \cdot V_{cb} \tag{6.77}$$

$$I_c = h_{fb} \cdot I_e + h_{ob} \cdot V_{cb} \tag{6.78}$$

General expressions for Amplifier performance parameters for current gain, input impedance, and voltage gain and output impedance can be adopted for CB Transistor Amplifier with an addition of 'b' for second subscript in the expressions as shown in the following equations:

$$\text{Current gain} \quad A_I = \frac{-h_{fb}}{1 + h_{ob} \cdot Z_L} \tag{6.79}$$

Input impedance $Z_{in} = h_{ib} + A_{I(cb)}\, h_{rb}\, Z_L$.

$$\text{Voltage gain} = A_{I\,cb} \cdot \frac{Z_L}{Z_{ib}} \tag{6.80}$$

$$\text{Output admittance} \quad Y_o = h_{ob} - \frac{h_{fb} \cdot h_{rb}}{R_S + h_{ib}}. \tag{6.81}$$

EXAMPLE 6.6

Find all the quantities A_I, Z_i, A_V and Z_o for the following CB Transistor Amplifier (Fig. 6.33): $h_{fb} = -0.99$, $h_{ib} = 20\ \Omega$, $h_{rb} = 2 \times 10^{-4}$, $h_{ob} = 0.5 \times 10^{-6}$ Siemens.

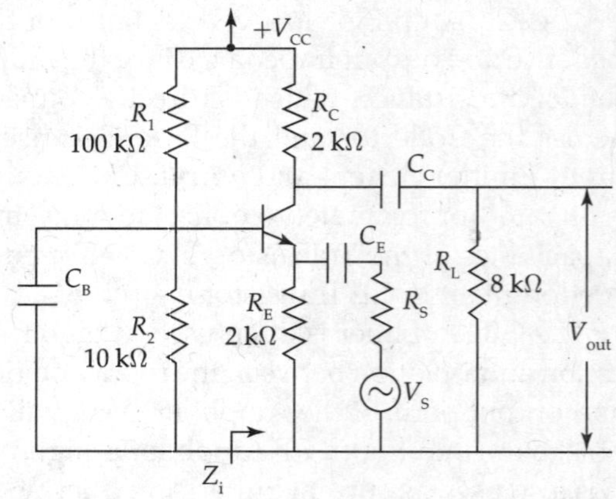

FIG. 6.33 *Common base transistor amplifier*

Solution: Current gain of CB Amplifier A_{Icb},

$$A_{Icb} = \frac{h_{fb}}{1 + h_{ob} \times Z_L}\ ; h_{ob} = 0.5 \times 10^{-6}$$

$$Z_L = Z_L' = \frac{R_L \cdot R_C}{R_L + R_C} = \frac{2 \times 10^3 \times 8 \times 10^3}{2 \times 10^3 + 8 \times 10^3}$$

$$= \frac{16 \times 10^3}{10} = 1.6 \times 10^3 = 1.6\ \text{k}\Omega$$

$$h_{ob} \cdot Z_L' = 0.5 \times 10^{-6} \times 1.6 \times 10^3 = 8 \times 10^{-4}$$

$$1 + h_{ob} \cdot Z_L' = 1 + 8 \times 10^{-4} = 1.0008$$

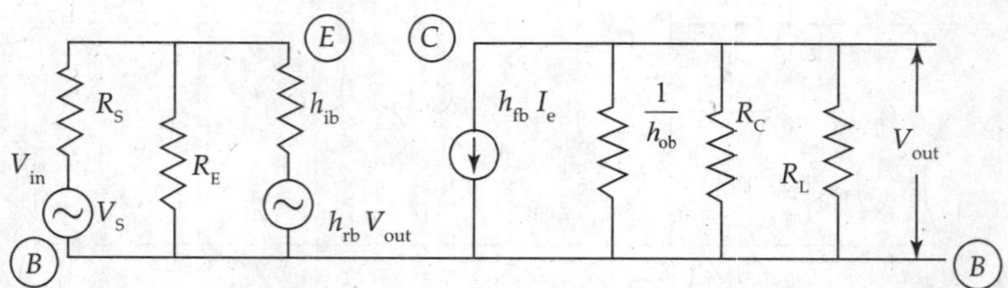

FIG. 6.34 *AC equivalent circuit of common base transistor amplifier using h-parameters*

$$\text{Current gain } A_I = \frac{0.99}{1.0008} = 0.989$$

$$Z_{in} = h_{ib} + A_I \cdot h_r \cdot Z_L$$

$$= 20 + 0.989 \times 2 \times 10^{-4} \times 1.6 \times 10^3$$

$$= 20.31\ \Omega \equiv 20\ \Omega$$

$$A_V = A_I \cdot \frac{Z_L}{Z_{in}} = \frac{0.989 \times 1.6 \times 10^3}{20.31} = 779$$

$$Y_o = h_{ob} + \frac{h_{fb} \times h_{rb}}{R_S + h_{ib}} = 0.5 \times 10^{-6} + \frac{0.989 \times 2 \times 10^{-4}}{1 \times 10^3 + 0.02 \times 10^3}$$

$$= \frac{1.978 \times 10^{-4}}{1.02 \times 10^3} = 1.940 \times 10^{-7}$$

$$Z_o = \frac{1}{Y_o} = 1.44 \times 10^6 = 1.44\ \text{M}\Omega$$

From the above calculations for the CB Transistor Amplifier, the Amplifier current gain is less than 1. The input impedance Z_{in} is only a few ohms and the output impedance Z_{out} is very high and is of the order of mega ohms. The reasons for such behaviour of CB Amplifiers are well discussed while discussing the CB Transistor characteristics. CB Transistor Amplifier is a Non-Inverting Amplifier.

We have to remember a few facts about the Transistor device and circuit analysis.

- The Transistor parameters are different among units of the same type.
- The Transistor parameters given in the manufacturer's data sheets will not be equal to those that are obtained by measurement. As an example, beta value given in the data sheet for a particular Transistor will be different from measured value. This also changes for different Transistors from the same make.
- The above two reasons suggest us that the simplifying assumptions made in the analysis of the circuits do not have much effect on the design accuracy.
- The simplified circuit models for the Transistors and the Amplifiers provide us with good understanding of the circuit functions.

EXAMPLE 6.7

For a single-stage Transistor Amplifier, $R_S = 2$ kΩ and $R_L = 5$ kΩ. The *h*-parameter values are $h_{fb} = 0.98$, $h_{ib} = 21\ \Omega$, $h_{rb} = 2.9 \times 10^{-4}$ and $h_{ob} = 0.49$ μA/V. Find A_I, A_V, A_{VS}, R_i and R_o for CB Transistor configuration (JNTU, Nov. 2006).

Solution:

$$\text{Current gain } A_I = -\frac{h_{fb}}{1 + h_{ob} \cdot Z_L}$$

$$= \frac{0.98}{1 + 0.49 \times 10^{-6} \times 5 \times 10^3} = 0.98$$

$$\text{Input impedance} \quad Z_{in} = h_{ib} + A_I \cdot h_{rb} \cdot Z_L$$

$$= 21 + 0.98 \times 2.9 \times 10^{-4} \times 5 \times 10^3$$

$$= 21 + 1.421 = 22.421\,\Omega$$

$$\text{Voltage gain} \quad A_V = A_I \cdot \frac{Z_L}{Z_{in}} = 0.98 \times \frac{5 \times 10^3}{22.421}$$

$$= \frac{4.9 \times 10^3}{22.421} = 218.45$$

$$\text{Output admittance} = Y_o = h_{ob} - \frac{h_{fb} \cdot h_{rb}}{R_S + h_{ib}}$$

$$= 0.49 \times 10^{-6} + \frac{0.98 \times 2.9 \times 10^{-4}}{(1 \times 10^3 + 21)}$$

$$\therefore \; Y_o = 0.49 \times 10^{-6} + \frac{98 \times 10^{-6} \times 2.9}{1021} = 10^{-6}\left[0.49 + \frac{98 \times 2.9}{1021}\right]$$

$$= 10^{-6}\left[0.49 + 0.278\right] = 0.768 \times 10^{-6}$$

$$\therefore \; \text{Output impedance} \quad Z_o = \frac{1}{Y_o} = \frac{1}{0.768 \times 10^{-6}} = \frac{1000 \times 10^6}{0.768} = 1.3 \times 10^6\ \Omega.$$

$$A_{VS} = \frac{A_V \times Z_{in}}{R_S + Z_{in}} = \frac{218.45 \times 22.421}{1000 + 22.421} = \frac{4897.87}{1022.421} = 4.79$$

6.12 COMMON COLLECTOR TRANSISTOR AMPLIFIER ANALYSIS

Common Collector Transistor Amplifier configuration is shown in Fig. 6.35. The input voltage is applied between the Base and the Collector terminals of the Transistor. Output voltage is available between the Emitter and the Collector terminals across the load resistance 'R_E or R_L'. Collector terminal is common to both the input and output voltages. So this Amplifier configuration is known as CC Amplifier.

Output current which is the Emitter current in CC Amplifier flows from Emitter terminal to the Collector lead through the load resistance 'R_E' as shown in Fig. 6.35. So the output voltage at the Emitter is positive going, while the input voltage is positive going and the output voltage is negative going, while the input voltage is negative going. Thus, the output voltage will be in phase with the input voltage or follows the input voltage. So the CC Amplifier is also known as Emitter follower. Emitter follower circuit is a *non-inverting* Amplifier.

The CC Amplifier circuit is also known as the Emitter follower or voltage follower, because the input and load voltages follow each other so closely. The output voltage is nearly *identical* to the input voltage, lagging behind only about 0.7 V (forward bias voltage of the conducting

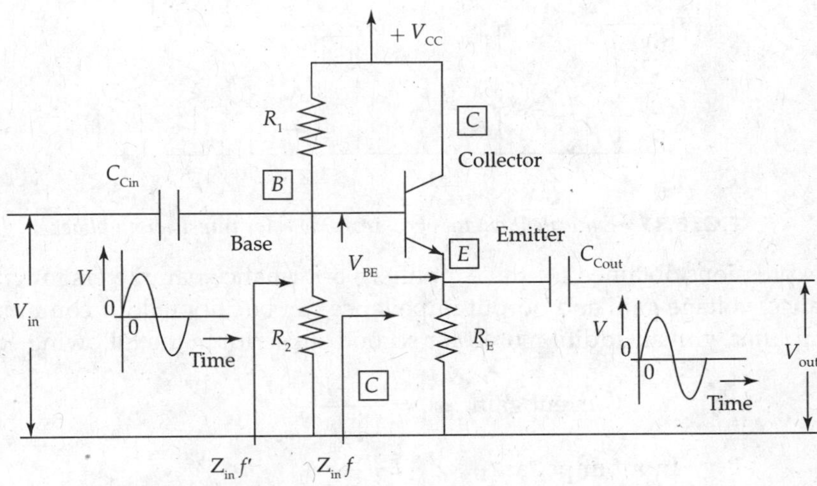

FIG. 6.35 *Common collector transistor amplifier (emitter follower)*

input junction of the Transistor). It is used as a buffer Amplifier, because of the unity voltage gain (0 dB) and the high input impedance associated with low output impedance of the Amplifier, which will be clear with the following worked out examples. Hence, CC Amplifier is used as an impedance matching device between low impedance loads and signal sources. It is also used in digital gate circuit implementations.

Output current is the Emitter current $I_E = I_C + I_B$.

Input current is the Base current I_B.

$$\text{Current gain } \; A_I = \frac{I_E}{I_B} = \frac{I_C + I_B}{I_B} = (\beta + 1).$$

The current gain of a CC Amplifier is equal to $(\beta + 1)$. The voltage gain is approximately equal to 1, as the output is fed back to the input port and the effective input signal decreases (due to negative feedback), which can be seen from the equivalent circuit of Fig. 6.36.

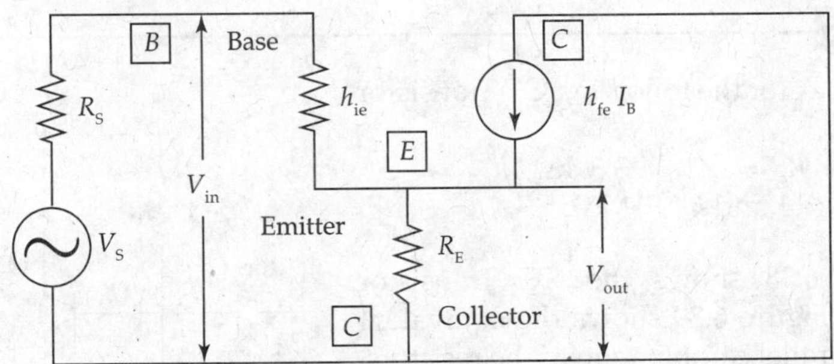

FIG. 6.36 *h-parameter model AC equivalent circuit for common collector amplifier (emitter follower)*

h-parameter equivalent circuit of CC Amplifier

Equations for h-parameter equivalent circuit for CC Transistor Amplifier:

$$V_{bc} = h_{ic} \cdot I_b + h_{rc} \cdot V_{ec} \tag{6.82}$$

$$I_e = h_{fc} \cdot I_b + h_{oc} \cdot V_{ec} \tag{6.83}$$

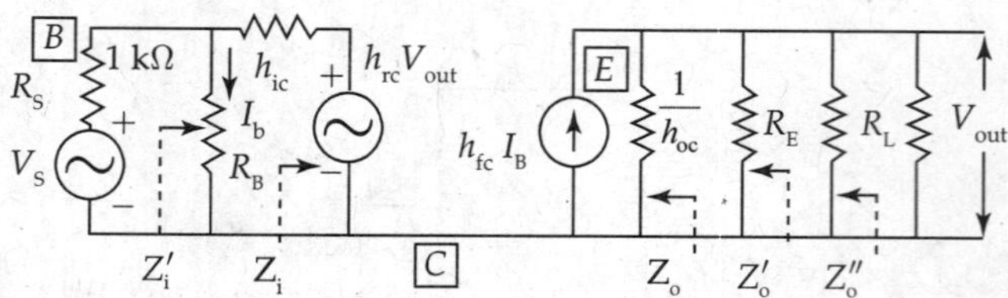

FIG. 6.37 *Equivalent circuit of common collector transistor amplifier*

The general expressions obtained for the Amplifier performance parameters for current gain, input impedance, voltage gain and output impedance can be adopted for common Collector Transistor Amplifier with an addition of 'c' for second subscript in the following equations:

$$\text{Current gain} \quad A_I = \frac{-h_{fc}}{1 + h_{oc} \cdot Z_L} \tag{6.84}$$

$$\text{Input impedance} \quad Z_{in} = h_{ic} + A_{I\,(cc)} \cdot h_{rc} \cdot Z_L \tag{6.85}$$

$$\text{Voltage gain} = A_{I\,cc} \cdot \frac{Z_L}{Z_{ic}} \tag{6.86}$$

$$\text{Output admittance} \quad Y_o = h_{oc} - \frac{h_{fc} \cdot h_{rc}}{R_S + h_{ic}} \tag{6.87}$$

One of the applications of CC Amplifiers is *Darlington pair*. A pair of Emitter follower circuits is connected as Darlington pair. The Emitter of one Transistor feeds current to the Base of the second Transistor in CC Amplifier configuration. Such a combination of special Transistors has an overall current gain equal to the product (multiplication) of their individual CC current gains ($\beta + 1$) or approximately equal to the product of the betas of the two Transistors. So the Darlington pair has current gain equal to 'β^2', if the current gain 'β' of the two Transistors is equal. Otherwise, the current gain of the Darlington pair becomes '$\beta_1 \times \beta_2$'. So Darlington pair is known as 'super beta Transistor'

EXAMPLE 6.8

Find A_I, Z_i, A_V and Z_0 for the following CC Transistor Amplifier circuit.

Data: $R_S = 1\ \text{k}\Omega$, $R_E = Z_L = 3.3\ \text{k}\Omega$, $h_{fc} = -51$, $h_{oc} = 25 \times 10^{-6}$, $h_{rc} = 1$ and $h_{ic} = 1\ \text{k}\Omega$.

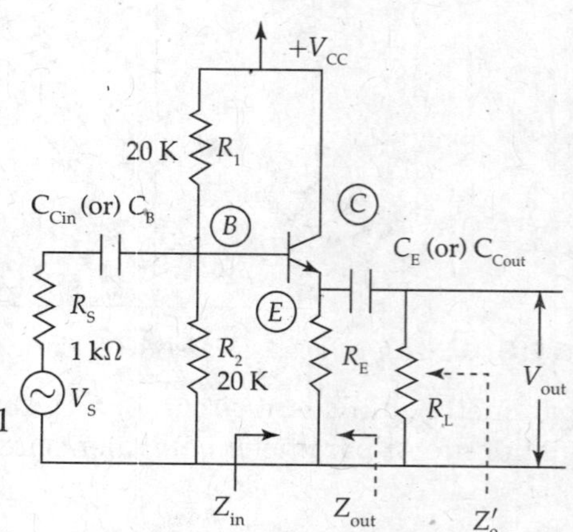

FIG. 6.38 *Common collector transistor amplifier*

Solution: Figure 6.38 shows the CC Transistor Amplifier circuit. Figure 6.39 shows the equivalent circuit. Z_L is the parallel combination of the resistors. 'R_E' and 'R_L' are equal to 3.3 kΩ.

$$A_I = -\frac{h_{fc}}{1 + h_{oc} \cdot Z_L} = -\frac{-51}{1 + 25 \times 10^{-6} \times 3.3 \times 10^3} = 47.11$$

$$Z_{in} = h_{ic} + A_I \cdot h_{rc} \cdot Z_L$$

$$= 10^3 + 47 \times 1 \times 3.3 \times 10^3 = 157.5\ \text{k}\Omega$$

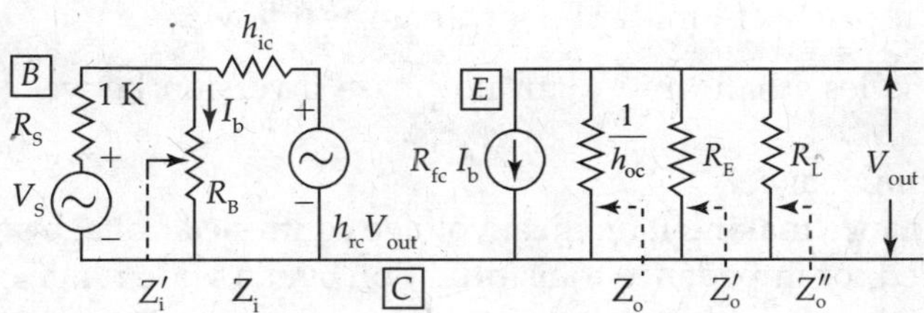

FIG. 6.39 *h-parameter equivalent circuit of common collector transistor amplifier*

$$Y_o = h_{oc} - \frac{h_{fc} \cdot h_{rc}}{R_S + h_{ic}}$$

$$\therefore \quad Y_o = 25\times10^{-6} - \left(\frac{51\times1}{10^3+10^3}\right) = 25\times10^{-6} + 25.5\times10^{-3} \cong 25.5\times10^{-3}$$

$$\therefore \; Z_{out} = \frac{1}{Y_o} = \frac{1}{25.5\times10^{-3}} = 39.2\,\Omega \text{ (low output impedance)}$$

$$Z_o' = \frac{Z_{out}\times R_L}{Z_{out}+R_L} = \frac{39.2\times3.3\times10^3}{39.2+(3.3\times10^3)} = 39\,\Omega$$

$$R_B = R_1 \parallel R_2 = 10\times10^3$$

$$Z_{in}' = [R_1 \parallel R_2] \| Z_{in} = 10\times10^3 \parallel 157\times10^3$$

$$\therefore \quad Z_{in}' = \frac{10\times10^3\times157\times10^3}{[(10\times10^3)+(157\times10^3)]}$$

$$= \frac{1570\times10^3}{167} = 9.4\times10^3 = 9.4\,\text{k}\Omega$$

Voltage gain $A_V = A_I \times \frac{Z_L}{Z_{in}} = \frac{47.11\times3.3\times10^3}{156.4\times10^3} = 0.99$

6.13 EMITTER FOLLOWER TRANSISTOR AMPLIFIER ANALYSIS

Emitter Follower is a CC Amplifier, since the Collector is at AC ground as can be seen from the circuit shown in Fig. 6.40. Output voltage across resistor 'R_E' is almost equal to or slightly less than the input Base-ground voltage (Base-Collector voltage). Emitter voltage follows the changes in input signal voltage. So the circuit is known as Emitter Follower. Feedback factor 'β' is unity as the voltage across resistor 'R_E' is entirely feedback to input port of the Amplifier.

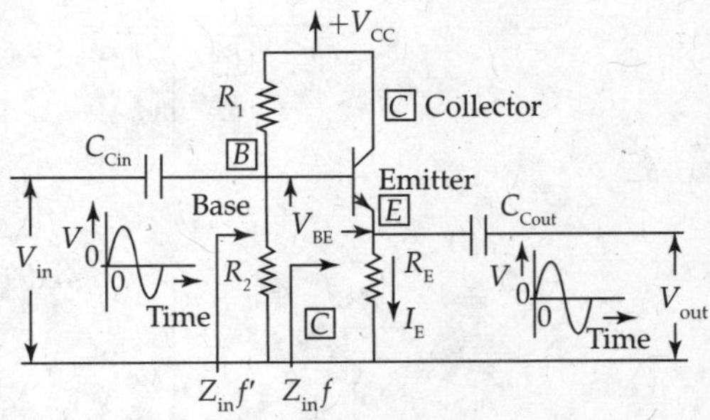

FIG. 6.40 *Emitter follower (common Collector Transistor Amplifier)*

The characteristic features of Emitter Follower are as follows:.

- The voltage gain is less than unity with no phase inversion between the input and the output signals.
- It has high input impedance.
- Low output impedance making it an ideal voltage controlled voltage source.
- It is commonly used for impedance transformation over a wide range of frequencies.
- The circuit has relatively high current gain and power gain, but voltage gain close to unity.

h-parameter model AC equivalent circuit of Emitter Follower

h-parameter model AC equivalent circuit of Emitter follower is drawn with an assumption that $h_{oe} = h_{oc} \cong 0$ so that $(1/h_{oc}) = \infty$, i.e., an open circuit and so the circuit component $1/h_{oc}$ parallel to the output current source is omitted in the equivalent circuit of Fig. 6.41. Further in the following analysis, the effect of R_1 and R_2 is not considered. The effect of R_1 and R_2 is to reduce the input impedance $Z_{in}(f)$ to $Z'_{in}(f)$.

From the CC Transistor Amplifier circuit in Fig. 6.40, because of the effects of R_1 and R_2 input impedance $Z'_{in}(f) = Z_{in}(f) \,||\, (R_1 \,||\, R_2)$

$$-V_{in} + h_{ie} \cdot I_B + I_E \cdot R_E = 0$$

$$\therefore \quad V_{in} = h_{ie} \cdot I_B + (1 + h_{fe}) \cdot I_B \cdot R_E \quad \text{using } [I_E = (1 + h_{fe}) \cdot I_B]$$

$$\therefore \quad Z_{in} = \frac{V_{in}}{I_{in}} = \frac{V_{in}}{I_B} = \frac{[h_{ie} + (1 + h_{fe}) R_E] I_B}{I_B}$$

$$= [h_{ie} + (1 + h_{fe}) R_E]$$

The input impedance Z_{in} has been enhanced by an amount $[(1 + h_{fe}) R_E]$

$$V_{in} = h_{ie} \cdot I_B + (1 + h_{fe}) \cdot I_B \cdot R_E$$

$$\text{But } V_{out} = I_E \cdot R_E = (1 + h_{fe}) \cdot I_B \cdot R_E$$

$$\therefore \quad \text{Voltage gain} \quad A_V = \frac{V_{out}}{V_{in}}$$

$$= \frac{[(1 + h_{fe}) \cdot R_E] \cdot I_B}{[h_{ie} + (1 + h_{fe}) \cdot R_E] \cdot I_B}$$

$$= \frac{[(1 + h_{fe}) \cdot R_E]}{[h_{ie} + (1 + h_{fe}) \cdot R_E]} < 1$$

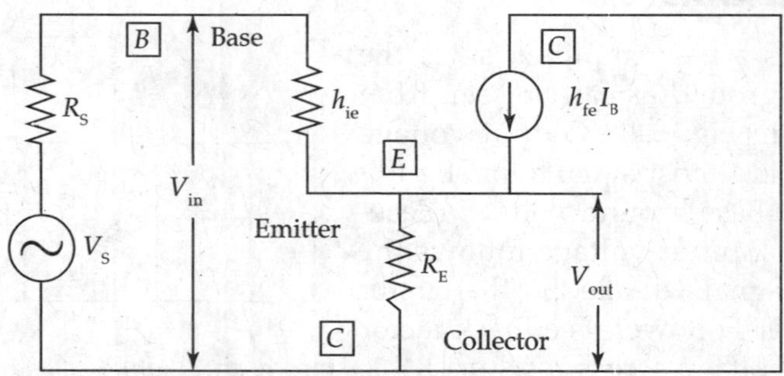

FIG. 6.41 *h-parameter model AC equivalent circuit for emitter follower (common collector amplifier)*

From the equation for $Z_{in} = [h_{ie} + (1 + h_{fe}) \times R_E]$, we understand that the input impedance of the CC Transistor Amplifier is increased to a very large value.

Further voltage gain A_V is less than unity because of the inherent feedback of the output voltage between the Emitter and Collector back to the input port. Output impedance Z_o across the output terminals is by definition $Z_o = \frac{V}{I}$, where V is the open circuit voltage across the output terminals and 'I' is the short circuit current.

When the Emitter is open, the open circuit voltage $V_o = V = V_{in} = I_B \cdot h_{ie}$.

The Short circuit current $I_{SC} = I = I_E$ and $I_{SC} = I_E = (1 + h_{fe}) \cdot I_B$.

$$\therefore \quad Z_o = \frac{V}{I} = \frac{h_{ie} \cdot I_B}{(1+h_{fe}) \cdot I_B} = \frac{h_{ie}}{(1+h_{fe})} \cong \frac{h_{ie}}{h_{fe}} = \frac{1}{g_m}$$

Then the output impedance Z_0 is $1/g_m$, a low value when compared to the input impedance Z_{in}, where $Z_{in} = [h_{ie} + (1+h_{fe}) \cdot R_E]$ that is relatively very large. One of the main applications of Emitter follower circuit is in voltage regulator circuits in power supply circuits.

EXAMPLE 6.9

Emitter Follower circuit:

For the following Emitter follower circuit shown in Fig. 6.42, $R_1 = R_2 = 20$ kΩ, $h_{ie} = 1$ kΩ, $h_{fe} = 100$, $R_E = 5$ kΩ, neglecting the effect of h_{0e}. Determine the input impedance Z_{in}, Z_{in}', output impedance Z_o and Z_o' and voltage gain A_V.

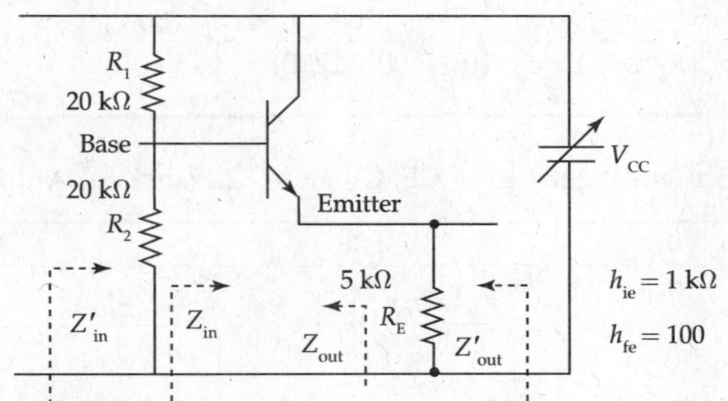

FIG. 6.42 *Common collector transistor amplifier (emitter follower circuit)*

Solution:

$$Z_{in} = h_{ie} + (1+h_{fe}) \cdot R_E = 1\,\text{k}\Omega + (1+100) \times 5 \times 10^3 = 506\,\text{k}\Omega$$

$$Z_{in}' = Z_{in} \,\|\, (R_1 \,\|\, R_2) = (506 \times 10^3) \,\|\, (10 \times 10^3) \cong 10 \times 10^3 = 10\,\text{k}\Omega$$

$$A_V = \frac{(1+h_{fe}) \cdot R_E}{h_{ie} + (1+h_{fe}) \cdot R_E} = \frac{(1+100) \times 5 \times 10^3}{1\,\text{K} + (1+100) \times 5 \times 10^3} = \frac{505 \times 10^3}{506 \times 10^3} \cong 1$$

$$Z_o = \frac{1}{g_m} = \frac{h_{ie}}{h_{fe}} = \frac{1 \times 10^3}{100} = 10\,\Omega = 10\,\Omega$$

$$Z_o' = Z_o \,\|\, R_E = 10\,\Omega \,\|\, 5 \times 10^3 \cong 10\,\Omega$$

Example 6.9 shows that the input impedance is very large, whereas the output impedance is very small for the CC Amplifier. This feature suggests that Emitter follower circuit can be used as buffer Amplifier for impedance matching between high impedance sources and low impedance loads for optimum power transfer. In such application it is known as buffer Amplifier and also known as voltage follower.

EXAMPLE 6.10

Calculate Current gain, Voltage gain, R_{IN} and R_{OUT} using the data below:
Data: $h_{fe} = 36$, $R_L = 2.5 \times 10^3$ and $h_{oe} = 2 \times 10^{-6}\ \Omega$; $h_{ie} = 1200\ \Omega$, $h_{re} = 0$ and $R_S = 500\ \Omega$.

Solution:

$$A_I = -\frac{h_{fe}}{[1+h_{oe}\cdot Z_L]} = \frac{36}{1+2\times10^{-6}\times2.5\times10^3} = \frac{36}{1+5\times10^{-3}} \equiv 36$$

Determination of Current gain, A_I,

$$A_V = \frac{A_I\cdot Z_L}{Z_{in}} = \frac{36\times2.5\times10^3}{1.2\times10^3} = \frac{90}{1.2} = 75$$

Determination of Voltage gain, A_V,
Determination of Input Resistance, R_{in}
$R_{in} = Z_{in} = h_{ie} + A_I \times h_{re}\, Z_L$ is approximately $= h_{ie}$ as $h_{re} = 0$; Therefore, $R_{in} = h_{ie} = 1200\ \Omega$.
Determination of output Resistance, R_0,

$$R_o = \frac{R_S + h_{ie}}{h_{oe}(R_S + h_{ie})} = \frac{500+1200}{2\times10^{-6}(500+1200)} = \frac{1700}{3.4\times10^{-3}} = 500\times10^3$$

Comparison of performance features of CE, CB and CC Transistor Amplifiers

Parameter	CE	CB	CC
A_I	$\dfrac{-h_{fe}}{1+h_{oe}\cdot Z_L}$	$\dfrac{-h_{fb}}{1+h_{ob}\cdot Z_L}$	$\dfrac{-h_{fc}}{1+h_{oc}\cdot Z_L}$
Z_i	$h_{ie} + A_{Ice}\, h_{re}\, Z_L$	$h_{ie} + A_{Icb} + h_{rbZL}$	$h_{ie} + A_{Icc}\cdot h_{rc} Z_L$
A_V	$A_{Ice}\cdot\dfrac{Z_L}{Z_{ie}}$	$A_{Icb}\cdot\dfrac{Z_L}{Z_{ib}}$	$A_{I\,cc}\cdot\dfrac{Z_L}{Z_{ic}}$
Y_o	$h_{oe} - \dfrac{h_{fe}\cdot h_{re}}{R_s + h_{ie}}$	$h_{ob} - \dfrac{h_{fb}\cdot h_{rb}}{R_s + h_{ib}}$	$h_{oc} - \dfrac{h_{fc}\cdot h_{rc}}{R_s + h_{ic}}$
Merits	Inverting Amplifier with reasonable voltage and current gain	Non-inverting Amplifier with high Z_0	Non-inverting Amplifier with large A_I. Excellent as an impedance transformer
Demerits	Non-ideal Amplifier	Low Z_i loads previous stage Low A_I.	$A_V < 1$.

Comparison of CE, CB, CC configurations of Transistor

CE Transistor is an inverting voltage Amplifier, whereas, CB and CC configurations form non-inverting Amplifiers.

CE Transistor configuration:

- Input impedance h_{ie} of the order of 1 kΩ.
- Output impedance $1/h_{oe}$ of the order of 40 kΩ.
- Neither a true current Amplifier nor a voltage Amplifier, but a bit of both.

CB Transistor configuration:

- Input impedance h_{ib} of the order of a few ohms 10–20 Ω.
- Output impedance $1/h_{ob}$ of the order of 2 MΩ.
- Reasonable voltage gain.
- Current gain h_{fb} is less than unity.
- Almost ideal current-controlled current generator, since its input impedance is low and can be connected to a current source. Since its output impedance is high it can act as a current source i.e., in effect a current-controlled current source.

CC Transistor configuration:

- Current gain of $h_{fc} = (1 + h_{fe})$.
- Voltage gain is less than unity (Due to voltage series negative feedback).
- Due to unity feedback factor has very high input impedance and very low output impedance and can act as a voltage controlled voltage source.

Although the active devices BJTs and FETs have different Bases of physical operation, once their circuit models replace these devices, their frequency response and other features can be analysed simultaneously.

6.14 FREQUENCY RESPONSE OF RC-COUPLED CE TRANSISTOR AMPLIFIER

Determination of Frequency response and Amplifier bandwidth:

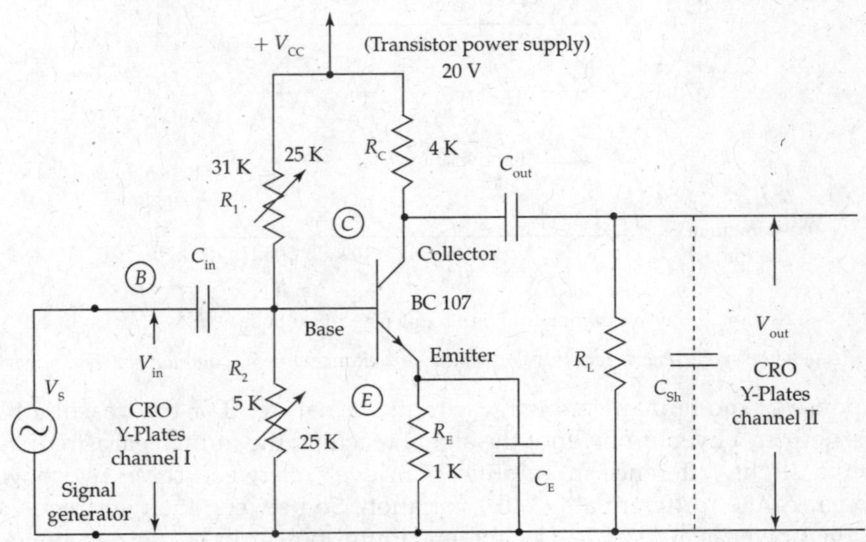

FIG. 6.43 *CE transistor amplifier circuit to obtain frequency response*

To determine the frequency response of an Amplifier,

- A function generator (signal generator) is connected to the input terminals of the Amplifier. The input signal is a sine wave applied from a signal generator.
- Sine wave signal is also connected to channel I of a CRO to measure the frequency and amplitude V_{in} of the input signal from the observed waveform on the screen of the CRO.
- The output signal of the Amplifier at the output port is connected to the second channel (channel II) of the CRO to measure the frequency and amplitude V_{out} from the observed waveform of output signal on the screen of the CRO.
- The voltage gain A_V of the Amplifier is $\frac{V_{out}}{V_{in}}$.
- The magnitude of V_{in} from the signal generator is maintained constant say 200 mV.
- By changing the frequency of the input signal in convenient steps, the output voltage V_{out} is measured.

The observations are tabulated as follows:

S. No.	Frequency	V_{in} (mV)	V_{out} (V)	Voltage gain, A_V	Voltage gain (dB)

Frequency response of an Amplifier

Frequency response characteristic is plotted on a Semi-log graph sheet with x-axis representing logarithm of frequency and y-axis representing voltage gains A_V for different frequencies of input signal from the observations already recorded in the table (Fig. 6.44).

It will be observed from the frequency response plot of the Amplifier that starting from DC frequency, the gain increases with frequency (Low-frequency region), enters the knee region, remains almost flat over a range of frequencies (Midrange frequencies) and starts falling off with frequency at High-frequency region.

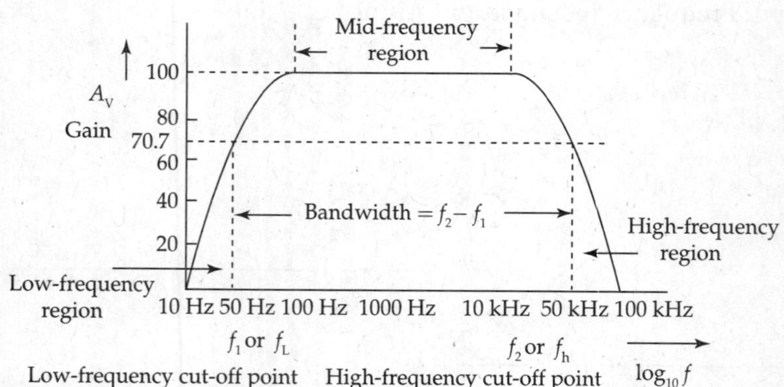

FIG. 6.44 *Frequency response characteristic of resistance capacitance coupled common emitter transistor amplifier*

The response of the human ears is logarithmic in nature. (For this reason only decibel (dB) gain is defined.) Even if power of the signal reaching the human ear is reduced to half, it feels only a slight difference in amplification levels. Between these two power levels, the human ear cannot differentiate (3-dB) variation. So between the two frequencies f_1 (f_L) and f_2 (f_h), the power is at least half of the maximum power in the mid-frequency region. Frequency f_1 is known as the lower cut-off frequency and f_2 is known as the upper cut-off frequency. The range of frequencies between them ($f_2 - f_1$) or ($f_h - f_L$) is called as bandwidth of the Amplifier. Between these two cut-off frequencies or corner frequencies, the response of the Amplifier is considered to be uniform or constant.

In voltage relationships, the 3-dB point refers to $1/\sqrt{2}$ of A_V mid (0.707 A_{Vmid}), where A_{Vmid} is the maximum gain in the flat region of the response characteristic. This flat response region is known as mid-frequency region of the Amplifier response.

From a typical Amplifier frequency response characteristic of the Amplifier of Fig. 6.43 maximum gain A_m in the mid-frequency region is 100. According to the definition for Amplifier bandwidth, low-frequency cut-off point f_1 is 50 Hz where the Amplifier gain is $0.707\ A_m = 0.707 \times 100 = 70.7$. Extending this same line to high-frequency end of the frequency response, high-frequency cut-off point $f_2 = 50$ kHz.

Amplifier bandwidth $B = (f_2 - f_1)$

Bandwidth = $[50 - 0.05] \times 10^3 = 49.95$ kHz.

Frequency response characteristic with gain in dB

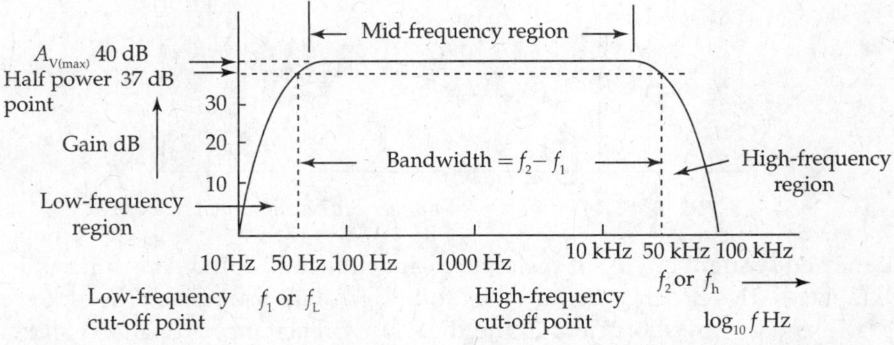

FIG. 6.45 *Frequency response characteristic of RCC amplifier voltage gain in dB (decibels)*

Reasons for fall of gain at low frequency (The combined effect of C_{in}, C_E and C_{out}):
As the frequency increases, reactance of these capacitances decreases and in the flat region and beyond, they are virtual short circuits. Up to the flat frequency region, the equivalent of a CE Amplifier acts as a high pass filter with f_1 as cut-off frequency, since high pass filter has stop band region up to f_1, beyond which it has the passband.

Reasons for fall of response at high frequencies:
R_L is the load resistance in the circuit of the appliance connected to the Amplifier, which has to receive only the AC output from the Amplifier. Every appliance will have two terminals with a potential difference across them and so possesses a capacitance. This is represented as a capacitance C_{Sh}, across the resistance R_L. R_L together with the capacitance across it could be the input impedance shunted by its input capacitance of a two-port network connected across the output terminals of the Amplifier. This represents a low pass filter with a cut-off frequency of f_h or f_2 and up to f_2 it is the passband and beyond which it has stop band, i.e., up to f_2, all higher frequencies.

EXAMPLE 6.11

Low-frequency cut-off point $f_1 = 500$ Hz, and high-frequency cut-off point $f_2 = 20.5$ kHz is identified on the frequency response of an Amplifier. Calculate the Amplifier bandwidth.

Solution: Amplifier bandwidth $B = (f_2 - f_1)$
High-frequency cut-off point $f_2 = 20.5$ kHz
Low-frequency cut-off point $f_1 = 500$ Hz $= 0.5$ kHz
Bandwidth $= [20.5 - 0.5] \times 10^3 = 20$ kHz

6.15 RESISTANCE CAPACITANCE COUPLED TRANSISTOR AMPLIFIER

Equivalent circuits for mid-, low- and high-frequency regions:
The exact equivalent circuit for a simplified CE Transistor Amplifier model resistance capacitance coupled Amplifier shown in Fig. 6.43 is shown in Fig. 6.46.

In the equivalent circuit shown in Fig. 6.46, it is assumed that $h_{re} = h_{oe} = 0$.

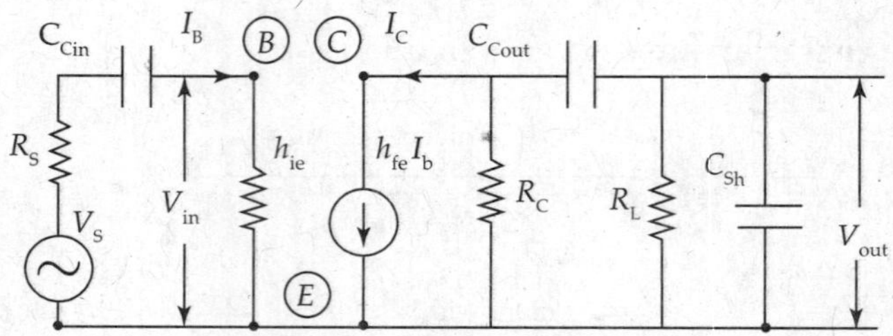

FIG. 6.46 *CE transistor amplifier equivalent circuit*

Mid-frequency equivalent circuit of resistance capacitance coupled Amplifier (Fig. 6.47):
At mid-frequencies, the circuit is redrawn as follows in Fig. 6.47. Since the series coupling capacitances C_C and C_{out} are short circuits and the shunt capacitance is open circuit in the mid-frequency range of Amplifier response, they are not considered in the equivalent circuit.

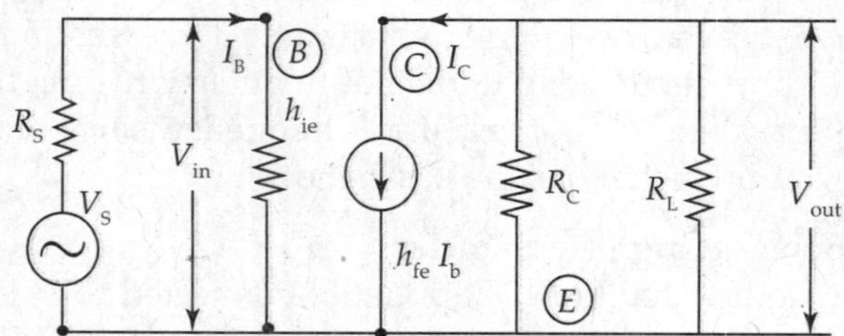

FIG. 6.47 *CE Transistor Amplifier mid frequency equivalent circuit for resistance capacitance coupled Transistor Amplifier*

Voltage gain A_{Vmid} can be calculated as shown below.

$$V_{out} = -h_{fe} \cdot I_b (R_C \parallel R_L); \quad V_{in} = h_{ie} \cdot I_b$$

$$\text{Voltage gain} \quad A_V = \frac{V_{out}}{V_{in}}$$

$$A_{Vmid} = -\frac{h_{fe}}{h_{ie}}(R_C \parallel R_L); \quad A_{Vmid} = -g_m(R_C \parallel R_L) \tag{6.88}$$

Low-frequency equivalent circuit of resistance capacitance coupled Amplifier (Fig. 6.48): At low frequencies, the coupling capacitances have effect in producing some signal loss for signal transmission through them. This can be observed by the fall in gain in the low-frequency region, from the maximum level of gain at the mid-frequency region, in the frequency response characteristic shown in Fig. 6.44. So output-coupling capacitor at the Collector terminal of the Transistor is shown in the low-frequency equivalent circuit of Resistance capacitance coupled Transistor Amplifiers.

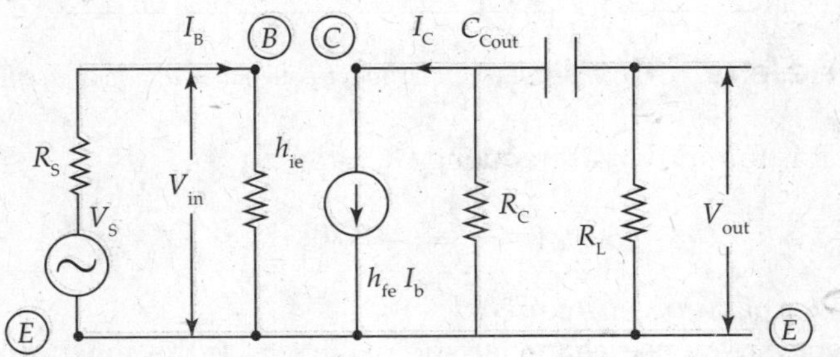

FIG. 6.48 *CE Transistor Amplifier low frequency equivalent circuit for resistance capacitance coupled Transistor Amplifier*

This circuit of Fig. 6.48 for the analysis can be redrawn as follows:

$$V_{out} = \frac{-h_{fe}\cdot I_b \cdot R_C \times R_L}{\left[R_C + R_L - \dfrac{j}{\omega C_C}\right]}; \quad V_{out} = \frac{-h_{fe}\cdot I_b \cdot R_C \cdot R_L}{(R_C + R_L)\cdot\left(1 - \dfrac{j}{\omega C_C (R_C + R_L)}\right)}$$

$$V_{in} = h_{ie}\cdot I_b$$

$$\therefore \text{ Voltage gain} = \frac{V_{out}}{V_{in}} = A_{VLow} = \frac{-h_{fe}\cdot I_b (R_C \parallel R_L)}{h_{ie}\cdot I_b\left(1 - \dfrac{j}{\omega C_C (R_C + R_L)}\right)} \tag{6.89}$$

$$A_{VLow} = \frac{g_m (R_C \parallel R_L)}{\left(1 - \dfrac{jf_1}{f}\right)} \tag{6.90}$$

$$\text{where } f_1 = \frac{1}{2\pi\, C_C (R_C + R_L)} \tag{6.91}$$

$$A_{Vlow} = \frac{A_{Vmid}}{\left(1 - \dfrac{jf_1}{f}\right)} \tag{6.92}$$

$$|A_{VLow}| = \frac{A_{Vmid}}{\sqrt{1 + \left(\dfrac{f_1}{f}\right)^2}} \tag{6.93}$$

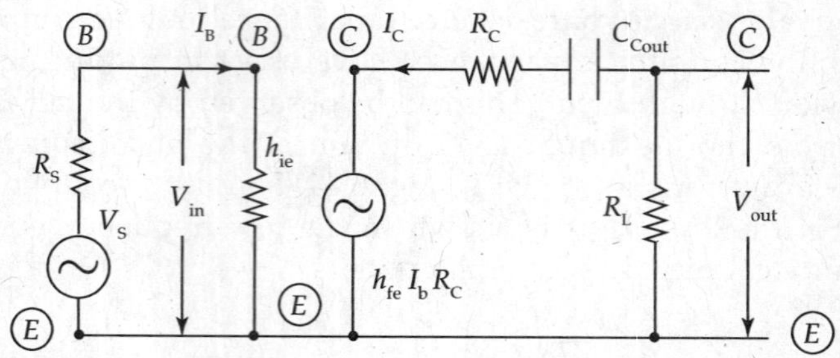

FIG. 6.49 *CE transistor ampifier low frequency equivalent circuit*

From Eq. (6.93), it is found that at frequency '$f = f_1$'

$$|A_{VLow}| = \frac{A_{Vmid}}{\sqrt{2}} = 0.707 A_{Vmid} \tag{6.94}$$

Thus, f_1 is the lower half power frequency.

From the frequency response characteristic of Fig. 6.44, the maximum gain in the mid-frequency region is assumed to be 100. $A_{Vmid} = A_{max} = 100$.

Gain in decibels at mid-frequency region = 20 $\log_{10}100 = 20 \times 2 = 40$ dB.

At low-frequency cut-off point f_1 or f_L,

$$\text{Gain} \quad A_{VLow} = \frac{A_{Vmid}}{\sqrt{2}} = 0.707 A_{Vmid} = 0.707 \times 100 = 70.7.$$

Gain in decibels at f_1

$$20 \log_{10} \frac{A_{Vmid}}{\sqrt{2}} = 20 \log_{10} 70.7 = 37 \text{ dB}.$$

So the fall in gain is 40 – 37 = 3 dB.

High-Frequency equivalent circuit of RC-Coupled Transistor Amplifier:

At high-frequency region, the series coupling capacitors behave as effective short circuits. So they are not considered in the high-frequency equivalent circuit. But the impedance of the reactances of shunt capacitances C_{Sh} which is a combination of stray wiring capacitance C_W and the junction capacitances C_{bc} due to the output junction and $C_{be'}$ the input junction capacitance of the Transistor will be low at high-frequency region. So this reduced reactance of the shunt capacitance C_{Sh} will be in parallel with the load resistance and cause reduction in the effective

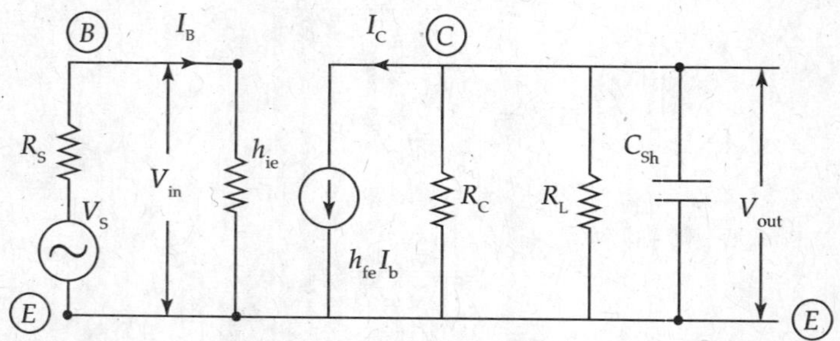

FIG. 6.50 *High frequency equivalent circuit of CE transistor amplifier*

load resistance of the Amplifier. This reduction in load resistance reduces the Amplifier gain, because the gain of the Amplifier is proportional to the load resistance in Eq. (6.46).

High-frequency equivalent circuit of the Amplifier is further simplified for the simplicity of analysis, by replacing the current source and R_C shown in Fig. 6.50 with voltage source $h_{fe} I_B R_L'$ and series resistance R_L', where R_L' is the parallel combination of R_C and R_L is the simplified equivalent circuit shown in Fig. 6.51:

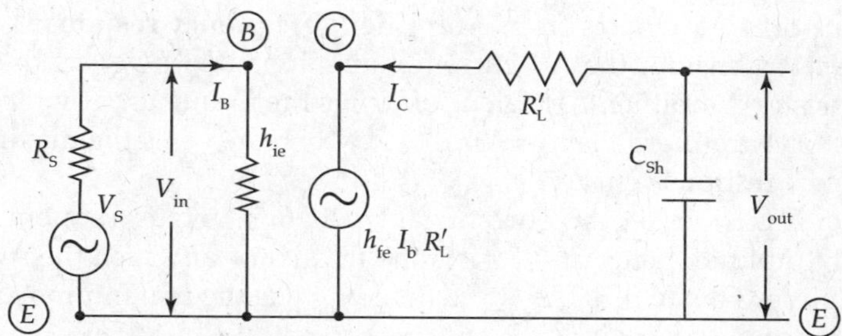

FIG. 6.51 *High frequency equivalent circuit of CE transistor amplifier (with voltage source)*

$$V_{out} = \frac{-h_{fe} \cdot i_b \cdot R_L' \times \left(\frac{1}{j\omega C_{Sh}}\right)}{R_L' + \left(\frac{1}{j\omega C_{Sh}}\right)}$$

$$V_{out} = \frac{-h_{fe} \cdot I_b \cdot R_L'}{1 + R_L' j\omega C_{Sh}}$$

$$V_{in} = h_{ie} \cdot I_{in}$$

$$\therefore \quad A_{Vhigh} = \frac{V_{out}}{V_{in}} = \frac{-h_{fe} \cdot I_b \cdot R_L'}{h_{ie} \cdot I_b (1 + R_L' j\omega C_{Sh})} \tag{6.95}$$

$$A_{Vhigh} = \frac{g_m \cdot R_L'}{\left(1 + \frac{j \cdot f}{f_2}\right)} = \frac{A_{Vmid}}{\left(1 + \frac{j \cdot f}{f_2}\right)}, \tag{6.96}$$

$$\text{where} \quad f_2 = \frac{1}{2\pi R_L' C_{Sh}} \tag{6.97}$$

$$|A_{Vhigh}| = \frac{A_{Vmid}}{\sqrt{1 + \left(\frac{f}{f_2}\right)^2}} \quad \text{at } f = f_2. \tag{6.98}$$

$$A_{Vhigh} = \frac{A_{Vmid}}{\sqrt{2}} = 0.707 A_{Vmid}. \tag{6.99}$$

So f_2 is the upper half power frequency. And the bandwidth is $(f_2 - f_1)$ but $f_1 \ll f_2$ so bandwidth $\cong f_2$. So f_1 is dependent on the series combination of R_C and R_L together with C_C and f_2 is dependent on the parallel combination of R_C and R_L together with C_{Sh}. At high-frequency cut-off point also, the gain is 3 dB down from the gain at the mid-frequency value. When $A_{mid} = 100$, the gain $A_{Vhigh} = 0.707 \times A_{Vmid} = 0.707 \times 100 = 70.7$.

Mid-band gain is 40 dB and gain at $f_2 = 37$ dB as frequency response shown in Fig. 6.44. In the first case, if C_C becomes large; f_1 will be reduced while C_{Sh} is large f_2 will be reduced. Larger the value of C_C smaller is f_1 and smaller the value of C_{Sh} larger is the f_2. So for maximum bandwidth C_C should be as large as possible and C_{Sh} should be made as small as possible.

Alpha and Beta cut-off frequencies from Amplifier's frequency response

Alpha cut-off frequency: Consider the AC current gain, alpha ('h_{fb}'), versus frequency response curve of a CB Transistor Amplifier. Alpha cut-off (f_α) is the frequency at which the AC current gain of CB Transistor Amplifier falls to 0.707 $h_{fb(max)}$, where $h_{fb(max)}$ is the maximum AC current gain on the frequency response curve.

Beta cut-off frequency: Consider the AC current gain, beta (h_{fe}), versus frequency response curve of a CE Transistor Amplifier. Beta cut-off (f_β) is the frequency at which the AC current gain of CE Transistor Amplifier falls to 0.707 $h_{fe(max)}$, where $h_{fe(max)}$ is the maximum AC current gain on the frequency response curve.

Beta cut-off frequency is normally lower than the alpha cut-off frequency for a Transistor because the current gain of CB Transistor is very low, causing large bandwidth. From the more advantageous features of CE Transistor Amplifier such as large voltage and current gains, CE Transistor Amplifiers find more practical applications. So beta cut-off frequency f_β normally appears in the Transistor specifications as its performance measure.

Gain bandwidth product: Important specification of a Transistor is its gain bandwidth product. Another parameter of interest is f_T. f_T is the frequency at which short circuit CE Transistor current gain is unity. It will be observed from experimental results that the frequency f_T will be very large.

EXAMPLE 6.12

Calculate current gain, voltage gain, R_{in} and R_{out} using the following data: Data: $h_{fe} = 36$, $R_L = 2.5 \times 10^3$ and $h_{oe} = 2 \times 10^{-6}$; $h_{ie} = 1200\ \Omega$, $h_{re} = 0$ and $R_S = 500\ \Omega$.

Solution: Determination of Current gain A_I:

$$A_I = -\frac{h_{fe}}{[1 + h_{oe} \cdot Z_L]} = \frac{36}{1 + 2 \times 10^{-6} \times 2.5 \times 10^3} = \frac{36}{1 + 5 \times 10^{-3}} \equiv 36$$

Determination of Voltage gain A_V:

$$A_V = \frac{A_I \cdot Z_L}{Z_{in}} = \frac{36 \times 2.5 \times 10^3}{1.2 \times 10^3} = \frac{90}{1.2} = 75$$

Determination of Input Resistance R_{in}:

$R_{in} = Z_{in} = h_{ie} + A_I h_{re} Z_L$ is approximately $= h_{ie}$ as $h_{re} = 0$. Therefore, $R_{in} = h_{ie} = 1200\ \Omega$.

Determination of output Resistance R_0:

$$R_o = \frac{R_S + h_{ie}}{h_{oe}(R_S + h_{ie})} = \frac{500 + 1200}{2 \times 10^{-6}(500 + 1200)} = \frac{1700}{3.4 \times 10^{-3}} = 500 \times 10^3.$$

SUMMARY

1. Transistor amplifier circuits have two sets of voltages, DC & AC, where – (1) DC voltages fix up quiescent operating conditions for applications such as *linear* or small signal amplifiers (described in this chapter) (2) AC signal voltages for amplification.
2. DC voltage variations, due to small signal voltages, are addressed in detail analyzing signal waveforms at input and output ports of the amplifiers.
3. An amplifier is designed using Common Emitter Transistor to handle small signal (class-A) operations. Once you get familiar with the circuit design, the concepts of frequency response and bandwidth of the amplifier are discussed.
4. Voltage gains obtained at *low, mid* and *high frequency* regions are analyzed 'mathematically', with the help of *h-parameter equivalent* circuits at various frequency ranges for Resistance Capacitance Coupled (RCC) amplifiers (RCC amplifiers are also described in detail Chapter 10).
5. Concepts of circuit design and frequency response of an amplifier form the foundation for our advanced topics in subsequent chapters on large signal (power) amplifiers and high frequency amplifiers. This foundation is a key to achieving competency in Electronic Circuit Analysis.

Questions for Practice

1. Discuss the classification of Amplifiers Based on frequency range, type of coupling, power delivered and signals handled? (JNTU, Nov. 2006)
2. Using the *h*-parameters model, derive the expressions for current gain A_I, input impedance, and Voltage gain A_V and output impedance of CE Transistor Amplifier? (JNTU, Feb. 2008)
3. (a) Define the *h*-parameters of a Transistor in a small signal Amplifier? What are the benefits of *h*-parameters? (JNTU, Nov. 2008)
   (b) In a single-stage CE Transistor Amplifier, $R_S = 1\ k\Omega$ and $R_L = 1.2\ k\Omega$, using $h_{fe} = 50$, $h_{oe} = 25 \times 10^{-6}$ A/V, $h_{re} = 2.5 \times 10^{-4}$, $h_{ie} = 1100\ \Omega$, find A_I, A_V, Z_i and Z_o.
4. With the help of necessary equations, discuss variations of A_I, R_i, R_o and A_V with R_S and R_L in CE Transistor Amplifier? (JNTU, Nov. 2007)
5. (a) What are the *h*-parameters? Why are they called so? Define them and what are the benefits of *h*-parameters?
   (b) In a single-stage CE Transistor Amplifier with an un-bypassed Emitter resistance, $R_o = 1\ k\Omega$, $R_1 = 50\ k\Omega$, $R_2 = 2\ k\Omega$, $R_E = 270\ \Omega$ and $R_L = 1.2\ k\Omega$. Find A_I, A_V, R_i and R_o. The *h*-parameters given are $h_{ie} = 1100\ \Omega$ and $h_{fe} = 50$ (JNTU, Nov. 2008)
6. (a) Draw the low-frequency *h*-parameter equivalent circuit of Transistor Amplifier and explain the significance of each parameter (JNTU, Nov. 2008)
   (b) In a single-stage CE Transistor Amplifier $R_S = 1\ k\Omega$, $R_1 = 50\ k\Omega$, $R_2 = 2\ k\Omega$, $R_C = 1\ k\Omega$, $R_L = 1.2\ k\Omega$, $h_{fe} = 50$ and $h_{ie} = 1.1\ k\Omega$. Find A_I, R_i, R_o and voltage gain.

7. Give typical values of h-parameters for CC Transistor configuration. (May 2005)
8. Draw the circuit of CE Transistor Amplifier and its equivalent circuit. Discuss its characteristics. (JNTU, Feb. 2008)
9. Draw the circuit diagram of Emitter follower and its equivalent circuit and derive the expression for voltage gain. (May 2004)
10. Draw the circuit of CB Transistor Amplifier and its h-parameter equivalent circuit. Discuss the characteristics of CB Amplifier. (JNTU, Nov. 2007)
11. The maximum voltage gain of a Transistor Amplifier is 100. Low-frequency cut-off frequency occurs at 1000 Hz and high-frequency cut-off occurs at 501 kHz. Determine the bandwidth of the Amplifier circuit and draw the frequency response marking the values of gain at the salient points.
12. Compare the CC and CE configurations with respect to R_i, R_o, A_V and A_I. (JNTU, Dec. 2004)
13. Draw a typical CE Amplifier and explain the function of each component. (JNTU, Nov. 2004)
14. Draw the equivalent circuit for CE Amplifier (1) with a bypassed Emitter resistor (2) with un-bypassed Emitter resistor. Briefly explain each stage. (Nov. 2004)
15. Explain the method of graphical determination of fixing the quiescent or DC-operating point of a Transistor Amplifier. Discuss the specifications that are considered in the design of a Transistor Amplifier and any precautions that are to be considered.
16. Draw the Transistor Amplifier circuit using potential divider biasing arrangement with $V_{CC} = +12$ V, $R_C = 0.8$ kΩ and $R_S = 0.2$ kΩ. To obtain maximum output signal swings, select suitable DC-operating point and calculate I_{CQ} and V_{CEQ}.
17. A Silicon Transistor is used in a CE Transistor Amplifier circuit with potential divider biasing arrangement with $V_{CC} = 18$ V and $R_C = 1.2$ kΩ. Transistor $\beta = 100$. So as to obtain symmetrical swings of output signals, the quiescent operating point is chosen with $I_{CQ} = 4.5$ mA and $V_{CEQ} = 9$ V. Calculate the values of R_1, R_2 and R_E assuming a stability factor $S = 10$.
18. DC equivalent circuit of CE Transistor Amplifier with Collector to Base bias arrangement has $V_{CC} = 18$ V, with component values of $R_C = 3.4$ kΩ, $R_E = 0.6$ kΩ, the Collector to Base resistance $R_B = 300$ kΩ, $V\gamma = 0.7$ V and beta of the Transistor $\beta = 100$. Calculate the magnitudes of Collector voltage, Emitter voltage and Base voltage.
19. Draw the CE Transistor Amplifier circuit with potential divider bias arrangement having $V_{CC} = 14$ V, $R_1 = 21$ kΩ, $R_2 = 7$ kΩ, $V_{BE} = 0.5$ V, $R_C = 0.8$ kΩ and $R_E = 0.6$ kΩ. Draw the DC equivalent circuit. Calculate the Base voltage, Emitter voltage, Collector voltage and V_{CE}. Comment on the results for the operation of the Amplifier circuit. Design suitable values of the capacitors to be used in the circuit considering the lowest frequency of the signal to be 116 Hz.
20. Discuss the main differences and similarities between hybrid pie and h-parameter models of BJT.
21. Discuss the concept of AC load line and DC load line.
22. Derive the expressions for feedback conductance, output conductance, Base spread resistance tranconductance, input conductance and diffusion capacitance of hybrid pie CE Amplifier.

23. Derive expressions for voltage gain and current gain of CE Amplifier. Explain how these are affected by swamping resister?
24. Disuses the limitations and effect of early effect on the hybrid pie model of BJT.

Multiple Choice Questions

1. An Amplifier ______________
   (a) performs signal processing of weak signals into larger signal amplitudes
   (b) increases the signal power
   (c) Introduces propagation delay
   (d) is a digital logic converter
2. For Amplifier applications, a BJT is operated in ______________
   (a) active mode
   (b) cut-off mode
   (c) saturation mode
   (d) both saturation and cut-off modes
3. For switching application, BJT is to be operated in ______________
   (a) active mode
   (b) cut-off mode
   (c) saturation mode
   (d) both saturation and cut-off modes
4. In active mode of operation of a BJT ______________
   (a) Emitter–Base junction is reverse biased and Collector to Base junction is reverse biased
   (b) Emitter to Base junction is forward biased and Collector to Base junction is reverse biased
   (c) Emitter to Base junction is forward biased and CB junction is forward biased
   (d) Emitter to Base junction is reverse biased and CBJ is forward biased
5. If a Transistor is operating with both junctions, forward biased but Collector to Base junction bias is greater than Emitter to Base junction forward bias its operating point is in the ______________
   (a) forward active mode
   (b) reverse active mode
   (c) reverse interactive mode
   (d) forward saturation mode
6. Substantial voltage gain and current gain are obtained in ______________
   (a) common Emitter Transistor Amplifier
   (b) common Base Transistor Amplifier
   (c) common Collector Transistor Amplifier
   (d) none of these

7. A common Base Transistor Amplifier is ______________
   (a) a voltage Amplifier
   (b) having a high-frequency response
   (c) an Emitter follower
   (d) associated with voltage and current gains
8. A common Base Amplifier finds application ______________
   (a) as a unity gain current Amplifier or current buffer
   (b) to provide substantial voltage gain and current gain
   (c) as a voltage buffer
   (d) used in multistage Amplifiers
9. Common Collector Transistor Amplifier is ______________
   (a) a current buffer
   (b) a voltage buffer Amplifier
   (c) used to provide both voltage gain and current gain
   (d) none of these
10. In a common Emitter Transistor Amplifier, the maximum supply voltage is limited by ______________
    (a) Avalanche breakdown of Base to Emitter junction
    (b) Zener breakdown of Base to Emitter junction
    (c) Collector to Base breakdown with Emitter open
    (d) Collector to Emitter breakdown with Base open
11. Introduction of a resister in Emitter of a CE Amplifier stabilises the variations in operating point against variations in ______________
    (a) only temperatures
    (b) only beta
    (c) the early effect
    (d) both temperature and beta
12. In low-frequency response of an Amplifier, ______________
    (a) coupling and bypass capacitors are treated as open circuits
    (b) coupling and bypass capacitors are to be included
    (c) the stray and Transistor capacitances treated as open circuits
    (d) the stray and Transistor capacitances treated as short circuits
13. In the high-frequency response of Amplifier, ______________
    (a) the coupling and bypass capacitors are treated as short circuits
    (b) the coupling and bypass capacitors are treated as open circuits
    (c) the Transistor and bypass capacitors must be taken into account
    (d) the stray and bypass capacitances are treated as open circuits
14. The largest bandwidth is provided by ______________
    (a) common Transistor Amplifier
    (b) common Collector Transistor Amplifier
    (c) common Base Transistor Amplifier

15. Low-frequency response of an Amplifier depends upon ______________
    (a) coupling and bypass capacitors
    (b) parasitic capacitances in the device
    (c) stray capacitances
    (d) biasing arrangement
16. In the mid-band range of frequency response of an Amplifier, ______________
    (a) the coupling and bypass capacitors act as open circuits
    (b) the coupling and bypass capacitors act as short circuits
    (c) stray and Transistor capacitances act open circuits
    (d) stray and Transistor capacitances act as short circuits
17. The units of hybrid parameters are
    (a) Siemens
    (b) Ohms
    (c) constant
    (d) a mixture of units

Answers to Multiple-Choice Questions

1. (both a and b)	2. (a)	3. (d)	4. (b)	5. (b)	6. (a)
7. (b)	8. (a)	9. (b)	10. (d)	11. (d)	12. (b and c)
13. (a and c)	14. (b)	15. (a)	16. (b and c)	17. (d)	

Chapter 7

FEEDBACK AMPLIFIERS

Learning Objectives

- Fundamental concepts of feedback in Amplifiers.
- Different types of feedback in Amplifiers.
- Analysis of negative Feedback Amplifiers using FET and BJT
- Transistor circuit operations for Feedback Amplifier concepts.
- Problem solving to understand the working of Feedback Amplifiers.

7.1 INTRODUCTION

In this age of automation when man is going on to moon, space stations for research, communication satellites in space, computation, inter continental ballistic missiles and remote control of energy systems by wireless control using digital signal processors and so on, the 'processes of signal feedbacks' play an important role.

As a day-to-day example of biological systems, if one is driving a car, the eyes sense the position and speed of the car. The input signals sent by the eyes to the human brain modify the biological inputs to the steering wheel by the human system by applying a brake or accelerates or decelerates the car engine. Using the data from the feedback, the process of driving the car is controlled.

Similarly in an Amplifier circuit, sampling a part of the output voltage or current feeding back to the input port through a network can control the process of amplification. An error generated between the input signal and feedback output signal to the input port of the Amplifier produces desired response. This is similar to 'closed loop control system'.

7.2 FUNDAMENTAL CONCEPTS OF FEEDBACK AMPLIFIER CIRCUITS

Basically, there are two types of signal feedbacks in electronic systems:

1. Positive feedback
2. Negative feedback

Harold S. Black at Bell telephone laboratories in United States of America has invented the concept of introduction of 'negative feedback' to obtain an Amplifier with stable gain while working with speech transmission over long distance telephone lines.

- If the signal feedback from the output is subtracted from the input in an Amplifier, the circuit is known as 'negative Feedback Amplifier'. It is generally known as Feedback Amplifier.
- So, a Feedback Amplifier consists of
  1. Basic or internal Amplifier, which is a simple Amplifier.
  2. Feedback network depending on the type of Feedback Amplifier.

The inherent working principles of negative feedback Amplifier are explained using the two circuits in Figs. 7.1 and 7.2.

7.2.1 Common Emitter Transistor Amplifier (Fig. 7.1)

Consider the common emitter Transistor Amplifier as the basic Amplifier. The Amplifier has an external signal voltage V_S and V_{out} as the output voltage response. As long as there are no other disturbances, signal voltage V_S will act as the effective input signal voltage V_{in} across the input port (base to emitter loop) to the Amplifier. Then the gain of the Amplifier A is the ratio of output voltage V_{out} to the input voltage V_{in}.

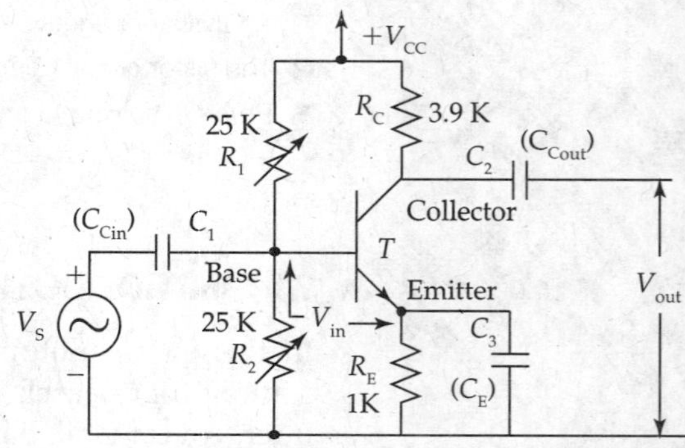

FIG. 7.1 *Common emitter transistor amplifier*

EXAMPLE 7.1

Calculate the gain A of the amplifier when the input signal voltage $V_{in} = 20$ mV and output response voltage $V_{out} = 1.8$ V.

Solution: Amplifier gain $A = \dfrac{V_{out}}{V_{in}}$

Effective input signal voltage $V_{in} = 20$ mV. Output response voltage of the amplifier $V_{out} = 1.8$ V. Amplification of the signal

$$A = \frac{V_{out}}{V_{in}} = \frac{1.8}{20 \times 10^{-3}} = \frac{1.8 \times 10^3}{20} = 90.$$

7.2.2 Negative Feedback Amplifier Circuit (Fig. 7.2)

- Feedback network consists of two resistors R_3 and R_4, which are connected at the output port of the Amplifier.
- Total output voltage V_{out} is across the series combination of R_3 and R_4.
- A portion of the output voltage sampled across R_4 is fed back to the input port of the Amplifier. The feedback voltage is V_F
- The feedback arrangement is such that the external signal V_S and the feedback signal V_F are opposing to one another.
- Then the feedback voltage is subtracted from the signal V_S. So, the effective input signal – V_{in} becomes equal to $V_S - V_F$
- Thus, the effective input signal gets reduced.
- The reduction in the effective input signal reduces the Amplifier gain.
- Gain of negative Feedback Amplifier will thus be lower than the normal Amplifier.

Figure 7.2 shows various parts of Feedback Amplifier. These concepts are further explained using the 'block diagram' model of a single-loop negative Feedback Amplifier in Fig. 7.3.

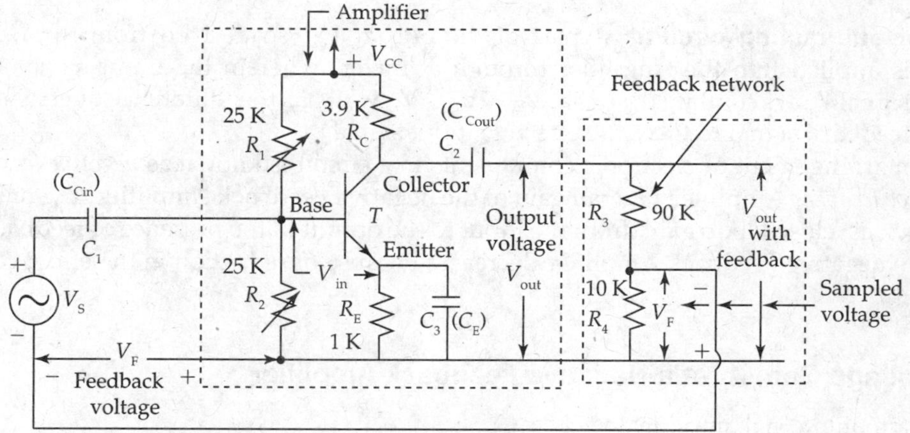

FIG. 7.2 *Various blocks in feedback amplifier circuit*

7.3 NEGATIVE FEEDBACK AMPLIFIER

When the feedback voltage is in phase opposition with the input source voltage, the type of feedback is negative feedback. The resulting Amplifier is a negative Feedback Amplifier.

Essential elements of a single-loop negative Feedback Amplifier shown in Fig. 7.3 are

Basic or internal Amplifier: V_{in} is the input signal and V_{out} is the output signal of the internal Amplifier. The ratio of the output voltage to the input voltage is called as forward gain or 'open loop gain' A (voltage gain without any feedback arrangement). The device features and Amplifier operation can be modified using negative feedback.

Sampling network (Voltage sampler or Current sampler): Voltage or current is sampled at the output port of the basic Amplifier by means of suitable sampling elements and fed to 'feedback network'.

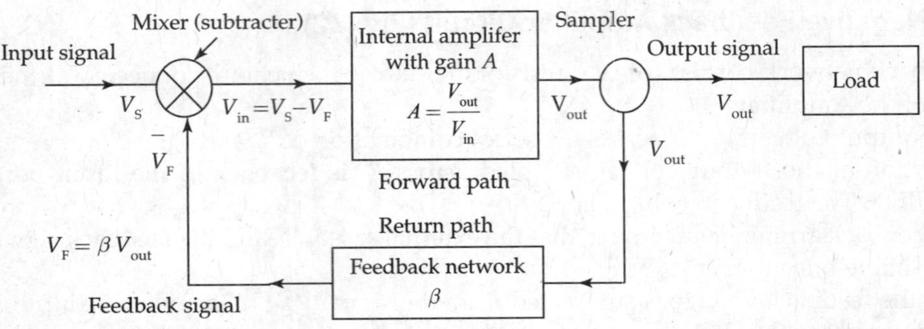

FIG. 7.3 *Block diagram of an ideal single loop negative feedback amplifier*

Feedback network: It contains passive components such as resistors or capacitors or inductors. (R_E in the emitter path of Transistor Amplifiers and R_S in the source path of FET Amplifiers explain this) Negative feedback arrangement is used in the feedback network. The sampled parameter is connected from the output port to the input port as feedback signal as shown in Fig. 7.3, through the feedback network.

Mixing or subtracting circuit: A part of the output V_{out} so derived from the feedback network is applied into the Amplifier through a 'mixer' wherein the input signal V_S and feedback signal V_F are combined, so that $V_{in} = V_S - V_F$, which is the 'difference or error signal'. Thus the feedback signal reduces the effective input signal.

Such an arrangement of electronic components in a circuit is known as *negative feedback or degenerative feedback* Amplifier. Overall gain of the negative Feedback Amplifier A_{NFB} or simply A_F is known as 'closed loop gain' from the input to the output in the presence of feedback. The gain of the negative Feedback Amplifier decreases because of reduction in the effective input signal V_{in}.

7.3.1 Voltage Gain A_F of a Negative Feedback Amplifier

Voltage gain of internal amplifier

$$A = \frac{\text{output voltage}}{\text{input voltage}} = \frac{V_{out}}{V_{in}} \tag{7.1}$$

Output voltage of internal amplifier

$$V_{out} = A \cdot V_{in} \tag{7.2}$$

For negative feedback

$$V_{in} = [V_S - V_F], \tag{7.3}$$

where $\beta = V_F / V_{out}$, β is known as reverse transmission factor or 'feedback factor' or 'feedback ratio'.

$$\therefore \quad V_F = \beta \cdot V_{out} \tag{7.4}$$

From Eqs. (7.3) and (7.4), we get

$$V_{in} = [V_S - \beta \cdot V_{out}] \tag{7.5}$$

From the equation $V_{out} = A \cdot V_{in}$ and Eq. (7.5),

$$V_{out} = A \cdot [V_S - \beta \cdot V_{out}] \tag{7.6}$$

Rearranging the terms in Eq. (7.6),

$$V_{out}[1 + A\beta] = A \cdot V_S \tag{7.7}$$

Therefore, voltage gain of negative feedback amplifier

$$A_{NFB} = A_F = \frac{V_{out}}{V_S}$$

$$A_F = \frac{V_{out}}{V_S} = \frac{A}{[1 + A\beta]}. \tag{7.8}$$

The magnitude of feedback introduced into the Amplifier is expressed in decibels (db).

Taking log on both sides of Eq. (7.8), we get

A_{NFB} in (db) = A in (db) – [1 + $A\beta$] in (db) (db is decibels)

A_{NFB} in (db) = A in (db) + N (db), for negative Feedback Amplifier, N is negative.

$$N = \text{Amount (db) of feedback} = 20\log\left|\frac{A_{NFB}}{A}\right| = 20\log\left|\frac{1}{[1 + A\beta]}\right|. \tag{7.9}$$

7.3.2 Concept of Positive Feedback

In case of the Positive Feedback signal V_F is in phase with V_S, then $V_{in} = [V_S + V_F]$ and the type of feedback is said to be 'positive feedback'.

$$A_{PFB} = \left|\frac{A}{1 - A\beta}\right|. \tag{7.10}$$

In this case, the gain with positive feedback increases. If loop gain $A\beta = 1$, the gain becomes ∞ (infinity). Then the circuit cannot be used as a normal Amplifier. The concept of 'positive feedback' is used in the design of electronic oscillator circuits.

EXAMPLE 7.2

If we consider the Voltage Series Feedback Amplifier of Fig. 7.2 with open loop gain of the internal Amplifier $A = 90$ and resistor $R_3 = 90$ kΩ and $R_4 = 10$ kΩ, calculate the values of feedback factor, loop gain and gain of the Feedback Amplifier.

Solution: Gain of internal amplifier $A = 90$

Feedback factor $\beta = \frac{V_F}{V_{out}} = \frac{R_4}{R_3 + R_4} = \frac{10 \times 10^3}{[90 \times 10^3 + 10 \times 10^3]} = \frac{10}{100} = 0.1$

Loop gain $A\beta = 90 \times 0.1 = 9$

Gain of feedback amplifier $A_F = \frac{A}{[1 + A\beta]} = \frac{90}{[1 + 9]} = \frac{90}{10} = 9$

These calculations show that gain of negative feedback amplifier reduces very much.

EXAMPLE 7.3

Internal Amplifier has a voltage gain A of 60 dB and uses 0.1 of the output voltage in the feedback. Calculate the gain of the negative Feedback Amplifier in db.

Solution: Open loop gain of internal amplifier $A = 60$ dB
Gain A in db $= 20 \log_{10} \cdot A = 60$ dB
Feedback factor $\beta = 0.1$

$$\therefore \quad \log_{10} A = \frac{60}{20} = 3$$

Gain A = antilog 3 = 1000
Gain of negative feedback amplifier

$$A_F = \frac{A}{[1 + A\beta]} = \frac{1000}{1 + 1000 \times 0.1}$$

$$= \frac{1000}{101} = 9.9.$$

Gain of the feedback amplifier A_F in db = 20 log (9.9) = 19.9 dB.

EXAMPLE 7.4

The open loop gain or forward path gain A of an Amplifier is 50,000. When negative feedback is introduced into the Amplifier, gain decreases to 10,000. Calculate the feedback factor and express the gain with negative feedback in db.

Solution: Open loop gain of the amplifier $A = 50{,}000$
Gain of the negative feedback amplifier $A_F = 10{,}000$

$$A_F = \frac{A}{1 + A\beta}$$

$$1 + A\beta = \frac{A}{A_F}$$

$$A\beta = \frac{A}{A_F} - 1 = \frac{A - A_F}{A_F}$$

$$\therefore \quad \beta = \frac{(A - A_F)}{A \cdot A_F} = \frac{1}{A_F} - \frac{1}{A}$$

$$\beta = \frac{1}{10 \times 10^3} - \frac{1}{50 \times 10^3} = \frac{40 \times 10^3}{500 \times 10^6} = 8 \times 10^{-5}$$

$$\text{Gain of feedback amplifier in db} = 20 \log_{10} 10{,}000 = 20 \log 10^4 = 80 \text{ dB}$$

$$\text{Feedback in db} = 20 \log_{10} \frac{A_F}{A} = 20 \log_{10} \frac{10{,}000}{50{,}000}$$

$$= 20 \log 0.2 = -13.98 \text{ dB}.$$

EXAMPLE 7.5

An Amplifier has a voltage gain of 100, feedback ratio, if 0.01, find the voltage gain with feedback, the output voltage of the Feedback Amplifier for an input voltage of 1 mV, and feedback voltage. (BEDC I B.Tech. EEE May/June 2008, set-4)

Solution: Voltage gain $A = 100$, feedback ratio $\beta = 0.01$.
Gain of feedback amplifier

$$A_F = \frac{A}{[1+A\beta]} = \frac{100}{[1+100\times 0.01]} = \frac{10}{[1+1]} = \frac{10}{2} = 5$$

Output voltage of feedback amplifier $V_{out} = A_F \cdot V_{in} = 5 \times 1\text{ mV} = 5\text{ mV}$
Feedback factor $\beta = 0.01$
Feedback voltage $V_F = \beta V_{out} = 0.01 \times 5 \times 10^{-3} = 0.05$ mV.

7.4 MERITS (GENERAL CHARACTERISTICS) OF NEGATIVE FEEDBACK AMPLIFIERS

It is already derived that the gain of negative feedback amplifier decreases. It is given by

$$A_{NFB} = \left|\frac{A}{1+A\beta}\right|$$

$$\text{If } |1+A\beta| \gg 1, \quad |A_{NFB}| < A. \tag{7.11}$$

7.4.1 Merits of *Negative Feedback Amplifiers*

a. Reduction of frequency and phase (waveform) distortion
(Waveform distortion due to the non-linearity of the active device characteristics)

Voltage gain of negative feedback amplifier

$$A_{NFB} = \frac{A}{[1+A\beta]}.$$

If the gain of the amplifier is very large so that

$$A\beta \gg 1, \quad A_{NFB} = \frac{1}{\beta}. \tag{7.12}$$

Now the gain A_{NFB} of the Feedback Amplifier is independent of the gain A of the internal Amplifier in the forward path and purely depends upon β of the feedback network. The sacrifice in gain brings many advantages for a better performance of the negative Feedback Amplifiers.

So, any changes in the following features do not have any effect on the gain of negative Feedback Amplifiers.

(1) Active device parameters; (2) Aging of the active device; (3) Temperature changes; (4) Supply voltage fluctuations; (5) If the feedback network contains only resistive elements, the gain becomes independent of frequency. So, there is substantial reduction in the frequency

and phase shift distortion; (6) Capacitors and resistors with high precision can also be used in the feedback path to stabilise the response of negative Feedback Amplifiers.

Thus, the stabilisation of Amplifier gain is achieved. This is the most desirable characteristic of Amplifier systems and in turn the practical electronic systems, because of the presence of Amplifier circuits in most of the electronic gadgets

b. Improvement in stability of gain of 'negative Feedback Amplifier'

Gain of the negative feedback amplifier

$$A_{NFB} = \left|\frac{A}{1+A\beta}\right|.$$

Differentiating the above equation with respect to A, we get

$$\frac{dA_{NFB}}{dA} = \frac{[1+A\beta]\cdot 1-(A\beta)}{[1+A\beta]^2} = \frac{1}{(1+A\beta)^2}.$$

Multiplying the numerator and denominator of RHS of the above equation by A, we get

$$\frac{dA_{NFB}}{dA} = \frac{A}{A(1+A\beta)^2} = \frac{A}{(1+A\beta)}\cdot\frac{1}{A(1+A\beta)} = \frac{A_{NFB}}{A(1+A\beta)} \tag{7.13}$$

$$\therefore \quad \left|\frac{dA_{NFB}}{A_{NFB}}\right| = \left|\frac{1}{[1+A\beta]}\right|\cdot\frac{dA}{A} \tag{7.14}$$

$$\text{Sensitivity } S \text{ is defined as} = \left|\frac{1}{[1+A\beta]}\right| = \frac{A_{NFB}}{A} \tag{7.15}$$

D is known as Desensitivity factor or Transfer gain T. D is also known as sacrifice factor. Further D is defined as

$$D = [1+A\beta] \tag{7.16}$$

$D = [1 + A\beta]$ is known as desensitivity factor or 'Return difference'.

$$\therefore \quad D = \frac{1}{S} = \frac{A}{A_{NFB}}. \tag{7.17}$$

Fractional change in gain A is reduced by the 'Desensitivity factor' D. Changes in output voltages or currents are feedback inputs and as such error changes in opposite direction affecting the output to be independent of changes in the basic Amplifier gain. So, overall stability is improved in a negative Feedback Amplifier against parameter changes caused due to active devices and temperature variations.

EXAMPLE 7.6

A negative Feedback Amplifier has open loop gain A_V of 160 dB and with negative feedback the gain is $A_{VF} = 80$ dB. Calculate the Desensitivity (sacrifice) factor D.

Solution: By definition,

$$\text{Desensitivity factor } D = \frac{A}{A_F} = [1+A\beta] = \frac{\text{Gain of basic amplifier}}{\text{Gain of feedback amplifier}}$$

Taking log on both sides of the above equation, we get

$$20\log_{10}D = 20\log_{10}A - 20\log_{10}A_F$$

$$\therefore\ 20\log_{10}D = 160 - 80 = 80 \text{ dB}$$

$$\log_{10}D = \frac{80}{20} = 4$$

$$D = \text{Anti}\log_{10} 4 = 10{,}000.$$

---

c. Reduction in non-linear distortion

Waveform distortion is produced when the Amplifier gain is not uniform for large input signal swings extending into non-linear region of the device characteristics. Assume the presence of second harmonic terms (Dominant component of harmonic distortion due to non-linear amplitude response in power Amplifiers) when a sinusoidal input signal is applied to large signal Amplifiers as in power Amplifiers. Second harmonic distortion B_2 is assumed in the output of internal Amplifier in the absence of feedback for analysis.

If negative feedback is applied, second harmonic signal B_{2F} is generated in the output of the active device. The feedback network receives sampled output and applies a feedback voltage of magnitude $-\beta \cdot B_{2F}$ at the input port of the internal Amplifier. This results in finding $-A\beta \cdot B_{2F}$ in the output. Now, the output voltage V_{out} contains two components: (1) The original distortion component B_2 when no feedback is introduced in the circuit and (2) $-A\beta \cdot B_{2F}$ is due to the introduction of negative feedback into the Amplifier forming a closed loop. Thus there is reduction of second harmonic distortion by a factor D in the negative Feedback Amplifier as explained below (Eq. 7.18).

Output signal of second harmonic distortion in the presence of feedback = B_{2F}

$$B_{2F} = (B_2 - A\beta \cdot B_{2F})$$

Rearranging the terms in the above equation, we get

$$B_{2F}(1 + A\beta) = B_2$$

$$\therefore\ B_{2F} = \frac{B_2}{[1 + A\beta]} = \frac{B_2}{D}, \quad \text{where } D = [1 + A\beta]. \tag{7.18}$$

Thus, the second harmonic distortion content B_2 in the output signal of a normal power Amplifier becomes B_{2F} due to the introduction of negative feedback into the Amplifier and reduces by a factor $[1 + A\beta]$ as shown in Eq. (7.18).

EXAMPLE 7.7

An Amplifier has a gain of 100 and a distortion of 8%. What is the effect of introducing negative feedback with a feedback factor of 0.05? (JNTU I B.Tech. EEE BEDC, May/June 2008, set-3)

Solution: Gain of an amplifier $A = 100$, feedback factor $\beta = 0.05$.
Gain of negative feedback amplifier

$$A_F = \frac{A}{[1 + A\beta]}$$

$$\therefore \quad A_F = \frac{A}{1+A\beta} = \frac{100}{[1+100\times 0.05]} = \frac{100}{6} = 16.25.$$

Distortion $B = 8\%$
With negative feedback amplifier distortion reduces

$$\text{Reduced distortion with feedback} \quad B_F = \frac{B}{[1+A\beta]}$$

$$\therefore \quad B_F = \frac{B}{[1+A\beta]} = \frac{8\%}{[1+100\times 0.05]} = \frac{8\%}{6} = 1.3\% \quad \text{after negative feedback.}$$

---

d. Reduction of Noise
If there is a noise source inside a system, application of negative feedback reduces noise by a factor $D = [1 + A\beta]$. The performance of the Amplifier is improved. In this connection, it is to be mentioned that signal to noise ratio will not have any effect. Further if the noise is extraneous to Amplifier, design modification has to be made to reduce noise. The hum introduced by a poorly designed direct current supply decreases considerably by using negative feedback.

e. Input impedance Z_{in} (FB)
The input impedance Z_{in}(FB) of a Feedback Amplifier depends upon the type of feedback connection in an Amplifier.

- Shunt feedback decreases input impedance by a factor $D = [1 + A\beta]$.
  Input impedance with shunt feedback

$$Z_{in}(\text{sh FB}) = \frac{Z_{in}}{D} = \frac{Z_{in}}{[1+A\beta]}. \tag{7.19}$$

- Series feedback increases the input impedance by a factor $D = [1 + A\beta]$.
  Input impedance with series feedback

$$Z_{in}(\text{ser FB}) = Z_{in} \times [1+A\beta]. \tag{7.20}$$

f. Output impedance $Z_{out}(f)$
The output impedance $Z_{out}(f)$ of a Feedback Amplifier depends upon the type of feedback connection in an Amplifier.

- The output impedance decreases by a factor D with Voltage Shunt feedback.
  Output impedance with shunt feedback

$$Z_{out}(\text{sh FB}) = \frac{Z_{out}}{D} = \frac{Z_{out}}{[1+A\beta]}. \tag{7.21}$$

- The output impedance increases with current feedback by a factor D.
  Output impedance with current feedback

$$Z_{out}(\text{ser FB}) = Z_{out} \times [1+A\beta]. \tag{7.22}$$

EXAMPLE 7.8

Calculate the gain of a negative Feedback Amplifier with internal Amplifier gain $A = 500$ and feedback factor $\beta = 0.01$. If the output impedance of the internal Amplifier $Z_{out} = 80$ kΩ. Determine the output impedance of the Feedback Amplifier. (Basic Electronics, II B.Tech. II-Sem; Mech. I, April/May 2008, set-1)

Solution: Gain of internal amplifier $A = 500$ and feedback factor $\beta = 0.01$.
Gain of the feedback amplifier

$$A_F = \frac{A}{[1 + A\beta]}$$

$$A_F = \frac{500}{[1 + 500 \times 0.01]} = \frac{500}{6} = 83.33$$

Output impedance of internal amplifier $Z_{out} = 80$ kΩ
Output impedance of shunt feedback amplifier Z_{out} (vol FB)

$$Z_{out}(\text{vol FB}) = \frac{Z_{out}}{[1 + A\beta]} = \frac{80 \times 10^3}{[1 + 500 \times 0.01]} = \frac{80 \times 10^3}{6} = 13.33\ \text{k}\Omega$$

Output impedance of series feedback amplifier Z_{out} (cur FB)

$$Z_{out}(\text{cur FB}) = Z_{out}[1 + A\beta] = 80 \times 10^3 \times [1 + 500 \times 0.01] = 80 \times 10^3 \times 6 = 480\ \text{k}\Omega.$$

EXAMPLE 7.9

An Amplifier without feedback has a gain of 1200, input impedance $Z_{in} = 1.5$ kΩ and output impedance Z_{out} of 50 kΩ. Determine the values of input and output impedances negative Feedback Amplifier with a feedback factor β of 0.01. (Basic Electronics, II B.Tech. II-sem; Mech. Engg. April/May 2008, set-2)

Solution: Amplifier gain $A = 1200$, feedback factor $\beta = 0.01$.
Gain of the feedback amplifier

$$A_{NFB} = \frac{A}{[1 + A\beta]} = \frac{1200}{[1 + 1200 \times 0.01]} = \frac{1200}{13} = 92.3$$

Input impedance of internal amplifier = $Z_{in} = 1.5$ kΩ
Input impedance of feedback amplifier Z_{in} (ser FB) = Z_{in} $(1 + A\beta)$

$$\therefore\ Z_{in}(\text{ser FB}) = 1.5 \times 10^3 \times [1 + 1200 \times 0.01] = 1.5 \times 10^3 \times 13 = 19.5\ \text{k}\Omega$$

$$Z_{in}(\text{sh FB}) = \frac{Z_{in}}{[1 + A\beta]} = \frac{1.5 \times 10^3}{[1 + 1200 \times 0.01]} = \frac{1500}{13} = 115\ \Omega$$

Output impedance of internal amplifier = $Z_{out} = 50$ kΩ
Output impedance of feedback amplifier Z_{out} (vol FB)

$$Z_{out}(\text{vol FB}) = \frac{Z_{out}}{[1 + A\beta]} = \frac{50 \times 10^3}{[1 + 1200 \times 0.01]} = \frac{50 \times 10^3}{13} = 3.846\ \text{k}\Omega.$$

EXAMPLE 7.10

The gain of internal Amplifier $A = 3900$ and feedback factor $\beta = 0.01$ in a Voltage Series negative Feedback Amplifier. Calculate the gain A_{FB} of the Feedback Amplifier and desensitivity factor D, Sensitivity factor S. For the internal Amplifier, the input resistance $Z_{in} = 1{,}200\ \Omega$ and output impedance $Z_{out} = 5\ k\Omega$. Calculate the input and output impedances of the Feedback Amplifier.

Solution: Gain $A = 3900$, feedback factor $\beta = 0.01$, desensitivity factor $D = [1 + A\beta] = [1 + 3900 \times 0.01] = 40$.

$$\text{Sensitivity factor}\quad S = \frac{1}{[1+A\beta]} = \frac{1}{[1+3900\times 0.01]} = \frac{1}{40} = 0.025$$

$$\text{Gain of the feedback amplifier}\quad A_{FB} = \frac{A}{[1+A\beta]} = \frac{3900}{40} = 97.5$$

$$\text{Input impedance}\quad Z_{in}(FB) = Z_{in}\cdot[1+A\beta] = 1200\times 40 = 48\ k\Omega$$

$$\text{Output impedance}\quad Z_{out}(FB) = \frac{Z_{out}}{[1+A\beta]} = \frac{5000}{40} = 125.$$

EXAMPLE 7.11

The basic Amplifier has a gain of -1000 and $\beta = -0.10$. If due to temperature change, the Amplifier gain changes by 10%. Calculate the percentage change in the gain of Amplifier with feedback.

Solution: The negative feedback introduced into the Amplifier reduces the amount of changes in gain due to temperature changes considerably.

$$\left|\frac{dA_F}{A_F}\right| = \left|\frac{1}{A\beta}\right|\left|\frac{dA}{A}\right|$$

$$\left|\frac{dA_F}{A_F}\right| = \frac{1}{(-1000)(-0.10)}\times 10\% = \frac{10}{100}\% = 0.1\%.$$

This is one of the major advantages of negative Feedback Amplifiers while there is reduction in gain of the Amplifier (a disadvantage).

EXAMPLE 7.12

An Amplifier has an open loop gain of 1000 with an input signal voltage V_{in} of 10 mV and second Harmonic distortion B_2 is 10%. If 40 dB negative Voltage Series Feedback is applied to develop same output signal voltage with the same distortion content $\beta = 0.1$. Determine (a) Desensitivity factor $[1 + A\beta]$; (b) Gain with feedback; (c) Second Harmonic distortion; (d) Required input voltage.

Solution:

a. Magnitude of feedback = 40 dB and $V_{in} = 10$ mV

$$\text{Feedback} = 40\text{ dB} = 20\log_{10}\frac{1}{1+A\beta}$$

$$\Rightarrow\ 40 = -20\log_{10}(1+A\beta)$$

$$\Rightarrow\ \log_{10}(1+A\beta) = 2$$

$$\therefore\ \text{Desensitivity factor}\quad D = [1+A\beta] = \text{anti}\log 2 = 100.$$

b. Open loop gain $A = 1000$ and feedback factor $\beta = 0.01$.

$$\text{Voltage gain of feedback amplifier}\quad A_F = \frac{A}{[1+A\beta]} = \frac{1000}{100} = 10.$$

c. Second harmonic distortion content $B_2(\text{FB}) = \dfrac{B_2}{D} = \dfrac{B_2}{[1+A\beta]}$

$$\therefore\ B_2(\text{FB}) = \frac{B_2}{[1+A\beta]} = \frac{10\%}{100} = 0.1\%$$

Therefore, second harmonic distortion has reduced to very small magnitude.

d. Required magnitude of input voltage V_S

$$V_S = \frac{V_{out}}{A_F} = \frac{A\cdot V_{in}}{A_F} = \frac{1000\times 10\times 10^{-3}}{10} = 1\text{ V}.$$

---

g. Improvement in Frequency Response (bandwidth) of the Amplifier

$$A_{HF} = \frac{A_m}{\left[1+j\left(\frac{f}{f_2}\right)\right]} \tag{7.23}$$

where A_m is the mid-frequency gain without feedback and A_{HF} = High-frequency gain without feedback.

In negative feedback amplifier 3-dB bandwidth (passband) stretches or increases because of reduction in gain with feedback (Gain × Bandwidth = constant).

The overall gain of negative feedback amplifier A_{FB} (HF) becomes

$$A_{FB}(\text{HF}) = \frac{A_{HF}}{[1+\beta A_{HF}]} \tag{7.24}$$

$$A_{FB}(\text{HF}) = \frac{\dfrac{A_m}{\left[1+j\left[\frac{f}{f_2}\right]\right]}}{\left[1+\beta\cdot\dfrac{A_m}{\left[1+j\frac{f}{f_2}\right]}\right]} = \frac{A_m}{\left[1+j\frac{f}{f_2}+\beta A_m\right]} = \frac{A_m}{\left[1+A_m\beta+j\frac{f}{f_2}\right]}. \tag{7.25}$$

Dividing both numerator and denominator of Eq. (7.25) by $[1 + A_m \cdot \beta]$, we get

$$A_{FB}(HF) = \frac{\left[\dfrac{A_m}{1+A_m\beta}\right]}{\left[\dfrac{1+A_m\beta+j\dfrac{f}{f_2}}{1+A_m\beta}\right]} = \frac{A_m(FB)}{\left[1+j\dfrac{f}{f_2[1+A_m\beta]}\right]} \tag{7.26}$$

$$\therefore \quad A_{FB}(HF) = \frac{A_m(FB)}{\left[1+j\dfrac{f}{f_2(FB)}\right]}, \quad \text{where } f_2(FB) = f_2[1+A_m\beta]. \tag{7.27}$$

From Eq. (7.27), high-frequency cut-off point, $f_2(FB) = f_2\,[1 + A_m \cdot \beta]$. Normally, the low-frequency cut-off point occurs at very low frequency. So, the Bandwidth of 'Negative Feedback Amplifier' increases by the factor $[1 + A_m \cdot \beta]$.
(The gain bandwidth product of an Amplifier is constant. Reduction in gain of a negative Feedback Amplifier causes an increase in the Amplifier bandwidth.)

- From Eq. (7.26), mid-frequency amplification with feedback $A_m(FB)$ is equal to the mid band amplification without feedback A_m divided by $[1 + \beta A_m]$.

$$A_m(FB) = \frac{A_m}{1+\beta A_m} \tag{7.28}$$

- The upper 3-dB frequency with feedback $f_2(FB)$ equals to the corresponding 3-dB frequency without feedback f_2 multiplied by the factor $[1 + \beta A_m]$.
- Similarly, it can be shown that lower −3 dB frequency with feedback $f_1(FB)$ is *decreased by the same factor* $[1 + \beta A_m]$

$$f_1(FB) = \frac{f_1}{1+\beta A_m}. \tag{7.29}$$

h. Gain Bandwidth product with feedback

$$A_m(FB)\cdot f_2(FB) = \frac{A_m}{1+\beta A_m}\cdot f_2(1+\beta A_m) = A_m \cdot f_2 \tag{7.30}$$

Thus, the gain bandwidth products with feedback and without feedback for Amplifier circuits are the same.

Using 'negative feedback' in Amplifier circuits, the lower half-power frequency is decreased and higher half-power frequency is increased, which is shown in the 'Frequency response of Amplifier circuits for both the situations in Fig. 7.2. The low-frequency cut-off f_1 is very much less than the high-frequency cut-off f_2 for Amplifiers. Hence, the bandwidth of the Amplifiers is usually decided by the high-frequency response of the Amplifier circuits. With negative feedback, the Amplifier gain is reduced and hence the Amplifier bandwidth (BW) is increased.

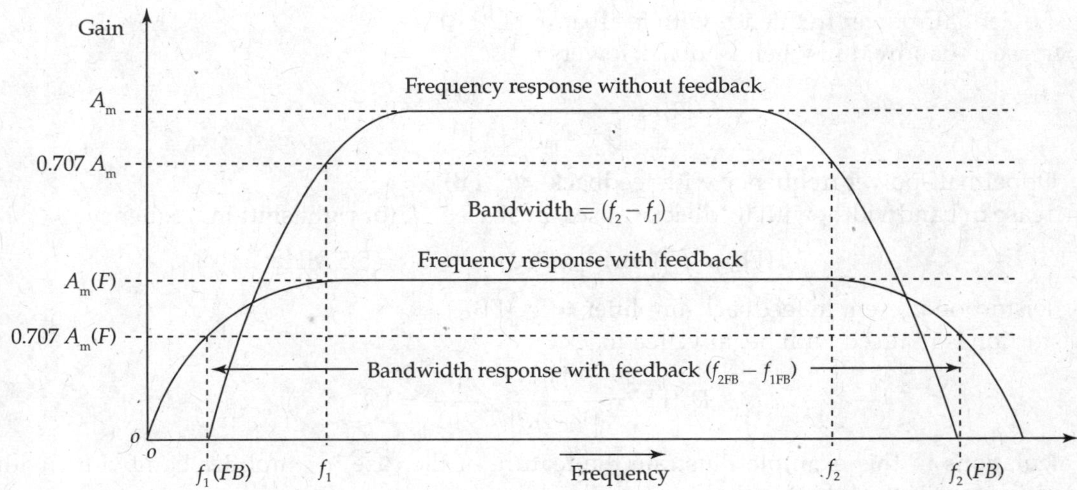

FIG. 7.4 *Frequency response of amplifiers with and without negative feedback*

Advantages of negative Feedback Amplifier (1) There is reduction in frequency and phase shift distortion. (2) Stability of the gain is improved. (3) Non-linear distortion is reduced. (4) Noise is reduced. (5) Input and output impedances of various Amplifiers can be modified. (6) Bandwidth stretching occurs due to increase in the 3 dB Bandwidth of the Amplifier. But the gain-bandwidth product of the Amplifier is unaltered.

Disadvantages (1) Amplifier gain gets reduced with negative feedback. (2) The advantages overweigh the disadvantages.

So, negative feedback is extensively used in Amplifiers, complex control systems, Robotics, Mechatronics, econometrics and modelling.

EXAMPLE 7.13

An RC-coupled Amplifier has mid-band voltage gain A_m of 1000. The half-power frequencies are $f_1 = 500$ Hz, $f_2 = 550$ kHz and distortion content D of 10% in the Amplifier without feedback. Find the Amplifier voltage gain A_{VF}, lower cut-off frequency with feedback f_1(FB) and upper cut-off frequency with feedback f_2(FB) and distortion with feedback D_F (FB) when negative feedback is introduced into the Amplifier with a feedback factor β of 0.009.

Solution: Open loop gain $A_m = 1000$, feedback factor $\beta = 0.009$, lower cut-off frequency $f_1 = 500$ Hz, upper cut-off frequency $f_2 = 550$ kHz, distortion $D = 10\%$.
Factor $[1 + A_m \cdot \beta] = [1 + 1000 \times 0.009] = [1 + 9] = 10$

a. Voltage gain with feedback

$$A_F = \frac{A_m}{[1+A_m\beta]} = \frac{1000}{[1+1000\times 0.009]}$$

$$= \frac{1000}{10} = 100.$$

b. Lower half-power frequency with feedback = f_1 (FB)
Increase in bandwidth, when feedback lowers f_1

$$\therefore \quad f_1(\text{FB}) = \frac{f_1}{[1+A_m\beta]} = \frac{500}{10} = 50 \text{ Hz.}$$

c. Upper half-power frequency with feedback = f_2 (FB)
Increase in bandwidth, with feedback causes, increase in f_2 (or right shift in frequency f_2)

$$\therefore \quad f_2(\text{FB}) = f_2[1+A\beta] = 550\times10^3\times10 = 5.5 \text{ MHz.}$$

d. Distortion present in feedback amplifier = D_F (FB)
Distortion is reduced with negative feedback

$$\therefore \quad D_F(\text{FB}) = \frac{D_F}{[1+A\beta]} = \frac{10\%}{10} = 1\%.$$

Calculations in this example illustrate the feature of increase in amplifier bandwidth with negative feedback at the sacrifice in amplifier gain.

EXAMPLE 7.14

An Amplifier without feedback has voltage gain A = 5000 with lower cut-off frequency f_L = 2.5 Hz upper cut-off frequency f_H = 100 kHz. Output resistance R_o = 2000 Ω. If negative feedback is applied to make the output resistance R_{out}(FB) = 600 Ω. Calculate the values of feedback factor β, voltage gain A_{FB} and half-power frequencies of the Feedback Amplifier. (Basic Electronics, II B.Tech. II-Sem. Mech. April/May 2008)

Solution:

a. Calculation of gain of the feedback amplifier A_{FB}
Output resistance of internal amplifier R_0 = 2000 Ω.
Output resistance with feedback R_{out} (FB) = 600 Ω.

$$R_{out}(\text{FB}) = \frac{R_{out}}{[1+A\beta]}$$

$$\therefore D = [1+A\beta] = \frac{R_{out}}{R_{out}(\text{FB})} = \frac{2000}{600} = \frac{20}{6} = 3.33.$$

Gain A of internal amplifier A = 5000.
Therefore, voltage gain of the feedback amplifier

$$A_{FB} = \frac{A}{[1+A\beta]} = \frac{5000\times6}{20} = 1500.$$

b. Calculation of the feedback factor β

$$[1 + A\beta] = 3.33$$

$$\therefore \quad A\beta = [3.33 - 1] = 2.33$$

Gain of internal amplifier A = 5000.

$$\text{Feedback factor} \quad \beta = \frac{2.33}{5000} = 0.466\times10^{-3}.$$

c. Upper cut-off frequency f_H (FB) of the feedback amplifier increases by a factor $(1 + A\beta)$

$$f_H\text{ (FB)} = f_H(1 + A\beta) = 100 \times 10^3 \times 3.33 = 333\text{ kHz.}$$

d. Lower cut-off frequency f_L (FB) of the amplifier decreases by a factor $[1 + A\beta]$

$$\therefore \quad f_l\text{(FB)} = \frac{f_L}{[1+A\beta]} = \frac{2.5}{3.33} = 0.75\text{ Hz.}$$

Classification of negative Feedback Amplifiers

	Type of feedback amplifier	Output connection	Input connection
1	(Voltage Amplifier) Voltage Series Feedback Amplifier	Shunt	Series
2	(Transresistance Amplifier) Voltage Shunt Feedback Amplifier	Shunt	Shunt
3	(Transconductance Amplifier) Current Series Feedback Amplifier	Series	Series
4	(Current Amplifier) Current Shunt Feedback Amplifier	Series	Shunt

The above classification is based on the principles of working of the Amplifiers.

1. Sampling of voltage or current at the output port:

- *Voltage*: Voltage can be sampled only by connecting a variable resistance, a potentiometer across the output terminals (output port of the Amplifier).

  Example: Voltage Series Feedback and Voltage Shunt Feedback Amplifiers

- *Current:* Current can be sampled by breaking the circuit at the output port and adding the feedback network in series. By adding the feedback network automatically couples the output into the input port.

  Example: Current series Feedback Amplifier and Current Shunt Feedback Amplifiers

2. Type of feedback signal connection into the input port:

- If the feedback quantity is to be connected in *series* with the input source, the quantity should always be voltage, i.e. it is V_F irrespective of whether the parameter sampled is voltage or current.

  Example: Voltage Series Feedback and current series Feedback Amplifiers

- On the other hand, if it is to be connected into the input port in *shunt* to input source, then it should always be current, i.e. it is I_F irrespective of voltage or current sampling being done at the output port.

  Example: Voltage Shunt feedback and Current Shunt Feedback Amplifiers

  We can neither connect in series a current sample to a current source nor can connect a voltage sample across a voltage source.

- So, invariably series feedback requires V_F and shunt feedback requires I_F

7.5 VOLTAGE AMPLIFIER (VOLTAGE SERIES FEEDBACK AMPLIFIER)

7.5.1 Block Diagram Configuration of Voltage Series Feedback Amplifier (Shunt-Series Type Amplifier)

- In Voltage Series Feedback Amplifiers, output voltage V_{out} is sampled and connected in series with the external input signal V_S at the input port of the Amplifier (Fig. 7.2).
- The feedback voltage V_F is applied through a two-port feedback network so as to cause negative feedback.
- Effective input signal to the internal Amplifier reduces and reduces the overall gain of the Feedback Amplifier.
- When there is no feedback process, feedback voltage $V_F = 0$. Then the voltage gain A_V of the simple Amplifier stage is the simple ratio of the output voltage V_{out} to the input voltage V_{in}.
- From Eq. (7.31), it is seen that, when a feedback signal V_F is connected in series opposition with the input signal, the voltage gain of the Voltage Series Feedback Amplifier A_F is reduced by a factor $(1 + A\beta)$

$$\text{Effective input voltage} \quad V_{in} = (V_S - V_f)$$

$$V_{out} = A \cdot V_{in} = A \cdot [V_S - V_f]$$

$$\therefore \quad V_{out} = A \cdot [V_S - \beta \cdot V_{out}], \quad V_{out}[1 + A\beta] = A \cdot V_S$$

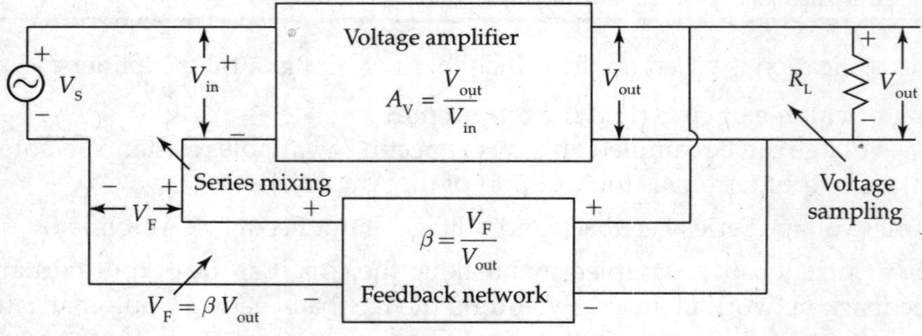

FIG. 7.5 *Voltage amplifier with voltage series feedback topology (voltage sampling and series mixing configuration)*

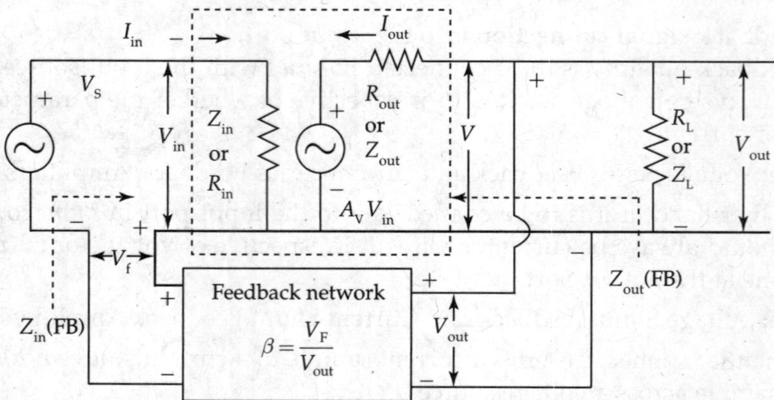

FIG. 7.6 *Amplifier equivalent circuit embedded into voltage series feedback amplifier block diagram*

$$\therefore \quad A_{FB} = A_F = \frac{V_{out}}{V_S} = \frac{A}{[1 + A\beta]}. \tag{7.31}$$

7.5.2 Effects of Feedback on the Amplifier Characteristics

h-parameter equivalent circuit of voltage series Feedback Amplifier to calculate input impedance Z_{in} (FB) and output impedance Z_{out} (FB) is shown in Fig. 7.6.

Input Impedance Z_{in}(FB)
From Fig. 7.6

$$\text{Input current} \quad I_{in} = \frac{V_{in}}{Z_{in}} = \frac{[V_S - V_f]}{Z_{in}} = \frac{V_S - \beta \cdot V_{out}}{Z_{in}}$$

$$\therefore \quad I_{in} \cdot Z_{in} = [V_S - \beta \cdot V_{out}] = [V_S - \beta A \cdot V_{in}] = [V_S - A\beta \cdot Z_{in} \cdot I_{in}]$$

$$\text{Rearranging the terms} \quad Z_{in} \cdot I_{in} \cdot [1 + A\beta] = V_S$$

$$\therefore \quad \text{Input impedance with feedback } Z_{in}(\text{FB}) = \frac{V_S}{I_{in}} = Z_{in} \cdot [1 + A\beta]\,\Omega. \tag{7.32}$$

Input impedance Z_{in} (FB) of voltage series feedback amplifier = Z_{in} $[1 + A\beta]$
The input impedance of the Voltage Series Feedback Amplifier Z_{in} (FB) increases by a factor $(1 + A\beta)$.

Output Impedance Z_{out}(FB)
Z_{out} (FB) is the impedance looking at the output port of the Amplifier with R_L disconnected and signal V_S is made zero. Therefore, source current $I_S = 0$.
Applying a voltage V across the output terminals, the resulting current I is calculated.
Output impedance of Feedback Amplifier

$$Z_{out}(\text{FB}) = \frac{V}{I}. \tag{7.33}$$

From the equivalent circuit diagram of Fig. 7.6

$$V = I \cdot Z_{out} + A_V \cdot V_{in} \quad \text{at the output port of the amplifier.} \tag{7.34}$$

When $V_S = 0$,

$$V_{in} = [V_S - V_F] = -V_F = -\beta \cdot V_{out} = -\beta V \quad \text{at amplifier input port.} \tag{7.35}$$

Therefore, Eq. (7.34) becomes

$$V = I \cdot Z_{out} - A_V \cdot \beta V \tag{7.36}$$

using $V_{in} = -\beta V$ in Eq. (7.34).

$$\therefore \quad V[1 + A_V \cdot \beta] = I \cdot Z_{out} \tag{7.37}$$

From Eq. (7.37),

$$Z_{out}(\text{FB}) = \frac{V}{I} = \frac{Z_{out}}{[1 + A_V \cdot \beta]}\,\Omega. \tag{7.38}$$

By definition, the output impedance of the Voltage Series Feedback Amplifier is reduced by a factor $(1 + A_V \cdot \beta)$, where $(1 + A_V \cdot \beta)$ is called as 'Desensitivity Factor – D'.

Equation (7.39) is the output terminal impedance Z'_{out}(FB) of feedback Amplifier including Z_L, where $Z'_o = Z_o \parallel Z_L$ is the output terminal impedance without feedback.

$$Z'_{out}(FB) = \frac{Z'_o}{1 + A_V \cdot \beta}. \tag{7.39}$$

7.5.3 Emitter Follower

Voltage Series Feedback Amplifier circuit (Fig. 7.7) familiar to us is 'Emitter Follower' in Transistor Amplifier circuits. It uses the process of Voltage Series Feedback in Amplifiers.

- The output voltage developed across R_E is fully applied back to input port so that this feedback voltage V_F is in series opposition with the input voltage V_S.
- Effective input signal voltage V_{in} is the difference of the two voltages V_S and V_F
- Output voltage V_{out} is across the emitter and ground terminals.
- Output and input voltages are in phase with unity gain as analysed below.

Various features of the Emitter Follower circuit

- The emitter voltage follows the variations in the input signal voltage and hence the Amplifier circuit is known as 'Emitter Follower circuit'. The feedback factor

$$\beta = \frac{V_F}{V_{out}} \cong 1. \tag{7.40}$$

Expression for voltage gain A_F of emitter follower circuit is

$$V_{out} = I_E \cdot R_E = [1 + h_{fE}] I_B \cdot R_E$$

$$\text{But } V_S = [h_{iE} + (1 + h_{fE}) R_E] I_B$$

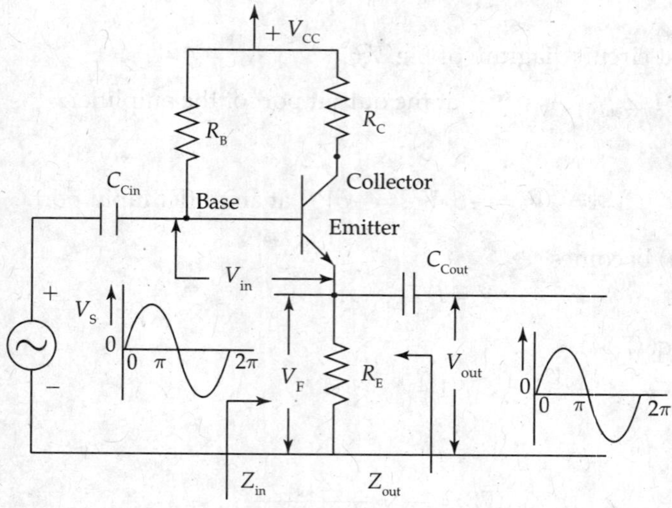

FIG. 7.7 *Voltage series feedback amplifier (emitter follower)*

$$\therefore \text{ Voltage gain } \quad A_F = \frac{V_{out}}{V_S} = \frac{[1+h_{fE}]I_B \cdot R_E}{[h_{iE}+(1+h_{fE})R_E]I_B} \cong 1 \quad [\because h_{iE} \ll [1+h_{fE}]R_E] \qquad (7.41)$$

- Input impedance Z_{in} (FB) (Eq. (7.32)) is very high.
- Output impedance Z_{out}(FB) (Eq. (7.38)) is very low.

Applications of Emitter Follower Circuit

- Emitter Follower circuit is used in the front end of various electronic test instruments.
- Emitter Follower circuit is used as a 'unity gain buffer Amplifier', while cascading Amplifiers as in multistage Amplifiers.

7.5.4 FET Source Follower Circuit

Source follower circuit is one type of Voltage Series Feedback Amplifier circuit (Fig. 7.8). It explains the operation with Voltage Series Feedback arrangement (Common Drain Amplifier to illustrate Voltage Series Feedback Amplifier operation).

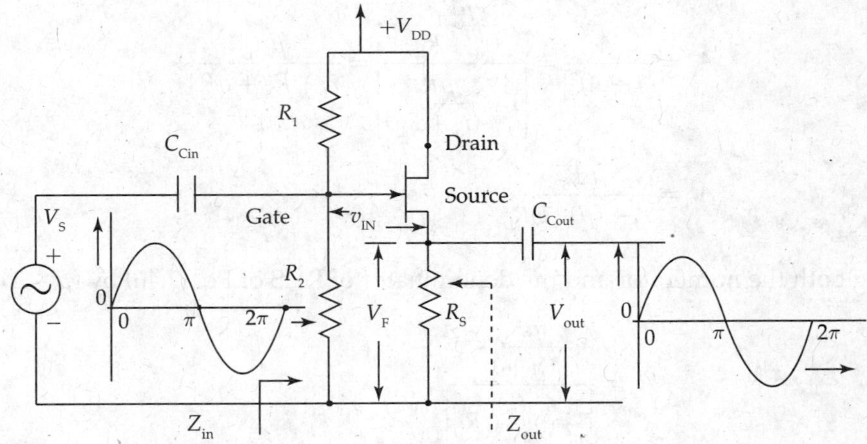

FIG. 7.8 *Voltage series feedback amplifier (source follower)*

Output voltage is across the source and the common terminals. Output voltage V_{out} across R_L or R_S is sampled and feedback to the input so that V_F is in series opposition with the input signal voltage V_S so that $V_{in} = (V_S - V_F)$. So, gain of Feedback Amplifier is reduced.

Voltage gain of source follower circuit is unity. It is a non-inverting Amplifier. There is no phase shift between output and input voltages. But there will be increase in input impedance and much decrease in output impedance. These properties are of much advantage in using the source follower as buffer Amplifier. Output voltage at the source terminal follows the input signal voltage variations. So, the circuit is known as 'source follower'. The circuit analysis follows:

From the nature of feedback arrangement shown in the source follower circuit of Fig. 7.8, feedback factor β is expressed by

$$\beta = \frac{V_F}{V_{out}} = 1. \qquad (7.42)$$

Voltage gain of FET amplifier

$$A = g_m \cdot R_L \tag{7.43}$$

Gain can also be expressed as

$$A = \frac{\mu R_L}{r_d + R_L}, \tag{7.44}$$

where R_L is the load resistance and r_d is the drain resistance of FET device.

Gain of feedback amplifier = A_F

Using the gain Eq. (7.43)

$$A_F = \frac{A}{[1 + A\beta]} = \frac{g_m \cdot R_L}{[1 + g_m \cdot R_L]} \tag{7.45}$$

using the value of $\beta = 1$ for source follower circuit.

Using the gain Eq. (7.44)

$$A_F = \frac{A}{[1 + A\beta]} = \frac{\dfrac{\mu R_L}{(r_d + R_L)}}{\left[1 + \dfrac{\mu R_L}{(r_d + R_L)}\right]} = \frac{\mu R_L}{r_d + R_L + \mu R_L}$$

$$\therefore \quad A_F = \frac{\mu R_L}{r_d + (\mu + 1)R_L}. \tag{7.46}$$

Dividing both the numerator and the denominator of RHS of Eq. (7.46) by $(\mu + 1)$, we get

$$A_F = \frac{\dfrac{\mu R_L}{(\mu + 1)}}{\left[\dfrac{r_d}{(\mu + 1)} + R_L\right]}$$

Generally load resistance $R_L \gg \dfrac{r_d}{(\mu + 1)}$

$$\text{Then} \quad A_F \cong \frac{\mu}{(\mu + 1)} \cong 1. \tag{7.47}$$

So the gain of source follower is unity.

Source follower circuit is used as unity gain amplifier known as buffer amplifier while acting as impedance transformer in multistage amplifier circuits.

EXAMPLE 7.15

Source follower has R_L = 2.5 kΩ. FET has $g_m = 400 \times 10^{-6}$ mhos, $\mu = 20$ and r_d = 50 kΩ. Calculate Gain A, Feedback Amplifier gain A_F, Z_{out}(FB), Z_{in}(FB) and feedback factor β for the source follower circuit (Voltage Series Feedback Amplifier).

Solution: Gain of FET amplifier

$$A = \frac{\mu R_L}{[r_d + R_L]} = \frac{20 \times 2.5 \times 10^3}{[50 \times 10^3 + 2.5 \times 10^3]}$$

$$= \frac{50 \times 10^3}{52.5 \times 10^3} \cong 0.95$$

For source follower feedback factor $\beta = 1$

Desensitivity factor $D = [1 + A\beta] = [1 + 0.95 \times 1] = 1.95$

The input resistance of FET device Z_{in} is very high.

Input impedance of source follower Z_{in} (FB) = Z_{in} $[1 + A\beta]$

Input impedance of feedback amplifier Z_{in} (FB) = Z_{in} $[1 + A\beta]$ is very large.

Output impedance of FET amplifier $Z_{out} = r_d \| R_L$

$\therefore$ Output impedance $Z_{out} = 50 \| 2.5 = 2.38\ k\Omega$

Output impedance of source follower $Z_{out}(FB) = \frac{Z_{out}}{[1 + A\beta]}$

$$\therefore\ Z_{out}(FB) = \frac{2.38\ k\Omega}{1.95} = 1220\ \Omega.$$

So, the output impedance of source follower reduces to a very small value. These calculations show that source follower can be used as impedance transformer and buffer amplifier.

7.5.5 Voltage Series Feedback Amplifier Circuit Using a BJT

In the 'Voltage Series Feedback Amplifier' circuit of Fig. 7.9, forward bias to the emitter junction and reverse bias to the collector junction of the Transistor are provided by the supply voltage V_{CC} and R_1 and R_2 so that the Transistor T acts as a basic Amplifier. Emitter resistor R_E stabilises the bias.

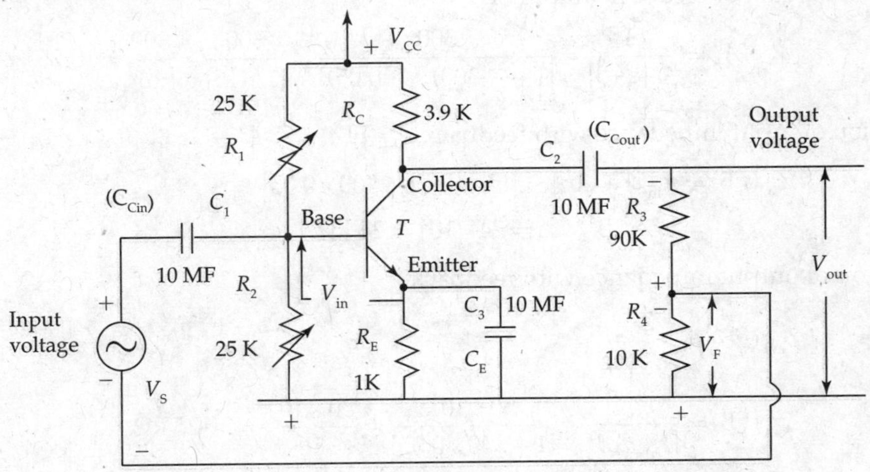

FIG. 7.9 *Voltage series feedback amplifier (practical circuit)*

The decoupling capacitor C_E keeps the emitter of the Transistor at AC ground. The output voltage V_{out} is sampled by using a potential divider using the two resistors R_3 and R_4. The voltage across the resistor R_4 provides the feedback voltage V_F.

$$\text{Feedback voltage} \quad V_F = \left[\frac{V_{out} \times R_4}{(R_3 + R_4)}\right] \tag{7.48}$$

$$\text{Feedback factor} \quad \beta = \frac{V_F}{V_{out}} = \frac{R_4}{(R_3 + R_4)}. \tag{7.49}$$

If $R_3 = 90$ kΩ and $R_4 = 10$ kΩ, then

$$\beta = \frac{R_4}{(R_3 + R_4)} = \frac{10\text{ k}\Omega}{[90\text{ k}\Omega + 10\text{ k}\Omega]}$$

$$= \frac{10 \times 10^3}{100 \times 10^3} = \frac{1}{10} = 0.1.$$

If the voltage gain of the Amplifier A_V without feedback is much greater than one, the gain of the negative Feedback Amplifier A_{NFB} is equal to $[\beta]^{-1}$. Then the gain of the feedback Amplifier $A_F = 10$, using the value of $\beta = 0.1$ as calculated previously.

From the circuit (Fig. 7.9), it is clear that V_F is connected in series opposition with V_S (signal voltage) thus contributing 'Voltage Series negative feedback' in the Amplifier circuit. The one disadvantage with the circuit is that the 'signal source' cannot be grounded and should be isolated from the ground.

EXAMPLE 7.16

Calculate the gain, input impedance and output impedance of Voltage Series Feedback Amplifier having gain $A = -300$, $R_{in} = 1.5$ kΩ and $R_{out} = 50$ kΩ, $\beta = 0.05$.

Solution:

a. Calculation of gain of feedback amplifier:

$$A_F = \frac{A}{[1 + A\beta]} = \frac{300}{[1 + (-300) \times (-0.05)]} = \frac{300}{[1 + 15]} = \frac{300}{16} = 18.75.$$

b. Calculation of input impedance with feedback Z_{in} (FB) $= Z_{in}$ $[1 + A\beta]$

$$Z_{in}(\text{FB}) = R_{in}[1 + A\beta] = 1.5 \times 10^3 [1 + 300 \times 0.05]$$

$$= 1.5 \times 10^3 \times 16 = 24 \times 10^3 = 24\text{ k}\Omega.$$

c. Calculation of output impedance with feedback

$$Z_{out}(\text{FB}) = \frac{Z_{out}}{[1 + A\beta]}$$

$$Z_{out}(\text{FB}) = \frac{R_{out}}{[1 + A\beta]} = \frac{50 \times 10^3}{[1 + 300 \times 0.05]} = \frac{50 \times 10^3}{16} = 3.125\text{ k}\Omega.$$

7.6 VOLTAGE SHUNT FEEDBACK AMPLIFIER (TRANSRESISTANCE AMPLIFIER)

7.6.1 Voltage Shunt Feedback Amplifier Block Diagram (Shunt–Shunt Feedback Amplifier)

The output voltage is sampled and the feedback output signal from the feedback network is connected in shunt across input signal and input port of the Amplifier. Voltage Shunt Feedback Amplifier is also called as 'shunt–shunt Feedback Amplifier' from the way the process goes on in the Amplifier circuit.

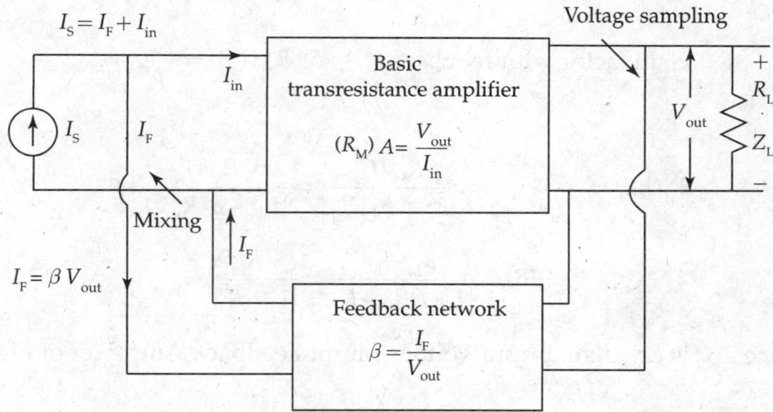

FIG. 7.10 *Block diagram of voltage shunt feedback amplifier (voltage sampling and shunt mixing topology)*

In Voltage Shunt Feedback Amplifier or Transresistance Amplifier, sampling parameter at the output port is the voltage V_{out}. The feedback connection at the input port of the Amplifier is shunt type and the feedback parameter is current I_F. In this way, the Amplifier gain R_M is the ratio of output voltage V_{out} to the input current I_{in}.

Voltage Shunt Feedback Amplifier is represented as equivalent circuit in Fig. 7.11.

$$\text{Transresistance } 'R_M' \text{ of feedback amplifier} \quad A_F = \frac{V_{out}}{I_S} = \frac{A \cdot I_{in}}{[I_{in} + I_f]} \tag{7.50}$$

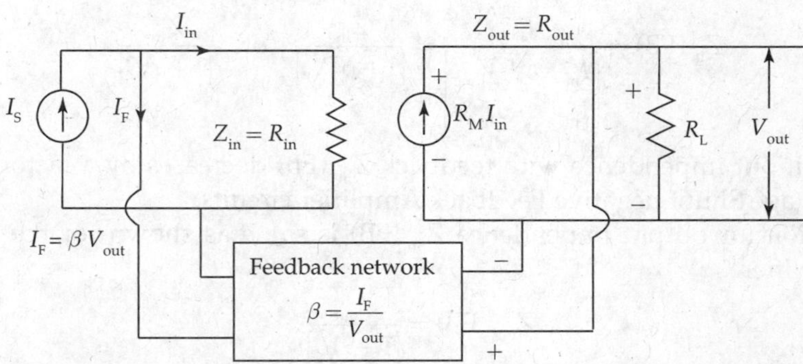

FIG. 7.11 *Equivalent circuit representing current controlled voltage source embedded into voltage shunt feedback amplifier block diagram*

$$A_F = \frac{A \cdot I_{in}}{[I_{in} + \beta \cdot V_{out}]} = \frac{A \cdot I_{in}}{[I_{in} + A\beta \cdot I_{in}]} = \frac{A \cdot I_{in}}{I_{in}[1 + A\beta]} = \frac{A}{[1 + A\beta]}$$

$$\therefore \quad A_F = \frac{A}{[1 + A\beta]} \tag{7.51}$$

$$\beta = \frac{I_F}{V_{out}} \text{ mhos} \tag{7.52}$$

$$\text{Gain factor} \quad A \text{ or } R_M = \frac{V_{out}}{I_{in}} \tag{7.53}$$

$$\text{Gain factor with feedback} \quad A_F \text{ or } R_M(\text{FB}) = \frac{V_{out}}{I_S}. \tag{7.54}$$

But $I_S = I_{in} + I_F$

$$R_M(\text{FB}) = \frac{V_{out}}{(I_{in} + I_F)} = \frac{R_M \cdot I_{in}}{[I_{in} + \beta \cdot V_{out}]} = \frac{R_M \cdot I_{in}}{[I_{in} + \beta \cdot R_M \cdot I_{in}]}$$

$$\therefore \quad R_M(\text{FB}) = \frac{R_M \cdot I_{in}}{[1 + \beta R_M] \cdot I_{in}} = \frac{R_M}{[1 + \beta R_M]}. \tag{7.55}$$

Input impedance Z_{inF} is calculated from Voltage Shunt Feedback Amplifier of Fig. 7.10.

$$\text{Input impedance} \quad Z_{in} = \frac{V_{in}}{I_{in}} \tag{7.56}$$

$$\text{Input impedance with feedback} \quad Z_{in}(\text{FB}) = \frac{V_{in}}{I_S} \tag{7.57}$$

But $I_S = I_{in} + I_F$

$$\therefore \quad Z_{in}(\text{FB}) = \frac{V_{in}}{(I_{in} + I_F)} = \frac{V_{in}}{(I_{in} + \beta \cdot V_{out})} \tag{7.58}$$

Dividing both the numerator and the denominator of Eq. (7.58), we get

$$Z_{in}(\text{FB}) = \frac{\dfrac{V_{in}}{I_{in}}}{\left[\dfrac{I_{in}}{I_{in}} + \dfrac{\beta \cdot V_{out}}{I_{in}}\right]} = \frac{Z_{in}}{[1 + \beta R_M]} \quad \text{or} \quad \frac{Z_{in}}{[1 + A\beta]}. \tag{7.59}$$

Therefore, the input impendence with feedback Z_{in} (FB) decreases by a factor $[1 + \beta \cdot R_M]$ or $[1 + A\beta]$ in Voltage Shunt negative Feedback Amplifier circuits.

The derivation for output impendence Z_{out}(FB) is same as shown for the Voltage Series Feedback Amplifier

$$Z_{out}(\text{FB}) = \frac{Z_{out}}{[1 + A\beta]}. \tag{7.60}$$

This type of Feedback Amplifier is called Transresistance Amplifier or I to V (Current to voltage) converter as shown in Fig. 7.12.

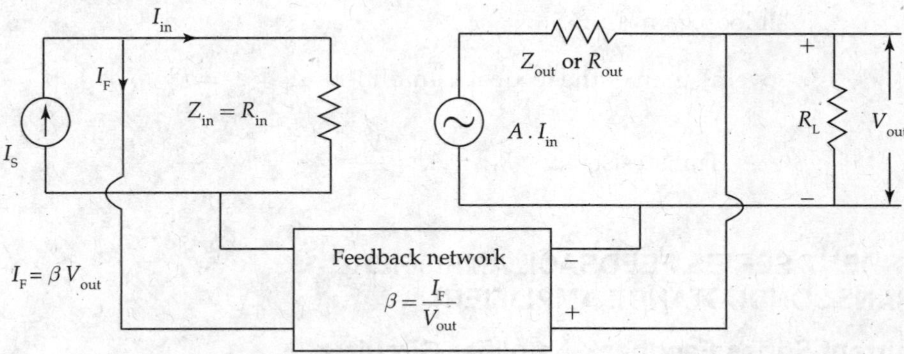

FIG. 7.12 *Thevinin's equivalent network embedded in voltage shunt feedback amplifier block diagram*

7.6.2 Voltage Shunt Feedback Amplifier Circuit

Voltage Shunt Feedback Amplifier in Fig. 7.13 is a Common Emitter Amplifier with the input signal V_S and the output voltage V_{out}. The resistor R_F is connected between the collector terminal and the Base terminal of the Transistor, sampling the output voltage and linking to the input port for feedback signal mixing with the input current. Feedback current I_F through the shunt resistor connecting the output and the input ports of the Feedback Amplifier is as follows:

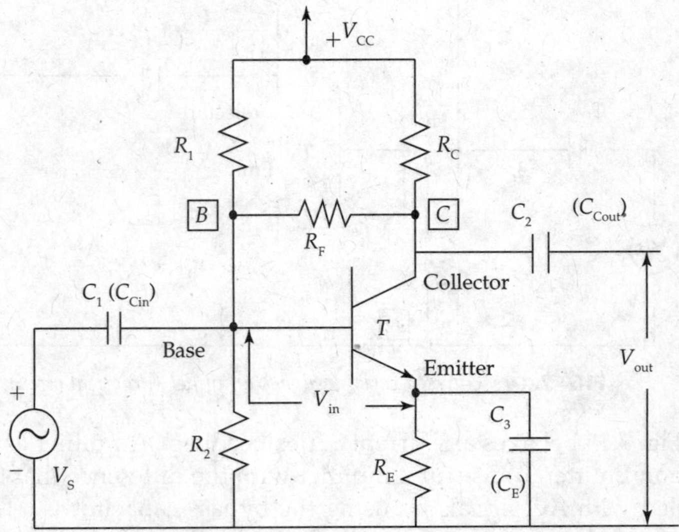

FIG. 7.13 *Voltage shunt feedback amplifier (practical circuit)*

$$I_F = \frac{[V_{in} - V_{out}]}{R_F} \cong -\frac{V_{out}}{R_F}, \quad V_{out} \rangle\rangle V_{in} \tag{7.61}$$

Since open loop gain A is very large

$$\therefore \quad I_F = \beta \cdot V_{out} \quad \text{considering } \beta = \frac{1}{R_F}$$

If loop gain $A\beta \gg 1$

overall gain of the feedback amplifier $A_F \cong \frac{1}{\beta} = R_F$ (7.62)

$$\text{Transresistance} \quad R_M = A_F = \frac{V_{out}}{I_S} = \frac{1}{\beta} = R_F. \tag{7.63}$$

7.7 CURRENT SERIES FEEDBACK AMPLIFIER (TRANSCONDUCTANCE AMPLIFIER)

7.7.1 Current Series Feedback Amplifier Circuit (SERIES–SERIES Type Amplifier)

Figure 7.14 represents a Current series feedback (Series–Series) Amplifier circuit, which is also known as Transconductance Amplifier.

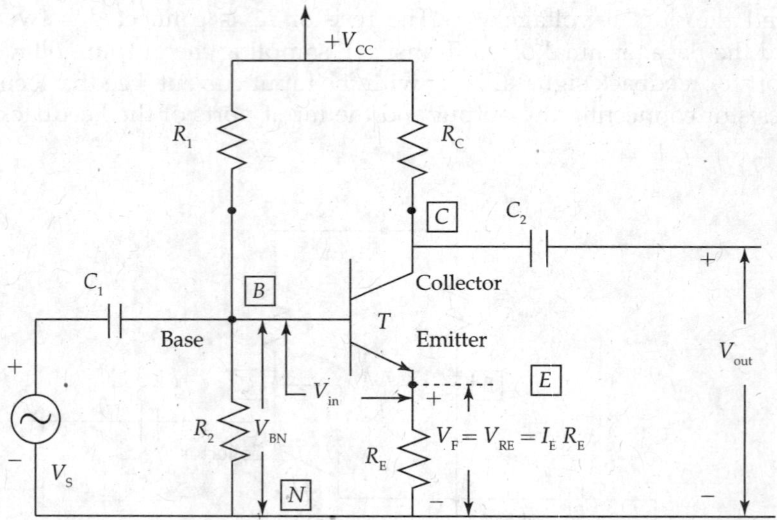

FIG. 7.14 *Current series feedback amplifier (practical circuit)*

The circuit in Fig. 7.14 behaves as Current Series Feedback Amplifier as the following. The circuit is a Common Emitter Transistor Amplifier with the difference that the Emitter resistor R_E is not decoupled with AC signals by using the bypass capacitor C_E. This means that the Emitter is not grounded even for AC signals. Since the Resistor R_E is common for both the input and the output circuits, it contributes or provides feedback. Since the output current flows through R_E, the sampling of the parameter is the 'current sampling' at the output port. This current I_E flowing through the Emitter resistor R_E develops a voltage $I_E \cdot R_E$ with the Emitter terminal positive compared to the Ground terminal N as shown in Fig. 7.14.

The voltage across R_E that is $I_E\, R_E$ known as V_{RE} is the feedback voltage, V_F, that opposes the voltage, V_{BN} across R_2 and feedback to the input port of the Amplifier. The actual signal input V_{in} to the Amplifier will be $V_{BN} - I_E \cdot R_E$, or $(V_S - I_E \cdot R_E)$ that is the input signal voltage V_S minus the feedback voltage V_F. Thus the sampling is series sampling of the current at the output port and coupling is series type of feedback voltage at the input port of the Amplifier for introducing negative feedback into the circuit. Hence the circuit is a Current Series Feedback Amplifier. The circuit is also known as series sampled-series Feedback Amplifier.

$$\text{Feedback factor} \quad \beta = \frac{V_F}{I_{out}} = \frac{I_E \cdot R_E}{I_{out}} = \frac{I_{out} \cdot R_E}{I_{out}} = R_E \tag{7.64}$$

$$\text{Gain factor with feedback} \quad g_m(F) = \frac{I_{out}}{V_S} \tag{7.65}$$

$$\therefore \quad g_m(F) = \frac{g_m}{[1 + g_m \cdot \beta]} \cong \frac{g_m}{g_m \cdot \beta} = \frac{1}{\beta} = \frac{1}{R_E} \tag{7.66}$$

$$\text{Gain of feedback amplifier} \quad A_F = -g_m(F) \cdot R_L = -g_m(F) \cdot R_C \tag{7.67}$$

$$\therefore \quad A_F = -g_m(F) \cdot R_C = -\frac{R_C}{R_E} \tag{7.68}$$

This can also be verified from gain expression using h-parameters

$$A_V = A_I \frac{Z_L}{Z_{in}} = -h_{fE} \frac{R_C}{[h_{iE} + (1 + h_{fE})R_E]} \cong -\frac{h_{fE} \cdot R_C}{(1 + h_{fE})R_E} = -\frac{R_C}{R_E} \tag{7.69}$$

Also it can be derived that

$$Z_{in}(F) = Z_{in}(1 + g_m \cdot \beta) = Z_{in}(1 + g_m \cdot R_E) \tag{7.70}$$

$$Z_{out}(F) = Z_{out}(1 + g_m \cdot \beta) = Z_{out}(1 + g_m \cdot R_E). \tag{7.71}$$

The block diagram in Fig. 7.15 explains the following features in the circuit.
(1) The concept of forward path Amplifier known as Transconductance Amplifier by context of the input and the output signals to the Amplifier. (2) The method of sampling the current

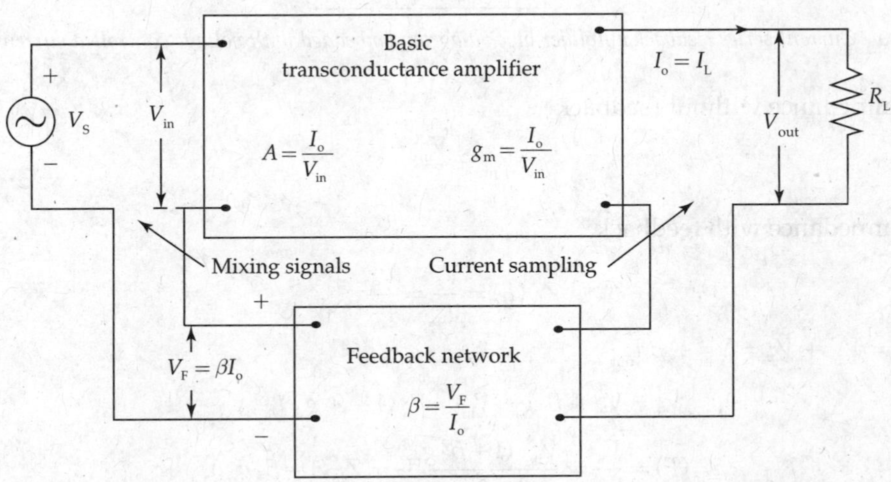

FIG. 7.15 *Transconductance amplifier (current series feedback amplifier) (series-series connected topology)*

at the output port. (3) The method of feeding the signal voltage V_F in series with V_S to cause mixing of the signals to provide negative feedback. Thus the figure explains the basic principles of Current Series Feedback Amplifier, which is also known as Transconductance Amplifier.

Gain factor without feedback

$$g_m = \frac{I_{out}}{V_{in}}. \tag{7.72}$$

Gain factor with feedback

$$g_m(F) = \frac{I_{out}}{V_S} \tag{7.73}$$

$$V_F = \beta \cdot I_{out} \quad \text{and} \quad V_S = V_F + V_{in}$$

$$\therefore \quad A_F = \frac{I_{out}}{V_S} = \frac{I_{out}}{(V_F + V_{in})}$$

$$A_F = \frac{I_{out}}{\beta \cdot I_{out} + \dfrac{I_{out}}{g_m}} = \frac{I_{out} \cdot g_m}{I_{out}(1 + g_m \cdot \beta)} = \frac{g_m}{(1 + g_m \cdot \beta)} \tag{7.74}$$

$$I_{out} = g_m V_{in}$$

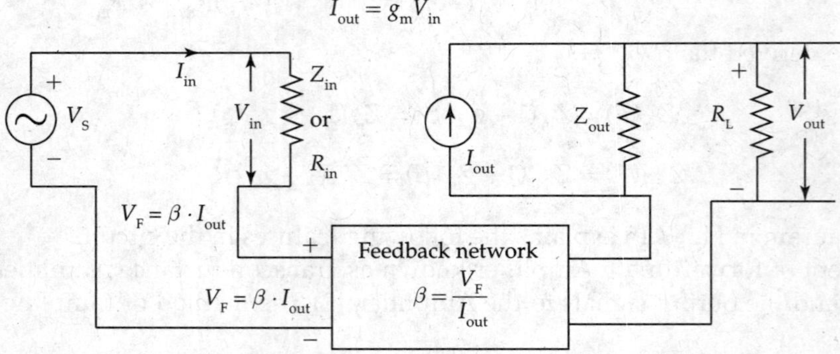

FIG. 7.16 *Current series feedback amplifier block diagram embedded with voltage controlled current source*

Input impedance without feedback

$$Z_{in} = \frac{V_{in}}{I_{in}}$$

Input impedance with feedback

$$Z_{in}(F) = \frac{V_S}{I_S} = \frac{V_S}{I_{in}} \tag{7.75}$$

But $V_S = V_{in} + V_F = V_{in} + \beta \cdot I_{out}$

$$\therefore \quad V_S = V_{in} + \beta \cdot g_m \cdot V_{in} = V_{in}(1 + \beta \cdot g_m) \tag{7.76}$$

$$Z_{in}(F) = \frac{V_S}{I_{in}} = \frac{V_{in} \cdot (1 + \beta \cdot g_m)}{I_{in}} = Z_{in}(1 + \beta \cdot g_m). \tag{7.77}$$

7.8 CURRENT SHUNT FEEDBACK AMPLIFIER CURRENT (SERIES-SHUNT) AMPLIFIER

The output current is sampled in this Feedback Amplifier and applied in shunt to input of the Amplifier. It is also called series derived shunt fed feedback.

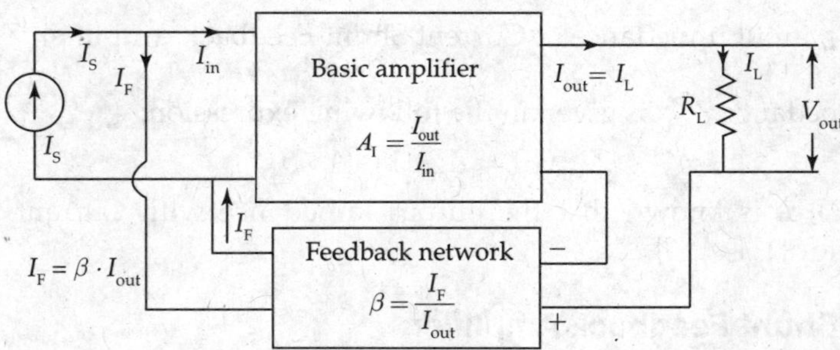

FIG. 7.17 *Block diagram of current shunt feedback amplifier (current sampling and shunt mixing arrangement)*

$$A_I = \frac{I_{out}}{I_{in}} = \text{current gain factor without feedback}$$

$$A_{IF} = \frac{I_{out}}{I_S} = \text{current gain with feedback}$$

$$I_S = (I_{in} + I_F) = (I_{in} + \beta \cdot I_{out}) \quad \text{using } I_F = \beta \cdot I_{out}$$

$$\therefore \quad A_{IF} = \frac{I_{out}}{I_S} = \frac{I_{out}}{I_{in} + \beta \cdot I_{out}} = \frac{\dfrac{I_{out}}{I_{in}}}{\dfrac{(I_{in} + \beta \cdot I_{out})}{I_{in}}} = \frac{A_I}{(1 + A_I \cdot \beta)} \tag{7.78}$$

Input Impedance with Feedback Z_{inF}

$$Z_{in} = \frac{V_{in}}{I_{in}} \quad \text{and} \quad Z_{inF} = \frac{V_{in}}{I_S} \tag{7.79}$$

$$\text{where} \quad I_S = [I_{in} + I_F] = [I_{in} + \beta \cdot I_o] \tag{7.80}$$

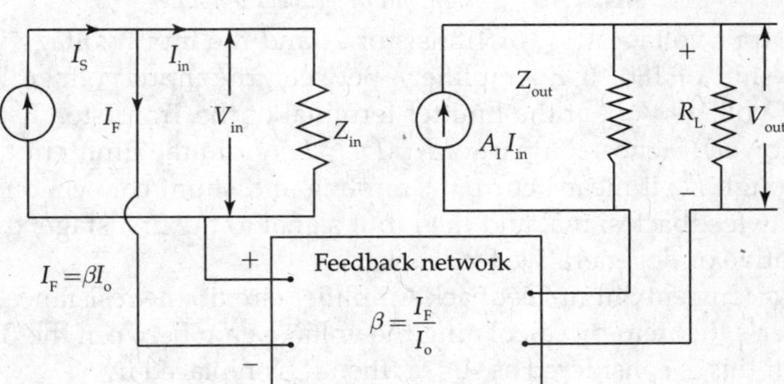

FIG. 7.18 *Equivalent circuit to determine input Impedance with feedback current shunt feedback amplifier – equivalent circuit*

$$Z_{in}(F) = \frac{V_{in}}{I_S} = \frac{V_{in}}{[I_{in} + \beta \cdot I_0]} = \frac{V_{in} / I_{in}}{[1 + \beta \cdot A_I]}$$

$$\therefore \quad Z_{in}(F) = \frac{Z_{in}}{[1 + A_I \cdot \beta]}. \tag{7.81}$$

From Eq. (7.81), input impedance of Current Shunt Feedback Amplifier, Z_{inF}, is reduced by the factor $[1 + A_I \cdot \beta]$.

The output impedance Z_{outF} is given in the following expression:

$$Z_{out}(F) = Z_{out}[1 + A_I \cdot \beta]. \tag{7.82}$$

From Eq. (7.82), it is known that the output impedance with Current Shunt Feedback increases by a factor $[1 + A_I \cdot \beta]$.

7.8.1 Current Shunt Feedback Amplifier

Current Shunt Feedback Amplifier (Fig. 7.19) is a two-stage Amplifier with the Emitter resistance R_{E2} of the second stage Amplifier Unbypassed and a resistor R_F is connected between the Emitter E_2 of Transistor T_2 and the Base B_1 of the Transistor T_1 to provide current sampling and shunt feedback arrangement as explained below.

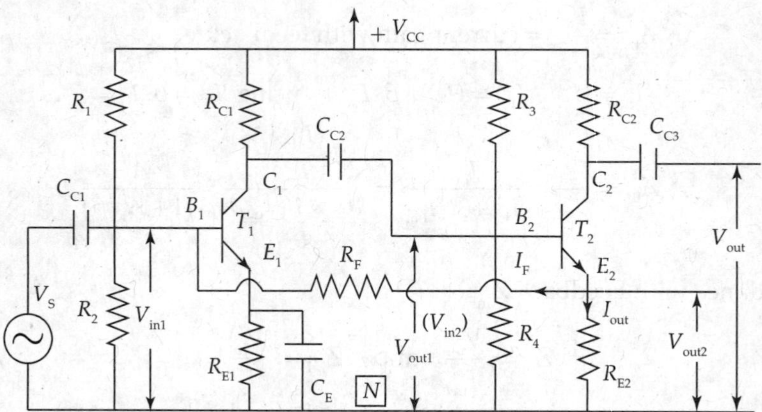

FIG. 7.19 *Current shunt feedback amplifier*

Between the output voltage $V_{out\text{-}1}$ of Transistor T_1 and the input voltage $V_{in\text{-}1}$ of Transistor T_1, there is phase shift of 180° (*CE* Amplifier). Between the input voltage $V_{in\text{-}2}$ of Transistor T_2 and the output voltage $V_{out\text{-}2}$ at the Emitter terminal of the Transistor T_2 there is no phase shift (Emitter Follower). Hence, if a part of I_{C2} ($I_{C2} \cong I_{E2}$) from the Emitter of T_2 is feedback at the Base of T_1 through R_F; it means current sampling and shunt connection with phase shift of 180° between the feedback signal and the input signal to the first stage, thus satisfying the condition for negative or degenerative feedback.

In the analysis of Current Shunt Feedback Amplifier circuit, the resistance R_F is replaced by applying the Miller's theorem by calculating the voltage gain between the Base of T_1 and the Emitter of T_2 and if this is considered as $A_{VB1.E2}$; then R_F is replaced by

$$R_{F1} = \frac{R_F}{1 - A_{VB1.E2}}. \tag{7.83}$$

The resistor R_F is across the two points B_1 and E_2 in the Amplifier circuit shown in Fig. 7.16. So, the feedback current I_F is according to the following equation

$$I_F = \frac{V_{\text{in-1}} - V_{\text{out-2}}}{R_F} \cong \frac{-V_{\text{out-2}}}{R_F} \qquad \because \; V_{\text{out-2}} \gg V_{\text{in-1}}$$

$$\text{But } V_{\text{out-2}} = -I_{\text{out}} \cdot R_{E2} \qquad I_F = \frac{I_{\text{out}} \cdot R_{E2}}{R_F} = \beta \cdot I_{\text{out}}$$

$$\therefore \; \text{Feedback factor} \quad \beta = \frac{R_{E2}}{R_F}.$$

Current gain with feedback = A_{IF}

It is already known that $A_F = (\beta)^{-1}$, when $A\beta \gg 1$.

Therefore, the current gain of the Feedback Amplifier, A_{IF}, where

$$A_{IF} = \frac{A}{1 + A\beta} \cong \frac{1}{\beta} = \frac{R_F}{R_{E2}}. \tag{7.84}$$

The output resistance R_{outF} with feedback is the parallel combination of the Emitter resistor R_{E2} of the second Transistor and the equivalent Miller resistance of feedback resistor at the output port $R_{F2} = R_F$

$$R_{\text{outF}} = \frac{R_{E2} \cdot R_F}{R_{E2} + R_F}. \tag{7.85}$$

On similar lines the input resistance with feedback R_{inF} is the parallel combination of the effective Miller resistance R_{F1} of the feedback resistor R_F at the input port of the first stage of the Amplifier and the input resistance h_{iE} of the first Transistor T_1:

The expression for $R_{F.1}$ is repeated here for convenience.

$$R_{F1} = \frac{R_F}{1 - A_{VB1.E2}} \qquad \therefore \; R_{\text{in}}(F) = \left[R_{F1} \| h_{iE}\right]. \tag{7.86}$$

Type of feedback				
Parameter mixing At input port	**Voltage series**	**Voltage shunt**	**Current series**	**Current shunt**
$Z_{\text{out}}(F)$ Output impedance	Decreases	Decreases	Increases	Increases
$Z_{\text{in}}(F)$ Input impedance	Increases	Decreases	Increases	Decreases
Type of amplifier	Voltage amplifier	Transresistance amplifier	Transconductance amplifier	Current amplifier
Stabilises amplifier performance	A_{VF}	R_{MF}	G_{MF}	A_{IF}
Amplifier bandwidth	Increases	Increases	Increases	Increases
Distortion	Decreases	Decreases	Decreases	Decreases
Feedback signal At input port	Voltage	Current	Voltage	Current
Sample signal At output port	Voltage	Voltage	Current	Current

Note: (1) For series comparison, $Z_{iF} > Z_{in}$ (2) For shunt mixing, $Z_{iF} < Z_{in}$

EXAMPLE 7.17

An Amplifier without feedback produces an output voltage V_{out} at the fundamental frequency of the signal of 18 V with 100% second harmonic distortion, when the input is 0.02 V. If 1% of the output is feedback to the input in Voltage Series Feedback Amplifier, calculate: (a) Output voltage, if the fundamental frequency is maintained at 18 V; (b) Desensitivity factor, if second harmonic distortion content is reduced to 10%; (c) Required input voltage to maintain 18 V for the Feedback Amplifier.

Solution:

a. Calculation of output voltage V_{out}

$$|A| = \frac{18}{0.02} = \frac{1800}{2} = 900 \qquad \beta = \frac{V_F}{V_{out}} = \frac{1}{100} = 0.01$$

$$A_F = \frac{A}{1+A\beta} = \frac{900}{1+900(0.01)} = \frac{900}{1+9} = \frac{900}{10} = 90$$

$$V_{out} = A_F \cdot V_{in} = 90 \times 0.02 = 1.8 \text{ V.}$$

b. Calculation of desensitivity factor $D = [1 + A\beta]$

$$B_{2F} = \frac{B_2}{[1+A\beta]}$$

$$\therefore \quad D = [1+A\beta] = \frac{B_2}{B_{2F}} = \frac{100}{10} = 10.$$

c. Calculation of required input for FB amplifier with $V_{out} = 18$ V

$$A_F = \frac{A}{[1+A\beta]} = \frac{900}{10} = 90$$

$$\therefore \quad V_{in}(\text{FB}) = \frac{V_{out}}{A_F} = \frac{18}{90} = 0.2 \text{ V.}$$

EXAMPLE 7.18

A negative Feedback Amplifier has open loop mid-band gain A_m of 1000. The 3-db frequencies f_1 and f_2 before the introduction of negative feedback into the Amplifier are 500 Hz and 5.5 MHz, respectively. The feedback factor $\beta = 9 \times 10^{-3}$. Calculate the parameters: (a) Gain with feedback $A_m(F)$, (b) Lower half-power frequency after feedback $f_1(F)$, (c) Upper half-power frequency after feedback $f_2(F)$, (d) Gain bandwidth product after feedback and (e) Gain bandwidth product before feedback.

Solution:

a. Open loop mid-band gain of feedback amplifier $A_m = 1000$
Feedback factor $\beta = 9 \times 10^{-3}$

$$\therefore \quad [1 + A_m \cdot \beta] = [1 + 1000 \times 9 \times 10^{-3}] = 10$$

Mid-band gain of feedback amplifier $$A_m(F) = \frac{A_m}{[1+A_m \cdot \beta]} = \frac{1000}{10} = 100.$$

b. Lower cut-off frequency after feedback f_1 (F)

$$f_1(F) = \frac{f_1}{[1 + A_m \cdot \beta]} = \frac{500}{10} = 50 \text{ Hz.}$$

c. Upper half-power frequency after feedback f_2 (F)

$$\therefore \quad f_2(F) = f_2 [1 + A_m \cdot \beta] = 5.5 \times 10^6 \times 10 = 55 \times 10^6 = 55 \text{ MHz.}$$

d. Bandwidth BW (F) of the negative feedback amplifier = 55 MHz
Gain Bandwidth product of feedback amplifier = A_m $(F) \times BW$ (F)

$$\therefore \quad A_m (F) \times BW (F) = 100 \times 55 = 5500 \text{ MHz.}$$

e. Bandwidth BW of amplifier before feedback = $A_m \times BW$

$$\therefore \quad A_m \times BW = 1000 \times 5.5 = 5500 \text{ MHz}$$

Thus it is clear that 'Gain Bandwidth' product of amplifiers is same.

The bandwidth of negative Feedback Amplifier has increased considerably at the sacrifice of gain A_F for negative Feedback Amplifier, satisfying the basic fact that the product of gain and bandwidth of Amplifiers is constant. At the same time, the distortion content is also reduced with negative feedback introduced into the Amplifier circuits.

7.9 VOLTAGE AND CURRENT SERIES FEEDBACK AMPLIFIERS (PRACTICAL CIRCUIT)

Aim:

1. To study the voltage and current series feedback configurations of Transistor Amplifiers.
2. To study the effects of feedback on voltage gains and bandwidths.
3. To observe the waveforms at various points in the Amplifier circuits and discuss on their use in practical circuits.

Circuit:

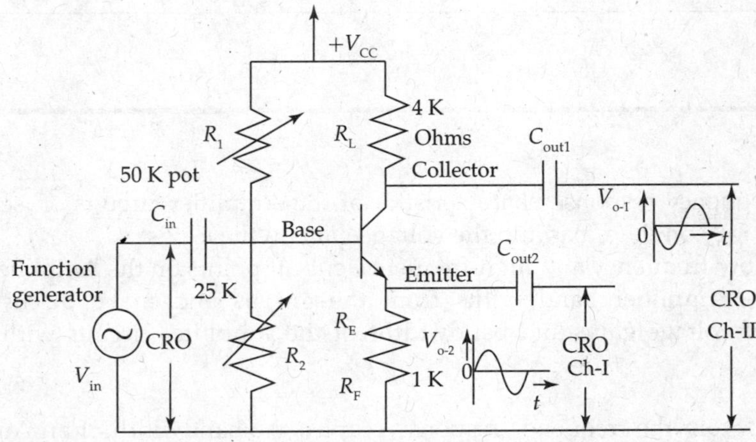

FIG. 7.20 *Voltage and current series feedback amplifier circuits*

Apparatus:

1. Voltage Series and Current Series Feedback Amplifier PCB Chassis
2. Function generator
3. Dual beam cathode ray oscilloscope
4. Transistor power supply

Observations:

1. Connections are made as per the circuit diagram.
2. **Voltage series feedback** Voltage V_{in} is the input signal voltage from the function generator and voltage at the Transistor Emitter point $V_{o\text{-}2}$ is the output voltage for Voltage Series Feedback circuit.
   - Observe the voltages V_{in} and $V_{o\text{-}2}$ and calculate the voltage gain.
   - Varying the input signal frequency of function generator, measure the input and output voltages. Tabulate them and calculate the voltage gains at each frequency.
   - Draw the frequency response characteristic and determine the bandwidth.
3. Draw the waveforms V_{in} and $V_{o\text{-}2}$ and observe that the two voltages are in-phase with one another and further the voltage gain is approximately less than unity.
4. **Current series feedback** Voltage V_{in} is the input signal voltage from the function generator and voltage at the Transistor Collector point $V_{o\text{-}1}$ is the output voltage for current series feedback circuit.
   - Observe the voltages V_{in} and $V_{o\text{-}1}$ and calculate the voltage gain.
   - Varying the input signal frequency of function generator, measure the input and output voltages. Tabulate them and calculate the voltage gains at each frequency.
   - Draw the frequency response characteristic and determine the bandwidth.
5. Draw the waveforms V_{in} and $V_{o\text{-}1}$ and observe that the two voltages are 180° out of phase with one another.
6. Connect the capacitor $C_{out\text{-}2}$ in parallel with R_E and draw the frequency response of the normal Amplifier without any feedbacks in the circuit.

S. No.	Input signal frequency	V_{in} (input voltage)	$V_{o\text{-}1}$ (OR) $V_{o\text{-}2}$ (output voltage)	(Voltage gain A_V) Output voltage/ input voltage	20 $\log_{10} A_V$ (dB)

Graphs:

1. Draw the frequency response characteristics of the Amplifier circuits on semi-log graph sheets with $\log_{10}^{f}$ on the x axis and the voltage gain on the y axis.
2. Identify the low-frequency and high-frequency cut-off points on the Amplifier responses.
3. Calculate the Amplifier bandwidths from the graphs and make discussions on the magnitude of voltage gains and bandwidths of the Amplifier circuits with and without feedback.

Calculations: From the frequency response graphs, the bandwidth of an Amplifier is the difference between high-frequency cut-off point f_2 and the low-frequency cut-off point f_1.

SUMMARY

1. The performance of Transistor Amplifiers is improved with some features by using negative feedback into the standard Amplifiers.
2. By using a frequency selective feedback network, the frequency response of an amplifier can be modified to the desired characteristic.
3. Negative feedback is effective in reducing distortion in Amplifiers.
4. The stability of operation of feedback Amplifiers is improved by the feedback network. Further, any drifts or changes in the parameters of active device do not have any effect on the performance of negative feedback Amplifiers as the gain of negative feedback amplifier $A_F = (\beta)^{-1}$.
5. The feedback voltage is 180° out of phase with the input signal for providing negative feedback in amplifier circuits. That is the reason why Common Emitter Transistor Amplifiers or Common Source FET Amplifiers find their use in negative feedback Amplifiers.
6. The four types of negative feedback Amplifiers: (1) Voltage Series feedback Amplifier; (2) Current Series shunt feedback Amplifier; (3) Voltage shunt feedback Amplifier; (4) Current shunt feedback Amplifiers are analysed with transistor circuits in detail.

Questions for Practice

1. How negative feedback does stabilises the quiescent operating point of an amplifier?
2. Discuss the effect on reduction of 'Harmonic Distortion' with negative feedback.
3. Draw the block diagrams for the four types of negative feedback amplifiers illustrating the sampling of signals at the output port and mixing of signals at the input port.
4. Mention the magnitudes of input and output impedances of negative feedback amplifiers and how they are modified from the input and output impedances of normal amplifiers.
5. Draw the Voltage Series Feedback Amplifier circuit. Substantiate the method of sampling and the nature of feedback. Derive the expressions for voltage gain, current gain, input impedance and output impedance of the circuit.
6. Draw the Voltage Shunt Feedback Amplifier circuit. Substantiate the method of sampling and the nature of feedback. Derive the expressions for voltage gain, current gain, input impedance and output impedance of the circuit.
7. Draw the current series Feedback Amplifier circuit. Substantiate the method of sampling and the nature of feedback. Derive the expressions for voltage gain, current gain, input impedance and output impedance of the circuit
8. Draw the Current Shunt Feedback Amplifier circuit. Substantiate the method of sampling and the nature of feedback. Derive the expressions for voltage gain, current gain, input impedance and output impedance of the circuit.

9. Using a block diagram illustrate the concept of negative feedback and further explain the concept of Current Series feedback in Amplifiers. Explain these concepts using appropriate Transistor Amplifier circuit or FET Amplifier circuit.
10. Using a block diagram illustrate the concept of negative feedback and further explain the concept of Current Shunt feedback in Amplifiers. Explain these concepts using appropriate Transistor Amplifier circuit or FET Amplifier circuit.
11. Draw the circuit of Emitter Follower. Discuss the nature of feedback that exists in the circuit. Derive the expressions for feedback factor β, voltage gain, input and output impedances. Discuss the circuit application as unity gain buffer Amplifier.
12. Draw the circuit of source follower. Discuss the nature of feedback that exists in the circuit. Derive the expressions for feedback factor β, voltage gain, input and output impedances. Discuss the circuit application as unity gain buffer Amplifier.

Multiple Choice Questions

1. The loop gain of Feedback Amplifier is ____________
   (a) β (b) $A\beta$ (c) $\beta \cdot V_{OUT}$ (d) $(1 + A\beta)$
2. Desensitivity factor D is ____________
   (a) $(1 - A\beta)$ (b) $|1 + A\beta|$ (c) $A\beta$ (d) $\dfrac{1}{(1+A\beta)}$
3. Negative feedback results in ____________
   (a) oscillations and gain becomes infinity (b) reduction of voltage gain
   (c) alteration of gain-bandwidth product (d) increase in nonlinear distortion
4. Positive feedback in amplifier ____________
   (a) initiates oscillations to occur (b) reduces voltage gain
   (c) does not alter the gain-bandwidth product (d) reduction in distortion
5. Transconductance amplifier is ____________
   (a) series–series feedback configuration
   (b) shunt–series feedback configuration
   (c) shunt–shunt configuration
   (d) series–shunt configuration
6. A transresistance amplifier belongs to the following topology
   (a) series–series feedback (b) shunt–shunt feedback
   (c) shunt–series feedback (d) series–series feedback
7. A series–shunt Feedback Amplifier is ____________
   (a) voltage amplifier (b) current amplifier
   (c) transresistance amplifier (d) transconductance amplifier
8. A shunt–series Feedback Amplifier is ____________
   (a) voltage amplifier (b) current amplifier
   (c) transresistance amplifier (d) transconductance amplifier

9. Common collector Transistor Amplifier is an example of ____________
   (a) voltage shunt feedback
   (b) voltage series feedback
   (c) current series feedback
   (d) current shunt feedback

10. CE Transistor Amplifier with unbypassed emitter resistor is an example for ____________
    (a) current series feedback
    (b) current shunt feedback
    (c) voltage series feedback
    (d) voltage shunt feedback

11. The basic Emitter Follower circuit is an example of ____________
    (a) shunt–series topology
    (b) series–shunt topology
    (c) series–series topology
    (d) shunt–shunt topology

12. Negative feedback in Amplifiers causes ____________
    (a) reduction in gain of the feedback amplifier
    (b) increase in gain of the feedback amplifier
    (c) no change in the gain of the feedback amplifier
    (d) oscillations occur in the feedback amplifier

13. Negative feedback in amplifiers causes ____________
    (a) reduction in noise of the feedback amplifier
    (b) increase in noise of the feedback amplifier
    (c) no change in the inherent noise of the feedback amplifier
    (d) oscillations occur in the feedback amplifier

14. Negative feedback in amplifiers causes ____________
    (a) reduction in stability of the feedback amplifier
    (b) increase in stability of the feedback amplifier
    (c) no change in the inherent stability of the feedback amplifier
    (d) oscillations occur in the feedback amplifier

15. Negative feedback in amplifiers causes ____________
    (a) reduction in bandwidth of the feedback amplifier
    (b) increase in bandwidth of the feedback amplifier
    (c) no change in the bandwidth of the feedback amplifier
    (d) oscillations occur in the feedback amplifier

16. Positive feedback is used in ____________
    (a) amplifiers
    (b) oscillators
    (c) tuned amplifiers
    (d) video amplifiers

Answers to Multiple-Choice Questions

1. (b)	2. (b)	3. (b)	4. (a)	5. (a)	
6. (c)	7. (b)	8. (a)	9. (a)	10. (c)	
11. (a)	12. (a)	13. (a)	14. (b)	15. (b)	16. (b)

Chapter 8

OSCILLATORS

Learning Objectives

- Fundamental principles of oscillator circuits using positive feedback
- Working principles of
    1. Low-frequency oscillators (RC oscillators)
    2. High-frequency oscillators (LC oscillators)
    3. Crystal oscillators (Stable frequency oscillators)
    4. UJT oscillator

8.1 INTRODUCTION

- Electronic oscillator circuits are generators of periodic signal waveforms without having any external input signal Source.
- During the process of generation of AC signals, oscillator draws DC power from supply Source (raw energy) and converts it into AC power.
- Oscillator circuits can be designed to produce AC signals of desired frequency and wave shape.
    - Today modern electronic communication systems essentially use 'sinusoidal oscillators'. Both the Amplifiers and oscillators find applications in radio and TV receivers, communication equipment, radar, biomedical instrumentation, cell phones, computers and so on.

8.1.1 Classification of Oscillators

Output signal waveform

1. Sinusoidal oscillators,
2. Non-sinusoidal oscillator such as Relaxation oscillators,
3. Multivibrators, Schmitt trigger, saw-tooth generators.

Frequency of output voltage

1. Fixed frequency oscillators,
2. Variable frequency oscillators.

Frequency band

1. Audio Frequency oscillators 20 Hz to 20 kHz,
2. RF oscillators 30 kHz to 30 MHz,
3. VHF oscillators 30 MHz to 300 MHz,
4. UHF oscillators 300 MHz to 3 GHz,
5. Microwave frequency oscillators above 3 GHz.

Principles of generating oscillations

1. Oscillators using positive feedback,
2. Negative resistance oscillators,
3. Crystal controlled oscillators.

Frequency determining circuit components

1. RC oscillators (Low-Frequency oscillators),
2. LC oscillators (High-Frequency oscillators).

Name of the inventor of oscillator circuits

1. Hartley oscillator,
2. Colpitts oscillator.

8.2 FUNDAMENTAL CONCEPTS OF SINUSOIDAL OSCILLATORS

Sinusoidal oscillator circuits consist of the following main sections:

1. Internal or basic Amplifier (using Transistors or Vacuum Tubes),
2. Feedback circuit (feedback arrangement),
3. Frequency determining circuit components (LC or RC elements).

Basic Amplifier
Electronic circuits along with active devices (BJT, FET, Vacuum Tubes) that produce increased output signals are known as *Amplifiers*.

$$\text{Voltage gain of Amplifier} \quad A = \frac{V_{\text{out}}}{V_{\text{in}}}. \tag{8.1}$$

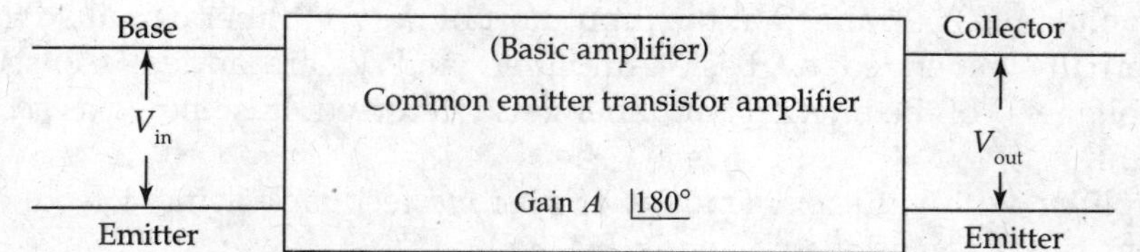

FIG. 8.1 *Block diagram of transistor amplifier*

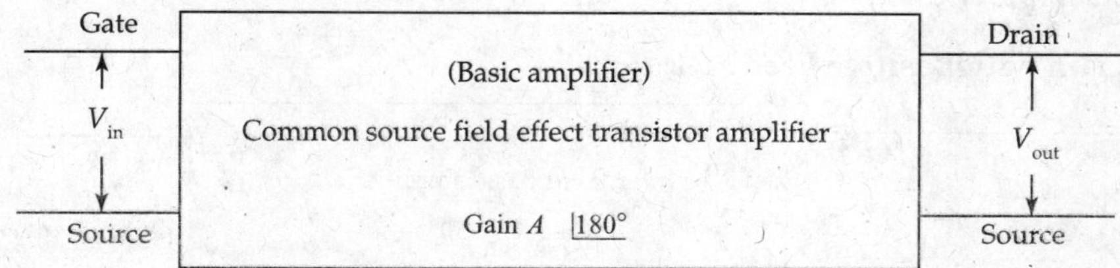

FIG. 8.2 *Block diagram of FET amplifier*

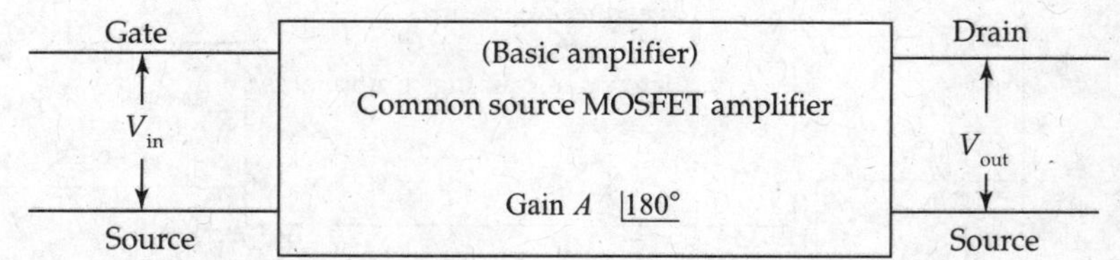

FIG. 8.3 *Block diagram of MOSFET amplifier*

Feedback Circuit

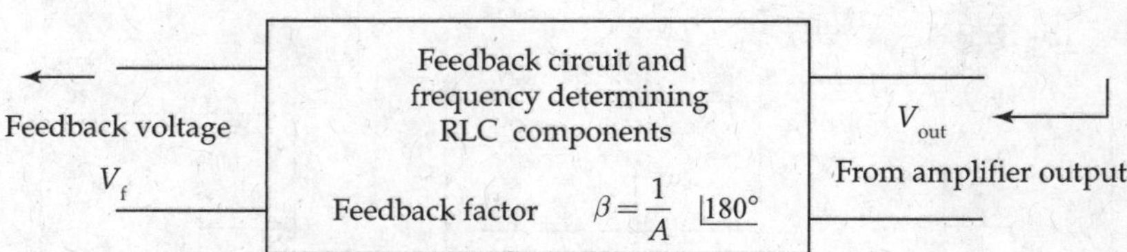

FIG. 8.4 *Block diagram of feedback circuit*

V_{out} = Output voltage from the output port of the Amplifier to the input port of β network.
V_f = Output voltage of feedback network serving as feedback voltage V_f to be connected to the input port of the Amplifier (refer to Fig. 8.5). The components of feedback network produce some attenuation of the signal passing through it.

$$\text{Feedback factor} \qquad \beta = \frac{V_f}{V_{out}} = \frac{V_f}{V_o} \tag{8.2}$$

$$\therefore \quad \text{Feedback voltage} \qquad V_f = \beta \cdot V_{out} = A \cdot \beta \cdot V_{in}. \tag{8.3}$$

Feedback network is designed and assembled to produce 180° phase shift (if the internal Amplifier is an inverting Amplifier).

- Feedback voltage V_f is arranged at the input port of the Amplifier circuit so that V_S (noise signal internally generated by basic Amplifier) and V_f are added. Consequently, the effective voltage v_{in} of the internal Amplifier gets reinforced. It is known as *positive feedback* arrangement.
  - An Amplifier with sufficient amount of positive feedback using a feedback network forms one type of an oscillator circuit (Fig. 8.5).

Oscillator circuit components depend upon frequency and output signal wave shape requirements in practical systems.

Amplifier Gain with Positive Feedback

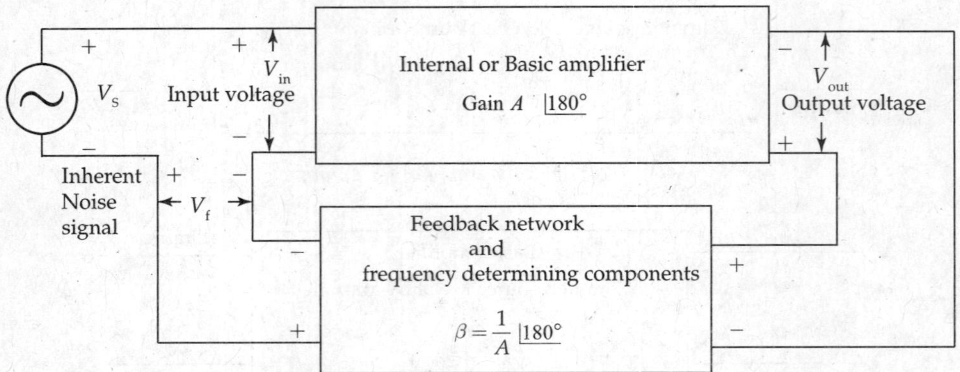

FIG. 8.5 *Block diagram of an oscillator circuit*

From the block diagram of an oscillator circuit (Fig. 8.5), Effective input voltage to the internal Amplifier [$V_{in} = V_S + V_f$].

From Fig. 8.6, feedback voltage [$V_f = A\beta\, V_{in}$]. $A\beta$ is known as *loop gain* in the oscillator circuit. Therefore, $V_{in} = V_S + A\beta{\cdot}V_{in}$ (refer Fig. 8.5).

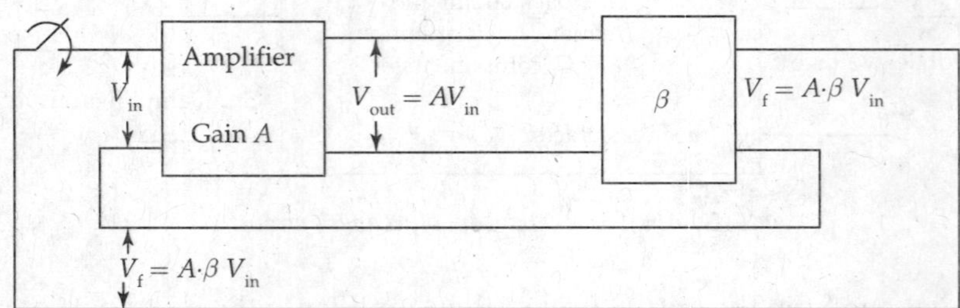

FIG. 8.6 *Illustration of working principle of an oscillator*

Rearranging the terms in the above equation, $V_{in}\,[1 - A\beta] = V_S$.

Gain of the feedback Amplifier $A_{pfb} = \dfrac{V_{out}}{V_S}$. (8.4)

Using $V_S = V_{in}\,[1 - A\beta]$ and $V_o = A\,V_{in}$, in Eq. (8.4),

Voltage gain of feedback Amplifier $A_{pfb} = \dfrac{V_o}{V_S} = \dfrac{A \cdot V_{in}}{V_{in}[1 - A\beta]} = \left[\dfrac{A}{[1 - A\beta]}\right]$. (8.5)

Oscillator circuits use *positive feedback* for initiation and maintaining the oscillations.

8.2.1 Bharkhausen Conditions for Oscillations

When the loop gain $A\beta$ is made equal to unity in Eq. (8.5), $A_{pfb} = \infty$. Then the circuit oscillates. The condition $[A\beta = 1]$ is known as *Bharkhausen conditions (Criteria) for oscillations*. Infinite gain at the start of oscillations is controlled by non-linearity of active device, so that the output signal of oscillator is stable.

In real systems, no external input signal is applied to an *oscillator.* Only the condition $A\beta = 1 \angle 360°$ or $\angle 0°$ must be satisfied to start and maintain the self-sustained oscillations.

They are implemented using the following principles in the oscillator circuits:

The content or magnitude of attenuation 'β' produced in the feedback network is compensated by the gain $[A = 1/\beta]$ contributed by the internal Amplifier so that $[A\beta = 1]$.
Oscillator output signal undergoes a total phase shift of 360° or 0°. The signal undergoes a phase shift of 180° in the 'internal Amplifier' and 180° phase shift in 'feedback network'.

- The introduction of 180° phase shift in the signal feedback path will be in different forms in various oscillator circuits. For example in RC phase-shift oscillator circuit 180° phase shift in the feedback network will be introduced by three identical RC elements, each RC section contributing 60° phase shift. In the Tuned Drain oscillator circuit, 180° phase shift is produced by transformer action.

Frequency determining network

- Frequency-determining circuit uses different circuit elements in various oscillator circuits. Transistor RC phase-shift oscillator circuits use 3-RC sections and R_L (load resistance) to determine the frequency of oscillations.
- Tuned Collector oscillator circuit, Colpitts oscillator and other LC oscillator circuits use *L* and *C* elements to determine the frequency of oscillations.
- *R* and *C* elements are used in low-frequency oscillator such as audio frequency oscillators. *L* and *C* elements are used in High-frequency oscillator circuits.

EXAMPLE 8.1

Mention the expression for gain (A_{pfb}) of an Amplifier with positive feedback. If the feedback factor $\beta = 0.02$, in an oscillator circuit, calculate open loop gain '*A*' of internal Amplifier and gain A_F with feedback satisfying Bharkhausen condition for oscillations.

Solution: Gain of the amplifier with positive feedback $A_{pfb} = \dfrac{A}{(1 - A\beta)}$.

Given $\beta = 0.02$. To satisfy Bharkhausen condition for oscillations $A\beta = 1$.

$$\therefore \text{ Required gain } \quad A = \frac{1}{\beta}$$

$$\text{Gain} \quad A = \frac{1}{\beta} = \frac{1}{0.02} = 50$$

$$A_{pfba} = \frac{A}{(1 - A\beta)} = \frac{50}{(1 - 50 \times 0.02)} = \infty.$$

8.2.2 General Concept of an Oscillator Action

- Every electronic oscillator has an active device such as a Transistor to convert the raw DC power to AC power and associated circuit components to provide positive feedback and frequency selection.
- When the DC power is switched on, random movement of current carriers through the active device produces 'noise signal' voltage having energy at all frequency components right from 0 Hz to infinite Hz. Such white noise input signal is amplified and fed back through a network for oscillator operation at the designed frequency.
- The feedback network selects the desired frequency component and provides required magnitude of positive feedback to the Transistor to function as an oscillator.
- The cycles of events repeat till the Bharkhausen conditions Criteria for oscillations $[A\beta = 1]$ is satisfied and the oscillator circuit generates an AC signal at desired frequency and amplitude.

8.3 TRANSISTOR RC PHASE-SHIFT OSCILLATOR

- RC phase-shift oscillators are used to generate AC signals of audio frequency range.
- Frequency of oscillations (in an oscillator using LC elements) $f_0 = 1/2\pi\sqrt{LC}$. For low-frequency signal generation, required values of L and C are large. Inductor will be bulky. So, RC elements will be used in low-frequency oscillator circuits.

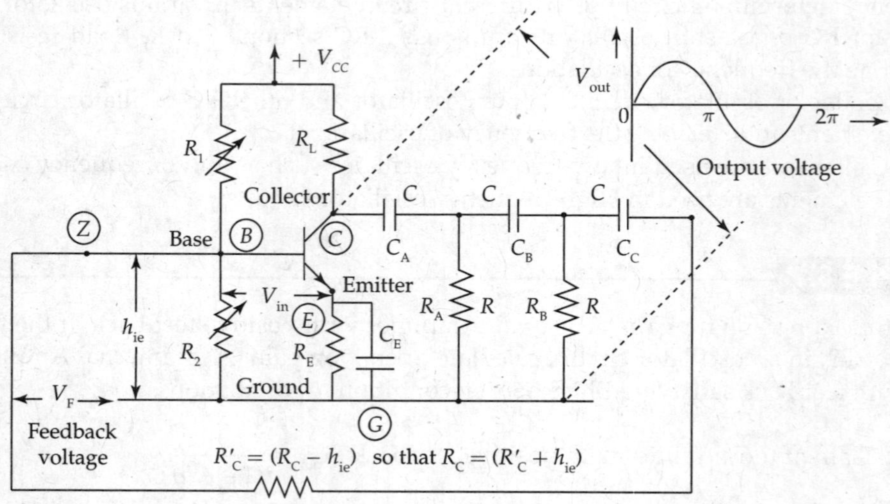

FIG. 8.7 *Transistor R-C phase shift oscillator circuit*

Understanding the circuit layout:

1. The combination of Transistor, R_1, R_2, R_L, R_E and V_{CC} function as internal (basic) Amplifier to provide voltage amplification. Output voltage V_{out} of the CE Transistor Amplifier will be 180° out of phase with the effective input voltage V_{in}.
2. Three identical RC elements are used as feedback network to provide signal path from the output port (Collector to Emitter path) of the Transistor Amplifier to its input port (Base to Emitter path).

3. The three RC elements produce 180° phase shift for the signal V_{out} moving through it. It also produces an attenuation of magnitude β. Feedback factor $\left[\beta = \frac{V_f}{V_{out}} = \frac{1}{A}\right]$.
4. Overall phase shift around the loop becomes 360° or 0°. Thus, the output and input voltages will be in phase at the input port of the Transistor to increase the effective input signal. Then the circuit begins to work as an oscillator satisfying 'Bharkhausen conditions for oscillation'.
5. Output voltage of the oscillator circuit will be a 'sinusoidal signal' at the designed frequency decided by the feed back network.

8.3.1 RC Phase-shift Oscillator Circuit Working

Considering $R_A = R_B = R_C = R$ and $C_A = C_B = C_C = C$ in RC phase-shift oscillator circuit of Fig. 8.7, the circuit in Fig. 8.8 is considered for the analysis of the circuit. When the DC Source is switched on, the random movement of charge carriers through the active device, the Transistor (BJT or JFET in general) and the circuit components produce a noise signal V_n (white noise containing signal frequencies from 0 Hz to infinite Hz).

- Noise signal V_n is of the order of a few Pico-volts at the input port of the Transistor.
- V_n acts as the virtual input signal V_{in}. V_n is amplified with gain A. Let the amplified voltage be V_0 at the output port of the Transistor. V_0 is larger than V_{in} and is 180° out of phase with V_{in}.
- The output voltage, V_0, is applied to the input port of the feedback network consisting of three RC sections.
- The three RC sections produce 180° phase shift, so that the signal V_0 passing through feedback network undergoes 180° phase shift.
- The output voltage of the feedback network is connected to the input port of the Transistor as fed back voltage V_f.
- So there is total 360° phase shift for V_{in} to come back as V_f and satisfies the condition for Positive feedback (one of the conditions of 360° phase shift for Bharkhausen condition of oscillations). The two signals are in phase and their instantaneous amplitudes get added.
- At this time, the effective input signal V_{in} increases. This cycle of events repeats and the output voltage goes on increasing (Fig. 8.9) unbounded till the setting in of the non-linearity

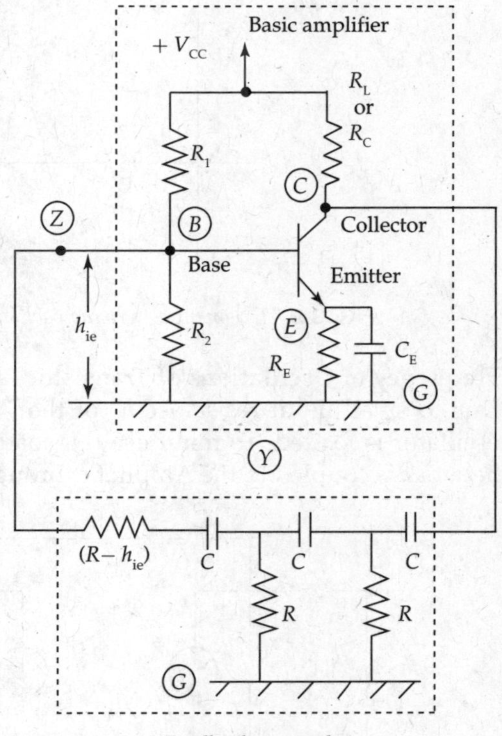

FIG. 8.8 *Transistor R-C Phase shift oscillator circuit various blocks in an oscillator*

of the active device that clamps the output voltage to a constant desired (designed) output voltage V_0.

- Amplitude limiting process is also achieved by the increase in forward bias by the exponentially growing signal together with coupling capacitor and the biasing resistance just like the bias stabilisation through R_E. Increase and decrease of oscillating signal is prevented by necessary bias changes.
- Feedback network also attenuates the voltage V_0, by a factor of $[\beta = 1/A]$ by the time it is connected as V_f to the input port of the Transistor to provide positive feedback.
- The Amplifier amplifies the reinforced input voltage with a gain $[A = 1/\beta]$, so as to satisfy the second Bharkhausen Criteria for oscillations in the circuit, i.e., $[A\beta = 1]$.
- The amplitude of the output sine wave depends upon the supply voltage and the bias conditions of the Transistor.
- The frequency of the output sine wave is decided by the three RC sections in the feedback network and R_L according to the equation

$$f_0 = \frac{1}{2\pi RC\sqrt{6+4\left(\frac{R_L}{R}\right)}} = \frac{1}{2\pi RC\sqrt{6+4K}}\ \text{Hz}, \tag{8.6}$$

where $k = R_L/R$.

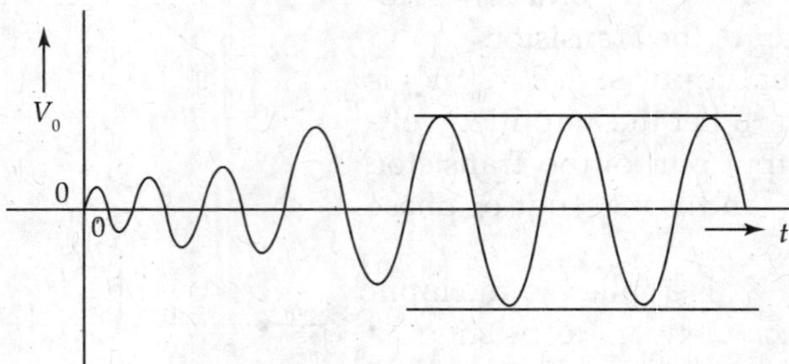

FIG. 8.9 *Exponentially rising oscillations clamped by nonlinearity of the active device*

Frequency of Oscillations of Transistor RC Phase-shift Oscillator

Due to small input resistance h_{ie} of the Transistor; output port of feedback network in the oscillator is loaded. To minimise this loading effect, *voltage shunt feedback* is used. Feedback network is coupled to the Amplifier through feedback resistor $R' = (R - h_{ie})$.

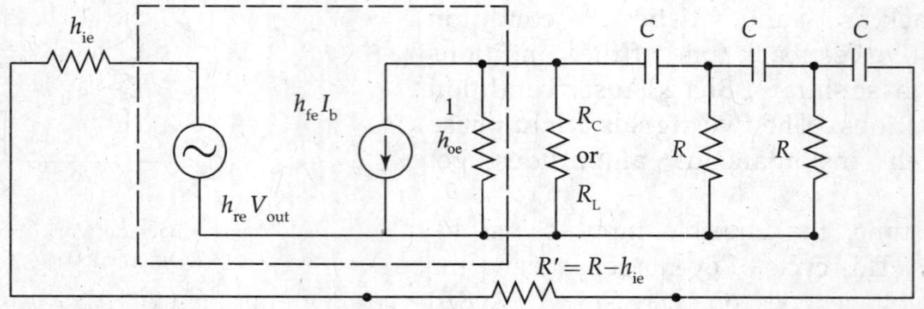

FIG. 8.10 *Equivalent circuit of transistor R-C phase shift oscillator*

Assumptions made in the h-parameter equivalent circuit (Fig. 8.10)

1. As h_{re} of the Transistor is very small, the Source signal '$h_{re}V_o$' can be omitted.
2. Since h_{oe} of the Transistor is very small and $\frac{1}{h_{oe}} = 40\ \text{k}\Omega \rangle\rangle R_C$ or R_L, the effect of h_{oe} is neglected in the equivalent circuit.
3. The current Source $h_{fe}I_b$ is replaced by equivalent Thevenin's voltage, $h_{fe}I_bR_C$ or $h_{fe}I_bR_L$ and its Source resistance is R_C or R_L.
4. For unity loop gain, $I_b = I_3$.
5. $R' + h_{ie} = R$ or $R' = R - h_{ie}$.

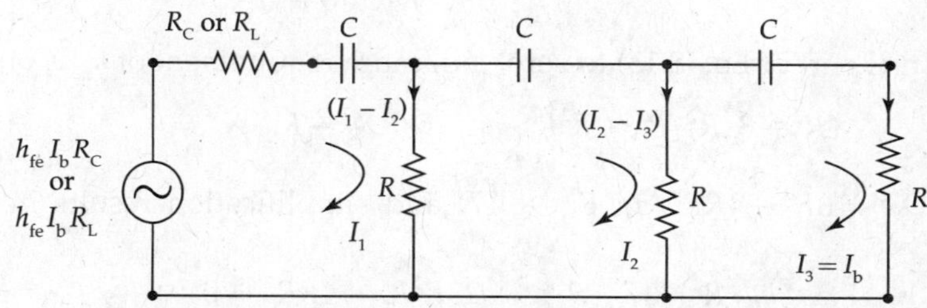

FIG. 8.11 *Modified equivalent circuit of R-C phase shift oscillator*

Using the above assumptions, the equivalent circuit in Fig. 8.10 is modified as in Fig. 8.11. From Fig. 8.11, the KVL mesh equations are as follows:

$$(R_C + R - jX_C)I_1 - RI_2 + h_{fe}R_CI_b = 0 \tag{8.7}$$

$$-RI_1 + (2R - jX_C)I_2 - RI_b = 0 \tag{8.8}$$

$$0 - RI_2 + (2R - jX_C)I_b = 0 \tag{8.9}$$

As I_1, I_2 and I_3 (or I_b) are non-vanishing, determinant formed by coefficient = 0. (8.10)

$$\Delta \cong \begin{vmatrix} R + R_C - jX_C & -R & +h_{fe}R_C \\ -R & 2R - jX_C & -R \\ 0 & -R & 2R - jX_C \end{vmatrix} = 0 \tag{8.11}$$

By simplifying the above determinant, we get

$$R^3 + R^2 \cdot R_C(3 + h_{fe}) - 5R \cdot X_C^2 - R_CX_C^2 - 6j \cdot R^2X_C - j \cdot 4 \cdot R \cdot R_C \cdot X_C + jX_C^3 = 0 \tag{8.12}$$

$$R^3 + R^2 \cdot R_C(3 + h_{fe}) - 5R \cdot X_C^2 - R_C \cdot X_C^2 + j \cdot \left[X_C^3 - 6R^2X_C - 4X_CRR_C\right] = 0 \tag{8.13}$$

Equating the imaginary part of Eq. (8.13) to zero, we get

$$X_C^3 - X_C \cdot (6R^2 + 4 \cdot R \cdot R_C) = 0 \tag{8.14}$$

Cancelling X_C throughout and carrying out negative term to other side

$$X_C^2 = 6R^2 + 4RR_C$$

$$\therefore \quad \frac{1}{\omega^2 \cdot C^2} = 6R^2 + 4R \cdot R_C$$

$$\omega^2 C^2 = \frac{1}{(6R^2 + 4R \cdot R_C)} \Rightarrow \omega^2 = \frac{1}{C^2(6 \cdot R^2 + 4 \cdot R \cdot R_C)}$$

$$\omega = \frac{1}{RC\sqrt{6 + 4\frac{R_C}{R}}} = \frac{1}{RC\sqrt{6+4K}}, \tag{8.15}$$

where $K = \frac{R_C}{R}$.

$$\therefore \text{ Frequency } \quad f = \frac{1}{2\pi R \cdot C\sqrt{6+4K}}. \tag{8.16}$$

Equating the real part of Eq. (8.13) to zero, minimum requirement of h_{fe} is obtained.

$$R^3 + 3 \cdot R^2 \cdot R_C + R^2 R_C \cdot h_{fe} - 5 \cdot RX_C^2 - R_C \cdot X_C^2 = 0 \tag{8.17}$$

Substituting $X_C^2 = 6R^2 + 4RR_C$ in Eq. (8.17) and simplification results in the following equation:

$$h_{fe} \cdot R^2 \cdot R_C - 29 \cdot R^3 - 23R^2 R_C - 4RR_C^2 = 0. \tag{8.18}$$

From Eq. (8.18),

$$h_{fe} R^2 R_C = 29R^3 + 23R^2 R_C + 4RR_C^2$$

$$h_{fe} = \frac{29R^3 + 23R^2 \cdot R_C + 4RR_C^2}{R^2 \cdot R_C} = 29 \cdot \frac{R}{R_C} + 23 + 4 \cdot \frac{R_C}{R} \tag{8.19}$$

Since $K = \frac{R_C}{R} = \frac{R_L}{R}$

$$h_{fe} = \frac{29}{K} + 23 + 4K. \tag{8.20}$$

For a maximum or minimum value of h_{fe}

$$\frac{d \cdot h_{fe}}{d \cdot K} = \frac{-29}{K^2} + 4 = 0 \quad \text{(differentiating Eq (8.20) w.r.t } K)$$

$$\therefore -29 + 4K^2 = 0$$

$$K^2 = \frac{29}{4} = 7.25 \text{ and hence } K = 2.7.$$

Substituting the value of $K = 2.7$ in Eq. (8.20),

$$h_{fe} = \frac{29}{2.7} + 23 + 4(2.7) = 44.54. \tag{8.21}$$

For a Transistor RC phase-shift oscillator, the minimum value of h_{fe} is 44.54 to sustain oscillations. (8.22).

If $R_C = R_L = R$ is substituted in Eq. (8.19), the minimum required value of h_{fe} of the Transistor is 56. (8.23).

- RC phase-shift oscillators are capable of generating frequencies of a few Hz to several kHz and are particularly suitable as audio frequency oscillators.
- By varying the elements of three RC networks, variable frequency operation can be achieved. A change in the value of resistor R causes change in input impedance. So, changes in values of capacitor C are preferred.
- Changes in component values are done while maintaining the phase shift β and $A\beta$ constant. To keep distortion low, RC phase-shift oscillator operates in Class A.
- Three RC networks are chosen in this circuit so that each section produces 60° phase shift and a total phase shift of 180° around the loop.
- Even though RC networks produce a phase shift of 90°, two-section phase-shift network cannot be used since it is required to make $R = 0$ which means infinite attenuation and because of losses in capacitors it is impracticable to construct the RC oscillator using two sections.
- Since reduction in h_{fe} causes difficulties in internal phase shift at high frequencies, RC phase-shift oscillators are not used in high-frequency applications.
- Main drawbacks: it is difficult to start oscillation due to small feedback arrangement and the frequency stability is relatively low compared to a Wien Bridge oscillator.

EXAMPLE 8.2

Transistor RC phase-shift oscillator circuit contains $R_A = R_B = R_C = R = 3.3\ \text{k}\Omega$, $R_L = 3.9\ \text{k}\Omega$ and $C_A = C_B = C_C = 0.01\ \mu\text{F}$. Calculate the frequency of oscillations of the Transistor RC phase-shift oscillator using the circuit component values.

Solution:

$$\frac{R_L}{R} = \frac{3.9\times 10^3}{3.3\times 10^3} = 1.18$$

$$f_0 = \frac{1}{2\pi\cdot R\cdot C\cdot\sqrt{6+4\cdot K}} = \frac{1}{2\pi\cdot 3.3\times 10^3\times 0.01\times 10^{-6}\sqrt{6+4\times 1.18}}$$

$$\therefore\ f = \frac{10^8}{2\pi\times 3.3\times 10^3\times 3.274} = \frac{10^8}{67.88\times 10^3} = \frac{100\times 10^3}{67.88} = 1.47\ \text{kHz}.$$

EXAMPLE 8.3

Find the value of 'C' of the frequency determining network and h_{fe} of the Transistor for RC Transistor Phase-shift oscillator. Data: Oscillator Frequency = 50 kHz. $R_C = 20\ \text{k}\Omega$ and $R = 6.8\ \text{k}\Omega$.

Solution:

Frequency of oscillations $$f_0 = \frac{1}{2\pi\cdot R\cdot C\sqrt{6+4\dfrac{R_C}{R}}}$$

$$f_0 = 50\times 10^3 = \frac{1}{2\pi\times 6.8\times 10^3\times C\sqrt{6+4\dfrac{R_C}{R}}}$$

$$50\times10^3 = \frac{1}{6.28\times6.8\times10^3\times C\sqrt{6+4\times2.94}}$$

$$\therefore \text{ Capacitance } C = \frac{1}{6.28\times6.8\times10^3\times50\times10^3\sqrt{17.76}}$$

$$= \frac{1}{2135.2\times10^6\times4.214}$$

$$\therefore C = \frac{1\times10^{-6}}{8997.73} = 111.1\times10^{-12} = 111\text{ pF}$$

Calculation of h_{fe} the transistor

$$h_{fe} = 23+\frac{29}{K}+4\cdot K = 23+\frac{29}{2.94}+4\times2.9$$

$$\therefore h_{fe} = 23+9.86+11.76 = 44.62.$$

8.4 FET–RC PHASE-SHIFT OSCILLATOR

- The combination of the circuit features, the supply voltage V_{DD}, R_D, R_1, R_2 and R_S–C_S provide stabilised DC Bias to the FET device in the circuit to act as basic amplifier.
- Three cascaded RC networks follow the basic FET Amplifier at Drain and Source.
- The output of the last section is returned to the input, constituting *voltage series feedback.* Voltage series feedback is used, because of the very high input resistance of the FET device.
- Feedback circuit with three RC sections determines the frequency of oscillations and also provide 180° phase shift to provide positive feedback.

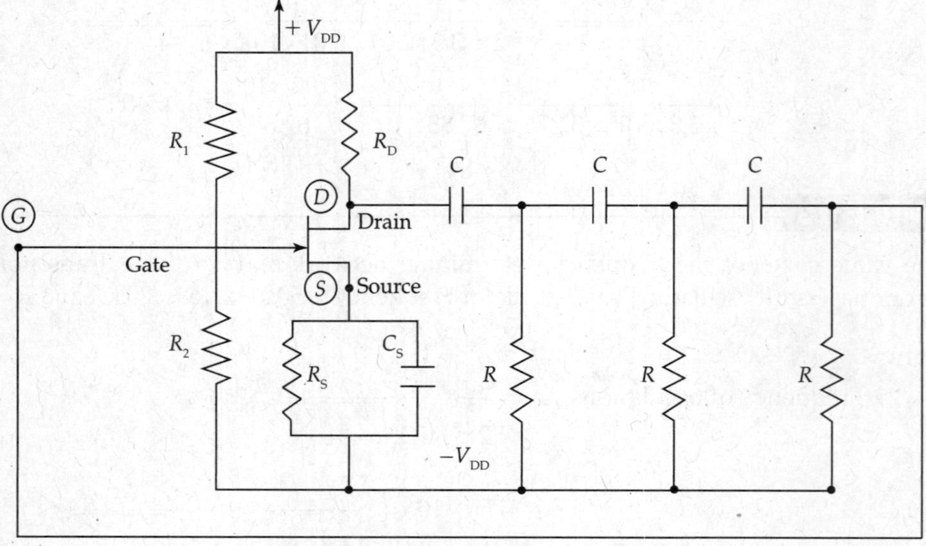

FIG. 8.12 *FET RC – phase shift oscillator*

- Phase shift of each RC network is $\phi = \tan^{-1}(1/\omega CR)$. Values of R and C elements are so chosen to give each RC section a phase shift of 60° and the three RC networks provide an overall phase shift of 180° at a specifically designed frequency f_0.
- CS FET Amplifier introduces a phase shift of 180° and the total phase shift is 360° around the total loop. The FET parameters of interest are g_m and r_d.

$$\text{Voltage gain} \quad A = g_m R_L, \tag{8.24}$$

where $R_L = \dfrac{R_0 r_d}{R_0 + r_d}$.

- Analysing the circuit, it can be found that feedback network attenuates the output voltage V_o and $\beta = 1/29$. So, the Amplifier has to provide a voltage gain $A = 29$ to satisfy the Bharkhausen condition of oscillations $[A\beta = 1]$ (loop gain) is satisfied.

$$\therefore \quad \text{Voltage gain} \quad A = 29 \text{ and feedback factor } \beta = \frac{1}{A} = \frac{1}{29} \tag{8.25}$$

A FET with $\mu > 29$ is to be used so as to keep $|A\beta|$ not less than unity and to satisfy the Bharkhausen condition for oscillation.

$$\omega = \frac{1}{\sqrt{6}RC}$$

$$\therefore \quad f_0 = \frac{1}{2\pi RC\sqrt{6}}. \tag{8.26}$$

Advantages

1. Oscillator produces sinusoidal output voltage without having distortion.
2. The circuit is simple with passive components RC elements that are cheap.
3. The signals can be produced over a wide frequency range from a few Hz to a few kHz.

Disadvantages

1. As the magnitude of feedback is small, starting of oscillations is difficult and also the output amplitude of the oscillator is small.
2. LC elements cannot be used in the frequency determination network, because at low frequencies, the component value of L is very high and becomes bulky.

EXAMPLE 8.4

A FET RC phase-shift Oscillator circuit uses three RC sections in the feedback Network containing $R = 10$ kΩ, $C = 0.05$ μF. Calculate the frequency of oscillations.

Solution: Frequency of oscillations f for FET RC phase-shift oscillator

$$f = \frac{1}{2 \cdot \pi \cdot R \cdot C \cdot \sqrt{6}}.$$

Substituting the values of $R = 10$ kΩ and $C = 0.05$ μF in the above equation,

$$f = \frac{1}{2\pi \times 10 \times 10^3 \times 0.05 \times 10^{-6} \times \sqrt{6}}$$

$$= \frac{10^4}{\pi\sqrt{6}} = \frac{10^4}{7.6914} = 0.13 \times 10^4 = 1.3 \text{ kHz.}$$

EXAMPLE 8.5

Design RC phase-shift oscillator using FET device having $g_m = 5000$ μs, $r_d = 50$ kΩ. Frequency determining network has $R = 10$ kΩ. Calculate the required value of C if the circuit has to oscillate at 2 kHz. Calculate the Drain circuit resistance R_D.

Solution: Gain of FET amplifier $A = g_m \cdot R_L$
Assuming gain to be greater than 29, consider gain $A = 50$.

$$\therefore\ R_L = \frac{A}{g_m} = \frac{50}{5000 \times 10^{-6}} = 10 \text{ k}\Omega$$

$$R_L = \frac{R_D \cdot r_d}{R_D + r_d}$$

$$\therefore\ R_D = \frac{R_L \cdot r_d}{[r_d - R_L]} = \frac{10 \times 10^3 \times 50 \times 10^3}{[50 \times 10^3 - 10 \times 10^4]}$$

$$\therefore \text{ Drain resistance} \quad R_D = \frac{500 \times 10^6}{40 \times 10^3}$$

$$= 12.5 \times 10^3\ \Omega$$

$$\text{Frequency of oscillation} \quad f_0 = \frac{1}{2\pi R \cdot C\sqrt{6}}$$

$$\therefore \text{ Capacitance} \quad C = \frac{1}{2\pi f \cdot R\sqrt{6}}$$

$$= \frac{1}{2 \cdot \pi \cdot 2 \times 10^3 \times 10 \times 10^3 \sqrt{6}} = 3.25 \text{ nF.}$$

8.4.1 Transistor RC Phase-shift Oscillator Circuit to Verify the Design

Aim:
To conduct an experiment to identify the various blocks of an RC phase-shift oscillator and observe the output signal waveform and measure its amplitude and frequency. To verify the theoretical and practical values of frequency of oscillations of the circuit.

Apparatus:

- DC power supply (0–30 V),
- RC phase-shift oscillator circuit board,
- Cathode ray oscilloscope.

Transistor RC phase-shift oscillator circuit:

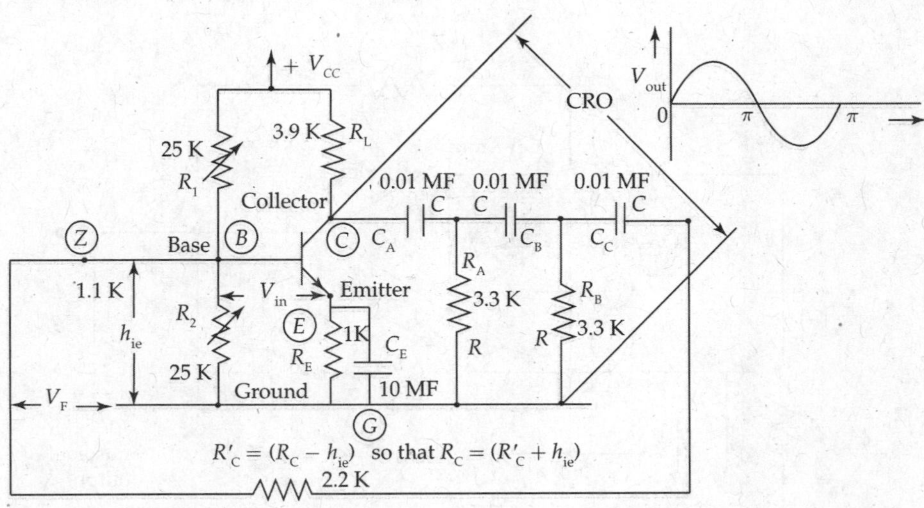

FIG. 8.13 *Transistor R-C phase shift oscillator circuit*

Procedure:

1. Trace the oscillator circuit and identify the various components that are necessary for the working of the oscillator.
2. Apply suitable magnitude of DC voltage from DC power supply unit to + V_{CC} and common terminal nodes of the oscillator circuit depending upon the type of Transistor used in the circuit and its specifications.
3. Connect the *Y*-plates probe of a CRO to the Collector terminal of the Transistor and the common terminal to observe the output voltage on the screen of CRO.
4. Adjust the biasing resistors R_1 and R_2 so that a good sine wave signal (without any distortion) appears at the oscillator output port.
5. Measure the amplitude and frequency of the output signal using CRO.
6. Draw the observed output waveform on a graph paper.
7. Calculate the theoretical frequency of oscillations from the circuit components in the Circuit and verify it with the practical frequency of the observed signal.
8. Measure the feedback voltage and calculate feedback ratio.

8.5 WIEN BRIDGE OSCILLATOR CIRCUIT USING OPERATIONAL AMPLIFIER

Wien Bridge oscillator generates low-frequency sine wave voltages. The oscillator circuit uses a network proposed by Max Wien in 1891 and later developed by William Hewlett of Stanford University (USA) during the year 1939. Hewlett and Packard developed Wien Bridge sine wave oscillator. It is a widely used low-frequency oscillator covering a wide range of frequencies.

Wien Bridge oscillator circuit consists of

1. Operational Amplifier as an internal or basic Amplifier,
2. Wien Bridge (Balanced Bridge) as the feedback network.

8.5.1 Wien Bridge Oscillator: Component Details

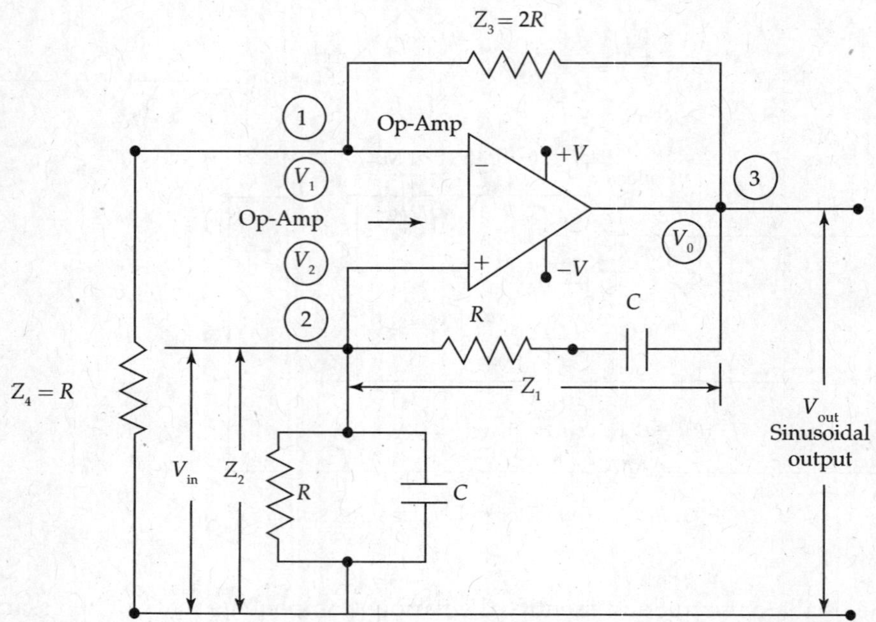

FIG. 8.14 *Wien bridge oscillator circuit using operational amplifier*

Understanding the circuit layout:

Internal Amplifier is an operational Amplifier with gain A in the above circuit. Output voltage V_{out} is gain 'A' times the differential input $(V_1 - V_2)$ voltage between positive and negative input terminals of the Operational Amplifier shown in Fig. 8.15.

$$V_{out} = A(V_1 - V_2).$$

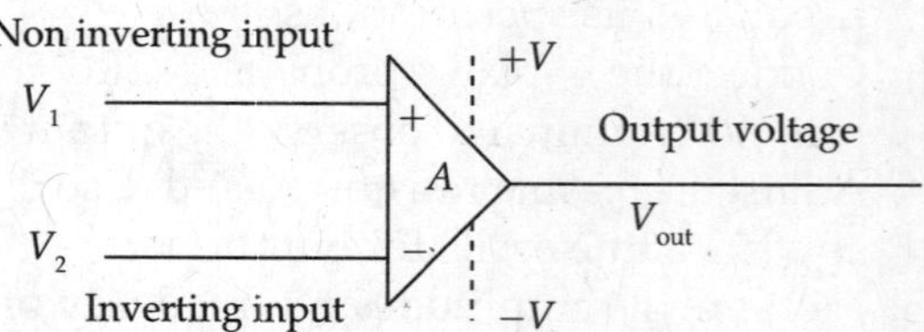

FIG. 8.15 *Operational amplifier*

All voltages are measured with respect to ground and the ground terminals with respect to input and output are not shown for convenience.

$$\text{Gain of operational amplifier} \quad A = \frac{V_{out}}{(V_1 - V_2)}. \tag{8.27}$$

Operational Amplifier is a high-gain Amplifier. Operational Amplifiers are used in amplification, instrumentation circuits and waveform generator circuits and so on by adding feedback circuits to control overall response characteristics of the Amplifier.

Operational Amplifiers were first used in 'Analog Computers' to perform mathematical operations to solve differential equations and so on before the popularity of present day computer systems. Various types of Operational Amplifiers in Integrated circuit (IC) form are available with very large input resistance R_{in}, Very low output resistance R_o and voltage gain A_V of the order of 10^5 or more.

Design of Wien Bridge Oscillator

1. Select an IC operational Amplifier having suitable parameters for application.
2. Feedback network in Wien bridge oscillator contains four resistors and two capacitors. They are grouped as four circuit elements Z_1, Z_2, Z_3 and Z_4 around the 'operational amplifier'.
3. Feedback Impedance Z_1 (Z_f) is a series combination of R and C (Fig. 8.16):

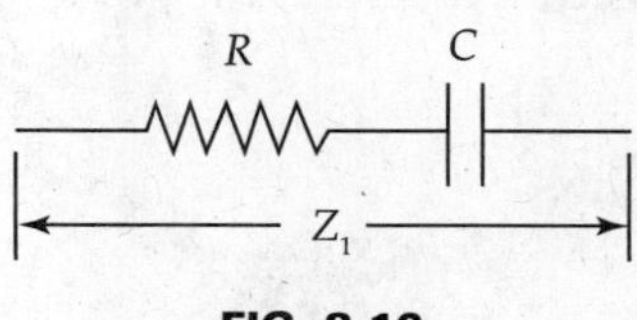

FIG. 8.16

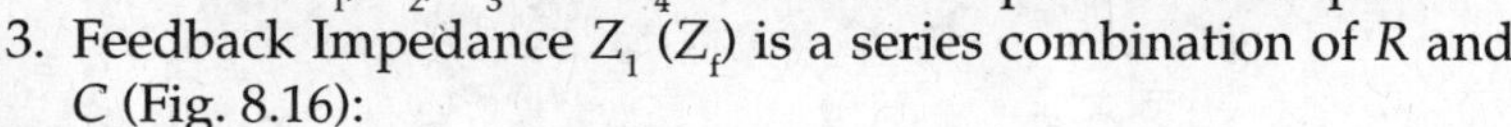

$$Z_1 = R + \frac{1}{j\omega C} = \frac{(1 + j\omega CR)}{j\omega C}. \qquad (8.28)$$

4. Z_2 is a parallel combination of R and C (Fig. 8.3):

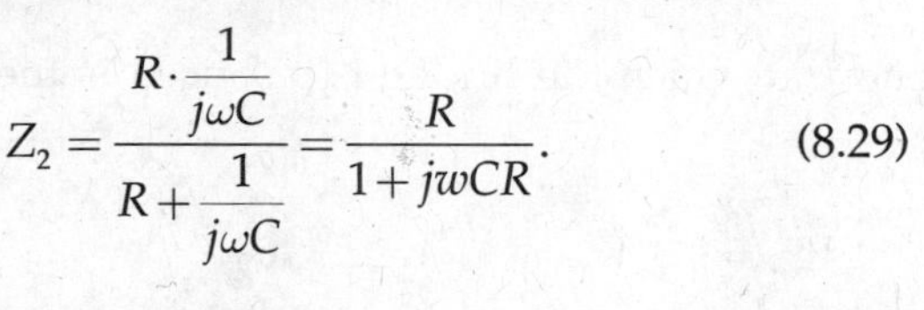

$$Z_2 = \frac{R \cdot \dfrac{1}{j\omega C}}{R + \dfrac{1}{j\omega C}} = \frac{R}{1 + jwCR}. \qquad (8.29)$$

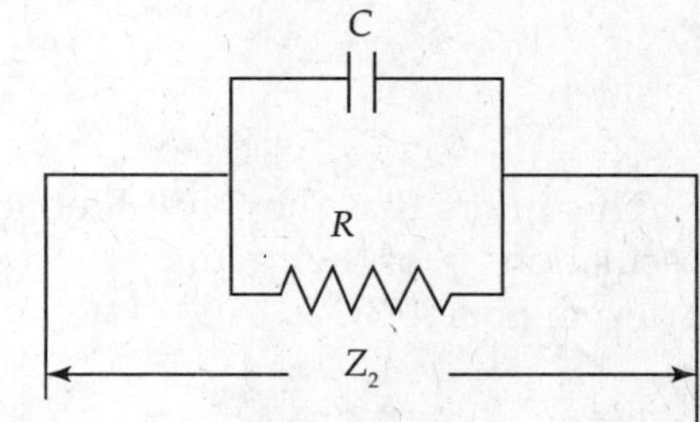

FIG. 8.17 *Impedance* Z_2

5. $Z_3 = 2R$ and $Z_4 = R$ (Fig. 8.14) (8.30)

Expressions for Frequency of Oscillations, Gain and Feedback Factor

From the circuit of Fig. 8.14,

$$V_{in} = \frac{V_{out} \times Z_2}{Z_1 + Z_2} \qquad (8.31)$$

$$V_{in} = \frac{V_{out} \times \dfrac{R}{1 + j\omega \cdot CR}}{\dfrac{(1 + j\omega \cdot CR)}{j\omega \cdot C} + \dfrac{R}{1 + j\omega \cdot CR}} = \frac{V_{out} \times \dfrac{R}{(1 + j\omega \cdot C \cdot R)}}{\dfrac{(1 + j \cdot \omega \cdot C \cdot R)^2 + j \cdot \omega \cdot C \cdot R}{j \cdot \omega \cdot C \cdot (1 + j \cdot \omega \cdot C \cdot R)}}$$

$$V_{in} = V_{out} \times \frac{j \cdot \omega \cdot CR}{\left[(1 + j \cdot \omega \cdot C \cdot R)^2 + j \cdot \omega \cdot C \cdot R\right]}$$

$$V_{in} = \frac{V_{out} \times j \cdot \omega \cdot C \cdot R}{\left[(1 - \omega^2 \cdot C^2 \cdot R^2 + 2j \cdot \omega \cdot C \cdot R) + j \cdot \omega \cdot C \cdot R\right]}$$

$$= \frac{V_{out} \times j \cdot \omega \cdot C \cdot R}{(1 - \omega^2 \cdot C^2 \cdot R^2 + 3 \cdot j \cdot \omega \cdot C \cdot R)}$$

$$\therefore \; V_{in} = \frac{V_{out}}{\left[\dfrac{-j}{\omega \cdot C \cdot R} + j \cdot \omega \cdot C \cdot R + 3\right]} = \frac{V_{out}}{\left[3 + j \cdot \left(\omega \cdot C \cdot R - \dfrac{1}{\omega \cdot C \cdot R}\right)\right]} \qquad (8.32)$$

$$\therefore \text{ Voltage gain } \; A = \frac{V_{out}}{V_{in}} = \left[3 + j \cdot \left(\omega \cdot C \cdot R - \frac{1}{\omega \cdot C \cdot R}\right)\right]. \qquad (8.33)$$

To satisfy the Bharkhausen condition for oscillations, total phase shift around the circuit is zero. Therefore, the *j* term in the expression (8.33) must be zero:

$$\omega RC - \frac{1}{\omega CR} = 0 \tag{8.34}$$

$$\omega^2 = \frac{1}{C^2 \cdot R^2}$$

$$\therefore\ \omega = \frac{1}{C \cdot R}$$

Hence, frequency of oscillations is given by

$$f = \frac{1}{2 \cdot \pi \cdot R \cdot C}. \tag{8.35}$$

This infers that the 'Null Frequency' at which the Bridge is balanced is the frequency of oscillations '*f*' of the circuit.

From Eq. (8.33),

$$V_{out} = +3 \cdot V_{in}. \tag{8.36}$$

Therefore, the voltage gain of the amplifier is

$$A = \frac{V_{out}}{V_{in}} = 3. \tag{8.37}$$

Therefore, voltage gain '*A*' of internal amplifier = 3.

To satisfy the loop gain condition $A\beta = 1$, the signal attenuation in the feedback network is of magnitude = 1/3. Attenuation in the feedback network $\beta = 1/3$ is compensated by the internal Amplifier having a minimum voltage gain of $A = 3$ making

$$A\beta = 3 \times \frac{1}{3} = 1. \tag{8.38}$$

To sustain oscillations in the circuit, the voltage gain is

$$A = \frac{V_{out}}{V_{in}} = +3. \tag{8.39}$$

Hence, gain of Amplifier in a Wien bridge oscillator must be at least '3' for the oscillations to occur. This is true for the Wien bridge oscillator using BJT and FET Amplifiers also:

$$V_{in} = -\frac{V_{out} \times Z_4}{Z_3 + Z_4}. \tag{8.40}$$

To make $V_{in} = +\frac{1}{3} V_{out}$ in Eq. (8.40), $Z_3 = 2\ Z_4$.

If $Z_4 = R$, $Z_3 = 2R$. If $R = 5$ kΩ, then $Z_4 = 5$ kΩ and $Z_3 = 10$ kΩ.

If $R = 5$ kΩ and $C = 0.01$ microfarads (μF),

$$\text{Frequence of oscillations} \quad f = \frac{1}{2\pi \times 5 \times 10^3 \times 0.01 \times 10^{-6}} = 3.185 \text{ kHz}.$$

Working Principle of Operation of Wien Bridge Oscillator

The details of Wien Bridge oscillator (Fig. 8.14) circuit are explained up to this point. Some more details are shown in Fig. 8.18. Output voltage is fed back to the input port of the operational amplifier by two paths.

- A portion of output voltage is fed to the non-inverting input (+) terminal of the op-Amp through the impedance Z_1, which is a series combination of resistor 'R' and Capacitor 'C'.
- Second portion of the output voltage is fed back to the inverting terminal (–) through the impedance Z_3 of magnitude $2R$ (part of voltage divider network consisting of Z_3 and Z_4 impedances). At the resonant frequency established by the Wien Bridge, Bharkhausen criteria for oscillations will be satisfied and the circuit works as an oscillator producing sine wave output voltages, whose frequencies are determined by Eq. (8.35).

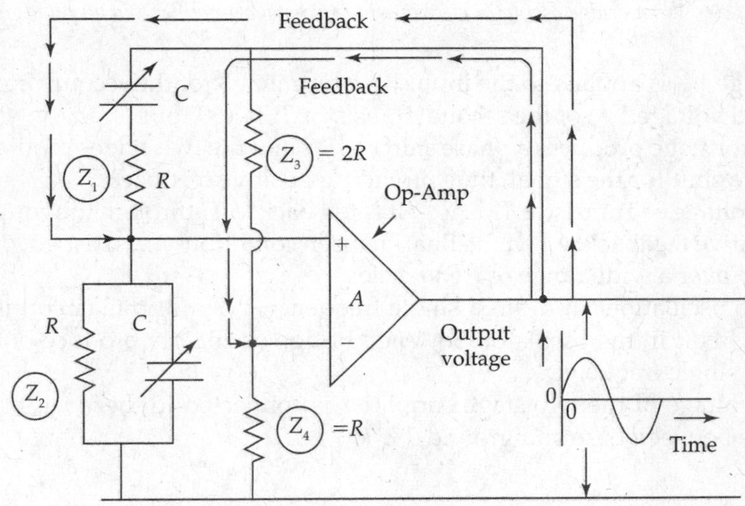

FIG. 8.18 *Wien bridge oscillator circuit working*

Wien Bridge Oscillator Circuit using Transistors

Wien Bridge oscillator circuits in Figs. 8.19 and 8.20 consist of

- Two-stage RC-coupled Amplifier,
- RC Bridge circuit to provide positive feedback path and selection of single frequency output signal voltage.

Wien Bridge Oscillator Circuit Operation

- When the DC Source is switched on, the random movement of charge carriers through the Transistor (BJT or JFET in general) and the circuit components produce a noise signal V_n at the Base B_1 Amplifier oscillator Transistor T_1 (White noise signal at B_1 contains signal frequencies from 0 Hz to infinite Hz).
- V_n is of the order of a few Pico volts at the input port of the Transistor T_1.
- V_n acts as virtual input signal V_{in}. V_n is amplified with gain A. Let the amplified voltage be V_{01} at output port of Transistor T_1. V_{01} is larger than V_{in} and is 180° out of phase with V_{in}.

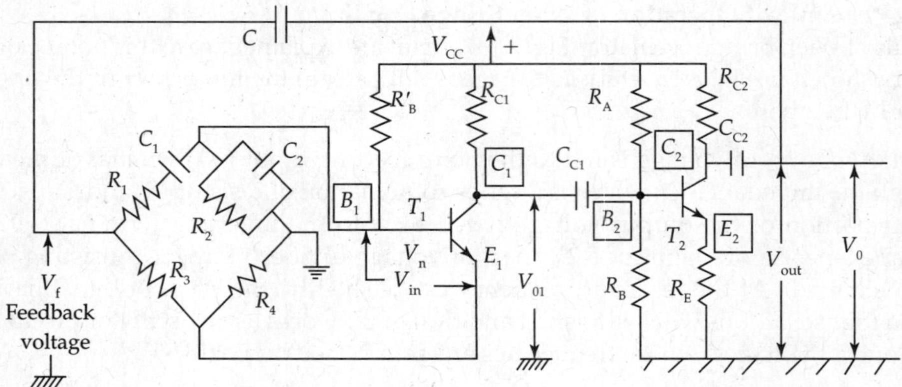

FIG. 8.19 *Wien bridge oscillator circuit using two-stage amplifier and a bridge circuit*

- Output voltage V_{01} is applied to the input Base terminal B_2 of the second Transistor T_2.
- The amplified voltage V_{out} of the second Transistor T_2 is 180° out of phase with V_{01}.
- Each Transistor stage produces a phase shift of 180° and the two stages produce 360° or zero degrees phase shift for the signal, thus providing positive feedback.
- The output voltage of Transistor T_2 is V_{out}. It is fed back to T_1 through the coupling capacitor 'C'. With positive feedback, when the Bharkhausen conditions are satisfied, oscillations will be developed over a wide range of frequencies.
- But, to obtain oscillations at desired single frequency, Wien Bridge circuit is incorporated as feedback circuit in the oscillator. So, Wien bridge oscillators produce output signals at highly stable single frequency.
- The output voltage of the two-stage Amplifier is connected to the Wien bridge circuit as input voltage between Base and ground.

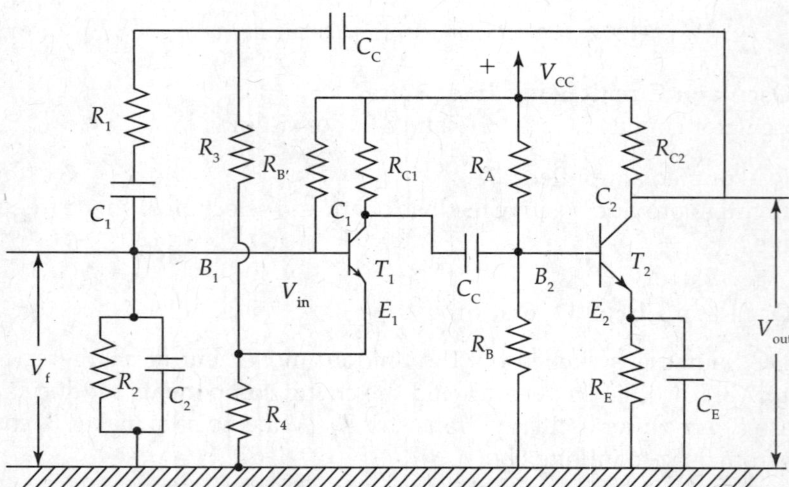

FIG. 8.20 *Wien bridge oscillator circuit for derivation of required gain a frequency of oscillations 'f' and feedback factor*

- Wien Bridge consists of four arms with four resistors and two capacitors grouped as (1) Series Combination of R_1 and C_1, (2) Parallel combination of R_2 and C_2, (3) R_3 in the third arm and (4) R_4 in the fourth arm.
- The Bridge gets balanced at a frequency where total phase shift is 0°.
- The resistors R_3 and R_4 form the voltage divider for the feedback voltage applied to the Emitter terminal of the Transistor T_1. Feedback voltage $V_f = \dfrac{V_0 \times R_4}{(R_3 + R_4)}$.
- The feedback signal will be present across the Base bias resistor R_2.

8.5.2 Frequency of Oscillations of the Wien Bridge Circuit

The balancing conditions in the Wien bridge are

$$\frac{R_3}{R_4} = \frac{\left[R_1 + \dfrac{1}{j \cdot \omega \cdot C_1}\right]}{\left[\dfrac{R_2}{1 + j \cdot \omega \cdot R_2 \cdot C_2}\right]}$$

$$R_2 \cdot R_3 = R_4 \cdot \left[R_1 - \frac{j}{\omega \cdot C_1}\right][1 + j \cdot \omega \cdot R_2 \cdot C_2]$$

$$R_2 \cdot R_3 = R_1 \cdot R_4 - \frac{j \cdot R_4}{\omega \cdot C_1} + j \cdot \omega \cdot R_1 \cdot R_2 \cdot R_4 \cdot C_2 + \frac{R_2 \cdot R_4 \cdot C_2}{C_1}. \quad (8.41)$$

Equating the real part to zero in Eq. (8.41), we get

$$R_2 \cdot R_3 - R_4 \cdot R_1 - \frac{R_2 \cdot R_4 \cdot C_2}{C_1} = 0$$

$$\frac{C_2}{C_1} = \frac{R_3}{R_4} - \frac{R_1}{R_2}.$$

If $C_1 = C_2 = C$ and $R_1 = R_2 = R$, then $R_3/R_4 = 2$, i.e. $R_3 = 2 \cdot R_4$.

A ratio of R_3 to R_4 greater than 2 provides sufficient loop gain for the circuit to oscillate at calculated frequencies.

Equating the imaginary part to zero in Eq. (8.41), we get

$$\frac{R_4}{\omega \cdot C_1} = \omega \cdot C_2 \cdot R_1 \cdot R_2 \cdot R_4$$

$$\therefore \omega^2 = \frac{1}{R_1 \cdot R_2 \cdot C_1 \cdot C_2}.$$

If $C_1 = C_2 = C$ and $R_1 = R_2 = R$,

Frequency of oscillation (at bridge balance) $f_0 = \dfrac{1}{2\pi RC}$.

For all frequencies other than the frequency of oscillations f_0, the bridge will be unbalanced and the circuit will not function as an oscillator. Wien bridge oscillator circuit uses positive feedback through R_1, C_1, R_2, C_2 to Transistor T_1 and negative feedback to the voltage divider to the input of T_1.

Advantages

1. The overall gain of the circuit is high, because of the two-stage Amplifier.
2. Frequency of oscillations can be changed by varying C_1 and C_2 or by using variable resistors.
3. Good frequency stability.
4. By replacing R_2 with a Thermistor, good amplitude stability can be achieved.
5. Stable and pure sine wave output waveform.
6. Absence of inductors and transformers makes the circuit suitable for VLSI technology.
7. No interference from external magnetic fields as no inductors are used in *RC* oscillator circuits.

Disadvantage

More number of circuit components.

EXAMPLE 8.6

In the Wien bridge oscillator circuit, if the RC network consists of resistors 200 kΩ and capacitors of 300 pF, find its frequency of oscillation. (JNTU, Nov 2003)

Solution: Frequency of oscillation $f_0 = \dfrac{1}{2\pi RC}$ Hz

Data: $R = 200$ kΩ and $C = 300$ pF

Substituting the components values in the equation for frequency of oscillations f_0, we get

$$f_0 = \frac{1}{2\pi \times 200 \times 10^3 \times 300 \times 10^{-12}}$$

$$= \frac{100 \times 10^3}{12 \times \pi} = 2.65 \times 10^3 \text{ Hz.}$$

8.6 LC OSCILLATORS (HIGH-FREQUENCY OSCILLATORS)

- LC Oscillators produce periodic sinusoidal voltages at High frequency.
- LC oscillators use a Transistor to function as an Amplifier initially to amplify the random noise signal. Common Emitter Transistor Amplifier or Common Source FET Amplifiers used in the circuit contributes 180° phase shift during amplifying action.
- The amplified signal is feedback through an LC network to select the desired signal and provide another 180° phase shift to cause positive feedback to the Transistor input to function as an oscillator. The cycles of events repeat till the Bharkhausen conditions (criteria) for oscillations [$A\beta = 1$] are satisfied and the designed oscillator circuit generates an AC signal at desired frequency and amplitude.

Colpitts Oscillator Circuit

Colpitts oscillator has basic amplifier and feedback network containing two Capacitors C_1 (Z_1) and C_2 (Z_2) and one Inductor L (Z_3).

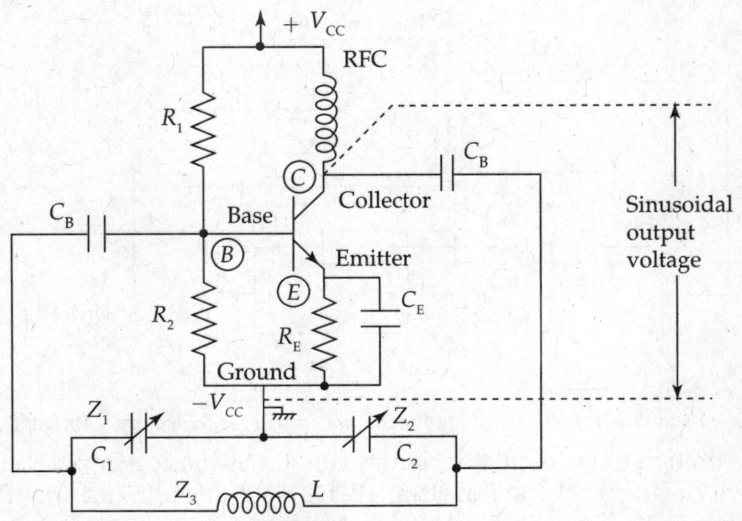

FIG. 8.21 *Colpitts oscillator*

Hartley Oscillator Circuit

Hartley oscillator has basic amplifier and feedback network containing two Inductors L_1 (Z_1) and L_2 (Z_2) and one Capacitor C (Z_3).

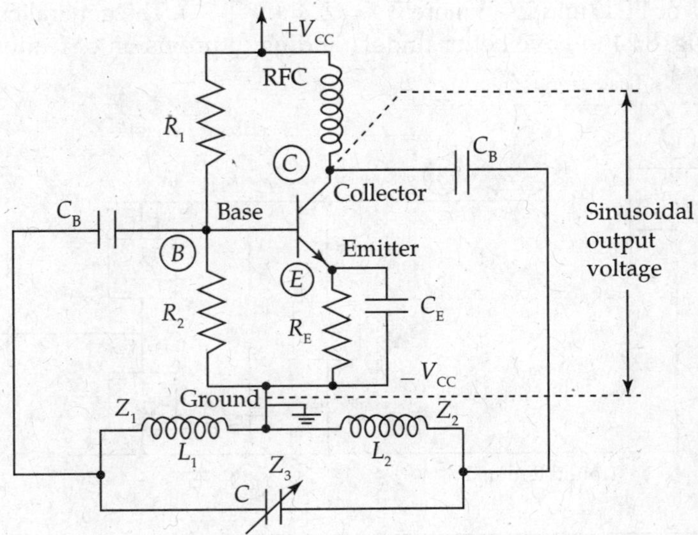

FIG. 8.22 *Hartley oscillator*

8.6.1 General Configuration of LC Oscillators

Observing the Colpitts and Hartley oscillator circuit layouts of Figs. 8.21 and 8.22, general configuration of circuit layout for LC oscillators could be as shown in Fig. 8.23. This configuration is also common to some other Tuned or Resonant oscillator circuits.

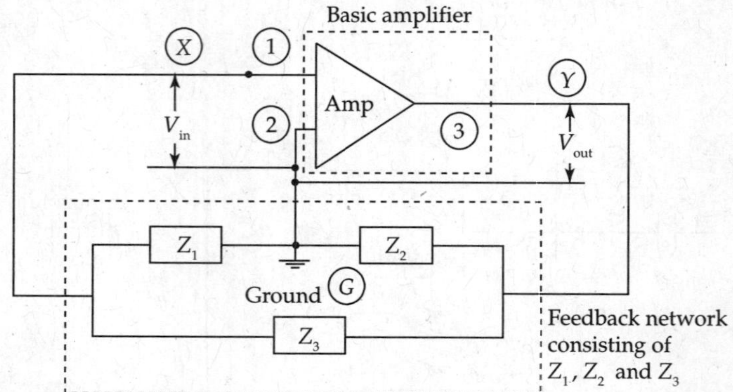

FIG. 8.23 *Circuit to illustrate the nature of the components in the feedback network of LC oscillators*

General configuration of LC oscillator circuits contain the basic Amplifier circuit that may contain the active devices such as Transistor (BJT), Field effect Transistor (FET), MOSFET or operational Amplifier. Terminals 1 and 2 are the input terminals of input voltage V_{in} and terminals 3 and 2 are output terminals (common terminal is 2) for output voltage V_{out}.

The general expression for gain 'A' of the Amplifier is expressed as

$$A = \frac{-V_{out}}{V_{in}} = -g_m Z_L. \tag{8.42}$$

Negative sign in the above expression indicates that Amplifier introduces 180° phase shift.

From the circuit in Fig. 8.23, Z_L is the parallel combination of the output resistance R_o of the active devices (BJT or FET) and Z'_L, where $Z'_L = Z_2 \,||\, (Z_1 + Z_3)$. The general circuit in Fig. 8.23 is redrawn as in Fig. 8.24 to have better understanding of provision of feedback 'V_f' to input

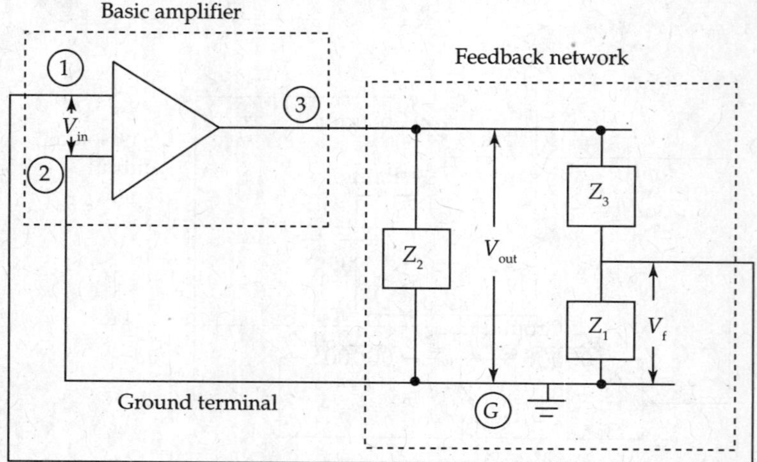

FIG. 8.24 *General configuration of LC oscillator circuits*

pair of terminals 1 and 2 from output pair of terminals 3 and 2 using the feedback network containing Z_1, Z_2 and Z_3.

Output voltage V_{out} is across Z_2 or $(Z_1 + Z_3)$. Feedback voltage V_f to the input terminals 1 and 2 is the voltage across Z_1.

From the circuit in Fig. 8.24, Z_L is the parallel combination of the output resistance 'R_o' of the active devices (BJT or FET) and Z_L where

$$Z_L' = \frac{Z_2(Z_1 + Z_3)}{(Z_1 + Z_2 + Z_3)} \tag{8.43}$$

$$\text{and } Z_L = R_o \| Z_L'$$

$$\text{Gain} \quad A = -g_m \cdot Z_L = \frac{-g_m \cdot R_o \cdot \dfrac{Z_2(Z_1 + Z_3)}{(Z_1 + Z_2 + Z_3)}}{R_o + \dfrac{Z_2(Z_1 + Z_3)}{(Z_1 + Z_2 + Z_3)}} \tag{8.44}$$

$$\therefore \ A = \frac{-g_m \cdot R_o [Z_2(Z_1 + Z_3)]}{R_o(Z_1 + Z_2 + Z_3) + Z_2(Z_1 + Z_3)} \tag{8.45}$$

$$\text{Feedback factor} \quad \beta = \frac{V_f}{V_{out}} = \frac{Z_1}{Z_1 + Z_3}. \tag{8.46}$$

Substituting the values of A and β from Eqs. (8.45) and (8.46) in the Bharkhausen condition for oscillations, $A\beta = 1$.

Output resistance R_0 of the active devices if finite,

$$A\beta = -g_m Z_L \beta = \frac{-g_m R_o Z_2(Z_1 + Z_3)}{R_o(Z_1 + Z_2 + Z_3) + Z_2(Z_1 + Z_3)} \times \frac{Z_1}{(Z_1 + Z_3)} = 1 \tag{8.47}$$

$$\text{Loop gain} \quad A\beta = \frac{-g_m \cdot R_o(Z_1 \cdot Z_2)}{R_o(Z_1 + Z_2 + Z_3) + Z_2(Z_1 + Z_3)} = 1 \tag{8.48}$$

In LC oscillator circuits, Z_1, Z_2 and Z_3 are reactances.

$$\text{Assume} \quad Z_1 = j \cdot x_1, \ Z_2 = j \cdot x_2 \text{ and } Z_3 = j \cdot x_3 \tag{8.49}$$

and substitute them in Eq. (8.48),

$$A\beta = \frac{-g_m \cdot R_o (j \cdot x_1) \cdot (j \cdot x_2)}{R_o(j \cdot x_1 + j \cdot x_2 + j \cdot x_3) + j \cdot x_2(j \cdot x_1 + j \cdot x_3)} = 1$$

$$\therefore \ A\beta = \frac{g_m \cdot R_o \cdot x_1 \cdot x_2}{j \cdot R_o(x_1 + x_2 + x_3) - \ x_2(x_1 + x_3)} = 1. \tag{8.50}$$

To satisfy the Bharkhausen condition for oscillations $A\beta = 1$, the phase shift is zero. Therefore, the j term or the reactive term of Eq. (8.50) is made zero.

Output resistance R_0 of the active devices if finite,

$$\therefore \ (x_1 + x_2 + x_3) = 0 \tag{8.51}$$

$$A\beta = \frac{-g_m R_o x_1 x_2}{x_2(x_1 + x_3)} = 1 \tag{8.52}$$

$$\text{i.e.,} \ \frac{-g_m \cdot R_o \cdot x_1}{(x_1 + x_3)} = 1 \tag{8.53}$$

$$\therefore \ \frac{g_m \cdot R_o \cdot x_1}{x_2} = 1 \ \ [\text{using } (x_1 + x_3) = -x_2 \text{ from Eq.(8.53)}]. \tag{8.54}$$

- Equation (8.54) suggests that x_1 and x_2 are similar types of reactance.
- So, both x_1 and x_2 are inductors in Hartley oscillator circuit (Fig. 8.22) and both are capacitors in Colpitts oscillator Circuit (Fig. 8.21).
- From Eq. (8.51), it is also known that $(x_1 + x_2) = -x_3$ so that when x_1 and x_2 are inductors in Hartley Oscillator circuit, x_3 is a capacitor as shown in Fig. 8.22.
- Similarly, in Colpitts oscillator circuit when x_1 and x_2 are capacitors, x_3 is an inductor in the circuit (refer Fig. 8.21).

S. No.	Type of tunable oscillator	Types of components for X_1, X_2 and X_3
1	Colpitts oscillator	x_1, x_2 are both capacitive and x_3 inductive
2	Clapp oscillator	x_1, x_2 are both capacitive and x_3 having L and C_3 in series
3	Hartley oscillator	x_1, x_2 are both inductive and x_3 is capacitive
4	Pierce Crystal oscillator	x_1, x_2 are both inductive and x_3 is capacitive determined by Crystal
5	Tuned gate/Tuned drain or TPTG oscillator (Miller Crystal oscillator)	x_1, x_2 parallel tuned circuit act as net L and x_3 capacitive

8.6.2 Frequency of Oscillations '*f*' for Hartley and Colpitts Oscillator Circuits

Using Eq. (8.51),

$$(x_1 + x_2) = -x_3. \tag{8.55}$$

In Hartley oscillator circuit,

$$x_1 = \omega \cdot L_1, \ \ x_2 = \omega \cdot L_2 \ \text{ and } \ x_3 = -\frac{1}{\omega \cdot C}$$

$$\omega \cdot L_1 + \omega \cdot L_2 = \frac{1}{\omega \cdot C}$$

$$\omega^2 (L_1 + L_2) = \frac{1}{C}$$

$$\therefore \ \omega^2 = \frac{1}{C(L_1 + L_2)}$$

$$\therefore \text{ Frequency of oscillations } \quad f = \frac{1}{2\pi\sqrt{(L_1 + L_2)C}} = \frac{1}{2\pi\sqrt{L_{eq}C}} \text{ Hz.}$$

On similar lines frequency of oscillations 'f' for Colpitts oscillator

$$f = \frac{1}{2\pi\sqrt{L \cdot \left[\dfrac{C_1 \cdot C_2}{C_1 + C_2}\right]}} = \frac{1}{2\pi\sqrt{L \cdot C_{eq}}} \text{ Hz.}$$

8.7 COLPITTS OSCILLATOR USING FET

Colpitts oscillator circuit in Fig. 8.25 uses common Source FET Amplifier. It is an example of resonant circuit (LC) oscillators. It uses split capacitor in the tuned circuit.

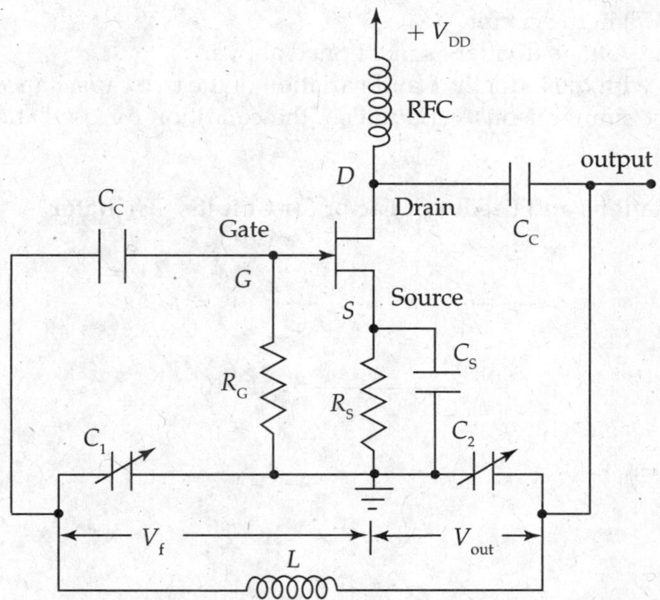

FIG. 8.25 *Colpitts oscillator using JFET device*

DC Biasing Circuit

- Radio frequency coil (RFC) (Inductive reactance ωL) acts like
  - DC short to connect the DC supply voltage V_{DD} to the Drain terminal of FET and
  - Large impedance for AC signals that prevents high frequency output oscillations to reach power supply.
- Drain supply voltage V_{DD} in association with Gate resistor (R_G) and parallel combination of R_S and C_S at source terminal (Self-Bias) provide the necessary stabilised biasing voltages to the Field Effect Transistor.

Basic Amplifier Action in the Oscillator

- Common Source FET Amplifier introduces a phase shift of 180° to the inherently generated noise signal due to the randomly moving charge carriers through the device.
- Another 180° phase shift is provided by the capacitive feedback for introducing positive feedback for oscillator action during the signal passage through the feedback circuit.
- The overall phase shift is 360° or 0° to the signals for satisfying the 'Bharkhausen condition' to start and maintain the oscillations and satisfy $[A\beta = 1]$.

Oscillator Action and Frequency Selection

- The feedback network has two variable capacitors C_1 and C_2 in series whose centre is grounded and are shunted by an inductor L.
- The capacitors and inductor constitute a resonant circuit in the feedback loop (one of the conditions of Bharkhausen conditions for oscillations) to determine the frequency of oscillations of the circuit.
- The voltage across the capacitor C_2 is the oscillator's output voltage V_{out}.
- The voltage across the capacitor C_1 is the fed back voltage (V_F) to the input port of the active device (BJT or JFET) in the circuit.
- C_1 and C_2 form the voltage divider as mentioned above.
- The tuned circuit with the inductor L and variation of the two capacitances C_1 and C_2 using ganged tuning for simultaneous variation set the condition for oscillations and oscillator frequency.

Frequency of Oscillations and Feedback Factor for Colpitts Oscillator

Using Eq. (8.51),

$$x_1 = -\frac{j}{\omega \cdot C_1}; \quad x_2 = -\frac{j}{\omega \cdot C_2}; \quad x_3 = j \cdot \omega \cdot L$$

$$-\frac{j}{\omega \cdot C_1} - \frac{j}{\omega \cdot C_2} + j \cdot \omega \cdot L = 0$$

At resonant frequency,

$$\omega \cdot L = \frac{1}{\omega \cdot C}.$$

$$\text{Then} \quad \omega \cdot L = \frac{1}{\omega}\left[\frac{1}{C_1} + \frac{1}{C_2}\right].$$

$$\therefore \quad \omega^2 = \frac{1}{L}\left[\frac{C_1 + C_2}{C_1 \cdot C_2}\right]$$

$$\omega^2 = \frac{1}{L \cdot C_{eq}}, \quad \text{where } C_{eq} = \left[\frac{C_1 \cdot C_2}{C_1 + C_2}\right].$$

$$\therefore \text{ Frequency of oscillations} \quad f_0 = \frac{1}{2\pi\sqrt{L \cdot C_{eq}}} \tag{8.56}$$

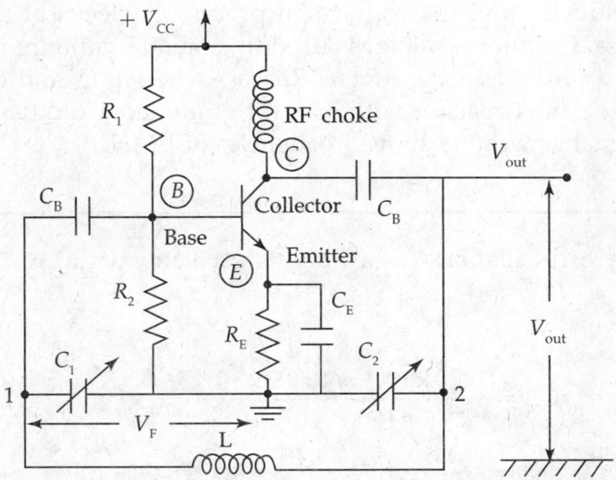

FIG. 8.26 *Colpitts oscillator circuit*

$$\text{Feedback factor} \quad \beta = \frac{V_f}{V_0} = \frac{I \cdot x_{C_1}}{I \cdot x_{C_2}} = \frac{C_2}{C_1}$$

From the condition $A\beta = 1$,

$$\text{Gain} \quad A = \frac{1}{\beta}$$

$$\therefore \quad A = \frac{C_1}{C_2}.$$

In a Transistor Colpitts oscillator, at increased frequencies of oscillations, the internal capacitance C_e (junction capacitance between Base and Emitter) is across C_1 at the input port of the Transistor and other stray capacitances C_s of Transistors come in parallel to C_2 at the output port of the Transistor. The internal capacitance changes due to variations in operating temperature or in replacing the active device affect the necessary phase relations. So, the frequency drifts or varies and may not be stable over a period of time. This is the disadvantage of Colpitts oscillator.

Analysis of Transistor oscillators gets complicated because the low input impedance Z_{in} (due to very low value of h_{ie}) of the Transistor shunts Z_1 (x_1) in the circuit. Also as the oscillators operate at very high frequencies (RF range of LC oscillators), the hybrid-π model of the Transistor is to be considered.

Circulating current is same for both the capacitors.The conditions for sustained oscillation

$$h_{fe} = \frac{x_1}{x_2} = \frac{C_2}{C_1}. \tag{8.57}$$

In general, Collector current flows continuously leading to Class-A operation. If it is desired to have high efficiency and large power output, Class-C operation is preferred and in that class of operation even though the Collector current flows in the shape of pulses, output of the tuned circuit becomes sinusoidal due to the sequence of charging and discharging of capacitor through inductor and the conversion process of electrical energy into magnetic energy and vice versa. Colpitts oscillator is used to generate radio frequencies. It is used as a local oscillator in a *Super Heterodyne Radio Receiver* wherein C_1 and C_2 are provided by a split-stator adjustable gang capacitor with its rotor grounded. Colpitts oscillator circuits are normally used as signal generators above frequencies of 1 MHz.

EXAMPLE 8.7

Determine frequency of oscillations for a Colpitts oscillator with $L = 10\ \mu\text{H}$, $C_1 = 1000$ pF, $C_2 = 10{,}000$ pF.

Solution:

$$C_{eq} = \frac{1000\times10^{-12}\times10000\times10^{-12}}{11000\times10^{-12}}$$

$$= \frac{10^7}{11000}\times10^{-12} = \frac{10^4}{11}\times10^{-12} = 909 \text{ pF}$$

$$f_0 = \frac{1}{2\pi\sqrt{L\cdot C_{eq}}} = \frac{1}{2\pi\sqrt{10\times10^{-6}\times909\times10^{-12}}}$$

$$= \frac{1}{2\pi\sqrt{10\times909\times10^{-18}}} = \frac{10^9}{2\pi\sqrt{9090}}$$

$$\therefore \text{ Frequency } f_0 = \frac{10^9}{2\pi\times95.34} = \frac{10^3\times10^6}{599} = 1.67 \text{ MHz.}$$

8.8 CLAPP OSCILLATOR

In the Colpitts oscillator circuits, the internal capacitances and the stray capacitances shunt the external capacitors C_1 and C_2 in the tuned circuit, because the two capacitors are directly across the input and output ports of the active device, the Transistor (as one of the ends of the two capacitors C_1 and C_2 are connected to the ground). It is known that the internal capacitances change with changes in temperature and quiescent operating point of Transistor bias conditions and as such the frequency of Colpitts oscillator is not stable and accurate over the desired frequency range.

To overcome such difficulty, a small variable capacitor C_3 is added in series with inductor L and the values of the (fixed) capacitors C_1 and C_2 are so chosen to be higher than C_3 in the Clapp oscillator circuit (Fig. 8.27) developed by James Kilton Clapp during 1948. It is a *modified version of Colpitts oscillator*.

The value of capacitor C_3 cannot be made too small as its value dominates inductive reactance and the circuit cannot oscillate. Clapp oscillator circuit improves frequency stability.

Circulating current in the tank circuit flows in series through the three capacitors, where

$$C_{eq} = \frac{C_1C_2C_3}{C_1C_2 + C_2C_3 + C_3C_1} \Rightarrow C_{eq} = \left(\frac{1}{\frac{1}{C_1}+\frac{1}{C_2}+\frac{1}{C_3}}\right).$$

Frequency of oscillation $f = \frac{1}{2\pi\sqrt{LC_{eq}}}$.

As C_3 is very much small compared to C_1 and C_2

$$\therefore\ C_{eq} = C_3 \text{ and} \ll \frac{C_1C_2}{C_1 + C_2}$$

$$\therefore\ f = \frac{1}{2\pi\sqrt{LC_3}} \tag{8.58}$$

For a Clapp oscillator, as the frequency of operation depends upon C_3 than C_1 and C_2, it is more stable and accurate than Colpitts oscillator and that is the reason why a Clapp oscillator is more preferred than Colpitts oscillator.

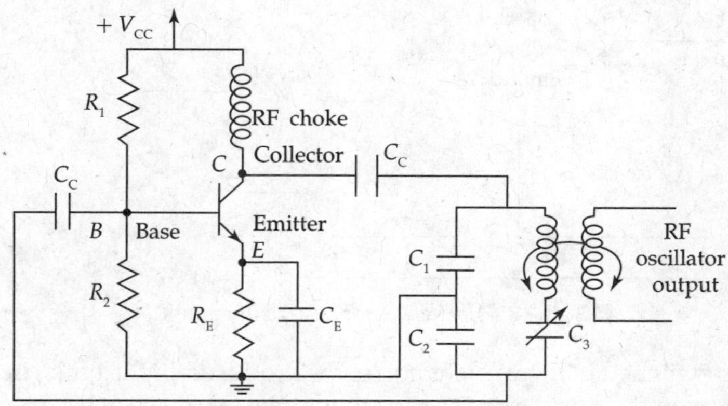

FIG. 8.27 *Basic circuit of clapp oscillator*

EXAMPLE 8.8

If a 50 pF capacitor is in series with 20 μH inductor of a Clapp oscillator with C_1 = 1000 pF, C_2 = 10,000 pF. Determine the frequency of oscillation.

Solution: Here $C_3 = 50\text{ pF} \ll (C_1\, C_2) / (C_1 + C_2)$

Frequency of oscillation of Clapp Oscillator circuit is

$$f = \frac{1}{2\pi\sqrt{LC_3}} = \frac{1}{2\pi\sqrt{20\times10^{-6}\times50\times10^{-12}}} \cong 5\text{ MHz}.$$

8.9 HARTLEY OSCILLATOR CIRCUIT

- Hartley circuit is analogous to that of a Colpitts oscillator circuit. It uses common Emitter Transistor in the basic Amplifier (Fig. 8.28).
- Here the inductors L_1 and L_2 are tapped at middle and there is a capacitor C across it.
- The tuned circuits determine the resonant frequency of the oscillator and the magnitude of feedback for maintaining oscillations through positive feedback.

- Radio frequency coil (RFC) (ωL) at the Collector terminal acts as DC short and has a high impedance for high-frequency oscillations and so high-frequency signals are blocked from reaching the power supply.
- The supply voltage V_{CC}, parallel combination of R_E and C_E and potential divider R_1 and R_2 provide the necessary stabilised bias to the Transistor.
- The phase-shift network consists of the two inductors L_1 and L_2.

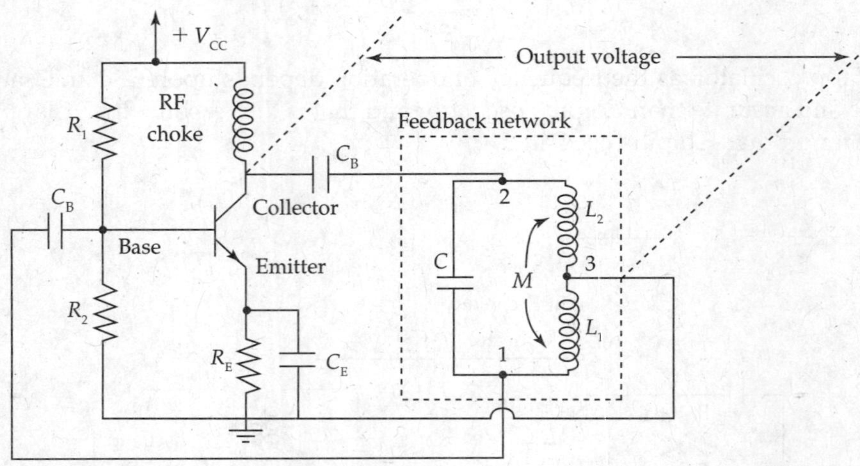

FIG. 8.28 *Hartley oscillator circuit using BJT*

When the circuit is switched ON, transient current is produced in the tank circuit and damped oscillations are set up in the tuned circuit. Referring to the Hartley oscillator circuit of Fig. 8.28, at any instant the voltages with respect to terminal 3 which is at ground, terminal 1 is positive and terminal 2 is negative and vice versa thus containing a phase shift 180° with the common Emitter Transistor Amplifier. The output voltage is across the inductor L_1. The coil L_2 is inductively coupled to coil L_1. Thus the induced voltage across L_1 forms the feedback voltage V_f, through C_b to the Base of the Transistor (input port of the Transistor). Common Emitter operation of the Transistor results in a phase shift of 180° and another 180° phase shift among the voltages across L_1 and L_2. The total phase shift around the loop is 360° or 0° and satisfies the Bharkhausen conditions (criteria) of oscillations.

Here $x_1 = \omega(L_1+M)$, $x_2 = \omega\,(L_2+M)$ and $x_3 = -1/\omega C$.

$$x_1 + x_2 + x_3 = 0$$

$$\therefore\ \omega\cdot(L_1 + L_2 + 2M) = \frac{1}{\omega\cdot C}$$

$$\omega^2 = \frac{1}{C\cdot(L_1 + L_2 + 2M)} = \frac{1}{C\cdot L_{eq}}, \qquad (8.59)$$

where $L_{eq} = (L_1 + L_2 + 2M)$

$$\therefore\ f_0 = \frac{1}{2\pi\sqrt{C\cdot L_{eq}}} \qquad (8.60)$$

$$\text{Feedback factor} \quad \beta = \frac{V_f}{V_{out}} = \frac{x_{L1}}{x_{L2}} = \frac{2\pi \cdot f \cdot L_1}{2\pi \cdot f \cdot L_2} = \frac{L_1}{L_2}. \tag{8.61}$$

The frequency of oscillation f_0 is calculated from the already familiar expression. Starting condition for oscillations is that $A > 1/\beta$; Gain $A = \frac{r_c}{r_e'} > \frac{L_2}{L_1}$.

Hartley oscillator is used as RF oscillator. Frequency of oscillation can be changed by making the core movable (varying the inductance) or by varying the capacitance. It is used in Super heterodyne Radio receivers.

Disadvantage of Hartley oscillator is that it cannot be used as low-frequency oscillator since the value of inductors becomes large and size of inductors becomes bulky. Hartley Oscillator circuit using a JFET device is similar in circuit operation, which is shown in Fig. 8.29.

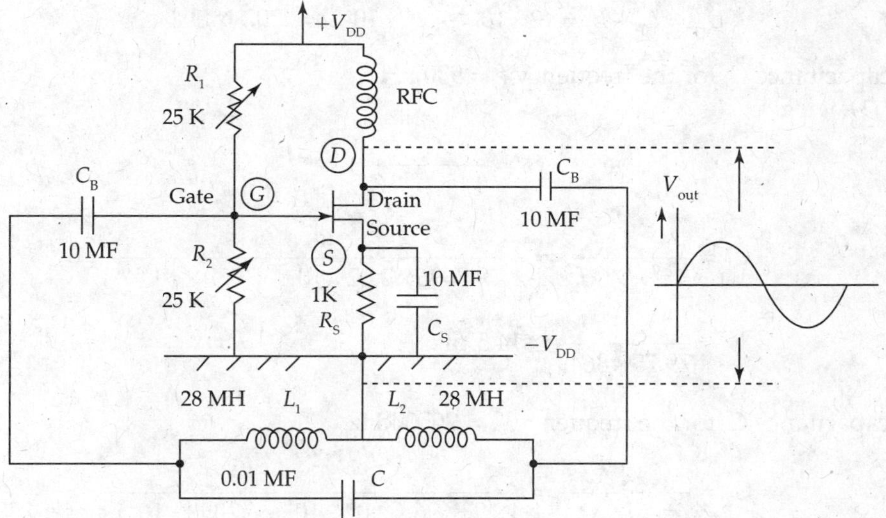

FIG. 8.29 *Hartley oscillator circuit using JFET device*

EXAMPLE 8.9

Determine the frequency of oscillations of a Transistor Hartley oscillator with $L_1 = 100$ μH, L_2 =1.5 MH, Mutual inductance $M = 100$ μH and $C = 150$ pF.

Solution: $L_{eq} = L_1 + L_2 + 2M$

$$L_{eq} = \left[(100\times10^{-6})+(1.5\times10^{-3})+(2\times100\times10^{-6})\right]$$

$$= \left[(100+1500+200)\times10^{-6}\right]$$

$$= 1800\times10^{-6} = 1.8 \text{ mH}$$

$$\text{Frequency of oscillations } f_0 = \frac{1}{2\pi\sqrt{L_{eq}\cdot C}} = \frac{1}{2\pi\sqrt{1.8\times10^{-3}\times150\times10^{-12}}}$$

$$\therefore\ f_0 = \frac{10^7}{2\pi\sqrt{27}} = \frac{10 \times 10^6}{32.6} = 306.7 \text{ kHz.}$$

EXAMPLE 8.10

In a Transistorised Hartley oscillator, the two inductances are 2 mH and 20 μH while the frequency is to be changed from 930 kHz to 2050 kHz. Calculate the range over which the capacitor is to be varied. (May/June 2006, set- 4)

Solution: In the Hartley oscillator frequency of oscillations

$$f = \frac{1}{2\pi\sqrt{L_{eq} \cdot C}} \text{ Hz} \tag{1}$$

$$L_{eq} = L_1 + L_2 = (2 \times 10^{-3} + 20 \times 10^{-6}) = 2.02 \text{ mH}$$

Value of capacitance C_1 for the frequency $f_1 = 930$ kHz
From Eq. (1), we get

$$f^2 = \frac{1}{4\pi^2 \cdot L_{eq} \cdot C}$$

$$\therefore\ C_1 = \frac{1}{4\pi^2 \cdot L_{eq} \cdot f_1^2} = \frac{1}{4 \times 9.8696 \times 2.02 \times 10^{-3} \times (930 \times 10^3)^2}$$

$$= \frac{10^{-5}}{79.75 \times 8649} \cong 14.5 \text{ pF.}$$

Value of capacitance C_2 for the frequency $f_2 = 2050$ kHz

$$\therefore\ C_2 = \frac{1}{4\pi^2 \cdot L_{eq} \cdot f_2^2} = \frac{1}{4 \times 9.8696 \times 2.02 \times 10^{-3} \times (2050 \times 10^3)^2}$$

$$= \frac{10^{-5}}{79.75 \times 42025} \cong 3 \text{ pF.}$$

Required range of variation of the capacitor = 3–14.5 pF.

8.9.1 Practical Working of Hartley Oscillator Circuit

Aim:
To conduct an experiment to identify the various blocks of a Hartley oscillator circuit and observe the output signal waveform and measure its amplitude and frequency.

Apparatus:
(1) DC power supply (0–30 V), (2) Hartley oscillator circuit board and (3) Cathode ray oscilloscope.

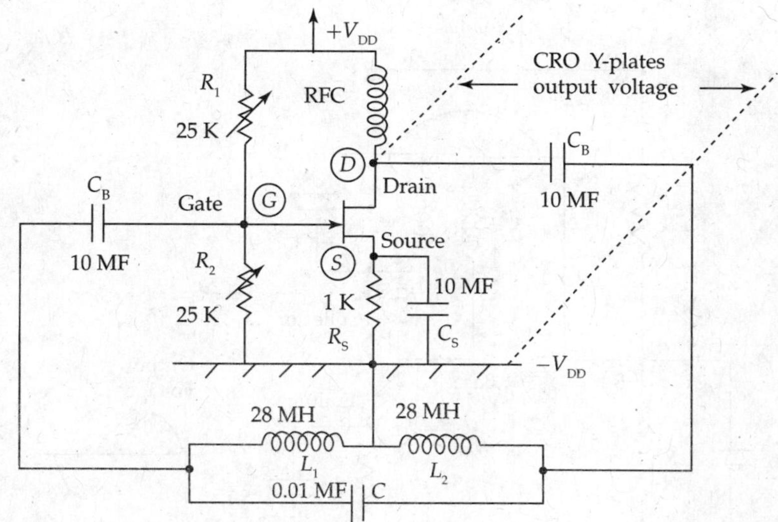

FIG. 8.30 *Hartley oscillator circuit using JFET device*

Procedure:

1. Trace the Hartley oscillator circuit and identify the various components that are necessary for the working of the oscillator.
2. Apply suitable magnitude of DC voltage from DC power supply unit to + V_{DD} and common terminal nodes of the oscillator circuit depending upon the type of field effect Transistor used in the circuit and its specifications.
3. Connect the *Y*-plates probe of a CRO to the Drain terminal of the FET device and the common terminal to observe the output voltage on the screen of CRO.
4. Adjust the biasing resistors R_1 and R_2 so that a good sine wave signal (without any distortion) appears at the oscillator output port.
5. Measure the amplitude and frequency of the output signal using CRO.
6. Draw the observed output waveform on a graph paper.
7. Calculate the theoretical frequency of oscillations from the circuit components in the circuit and verify it with the practical frequency of the observed signal.
8. Hartley oscillator is a high-frequency oscillator.

8.10 TUNED COLLECTOR OSCILLATOR

In the tuned Collector oscillator circuit (Fig. 8.31), components R_1, R_2-C_2, R_E, C_E and V_{CC} determine the necessary DC-operating conditions of the circuit. C_2 provides bypass path for AC signals around R_2 and C_E provides bypass path for AC signals around R_E. The Tuned circuit consisting of L_P and C_T is connected in to the Collector path of the Transistor in the circuit. Tuned circuit is the load impedance. Output voltage across the tuned circuit is fed back to the input port through the Radio Frequency transformer (L_S and L_P).

The secondary winding of the transformer (L_S) is so chosen as to make the secondary inducted voltage V_f fed in phase to the input (Positive feedback) (180° phase shift is introduced during feedback through transformer action.) so as to allow oscillations to setup satisfying the

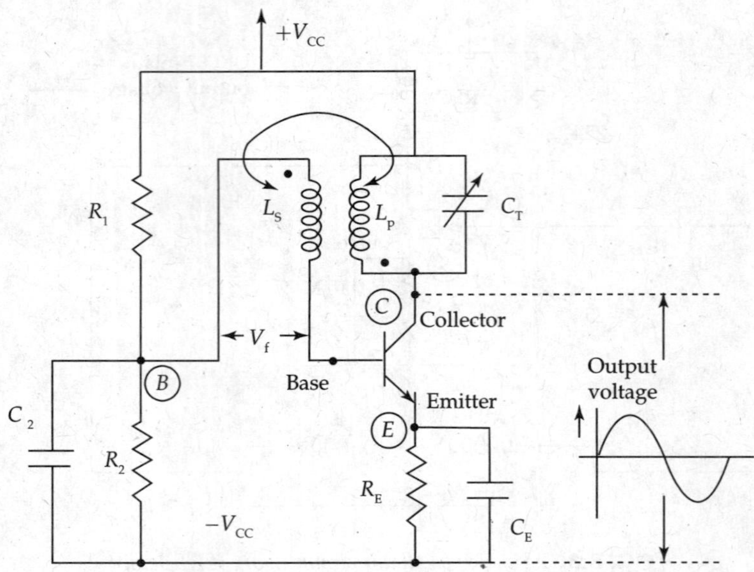

FIG. 8.31 *Tuned collector oscillator*

Bharkhausen criteria and sustain the oscillations in the circuit. The parallel tuned circuit with L_P and C_T determines the frequency of oscillations f_0 of the circuit.

$$\text{Frequency of oscillations} \quad f_0 = \frac{1}{2\pi\sqrt{L_P C_T}}. \tag{8.62}$$

8.11 TUNED DRAIN OSCILLATOR CIRCUIT

Tuned Drain Oscillator circuit is shown in Fig. 8.32.

In the tuned Drain oscillator circuit, the components R_1, R_2, C_2, R_S, C_S and V_{DD} determines the necessary stabilised bias operating conditions of the circuit. The tuned circuit consisting

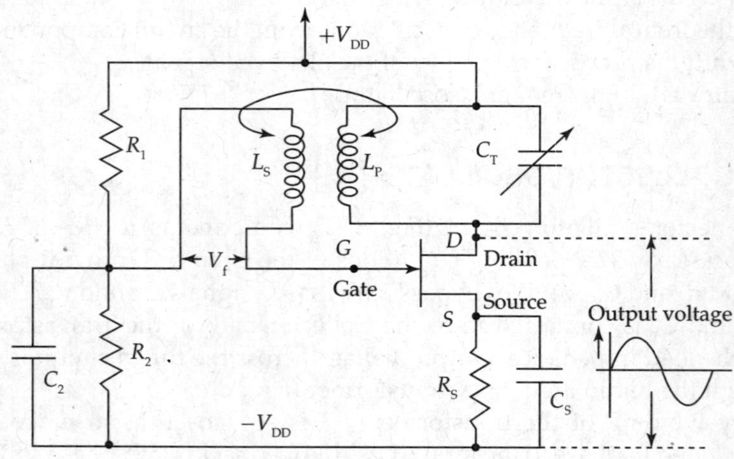

FIG. 8.32 *Tuned drain oscillator*

of L_P and C_T is connected to the Drain path of the Junction Field Effect Transistor (JFET) in the circuit. The output voltage V_0 that is produced across the parallel tuned circuit is fed back into the input port through RF transformer secondary coil L_S. The secondary winding direction of the transformer is so chosen as to make the feedback voltage V_f to be in phase with the effective input for providing positive feedback satisfying the 'Bharkhausen criteria' by an additional phase shift produced by the transformer action. Thus the circuit produces the desired sinusoidal output voltage at the desired frequency and amplitude.

The frequency of oscillation is $f_0 = \dfrac{1}{2\pi\sqrt{L_P C_T}}$.

EXAMPLE 8.11

The tank circuit of Tuned Drain oscillator has $L = 50\ \mu\text{H}$ and $C = 200$ pF. Calculate the frequency Oscillation.

Solution:

Frequency of oscillations $f_0 = \dfrac{1}{2\pi\sqrt{LC}}$

$$= \frac{1}{2\pi\sqrt{50\times10^{-6}\times200\times10^{-12}}} = \frac{10\times10^{6}}{2\pi} = 1.6 \text{ MHz}.$$

8.11.1 Tuned Gate and Tuned Base Oscillator circuits

Tuned Gate (/Base) Oscillator consists of parallel tuned circuit of L_P and C elements between Gate (/Base) and ground. Required positive feedback to initiate and maintain oscillations in the circuit is provided by RF transformer of mutually coupled coils L_S and L_P.

The frequency of oscillations is given as

$$f_0 = \frac{1}{2\pi\sqrt{L_P C}}. \tag{8.63}$$

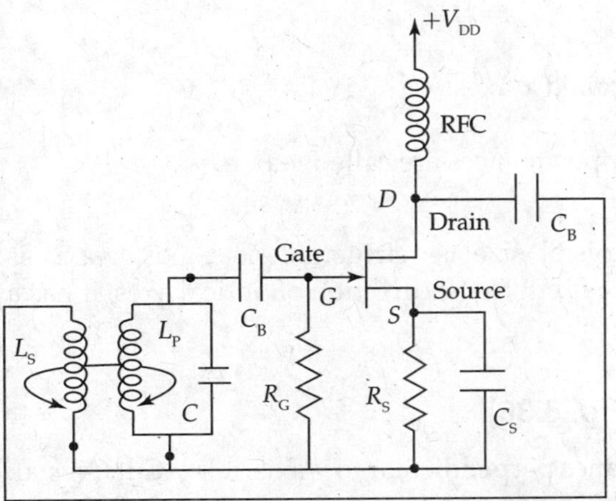

FIG. 8.33 *Tuned gae oscillator (armstrong oscillator)*

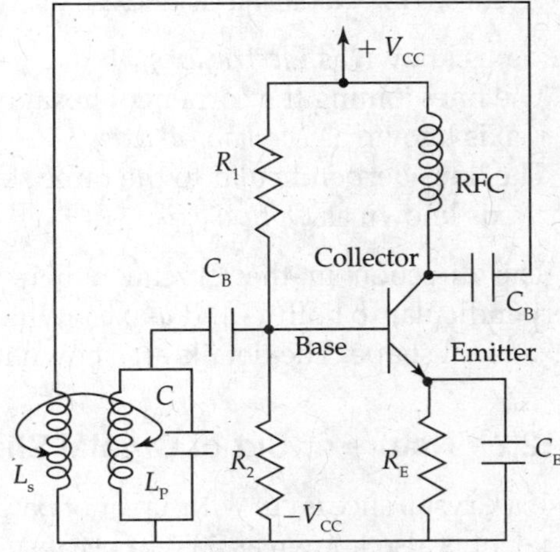

FIG. 8.34 *Tuned base oscillator*

8.12 CRYSTAL OSCILLATORS

8.12.1 Features of Quartz Crystals

Quartz or piezoelectric Crystals are used in electronic oscillator circuits as substitute for series or parallel resonant circuits for maintaining stable fixed frequency signal generation. Crystals used in oscillator circuits consist of a Crystal slice having suitable dimensions (Based on the Crystal frequency) embedded between two conducting electrodes enclosed in a protective box with external leads for electrical connections. Quartz Crystals exhibit piezoelectric effect.

The phenomenon of piezoelectric effect means that when an AC voltage is applied across the faces of a Crystal, they vibrate at a frequency of the applied voltage and mechanical distortion occurs in the Crystal shape. Conversely when the Crystals are mechanically stressed across the faces, then a proportional AC voltage is developed across the opposite faces of the Crystal. This phenomenon is known as *Piezoelectric Effect*.

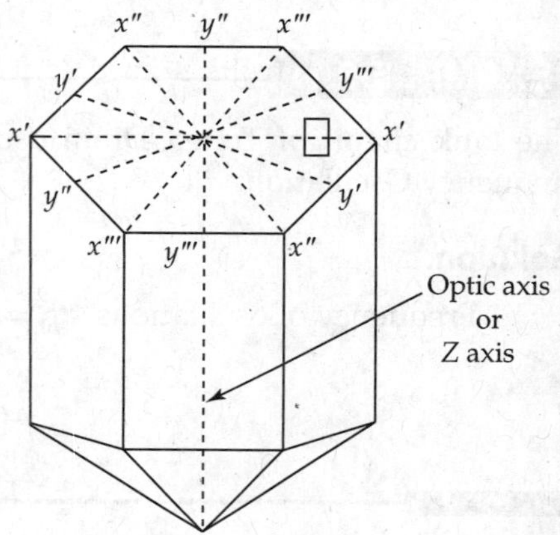

FIG. 8.35 *Quartz crystal axes, x-axis, y-axis, z-axis*

Some Crystal materials such as Rochelle salt, Quartz and Tourmaline exhibit piezoelectric property. Rochelle Salt even though exhibits excellent piezoelectric phenomena is mechanically weakest, easily breaks and effected by heat and moisture. Crystals made up of Tourmaline are mechanically strongest but exhibit least piezoelectric phenomena and are very costlier. They are used at high frequencies. Quartz Crystals have properties intermediate in between Rochelle salts and Tourmaline. With very high values of Q, quartz Crystals are readily available in nature and are cheap. Crystals having values of Q of the order of 100,000 have resonant frequencies ranging from a few kilo Hertz to many mega Hertz. It has got several applications in RF oscillators and filters, communication Transmitters and receivers, digital clocks and in time standards etc.

Quartz Crystal looks like a hexagonal prism with pyramids at both ends. A Crystal has three axes, viz. x-axis, y-axis and z-axis.

- x-axis known as *Electric axis*:
  The lines joining the corners of hexagon are called x-axes.
- y-axis known as *Mechanical axis*:
  The lines perpendicular to the three sets of opposite faces are called y-axes.
- z-axis known as *Optical axis*:

The direction in the Crystal, where the Crystal size is refrigerant, is z-axis. z-axis is perpendicular to both x- and y-axes. When a slice is cut to the direction of optical axes, it has a hexagonal shape. The details are shown in Fig. 8.35.

8.12.2 Nature of Cut of Crystal Slices (Fig. 8.36)

- If a Crystal slice is cut with a pair of parallel planes perpendicular to y-axis, it is y-Cut Crystal. y-Cut is also known as 30° cut because the angle between adjacent x- and y-axes is 30°.
- The Crystal slices cut perpendicular to x-axis are called as x-cut Crystals.

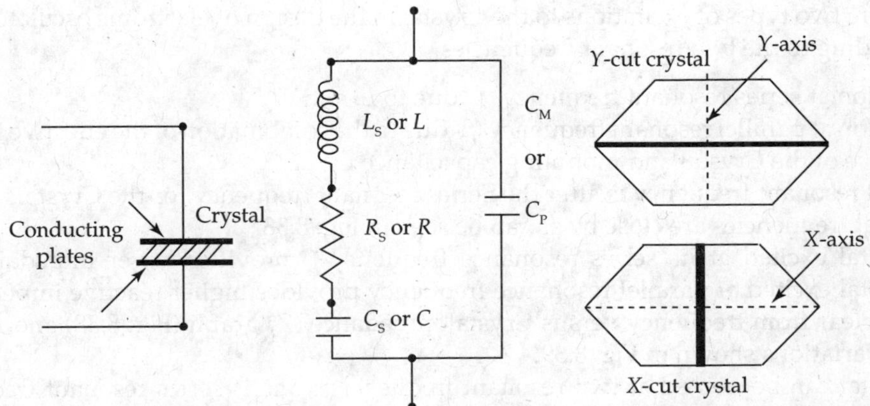

FIG. 8.36 *Electrical equivalent circuit of a quartz crystal*

Some manufactures cut the slices in different cuts known as AT, BT, CT and GT. A thin slice of Crystal is sandwiched between two thin-plated electrodes and when an AC Source is applied, it vibrates such that AC current has maximum at resonant frequency.

Frequency of oscillation of the electro-mechanical system depends upon the mass, thickness of slice, mode of vibration and on Crystal mounting. Crystal has a very high Q. Electrical equivalent circuit of a quartz Crystal is a tuned circuit shown in Fig. 8.36.

8.12.3 Frequency Stability of Crystal Oscillators

The electrical equivalent circuit of Crystal is shown in Fig. 8.37. It has reactive elements as explained below.

Inductance 'L_S' or L is analogous to the mass of Crystal. Capacitance 'C_S' or C represents compliance (reciprocal of stiffness) in Pico farads. Resistance 'R_S' represents friction. (Crystal loses) 'C_P' or C_M represents self-capacitance of the total Crystal assembly which has a Crystal slice as dielectric between two electrodes (Conducting planes). C_P or C_M is higher than C_S. It is also known as mounting capacitance. It will be in the order of Pico farads.

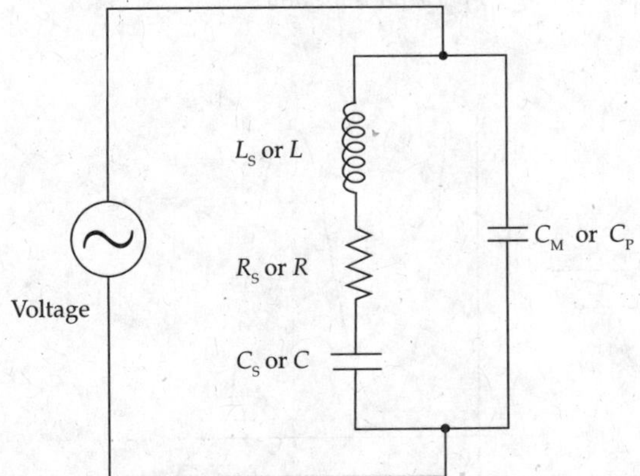

FIG. 8.37 *Electrical equivalent circuit of quartz crystal with excitation*

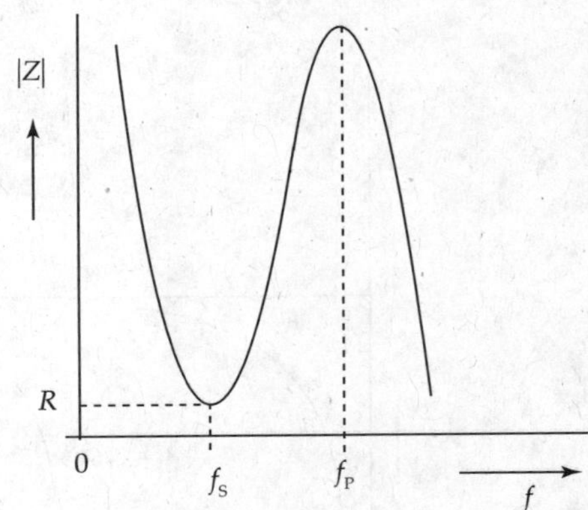

FIG. 8.38 *Variation of crystal impedance $|Z|$ with frequency f*

There are two types of excitations to the Crystal in the design of electronic oscillator circuits corresponding to the two resonant frequencies.

- Excitation at series resonant frequency 'f_S' due to L_S and C_S.
- Excitation at parallel resonant frequency 'f_P' due to the combination of the effective inductive reactance of the Crystal and mounting capacitance C_M or $C_{P'}$
- Parallel resonant frequency is after the Series resonant frequency for the Crystals. Both the resonant frequencies are close by as can be seen in Fig. 8.38.
    - Crystal excited at its series resonance frequency f_S provides lower impedance path. Crystal excited at parallel resonance frequency provides higher reactive impedance, as it is clear from frequency versus Crystal impedance (*Z*) graph (Fig. 8.38) and reactance (*X*) variations shown in Fig. 8.39.
    - Crystal can be excited at two resonant frequencies: one is series resonant frequency f_S and the other is a parallel resonant frequency f_P as shown in Figs. 8.38, 8.39 and 8.40.

The resonant frequencies and the quality factor '*Q*' of the Crystal depend upon the Crystal slice dimensions, how the Crystal surfaces are oriented with respect to its axes and how the

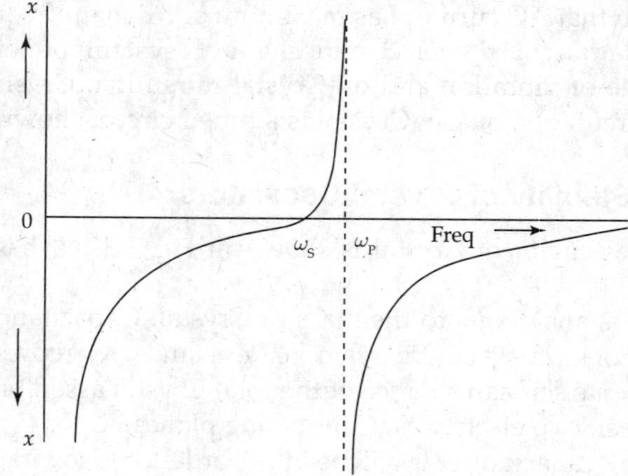

FIG. 8.39 *Frequency versus reactance graph o f crystal impedance*

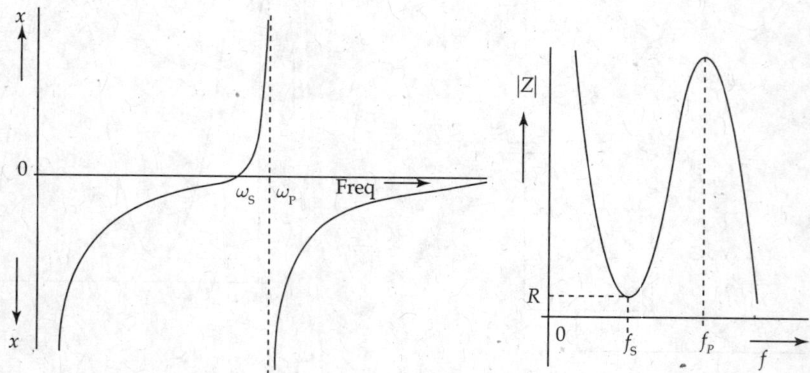

FIG. 8.40 *Crystal reactance (X) and impedance |Z| variations with frequency f*

device is mounted. Commercially available Crystals have the values of 'Q' ranging from several thousands to several hundred thousands. The Crystals vibrating frequencies range from a few kilo Hertz to a few Mega Hertz.

Crystal reactance (X) and impedance (Z) versus frequency curves are shown in Fig. 8.40. Considering the equivalent circuit of the Crystal, its impedance Z is given by Eq. (8.64). When R is negligible

$$Z=\frac{\frac{1}{j\omega C_P}\left(j\omega L+\frac{1}{j\omega C}\right)}{\frac{1}{j\omega C_P}+j\omega L+\frac{1}{j\omega C}}=\frac{\frac{1}{\omega C_P}\left(\omega L-\frac{1}{\omega C}\right)}{j\left(\omega L-\frac{1(C+C_P)}{\omega C C_P}\right)}. \tag{8.64}$$

Multiply both numerator and denominator by $j\omega/L$.

$$Z=\frac{\frac{j\cdot 1}{\omega C_P}\left(\omega L-\frac{1}{\omega C}\right)\frac{\omega}{L}}{-\left(\omega L-\frac{1(C+C_P)}{\omega C C_P}\right)\frac{\omega}{L}}=-\frac{j}{\omega C_P}\frac{\left(\omega^2-\frac{1}{LC}\right)}{\left(\omega^2-\frac{1}{L\frac{CC_P}{(C+C_P)}}\right)}.$$

$$Z=-\frac{j}{\omega C_P}\frac{(\omega^2-\omega_S^2)}{(\omega^2-\omega_P^2)},\text{ where } \omega_S^2=\frac{1}{LC}\text{, i.e., } f_S=\frac{1}{2\pi\sqrt{L_S C_S}}\text{ or } f_S=\frac{1}{2\pi\sqrt{LC}}.$$

$$\omega_P=\frac{1}{\sqrt{L\frac{CC_P}{(C+C_P)}}}=\frac{1}{\sqrt{LC_{EQ}}},\text{ where } C_{eq}=\frac{CC_P}{(C+C_P)},\ f_P=\frac{1}{2\pi\sqrt{LC_{eq}}} \tag{8.65}$$

$$\frac{\omega_P^2}{\omega_S^2}=\frac{C+C_P}{C_P}=1+\frac{C}{C_P}$$

$$\therefore\ \frac{\omega_P}{\omega_S}=\sqrt{1+\frac{C}{C_P}} \tag{8.66}$$

$$\frac{f_P}{f_S}=\sqrt{1+\frac{C}{C_P}}$$

As the ratio C/C_P is very small, the two frequencies f_S and f_P are very close. The separation between them is only a few Hz. Hence, Crystal oscillators generate stable frequency signals.

From the expression for the impedance 'Z'

$$\text{When } \omega=\omega_S,\ Z=0 \tag{8.67}$$

$$\text{When } \omega=\omega_P,\ Z=\infty \tag{8.68}$$

Thus, a quartz Crystal has two resonant frequencies. Between these two frequencies f_P occurs at a frequency higher than f_S and the difference $(f_P - f_S)$ is very small. Both the frequencies f_S and f_P set the lower and upper frequency limits of the Crystal oscillator. Between the two frequencies f_P and f_S the Crystal is inductive and it can replace the inductor in COLPITTS oscillator circuit and Tuned Drain Tuned Gate (TDTG) oscillator circuits.

From the Crystal equivalent circuit,

$$f = \frac{1}{2\pi\sqrt{LC}} \tag{8.69}$$

$$Q = \omega L/R.$$

Advantages

1. The Q factor of a Crystal is very high of the order 10^6 compared to that of an LC circuit and as such the frequency of Crystal is highly stable. Thus, f_r/Q = Bandwidth virtually becomes zero. The circuit frequency depends upon the Crystal resonance frequency alone and nothing else.
2. By changing the Crystal with another Crystal, different oscillator frequencies can easily be achieved.
3. As the frequency of Crystal slightly drifts at an ambient temperature, they are often enclosed in temperature-controlled oven so as to achieve good frequency stability. The frequency drift can be made less than 1 part in 10^6.
4. The rate of charge of phase shift θ with angular frequency ω is $d\theta/d\omega$, which is large; frequency charge is very small even if the phase shift of the circuit changes due to variations in stray capacitance.
5. Crystal oscillator does not need a separated tuned circuit.

Disadvantages

1. As the Crystal has a very large Q, the bandwidth of Crystal oscillator is very small and is of the order of few Hz, the Crystal vibrates at resonant frequency and does not vibrate at all at the other frequencies.
2. If excessive power is applied, oscillator waveform will be distorted. Overheating causes frequency drift and the Crystal being fragile is likely to fail. A Crystal oscillator is used in low-power circuits.

Crystal oscillators are used in frequency synthesizers, which have revolutionised the communication equipment. It is now possible to have complete systems economically.

They are used to generate a chromatic sub-carrier or colour sub-carrier in TV receivers. They got wide applications in Microprocessors, Microcontrollers, embedded systems, electronic clocks and watches, frequency and time standards, computer-clock pulse generators in Radio and Communication equipment.

Crystal behaves as a series resonant circuit (Fig. 8.38) at frequency f_S, where the impedance offered by the Crystal is smallest and the amount of positive feedback is large. A Crystal-controlled oscillator using this property is shown in Fig. 8.41.

8.12.4 Pierce Crystal Oscillator Circuit using BJT (Fig. 8.41)

- The combination of circuit features V_{CC}, R_1, R_2, R_E, C_E, $R\,f\,C$ provides stabilised DC Bias conditions for the oscillator circuit.
- Reactance $X_L = \omega L$ of RFC provides short circuit path for the DC Bias V_{CC}, simultaneously preventing high-frequency signal (by offering large impedance path to AC signal) from entering the DC supply.

- Crystal is connected between the Collector and the Base of the Transistor.
- The feedback voltage V_f fed back from the output of the Transistor to the input port is maximum, when the Crystal is excited at its series resonance frequency. The Crystal offers minimum or least impedance path from the Collector to the Base of the Transistor.
- The coupling capacitor C_C offers low reactance path for the generated Radio Frequency signal and does not allow any DC from the output port to the input port.
- This type of Pierce Crystal oscillator operates at stable frequency set by the series resonance frequency f_S of the Crystal. This capacitor is also used to fine tune the Crystal frequency slightly about the resonance frequency of the Crystal.

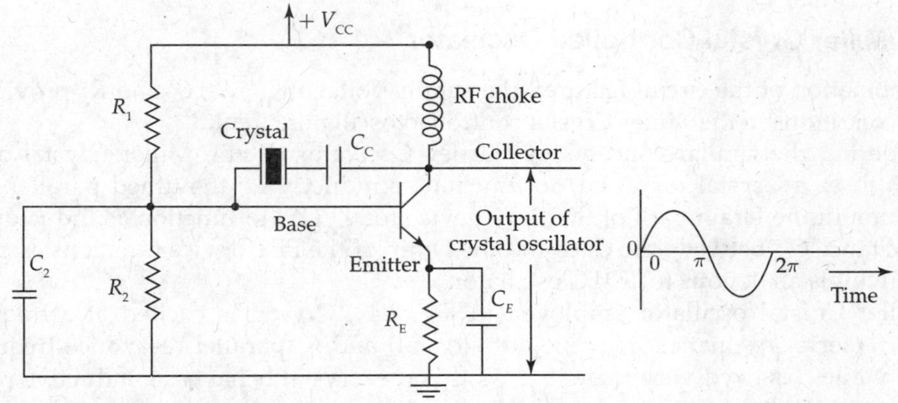

FIG. 8.41 *Pierce crystal oscillator circuit using a crystal excited at series resonant frequency f*

8.12.5 Pierce Crystal Oscillator Circuit using FET (Fig. 8.42)

Frequency of oscillations of *Pierce Crystal Oscillator* is determined by series resonant frequency f_S of Crystal. Frequency stability is good as the difference between the series and the parallel resonant frequencies of crystal operations is negligibly small, which is of the order of a few Hz. The operating frequency is not effected due to changes in supply voltage, device parameters, etc.

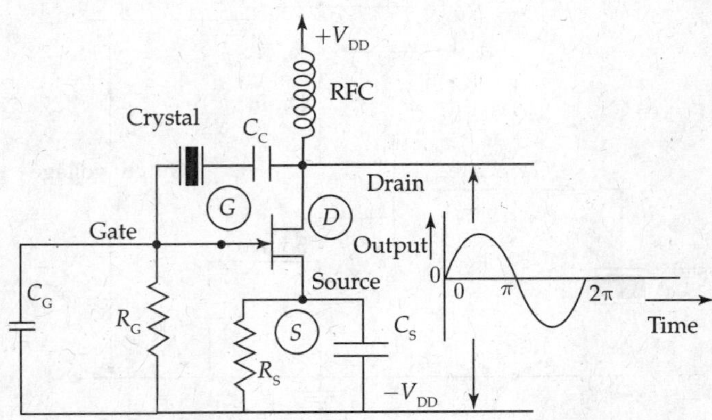

FIG. 8.42 *Pierce crystal oscillator using JFET*

The Pierce Crystal oscillator circuit shown in Fig. 8.42 uses JFET device and the biasing circuit of R_G, R_S and C_S. The reactance $X_L = \omega L$ of the RFC provides short circuit path for the DC Bias V_{CC}, simultaneously preventing the high-frequency signal (by offering large impedance path to AC signal) from entering into the DC supply. The coupling capacitor C_C offers low reactance path for the generated Radio Frequency signal and does not allow any DC from the output port to the input port.

Crystal excited at the series resonance provides least impedance path for maximum feedback voltage V_f from the Drain to the Gate circuit of the Field effect Transistor. When the positive feedback satisfies the Bharkhausen criterion for Oscillations, the circuit produces AC signal voltages, whose frequency is determined by *series resonance frequency of the Crystal.*

8.12.6 Miller Crystal Controlled Oscillator

The Combination of the circuit features, the supply voltage V_{DD}; R_S, C_S and R_G provide stable DC Bias conditions in the Miller Crystal controlled oscillator circuit.

Considering the similar concepts of Hartley Oscillator circuit, Miller Crystal oscillator (Fig. 8.43) uses a Crystal for Z_1 (x_1) to function as inductor L_1; the tuned parallel L and C combination in the Drain path of the JFET device for Z_2 (X_2) to function as inductor L_2; and the capacitance C_{gd} between the Gate and the Drain of the FET device functions as capacitor C. This circuit is analogous to TDTG oscillator.

In Miller Crystal oscillator employed in Fig. 8.43, Crystal is excited at a frequency f_0 between f_S (series resonance frequency of Crystal) and f_P (parallel resonance frequency of Crystal), while f_0 is fixed very close to f_P, so that the Crystal behaves as inductive reactance x_1 (L_1). Tuned parallel LC network in the Drain section is tuned to the frequency f_0 to which the Crystal is excited. Once the DC Source is switched ON and the Bharkhausen conditions of oscillations are satisfied, the output voltage V_{out} is sine wave at the frequency f_0, the oscillation frequency of the circuit. As the difference between the resonance frequencies of

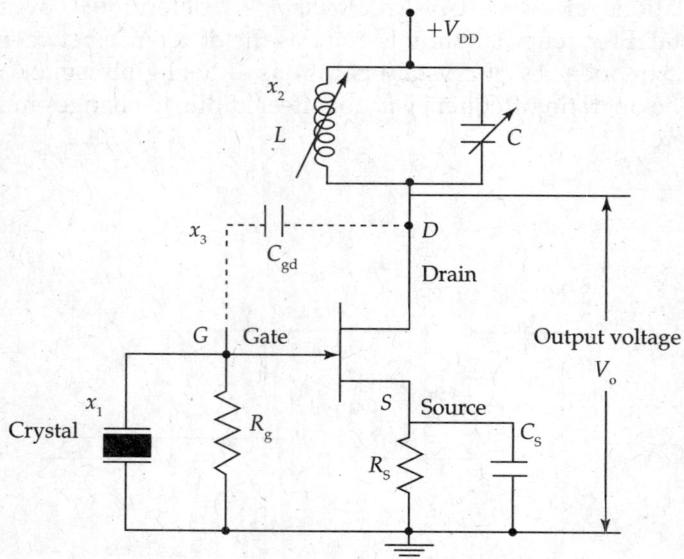

FIG. 8.43 *Miller crystal controlled oscillator*

the Crystal '$f_P - f_S$' is only a few Hz, the circuit produces stable frequency of oscillations. So, the Crystal oscillator circuits find their applications in radio transmitter circuits, so as to maintain constant radio station signal for fixed location tuning in receiver equipments.

8.12.7 Modified Colpitts Oscillator Circuit using Crystal

Other forms of Crystal oscillator circuits with Crystal excited at f_P

Modified Crystal oscillator circuit shown in Fig. 8.44 uses a Crystal in the parallel mode. The feedback is provided through Potential divider formed by C_1 and C_2 in series. The Transistor itself operates in the common Base mode since Base is grounded through C_B. Thus, it is equivalent to a Colpitts oscillator with a common Base nominal Amplifier. Here, the conventional inductor is replaced with a Crystal.

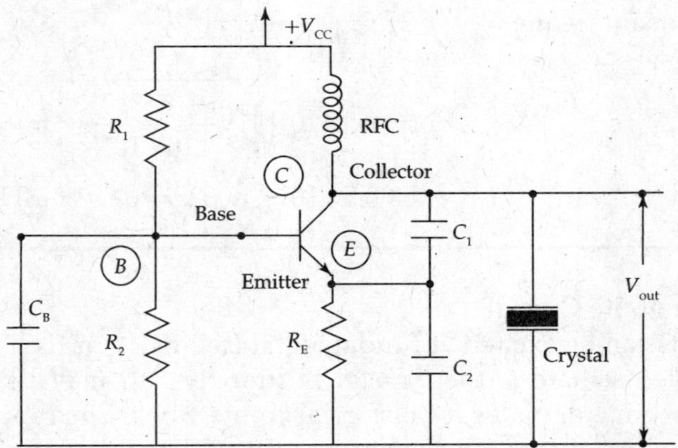

FIG. 8.44 *Modified colpitts oscillator (pierce oscillator) using crystal excited at f_P (the crystal excited at f_P behaves as an inductance)*

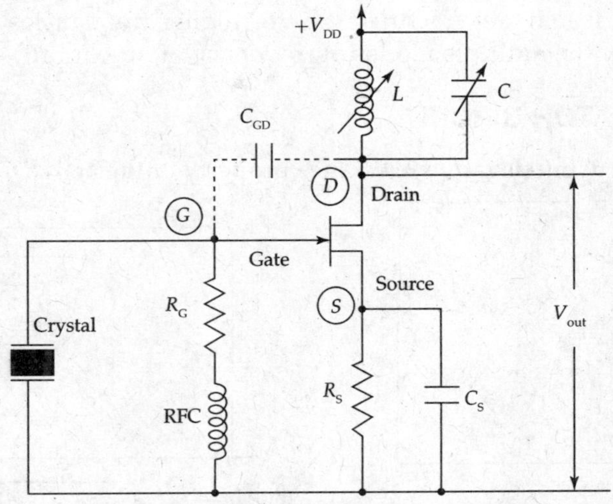

FIG. 8.45 *Crystal oscillator with the crystal excited at f_P*

EXAMPLE 8.12

In a Crystal oscillator, the Crystal parameters are $L_s = 2H$, $C_S = 0.05$ pF, $R = 2000\ \Omega$, $C_P = 10$ pF. Calculate series and parallel resonant frequencies.

Solution: Series resonant frequency $f_S = \dfrac{1}{2\pi\sqrt{L_S C_S}} = \dfrac{1}{2\pi\sqrt{2\times 0.05\times 10^{-12}}}$

$$f_S = \frac{1}{2\pi\sqrt{10\times 10^{-14}}} = \frac{10^7}{2\pi\sqrt{10}} = \frac{5\times 10^6}{\pi\sqrt{10}} = 0.5 \text{ MHz}$$

Using $\dfrac{f_P}{f_S} = \sqrt{\left[1+\dfrac{C_S}{C_P}\right]}$,

Parallel resonant frequency $f_P = f_S \cdot \sqrt{\left[1+\dfrac{C_S}{C_P}\right]}$ Hz

$$\therefore\ f_P = 0.5\times 10^6 \sqrt{\left[1+\frac{0.05\times 10^{-12}}{10\times 10^{-12}}\right]} = 0.5\times 10^6 \cdot \sqrt{\frac{10.05}{10}}$$

$$= 0.5\times 10^6 \times 1.0025 = 0.5012 \text{ MHz.}$$

Modes of Operation of the Crystal

Piezoelectric Crystals can be excited at fundamental frequency or its harmonics. For using Crystals in electronic oscillators, the Crystal is suitably cut, may be *x*-cut or *y*-cut and mounted between two metal plates for electrical connections. The fundamental frequency of electro-mechanical resonance (Piezoelectric effect) depends upon the Crystal slice dimensions – nature of the Crystal cut and the thickness of the Crystal slice. There is an upper limit on the fundamental frequency of excitation of a Crystal. As the thickness of the Crystal is inversely proportional to the frequency, for higher frequencies the Crystal becomes so thin that it may get fractured. So, for higher frequencies above 20 MHz, the Crystals are excited at the multiple modes of frequency of operation

8.13 UJT OSCILLATOR CIRCUIT

UJT oscillator circuit using negative resistance property of the active device (UJT)

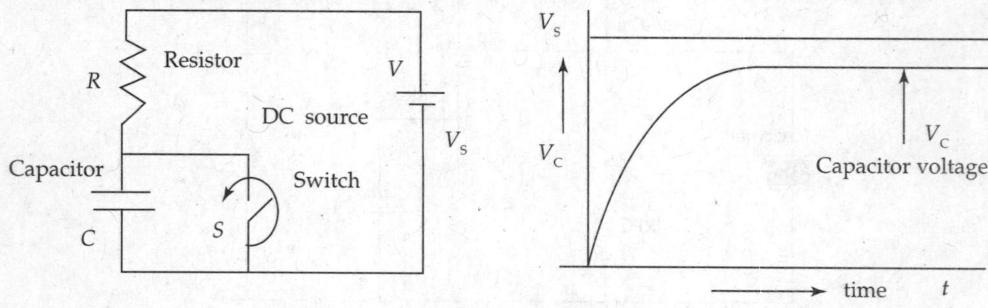

FIG. 8.46 *Illustration of charging a capacitor through 'R' to the supply voltage V_S*

The basic concepts follow for understanding the principle of working of UJT oscillator.

Saw tooth waveform can be generated by connecting a DC Source across R-C combination and adding a switch across the capacitor.

As long as the switch 'S' is open, the capacitor goes on charged exponentially reaching the Source voltage V_S in infinite time as shown in Fig. 8.47. The time taken to reach 63.7% of the final voltage V_S is attained with charging time constant $\tau = RC$. The expression for the charging voltage across the capacitor $V_C(t)$ is

$$V_C(t) = V_S(1 - e^{-t/RC}) = V_S(1 - e^{-t/\tau}) \tag{8.70}$$

When $t = 0$, $V_C(t) = 0$ and when $t = \infty$, $V_C(t) = V_S$; In between it charges exponentially. When $t = \tau$, the voltage across the capacitor $V_C(t) \approx 63.7\%$ of V_S.

When switch 'S' is closed between time $t = 0$ and ∞, Capacitor discharges to $V_C(t) = 0$; when the switch is again opened after $V_C(t) = 0$, it starts all over again and the waveform will be as shown in Fig. 8.47.

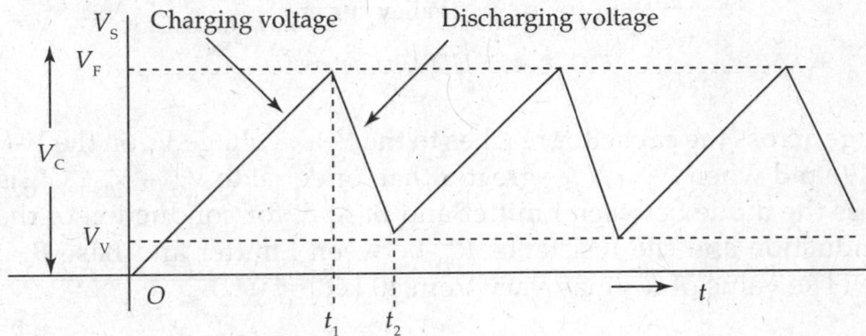

FIG. 8.47 *Output voltage across the capacitor V_C with the switch operations*

When the switch is closed at constant intervals and opened again, a near saw tooth waveform results. If the switch 'S' is replaced by a voltage-sensitive switch as Unijunction Transistor (UJT), the output voltage of the UJT oscillator circuit using the above principles of operation will be a repeating Saw Tooth voltage waveform.

The UJT relaxation oscillator consists of UJT (Unijunction Transistor), Resistor (*R*) Capacitor (C) combination, Resistors R_1 and R_2 and a power supply voltage V_{BB}.

When V_{BB} is switched on, the voltage across the capacitor is zero initially. (The voltage across the capacitor cannot change instantaneously.) ($V_C = Q_C = I{\cdot}T/C$, when $T = 0$, $V_C = 0$).

As discussed earlier, the capacitor *C* starts charging through resistor *R* with a time constant $\tau_1 = RC$ during the time t_1. The voltage across capacitor increases exponentially, during '0 to *A*' portion on the UJT Characteristics of Fig. 8.49.

The voltage across the capacitor,

$$V_C = V_{BB}\left[1 - e^{-t/RC}\right]. \tag{8.71}$$

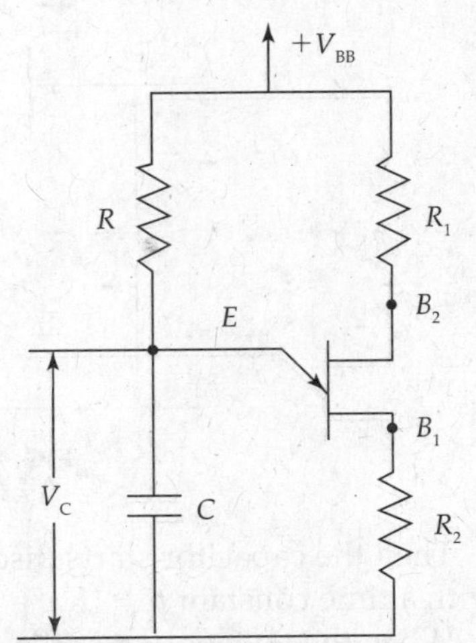

FIG. 8.48 *UJT relaxation oscillator*

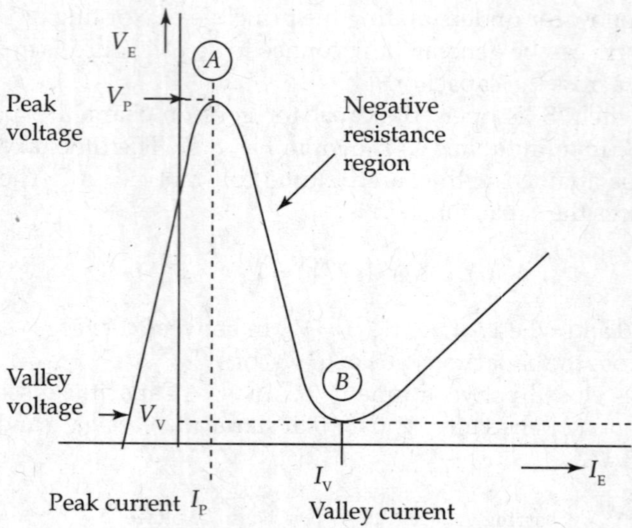

FIG. 8.49 *UJT Characteristic*

When the voltage across the capacitor reaches to the Peak voltage V_P on the V-I characteristic of UJT (Fig. 8.49) and when $V_P = V_E$ is greater than or equal to $V_D + V_{RB1}$ (V_D is equal to cut-in voltage across the diode between Emitter and Base B_1 for conduction of the device); UJT comes into conduction and the resistance R_{B1} between Emitter and Base B_1 decreases to a very low value (The value of R_{B1} may vary from 40 kΩ to 50 Ω).

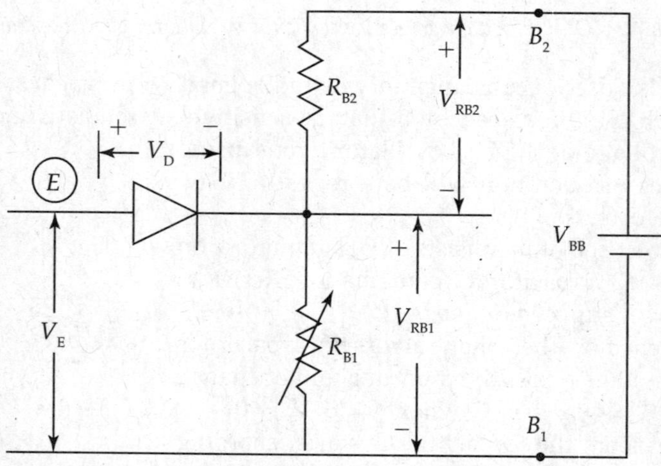

FIG. 8.50 *Working principle of UJT oscillator*

Then the capacitor starts discharging between Emitter E and Base B_1 through R_2 for time t_2 with a time constant $t_2 = (R_{B1} + R_2)C$.

When the voltage across the capacitor V_C moves back to a minimum voltage $V_{min} = V_V$, UJT will be switched-off.

Then the charging and discharging cycles of the capacitor repeat with a frequency determined by the time constant $\tau = (R + R_{B1} + R_2)\,C \cong RC$, because R is much larger than R_{B1} (when UJT conducts) and R_2.

Pulse repetition frequency

$$f = \frac{1}{\tau}\ \text{Hz}. \tag{8.72}$$

Negative resistance region of the device characteristic is used to fix up the DC-operating conditions so that UJT works as on oscillator.

The final expression for the time period of oscillations '*t*' is given as

$$\text{Time period } t = 2.3RC\log_{10}\frac{1}{1-\eta}. \tag{8.73}$$

Here Intrinsic Stand-off ratio η

$$\eta = \frac{R_{B1}}{R_{B1}+R_{B2}}. \tag{8.74}$$

(h has a typical value between 0.5 and 0.8).

If the charging Resistor '*R*' is replaced by a constant current Source, a linear Saw tooth waveform as shown in Fig. 8.47 results.

Variable frequency operation can be achieved using adjustable '*R*' and including multiple capacitors for charging and discharging purposes.

This capacitor voltage V_C may be used as the sweep voltage to be connected to *x*-plates in a Cathode Ray tube of a CRO.

EXAMPLE 8.13

A UJT relaxation oscillator has R = 50 kΩ, C = 0.1 μF and Intrinsic stand-off ratio h = 0.5. Determine the frequency of the saw tooth voltage.

Solution:

$$\text{Frequency } f = \frac{1}{2.3RC\cdot\log_{10}\dfrac{1}{(1-\eta)}}$$

$$= \frac{1}{2.3\times 50\times 10^3\times 0.1\times 10^{-6}\cdot\log_{10}\dfrac{1}{(1-0.5)}}$$

$$= \frac{1}{11.5\times 10^{-3}\cdot\log_{10}\dfrac{1}{0.5}}$$

$$= \frac{1}{11.5\times 10^{-3}\cdot\log_{10}2} = \frac{10^3}{11.5\times 0.3} = 290\ \text{Hz}.$$

SUMMARY

Following concepts in oscillator circuits are discussed:

1. Positive feedback concept with a block diagram.
2. Influence of Bharkhausen (Criteria) conditions for oscillations in various circuits.
3. Low-frequency oscillators such as RC phase-shift oscillators using Transistors, FET devices and Wien Bridge circuits with practical layout.
4. High-frequency oscillator circuits such as Colpitts oscillator, Hartley oscillator and Clapp Oscillator.
5. Working of various Crystal Oscillator circuits is explained keeping in vies of how they could produce stable high-frequency oscillations.
6. Finally UJT relaxation oscillator having much role in CRO instruments is discussed. Another feature of the circuit is the role of negative resistance property of the device for oscillator function

Questions for Practice

1. What are the essential constituents of an oscillator? State the Bharkhausen conditions for oscillations?
2. Draw the diagram of RC Phase-shift oscillator using BJT and explain its operation and obtain the expression for frequency of oscillation and minimum value of h_{fe} required for BJT to oscillate. Discuss the various types of feedbacks in the circuit.
3. Draw the diagram of RC Phase-shift oscillator using JFET, explain its operation and obtain the expression for frequency of oscillation and minimum value of gain required for JFET to oscillate. Discuss the various types of feedbacks in the circuit.
4. Draw the circuit and explain the working of a Wien Bridge oscillator using two stages of Transistor Amplifiers. Derive the expression for frequency of oscillations.
5. Draw the circuit and explain the working of a Wien-Bridge oscillator using Operational Amplifier. Derive the expression for frequency of oscillations.
6. Draw the circuit diagram of 'Tuned Collector oscillator'. Identify the various components for producing oscillations and maintaining stable output signal. Explain the method of measuring the frequency of the output signal using a CRO.
7. Draw the circuit diagram of 'Tuned Drain oscillator'. Identify the various components for producing oscillations and maintaining stable output signal. Explain the method of measuring the frequency of the output signal using a CRO.
8. Draw the circuit diagram of 'Colpitts oscillator'. Identify the various components for producing oscillations and maintaining stable output signal. Explain the method of measuring the frequency of the output signal using a CRO.

9. Draw the circuit diagram of 'Hartley oscillator'. Identify the various components for producing oscillations and maintaining stable output signal. Explain the method of measuring the frequency of the output signal using a CRO.
10. Why Clapp oscillator is preferred over Colpitts circuit?
11. Draw the circuit diagram of 'Crystal controlled oscillator'. Identify the various components for producing oscillations and maintaining highly stable output signal. Explain the method of measuring the frequency of the output signal using a CRO.
12. Define frequency stability in Crystal oscillator circuits and why is it necessary. Explain its significance referring to the reception of various radio station signals at the same spot on radio receivers.
13. Draw the frequency versus impedance curve of a quartz Crystal.
14. A quartz Crystal has the following constants. L = 50 MH, C_1 = 0.02 pF, C_2 = 12 pF, R = 500 Ω. Find the values of series and parallel resonant frequencies. If the external capacitance across the Crystal changes from 5 pF to 6 pF, find the change in the frequency of oscillations. (May/June 2006, set-3)

Multiple Choice Questions

1. Output signal of Hartley oscillator is ______________.
   (a) square wave
   (b) triangular wave
   (c) sine wave
   (d) non-sinusoidal signal.
2. Feedback factor of Wien Bridge Oscillator using Op-Amp ______________.
   (a) 44.5
   (b) 1/3
   (c) 2
   (d) 5
3. The oscillator having highest frequency stability is ______________.
   (a) RC oscillators
   (b) LC oscillators
   (c) crystal oscillators
   (d) relaxation oscillators
4. The amplitude stability of an oscillator can be achieved ______________.
   (a) $|A\beta|$ drops below unity
   (b) $|A\beta|$ slightly higher than unity
   (c) $|A\beta|$ is higher than unity
   (d) dA/dV_0 must be a large negative number
5. In RC phase-shift oscillator circuit ______________.
   (a) feedback network provides a phase shift of 180°
   (b) the total phase shift around the loop is 360°
   (c) $A\beta = -1$
   (d) the amplifier gain has to be a positive number
6. A wide range of oscillations in the audio range is obtained ______________.
   (a) phase-shift oscillator
   (b) Wien bridge oscillator
   (c) Hartley oscillator
   (d) Colpitts oscillator

7. The feedback network in Wien bridge oscillators is a ______________.
   (a) ladder network (b) high pass network
   (c) low pass network (d) band pass network

8. To maintain sustained oscillations in a Wien bridge oscillator, the gain should be
   (a) slightly higher than 3 (b) 43.5
   (c) 29 (d) equal to unity

9. Circuit elements in the equivalent circuit of a Crystal ______________.
   (a) resistance and capacitance
   (b) resistance and inductance
   (c) resistance, capacitance and inductance
   (d) resistance

10. Conditions for oscillations in sinusoidal oscillators ______________.
    (a) Bharkhausen conditions (b) Lenz' law
    (c) Faraday's law (d) Nyquist criteria

11. The type of feedback used in LC oscillator circuits ______________.
    (a) negative feedback (b) positive feedback
    (c) none of the above

12. Which sinusoidal oscillator is preferred for microwave frequencies?
    (a) RC phase-shift oscillators
    (b) LC oscillators
    (c) oscillators using negative resistance devices
    (d) all of the above

Answers to Multiple-Choice Questions

1. (c)	2. (b)	3. (c)	4. (b & d)	5. (a)
6. (b)	7. (b)	8. (a)	9. (c)	10. (a)
11. (b)	12. (b)			

Chapter 9

FET AND MOSFET AMPLIFIERS

Learning Objectives

- Design and analysis of FET and MOSFET Amplifiers
- CS, CG and CD FET Amplifiers
- CS, CG, CD MOSFET Amplifiers

9.1 AMPLIFIER GAIN USING DECIBELS

Amplifiers using Vacuum Tube, BJT, FET and MOSFET amplify signals. One simple application is Audio Amplifier in Public Address system (Fig. 9.1). Microphone converts Audio signal into electrical signal, which is amplified by the Amplifier. Its output is connected to a loud speaker. Final output is an audio signal. Amplified sound from the speaker is utilised in public gatherings, large auditoriums or conference rooms.

During the process of amplification, the information contained in the output signal should be an exact replica of the input signal,

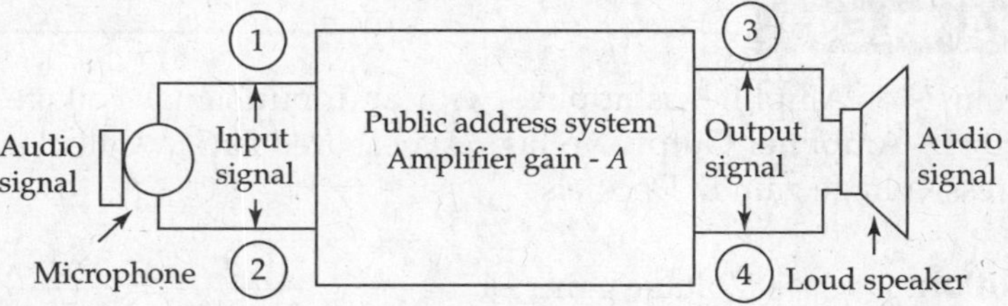

FIG. 9.1 *Audio amplifier system*

without adding new or, deleting or distorting existing. There is a necessity to maintain linearity in amplification.

Amplifier performance is expressed as voltage gain, current gain, power gain, frequency responses and bandwidth. They are measured as input and output voltages and currents, shown in Eqs. (9.1), (9.2) and (9.3), for comparison and design implementation.

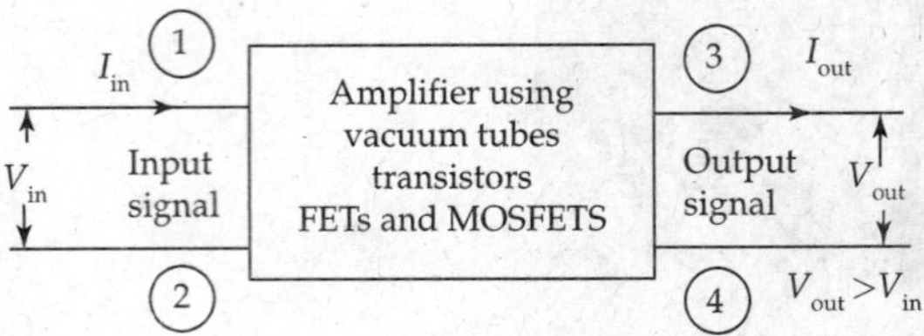

FIG. 9.2 *Amplifier concept*

$$\text{Amplifier Gain (Amplification)} \quad A_V = \frac{\text{Output voltage } (V_{out})}{\text{Input voltage}(V_{in})} = \frac{V_{out}}{V_{in}} \tag{9.1}$$

$$\therefore \text{ Output voltage} \quad V_{out} = AV_{in} \tag{9.2}$$

$$\text{Current gain} \quad A_i = \frac{\text{Output load current } I_L (I_{out})}{\text{Input signal current } (I_{in})} = \frac{I_{out}}{I_{in}} \tag{9.3}$$

$$\text{Power gain} \quad A_P = \frac{\text{AC signal output power across the load}}{\text{Input AC signal power}}$$

$$A_P = \frac{V_{out} \cdot I_{out}}{V_{in} \cdot I_{in}} = A_V A_i \tag{9.4}$$

$$\text{Voltage gain in decibels (dB)} = 20 \log_{10} |A_V| \text{dB} \tag{9.5}$$

$$\text{Current gain in decibels (dB)} = 20 \log_{10} |A_i| \text{dB} \tag{9.6}$$

$$\text{Power gain in decibels (dB)} = 10 \log_{10} |A_P| \text{dB} \tag{9.7}$$

Amplifiers are used in various applications such as Radio, TV and Home Theatre, Telephones, Cell phones, Video Conference systems, Computer communications and Satellite communications, etc., visible in all walks of life. Output power levels of various Amplifiers range from fraction of a watt to hundreds of watts. They are expressed in Decibel units for estimation and comparison.

If Amplifier's output signal is without any change, except increased signal amplitude, Amplifier is an *ideal* or *linear Amplifier*. Non-linear characteristics of electronic devices such as *BJT, FET or Vacuum Tube* may introduce changes in some features of an electrical signal, as it passes through an Amplifier. Undesirable changes in the signal are considered as 'Distortion', which is discussed in Chapter 11.

EXAMPLE 9.1

A Transistor Amplifier is applied with an input signal voltage of 50 mV at a frequency of 1,000 Hz. Amplifier Output voltage is measured as 5 V. Calculate Amplifier voltage gain and express voltage gain in Decibels.

Solution: Voltage gain $A_V = \dfrac{V_{out}}{V_{in}} = \dfrac{5}{50 \times 10^{-3}} = \dfrac{5 \times 10^3}{50} = \dfrac{5000}{50} = 100$

Voltage gain in decibels dB $= 20 \log_{10} |A_V| = 20 \log_{10} 100 = 20 \times 2 = 40$ dB.

9.2 BASIC CONCEPTS OF FET AMPLIFIER

Gate supply voltage V_{GG} is based on the class of Amplifier operation. For Class-A operation

$$V_{GG} = V_{GS} = \frac{V_P}{2} = \frac{-4\text{ V}}{2} = -2\text{ V}, \text{ where}$$

$$\text{Pinch-off voltage} \quad V_P = -4\text{ V} \quad \text{(for BFW10 FET device)} \tag{9.8}$$

Voltage across load resistance $R_L = I_D \cdot R_L = V_{RL}$ is about half the supply voltage V_{DD} to satisfy Class-A operation.

Gate–Channel junctions of FET are reverse biased. So, its input resistance R_{in} is very large.

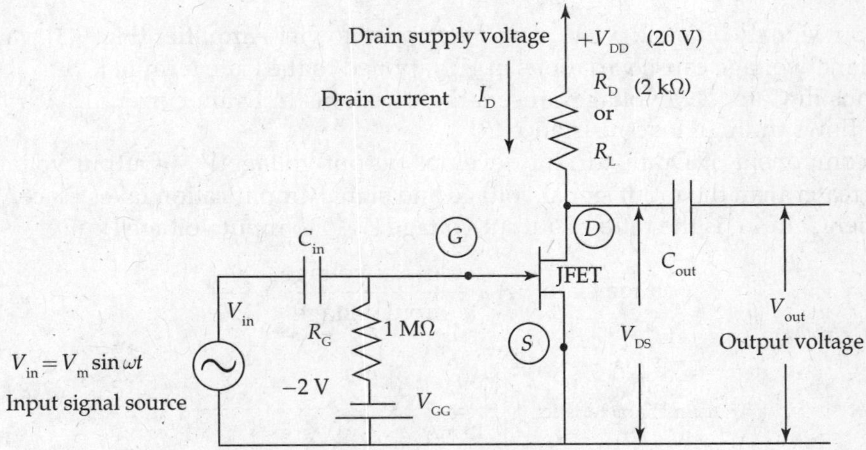

FIG. 9.3 *Basic JFET amplifier*

DC load line on FET output characteristics to establish DC conditions in the Amplifier

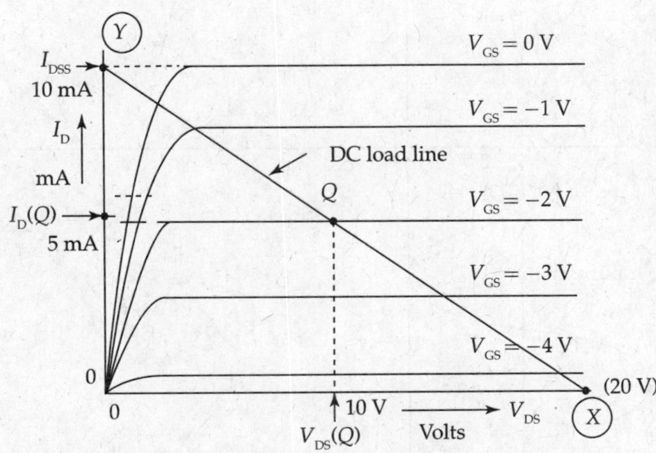

FIG. 9.4 *JEFT output characteristics with DC load line and quiscent operating point 'Q' for Class-A operation of amplifier*

DC Load line equation from the Amplifier (Fig. 9.3) (Discussed in Chapter 6).

$$[V_{DD} = V_{DS} + I_D R_L] \quad \text{(DC Load line equation)} \tag{9.9}$$

$$\therefore \quad [V_{DS} = V_{DD} - I_D R_L]$$

- Coordinates of point Y of DC load line are $V_{DS} = 0$ V and $I_D = (V_{DD}/R_L)$
- Coordinates of point X of the line are $V_{DS} = V_{DD}$ volts and $I_D = 0$ mA.
- Joining the two points X and Y, the *DC load line* is drawn on output characteristics.
- Drain current $I_D(Q)$ and voltage $V_{DS}(Q)$ are obtained from quiescent operating point (Q) for Class-A amplifier operation.

Class-A Amplifier operation

- If an input signal voltage $V_{in} = V_m \sin \omega t$ is applied to FET Amplifier (Fig. 9.3), variations in input signal voltage cause variations in V_{GG} applied to the Gate terminal.
- Variations in Gate (DC) voltage cause large variations in Drain current I_D. Varying Drain current flows through load resistance 'R_L'.
- Signal component of Drain current develops output voltage V_{out}. Output voltage will be much greater than the input signal voltage and signal amplification takes place.
- Amplifier Gain 'A' is the ratio of output voltage (V_{out}) to input voltage (V_{in}):

$$\text{Voltage Gain} \quad A = \frac{\text{Output voltage}}{\text{Input voltage}} = \frac{V_{out}}{V_{in}}.$$

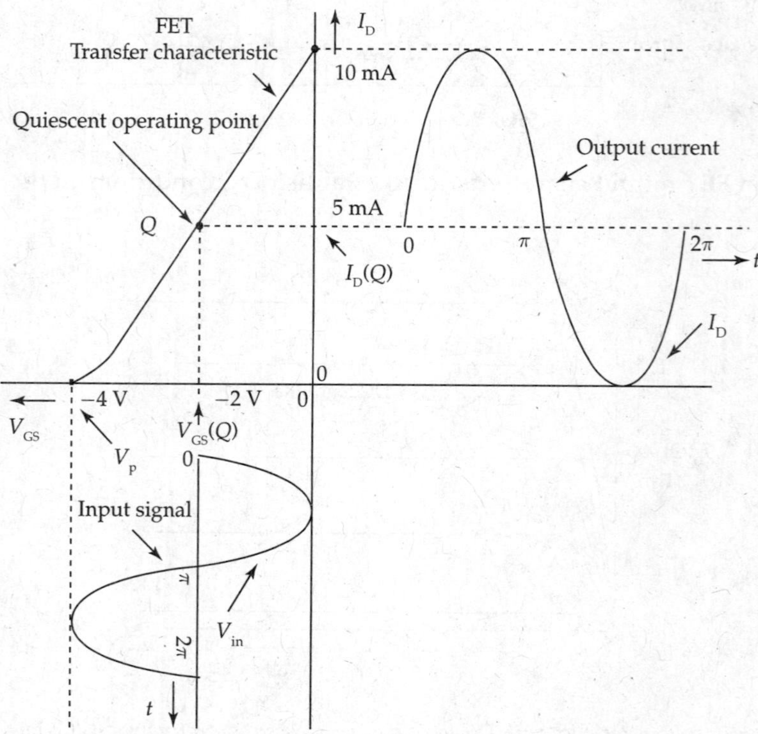

FIG. 9.5 *Class-A operation of FET amplifier using transfer characteristic of FET*

9.3 COMMON SOURCE FET AMPLIFIER

CS FET Amplifier (Fig. 9.6) provides both voltage and current gains. Maximum voltage gain of FET Amplifiers is limited to 'μ' of FET device. FET Amplifiers are useful with high output impedance signal sources to get large current gain. Voltage gain is usually less than that available from a CE Transistor Amplifier.

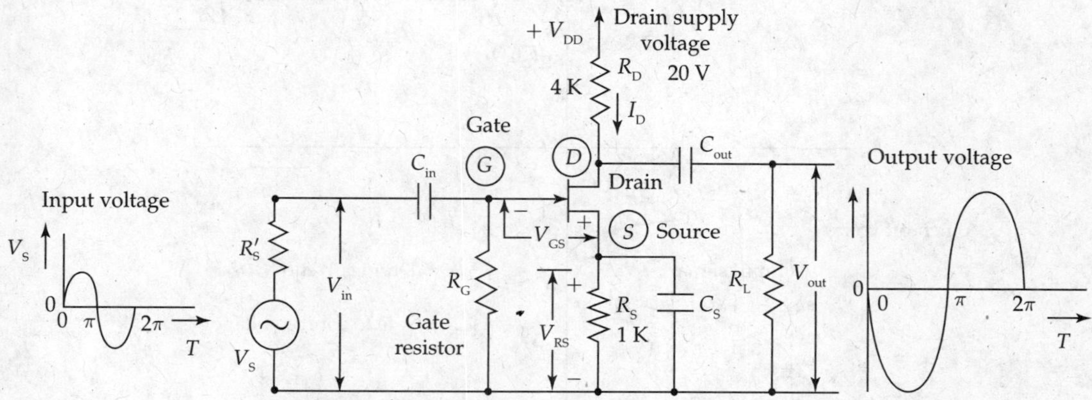

FIG. 9.6 *Single-stage field effect transistor amplifier*

- FET Amplifier circuit uses the self-bias arrangement using R_S and R_G to fix the DC operating conditions for Class-A operation. DC bias V_{GS} is equal to the voltage drop across the resistor R_S due to the flow of the quiescent component of Drain current I_D $[V_{GS} = V_{RS} = I_D(Q)R_S]$. Drain current flows from upper terminal of Source through R_S to common terminal. So V_{RS} provides reverse bias to Gate terminal.
- Gate to Source junction is reverse biased. So, the Gate current is zero resulting in zero voltage drops across R_G (Gate resistor). Input signal voltage V_S superimposes on the DC bias V_{GS}. Signal variations results in variations in Drain current I_D.
- Various signals and DC voltages in the Amplifier are shown in Fig. 9.7 using
(1) Transfer characteristic of a FET device; (2) DC bias ($v_{GS}(Q) = -2$ V) between the Gate and Source terminals of the FET; (3) Input signal $V_{in} = 2 \sin \omega t$ (Maximum amplitude of Signal = 2 V); (4) The input signal superimposes on the DC bias.
Effective input signal voltage $v_{GS} = V_{GS}(Q) + V_{in} = V_{GS}(Q) + V_m \cdot \sin \omega t$

$$\therefore \quad v_{GS} = V_{GS}(Q) + 2 \sin \omega t = -2 + 2 \sin \omega t.$$

- At '0' instance of the input signal voltage, $V_{in} = 0$ V and $V_{GS}(Q) = -2$ V

$$\therefore \quad v_{GS} = V_{GS}(Q) + 2 \sin \omega t = -2 + 0 = -2 \text{ V}$$

Corresponding value of Drain current $I_D = 2$ mA
- At $\pi/2$ instance of the input signal voltage, $V_{in} = 2$ V

$$\therefore \quad v_{GS} = V_{GS}(Q) + 2 \sin \omega t = -2 + 2 = 0 \text{ V}$$

Corresponding value of Drain current $I_D = 4$ mA

The increase in Drain current from 2 to 4 mA is due to the decrease in the reverse bias between the Gate and the source from –2 to 0 V.

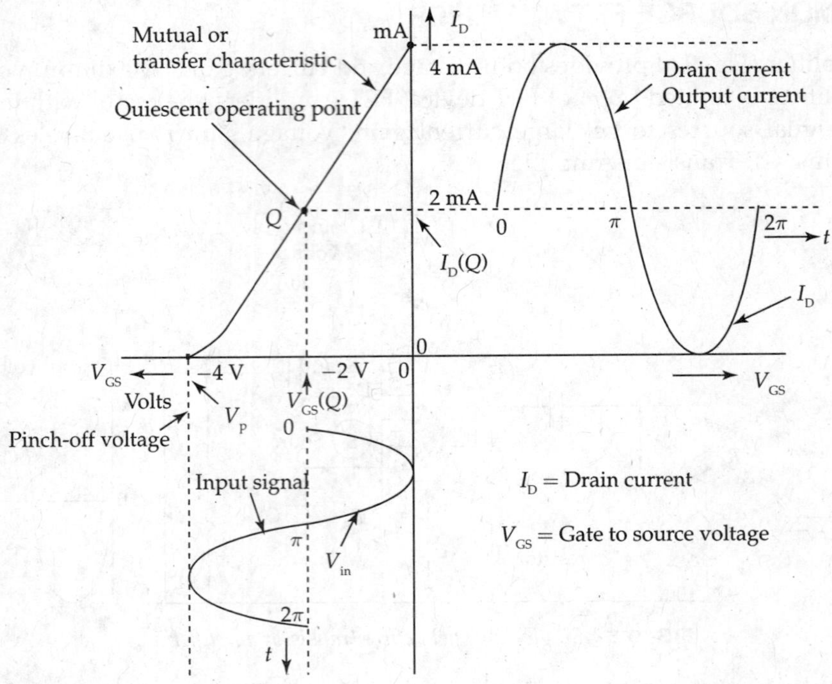

FIG. 9.7 *Class A operation of JFET amplifier using transfer characteristic of JFET*

- During the interval '0 to $\pi/2$' of input signal voltage, effective input signal voltage decreases from −2 to 0 V. This causes a decrease in reverse bias to Gate junctions. Such variations in DC reverse bias cause sinusoidal variations in Drain current from 2 to 4 mA.
- At 'π' instance of the input signal voltage, $V_{in} = 0$ V and $V_{GS}(Q) = -2$ V

$$\therefore \quad v_{GS} = V_{GS}(Q) + 2\sin\omega t = -2 + 0 = -2 \text{ V}$$

Corresponding value of Drain current $I_D = 2$ mA

The reverse bias to Gate junctions varies from '0' to −2 V during interval $\pi/2$ to π. Such sinusoidal variations in reverse bias to Gate junctions decrease Drain current from 4 to 2 mA.

- At $3\pi/2$ instance of the input signal voltage, $V_{in} = -.2$ V

$$\therefore \quad v_{GS} = V_{GS}(Q) + 2\sin\omega t = -2 - 2 = -4 \text{ V}$$

Corresponding value of Drain current $I_D = 0$ mA

Decrease in Drain current from 2 to 0 mA is due to increase in reverse bias to Gate junctions from –2 to –4 V. These variations in Drain current are also sinusoidal.

- At '2π' instance of the input signal voltage, $V_{in} = 0$ V.

$$\therefore \quad v_{GS} = V_{GS}(Q) + 2\sin\omega t = -2 + 0 = -2 \text{ V}$$

Corresponding value of Drain current $I_D = 2$ mA

Increase in Drain current from 0 to 2 mA is due to decrease in reverse bias to Gate junctions from –4 to –2 V during the interval from $3\pi/2$ to 2π. These variations in Drain current from 0 to 2 mA are also sinusoidal as discussed previously.

Various signal waveforms in FET Amplifier are shown in Figs. 9.7 and 9.8. One of the main advantages of the FET Amplifiers is that signal sources need not supply power to FET Amplifiers. High input impedance of FET device draws negligible current and in turn negligible power from the signal sources.

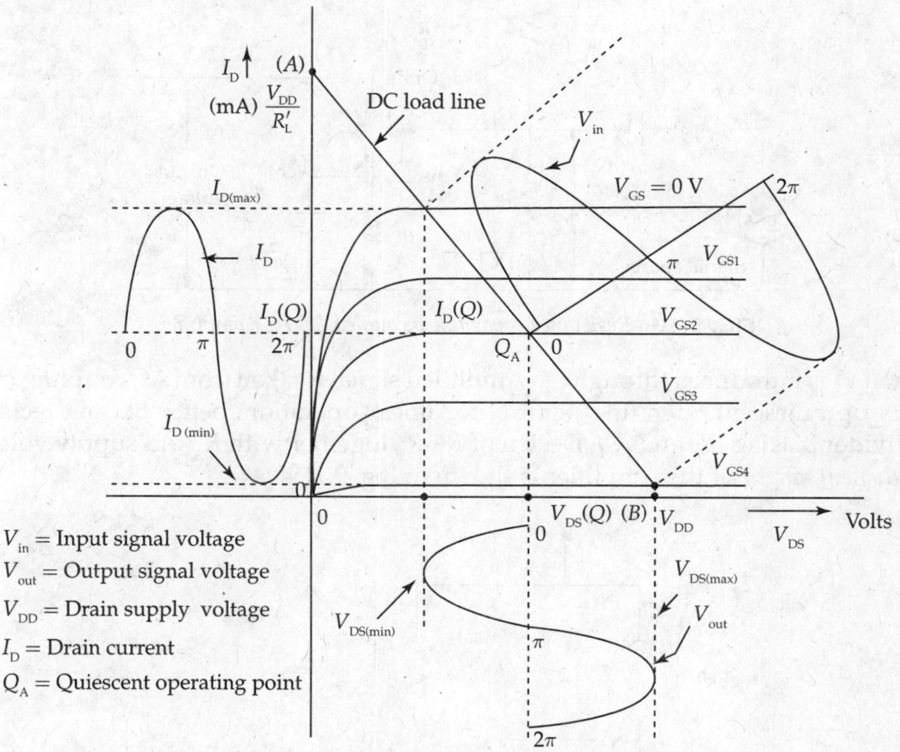

FIG. 9.8 *Signal variations of I_D, V_{in} and V_{out} in JFET amplifier*

Figure 9.9 shows the small signal AC equivalent circuit of CS FET Amplifier.

Varying Drain current flows through effective load resistance R'_L (R'_L is the parallel combination of the two resistors R_D and R_L). This is shown in the AC equivalent circuit of FET Amplifier (Fig. 9.9). There will be 180° phase shift between output voltage (V_{out}) and externally applied input signal voltage (V_S) for s CS FET Amplifier (Fig. 9.8). Hence, CS FET Amplifier is known as *Inverting Amplifier*.

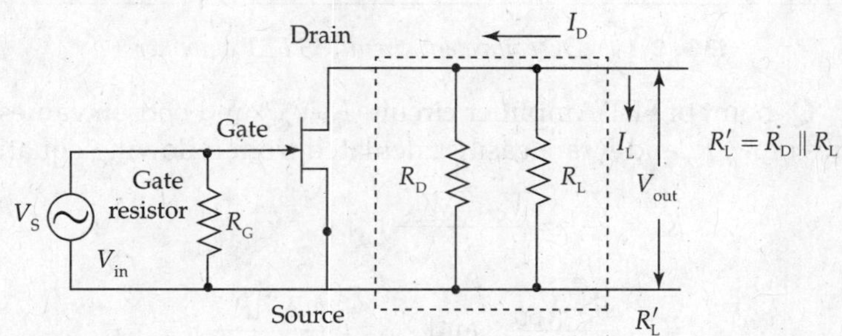

FIG. 9.9 *AC equivalent circuit for single-stage JFET amplifier*

9.4 RESISTANCE CAPACITANCE COUPLED FET AMPLIFIER

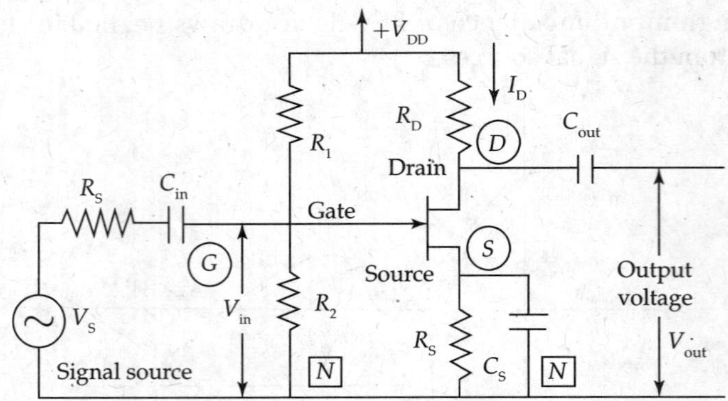

FIG. 9.10 *Resistance capacitance coupled CS FET amplifier*

Input signal V_S is introduced through C_{in}. Amplified signal is taken from AC coupling capacitor C_{out}. To fix up quiescent operating point Q for linear operation, better biasing technique is voltage divider biasing (R_1 and R_2). Design of R_1, R_2, together with R_S and supply voltage V_{DD} in *DC equivalent circuit* of the Amplifier is shown in Fig. 9.11.

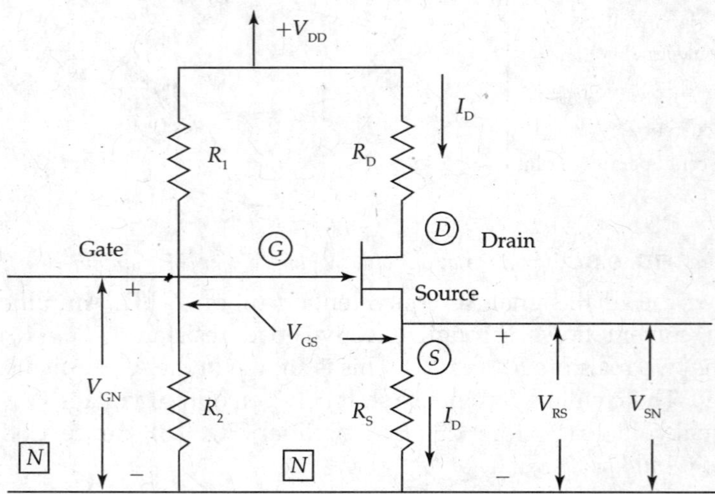

FIG. 9.11 *DC equivalent circuit CS FET amplifier*

For a specified 'Q' point of FET Amplifier circuit (I_D, V_{GS}) and chosen values of V_{GN} and R_G, the required values of R_1, R_2 and R_S are easily calculated from following equations:

$$R_S = \frac{V_{SN}}{I_D} = \frac{V_{GN} - V_{GS}}{I_D} \tag{9.10}$$

$$R_1 = \frac{R_G \cdot V_{DD}}{V_{GN}} \quad \text{and} \quad R_2 = \frac{R_G \cdot R_1}{(R_1 - R_G)} \tag{9.11}$$

The effect of any shift in V_{GS} is reduced by making $|V_{SN}|$ large compared to $|V_{GS}|$.

$$V_{GN} = \frac{V_{DD} \cdot R_2}{(R_1 + R_2)} = \frac{V_{DD} \cdot R_1 R_2}{(R_1 + R_2)R_1} = \frac{V_{DD} \cdot R_G}{R_1}, \text{ where } R_G = \frac{R_1 R_2}{(R_1 + R_2)}. \quad (9.12)$$

EXAMPLE 9.2

JFET is to be operated at a quiescent point defined by $I_D = 4$ mA, $V_{DS} = 8$ V and $V_{GS} = -2$ V. Design an appropriate biasing circuit with $V_{DD} = 30$ V.

Solution: Using the voltage divider bias circuit, R_1 and R_2 assume $V_{GN} = +12$ V, so that $V_{GN} - V_{GS} = V_{SN}$ is large compared to V_{GS} for stability, i.e., the effect of any shift in V_{GS} is reduced.

Assume $R_G = 1$ MΩ to keep the input resistance high.

Then $V_{SN} = I_D R_S = V_{GN} - V_{GS} = 12 - (-2) = 14$ V

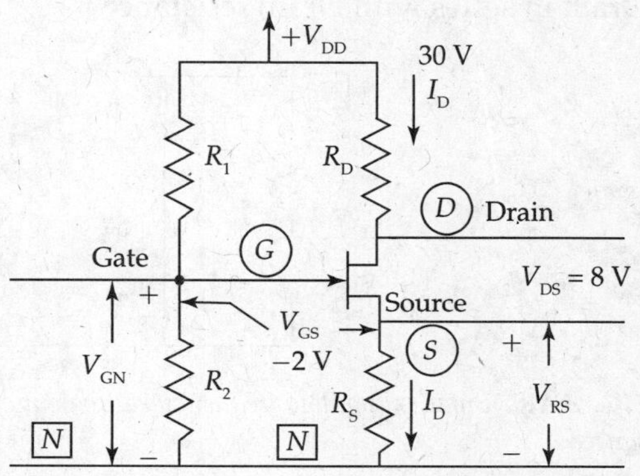

FIG. 9.12 *DC equivalent circuit CS FET amplifier*

$$R_S = \frac{V_{SN}}{I_D} = \frac{(V_{GN} - V_{GS})}{I_D} = \frac{(12 - (-2))}{4 \text{ mA}} = \frac{14}{4 \times 10^{-3}} = 3.5 \text{ k}\Omega$$

$$R_1 = \frac{R_G \cdot V_{DD}}{V_{GG}} = \frac{1 \times 10^6 \times 30}{12} = 2.5 \text{ M}\Omega$$

$$R_2 = \frac{R_G \cdot R_1}{(R_1 - R_G)} = \frac{1 \times 10^6 \times 2.5 \times 10^6}{(2.5 - 1.0) \times 10^6} = \frac{2.5 \times 10^6}{1.5} = 1.67 \times 10^6 = 1.67 \text{ M}\Omega$$

Summing voltages around the Drain loop yields $\Sigma V = 0$, i.e., $V_{DD} - I_D R_D - V_{DS} - I_D R_S = 0$

For $V_{DS} = 8$ V (from the data of the given *Q*-point (Quiescent operating point))

$$R_D = \frac{(V_{DD} - V_{DS} - I_D R_S)}{I_D} = \frac{30 - 8 - 14}{4 \times 10^{-3}} = 2 \times 10^3 = 2 \text{k}\Omega.$$

Thus the magnitudes of component values of R_1, R_2, R_S and R_D are estimated based on the given data of the DC equivalent circuit of common Source JFET Amplifier.

EXAMPLE 9.3

JFET amplifier has $V_{DD} = 20$ V, quiescent operating point is located at $V_{DS}(Q) = 10$ V and $I_D(Q) = 4$ mA, Voltage $V_{RS} = 2$ V. Calculate load resistance R_D.

Solution: Load line equation $[V_{DD} - V_{DS}(Q) - V_{RS}] = I_D(Q) R_D$

$$\therefore \quad R_D = \frac{[V_{DD} - V_{DS}(Q) - V_{RS}]}{I_D(Q)} = \frac{(20 - 10 - 2)}{4 \times 10^{-3}} = 2 \times 10^3 \ \Omega.$$

Small Signal Low-frequency (LF) Model of FET Amplifier
For small signal LF operations, FET, MOSFET and BJT Amplifiers respond in a linear fashion. Following equivalent circuits explain small signal LF response of FET Amplifier.

FET device parameters μ, g_m and r_d suggest AC equivalent circuit similar to equivalent circuit for BJT (Fig. 9.13). Irrespective of CS, CG and CD configurations of FET amplifiers, FET can be represented by derived voltage Source 'μV_{GS}' with shown polarity between Source and Drain in series with Drain resistance r_d.

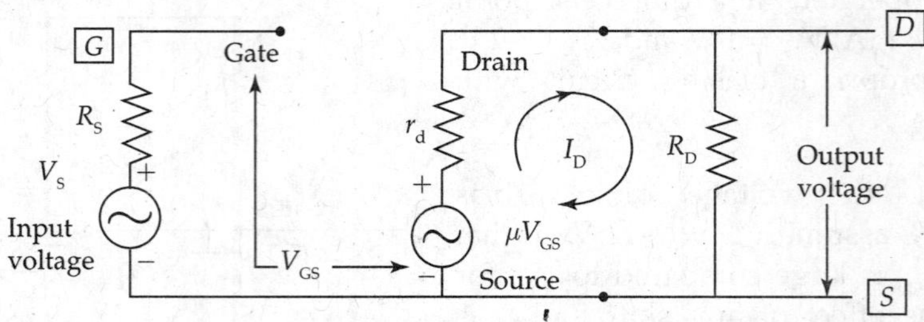

FIG. 9.13 *Small signal low frequency equivalent circuit of FET amplifier when FET is represented with voltage source*

From Fig. 9.13,

$$\text{Drain current} \quad I_D = \frac{\mu \cdot v_{GS}}{[r_d + R_D]} \text{ mA} \tag{9.13}$$

$$\text{Output voltage} \quad V_{out} = -I_d \cdot R_D = -\frac{\mu \cdot v_{GS}}{[r_d + R_D]} \times R_L \text{ volts} \tag{9.14}$$

Normally drain resistance of FET will be much greater than R_D
Therefore $(r_d + R_D) \cong r_d$.

$$\therefore \text{ Output voltage} \quad V_{out} = -\frac{\mu \cdot v_{GS}}{r_d} \times R_D = -\frac{g_m \cdot r_d \cdot v_{GS}}{r_d} \times R_D = -g_m \cdot R_D \times v_{GS} \text{ volts}$$

$$\text{Voltage gain of FET amplifier} = \frac{V_{out}}{v_{GS}} = -g_m \cdot R_D \tag{9.15}$$

FET Amplifier analysis using FET as a current Source (Fig. 9.14)

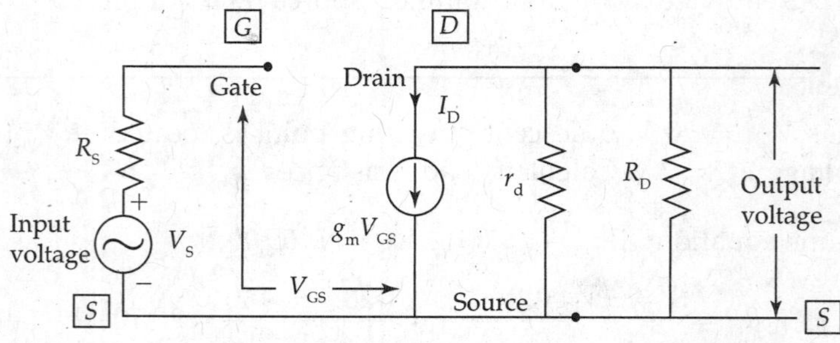

FIG. 9.14 *Small signal low frequency equivalent circuit of FET amplifier when FET is represented as a current source*

From Fig. 9.14,

$$\text{Output voltage} \quad V_{out} = -g_m \cdot v_{GS} \left[\frac{r_d \cdot R_L}{r_d + R_D}\right] \text{volts}$$

$$\text{Voltage gain of the amplifier} \quad A = \frac{V_{out}}{v_{GS}} = -g_m \cdot \left[\frac{r_d \cdot R_D}{r_d + R_D}\right]$$

$$\therefore \text{ Gain } A \cong -\frac{g_m \cdot r_d \cdot R_D}{r_d} \text{ (using } (r_d + R_D) \cong r_d \text{, as } r_d \gg R_D))$$

Hence, amplifier gain $A \cong -g_m \cdot R_D$.

Resistance capacitance coupled CSFET Amplifier with additional load R_L

Class-A Amplifier with low frequency small signal operation is considered.

Input signal V_{in} is applied through capacitor (C_{in}). R_G is the Gate resistor to provide high input impedance as well as a leakage path for the capacitance to discharge. R_G also provides connecting path for DC bias developed across the Source resistance R_S to be applied to input terminal. R_S provides the necessary bias and C_S provides bypass path for AC signals (around R_S) to make the Source at AC ground potential (eliminates signal feedback), justifying the common Source operation of FET Amplifier (Fig. 9.15).

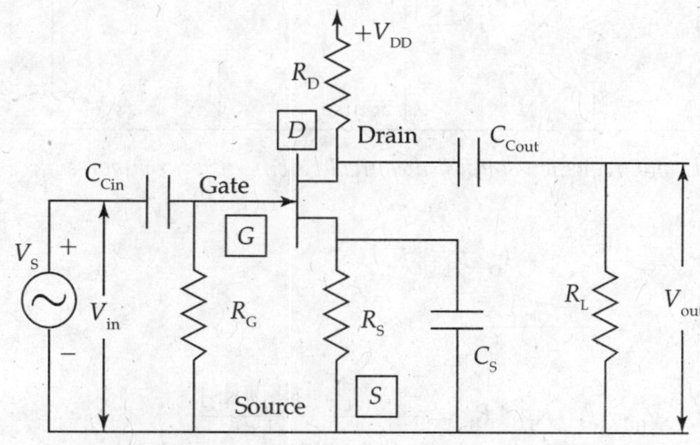

FIG. 9.15 *CS FET amplifier with external load R_L*

AC output signal voltage V_{out} is developed across Drain circuit resistance R_D.

Figure 9.16 shows small signal LF AC equivalent circuit of CS FET Amplifier. Equivalent circuit is drawn using the following norms. (1) FET device is replaced by a controlled voltage Source of 'μV_{GS}' with indicated polarity in series with Drain resistance 'r_d' (between Source and Drain terminals). (2) Gate terminal is left free and can be connected to points depending on actual components in the circuit. (3) Other components are replaced as they are from the actual circuit.

Circuit design will be such that $X_{Cin} \ll R_G$ and $X_{Cout} \ll R_L$ and $X_{CS} \ll R_S$. Then above circuit gets simplified as shown in Fig. 9.17.

Voltage gain A_V, Z_{in} and Z_{out} can be calculated using circuit theory principles. From Fig. 9.17, voltage $V_{GS} = V_S$ at Amplifier input port. Using Kirchoff's Voltage Law to the loop 'S D S' from Source to Drain and then back to Source at output port of Amplifier:

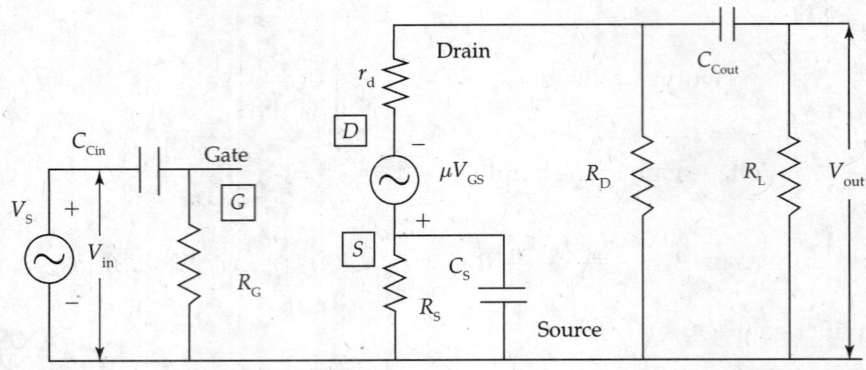

FIG. 9.16 *Small signal low frequency equivalent circuit of common source FET amplifier*

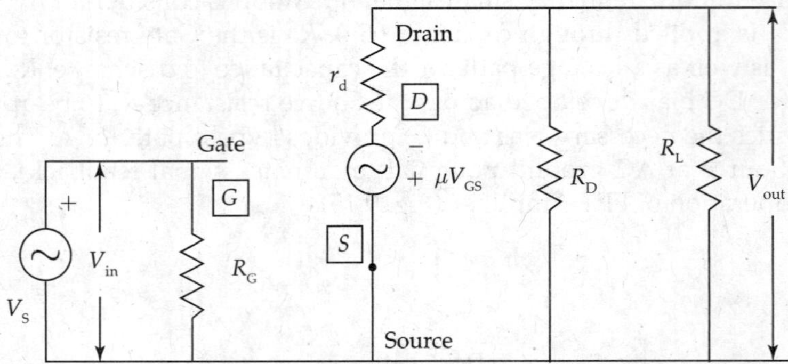

FIG. 9.17 *Small signal low frequency simplified equivalent circuit for common source field effect transistor amplifier*

$$\mu \cdot V_{GS} + I_D \cdot r_d + I_D \cdot R_D \parallel R_L = 0$$

$$\text{or } I_D = \frac{-\mu \cdot V_{GS}}{r_d + R_D \parallel R_L}$$

$$\therefore \quad V_0 = I_D \cdot R_D \parallel R_L = \frac{-\mu \cdot V_{GS} \cdot [R_D \parallel R_L]}{r_d + R_D \parallel R_L}$$

$$\therefore \quad A_V = \frac{-\mu \cdot [R_D \parallel R_L]}{r_d + [R_D \parallel R_L]} \cong \frac{-\mu \cdot R_L'}{r_d + R_L'} \quad \text{when } R_D \parallel R_L = R_L'$$

$$Z_{in} = R_G$$

$$Z_{out} = r_d \parallel R_D \parallel R_L$$

For alternate approach to Amplifier gain calculation, FET is replaced by Norton's equivalent circuit (Fig. 9.18):

$$V_{out} = V_0 = -g_m \cdot V_{GS} \times (r_d \parallel R_D \parallel R_L)$$

$$\text{Amplifier Gain} \quad A_V = \frac{V_0}{V_{GS}} = -g_m \times (r_d \parallel R_D \parallel R_L) = -g_m \times \text{Req}$$

where $\text{Req} = (r_d \parallel R_D \parallel R_L \cong g_m \times R_L)$

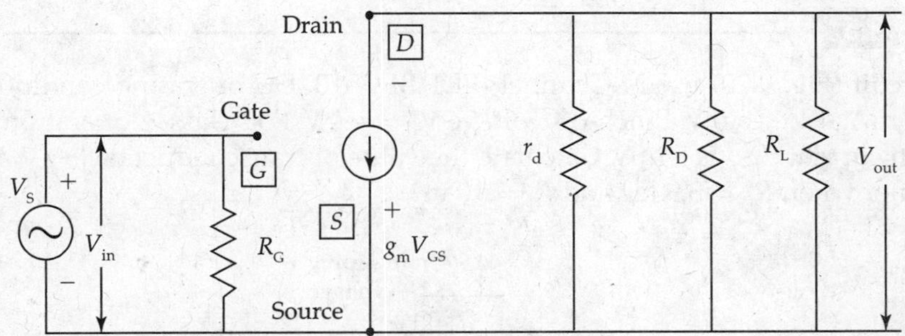

FIG. 9.18 *Small signal low frequency equivalent circuit with current source for FET in common source FET amplifier circuit*

$$\text{Input impedance } Z_{in} = R_G \text{ and } Z_{out} = R_{eQ} = [r_d \,||\, R_D \,||\, R_L] \tag{9.16}$$

For maximum voltage gain A_V, it requires that $r_d \ll R_D \,||\, R_L$. Maximum value of voltage gain will be approximately equal to μ (maximum theoretical gain) from FET Amplifier circuits. CS FET Amplifier is popular circuit in applications.

EXAMPLE 9.4

Common Source FET Amplifier has $r_d = 36$ kΩ, $\mu = 50$, $R_D = 4$ kΩ. Calculate the voltage gain A_V for the Common Source FET Amplifier in Fig. 9.19.

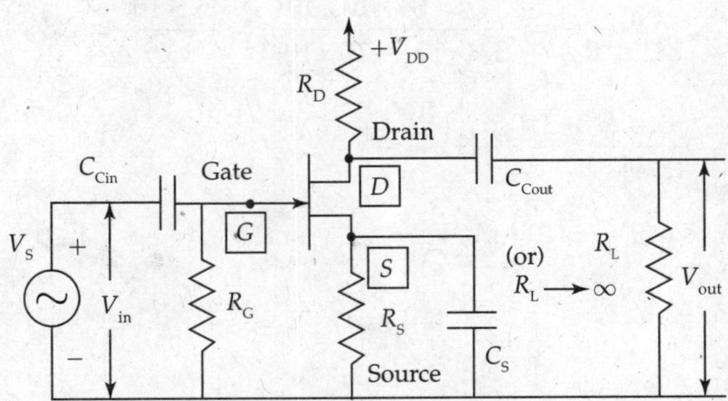

FIG. 9.19 *Common source RC-coupled FET amplifier*

Solution: From Fig. 9.19, Load resistance R_L is very large. So, the parallel combination of the load resistance R_L and the Drain circuit resistance R_D will be equal to R_D

$$\text{Voltage Gain } A_V = \frac{\mu \cdot R_D}{(r_d + R_D)} \quad \text{when } R_D \ll R_L$$

$$A_V = \frac{50 \times 4 \times 10^3}{36 \times 10^3 + 4 \times 10^3} = \frac{50 \times 4 \times 10^3}{40 \times 10^3} = 5.$$

EXAMPLE 9.5

Amplifier circuit (Fig. 9.20) use N-Channel JFET BFW-10. DC operating conditions are such that $I_{DSS} = 10$ mA at $V_{GS} = 0$ V. Pinch-off voltage $V_P = -4$ V. For Class-A operation $V_{GS} = -2$ V and the Drain current 'I_{DQ}' is 5 mA. Calculate the value of transconductance 'g_m' and the gain of the Amplifier when $R_D = 3.9$ kΩ and $R_L = 10$ kΩ.

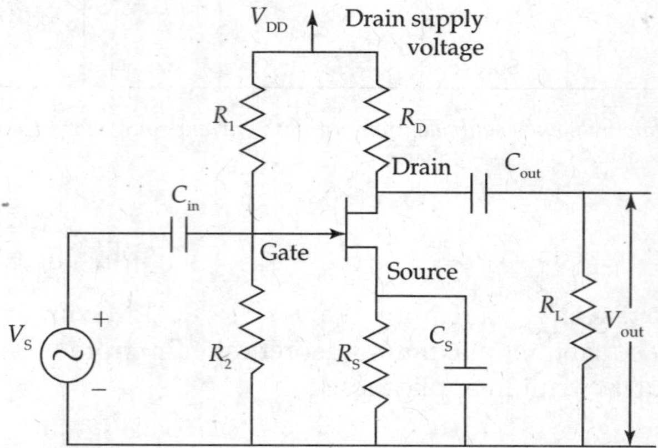

FIG. 9.20 *Single-stage RC-coupled FET amplifier*

Solution: The effective load resistance R_{Leq} for AC signals is the parallel combination of $R_D = 3.9$ kΩ and $R_L = 10$ kΩ.

$$R_{Leq} = \frac{R_D \cdot R_L}{(R_D + R_L)} = \frac{3.9\times10^3 \times 10\times10^3}{(3.9\times10^3 + 10\times10^3)} = \frac{39\times10^3}{13.9} = 2.8\times10^3$$

$$g_m = \frac{\Delta \cdot I_D}{\Delta \cdot V_{GS}} = \frac{2\cdot I_{DSS}}{V_P}\left[1 - \left|\frac{V_{GS}}{V_P}\right|\right]$$

Substituting the values of $I_{DSS} = 10$ mA, $V_P = -4$ V and $V_{GS} = -2$ V in the above equation

$$g_m = -\frac{2\cdot I_{DSS}}{V_P}\left[1 - \left|\frac{V_{GS}}{V_P}\right|\right] = -\frac{2\times10\times10^{-3}}{4}\left[1 - \frac{2}{4}\right] = -\frac{5\times10^{-3}}{2} = -2.5\times10^{-3} \text{ mhos}$$

Voltage gain of the FET Amplifier $A_V = -g_m \cdot R_{Leq}$

Hence, Voltage $A_V = -2.5 \times 10^{-3} \times 2.8 \times 10^3 = 7.0$.

9.5 COMMON GATE FET AMPLIFIER ANALYSIS

In Common Gate Amplifier, Gate is grounded and driving signal source (V_S) is connected to the Source terminal. Load resistance R_D in Drain path and external load resistance R_L are connected between Drain and Gate electrodes as shown in Fig. 9.21. Common Gate FET Amplifier circuit with self-bias arrangement is shown in the Fig. 9.22. Self-Bias is obtained using R_S.

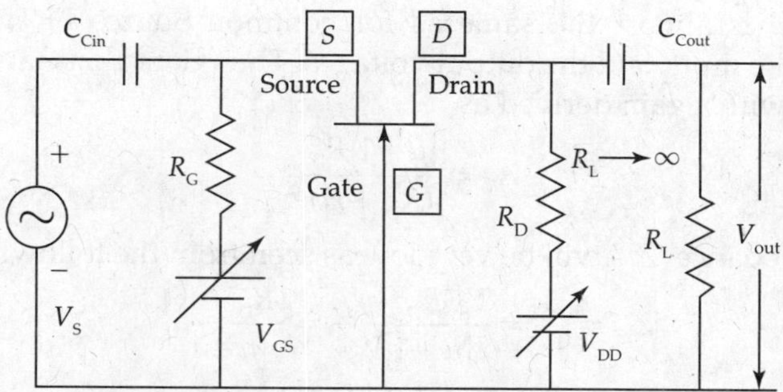

FIG. 9.21 *Common gate FET amplifier circuit*

When $C_{in} = C_{out} = C_S$ are all so large as to make their reactance negligible compared to the resistances R_{in}, R_{out} and R_S the equivalent circuit in Fig. 9.23 is drawn.

From the equivalent circuit of Fig. 9.23, it is clear that $V_{GS} = V_S$. On the loop D-S-G-D in Fig. 9.23,

$$-V_S + \mu \cdot V_{GS} + I_D(R_D + r_d) = 0$$

But $V_{GS} = -V_S$

$$\therefore \quad -V_S - \mu \cdot V_S + I_D \cdot (R_D + r_d) = 0$$

$$\therefore \quad I_D = \frac{[\mu + 1]V_S}{[(R_D + r_d)]}$$

But $V_0 = I_D \cdot R_D = \dfrac{[\mu + 1]V_S \cdot R_D}{(R_D + r_d)}$

$$\therefore \quad \text{Voltage Gain } A_V = \frac{V_0}{V_S} = \frac{[\mu + 1]V_S \cdot R_D}{(R_D + r_d)V_S} = \frac{[\mu + 1]R_D}{(R_D + r_d)}$$

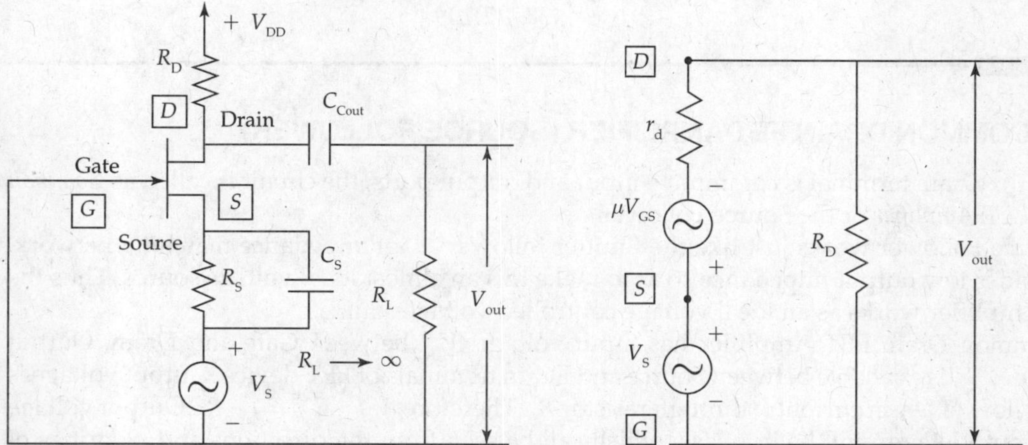

FIG. 9.22 *Common gate amplifier with self bias* **FIG. 9.23** *Common gate FET amplifier equivalent circuit*

Voltage Gain (A_V) is almost the same as for common Source FET Amplifier with 'no inversion' between the input and the output voltages. Thus Common Gate Amplifier is a non-inverting Amplifier with a gain derived as

$$A_V = \frac{[\mu+1]R_D}{(R_D + r_d)}. \tag{9.17}$$

But the input impedance 'Z_{in}' will be very low as seen from the following equation

$$Z_{in} = \frac{V_S}{I_D} = \frac{V_S(R_D + r_d)}{[\mu+1]V_S} = \frac{(R_D + r_d)}{[\mu+1]}$$

$$\text{If } r_d \gg R_D, \text{ then } Z_{in} = \frac{r_d}{[\mu+1]} \cong \frac{1}{g_m}.$$

EXAMPLE 9.6

For a common Drain FET Amplifier circuit, $\mu = 50$, $R_D = 2$ kΩ and $r_d = 38$ kΩ, calculate the voltage gain A_V (Fig. 9.24). (R_S functions as R_D. Therefore $R_S = R_D = 2$ kΩ)

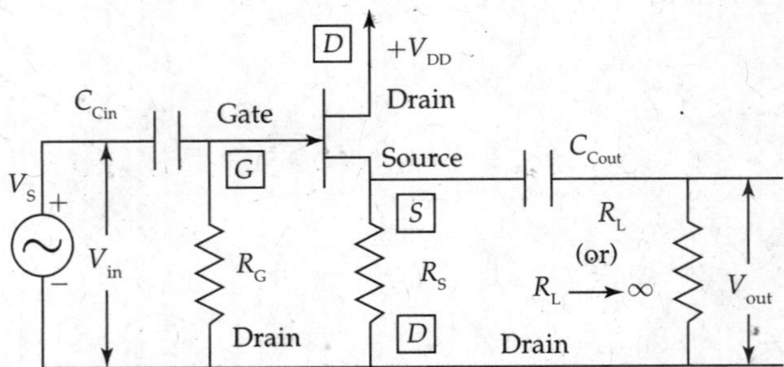

FIG. 9.24 *Common drain FET amplifier circuit*

Solution: The voltage gain of common Gate FET Amplifier A_V

$$A_V = \frac{(\mu+1)R_D}{r_d + R_D} = \frac{(50+1)2\times10^3}{(38+2)10^3} \equiv 2.5.$$

9.6 COMMON DRAIN FET AMPLIFIER (SOURCE FOLLOWER)

When the Drain terminal is common to input and output ports, the circuit is called as *Common Drain FET Amplifier* or the Source Follower.

Source Follower works just like the Emitter follower as an impedance matching network. It provides low output impedance so as to make this an almost ideal voltage source. Thus the total Amplifier works as an ideal voltage controlled voltage source.

Common Drain FET Amplifier has input voltage 'V_{in}' between Gate and Drain, Output voltage 'V_{out}' is available between Source and Drain terminals of FET device. Output voltage is due to flow of Drain current I_d through resistor R_S. Therefore, $V_S = V_{out} = I_d \cdot R_S$. Output voltage and input voltages will be in phase, which can be seen from the directions and polarities of input and output voltages.

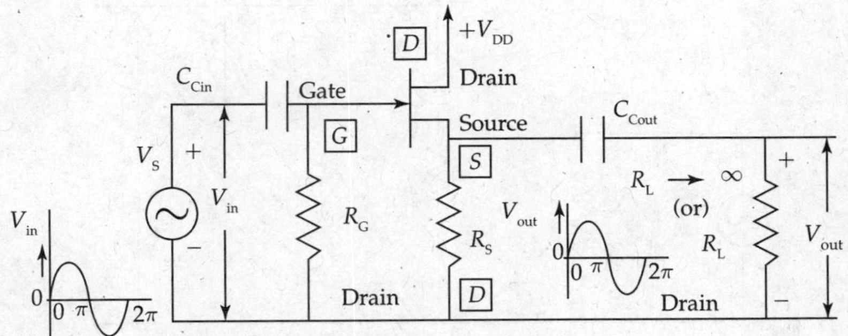

FIG. 9.25 *Common drain FET amplifier (source follower)*

Source voltage (output voltage) follows the variations in input signal. So, CD Amplifier circuit is also known as Source Follower circuit. Since the output voltage is almost equal to input voltage, the gain is approximately unity.

From Fig. 9.26, $V_{GS} = V_{in} - I_d R_S$

$$-\mu V_{GS} + (r_d + R_S) I_d = 0$$

$$\therefore \quad -\mu(V_{in} - I_d R_S) + (r_d + R_S) I_d = 0$$

$$-\mu V_{in} + I_d [r_d + R_S(\mu + 1)] = 0$$

$$\text{or} \quad I_d = \frac{\mu V_{in}}{[r_d + R_S(\mu + 1)]}$$

$$\text{Output Voltage} \quad V_{out} = I_d \cdot R_S = \frac{\mu \cdot V_{in} \cdot R_S}{[r_d + R_S(\mu + 1)]}$$

$$V_{out} \text{ can be written as } V_{out} = \frac{\left[\dfrac{\mu}{(\mu + 1)}\right] V_{in} \cdot R_S}{\dfrac{r_d}{(\mu + 1)} + R_S}$$

Equivalent circuits for CDFET amplifier

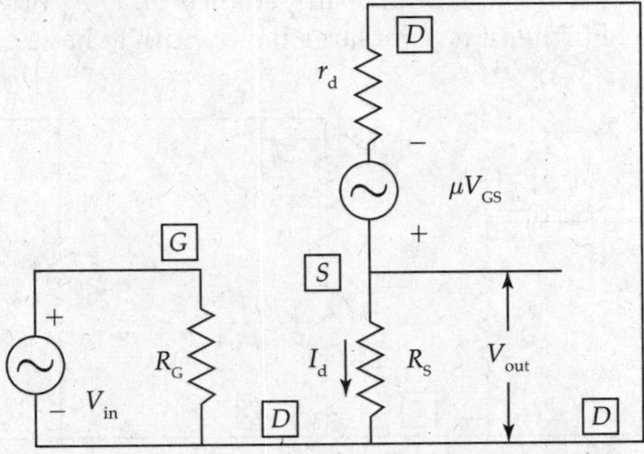

FIG. 9.26 *Common drain FET amplifier equivalent circuit*

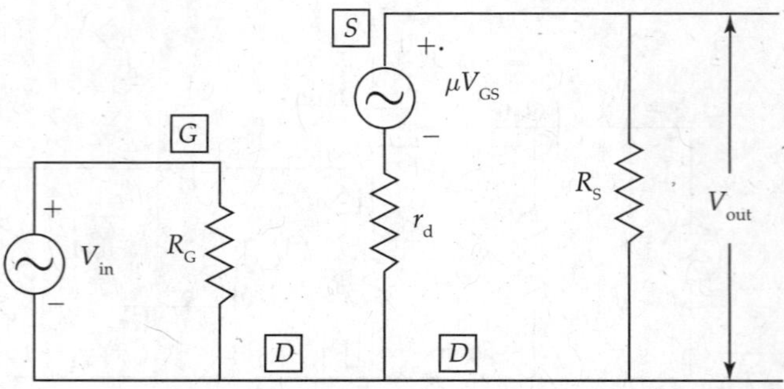

FIG. 9.27 *Common drain FET amplifier equivalent circuit reoriented for analysis*

$$\therefore \text{ Voltage Gain } \quad A_V = \frac{V_{out}}{V_{in}} = \frac{[\mu/(\mu+1)]R_S}{[r_d/(\mu+1)]+R_S} \cong \frac{\mu}{(\mu+1)}; \text{ if } R_S \gg \frac{r_d}{(\mu+1)}$$

$$\text{Also Voltage Gain } \quad A_V = \frac{\mu R_S}{[r_d+(\mu+1)R_S]}$$

$$\therefore \text{ Voltage Gain } \quad A_V = \frac{\mu}{(\mu+1)} \cong 1.$$

Therefore, the voltage gain A_V of a Common Drain FET Amplifier is close to unity, similar to the Emitter Follower circuit.

From the equation of Drain current

$$I_d = \frac{[\mu/(\mu+1)]V_{in}}{[r_d/(\mu+1)]+R_S}.$$

We can draw a circuit as shown in the Fig. 9.28. The circuit represents a controlled Source with voltage $\mu \cdot V_{in}/(\mu + 1)$ in series with an impedance $r_d/(\mu + 1)$ driving a load resistance R_S. Thus the controlled Source impedance or the so-called output impedance 'Z_{out}' or the output resistance 'R_{out}' of the Source follower is $r_d/(\mu + 1) \cong (1/g_m)$. It is interesting to note that the output impedance of Source follower is same as the input impedance of Common Base Transistor Amplifier that is $1/g_m$. Thus Source follower acts as a unity voltage gain Non-Inverting Amplifier with a very low Source impedance of $1/g_m$ or acts as an ideal voltage controlled voltage source. Further it has got large bandwidth to have good frequency response

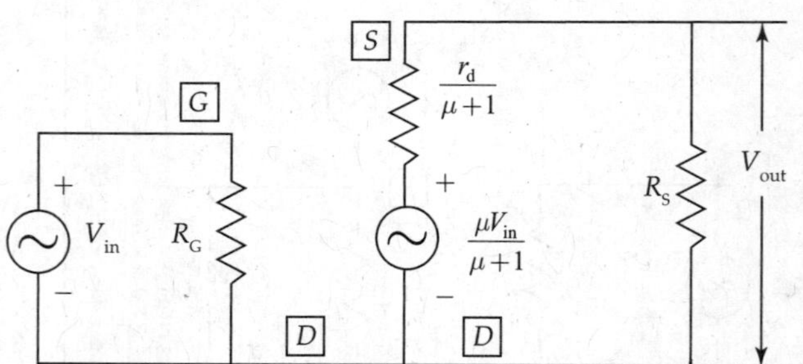

FIG. 9.28 *Common drain FET amplifier equivalent circuit after simplification*

(from the concept of the product of gain and Bandwidth of Amplifiers is constant) and stable operation due to inherent 'negative feedback' in the Amplifier operation.

Further, the high input impedance 'Z_{in}' of the Source follower circuit is used in measuring instruments when loading on the input signal sources is to be minimised. One of the applications is at the input stages of Amplifiers used in CRO.

Unity gain of the Source follower and impedance transformation feature, i.e., the very low output impedance 'Z_{out}' and high input impedance 'Z_{in}' feature of Source follower circuit is used as 'Unity Gain Buffer Amplifier' in instrumentation circuits.

EXAMPLE 9.7

For Common Drain Amplifier shown in Fig. 9.29, Transconductance $g_m = 2.5$ mA, $r_d = 25$ kΩ. Calculate R_G, R_0 and A_V.

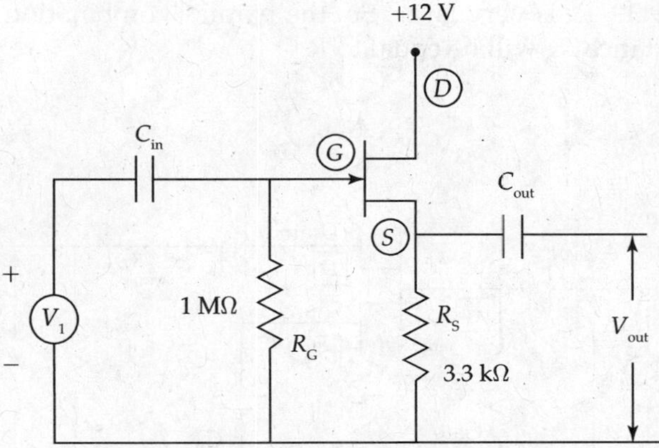

FIG. 9.29 *Common drain FET amplifier*

Solution: Data: $g_m = 2.5$ ms $= 2.5 \times 10{-}3$ s, $R_S = 3.3$ kΩ = 3300 Ω and $r_d = 25$ kΩ $= 25 \times 10^3$ Ω

To calculate R_G, R_0 and A_V

Data: Gate resistor $R_G = 1$ MΩ.

To calculate Output resistance R_0

We know that for $r_d \geq 10\, R_S$

$$R_0 = R_S \parallel \frac{1}{g_m} = \frac{\left[R_S \times \frac{1}{g_m}\right]}{\left[R_S + \frac{1}{g_m}\right]} = \frac{R_S}{[R_S \cdot g_m + 1]}$$

$$\therefore \quad R_0 = \frac{3300}{(3300 \times 2.5 \times 10^{-3} + 1)} = 356.756\ \Omega.$$

Calculation of Voltage gain A_V considering r_d (drain resistance)

$$A_V = \frac{g_m (r_d \parallel R_S)}{1 + g_m (r_d \parallel R_S)} = \frac{[(2.5\text{ ms})(25\text{ k}\Omega \parallel 3.3\text{ k}\Omega)]}{[1 + (2.5\text{ ms})(25\text{ k}\Omega \parallel 3.3\text{ k}\Omega)]} = \frac{7.366}{(1 + 7.366)} = 0.88$$

Calculation of Voltage gain A_V without considering r_d (drain resistance)

$$A_V = \frac{g_m \cdot R_S}{1 + g_m \cdot R_S} = \frac{[(2.5\times10^{-3})\times3300]}{[1+(2.5\times10^{-3}\times3300)]} = \frac{8.25}{1+8.25} = 0.89.$$

Examples worked out to show the different magnitudes of gain for common source, common Gate and common Drain FET Amplifier circuits

EXAMPLE 9.8

Common Source FET Amplifier has $r_d = 36\text{ k}\Omega$, $\mu = 50$, $R_D = 4\text{ k}\Omega$. Calculate voltage gain (A_V) using the given data for CS FET Amplifier. If $R_S = 500\ \Omega$, calculate the value of decoupling (bypass capacitor) C_S at lowest frequency of input signal $f_S = 64$ Hz.

Solution: From Fig. 9.30, R_L is very large. So, the parallel combination of load resistance R_L and Drain circuit resistance R_D will be equal to R_D

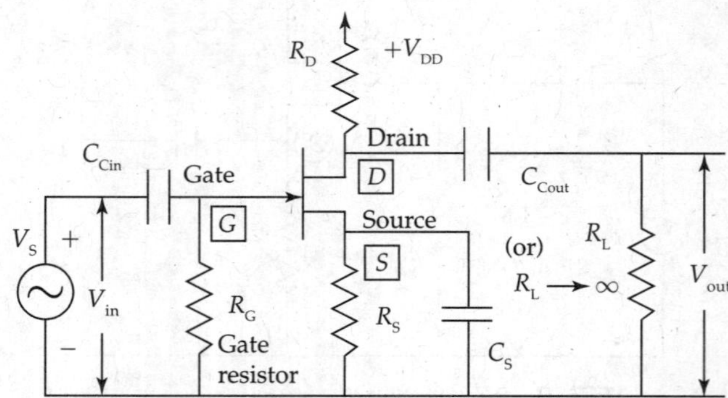

FIG. 9.30 *Common source RC-coupled FET amplifier*

$$\text{Voltage gain} \quad A_V = \frac{\mu R_D}{r_d + R_D} \text{ when } R_D \ll R_L$$

$$A_V = \frac{50\times4\times10^3}{[36\times10^3 + 4\times10^3]}$$

$$= \frac{200\times10^3}{40\times10^3} = 5$$

$$\text{Reactance of bypass capacitor} \quad X_{CS} = \frac{1}{10}\times R_S$$

$$X_{CS} = \frac{1}{2\pi f_S \times C_S} = \frac{1}{10}\times R_S$$

$$\therefore \quad C_S = \frac{10}{2\pi f_S \times 500} = \frac{10}{6.28\times64\times500}$$

$$= \frac{10}{0.2\times10^6} = \frac{100\times10^{-6}}{2} = 50\ \mu\text{F}.$$

EXAMPLE 9.9

For a Common Gate FET Amplifier circuit shown in Fig. 9.31, $\mu = 50$, $R_D = 2\text{ k}\Omega$ and $r_d = 38\text{ k}\Omega$, calculate the voltage gain A_V.

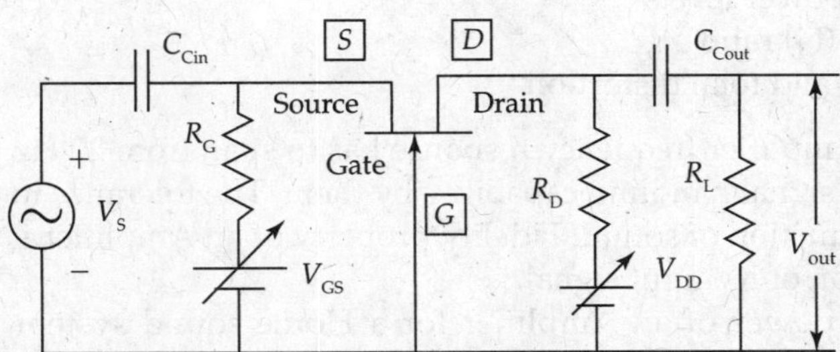

FIG. 9.31 *Common gate FET amplifier circuit*

Solution: The voltage gain of common Gate FET Amplifier A_V

$$A_V = \frac{(\mu+1)R_D}{r_d + R_D} = \frac{(50+1)\,2\times10^3}{(38+2)10^3} \equiv 2.5.$$

EXAMPLE 9.10

The common Drain FET Amplifier circuit is shown in Fig. 9.32. The FET device has $\mu = 50$, $r_d = 46\text{ k}\Omega$, $g_m = 2$ millimhos. $R_S = 4\text{ k}\Omega$ and $R_G = 1\text{ m}\Omega$ in the circuit. Calculate the voltage gain A_V for the Amplifier and the output resistance of the Amplifier.

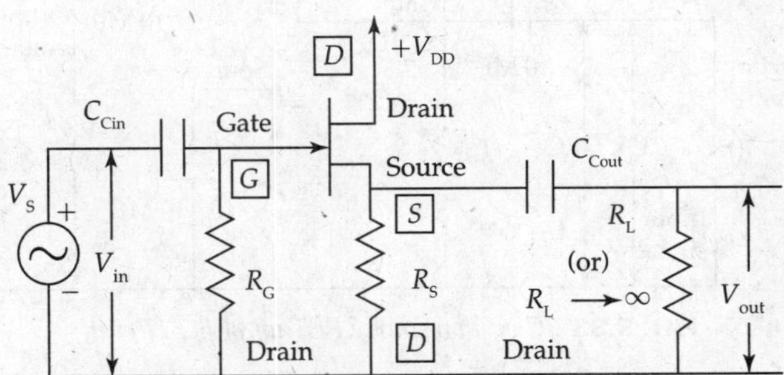

FIG. 9.32 *Common drain FET amplifier circuit*

Solution: The voltage gain A_V of the Common Drain FET Amplifier circuit is

$$A_V = \frac{\mu R_S}{(\mu+1)R_S + r_d} = \frac{50\times4\times10^3}{(50+1)4\times10^3 + 46\times10^3} = \frac{200}{250} = 0.8 < 1$$

$$\text{Output resistance } R_{out} = \frac{1}{g_m} = \frac{1}{2\times10^{-3}} = 500\ \Omega.$$

9.7 FREQUENCY RESPONSE OF SINGLE-STAGE AMPLIFIER

Main properties for an Amplifier

1. Frequency response and useful range of uniform gain given by 3-dB bandwidth
2. Maximum output power levels
3. Signal (S) to Noise (N) ratio
4. Types of distortion and total distortion

Audio frequency Amplifier frequency response has to span from 20 Hz to 20 kHz to cover both voice and music signal handling capability by them. The tolerance in distortion is about 0.01% for good reproduction of sound. Fidelity property of an Amplifier is equally important in faithful reproduction of its input signal.

Maximum output power of an Amplifier for a Home sound system could be 10 W to provide pleasant level of sound from the loud speakers in a room, as the efficiency of acoustic power delivery from a loud speaker could be very low. Power levels for public address system could be very large depending upon the area of coverage. The useful range for faithful reproduction of such signals is known as Amplifier Bandwidth.

Frequency response for an Amplifier stage can be determined as following:
From Fig. 9.33

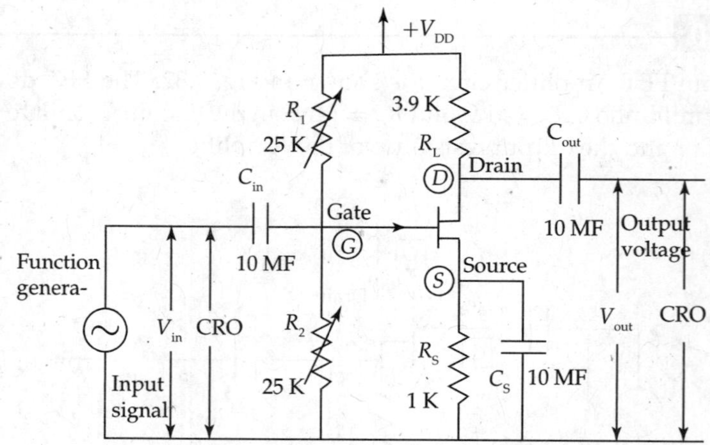

FIG. 9.33 *Common source FET amplifier circuit*

1. Input signal voltage to the Amplifier is kept at constant voltage. For CS FET Amplifier, the input signal may be kept around 1 V (based on Class-A type DC bias).
2. Frequency of the input signal is varied over a desired frequency range of prediction of the use of an Amplifier or its performance.
3. Magnitudes of output voltages are measured at various frequencies.
4. Gains of the Amplifier are calculated at each frequency of interest.

S. No	Frequency of input signal	Input voltage	Output voltage	Voltage gain	Voltage gain $20 \log_{10} A$ (dB)

Graph between the input signal frequency on the x-axis and voltage gain 'A' on the y-axis is known as frequency response characteristic of the Amplifier.

1. Amplifier gain will remain uniform or constant over a moderate range of frequencies of the input signal. It is known as the *mid-frequency region* of the Amplifier.
2. Amplifier gain falls off or decreases at both low- and high-frequency regions of frequency band over which Amplifier performance has to be predicted from expt.
   The amplitude response of an Amplifier is divided into the following three regions.
   - Low-frequency region
   - Mid-frequency region
   - High-frequency region

1. Consider the maximum voltage gain 'A_m' of an Amplifier to be 100.
2. Identify a point 0.707 A_m, on the y-axis.
3. Later, it can be seen that the power at this point is half that at the constant gain region. Draw a dotted line horizontally from this half-power point onto the frequency response curve. This line intersects the characteristic at two points f_1 and f_2 (with Voltage amplification or gain of 70.7) as shown in Fig. 9.34.
4. Frequency f_1 is known as the lower cut-off frequency.f_2 is the upper cut-off frequency.
5. The band of frequencies between f_1 and f_2 that is the region between f_1 and f_2 is known as the Amplifier bandwidth.
6. This region with uniform gain is known as 'mid-frequency region'.

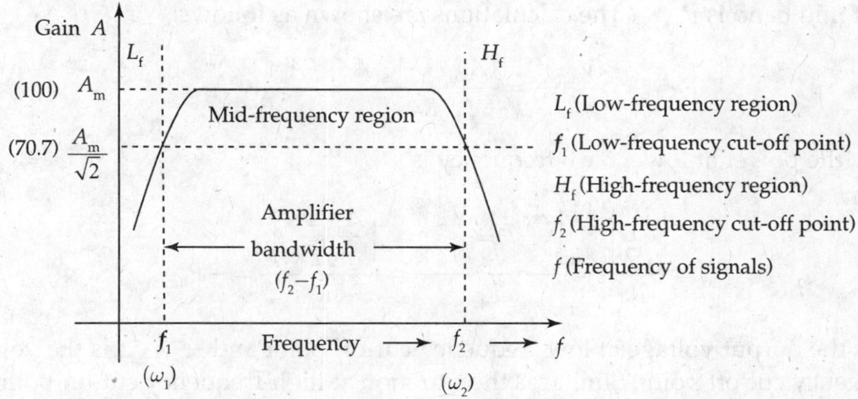

FIG. 9.34 *Frequency response of an amplifier showing bandwidth calculation*

Figure 9.35 shows the method of calculation of bandwidth from the frequency response characteristic of an Amplifier when the Amplifier gain is calculated in terms of decibels.

Gain in decibels = 20 $\log_{10} A_V$, where the voltage gain of an Amplifier is A_V.

f_1 or f_L and f_2 or f_h are also known as 3-dB frequencies or 1/2 power frequencies, where A_m or A_{max} is the maximum value of gain obtained from the frequency response characteristic. So fall in gain at cut-off frequencies is as follows:

$$\text{Gain at cut-off frequencies} \quad f_1 \text{ or } f_2 = \frac{A_m}{\sqrt{2}}$$

$$20 \log_{10} \frac{1}{\sqrt{2}} = 20 \log_{10} 2^{-1/2} = -3.0 \text{ dB}$$

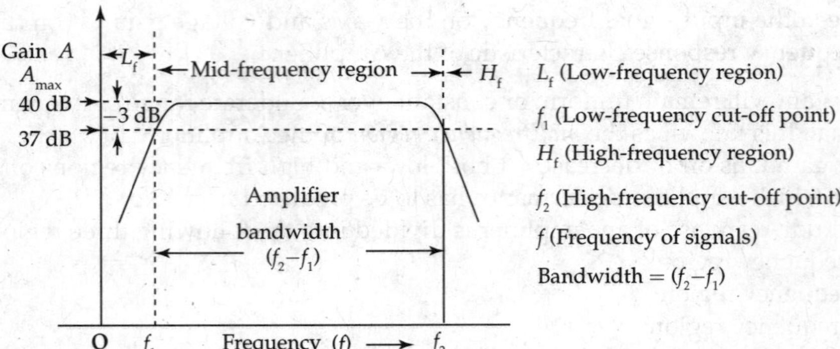

FIG. 9.35 *Frequency response of a FET amplifier showing the calculations for amplifier bandwidth using gain in dB*

So the lower cut-off (f_1) and upper cut-off (f_2) frequencies are marked on the frequency response characteristic in Fig. 9.35, with a gain –3 dB lower than the gain at mid-range frequencies, where the gain is maximum and uniform.

The maximum gain in this case is considered as 40 dB (40 dB corresponds to voltage gain 100 considered in Fig. 9.34 and so the gain at the cut-off frequency points will be 37 dB which will be lower by 3 dB from the maximum gain.

f_1 and f_2 are known as corner, break or half-power frequencies or roll-off frequencies.

Power at mid band is $P_{0(mid)}$. The calculations are shown as follows:

$$P_{0(mid)} = \frac{(V_{out})^2}{R_L} = \frac{(A_{V(mid)} \cdot V_{in})^2}{R_L} = \frac{(A_m \cdot V_{in})^2}{R_L} \tag{9.18}$$

Similarly the power at lower 3 dB frequency is

$$P_{0(3\text{-dB})} = \frac{\left(\frac{V_L}{\sqrt{2}}\right)^2}{R_L} = \frac{\left(\frac{A_{V(mid)} \cdot V_{in}}{\sqrt{2}}\right)^2}{R_L} = \frac{\left(\frac{A_m \cdot V_{in}}{\sqrt{2}}\right)^2}{R_L} = \frac{1}{2} P_{0(mid)}, \tag{9.19}$$

where V_L is the output voltage at low-frequency cut-off point and $A_m / \sqrt{2}$ is the voltage gain at low frequency cut-off point. Similar is the situation at high-frequency cut-off point.

If we examine the Amplifier response curves of Figs. 9.34 and 9.35, Amplifier gain is uniform over middle range of frequencies of signals. Hence, small signal low frequency equivalent circuit for BJT or JFET Amplifier circuits contains only resistive elements and all series and shunt capacitor elements need not be considered because their reactance effects are negligible in linear operation of Amplifiers in the mid-frequency region.

Mid-frequency equivalent circuit of FET Amplifier

Coupling and the bypass capacitors function as effective short circuits in mid-frequency region (Fig. 9.36). So voltage gain in mid-frequency region is almost uniform (Fig. 9.34). All the frequency components in the mid-frequency region undergo uniform amplification. This suggests that frequency distortion is zero. Amplifier response over the region defined by its bandwidth (Fig. 9.33) is considered to be at uniform level for the human ear.

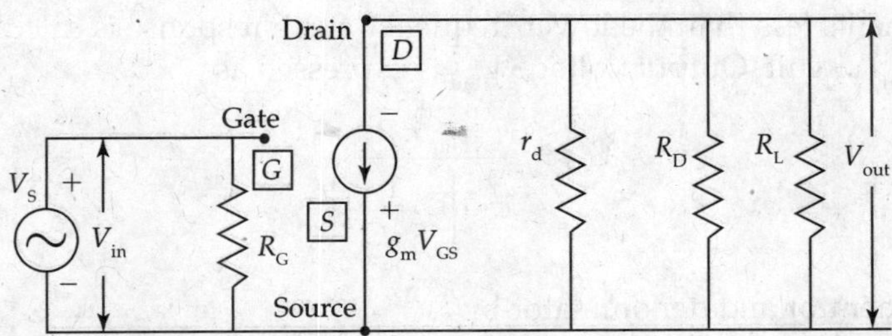

FIG. 9.36 *Small signal low frequency equivalent circuit with current source for FET in common source FET amplifier circuit for mid-frequency range*

Low-frequency and High-frequency regions

The region on the frequency response characteristic below 'f_1' is Low-frequency region 'L_f'. The region on the frequency response characteristic beyond 'f_2' is known as High-frequency region 'H_f'. Concept of Half-Power Points at f_1 and f_2 is from the following expressions. The signal power at mid-range frequency region can be calculated as

$$\text{Output Power} \quad P_{\text{out(mid-freq)}} = \frac{V_{\text{out}}^2}{R_L} = \frac{(A_m \cdot V_{\text{in}})^2}{R_L} \text{ watts} \tag{9.20}$$

The signal power levels at f_1 and f_2 can be calculated as

$$P_{\text{out}} \text{ at } f_1 \text{ or } f_2 = \frac{(V_{\text{out}}^2)}{R_L} = \frac{\left(\dfrac{A_m \cdot V_{\text{in}}}{\sqrt{2}}\right)^2}{R_L} = \frac{(A_m \cdot V_{\text{in}})^2}{2R_L} = \frac{1}{2} P_{\text{out(mid-freq)}} \tag{9.21}$$

Equation (9.21) shows that the power at lower cut-off frequency f_1 and upper cut-off frequency point f_2 is half the power at mid-frequency region. Thus, f_1 and f_2 are considered as half-power points. Reduction in power (decibels) $= 10 \log_{10}(1/2) = -3$ dB at half-power frequencies. Therefore, they are also considered as –3 dB points. Amplifier response between the two half-power points is defined as Amplifier Bandwidth (Useful Specification).

Effect of coupling capacitors on low frequency response of an Amplifier (High pass circuit)

Circuit components C_C at input and output terminals of active devices are responsible for reduction in gain at low-frequency region of frequency response. Equivalent resistance–capacitance circuit is shown in Fig. 9.37. This R-C circuit behaves as a High Pass filter allowing the signals having frequencies above f_1 and causing a reduction in gain for the signals with

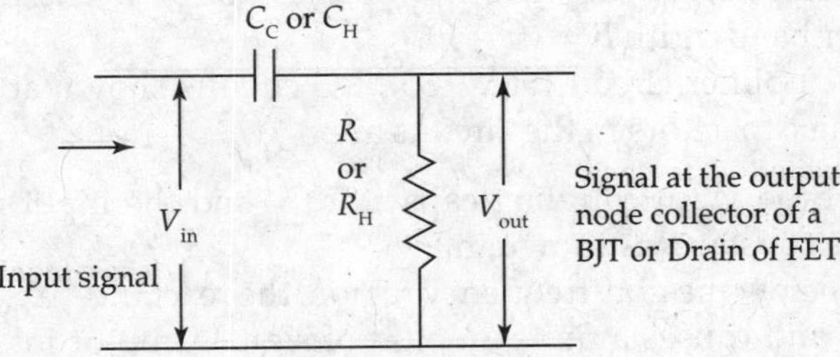

FIG. 9.37 *High pass circuit containing R-C_C elements to cause reduction of signal strength in low-frequency region below f_1 (low-frequency cut-off)*

frequency components less than the lower frequency f_1. LF response is calculated using High pass R_H and C_C (C_H) circuit. Output voltage V_{out} is expressed as;

$$V_{out} = \frac{V_{in} \times R}{R + \left(\frac{1}{J\omega \cdot C_C}\right)}$$

Dividing both numerator and denominator by R

$$\text{Output voltage } V_{out} = \frac{V_{in}}{1 + \left(\frac{1}{J\omega \cdot C_C \cdot R}\right)}$$

$$\therefore \text{ Voltage Gain } A = \frac{V_{out}}{V_{in}} = \frac{1}{\left[1 + \frac{1}{J\omega \cdot C_C \cdot R}\right]} = \frac{1}{\left[1 - \frac{J}{2\pi f \cdot C_C \cdot R}\right]}$$

$$\text{Assuming lower cut-off frequency } f_1 = \frac{1}{2\pi \cdot C_C \cdot R}$$

$$\text{At lower cut-off frequency, Gain } A = \frac{1}{\left[1 - J\frac{f_1}{f}\right]}$$

The magnitude $|A_L| = \dfrac{1}{\sqrt{1 + \left(\frac{f_1}{f}\right)^2}}$, Phase lead angle $\theta_1 = \arctan\left(\dfrac{f_1}{f}\right)$.

Substituting $f_1 = f$ in the above equation, we get

$$|A_L| = \frac{1}{\sqrt{1 + \left(\frac{f_1}{f}\right)^2}} = \frac{1}{\sqrt{2}} = 0.707.$$

This shows that the magnitude at the cut-off frequency f_1 = 0.707 f. From this relation, similar expression for high-frequency cut-off point f_2 = 0.707 f. These relations are used to calculate the Amplifier bandwidth $B = (f_2 - f_1)$.

As shown in FET Amplifier circuit (Fig. 9.35), RC circuits shown at the input and output ports with the coupling capacitors in the circuits are:

1. The combination of the input coupling capacitor C_{in} and the input resistance 'R_{in}' or 'R_H' form the input time constant, as shown in Fig. 9.37.
2. At low frequencies below the mid-frequency region, the reactance 'X_{C1}' of the input coupling capacitor increases and causes a finite amount of signal drop or loss across it. Hence, the signal amplitude available to the effective input terminals of FET devices is actually less. It causes a reduction in output signal amplitudes and overall Amplifier gains.

At $f = 200$ Hz and $C_{in} = 10\,\mu F$, $X_{Cin} = \dfrac{1}{2\pi \cdot f \cdot C_{in}} = 75\,\Omega$

At $f = 1000$ Hz and $C_{in} = 10\,\mu F$, $X_{Cin} = \dfrac{1}{2\pi \cdot f \cdot C_{in}} = 15\,\Omega$.

Variations in the reactance of coupling capacitors at two lower frequencies suggest the variations in gain at low frequencies. Coupling capacitors allow the high-frequency signals and attenuates low-frequency signals, because of loss of signal at low frequencies due to voltage drop across the coupling capacitors.

$$\text{Voltage amplification } |A_V| (\text{at } f_1 = f) = \sqrt{1 + \left(\frac{f_1}{f}\right)^2} = \frac{1}{\sqrt{2}}$$

At the frequency $f = f_1$ and reactance $X_{Cin} = R_{in}$.

From the previous expression, Voltage Gain $A_V = 1/\sqrt{2} = 0.707$.

With this basis, the gain is 0.707 times the gain A_m at mid-frequency region. This drop in signal level corresponds to a decibel reduction of $20\log_{10}(1/\sqrt{2}) = -3$ dB. Accordingly, f_1 or f_L is also known as lower 3 dB frequency or –3 dB point or half-power point. At low-frequency cut-off point f_1, the magnitude of the resistance 'R' will be equal to reactance X_C of C_C.

Voltage gain at the mid-frequency region for a FET Amplifier = AVM = $g_m \cdot R_L$.

Effect on Low frequency response due to coupling capacitors C_{in} and C_{out} (Fig. 9.38)

Similarly, the combination of output coupling capacitance 'C_{out}' and output resistance forms the output time constant of the Amplifier circuits similar to that as shown in Fig. 9.37 and causes reduction in gain at low-frequency region as shown (Fig. 9.38).

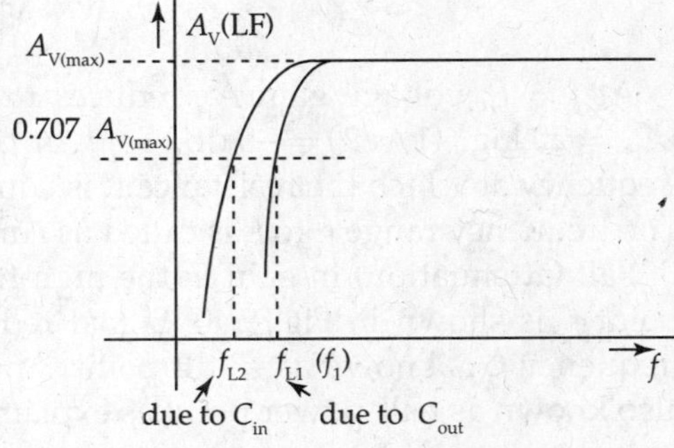

FIG. 9.38 *Reduction in gain at low-frequency region of amplifier frequency response*

Effect of Shunt Capacitances on the High Frequency Response of an Amplifier

High-Frequency response of an Amplifier (Low Pass Circuit)

In the high-frequency region (above the mid-band frequency region), an Amplifier stage can often be approximated by the simple *low pass circuit* shown in Fig. 9.39.

This circuit can be considered as a voltage divider with the input voltage excitation 'V_{in}' across the series combination of Resistor R_L or R and the Capacitor 'C_L' or 'C_{SH}' impedances and the output voltage is across the Capacitor 'C_{SH}'.

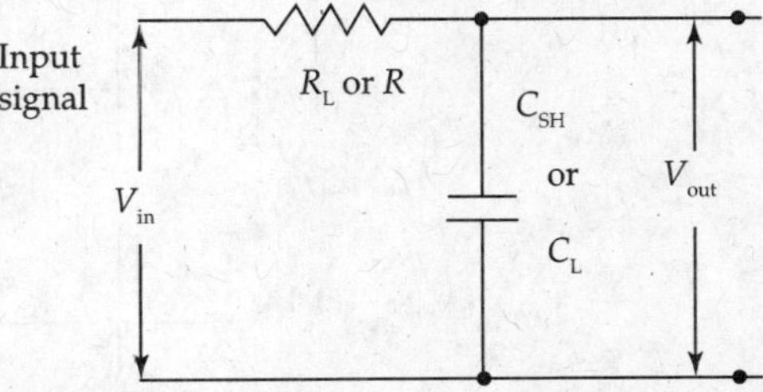

FIG. 9.39 *Low pass circuit containing shunt capacitor C_L or C_{SH} reducing the gain at beyond f_2 (high-frequency cut-off point) in the high-frequency region of the frequency response of the amplifier*

Shunt capacitance 'C_{SH}' consists of junction capacitances of active devices (BJT or FET) and stray wiring capacitances that add together to determine the fall in gain (attenuation) at high-frequency region of frequency response of the Amplifier (Fig. 9.34).

$$V_{out} = \frac{V_{in} \times \left(\frac{1}{J\omega C_{SH}}\right)}{R + \left(\frac{1}{J\omega C_{SH}}\right)};$$

$$\therefore \text{ Gain at high frequencies } A_H = \frac{V_{out}}{V_{in}} = \frac{1}{1 + J\omega \cdot C_{SH} \cdot R}$$

Assuming the higher cut-off frequency $\omega_2 = 1/C_{SH}R$, so that $f_2 = 1/(2\pi \cdot C_{SH} \cdot R)$, the above expression for gain 'A_H' at high frequencies can be written as

$$A_H = \frac{1}{1 + J\frac{\omega}{\omega_2}} = \frac{1}{1 + J\frac{f}{f_2}}.$$

$$\text{At frequency } f = f_H, \text{ Gain } |A_H| = \frac{1}{\sqrt{1 + \sqrt{\left[\frac{f}{f_2}\right]^2}}} = \frac{1}{\sqrt{2}} \text{ and } \theta_H = \tan^{-1}\left(\frac{f}{f_2}\right).$$

At $f = f_2$, voltage gain A_H reduces to 0.707 times the gain at mid-frequency region. So $|A_H| = 20 \log_{10}(1/\sqrt{2}) = -3$ dB. So, f_2 or f_H is also called as –3 dB point. It also represents the frequency at which the resistance R is equal to the capacitive reactance $X_{CSH} = 1/(2\pi f_2 \cdot C_{SH})$. The frequency range f_1 to f_2 is called as Amplifier Bandwidth.

Fall (attenuation) in gain in the high-frequency region beyond the high-frequency cut-off point f_2 is shown in Fig. 9.40. Signal reduction at f_2 is $-10\log_{10}^2 = -3$ dB $= -3$ dB. So the frequency f_2 is known as –3 dB point. And calculated at f_2, in terms of power reduction, it is also known as half-power point as explained from Eqs. (9.20) and (9.21).

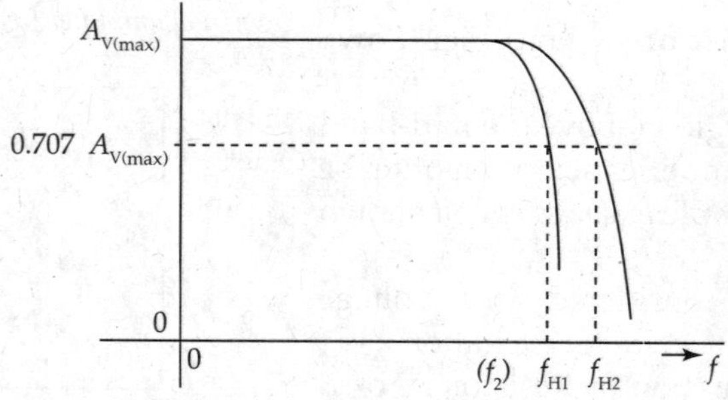

FIG. 9.40 *Reduction in gains at high-frequency regions of the frequency response curve of an amplifier due to C_{SH}*

Total Frequency Response of an Amplifier Total frequency response of an Amplifier shown in Fig. 9.34 is due to combined effects of low pass and high pass circuits (Fig. 9.41).

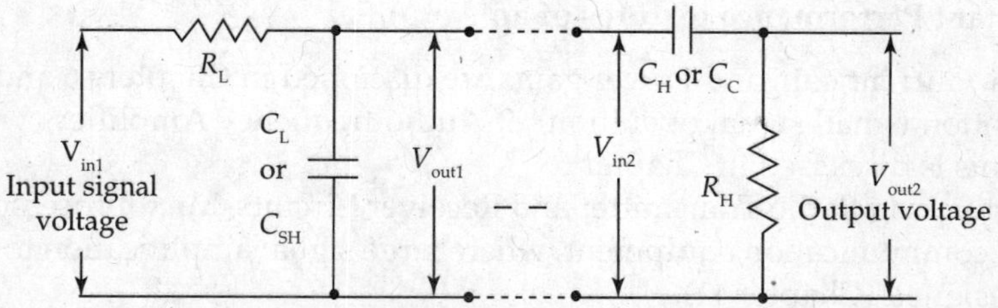

FIG. 9.41 *Total frequency response shown in Fig. 9.34 is due to the combination of series and shunt capacitors (combination of low pass and high pass action of capacitors in amplifier circuit)*

9.8 BASIC CONCEPTS OF MOSFET AMPLIFIERS

Before discussing the concepts of MOSFET amplifiers, some more details of MOSFET devices are considered here.

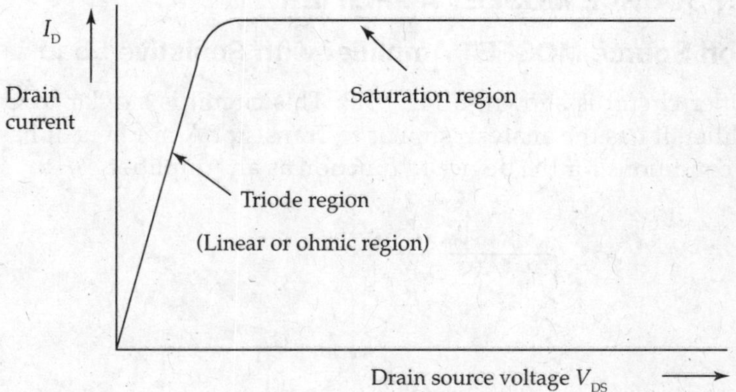

FIG. 9.42 *Output characteristic of MOSFET*

Consider the typical characteristic between Drain–Source voltage V_{DS} and Drain current I_D. When the biasing conditions for MOSFET are such that MOSFET device operates in the linear region of output characteristics, MOSFET works like a resistor. MOSFET is turned ON when the Gate voltage V_G is greater than Threshold voltage (V_{TH}). Then the Drain current (I_D) flow is observed as seen in the output characteristic of the device. Increase in Gate voltage increases or strengthens the channel current between Source and Drain which further increases Drain current. MOSFET can be used as variable resistor based on the magnitude of Gate voltages and Threshold voltage for the device to work in *Triode (linear or ohmic) region*. Transconductance g_m is the important performance parameter of the MOSFET device to indicate the level of control of Gate voltage on the Drain current

$$\text{Transconductance} \quad g_m = \frac{\partial I_D}{\partial V_{GS}} | V_{DS} = \text{Constant}$$

If the bias conditions for MOSFET are in **saturation region** of the output characteristic, MOSFET works like a Current Source in parallel with a resistor. On the total output curve, the MOSFET behaves like a resistor and it is considered as *output resistance* of MOSFET device.

Various important Performance features of an Amplifier

- Voltage gain, Current gain and Power gains are discussed in Chapters 6 and 12.
- Linear operation (small signal operation) of Audio frequency Amplifiers with speech and Music systems is discussed in Chapter 6.
- Power Amplifiers in Radio Transmitter and Receiver Circuits, Microprocessors, Computers and various communication equipment, where large signal amplification to desired power levels are designed (Chapter 11).
- Various design criteria such as maximum supply voltages and Power dissipation levels in the circuits along with signal amplitude operations are considered at various stages in the study of amplifier Chapters 5 and 6 and in the previous sections of this chapter.
- Knowledge of input and output impedance levels of different types of amplifier configurations is important in the design of various interfacing circuits; Cascading and Cascoding amplifier configurations are discussed in multistage Amplifiers.

9.9 COMMON SOURCE MOSFET AMPLIFIER

9.9.1 Common Source MOSFET Amplifier with Resistive Load

MOSFET Amplifier circuit is shown in Fig. 9.43. This circuit is similar to Common Emitter Transistor Amplifier. It has the analysis similar to Transistor Amplifiers. It has both AC signal and DC biasing conditions for the device to function as an Amplifier.

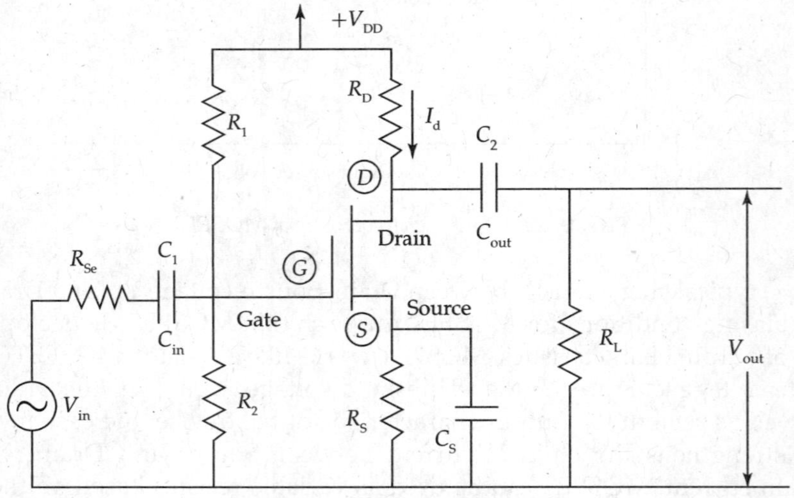

FIG. 9.43 *Common source MOSFET amplifier circuit*

DC equivalent circuit of the MOSFET Amplifier

For DC voltages in the Amplifier circuit, the two capacitors C_{in} and C_{out} behave as open circuits. So they are not considered in the DC equivalent circuit (Fig. 9.44). V_{DD} is the Drain Supply voltage. Resistors R_1, R_2, R_D and R_S fix up the various DC voltages (in association with V_{DD}) for the MOSFET

$$\text{Gate voltage} \quad V_G = V_{DD}\frac{R_2}{R_1 + R_2}$$

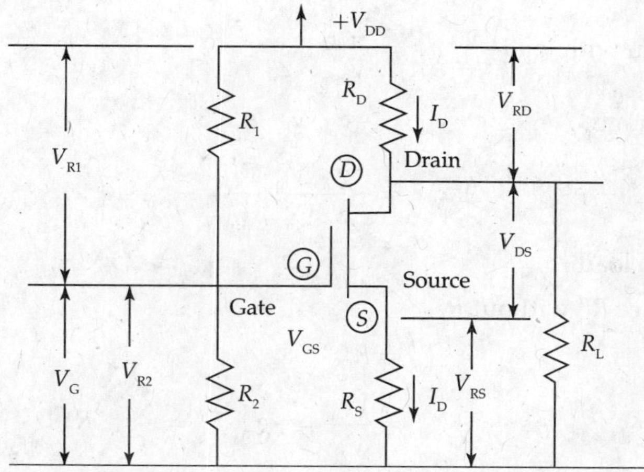

FIG. 9.44 *DC equivalent circuit for CS FET amplifier*

$$\text{Drain current} \quad I_D = \frac{[V_G - V_{GS}]}{R_S}$$

$$\text{Transconductance} \quad g_m = 2K[V_{GS} - V_T], \text{ where } V_{TH} \text{ is the Threshold voltage.}$$

DC Load line equation: $[V_{DD} - V_{DS}] = I_D [R_D + R_S]$. Construction of DC and AC load lines on MOSFET device output characteristics is similar for Transistor Amplifier analysis. *Q*-point for Class-A Amplifier is fixed at middle of DC load line.

The input signal V_{in} superimposes on DC bias V_{GS}. Effective input signal Variations cause variations in Drain current. Output voltage is developed across R_D due to the flow of signal Drain current through it. $V_{out} = I_D R_D$ Volts.

Small Signal Model of Common Source MOSFET Amplifier

Analysis of Enhancement mode common Source MOSFET Amplifier (Fig. 9.45)

Input resistance	$R_{in} = R_G = R_1 \parallel R_2$
Output resistance	$R_0 = R_D$
Total load resistance	$R'_L = R_D \parallel R_L$

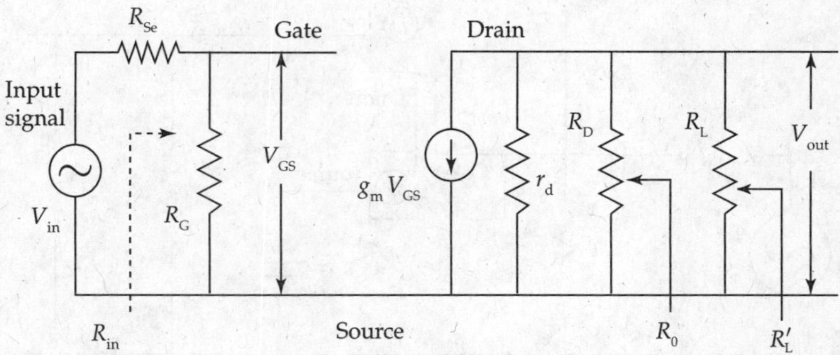

FIG. 9.45 *Small signal AC equivalent circuit for common source MOSFET amplifier*

Voltage gain with load $A_{V(WL)} = \frac{V_{out}}{V_{in}}$

Output voltage $V_{out} = -g_m \cdot V_{in} \cdot R'_L$

$$\therefore \quad A_{V(WL)} = \frac{-g_m \cdot V_{in} \cdot R'_L}{V_{in}} = -g_m \cdot R'_L \tag{9.22}$$

Voltage gain with no load $A_{V(NL)}$

Effective load resistace R''_L without R_L

$$R''_L = R_D$$

$$A_{V(NL)} = \frac{V_{out}}{V_{in}} = -g_m \cdot R''_L = -g_m \cdot R_D \tag{9.23}$$

$$\text{Current gain} \quad A_I = \frac{I_0}{I_{in}} = \frac{R_{in}}{R_L} \cdot A_V(\text{with load})$$

$$\text{Power gain} \quad A_P = \frac{P_{out}}{P_{in}} = A_V(\text{with load}) \times A_I \tag{9.24}$$

Common Source MOSFET Amplifier with Diode as load resistance R_D

In VLSI technology, manufacture of resistors to the desired values and sizes is a tough process. So, Diode connected NMOS transistor is used to simulate the resistance load functions in Amplifier circuits and in general in electronic circuit design Gate voltage and V_{DD} are designed such that Drain current I_D at Q-point is in Triode region of MOSFET output characteristics. Then the design can be made for the device to work with specified load resistance R_D. Analysis and Performance are similar in all respects to CS amplifier with Resistive Load.

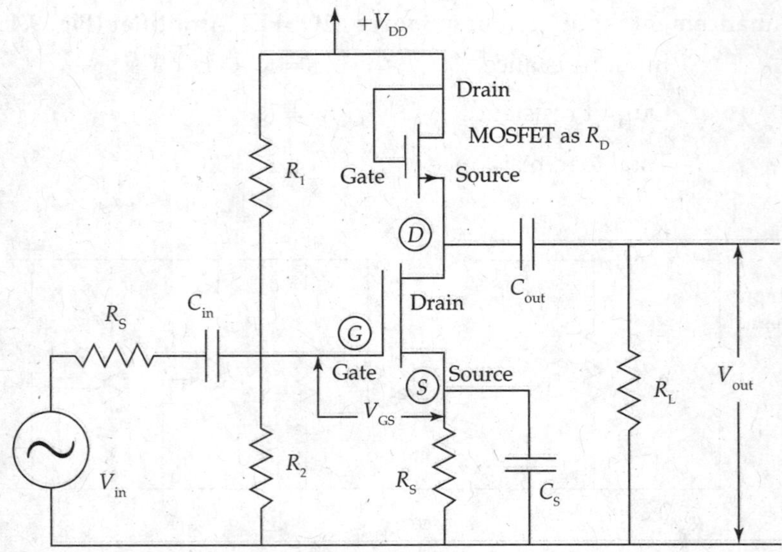

FIG. 9.46 *Common source MOSFET amplifier circuit with diode connected (MOSFET) as load resistor R_D*

Common Source MOSFET Amplifier with Current Source as load resistance R_D

For Amplifiers, which need large values of Voltage gain, increase in load resistance has limitations due to reduction in Drain to source voltage of MOSFET. Gate voltage and V_{DD} are designed such that the Drain current for MOSFET is in Saturation region at desired point to simulate *current Source with parallel resistance combination*. It ultimately works as desired Load resistance to realise required values of voltage Gains from the Amplifiers.

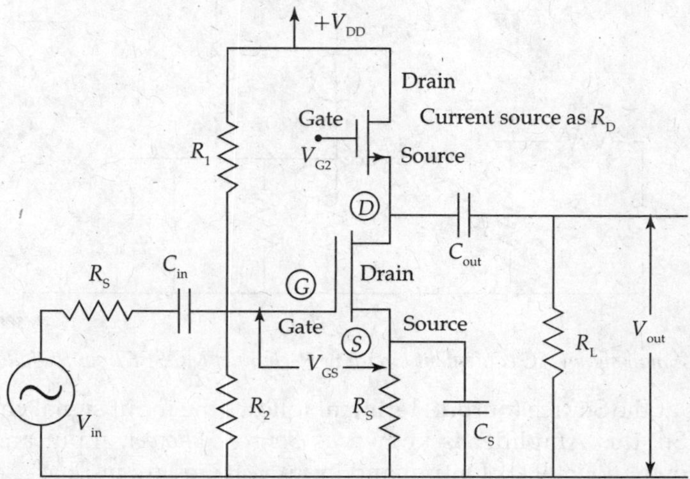

FIG. 9.47 *Common source MOSFET amplifier circuit with current source (MOSFET) as load resistor R_D*

9.10 SOURCE FOLLOWER USING MOSFET

(Common Drain MOSFET Amplifier)

When the Drain terminal is common to input and the output ports of MOSFET Amplifier, the circuit is 'Common-Drain MOSFET Amplifier' ('Source Follower'). This configuration is similar to Emitter follower or common-Drain FET Amplifier. Source follower acts as an impedance matching network and buffer amplifier.

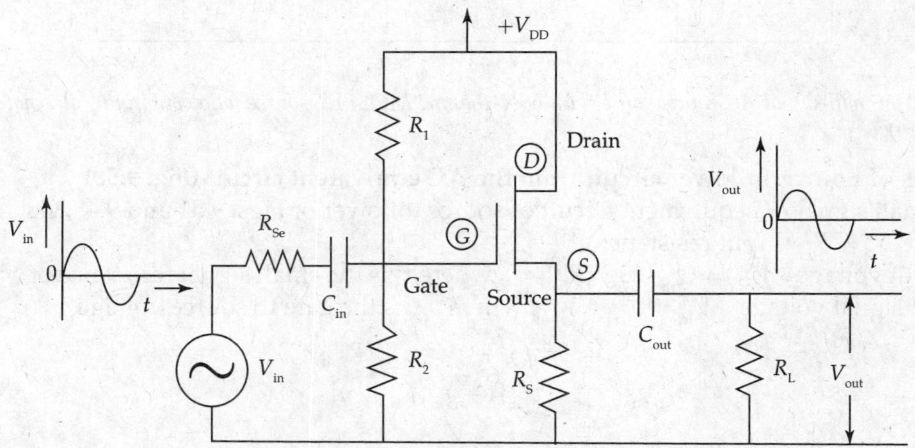

FIG. 9.48 *Source follower circuit using MOSFET*

In Source follower circuit, output voltage is taken off at the Source wrt ground (Common Drain terminal of MOSFET). Output signal is developed by the flow of Drain current I_D through R_S (load resistor). Drain terminal is directly connected to V_{DD}.

From the small signal AC equivalent circuit (Fig. 9.49), it is known that V_{DD} becomes signal ground. So, the circuit is known as a 'Common Drain Amplifier'.

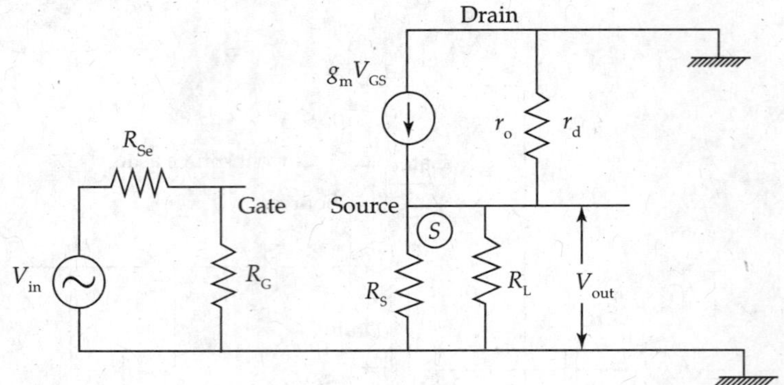

FIG. 9.49 *Small signal AC equivalent circuit of N-channel MOSFET source follower circuit*

Output voltage at the Source (output) terminal follows the input signal changes at the Gate (input) terminal. So, this Amplifier is known as *Source Follower.* Input and output voltage waveforms are shown in Fig. 9.48. Output and input voltages are in-phase.

Alternate small signal AC equivalent circuit for Source follower using MOSFET (Fig. 9.50)

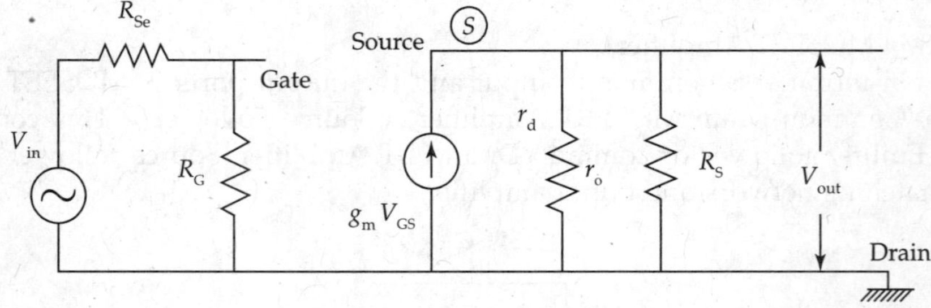

FIG. 9.50 *Small signal AC equivalent circuit of N-channel MOSFET source follower circuit (alternate circuit to Fig. 9.49)*

Analysis of Source follower circuit from the AC equivalent circuit (Fig. 9.50)
From small signal AC equivalent circuit of source follower of Figs. 9.51 and 9.48, Gate resistor $R_G = R_1 \parallel R_2 = R_{in}$ (Input resistance)

Output voltage $V_{out} = (g_m \cdot v_{GS})\,(R_S \parallel r_o)$, where r_o is the small signal resistance of MOSFET.

Input signal voltage $V_{in} = v_{GS} + V_{out}$, where v_{GS} is the gate to source voltage.

$$V_{in} = v_{GS} + (g_m \cdot v_{GS})(R_S \parallel r_o)$$

$$\therefore \quad V_{in} = v_{GS}[1 + g_m (R_S \parallel r_o)]$$

$$\therefore \quad v_{GS} = \frac{V_{in}}{[1 + g_m (R_S \parallel r_o)]}$$

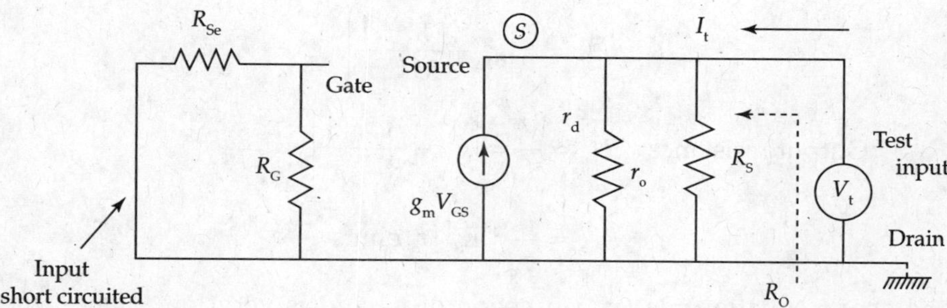

FIG. 9.51 *Small signal AC equivalent circuit of N-channel MOSFET source follower circuit to determine output resistance R_{out}*

Dividing both numerator and denominator of RHS by g_m

$$v_{GS} = \frac{V_{in} / g_m}{\left[\frac{1}{g_m} + R_S \parallel r_o\right]}$$

$$\therefore \quad V_{in} = g_m \cdot v_{GS}\left[\left(\frac{1}{g_m} + R_S \parallel r_o\right)\right]$$

$$\text{Voltage gain} \quad A_V = \frac{V_{out}}{V_{in}} = \frac{g_m \cdot v_{GS}\,[R_S \parallel r_o]}{g_m \cdot v_{GS}\left[\left(\frac{1}{g_m} + R_S \parallel r_o\right)\right]} \cong 1 \qquad (9.25)$$

From Eq. (9.25), voltage gain is slightly less than unity.

As the voltage gain is positive, output and input signals are in phase. Hence the Amplifier circuit configuration acts as a 'source-follower'.

As the Transconductance g_m of a BJT is larger than that of a MOSFET, voltage gain of Emitter follower is closer to unity than that of MOSFET Source follower.

Input resistance R_{in}

$R_{in} = R_G = R_1 \parallel R_2 =$ Thevinin equivalent of bias resistors in circuit (Fig. 9.51).

Output resistance R_{out}

To calculate R_{out}, all independent sources are set to zero and a test voltage V_t is applied at output terminals. This simplified circuit is in Fig. 9.51.

Output resistance $R_{out} = V_t/I_t$ from circuit of Fig. 9.51.

$$\text{Using KCL at the output circuit} \quad I_t + g_m \cdot v_{GS} = \frac{V_t}{R_S} + \frac{V_t}{r_0}.$$

$[I_t - V_t \cdot g_m] = \dfrac{V_t}{R_S} + \dfrac{V_t}{r_o}$ using ($v_{GS} = -V_t$) in the above equation.

$$\therefore \quad I_t = V_t \left[g_m + \frac{1}{R_S} + \frac{1}{r_o} \right]$$

$$\therefore \text{ Output resistance } \quad R_{out} = \frac{V_t}{I_t} = \left[g_m + \frac{1}{R_S} + \frac{1}{r_o} \right]$$

$$\therefore \quad R_{out} = \frac{1}{g_m} \parallel R_S \parallel r_o \text{ ohms}$$

From the above equation, R_{out} is a function of g_m and is very low. Hence, the Source follower behaves as an ideal voltage source. It can drive another circuit without loading effect.

Current Gain Simplified Norton equivalent circuit (Fig. 9.52) to determine current gain A_i.

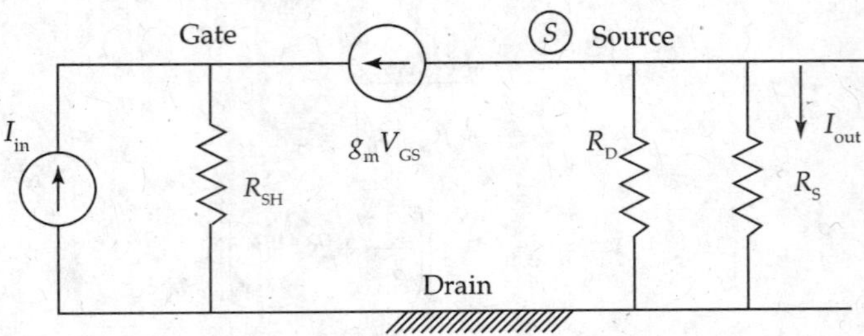

FIG. 9.52 *Small signal AC equivalent circuit of N-channel MOSFET source follower circuit to determine current gain*

From the equivalent circuit in Fig. 9.52,

$$\text{Output current} \quad I_{out} = -g_m \cdot v_{GS} \cdot \frac{R_D}{(R_D + R_L)}$$

$$I_{in} + g_m \cdot v_{GS} + \frac{v_{GS}}{R_S} = 0$$

$$\therefore \quad v_{GS} \left[g_m + \frac{1}{R_S} \right] = -I_{in}$$

$$I_{in} = v_{GS} \left[\left(\frac{g_m \cdot R_S + 1}{R_S} \right) \right]$$

$$\text{Current gain} \quad A_i = \frac{I_{out}}{I_{in}} = g_m \left(\frac{R_D}{R_D + R_L} \right) \left(\frac{R_S}{1 + g_m \cdot R_S} \right) \cong 1.$$

9.11 COMMON GATE MOSFET AMPLIFIER

Analysis for Common Gate MOSFET Amplifier is similar to the previous analysis of JFET CG Amplifier. Likewise, it is a non-inverting Amplifier with relatively low output resistance. CG MOSFET Amplifier is shown in Fig. 9.53. The DC operating conditions can be designed similar to Common-Source FET Amplifier.

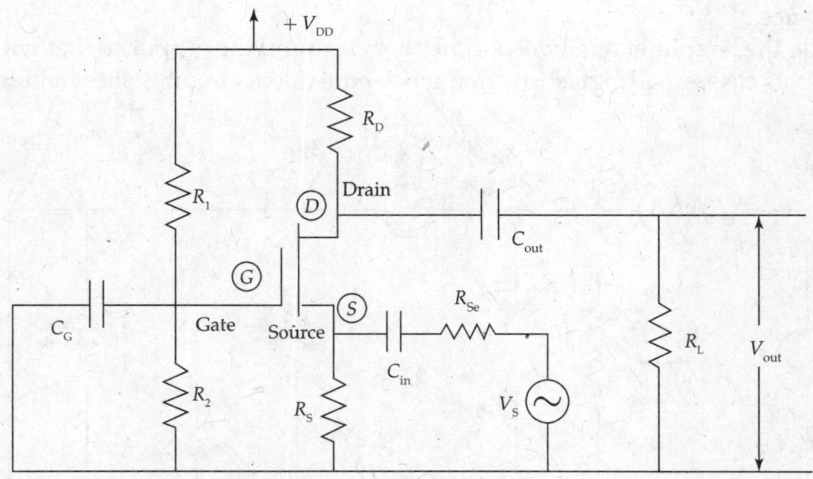

FIG. 9.53 *Common gate MOSFET amplifier circuit*

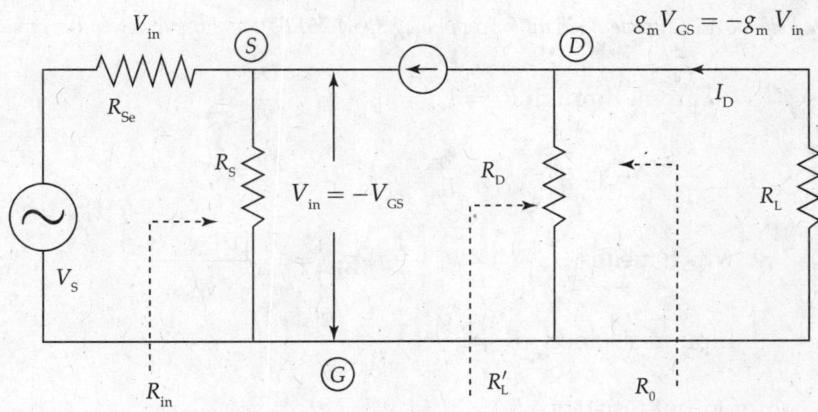

FIG. 9.54 *Small signal AC equivalent circuit for common gate MOSFET amplifier*

Voltage gain with load A_V(WL)

Effective load reistance $R_L' = R_D \parallel R_L$

Drain current $I_D = g_m \cdot V_{GS}$

Input voltage $V_{in} = -V_{GS}$

Output voltage with load $V_{out} = -I_D \cdot R_L' = -g_m \cdot V_{GS} \cdot R_L'$

Voltage gain with load $A_V = \dfrac{V_{out}}{V_{in}} = \dfrac{-g_m \cdot V_{GS} \cdot R_L'}{-V_{GS}} = g_m \cdot R_L'$

No load voltage gain $A_V(\text{NL}) = \dfrac{V_{out}}{V_{in}} = g_m \cdot R_D$

(when there is no load resistance $R_L = \infty$ ohms. Then $R_L' = R_D$.

Input Resistance

By considering the Amplifier as a two-port network, ground the output so that output voltage $V_{out} = 0$ V. In this case, small signal low frequency equivalent circuit is shown in Fig. 9.55

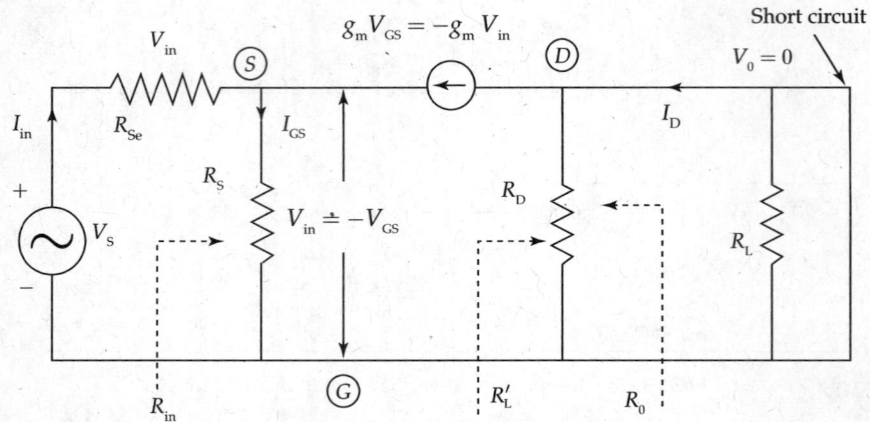

FIG. 9.55 *Small signal AC equivalent circuit for common gate MOSFET amplifier to determine input resistance R_{in}*

$$\text{Current through } R_S = I_{RS} = I_{in} + I_d = \frac{V_{in}}{R_S}$$

$$\therefore \quad I_{in} = \frac{V_{in}}{R_S} - I_d = \frac{V_{in}}{R_S} - g_m \cdot v_{GS} = \frac{V_{in}}{R_S} + g_m \cdot V_{in}$$

$$\text{which means} \qquad I_{in} = V_{in}\left[\frac{1}{R_S} + g_m\right] = V_{in}\frac{[1 + g_m \cdot R_S]}{R_S}$$

$$\text{Input resistance} \quad R_{in} = \frac{V_{in}}{I_{in}} = \frac{R_S}{[1 + g_m \cdot R_S]}$$

$$\text{Output resistance } R_o = r_o \,||\, R_D = R_D.$$

Common gate MOSFET has very low input resistance R_{in} and large Output resistance R_{out}

As the output voltage for both CS and CG Amplifiers are taken at Drain terminal, the output resistance for CG Amplifier will be equal to the output resistance for CS Amplifier.

Advantages: Desirable high frequency response.

Disadvantages: Relatively low input resistance

Output resistance $R_{out} = r_o \,||\, R_D \cong R_D$

Output resistance will be very large

$$\text{Current gain} \quad A_i = \frac{I_{out}}{I_{in}} = A_{V(WL)} \times \frac{R_{in}}{R_L}$$

Common gate amplifier has got no current gain

$$\text{Power gain} \quad A_P = \frac{P_{out}}{P_{in}} = A_i \cdot A_{V(WL)}$$

$$\text{Output voltage with load } V_{O(WL)} = A_{V(WL)} \cdot V_{in}$$

$$\text{where } V_{in} = V_S \frac{R_{in}}{R_{SE} + R_{in}}$$

Comparison of Basic MOSFET Amplifiers

S. No	Parameter	CS amplifier	Source follower	CG amplifier
1	Voltage gain	$A_V > 1$	$A_V \cong 1$	$A_V > 1$
2	Current gain	–	–	$A_i > 1$
3	Input resistance	R_{TH}	R_{TH}	Low
4	Output resistance	Moderate to high	Low	Moderately high

SUMMARY

1. In all types of Amplifiers, the input coupling capacitor C_{in} couples the AC input signal and at the same time blocks DC voltage that may be present in the signal Source so that it will not effect the DC operating conditions of the Amplifiers.
2. Input signal gets superimposed on DC biasing voltages in the input circuit causing variations in DC biasing voltages. Resulting increased values of output currents (for example Drain current for FET) of Amplifiers. Output currents flowing through load resistances develop output voltages that are used in various circuits.
3. If voltage gains are observed for different values of input signal frequencies and plotted on a Semi-log or a linear graph paper the resulting characteristic of the Amplifier is known as the *Frequency response of the Amplifier*.
4. Frequency response characteristic of the Amplifier is divided into three operating regions: (1) low-frequency region, (2) mid-frequency region and (3) high-frequency region.
5. Once the low-frequency and high-frequency cut-off points are known from the frequency response characteristic, Bandwidth 'BW' is the difference in High-frequency cut-off point and Low-frequency cut-off point.
6. Frequency responses of various types of Amplifiers reflect in the specifications of Amplifier (System) Bandwidths.
7. Power capabilities of various types of Amplifiers are analysed in the chapter of "Large signal" or power Amplifiers.
8. Frequency responses and power levels of Amplifiers decide their practical utility.

Questions for Practice

1. What is square law distortion? What is its effect in FET Amplifiers? Compare the important characteristics of CS, CG and CD FET Amplifiers. (JNTU, May 2005)

2. Draw the circuit diagram and low frequency equivalent circuit of common Source FET Amplifier. Derive the expression for voltage gain. (JNTU, Feb. 2008)
3. Sketch common Source Amplifier using JFET and draw its equivalent circuit. Derive the expression for voltage gain. (May 2004)
4. Derive the expression for voltage gain of CD FET Amplifier. (May 2004)
5. Write short notes on bandwidth of an Amplifier. (May 2005)
6. Draw the circuit diagram of low frequency equivalent circuit of Common Source FET Amplifier and derive the expression for voltage gain. (JNTU, Feb. 2008)
7. Draw the circuit of a single-stage FET Amplifier. Draw the mid-frequency equivalent circuit. Derive the expression for voltage gain. Data: $g_m = 2$ millimhos; $r_d = 200$ kΩ; $R_D = 5$ kΩ. R_L is very large and does not affect the load. Calculate voltage gain.
8. Draw the circuit of CS FET Amplifier with Gate resistor $R_G = 1$ mΩ, source resistor $R_S = 0.5$ kΩ, load resistance $R_D = 1.5$ kΩ and $V_{CC} = 20$ V. Pinch-off voltage 'V_P' of selected FET device is equal to −3.6 V. Design the Gate to Source bias voltage to be half the pinch-off voltage for Class-A Amplifier circuit. Calculate the magnitudes of Source voltage and Drain voltages. Calculate input resistance when the current through Gate resistor 'R_G' is zero.
9. Consider common Source FET Amplifier circuit with potential divider bias circuit. Draw the waveforms at salient points in the circuit and discuss the feature of 180° phase shift between the input and output signal voltages.
10. Draw the circuit of a single-stage FET Amplifier. Draw the mid-frequency equivalent circuit and derive the expression for voltage gain. Data: $g_m = 2$ milli mhos; $r_d = 200$ kΩ; $R_D = 5$ kΩ. R_L is very large and does not affect the load. Calculate voltage gain.
11. Draw the CG FET Amplifier circuit. Show the waveforms at different points in the circuit with qualitative explanation. Derive the expressions for A_V, Z_{in} and Z_{out}.
12. Draw the CD Drain FET Amplifier circuit. Show the waveforms at different points in the circuit with qualitative explanation. Derive the expressions for A_V, Z_{in} and Z_{out}.
13. Discuss the reasons for the absence of Thermal runaway in FET devices.
14. Draw the mid-frequency equivalent circuit of a FET Amplifier circuit and derive the expression for voltage gain mentioning the reasons for not including the effect of various capacitances in the equivalent circuit?
15. For Source Follower circuits discuss the various features in detail that made it suitable as 'Unity Gain Buffer Amplifier Circuits'. Mention a few practical applications.

Multiple Choice Questions

1. The FET configuration used in the unity gain buffers is ____________.
   (a) CG (b) CD (c) CS (d) none of the above
2. FETs are seldom used in small signal Amplifiers because ____________.
   (a) they have lower voltage gain
   (b) they have lesser voltage gain
   (c) they have a wider spread in device characteristics
   (d) none of the above

3. FETs are having faster on and off times due to ________________
   (a) absence of barrier potential
   (b) presence of threshold voltage
   (c) larger interelectrode capacitances
   (d) absence of storage charges as they are unipolar

4. The most popular configuration in measuring instruments like voltmeters and oscilloscopes having a high input impedance is ________________.
   (a) CS (b) CD (c) CG (d) cascode

5. FETs are used in mixers of TV and FM receivers for the reason ________________.
   (a) they have got no drift (b) they have low noise margins
   (c) they have square law distortion (d) they have high voltage gains

6. In the Norton small signal equivalent of FET Amplifier ________________.
   (a) gate to source connection is open circuited
   (b) a capacitor is connected between Gate and Drain
   (c) voltage source is shown
   (d) drain resistance is shown in shunt with a current Source between Gate and Drain

7. In Thevinin small signal equivalent circuit for a FET Amplifier ________________.
   (a) gate and source are open circuited
   (b) presence of current source
   (c) a capacitor connected between Gate and Drain
   (d) the drain resistance shown in series with voltage Source between Drain and source

8. The current buffer having current gain less than unity is ________________.
   (a) CS (b) CG (c) CD (d) cascode

9. The most frequently used FET stage having voltage and current gain is ________________.
   (a) CS (b) CG (c) CD (d) cascode

10. A JFET device can operate on ________________.
    (a) enhancement mode (b) depletion mode
    (c) both depletion and enhancement modes (d) none of the above

11. N-channel FETs are used in practical circuits, when compared to P-channel FETS because ________________.
    (a) mobility of electrons is greater than that of holes
    (b) better switching times
    (c) higher input impedance
    (d) high gain

12. N-channel MOSFET devices are better than P-channel MOSFETS because ________________.
    (a) TTL compatibility (b) noise immunity
    (c) operate faster (d) easy operation

Answers to Multiple-Choice Questions

1. (b)	2. (b & c)	3. (c)	4. (b)	5. (c)	6. (a & d)
7. (a & d)	8. (b)	9. (b)	10. (a)	11. (b)	12. (c)

Chapter 10

MULTISTAGE (CASCADED) AMPLIFIERS

Learning Objectives

Many performance features of practical amplifiers may not be met with *single-stage Amplifiers*. Considering the various aspects of single-stage amplifiers, the following aspects of *Multistage (Cascaded) Amplifier* are discussed:

- *Multistage (Cascaded) Amplifier* gain and Bandwidth
- Advantages of different coupling schemes of Multistage Amplifiers
- Various schemes of *CASCODE Amplifiers*
- Cascading amplifier design to obtain 'Darlington Pair circuits'
- Concepts and design aspects of *differential amplifiers.*

10.1 CONCEPTS OF CASCADED (MULTISTAGE) AMPLIFIERS

10.1.1 Introduction

The gain of a *single-stage Amplifier* may not be adequate for certain applications and the input and output impedances also may not be of required magnitudes. Such limitations of a single-stage Amplifier necessitated the development of *Cascaded or Multistage Amplifiers.*

In Multistage Amplifiers, multiple amplifiers are interconnected to obtain desired performance features for the total Amplifier. In this process, the output of stage (amplifier) 1 is connected to the input of stage (amplifier) 2, whose output voltage is applied to the following next (third) stage and so on, till the overall performance of the resulting amplifier meets the desired performance requirements. Such method of interconnecting more than one Amplifiers in stages is known as *Cascading of Amplifiers*. The resulting overall Amplifier is known as *Multistage or Cascaded Amplifier.*

While cascading, the first (input) stage has to provide high input impedance when fed from a Source. The intermediate stages are intended to provide the necessary voltage gain. Final stage has to provide a low output resistance to avoid losses when the load is of low impedance. In multistage Amplifiers, the primary characteristics of interest are large amplification (gain), high input impedance, low output impedance and improved frequency response.

Concept of multistage (cascaded) Amplifiers using a block diagram (Fig. 10.1)
In Fig. 10.1, output voltage of the first Amplifier in stage 1 is fed as input to the second Amplifier at stage 2. Output voltage of the second Amplifier is fed to the load resistance R_L in the two-stage Amplifier. Sometimes the output voltage of stage 2 Amplifier may be connected to stage 3 Amplifier and so on until the required characteristics of signal amplitude, output and input resistances and the frequency responses are achieved.

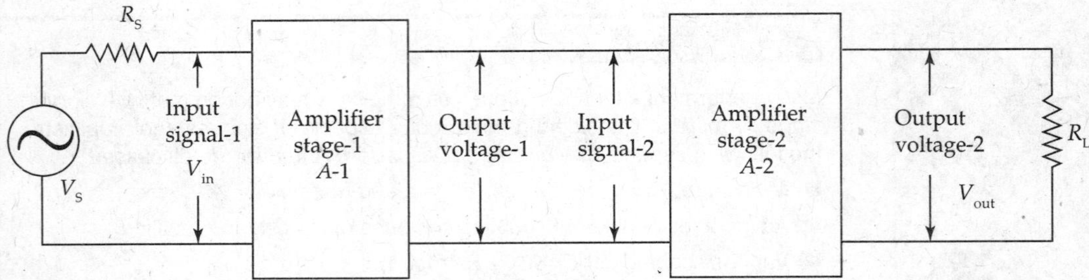

FIG. 10.1 *Two amplifier stages connected in cascade to form a multistage amplifier*

- Fundamental building blocks of electronic circuits in modern VLSI technology are Transistors. Advanced technologies such as carbon nano tubes are used for the semi-conductor channel in MOSFET Transistors, to further increase the density of Transistors in IC fabrication, satisfying Moore's law.
- The latest technologies and engineering marvels of the 21st century further reduce the size of electronic gadgets for the use of citizens in this electronic government age to lead comfortable life.
- Multistage Amplifiers are used in a variety of communication systems such as optical, satellite, and cellular and mobile communications.
- At the same time, certain constraints – such as (i) inherent noise generation in active devices, (ii) bandwidth considerations and (iii) stability of Amplifier operations – restrict the maximum gains available with each amplifier stage and the *maximum number of amplifier stages from seven to nine.*

10.1.2 Classification of Multistage Amplifiers

Multistage Amplifiers can be classified in different ways:

1. Based on the type of active device (Transistors) used in each amplifier stage and
2. Based on the type of coupling between the amplifier stages.

Classification based on the active device in each amplifier stage

There are nine types of possible combinations of connecting two BJT amplifier stages in cascade. Out of which, six combinations are used in cascading taking into the characteristics of individual stages. Out of the six combinations, the following four configurations are more popular.

1. CE Transistor Amplifier + CE Transistor Amplifier configurations
2. CE Transistor Amplifier + CC Transistor Amplifier configurations
3. CE Transistor Amplifier + CB Transistor Amplifier configurations (Cascade)
4. CC Transistor Amplifier + CC Transistor Amplifier configurations.

In addition to the above configurations, compound devices such as Darlington pair can also be used in multistage Amplifiers.

Similar to the various classifications used in BJTs, FET stages can also be classified as

1. CS FET Amplifier + CS FET Amplifier configurations
2. CS FET Amplifier + CD FET Amplifier configurations
3. CS FET Amplifier + CG FET Amplifier configurations (Cascode)
4. CD FET Amplifier + CD FET Amplifier configurations.

10.2 DIFFERENT COUPLING SCHEMES USED IN AMPLIFIERS

The process of transferring signal energy between Amplifier circuits is known as coupling between amplifier stages. Classification of multistage Amplifiers is decided from the type of inter-stage coupling component between amplifier stages:

1. Resistance capacitance (RC)-coupled Amplifier,
2. Transformer-coupled Amplifier and
3. Direct-coupled Amplifier.

10.2.1 Resistance Capacitance Coupled Amplifier (RCC Amplifier)

Figure 10.2 shows a typical two-stage resistance capacitance coupled Amplifier. RCC Amplifier is one of the more important circuits, which is popularly used. It has uniform gain over a wide range of frequencies. It is used in Audio and Video Amplifiers.

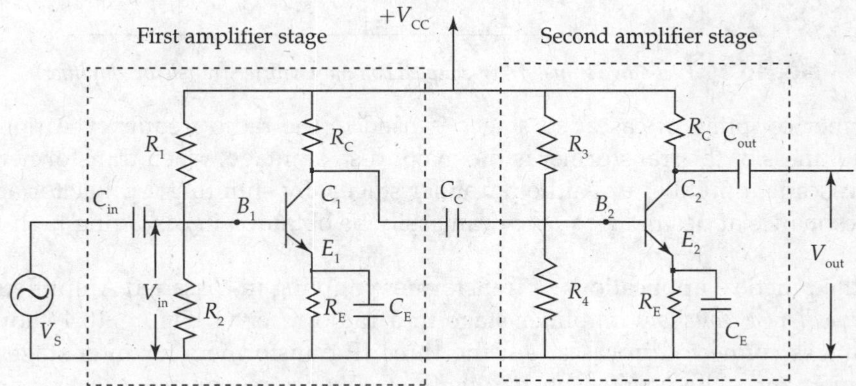

FIG. 10.2 *Two stage resistance capacitance coupled amplifier (example for a cascaded connection of amplifiers) multi stage amplifier*

Coupling between two stages is capacitive coupling. A signal developed in one stage is coupled to the next stage using capacitors to get improved overall gain equal to the cascaded amplifiers. The coupling allows only AC component of the signal to pass from one stage to the next stage, while the individual stages are isolated in respect of DC. The drawback of this coupling is limitation of low-frequency response of overall Amplifier.

10.2.2 Transformer-coupled Amplifier (Fig. 10.3)

When the coupling element between two amplifier stages to be cascaded is a transformer, the cascaded Amplifier is known as 'Transformer-coupled Amplifier'.

The following are the three functions of the transformer in the Amplifier circuit:

1. The transformer couples or transfers the AC output voltage (energy) of one amplifier stage to the input stage of the following Amplifier.
2. The transformer isolates the DC conditions of one amplifier stage to the following stage so that the DC biasing conditions for the active device are not disturbed.
3. Impedance matching between the output impedance of one amplifier stage and the input impedance of the following connecting amplifier stage for achieving maximum power transfer conditions.

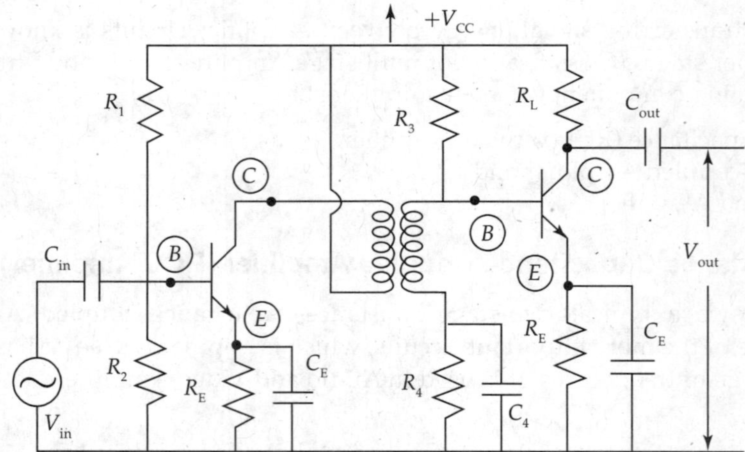

FIG. 10.3 *Two-stage transformer-coupled common emitter transistor amplifier*

Transformer coupling of cascaded stages is used in the radio frequency Amplifiers. The cost and bulkiness of the transformer is the major disadvantage, when transformer coupling is used to cascade a number of Audio amplifier stages. By shunting a capacitor across each winding, resonance at desired frequency can easily be obtained in cascading high-frequency Amplifiers.

One of the practical applications of transformer coupling in cascaded Amplifiers is found in IF (*Intermediate Frequency*) amplifier stage in a radio receiver. Figure 10.4 shows two IF amplifier stages connected in cascade using three IF transformers for inter-stage coupling. IF transformers serve the purpose of shaping the frequency response of the Amplifier and in association with capacitors they form tuned circuits to improve the selectivity of the Amplifier to the required values.

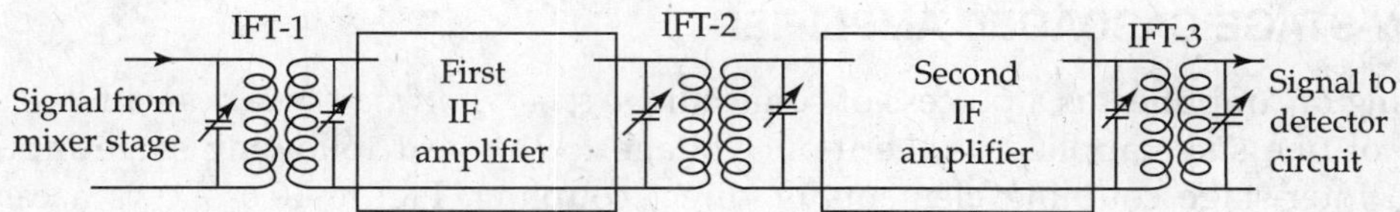

FIG. 10.4 *Transformer coupling in IF amplifier stage using IF transformers in a radio receiver*

10.2.3 Direct-coupled Amplifier

The direct-coupled Amplifier circuit is shown in Fig. 10.5. It is also known as DC Amplifier circuit. In this Amplifier, output voltage of one amplifier stage is directly connected to the input of the following amplifier stage without using any reactive elements such as inductors or capacitors.

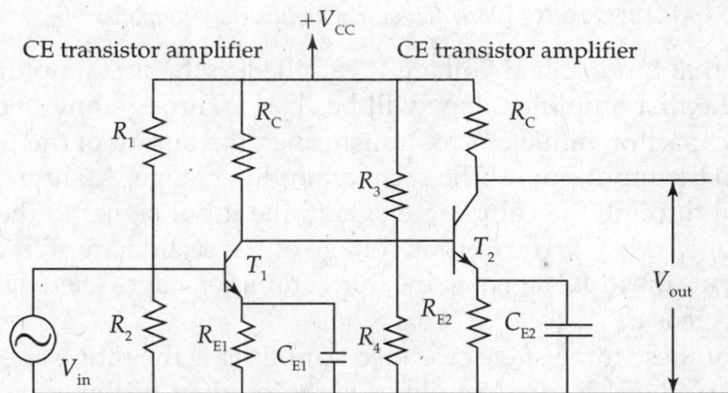

FIG. 10.5 *Cascaded amplifier using direct coupling direct coupled amplifier*

No reactive elements are involved in inter-stage coupling. So, it can be considered as a broadband Amplifier with flat response starting from 0 Hz. But the junction capacitances of semiconductor devices such as Transistors, FETs and MOSFETs affect the high frequency response of DC Amplifiers.

The DC-coupled Amplifier allows the DC component from the Source and amplifies signals ground to zero frequency. It even amplifies very slow changes in the input voltage. It is more popular in biomedical electronic instrumentation, Siesmo-graphic instruments. This has got a severe disadvantage like unwanted drift and it even responds to changes in supply voltages and Transistor parameter variations due to temperature. The design of direct-coupled multistage Amplifier is complicated.

Due to the absence of inductors and capacitors, direct-coupled cascaded Amplifiers are used in ICs and operational Amplifiers.

Advantages

1. Simple circuit without any inter-stage coupling elements.
2. DC Amplifier circuits are compact as no capacitors are required.
3. Amplifies DC and low-frequency signals with its flat frequency response starting from 0 Hz.
4. Fabrication of DC Amplifiers in IC form is compact and simple.
5. Reduction in circuit elements reduces the Amplifier construction cost.

10.3 *N*-STAGE CASCADED AMPLIFIER

Cascading of Amplifiers is a process of connecting *a set of Amplifiers in series* with the output voltage of one stage applied as the input voltage to the next following stage input port through inter-stage coupling elements or direct coupling. Figure 10.6 shows a cascaded Amplifier with three stages of Amplifiers having gains A_1, A_2 and A_3.

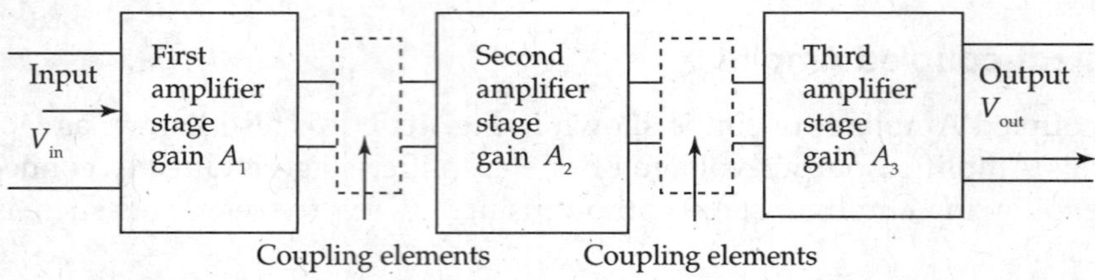

FIG. 10.6 *Block diagram of a multistage amplifier*

Consider an external input signal voltage V_{in} applied to the first amplifier stage. Then the output voltage of the first amplifier stage will be $A_1 \cdot V_{in}$. Through any one type of coupling elements such as a capacitor, inductor or a transformer, the output of the first amplifier stage will be connected to the input port of the second amplifier stage. Assuming no loss of signal during transmission through the coupling element, the input signal to the input port of the second amplifier stage is $A_1 \cdot V_{in}$. The output voltage of the second amplifier stage is $A_1 \cdot A_2 \cdot V_{in}$. This signal again forms the input signal to the third amplifier stage. Then the output signal V_{out} of third amplifier stage is $A_1 \cdot A_2 \cdot A_3 \cdot V_{in}$.

Now the gain A of these three-stage cascaded Amplifier is the ratio of V_{out} to V_{in}.

The total gain A of the cascaded Amplifier is the product of the gains of the individual stages, which is $A_1 \cdot A_2 \cdot A_3$.

Gain calculations of *N*-stage cascaded Amplifier (Fig. 10.7)

The concept of cascading of Amplifiers can thus be generalised to *N*-amplifier stages.

Then the total or overall or effective gain A of the N number of amplifier stages in the cascaded system is

$$A_E = A_1 \cdot A_2 \cdot A_3 \cdot A_4 \ldots A_N \tag{10.1}$$

Thus the overall voltage gain of a multistage Amplifier is the product of the gains of individual stages of Amplifiers. Overall gain of the multistage Amplifier is much larger than the individual stage gains at the expense of bandwidth of the overall Amplifier. (However, the gain of each stage is to be determined under loaded conditions. The inter-stage loading of each stage must be considered in Transistor Amplifiers.)

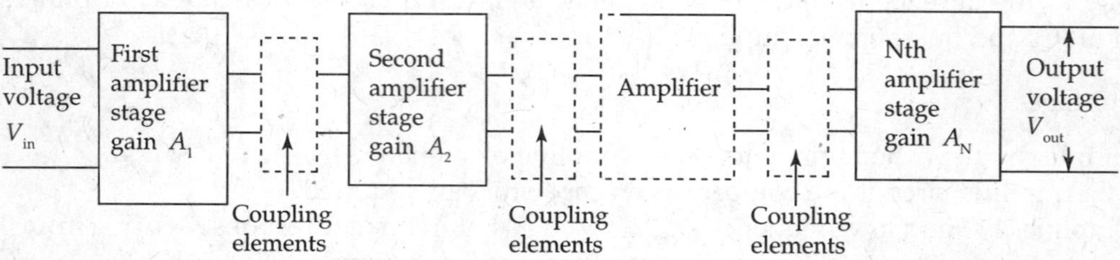

FIG. 10.7 *Block diagram of a multistage amplifier (N-stage amplifier)*

The resultant phase shift between input and output voltages of the multistage Amplifier is the sum of the individual Amplifier voltage phase shifts. If $\theta_1, \theta_2, \ldots, \theta_n$ are the phase shifts among the input and output voltages of each stage, then the phase shift θ between the output voltage of the last stage Amplifier and the input voltage of the first stage Amplifier of the cascaded Amplifier is the sum of the individual Amplifier signal phase shifts. It can be represented as

$$\theta = \theta_1 + \theta_2 + \theta_3 + \cdots + \theta_n. \tag{10.2}$$

Gain of N-stage cascaded amplifier calculated in decibels is

$$20 \log_{10} A = 20 \cdot \log_{10} A_1 \cdot A_2 \cdot A_3 \cdot A_4 \ldots A_N, \tag{10.3}$$

which result in

$$20 \log_{10} A = 20 \log_{10} A_1 + 20 \log_{10} A_2 + 20 \log_{10} A_3 + \cdots + 30 \log_{10} A_N. \tag{10.4}$$

Equation (10.4) shows that the gain A expressed in decibels (dB) of a multistage Amplifier is the sum of the individual Amplifier gains calculated in decibels (dB). Calculations are simple with voltage gains expressed in decibel units. This concept is explained in the worked out Example 10.4.

From the Amplifier configuration in Fig. 10.6,

- Input resistance R_{in} (total) of the multistage Amplifier is the input resistance R_{in1} of the first amplifier stage.

$$R_{in} \text{ (total)} = R_{in1}. \tag{10.5}$$

- Output resistance of the multistage Amplifier R_{out} (total) is equal to the output resistance of the last stage R_{outN} of the multistage Amplifier:

$$R_{out} \text{ (total)} = R_{outN}. \tag{10.6}$$

EXAMPLE 10.1

Three-stage multistage Amplifier has its first stage Amplifier with voltage gain $A_1 = 1000$, second amplifier stage gain with gain $A_2 = 100$ and third amplifier stage with gain $A_3 = 10$. Calculate the voltage gain A_E of cascaded Amplifier in decibels.

Solution: Voltage gain of first amplifier stage in decibels $= 20 \log_{10} A_1 = 20 \log_{10} 1000 = 60$ dB.
Voltage gain of the second amplifier stage in decibels $= 20 \log_{10} A_2 = 20 \log_{10} 100 = 40$ dB.
Voltage gain of the third amplifier stage in decibels $= 20 \log_{10} A_3 = 20 \log_{10} 10 = 20$ dB.
The effective voltage gain A_E of the cascaded Amplifier is the sum of the three individual amplifier stage gains expressed in decibels.

$$\therefore \quad A_E = 60 + 40 + 20 = 120 \text{ dB}.$$

EXAMPLE 10.2

A multistage Amplifier has effective voltage gain A_V of 60 decibels (dB). Calculate the magnitude of voltage gain A_V. If the input signal voltage to the Amplifier is 0.1 V, calculate the magnitude of the output voltage.

Solution: Voltage gain A_V of the Amplifier in decibels $= 20 \log_{10} A_V = 60$ dB.

Therefore, $\log_{10} A_V = 3$. Hence, $A_V = 1000$.

Voltage gain $A_V = \dfrac{V_{out}}{V_{in}} = 1000.$

Output voltage $= V_{out} = 1000;\cdot V_{in} = 1000 \times 0.1 = 100$ V.

Derivation of expression for high-frequency cut-off point f_2 for a multistage Amplifier

For a single-stage Amplifier using any type of an active device (amplifying device may be a BJT or JFET or MOSFET), the voltage gain in the high-frequency region of the frequency response characteristic of an Amplifier is given by the following equation, assuming the normalised mid-band gain (A_m) of the Amplifier to be unity.

$$A_{HF}^{1} = \frac{1}{\sqrt{1+\left[\frac{f}{f_{HF}}\right]^2}} \quad \text{or} \quad A_{HF}^{1} = \frac{1}{\sqrt{1+\left[\frac{f}{f_2}\right]^2}}, \tag{10.7}$$

where the superscript to the letter A_{HF} denotes the number of amplifier stages in the cascaded Amplifier and f_2 is denoted as f_{HF} also. So, A_{HF}^{1} is the gain of the first amplifier stage in the high-frequency region, A_{HF}^{2} is the gain of the second amplifier stage in the high-frequency region and so on, so that A_{HF}^{n} is the gain of the nth amplifier stage in the high-frequency region. The overall gain of N-stage Amplifier containing N-amplifier stages is given as A_{HF}^{N}.

Similarly, f_2^1 is the high-frequency cut-off point for the first stage Amplifier, f_2^2 is the high-frequency cut-off of the second stage Amplifier and so on. Finally, f_2^n is the high-frequency cut-off point of N-stage Amplifier. Hence, f_2^N is the high-frequency cut-off of N-stage cascaded Amplifier.

In the beginning of this chapter, it is observed mathematically that the gain A_{HF}^{N} of N-stage cascaded Amplifier is the product of the individual stage voltage gains. This is represented as

$$A_{HF}^{N} = A_{HF}^{1} \cdot A_{HF}^{2} \cdot A_{HF}^{3} \cdot A_{HF}^{4} \ldots A_{HF}^{n}. \tag{10.8}$$

Using the above concept and the following expressions for the voltage gains of individual amplifier stages, assuming identical amplifier stages, the upper cut-off frequency points f_2 for all individual amplifier stages will be the same f_2:

$$\text{i.e.,}\quad f_2^1 = f_2^2 = f_2^3 = f_2^n = f_2.$$

HF gain of the first amplifier stage

$$A_{HF}^{1} = \frac{1}{\sqrt{1+\left[\frac{f}{f_2}\right]^2}}. \tag{10.9}$$

HF gain of the second amplifier stage

$$A_{HF}^{2} = \frac{1}{\sqrt{1+\left[\frac{f}{f_2}\right]^2}}. \tag{10.10}$$

HF gain of the third amplifier stage

$$A_{\text{HF}}^{3} = \frac{1}{\sqrt{1+\left[\frac{f}{f_2}\right]^2}}.$$

HF gain of the *N*th amplifier stage

$$A_{\text{HF}}^{n} = \frac{1}{\sqrt{1+\left[\frac{f}{f_2}\right]^2}}.$$

Therefore, HF gain of overall *N*-stage cascaded amplifier at high-frequency cut-off point is

$$A_{\text{HF}}^{N} = \frac{1}{\sqrt{1+\left[\frac{f}{f_2}\right]^2}} \cdot \frac{1}{\sqrt{1+\left[\frac{f}{f_2}\right]^2}} \cdot \frac{1}{\sqrt{1+\left[\frac{f}{f_2}\right]^2}} \cdots \frac{1}{\sqrt{1+\left[\frac{f}{f_2}\right]^2}} = \frac{1}{\sqrt{2}} \tag{10.11}$$

$$A^{N} = \left[\frac{1}{\sqrt{1+\left[\frac{f_2^N}{f_2}\right]^2}}\right]^N = \frac{1}{\sqrt{2}}. \tag{10.12}$$

Bandwidth shrinkage (reduction) factor due to increase in gain in multistage Amplifiers

At high-frequency cut-off point $f = f_2^N$ of multistage Amplifier, normalised gain is $\frac{1}{\sqrt{2}}$.

$$\therefore \quad A^{N} = \left[\frac{1}{\sqrt{1+\left(\frac{f_2^N}{f_2}\right)^2}}\right]^N = \frac{1}{\sqrt{2}}, \ldots, \frac{f_2^N}{f_2} = \sqrt{2^{1/N}-1}. \tag{10.13}$$

Therefore, high-frequency cut-off point of the multistage Amplifier f_2^N or the bandwidth of the cascaded Amplifier is given by the following expression:

$$f_2^N = f_2 \cdot \sqrt{2^{1/N}-1}.$$

Effect on Bandwidth due to increase in gain of a multistage Amplifier

$$\therefore \quad f_2^N = f_2 \cdot \sqrt{2^{1/N}-1} \tag{10.14}$$

$$\frac{f_2^N}{f_2} = \sqrt{2^{1/N}-1}.$$

This equation indicates that the bandwidth of the multistage Amplifier is reduced by the factor $\sqrt{2^{1/N}-1}$. The reduction in bandwidth is considered as 'Shrinkage' in Bandwidth.

The bandwidth reduction factor $\sqrt{2^{1/N}-1}$ for *N*-stage multistage Amplifier can be used to calculate the factor by which bandwidth of certain number of cascaded Amplifiers can be calculated. Shrinkage factor values for nine Amplifiers are given in Table 10.1.

Table 10.1 Bandwidth reduction factor $\sqrt{2^{1/N}-1}$ of cascaded (multistage) amplifiers

Number of stages (N)	2	3	4	5	6	7	8	9
$\frac{f_2^N}{f_2}$	0.6435	0.5098	0.4349	0.3856	0.3499	0.3226	0.3008	0.2829

On similar lines, Low-Frequency Cut-off Point f_1^N of multistage Amplifier can be derived as

$$\frac{f_1^N}{f_1} = \frac{1}{\sqrt{2^{1/N}-1}}. \tag{10.15}$$

EXAMPLE 10.3

Consider a two-stage Amplifier with identical amplifier stages. If the bandwidth of each individual stage is 20 kHz, calculate the bandwidth of the overall Amplifier.

Solution: For a two-stage Amplifier $N = 2$.
Expression for bandwidth of N-stage Amplifier

$$f_2^N = f_2 \cdot \sqrt{2^{1/2}-1} = f_2 \cdot \sqrt{1.414-1} = f_2 \cdot \sqrt{0.414} = 0.64 f_2.$$

If the bandwidth of each stage of a two-stage cascaded Amplifier is approximately equal to $f_2 = 20$ kHz, then the bandwidth of the cascaded Amplifier is 12.8 kHz.
This shows that the bandwidth of a cascaded or a multistage Amplifier is reduced. Further, it can be seen that further increase in the number of stages increases the gain at the sacrifice of the bandwidth of the overall cascaded Amplifier system.

EXAMPLE 10.4

If the low-frequency cut-off point f_1 of an amplifier stage of multistage Amplifier consisting of two identical stages is 100 Hz, calculate the low-frequency cut-off point of the cascaded Amplifier.

Solution: The low-frequency cut-off point f_1^N of the multistage Amplifier can be obtained as

$$\frac{f_1^N}{f_1} = \frac{1}{\sqrt{2^{1/N}-1}}.$$

If the low-frequency cut-off point f_1 of an amplifier stage of multistage Amplifier consisting of two identical stages is 100 Hz, then

$$f_1^N = \frac{1}{\sqrt{2^{1/N}-1}} \cdot f_1 = \frac{1}{\sqrt{2^{1/2}-1}} \cdot f_1$$

$$= \frac{1}{0.6435} \cdot f_1 = 1.554 \times 100 = 155.4 \text{ Hz}.$$

EXAMPLE 10.5

Mention the equations for the overall cut-off frequencies f_1 and f_2 of cascaded or multistage Amplifier circuit having identical amplifier stages. If five identical Resistance Capacitance-coupled amplifier stages having their lower cut-off frequencies as $f_1 = 200$ Hz and the upper cut-off frequencies as $f_2 = 36$ MHz are cascaded, calculate the effective cut-off points f_1^5 and f_2^5 of the cascaded Amplifier.

Solution: The high-frequency cut-off point of the multistage Amplifier f_2^N or the bandwidth of the cascaded Amplifier is given by

$$f_2^N = f_2 \cdot \sqrt{2^{1/N} - 1}, \tag{10.16}$$

where N is the number of stages of a cascaded Amplifier and the high-frequency cut-off point of an Amplifier is expressed as f_2.

In the above example, the number of stages = 5.

$$\begin{aligned} f_2^5 &= f_2 \cdot \sqrt{2^{1/5} - 1} \\ &= 36 \times 10^6 \sqrt{2^{1/5} - 1} \\ &= 36 \times 10^6 \times \sqrt{1.1486 - 1} \\ &= 36 \times 10^6 \times \sqrt{0.1486} \\ &= 36 \times 10^6 \times 0.3856 = 13.87 \times 10^6 \text{ Hz.} \end{aligned}$$

Thus, the high-frequency cut-off point is drastically reduced to 13.87 MHz.

Table 10.2 shows that the bandwidth of a multistage Amplifier decreases with increase in the number of stages in cascade.

Table 10.2

Number of amplifiers in cascade	High-frequency cut-off point f_2^N
2	$f_2^2 = 0.6435\, f_2$
3	$f_2^3 = 0.5098\, f_2$
4	$f_2^4 = 0.4349\, f_2$
5	$f_2^5 = 0.3856\, f_2$
6	$f_2^6 = 0.3499\, f_2$
7	$f_2^7 = 0.3226\, f_2$
8	$f_2^8 = 0.3008\, f_2$
9	$f_2^9 = 0.2829\, f_2$

Low-frequency cut-off point f_1^N of multistage Amplifier is given as

$$\frac{f_1^N}{f_1} = \frac{1}{\sqrt{2^{1/N} - 1}}. \tag{10.17}$$

$$f_1^5 = \frac{1}{\sqrt{2^{1/5} - 1}} \cdot f_1$$

$$f_1^5 = 2.5933\, f_1$$

$$f_1^5 = 2.5933 \times 200 = 318 \text{ Hz}$$

The above calculations show that there is an increase in frequency of the low-frequency cut-off point and reduction in the frequency of the high-frequency cut-off point. Thus, there is in effect reduction in bandwidth of the multistage Amplifier. This process is considered as *shrinkage in bandwidth for multistage Amplifier* in comparison with the individual Amplifiers of the overall system (Table 10.3).

Table 10.3

Number of amplifiers in cascade	Low-frequency cut-off point f_1^N
2	$f_1^2 = 1.554\, f_1$
3	$f_1^3 = 1.9646\, f_1$
4	$f_1^4 = 2.2993\, f_1$
5	$f_1^5 = 2.5933\, f_1$
6	$f_1^6 = 2.8579\, f_1$
7	$f_1^7 = 3.0998\, f_1$
8	$f_1^8 = 3.3244\, f_1$
9	$f_1^9 = 3.5348\, f_1$

10.4 CASCADED RC-COUPLED BJT AMPLIFIERS

Figure 10.8 shows two-stage CE Transistor Amplifier. Supply voltage V_{CC}, potential divider networks containing R_1, R_2 elements and the resistors R_C and R_E fix up the DC operating conditions of the Amplifier circuits. For linear operation of Amplifiers, quiescent or DC operating conditions are fixed so that the Quiescent operating point is located at the middle of the DC load line (drawn on the Transistor output characteristics) for maximum signal conditions without distortion and symmetrical signal swings.

External input signal V_S is applied for amplification through the input coupling capacitor C_{in} and the amplified output voltage V_{out2} is obtained across the load resistance R_L.

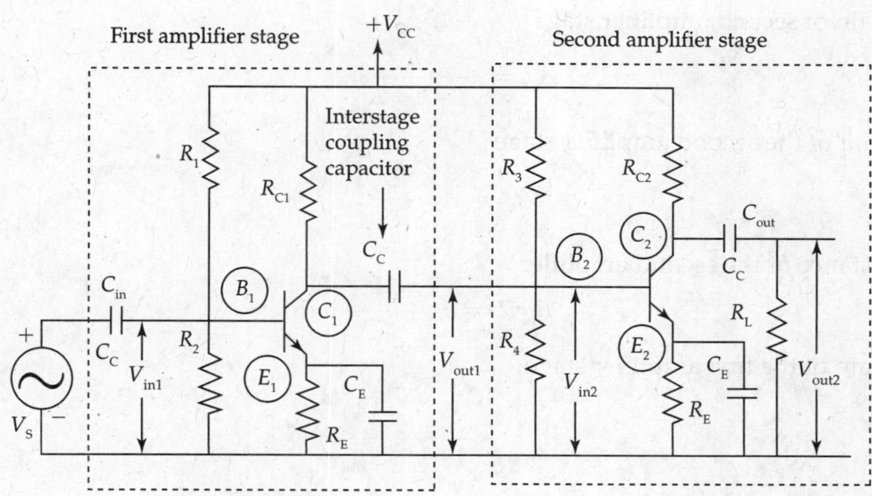

FIG. 10.8 *Cascaded RC coupled BJT amplifiers*

Amplifier circuit operation

- Input signal V_{in1} is applied at the Base B_1 of the Transistor in the first amplifier stage through the input coupling capacitor C_C. First amplifier stage amplifies it with voltage gain A_{V1} and the current gain A_{I1}.
- Output voltage V_{out1} of the first amplifier stage is fed as input voltage V_{in2} to the input terminal B_2 of the Transistor in the second amplifier stage through the inter-stage coupling capacitor C_C. The amplified voltage by the second Amplifier is the output voltage V_{out2}.
- The required output voltages with gain A_{Vn} for the multistage Amplifier can thus be achieved using suitable design.
- Effective load resistance R_{L2} of the second amplifier stage is the parallel combination of R_{C2} and R_L.
- Effective load resistance of the first amplifier stage is R_{L1}, which is the parallel combination of R_{C1} and the input resistance R_{in2} of the second amplifier stage

AC equivalent circuit of Two-stage Transistor Amplifier connected in cascade (Fig. 10.9)

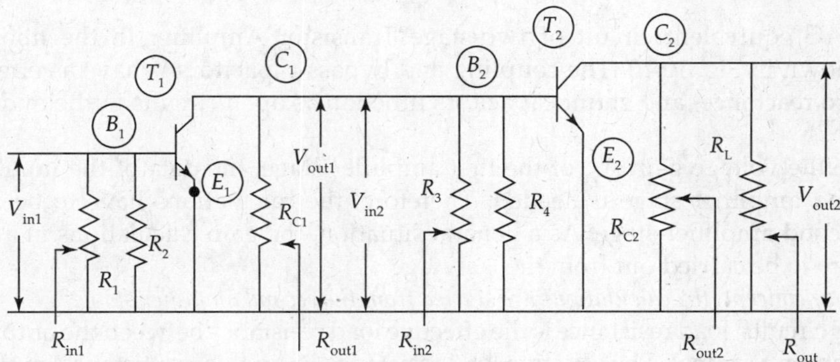

FIG. 10.9 *AC equivalent circuit of cascaded RC coupled BJT amplifiers*

Current gain of second amplifier stage

$$A_{I2} = -\frac{h_{fe}}{1 + h_{oe} \cdot R_{L2}}. \tag{10.18}$$

Voltage gain of the second amplifier stage

$$A_{V2} = A_{I2} \cdot \frac{R_{L2}}{R_{in2}}. \tag{10.19}$$

Input resistance of the second amplifier

$$R_{in2} = R_{ie}(2) = h_{ie} + h_{re} \cdot A_{I2} \cdot R_{L2}. \tag{10.20}$$

Current gain of the first amplifier stage

$$A_{I1} = -\frac{h_{fe}}{[1 + h_{oe} \cdot R_{L1}]}. \tag{10.21}$$

Voltage gain of the first amplifier stage

$$A_{V1} = A_{I1} \cdot \frac{R_{L1}}{R_{in1}}, \tag{10.22}$$

where $R_{L1} = R_{C1} \parallel R_{in2}$.
Input resistance of the first amplifier stage

$$R_{in1} = R_{ie}(1) = h_{ie} + h_{re} \cdot A_{I1} \cdot R_{L1}. \tag{10.23}$$

The worked out Example 10.6 explains in detail the levels of voltage and current gains and the impedance values for the (cascaded) multistage Amplifier.

EXAMPLE 10.6

The following two-stage CE + CE Transistor Amplifier of Fig. 10.9 shows component values for a designed multistage Amplifier. The two Transistors are BC107. The hybrid (h) parameters are as follows: input resistance $h_{ie} = 1.1$ kΩ, Current gain $h_{fe} = \beta = 100$, $h_{oe} = 25 \times 10^{-6}$ mhos and $h_{re} = 2.4 \times 10^{-4}$. Calculate the parameters (a) Input resistance R_{in}; (b) output impedance; (c) current gain A_I; (d) Voltage gain A_V for each stages. Assume that the two Transistors have same h-parameters.

Solution: AC equivalent circuit of two-stage Transistor Amplifier in the mid-frequency region is shown in Fig. 10.10. The coupling and bypass capacitors behave as effective short circuits (zero reactance) and shunt capacitors function as open circuits in the mid-frequency region.
To calculate the voltage gain A_{V1} of the first amplifier stage, the data of the input resistance of the second amplifier stage is needed. Therefore, the calculations have to be carried out from the second amplifier stage. As a general situation, the gain calculations in a multistage Amplifier are to be carried out from the last stage.
From the above concept, the calculations are started from the second amplifier stage.
In Amplifier circuits, load resistance is the effective load resistance between the output terminal and the common terminal. The effective load resistance R_{L2} in this multistage Amplifier circuit is the parallel combination of R_{C2} and R_L:

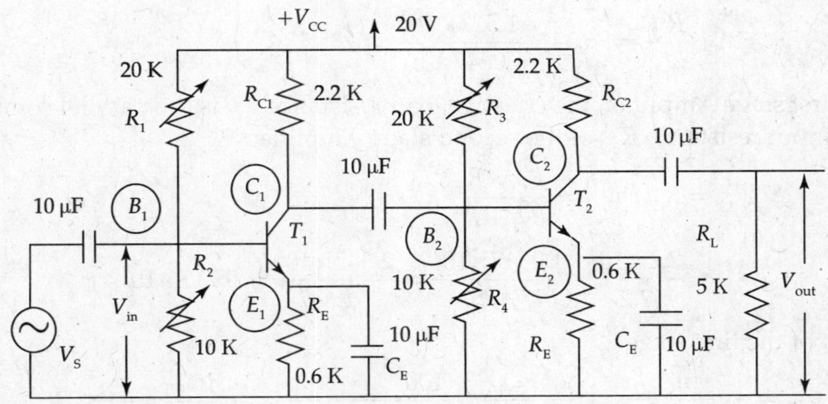

FIG. 10.10 *Two stage common emitter transistor RCC amplifier multi stage amplifier*

$$R_{L2} = R_{C2} \parallel R_L = \frac{R_{C2} \times R_L}{R_{C2} + R_L} = \frac{2.2 \times 10^3 \times 5 \times 10^3}{(2.2+5) \times 10^3} = \frac{11 \times 10^3}{7.2} = 1.527 \times 10^3.$$

Transistor parameters: $h_{fe} = 100$, $h_{ie} = 1.1\ \text{k}\Omega$, $h_{oe} = 25 \times 10^{-6}$ mhos, $h_{re} = 24 \times 10^{-4}$.

Current gain: $A_{I2} = -\dfrac{h_{fe}}{1 + h_{oe} \cdot R_{L2}} = -\dfrac{100}{1 + 25 \times 10^{-6} \times 1.527 \times 10^3} = -\dfrac{100}{1.038} = -96.34$

Voltage gain $A_{V2} = A_{I2} \cdot \dfrac{R_{L2}}{R_{in2}}$

Input resistance R_{in2} of the second stage Amplifier:

$$R_{in2} = R_{ie} = h_{ie} + h_{re} \times A_{I2} \cdot R_{L2}$$

$$\therefore\ R_{in2} = 1.1 \times 10^3 - 2.4 \times 10^{-4} \times 96.34 \times 1.527 \times 10^3$$

$$= 1100 - 35.3 = 1067.7\ \Omega = 1.0647 \times 10^3\ \Omega.$$

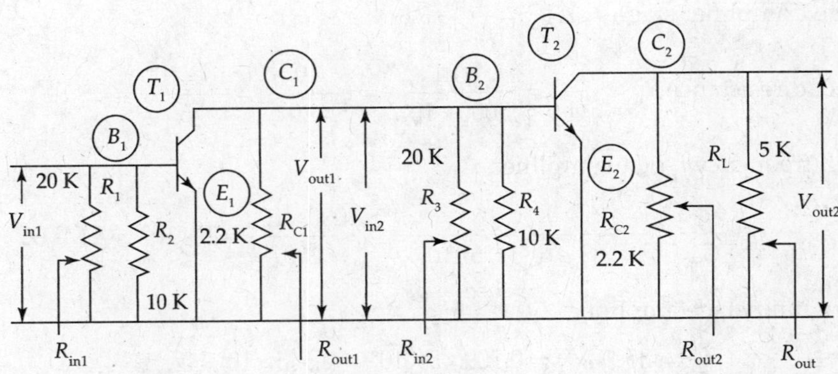

FIG. 10.11 *AC equivalent circuit of two-stage transistor amplifiers connected in cascade*

Voltage gain $A_{V2} = A_{I2} \cdot \frac{R_{L2}}{R_{in2}} = 96.34 \times \frac{1.527 \times 10^3}{1.0647 \times 10^3} = 138.17.$

But for the first stage Amplifier, the effective load resistance R_{L1} is the parallel combination of R_{C1} and the input resistance R_{in2} of the second stage Amplifier.

$$R_{L1} = R_{C1} \parallel R_{in2} = \frac{R_{C1} \times R_{in2}}{R_{C1} + R_{in2}}$$

$$= \frac{2.2 \times 10^3 \times 1.0647 \times 10^3}{(2.2 + 1.0647) \times 10^3} = \frac{2.342 \times 10^3}{3.2647} = 0.717 \times 10^3.$$

Current gain of the first stage

$$A_{I1} = -\frac{h_{fe}}{1 + h_{oe} R_{L1}} = -\frac{100}{1 + 25 \times 10^{-6} \times 0.717 \times 10^3} = -\frac{100}{1.018} = -98.23.$$

Input resistance of the first amplifier stage

$$R_{in1} = h_{ie} + h_{re} \cdot A_{I1} \cdot R_{L1}.$$

Substituting the various quantities into the equation of R_{in1}

$$R_{in1} = 1.1 \times 10^3 - 2.4 \times 10^{-4} \times 98.23 \times 0.717 \times 10^3$$

$$= 1100 - 16.9 = 1083.1\,\Omega$$

$$A_{V1} = A_{I1} \cdot \frac{R_{L1}}{R_{in1}} = -98.23 \times \frac{0.717 \times 10^3}{1083.1 \times 10^3} = -65.03.$$

EXAMPLE 10.7

Multistage Common Source (CS) FET Amplifier has three stages of Amplifiers. The effective mid-band gain of Amplifier is $A_E^3 = 1000$ for each amplifier stage, the overall shunt capacitance $C_{Sh} = C_2 = 16$ pF and g_m of active device FET = 5000 μmhos. Calculate the gain of the individual Amplifiers, load resistance and effective bandwidth of the multistage Amplifier. Assume that the individual Amplifiers are identical.

Solution: Mid-band gain of individual amplifiers = $1000^{1/3} = 10$.
Gain of CS FET amplifier stage = $A_{mid} = -g_m \cdot R_L$

$$\therefore \text{Load resistance } R_L = \frac{A_{mid}}{g_m} = \frac{10}{5000 \times 10^{-6}} = \frac{10}{5 \times 10^{-3}} = \frac{10 \times 10^3}{5} = 2 \times 10^3\ \Omega.$$

Upper cut-off frequency f_2 of an amplifier

$$f_2 = \frac{1}{2\pi \cdot R_L \cdot C_{Sh}} = \frac{1}{2\pi \times 2 \times 10^3 \times 16\pi 10^{-12}} \cong \frac{1000 \times 10^6}{200} = 5 \times 10^6 = 5\text{ MHz}.$$

Bandwidth of multistage amplifier

$$f_2^3 = 0.509 \times f_2 = 0.509 \times 5 \times 10^6 = 2.545 \times 10^6\text{ Hz}.$$

10.5 CASCADED RC-COUPLED FET AMPLIFIERS

Figure 10.12 shows a two-stage RC-coupled JFET Amplifier.

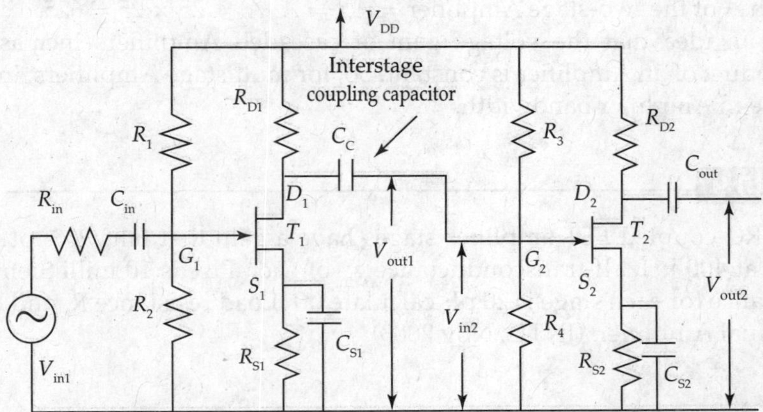

FIG. 10.12 *Cascaded R-C coupled FET amplifiers*

- The amplifier stages use two FET devices T_1 and T_2.
- The multistage Amplifier uses Class-A operation with biasing resistors R_1, R_2, R_{S1}, for the first amplifier stage and R_3, R_4, R_{S2} combinations for the second amplifier stage.
- Input signal to the first stage is V_{in1}. It will be amplified by the Transistor T_1 with gain $A_1 = -g_m \cdot R_{L1}$.
- Load R_{L1} for the first amplifier stage is the parallel combination of R_{D1} and the input resistance R_{in2} of the second amplifier stage.
- Input resistance of the second CS FET amplifier stage is very high, and hence, there is no loading effect phenomena with FET Amplifiers.
- Output signal V_{out1} between the Drain and the Source of the first Transistor is fed as the input signal V_{in2} for input port of the second amplifier stage.
- Input signal V_{in2} varies the DC biasing conditions that exist for Transistor T_2 and the varying DC voltage is amplified and appears as the output voltage V_{out2} with gain $A_2 = g_m \cdot R_{L2} = g_m \cdot R_{D2}$.
- For the multistage Amplifier, the overall gain $A_{1,2}$ is the product of the individual voltage gains A_1 and A_2. Therefore, the output voltage $V_{out2} = A_1 \cdot A_2 \cdot V_{in1}$.

EXAMPLE 10.8

If transconductance of each JFET device is $g_m = 5 \times 10^{-3}$ mhos and $R_{D1} = R_{D2} = 5 \times 10^3\ \Omega$ in a two-stage FET Amplifier of Fig. 10.12, calculate the gains of each stage and the gain of the overall Amplifier.

Solution: For the active devices, JFETs the input resistances are very high, as the Gate to Source junctions of JFET devices are reverse biased. So, R_{in2} is very high, and hence, the load resistance of the first amplifier stage $R_{L1} = R_{D1} = 5\ \text{k}\Omega$.
Therefore, the gain of the first stage Amplifier is

$$A_1 = -g_m \cdot R_{L1} = 5 \times 10^{-3} \times 5 \times 10^3 = 25.$$

On similar lines the gain of the second Amplifier is

$$A_2 = g_m \cdot R_{D2} = 5 \times 10^{-3} \times 5 \times 10^3 = 25.$$

Hence, the gain A of the two-stage Amplifier $= A_{1.2} = A_1 \cdot A_2 = 25 \times 25 = 625$.
This gives us an idea that the voltage gain of cascaded Amplifiers increases. The gain-bandwidth product of an Amplifier is constant. So, for multistage Amplifiers, increase in gain causes decrease in Amplifier bandwidth.

EXAMPLE 10.9

Two identical RC-coupled FET amplifier stages have a gain that falls 90% of the mid-band gain value A_M at 400 kHz. If transconductance g_m of each FET is 10 milli Siemens and total output capacitance for each stage is 20 pF, calculate (a) Load resistance R_L and (b) Mid-band gain of individual Amplifier. (JNTU, Nov 2003)

Solution:

a. Gain of the multistage amplifier $A = \dfrac{A_M}{\sqrt{1 + \left[\dfrac{f}{f_H^N}\right]^2}}$.

From the given data, $\dfrac{A}{A_M} = 0.9$ at frequency $f = 400$ kHz.

$$\therefore \quad \frac{1}{\sqrt{1 + \left[\dfrac{f}{f_H^N}\right]^2}} = 0.9$$

$$\left[\frac{f}{f_H^N}\right]^2 = 0.235$$

$$\therefore \quad \frac{f}{f_H^N} = \sqrt{0.235} = 0.4848.$$

High-frequency cut-off of multistage amplifier $= f_H^N$

$$f_H^N = \frac{f}{0.4848} = \frac{400 \times 10^3}{0.4848} = 825.08 \text{ kHz}$$

Upper cut-off frequency of individual another stages $= f_H^1 = f_H^2 = f_H$

$$f_H = \frac{f_H^N}{\sqrt{2^{1/N} - 1}} = \frac{f_H^N}{\sqrt{2^{1/N} - 1}}$$

$$= \frac{825.08 \times 10^3}{\sqrt{2^{\frac{1}{2}} - 1}} = \frac{825.08 \times 10^3}{0.643} = 1282.32 \text{ kHz}$$

$$f_H = \frac{1}{2 \times \pi \times R_L \times C_0} = 1282.32 \text{ kHz}.$$

Given data: Output capacitance $C_0 = 20\times10^{-12}$ F

$$\therefore \text{ Load resistance } R_L = \frac{1}{2\times\pi\times20\times10^{-12}\times1282.32\times10^3}$$

$$= \frac{100\times10^3}{16.12} = 6.2 \text{ k}\Omega.$$

b. Mid-band gain of individual stages $A_M^1 = A_M^2 = -g_m \times R_L$
Given data: $g_m = 10$ milli Siemens and $R_L = 6.2$ kΩ

$$\text{Mid-band gain } A_M^1 = A_M^2 = 10\times10^{-3}\times6.2\times10^3 = 62.$$

EXAMPLE 10.10

Three identical non-interacting amplifier stages have an overall gain of 0.3 dB down at 20 kHz compared to mid-band gain. Calculate the upper cut-off frequencies of the individual stages. (JNTU, RR21041)

Solution: Multistage amplifier gain

$$\text{Gain } A = \frac{A_M}{\sqrt{\left[1+\left[\frac{f}{f_H^N}\right]^2\right]}}$$

$$\therefore \frac{A}{A_M} = \left[1+\left[\frac{f}{f_H^N}\right]^2\right]^{-1/2}$$

$$20\log_{10}\frac{A}{A_M} = -20\times\frac{1}{2}\cdot\log_{10}\left[1+\left[\frac{f}{f_H^N}\right]^2\right] = -0.3 \text{ dB}$$

$$\therefore \log_{10}\left[1+\left[\frac{f}{f_H^2}\right]^2\right] = \frac{0.3}{10} = 0.03$$

$$\left[1+\left[\frac{f}{f_H^N}\right]^2\right] = \text{anti}\log_{10} 0.03 = 1.072$$

$$\left[\frac{f}{f_H^N}\right]^2 = 1.072 - 1.0 = 0.072$$

$$\left[\frac{f}{f_H^N}\right] = \sqrt{0.072} = 0.2683$$

$$f_H^N = \frac{f}{0.2683} = \frac{50\times10^3}{0.2683} = 186.36 \text{ kHz.}$$

Given data: $f = 50 \times 10^3$ Hz
Number of stages in the amplifier = 3
Therefore, upper cut-off frequency of individual stages is given by

$$f_H = \frac{f_H^N}{\sqrt{2^{1/N} - 1}}$$

$$\text{Hence,}\quad f_H = \frac{186.36 \times 10^3}{\sqrt{2^{1/3} - 1}} = 365.264 \text{ kHz.}$$

10.6 FREQUENCY RESPONSE CHARACTERISTIC OF RC-COUPLED AMPLIFIER

The *frequency response* characteristic of an Amplifier is a graph between Amplifier gain and signal frequency. Typical frequency response characteristic of an Amplifier is shown in Fig. 10.14. It can be obtained from a typical Amplifier circuit in Fig. 10.13.

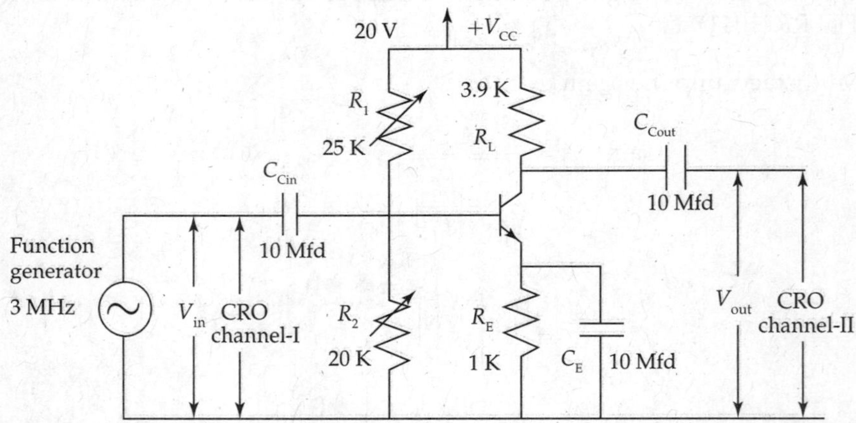

FIG. 10.13 *Common emitter transistor amplifier practical circuit variable gain amplifier circuit*

Amplifier bandwidth calculations are shown in Fig. 10.14 by assuming maximum gain of the Amplifier $A_{max} = 100$.

Calculation of amplifier Bandwidth using gain:

$$A_{max} = 100$$

$$\text{Gain at cut-off points}\quad \frac{A_{max}}{\sqrt{2}} = 0.707 \times 100 = 70.7.$$

If the upper cut-off frequency $f_2 = 3.3$ kHz and lower cut-off frequency $f_1 = 0.3$ kHz, Amplifier bandwidth BW = $(f_2 - f_1)$ = (3.3 – 0.3) kHz = 3 kHz.

Calculation of bandwidth from the frequency response characteristic of an Amplifier: When the Amplifier gain is calculated in terms of decibels (Fig. 10.15)

$$\text{Gain in db} = 20 \log 10\ A_V, \tag{10.24}$$

where the voltage gain of an Amplifier is A_V.

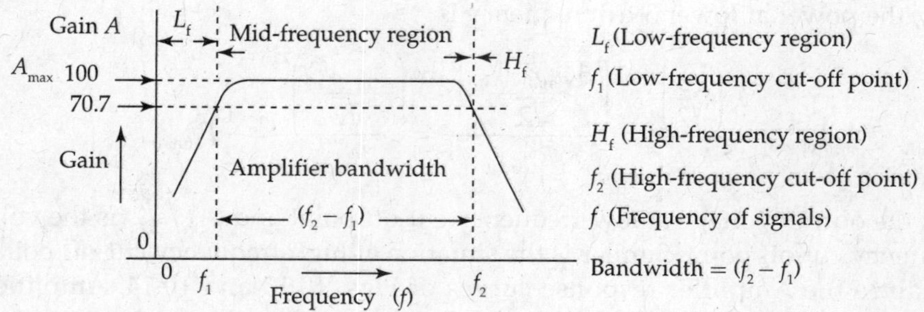

FIG. 10.14 *Frequency response of a single-stage amplifier showing the calculation of amplifier band width using gain A_{max} = 100 gain at cut-off points = A_{max} = 0 707 × 100 = 70 7*

$$\text{Gain at cut-off frequencies } f_1 \text{ or } f_2 = \frac{A_m}{\sqrt{2}} \tag{10.25}$$

f_1 (or f_L) and f_2 (or f_h) are also known as 3-db frequencies or 1/2 power frequencies.

$$20\log_{10}\frac{1}{\sqrt{2}} = 20\log_{10}2^{-1/2} = -3.0 \text{ dB}, \tag{10.26}$$

where A_m or A_{max} is the maximum value of gain obtained from the frequency response characteristic. So, fall in gain at cut-off frequencies is given below.

At the half power points or lower and upper cut-off frequency points f_1 and f_2, respectively, gain is lower by 3 dB from the gain in db at the maximum value of gain. The maximum gain in this case is considered as 40 dB (corresponding to A_{max} = 100) and so the gain at the cut-off frequency points will be 37 dB which will be lower by 3 dB from the maximum gain of 40 dB in the mid-frequency region.

The frequencies f_1 and f_2 are known as corner, break or half power frequencies.

Power at mid-band is

$$P_{o(mid)} = \frac{(V_{out})^2}{R_{Load}} = \frac{(V_L)^2}{R_{Load}} = \frac{(A_{V(mid)} V_{in})^2}{R_{Load}} = \frac{(A_m \cdot V_{in})^2}{R_{Load}}.$$

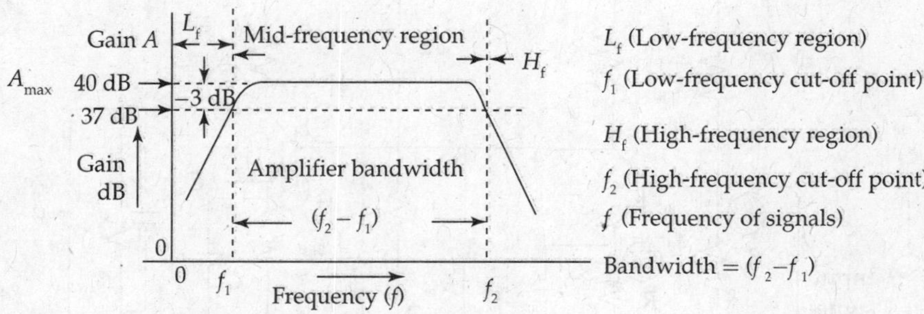

FIG. 10.15 *Frequency response of a single-stage amplifier showing the calculation of amplifier band width using gain in decibels (db)*

Similarly, the power at lower 3-dB frequency is

$$P_{o(3\text{ dB})} = \frac{\left(\frac{V_L}{\sqrt{2}}\right)^2}{R_{Load}} = \frac{\left(\frac{A_{V(mid)} \cdot V_{in}}{\sqrt{2}}\right)^2}{R_{Load}} = \frac{\left(\frac{A_m \cdot V_{in}}{\sqrt{2}}\right)^2}{R_{Load}} = \frac{1}{2} P_{o(mid)}, \tag{10.27}$$

where V_L is the output voltage at low-frequency cut-off point and $A_m/\sqrt{2}$ is the voltage gain at low-frequency cut-off point. Similar is the situation at high-frequency cut-off point.

If we analyse the Amplifier response curves of Figs. 10.13 and 10.14, Amplifier gain is uniform over middle range of frequencies of the signals. It suggests that the small signal low frequency equivalent circuit for BJT or JFET Amplifier circuits contain only resistive elements in the mid-frequency region. All the series and shunt capacitor elements in an Amplifier circuit shown in the circuit of Fig. 10.13 need not be considered because their reactance effects are negligible in linear operation of Amplifiers in mid-frequency or intermediate frequency region.

Typical shape or nature of variation of voltage gain in an Amplifier frequency response curve for multistage Amplifiers will also be similar in nature but there will be increase in gain and simultaneous reduction in bandwidth to a magnitude of shrinkage factor.

10.7 EQUIVALENT CIRCUITS OF CASCADED RC-COUPLED TRANSISTOR AMPLIFIERS

Let us consider a general multistage Amplifier circuit containing two amplifier stages shown in Fig. 10.15 for the discussion of the equivalent circuits under the three types of frequency ranges of operation.

- Mid-frequency region,
- Low-frequency region and
- High-frequency region.

There are various capacitances in the Amplifier circuit. They are coupling capacitors, bypass capacitors and Junction capacitances about the Transistors which are shunt capacitances in the circuit shown in Fig. 10.16.

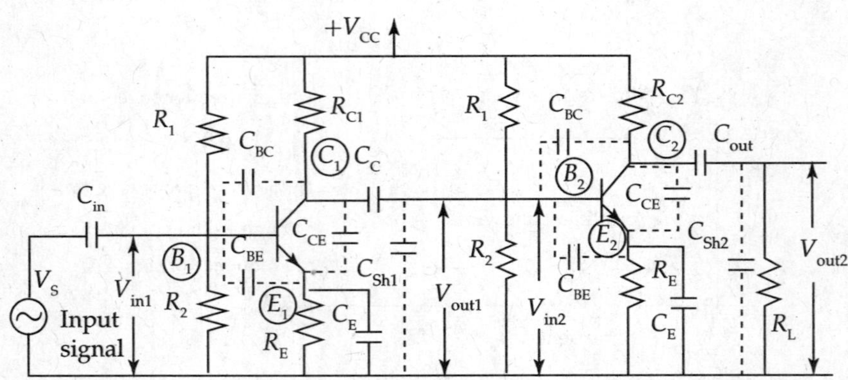

FIG. 10.16 *Two stage transistor resistance capacitance coupled amplifier*

- Series and bypass capacitances affect or reduce the Amplifier gain (because of their high reactance at low frequencies) when the AC signals pass through them.
- Similarly, the junction capacitances in the Transistors in association with the shunt capacitances form C_{Sh1} and C_{Sh2} to affect or reduce the gain of the Amplifier in high-frequency operation. The reduced reactance of C_{Sh} comes in parallel with output resistances at the respective ports of amplifier stage-1 or amplifier stage-2.

AC Equivalent Circuit

AC equivalent circuit is shown in Fig. 10.17 with all the capacitances in the circuit, except the parallel combination of R_E and C_E that work as short circuits. The DC Source V_{CC} is replaced by its internal resistance, and under ideal conditions, it is taken as zero ohms. Hence, the DC Source is replaced by short circuit between the Resistors (R_{C1} and R_{C2}) and the ground in the AC equivalent circuit.

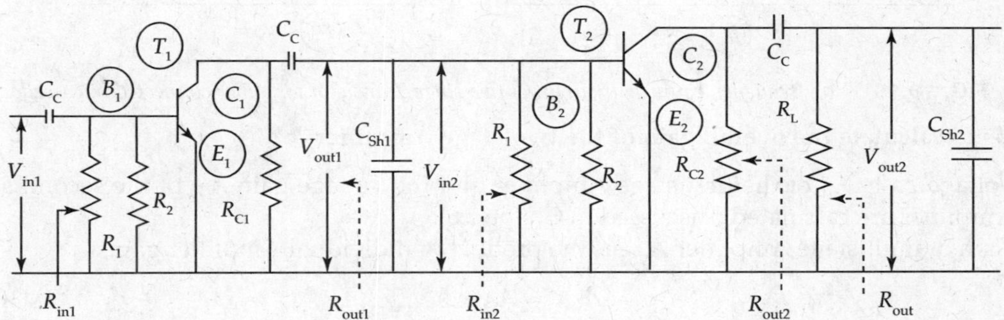

FIG. 10.17 *AC equivalent circuit of two stage transistor amplifier connected in cascade*

Mid-Frequency Circuit of Cascaded Amplifier using Transistor *h*-parameters (Fig. 10.18)

The analysis of two-stage Amplifier is carried out for small signal operation by replacing the two Transistors with their *h*-parameter equivalent circuits. The effects of series and shunt capacitances on the frequency response are negligible in the intermediate or mid-frequency region. So, the capacitors are not shown in the *h*-parameter equivalent circuit of the Amplifier for mid-frequency region.

In the equivalent circuit of Fig. 10.18, the voltage Sources $h_{re1} \cdot V_{B1E1}$ in series with h_{ie1} and $h_{re2} \cdot V_{B2E2}$ in series with h_{ie2} are neglected, because h_{re} of a Transistor is negligibly small. Further level of simplification of the circuit is shown in Fig. 10.19.

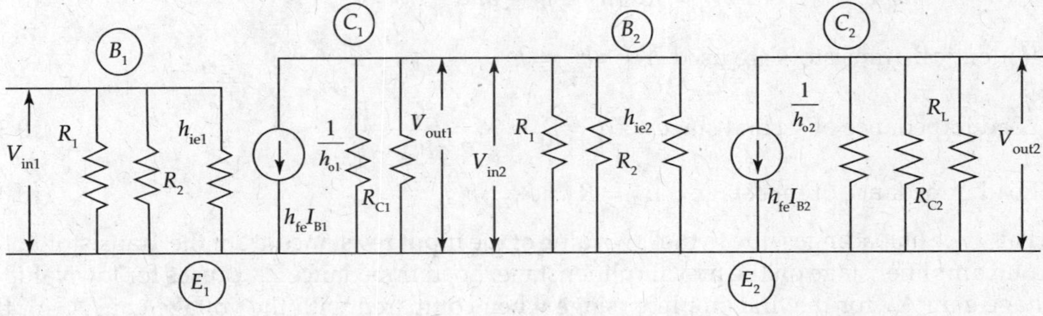

FIG. 10.18 *Mid frequency equivalent circuit of cascaded RC coupled transistor amplifier*

- By neglecting large values of h_{o1} and h_{o2}, because of their larger values in parallel with lower values of resistances in the output circuits.
- Representing the parallel combination of resistors R_1 and R_2 to be equal to R_B.

Simplified h-parameter model circuit of two-stage RC-coupled Amplifier is shown in Fig. 10.19.

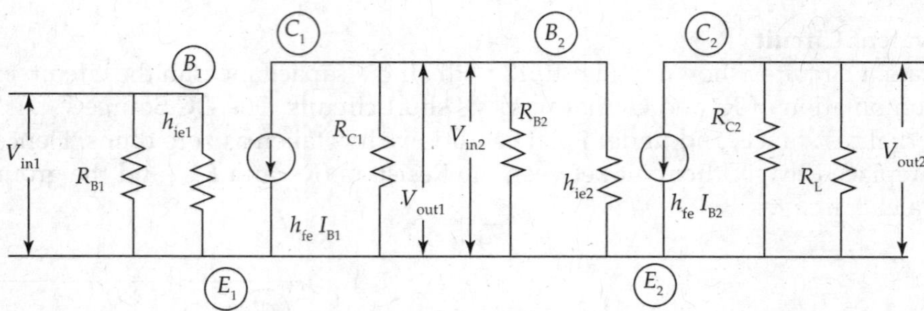

FIG. 10.19 *Mid frequency equivalent circuit of two stage Transistor amplifier connected in cascade*

Steps to calculate the overall gain of the two-stage Amplifier

- Voltage gain A_{V1} of the first stage Amplifier and the voltage gain A_{V2} of the second stage Amplifier are calculated (discussed in Chapter 6).
- Gain of multistage Amplifier $A_{V1.2}$ is the product of individual Amplifier gains.

$$A_{V1.2} = A_{V1} \times A_{V2} \tag{10.28}$$

Equation (10.28) shows considerable increase in gain for the cascaded amplifier stages.

- Further it can be observed that the load resistance of the first amplifier stage is the parallel combination of the load resistance and output resistance of the first amplifier stage and the input resistance of the second amplifier stage.
- Hence, the gain of the second amplifier stage A_{V2} is calculated first and then the gain A_{V1} of the first stage is determined to calculate the gain of the multistage Amplifier.

$$\text{Voltage gain} \quad A_V = -\frac{A_I \cdot Z_L}{Z_{in}} \cong -\frac{h_{fe} \cdot Z_L}{h_{ie}}. \tag{10.29}$$

Approximations taken in Eq. (10.29) are

$$Z_{in} = R_B \parallel h_{ie} \equiv h_{ie} \quad \text{and} \quad A_I = h_{fe}. \tag{10.30}$$

If identical Transistors are used $h_{ie1} = h_{ie2} = h_{ie}$.

$$\text{Load impedance of second stage} \quad Z_{L2} = R_{C2} \parallel R_L = \frac{R_{C2} \times R_L}{R_{C2} + R_L} \tag{10.31}$$

$$\text{Load impedance of first stage} \quad Z_{L1} = R_{C1} \parallel R_{B2} \parallel h_{ie2}. \tag{10.32}$$

This type of loading is due to the low value of the input resistance h_{ie2} of the Transistor in the second amplifier stage on the first amplifier stage. Load impedance Z_{L1} causes for low value of voltage gain A_{V1} for the first amplifier stage when compared with the voltage gain A_{V2} of the second amplifier stage.

Once the values for load impedances Z_{L1} and Z_{L2} and the input impedances are determined at desired cross sections of the Amplifier, individual Amplifier voltage gains and the overall gain of the cascaded or multistage Amplifier are calculated.

Miller's theorem to calculate the effect of feedback impedance

Consider a four-terminal network, with feedback impedance Z_f connected between input and output ports, as shown in Fig. 10.20.

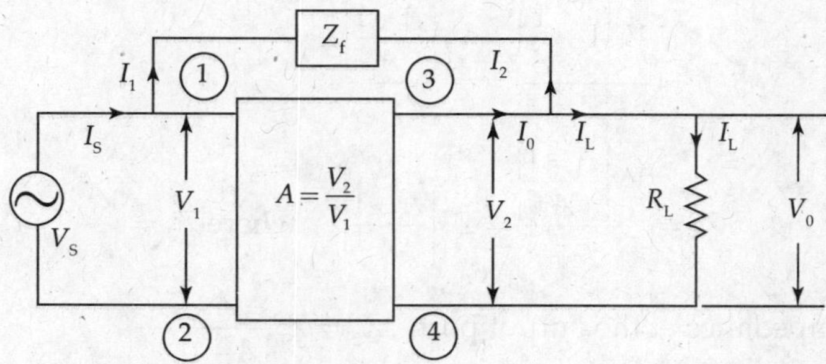

FIG. 10.20 *Four-terminal network with impedance Z_f connecting input and output ports to illustrate miller's theorem*

Figure 10.21 shows Miller's equivalent circuit with impedances Z_1 connected at the input port and Z_2 at the output port to the four-terminal networks. The values of the impedances are calculated as shown below.

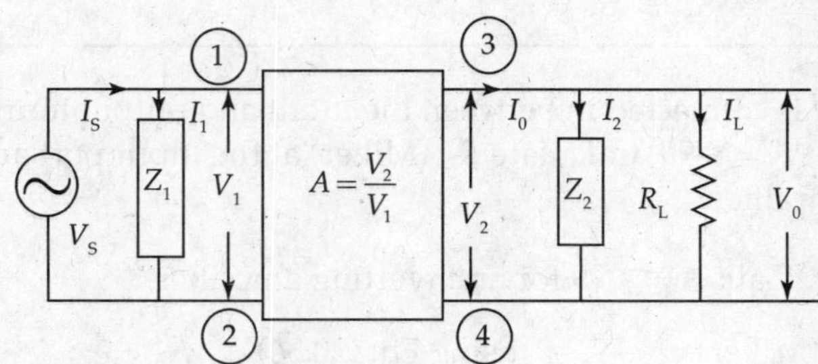

FIG. 10.21 *Miller's equivalent circuit with impedances Z_1 and Z_2*

At the input port side,

$$I_1 = \frac{V_1 - V_2}{Z_f}$$

$$= \frac{V_1[1 - V_2/V_1]}{Z_f} = \frac{V_1[1-A]}{Z_f}$$

$$\therefore \; Z_f = \frac{V_1[1-A]}{I_1} = Z_1[1-A],$$

$$\text{where } Z_1 = \frac{V_1}{I_1}$$

$\therefore$ Equivalent impedance at the input port $Z_1 = \dfrac{Z_f}{[1-A]}$. (10.33)

At the output port side,

$$I_2 = \frac{V_2 - V_1}{Z_f} = \frac{V_2\left[1-\dfrac{V_1}{V_2}\right]}{Z_f}$$

$$= \frac{V_2\left[1-\dfrac{1}{A}\right]}{Z_f} = \frac{V_2\left[\dfrac{A-1}{A}\right]}{Z_f}$$

$$\therefore \quad Z_f = \frac{V_2\left[\dfrac{A-1}{A}\right]}{I_2} = Z_2\left[\frac{A-1}{A}\right], \quad \text{where } Z_2 = \frac{V_2}{I_2}$$

$\therefore$ Equivalent impedance at the output port $Z_2 = Z_f\dfrac{A}{[A-1]}$. (10.34)

It can be written as $Z_2 = Z_f\dfrac{1}{\left[\dfrac{A-1}{A}\right]} = Z_f \cdot \dfrac{1}{\left[1-\dfrac{1}{A}\right]}$. (10.35)

Because gain is negative. Normally the amplifier gain is very large. Therefore, $Z_2 \cong Z_f$ at the output port of the four-terminal network.

EXAMPLE 10.11

A resistor $R_f = 100\ \text{k}\Omega$ is connected in between the input and output terminals of an inverting Amplifier having gain $A = -99$. Calculate R_{in} (Miller) at the input port and R_{out} (Miller) at the output port of the Amplifier.

Solution: Given data: Gain $A = -99$ for an inverting amplifier

$$R_{in}\,(\text{Miller}) = \frac{R_f}{[1-A]} \quad \text{using Eq. (10.36)}$$

$$\therefore \; R_f = \left[\frac{100\ \text{k}\Omega}{(1-(-99))}\right] = \left[\frac{100\times10^3}{100}\right] = 1\ \text{k}\Omega.$$

$$R_{out}\,(\text{Miller}) = R_f \cdot \frac{A}{[A-1]} \quad \text{using Eq. (10.34)}$$

$$R_{out}\,(\text{Miller}) = 100\times10^3\left[\frac{-99}{-99-1}\right]$$

$$= (100\times10^3)\frac{99}{100} = 100\times0.99\times10^3$$

$$= 99\times10^3\ \Omega.$$

In general, R_{out} (Miller) is approximately equal to $R_f = 100\ \text{k}\Omega$.

EXAMPLE 10.12

A 10 pF capacitor is connected across the output and input terminals of an inverting Amplifier of gain $A = -99$. What is C_{in} (Miller) and C_{out} (Miller)?

Solution:

$$Z_{in}(\text{Miller}) = \frac{Z_f}{[1-A]} \quad \text{using Eq. (10.33)}$$

For capacitive reactance,
$$\frac{1}{\omega \cdot C_{in}(\text{Miller})} = \frac{1}{\omega \cdot C_f(1-A)} \tag{10.36}$$

$$\because \quad C_{in}(\text{Miller}) = C_f(1-A)$$

$$C_{in}(\text{Miller}) = 10 \text{ pF } [1-(-99)] = 10 \text{ pF} \times 100 = 1000 \text{ pF.}$$

$$Z_{out}(\text{Miller}) = Z\frac{A}{[A-1]} \quad \text{using Eq. (10.34)}$$

For capacitive reactance,
$$\frac{1}{\omega \cdot C_{out}(\text{Miller})} = \frac{1}{\omega \cdot C_f}\left[\frac{A}{A-1}\right]$$

$$\therefore \quad C_{out}(\text{Miller}) = C_f\left[\frac{A-1}{A}\right] \tag{10.37}$$

$$C_{out}(\text{Miller}) = 10 \text{ pF}\left[\frac{-99-1}{-99}\right] = 10 \text{ pF}\left[\frac{100}{99}\right] = 10 \times 1.01 = 10.1 \text{ pF.}$$

Cascading of different types of Amplifiers

Different types of electronic amplifier stages are used to accomplish different goals of responses. The multistage Amplifiers may be a combination of

- CE (Common Emitter) Transistor amplifier stage to obtain voltage gain, CB (Common Base) Transistor amplifier stage to act as a current buffer and CC (Common Collector) Transistor amplifier stage to act as a voltage buffer using Bipolar Junction Transistors of NPN or PNP type.
- CS (Common Source) FET amplifier stage to obtain voltage gain, CG (Common Gate) FET amplifier stage to act as current buffer and CD (Common Drain) FET amplifier stage to act as voltage buffer using the Field Effect Transistor (JFET or MOSFET) family devices.
- Further the amplifier stages may be one of the basic or main groups of (1) *Voltage*, (2) *Current*, (3) *Transconductance* and (4) *Transresistance Amplifiers*.

Normally, the requirements of a multistage Amplifier are met from the combination of simple amplifier stages connected as follows:

1. Input-side amplifier stage to meet high input resistance of the overall multistage Amplifier.
2. Middle-tier Amplifiers consisting of one or more amplifier stages to obtain required high gain of total multistage Amplifier using CE Transistor Amplifier or CS FET Amplifier.
3. Output-side amplifier stage to meet low output resistance of the total multistage Amplifier to provide maximum power transfer to the industrial or practical (real time) loads. The output stage may be Emitter follower or Source follower stages with low output resistance and unity gain.

10.8 (CE + CC) TRANSISTOR AMPLIFIER

Cascading of CE Transistor Amplifier and CC Transistor Amplifier (Emitter follower) circuits is shown in Fig. 10.22.

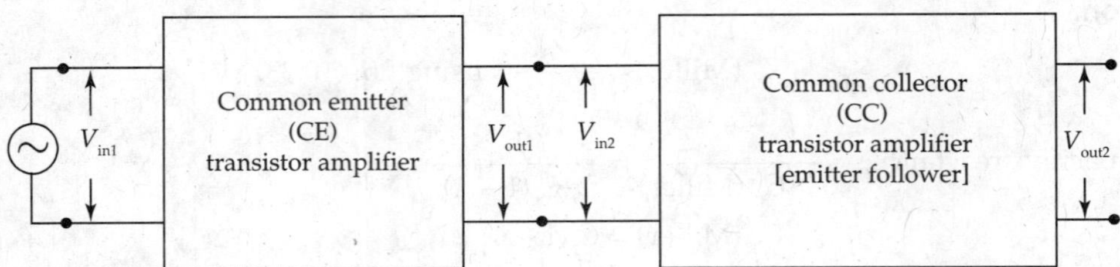

FIG. 10.22 *Cascading of CE transistor amplifier and emitter follower circuits*

Common Emitter Transistor Amplifier has moderately large values of voltage and current gains and their input resistance R_{in} and output resistances R_{out} are in the ranges of kilo ohms depending upon the biasing conditions of the Transistor, whereas the 'Emitter follower circuit' has unity voltage gain, Very large input resistance R_{in} and Very low value of output resistance R_{out}. Cascading of CE Transistor amplifier stage and Emitter follower stage into a multistage Amplifier modifies the system parameters, and the design of so connected multistage Amplifier can be carried out for the practical specifications with the desirable features of both the configurations.

Whenever the output of a CE Transistor Amplifier has to be used with low values of load resistance, a CC Amplifier is used as a buffer Amplifier circuit between the output port of CE Transistor Amplifier and load resistance. Such cascaded Amplifier arrangement is shown in Fig. 10.23.

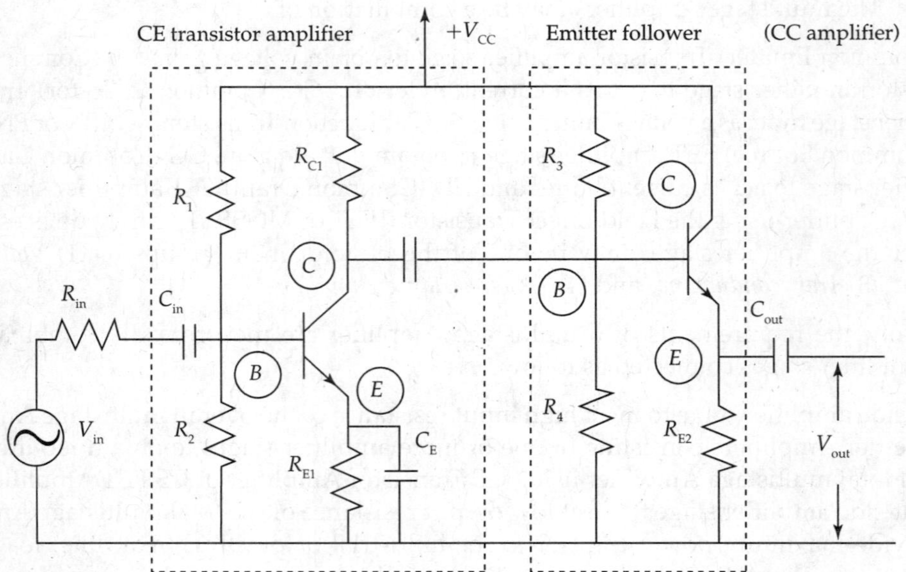

FIG. 10.23 *CE transistor amplifier and CC transistor amplifier (emitter follower) in cascade*

10.9 (CS + CD) FET AMPLIFIER

The block diagram of cascaded CS FET Amplifier and CD FET Amplifier circuit is shown in Fig. 10.24.

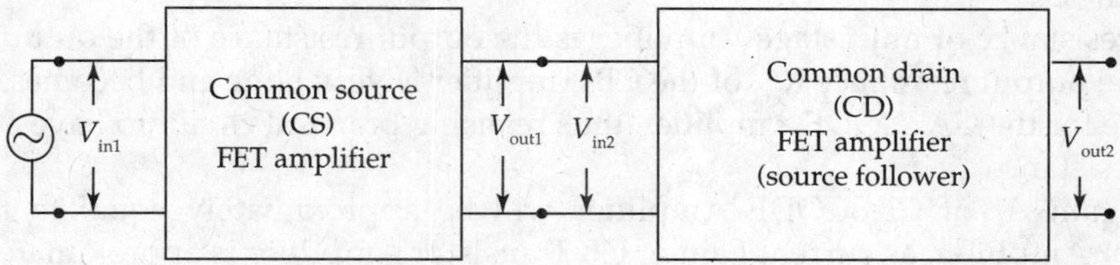

FIG. 10.24 *Cascading of CS FET amplifier and source follower circuits*

Common Source FET Amplifier has moderately large values of voltage and current gains and very high input resistance R_{in} in the ranges of mega ohms and few kilo ohms of output resistance R_{out}, whereas the Source follower circuit has unity voltage gain, Very large input resistance R_{in} and Very low value of output resistance R_{out}. Cascading of CS FET amplifier stage and Source follower stage (CD FET Amplifier) into a multistage Amplifier modifies the system parameters and the design of so connected multistage Amplifier can be carried out for the practical specifications with the desirable features of both the configurations (Fig. 10.25).

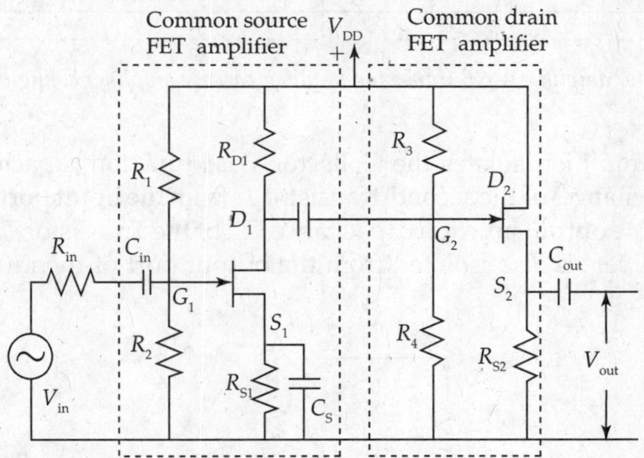

FIG. 10.25 *CS FET amplifier and source follower (CD) FET amplifier circuits in cascade*

10.10 CASCODE (CE + CB) AMPLIFIER

CASCODE Amplifier consists of CE Transistor Amplifier driving CB Transistor Amplifier circuit as shown in Fig. 10.26.

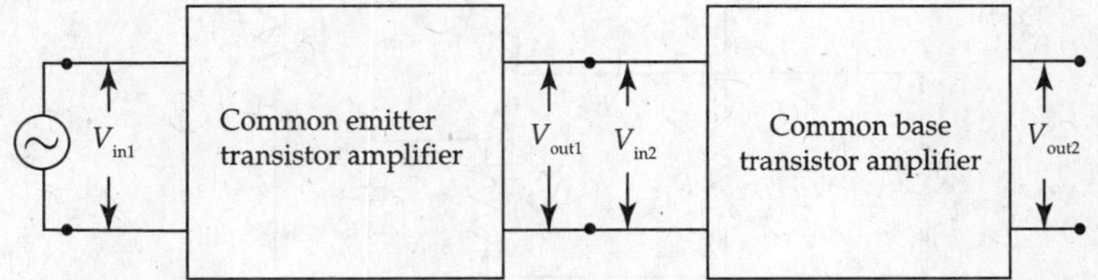

FIG. 10.26 *(CE transistor amplifier and CB transistor amplifier) cascode amplifier*

- It has both the advantages of CE Transistor Amplifier and CB Transistor Amplifier. It has larger bandwidth with good isolation between input and output stages.
- The merit of CASCODE Amplifier is that it provides a high voltage gain over a wide range of frequencies.
- Output resistance of multistage Amplifier is the output resistance of the output amplifier stage. The output resistance R_{out} of the CB Amplifier is very large and becomes the output resistance for the CASCODE Amplifier, thus realising practical circuit to have a very large output resistance.
- Voltage gain A_V of CASCODE Amplifier will be approximately equal to that of CE Transistor Amplifier, as current gain of CB Transistor Amplifier is approximately 1.

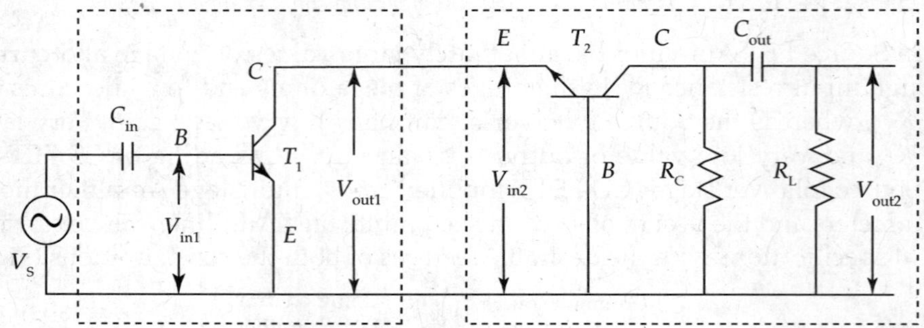

FIG. 10.27 *CE transistor amplifier + CB transistor amplifier cascode amplifier configuration without biasing circuits*

The CB Transistor Amplifier isolates the Collector–Base Junction capacitance C_{BC} or C_C (the output junction capacitance) of the second Transistor T_2 from the input port as the Base terminal is at signal ground. The output junction capacitance C_{BC} of the Transistor T_1 associated with the 'CE Transistor Amplifier' is also isolated from the output port of the multistage 'CASCODE

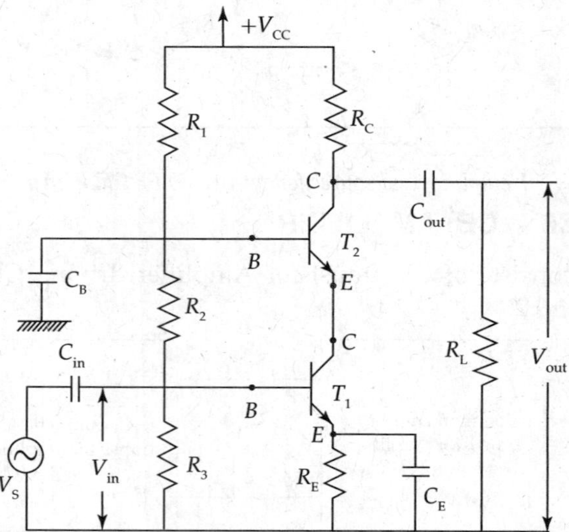

FIG. 10.28 *CASCODE amplifier CE amplifier + CB amplifier configuration*

Amplifier' and so the 'Miller capacitance multiplication effect' is reduced and that further reduces the load resistance of CE Amplifier. This extends the upper cut-off frequency.

Bandwidth of CASCODE Amplifier is very wide, because the reduction in output signal in the high frequency response region is shifted farther and farther and wide bandwidth is realised. So, the 'CASCODE Amplifiers' are used in RF amplifier stages. 'CASCODE Amplifier' has advantages of high-speed working and high voltage Amplifier applications.

Following h-parameter equivalent circuit can be used to calculate various gains:

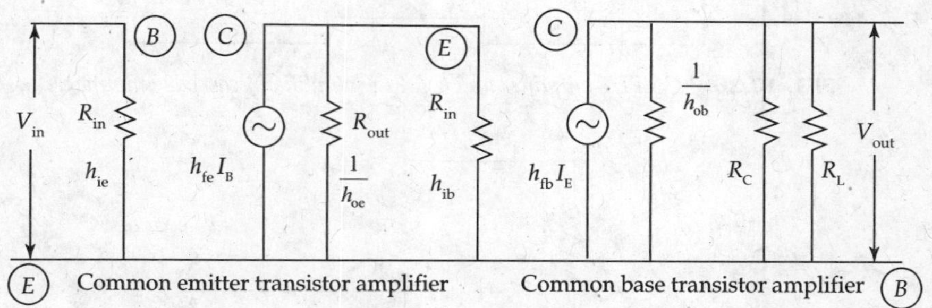

FIG. 10.29 *Cascade amplifier circuit using CE and CB in cascade*

Effective load resistance of the CASCODE Amplifier is R'_L, which is the parallel combination of R_C and R_L.

$$\text{Voltage gain} \quad A_V = -g_m \cdot R'_L .$$

Advantages

1. The output resistance of CASCODE Amplifier is very high and acts close to an ideal current Source.
2. Excellent high frequency response with larger bandwidth.
3. It has the combined properties of wide bandwidth of CB circuit (due to the absence of Miller effect) and high input impedance of CE Amplifier.
4. It provides large voltage gain.
5. As there is no direct coupling between the input and the output, there is improved input and output isolations due to the absence of reverse transfer.

Disadvantage

Relatively high supply voltage is required as two Transistor stages are used in series.

Applications

1. It is used in RF Amplifiers using tuned circuits connected at input and output ports to realise synchronously tuned or stagger tuned Amplifier.
2. It is used in current mirrors to create relatively constant current Source while designing integrated circuits.
3. It is used as a modulator in amplitude modulation by connecting RF signal at the input and audio signal at the output.
4. Widely used in TV tuners as cascading provides higher bandwidth.
5. Popularly used in front end of VHF receivers.

10.11 CASCODE (CS + CG) AMPLIFIER

Figure 10.30 shows a CASCODE circuit containing CS FET Amplifier and CG FET Amplifier circuits. The schematic diagram of a MOSFET CASCODE Amplifier is shown in Fig. 10.31.

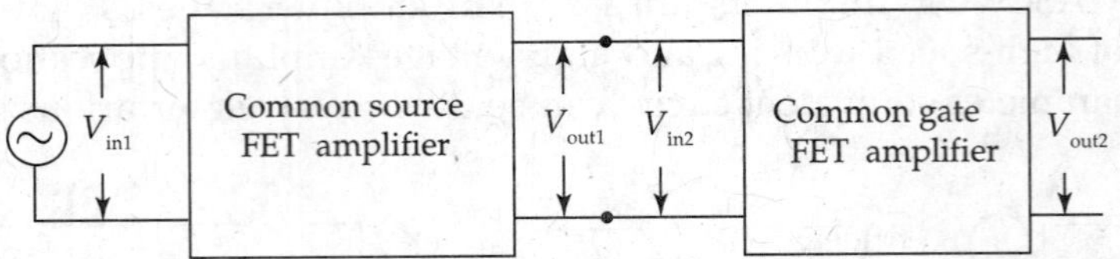

FIG. 10.30 *(CS FET amplifier and CG FET amplifiers) cascode amplifier*

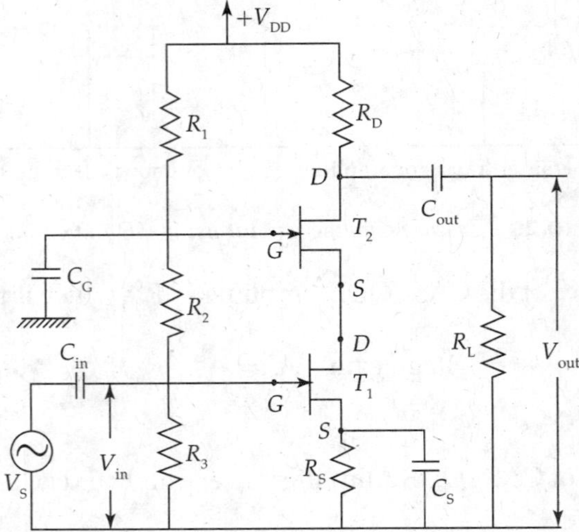

FIG. 10.31 *CASCODE amplifier using FET devices CS FET amplifier + CG FET amplifier*

The Drain current of a CS Amplifier is fed to the Source of another MOSFET Amplifier connected in common Gate Mode (Current Amplifier). The miller effect of CS Amplifier due to the capacitor C_{GD} is reduced to a very small quantity. As a result, the bandwidth of CASCODE Amplifier is very high when compared to the bandwidth of CS Amplifier when connected alone. The current buffer is associated with low input resistance and high bandwidth. This is also one of the reasons for the wider bandwidth of CASCODE Amplifier. The primary advantages of CASCODE Amplifier, the loaded Source is completely decoupled. It has high input resistance and wider bandwidth.

10.12 (CC + CE) TRANSISTOR AMPLIFIER

The phenomenon of 'Increase in Capacitance at the input port due to the Miller Effect' in the CE Transistor Amplifier circuit is absent in CC Transistor Amplifier circuit. So, Emitter follower Circuits have large bandwidth. Cascading two single-stage Amplifiers such as 'Common Collector Amplifier' (Emitter Follower) circuit and 'Common Emitter Transistor Amplifier' circuit, the resulting multistage Amplifier provides 'large gain with excellent high frequency response'.

Emitter Follower and CE Transistor Amplifiers connected in cascade for realising improved performance characteristics as discussed below are shown in Fig. 10.32. Figure 10.33 shows the (CC + CE) Transistor Amplifier configuration.

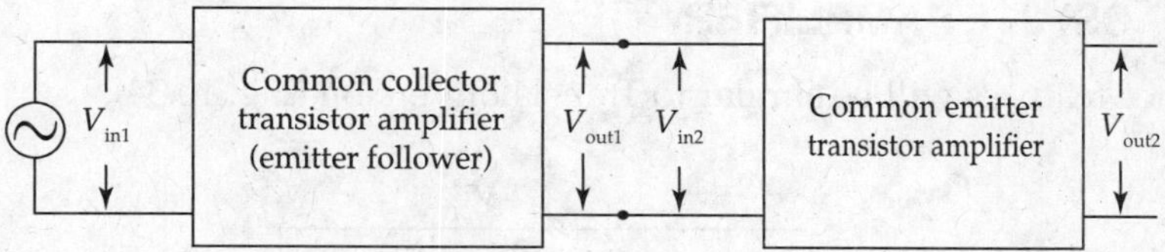

FIG. 10.32 *CC transistor amplifier and CE transistor amplifiers*

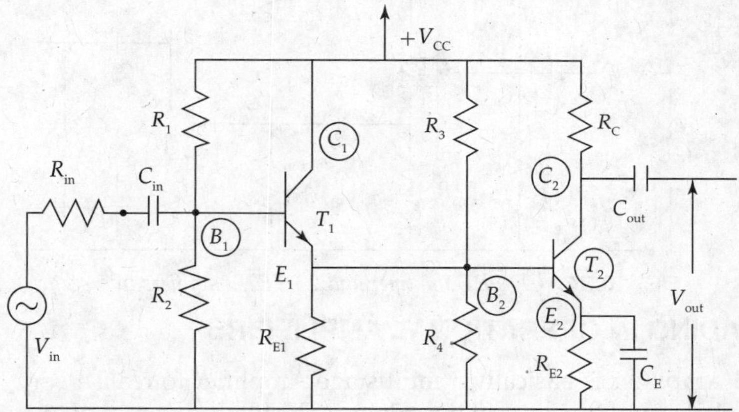

FIG. 10.33 *CC transistor amplifier and CE transistor amplifiers in cascade*

The Transistor T_2 of the CE Transistor Amplifier provides the voltage gain. Due to 'Miller Capacitance Multiplication Effect' present at the input port of CE Transistor Amplifier, the effective input capacitance between the Base and the Ground of the second Transistor T_2 increases and will be large. Time constant τ at that port is reduced to a small value because it is the product of the large Miller capacitance with very low output resistance (of Emitter Follower circuit with the Transistor T_1). So, there is no reduction in signal in the High-frequency region of the multistage Amplifier. Thus, the wide bandwidth response characteristic of the Emitter follower and the large gain of the CE Transistor Amplifier are realised with the multistage CC–CE Amplifier configuration.

Advantages

1. The CC + CE cascade has an excellent high frequency response when compared with a CASCODE.
2. The frequency response is also superior to that of a CC + CC circuit.
3. It is a modified version of Darlington configuration and can be used as a high performance voltage follower circuit.
4. Even though the upper 3-dB frequency is not as high as that of a CASCODE, the mid-frequency gain is higher that results in improved gain bandwidth product.

Applications

It is used in RF Amplifiers.

Analogous to the above configuration, a sister configuration (CD + CG) is also available using MOSFET devices.

10.13 (CD + CS) JFET AMPLIFIER

The performance features will be similar to Transistor version (Fig. 10.34).

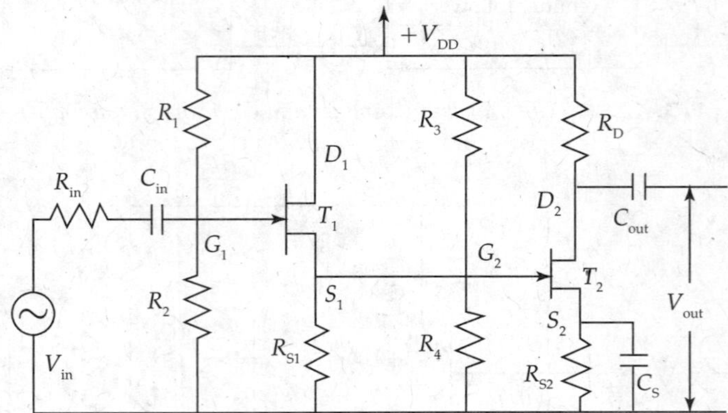

FIG. 10.34 *CD FET amplifier and CS FET amplifiers in cascade*

10.14 CASCADING IN OPERATIONAL AMPLIFIERS

The operational Amplifier is basically a multistage Amplifier to realise very high gain, very high input impedance and low output impedance. The block diagram of an op-amp is shown in Fig. 10.35.

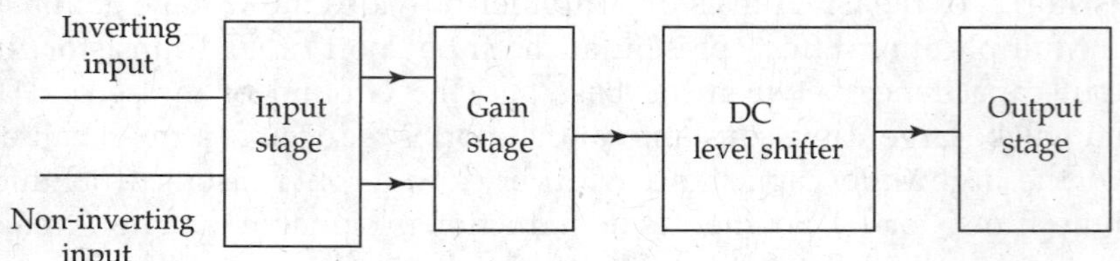

FIG. 10.35 *Block diagram of a two-stage operational amplifier (op-amp)*

The input stage is a dual input, balanced output differential Amplifier. It provides most of the voltage gain. It also establishes the input resistance of the operational Amplifier. The input stage is cascaded with a second stage usually a differential Amplifier. Both the stages are directly coupled. The second stage provides necessary gain and employs a dual input unbalanced differential Amplifier to provide the additional voltage gain. A level shifting arrangement known as DC level shifter employs an Emitter follower using a constant current Source. The level shifting is necessary to bias the final stage. It can be a part of either the second stage or the output stage. It may be even a separate block in between the stages shown in Fig. 10.35. The output stage will be a complementary symmetry push pull Amplifier or a current driver to increase the swing in thee output voltage and to enhance the current supplying capacity of the operational Amplifier. It provides low impedance and capable of driving smaller loads.

As direct coupling is used, the op-amp behaves like an ideal Amplifier with a very large gain that remains constant over a wide band of frequency starting down from a zero frequency to a very large frequency. *Operational Amplifier behaves like a multistage Amplifier.* It employs precision-engineered complex circuitry under the domain of microelectronics. It is sufficient to know at this stage that the *op-amp* is a multistage Amplifier with low off-set voltage and current, high CMRR, high input impedance, relatively large gain and larger gain bandwidth product and popularly available in IC form requiring fewer external components depending upon the type of application.

10.15 DARLINGTON PAIR (COMPOUND TRANSISTOR CONFIGURATIONS)

Composite Transistor Amplifiers are special purpose multistage Amplifiers considered as a single unit with special purposes. These configurations are popularly available in discrete Amplifiers as well as in Amplifiers using ICs. The various types of composite Transistor configurations are

1. (CC + CC) pair (Darlington configuration),
2. (CC + CB) pair and
3. (CC + CE) pair.

10.15.1 Darlington (CC + CC) Pair

The Darlington pair configuration shown in Fig. 10.36 is named after its inventor Sidney Darlington, an engineer of Bell Telephone laboratories, USA. It is a compound device using two Transistors in a single unit, wherein the Collector terminals are tied together; the Emitter current of one Transistor is connected to the Base of the second Transistor. Only three external connections Base, Emitter and Collector are made available for application use. The configuration used is (CC + CC) Amplifier configuration or two Emitter follower circuits in one module.

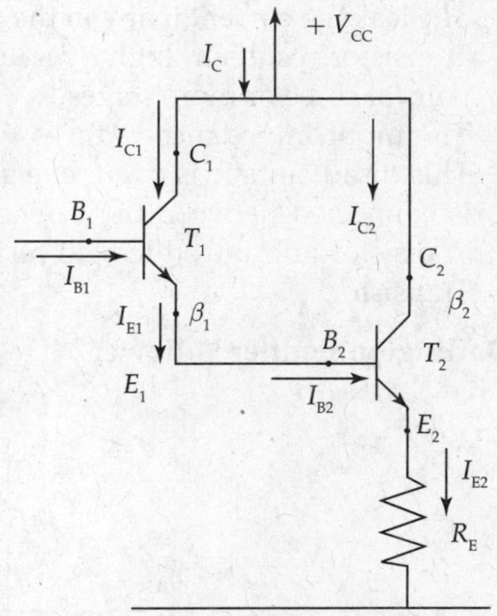

FIG. 10.36 *Darlington pair using two NPN transistors in CC + CC model*

Analysis:

$$I_{C1} = \beta_1 \cdot I_{B1} \tag{10.38}$$

$$I_{B2} = I_{E1} = I_{B1} + I_{C1} = I_{B1} + \beta_1 \cdot I_{B1} = [1+\beta_1] \cdot I_{B1} \tag{10.39}$$

$$I_{C2} = \beta_2 \cdot I_{B2} \tag{10.40}$$

$$\text{But} \quad I_C = [I_{C1} + I_{C2}] = \beta_1 \cdot I_{B1} + \beta_2 \cdot I_{B2} = \beta_1 \cdot I_{B1} + \beta_2 \cdot [1+\beta_1] \cdot I_{B1} \tag{10.41}$$

$$\therefore \quad I_C = I_{B1}[\beta_1 + \beta_2 + \beta_1 \cdot \beta_2] \tag{10.42}$$

$$\text{Overall composite gain} \quad [\beta_1 + \beta_2 + \beta_1 \cdot \beta_2] \approx \beta_1 \cdot \beta_2. \tag{10.43}$$

If β of each Transistor = 100, the overall current gain = 10,200 from Eq. (10.43), which is approximately equal to 10,000.

The two Transistors need not necessarily be of NPN type. They can be PNP type Transistors also, which are used popularly in discrete circuit designs. But good quality PNP Transistors are rarely available. An alternative compound configuration in which one Transistor is PNP and another Transistor is NPN known as feedback pair is more popular in integrated circuits. Feedback pair is shown in Fig. 10.37.

The overall current gain of the feedback pair = $\beta_N \cdot \beta_P$ where β_N is the beta (β) parameter of NPN Transistor and β_P is the beta (β) parameter of PNP Transistor.

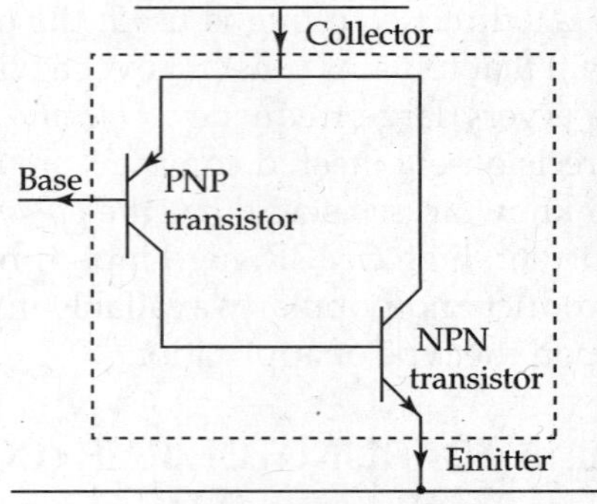

FIG. 10.37 *Feedback pair (Modified Darlington pair)*

Advantages

- As the overall current gain is very large, they are used for large current applications to enhance the current supplying capacity of voltage regulators.
- High input impedance.
- They can be used as voltage follower with high current gain, high input resistance and overall voltage gain very close to unity.

Disadvantages

- The leakage current if any in the first Transistor in the Amplifier is amplified by the second Transistor results in high overall leakage current. This limits the usage of the Darlington pair for three or more stages.
- The input impedance reduces when biasing arrangement is made in a Darlington pair. This disadvantage is overcome by adopting bootstrapping technique, wherein a capacitor is connected between the Collector of the first Transistor and the Emitter of the second Transistor and introducing a resistor in between the Base and the Collector circuit of the Transistor.

Darlington Emitter Follower

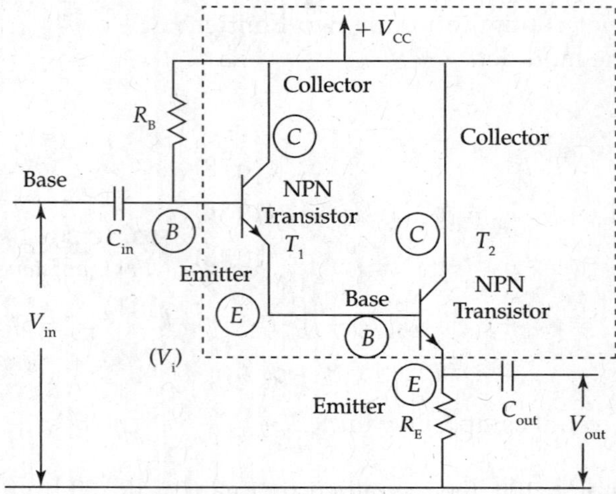

FIG. 10.38 *Darlington emitter follower*

Some examples of NPN Darlington Transistors are BSP 29, BCV 49 (Philips) (High Current: 1 A, High DC current gain: Min = 1000 and Max = 20,000, Low voltage: 80 V). BCV 26 and BCV 46 are PNP Darlington Transistors (Philips) (High current: 500 mA, High DC current gain: 10,000, Low voltage: 60 V).

Analysis of Darlington Emitter Follower

Figure 10.38 shows Darlington Emitter follower circuit and Fig. 10.39 shows the small signal equivalent circuit of Darlington Emitter Follower.

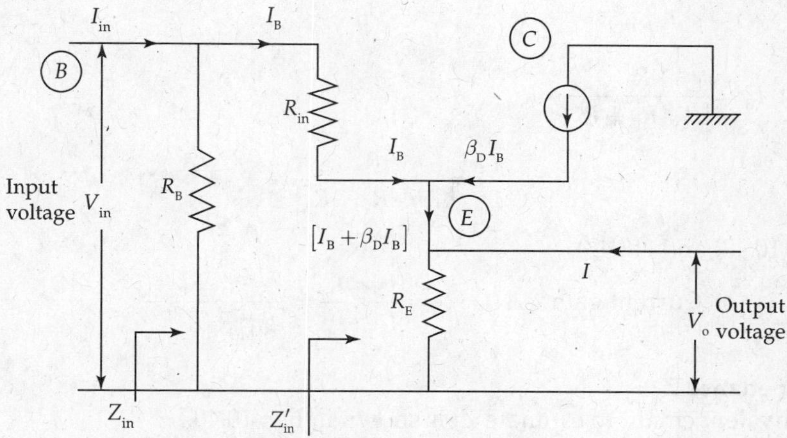

FIG. 10.39 *Small signal equivalent circuit of darlington emitter follower*

Derivation for current gain A_1 of Darlington Emitter Follower

Current flowing through the input resistance R_{in} if the Transistor T_1

$$I_B = \frac{[V_{in} - V_o]}{R_{in}}. \tag{10.44}$$

Current of $I_B + \beta_D \cdot I_B$ flows through the Emitter resistor R_E where β_D is the overall β of the Darlington pair:

$$\therefore \quad [I_B + \beta_D \cdot I_B] \cdot R_E = V_o. \tag{10.45}$$

Equation (10.44) can be written as

$$V_{in} = V_o + I_B \cdot R_{in}.$$

Substituting Eq. (10.45) in the above equation,

$$V_{in} = [I_B + \beta_D \cdot I_B] \cdot R_E + I_B \cdot R_{in} = I_B \cdot [(1 + \beta_D) \cdot R_E + R_{in}]$$

But $\beta_D \gg 1$. Therefore, $\beta_D \cdot R_E \gg R_{in}$

$$\therefore \quad V_{in} = I_B [\beta_D \cdot R_E]$$

Therefore, The impedance looking into the base of the Transistor $T_1 = Z'_{in}$:

$$Z'_{in} = \frac{V_{in}}{I_B} = \beta_D \cdot R_E.$$

Therefore, Input impedance

$$Z_{in} = R_B \parallel Z'_{in} = R_B \parallel \beta_D \cdot R_E \tag{10.46}$$

is very high. Hence, Darlington pair is a High input resistance circuit.

AC current gain

Current through R_E is I_o

$$I_o = I_B + \beta_D \cdot I_B \approx \beta_D \cdot I_B$$

$$\therefore \quad \frac{I_o}{I_B} = \beta_D. \tag{10.47}$$

Base current $I_B = \dfrac{R_B}{[\beta_D \cdot R_E + R_B]} \times I_{in}$

$$\therefore \quad \frac{I_B}{I_{in}} = \frac{R_B}{[\beta_D \cdot R_E + R_B]}. \tag{10.48}$$

Using Eqs. (10.47) and (10.48),

$$\text{Current gain} \quad A_I = \frac{I_o}{I_{in}} = \frac{I_o}{I_B} \times \frac{I_B}{I_{in}} = \frac{\beta_D \cdot R_B}{[\beta_D \cdot R_E + R_B]}. \tag{10.49}$$

AC output impedance Z_o

Simplified equivalent circuit to estimate Z_o is shown in Fig. 10.40.

Make Source voltage zero in the equivalent circuit of Fig. 10.40 and apply a test voltage *V* at the output terminals to find *I*. Then the output impedance is the ratio of *V* to *I*. The method of finding the output impedance is shown in Fig. 10.41.

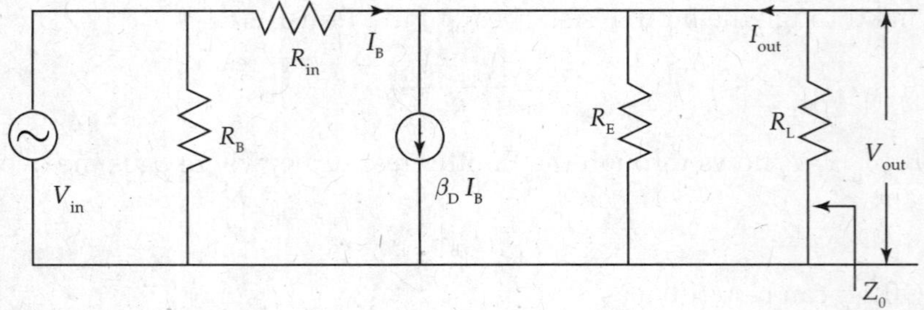

FIG. 10.40 *Equivalent circuit to estimate the AC output impedance Z_o*

$$I = \frac{V}{R_E} + \frac{V}{R_{in}} - \beta_D \cdot I_B.$$

But $I_B = V/R_{in}$.

$$\therefore \quad I = V\left[\frac{1}{R_E} + \frac{1}{R_{in}} - \frac{\beta_D}{R_{in}}\right]$$

$$\text{Hence,} \quad Y_o = \frac{I}{V} = \left[\frac{1}{R_E} + \frac{1}{R_{in}} - \frac{\beta_D}{R_{in}}\right]$$

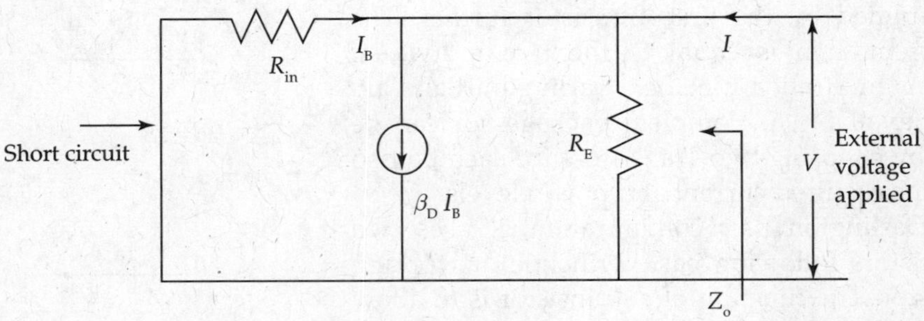

FIG. 10.41 *Method of determination of Z_0*

$$\therefore\ Z_o = \frac{1}{Y_o} = \left[R_E \middle\| R_{in} \middle\| \frac{R_{in}}{\beta_D} \right] \approx \frac{R_{in}}{\beta_D}. \tag{10.50}$$

Voltage gain A_V

Simplified equivalent circuit is shown in Fig. 10.42 to calculate voltage A_V:

$$V_{in} = I_B \cdot R_{in} + I_B \cdot R_E + I_B [\beta_D \cdot R_E]$$

$$V_o = I_B \cdot R_E + I_B \cdot \beta_D \cdot R_E$$

$$\therefore \text{ Voltage gain } \quad A_V = \frac{V_o}{V_{in}} = \frac{R_E[1+\beta_D]}{R_{in} + R_E[1+\beta_D]} \approx 1. \tag{10.51}$$

From the above calculations, voltage gain of a Darlington pair is approximately 1.

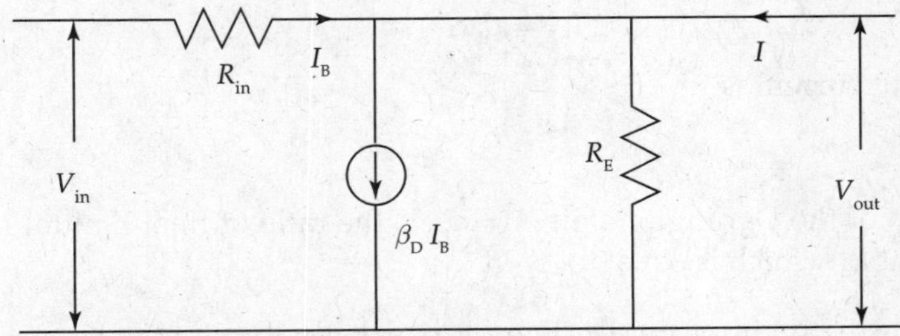

FIG. 10.42 *Method of determination of voltage gain A_V*

10.16 HIGH INPUT RESISTANCE TRANSISTOR CIRCUITS (DARLINGTON PAIR)

One of the popular high input resistance Transistor Amplifier circuits is a Darlington pair. Darlington pair consists of two Transistors that behave like a single Transistor (known as super beta Transistor) with high current gain and very high input resistance.

It is used to amplify weak signals so that another transducer or a circuit can detect them. The first Transistor's Emitter feeds into the second Transistor's Base and as a result the input signal V_{in1} is amplified and appears across the output of the second Transistor as V_{out2}. The ratio of the output voltage V_{out2} to the input voltage V_{in1} is the voltage gain of the Amplifier and it represents the magnitude by which the weak signal is amplified.

Darlington Transistor circuits have high current gains and power handling capabilities. Darlington pair configuration is shown in Fig. 10.43. The second Transistor amplifies the current amplified by the first Transistor further. The overall current gain is equal to the two individual current gains multiplied together. Darlington Pairs are completely available as complete packages or can be assembled by choosing two Transistors for each pair so as to meet the required current and power levels.

In this Darlington pair configuration, it uses two NPN Transistors following basic principle of working of a Transistor. The function of a Transistor is to allow the small amount of input Base current I_B that enters its Base terminal to control the amount of Collector current I_C, flowing from its Collector terminal to its Emitter terminal. Thus the low power in the input port of the Transistor controls a higher power in its output port to meet the practical applications.

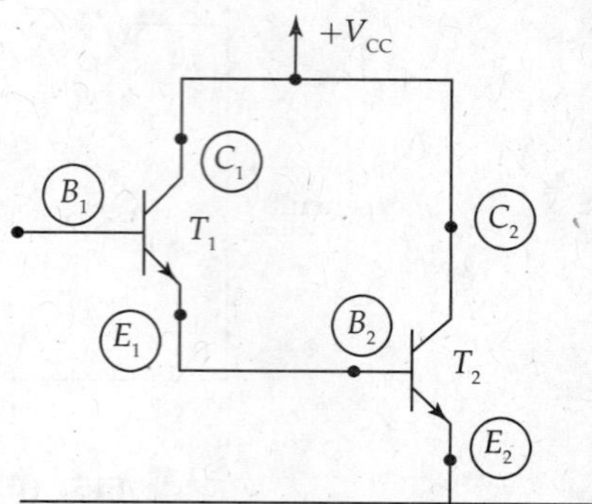

FIG. 10.43 *Darlington pair using NPN transistors*

The function of the Darlington pair acting as *super beta Transistor* having β as the product of the beta values β_1 and β_2 of the two Transistors ($\beta = \beta_1 \cdot \beta_2$) is explained from the following concepts.

The input current to the first Transistor T_1 is the Base current I_{B1}.

The amplified current of the Transistor T_1 is I_{E1}, which is equal to $\beta_1 \cdot I_{B1}$. (10.52)

This Emitter current I_{E1} is fed to the input port of the second Transistor. Therefore, the input current to the second Transistor T_2 is $I_{B2} = \beta_1 \cdot I_{B1}$. The second Transistor T_2 amplifies this Base current I_{B2}.

The Emitter current of the Darlington pair is

$$I_{E2} = \beta_1 \cdot I_{B2} \cdot \tag{10.53}$$

The final output current is

$$I_{E2} = \beta_1 \cdot \beta_2 \cdot I_{B1} \cdot \tag{10.54}$$

Total current gain of the Darlington pair $A = h_{FE}$ is the ratio of final Emitter current I_{E2} to the input Base current I_{B1}, which is equal to $\beta_1 \cdot \beta_2 \cdot$ (10.55)

So, the total current gain of the Darlington pair is the product of the individual current gains $\beta_1 \cdot \beta_2$. Total current gain of the Darlington pair is also known as β_D.

Therefore, the super beta of Darlington pair is $\beta_D = \beta_1 \cdot \beta_2$.

Current gain of Darlington pair is also represented as $h_{FE} = h_{FE1} \cdot h_{FE2}$, where the individual current gains of the two Transistors are h_{FE1} and h_{FE2}. If $h_{FE1} = 100$ and $h_{FE2} = 100$, then the total current gain = 10,000.

The maximum Collector current $I_{C(max)}$ from the Darlington pair is the maximum Collector current $I_{C(max)}$ for the second Transistor as seen from the currents flowing through the individual Transistors. Darlington pair is used in power output stages.

To turn on the Darlington pair Transistor, there must be 0.7 V across both the Base–Emitter junctions that are connected in series inside the Darlington pair. As can be seen in Fig. 10.43, $V_{BE1} = 0.7$ V and $V_{BE2} = 0.7$ V and the total required bias voltage is 1.4 V for the Darlington pair to conduct. As the current gains of Darlington pair Transistors is very high, the pair is highly sensitive to small magnitudes of currents. Darlington pair on a chip occupies less space than two individual discrete devices. Sidney Darlington, an Engineer at Bell Laboratories (USA) in the 1950s, is the pioneer to combine the two Transistors on a single chip.

10.16.1 Darlington Emitter Follower

Second aspect of Darlington Pair is considered as Emitter follower circuit configuration (Fig. 10.44). Common Collector (CC) Transistor Amplifier popularly known as Emitter follower has its important applications as *buffer Amplifier* in interfacing two circuits having high output resistance, R_{out} and low resistance loads R_L in real-world applications. They function as *impedance transformers*.

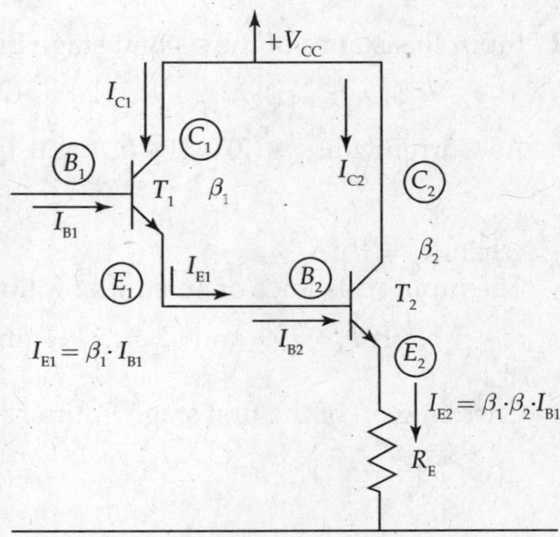

FIG. 10.44 *Darlington pair with emitter resister R_E*

To realise Amplifier circuit with high input resistances, *Darlington pair circuits* are used. The input resistance R_{in} of an Emitter follower circuit is a function of the current gain β of the Transistor. Hence, the maximum input resistance of an Emitter follower circuit is limited by the Beta value of the Transistor as evident from the equation

$$R_{in} = \beta R_E.$$

The Darlington Emitter follower circuit shown in Fig. 10.45 is a cascaded Amplifier of two Emitter follower circuits. It is already familiar that in Cascaded Amplifiers, the Amplifier analysis has to be started from the last amplifier stage. Therefore, the circuit analysis starts from the second Amplifier (Emitter follower) stage as shown below.

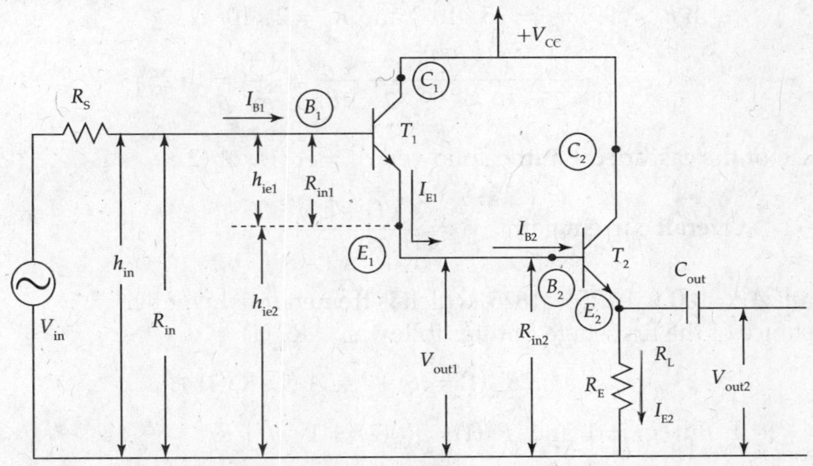

FIG. 10.45 *Darlington emitter follower circuit*

1. Current gain of Emitter follower (2) $= A_I(2) = \dfrac{h_{fc}}{[1 + h_{oc} \cdot R_E]}$ (10.56)

$$\text{As } h_{oc} \cdot R_E \leq 0.1 \qquad [1 + h_{oc} \cdot R_E] \cong 1 \tag{10.57}$$

$$\therefore \; A_J(2) = h_{fc} = [1 + h_{fe}] \tag{10.58}$$

If $h_{fe} = 99$, current gain $A_I(2) = 100$.

2. Input Resistance of the second stage Emitter follower (2) = R_{in} (2)

$$\therefore \; R_{in}(2) = h_{ie} + h_{rc} \cdot A_I(2) \cdot R_E \tag{10.59}$$

As Current gain $A_I(2) = (1 + h_{fe})$ and $h_{ie} \langle\langle (1 + h_{fe})R_E$ and $h_{rc} \equiv 1$

$$R_{in}(2) \cong (1 + h_{fe})R_E = (1 + h_{fe})R_L \tag{10.60}$$

when $R_E = R_L$.
The input resistance of an emitter follower circuit is very high from Eq. (10.60).

If $h_{fe} = 99$ and $R_E = 2\ \text{k}\Omega$, then $R_{in}(2) = (1 + 99) \times 2 \times 10^3 = 200\ \text{k}\Omega$.

3. Current gain of the first stage Emitter follower

$$A_I(1) = \frac{h_{fc}}{[1 + h_{oc} \cdot R_L(1)]}. \tag{10.61}$$

But the load resistance of the first Emitter follower stage $R_L(1) = R_{in}(2)$ and $h_{fc} = (1 + h_{fe})$.

$$\therefore \; A_I(1) = \frac{[1 + h_{fe}]}{[1 + h_{oe}(1 + h_{fe}) \cdot R_E]} \tag{10.62}$$

$$\text{As } h_{oe} \cdot R_E \ll 0.1$$

$$\text{Current gain} \quad A_I(1) \cong \frac{[1 + h_{fe}]}{[1 + h_{0e} \cdot h_{fe} \cdot R_E]} \tag{10.63}$$

$$\text{If } h_{fe} = 99,\ h_{oe} = 25 \times 10^{-6} \text{ and } R_E = 2 \times 10^3\ \Omega$$

$$A_I(1) = \frac{[1 + 99]}{[1 + 25 \times 10^{-6} \times 99 \times 2 \times 10^3]} = \frac{100}{6} = 16.25.$$

Current gain of the cascaded Emitter follower $A_I = A_I(1) \times A_I(2)$ (10.64)

$$\text{Overall current gain} \quad A_I = \frac{(1 + h_{fe})^2}{[1 + h_{oe} \cdot h_{fe} \cdot R_E]}. \tag{10.65}$$

Current gain $A_I = 100 \times 16.25 = 1625$, which is tremendously higher.

4. Input resistance of the first stage Emitter follower = R_{in} (1)

$$R_{in}(1) = h_{ie} + h_{rc} \cdot A_I(1) \cdot R_L(1)$$

$$\text{But } h_{rc} = 1 \text{ and } R_L(1) = R_{in}(2) = [1 + h_{fe}] \cdot R_E. \tag{10.66}$$

Using Eqs. (10.64), (10.65) and (10.66), we get

$$\therefore \; R_{in}(1) = \frac{[1 + h_{fe}]^2 \cdot R_E}{[1 + h_{oe} \cdot h_{fe} \cdot R_E]}. \tag{10.67}$$

From the practical values, $h_{fe} = 99,\ R_E = 2\ \text{k}\Omega,\ h_{oe} = 25\times10^{-6}$

$$\therefore R_{in}(1) = \frac{[1+99]^2 \times 2\times10^3}{[1+25\times10^{-6}\times99\times2\times10^3]} = \frac{10^4\times20\times10^2}{6} \cong 3.3\ \text{M}\Omega.$$

EXAMPLE 10.13

Explain how the input impedance of Darlington Emitter follower (Fig. 10.46) is higher than that of a single-stage Emitter follower circuit.

Solution: Circuit of Darlington Emitter follower (Fig. 10.46):

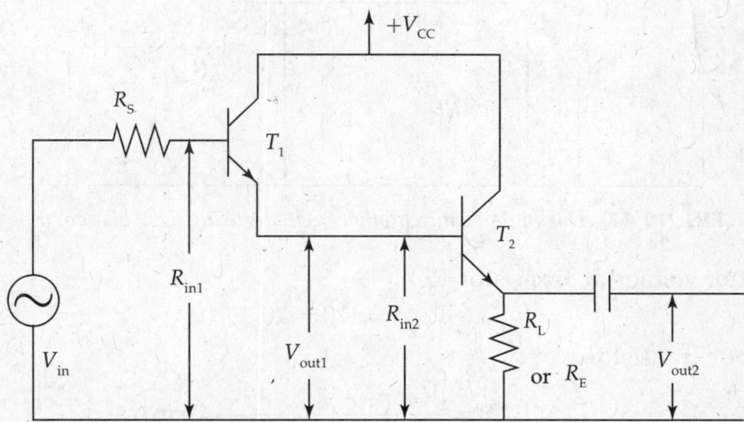

FIG. 10.46 *Darlington emitter follower circuit*

a. The input resistance of the second stage Emitter follower $= R_{in2}$

$$R_{in2} = h_{ie} + (1+h_{fe})R_L.$$

b. The input resistance R_{in1} of the first stage Emitter follower

$$R_{in1} = h_{ie} + A_{I1}\cdot R_{in2} = \frac{(1+h_{fe})^2\cdot R_E}{(1+h_{oe}\cdot h_{fe}\cdot R_E)}$$

This equation for R_{in1} is valid for the condition that $h_{oe}\cdot h_{fe} < 0.1$.

If $R_E = R_L = 4.5\ \text{k}\Omega,\ h_{ie} = 1\ \text{k}\Omega,\ h_{re} = 2.5\times10-6,\ h_{oe} = 25\times10-6$ mhos, $h_{fe} = 49$

$$R_{in2} = h_{ie} + (1+h_{fe})R_L = 1\times10^3 + (1+49)\times4.5\times10^3 = 226\ \text{k}\Omega$$

R_{in2} is the input resistance for the second stage Emitter follower

$$R_{in1} = h_{ie} + A_{I1}\cdot R_{in2} = \frac{(1+h_{fe})^2\cdot R_E}{(1+h_{oe}\cdot h_{fe}\cdot R_E)}$$

$$= (1+49)^2\times5\times10^3 = 12.5\times10^6\ \Omega.$$

These calculations clearly show that a two-stage Emitter follower has very high input impedance than that of a single-stage Emitter follower.

EXAMPLE 10.14

Calculate the value of Bias resistance R_B in the Darlington Emitter follower circuit using fixed biasing circuit shown in Fig. 10.47. $\beta_1 = 99$, $\beta_2 = 99$ and $R_E = 2000\ \Omega$.

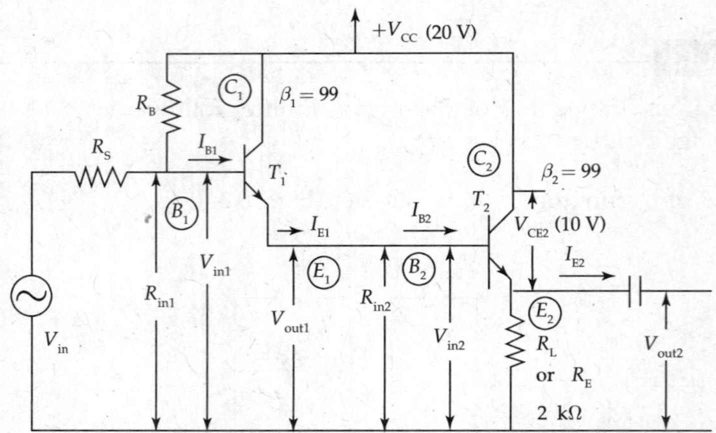

FIG. 10.47 *Darlington emitter follower circuit using fixed bias circuit*

Solution: Emitter voltage at Transistor T_2

$$V_{E2} = (V_{CC} - V_{CE2}) = (20 - 10) = 10\text{ V}.$$

Emitter current of Transistor

$$T_2 = I_E(2) = \frac{[V_{CC} - V_{CE2}]}{R_E} = \frac{10}{2000} = 5\text{ mA}.$$

Base current

$$I_B(2) = \frac{I_E(2)}{(1+\beta_2)} = \frac{5\times 10^{-3}}{100} = 50\times 10^{-6} = 50\ \mu\text{A}.$$

Emitter current of Transistor

$$T_1 = I_E(1) = I_B(2) = 50\ \mu\text{A}.$$

Base current of T_1 $$I_B(1) = \frac{I_E(1)}{(1+\beta_1)} = \frac{50\times 10^{-6}}{(1+99)} = \frac{50\times 10^{-6}}{100} = 0.5\ \mu\text{A}.$$

Voltage across R_B $$V_{RB} = [V_{CC} - V_{BE}(1) - V_{BE}(2) - V_{RE}]$$

$$V_{RB} = [20 - 0.7 - 0.7 - 5\times 10^{-3}\times 2\times 10^{3}] = [20 - 11.4] = 8.6\text{ V}$$

Resistor R_B $$\frac{V_{RB}}{I_B(1)} = \frac{8.6}{0.5\times 10^{-6}} = 17.2\times 10^{6} = 17.2\ \text{M}\Omega.$$

(CC + CE) Composite Pair

This configuration of composite Transistor is a development over Darlington pair. The Collector terminals are not tied together, but are connected in parallel resulting in reduced output resistance. The frequency response is superior to that of CC configuration. For connecting equivalent Common Emitter stages, the CC + CE combination is preferred over CC + CC combination. The (CC + CE) configuration is shown in Fig. 10.48.

The Emitter current of Transistor T_1 drives the Base of the Transistor T_2. The overall composite gain is $\beta_1 \cdot \beta_2$. When the Collector terminal of the second Transistor is returned to the same power supply, CC + CE pair becomes a Darlington pair.

Apart from the above configurations, CC + CB pair is used in Emitter-coupled Amplifier to realise larger bandwidth in high-frequency Amplifiers.

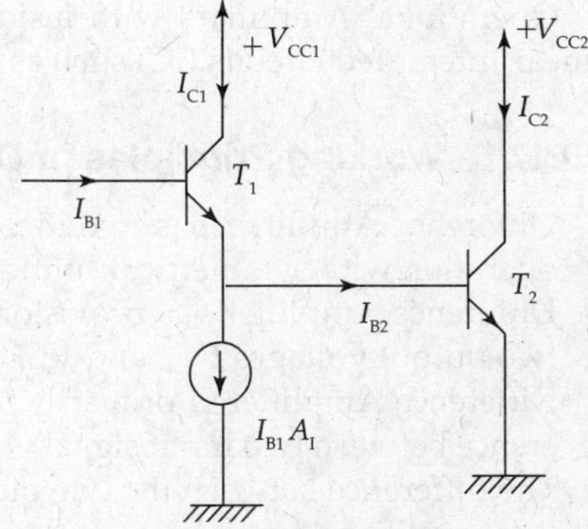

FIG. 10.48 *CC + CE composite transistor*

10.17 DIFFERENCE AMPLIFIERS

Basically computers were classified as

1. Analog computers and
2. Digital computers.

A difference Amplifier is also known as *differential Amplifier*. Differential Amplifier is one of the basic building blocks of an operational Amplifier used mostly in analog computers in previous days to solve differential equations used in computations and electronic instrumentation Amplifiers.

The Amplifier (Fig. 10.49) circuit amplifies the difference of the two input voltages applied to the two Transistors. Hence, it is known as *Difference Amplifier*. The differential pair is also known as *Emitter-Coupled Pair*.

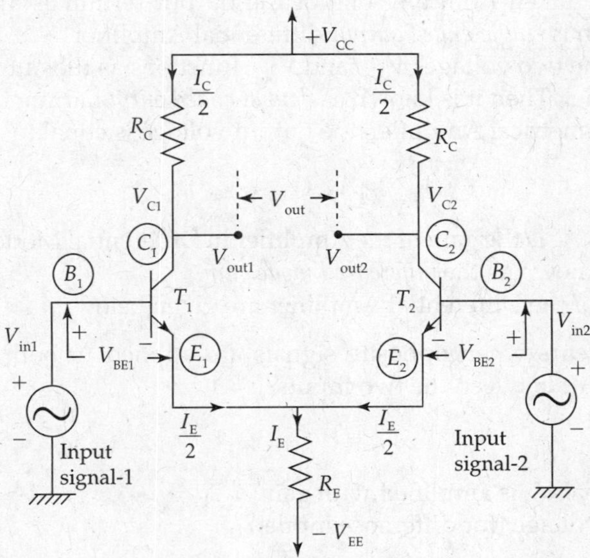

FIG. 10.49 *BJT difference amplifier*

Operational Amplifiers with inside differential Amplifier circuits are popularly used in linear Integrated circuits (IC) such as **μA 741** (Operational Amplifier).

10.17.1 Working Principles of Difference Amplifier (Fig. 10.49)

- Difference Amplifier has two identical Transistors, T_1 and T_2, connected in CE Transistor operation with symmetrical configuration.
- Difference Amplifier has provision to connect two input voltages V_{in1} and V_{in2} and obtain two output voltages V_{out1} and V_{out2}.
- Difference Amplifier is primarily used to amplify the differential signal, which is the difference between two input signal voltages and produces two output voltages V_{out1} and V_{out2}.
- The difference between the two output voltages is taken as single output voltage V_{out} from the Amplifier.
- The circuit is designed for equal biasing voltages V_{BE1} and V_{BE2}, so that biasing voltage becomes $V_{BE} = 0.7$ V for the two identical Transistors.
- The two Emitters are connected together and resulting DC bias current I_E through R_E will be shared equally by the two Transistors T_1 and T_2.
- Each Transistor shares 0.5 I_E to contribute to total Emitter current I_E through the Emitter resistor R_E connecting the two Emitters.
- The two Collector currents I_{C1} and I_{C2} are equal to 0.5 I_C.
- Each Transistor Collector current $I_C = 0.5\ I_E$.
- The total current I_E is the sum of the two DC Collector currents of each Transistor.
- The two Collector resistances R_{C1} and R_{C2} are set to equal value R_C.
- Then the two Collector currents will be equal resulting equal magnitudes of DC Collector voltages V_{C1} and V_{C2}.
- Output voltages V_{out1} and V_{out2} are developed at the two Collector points, when the input signal voltages are applied.
- The output can be taken from any one of the output terminals and ground. Then the Amplifier operation is *single-ended output* differential Amplifier.
- The difference of the two voltages V_{out1} and V_{out2} functions as the output voltage V_{out} of the differential Amplifier. Then it is known as *double-ended output* arrangement.
- For a perfectly symmetrical Amplifier, the output voltage is equal to

$$V_{out} = A_D \cdot (V_{in1} - V_{in2}).$$

In the above equation, A_D is the gain of the Amplifier in Differential Mode operation of the two input voltages. A_D is known as the *differential mode gain*.

The main features of the Differential Amplifier are given below:

- Very large gain occurs when opposite signals are applied to both the input terminals. Difference voltage V_D between the two inputs

$$V_D = [V_{in1} - V_{in2}] \tag{10.68}$$

then difference signal V_D is amplified with gain A_D.

- Amplified output voltage (for difference inputs)

$$V_{out}(D) = A_D\ [V_{in1} - V_{in2}]. \tag{10.69}$$

- Very small gain occurs when common type signals are applied to the two input terminals. Average of the sum of the two input signals

$$V_C = \frac{[V_{in1} + V_{in2}]}{2}. \tag{10.70}$$

Then the common signal is amplified with gain A_C.

- Amplified output voltage (for common inputs)

$$V_{out}(C) = A_C\left[\frac{V_{in1} + V_{in2}}{2}\right]. \tag{10.71}$$

- The overall operation is to amplify the difference signals, while rejecting the common signal at the two inputs.
- Overall output voltage $V_{out} = V_{out}(D) + V_{out}(C)$. It is sum of the two types of output voltages that occur due to difference input signals and common mode input signals.

$$V_{out} = \left[A_D[V_{in1} - V_{in2}] + A_C\left[\frac{V_{in1} + V_{in2}}{2}\right]\right]. \tag{10.72}$$

- Noise or any unwanted signal is generally common to both the input terminals of the Amplifier. The differential connection in the Amplifier causes attenuation (due to cancellation) of the noise (unwanted) input signals. This operating feature is known as common mode rejection.
- Since the amplification of the opposite signals is much greater than that of the common input signals, the Amplifier provides a common mode rejection. It is described by a parameter known as common mode rejection ratio.

$$\text{Common Mode Rejection Ratio (CMRR)} \quad A_{RR} = \frac{A_D}{A_C} \tag{10.73}$$

$$\text{CMRR in dB} = 20\log_{10}\frac{A_D}{A_C}. \tag{10.74}$$

- Typical values of CMRR are in between 100 and 120 dB. Normally differential amplifiers with larger values of CMRR are used. It measures how well the differential amplifier attenuates or rejects the common mode signals. The amplifier is virtually free from interfering signals. Signal to noise ratio will be improved by a factor of CMRR.
- Interference, static, induced voltages, etc. drive a Differential Amplifier in the *common mode operation*.
- A *common mode input signal* is used to test a Difference Amplifier to see how well the sections are working.
- Internal circuit of Operational Amplifiers use Differential Amplifiers in Cascade.
- As no coupling or bypass capacitors are used in Differential Amplifiers, they are simple Cascaded Direct-Coupled (DC) Amplifiers capable of amplifying signals with frequencies as low as 0 Hz (DC is nothing but AC with zero frequency).
- Transistors in IC circuits using Differential Amplifiers will be almost at the same temperature. So there will be almost no drift in cascaded differential Amplifiers.

10.17.2 JFET Difference Amplifier (Fig. 10.50)

Difference Amplifier using JFET is similar to BJT Differential Amplifier. It is also known as Source-coupled pair. Simple process of fabrication of JFET in IC version and very high input resistance of FET devices make the application of FET differential Amplifiers more popular.

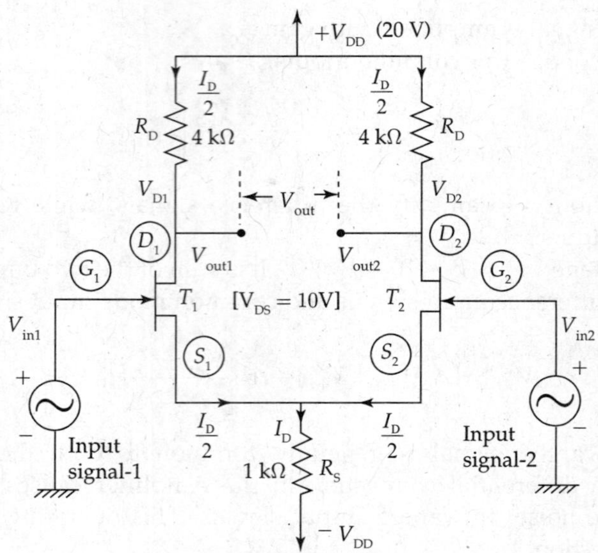

FIG. 10.50 *JFET difference amplifier*

For Double-ended Amplifier operation

- When two input voltages V_{in1} and V_{in2} (which are out of phase to one another) are applied, the *effective input signal* $V_{E(in)} = V_{in} = [V_{in1} - V_{in2}]$.
- Then two output voltages V_{out1} and V_{out2} will be developed at the two Drain terminals of the FET devices with gain A_D.
- The effective output voltage

$$V_{E(out)} = V_{out} = [V_{out1} - V_{out2}]. \tag{10.75}$$

Voltage gain of Double-ended Difference amplifier

$$\text{Voltage Gain } A_D = \left[\frac{V_{out1} - V_{out2}}{V_{in1} - V_{in2}}\right] = -g_m \cdot R_D, \tag{10.76}$$

which is equal to gain of CS FET Amplifier (using one FET) Voltage gain of Single-ended Amplifier A_S (with single input signal).

$$\text{Voltage gain} \quad A_S = -\frac{g_m \cdot R_D}{2}. \tag{10.77}$$

DC Drain currents $I_D^1 = I_D^2 = I_D/2$ through each FET device where I_D^1 is the Drain current, through FET device T_1 and I_D^2 is the Drain current through FET device T_2 when the two FET devices are identical matched pair $I_D^1 = I_D^2 = I_D/2$.

DC drain voltages are at the Drain of the first FET device

$$V_{D1} = [V_{DD} - I_D^1 \cdot R_D] = \left[V_{DD} - \frac{I_D \cdot R_D}{2}\right] \quad (10.78)$$

At the Drain of the second FET device,

$$V_{D2} = [V_{DD} - I_D^2 \cdot R_D] = \left[V_{DD} - \frac{I_D \cdot R_D}{2}\right]. \quad (10.79)$$

EXAMPLE 10.15

For the JFET Difference Amplifier circuit in Fig. 10.50 $I_{DSS} = 4$ mA and pinch-off voltage $V_P = -4$ V. Calculate (a) DC output voltages; (b) Gain of single-ended Amplifier and (c) Gain of Double-ended Amplifier.

Solution: Current through R_S

$$I_D = \left[\frac{V_{DD} - V_{DS}}{R_D + R_S}\right]$$

$$= \left[\frac{(20-10)}{(4+1)\times 10^3}\right] = \left[\frac{10}{5\times 10^3}\right] = 2 \text{ mA.}$$

a. DC voltages at the Drain terminals of FET device

$$V_{D1} = V_{D2} = \left[V_{DD} - \frac{I_D \cdot R_D}{2}\right]$$

$$= \left[20 - \frac{2\times 10^3 \times 4\times 10^3}{2}\right]$$

$$= [20 - 4 \text{ V}] = 16 \text{ V.}$$

b.

$$g_m = \left[\frac{2I_{DSS}}{|V_P|} \cdot \sqrt{\frac{I_D}{I_{DSS}}}\right]$$

$$= \left[\frac{2\times 4\times 10^{-3}}{4} \cdot \sqrt{\frac{2\times 10^{-3}}{4\times 10^{-3}}}\right] = 1.414 \text{ milli mhos}$$

∴ Gain of Single-ended Amplifier $A_S = -\dfrac{g_m \cdot R_D}{2}$

$$= -\frac{1.414\times 10^{-3} \times 4\times 10^3}{2} = 2.828.$$

c. Gain of Double-ended Amplifier

$$A_D = -g_m \times R_D = 1.414\times 10^{-3} \times 4\times 10^3 = 5.656.$$

The Difference Amplifier circuit in Fig. 10.50 can be operated in any of the following three types of input signal voltage combinations.

1. Single-ended Difference Amplifier

If one input signal V_{in} is applied to one of the input terminals of the two Transistors, while the second input terminal of the other Transistor is grounded, the electronic Amplifier is known as *single-ended Difference Amplifier* which is shown in Fig. 10.51.

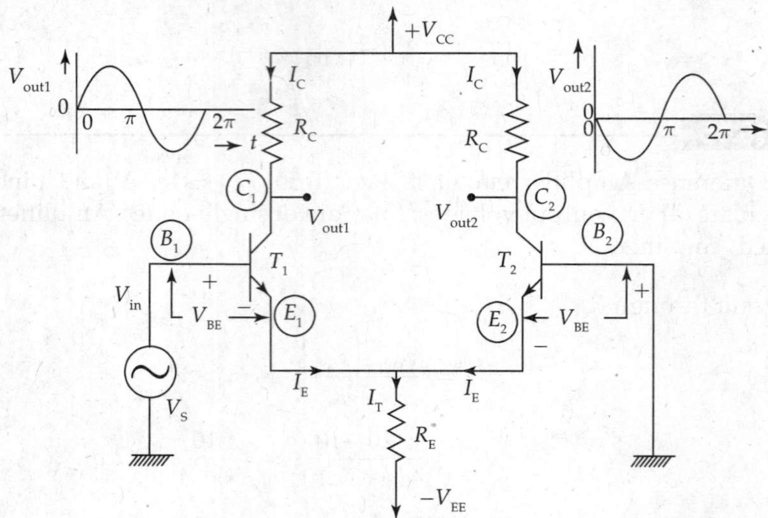

FIG. 10.51 *Single-ended difference amplifier*

In this Amplifier configuration, a single input signal is applied. But, due to CE connection of the two Transistors, the input signal operates the two Transistors into conduction, which results in two output voltages V_{out1} and V_{out2}. Output voltage can be taken from any of the output terminals and ground. Then it is considered as *Single-ended output Differential Amplifier.*

Signal waveforms at different points in the Amplifier circuit are shown in Fig. 10.51. Transistor T_1 acts as CE Transistor Amplifier. Therefore, amplified output voltage V_{out1} of Transistor T_1 is 180° out of phase with input signal voltage. Transistor T_2 functions as *Common Base Transistor Amplifier*. The amplified output voltage V_{out2} will be in phase with the input signal V_{in}.

- **When only one output terminal is available at the Collector terminal of the Transistor** T_1, the output voltage V_{out} will be 180° out of phase to the input signal applied to Base-1 of the Transistor T_1.
- **When only one output terminal is available at the Collector terminal of the Transistor** T_2, the output voltage V_{out} will be in phase to the input signal applied to Base-1 of the Transistor T_1.

2. Double-ended Difference Amplifier

When two equal input signals V_{in1} and V_{in2} of opposite polarity are applied to the two inputs of the Differential Amplifier, the electronic Amplifier is known as *Double-ended Difference Amplifier.* Typical Amplifier configuration is shown in Fig. 10.52.

The differential mode signals are amplified. The difference between the two equal and opposite polarity input signals is double the magnitude of each signal. Then Amplifier

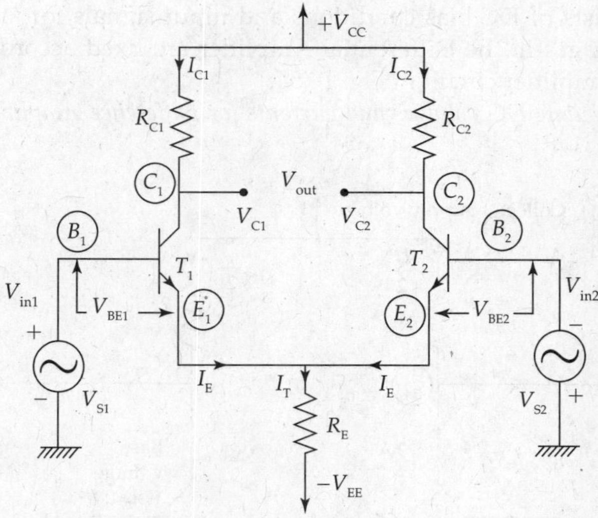

FIG. 10.52 *Double-ended difference amplifier*

provides larger gain. The output voltage is taken between the two output terminals. Then it is known as *Double-ended output Differential Amplifier.*

3. Common Mode Operation of the Difference Amplifier (Fig. 10.53)

When the same input signal is applied to both the input terminals of the two Transistors of the Differential Amplifier, the electronic Amplifier is considered to be in *Common Mode Operation* of the Amplifier. Then the input signals to the two Transistors are in phase and equal in magnitude. The common mode input signals get cancelled or not amplified by the Differential Amplifier because it is designed to amplify only the *difference signals.*

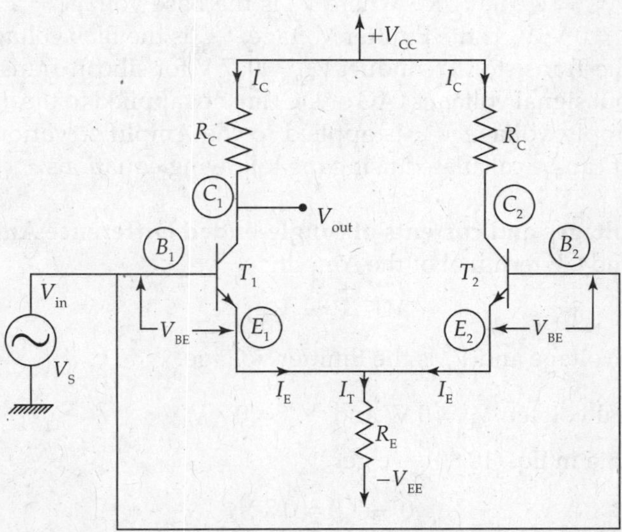

FIG. 10.53 *Common mode operation of difference amplifier*

An Amplifier consists of DC bias conditions and input signals for amplification. Various levels of DC bias voltages in the Differential Amplifier are fixed according to the following equations using the Amplifier circuit in Fig. 10.54.

Design equations to calculate DC voltages and currents for Difference Amplifier V_{CC} is the Collector supply voltage.

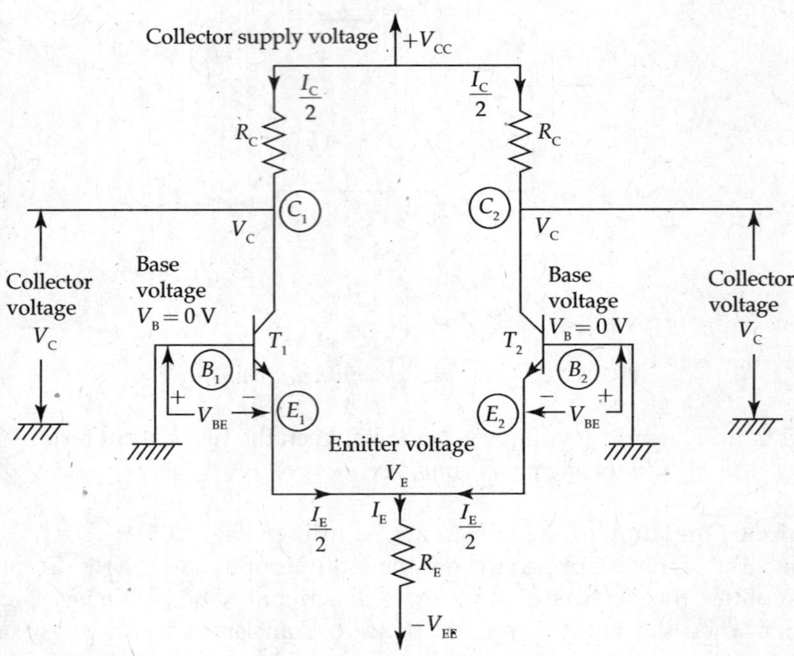

FIG. 10.54 *DC bias voltages in BJT difference amplifier*

Collector voltage $V_C = V_{CC} - I_C \cdot R_C$, where V_B is the Base voltage = External input signal voltages (AC or DC) = 0 V; V_E is the Emitter Voltage; V_{BE} is the Bias voltage between the Base and the Emitter for the Transistor to conduct $V_{BE} = 0.7$ V for silicon Transistors.

When external input signal voltages (AC or DC) are not applied to the differential Amplifier and the Collector supply voltage V_{CC} is applied to the Amplifier various DC voltages and currents in the circuit can be calculated using the following equations.

Equations for DC voltages and currents of Single-ended Difference Amplifier (Fig. 10.54)
Between each Base and CE terminal of the Amplifier

$$V_B - V_E = V_{BE}, \tag{10.80}$$

where V_B is the Base Voltage and V_E is the Emitter voltage.

Data external signal voltage $V_E = 0$ V and $V_{BE} = 0.7$ V. (10.81)

Substituting the data in Eq. (10.80), we get

$$(0 - V_E) = 0.7 \text{ V}$$

$$\therefore \text{ Emitter voltage } \quad V_E = -0.7 \text{ V}. \tag{10.82}$$

$$\text{DC bias Emitter current} \quad I_E = \left[\frac{[V_E - (-V_{EE})]}{R_E}\right] = \left[\frac{V_E + V_{EE}}{R_E}\right] = \left[\frac{-0.7 + V_{EE}}{R_E}\right]$$

$$\therefore \quad I_E = \left[\frac{V_{EE} - 0.7}{R_E}\right]. \tag{10.83}$$

When matched pair Transistors T_1 and T_2 are used in the circuit, the two collector currents I_{C1} and I_{C2} are equal to I_C.

$$\therefore \; I_{C1} = I_{C2} = I_C. \tag{10.84}$$

From the Transistor configuration when the two Emitter terminals are connected together, Collector currents of each Transistor

$$I_C = \frac{I_E}{2}. \tag{10.85}$$

Collector voltages V_{C1} and V_{C2} of the two Transistors are equal and $= V_C \cdot V_{C1} = V_{C2} = V_C$.

$$\text{Collector voltage} \quad V_C = (V_{CC} - I_C \cdot R_C). \tag{10.86}$$

EXAMPLE 10.16

Calculate the DC bias voltages and currents in the Difference Amplifier circuit of Fig. 10.55.

Solution: From Eq. (10.81), Emitter voltage $V_E = V_{BE} = 0.7$ V.

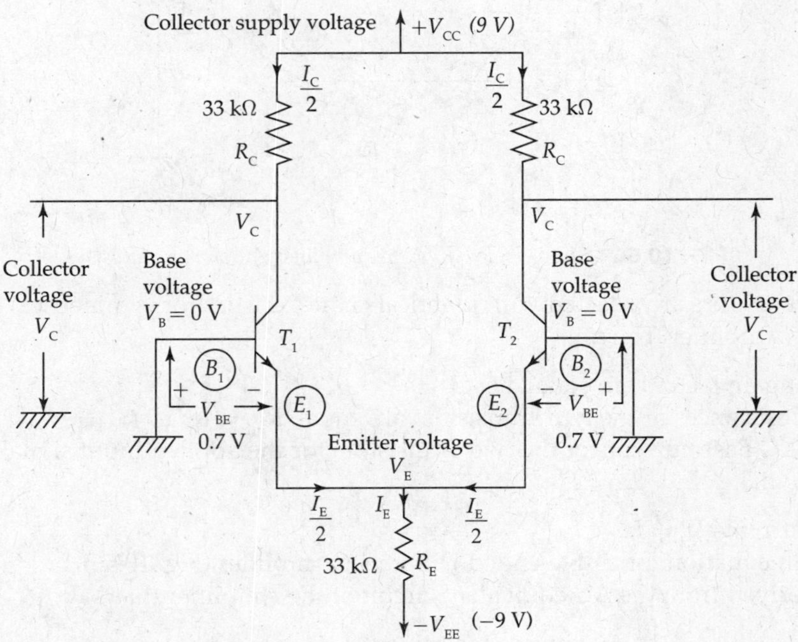

FIG. 10.55 *DC bias voltages in BJT difference amplifier*

Using Eq. (10.86), $I_E = \dfrac{[V_{EE} - 0.7]}{R_E} = \dfrac{(9 - 0.7)}{3\ \text{k}\Omega}$

$$= \frac{8.3 \times 10^{-3}}{3} = 2.766\ \text{mA}.$$

Collector current $I_C = \dfrac{I_E}{2} = 1.383\ \text{mA}$

Collector voltage $V_C = [V_{CC} - I_C \cdot R_C] = [9 - 1.383 \times 10^{-3} \times 3 \times 3 \times 10^3]$

$$= (9 - 4.56) = 4.44\ \text{V}.$$

10.17.3 AC Signal Voltage Gain of Single-ended Difference Amplifier

For single-ended Amplifier operation of a Difference Amplifier, one input signal V_{in1} is applied to the Transistor T_1. The input terminal of the Transistor T_2 is connected to ground terminal so that $V_{in2} = 0$ (Fig. 10.56).

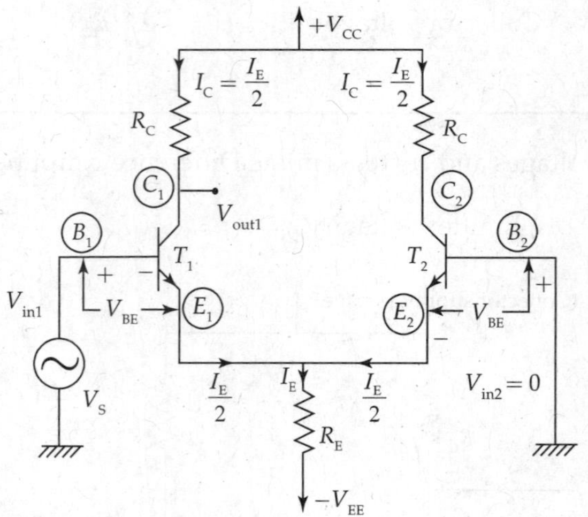

FIG. 10.56 *Single-ended difference amplifier AC signal analysis*

The two Transistors are selected with identical characteristics for symmetrical operation of the Difference Amplifier. Then,

- The current gain factors are equal, i.e., $\beta_1 = \beta_2 = \beta$.
- The input resistances of the two Transistors are equal. $R_{in1} = R_{in2}$.
- The input AC Base currents of the two Transistors for the applied input signal voltage V_{in1} are also equal, i.e., $I_{B1} = I_{B2} = I_B$.

Collector currents $I_{C1} = I_{C2} = I_C$

Voltage gain equation for single-ended Difference Amplifier (Fig. 10.56)

It can be derived from the AC equivalent circuit of the Amplifier that

$$\text{AC input base current} \quad I_B = \frac{V_{in}}{2 \cdot R_{in}} \tag{10.87}$$

$$\text{Collector currents} \quad I_C = \beta \cdot I_B = \frac{\beta \cdot V_{in}}{2 \cdot R_{in}}$$

$$\text{Output voltage} \quad V_{out} = I_C \cdot R_C = \frac{\beta \cdot V_{in} \cdot R_C}{2 \cdot R_{in}}. \tag{10.88}$$

But the transistor input resistance $R_{in} = \beta \cdot R_E$

$$\therefore \; V_{out} = \frac{\beta \cdot V_{in} \cdot R_C}{2 \cdot \beta \cdot R_E} = \frac{V_{in} \cdot R_C}{2 \cdot R_E}, \tag{10.89}$$

where the emitter diode resistance

$$R_E = \frac{V_T}{I_C(Q)} \tag{10.90}$$

where V_T is the Voltage equivalent of temperature and V_T = 26 mV at 27°C; $I_C(Q)$ is the quiescent component of collector current.

$$\text{Voltage gain} \quad A_V = \frac{V_{out}}{V_{in}} = \frac{R_C}{2 \cdot R_E}. \tag{10.91}$$

EXAMPLE 10.17

Calculate the DC currents, Output voltage and voltage gain for the *Single-ended Difference Amplifier* with an input voltage of 5 mV (Fig. 10.57).

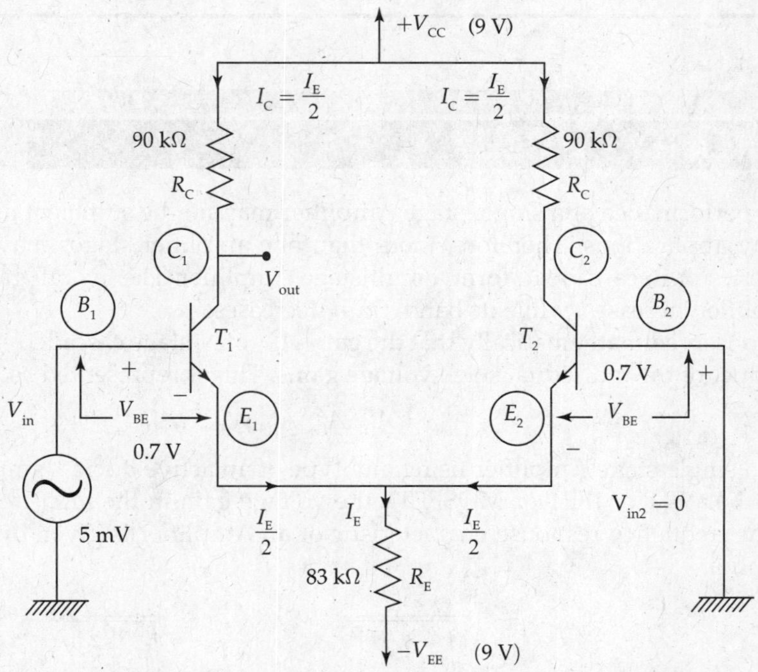

FIG. 10.57 *Single-ended difference amplifier for gain calculations using AC signal analysis*

Solution:

$$\text{DC Emitter current } I_E = \left[\frac{V_{EE} - V_{BE}}{R_E}\right] = \left[\frac{9-0.7}{83\times10^3}\right]$$

$$= \frac{8.3\times10^{-3}}{83} = 100\ \mu\text{A}.$$

$$\text{DC Collector current } \quad I_C = \frac{I_E}{2} = \frac{100\ \mu\text{A}}{2} = 50\ \mu\text{A}.$$

$$\text{DC Collector voltage } \quad V_C = [V_{CC} - I_C\cdot R_C]$$

$$= [9 - 50\times10^{-6}\times90\times10^3] = (9-4.5) = 4.5\ \text{V}.$$

$$\text{Emitter Diode resistance } \quad r_e = \frac{V_T}{I_C(Q)} = \frac{26\ \text{mV}}{50\times10^{-6}}$$

$$= \frac{26\times10^{-3}\times10^6}{50} = 520\ \Omega.$$

Using Voltage equivalent of Temperature V_T (at 300°K) = 26 mV (already known)

$$\text{AC Voltage gain } \quad A_V = \frac{V_{out}}{V_{in}} = \frac{R_C}{2\cdot r_e} = \frac{90\times10^3}{2\times520} = 86.5.$$

$$\text{Input voltage } \quad V_{in} = 5\ \text{mV}$$

$$\text{Output voltage } \quad V_{out} = A_V\cdot V_{in} = 86.5\times5\times10^{-3} = 0.4325\ \text{V}.$$

SUMMARY

- The performance of a single-stage Amplifier may not be sufficient and suitable for many applications. Therefore, more than one amplifier stages may be connected in series or in cascade to form a multistage Amplifier. The overall gain of cascaded amplifier increases, while its bandwidth decreases.
- It is observed mathematically that the gain A_{HF}^{N} of *N*-stage cascaded Amplifier is the product of the individual stage voltage gains. This is represented as the following.

$$A_{HF}^{N} = A_{HF}^{1}\, A_{HF}^{2}\, A_{HF}^{3}\, A_{HF}^{4} \ldots A_{HF}^{n}.$$

- For a single-stage Amplifier using any type of an active device (amplifying device may be a BJT or JFET or MOSFET), the voltage gain in the high-frequency region of the frequency response characteristic of an Amplifier is given by the following equation:

$$A_{HF}^{1} = \frac{1}{\sqrt{1+\left[\frac{f}{f_{HF}}\right]^2}} \quad \text{(or)} \quad A_{HF}^{1} = \frac{1}{\sqrt{1+\left[\frac{f}{f_2}\right]^2}}.$$

- The Low-frequency cut-off point f_1^N of the multistage Amplifier of the cascaded Amplifier is given by the expression $f_1^N = \frac{f_1}{\sqrt{2^{1/N}-1}}$, where N is the number of stages of a cascaded Amplifier and f_1 is the *low-frequency* cut-off point of an Amplifier.
- The high-frequency cut-off point f_2^N of the multistage Amplifier or the bandwidth of the cascaded Amplifier is given by the expression. $f_2^N = f_2 \cdot \sqrt{2^{1/N}-1}$, where N is the number of stages of a cascaded Amplifier and the high-frequency cut-off point of an Amplifier is expressed as f_2.
- The input stage of a multistage Amplifier should have high input resistance in order to avoid power Drain from the signal source. The middle stages of cascaded Amplifier provide the bulk gain. The output stage of a multistage Amplifier need to have low output resistance whenever it is required to feed low impedance loads.
- A Darlington pair is a cascaded Emitter follower. Its voltage gain is unity. But its current gain is very large. Current gain of Darlington pair using Transistors is the product of individual Transistor gains.
- CASCODE Amplifier configuration has CS FET Amplifier followed by CG Amplifier. It has increased bandwidth due to CG FET Amplifier and very high input resistance, because the front stage is a CS FET Amplifier.
- CASCODE Amplifier configuration has CE Transistor amplifier stage feeding CB Transistor amplifier stage.
- Difference Amplifiers are usually either Emitter-Coupled pair using BJTs or Source-Coupled pair using FETS.

Questions for Practice

1. If there are two amplifier stages in cascade with each Amplifier having a voltage gain of 100, calculate the gain of the cascaded Amplifier in decibels.
2. Explain how the loading effect in early amplifier stages of a multistage Amplifier using Transistors causes in reduction in gain. Further explain whether such effect come into picture in multistage Amplifiers using MOSFET devices.
3. Calculate the magnitude of the output voltage of a multistage Amplifier having a gain of 40 dB, when the input voltage to the Amplifier is 50 mV.
4. The voltage gain of three-stage cascaded Amplifier is 120 dB. If the first and the third amplifier stages have gains of 100, calculate the gain of the second amplifier stage.
5. Draw two frequency response characteristics for a single-stage amplifier and a Multistage Amplifier on a single semi-log graph paper and give the comments on variations in gain, bandwidth and cut-off frequencies.
6. The gain of an RC-coupled two-stage FET Amplifier falls by 90% of the mid-band value at 400 kHz. If g_m of each FET is 10 mA/V and total output capacitance for each stage is 20 pF, calculate the required load resistance R_L and the stage mid-band gain.
7. Write a short note on bandwidth of Amplifiers. (JNTU, May 2005)

8. Obtain the theoretical expressions for f_{1n} and f_{2n} when n-stages of identical Amplifiers are cascaded. (JNTU, March 2006)
9. Three identical non-interacting amplifier stages connected in cascade have an overall gain of 0.3 dB down at 50 kHz compared to mid-band. Calculate the upper cut-off frequency of the individual stages. (JNTU, March 2006; Feb 2008)
10. Draw the circuit diagram of a differential Amplifier using BJTs. Describe common mode and differential modes of working.
11. What is CASCODE Amplifier? Explain.
12. Discuss the frequency response characteristics of RC-Coupled Amplifier. Derive the general expressions for voltage gains at middle, low and high frequencies.
13. Draw the circuit diagram of two stages RC-coupled Amplifier using BJT and its equivalent circuit. Derive the expression for its overall voltage gain.
14. Draw the circuit diagram of two stages RC-coupled Amplifier using FET and its equivalent circuit. Derive the expression for its overall voltage gain.
15. What are the merits and demerits of DC Amplifier in comparison with RC Amplifiers?
16. Explain the importance of CMRR (Common Mode Rejection Ratio).

Multiple Choice Questions

1. In Multistage Amplifiers, the total voltage gain is usually realised by one or more ________________.

   (a) CE stages (b) CB stages (c) CC stages (d) CG stages

2. The impedance buffering action of the following configuration can be employed to extend the high frequency response of amplifiers and to speed up the operation of digital circuits.

   (a) CE (b) CS (c) CD (d) none of these

3. Following statements are made with respect to features of Multistage Amplifiers.
   1. It consists of an input stage having low input resistance.
   2. It consists of an output stage having low output resistance.
   3. One or more intermediate stages to realise the bulk of gain.
   4. It consists of an input stage having high input resistance.
   5. It consists of an output stage having high output resistance.

   Of the above statements, the true statements are ________________.

   (a) 1, 2, 3 (b) 1, 2, 4 (c) 1, 2, 5 (d) 3, 2, 5

4. Two stages of a cascaded amplifier have individual upper cut-off frequencies of 5 MHz and 3.33 MHz. The best approximation of the upper cut-off frequencies of the Cascaded amplifier is ________________

   (a) geometric mean of the two (b) arithmetic mean of the two
   (c) 3.33 MHz (d) 5 MHz

5. The following statements are made with reference to Cascaded amplifiers.
   1. It is CB followed by CE
   2. Increased output resistance
   3. Bandwidth decreases
   4. Output stage having low output resistance

   Out of the above statements, the true statements are ______________.

   (a) 1, 4 (b) 2, 3, 4 (c) 3, 4 (d) 1, 4

6. A Darlington transistor amplifier configuration is ______________.

   (a) CE–CB pair (b) CC–CC pair
   (c) CE–CC pair (d) CE–CE pair

7. The individual high cut-off frequencies of a two-stage amplifier is 10.2 MHz, the overall high-frequency cut-off frequency of the cascaded amplifier is ______________.

   (a) 10.2 MHz (b) 200 kHz (c) 392 kHz (d) 6.56 MHz

8. In Multistage Amplifier, the coupling capacitor ______________.

   (a) limits the low frequency response
   (b) limits the high frequency response
   (c) does not affect the frequency response
   (d) blocks DC without affecting frequency response

9. CASCODE amplifier stage is equivalent to ______________.

   (a) a CE stage followed by a CB stage
   (b) a CB stage followed by a CE stage
   (c) an emitter follower by a CB stage
   (d) a CB stage followed by a CC stage

10. Match the Following:

1. Combination of BJT and FET stages	(a) CASCODE
2. Super Beta transistor	(b) CASCADE
3. Combination of NPN and PNP	(c) DARLINGTON
4. CC–CE act as a PNP transistor	(d) FEEDBACK PAIR
5. A CE (CS) transistor followed by a CB (CG) transistor	(e) BIFET

Answers to Multiple-Choice Questions

1. (a)	2. (c)	3. (a)	4. (c)	5. (b)
6. (b)	7. (d)	8. (a)	9. (a)	10. [e, c, d, b, a]

Chapter 11

LARGE SIGNAL (POWER) AMPLIFIERS

Learning Objectives

To get familiarity of the concepts and working principles of Power Amplifiers

- Different classes of operation of Amplifiers
- Power conversion capabilities and applications
- Merits and demerits
- Problems with distortion in Amplifiers and remedies
- Push-Pull power Amplifiers
- Advanced Power Amplifiers

11.1 CLASS-A, CLASS-B AND CLASS-C AMPLIFIERS

Power Amplifiers essentially

- Operate as *large signal* Amplifiers with modest amount of *voltage gain* and substantial amount of *current gain;*
- Convert as much *DC input power* as possible into *AC signal output power;*
- Find applications in Radio receiver, Public address systems, Stereo Amplifiers, Home Theatres, TV, Radio, Communication equipment Cell Phones and so on.

Classification of Power Amplifiers based on Class of operation

(1) Class-A Power Amplifier; (2) Class-B Power Amplifier; (3) Class-AB Power Amplifier; and (4) Class-C Power Amplifier.

11.1.1 Class-A Amplifier

Transistors and Vacuum Tubes in Class-A Amplifier conduct continuously for entire cycle (0° to 360°) of input signal, which means that 100% of signal is used. Distortion is less and has good fidelity in amplifier response.

Class-A Operation of Amplifier (Figs. 11.1, 11.2 and 11.3)
V_{DD} is the Drain Supply voltage. DC bias $V_{GS}(Q) = -2$ V (It is half the pinch-off voltage $V_P = -4$ V for BFW10 (FET).). *Gate resistor* R_G connects the DC Source V_{GS} to Gate terminal.

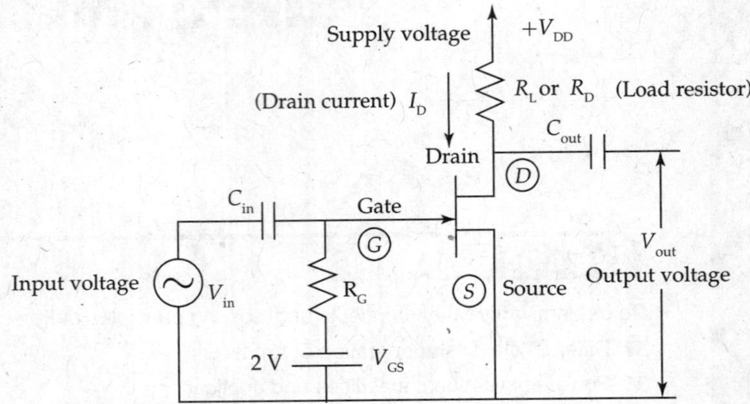

FIG. 11.1 *Class-A amplifier using FET device*

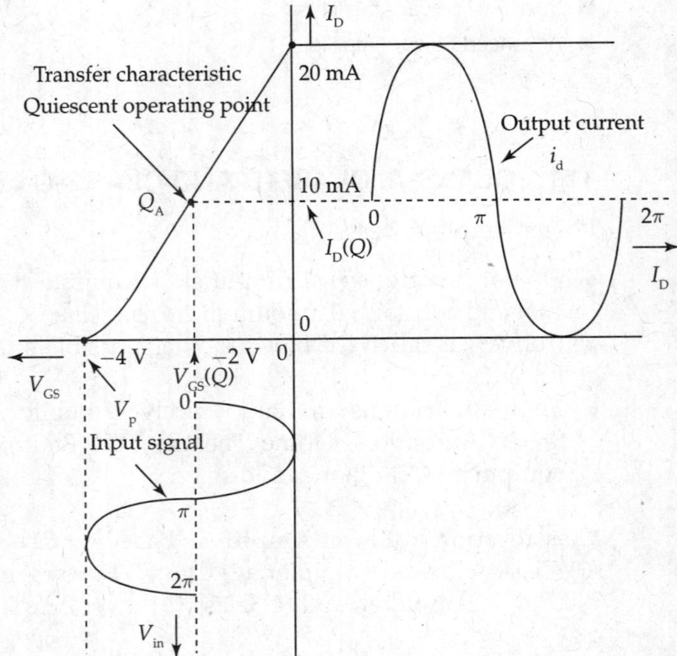

FIG. 11.2 *Class-A amplifier operation using mutual characteristic of FET device*

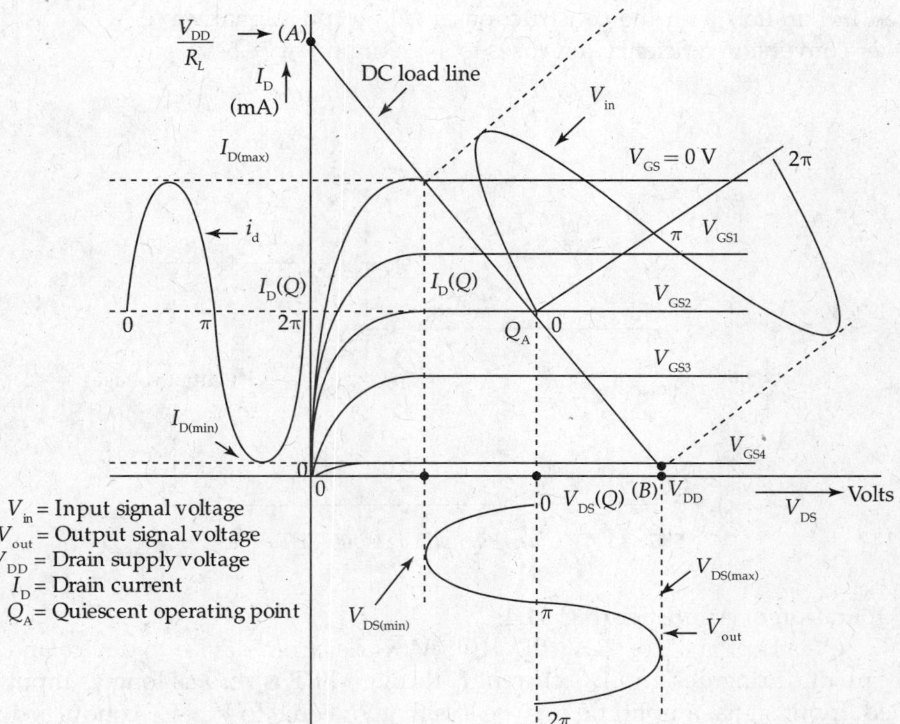

FIG. 11.3 *Signal variations of I_D, V_{in} and V_{out} in FET amplifier*

Input signal V_{in} is a sine wave. (V_{in} should be less than or equal to decided DC Bias $V_{GS}(Q)$). Input signal superimposes on DC bias $V_{GS}(Q)$ and vary DC voltage, during the interval '0 to 2π' of signal variations. They cause variations in *Drain current* i_d (Fig. 11.3). All currents (DC and AC) flow for total time from 0° to 360° of signals. *Output Voltage* 'V_{out}' exists throughout the cycle.

Waveforms in Class-A Amplifier Various signals are explained using (1) Mutual (Transfer) characteristics (Fig. 11.2) and (2) Output (Drain) characteristics (Fig. 11.3) of FET device. DC load line AB is drawn on Transistor (FET) output characteristics and Quiescent (*Q*) operating point is selected at the middle of DC load line for Class-A operation. Signal waveforms can be observed on CRO.

General Features of Class-A Power Amplifiers

DC current flows through the Transistor even when AC signal is not applied. So, power dissipation by the device and resistive components in signal path is more. Useful output power becomes less. It results in low power conversion efficiency (Amplifier efficiency η is the ratio of AC output power to DC input power) with a maximum theoretical value of 25%.

11.1.2 Class-B Amplifier

Class-B Amplifiers amplify only one-half waves (0° to 180°) of input signal wave. So, distortion will be more. Class-B operation is used in push-pull Amplifiers using two Transistors (connected in parallel) with the construction of full output signal wave, which is discussed later. Power conversion efficiency increases to a maximum of 78.54%.

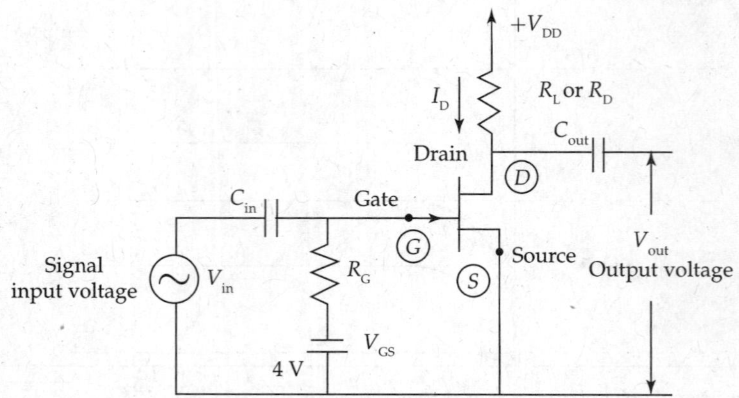

FIG. 11.4 *Class-B amplifier using JFET device*

Class-B Operation of Amplifier (Fig. 11.4)

DC Bias $V_{GS}(Q)$ is kept at V_P of –4 V (BFW10). (V_P varies from device to device and it has to be taken care of during design.) DC current I_D through FET is zero as long as input signal is not applied. Input signal amplitude may be less than or equal to V_P. FET conducts during the interval '0 to π' (Fig. 11.5) of input signal. Effective input signal V_{GS} is superimposed version of

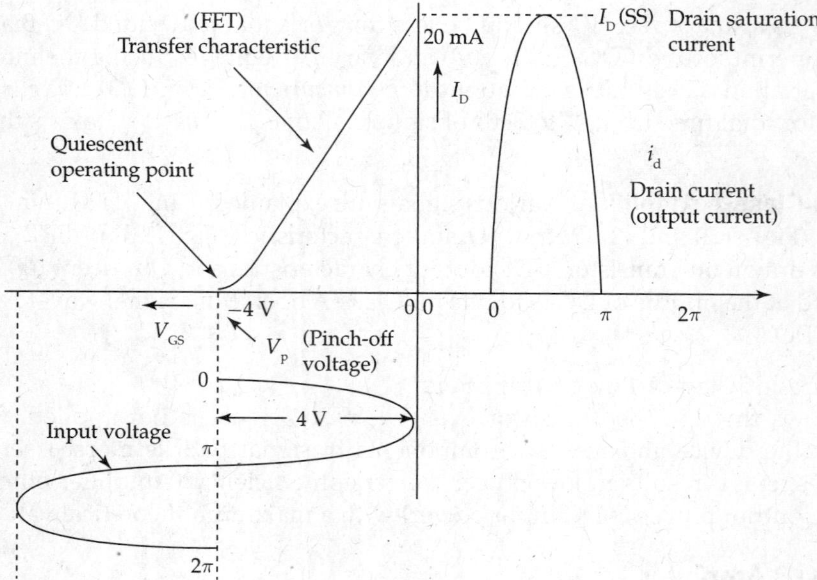

FIG. 11.5 *DC bias and signal waveforms for Class-B operation of amplifiers using the 'mutual characteristic' of JFET device*

input signal V_{in} over DC Bias V_P. Net voltage is below V_P causing Drain current i_d to flow for a period of '0 to π'. But, during the negative half cycle of input signal voltage, effective input voltage V_{GS} is beyond cut-off voltage (V_P). FET will not conduct causing i_d to zero during the time interval 'π to 2π'.

Waveforms in Class-B Amplifier DC voltage levels, input signal voltage swings and resulting amplified signal waveforms are shown in Fig. 11.5. Waveforms can be observed on CRO screen.

General features of Class-B power Amplifiers

1. Input signal amplitude is larger than that of Class-A operation. So, there is an increase in output power. Class-B operation uses 50% of signal. Power dissipation by active device is reduced. So, there is an increase in power and efficiency.
2. When an input signal is not applied, current through the device is zero. Power dissipation by the devices is zero under no signal conditions or standby operation. Class-B Amplifiers are used in *satellite systems* to save power.

11.1.3 Class-C Amplifier

Class-C Amplifier Operation

In Class-C Amplifier, DC bias V_{GS} is set to be greater than V_P. Output current will be in the form of pulses, since the device conducts for a time period less than 180°. Normal device conduction intervals are about 60° to 120° in design. (Full sine wave output is realised by using *Parallel Tuned Circuit* as load.) Separate type of biasing circuit is used, because the biasing voltage at input port is zero. No DC current flows through the circuit. Power dissipation in the active device is reduced. So, AC Power output is increased. Power conversion efficiency is high.

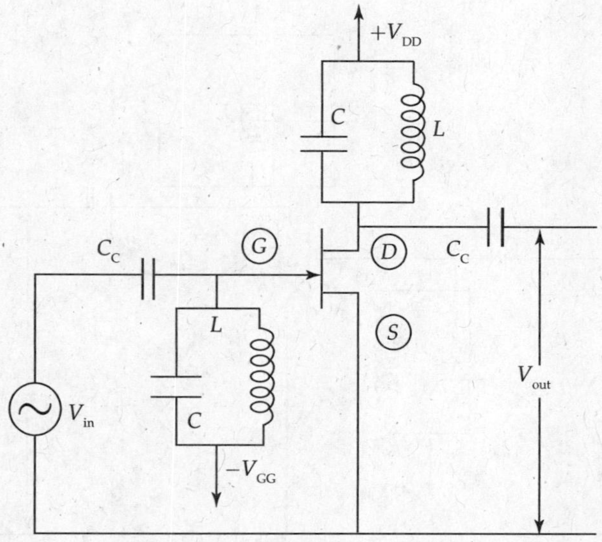

FIG. 11.6 *Class-C amplifier*

General features of Class-C Power Amplifiers

Class-C Power Amplifier using BJT provides modest amount of voltage gain and substantial amount of current gain. It absorbs little Power from signal Source (as input current is zero) and delivers a large amount of Power to the load with maximum efficiency about 90%. It has maximum input signal drive and reduced distortion

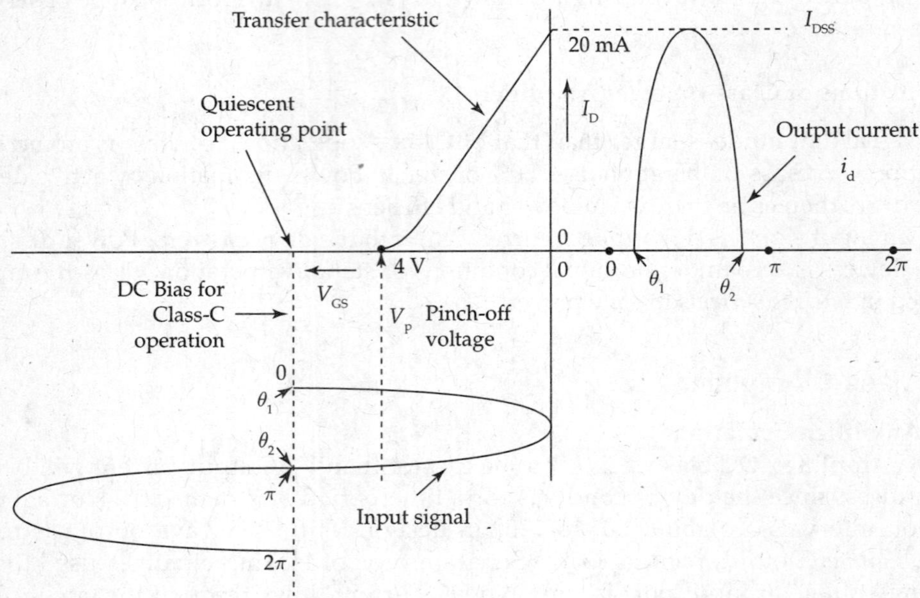

FIG. 11.7 *DC bias and signal waveforms for Class-C FET amplifier*

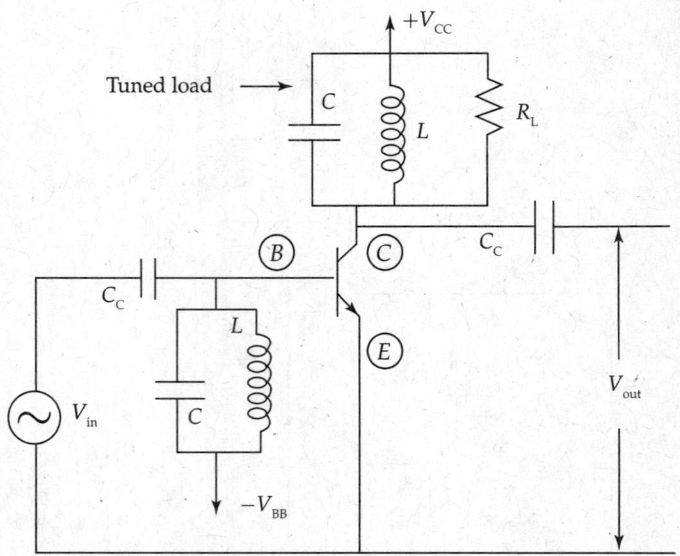

FIG. 11.8 *Class-C power amplifier using a transistor (BJT)*

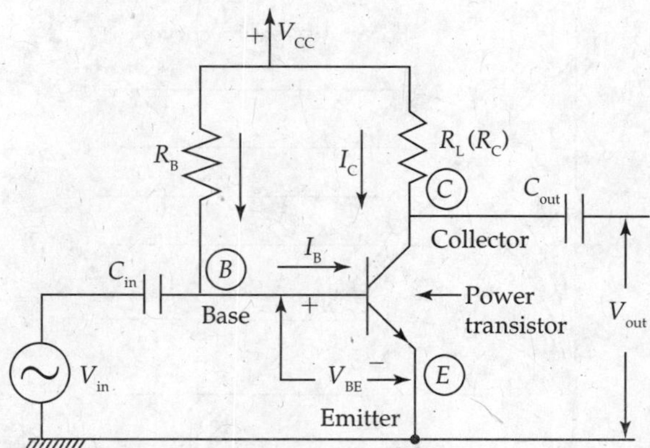

FIG. 11.9 *Class-A power amplifier*

11.2 CLASS-A POWER AMPLIFIER

11.2.1 Series-Fed Class-A Power Amplifier Circuit (Fig. 11.9)

Load resistance (R_L) and power supply are in the same series signal path of Amplifier circuit. So, it is known as *Series-fed Class-A Amplifier*.

Various factors for the selection of Power Transistors

(1) Application and use; (2) Magnitude of AC signal output power; (3) Output (Collector) circuit efficiency (for the desired Class of operation of Amplifier) to provide desired value of output power in watts to the desired load; (4) Predicting the DC power input to the Power Amplifier considering above factors; (5) Selection of active device (BJT or FET or a Vacuum Tube) depending on the output power level required in a practical system, for example in Radio and TV Transmitter circuits' output power requirements are large; (6) Maximum Power dissipation rating of the active device so as to withstand the Power dissipation in the active device with suitable Heat sink and cooling system.

Once the Transistor is selected, determine the DC operating conditions of Amplifier from quiescent operating point '*Q*' on Transistor output characteristics (Fig. 11.10).

Location of Quiescent operating point *Q* in Class-A Amplifier

1. Note down the Power dissipation rating P_D of Transistor from data manuals.
2. Using the expression $P_D = V_{CE} \cdot I_C$, calculate different values of Collector current I_C for different selected values of V_{CE} (available on the output characteristics).
3. Draw the *Power dissipation curve* (using the calculations made in the above step) on the Transistor output characteristics. *It will have the shape of Hyperbola*.
4. Draw the DC load line (as explained below) tangential to Power dissipation curve or a little below to it to ensure the Transistor operation within safe limits.
5. Using the *DC load line Eq. (11.1)*, coordinates of point *A* on current (I_C) axis and coordinates of point *B* on voltage (V_{CE}) axis (Fig. 11.10) are calculated:

$$[V_{CC} - V_{CE}] = (I_C \times R_L) \quad \text{(DC load line equation)} \tag{11.1}$$

6. One coordinate of point *A* is $V_{CE} = 0$ V (from location of point *A*). Other coordinate is $I_C = V_{CC}/R_C$ from DC load line equation, substituting $V_{CE} = 0$ in it.

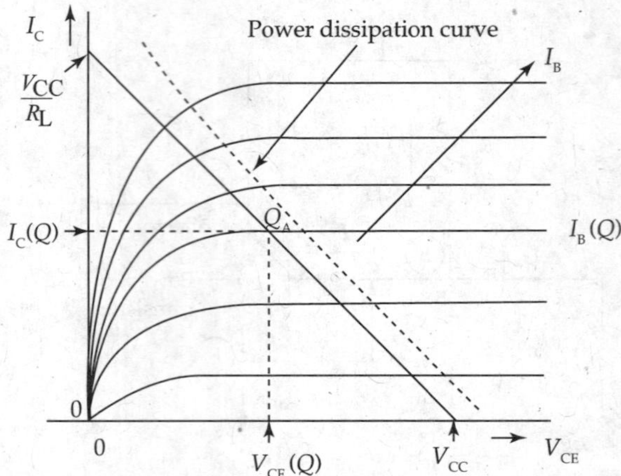

FIG. 11.10 *DC load line, Q-point and power dissipation curve on typical transistor characteristics*

7. One coordinate of point *B* is $I_C = 0$ mA (from location of point *B*).Other coordinate is $V_{CE} = V_{CC}$ from DC load line equation, substituting $I_C = 0$ mA in it.
8. DC load line is drawn (on the output characteristics) by joining the two points *A* and *B* fixed from the previous calculations.
9. Quiescent operating point Q_A is fixed at the middle point on DC load line.
10. Intersection of Transistor output characteristic with DC load line at 'Q_A' is identified.
11. DC Bias value of current $I_B(Q)$ on the identified characteristic with DC load line, $I_C(Q)$ and $V_{CE}(Q)$ determine the operating point Q_A for Class-A operation.
12. Intersection point of DC load line with Transistor output characteristic for $I_B = 0$ mA is the Quiescent operating point (Q_B) for Class-B operation. Location and the Concept of *Q*-point Q_B will be used in Class-B Amplifier in later sections.

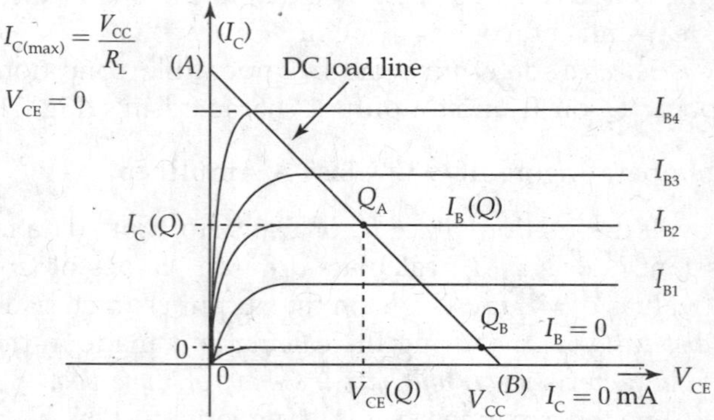

FIG. 11.11 *DC load line on transistor output characteristics*

Design equations for DC Bias operating conditions for fixed bias circuit (Fig. 11.9)
The DC bias currents are set by V_{CC}, R_B and R_C.

$$\text{DC bias current} \quad I_B(Q) = \frac{(V_{CC} - V_{BE})}{R_B} \tag{11.2}$$

$$I_C(Q) = \frac{(V_{CC} - 0.7\ \text{V})}{R_B} \quad \text{if } V_{BE} = 0.7\ \text{V} \tag{11.3}$$

$$I_C(Q) = \beta \times I_B(Q) \tag{11.4}$$

$$V_{CE} = [V_{CC} - I_C \cdot R_C] \tag{11.5}$$

AC signal operation

Signal waveforms and the signal swings (Fig. 11.12) are used to calculate DC Power input [P_{in}(DC)], AC Power output [P_{out}(AC)], Power conversion efficiency (η) and Power dissipation (P_D) by the active device (and resistors in the signal paths).

AC Signal waveforms and DC voltages

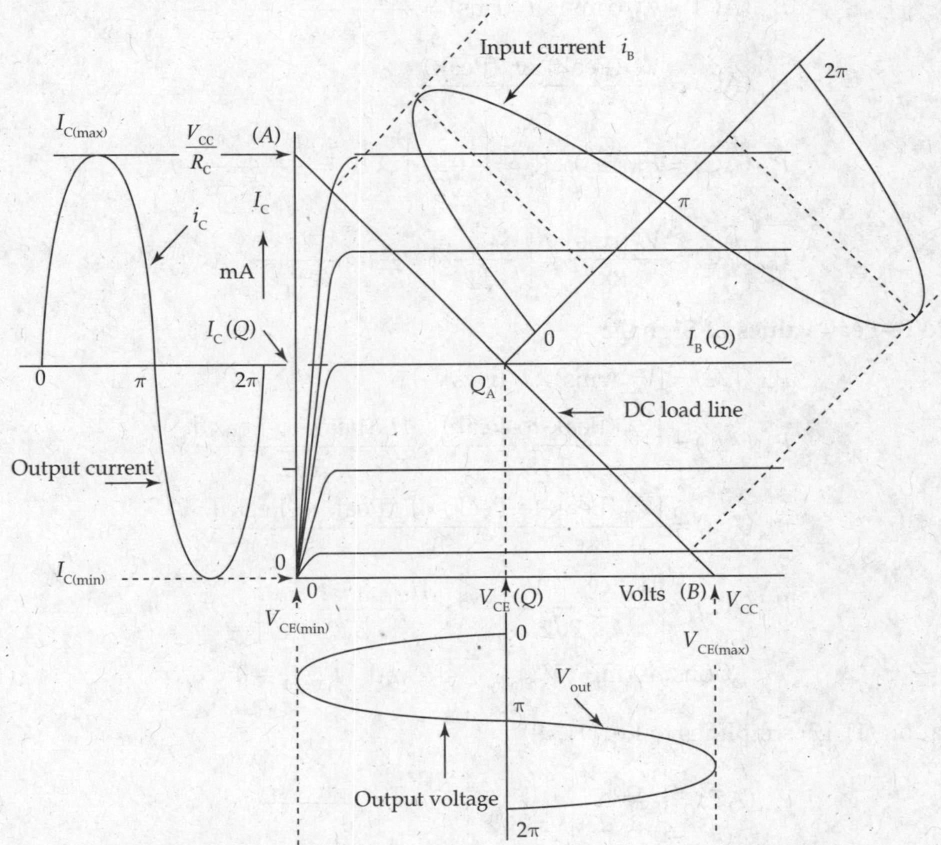

FIG. 11.12 *Signal waveforms of I_B, I_C and V_{out} in a Class-A amplifier*

11.2.2 Power conversion efficiency of Class-A Amplifier

DC input power to Amplifier (drawn from Power supply V_{CC})

DC input power is the product of DC supply voltage V_{CC} and Collector current $I_C(Q)$:

$$P_{in}(\text{DC}) = V_{CC} \times I_C(Q). \tag{11.6}$$

Even if an AC signal is applied, average current drawn from the supply remains same. So, $P_{in}(DC) = [V_{CC} \cdot I_C(Q)]$ represents DC power input to series-fed Class-A Amplifier.

AC output power can be calculated in a number of ways as shown below:

1. RMS values of current and voltage Signals:

$$P_{out}(AC) = V_{CE}(\text{rms}) \times I_C(\text{rms}) \tag{11.7}$$

$$P_{out}(AC) = I_C^2(\text{rms}) \times R_C \tag{11.8}$$

$$P_{out}(AC) = \frac{V_C^2(\text{rms})}{R_C} \tag{11.9}$$

2. Peak Signals of current and voltage variations in the Amplifier circuit:

$$P_{out}(AC) = V_{CE}(\text{rms}) \times I_C(\text{rms}) = \frac{V_{CE}(\text{Peak})}{\sqrt{2}} \times \frac{I_C(\text{Peak})}{\sqrt{2}} \tag{11.10}$$

$$\therefore \; P_{out}(AC) = \frac{V_{CE}(\text{Peak}) \times I_C(\text{Peak})}{2} \tag{11.11}$$

$$P_{out}(AC) = I_C^2(\text{rms}) \times R_C = \left[\frac{I_{C(\max)}}{\sqrt{2}}\right]^2 \times R_C = \frac{I_{C(\max)}^2 \times R_C}{2} \tag{11.12}$$

$$P_{out}(AC) = \frac{V_C^2(\text{rms})}{R_C} = \left(\frac{V_{CE(\max)}}{\sqrt{2}}\right)^2 \cdot \frac{1}{R_C} = \frac{V_{CE(\max)}^2}{2R_C} \tag{11.13}$$

3. Peak-To-Peak values of Signals:

$$P_{out}(AC) = [V_{CE}(\text{rms}) \times I_C(\text{rms})] \tag{11.14}$$

$$P_{out}(AC) = \frac{[V_{CE}(\text{Peak-to-Peak})]}{2\sqrt{2}} \times \frac{[I_C(\text{Peak-to-Peak})]}{2\sqrt{2}} \tag{11.15}$$

$$\therefore \; P_{out}(AC) = \frac{[V_{CE}(\text{Peak-to-Peak}) \times I_C(\text{Peak-to-Peak})]}{8} \tag{11.16}$$

$$P_{out}(AC) = \left[\frac{[V_{CE(\max)} - V_{CE(\min)}}{2\sqrt{2}}\right] \times \left[\frac{[I_{C(\max)} - I_{C(\min)}}{2\sqrt{2}}\right] \tag{11.17}$$

$$\text{Considering} \quad V_{CE(\min)} \cong 0 \quad \text{and} \quad I_{C(\min)} \equiv 0 \tag{11.18}$$

Equation (11.17) simplifies to Eq. (11.19)

$$P_{out}(AC) = \left[\frac{V_{CE(\max)}}{2\sqrt{2}}\right] \times \left[\frac{I_{C(\max)}}{2\sqrt{2}}\right] = \frac{[V_{CE(\max)} \cdot I_{C(\max)}]}{8} \tag{11.19}$$

4. Other forms of output AC Power calculations:

$$P_{out}(AC) = I_C^2(\text{rms}) \cdot R_C = \left[\frac{I_C(\text{Peak-to-Peak})}{2\sqrt{2}}\right]^2 \cdot R_C \tag{11.20}$$

$$\therefore \; P_{out}(AC) = \frac{I_C^2(\text{Peak-to-Peak})}{8} \times R_C \tag{11.21}$$

$$P_{out}(AC) = \frac{V_{CE}^2(\text{rms})}{R_C} = \left[\frac{V_{CE}(\text{Peak-to-Peak})}{2\sqrt{2}}\right]^2 \times \frac{1}{R_C} \tag{11.22}$$

$$\therefore \quad P_{out}(AC) = \frac{V_{CE}^2(\text{Peak-to-Peak})}{8R_C} \tag{11.23}$$

Power dissipation Power dissipation $= P_D = [P_{in}(DC) - P_{out}(AC)]$

Transistor Collector Circuit (power conversion) Efficiency

$$\%\eta = \frac{\text{AC power output to the load}}{\text{DC power input to the amplifier}} \times 100$$

$$= \frac{P_{out}(AC)}{P_{in}(DC)} \times 100 \tag{11.24}$$

Maximum theoretical efficiency for Class-A series-fed Amplifier
For the voltage swing,

$$\text{Maximum } V_{CE}(\text{Peak-to-Peak}) = V_{CC}$$

For the current swing,

$$\text{Maximum} \quad I_C(\text{Peak-to-Peak}) = \frac{V_{CC}}{R_C} \tag{11.25}$$

Using the maximum voltage and current swings,

$$\text{Maximum} \quad P_{out}(AC) = \frac{V_{CC} \times \left(\dfrac{V_{CC}}{R_C}\right)}{8} = \frac{V_{CC}^2}{8R_C} \tag{11.26}$$

Maximum Power input can be calculated using the DC Collector current $I_C(Q)$ set to half the maximum value $I_{C(max)}$ for Class-A Amplifier (Fig. 11.12).

$$P_{in}(DC)(\max) = V_{CC} \times I_C(Q) = V_{CC} \times \left(\frac{I_{C(max)}}{2}\right) \tag{11.27}$$

$$P_{in}(DC)(\max) = V_{CC} \times \left[\frac{V_{CC}/R_C}{2}\right] = \frac{V_{CC}^2}{2R_C}, \tag{11.28}$$

where $I_{C(max)} = \dfrac{V_{CC}}{R_C}$.

$$\text{Maximum} \quad \%\eta = \frac{P_{out}(AC)(\max)}{P_{in}(DC)(\max)} \times 100\% \tag{11.29}$$

$$\text{Maximum} \quad \%\eta = \frac{V_{CC}^2/8R_C}{V_{CC}^2/2R_C} \times 100\% = 25\%. \tag{11.30}$$

EXAMPLE 11.1

Calculate (a) DC input power, (b) output signal AC Power and (c) Collector circuit conversion efficiency 'η' of the Amplifier circuit in Fig. 11.13 for an input signal voltage V_{in} that causes variations in the input Base current I_B of 10 mA peak.

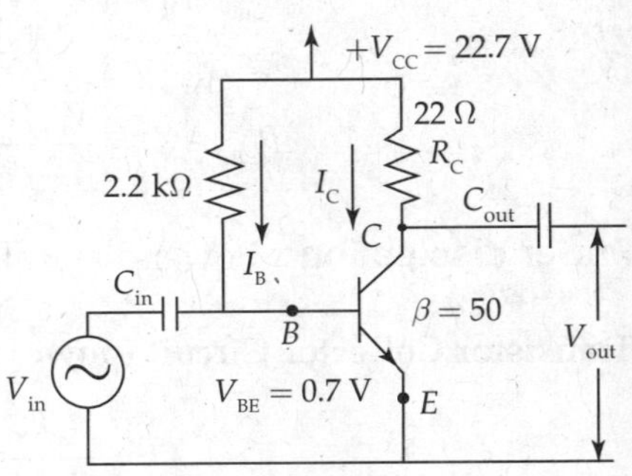

FIG. 11.13 *Class-A power amplifier with resistive load R_C*

Solution: Parameters at Quiescent operating point Q

$$I_B(Q) = \frac{[V_{CC} - V_{BE}]}{R_B} = \frac{[22.7\text{ V} - 0.7\text{ V}]}{2.2\text{ k}\Omega} = 10\text{ mA}$$

$$I_C(Q) = \beta \cdot I_B = 50(10\text{ mA}) \cong 0.5\text{ A}$$

$$V_{CE}(Q) = [V_{CC} - I_C \cdot R_C] = [22.7\text{ V} - (0.5) \times (22)] = 11.7\text{ V}$$

Q-Point parameters are $V_{CE}(Q) = 11.7$ V and $I_C(Q) = 0.5$ A

AC Signals:

$$I_C(\text{Peak}) = \beta \times I_B(\text{Peak}) = [50 \times (10\text{ mA Peak})] = 500\text{ mA (Peak)}$$

Output power $$P_{out}(\text{AC}) = \frac{[I_C^2(\text{Peak}) \times R_C]}{2} = \frac{(0.5)^2 \times 22}{2} = 2.75\text{ W}$$

Input DC Power:

DC input power $$P_{in}(\text{DC}) = V_{CC} \times I_C(Q) = (22.7\text{ V}) \times (0.5\text{ A}) = 11.35\text{ W}$$

Power dissipation $$P_D = [P_{in}(\text{DC}) - P_0(\text{AC})] = [11.35 - 2.75] = 8.6\text{ W}$$

Collector Circuit efficiency:

$$\%\text{ Efficiency} = \%\eta = \frac{P_{out}(\text{AC})}{P_{in}(\text{DC})} \times 100\% = \frac{2.75\text{ watts}}{11.35\text{ watts}} \times 100 = 24.22\%.$$

11.3 TRANSFORMER-COUPLED AUDIO POWER AMPLIFIER

Disadvantage of DC Power dissipation in resistive load in series-fed Power Amplifier is overcome by using a Transformer to couple the output signal Power to the load.

Amplifier Operation

- Input signal (V_{in}) variations cause variations in Transistor biasing voltages. They cause variations in output current and voltages in the Amplifier circuit (Fig. 11.14).
- Optimum Power transfer is obtained by using a Transformer between the (high impedance) output circuit of the Transistor and low impedance load (by providing impedance matching by transformer action).

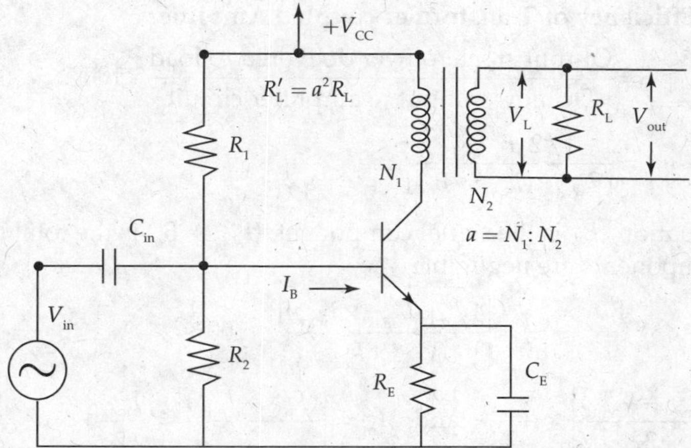

FIG: 11.14 *Transformer-coupled Class-A transistor power amplifier*

11.3.1 Merits of Using Transformer in the Amplifier in Place of R_L

- If load resistance (R_C) R_L is connected directly in Amplifier output circuit (Fig. 11.9), DC Collector current passes through load resistance R_L. Collector current causes Power dissipation (P_D) and heats the resistor. It represents a waste of power, since AC signal component of Power only is used for practical applications.
- Output Transformer is used (in place of R_L) to couple actual load into Amplifier output circuit (Fig. 11.14). Such circuit is known as Transformer-coupled Power Amplifier.
- DC power supply path to Transistor and signal power path through load resistance R_L are separate. Hence, it is known as Shunt feed Amplifier.
- Power transfer through impedance matching is done by using primary winding having number of turns N_1 greater than secondary winding with number of turns N_2.
  - o Amplifier design starts from final load R_L (for example, loudspeaker's impedance of 8 Ω) into output circuit of active device for maximum Power transfer (considering the turns ratio and efficiency of output Transformer) is considered.
  - o Hence, voltage and current levels can be changed by transformer turns ratio.
  - o Load resistance R_L is connected across Transformer secondary winding. It appears as reflected resistance R'_L at the Transformer primary.

$$\frac{R_L}{R'_L} = \frac{V_2 / I_2}{V_1 / I_1} = \frac{V_2}{V_1} \times \frac{I_1}{I_2} = \frac{N_2}{N_1} \times \frac{N_2}{N_1} = \left(\frac{N_2}{N_1}\right)^2$$

If we define Transformer Turns ratio $a = \dfrac{N_1}{N_2}$,

$$\frac{R'_L}{R_L} = \left(\frac{N_1}{N_2}\right)^2 = a^2$$

$$\therefore \quad R'_L = a^2 \cdot R_L.$$

Collector Circuit Efficiency of Transformer-coupled Amplifier

$$\eta = \frac{\text{Output signal power delivered to load}}{\text{DC power input to amplifier circuit}} \times 100\% \tag{11.31}$$

$$\eta = \frac{1/2 \cdot B_1^2 \cdot R_L'}{[V_{max} \times (I_{max} + B_0)]} \tag{11.32}$$

In the above equation, B_1^2 is the signal component, $(I_{max} + B_0)$ is the total DC component. If distortion components are negligible,

$$\eta = \left[\frac{1/2[(V_{max} \times I_{max})]}{(V_{max} \times I_{max})}\right] \times 100\%$$

$$\left(\text{where } V_m = \left[\frac{(V_{max} - V_{min})}{2}\right] \cong \frac{V_{max}}{2} \text{ and } I_m = \left[\frac{(I_{max} - I_{min})}{2}\right] \cong \frac{I_{max}}{2}\right)$$

$$\eta = \frac{100\%}{2} = 50\% \tag{11.33}$$

Maximum theoretical efficiency of a *Transformer-coupled Power Amplifier* = 50%. This is twice the efficiency of Series-fed Class-A amplifier with resistive load.

DC Load Line DC winding resistance determines DC Load line. DC resistance is very small (ideally 0 Ω). So, DC load line is a vertical line (Fig. 11.15) at $V_{CE}(Q) = V_{CC}$.

Quiescent Operating Point (Q) *Q*-Point is set at the intersection of DC load line and Base current set by the biasing circuit for Class-A operation.

AC Load Line To carry out the AC analysis, it is necessary to calculate the AC load resistance seen looking into the primary side of the Transformer, as the slope of AC load line depends on the reflected load resistance, $R_L' = a^2 R_L$, where $a = N_1/N_2$.

Draw the AC Load line through the operating point with a slope equal to $-1/R_L'$

$$Q\text{-point is located at } I_C(Q) = \frac{V_{CC}}{R_{AC}} = \frac{V_{CC}}{R_L'} = \frac{V_{CC}}{a^2 \times R_L}. \tag{11.34}$$

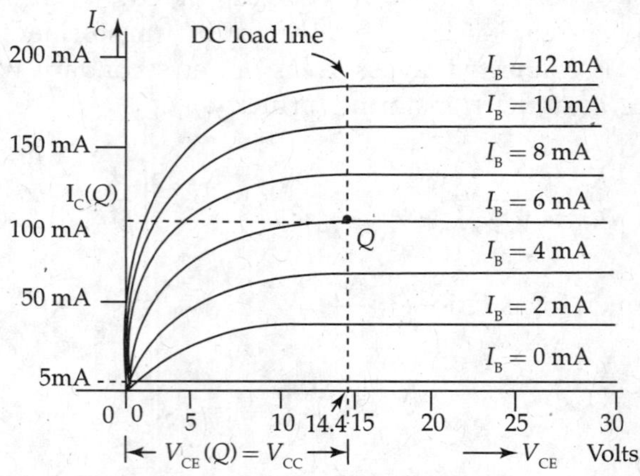

FIG. 11.15 *Transformer-coupled amplifier: DC load line construction*

Collector Current Signal Swing and AC Power output

From the signal variations in Fig. 11.16, the peak-to-peak voltage signal swing is

Output voltage swing = V_{CE}(peak-to-peak) = $[V_{CE(max)} - V_{CE(min)}] = 2V_m$

Output current swing = I_C(peak-to-peak) = $[I_{C(max)} - I_{C(min)}] = 2I_m$ and $V_{CE}(Q) = V_{CC}$

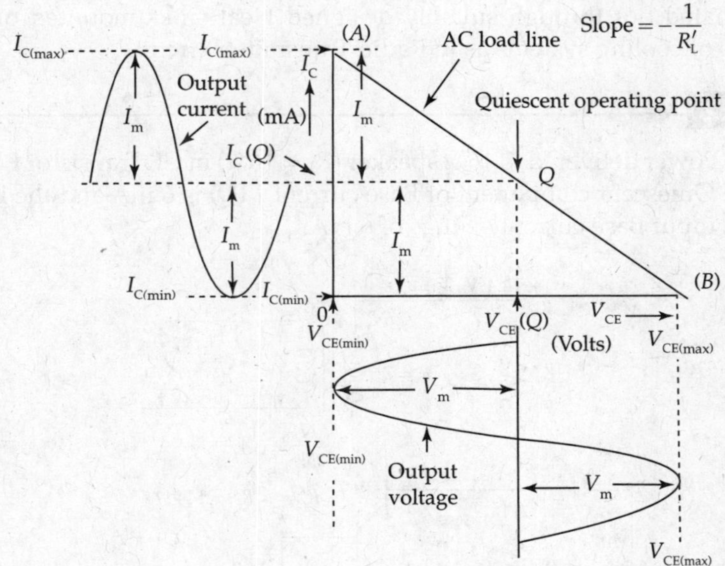

FIG. 11.16 *Output voltage and current swings of transformer-coupled Class-A power amplifier*

Collector Circuit Efficiency

AC power P_{out}(AC) delivered to the load

$$P_{out}(AC) = \frac{V_m}{\sqrt{2}} \cdot \frac{I_m}{\sqrt{2}} = \frac{(V_{CE(max)} - V_{CE(min)})}{2\sqrt{2}} \times \frac{(I_{C(max)} - I_{C(min)})}{2\sqrt{2}} \tag{11.35}$$

$$\therefore \quad P_{out}(AC) = \frac{(V_{CE(max)} - V_{CE(min)}) \times (I_{C(max)} - I_{C(min)})}{8} \tag{11.36}$$

As $[V_{CE(max)} - V_{CE(min)}] = 2\,V_{CC}$ when $V_{CE(min)} \cong 0$ and also $[I_{C(min)}] \cong 0$,

$$P_{out}(AC) = \frac{[2V_{CC} \times I_{C(max)}]}{8} = \frac{[V_{CC} \times I_{C(max)}]}{4} \tag{11.37}$$

DC input power P_{in}(DC) from supply voltage V_{CC}

$$P_{in}(DC) = [V_{CC} \times I_C(Q)] = \left[\frac{V_{CC} \times I_{C(max)}}{2}\right], \tag{11.38}$$

$$\text{where } I_C(Q) = \left[\frac{I_{C(max)}}{2}\right] \tag{11.39}$$

$$\%\eta = \frac{P_{out}(AC)}{P_{in}(DC)} = \left[\frac{[V_{CC} \times I_{C(max)}]/4}{[V_{CC} \times I_{C(max)}]/2}\right] \times 100 = \left[\frac{2 \times [V_{CC} \times I_{C(max)}]}{4 \times [V_{CC} \times I_{C(max)}]}\right] \times 100 = 50\%. \tag{11.40}$$

Only Power loss considered here is that dissipated by the Power Transistor.
Power dissipation in the Transistor,

$$P_D = [P_{in}(DC) - P_{out}(AC)] \text{ watts} \tag{11.41}$$

This dissipated Power in the Power Transistor produces Heat in the Transistor. Heat from Transistor is radiated out through suitably designed Heat sinks mounted on Transistor or some other types of cooling systems as in Radio Transmitter circuits.

EXAMPLE 11.2

Calculate the AC Power delivered to 16 Ω speaker ($R_L = 16\ \Omega$) in CE Transistor Power Amplifier circuit, when the Quiescent component of Base current $I_B(Q)$ is 6 mA and the input signal V_{in} resulting in peak input Base current swings of 6 mA.

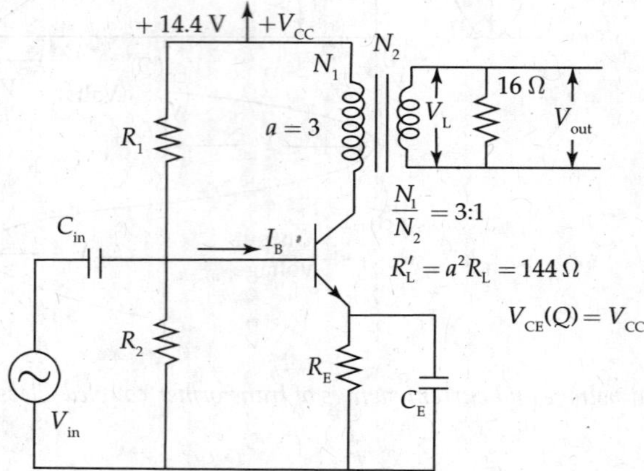

FIG. 11.17 *Transformer-coupled Class-A power amplifier*

Solution: Draw Vertical DC Load line on the output characteristics of the active device considering the primary winding resistance of Transformer as approximately zero ohms. R_E is also assumed to be a very small resistance.

$V_{CE}(Q) = 14.4$ V and $I_C(Q) = 100$ mA

Turns ratio 'a' (of output Transformer) = 3 and $R_L = 16\ \Omega$

Reflected load resistance into output circuit of the Transistor,

$$R'_L = 9 \times 16 = 144\ \Omega$$

Let us draw an AC load line connecting points X and Y with coordinates for point X as $V_{CE} = 0$ V (known from the location of the point X on the load line) and

$$I_C = \frac{V_{CE}}{R'_L} = \frac{14.4\text{ V}}{144\ \Omega} \cong 100\text{ mA.}$$

Coordinates of point Y are $I_C = 0$ mA, known from the location of the point Y and $V_{CE} = V_{CE}(Q)$ that can be calculated from DC load line equation of circuit. AC load line is drawn connecting points X and Y with $R'_L = 144\ \Omega$ with a slope $(-1/144)$.

Now another AC load line is drawn parallel to the line X-Y passing through 'Q' point to obtain the actual output signal swings for given external input excitation of the Amplifier,

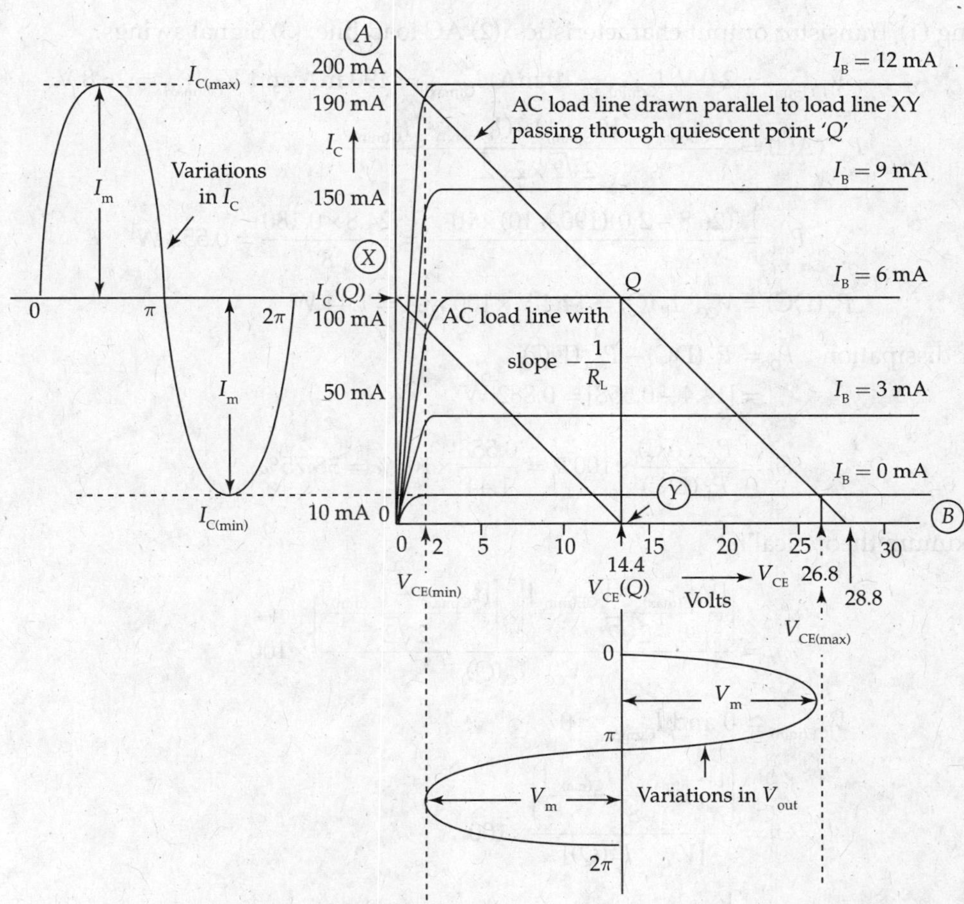

FIG. 11.18 *Signal waveforms in transformer-coupled amplifier*

resulting in the input Base current of 6 mA (assumed in the problem). All these signal swings are about the quiescent operating point Q.

For AC Load line connecting points A and B, coordinates of Point B are $V_{CE} = 28.8$ V and $I_C = 0$ mA.

Coordinates of Point A are $I_C = 200$ mA and $V_{CE} = 0$ V.

Effective AC resistance at primary $R'_L = a^2 R_L$

$$\text{At Point } A, \quad I_C = \frac{V_{CE} \text{ at point } B}{144\ \Omega} \cong \frac{28.8}{144} = 200 \text{ mA.}$$

At the quiescent operating point 'Q',

$$I_C(Q) = \frac{V_{CE}(Q)}{R'_L} = \frac{14.4 \text{ V}}{144\ \Omega} = 100 \text{ mA.}$$

At Point A on the AC load line,

$$[I_C(Q) + I_C] = [100 + 100] = 200 \text{ mA.}$$

Using (1) Transistor output characteristics, (2) AC load line, (3) Signal swings,

$$V_{CE(min)} = 2.0 \text{ V}, I_{C(min)} = 10 \text{ mA}, I_{C(max)} = 190 \text{ mA and } V_{CE(max)} = 26.8 \text{ V}.$$

$$P_{out}(AC) = \frac{(V_{CE(max)} - V_{CE(min)}) \times (I_{C(max)} - I_{C(min)})}{2\sqrt{2} \times 2\sqrt{2}}$$

$$P_{out} = \frac{(26.8 - 2.0)(190 - 10) \times 10^{-3}}{8} = \frac{24.8 \times 0.180}{8} = 0.558 \text{ W}$$

$$P_{in}(DC) = V_{CC} \cdot I_C(Q) = 14.4 \text{ V} \times 100 \text{ mA} = 1.44 \text{ W}$$

Power dissipation $P_D = P_{in}(DC) - P_{out}(DC)$

$$= [14.4 - 0.558] = 0.882 \text{ W}$$

$$\%\eta = \frac{P_{out}(AC)}{P_{in}(DC)} \times 100\% = \frac{0.558}{1.44} \times 100\% = 38.75\%$$

Maximum theoretical η:

$$\%\eta = \frac{\left[\frac{[V_{CE(max)} - V_{CE(min)}]}{2\sqrt{2}}\right] \times \left[\frac{[I_{C(max)} - I_{C(min)}]}{2\sqrt{2}}\right]}{V_{CC} \cdot I_C(Q)} \times 100$$

$$V_{CE(min)} \cong 0 \text{ and } I_{C(min)} \cong 0$$

$$\therefore \quad \%\eta = \frac{\left(\frac{[V_{CE(max)} \times I_{C(max)}]}{8}\right)}{[V_{CC} \times I_C(Q)]} \times 100$$

$$V_{CE} = \frac{V_{CE(max)}}{2} \quad \text{and} \quad I_C(Q) = \frac{I_{C(max)}}{2}$$

$$\%\eta = \frac{\left(\frac{V_{CE(max)} \times I_{C(max)}}{8}\right)}{\left(\frac{V_{CE(max)}}{2} \times \frac{I_{C(max)}}{2}\right)} \times 100 = 50\%.$$

11.4 CLASS-A PUSH-PULL AMPLIFIER

11.4.1 Parallel Operation of Amplifiers

Increase in output power can be obtained by connecting two Amplifiers (Fig. 11.19) in parallel with a common load R_L. Such operation is considered as parallel Amplifier.

The two Collector currents i_{C1} and i_{C2} flow through R_L in the same direction. Hence, output current is sum of two Collector currents, which increases output power. Each Amplifier supplies half load current (I_L) and increases output power. Simultaneously distortion increases. Increase in distortion content in output signal is an unwanted phenomenon.

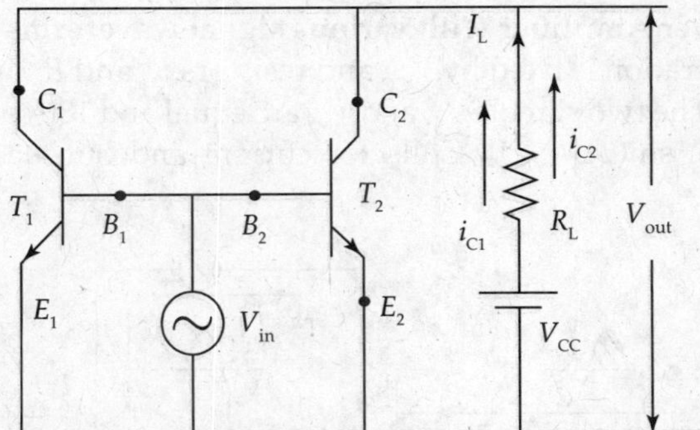

FIG. 11.19 *Parallel operation of amplifiers*

11.4.2 Push-Pull Amplifier Circuit (Fig. 11.20)

Push-Pull Amplifiers work on the principle of operation of parallel connection of two Amplifiers with common load R_L.

- The two Base terminals B_1 and B_2 of Transistors T_1 and T_2 are connected through the secondary winding of the input Transformer. The two Emitter terminals E_1 and E_2 are connected to form the common terminal. Input signal is applied (at primary of input Transformer) between Base and Emitter of CE Transistor Amplifiers.
- The two Collector terminals C_1 and C_2 are connected together through the primary winding of the output Transformer. Common Emitter and Collector terminals form output port for biasing arrangements, main DC Source and output signal operations. Output Power is obtained at output Transformer secondary with R_L.
- Load resistance is connected at the secondary winding of the output Transformer. By suitable design of *turns ratio* of output Transformer, *optimum Power transfer* is obtained from the Transistor output circuit to load resistance R_L.
- The two Transistors should be a *matched pair* with identical characteristics so that certain advantages of push-pull operation can be derived.

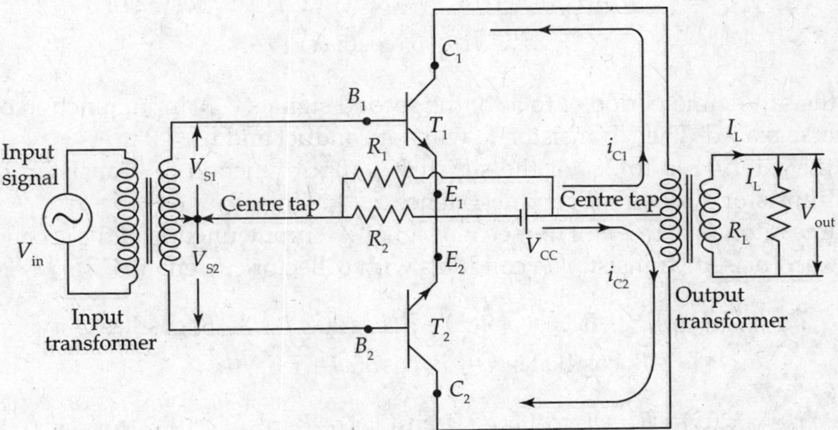

FIG. 11.20 *Push-pull Class-A power amplifier*

Class-A Push-Pull Power Amplifier with various signal waveforms (Fig. 11.21)
Q-point for Class-A operation is fixed by V_{CC} and resistors R_1 and R_2. Input signal is applied to the input Transformer. The two signals V_{S1} and V_{S2} are equal and 180° out of phase [$V_{S1} = V_{in(max)} \sin(\omega t)$ and $V_{S2} = V_{in(max)} \sin(\omega t + \pi)$]. Collector current and output voltage waveforms are shown in Fig. 11.21.

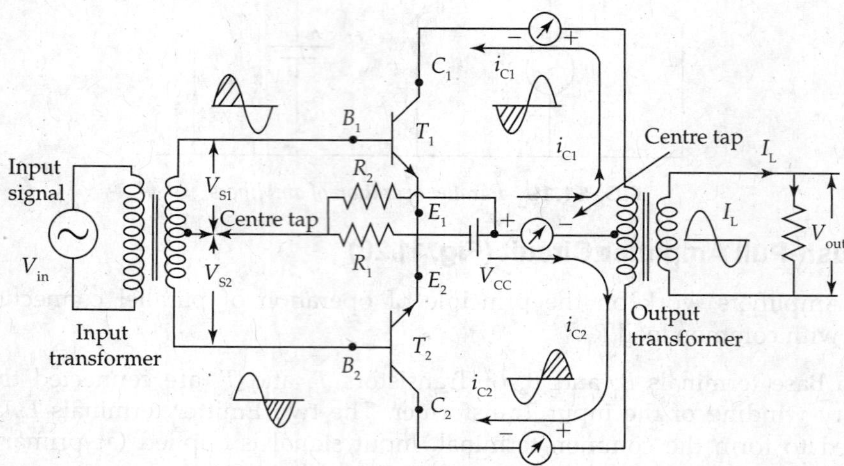

FIG. 11.21 *Push-pull power amplifier*

A number of circuit arrangements exist to obtain push-pull operation, but Transformer with centre-tapped secondary winding arrangement to produce the phase-inverted signals (phase splitter) is a simple way to obtain push-pull operation in this Amplifier circuit.

11.4.3 Circuit Operation of Push-Pull Amplifier

- During the interval '0 to π' of input signal (V_{S1}), input junction of Transistor T_1 is forward biased. Then Transistor T_1 conducts with collector current 'i_{C1}':

$$i_{C1} = I_C(Q) + B_0 + B_1 \cos(\omega t) + B_2 \cos(2\omega t) + B_3 \cos(3\omega t) + B_4 \cos(4\omega t) + B_5 \cos(5\omega t) + \cdots. \tag{11.42}$$

- During the same time period (0 to π) of the second signal (V_{S2}), input junction of Transistor T_2 is reverse biased. Then Transistor T_2 will not conduct and $i_{C2} \cong 0$ mA.
- During the interval 'π to 2π' of the signal V_{S1}, input junction of Transistor T_1 is reverse biased. Transistor T_1 does not conduct. Hence, $i_{C1} \cong 0$ mA.
- At the same interval 'π to 2π' of the second signal V_{S2}, input junction of the second Transistor T_2 is forward biased. Transistor T_2 conducts with collector current 'i_{C2}'.

$$i_{C2} = I_C(Q) + B_0 + B_1 \cos(\omega t + \pi) + B_2 \cos 2(\omega t + \pi) + B_3 \cos 3(\omega t + \pi) + B_4 \cos 4(\omega t + \pi) + B_5 \cos 5(\omega t + \pi) + \cdots. \tag{11.43}$$

$$i_{C2} = I_C(Q) + B_0 - B_1 \cos(\omega t) + B_2 \cos(2\omega t) - B_3 \cos(3\omega t) + B_4 \cos(4\omega t) - B_5 \cos(5\omega t) + \cdots. \tag{11.44}$$

The two output currents i_{C1} and i_{C2} (which are 180° out of phase with each other) combine or add together. Addition of output currents results in more Power than one Amplifier circuit can provide.

Output current through load resistance is proportional to the difference of the two collector currents i_{C1} and i_{C2}.

$$\therefore \quad I_L \propto (i_{C1} - i_{C2}) \tag{11.45}$$

Using the expressions for i_{C1} and i_{C2} in Eq. (11.45), resultant load current

$$I_L = 2B_1 \cos(\omega t) + 2B_3 \cos(3\omega t) + 2B_5 \cos(5\omega t) + \cdots. \tag{11.46}$$

Equation (11.46) is obtained by considering the two Transistors to have identical characteristics. Then the two Transistors are known as *matched pair.*

- Load current (Eq. (11.46)) does not contain even harmonic terms. Hence, the even harmonic distortion content is cancelled in output power for Push-Pull Amplifiers.
- There is simultaneous (1) reduction in distortion content and (2) increase in output power (Half of the signal amplification is done by Transistor T_1 and the other half of the signal amplification is achieved by Transistor T_2 in push-pull Amplifiers.
- Additional advantage is reduction in size and weight of output Transformer due to considerable reduction of core losses in it. It is due to the flow of the two collector currents in opposite directions through the two half windings of its primary windings of output transformer (Fig. 11.21).
- Output Transformer couples the amplified output signal to load resistance R_L, while providing maximum power transfer from the Transistor.

Three point method of analysis to determine second harmonic distortion and signal Power

Dynamic transfer characteristic of an active device for large signal operation is non-linear.

Collector current I_{out} (i_C) and input signal V_{in} are related by

$$I_{out} = i_C = A_1 \cdot V_{in} + A_2 \cdot V_{in}^2. \tag{11.47)}$$

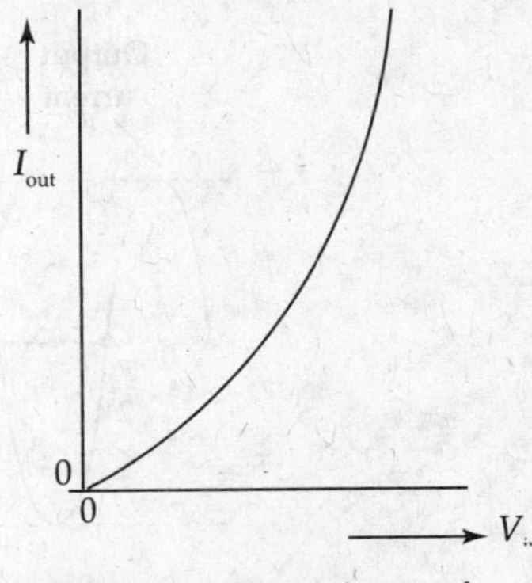

Assume input signal

$$V_{in} = V_m \cdot \cos(\omega t) \tag{11.48}$$

Substituting the value of V_{in} in Eq. (11.47)

$$i_C = A_1 \cdot V_m \cdot \cos(\omega t) + A_2 \cdot V_m^2 \cdot \cos^2(\omega t) \tag{11.49}$$

FIG. 11.22 *Dynamic transfer curve*

$$i_C = A_1 \cdot V_m \cdot \cos(\omega t) + A_2 \cdot V_m^2 \left[\frac{(1+\cos(2\omega t)}{2}\right]$$

$$i_C = A_1 \cdot V_m \cdot \cos(\omega t) + \frac{A_2 \cdot V_m^2}{2} + \frac{A_2 \cdot V_m^2}{2} \cdot \cos(2\omega t) \tag{11.50}$$

But the instaneous value of the output current

$$i_C = I_C(Q) + i_C, \tag{11.51}$$

where $I_C(Q)$ is the magnitude of average or DC component of output current i_C.

Substituting the value of i_c from Eq. (11.50) in Eq. (11.51), we get

$$i_C = I_C(Q) + \frac{A_2 \cdot V_m^2}{2} + A_1 \cdot V_m \cdot \cos(\omega t) + \frac{A_2 \cdot V_m^2}{2} \cos(2\omega t). \tag{11.52}$$

Equation (11.52) can be written as

$$i_C = I_C(Q) + B_0 + B_1 \cdot \cos(\omega t) + B_2 \cdot \cos(2\omega t), \tag{11.53}$$

$$\text{where } B_0 = \frac{A_2 \cdot V_m^2}{2}, \quad B_1 = A_1 \cdot V_m \quad \text{and} \quad B_2 = \frac{A_2 \cdot V_m^2}{2}. \tag{11.54}$$

Total DC content in output current $[I_{DC} = I_C(Q) + B_0]$. (11.55)

(1) B_1 is the amplitude of fundamental frequency (same frequency of input signal) of output signal. It contributes output power. (2) B_2 is the amplitude of second harmonic component. Frequency of second harmonic is twice that of sinusoidal excitation. (3) Output signal contains frequency components that are integer multiples of fundamental frequency component. Those multiple frequency components are known as higher order Harmonics. They are considered as *Harmonic Distortion.*

Voltage and current signal waveforms in Fig. 11.23 are used for the analysis of power and distortion in Amplifier using graphical method by Three point method of Analysis.

The three constants in Eq. (11.53) are determined by the measured values of $I_{C(max)}$, $I_{C(min)}$ and $I_C(Q)$ from the Transistor output characteristics and dynamic load line in Fig. 11.23.

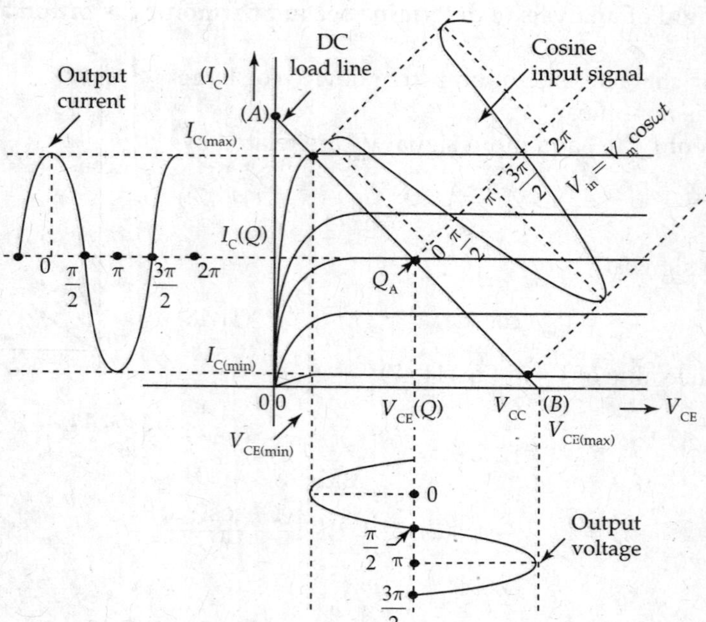

FIG. 11.23 *Output voltage and current swings of push-pull amplifier for three-point method of analysis*

From the waveforms in Fig. 11.23,

$$\text{at } \omega t = 0, \quad \text{output current } i_C = I_{C(\max)}, \tag{11.56}$$

$$\text{at } \omega t = (\pi/2), \quad \text{output current } i_C = I_C(Q), \tag{11.57}$$

$$\text{at } \omega t = \pi, \quad \text{output current } i_C = I_{C(\min)}. \tag{11.58}$$

Substituting the values from Eqs. (11.56)–(11.58) in Eq. (11.53), we get

$$I_{C(\max)} = I_C(Q) + B_0 + B_1 + B_2, \tag{11.59}$$

$$\therefore \quad I_C(Q) = I_C(Q) + B_0 - B_2 \tag{11.60}$$

$$I_{C(\min)} = I_C(Q) + B_0 - B_1 + B_2. \tag{11.61}$$

From Eq. (11.60),

$$B_0 = B_2. \tag{11.62}$$

Subtracting Eq. (11.61) from Eq. (11.59)

$$B_1 = \frac{I_{C(\max)} - I_{C(\min)}}{2}. \tag{11.63}$$

From Eqs. (11.59), (11.60) and (11.63),

$$I_{C(\max)} = I_C(Q) + 2B_2 + \frac{[I_{C(\max)} - I_{C(\min)}]}{2} \tag{11.64}$$

$$\therefore \quad 2B_2 = I_{C(\max)} - I_C(Q) - \frac{[I_{C(\max)} - I_{C(\min)}]}{2} \tag{11.65}$$

$$B_0 = B_2 = \frac{[I_{C(\max)} + I_{C(\min)} - 2I_C(Q)]}{4} \tag{11.66}$$

Distortion

Distortion is contributed by the second harmonic component B_2.

Distortion factor D_2 is defined as follows:

$$D_2 = \left|\frac{B_2}{B_1}\right|, \tag{11.67}$$

$$\% \text{ Distortion factor} = D_2 \times 100 = \left|\frac{B_2}{B_1}\right| \times 100. \tag{11.68}$$

Similarly, the higher order harmonic distortions are expressed as

$$D_2 = \left|\frac{B_2}{B_1}\right| \tag{11.69}$$

$$\text{and} \quad D_3 = \left|\frac{B_3}{B_1}\right|. \tag{11.70}$$

Total harmonic distortion D (THD) in the power amplifier is

$$D=\sqrt{[D_2^2+D_3^2+\cdots]}. \tag{11.71}$$

Power output P_1 due to the fundamental frequency component B_1 is

$$P_1=\frac{B_1^2\cdot R_L}{2}\text{(watts)}, \tag{11.72}$$

where R_L is the load resistance.

Total power output P is calculated using the first and second harmonic amplitude contributions B_1 and B_2, respectively.

$$P=\frac{(B_1^2+B_2^2)R_L}{2}=P_1(1+D_2^2)$$

$$P=P_1(1+D^2),$$

where total distortion $D=\sqrt{D_2^2}$ watts Eq. (11.72) due to the second harmonic.

This analysis can be extended to other active devices like Field effect Transistors.

Five-point method of analysis

To determine higher order harmonic distortion and signal Power

For Power Amplifiers with large magnitudes of input signal voltages, higher order harmonics beyond second harmonic distortion would be introduced. Output Collector current i_C is expressed by the following equation, with significant amplitudes up to fourth harmonic term with B_4:

$$i_C=I_C(Q)+B_0+B_1\cos(\omega t)+B_2\cos(2\omega t)+B_3\cos(3\omega t)+B_4\cos(4\omega t). \tag{11.73}$$

Five terms B_0, B_1, B_2, B_3 and B_4 in Eq. (11.73) are calculated by finding five values of output Collector current at five instances of input signal of cosine function using the dynamic transfer characteristic between the input cosine signal and output current.

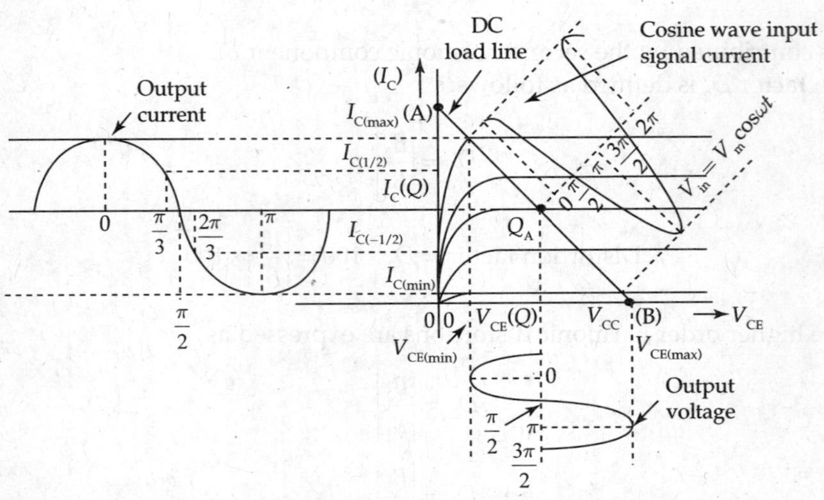

FIG. 11.24 *Class-A amplifier voltage and current signal waveforms*

Assuming the input signal of cosine wave, $V_{in} = V_m \cdot \cos(\omega t)$
Resulting output Collector current $i_C = I_{C(max)}\cos(\omega t)$
From the waveforms in Fig. 11.24, currents for different values of weight are

$$\text{at } \omega t = 0, \quad i_C = I_{C(max)}, \tag{11.74}$$

$$\text{at } \omega t = \frac{\pi}{3}, \quad i_C = I_{C(1/2)}, \tag{11.75}$$

$$\text{at } \omega t = \frac{\pi}{2}, \quad i_C = I_C(Q), \tag{11.76}$$

$$\text{at } \omega t = \frac{2\pi}{3}, \quad i_C = I_{C(-1/2)}, \tag{11.77}$$

$$\text{at } \omega t = \pi, \quad i_C = I_{C(min)}. \tag{11.78}$$

Substituting the condition of Eq. (11.74) in Eq. (11.73), we get

$$I_{C(max)} = I_C(Q) + B_0 + B_1 + B_2 + B_3 + B_4. \tag{11.79}$$

Substituting the condition of Eq. (11.75) in Eq. (11.73), we get

$$I_{C(1/2)} = I_C(Q) + B_0 + \frac{B_1}{2} - \frac{B_2}{2} - B_3 - \frac{B_4}{2}. \tag{11.80}$$

Substituting the condition of Eq. (11.76) in Eq. (11.73), we get

$$I_C(Q) = I_C(Q) + B_0 - B_2 + B_4. \tag{11.81}$$

Substituting the condition of Eq. (11.77) in Eq. (11.73), we get

$$I_{C(-1/2)} = I_C(Q) + B_0 - \frac{B_1}{2} - \frac{B_2}{2} + B_3 - \frac{B_4}{2}. \tag{11.82}$$

Substituting the condition of Eq. (11.78) in Eq. (11.73), we get

$$I_{C(min)} = I_C(Q) + B_0 - B_1 + B_2 - B_3 + B_4. \tag{11.83}$$

Now use the following manipulations:

$$\text{Equations (11.79)} - \text{(11.83)} \Rightarrow I_{C(max)} - I_{C(min)} = 2(B_1 + B_3)$$

$$\Rightarrow (B_1 + B_3) = \left[\frac{(I_{C(max)} - I_{C(min)})}{2}\right] \tag{11.84}$$

$$\text{Equations (11.80)} - \text{(11.82)} \Rightarrow I_{C(1/2)} - I_{C(-1/2)} = B_1 - 2(B_3) \tag{11.85}$$

$$\text{Equations (11.84)} - \text{(11.85)} \Rightarrow 3B_3 = \left[\frac{I_{C(max)} - I_{C(min)}}{2}\right] - \left[I_{C(1/2)} - I_{C(-1/2)}\right] \tag{11.86}$$

$$\Rightarrow B_3 = \left[\frac{I_{C(max)} - I_{C(min)}}{6}\right] - \left[\frac{I_{C(1/2)} - I_{C(-1/2)}}{3}\right]. \tag{11.87}$$

Substituting Eq. (11.87) in Eq. (11.84), we get

$$B_1 = \left[\frac{I_{C(max)} - I_{C(min)}}{2}\right] - \left[\frac{I_{C(max)} - I_{C(min)}}{6}\right] + \left[\frac{I_{C(1/2)} - I_{C(-1/2)}}{3}\right]$$

$$\Rightarrow B_1 = \left[\frac{2I_{C(max)} - 2I_{C(min)}}{6}\right] + \left[\frac{I_{C(1/2)} - I_{C(-1/2)}}{3}\right]$$

$$\therefore \quad B_1 = \left[\frac{I_{C(max)} - I_{C(min)} + I_{C(1/2)} - I_{C(-1/2)}}{3}\right] \tag{11.88}$$

Subtracting the expression (11.80) for $I_{C(1/2)}$ from the expression (11.79) for $I_{C(max)}$, we get

$$I_{C(max)} - I_{C(1/2)} = \frac{B_1}{2} + \frac{3B_2}{2} + 2B_3 + \frac{3B_4}{2} \tag{11.89}$$

$$\Rightarrow \frac{3(B_2 + B_4)}{2} = (I_{C(max)} - I_{C(1/2)}) - \frac{B_1}{2} - 2B_3. \tag{11.90}$$

Substituting the values of B_1 from Eq. (11.88) and the value of B_3 from Eq. (11.87) in Eq. (11.90), we get

$$[I_{C(max)} - I_{C(1/2)}] - \left[\frac{I_{C(max)} - I_{C(min)} + I_{C(1/2)} - I_{C(-1/2)}}{6}\right] - \left[\frac{I_{C(max)} - I_{C(min)}}{3}\right] + \frac{2(I_{C(1/2)} - I_{C(-1/2)})}{3}$$

$$\Rightarrow \left[-1 - \frac{1}{6} + \frac{2}{3}\right] I_{C(1/2)} + \left[\frac{1}{6} - \frac{2}{3}\right] I_{C(-1/2)} + \left[+1 - \frac{1}{6} - \frac{1}{3}\right] I_{C(max)} + \left[+\frac{1}{6} + \frac{1}{3}\right] I_{C(min)} \tag{11.91}$$

$$\Rightarrow -\frac{3}{6}\left[I_{C(1/2)} + I_{C(-1/2)}\right] + \frac{3}{6}\left[I_{C(max)} + I_{C(min)}\right] \tag{11.92}$$

From Eqs. (11.90) and (11.92),

$$[B_2 + B_4] = -\frac{1}{3}\left[I_{C(1/2)} + I_{C(-1/2)}\right] + \frac{1}{3}\left[I_{C(max)} + I_{C(min)}\right] \tag{11.93}$$

Subtracting Eq. (11.83) from Eq. (11.81), we get

$$\left[I_C(Q) - I_{C(min)}\right] = [B_1 - 2B_2 + B_3] \tag{11.94}$$

$$\therefore \quad 2B_2 = \left[I_{C(min)} - I_{CQ}\right] + [B_1 + B_3]. \tag{11.95}$$

Substituting the values of $(B_1 + B_3)$ from Eq. (11.84) in Eq. (11.95), we get

$$2B_2 = \left[I_{C(min)} - I_C(Q)\right] + \frac{1}{2}\left[I_{C(max)} - I_{C(min)}\right] \tag{11.96}$$

$$\therefore \quad B_2 = \frac{1}{4}\left[I_{C(min)} + I_{C(max)} - 2I_C(Q)\right] \tag{11.97}$$

Substituting the values of B_2 from Eq. (11.97) in Eq. (11.93), we get

$$B_4 = -\frac{1}{3}\left[I_{C(1/2)} + I_{C(-1/2)}\right] + \frac{1}{3}\left[I_{C(max)} + I_{C(min)}\right] - \frac{1}{4}\left[I_{C(min)} + I_{C(max)} - 2I_C(Q)\right] \tag{11.98}$$

$$\therefore \quad B_4 = \frac{1}{12}\left[I_{C(max)} + I_{C(min)} - 4I_{C(1/2)} - 4I_{C(-1/2)} + 6I_C(Q)\right]. \tag{11.99}$$

From Eq. (11.81),

$$B_0 = (B_2 - B_4) \tag{11.100}$$

$$B_0 = \frac{1}{4}\left[I_{C(max)} + I_{C(min)} - 2I_C(Q)\right] - \frac{1}{12}\left[I_{C(max)} + I_{C(min)} - 4I_{C(-1/2)} - 4I_{C(1/2)} + 6I_C(Q)\right] \tag{11.101}$$

$$\Rightarrow B_0 = \frac{1}{12}\left[2I_{C(max)} + 2I_{C(min)} + 4I_{C(1/2)} + 4I_{C(-1/2)} - 12I_C(Q)\right] \tag{11.102}$$

$$\therefore\ B_0 = \frac{1}{6}\left[I_{C(max)} + I_{C(min)} + 2I_{C(1/2)} + 2I_{C(-1/2)} - 6I_C(Q)\right] \tag{11.103}$$

$$B_0 = \frac{1}{6}\left[I_{C(max)} + I_{C(min)} + 2I_{C(1/2)} + 2I_{C(-1/2)}\right] - I_C(Q) \tag{11.104}$$

Final values of the determined constants are as follows:

$$B_0 = \frac{1}{6}[I_{C(max)} + 2I_{C(1/2)} + 2I_{C(-1/2)} + I_{C(min)}] - I_C(Q). \tag{11.105}$$

$$B_1 = \frac{1}{3}[I_{C(max)} + I_{C(1/2)} - I_{C(-1/2)} + I_{C(min)}] \tag{11.106}$$

$$B_2 = \frac{1}{4}[I_{C(max)} - 2I_C(Q) + I_{C(min)}] \tag{11.107}$$

$$B_3 = \frac{1}{6}[I_{C(max)} - 2I_{C(1/2)} + 2I_{C(-1/2)} - I_{C(min)}] \tag{11.108}$$

$$B_4 = \frac{1}{12}[I_{C(max)} - 4I_{C(1/2)} + 6I_C(Q) - 4I_{C(-1/2)} + I_{C(min)}]. \tag{11.109}$$

11.5 CLASS-B PUSH-PULL AMPLIFIERS

For Class-B operation of push-pull Amplifier, biasing resistors R_1 and R_2 are omitted. DC bias to input junctions of two Transistors is zero. When input signal is not applied, both Transistors are in cut-off state. Then the output voltage is zero. When no signal is applied, DC power input and Power dissipation are zero.

When input signal V_{in} is applied, input Transformer produces two out of phase signal voltages V_{S1} and V_{S2} at its secondary winding. Input signal V_{S1} causes forward bias to Transistor T_1 during positive half cycle. It conducts during the interval 0 to π producing (push) output Collector current I_{C1} for only one-half cycle of the signal (Fig. 11.25).

Similarly, V_{S2} causes Transistor T_2 to conduct during the interval π to 2π producing (pull) output Collector current I_{C2}. Thus, it is clear that output currents through each Transistor flow for a time period of one-half cycles. To obtain complete output signal, push-pull actions of two Transistors are used. Such Amplifier is known as push-pull Amplifier.

Half of original input signal gets amplification by Transistor T_1. Other half of input signal gets amplified by second Transistor T_2. Due to combined operation of the two Transistors, output current at secondary winding of output Transformer is a complete sine wave producing output voltage across load resistance R_L, as shown in Fig. 11.25.

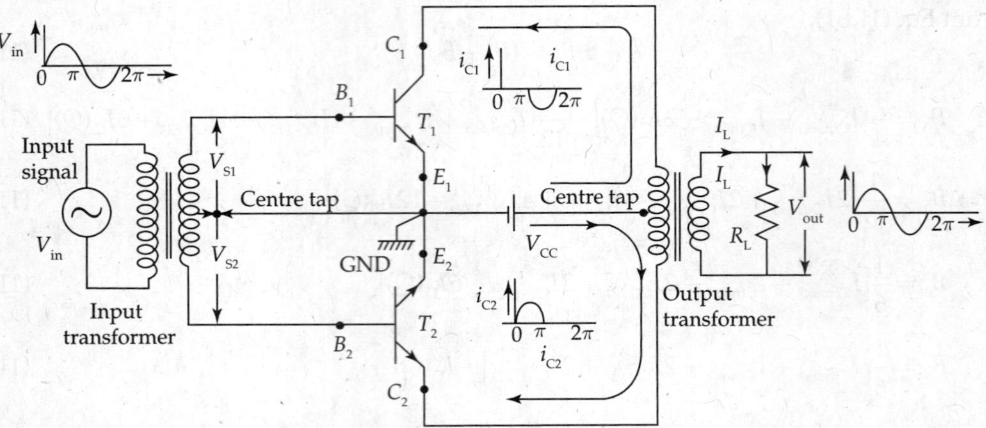

FIG. 11.25 *Class-B push-pull amplifier circuit with signal waveforms*

Calculations for output power and output circuit efficiency

Power conversion efficiency for Class-B push-pull Power Amplifier

$$\text{Collector circuit efficiency} \quad \eta = \frac{\text{AC output power}}{\text{DC input power}} = \frac{P_{AC}}{P_{DC}} \tag{11.110}$$

Collector circuit efficiency is equal to Class-B push-pull amplifier efficiency η

$$\text{AC output power} \quad P_{AC} = V_{rms} \cdot I_{rms} = \frac{V_m}{\sqrt{2}} \times \frac{I_m}{\sqrt{2}} = \frac{V_m \times I_m}{2} = \frac{V_m^2}{2R_L}, \tag{11.111}$$

where V_{rms} is the rms value of the output voltage V_{out}, I_{rms} is the rms value of the output current I_L, V_m is the maximum value of the output voltage V_{out}, and I_m is the maximum value of the load or output current I_L.

$$\text{Maximum AC output power} \quad P_{out(max)} = \frac{V_{CC}^2}{2R_L} \quad (\text{when } V_m = V_{CC})$$

During the signal amplification, DC input power $P_{DC} = (V_{CC} \times I_{DC})$, where I_{DC} is the average value of output Collector currents (in half sinusoid form):

$$\therefore \quad I_{DC} = \frac{I_m}{\pi}, \tag{11.112}$$

where I_{DC} is the average value of output collector currents, which are half sinusoids.

For each transistor during conduction,

$$P_{DC}(\text{input}) = (V_{CC} \times I_{DC}) = \frac{V_{CC} \times I_m}{\pi}. \tag{11.113}$$

Input power for two transistors during conduction,

$$P_{DC} = \frac{2 \times V_{CC} \times I_m}{\pi} \tag{11.114}$$

$$\text{Efficiency of Class-B push-pull amplifier} = \eta = \frac{\text{AC power output}}{\text{DC power input}} = \frac{P_{AC}}{P_{DC}} \quad (11.115)$$

$$\eta = \frac{P_{AC}}{P_{DC}} = \left[\frac{(V_m \times I_m)/2}{(2 \times V_{CC} \times I_m)/\pi}\right] = \left[\frac{V_m \times I_m}{V_{CC} \times I_m}\right] \times \frac{\pi}{4}. \quad (11.116)$$

Maximum amplitude of output voltage swing $V_m = V_{CC}$.
Substituting $V_m = V_{CC}$ in Eq. (11.116), we get

$$\text{Maximum effficiency} \quad \eta = \frac{\pi}{4} = 0.7854.$$

$$\text{Hence,} \quad \%\ \text{efficiency} = \%\eta = 0.7854 \times 100 = 78.54\% \quad (11.117)$$

% efficiency of Class-B push-pull power amplifier = 78.54%.

Maximum Power Dissipation in Transistors

In the absence of input signal to Amplifier, Power dissipation in the circuit is zero. AC output power increases with signal amplitude. This results in increase in average Power dissipation 'P_D' by Transistors during conduction. Maximum Power dissipation, $P_{D(max)}$, can be predicted as follows:

$$\text{Input power} = P_{DC}(\text{input}) = \frac{2V_{CC} \cdot I_m}{\pi} = \frac{2V_{CC} \cdot V_m}{\pi \cdot R_L} \left(\text{using} \quad I_m = \frac{V_m}{R_L}\right) \quad (11.118)$$

$$\text{AC output power} \quad P_{AC} = \frac{V_m \cdot I_m}{2} = \frac{V_m \cdot V_m}{2 \cdot R_L} = \frac{V_m^2}{2 \cdot R_L} \quad (11.119)$$

$$\text{Power dissipation by the transistors} \quad P_D = \left[\frac{2V_{CC} \cdot V_m}{\pi \cdot R_L} - \frac{V_m^2}{2R_L}\right] \quad (11.120)$$

P_D is zero when there is no input signal, i.e., when $V_m = 0$.
Increasing input signal V_{out} increases and reaches maximum value V_m when $V_m = V_{CC}$.
In this process, maximum Power dissipation in transistors reaches $P_{D(max)}$.
$P_{D(max)}$ can be obtained by differentiating Eq. (11.120) with respect to V_m and equate the resulting expression to zero:

$$\frac{dP_D}{dV_m} = \left[\frac{2V_{CC}}{\pi R_L} - \frac{2V_m}{2R_L}\right] = \left[\frac{2V_{CC}}{\pi R_L} - \frac{V_m}{R_L}\right]. \quad (11.121)$$

Now, equate Eq. (11.121) to zero to get the condition for $P_{D(max)}$ in the Transistors:

$$\text{i.e.,} \quad \left[\frac{2V_{CC}}{\pi R_L} - \frac{V_m}{R_L}\right] = 0$$

$$V_m = \frac{2V_{CC}}{\pi}. \quad (11.122)$$

Substituting this value of V_m from Eq. (11.122) in Eq. (11.120), we get

$$P_{D(max)} = \left[\frac{2V_{CC} \cdot 2V_{CC}}{\pi^2 \cdot R_L} - \frac{4V_{CC}^2}{2\pi^2 \cdot R_L}\right] = \left[\frac{8V_{CC}^2 - 4V_{CC}^2}{2\pi^2 \cdot R_L}\right] = \frac{2V_{CC}^2}{\pi^2 \cdot R_L} = 0.2\frac{V_{CC}^2}{R_L} \quad (11.123)$$

$\therefore$ Maximum AC output power $P_{AC(max)} = \dfrac{V_{CC}^2}{2R_L}$ when $V_m = V_{CC}$ (11.124)

$$\therefore \quad P_{D(max)} = 0.2\frac{V_{CC}^2}{R_L} = 0.2 \times 2P_{AC(max)} = 0.4P_{AC(max)} \quad (11.125)$$

Maximum power dissipation $P_D = 0.4P_{AC(max)}$.

Power Dissipation by Transistors

Power dissipated by the two Transistors ($P_D(2T)$) in push-pull Power Amplifiers is the difference between *input power* P_{DC} and *AC output signal power* P_{AC}:

$$P_D\,(2T) = [P_{DC}\,(\text{input power}) - P_{AC}\,(\text{output power})].$$

Power dissipated by each Transistor is then $P_{DC}(2T)/2$.

$\therefore$ Power dissipation by each Transistor $\dfrac{P_D(2T)}{2} = \dfrac{P_{D(max)}}{2} = 0.2 \cdot P_{AC(max)}$ (11.126)

Maximum Power dissipated by the two Transistors $= P_{D(max)}\,(2T) = 0.4 \cdot P_{AC(max)}$.

Advantages of push-pull Class-B Power Amplifier

(1) DC input power and Power dissipation in Transistors are zero, when the input signal is not applied. Hence, standby Power dissipation is zero. This feature is advantageous in Satellite communication Amplifiers. (2) Due to increase in input signal swing output AC Power increases. Amplifier efficiency is increased to a maximum value of 78.54%. (3) Even harmonics in the output signal having significant amplitude are eliminated, which causes a reduction in distortion content. (4) There is no DC saturation in the core of output Transformer. (5) Output Transformer makes optimum Power transfer to low impedance loads.

Disadvantages of push-pull Class-B Power Amplifier

(1) Crossover distortion occurs. (2) Use of Transformers increases the cost of Amplifier. (3) Poor high frequency response of Transformer causes reduction in bandwidth. (4) *Total harmonic distortion (THD)* is higher than standard THD of 1%.

Limitations of Class-B push-pull Amplifier using Transformers

(1) Frequency response and bandwidth are limited due to limitations of frequency response of Transformer. (2) Non-linearity of core material causes distortion in output response. (3) Bulkiness and cost limit the use of Transformer.

11.6 TRANSFORMER-LESS PUSH-PULL AMPLIFIER

Phase Inverter circuit using a single-stage Transistor feedback Amplifier

Transformer-less Class-B push-pull Amplifier circuit is realised by using a phase splitter circuit (Fig. 11.26) with two out of phase voltages. It is a single-stage Transistor feedback Amplifier.

(1) When the bypass capacitor across R_E is not present, the voltage V_{out} (2) across R_E is *in-phase to the original input signal*. (2) Output voltage V_{out} (1) at the Collector terminal of the Amplifier will be *180° out of phase to the input signal*.

Choosing R_E and R_L to be of equal values, two output voltages will be equal with 180° phase shift to one another. Such equal and out of phase voltages are used in place of input Transformer in push-pull Amplifier circuits to eliminate the disadvantages of Transformers. Transformer-less Amplifier design is used in VLSI technology.

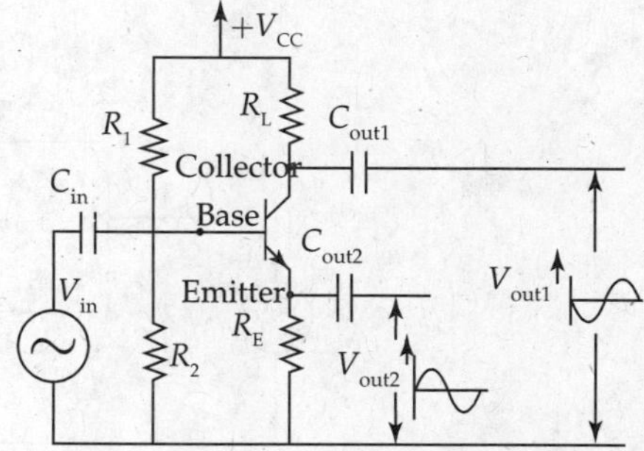

FIG. 11.26 *Phase inverter circuit to obtain two out of phase signals*

11.6.1 Class-B Transformer-less Push-Pull Amplifier

Figure 11.27 show Class-B Transformer-less push-pull Amplifier circuit. At the input port of push-pull Amplifier circuit, two 180° out of phase signals are obtained using a phase splitter circuit using a single-stage Transistor Amplifier.

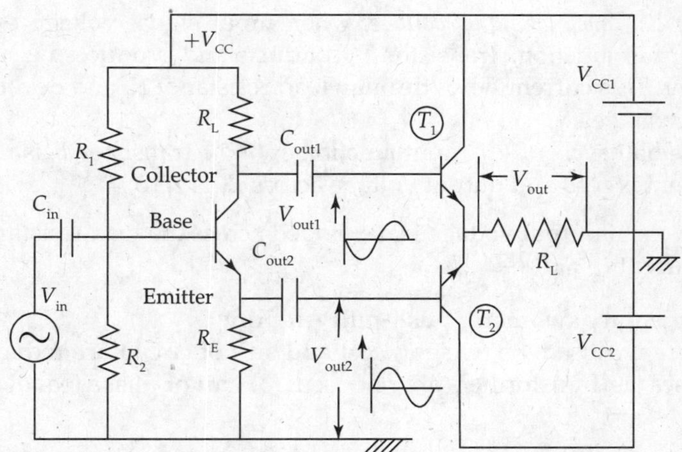

FIG. 11.27 *Transformer-less Class-B push-pull amplifier circuit*

Class-B complementary symmetry push-pull Amplifier (Fig. 11.28)

- Two identical Transistors (one NPN Transistor and one PNP Transistor) constituting complementary symmetry pair are used in Class-B output stage in such a fashion that both cannot conduct simultaneously. Symmetry in operation having equal biasing voltages to have identical signals. Full cycle of output signal across the load resistance (R_L) is obtained due to the contribution of individual half cycles from each Transistor conducting in alternate half cycles, using a single input signal V_{in}. No input and output Transformers are required.
- No phase splitting signals arrangement at the input port. Each Transistor circuit segment functionally works as Emitter follower circuit with the load resistance R_L in the output Emitter terminal circuit.

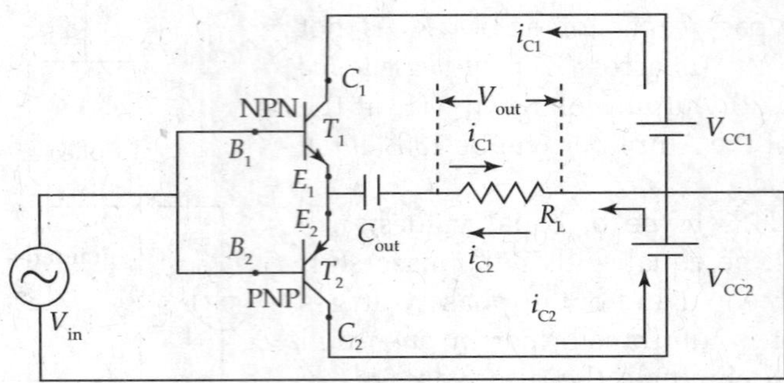

FIG. 11.28 *Complementary symmetry push-pull amplifier*

Circuit Operation

- During *positive half cycle of input signal* (V_{in}), when the signal voltage exceeds cut-in voltage of input junction, Transistor T_1 conducts and its Collector current i_{C1} flows through R_L. It develops output voltage across R_L.
- During positive half cycle of input signal V_{in}, input junction of PNP Transistor T_2 is reverse biased. So, the Transistor T_2 is in off (cut-off) state. Collector current i_{C2} through load R_L is zero during the interval '0 to π'.
- During *negative half cycle of input voltage*, when input signal voltage exceeds the cut-in voltage of the input junction, Transistor T_2 conducts and produces negative half cycle of Collector current. This current flows through load resistance R_L and develops negative half cycle of output voltage.
- During negative half cycle of V_{in}, input junction of NPN Transistor T_1 is reverse biased. So, the output current is zero and output voltage across R_L is zero.

The complete cycle of output voltage is developed across the load resistance R_L.
Finally, Load current $I_L = (i_{C1} - i_{C2})$.

Merits of complementary symmetry push-pull Amplifier

(1) Transformers are not used. So, weight, cost and size of circuit are reduced. (2) Amplifier bandwidth increases, as Transformers are not used. (3) Out of phase (equal amplitude) input signals are not necessary.

Demerits of push-pull Amplifier circuit

(1) Distortion increases due to the addition of higher harmonic components that flow in the same direction through the load resistance. (2) Additional distortion occurs due to crossover distortion. (3) Two Power supplies are used separately to provide the bias conditions for NPN and PNP Transistors. (4) Selection of NPN and PNP Transistors with identical characteristics is difficult.

Magnitudes of various Powers in Class-B complementary symmetry Power Amplifier

$$\text{DC power input} \quad P_{DC} = \frac{V_{CC} \times 2I_m}{\pi} \tag{11.127}$$

$$P_{DC} = \frac{2V_{CC} \cdot V_m}{\pi R_L} = \frac{2V_{CC}^2}{\pi R_L} \quad (\text{assuming } V_m = V_{CC}) \tag{11.128}$$

$$\text{AC power output} \quad P_{AC} = \frac{V_{CC}^2}{2R_L} \tag{11.129}$$

$$\text{Collector circuit power conversion efficiency} = \frac{P_{AC}(\text{output})}{P_{DC}(\text{input})} = \frac{V_{CC}^2 / 2R_L}{2V_{CC}^2 / \pi R_L} = \frac{\pi}{4} = 0.7854$$

$$\%\ \text{Power efficiency} \quad \eta = 0.7854 \times 100 = 78.54. \tag{11.130}$$

EXAMPLE 11.3

In a complementary symmetry push-pull Amplifier using PNP and NPN Transistors, maximum value of Collector current in each Transistor $I_{C(max)} = 2$ A. Break-down voltages for each Transistor $BV_{CE0} = 50$ V. Load resistance $R_L = 8\ \Omega$. Maximum Power dissipation in each Transistor $P_{D(max)} = 5$ W. Calculate V_{CC} and verify whether operating conditions are within Transistor specifications.

Solution: Data given: $I_{C(max)} = 2$ A and $R_L = 8\ \Omega$
From the equation $I_{C(max)} = V_{CC}/R_L$,

$$V_{CC} = I_{C(max)} \cdot R_L = 2 \times 8 = 16\ \text{V}.$$

Maximum signal (peak-to-peak) handling capacity of the amplifier = 2 V_{CC}.
Maximum output voltage swings may be possible up to 2 V_{CC}. (This can be obtained from the graphical analysis using the composite curves.)
Signal swings of amplitude, $2V_{CC} = 32$ V, is quite less than the break-down voltage levels $B_{CE0} = 50$ V for the output junctions of power Transistors.

$$\text{AC power output} = \frac{[V_{CC}]^2}{2R_L} = \frac{(16)^2}{2 \times 8} = 16\ \text{W}$$

$$P_{DC}(\text{input}) = \frac{2V_{CC} \cdot I_m}{\pi} = \frac{2 \times 16 \times 2}{3.14} \cong 20.4\ \text{W}.$$

Complementary symmetry Push-pull Amplifier circuit with biasing voltages (Fig. 11.29)

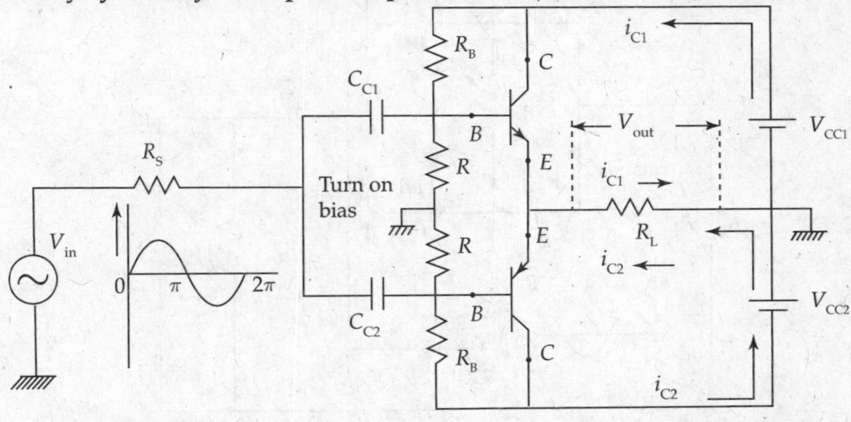

FIG. 11.29 *Complementary symmetry push-pull amplifier with biasing arrangement*

Complementary symmetry Push-pull Amplifier with single DC Source (Fig. 11.30)
Power and circuit efficiency η calculations are similar to Class-B Power Amplifier.

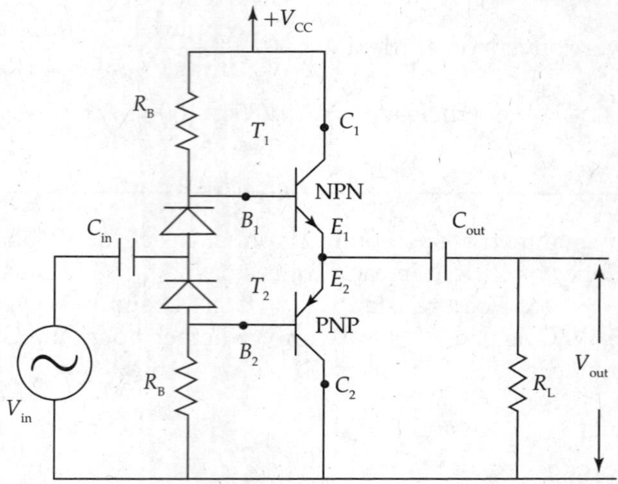

FIG. 11.30 *Complementary symmetry amplifier using a single DC source*

EXAMPLE 11.4

Class-B Power Amplifier (Fig. 11.31) has input voltage $V_{in} = 7.1$ V (rms), and Collector supply voltages $V_{CC1} = V_{CC2} = 22$ V. Amplifier feeds the output voltage V_{out} to a load resistance R_L of 5 Ω. Calculate (a) AC output power P_{out}(AC); (b) Input power P_{in}(DC); (c) Power dissipation by each Transistor; (d) Power conversion efficiency of the Amplifier circuit.

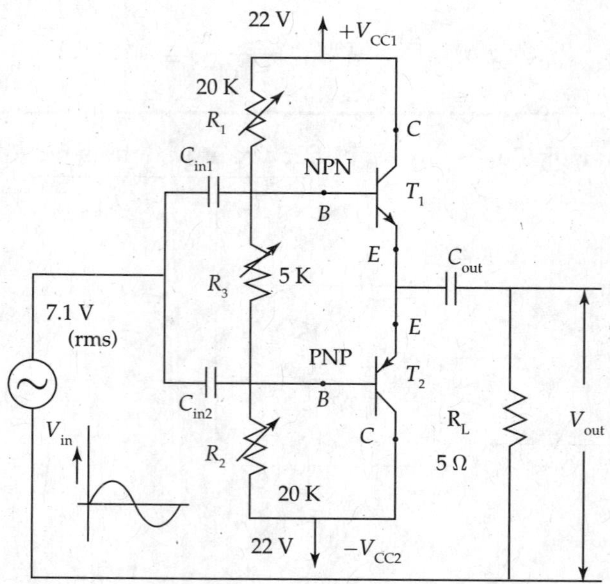

FIG. 11.31 *Class-B push-pull amplifier with variable gain*

Solution:

(a) Data given: $V_{in}\,(\text{rms}) = 7.1$ V and $R_L = 5\,\Omega$

$$V_{in}\,(\text{peak}) \cong \sqrt{2} \times 7.1 = 10 \text{ V}.$$

Amplifier works as Emitter follower, with voltage gain of unity.

∴ Load voltage (output voltage) $V_L\,(\text{peak}) = V_{in}\,(\text{peak}) = 10$ V.

AC power produced across the load $P_{out}\,(\text{AC}) = \dfrac{[V_L\,(\text{peak})]^2}{2 \times R_L} = \dfrac{(10)^2}{10} = 10$ W.

(b) Peak load current $I_L\,(\text{peak}) = \dfrac{V_L\,(\text{peak})}{R_L} = \dfrac{10}{5} = 2$ A

DC current supplied by the two power supplies to meet Peak load current,

$$I_{DC} = \frac{2I_L\,(\text{peak})}{\pi} = \frac{2 \times 2}{3.142} = \frac{4}{3.142} = 1.273 \text{ A}$$

$$\text{Input DC power} = P_{in}\,(\text{DC}) = V_{CC} \times I_{DC} = 22 \times 1.273 = 28 \text{ W}.$$

(c) Power dissipation by two Transistors $P_D = [P_{in}\,(\text{DC}) - P_{out}\,(\text{AC})]$

$$P_D = (28 - 10) = 18 \text{ W}$$

Power dissipation by each Transistor $= \dfrac{P_D}{2} = \dfrac{18}{2} = 9$ W.

(d) Power conversion efficiency $= \dfrac{P_{out}\,(\text{AC})}{P_{in}\,(\text{DC})} = \dfrac{10}{28} = 35.5\%$.

11.7 CROSSOVER DISTORTION

In Class-B Transistor Amplifier, *Q*-point (Q_B) is located at the intersection of $I_B = 0$ mA characteristic and DC load line on Transistor output characteristics (Fig. 11.11). On the input characteristics, bias is at cut-off, so that quiescent components of currents are zero.

Input current flows only when the input signal voltage is larger than built-in potential (V_B) for the Transistor input junction. Output current has some non-conduction intervals or dead zones in its signal waveforms. Transistor conduction process is shown in Fig. 11.32.

Illustration of crossover distortion (Fig. 11.32) in Class-B Push-Pull Amplifiers

- Input characteristic of Transistor (T_1) is between V_{BE1} and I_{B1}. Input characteristic of Transistor (T_2) is between V_{BE2} and I_{B2}. Input signal causes conduction during the intervals ($\theta_1 - \theta_2$) for positive half cycle and during the interval ($\theta_3 - \theta_4$) for negative half cycle. Non-conduction intervals in the input current due to cut-in voltages ($V\gamma$) of the two Transistors are called *dead zones* (Fig. 11.32).
- Absence of the signal during *dead zones* in the output voltage and current signals is known as *crossover distortion.*
- Crossover distortion is more significant, when the magnitude of the input signal is small.

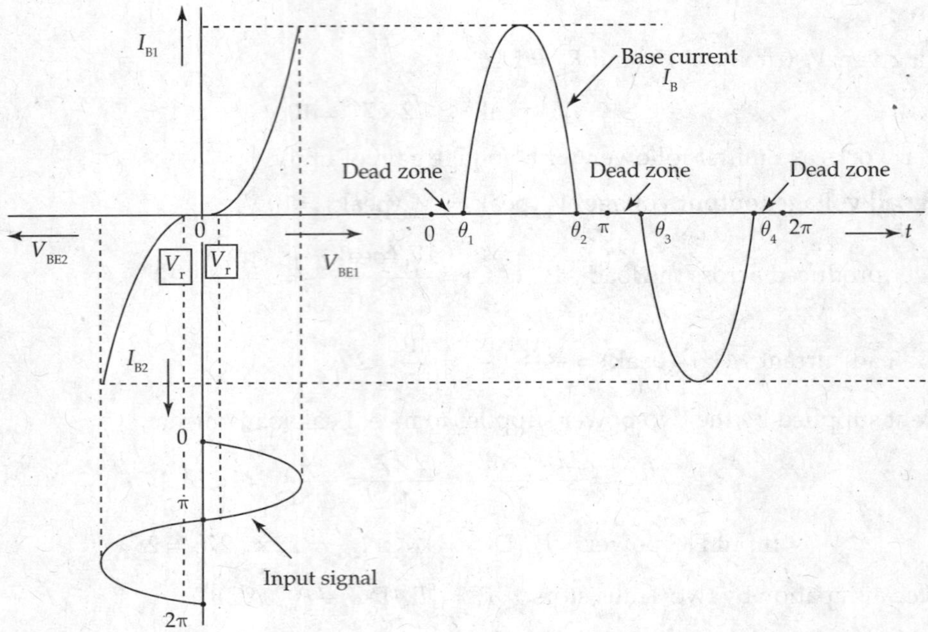

FIG. 11.32 *Crossover distortion in Class-B power amplifiers due to dead zones in the input and consequent output currents [V_r is cut-in voltage]*

- Providing a small initial forward bias (called as 'tickle biases') to the input junctions of the two Transistors can minimise this crossover distortion. So, the Transistor operation goes into Class-AB operation of Amplifier.
- Voltage across resistor R in the input circuits (Fig. 11.33) is designed to provide cut-in voltages ($V\gamma$) during both half cycles of input signal voltage. This provides Class-AB operation to minimise crossover distortion.

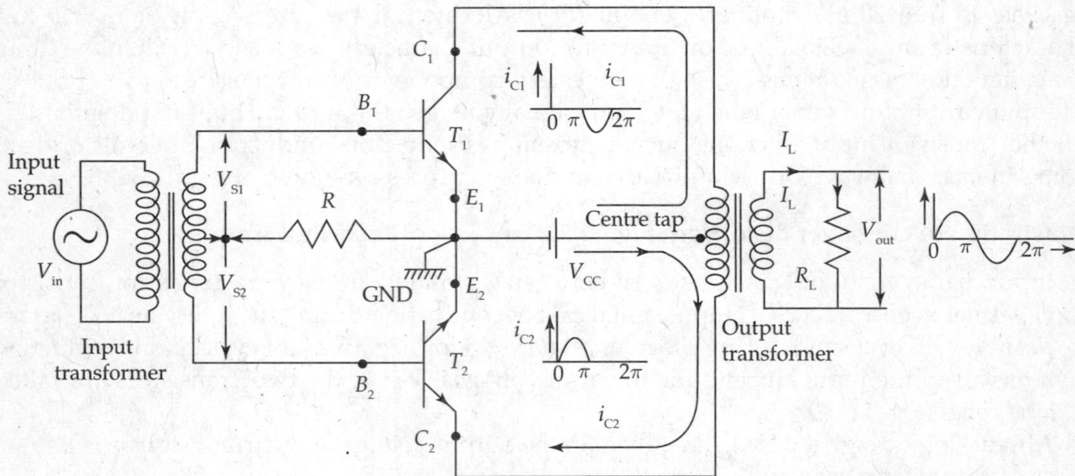

FIG. 11.33 *Class-AB operation to minimise crossover distortion in Class-B push-pull amplifier circuit*

- Measurement of Power in Decibels Let P_2 is the output power from an Amplifier at an instance and P_1 is the reference Power. One of the units of power the *Bel* is defined as the common logarithm of the *ratio of the two Powers P_1 and P_2* (Power ratio *P*) as defined above.

$$\text{Number of Bels } (N_B) \text{ of power ratio } \quad P = \log_{10}\frac{P_2}{P_1} \text{ (Bels).} \tag{11.131}$$

As the Power levels with the electronic Amplifiers are small, a smaller magnitude unit considered for the practical situations is *Decibels* (dB), which is one-tenth of *Bel*:

$$\text{Number of decibels } (N_{dB}) \text{ of power ratio } \quad P = 10\log_{10}\frac{P_2}{P_1} \text{ (dB).} \tag{11.132}$$

If Power P_2 is larger than Power P_1, Power ratio is a positive quantity (for Amplifiers). If the Power P_2 is less than P_1, the Power ratio is a negative quantity. Negative value of measured Power indicates reduction (attenuation) in the circuit under case study.

EXAMPLE 11.5

In a complementary symmetry Class-B Amplifier (Fig. 11.34), $V_{CC} = 16$ V and $R_L = 8\ \Omega$. For sinusoidal input signal, calculate (a) maximum output signal power, (b) DC input power, (c) power dissipation and (d) conversion efficiency.

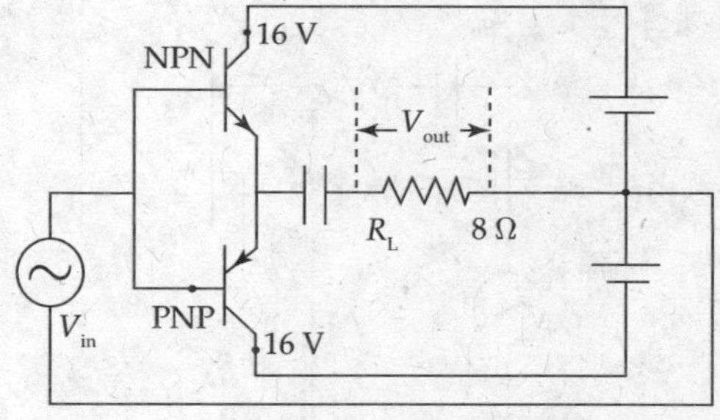

FIG. 11.34 *Class-B complementary symmetry amplifier*

Solution: Collector supply voltage = 16 V; Load resistance $R_L = 8\ \Omega$

$$\text{Maximum Collector current } \quad I_{max} = \frac{V_{CC}}{R_L}$$

$$= \frac{16}{8} = 2\text{ A}$$

$$\text{AC signal output power } \quad P_{out}(\text{AC}) = (I_{rms})^2 \cdot R_L$$

$$\therefore \quad P_{out}(\text{AC}) = \left(\frac{2}{\sqrt{2}}\right)^2 \times 8 = \frac{32}{2} = 16\text{ W}$$

DC Input Power to the Amplifier,

$$P_{in}(\text{DC}) = \frac{V_{CC}\cdot 2\cdot I_{max}}{\pi} = \frac{16\times 2\times 2}{\pi} = 20.4\text{ W}$$

Conversion efficiency or Collector circuit efficiency 'η' is the ratio of the AC Power output to the DC Power input to the Amplifier:

$$\%\text{ efficiency} = \frac{\text{AC power output}}{\text{DC power input}}$$

$$\frac{P_{out}(\text{AC})}{P_{in}(\text{DC})}\times 100 = \frac{16}{20.4}\times 100 = 78.43.$$

Reduction of *crossover distortion* in Class-B output stage
Crossover distortion of Class-B output stage will be reduced drastically by using a high gain Operational Amplifier in the overall negative feedback loop of Class-B stage. It reduces the plus (+) or minus (–) 0.7 V *Dead Band* between the zero crossings by an amount + or – 0.7 V/ A_0, where A_0 is the open loop gain (DC gain) of Operational Amplifier. At high frequencies, the zero crossover distortion is somewhat noticeable, because of the limitation imposed by the skew rate of the operational Amplifier. It causes the Transistors to turn between ON and OFF. Zero *crossover distortion* can be totally eliminated by special biasing techniques known as Class-AB biasing.

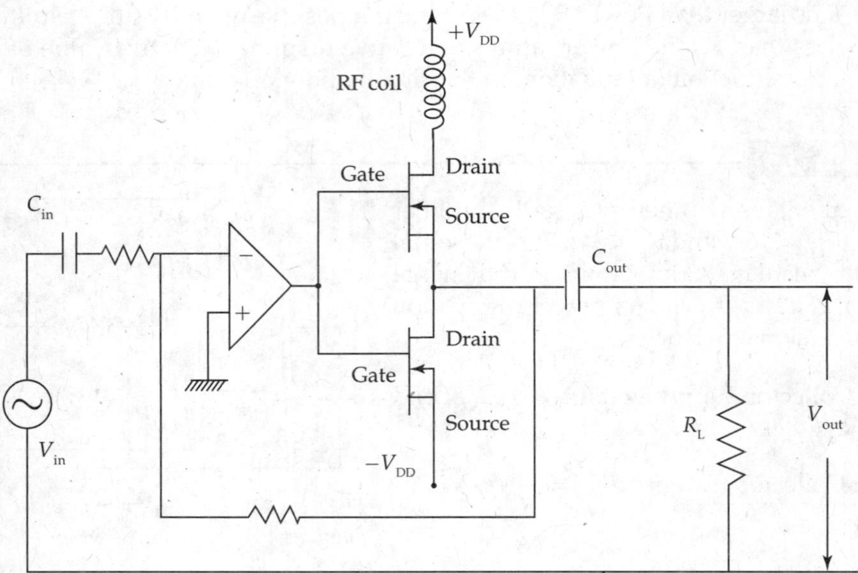

FIG. 11.35 *Buffered Class-B push-pull amplifier with an op-amp connected in negative feedback loop to reduce crossover distortion*

11.8 CLASS-AB POWER AMPLIFIER

Class-AB Amplifier is an intermediate Class of Amplifier between Class-A and Class-B. Conduction takes place for an interval slightly greater than half a cycle that is greater than 180° but very much less than 360°. Another Transistor is also employed to conduct for an interval slightly greater than that of a negative half cycle. Thus, the output currents of both the Transistors when combined in the connected load, the current flows continuously. During the interval between the zero crossings (*dead zone*) both the Transistors conduct which distinguishes from Class-B mode. The unwanted zero crossover distortion is totally eliminated.

Among Class-A, Class-AB and Class-B, the Class-AB stage is the most preferred stages. It is very popular in the design of discrete circuits as well as in integrated circuits.

Circuit Operation Two matched pairs of Transistors T_N (NPN) and T_P (PNP) are connected in complementary mode and biased as shown in Fig. 11.36.

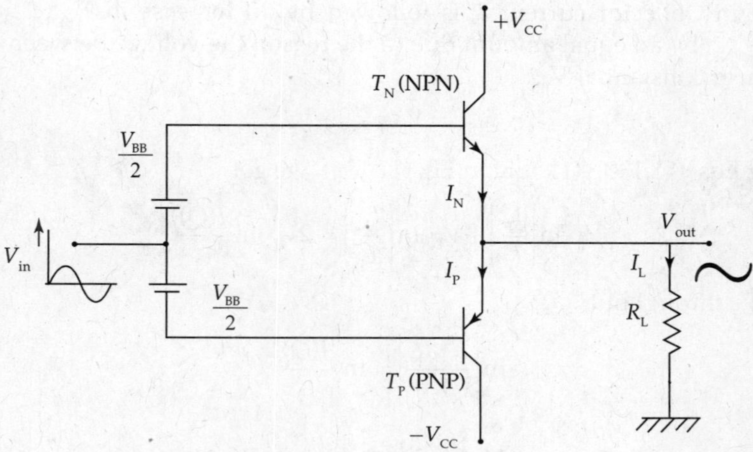

FIG. 11.36 *Class-AB power amplifier*

Under the conditions when the input voltage is zero, the output voltage is zero. Half of the bias voltage is connected across the Base–Emitter junctions of the Transistors. A small quiescent current $I(Q)$ is supplied by small forward bias V_{BB}.

$$\therefore \quad I(Q) = I_P = I_N = I_S \cdot e^{V_{BB}/2V_T},$$

where I_S is the saturation current.

$$\therefore \quad \frac{I_P}{I_S} = e^{V_{BB}/2V_T}$$

Taking natural logarithms on both sides,

$$\ln\left[\frac{I_P}{I_S}\right] = \frac{V_{BB}}{2V_T}$$

$$\therefore \quad \frac{V_{BB}}{2} = V_T \cdot \ln\left[\frac{I_P}{I_S}\right] \tag{11.133}$$

Similarly,

$$\frac{V_{BB}}{2} = V_T \cdot \ln\left[\frac{I_N}{I_S}\right] \tag{11.134}$$

$$\frac{V_{BB}}{2} = V_T \cdot \ln\left[\frac{I(Q)}{I_S}\right]. \tag{11.135}$$

When the input voltage is slightly made positive, the voltage at the Base of the Transistor T_N (NPN) increases turning the Transistor into ON state and the Transistor functions as an Emitter follower. As a result, the output voltage V_{out} also increases and it is positive:

$$V_{out} = V_{in} + \frac{V_{BB}}{2} - V_{BE(NPN)}.$$

The increased output voltage causes a current I_L to flow into the load resistor.

$$\therefore \quad I_N = I_P + I_L$$

An increase in Collector current I_P is followed by an increase in $V_{BE(NPN)}$ followed by a decrease in $V_{EB(PNP)}$ by an equal amount due to the reason the voltage between the Base of the Transistors is held constant at V_{BB}.

$$\therefore \quad V_{BB} = V_{BE(NPN)} + V_{EB(PNP)} \tag{11.136}$$

Substituting Eqs. (11.133)–(11.135) in Eq. (11.136), we get

$$V_T \cdot \ln\left(\frac{I_P}{I_S}\right) + V_T \cdot \ln\left(\frac{I_N}{I_S}\right) = 2V_T \cdot \ln\frac{I(Q)}{I_S}.$$

Cancelling V_T throughout

$$\ln\left[\frac{I_P \cdot I_N}{I_S^2}\right] = \ln\left[\frac{I(Q)}{I_S}\right]^2$$

$$\therefore \quad I_P \cdot I_N = [I(Q)]^2. \tag{11.137}$$

Therefore, this is important relation on which Class-AB operation depends. While I_P increases I_N decreases and vice versa and never I_P or I_N goes down to zero and always allowing some current to pass through the load except in the situation when V_{in} is zero.

For smaller values of input voltage going negative, the load current will be supplied by the PNP Transistor T_P acting as an Emitter follower.

Class-AB operation is an improvement over Class-B particularly when the input voltage is increased or decreased, either of the Transistors continues to conduct and there is a smooth transition in load sharing by the Transistors. Thus the crossover distortion is completely eliminated. Under DC operating conditions, each Transistor in Class-AB operation dissipates a small amount of DC Power dissipation $= V_{CC} \times I_C(Q)$.

Adding the DC Power dissipations in various Power relations modifies the equation derived earlier during Class-B operations:

$$\text{Power dissipation} \quad P_{DC} = \frac{2V_{CC} \times V_m}{\pi R_L} + 2V_{CC} \times I_C(Q)$$

$$P_{DC} = 2V_{CC}\left[\frac{V_m}{\pi R_L} + I_C(Q)\right] \tag{11.138}$$

$$P_{AC} = \frac{V_m^2}{2R_L} \tag{11.139}$$

$$\therefore \text{Efficiency} \quad \eta = \frac{P_{AC}}{P_{DC}} = \frac{V_m^2 / 2R_L}{2V_{CC}\left[\frac{V_m}{\pi R_L} + I_C(Q)\right]} \tag{11.140}$$

Maximum efficiency occurs when $V_m = V_{CC}$.

Therefore, maximum efficiency η is less than 78.54% due to the quiescent power dissipation in each transistor.

The Power dissipation in each Transistor $= 0.5\,[P_{DC} - P_{AC}]$.

The elimination of zero crossover distortion in Class-AB operation can be explained in another way by considering the output resistance of the Amplifier stage. The output resistance R_{OUT} is equal to equivalent parallel resistance of the two Amplifiers:

$$R_{OUT} = r_{e(NPN)} \parallel r_{e(PNP)}$$

where $r_{e(NPN)}$ is the small signal emitter resistance of NPN transistor $= V_T/I_N$, and $r_{e(PNP)}$ is the small signal equivalent resistance of PNP transistor $= V_T/I_P$.

$$\therefore \quad R_{OUT} = \frac{V_T}{[I_N + I_P]},$$

where I_P and I_N are governed by the relation $I_P\, I_N = [I(Q)]^2$.

When V_{in} is very very small, I_P and I_N are very small and output resistance R_{OUT} will be very large.

When V_{in} increases or decreases I_P and I_N also increase or decrease and output resistance will be small

This is the basic principle for the elimination of crossover distortion in Class-AB amplifiers.

Biasing of Class-AB output stages

The Class-AB output stages are biased using two semiconductor Diodes (Fig. 11.37). A current Source passing through the two diodes D_1 and D_2 provides the necessary bias. The two diodes are mounted in close thermal contact with the output Transistors to protect the Transistors against thermal runaway under quiescent conditions. I_{Bias} is chosen to be greater than the maximum anticipated Base drive for the NPN Transistors. It supplies the Base current of the NPN Transistor T_N to increase from $I_C(Q)/\beta_N$ to I_L/β_N, while sourcing current on the application of a small positive voltage V_{in}.

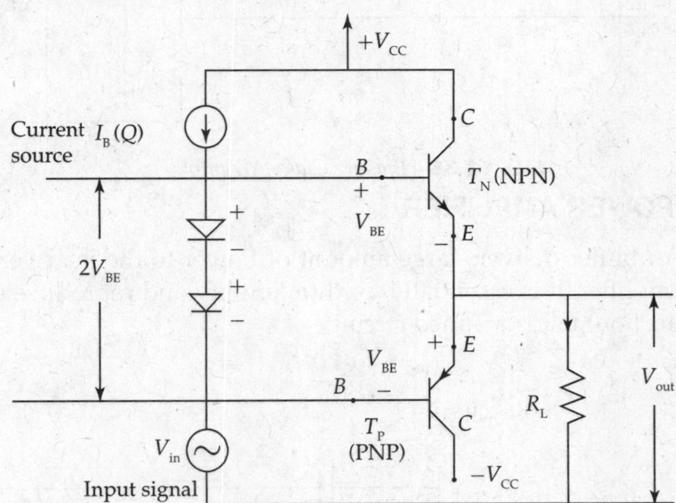

FIG. 11.37 *Biasing for Class-AB output amplifier stage*

Under these conditions, the effective input voltage is

$$V'_{in} = V_{in} + 2V_{BE} - V_{BE(NPN)}$$

$$V'_{in} = V_{in} + 2 \times 0.7 - 0.7 = V_{in} + 0.7\text{ V}.$$

When a small negative voltage is applied, the other PNP Transistor conducts.

Effective input voltage $V'_{in} = V_{in} + V_{BE(PNP)} = V_{in} + 0.7\text{ V}$. Whether it is PNP or NPN Transistor, there is always an off-set voltage of 0.7 V. This is the reason for the elimination of dead zone and zero crossover distortion. Only disadvantage is *Thermal runaway*, which will be minimised by mounting the Diodes in close vicinity and in thermal contact with output Transistors, in the case of discrete circuits.

Another popular circuit uses buffer stage and current Source for biasing Class-AB Amplifiers with feedback arrangement and gain using op-amp is shown in Fig. 11.38.

It eliminates zero crossover distortion. While using Class-AB output stages in IC form thermal shutdown protection is incorporated in the chip level.

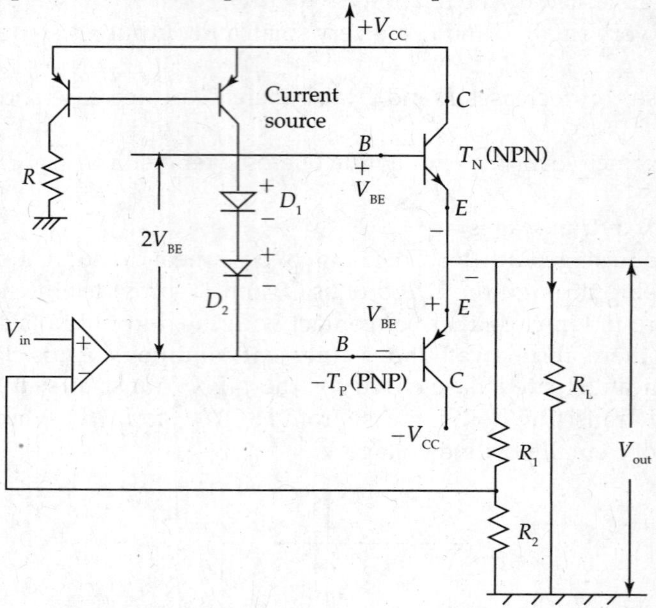

FIG. 11.38 *Biasing Class-AB amplifier*

11.9 CLASS-C POWER AMPLIFIER

A Class-C Power Amplifier delivers large amount of Power to the load very efficiently than a Class-B Power Amplifier. It is essentially used to amplify and reproduce sine wave signals (without much distortion) using a Tuned circuit.

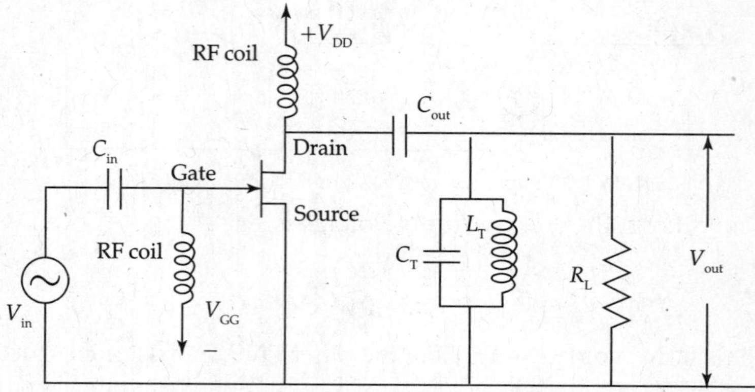

FIG. 11.39 *Single-stage Class-C power amplifier using JFET*

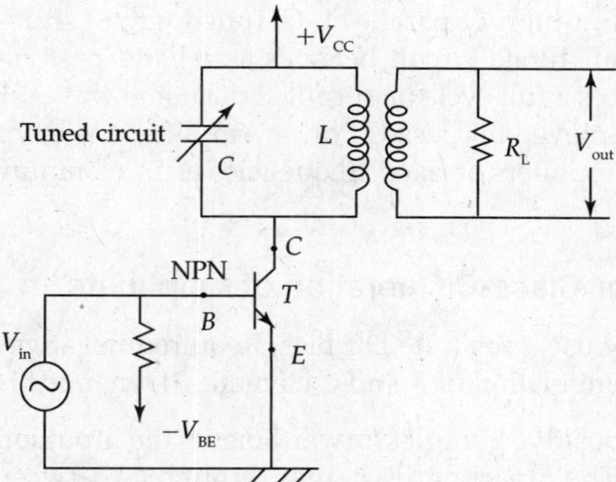

FIG. 11.40 *Basic circuit of Class-C amplifier*

Class-C Amplifier circuits using FET (Fig. 11.39) and BJT (Fig. 11.40)

In a Class-C Power Amplifier, *Q*-point is located beyond the cut-off by using sufficient negative bias V_{BE} (Fig. 11.40) to decide the Transistor conduction intervals. Collector current I_C will be in the form of pulse train (waveforms) shown in Fig. 11.41. This train of output current

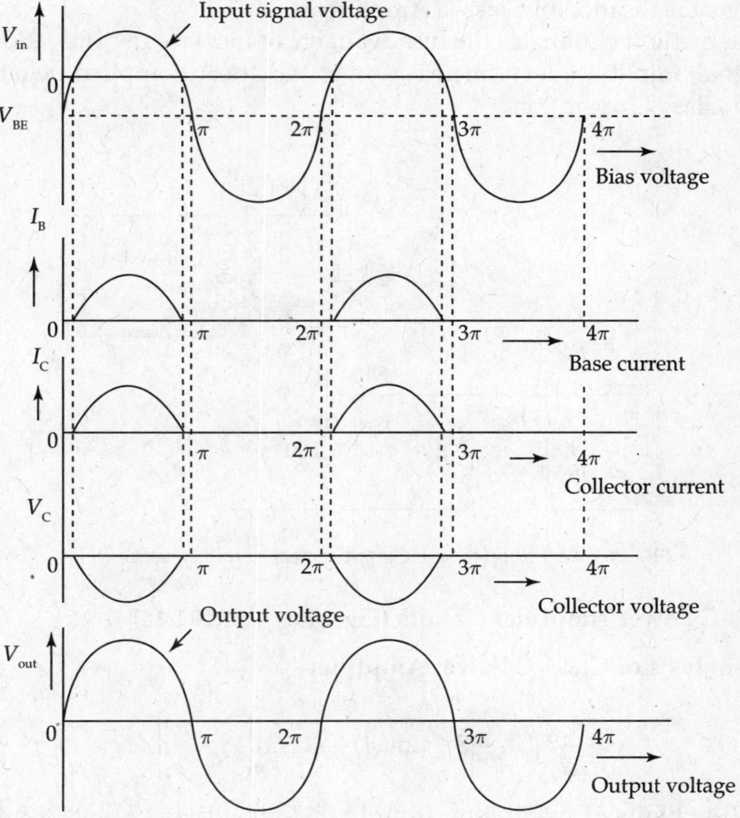

FIG. 11.41 *Signal waveforms at various points of the active device in Class-C amplifiers*

pulses is passed through a high Q parallel L-C Tuned circuit. It is tuned to the frequency of input sinusoidal signal. Tuned circuit behaves as a band pass filter with a narrow band characteristic and produces a full cycle of output signal (sine wave) at the resonant frequency. This basic principle is employed in Class-C Power Amplifier. Such type of operation limits the application of Class-C Amplifiers at fixed frequencies as in communication transmitters and receivers.

11.9.1 Signal Bias for Class-C Operation of Amplifiers

Class-C Amplifiers rarely use a separate DC bias. Required negative bias is derived from the input signal using the combination of R and C elements (Dynamic bias) (Fig. 11.42).

- When input signal is positive going, it forward biases the input junction of Transistor. Base current I_B charges C_B. The capacitor discharges through R_B between positive peaks of input signal. Component values of C_B and R_B are designed such that C_B could not discharge totally in this discharging interval. Therefore, average voltage V_C builds up across C_B, which acts as sufficient reverse bias V_{BE} for Class-C Amplifier.
- Main advantage of obtaining signal bias is that it automatically adjusts the deep reverse bias necessary for Class-C operation according to the amplitude level of the input signal voltage (V_{in}). Signal bias method maintains constant conduction angles that are small for higher efficiency realisation of Class-C Amplifiers.
- Signal biasing method eliminates the disadvantage of increase in conduction angles, when the input signal amplitude level increases over the already applied negative bias to the Transistor for Class-C operation.

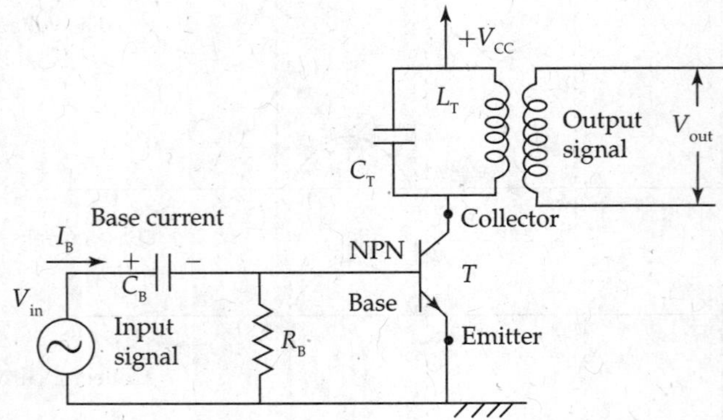

FIG. 11.42 *Single-stage Class-C power amplifier using BJT*

Push-Pull Class-C Power Amplifier circuits (Figs. 11.43 and 11.44)

Mathematical analysis of Class-C Power Amplifier

$$I_{av} = \frac{1}{2\pi} \cdot \int_{\left[\frac{\Pi}{2}-\frac{\theta}{2}\right]}^{\left[\frac{\Pi}{2}+\frac{\theta}{2}\right]} \left[I_P \cdot \sin(\omega t) - I_P \cdot \sin\left(\omega t - \frac{\theta}{2}\right)\right] d\omega t \tag{11.141}$$

By changing the limits of integration from $[\pi/2 - \theta/2]$ to $[\pi/2 + \theta/2]$ into the range of $[\pi/2 - \theta/2]$ to $[\pi/2]$

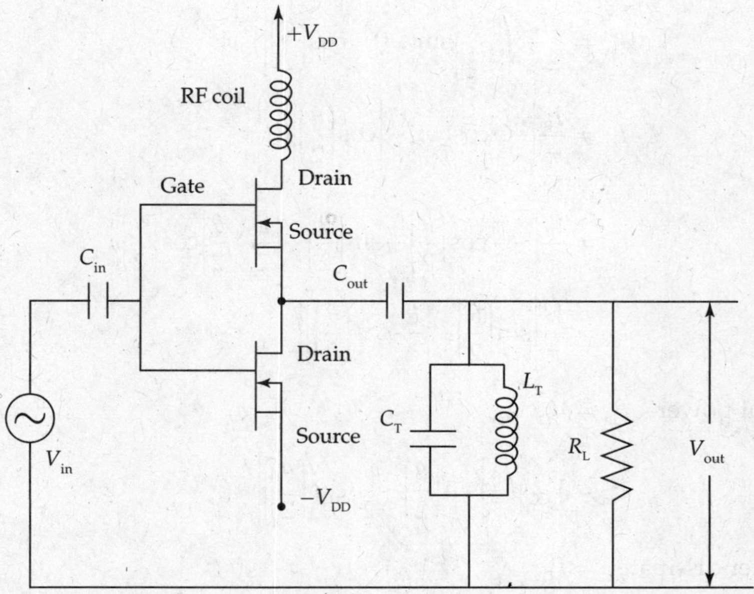

FIG. 11.43 *Complementary symmetry push-pull Class-C power amplifier using MOSFET devices*

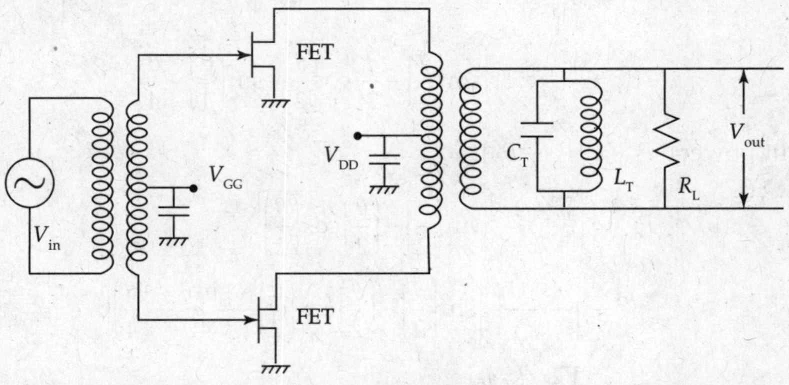

FIG. 11.44 *Transformer-coupled push-pull amplifier using FET devices*

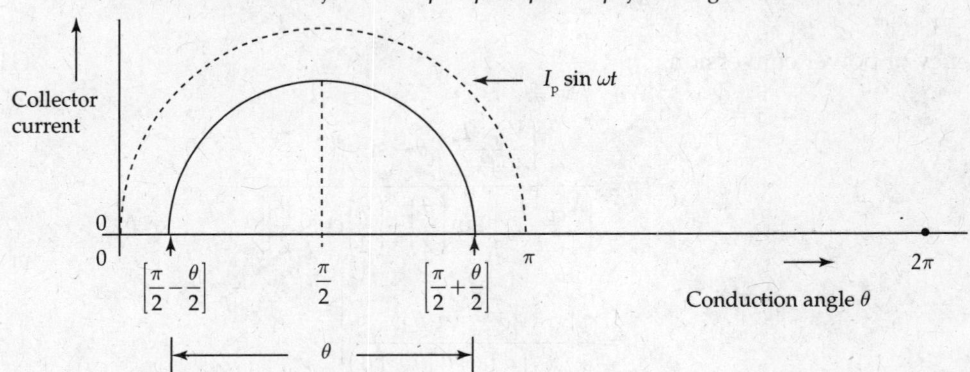

FIG. 11.45 *Conduction angle versus collector current*

$$I_{av} = \frac{2I_P}{2\pi}\int_{\left[\frac{\pi}{2}-\frac{\theta}{2}\right]}^{\pi/2}\left[\sin(\omega t) - \cos\left(\frac{\theta}{2}\right)\right] d\omega t$$

$$I_{av} = \frac{I_P}{\pi}\left[-\omega\cos(\omega t) - \left[\cos\left(\frac{\theta}{2}\right)\right]\omega t\right]_{\left[\frac{\pi}{2}-\frac{\theta}{2}\right]}^{\pi/2}$$

$$\therefore\ I_{av} = \frac{I_P}{\pi}\left[-\frac{\pi}{2}\cos\left(\frac{\theta}{2}\right) + \sin\left(\frac{\theta}{2}\right) + \left[\frac{\pi}{2} - \frac{\theta}{2}\right]\cos\left(\frac{\theta}{2}\right)\right].$$

$$I_{av} = \frac{I_P}{\pi}\left[\sin\left(\frac{\theta}{2}\right) - \left(\frac{\theta}{2}\cdot\cos\left(\frac{\theta}{2}\right)\right)\right] \quad (11.142)$$

Input power $P_{in} = V_{CC}\cdot I_{av}$ (11.143)

$$P_{in} = V_{CC}\left[\frac{I_P}{\pi}\left[\sin\left(\frac{\theta}{2}\right) - \frac{\theta}{2}\cdot\cos\left(\frac{\theta}{2}\right)\right]\right] \quad (11.144)$$

Average power dissipated $= P_D$

$$= \int_{\left[\frac{\pi}{2}-\frac{\theta}{2}\right]}^{\left[\frac{\pi}{2}+\frac{\theta}{2}\right]}\left[(V_{CC} - V_{CC}\cdot\sin(\omega t)\right] I_P\left[\sin(\omega t) - \cos\left(\frac{\theta}{2}\right)\right] d\omega t$$

$$\therefore\ P_D = \frac{V_{CC}}{\pi}\times I_P\left[\sin\left(\frac{\theta}{2}\right) - \frac{\theta}{2}\cdot\cos\frac{\theta}{2} + \sin\left(\frac{\theta}{4}\right) - \frac{\theta}{4}\right] \quad (11.145)$$

Output power $P_{out} = [P_{in} - P_D]$ (11.146)

$$P_{out} = \left[\frac{V_{CC}\cdot I_P}{\pi}\right]\left[\sin\left(\frac{\theta}{2}\right) - \left(\frac{\theta}{2}\right)\cos\left(\frac{\theta}{2}\right)\right].$$

$$-\left[\frac{V_{CC}\cdot I_P}{\pi}\right]\left[\sin\left(\frac{\theta}{2}\right) - \left(\frac{\theta}{2}\right)\cos\left(\frac{\theta}{2}\right) - \frac{\sin\theta}{4} + \left(\frac{\theta}{4}\right)\right]$$

$$\therefore\ P_{out} = \frac{V_{CC}\cdot I_P}{4\pi}[\theta - \sin\theta] \quad (11.147)$$

Efficiency of power conversion $\eta = \frac{P_{out}}{P_{in}}$ (11.148)

$$\eta = \frac{\left(\frac{V_{CC}\cdot I_P}{4\pi}\right)[\theta - \sin(\theta)]}{\left(\frac{V_{CC}\cdot I_P}{\pi}\right)\left[\sin\left(\frac{\theta}{2}\right) - \left(\frac{\theta}{2}\right)\cos\left(\frac{\theta}{2}\right)\right]}$$

$$= \frac{1}{4}\left[\frac{[\theta - \sin(\theta)]}{\sin\left(\frac{\theta}{2}\right) - \left(\frac{\theta}{2}\right)\cos\left(\frac{\theta}{2}\right)}\right] \quad (11.149)$$

From the expression for η, efficiency depends on conduction angle θ. If $\theta = \pi/2$ as in the case of Class-B amplifier,

$$\eta = 0.7854. \tag{11.150}$$

By using small conduction angles in Class-C amplifier, it can be shown that $\eta = 1$.
When θ is small, by trigonometric series,

$$\sin\theta = \theta - \frac{\theta^3}{3.2.1}$$

$$\sin\left(\frac{\theta}{2}\right) = \left[\left(\frac{\theta}{2}\right) - \frac{(\theta/2)^3}{3.2.1}\right] = \left[\frac{\theta}{2} - \frac{\theta^3}{48}\right]$$

$$\cos\left(\frac{\theta}{2}\right) = \left[1 - \left(\frac{\theta^2}{8}\right)\right].$$

Substituting these values in the expression for η, η becomes 1.

$$\%\eta = \frac{P_{\text{out}}}{P_{\text{in}}} \times 100 = 100\%. \tag{11.151}$$

Practical values of efficiency of a Class-C Power Amplifier are around 95%.

Conduction angle versus efficiency (Fig. 11.46)
Power conversion efficiency of Power Amplifiers decreases with increase in conduction angles of the active devices used in the Amplifiers.

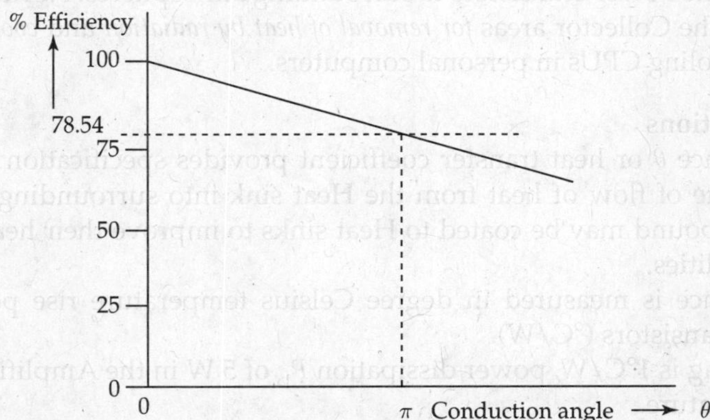

FIG. 11.46 *Active device conduction angle versus efficiency of power amplifiers*

Applications of Class-C Amplifiers
(1) RF Amplifiers (high-efficiency operation). (2) Cellular phones. (3) Radio and TV transmitters. (4) Harmonic multipliers for generating lower harmonics.

Requirements of Power Amplifiers
(1) Power Amplifiers have to be designed to drive low impedance loads efficiently. (2) They must deliver large amounts of Power efficiently, while dissipating low amounts of Power internally. (3) They should deliver signal Power at low levels of distortion. (4) Total harmonic distortion should be kept well within 0.1%.

11.10 THERMAL MODELLING AND HEAT SINKS

DC Power input = (AC output power + Power dissipation in the active device and load).

- Base current in the input circuit of Transistor Amplifiers is very small. So, Power dissipation at input circuit is negligible. But Collector current is larger in output circuits. Collector current flow produces larger Power dissipation at the Collector junction in Transistors. A portion of DC input power is dissipated due to large values of current flow through Transistors. It is known as *Power dissipation*.
- Due to the Power dissipation, Transistors become hot. Heat from Transistor can be radiated to surrounding air through Heat sinks attached to Collector nodes of Transistors. They work at maximum Power dissipation rating $P_{D(max)}$ with Heat sinks. This rating is important for Transistors operating as Amplifiers, because the Transistor will be in the ON condition for more time during Amplifier operation. Hence, maximum Collector current ($I_{C(max)}$) and $V_{CE(max)}$ ratings are more important during circuit design of Transistor Amplifiers.

  DC Power dissipation for class-A amplifier $P_D = V_{CE} \cdot I_{C(max)} = 0.5\, V_{CC} \cdot I_{C(max)}$

$$P_D = I_{C(max)} \cdot V_{CE} = I_{C(max)} \cdot \frac{1}{2} \cdot V_{CC},$$

where V_{CC} is the Collector supply voltage.

If the maximum Collector current is 2 A and Collector supply voltage is 20 V, then the Power dissipation by the Transistor $P_D = 2 \times 0.5 \times 20 = 20$ W.

Power Transistors have metal bodies. Metal Heat sinks are used to enhance thermal conductivity from the Power Transistors to surrounding atmospheres. Normally, *Heat sinks* are firmly attached to the Collector areas *for removal of heat by radiation* and *cooling the Transistors with fans* such as cooling CPUs in personal computers.

Thermal considerations

- Thermal resistance θ or heat transfer coefficient provides specification for Heat sinks. It describes the rate of flow of heat from the Heat sink into surrounding air. Sometimes a conductive compound may be coated to Heat sinks to improve their heat conduction and radiation capabilities.
- Thermal resistance is measured in degree Celsius temperature rise per watt of Power dissipation in Transistors (°C/W).
- If Heat sink rating is 1°C/W, power dissipation P_D of 5 W in the Amplifier results in a rise of 5°C in temperature.
- Semiconductors have negative temperature coefficient. When semiconductor devices get hotter, their resistance falls, resulting in larger currents to flow through them. Increase in current flows through Transistor materials increases heating effect yet further, so Transistors get hotter. Eventually, they may burn out.
- Assume that the average ambient temperature (Temperature of the surroundings of Transistor operation) is 27°C and the maximum operating temperature in tropical countries is of the order of 67°C. When the Power dissipation in the Transistor for example is 20 W. Then thermal resistance becomes the ratio of the difference in

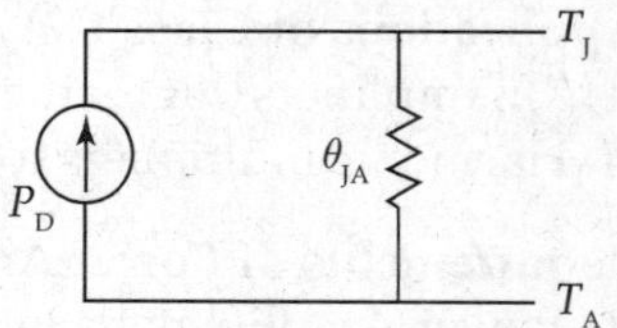

FIG. 11.47 *Electrical equivalent of thermal conduction process in a power transistor*

operating temperature and the surrounding temperature (similar to potential difference (voltage) across a resistor), divided by the maximum Power dissipation due to current flow in the Transistor. Thermal resistance $R_{TH} = \theta = 2°C/W$.

$$\text{Thermal resistance} = R_{TH} = \theta_{JA} = \frac{\Delta T}{P_D} = \frac{(T_J - T_A)}{P_D} \quad (°C/W)$$

$$\Delta T = (T_J - T_A),$$

where T_J is the Collector Junction temperature of the power transistor in °C, T_A is the Ambient Temperature of the surroundings in °C, and P_D is the Power dissipation = (1/2) · $V_{CC}\, I_{C(max)}$ watts.

$$\text{Ex: If } T_J = 67°C,\ T_A = 27°C,\ \Delta T = (T_J - T_A) = 67 - 27 = 40°C.$$

$$\text{If } I_{C(max)} = 2\text{ A and } V_{CC} = 20\text{ V, then } P_D = \frac{1}{2} \times 20 \times 2 = 20\text{ W.}$$

$$\text{Thermal resistance} \quad R_{TH} = \theta_{JA} = \frac{(T_J - T_A)}{P_D} = \frac{40}{20} = 2°C/W.$$

Thermal considerations between Power Transistor and heat sink (Fig. 11.48)

For a Power Transistor, θ_{JA} = [θ_{JC} + θ_{CA}], where θ_{JA} is the Thermal resistance between junction and ambient, θ_{JC} is the Thermal resistance between junction and the case of the Transistor, and θ_{CA} is the Thermal resistance between case and ambient.

The electrical equivalent of thermal conducting process when a Transistor is mounted over a heat sink is shown in Fig. 11.48.

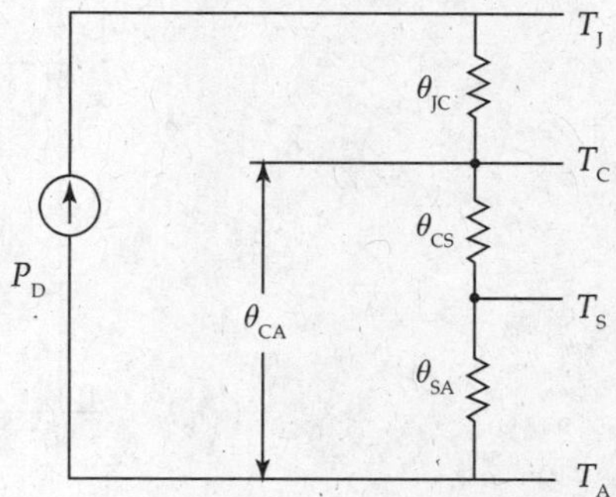

FIG. 11.48 *Equivalent circuit of thermal considerations for a transistor mounted over a heat sink*

$$[T_J - T_A] = [\theta_{JC} + \theta_{CS} + \theta_{SA}] \cdot P_D,$$

where T_J is the Junction temperature in °C, T_A is the Ambient temperature in °C, θ_{JC} is the Thermal resistance between junction and the Transistor case in °C/W, θ_{CS} is the Thermal resistance between case and heat sink in °C/W, and θ_{SA} is the Thermal resistance between heat sink and the ambient in °C/W.

$$[T_C - T_A] = [\theta_{CS} + \theta_{SA}] \cdot P_D,$$

where T_C is the Transistor case temperature in °C.

To prevent heating in Transistors and resulting damage to them, heat sinks are attached to the Transistors to radiate the heat and cool the active device.

- A fan removes heat dissipated in a Heat sink associated with central processing unit (CPU) in a computer. On similar lines, Heat sinks are available in different shapes depending upon the practical applications to radiate to the Ambient.
- Low Power Transistors are mounted on metal chassis for providing enough area for ventilating the generated heat on the mounted Transistors.

- Heat generation takes place at the Collector junction of the Transistors. Heat radiation is provided to atmosphere by providing suitable Heat sinks of different shapes on the tops of high power Transistors.
- There are many types of *Heat sinks* in shapes using Aluminium alloy sheets depending upon the device structure for radiation of heat to the surroundings.
- There are two types of heat sinks: (1) *Low Power Heat sinks* and (2) *High Power Heat sinks.* Power Transistors delivering a power of 3 W are called low power Transistors. They use low power Heat sinks, when the Power exceeds 1 W.
- When a Power Transistor is mounted over a Heat sink, an electrically insulating material (heat sinking compounds such as silicon compound, Zinc oxide compound, Beryllium compound with good thermal conductivity) is placed in between metal case and metallic heat sink. Insulating bushes are also provided. Metal case is the Collector of the Transistor, which is electrically connected.

Power dissipation versus temperature for a typical Power Transistor (Fig. 11.49)

Transistor manufacturers specify maximum Power dissipation of a Transistor $P_{D(max)}$ at a specified temperature T_{A0}, $T_{J(max)}$, thermal resistance θ_{JA} and θ_{JC} at 25°C along with derating curves between Power dissipation and temperature.

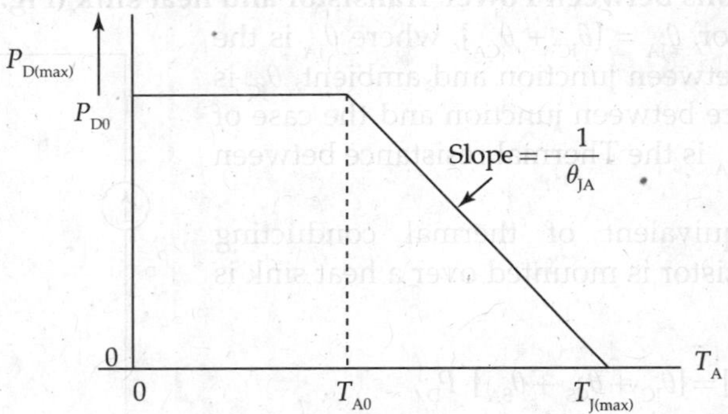

FIG. 11.49 *Power transistor derating curve: maximum allowed P_N versus ambient temperature*

From Fig. 11.49,

$$\theta_{JA} = \frac{(T_{J(max)} - T_{A0})}{P_{D0}} = \text{Inverse slope derating curve}$$

$$\therefore \quad P_{D0} = \frac{[T_{J(max)} - T_{A0}]}{\theta_{JA}}$$

The inverse slope of the derating curve limits the amount of heat removed from the transistor junction, when $T_A = T_{J(max)}$ heat transfer does not take place.

EXAMPLE 11.6

Determine the dissipated Power $P_{D(max)}$ for Power Transistor. Assume that it is mounted over a heat sink with θ_{CS} (Thermal resistance between case and sink) = 0.7°C/W. θ_{SA} (Thermal resistance between heat sink and ambient) = 1.8°C/W. For Power Transistor $P_{D(max)}$ at (case temperature of the Transistor) T_{C0} of 25°C is 30 W. It is to be derated by 0.24 W/°C. $\theta_{JC} = 4.167$°C/W approximately 4°C. $T_{J(max)} = 150$°C. Assume an ambient temperature $T_A = 35$°C.

Solution: From the thermal equivalent circuit of a Transistor mounted over heat sink,

$$[T_J - T_C] = [\theta_{JC} + \theta_{CS} + \theta_{SA}] \cdot P_D$$

$$(T_j - T_C) = (150 - 35) = 115 \text{ C}$$

$$(\theta_{JC} + \theta_{CS} + \theta_{SA}) = (4 + 0.7 + 1.8) = 6.5°\text{C/W}$$

$$\therefore \quad P_D = (115/6.5) = 17.69 \text{ W}$$

Case temperature $\quad T_C = T_A + P_D\,(\theta_{CS} + \theta_{SA}) = 35 + 17.69\,(0.7 + 1.8)$

$$\therefore \quad T_C = 79.22°\text{C}.$$

Applying derating factor for T_C,

$$P_{D(max)}\,|T_C = P_{D(max)} - 0.24(T_C - T_{C0})$$

$$\therefore \quad P_{D(max)}\,|T_C = 30 - 0.24(79.22 - 25) = 17 \text{ W}.$$

11.11 ADVANCED POWER AMPLIFIERS

Present day advanced researchers are aiming in realisation of highly efficient output stages with latest devices. Intelligent and novel designs led to the Classification of Power Amplifiers in an alphabetical fashion after Class-C as Class-D, Class-E, Class-F and Class-H and up to Class-T.

Particularly in battery-operated (portable equipment) transmitters, aero space, military and high-end audio equipment in automobiles; Power drain from the batteries has to be made much insignificant. This necessitates conserving Power by employing highly efficient Amplifiers using latest MOS devices and Monolithic ICs.

While employing conventional Class-AB output stages in high-end Audio Amplifiers, its efficiency is well within 78.54%. Using the latest Class of operation of Amplifiers, efficiencies as high as 95% or more can be realised with good Audio quality.

All the latest Class of output stages basically uses *either of the two principles.*

- *Small fraction of cycle is used by using switching principle such as PWM or delta sigma modulation.*
- *Modulating the supply rails in the final output stages.*

11.11.1 Class-D Amplifier

Concept of Class-D Amplifier using Vacuum Tubes existed 60 years back. It is basically an Audio Amplifier using switching principles by using PWM or delta sigma modulation. It is more popular in music systems, sub woofer systems, car stereos, high power (more than 1000 W) quality Amplifiers due to the reasons – *highest efficiency, reduced size and weight, reduced Power dissipation and less drain on the DC Source*. They do not need any external heat sinks, require less board space, portable and deliver good quality audio. Class-D Amplifier is basically a switching Amplifier, where the current flows through the switching Transistors (preferably MOSFETs) in the shape of *a train of narrow needle pulses*. Class-D Amplifier circuit is shown in Fig. 11.52.

Operation of Class-D Amplifier Circuit

Class-D Amplifier employs a comparator to drive MOSFET switches to make them fully ON and OFF with minimum losses in the device. NIV input to comparator is the audio signal and other input to NINV terminal is a triangular wave generated by an oscillator. When audio signal (reference voltage) is continuously changing, duty cycle also changes continuously. Resulting changes in output voltages depends on V_S (audio signal) and V_T (Triangular waveform signal).

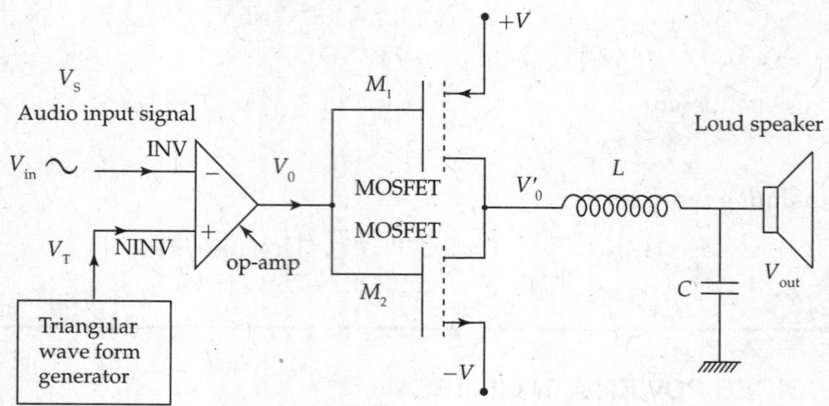

FIG. 11.50 *Class-D amplifier circuit using PWM*

When V_S is greater than V_T, comparator output goes negative. Comparator output is denoted by $-V_0$. when V_S is less than V_T, comparator output V_0 is positive.

1. When V_0 is made positive, Gate drive makes the MOSFET M_1 turn ON and lower MOSFET M_2 is turned OFF. Assuming a very small voltage drop in MOSFETs a voltage $V_0' = V_0 +$ that appears before the LC filter driving the loud speaker.
2. Similarly, when V_S is less than V_T, the Gate drive V_0 is negative and the lower MOSFET is turned ON, while the upper MOSFET is turned OFF causing a voltage $V_0' = V_0 -$ that appears before the LC filter driving the loud speaker.

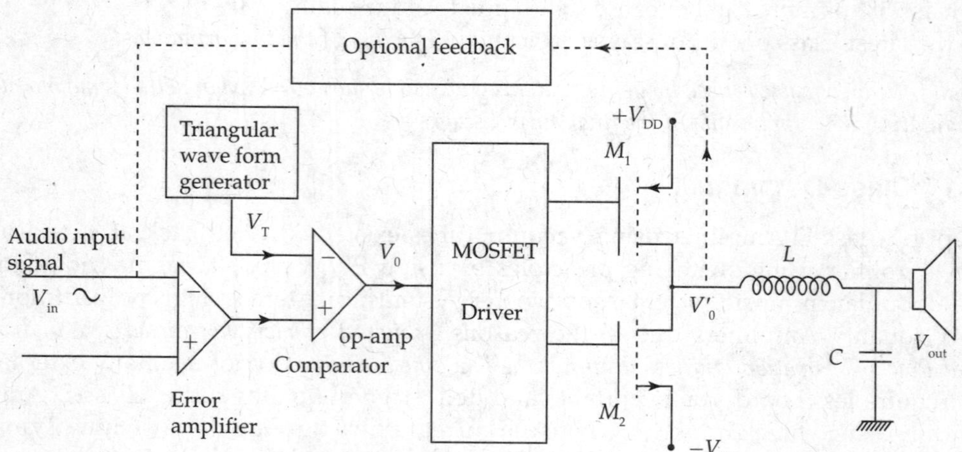

FIG. 11.51 *Class-D amplifier with feedback*

Active filters or cascaded LC filters are used to provide adequate filtering of unwanted spectral components (byproducts) of PWM. While choosing the frequency of triangular waveform generator, it should be greater than the cut-off frequency of LC filter.

Using negative feedback with *operational Amplifiers* or *current feedback Amplifiers (CFA)*, performance of Class-D Amplifier (Fig. 11.51) can be stabilised and improved.

The operation of Class-D Amplifier is similar to that employed in DC to DC converters (Chapter 14) except in connecting the audio signal in place of reference voltage.

While employing *PWM* technique several drawbacks occur in Class-D Amplifier like (1) Additional Distortion, (2) Harmonics causing *Electro Magnetic Interference* (*EMI*) in audiogram and (3) Difficulty in achieving full modulation.

These drawbacks can be eliminated by using another popular technique *Pulse Density Modulation (PDM)*. In PDM, a number of pulses in a given time slot (that is in a time window) are made proportional to the average value of the input audio signal. These individual pulses are quantised in multiples of modulated clock period. This process is known as *Delta Sigma modulation* or *Sigma Delta modulation*, which is a more popular technique in *Digital Power Supplies*.

11.11.2 Class-E Amplifier

It is primarily used as Tuned Power Amplifier. Class-E Amplifier output is passed through a tuned circuit to obtain damped signal pulse. It is invented by Nathane O Sokal and Elon D Sokal in 1972. Class-E Amplifier circuit is shown in Fig. 11.52.

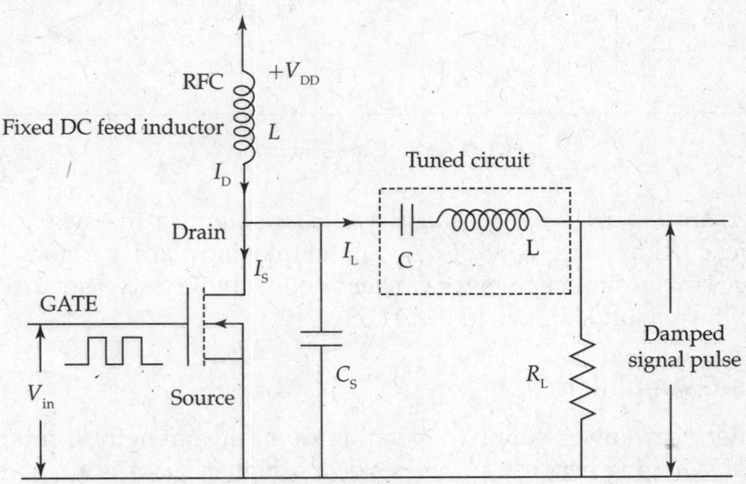

FIG. 11.52 *Class-E amplifier*

Tuned circuit selects fundamental frequency of the signal. When the switch is turned ON, voltage across switch is zero. During zero voltage switching (ZVS), zero Power loss occurs in switch. When switch is in ON state and switching waveform is zero, zero voltage derivatives switching waveform occurs (ZVDS). Now, there is no current flow through C_S. All current flows through LC circuit, which resonates at fundamental frequency of input signal. Current rises smoothly till the switch is OFF again. These ON and OFF actions are controlled by driving the Gate of MOSFET switch by using HF switching and PWM. Ideal efficiency of a Class-E Amplifier is 100%. Switches may be BJT, MOSFET, MESFET. These Amplifiers are used in Blue tooth Class-1 wireless connections, to increase the readable range of RFID, in VHF communication systems and where RF Power conversion efficiency of 100% is desirable.

11.11.3 Class-F Amplifier

Class-E and Class-F output stages essentially use perfect switching device and an impedance network. Class-F Amplifier is intelligent extension of fundamental Class-C Amplifier, where the output is tuned to harmonic of input signal. Harmonic resonator removes the odd harmonics particularly the third harmonic at output of Amplifier. Magnitude and phase of third harmonic control the flatness of Collector waveform.

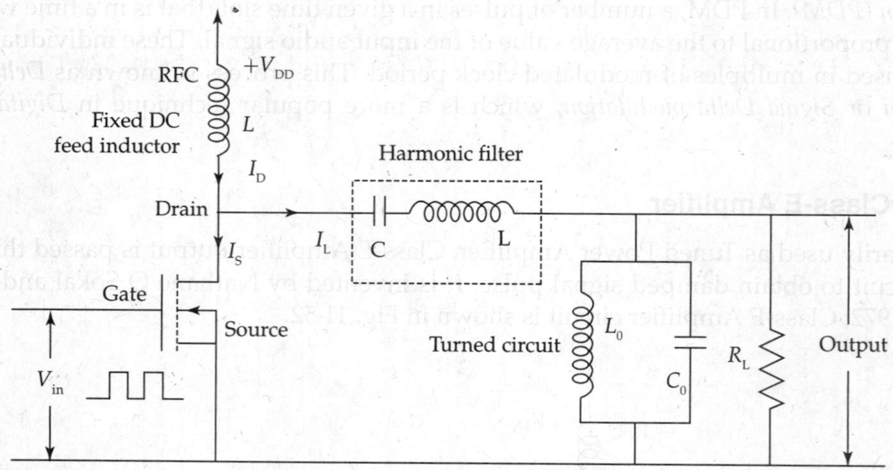

FIG. 11.53 *Class-F amplifier*

Due to lower drop of DC voltage and Power dissipation in the switch, Class-F Power Amplifier efficiency is as high as 88.36%, while employing third harmonic elimination. If second harmonic is also eliminated, peak efficiency of Amplifier increases to 88.5%.

Class-F Amplifier circuit is shown in Fig. 11.53.

11.11.4 Class-G Amplifier

Class-G Amplifier uses Power supply depending upon the strength of input audio signal. A musical signal swinging between lower peaks and higher peaks is applied at input of an Amplifier system. Assume $V_{CC}(1) = 5$ V. For a smaller swing in music voltage, Transistor T_2 connected to lower voltage $V_{CC}(1)$ conducts (because of Diode D_1). Then, voltage at the Collector of T_2 is equal to $[V_{CC}(1) - \text{Diode drop}] = 4.3$ V.

If the swing in the musical signal at higher peaks exceeds the above voltage of 4.3 V, Transistor T_1 connected to higher voltage $V_{CC}(2)$ also starts conducting and provides the necessary Power for larger swings in the musical input.

Thus the Power is optimised as per the input signal strength. The Power dissipation in the Transistors is minimised resulting higher efficiency. This type of Amplifier is more popular in the audio designs.

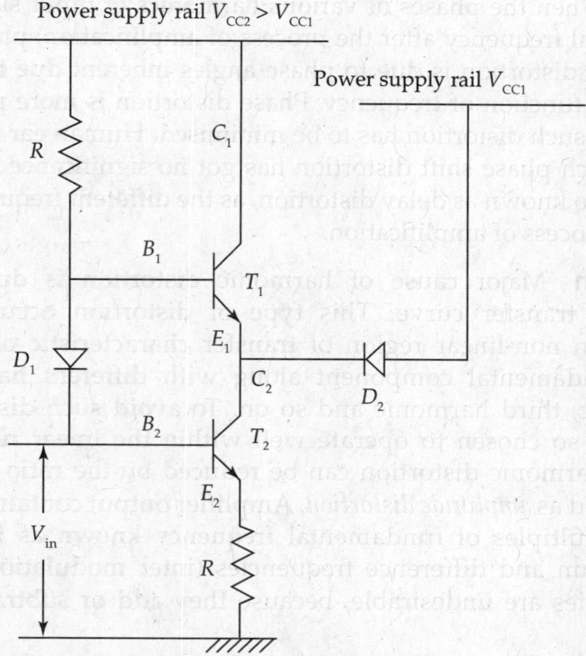

FIG. 11.54 *Class-G power amplifier*

11.12 DISTORTION IN AMPLIFIERS

Output signal of an Amplifier should be an exact replica of input signal, except of course with larger magnitudes. Any change in the output waveform is undesirable. Undesirable changes in the output are known as *distortion.*

Frequency distortion If the input signal consists of a spectrum of equal amplitude waveforms, and after amplification higher frequencies get attenuated, then frequency distortion is said to occur. Typical frequency distortion is shown in Fig. 11.55.

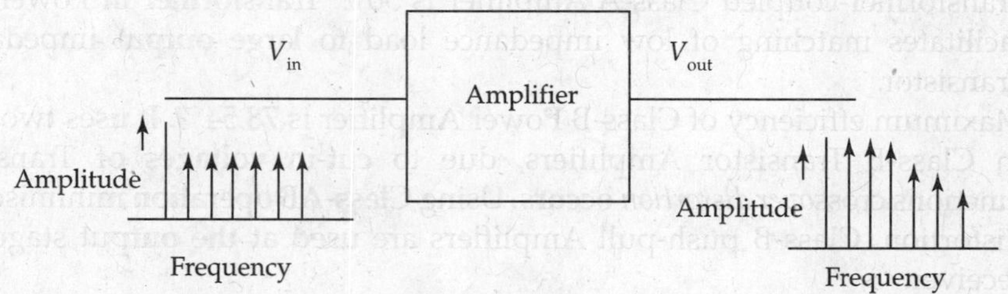

FIG. 11.55 *Frequency distortion in amplifiers*

This distortion occurs when the Cut-off frequency is less than the highest input signal frequency. Spectral components of input signals should be well within the bandwidths of Amplifier to ensure equal amplification. Otherwise individual components of input signal get amplified by different amounts due to attenuation or filtering action of Amplifier. Frequency distortion is due to changes in Amplifier gain with frequency.

Phase distortion When the phases of various harmonics of input signal get shifted with respect to fundamental frequency after the process of amplification, phase distortion is said to occur. This type of distortion is due to phase angles inherent due to complex gain A of Amplifier, which is a function of frequency. Phase distortion is more prominent in TV and video Amplifiers and such distortion has to be minimised. Human ear is not sensitive to the phase shift and as such phase shift distortion has got no significance in Audio Amplifiers. Phase distortion is also known as delay distortion, as the different frequency components are delayed during the process of amplification.

Harmonic distortion Major cause of harmonic distortion is due to the nature of hyperbolic dynamic transfer curve. This type of distortion occurs, when the input signal excitation is in non-linear region of transfer characteristic of amplifying device. Output contains fundamental component along with different harmonic components like second harmonic, third harmonic and so on. To avoid such distortion, input signal excitation should be so chosen to operate well within the linear region. By employing negative feedback, harmonic distortion can be reduced by the ratio $[1 + A\beta]$. Harmonic distortion is also called as *amplitude distortion*. Amplifier output contains newer frequencies, which are integer multiples of fundamental frequency known as harmonics. They are in addition to the sum and difference frequencies (inter modulation frequencies). Inter modulation frequencies are undesirable, because they add or subtract from the original signal.

SUMMARY

- Main *classification of Power Amplifiers* is based on angle of conduction of the Power Transistors: *Class-A (360°), Class-B (180°), Class-AB (slightly more than 180°)* and *Class-C (less than 180°)* and so on.
- The ratio of AC output power to DC input power is the *efficiency of operation of an Amplifier*. It is also known as *output circuit efficiency*.
- *Maximum efficiency* of *power conversion* in Class-A Amplifier is 25%, efficiency of Transformer-coupled Class-A Amplifier is 50%. Transformer in Power Amplifiers facilitates matching of low impedance load to large output impedance of the Transistor.
- Maximum efficiency of Class-B Power Amplifier is 78.54%. It uses two transistors. In Class-B Transistor Amplifiers, due to cut-in voltages of Transistor input junctions cross*over distortion* occurs. Using Class-AB operation minimises crossover distortion. Class-B push-pull Amplifiers are used at the output stage of a radio receiver.
- Output signal in Class-C Amplifiers exists for less than 180°. Efficiency of Class-C Power Amplifiers is around 100%, under ideal conditions. Class-C Amplifiers use *tuned circuits* as load to construct the total signal in the output circuit.
- Heat Sinks are used to keep junction temperature T_J of power Transistor below the maximum temperature $T_{J(max)}$ rating to avoid overheating and destruction of Transistor. Heat Sink transfers heat from output junctions of Power Transistors by conduction and radiation to surroundings.

Questions for Practice

1. Classify large signal Amplifiers Based on operating point. Distinguish between these Amplifiers in terms of conversion efficiency. (JNTU, Feb. 2008)
2. In series-fed Class-A Power Amplifier, explain the importance of the position of operating point on the output signal swing. Show that the Power conversion efficiency is 25%. (JNTU, Nov. 2007)
3. (a) Draw the circuit diagram of series-fed Class-A Power Amplifier. Explain the circuit operation with necessary details.
   (b) Derive the expression for its output power P_{out} in terms of load resistance R_L and Collector circuit efficiency.
4. Show that the Power conversion efficiency of a Transformer-coupled Power Amplifier is 50%. (JNTU, May–June 2005)
5. Compare the series-fed and Transformer-fed Class-A Power Amplifiers. Why is the conversion efficiency is doubled in a Transformer-coupled Class-A Power Amplifier? (JNTU, Feb. 2007)
6. Show that even harmonics are eliminated in Class-B push-pull connection. (JNTU, Nov. 2005)
7. What are the advantages and disadvantages of push-pull connection? Show that in Class-B push-pull Amplifier, maximum theoretical efficiency is 78.54%. (JNTU, Feb. 2008)
8. Compare and contrast push-pull and complementary configurations of Class-B Power Amplifiers. (JNTU, Feb. 2008)
9. What is crossover distortion? How does a Class-AB Power Amplifier avoid crossover distortion? (JNTU, Feb. 2008)
10. Derive an expression relating the total output power '*P*', fundamental Power 'P_1' and total harmonic distortion '*D*' in Power Amplifiers. If total distortion in Amplifier is 9%, calculate the contribution to the total Power. (JNTU, Mar. 2006)
11. Write short notes on requirement and types of heat sinks for dissipation in large Power Amplifiers. (JNTU, Feb. 2008)
12. Calculate the second harmonic distortion, if the output signal waveform of a push-pull Amplifier has measured values of $V_{CE(min)} = 1$ V, $V_{CE(max)} = 24$ V and $V_{CE}(Q) = 14$ V using an oscilloscope. (JNTU, Mar. 2006)
13. (a) What is thermal resistance? What is the unit of Thermal resistance?
    (b) Derive a relation to prove that increasing the effective surface area of the Transistor case the resistance of heat flow could be decreased. (JNTU, Nov. 2005)
14. (a) Explain crossover distortion in Class-B complementary symmetry Amplifier.
    (b) In complementary symmetry Class-B Power Amplifier circuit, $V_{CC} = 25$ V, load resistance $R_L = 16\ \Omega$ and $I_{max} = 2$ A. Determine the input power, output power and efficiency. (JNTU, Nov. 2005)
15. Discuss about different types of distortions in Amplifiers. (JNTU, Feb. 2008)

Multiple Choice Questions

1. Maximum conversion efficiency occurs in the ______.
   (a) Class-A Power Amplifier (b) Class-B Power Amplifier
   (c) Class-AB Power Amplifier (d) Class-C Power Amplifier
2. A Class-C Power Amplifier conducts ______.
   (a) for the entire cycle of input signal (b) for only one-half of input signal
   (c) for slightly more than half a cycle (d) less than half a cycle
3. Maximum efficiency of Class-A Amplifier that could be realised is ______.
   (a) 25% (b) 50% (c) 78.54% (d) more than 90%
4. Maximum efficiency of a Class-A inductively coupled Amplifier is ______.
   (a) 25% (b) 50% (c) 78.54% (d) more than 90%
5. Maximum possible efficiency of Class-A Transformer-coupled Amplifier is ______.
   (a) 25% (b) 50% (c) 78.54% (d) more than 90%
6. Crossover distortion is present in ______.
   (a) Class-A Power Amplifier (b) Class-B Power Amplifier
   (c) Class-AB Power Amplifier (d) Class-C Power Amplifier
7. Maximum efficiency of Class-B push-pull Amplifier is ______.
   (a) 25% (b) 50% (c) 78.54% (d) more than 90%
8. Out of the following Power Amplifier operations, the ______ is non-linear.
   (a) Class-A (b) Class-B (c) Class-C (d) Class-AB
9. Amplifier Class of operation using digital or pulse signals is ______.
   (a) Class-A (b) Class-B (c) Class-C (d) Class-D
10. Power Transistor capable of handling both high current and high voltage is ______.
    (a) BJT (b) FET (c) MOSFET (d) DMOS
11. Amplifier operation preferred between high fidelity and high efficiency is ______.
    (a) Class-A (b) Class-B (c) Class-AB (d) Class-C
12. Class of operation used in RF Power Amplifiers is ______.
    (a) Class-A (b) Class-B (c) Class-AB (d) Class-C

Answers to Multiple-Choice Questions

1. (d) 2. (d) 3. (a) 4. (b) 5. (b) 6. (b)

7. (c) 8. (c) 9. (d) 10. (d) 11. (c) 12. (d)

Chapter 12

HIGH FREQUENCY TRANSISTOR CIRCUITS

Learning Objectives

To get familiarity of high-frequency behaviour of transistor

- Concepts of Transistor behaviour at high frequencies.
- Hybrid-π circuits of Transistor for various configurations.
- Analysis of f_α, f_β, f_T, Bandwidth and Amplifier frequency response.
- High frequency equivalent circuits and analysis of FET and MOSFET.

12.1 TRANSISTOR AT HIGH FREQUENCY INPUT SIGNALS

For high frequency signal operation, junction capacitances of Transistors and terminal capacitances play dominant role. Separate high frequency equivalent circuits using hybrid-π model are developed for suitable analysis. This model can be used at low-frequency and high-frequency operations.

1. When input signals are of high frequency (HF), amplifier gain reduces due to following conditions:
   a. Impedances of internal capacitances tend to decrease, resulting in a close to short circuit at the output.
   b. Input capacitance increases (Miller effect) due to feedback capacitance from output to input, causing finite amount of energy feedback.
2. Electric current charge carriers take finite *transit time* to travel through a Transistor. Transistor response to low-frequency signals is considered to be instantaneous, as the transit time is

insignificant compared to the wavelength of low-frequency signals. For high-frequency signals, *transit time* of charge carriers is significant compared to their wavelengths, hence something to consider.

3. Practical applications such as wideband and Radio Frequency Amplifiers use high-frequency Transistors. MRF 313 is one such HF Transistor used in wideband amplifiers and oscillators in mobile and aircraft instruments. It has unity gain frequency $f_T \cong 200$ MHz.

Transistor behaviour at high-frequency signals is analysed better using Hybrid-π or Giacoletto model circuit shown in Fig. 12.1.

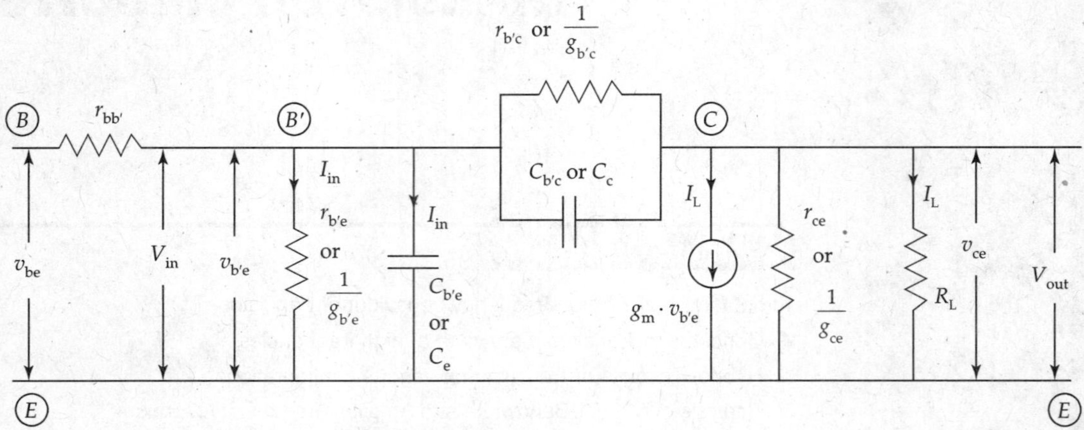

FIG. 12.1 *Hybrid-π equivalent circuit of high frequency transistor amplifier with load* R_L

12.2 HYBRID-π MODEL FOR HF TRANSISTORS

Hybrid-π model is used for HF analysis of CE Amplifier. Transistor input junction is forward-biased and its output junction is reverse-biased. Using this concept, circuit components of hybrid-π circuit model are derived. All the parameters (resistances and capacitors) are assumed to be independent of frequency.

12.2.1 Circuit Components of Hybrid-π Model of CE Amplifier (Fig. 12.1)

1. Base-spreading resistance $r_{bb'}$ (resistance of conducting path between *B* and *B'*)

Base-spreading resistance $r_{bb'}$ is the series Base resistance that represents the conducting path distributed from (external) Base point *B* to the internal (fictitious) Base point *B'* for active forward-bias to input junction of Transistor. Hence, $r_{bb'}$ is known as Base-spreading resistance. Typically, $r_{bb'} \cong 100\ \Omega$.

2. Dynamic Emitter junction resistance $r_{b'e}$ or r_π or input resistance h_{ie} of Transistor

In an Amplifier

- Transistor Emitter junction is forward-biased by V_{BE} and its Collector junction is reverse-biased by V_{CE}.
- Quiescent operating point *Q* might be fixed by forward-bias $V_{BE}(Q)$ and Base current $I_B(Q)$. When AC input signal is applied, it superimposes on DC bias $V_{BE}(Q)$.

- Effective input signal voltage becomes $v_{b'e}$.
- Variations in DC bias to the Emitter junction results in varying Base current i_B. There will be corresponding variations in output Collector current i_C of the Transistor.

$$\therefore \quad v_{b'e} = [r_{b'e} \times i_B] = [r_\pi \times i_B]$$

$$\text{Input resistance} \quad h_{ie} = \frac{v_{b'e}}{i_B} = r_{b'e} = r_\pi \ (\Omega)$$

$$\text{Conductance} \quad g_{b'e} = \frac{1}{r_{b'e}} = \frac{1}{r_\pi} \tag{12.1}$$

Typical value of $r_{b'e}$ is $\sim$1000 Ω.
Input resistance R_{in} at Transistor Base alone:

$$R_{in} = (r_{bb'} + r_{b'e}) \cong r_{b'e} = h_{ie}.$$

3. Base to Emitter capacitance $C_{b'e}$ or C_e or C_π (Emitter junction capacitance)
Assume NPN Transistor in CE configuration. Due to forward-bias $V_{BE'}$ electrons are injected from Emitter into Base region. These electrons, injected into P-material Base region, form excess minority carriers at Emitter junction in Base region. Excess minority carriers in the depletion region contribute to diffusion capacitance $C_{e'}$ $C_{b'e}$ or C_π (between Base B' and Emitter E). Due to forward-bias, width of depletion region about Emitter junction will be very small, making Emitter junction capacitance $C_e = \sim$100 pF. Small signal variations in $v_{b'e}$ cause variations in charge about the junction and causes small variations in Emitter junction capacitance $C_{b'e}$. It is specified as C_{ib} in the Transistor data.

4. Resistance $r_{b'c}$ (r_μ): Resistance of reverse-biased Collector junction
Depletion region width, about Collector junction, changes with variations in reverse-bias V_{CE} to Transistor output junction. Changes in depletion region cause variations in Base width. It is known as *Base width modulation* due to *Early effect*. Base width modulation causes the slope of minority carrier distribution to change. This in turn causes a slight increase in Collector current, which appears as slight upward slope in Transistor output characteristics. This small magnitude of feedback action between output and input ports is taken care by including *Base width modulation resistance* $r_{b'c}$ (r_μ) between the points B' and C. Resistance $r_{b'c}$ is $\sim$ 4 to 5 MΩ, playing insignificant role in amplifier analysis.

5. Capacitance $C_{b'c'}$ C_c or C_μ (output junction capacitance)
Capacitance $C_{b'c'}$ C_c or C_μ represent barrier capacitance of reverse-biased Collector junction. Due to reverse-bias, depletion region width about Collector junction will be large. Variations in reverse-bias cause variations in output junction capacitance. Hence, the output junction capacitance C_c is normally very small, about 1–5 pF. Output junction capacitance C_0 or C_c is specified as C_{0b} in a Transistor datasheet. Elements $r_{b'c}$ and $C_{b'c}$ between B' and C are due to reverse-bias to output junction that has some interaction to the input port.

6. Current generator $g_m\ v_{b'e}$
$v_{b'e}$ is the effective input signal voltage across Emitter junction, and its small changes cause variations in Base current, resulting in increased output Collector current i_C (depending upon Transistor β). Corresponding small-signal Collector current is proportional to $v_{b'e}$. This justifies the inclusion of current generator $i_C = g_m \cdot v_{b'e}$ across Collector C and Emitter E.

7. Conductance in Hybrid-π circuit
Parameter g_m is transconductance or mutual conductance of Transistors. Transconductance g_m represents small signal variations in Collector current for effective input signal voltage variations, about quiescent operating point Q. Therefore, for small signal operation (in linear region) about quiescent operating point g_m is given by

$$\text{Transconductance of the transistor} \quad g_m = \frac{|I_C(Q)|}{V_T} \text{ mA / V or mhos,} \tag{12.2}$$

where $I_C(Q)$ is the Quiescent component of collector current and V_T is the Voltage equivalent of temperature T:

$$V_T = \frac{kT}{e} = \frac{T}{11,600} = \frac{300}{11,600} \cong 26 \text{ mV} \quad \text{at temperature } T = 300°\text{K}. \tag{12.3}$$

8. Output resistance r_{ce} or r_o
Output resistance r_{ce} is the total conducting path resistance between Collector and Emitter, due to flow of Transistor currents. Typical value of r_{ce} is $\sim$ 40–80 kΩ.
9. g_{ce} Conductance between Collector and Emitter conducting areas (reciprocal of output resistance r_{ce} or r_o)
10. R_L (Load resistance). Normally, $R_L = 2$ kΩ for high-frequency Amplifiers
11. v_{ce} (output voltage)
12. I_L (Load current)
13. I_{in} (Input current)

Typical values for hybrid-π circuit parameters
(1) $g_m = 50$ milli mhos, (2) $r_{bb'} = 100\ \Omega$, (3) $r_{b'e} = 1$ kΩ, (4) $r_{b'c} = 4$ to 5 MΩ, (5) $r_{ce} = 40$ to 80 kΩ, (6) $C_c = 5$ pF, (7) $C_e = 100$ pF.

12.3 DETERMINATION OF HYBRID-π CIRCUIT PARAMETERS

1. Dynamic resistance of emitter Diode r_e

$$r_e = \frac{\partial V_E}{\partial I_E} = \eta \frac{V_T}{I_E} = \frac{V_T}{I_E} \tag{12.4}$$

when η is considered as unity.
2. Transconductance g_m at $I_C(Q)$ at quiescent operating point

$$g_m = \frac{I_C}{V_E} = \frac{I_C}{I_E} \cdot \frac{I_E}{V_E} = \frac{\alpha}{r_e}$$

$$g_m = \alpha_o \cdot \frac{\partial I_E}{\partial V_E} = \frac{\alpha_o}{r_e}$$

$$g_m = \frac{\alpha I_E}{V_T} \text{ or } \frac{|I_C|}{V_T} \tag{12.5}$$

Voltage equivalent of temperature $V_T = KT/e$, where K is the Boltzman constant joules/°K.

$$V_T = \frac{T}{11,600} = 26\text{ mV} \quad \text{at} \quad T = 300°\text{K}.$$

3. Resistance $r_{b'e}$

$$g_m = \frac{|I_C|}{V_T}$$

$$i_C = g_m \cdot v_{b'e} = g_m \cdot i_B \cdot r_{b'e}$$

$$v_{b'e} = i_B \cdot r_{b'e}$$

$$h_{fe} = \frac{i_C}{i_B} \mid v_{CE} = \text{constant}$$

$$\text{but} \quad i_C = g_m \cdot v_{b'e} = g_m \cdot i_B \cdot r_{b'e}$$

$$h_{fe} = \frac{i_C}{i_B} = \frac{g_m r_{b'e} \cdot i_B}{i_B} = g_m r_{b'e}$$

$$h_{fe} = g_m \cdot r_{b'e} \cdot \quad \left[\therefore \quad g_{b'e} = \frac{g_m}{h_{fe}}\right]$$

$$r_{b'e} = \frac{h_{fe}}{g_m} = \frac{h_{fe} V_T}{|I_C|} \tag{12.6}$$

4. Feedback conductance $g_{b'c}$

$$h_{re} = \frac{v_{b'e}}{V_{CE}} = \frac{(r_{b'e}) \cdot i_B}{(r_{b'e} + r_{b'c}) \cdot i_B}$$

$$h_{re}[r_{b'e} + r_{b'c}] = r_{b'e}$$

$$r_{b'e}[1 - h_{re}] = h_{re} \cdot r_{b'c} \tag{12.7}$$

Since $h_{re} \langle\langle 1$ or of the order of 10^{-4}. Equation (12.7) reduces to

$$r_{b'e} = h_{re} \cdot r_{b'e} \tag{12.8}$$

$$\therefore \quad g_{b'c} = h_{re} \cdot g_{b'e}.$$

Since h_{re} is almost insensitive to variations in current and temperature, $r_{b'c}$ depends on current (I_C) and temperature T to the same extent as $r_{b'e}$.

5. $r_{bb'}$: Base-Spreading resistance (Series Base resistance)

h_{ie} is defined as the input resistance of the Transistor with output short circuited. From the hybrid-π circuit, if output is short-circuited, $r_{b'c}$ comes in parallel with $r_{b'e}$.

$r_{b'c} \sim 4\ \text{M}\Omega$ and $r_{b'e} = 1\ \text{k}\Omega$ (typical values)

So, $r_{b'c} \gg r_{b'e}$. Hence $r_{b'e} \| r_{b'c} \cong r_{b'e}$. So, the input resistance $h_{ie} \cong r_{bb'} + r_{b'e}$

$$\therefore \quad r_{bb'} = h_{ie} - r_{b'e} \tag{12.9}$$

$$r_{bb'} = h_{ie} - \frac{h_{fe}}{g_m}$$

Since Base section is very thin, Base current I_B passes through a region of extremely small cross-section. Hence, resistance $r_{bb'}$ is large and may be of order of a few hundred ohms. But Collector and Emitter resistances are only a few ohms and may usually be neglected.

6. Output conductance

$$g_{ce} = (h_{oe} - g_m \cdot h_{re}). \tag{12.10}$$

7. Collector junction capacitance: C_C

Collector junction capacitance C_C is nothing but the output capacitance with input open circuited, i.e., with $I_E = 0$ and specified by manufacturers as C_{0b}. In the active region, the Collector junction being reverse-biased, Collector diffusion capacitance is negligibly small.

Typical value of $C_C \sim$ 1–5 pF.

8. Emitter junction capacitance: C_e

$$C_e \cong \frac{g_m}{2\pi f_T} \tag{12.11}$$

From Eq. (12.11) at frequency f_T, the current gain A_I of high frequency transistor amplifier with short circuited load is unity.

Typical value of $C_e = 100$ pF; frequency f_T is known as Transistor frequency.

Sequence to be followed in calculating the parameters of the hybrid-π equivalent circuit:

1. $g_m = \dfrac{|I_C|}{V_T}$, considering $I_C(Q)$ (12.12)

2. Input resistance $h_{ie} \cong r_{b'e} = \dfrac{h_{fe}}{g_m}$ or $g_{b'e} = \dfrac{g_m}{h_{fe}}$ (12.13)

3. $r_{bb'} = h_{ie} - r_{b'e} = h_{ie} - \dfrac{h_{fe}}{g_m}$ (12.14)

4. $r_{b'c} = \dfrac{r_{b'e}}{h_{re}}$ or $g_{b'c} = h_{re} \cdot g_{b'e}$ (12.15)

5. $g_{ce} = h_{oe} - g_m h_{re}$ or $g_{ce} = h_{oe} - (1+h_{fe}) g_{b'c}$ (12.16)

EXAMPLE 12.1

Calculate the components of hybrid-π equivalent circuit for $I_C(Q) = 2.6$ mA, $V_T = 26$ mV, $\beta = 100$, $h_{ie} = 1100\ \Omega$, $h_{re} = 2.0 \times 10^{-4}$, $h_{oe} = 30 \times 10^{-6}$ mhos, $f_T = 160$ MHz and $C_c = 5$ pF.

Solution:

Step 1: Calculation of g_m at collector current $I_C(Q) = 2.6$ mA

$$\text{Transconductance } g_m = \frac{|I_C(Q)|}{V_T} = \frac{2.6\times10^{-3}}{26\times10^{-3}} = 100\times10^{-3}\text{ mhos.}$$

Step 2: Calculation of $r_{b'e}$

$$r_{b'e} = \frac{h_{fe}}{g_m} = \frac{\beta}{g_m} = \frac{100}{100\times10^{-3}} = 10^3 = 1000\ \Omega.$$

Step 3: Calculation of Base-spreading resistance $r_{bb'}$

$$r_{bb'} = h_{ie} - r_{b'e} = 1100 - 1000 = 100\ \Omega.$$

Step 4: Calculation of $r_{b'c}$

$$r_{b'c} = \frac{r_{b'e}}{h_{re}} = \frac{1000}{2.0\times10^{-4}} = \frac{1000\times10^{4}}{2.0} = 5\times10^{6} = 5\ \text{M}\Omega.$$

Step 5: Calculation of g_{ce}

$$g_{ce} = (h_{oe} - g_m \cdot h_{re})$$

$$g_{ce} = (30\times10^{-6} - 100\times10^{-3}\times2.0\times10^{-4})$$

$$g_{ce} = (30\times10^{-6} - 20\times10^{-6}) = 10\times10^{-6}\ \text{mhos}$$

$$\therefore\quad r_{ce} = \frac{1}{g_{ce}} = \frac{1}{10\times10^{-6}} = 100\ \text{k}\Omega.$$

Step 6: Calculation of $C_c = C_{b'c}$

Collector junction capacitance $C_c = C_{b'c} = C_{ob} = 3$ pF

Step 7: Emitter junction capacitance C_e

$$C_{b'e} = C_e = \frac{g_m}{2\cdot\pi\cdot f_T} = \frac{100\times10^{-3}}{2\pi\times160\times10^{6}} \cong 100\ \text{pF}.$$

EXAMPLE 12.2

Given a germanium PNP Transistor, whose Base width is 10^{-4} cm at room temperature. For a DC Emitter current of 2 mA, find (a) Emitter diffusion capacitance and (b) f_T, assuming a diffusion constant of 47 cm³/s. (JNTU, Nov. 2006, 2007)

Solution: Assuming a room temperature of 27°C

$$T = 273 + 27 = 300^\circ\text{K}.$$

Assuming negligible base current $I_C = I_E = 2$ mA.

$$\text{Transconductance}\quad g_m = \frac{|I_C|}{V_T} = \frac{2\times10^{-3}}{26\times10^{-3}} = \frac{1}{13}\ \text{mhos}.$$

(a) Emitter diffusion capacitance $C_e = \dfrac{g_m \cdot W^2}{2\times D_B} = \dfrac{1\times(10^{-4})^2}{13\times2\times47} = 8.172$ pF

(b) $f_T = \dfrac{g_m}{2\times\pi\times C_e} = \dfrac{1}{13\times2\times\pi\times8.172\times10^{-12}} \cong 1500$ MHz.

EXAMPLE 12.3

Calculate the parameters of hybrid-π equivalent circuit of a high-frequency Transistor provided with the following data. $I_C(Q) = 5$ mA, $h_{ie} = 1$ kΩ, $h_{oe} = 4\times10^{-5}$ mhos, $h_{re} = 10^{-4}$, $h_{fe} = 100$, $C_{ob} = 2$ pF. (JNTU, Nov. 2003)

Solution:

(a) Transconductance $g_m = \frac{|I_C(Q)|}{V_T} = \frac{5\times10^{-3}}{26\times10^{-3}} = 0.192 = 192$ milli mhos

(b) Resistance $r_{b'e} = \frac{h_{fe}}{g_m} = \frac{100}{192\times10^{-3}} = 520\ \Omega$

(c) Base-spreading resistance $r_{bb'} = (h_{ie} - r_{b'e}) = (1000 - 520) = 480\ \Omega$

(d) $r_{b'c} = \frac{r_{b'e}}{h_{re}} = \frac{520}{10^{-4}} = 5.2\times10^{6}\ \Omega = 5.2\ \text{M}\Omega$

(e) $g_{ce} = [h_{oe} - (1+h_{fe})\cdot g_{b'c}]$

$$\therefore\ g_{ce} = \left[4\times10^{-5} - \frac{(1+100)}{(5.3\times10^{6})}\right]$$

$$= 10^{-5}(4-1.95) = 2.05\times10^{-5}\ \text{mhos}$$

(f) $r_{ce} = \frac{1}{(2.05\times10^{-5})} = 48.8\ \text{k}\Omega.$

EXAMPLE 12.4

A particular Transistor operates at $I_C = 1$ mA has $C_\mu = 1$ pF, $C_\pi = 9$ pF and Transistor $\beta = 150$. Find the values of ω_T and ω_β.

Solution:

$$g_m = \frac{|I_C|}{V_T} = \frac{1\times10^{-3}}{26\times10^{-3}} = \frac{1}{26}\ \text{mhos.}$$

(a) $\omega_T = \frac{g_m}{C_\pi + C_\mu} = \frac{1}{26[9+1]\times10^{-12}} = 3.85\times10^{9}$ rad/s

(b) $\omega_\beta = \frac{\omega_T}{\beta} = \frac{3.85\times10^{9}}{150} = 25.66\times10^{6}$ rad/s.

EXAMPLE 12.5

The following low-frequency parameters are known for a given Transistor. $I_C = 10$ mA, $V_{CE} = 10$ V and at room temperature $h_{ie} = 500\ \Omega$, $h_{0e} = 4\times10^{-5}$ mhos. And $h_{fe} = 100$ and $h_{re} = 10^{-4}$. At the same operating point, $f_T = 50$ MHz, $C_c = 3$ pF. Compute all the values of hybrid-π parameters. (JNTU, Nov. 2007)

Solution: Assuming g a room temperature of 27°C, Temperature $T = (27 + 273) = 300$°K, $V_T = 26$ mV

(a) Transconductance $g_m = \frac{|I_C|}{V_T} = \frac{10}{26} = 0.384$ Siemens

(b) Resistance $r_{b'e} = \dfrac{h_{fe}}{g_m} = \dfrac{100}{0.384} = 260.4\ \Omega$

(c) Base-spreading resistance $r_{bb'} = (h_{ie} - r_{b'e}) = 500 - 260.4 = 239.6\ \Omega$

(d) $r_{\mu} = r_{b'c} = \dfrac{r_{b'e}}{h_{re}} = \dfrac{260.4}{10^{-4}} = 2604\ \text{k}\Omega$

(e) $r_{ce} = r_0 = \dfrac{1}{(h_{oe} - g_m \cdot h_{re})} = \dfrac{1}{(4\times10^{-5} - 0.384\times10^{-4})} = 667\ \text{k}\Omega$

(f) $[C_c + C_e] = \dfrac{g_m}{2\cdot\pi\cdot f_T} = \dfrac{0.384}{(2\pi\times50\times10^6)} = 1222\ \text{pF}$

But $C_c = 3$ pF. Therefore, $C_e = 1219$ pF.

EXAMPLE 12.6

A Bipolar Junction Transistor (BJT) is operated at the quiescent operating point Q with Collector current $I_C(Q) = 2$ mA, $V_{CE}(Q) = 20$ V, $I_B(Q) = 20$ μA. Data given: $T = 300$°K, $f_T = 50$ MHz, $h_{ie} = 1400\ \Omega$, $h_{re} = 2.5 \times 10^{-4}$, $h_{oe} = 25$ μmhos, $C_c = 5$ pF. Calculate the parameters of the hybrid-π model for the BJT.

Solution:

(a) Transconductance $g_m = \dfrac{|I_C(Q)|}{V_T} = \dfrac{1}{13}$ mhos

(b) Transistor DC current gain $\beta_0 = \dfrac{I_C(Q)}{I_B(Q)} = \dfrac{2\times10^{-3}}{20\times10^{-6}} = 100$

(c) Resistance $r_{b'e} = \dfrac{\beta_0}{g_m} = \dfrac{100\times13}{1} = 1300\ \Omega$

(d) Base-spreading resistance $r_{bb'} = (h_{ie} - r_{b'e}) = (1400 - 1300) = 100\ \Omega$

(e) Resistor $r_{b'c} = \dfrac{r_{b'e}}{h_{re}} = \dfrac{1300}{2.5\times10^{-4}} = 5.2\times10^6\ \Omega$

(f) $g_{ce} = (h_{oe} - g_m \cdot h_{re}) = \left(2.5\times10^{-6} - \dfrac{1}{13}\times2.5\times10^{-4}\right)$

$= 5.66\times10^{-6}$ mhos.

$r_{ce} = \dfrac{1\times10^6}{5.66} = \dfrac{1000\times10^3}{5.66} = 176.6\ \text{k}\Omega$

(g) $[C_e + C_c] = \dfrac{g_m}{[2\cdot\pi\cdot f_T]} = \dfrac{1}{13\times6.28\times50\times10^6} = 245\ \text{pF}$

(h) $\therefore\ C_e = [245 - 5]\ \text{pF} = 240\ \text{pF}.$

EXAMPLE 12.7

The h-parameters of a Transistor at $I_C = 8$ mA, $V_{CE} = 10$ V and at room temperature are $h_{ie} = 1$ kΩ, $h_{oe} = 2 \times 10^{-5}$ A/V, $h_{fe} = 50$, $h_{re} = (2.5 \times 10^{-4})$. At the same operating point, $f_T = 60$ MHz and $C_{ob} = 2$ pF. Compute the values of hybrid-π parameters. (Nov. 2005)

Solution:

(a) Transconductance $g_m = \dfrac{|I_C(Q)|}{V_T} = \dfrac{8\times10^{-3}}{26\times10^{-3}} = \dfrac{4}{13}$ mhos

(b) Resistance $r_{b'e} = \dfrac{h_{fe}}{g_m} = \dfrac{\beta}{g_m} = \dfrac{50\times13}{4} = 162.5\,\Omega$

(c) Base-spreading resistance $r_{bb'} = (h_{ie} - r_{b'e}) = (1000 - 162.5) = 837.5\,\Omega$

(d) $r_{b'c} = \dfrac{r_{b'e}}{h_{re}} = \dfrac{162.5}{2.5\times10^{-4}} = 650\text{ k}\Omega$

(e) $C_c = C_{b'c} = C_{ob} = 2$ pF

(f) $C_e = C_{b'e} = \dfrac{g_m}{2\cdot\pi\cdot f_T} = \dfrac{4}{13\times6.28\times(60\times10^6)} = 816\times10^{-12} = 816$ pF.

12.4 CURRENT GAIN OF CE TRANSISTOR AMPLIFIER WITH RESISTIVE LOAD

Hybrid-π equivalent circuit of high-frequency Amplifier with resistive load R_L is shown in Fig. 12.2 to obtain its frequency response analysis.

At input and output circuits, effects of parallel combination of elements $r_{b'c}$ and $C_{b'c}$ (C_c) are explained using Miller's theorem. Emitter junction capacitance $C_{b'e}$ is considered as C_e and Collector junction capacitance $C_{b'c}$ is considered as C_c. Simplified equivalent circuit is shown in Fig. 12.3.

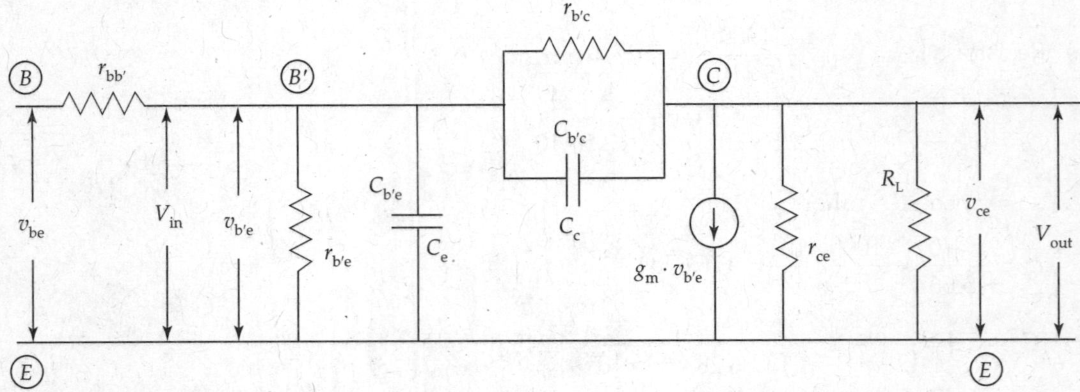

FIG. 12.2 *Hybrid-π equivalent circuit of high frequency transistor amplifier with load R_L*

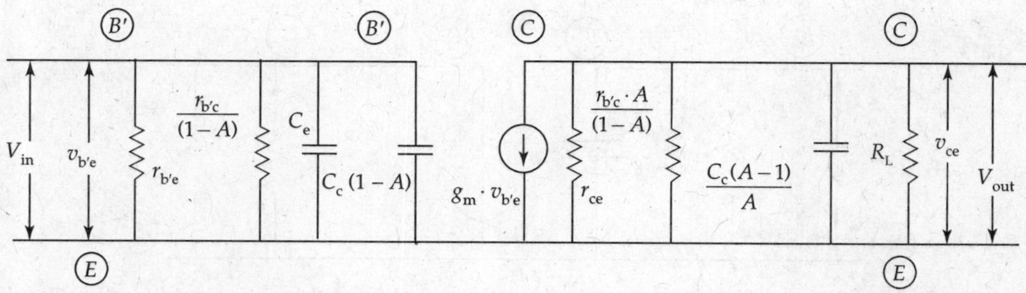

FIG. 12.3 *Simplified equivalent circuit using 'Millers theorem' for high-frequency arnplifier with load R_L*

Time constant of Transistor output circuit between Collector and Emitter

Typical values of various components are:

(a) $R_L = 2\ \text{k}\Omega$

(b) $r_{ce} = 40\ \text{k}\Omega$

(c) $r_{b'c} \cdot \dfrac{A}{(1-A)} \cong r_{b'c}$ for large values of amplifier gain A, $r_{b'c} \cong 4\ \text{M}\Omega$

(d) $C_c \cdot \dfrac{(A-1)}{A} \cong C_c$ for large values of amplifier gain A, $C_c = 5$ pF

(e) Effective resistance of the parallel combination of resistors r_{ce}, $r_{b'c}$ and R_L is equal to R_L (2 kΩ)

(f) Value of capacitance $C_c = 5$ pF

Therefore, output time constant,

$$T_{out} = T_o = R_L \times C_c = (2 \times 10^3) \times (5 \times 10^{-12}) = 10 \text{ ns.}$$

Time constant of Transistor input circuit between Base and Emitter.

Typical values of the components are

$$r_{b'e} = 1\ \text{k}\Omega$$

$$\frac{r_{b'c}}{(1-A)} = \frac{4\times 10^6}{(1+100)} \cong 40\ \text{k}\Omega,$$

where $A = -g_m \cdot R_L = -50 \times 10^{-3} \times 2\times 10^3 = -100$ (assuming, $g_m = 50 \times 10^{-3}$ mhos).

Therefore, parallel combination of $r_{b'e}$ and $r_{b'c}/(1-A)$ will be equal to $r_{b'e}$ only.

Only one resistor $r_{b'e}$ is considered in the input circuit after simplification (Fig. 12.4).

Value of total capacitance $C_T = [C_e + C_c\,(1-A)]$ where $C_M = C_c\,(1-A)$

Therefore, C_T or C_M is increased. (*C_M is Miller capacitance), as there is an increase in input capacitance due to Miller effect.*

Input time constant $T_{input} = T_{in} = r_{b'e} \times [C_e + C_c\,(1-A)]$

$$T_{in} = 1 \times 10^3 \times [100 \text{ pF} + 5 \text{ pF } (1 + 100)]$$

$$= 605 \times 10^{-9} = 605 \text{ nF (nano Farads)}$$

So, input time constant of Amplifier with resistive load is very high. Amplifier Bandwidth can be decided by input time constant.

During these calculations, it is clear that *output circuit time constant is negligible,* making the effect of C_c in output circuit negligible. Amplifier equivalent circuit after final simplifications is shown in Fig. 12.4.

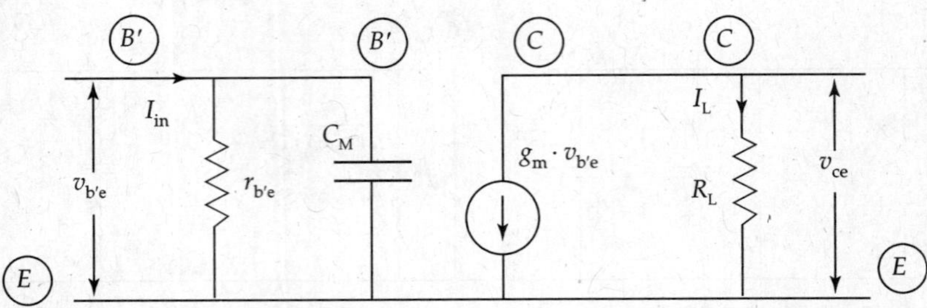

FIG. 12.4 *Final simplified circuit of high frequency transistor amplifier with resistive load for calculating the current gain A_I*

Expression for current gain A_I for high-frequency Amplifier with load R_L

$$\text{Current gain} \quad A_I = \frac{I_{load}}{I_{in}} = \frac{I_L}{I_{in}}, \tag{12.17}$$

where $I_L = g_m \cdot v_{b'e}$.

$$I_{in} = v_{b'e}\,[g_{b'e} + j \cdot w \cdot (C_e + C_M)] \tag{12.18}$$

$$\therefore \quad A_I = -\frac{I_L}{I_{in}} = -\frac{g_m \cdot v_{b'e}}{v_{b'e}[g_{b'e} + j \cdot 2 \cdot \pi \cdot f(C_e + C_M)}$$

$$= -\frac{g_m}{[g_{b'e} + j \cdot 2 \cdot \pi \cdot f(C_e + C_M)]}$$

$$A_I = -\frac{g_{b'e} \cdot h_{fe}}{[g_{b'e} + j \cdot 2 \cdot \pi \cdot f(C_e + C_M)]} \quad \text{using } (g_m = g_{b'e} \times h_{fe}) \tag{12.19}$$

Dividing both numerator and denominator of Eq. (12.19) by $g_{b'e}$, we get

$$A_I = -\frac{h_{fe}}{\left[1 + \frac{j \cdot 2 \cdot \pi \cdot f(C_e + C_M)}{g_{b'e}}\right]}$$

$$A_I = -\frac{h_{fe}}{\left[1 + j\left(\frac{f}{f_H}\right)\right]}, \quad \text{where } f_H = \frac{g_{b'e}}{2 \cdot \pi \cdot (C_e + C_M)} \tag{12.20}$$

$$|A_I| = \frac{h_{fe}}{\sqrt{1 + \left(\frac{f}{f_H}\right)^2}} \tag{12.21}$$

At frequency $f = 0$ Hz, Current gain is

$$A_I = h_{fe}. \tag{12.22}$$

At lower frequencies, $A_I = h_{fe}$.

From Eq. (12.21) at frequency ($f = f_H$)

$$\text{Current gain} \quad A_I = \frac{h_{fe}}{\sqrt{2}}. \tag{12.23}$$

From these values, amplifier response can be obtained:

$$\text{Comparing the values of } f_\beta = \frac{g_{b'e}}{2\cdot\pi\cdot(C_e + C_c)} \tag{12.24}$$

$$\text{and } f_H = \frac{g_{b'e}}{2\cdot\pi\cdot(C_e + C_M)} \tag{12.25}$$

Amplifier bandwidth f_H will be very low.

This is due to increased value of input capacitance due to 'Miller effect', where the increased capacitance is known as Miller capacitance.

12.5 SHORT CIRCUIT CURRENT GAIN A_I AND f_B OF CE TRANSISTOR AMPLIFIER

For hybrid-π circuit of CE Amplifier, with load R_L (as shown in Fig. 12.5), gain and bandwidth expressions are already derived for high frequencies. Figure 12.6 shows a hybrid-π circuit of CE Amplifier, with short circuited load R_L, input current I_{in} and load current I_L.

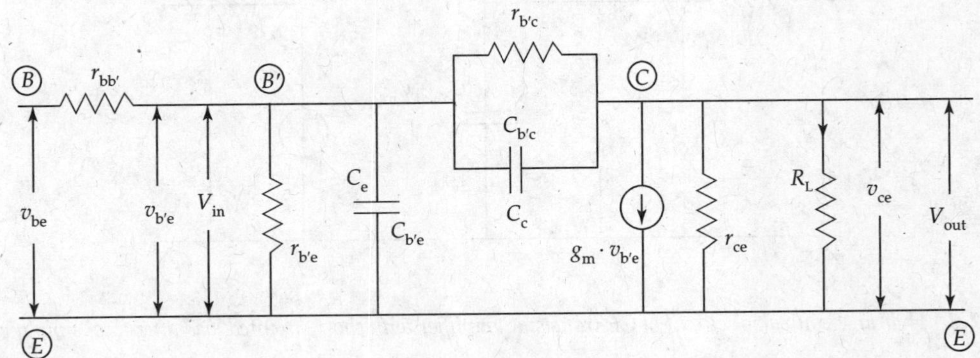

FIG. 12.5 *Hybrid-π equivalent circuit of CE transistor amplifier with load R_L*

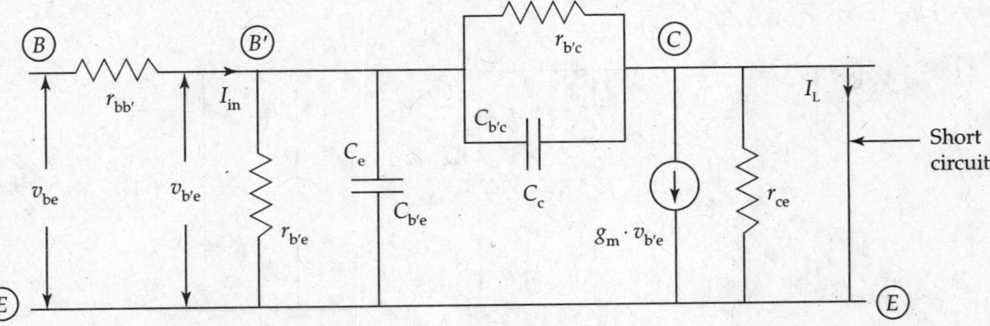

FIG. 12.6 *Hybrid-π equivalent circuit of CE transistor amplifier with short circuited load*

Equivalent circuit in Fig. 12.6 is simplified assuming negligible current flow to output circuit through $r_{b'c}$ (few mega ohms). r_{ce} is in shunt with a short circuit and eliminated in the output circuit. This simplified circuit is shown in Fig. 12.7.

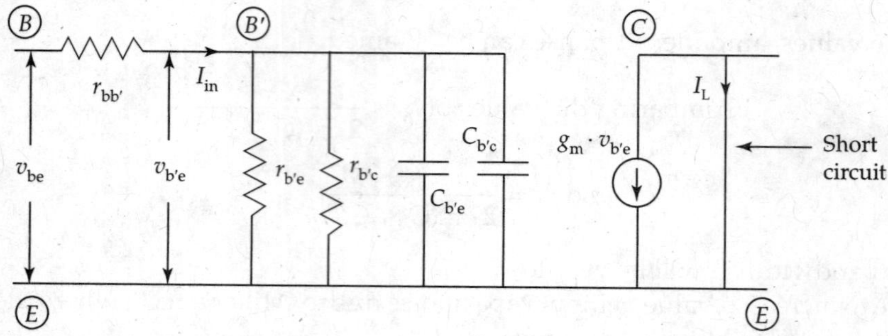

FIG. 12.7 *Hybrid-π equivalent circuit of CE transistor amplifier with short circuited load after simplification*

Typical values are $r_{b'e} = \sim 1$ kΩ and $r_{b'c} = \sim 4$ MΩ. Therefore, parallel combination of $r_{b'e}$ and $r_{b'c}$ will be equal to $r_{b'e}$ in the circuit shown in Fig. 12.7. Similarly, parallel combination of capacitors $C_{b'e}$ and $C_{b'c}$ becomes $[C_{b'e} + C_{b'c}]$, for *final simplification*. Final simplified circuit of CE Transistor amplifier with short circuited load resistance ($R_L = 0$ Ω) is shown in Fig. 12.8.

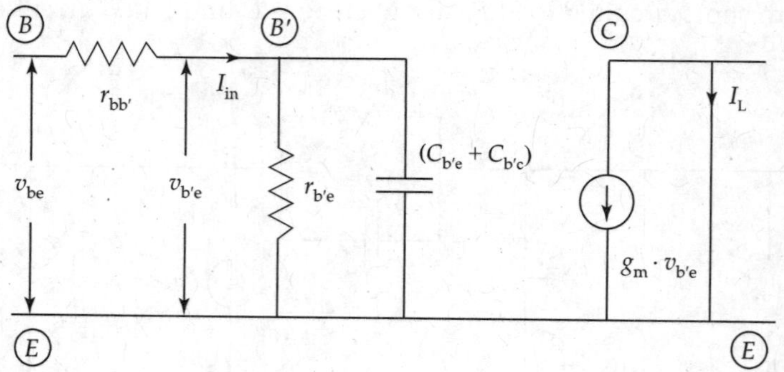

FIG. 12.8 *Hybrid-π equivalent circuit of CE transistor amplifier with short circuited load after final simplification*

For notation simplicity, Emitter junction capacitance $C_{b'e} = C_e$ and Collector junction capacitance $C_{b'c} = C_c$. *Including this nomenclature, hybrid-π circuit of HF amplifier appears as shown in Fig. 12.9.*

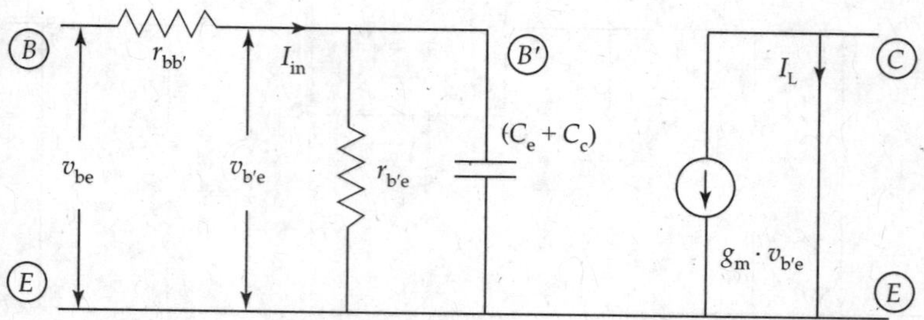

FIG. 12.9 *Hybrid-π equivalent circuit of CE transistor amplifier with short circuited load after final simplification*

For HF amplifier, current gain A_I is the ratio of load current I_L to input current I_{in}. Using the expression of current gain, expression for bandwidth frequency f_β is obtained as follows.

$$\text{Current gain} \quad A_I = \frac{I_{load}}{I_{in}} = \frac{I_L}{I_{in}} \tag{12.26}$$

$$\text{Load current} \quad I_L = -g_m \cdot v_{b'e} \tag{12.27}$$

$$\text{Input current} \quad I_{in} = v_{b'e}[g_{b'e} + j \cdot \omega (C_e + C_c)] \tag{12.28}$$

$$A_I = \frac{I_L}{I_{in}} = -\frac{g_m}{[g_{b'e} + j \cdot \omega \cdot (C_e + C_c)]} \quad \text{from Eqs. (12.27) and (12.28).} \tag{12.29}$$

Using the equation ($g_m = g_{b'e} \cdot h_{fe}$) in Eq. (12.29), we get

$$A_I = -\frac{g_{b'e} \cdot h_{fe}}{[g_{b'e} + j \cdot \omega \cdot (C_e + C_c)]}. \tag{12.30}$$

Dividing both numerator and denominator by $g_{b'e}$

$$A_I = -\frac{h_{fe}}{\left[1 + \frac{j \cdot 2 \cdot \pi \cdot f \cdot (C_e + C_c)}{g_{b'e}}\right]} \tag{12.31}$$

Equation (12.31) can be written as

$$A_I = -\frac{h_{fe}}{\left[1 + j \cdot \frac{f}{f_\beta}\right]}, \quad \text{where } f_\beta = \frac{g_{b'e}}{2 \cdot \pi \cdot (C_e + C_c)} \tag{12.32}$$

$$|A_I| = \frac{h_{fe}}{\sqrt{1 + \left[\frac{f}{f_\beta}\right]^2}} \tag{12.33}$$

$$A_I = h_{fe}, \quad \text{when } f = 0 \text{ Hz} \tag{12.34}$$

$$\text{At the frequency } f = f_\beta, \quad [A_I] = \frac{h_{fe}}{\sqrt{2}}. \tag{12.35}$$

If the maximum gain of Amplifier at mid frequencies is h_{fe}, f_β will be high-frequency cut-off point representing Amplifier bandwidth.

$$f_\beta = \frac{g_{b'e}}{2 \cdot \pi \cdot (C_e + C_c)} = \frac{g_m}{h_{fe}[2 \cdot \pi \cdot (C_e + C_c)]} \text{Hz} \tag{12.36}$$

Transition Frequency f_T and gain bandwidth product:

Frequency f at which current gain A_I becomes unity is known as f_T.
From Eq. (12.33),

$$A_I = \frac{h_{fe}}{\sqrt{1 + \left[\frac{f_T}{f_\beta}\right]^2}} = 1 \tag{12.37}$$

$$\therefore \frac{f_T}{f_\beta} = h_{fe} \tag{12.38}$$

$$\text{which means } f_T = (h_{fe} \times f_\beta) = (\text{gain} \times \text{bandwidth}) \tag{12.39}$$

$$f_T = \frac{h_{fe} \cdot g_m}{h_{fe}[2 \cdot \pi \cdot (C_e + C_c]}$$

$$\therefore f_T = \frac{g_m}{[2 \cdot \pi \cdot (C_e + C_c)]} \cong \frac{g_m}{2 \cdot \pi \cdot C_e} \quad (\text{as } C_e \gg C_c). \tag{12.40}$$

From Eq. (12.40), *product of gain and bandwidth* of Amplifier is constant.

Transition frequency f_T and gain bandwidth product

From Eq. (12.39), $f_T = h_{fe} f_\beta$

From Eq. (12.34), current gain $A_I = h_{fe}$, when $f = 0$ Hz (at low frequency).

From Eq. (12.35), f_β is the high-frequency cut-off point or Amplifier bandwidth.

Therefore, frequency f_T is gain bandwidth product for high-frequency Amplifier with a load condition of short circuit (Fig. 12.10).

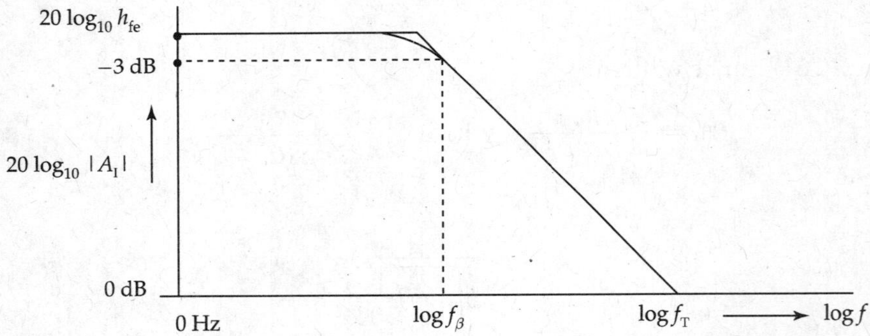

FIG. 12.10 *Frequency response of common emitter high frequency transistor amplifier with short circuited load ($R_L = 0$)*

In practice, various manufacturers specify the transition frequency f_T in relevant data sheets. Typical values range from 100 MHz to 10s of GHz. It can also be determined by following the below steps.

Transconductance g_m can be determined by the relation

$$g_m = \frac{|I_C(Q)|}{V_T} = \frac{|I_C(Q)|\text{mA}}{25\text{ mV}} \quad \text{at room temperature.} \tag{12.41}$$

C_μ can be obtained by conducting independent measurement of capacitance between base and collector at the desired reverse bias voltage.

Having known f_T, g_m and C_μ, the value of C_π can be determined by the relation

$$f_T = \frac{g_m}{2 \cdot \pi \cdot [C_\pi + C_\mu]}, \tag{12.42}$$

where $C_c = C_\mu$ and $C_e = C_\pi$.

If two Transistors have equal transition frequency f_T (low frequency current gain multiplied by upper 3-dB frequency), the Transistor with lower β is to be chosen for larger bandwidth.

Transition frequency is the upper bound frequency for which the hybrid-π model is valid. In practice, hybrid-π model is useful for analysis only up to $(1/3)f_T$. At frequencies above this range, Transistor modelling is quite complicated and has to consider the effects of parasitic elements and splitting r_x into number of parts.

- *Transition frequency* f_T is a function of Collector current I_C and V_{CE} (Fig. 12.11).
- Frequency f_T is a function of g_m and a small part of C_π, both directly proportional to I_C. This explains the lower f_T at lower currents.
- Low-frequency value of β_0 decreases with higher currents and $\omega_T = \beta_0 \cdot \omega_\beta$. This explains the decrease in *transition frequency* at high currents.
- Transition frequency is relatively constant, in between these regions.

Similar to the analysis of CE Amplifier with short circuit current gain, a CB Amplifier can also be analysed with similar short circuit current gain. Instead of $\beta_{\text{cut-off}}$ frequency, cut-off frequency known as $\alpha_{\text{cut-off}}$ frequency can be arrived using a similar expression. Alpha cut-off frequency f_α of a CB Amplifier has a wider frequency range than Beta cut-off frequency f_β of a CE Amplifier. This is the primary reason for using CB Amplifier in CE + CB Cascode, where high frequency response is improved better than a single CE Amplifier.

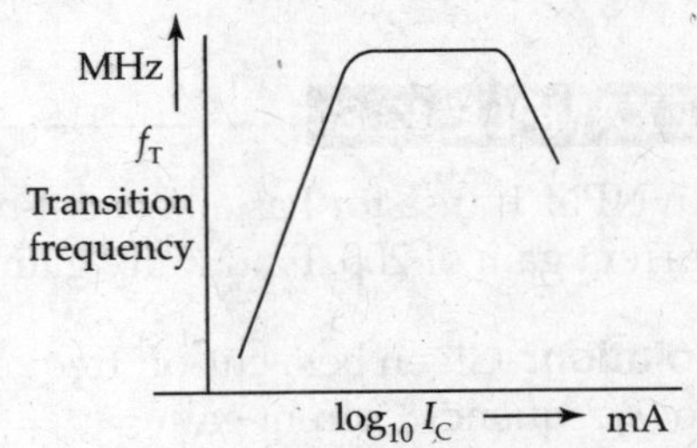

FIG. 12.11 *Typical variation of transition frequency f_T with collector current I_C*

Alpha cut-off frequency f_α for CB Transistor Amplifier at high frequencies

Current gain α (Alpha) of CB Amplifier varies when operating at high frequencies versus low frequencies, due to the differences in *transit times* of low- and high-frequency currents through the Amplifier. Journey time (transit time) through Transistor Base for diffusion process of charges depends on diffusion constant D_B and Base width W_B. These device parameters in turn predict the Alpha cut-off frequency f_α.

$$\text{Alpha cut-off frequency of } f_\alpha = \frac{D_B}{\pi \times W_B^2} \text{ MHz} \tag{12.43}$$

As base width W_B is of the order of a few microns, Alpha cut-off frequency f_α for common base transistor amplifier will be of the order of a few MHz.

By reorienting hybrid-π circuit of HF CE Amplifier into equivalent CB, it provides the following *hybrid-T model* with suitable circuit parameters (Fig. 12.12).

Various parameters for the hybrid-T model change their values from their equivalent hybrid-π parameters. Hence, current gain α_{hf} for HF CB Transistor also changes. Relation between parameters (Alpha) α, frequency f, Alpha-cut-off frequency f_α and α_{hf} is shown as follows.

$$\alpha_{hf} = \frac{\alpha}{\left[1 + j \cdot \frac{f}{f_\alpha}\right]} \tag{12.44}$$

At frequency $f = f_\alpha$, high-frequency transistor alpha $\alpha_{hf} = \frac{\alpha}{\sqrt{2}} = 0.707\alpha$.

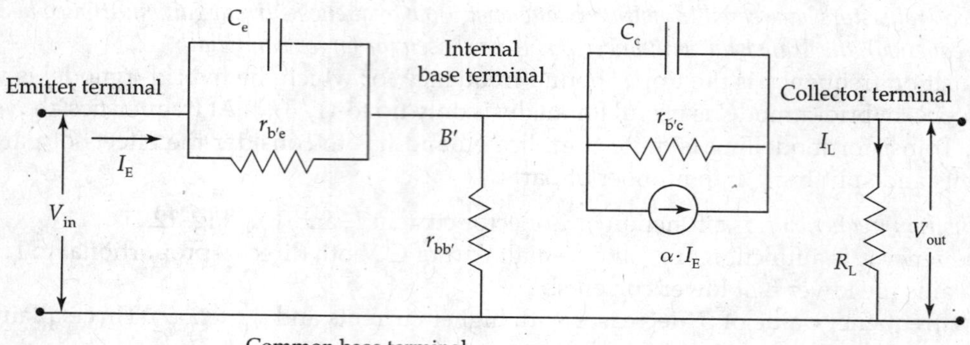

FIG. 12.12 *Hybrid-T model equivalent circuit of common base transistor at high frequencies*

EXAMPLE 12.8

An NPN Transistor has a beta cut-off frequency of 1 MHz and CE short circuit low frequency current gain of 200. Find unity gain frequency f_T and Alpha cut-off frequency.

Solution: Given beta cut-off frequency $f_\beta = 1$ MHz
Low frequency current gain $= \beta_0 = h_{fe} = 200$
From the relation, $f_T = \beta_0 \times f_\beta = 200 \times (1 \times 10^6) = 200$ MHz
From the definition of alpha cut-off frequency,

$$f_\alpha = h_{fe} \times f_\beta \left[\frac{C_\pi + C_\mu}{C_\pi}\right]$$

Assuming g, $C_\pi = 9$ pF and $C_\mu = 1$ pF

$$f_\alpha = 200 \times (1 \times 10^6)\left[\frac{(9+1)}{9}\right] = 222.22 \text{ MHz},$$

which is evidently greater than f_β

EXAMPLE 12.9

For the following measurements, $I_C = 5$ mA, $V_{CE} = 10$ V at room temperature. $h_{fe} = 100$, $h_{ie} = 600\ \Omega$, $A_{ie} = 10$ at 10 MHz, $C_c = 3$ pF. Find f_β, f_T, $r_{b'e}$, $r_{bb'}$ and C_e. (JNTU, Nov. 2003, 2005)

Solution:

(a) From the equation, $|A_{ie}| = \dfrac{h_{fe}}{\sqrt{1 + \left[\dfrac{f}{f_\beta}\right]^2}}$

Given $A_{ie} = 10$, $f = 10$ MHz, $h_{fe} = 100$.
Substituting these values in the above equation, we get

$$f_\beta = 1.005 \text{ MHz}.$$

(b) $f_T = h_{fe} \times f_\beta = 100 \times 1.005 \times 10^6 = 100.5$ MHz

$$g_m = \frac{|I_C|}{V_T} = \frac{5}{26} = 192 \text{ milli Siemens}$$

(c) $r_{b'e} = \dfrac{h_{fe}}{g_m} = \dfrac{100}{192 \times 10^{-3}} = 520.8\ \Omega$

(d) $r_{bb'} = (h_{ie} - r_{b'e}) = (600 - 520.8) = 79.2\ \Omega$

(e) $(C_c + C_e) = \dfrac{g_m}{2\pi f_T} = \dfrac{0.192}{6.28 \times (100.5 \times 10^6)} = 19.14$ pF

But $C_c = 3$ pF $\therefore C_e = 16.14$ pF.

EXAMPLE 12.10

Hybrid parameters of Transistor shown in circuit (Fig. 12.13) are transconductance $g_m = 50$ mA/V, $r_{bb'} = 100\ \Omega$, $r_{b'e} = 1\ \text{k}\Omega$, $r_{b'c} = 4\ \text{M}\Omega$, $r_{ce} = 80\ \text{k}\Omega$, $C_c = 3$ pF, $C_e = 100$ pF. Using Millers theorem and appropriate analysis, compute (a) upper 3-dB frequency of current gain A_I and (b) Voltage gain at frequency calculated above. (JNTU, Mar. 2006)

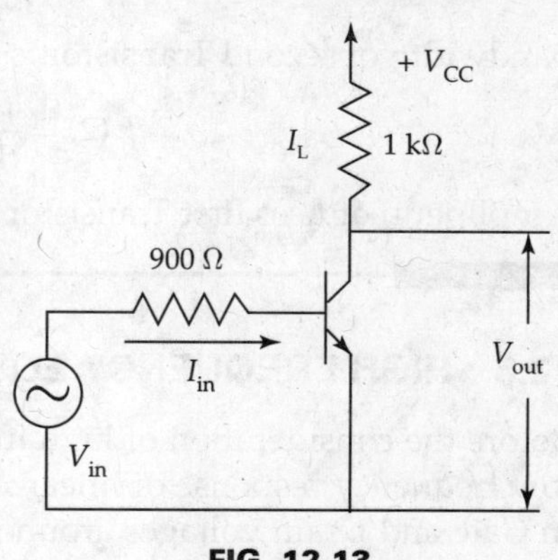

FIG. 12.13

Solution:

Step 1: Upper 3-dB frequency

$$f_\beta = \frac{g_{b'e}}{2 \cdot \pi \cdot [C_e + C_c]} = \frac{1}{r_{b'e} \times 2 \cdot \pi [C_e + C_c]}$$

$$\therefore\ f_\beta = \frac{1}{1 \times 10^3 \times 6.28\,[100 + 3] \times 10^{-12}}$$

$$= \frac{1000 \times 10^6}{6.28 \times 103} = 1.546 \text{ MHz}$$

Step 2: $f_T = h_{fe} \times f_\beta$

$$h_{fe} = g_m \cdot r_{b'e} = (50 \times 10^{-3}) \times (1 \times 10^3) = 50$$

$$f_\beta = (1.546 \times 10^6) \text{ Hz calculated above}$$

$$\therefore\ f_T = (h_{fe} \times f_\beta) = (50) \times (1.546 \times 10^6) = 77.3 \text{ MHz}$$

Step 3: Mid-band voltage gain $= \dfrac{(h_{fe} \times R_L)}{R_{in}}$

$$R_L = 1\ \text{k}\Omega \text{ and source resistance } r_S = 900\ \Omega$$

Input resistance $R_{in} = h_{ie} = (r_S + r_{bb'} + r_{b'e}) = (900 + 100 + 1000)\ \Omega = (2.0 \times 10^3)\ \Omega$

$$A_V = \frac{(1 \times 10^3) \times (50)}{2 \times 10^3} = \frac{50}{2} = 25$$

Step 4: $A_V(f_\beta)$ = Voltage gain at f_β, when $A_I = 1$

$$A_V(f_\beta) = A_I \times \frac{R_L}{R_{in}} = A_I \times \frac{R_L}{(r_S + h_{ie})} = \frac{1\times(1\times10^3)}{2.0\times10^3} = 0.5.$$

EXAMPLE 12.11

Calculate the bandwidths of following two Transistors. The first Transistor has Beta $\beta_1 = 100$ and the second Transistor has Beta $\beta_2 = 200$. If both Transistors have *transition frequency* f_T equal to 200 MHz, compare their performance.

Solution: $f_T = \beta_1 f_\beta(1)$ for the first Transistor.
Therefore, bandwidth of the first Transistor is

$$f_\beta(1) = \frac{f_T}{\beta_1} = \frac{200\times10^6}{100} = 2\times10^6 = 2\text{ MHz}$$

Bandwidth of second Transistor is

$$f_\beta(2) = \frac{f_T}{\beta_2} = \frac{200\times10^6}{200} = 1\times10^6 = 1\text{ MHz}$$

Amplifier that uses first Transistor with lower value of $\beta = 100$ has higher Bandwidth.

12.6 HIGH FREQUENCY EQUIVALENT CIRCUIT OF JFET

Before the consideration of high frequency response of JFET, we need to discuss small signal low frequency response of linear Amplifier. It represents the operation of device, as changes in Gate and Drain voltages around the operating point determine I_D, V_G and V_D. Incremental changes in total instantaneous current I_D and the incremental changes in Drain voltage V_D and Gate voltage V_G are governed by the following linear relationship:

$$I_D = g_m \times V_G + g_D \times V_D,$$

where g_m is the transconductance and g_D is the channel conductance.

Inter-electrode capacitances do not play any role in small incremental changes at low frequencies. Above relationship can be shown as a small signal linear low frequency equivalent circuit (Fig. 12.14). The input resistance R_{in} of the FET device is of the order of a few mega ohms, because of the reverse-biased operation between Gate and Source terminals. Hence, no component is shown in the input port of equivalent circuit for the FET device.

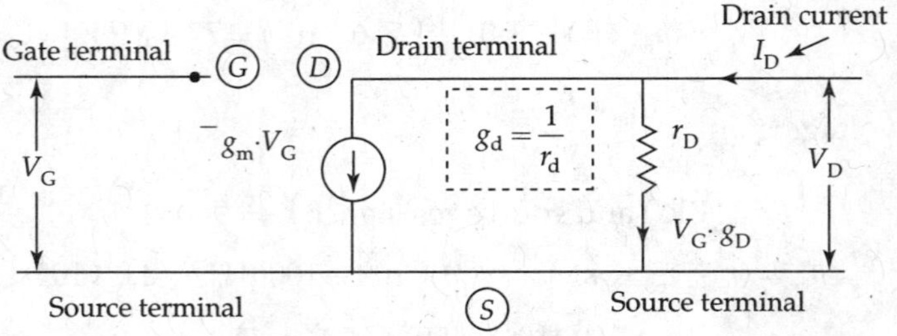

FIG. 12.14 *Small signal low frequency equivalent circuit for JFET*

Figure of merit of JFET is a measure of both gain and high frequency response. Analogous to that of a BJT, high frequency response of JFET is influenced by internal capacitances $-C_{GS}$ (capacitance between Gate and Source), C_{GD} (capacitance between Gate and Drain) and C_{DS} (capacitance between Drain and Source). In the small signal equivalent circuit in Fig. 12.14, these capacitors are introduced to show the HF equivalent circuit of JFET in Fig. 12.15.

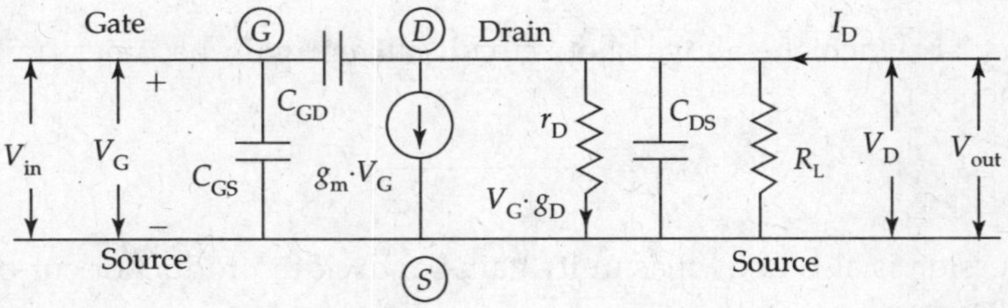

FIG. 12.15 *High frequency equivalent circuit for JFET showing junction capacitances*

Drain resistance r_D will be about tens of kilo ohms, whereas Drain circuit resistance and load resistances R_L in the Drain circuit will be about a few kilo ohms. So, Drain resistance r_D will not have any influence on the amplifier function.

High frequency equivalent circuit can further be simplified by splitting up C_{GD} into two components by using Millers theorem, with one component at input port and other component at output port of JFET Amplifier, which will be studied in further sections.

Short circuit current gain of a JFET

High frequency equivalent circuit of CS configuration employing short circuit between Drain and Source is shown in Fig. 12.16.

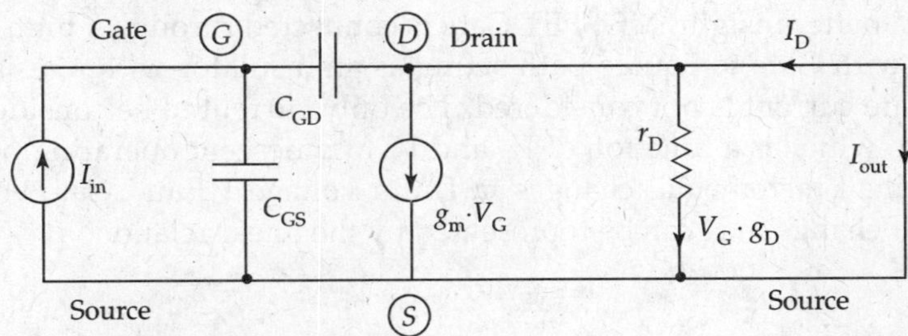

FIG. 12.16 *Short circuit current gain of JFET amplifier*

Sinusoidal gate current having an rms value of I_{in} is applied at input port between Gate and Source. I_{out} is output current, corresponding rms current through short circuit between Drain and Source.

Current flowing through internal capacitor C_{GD} can be neglected when compared to current $g_m \cdot V_G$, where V_G is the rms value of v_{gs}.

$$I_{in} = V_G \cdot j\omega \cdot [C_{GS} + C_{GD}] \tag{12.45}$$

$$I_{out} = V_G \cdot [g_m - j \cdot \omega \cdot C_{GD}] \tag{12.46}$$

$$\therefore \quad \text{Short circuit current gain } A_I(\text{sc}) = \left|\frac{I_{\text{out}}}{I_{\text{in}}}\right| = \frac{[g_m - j\cdot\omega\cdot C_{GD}]}{j\cdot\omega[C_{GS} + C_{GD}]}$$

$$A_I(\text{sc}) = \frac{g_m}{j\cdot\omega\cdot[C_{GS} + C_{GD}]} - \frac{C_{GD}}{[C_{GS} + C_{GD}]} \cong \frac{g_m}{j\cdot\omega\cdot[C_{GS} + C_{GD}]}$$

The frequency at which the above short circuit current gain becomes unity is known as Transition frequency f_T:

$$f_T = \frac{g_m}{2\cdot\pi\cdot[C_{GS} + C_{GD}]} \tag{12.47}$$

The above expression is also known as unity gain bandwidth product or cut-off frequency or Transition frequency.

Manufacturers specify *transition frequency* in data sheets. Based on prior discussion about BJT and the above formula, C_{GS} and C_{GD} can be estimated.

By using the method of *short circuit time constant*, individual high-frequency cut-offs of the FET amplifier can be estimated.

High frequency response is independent of physical constants and dimensions of JFET. It can be improved by *decreasing channel length*, which in turn determines the capacitance and increases g_m, resulting in improved bandwidth gain product. By employing semiconductors with high mobility charges (electrons), current travels with high velocity taking less *transit time*, thereby improving HF response.

12.7 HIGH FREQUENCY EQUIVALENT CIRCUIT OF MOSFET

Before analysing the HF response of a MOSFET, studying its small signal LF equivalent circuit provides a better insight. MOSFET Gate is connected through a high quality layer of Silicon Dioxide, with Gate to Source path acting as an insulator with a resistance of 10^{14} to $10^{15}\ \Omega$. Hence, Gate current is not considered. The only current to be considered is I_D, Drain to Source current, which is a function of V_D and V_G at quiescent operating point.

Considering small incremental changes in Drain voltage V_D and Gate voltage V_G, small signal component change in I_D can be represented by the linear relation

$$I_D = g_m\cdot V_G + g_D\cdot V_D,$$

where g_m is *transconductance* representing the control of input Gate voltage over Drain current and g_D is the output conductance representing the control of output voltage over Drain

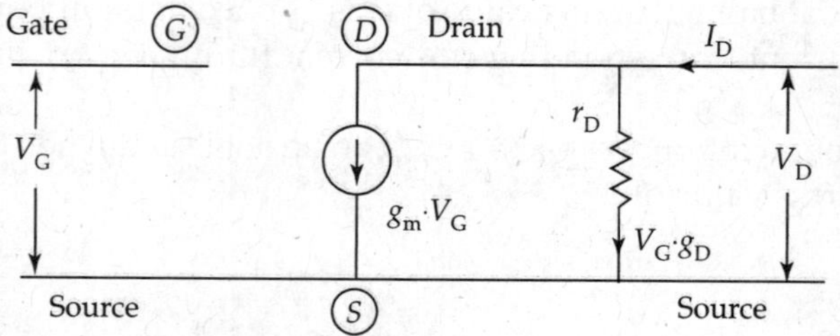

FIG. 12.17 *Small signal low frequency equivalent circuit for MOSFET*

current. In the discussion of small signal response, the capacitances inherent in the operation of the device and the parasitic capacitances have no influence. Small signal low frequency equivalent circuit of a MOSFET is shown in Fig. 12.17.

While considering the high frequency response of MOSFET, Gate to Source capacitance C_{GS}, Drain to Gate capacitance C_{DG}, parasitic capacitances C_{OS} and C_{OG} between Gate and Source and Gate and Drain are considered. They are introduced in the above equivalent circuit (Fig. 12.17) at appropriate places to form high frequency equivalent circuit (Fig. 12.18).

In addition to the above four capacitances, there are two more capacitances C_{tS} and C_{tD} which are considered. C_{tS} is the depletion layer capacitance between Source and the substrate and C_{tD}

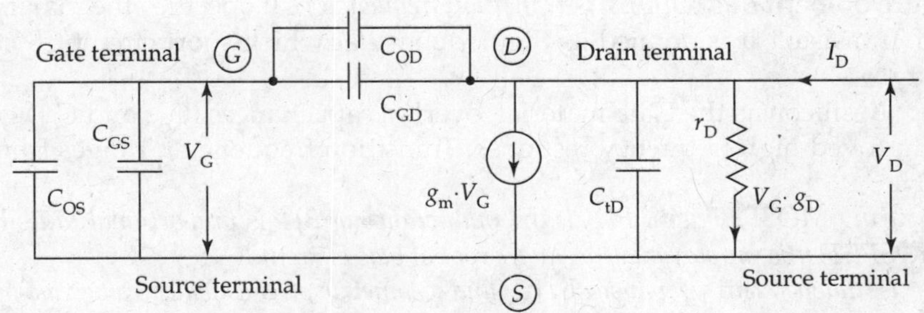

FIG. 12.18 *Complete high frequency equivalent circuit of MOSFET*

is capacitance between Drain and substrate. As the substrate is assumed to be connected to Source, C_{tS} is not shown in Fig. 12.16. Only five capacitances are added.

Short circuit current gain of a Common Source (CS) MOSFET

Equivalent circuit in Fig. 12.14 is slightly modified to calculate short circuit current gain of CS MOSFET. Overall, capacitances C_{OS} (between Gate and Source) and C_{OD} (between Gate and Drain) are lumped together in the Gate–Drain capacitance C_{GD}. Capacitance between Drain and substrate C_{tD} is larger and hence neglected. Equivalent circuit after applying a short circuit at the output port and an input signal applied at input port is shown in Fig. 12.19.

$$\text{Input current} \quad I_{in} = V_{GS} \cdot j \cdot \omega \cdot (C_{GS} + C_{GD}) \tag{12.48}$$

$$\text{Output current} \quad I_{out} = g_m \cdot V_{GS} \tag{12.49}$$

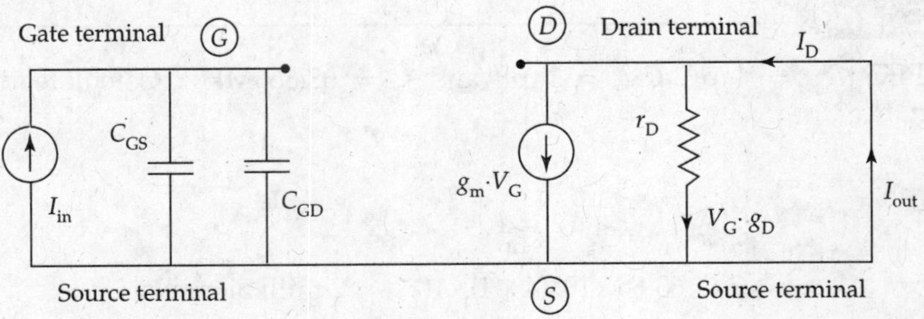

FIG. 12.19 *Simplified equivalent circuit of MOSFET to determine f_T*

$$\text{Short circuit current gain } A_I(\text{sc}) = \frac{|I_{\text{out}}|}{|I_{\text{in}}|} = \frac{g_m}{j\cdot\omega\cdot[C_{GS}+C_{GD}]} \tag{12.50}$$

Frequency at which the magnitude of short circuit current gain is unity is known as Transition frequency f_T:

$$\therefore\ f_T = \frac{g_m}{2\cdot\pi\cdot[C_{GS}+C_{GD}]} \tag{12.51}$$

From Eq. (12.49) for short circuit current gain, when $\omega = 0$, current gain of MOSFET reaches infinity. This happens in MOSFET, if extremely small leakage current present at the input is ignored. Figure of merit of MOSFET is *transition frequency* f_T. It specifies the maximum usable frequency of Transistor. It is defined as the frequency at which short circuit CS current gain becomes unity.

By using poly-silicon as the Gate material, overlap capacitances (C_{OD} and C_{OS}) get reduced, leading to improved high frequency response. Transition frequency f_T ranges from 100 MHz to several GHz.

While comparing MOSFET with BJT, transconductance of BJT is proportional to its bias current, whereas in MOSFET, it is proportional to square root of bias current.

In BJT, g_m is independent of physical size and geometry of the device, whereas in MOSFET, the transconductance depends on the physical size and geometry of the MOSFET. Even though, transconductance is much smaller in MOSFET compared to BJT, the MOSFET is more popular because it is small in size, cheap, has improved high frequency response and easy to implement in fabrication of integrated circuits.

The above concepts offer insight into HF operations of different transistor types.

EXAMPLE 12.12

A Transistor datasheet lists three parasitic capacitances of a JFET, as given below. Input capacitance $C_{iSS} = 6$ pF; Output capacitance $C_{oSS} = 4$ pF; Reverse transfer capacitance $C_{RSS} = 2$ pF. Find inter-electrode capacitances of the device, which influence high frequency response.

Solution:

$$C_{GS} = C_{iSS} - C_{RSS} = (6\text{ pF} - 2\text{ pF}) = 4\text{ pF}$$
$$C_{GD} = C_{RSS} = 2\text{ pF}$$
$$C_{DS} = C_{oSS} - C_{RSS} = 4\text{ pF} - 2\text{ pF} = 2\text{ pF}.$$

EXAMPLE 12.13

For a MOSFET, $C_{GS} = 4$ pF, $C_{GD} = 1$ pF and $f_T = 63.66$ MHz. Calculate the value of transconductance?

Solution:

$$g_m = 2\cdot\pi\cdot f_T \times (C_{GS} + C_{GD})$$
$$= 2\pi \times 63.66 \times 10^6 (4+1) \times 10^{-12} = 2\text{ milli Siemens.}$$

SUMMARY

1. Small signal HF Transistor amplifiers are analysed using hybrid-π parameters of Transistors such as $r_{bb'}$, $r_{b'e}$, $r_{b'c}$, r_{ce}, C_e and C_c.
2. In HF Transistor amplifier analysis, junction capacitances and transit time play a dominant role in their frequency response.
3. Alpha cut-off frequency Base width W_B and diffusion constant D_B – for the movement of charge carriers through Base region – are related by equation:

$$f_\alpha = \frac{D_B}{\pi \cdot W_B^2} \text{ MHz.}$$

4. Alpha cut-off frequency f_α, current gain α of LF CB Transistor and gain α_{hf} of HF CB Transistor are governed by the following equation:

$$\alpha_{hf} = \frac{\alpha}{\left[1 + j \cdot \frac{f}{f_\alpha}\right]}.$$

5. Cut-off f_β of a HF Transistor amplifier is the frequency at which short circuit current gain of CE Transistor falls by 3 dB.
6. HF cut-off f_β provides the bandwidth of HF Amplifier

$$f_\beta = \frac{g_m}{h_{fe}[2 \cdot \pi \cdot (C_e + C_c)]} \text{ Hz,}$$

where g_m is the transconductance, C_e is the Emitter junction capacitance and C_c is the Collector junction capacitance and $h_{fe} = \beta$.
7. Transition frequency f_T is the frequency at which the short circuit current gain A_I of a HF Amplifier is unity.
8. Current gain h_{fe}, f_T and f_β are related by the equation $f_T = h_{fe} \cdot f_\beta$.
9. Transition frequency f_T, transconductance g_m and the junction capacitances are related by the following equation:

$$f_T = \frac{g_m}{2 \cdot \pi \cdot [(C_e + C_c)]} = \frac{g_m}{2 \cdot \pi \cdot [(C_\pi + C_\mu)]} \text{ Hz.}$$

10. Transition frequency f_T measures the quality of performance of HF Transistors and it is known as 'figure of merit'.
11. Input time constant of an amplifier determines the frequency response of an amplifier.

Questions for Practice

1. Draw HF equivalent circuit of a Transistor using hybrid-π model and discuss the significance of each component in the circuit.

2. Derive the various expressions used in the determination of hybrid-π circuit indicating the sequence of calculations.
3. Discuss the roles of junction capacitances in the determination of Transistor performance at high frequencies.
4. Derive the expressions for transconductance and input conductance of CE Transistor Amplifier using HF model.
5. Derive expressions for feedback capacitance and Base-spreading resistance of CE Transistor amplifier using hybrid-π model.
6. Derive the expression for output conductance and diffusion capacitance of hybrid-π Equivalent circuit of CE Transistor amplifier.
7. (a) Prove that in hybrid-π model circuit, the diffusion capacitance is proportional to bias current?

   (b) In Giacolletto model of a Transistor at high frequencies, how does C_c vary with I_C and V_{CE}? How does C_e vary with I_C and V_{CE}? (JNTU, Feb. 2008)
8. Define f_β and f_T and derive the relation between f_β and f_T? (Nov. 2005)
9. Define the terms f_β, f_α and f_T from hybrid-π and hybrid-T model circuits?

Multiple Choice Questions

1. Transition frequency of BJT _______________.

   (1) is limited by C_μ (2) is limited by C_π

   (3) increases with increase in g_m (4) decreases with increase in g_m

   The correct statements are

   (a) 1 & 4 (b) 2 & 4 (c) 1, 2 & 3 (d) 1, 2 & 4
2. The internal capacitances of a BJT (or an FET) exhibit _______________.

   (a) low-pass characteristic (b) high-pass characteristic

   (c) band-pass characteristic (d) band-stop characteristic
3. At high frequencies, _______________.

   (1) external capacitance is large so that they are effectively short-circuited

   (2) external capacitance can be assumed to be open circuited

   (3) internal capacitance is small so that they are effectively open-circuited

   (4) internal capacitance can be treated as open circuits

   Out of the above statements, the correct statements are

   (a) 1, 2, 4 (b) 1, 2, 3 (c) 2, 4 (d) 2 & 3
4. Graphical method of determining cut-off frequencies is _______________.

   (a) open circuit time constant method (b) short circuit time constant method

   (c) bode plot (d) Miller capacitance method

5. Depletion capacitances of MOSFET ______________.
   (a) C_g & C_{sb} (b) C_{sb} & C_{db}
   (c) C_{sd} & C_{GD} (d) C_{db} & C_{GD}
6. The dominant high cut-off frequency of a CE amplifier ______________.
   (a) F_h introduced by C_{in}
   (b) F_h introduced by C_{out}
   (c) F_h introduced by β
   (d) is the lowest of above three cut-off frequencies
7. A CE–CB cascode amplifier ______________.
   (1) has all the characteristics of a CE amplifier
   (2) has all the characteristics of a CB amplifier
   (3) has a superior high frequency response of CB amplifier
   (4) has a lower high frequency response of CB amplifier
   The correct statements out of above are
   (a) 1 & 4 (b) 2 & 4 (c) 1, 2 & 4 (d) 1, 2 & 3
8. Source follower high frequency response ______________.
   (1) is limited by the Miller effect
   (2) is not limited by the Miller effect
   (3) better high frequency response compared to common–source amplifier
   (4) poorer frequency response compared to common source amplifier
   The correct statements out of the above are
   (a) 1, 2, 4, 3 (b) 3, 2 & 4 (c) 1, 2 & 3 (d) 1, 2 & 4
9. If the high-frequency cut-off point is not high as designed it can be increased ______________.
   (a) by choosing a transistor that has a higher frequency capability
   (b) by choosing an amplifier configuration that is not as frequency sensitive
   (c) by reducing the gain per stage to reduce the Miller effect
   (d) none of the above methods

Answers to Multiple-Choice Questions

1. (c)	2. (a)	3. (c)	4. (c)	5. (b)
6. (d)	7. (d)	8. (c)	9. (a, b & c)	

Chapter 13

TUNED AMPLIFIERS

Learning Objectives

To get familiarity of fundamental concepts, design aspects and applications of

- Various types of Tuned Amplifiers
- Stagger-Tuned Amplifiers
- Radio Frequency Amplifiers
- Wideband Amplifiers

13.1 INTRODUCTION

Tuned Amplifiers are used in Radio transmission and reception (550 kHz to 30 MHz), TV (54–88 MHz) VHF band-1, (174–216 MHz) VHF band-2, (472–890 MHz) UHF band and (88–108 MHz) in FM services. It has become so popular that small remote controlled Toys Aero Planes to Space communications to Inter Galaxies use Tuned Amplifiers.

Radio, TV, Cell phone, Broad Band Internet and Satellite Communication systems use Electromagnetic (EM) waves for transmission and reception.

In Radio communications (Fig. 13.1), EM Waves from different Radio Broadcast stations such as Visakhapatnam, Bombay, Ceylon, New Delhi, Hyderabad, Moscow, New York and so on travel through space and reach the receiving Antenna. Receiving Antenna picks up weak/strong signals before feeding into the Radio Receiver.

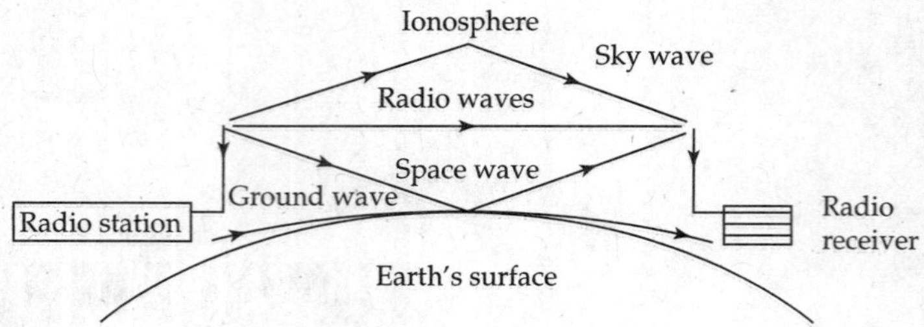

FIG. 13.1 *Block diagram of radio transmission and reception*

Received signal will be weak/strong due to the following factors:
(1) Mode of propagation of EM Waves for Medium wave or Short wave signals transmitted from various Radio Stations. (2) Distance from radio station.

Tuned Amplifiers have several functions to perform in a Radio (Transistor) Receiver:

- Selection of desired radio station signal among many broadcast station signals in the frequency band from 550 kHz to 30 MHz (The upper limit of 30 MHz for broadcast signals is fixed by the higher limit of frequency in sky wave propagation. Beyond 30 MHz radio signals will be lost into space without getting reflected by the ionosphere to the earth).
- Amplification of the received station signal containing a narrow passband of audio signals transmitted from selected radio station.

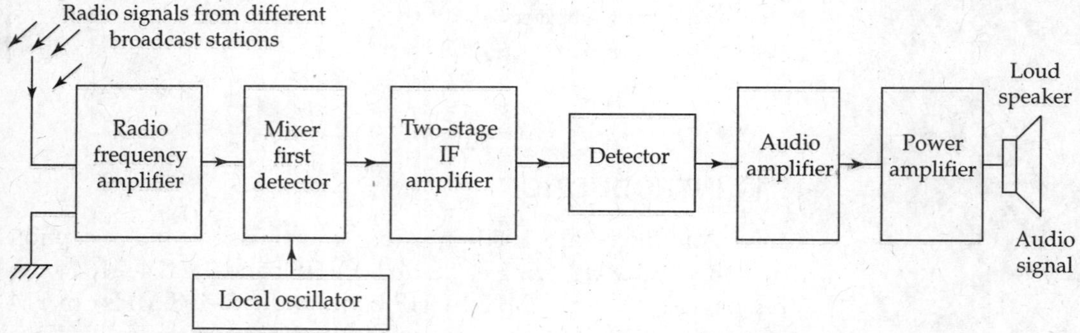

FIG. 13.2 *Radio receiver block diagram to illustrate the use of various amplifiers*

In AM radio receivers, the requirements of RF Amplifier and 10 kHz passband (containing audio signals) with uniform gain in IF (455 kHZ) Amplifiers around the selected station signals frequency are achieved by using *Tuned Amplifiers*.

13.2 BASIC CONCEPTS OF TUNED AMPLIFIERS

13.2.1 Single-tuned Transistor Amplifier

- Salient functions of Tuned Amplifier
  1. *Amplification at centre (resonant) frequency of passband of the Amplifier*
  2. *Narrow or Wideband response that depends upon the circuit application*
  3. *Selectivity and Image rejection features*

- Voltage gain of Tuned Amplifier is directly proportional to its load impedance Z_L. Resonant load L and variable C provide maximum Amplifier gain, at its resonant frequency (*centre frequency*) *of its bandwidth*. Amplifier gain $A = -g_m \cdot Z_L$. Load impedance will be large at its resonant frequency and falls off on either side of resonant frequency (Figs. 13.3 and 13.4).

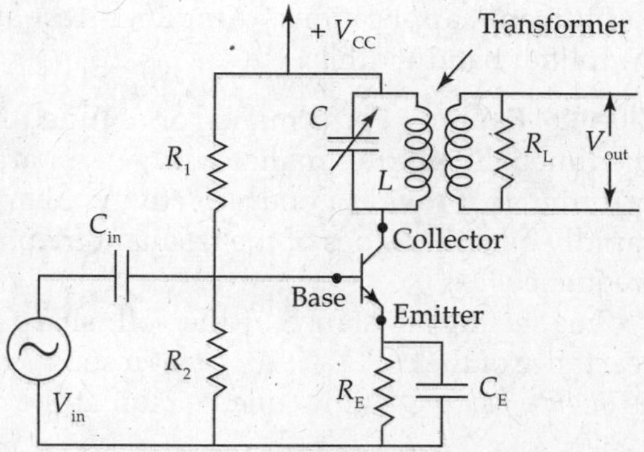

FIG. 13.3 *Tuned amplifier using BJT*

Resonant frequency of the tuned circuit

- When signals of various frequencies are present at the input of Tuned Amplifier, the Amplifier selects the desired frequency and strongly amplifies the signals at resonant frequency and rejects all other signals.
- *Frequency response* of an Amplifier depends upon the quality factor 'Q' of the tuned circuit. Q is equal to the ratio of the *resonant frequency* (f_r of the tuned circuit) to *bandwidth* (B of the Amplifier).

Frequency response curve and Bandwidth of Tuned Amplifier (Fig. 13.4)
Amplifier response (voltage gain A_V) is large, at resonant frequency f_0 and is sharply lower before and after. Reduction in gain depends upon quality factor Q of the circuit. Frequency response of Tuned Amplifier is similar to that of a *band pass filter.*

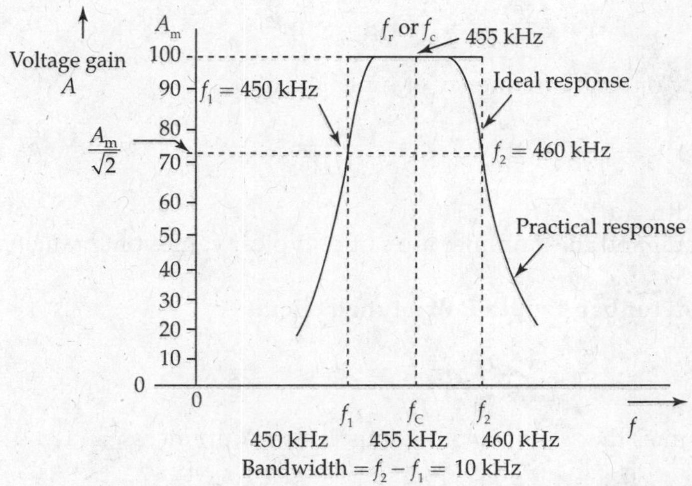

FIG. 13.4 *Frequency response characteristic of single-tuned amplifier for practical and ideal conditions*

Let A_m be the maximum value of Amplifier gain (at f_r). On the frequency response characteristic of the Amplifier, two *frequencies*, f_1 as lower cut-off and f_2 as upper cut-off, are identified where the voltage gains are 0.707 of A_m. Then *Amplifier bandwidth* is shown as $BW = (f_2 - f_1)$. Assuming $A_m = 100$, gain at f_1 and f_2 is 70.7. If $f_2 = 460$ kHz and $f_1 = 450$ kHz, the Amplifier bandwidth $BW = (f_2 - f_1) = 10$ kHz.

In many applications, Amplifier response improves depending on how narrow the Amplifier bandwidth is.

Quality Factor Performance of a Tuned Amplifier depends upon the quality factor Q of the tuned circuit. Resonance curve of a parallel tuned circuit should be as sharp as possible in order to provide good selectivity. Sharp resonance curve means that impedance falls rapidly on either sides of the resonant frequency, as the frequency is varied from the resonant frequency f_r.

Smaller the resistance of the coil, sharper is the resonance curve. The ratio of inductive reactance of the coil $X_L = 2\pi f_r L$ at resonance, to resistance R associated with the coil is known as *quality factor* Q of the tuned circuit at the desired frequency.

$$Q = \frac{X_L}{R} = \frac{2\pi \cdot f \cdot L}{R} = \frac{\omega_0 \cdot L}{R}$$

$$Q = \frac{\text{energy stored}}{\text{energy lost}}$$

$$Q = \frac{\text{reactive power}}{\text{resistive power}} = \frac{I_L^2 \cdot X_L}{I_L^2 \cdot R} = \frac{X_L}{R} = \frac{\omega_0 \cdot L}{R}. \tag{13.1}$$

Relation between quality factor Q and Bandwidth B

Quality of performance of Tuned Amplifier is known as *figure of merit*. Quality factor (Q) of the Tuned Amplifier is the ratio of (centre frequency) resonant frequency f_r of the tuned circuit to the bandwidth B of the Amplifier:

$$Q = \frac{f_r}{BW} = \frac{f_r}{B} \tag{13.2}$$

Resonant frequency or centre frequency of tuned circuit

$$f_r = \frac{1}{2\pi\sqrt{LC}} \tag{13.3}$$

Amplifier bandwidth $= BW = B$

Q of the tuned circuit and the amplifier are same. Typical values of Q range from 50–200.

General expression for bandwidth B of a tuned circuit

$$B = \frac{\omega_0}{Q} = \frac{\omega_0}{\omega_0 \cdot R \cdot C} = \frac{1}{RC} \tag{13.4}$$

Bandwidth determines the *selectivity* of various Tuned Amplifiers.

13.3 PERFORMANCE OF PARALLEL RESONANT CIRCUITS

To learn the working of Tuned Amplifiers, we need to understand the qualitative behaviour of passive L, C and R elements.

A specific circuit of L and C components, which responds to a specific frequency or a set of frequencies (within a narrow passband), is known as *parallel resonant circuit* (Fig. 13.5). This circuit is also known as a *Tuned Circuit*.

Assume the impedance of the Tuned circuit as Z_p or Z_T

$$Z_p = \frac{Z_1 \times Z_2}{Z_1 + Z_2}, \quad (13.5)$$

where $Z_1 = (R + j\omega L) = (R + jX_L)$.

For large values of Q of the coils, $Z_1 = (R + j\omega L) \cong j\omega L = jX_L$.

When R is very small

$$Z_2 = \frac{1}{j\omega C} = -j\frac{1}{\omega C} = -jX_C$$

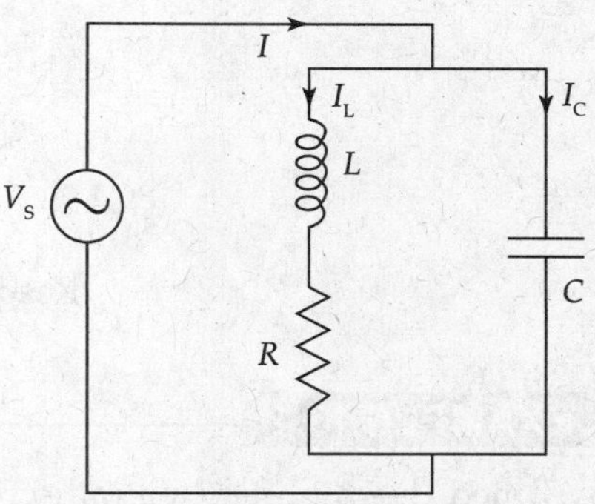

FIG. 13.5 *Parallel tuned circuit*

Further at resonance $\quad X_L = X_C \quad (13.6)$

$$\therefore \quad (Z_1 + Z_2) = (R + jX_L - jX_C) = R$$

$$\therefore \quad Z_p = Z_r = \frac{(jX_L) \times (-jX_C)}{(R + jX_L - jX_C)} = \frac{X_L \times X_C}{R} = \frac{\omega L}{\omega CR} = \frac{L}{CR}$$

$$\text{Circuit impedance at resonance} \quad Z_r = R_0 = \frac{L}{CR} \quad (13.7)$$

$$\text{Line or circuit current} \quad I = \frac{V}{Z_r} = \frac{V}{R_0}.$$

- As shown in Fig. 13.6, inductive reactance increases with increasing frequency (below the resonant frequency), whereas the capacitance reactance decreases (beyond the resonant frequency). Naturally, at a single frequency, reactances of the inductor and capacitor will equal and balance out. This frequency is known as *resonant frequency*. Hence, the load at resonance becomes resistive, known as *effective load resistance* $R_{L(eq)}$ of the tuned circuit.
- Maximum impedance occurs at resonance for the parallel resonant circuit, resulting in minimum current. The circuit impedance will be resistive at resonant frequency, resulting in the applied voltage and supply current to be in phase. This effect is called *parallel resonance*.

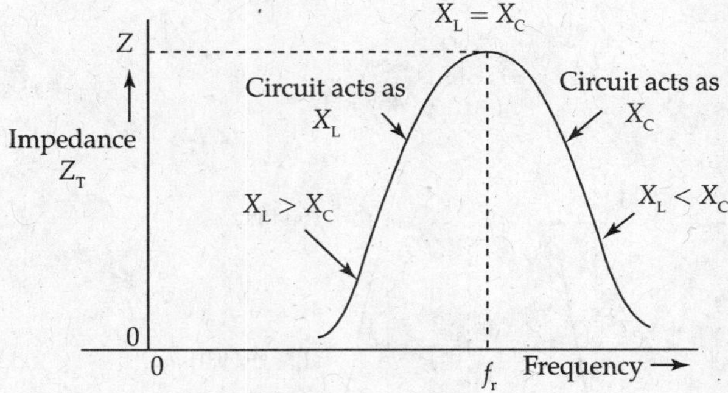

FIG. 13.6 *Frequency versus impedance curve of a parallel resonant circuit*

At resonant frequency, $f_r = f_0$, Inductive reactance X_L = Capacitive reactance X_C

$$\therefore \quad \omega_0 \cdot L = \frac{1}{\omega_0 C}$$

$$\text{Hence,} \quad \omega_0^2 = \frac{1}{LC}$$

$$\therefore \quad (2\pi f_0)^2 = \frac{1}{LC}$$

$$\therefore \text{ Resonant frequency } \quad f_r = \frac{1}{2\pi\sqrt{LC}}. \qquad (13.8)$$

EXAMPLE 13.1

A Tuned IF Amplifier in a radio receiver is designed to amplify only those frequencies that are within the passband of ±10 kHz of central frequency of 455 kHz. That is, f_r (f_c) = 455 kHz, f_1 = 450 kHz, f_2 = 460 kHz. Determine the bandwidth.

Solution: Bandwidth $B = (f_2 - f_1) = (460 - 450)$ kHz = 10 kHz.

- This means that, as long as the input signal is within the passband frequency range of 450–460 kHz, it will be amplified well. If the frequency of input signal goes out of this range, amplification will be drastically reduced or attenuated.
- The desired selectivity is evident from the passband in the response. Steepness in the response or sharp attenuation around the central or carrier frequency f_0 in the response shows *skirt selectivity* of the Amplifier.

13.3.1 Skirt Selectivity of Tuned Amplifiers

Tuned Amplifiers use LC circuits. In LC circuits, the charge flows back and forth between the inductor L and the capacitor C, several billion times similar to flywheel effect. The energy oscillates back and forth just like lashing of water between two levels in a water tank. As a result of this effect, LC circuit is also known as a *Tank Circuit*.

The response of an ideal Tuned Amplifier resembles that of a *Band Pass Filter*, with stop bands on either side as shown in Fig. 13.7. But in practical tuned circuits, it is very difficult

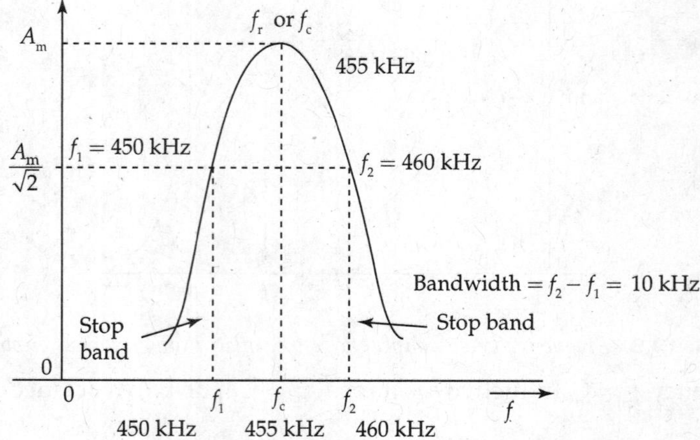

FIG. 13.7 *Frequency response of tuned amplifier (IF amplifier)*

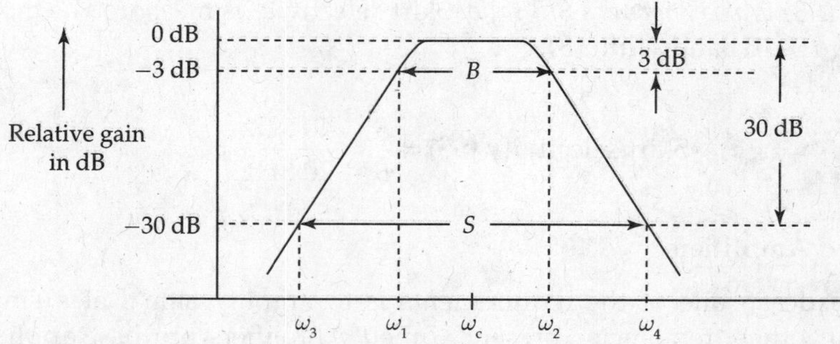

FIG. 13.8 *Frequency response of a tuned amplifier defining skirt selectivity*

to achieve such an ideal response with flat top and steep falls. The frequency response of a practical tuned circuit resembles the skirt of a lady, having a narrow passband with fall-offs on either side, centred on a carrier frequency, as shown in Fig. 13.8.

- Normal Passband B between 3-dB frequencies is $(\omega_2 - \omega_1)$. It is usually less than 5% of the central carrier frequency ω_0. (Passband of tuned Amplifiers is narrower than that of untuned Amplifiers.)
- S denotes the 30-dB bandwidth $(\omega_4 - \omega_3)$.
- *Skirt selectivity* (SS) of Tuned Amplifier is defined as the ratio of 30-dB Bandwidth (S) to 3-dB Bandwidth (B). It is also called S/B ratio, or *rejection quantity*.

$$\text{Skirt selectevity (SS)} = \frac{S}{B} = \frac{\text{30-dB bandwidth}}{\text{3-dB bandwidth}}.$$

Skirt selectivity has no units. For an ideal Tuned Amplifier, Skirt Selectivity is 1. But achieving such selectivity is not possible. Skirt Selectivity of 3 or less is preferable in most of the communication applications.

- In Fig. 13.9, Skirt Selectivity factor 'SS' is shown again with maximum Amplifier gain A_V at 100. Maximum gain on the top in dB $= 20 \log_{10} A_V = 20 \log_{10} 100 = 40$ dB.
- At 3 dB less from the top (maximum gain) in the graph, normal gain is 37 dB. Assuming gain of 3 dB, Bandwidth $B = (\omega_2 - \omega_1) = 20$ kHz.
- At 30 dB less from the top (maximum gain) in the graph, normal gain is 10 dB. Assuming gain of 30 dB, Bandwidth $S = (\omega_4 - \omega_3) = 40$ kHz.
- From the Bandwidth values (S and B), skirt selectivity can be calculated.

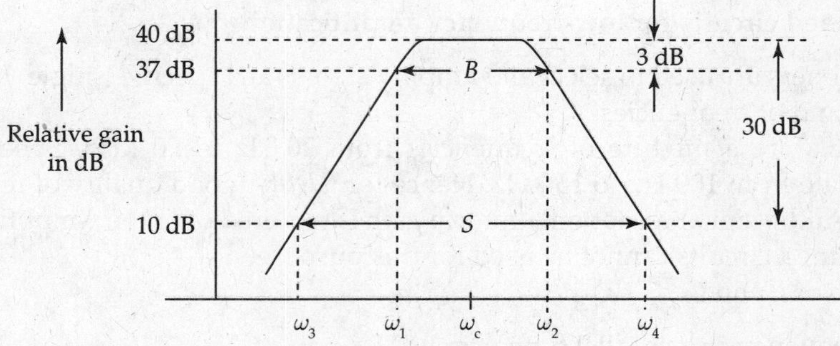

FIG. 13.9 *Frequency response of a tuned amplifier represented in dB*

Skirt Selectivity (SS) is also known as S/B ratio. Skirt selectivity is measured as the ratio of 30-dB bandwidth (S) to 3-dB bandwidth (B).

From Fig. (13.11),

$$\text{Skirt selectivity (SS)} = \frac{S}{B} = \frac{40\text{ kHz}}{20\text{ kHz}} = 2.$$

Merits of Tuned Amplifiers

1. In radio broadcast, one of the requirements is to amplify signal at a single frequency, and reject all other frequencies present. Tuned Amplifiers are best applicable for such applications.
2. **Good selectivity** A tuned circuit has the property of *selectivity*, i.e., it can select signal at a desired frequency for amplification, out of a large number of frequencies available. The circuit selects signal at resonant frequency and attenuates all other signal components. The circuit has high selectivity, if circuit quality Q is high.
3. Tuned circuits in Transistor Amplifiers help selection and efficient amplification of signal at a particular radio frequency. Such an Amplifier is a 'Tuned Amplifier' or 'Radio frequency Amplifier'.
4. Tuned circuit takes energy from the power Source. It has the property of storing energy in an oscillating state between the kinetic energy cycles – between magnetic field associated with current flow through the inductor L and potential energy of the electric field when the capacitor is charged – and back and forth. Such cycle of events occur at a particular frequency – *resonant frequency*. Therefore, output response of a tuned circuit is also an AC wave.
5. Even if the input exciting signal to a parallel resonant circuit is in *pulse form*, output is a continuous *sine wave.* So, tuned circuits have good use in Class-C and Class-D operations of Tuned Power Amplifiers, in order to improve efficiency of power conversion and reduce distortion.
6. **Reduced power loss/dissipation** Tuned parallel circuit consists of inductor L and capacitor C. Consequently, power loss in such a load in Amplifiers is quite low. Therefore, Tuned Amplifiers are highly efficient in power consumption with least amount of power dissipation.
7. Tuned Amplifiers are highly effective as narrow band Amplifiers with *bandwidth decided by the time constant* of the Tuned load circuit.

Demerits of tuned circuits for low-frequency amplification

1. Tuned Amplifiers are used to select and amplify signals at a specific (single) high frequency or narrow band of frequencies.
2. Audio Signals are a mixture of frequencies from 20 Hz to 20 kHz. Speech and music signals operate from 100 Hz to 15 kHz. Hence for *fidelity* (good quality of reproducing the original) of Audio signal reception, uniform gain is required over the Amplifier bandwidth. Therefore, tuned circuits cannot be used for this purpose.
3. Large values of L and C:

   Resonant frequency of a parallel tuned circuit $f_r = \dfrac{1}{2\pi\sqrt{LC}}$.

For low-frequency signal amplification, values of L and C are large, meaning the inductive and capacitance elements will be bulky and expensive. So, RC and Transformer-coupled Amplifiers are used for low-frequency applications.

13.3.2 Classification of Amplifiers

1. Small-signal Tuned Amplifier
2. Large-signal Tuned Amplifier

Depending upon the *type of coupling* used in the cascaded Tuned Amplifiers, they can also be classified as follows:

1. Capacitance-coupled and
2. Inductively coupled (transformer).

However, the normal classification is based on the type of *tuned stages*:

1. **Single-tuned Amplifier** The Amplifier is tuned to a certain desired frequency either at the input or at the output side of the Amplifier.
2. **Double-tuned Amplifier** Amplifiers use two tuned circuits to obtain sharp response.
3. **Synchronously tuned Amplifier** All of the tuned circuits in the Tuned Amplifier are tuned to the same frequency
4. **Stagger-tuned Amplifier** When primary and secondary tuned circuits of an RF transformer are tuned to slightly different frequencies, one staggered from the other, the Tuned Amplifier is known as *Stagger-Tuned* Amplifier. Stagger tuning allows more narrow band signals and thus *increases the passband* of the Amplifier.

13.4 SINGLE-TUNED CAPACITANCE-COUPLED (DIRECT-COUPLED) AMPLIFIER

Direct- or capacitance-coupled Tuned Amplifier (Fig. 13.10) consists of single-tuned circuit in the output (Collector) circuit. Output is taken through a coupling-capacitor to load resistance R_L or to the input port of a latter amplifying stage.

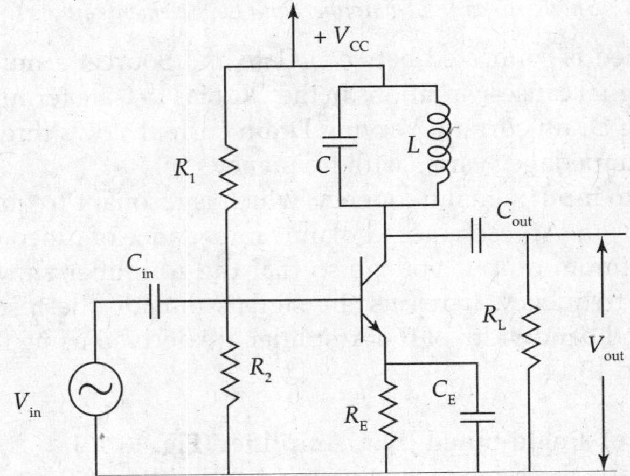

FIG. 13.10 *Single-tuned capacitance-coupled tuned amplifier*

Typical frequency response of the Amplifier is shown in Fig. 13.7. Amplifier gain is large, when the input signal matches the resonant frequency, because the impedance of the tuned circuit is high at the resonant frequency. At frequencies other than the resonant frequency, (i) reactances of the inductor coil and the capacitor no longer balance, (ii) the impedances are less than those at the resonant frequency f_r, and (iii) hence the voltage gains are reduced to low values. The Amplifier response has *Band Pass Filter* characteristic with *narrow passband*. Thus, inductor L and capacitor C of the tuned circuit decide the frequency response of the Amplifier.

13.4.1 Analysis of Single-stage Capacitance-coupled Tuned JFET Amplifier

Single-Tuned Amplifier works as the simplest band pass Amplifier. A single-tuned Amplifier circuit may utilise FET, BJT or operational Amplifier. JFET single-tuned Amplifier is shown in Fig. 13.11, with capacitance-coupling circuit. One of the functions of resistor R_D is to stabilise Amplifier gain.

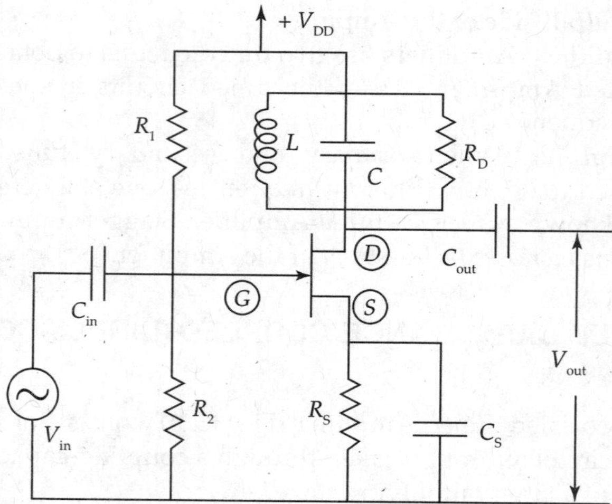

FIG. 13.11 *Single-tuned JFET amplifier with LCR elements in the tuned circuit*

Signal to be amplified is connected between Gate and Source terminals of JFET device of the Amplifier. Input signal causes variations in the DC bias to Gate terminal, and varying Gate voltage causes varying Drain current. Varying Drain current flows through the tuned circuit (Drain circuit) whose impedance varies with frequency.

LC circuit is tuned to input signal frequency, which is resonant frequency of tuned circuit, based on Amplifier design. At resonance, dynamic impedance of tuned circuit is large. Signal current develops maximum output voltage so that the Amplifier shows maximum gain at the selected resonant frequency. It rejects the signals outside the resonant frequency. The expressions for gain and bandwidth of the Amplifier are derived using the equivalent circuits in Figs. 13.12, 13.13 and 13.14.

AC equivalent circuit of single-tuned JFET Amplifier (Fig. 13.12)

To obtain mathematical expressions for gain and bandwidth of the Amplifier, JFET is also replaced with its equivalent circuit shown in Fig. 13.13.

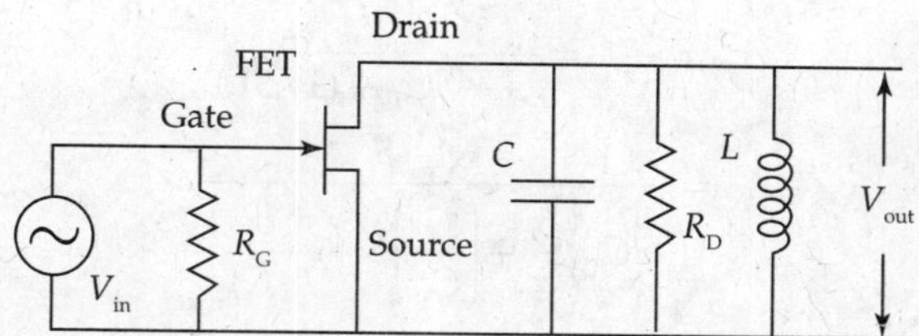

FIG. 13.12 *AC equivalent circuit of FET tuned amplifier*

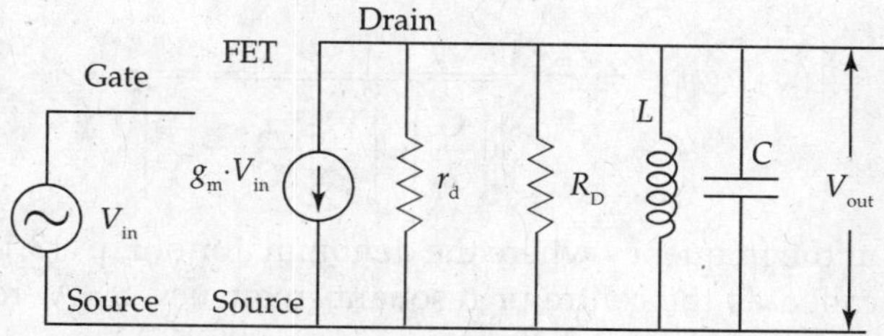

FIG. 13.13 *Equivalent circuit of FET tuned amplifier*

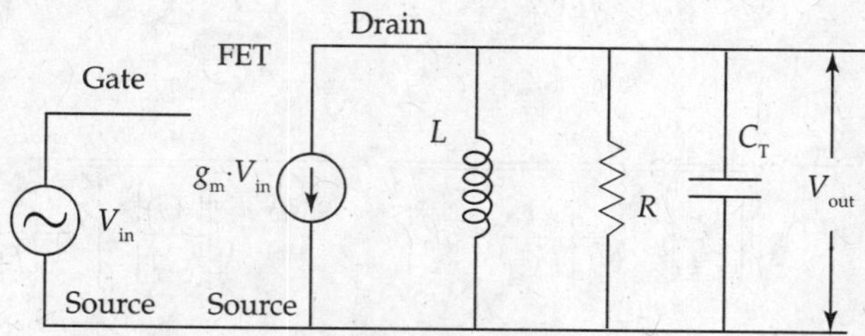

FIG. 13.14 *Simplified equivalent circuit of FET tuned amplifier*

Simplified equivalent circuit of JFET Single-tuned Amplifier (Fig. 13.14)
Resistance in the circuit

$$R = R_D \,||\, r_d. \tag{13.9}$$

Total capacitance in the output circuit

$$C_T = C + C_{DS}, \tag{13.10}$$

where C is the tuning capacitance, and C_{DS} is the Capacitance between Drain and Source of FET device, usually small and ignored.

Single-tuned FET Amplifier voltage gain is derived by using its simplified equivalent circuit of Fig. 13.14.

Derivation for voltage gain A_V of capacitive-coupled tuned amplifier

$$\text{Output voltage} \quad V_{out} = -\frac{g_m . V_{in}}{Y} = -\frac{g_m \cdot V_{in}}{G + SC_T + (1/SL)} \tag{13.11}$$

$$A_V = \frac{V_{out}}{V_{in}} = \frac{-g_m}{G + SC_T + (1/SL)} \tag{13.12}$$

$$A_V = -\frac{g_m \cdot S}{C_T\left[S^2 + \left(\frac{G}{C_T}\right)S + \frac{1}{L \cdot C_T}\right]} \tag{13.13}$$

Equation (5.13) represents second-order band pass function. Equation (5.13) is known as resonator expression.

The magnitude of the gain function $|A_V (j \cdot \omega)|$ is

$$|A_V (j \cdot \omega)| = -\frac{g_m}{C_T} \cdot \frac{\omega}{\sqrt{\left[\left(\frac{G}{C_T}\right)\omega\right]^2 + \left[\frac{1}{LC_T} - \omega^2\right]^2}}. \tag{13.14}$$

·Magnitude is high at the frequency where the denominator of Eq. (13.14) is minimum. This frequency is usually called as the centre or resonant frequency ω_0. At resonant frequency f_r centre frequency gain

$$\omega_0 = \frac{1}{\sqrt{LC_T}} \quad \text{or} \quad (\omega_0)^2 = \frac{1}{LC_T} \tag{13.15}$$

At $\omega = \omega_0$,

$$|A_V \cdot (j \cdot \omega)| = -\frac{g_m}{C_T} \cdot \frac{\omega}{\sqrt{\left[\left(\frac{G}{C_T}\right)\omega\right]^2 + \left[\frac{1}{LC_T} - \omega_0^2\right]^2}} = -\frac{g_m}{C_T} \cdot \frac{\omega}{\sqrt{\left[\left(\frac{G}{C_T}\right)\omega\right]^2 + \left[\frac{1}{LC_T} - \frac{1}{LC_T}\right]}}. \tag{13.16}$$

Therefore, on simplification of Eq. (13.16), we obtain

$$\text{Gain} \quad |A_V (j \cdot \omega)| = -\frac{g_m}{G} = -g_m \cdot R \quad \text{at } \omega = \omega_0. \tag{13.17}$$

Amplifier gain at centre frequency

$$|A_V (j \cdot \omega)| = -g_m \cdot R = -g_m \cdot (R_D \parallel r_d) \tag{13.18}$$

13.4.2 Gain at 'Half-Power Frequencies' and the Bandwidth

Frequencies at which function gain is down by 3 dB, from the centre frequency, are called as *half-power frequencies*. They are determined from the above equations of the previous section.

For the Amplifier circuit, assuming input resistance R_{in} and output resistance R_{out} to be equal, equations at half-power frequencies are

$$\left(\frac{-g_m}{G}\right)^2 \cdot \frac{1}{2} = \left(\frac{-g_m}{C_T}\right)^2 \frac{\omega^2}{\left[\left(\frac{G}{C_T}\right)\omega\right]^2 + (\omega_0^2 - \omega^2)^2}. \tag{13.19}$$

Rearranging the terms in Eq. (13.19):

$$\left[\frac{G}{C_T}\cdot\omega\right]^2+(\omega_0^2-\omega^2)^2=2\cdot\frac{G^2}{C_T^2}\cdot\omega^2 \tag{13.20}$$

$$\therefore\quad (\omega_0^2-\omega^2)^2=\left(\frac{G^2}{C_T^2}\right)\omega^2 \tag{13.21}$$

$$\Rightarrow\quad (\omega_0^2-\omega^2)=\frac{G}{C_T}\cdot\omega \tag{13.22}$$

Rearranging the terms, the following quadratic equation in the variable ω is obtained:

$$\omega^2+\frac{G}{C_T}\cdot\omega-\omega_0^2=0. \tag{13.23}$$

Two solutions of the above quadratic equation are given below:

$$\omega_1=-\frac{G}{2C_T}+\sqrt{\omega_0^2+\frac{G^2}{4{C_T}^2}} \tag{13.24}$$

$$\omega_2=-\frac{G}{2C_T}-\sqrt{\omega_0^2+\frac{G^2}{4{C_T}^2}} \tag{13.25}$$

Since frequency ω_2 cannot be negative for real systems

$$\omega_2=\frac{G}{2C_T}+\sqrt{\omega_0^2+\frac{G^2}{4{C_T}^2}}. \tag{13.26}$$

$$\therefore\quad BW=(\omega_2-\omega_1)=2\pi(f_2-f_1)=\frac{G}{C_T}=\frac{1}{RC_T} \tag{13.27}$$

$$\text{and}\quad \omega_0=\sqrt{\omega_1\cdot\omega_2}, \tag{13.28}$$

where ω_0 is the geometric mean of upper and lower 3-dB frequencies Quality factor.

$$Q=\frac{\omega_0}{\text{Bandwidth}}=\frac{\omega_0\times C_T}{G}$$

$$\text{Gain}\times\text{bandwidth product}=A_V\cdot(BW)=-\frac{g_m}{G}\cdot\frac{G}{C_T}=-\frac{g_m}{C_T}. \tag{13.29}$$

Thus, it is known that the product of Amplifier gain and bandwidth is constant.

$A_V(j\cdot\omega)$ can also be written as

$$A_V(j\cdot\omega)=\frac{-g_m\cdot R}{1+j\cdot\omega\cdot RC_T-j\left(\frac{R}{\omega L}\right)}=-\frac{g_m\cdot R}{\left[1+j\cdot\omega_0\cdot RC_T\cdot\left(\frac{\omega}{\omega_0}-\frac{\omega_0}{\omega}\right)\right]}. \tag{13.30}$$

Using the equations $Q = \omega_0 \cdot R \cdot C_T = \dfrac{R}{L \cdot \omega_0}$ and $|A_V(j \cdot \omega)| = -\dfrac{g_m}{G}$ into Eq. (13.30), we obtain

$$\frac{A_V(j \cdot \omega)}{A_V(j \cdot \omega_0)} = \frac{1}{1 + jQ\left(\dfrac{\omega}{\omega_0} - \dfrac{\omega_0}{\omega}\right)}. \tag{13.31}$$

Plot of relative magnitude, for a value of $Q = 10$, is shown in Fig. 13.15.

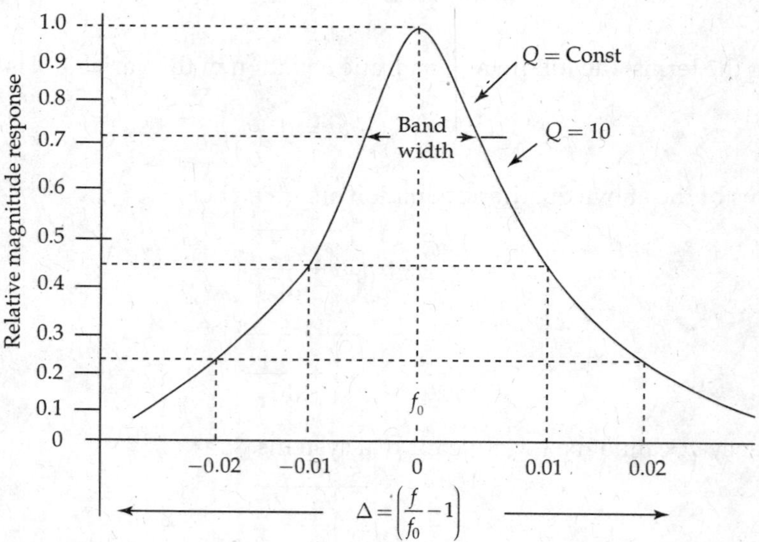

FIG. 13 15 *Typical response curve of single-tuned amplifer*

We may express the response in the passband of the narrow band Amplifier by putting

$$\frac{\omega}{\omega_0} = 1 + \Delta, \quad \text{where } \Delta = \left[\frac{f}{f_0} - 1\right] = \left[\frac{f - f_0}{f_0}\right]. \tag{13.32}$$

Δ is the fractional deviation of the actual frequency f from the resonant frequency f_0 which is evident from the above definition.

When $\Delta \langle\langle 1$, Eq. (13.32) reduces to

$$\frac{A_V(j \cdot \omega)}{A_V(j \cdot \omega_0)} = \frac{1}{[1 + j \cdot 2 \cdot Q \cdot \Delta]}, \tag{13.33}$$

where $A_V(j \cdot \omega_0)$ is the amplifier gain at resonance.

Typical response of single-tuned amplifier is shown in Fig. 13.14 using Eq. (13.33). On similar lines universal resonance curves can also be obtained.

Input impedance of FET Amplifier includes *negative resistance* for frequencies below resonance. Negative input resistance is due to *parasitic capacitance* C_{GD}. It acts as a *feedback loop* from output to input of the Amplifier, in turn causing the Amplifier to be *bilateral*. Feedback destabilises the Amplifier and forces undesirable oscillations, and in order to cancel the feedback effect, a *neutralisation circuit* can be added. A second coil is tightly coupled to the inductor, with unity turns ratio, and voltage induced will be opposite in phase to that of AC component. This voltage, by the additional inductor, is fed back through a *neutralisation capacitor* $C_N = C_{GD}$. Alternative approach is to employ CG or CB Amplifier or to use CASCODE and differential Amplifiers.

EXAMPLE 13.2

Design a single-stage capacitance-coupled FET Tuned Amplifier with the following specifications: $\omega_0 = 2\pi\ (10^6\ \text{Hz})$; $\omega_{3\text{-dB}} = 2\pi\ (10^4\ \text{Hz})$ and a mid-band voltage gain $A_V\ (j \cdot \omega_o) = -8$. The FET parameters are as follows: $g_m = 4 \times 10^{-3}$ mhos; Drain resistance $r_d = 18\ \text{k}\Omega$; $C_{GS} = 40$ pF, $C_{GD} = C_{DS} = 5$ pF, Tuning capacitance $C = 15$ nF.

Solution: From gain requirements $A_V\ (j \cdot \omega_0) = -g_m\ (R_D \,||\, r_d) = -8$.
Transconductance $g_m = 4 \times 10^{-3}$ mhos. Drain resistance $r_d = 18\ \text{k}\Omega$.

$$A_V = g_m \cdot R_D \,||\, r_d$$

$$\therefore\quad R_D \,||\, r_d = \frac{A_V}{g_m} = \frac{8}{4\times 10^{-3}} = 2\times 10^3\ \Omega.$$

From above equation, $\dfrac{R_D \times 18\times 10^3}{R_D + 18\times 10^3} = 2\times 10^3$

$$\therefore\quad \text{Load resistance in Drain circuit}\quad R_D = 2.25\ \text{k}\Omega.$$

From the Bandwidth specification,

$$\omega_{3\text{-dB}} = \frac{G}{C_T} = \frac{G_D + g_d}{C + C_{DS} + C_{GD}} = 2\pi(10^4)\ \text{radians}.$$

Substituting the values of C_T, G_D, g_d, C_{DS}, C_{GD} and $\omega_{3\text{-dB}}$
$C_T = 15$ nF.

$$\omega_0 = \frac{1}{\sqrt{LC_T}}$$

$$\therefore\quad L = \frac{1}{C_T \cdot \omega_0^2}$$

$$C_T = C + C_{GD} + C_{DS}$$

$$C_T = 15\times 10^{-9} + 5\times 10^{-12} + 5\times 10^{-12} \cong 15\times 10^{-9}\ \text{Farads}$$

$$L = \frac{1}{15\times 10^{-9} \times \left[2\pi \times 10^6\right]^2} = 10.6\ \mu\text{H}.$$

13.5 SINGLE-TUNED CAPACITANCE-COUPLED CE TRANSISTOR AMPLIFIER

Single-tuned capacitance-coupled CE Transistor Amplifier circuit is shown in Fig. 13.16. Signal to be amplified is connected at the input terminals of the transistor. The tank circuit is tuned to the input signal frequency as per design of the Amplifier.

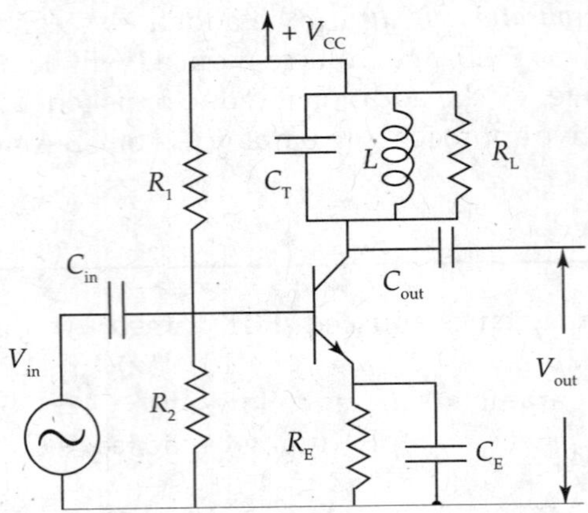

FIG. 13.16 *Single-tuned capacitance-coupled transistor amplifier*

At resonant frequency, the tuned circuit has large impedance and so the signal current develops large amplitude, with the signal amplitudes limited to a maximum of supply voltage V_{CC}. Tuning capacitance is C_T. Using the high frequency equivalent circuit of Transistor hybrid-π model (Fig. 13.17) and expressions for voltage gain A_V, bandwidths of the Amplifier are derived as follows.

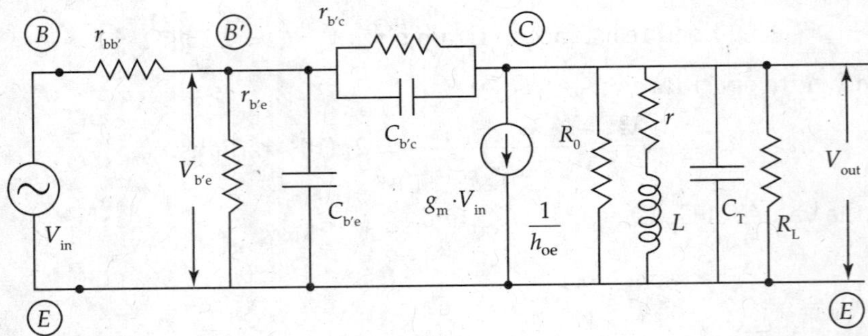

FIG. 13.17 *Hybrid-π equivalent circuit and tuned circuit of single-tuned capacitive-coupled CE transistor amplifier*

Circuit (Fig. 13.17) is further simplified as follows.

Using Miller's theorem, capacitor $C_{b'c}$ and resistor $r_{b'c}$, connecting the input and output ports of the Transistor, are replaced with equivalent components.

1. $C_{in}(M) = C_{b'c}(1 + A)$ – at input port, between B' (Base) and E (Emitter) terminals,
2. $C_{out}(M) = C_{b'c}\left[1 - \frac{1}{A}\right]$ – at output port between Collector and Emitter terminals.

Resistor $r_{b'c}$ is about 4 MΩ and it is very large compared to other resistances in the circuit. It has negligible effect on circuit performance. The equivalent circuit is shown in Fig. 13.18.

Base spread resistance $r_{bb'}$ is about 100 Ω and hence it is also ignored. R_P is the dynamic resistance at resonance for parallel-tuned circuit. The equivalent circuit in Fig. 13.18 is *further simplified* as Fig. 13.19 with the following assumptions:

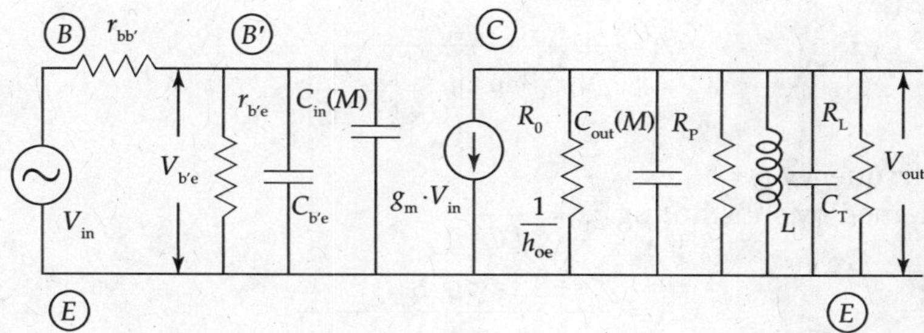

FIG. 13.18 *Hybrid-π equivalent circuit and tuned circuit of single-tuned capacitive-coupled CE transistor amplifier transfering the effects of $C_{b'c}$ and $r_{b'c}$ into input and output ports*

$$C_{in} = C_{b'e} + C_{in}(M) + C_{STRAY} \tag{13.34}$$

$$C_{in} = C_{b'e} + C_{b'c}(1+A) + C_{STRAY} \tag{13.35}$$

$$C_{eff}(\text{out}) = C = C_{TUNIING} + C_{b'c}\left[\frac{(1+A)}{A}\right] \text{ gets simplified to}$$

$$C = C_{TUNING} + C_{out}(M) \tag{13.36}$$

$$\text{Loaded resistance } R \text{ of the tuned circuit} = R = R_0 \| R_P \| R_L. \tag{13.37}$$

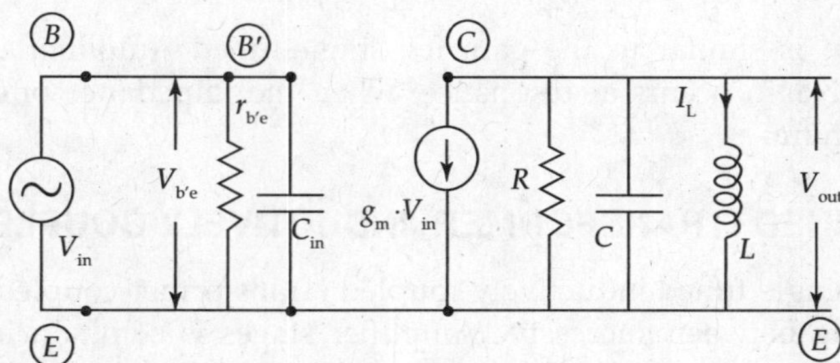

FIG. 13.19 *Simplified equivalent circuit and tuned circuit single-tuned capacitive-coupled CE transistor amplifier*

From the final simplified circuit (Fig. 13.19), Output voltage V_{out} is due to the flow of output Collector current '$g_m \cdot V_{in}$' through the output admittance Y_L, which is the parallel combination of R, L and C elements.

$$V_{out} = \frac{-g_m \cdot V_{in}}{\left(\frac{1}{R} + SC + \frac{1}{SL}\right)} \tag{13.38}$$

$$\text{Voltage gain } A = \frac{V_{out}}{V_{in}} = \frac{-g_m \cdot SLR}{[S^2 \cdot LCR + SL + R]} \tag{13.39}$$

$$A = \frac{-g_m \cdot S}{C\left[S^2 + S\left(\frac{1}{CR}\right) + \frac{1}{LC}\right]} \tag{13.40}$$

Substituting $S^2 = (j \cdot \omega)^2 = -\omega^2 = -\omega_0^2$ at f_0

$$A = \frac{-g_m \cdot S}{C\left[-\omega_0^2 + S\left(\frac{1}{CR}\right) + \frac{1}{LC}\right]} \tag{13.41}$$

at the resonant frequency f_0.

Gain A is obtained by substituting $\omega_0^2 = \frac{1}{LC}$

$$A = \frac{-g_m \cdot S}{C\left[-\frac{1}{LC} + S\left(\frac{1}{CR}\right) + \frac{1}{LC}\right]} \tag{13.42}$$

$$\therefore \quad A = \frac{-g_m \cdot S}{S/R} = -g_m \cdot R. \tag{13.43}$$

Later, we will see the advantages of Tuned Amplifiers in tuning out the parasitic capacitances in the circuit.

Another way of obtaining the expression for gain from the equivalent circuit (Fig. 13.19)

$$\text{Output current} \quad I_L = -g_m \cdot V_{b'e} = -g_m \cdot V_{in} \tag{13.44}$$

$$\text{Output voltage} \quad I_L \cdot Z = -g_m \cdot V_{in} \cdot Z, \text{ where } Z = R_0 \parallel R_P \parallel R_L = R. \tag{13.45}$$

$$\text{Voltage gain} \quad A_V = -g_m \cdot R. \tag{13.46}$$

This gain expression is similar to the gain for single-tuned Amplifier using FET device. Maximum Amplifier gain occurs at resonance, when the impedance offered by the tuned circuit is at its maximum.

13.6 SINGLE-TUNED (TRANSFORMER) INDUCTIVELY COUPLED AMPLIFIER

Figure 13.20 shows single-tuned inductively coupled (Transformer-coupled Amplifier).

Inductive coupling between successive Amplifier stages is common in radio frequency Amplifiers such as IF Amplifiers. *Coefficient coupling K* depends upon the mutual inductance between the coils. *Amplifier gain* and *bandwidth* are decided by Q of the tuned circuit. With the increase in gain, bandwidth of the Amplifier decreases. Inductive coupling in RF Amplifiers is used to *achieve maximum power transfer* from the Amplifier output to the load (such as a speaker) or the input of subsequent stage Amplifier (in the case of multistage Amplifiers).

Input signal V_{in} is applied at the input port between Base and Emitter terminals of the transistor. The amplified signal of the selected frequency appears as the output voltage V_{out}. Maximum output signal is obtained at resonant frequency, where the tuned circuit offers maximum impedance. Equivalent circuit is shown in Fig. 13.21.

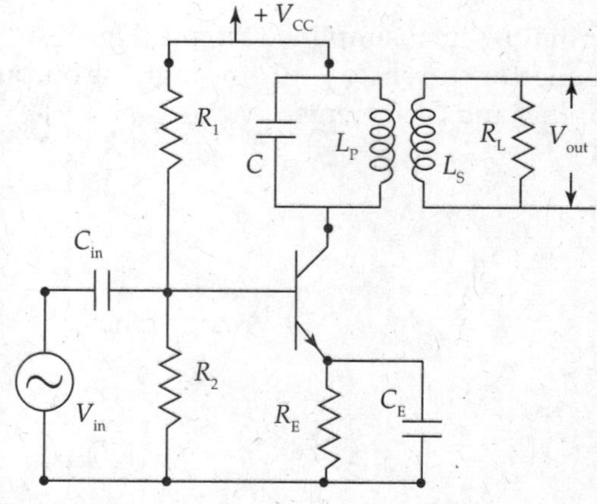

FIG. 13.20 *Single-tuned inductively coupled CE transistor amplifier*

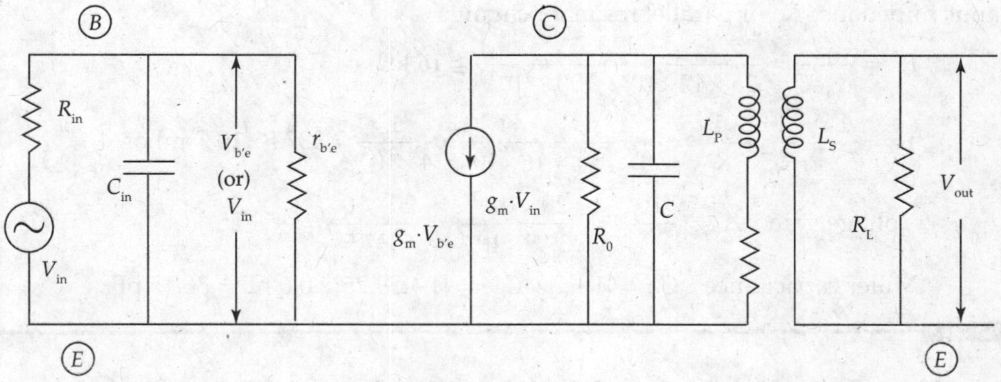

FIG. 13.21 *h-parameter equivalent circuit for single-tuned inductance coupled amplifier*

$$\text{Bandwidth } (B) = \frac{\omega_0}{Q} = \frac{\omega_0}{\omega_0 RC} = \frac{1}{RC}.$$

EXAMPLE 13.3

A common Emitter transistor Amplifier has a tuned circuit in the Collector, which resonates at 12 MHz (25 m band), with total tuning capacitance of 100 pF (Fig. 13.22). The Q-factor of the tuned circuit is 120. Output resistance R_0 of the transistor is 40 kΩ. Load resistance R_L is 4 kΩ. Transistor is biased at the Collector current of 500 μA. Reverse-biased output junction capacitance between Collector and Base is C_{CB} = 0.6 pF. Calculate the voltage gain of the Amplifier and Miller capacitance at its input terminals.

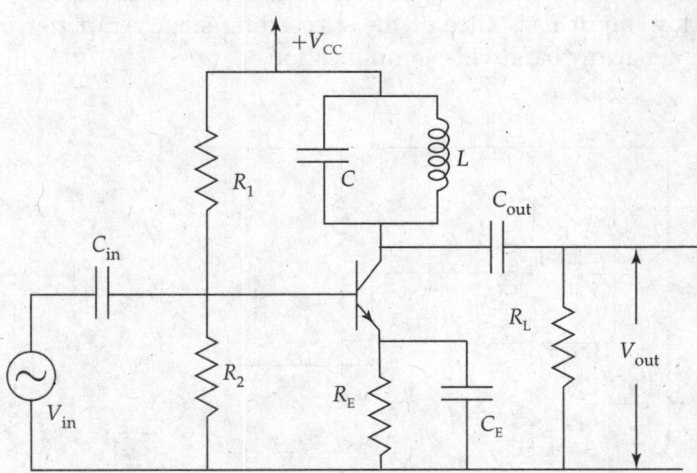

FIG. 13.22 *Single-tuned common emitter transistor amplifier*

Solution:

$$g_m = \frac{I_C}{V_T} = \frac{500 \times 10^{-6}}{25 \times 10^{-3}} = 20 \text{ milli mhos}$$

Dynamic impedance R_D of parallel resonant circuit

$$R_D = \frac{Q}{\omega_0 \cdot C} = \frac{120}{2\pi \times 12 \times 10^6 \times 100 \times 10^{-12}} \cong 16\ \text{k}\Omega$$

$$Y_0 = \frac{1}{R_0} + \frac{1}{R_D} + \frac{1}{R_L} = \frac{1}{40 \times 10^3} + \frac{1}{16 \times 10^3} \times \frac{1}{4 \times 10^3} = 337.5 \times 10^{-6}\ \text{mhos}$$

$$\text{Voltage gain} \quad A_V = -\frac{g_m}{Y_0} = \frac{20 \times 10^{-3}}{337.5 \times 10^{-6}} = \frac{20 \times 10^3}{337.5} = -59.25$$

$$\text{Miller capacitance} \quad C_M = [1 + A_V]C_{b'c} = [1 + 59.25] \times 0.6\ \text{pF} = 36.15\ \text{pF}.$$

13.6.1 Impedance Matching or Adjustment for Optimum Power Transfer

In cascaded capacitance-coupled Amplifiers, output of the first Amplifier is loaded by the low input resistance of the succeeding Amplifier. So, the gain of the first Amplifier will be reduced, making the bandwidth become large. They face difficulties in meeting the requirement of narrow bandwidth and large gain for radio frequency Amplifiers.

To achieve narrow bandwidth and optimum signal transfer between output stage and load resistance R_L or to input port of the following Amplifier stage, *two methods of coupling* are used: (1) Inductive tap coupling or autotransformer and (2) Capacitance tap coupling.

13.7 TAPPED SINGLE-TUNED CAPACITANCE-COUPLED AMPLIFIER (INDUCTIVE TAP BETWEEN AMPLIFIERS FOR OPTIMUM POWER TRANSFER)

The circuit in Fig. 13.23 is a two-stage Amplifier, which uses *inductive tap method* for transforming the low input resistance of the succeeding stage Amplifier into a reasonably high value, allowing narrow bandwidth amplification.

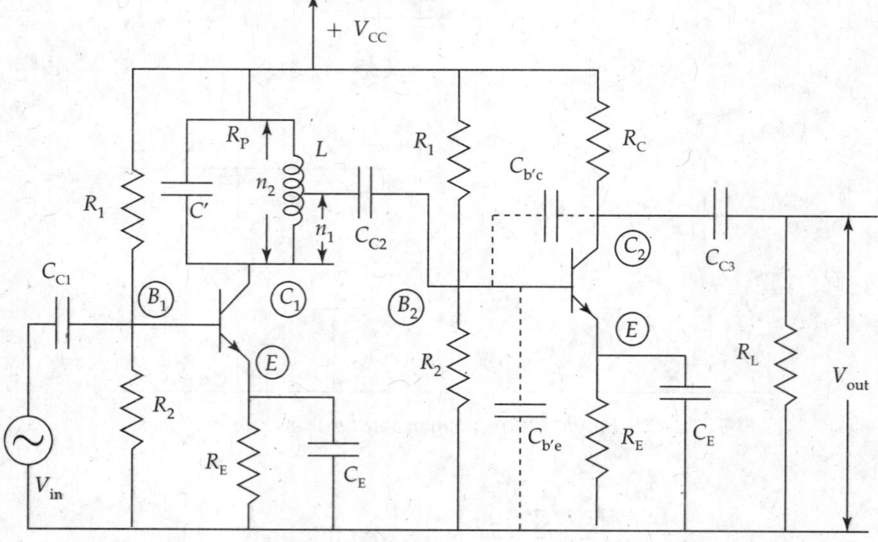

FIG. 13.23 *Inductance tap for optimum power transfer between amplifiers*

Inductor L (total number of turns n_2) with inductive tap can be considered as an autotransformer with inductance L_1 (number of turns n_1) and L_2 at the tap so that the turns ratio 'a' is the ratio of n_1 to n_2. Resistor R_P is in parallel with inductor L. Low input resistance of succeeding Amplifier has large reflected impedance into output circuit of the first Amplifier stage to satisfy the maximum power transfer condition. Then optimum power transfer takes place from the first Amplifier output to the next stage input port. At the same time narrow bandwidth is achieved. This type of inductance coupling uses large value of inductance with small capacitance.

High-frequency equivalent circuit for two-stage Amplifier illustrating inductive taps

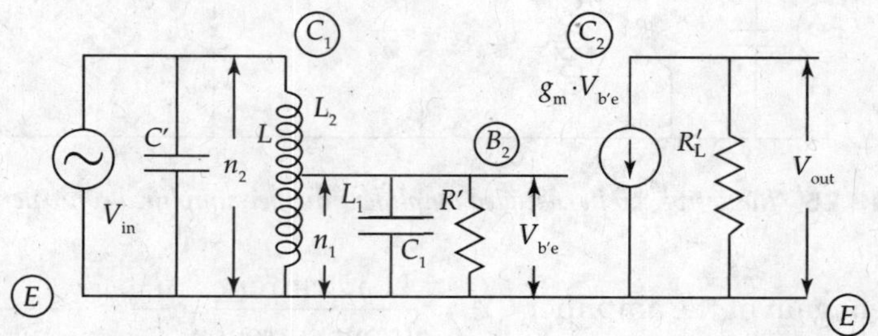

FIG. 13.24 *Inductive tap in cascaded amplifiers equivalent circuit*

From the circuit in Fig. 13.23, the Amplifier circuit has an autotransformer, as a tapped inductor with primary to secondary turns ratio 'a', where

$$a = \frac{n_1}{n_2} = \frac{V_{b'e}}{V_{in}} < 1 \tag{13.47}$$

$$R' = R_1 \parallel R_2 \parallel r_{b'e} \tag{13.48}$$

$$C_1 = C_{b'e} + C_{b'c}(1 + g_m \cdot R_L'), \tag{13.49}$$

$$\text{where } R_L' = R_C \parallel R_L = R_L \quad \text{if } R_C \gg R_L. \tag{13.50}$$

Using the properties of impedance transformation due to inductive tap and the following approximations, the equivalent circuit in Fig. 13.27 can be obtained as follows:

$$C_2 = C' + a^2 \cdot C_1 = C' + a^2 \, [C_{b'e} + C_{b'c}(1 + g_m \cdot R_L)]. \tag{13.51}$$

L inductance of autotransformer

$$R'' = \frac{R'}{a^2} = \frac{R_1 \parallel R_2 \parallel r_{b'e}}{a^2}. \tag{13.52}$$

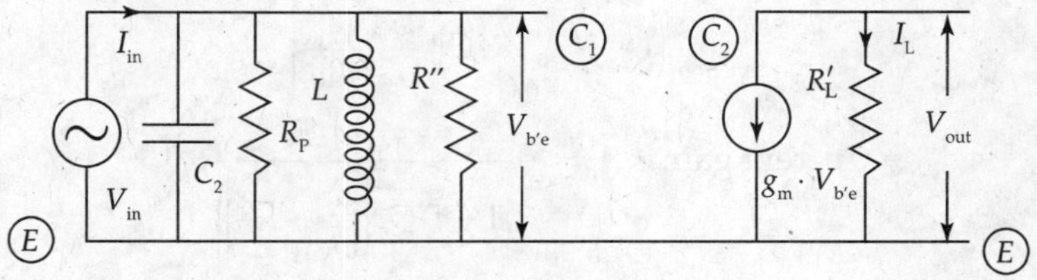

FIG. 13.25 *Inductive tap in cascaded amplifiers simplified equivalent circuit*

To obtain the Amplifier equivalent circuit in the familiar form derived earlier and the expression for Amplifier gain, make the following approximation.

$$R = R_P \parallel R' \tag{13.53}$$

$$V'_{b'e} = \frac{V_{b'e}}{a} = V_{in} \tag{13.54}$$

$$\text{and} \quad C_2 = C. \tag{13.55}$$

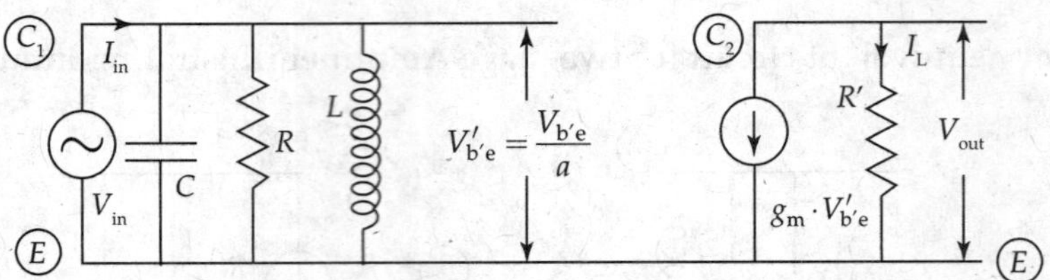

FIG. 13.26 *Inductive tap in cascaded ampliflers further simplified equivalent circuit*

$$\text{Current gain of the amplifier} \quad A_I = \frac{\text{Load current}}{\text{Input current}} = \frac{I_L}{I_{in}} = -\frac{g_m \cdot V_{b'e}}{I_{in}} \tag{13.56}$$

From the equivalent circuit in Fig. 13.29

$$\text{Input current} \quad I_{in} = V_{in}\left[\frac{1}{R} + j\left[\omega \cdot C - \frac{1}{\omega \cdot L}\right]\right]. \tag{13.57}$$

Using Eq. (13.54) in Eq. (13.57), we obtain

$$I_{in} = \frac{V_{b'e}}{a}\left[\frac{1}{R} + j\left[\omega \cdot C - \frac{1}{\omega \cdot L}\right]\right]$$

$$\therefore \quad A_I = -\frac{g_m \cdot V_{b'e} \cdot a}{V_{b'e}\left[\frac{1}{R} + j\left[\omega \cdot C - \frac{1}{\omega \cdot L}\right]\right]}$$

$$A_I = -\frac{g_m \cdot a \cdot R}{\left[1 + j \cdot R \cdot C\left[\omega - \frac{1}{\omega LC}\right]\right]} = -\frac{g_m \cdot a \cdot R}{\left[1 + j \cdot R \cdot C\left[\omega - \frac{\omega_0^2}{\omega}\right]\right]}, \tag{13.58}$$

where $\omega_0^2 = 1/LC$.
Equation (13.58) can be written as

$$A_I = -\frac{g_m \cdot a \cdot R}{[1 + j \cdot \omega_0 \cdot R \cdot C]\left[\frac{\omega}{\omega_0} - \frac{\omega_0}{\omega}\right]}$$

$$\text{Current gain} \quad A_I = -\frac{-g_m \cdot a \cdot R}{\left[1 + j \cdot Q_0\left[\frac{\omega}{\omega_0} - \frac{\omega_0}{\omega}\right]\right]},$$

where $Q_0 = \omega_0 RC$.

13.7.1 Function of Capacitance Tap Usage in Tuned Amplifiers

Alternative configuration to inductive tapping is *capacitance tap*. The main purpose of using *capacitance tap* is to reduce loading effect of R_L or lower input impedance of succeeding stage in Tuned Amplifier circuits, so as to maintain good selectivity and flat response in the passband.

Capacitance tap arrangement for optimum power transfer and to reduce loading effect
Impedance transformation ratio '*a*' for capacitance tap arrangement (Fig. 13.27)

Voltage gain $A_V = -\frac{g_m \cdot V_{b'e}}{Y} = -g_m aR,$ (13.59)

where $R = R_P \parallel (R_1 \parallel R_2) \parallel R_{b'e}$. (13.60)

Resonant frequency $f_0 = \frac{1}{2\pi\sqrt{LC}}$ (13.61)

and bandwidth $= (f_H - f_L) = \frac{1}{2\pi RC}$ (13.62)

where $C = C' + a^2[C_{b'e} + C_{b'c}(1 + g_m \cdot R_L)]$ (13.63)

and $a = \frac{(C_1 + C_2)}{C_1}$. (13.64)

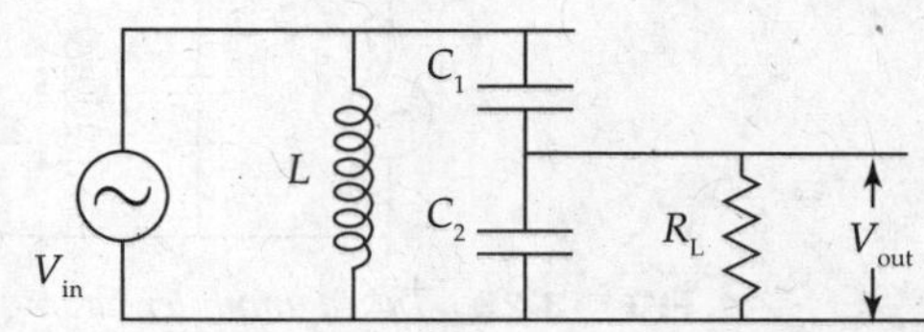

FIG. 13 27 *Capacitance tap for optimum power transfer*

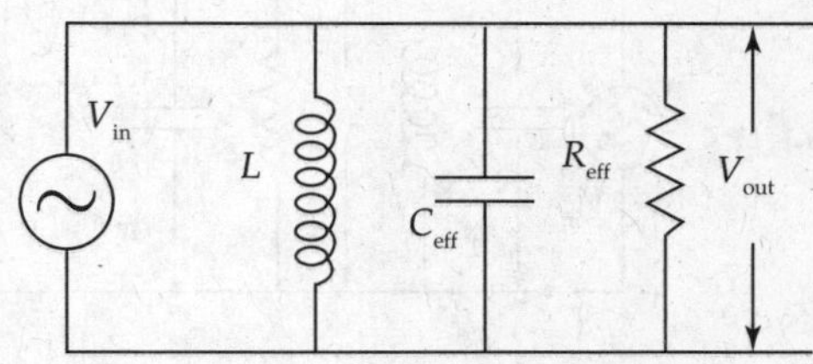

FIG. 13.28 *Capacitance tap for optimum power transfer using effective capacitance and resistance*

Effective resistance $R_{eff} = a^2 \cdot R_L$.
Effective tuning capacitance C_{eff} is the series combination of C_1 and C_2.

Tuning capacitance $C_{eff} = \frac{C_1 \cdot C_2}{C_1 + C_2}$ (13.65)

Considering the concept of effective capacitance C_{eff} and the effective resistance R_{eff}, the circuit is transformed into the equivalent circuit in Fig. 13.28:

$$\text{Amplifier bandwidth} = \frac{f_0}{Q} = \frac{f_r}{Q}. \quad (13.66)$$

13.8 AMPLIFIERS WITH MULTIPLE TUNED CIRCUITS

Selecting desired signals in the passband of a single-tuned Amplifier might not be sufficient in many practical situations. One such situation arises with IF Amplifiers of Radio and TV receivers. Improved selectivity is obtained by using additional tuned circuits. Figure 13.29 shows Tuned Amplifier circuit containing tuned circuits at both input and output ports.

Simplified equivalent circuit is shown in Fig. 13.30. The effect of $r_{b'c}$ is neglected because of its very large value of about 4 MΩ. The effect of $C_{b'c}$ (feedback capacitance) is transferred to input and output circuits using Miller theorem.

Analysis
Let the admittance at the output port $= Y_2$

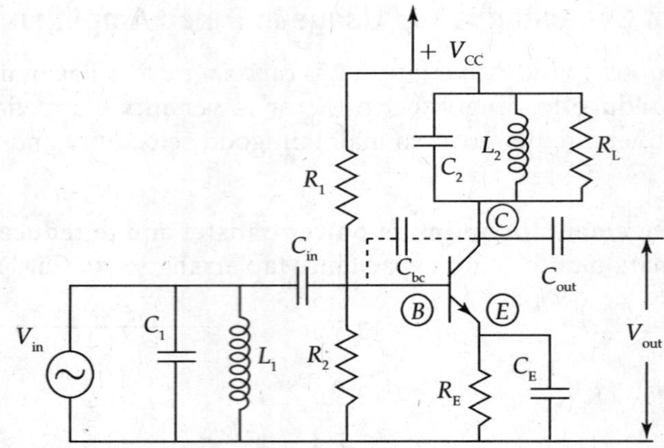

FIG. 13.29 *Tuned amplifier having tuned circuits at both collector and base circuits*

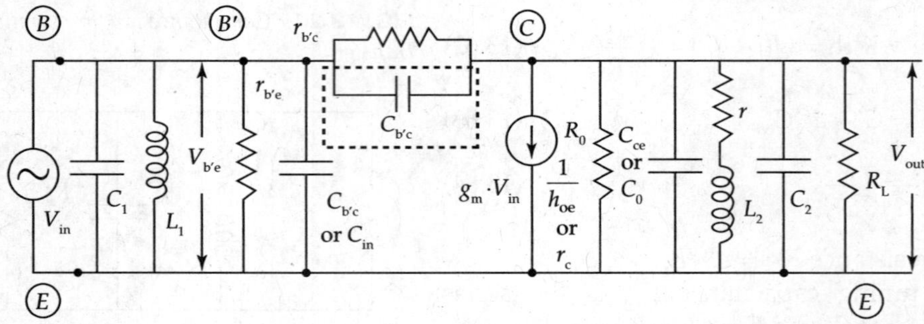

FIG. 13.30 *Equivalent circuit of tuned amplifier with tuned circuits at both input base and output collector circuits*

$$Y_2 = \frac{1}{r_0} + \frac{1}{R_L} + \frac{1}{R_2 + j\cdot\omega\cdot L_2} + j\cdot\omega\cdot(C_0 + C_2) \tag{13.67}$$

$$\frac{1}{R_{D2(eff)}} = \frac{1}{r_0} + \frac{1}{R_L} + \frac{(C_0 + C_2 + C_{b'c})R_2}{L_2}, \tag{13.68}$$

where $R_{D2(eff)}$ is the effective dynamic resistance at the output port of the Amplifier.

Effective Q_2-factor at the output circuit $= Q_{2(eff)} = \omega_2\,(C_0 + C_2 + C_{b'c})\,R_{D2(eff)}$.

Amplifier Voltage gain can be written as

$$A_V = \frac{g_m \cdot R_{D2(eff)}}{1 + j\cdot Y\cdot Q_{2(eff)}}, \quad \text{where } Y = \left[\frac{\omega}{\omega_0} - \frac{\omega_0}{\omega}\right] \tag{13.69}$$

$$Y_{in} = Y_1 + j\cdot\omega\cdot C_{b'c}\,[1 + g_m \cdot R_{D2(eff)}] \tag{13.70}$$

$$C_{in}(M) = C_{b'c}[1 + g_m \cdot R_{D2(eff)}] \tag{13.71}$$

$$\therefore \quad Y_{in} = Y_1 + j\cdot\omega\cdot C_{in}(M). \tag{13.72}$$

This expression for the input admittance Y_{in} is increased due to the feedback effect of Miller capacitance $C_{in}(M)$ caused by the inter electrode capacitance between the Base and the Collector of the transistor known as $C_{b'c}$.

At resonance, the dynamic resistance of the input tuned circuit $= R_{D1} = Q \cdot \omega_0 \cdot L_1$, where Q is the Q-factor of the input tuned circuit.

$$\frac{1}{R_{D1(eff)}} = \frac{1}{R_S} + \frac{1}{R_{D1}} + \frac{1}{r_{b'e}} \tag{13.73}$$

$$\text{Effective } Q_1\text{-factor} = Q_{1(eff)} = \frac{R_{D1(eff)}}{\omega_0 \cdot L_1} = \omega_0.[C_1 + C_{b'c} + C_{in}(M)] \cdot R_{D1(eff)}. \tag{13.74}$$

Since, Tuned Amplifier circuits work at radio frequencies, capacitors used in Tuned RF Amplifiers can be much smaller than those used in audio frequency Amplifiers.

Worked out example to show the influence of low input impedance on Effective Q

EXAMPLE 13.4

Referring to Fig. 13.30, the input tuned circuit has a Q-factor 'Q_1' of 120 at a frequency of 6 MHz. Inductance $L_1 = 4$ μH; Source resistance $R_S = 1$ kΩ; Current gain β of transistor is 200. Junction capacitance between the Base and the Emitter $C_{b'e} = 10$ pF and $r_{b'e} = 0.1$ kΩ. Calculate the effective Q-factor of the tuned circuit at the input port.

Solution: Dynamic resistance of the tuned circuit L_1 and C_1 is R_{D1}:

$$R_{D1} = Q_1 \cdot \omega_0 \cdot L_1 = 120 \times 2\pi \cdot 6 \times 10^6 \times 4 \times 10^{-6} = 18\ \text{k}\Omega.$$

Effective dynamic impedance $= R_{D1(eff)}$

$$\frac{1}{R_{D1(eff)}} = \frac{1}{R_S} + \frac{1}{R_{D1}} + \frac{1}{r_{b'e}}$$

$$\frac{1}{R_{D1(eff)}} = \frac{1}{1000} + \frac{1}{18 \times 10^3} + \frac{1}{100} = 11.05 \times 10^{-3} \text{ mhos}$$

$$R_{D1(eff)} = 90.5\ \Omega$$

$$\text{Effective } Q\text{-factor} = \frac{R_{D1(eff)}}{\omega_0 \cdot L} = \frac{90.5}{2\pi \times 6 \times 10^6 \times 4 \times 10^{-6}} = 0.6.$$

These calculations show that the effective Q of the tuned circuit at the input port is very much reduced. So, in practical applications the signal Source is applied through capacitance tap on the input tuned circuit to provide power transfer matching and reduced loading effect. Such arrangement is shown in Fig. 13.31.

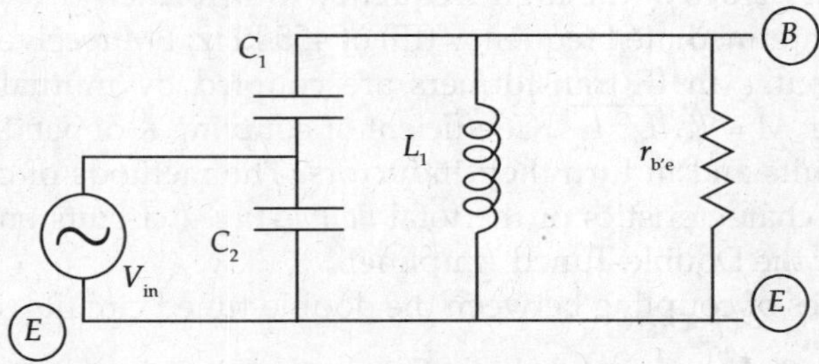

FIG. 13.31 *Capacitive tap for reducing loading effect at input port of common emitter transistor tuned amplifier*

13.9 DOUBLE-TUNED AMPLIFIER

In Amplifiers, there are two sets of conditions: (1) DC biasing is obtained through V_{CC}, R_1, R_2 and R_E for stable operation and (2) input AC signal is connected to the input port. Double-tuned Amplifier circuit contains two tuned circuits as shown in Fig. 13.32. One tuned circuit consists of inductance L_P and capacitor C_P, connected in parallel between Collector terminal and the positive terminal of V_{CC}. This acts as a tuned load. The second parallel resonant circuit consists of load resistance R_L in parallel to an inductor L_S and a capacitor C_S. Double-tuned circuit has two frequencies of resonance, to obtain wider bandwidths than those obtained from single-tuned Amplifiers.

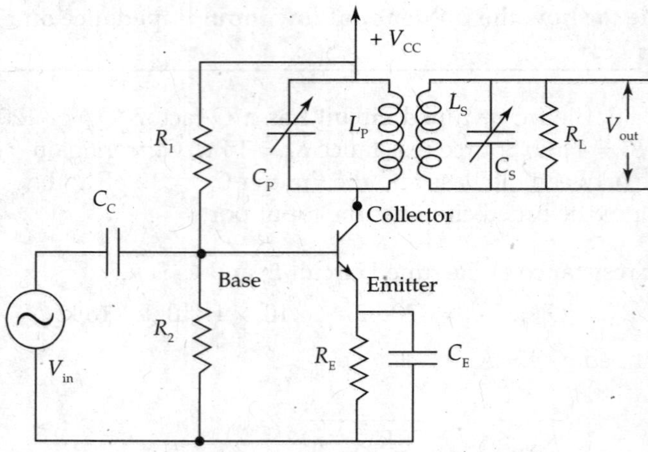

FIG. 13.32 *Transistor double tuned amplifier*

Double-Tuned Amplifiers are used mostly in Television, Radar and other communication receivers. They provide constant amplification of signals over a selected passband and reject the signals sharply outside the passband.

As a common example, IF Transformers in radio receiver circuits contain Double-Tuned circuits with *stagger tuning* to obtain the desired passband of 10 kHz. Tuning capacitances in the tuned circuits are of the order 50 to 120 pF. Q values range from 60 to 70. If a receiver contains single-stage IF Amplifier, there will be two IF Transformers. Whenever a radio receiver has two IF Amplifier stages for better selectivity, the IF Amplifier stages contain three IF Transformers. Each high-frequency transformer contains two tuned circuits, at primary and secondary sides of the high-frequency transformer.

AM receivers use Intermediate Frequency (IF) of 455 kHz; FM receivers use 10.7 MHz.

The two tuned circuits in IF transformers are coupled by mutual inductance M with coefficient of coupling $M = K\sqrt{L_P \cdot L_S}$. Coefficient of coupling K depends upon the proximity of the two tuned circuits and in turn their inductors. The methods of coupling between the two coils modify the characteristics of the total *double-tuned* circuits and in turn the output frequency response of the Double-Tuned Amplifier.

There are three types of coupling between the double-tuned circuits (Fig. 13.33):

1. **Critical-coupling or loosely coupled** For critical-coupling case, the primary and secondary tuned circuits are identical, and the frequency response is similar to a normal Amplifier response. Then, $KQ = 1$.

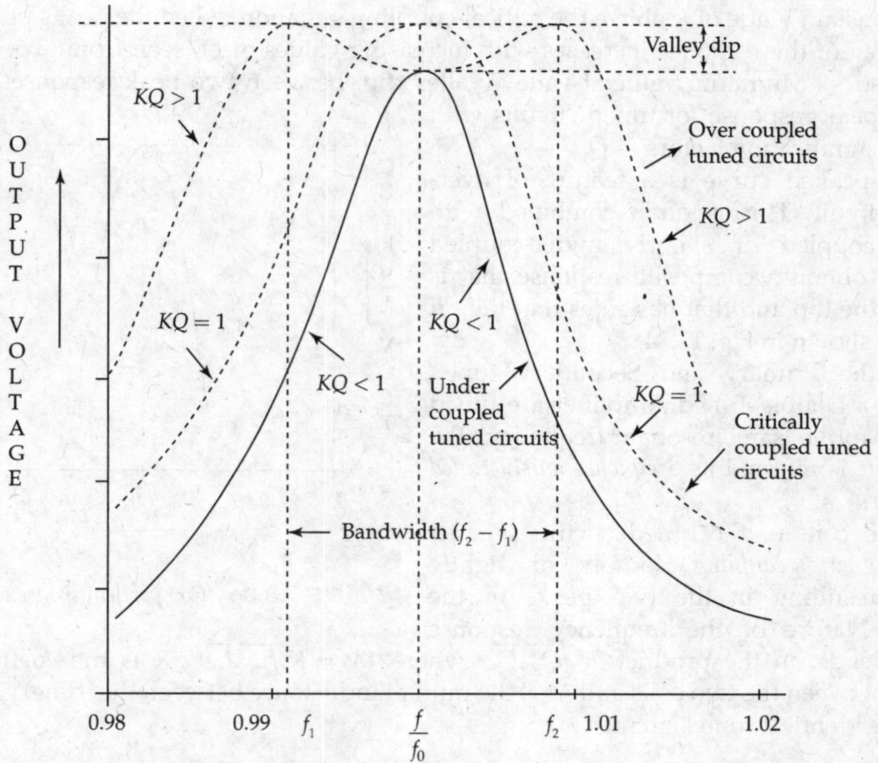

FIG. 13.33 *Frequency response of double tuned amplifier for different values of K and Q*

2. Under-coupled response occurs when $KQ < 1$.
3. **Over-coupled scenario** Q factors of primary and secondary tuned circuits differ appreciably, and response includes two peaks. Then, $KQ > 1$.

But these peaks do not appear immediately after the deviation from critical-coupling situation. The value of K where the double peaks of maximum response will appear in the output response characteristics, also known as *Transitional-Coupling factor 'K_t'*. The distance between these two peaks increases with increased values beyond the values of coupling factor K_C.

Bandwidth between the two peaks can be calculated as follows:

$$\left(\frac{f_2 - f_1}{f_0}\right) = \sqrt{(K^2 - K_t^2)},$$

where f_1 and f_2 are frequencies at the two peaks of frequency response of Double-Tuned Amplifier or high-frequency transformers used in Amplifier circuits. $(f_2 - f_1)$ is the bandwidth between the two peaks.

Approximate formula relating these factors is given as

$$\left(\frac{(f_2 - f_1)}{f_0}\right) \cong K. \tag{13.75}$$

For a constant value of K above the critical-coupling situation, when the peaks pop in, dip in the valley of the response increases with increasing values of Q (seen from experimental characteristics). Minimum value at valley (valley dip) between two peak responses is 1.414 times the peak response for tuned circuits with unequal magnification factors of Q.

Double-peaked curve is a feature of over-coupled circuit. This is often combined with critically coupled or slightly under-coupled circuits to obtain a composite response that is flat along the top and that has sides that fall off sharply as shown in Fig. 13.34.

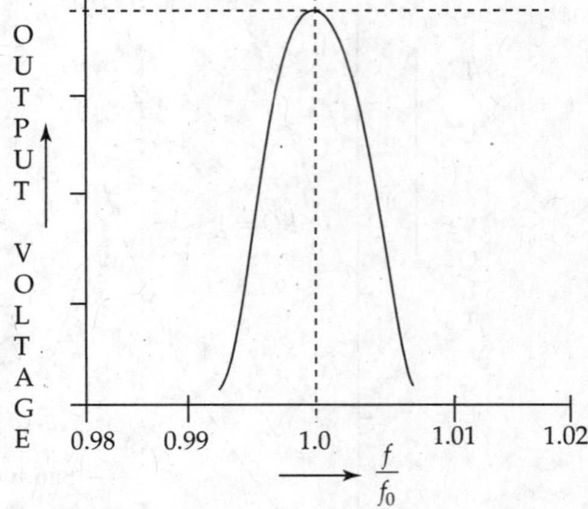

FIG. 13.34 *Composite response curve*

When the Primary and Secondary tuned circuits in a Double-Tuned Amplifier are tuned separately to the 'same resonant frequency', the transformer is referred as a *synchronously tuned transformer*.

The two coils of the tuned circuits interact using *magnetic coupling*, loosely or tightly coupled, resulting in the two peaks in the response. Nature of the frequency response curve depends on the product $K\sqrt{Q_P \cdot Q_S}$ where $(M = K\sqrt{L_1 \cdot L_2})$ (K is the coefficient of coupling between the two coils and M is the mutual inductance between the same).

Assume identical tuned circuits:

$$L_P = L_S = L, \quad Q_P = Q_S = Q \quad \text{and} \quad C_P = C_S = C.$$

Load Voltage $V_S = IZ_T$; Input Voltage $= IZ_{in}$ where Z_{in} is the input impedance as seen by Source I. (Internal impedance of Source R_0 is included in Z_{in}.)

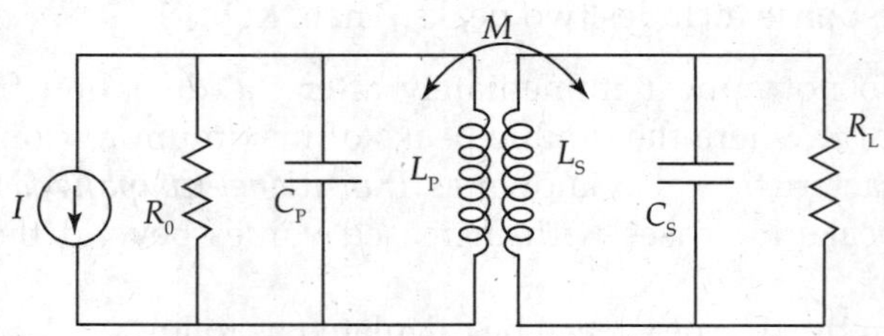

FIG. 13.35 *High-frequency transformer in circuit*

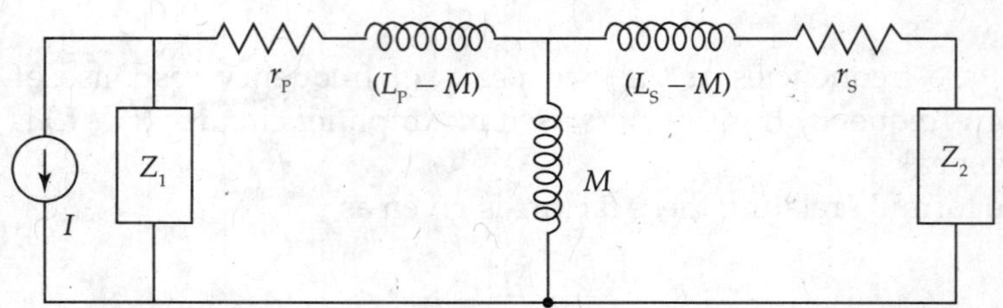

FIG. 13.36 *Equivalent circuit including HF Transformer*

Equivalent circuit of high-frequency transformer:

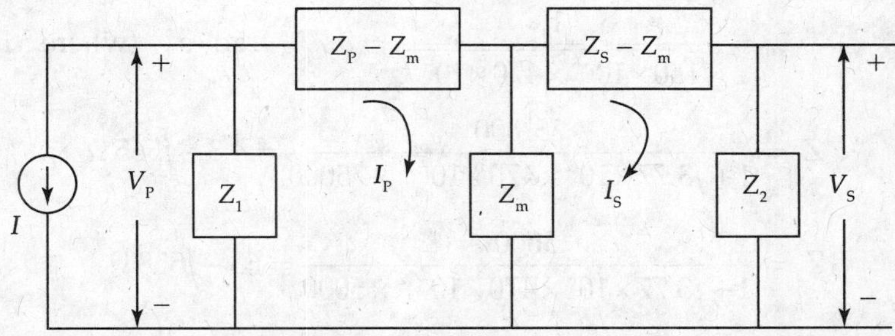

FIG. 13.37 *Synchronously tuned amplifier equivalent circuit including high-frequency transformer*

Block schematic for the transformer's equivalent circuit shown in Fig. 13.36.

$$Z_P = (r_P + j\cdot\omega\cdot L_P) \tag{13.76}$$

$$Z_S = (r_S + j\cdot\omega\cdot L_S) \tag{13.77}$$

$$Z_m = j\cdot\omega\cdot M \tag{13.78}$$

Load resistance R_L is in parallel with secondary tuning capacitor C_S in Fig. 13.35. This parallel impedance Z_2 is given by

$$Z_2 = \frac{R_L}{1 + j\cdot\omega\cdot C_S\cdot R_L}. \tag{13.79}$$

Likewise, at the primary, Source resistance R_0 appears in parallel with C_P and impedance Z_1 is

$$Z_1 = \frac{R_0}{1 + j\cdot\omega\cdot C_P\cdot R_0}. \tag{13.80}$$

Transfer impedance Z_T is the ratio of output voltage V_S to input current I:

$$Z_T = \frac{V_S}{I} \tag{13.81}$$

$$Z_T = \frac{Z_1\cdot Z_2\cdot Z_m}{(Z_P + Z_1)(Z_S + Z_2) - Z_m^2}. \tag{13.82}$$

It can be observed from the expression for Z_T that Z_T takes into account the damping effects of resistances at Source R_0 and load R_L.

EXAMPLE 13.5

High-frequency transformer has identical primary and secondary circuits for which $L_P = L_S = 150$ μH, $C_P = C_S = 470$ pF and Q-factor for each circuit alone (that is not coupled) is 85. The coefficient of coupling $K = 0.01$; Load resistance $R_L = 5$ kΩ; Constant current Source, feeding the transformer, has an internal resistance R_0 of 75 kΩ. Determine the transfer impedance Z_T at resonance.

Solution: Common resonant frequency ω_0 is given as

$$\omega_0 = \frac{1}{\sqrt{150\times10^{-6}\times470\times10^{-12}}} = 3.77\ \text{Mrad/s} \quad (\text{where } M = 10^6)$$

$$Z_1 = \frac{75,000}{1 + j3.77\times10^6\times470\times10^{-12}\times75000} = 4.3 - j565\ \Omega$$

$$Z_2 = \frac{5000}{1 + j3.77\times10^6\times470\times10^{-12}\times5000} = 63 - j558\ \Omega$$

At resonance, $Q = \frac{\omega_0 \cdot L}{r}$,

$$\therefore\quad r = \frac{\omega_0 \cdot L}{Q}.$$

$$f' = \frac{f - f_0}{f_0}$$

$$\Delta = \frac{\omega - \omega_0}{\omega_0} = \frac{(f - f_0)}{f_0}$$

$$r = \frac{\omega_0 \cdot L}{Q} = 6.7\ \Omega \tag{1}$$

$$\therefore\quad Z_P = Z_S = r + j\cdot\omega_0\cdot L, \tag{2}$$

where $r_P = r_S = r$ and $L_P = L_S = L$.
From Eqs. (1) and (2)

$$Z_P = Z_S = \omega_0 \cdot L \cdot \left[\frac{1}{Q} + j\right]$$

$$Z_P = Z_S = 6.65 + j565\ \Omega$$

$$Z_m = j\cdot 3.77\times10^6\ \text{M}$$

$$\text{where } M = K\sqrt{L_1 \cdot L_2} \tag{3}$$

$$L_1 = L_2 = 150\times10^{-6} \text{ and } K = 0.01$$

$$\therefore\quad M = 0.01\sqrt{150\times10^{-6}\times150\times10^{-6}}$$

$$M = 0.01\times150\times10^{-6} = 1.5\times10^{-6}$$

Hence, $Z_m = j\cdot\omega_0\cdot M = j\cdot 3.77\times10^6\times0.01\times150\times10^{-6}$

$$Z_m = j5.65\ \Omega$$

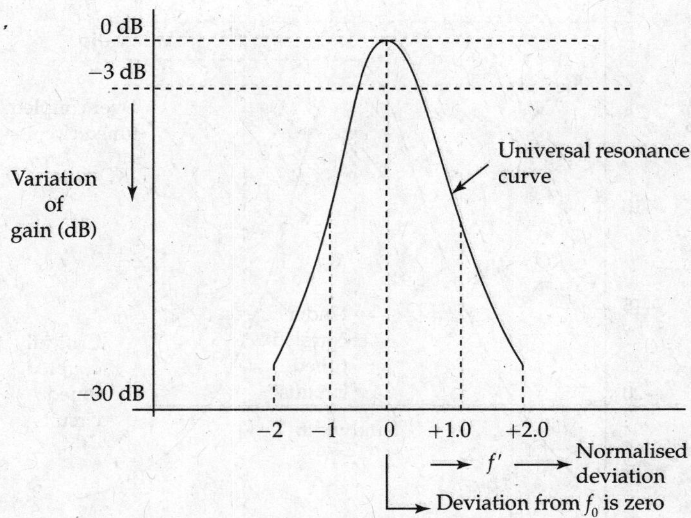

FIG. 13.38 *Universal resonance curve*

Let $\Delta = (Z_P + Z_1)(Z_S + Z_2) - Z_m^2$

$$\Delta = 791 + j80\ \Omega$$

Hence, $Z_T = \dfrac{Z_1 \cdot Z_2 \cdot Z_m}{\Delta} = -j \cdot 2.25 \times 10^3 + 43.8$

$$\therefore\ \ Z_T = 43.8 - j2.25 \times 10^3\ \Omega.$$

This example shows that at resonance, Z_T is almost entirely capacitance with maximum output voltage. For current of 1 mA (input current I), output voltage is approx $-j$ 2.25 V.
Z_T reactive part $= -j\,2.25 \times 10^3$.
$I \times Z_T$ reactive part $= -j\,2.25 \times 10^3 \times 1 \times 10^{-3} = -j\,2.25$ V.
Assuming that the primary is fed from a constant current Source, transfer impedance Z_T gives the variation of output voltage with frequency.

$$\text{Normalised deviation}\quad f' = \left[\frac{(f - f_0)}{f_0}\right] = \left[\frac{f}{f_0} - 1\right] \tag{13.83}$$

When $(f = f_0)$	$f' = 0$
When $(f = 2f_0)$	$f' = 1$
When $(f = -2f_0)$	$f' = -1$
When $(f = 3f_0)$	$f' = +2$
When $(f = -3f_0)$	$f' = -2$

$K\sqrt{Q_P \cdot Q_S} = KQ$, if $Q_P = Q_S$ curves drawn for various values of KQ in Fig. 13.39.

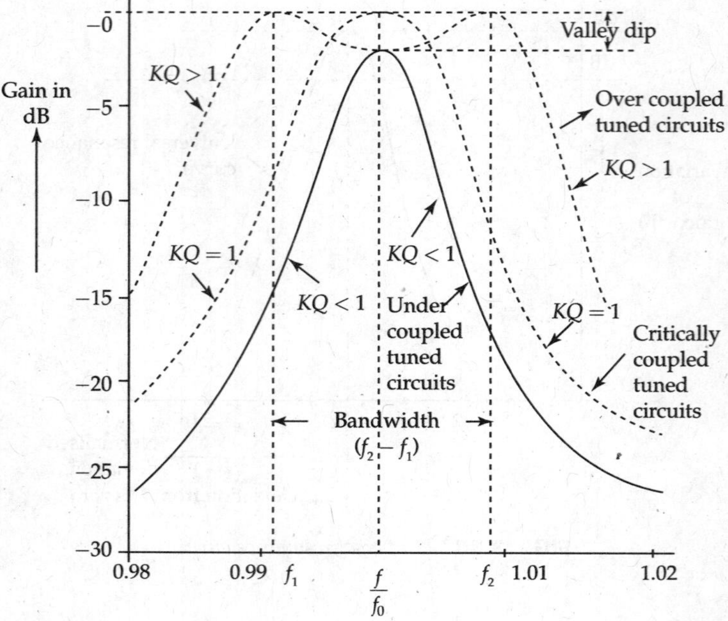

FIG. 13.39 *Frequency response of double tuned amplifier for different values of KQ*

13.10 APPLICATIONS OF TUNED AMPLIFIERS

(1) Radio transmitter and receivers, (2) TV transmitter and receivers, (3) Base stations of Cellular and mobile Communications, (4) Low-Noise Amplifiers in Cable and Satellite transmitter and receivers, (5) Military Communications, (6) Industrial Automation, (7) Medical Instrumentation, etc.

Radio Receiver

A radio receiver must perform a number of functions.

1. Receiver must select wanted radio signal, from all other radio signals that may be picked up by the antenna and reject the unwanted ones.
2. Receiver must amplify the desired signal to a usable level. Finally, the receiver must recover the signal information from the radio carrier and pass it on to the user.

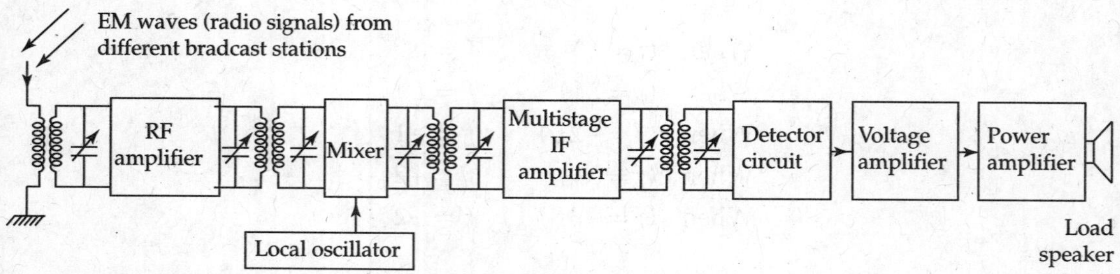

FIG. 13.40 *Block diagram of a radio receiver*

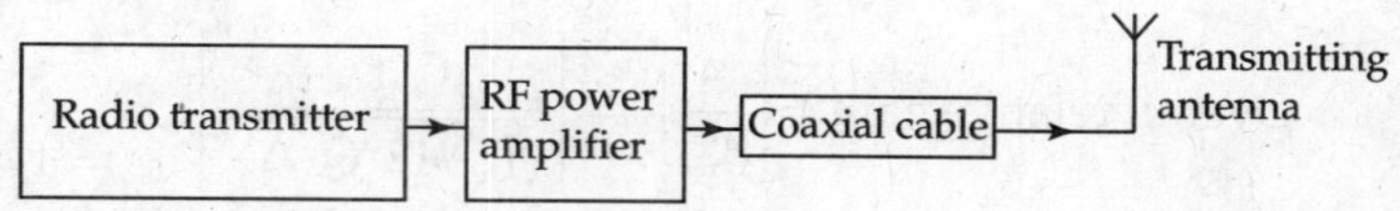

FIG. 13.41 *Tuned RF power amplifier in radio transmitter circuit*

Radio Transmitters

Figure 13.41 shows Tuned RF power Amplifier delivering power to a transmitting aerial through a coaxial cable as a matching network.

Tuned Amplifiers are used in various types of reliable radio communications for various applications such as telemetry systems, satellite communications, industrial remote control, security systems paging, mobile communications, factory automation and so on.

Salient features of Tuned Amplifiers

1. Maximum amplification at centre frequency of the desired passband of signals,
2. Variations in amplification around the centre frequency or resonant frequency and
3. Selectivity of the desired signals.

To realise good selectivity of Tuned Amplifiers, skirt selectivity of 3-dB or less is desirable in the communication applications, particularly in IF Amplifier stages.

Popular methods of tuning in multiple tuned circuits of Amplifiers are

1. Synchronous tuning and
2. Stagger tuning in communication receivers.

13.11 SYNCHRONOUSLY TUNED AMPLIFIER

In synchronous tuning, centre frequencies of non-interacting Tuned Amplifiers are tuned to the same frequency. It results in overall bandwidth lower than that of a single-tuned circuit. When *N*-identical non-interacting Tuned Amplifiers are cascaded (Fig. 13.42), it constitutes a synchronously Tuned Amplifier system using the same centre frequency f_0.

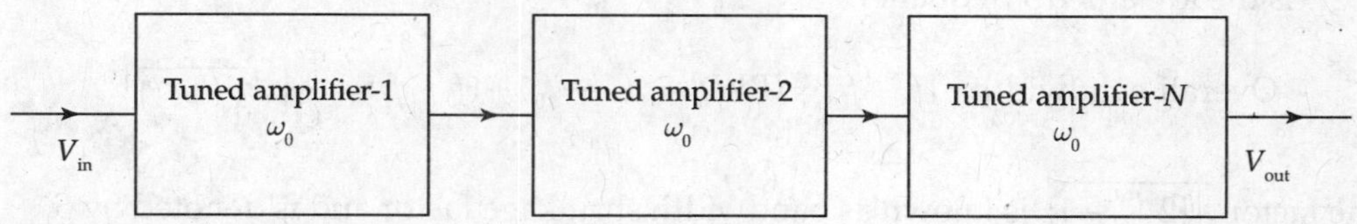

FIG. 13.42 *Block diagram of cascaded synchronously tuned amplifier with N-stages*

Relative gain of a single-stage amplifier:

$$\frac{A}{A_{res}} = \frac{1}{[1 + j \cdot 2 \cdot Q \cdot \Delta}$$

$$\left|\frac{A}{A_{res}}\right| = \frac{1}{\sqrt{[1 + (2 \cdot Q \cdot \Delta)^2]}}$$

When such *N*-stages are cascaded with the same centre frequency f_0 the overall relative gain is the product of individual relative gains of the identical stages.

$$\text{Overall relative gain} = \left|\frac{A}{A_{res}}\right|^N = \left[\frac{1}{\sqrt{[1+(2\cdot Q\cdot \Delta)^2}}\right]^N$$

$$\text{At 3-dB, the overall relative gain} = \frac{1}{\sqrt{2}}$$

$$\left[\frac{1}{\sqrt{[1+(2\cdot Q\cdot \Delta)^2}}\right]^N = \frac{1}{\sqrt{2}}$$

$$\left[1+(2\cdot Q\cdot \Delta)^2\right]^{N/2} = 2^{1/2}$$

$$\therefore \quad (2\cdot \Delta\cdot Q)^2 = 2^{1/N} - 1$$

$$\Rightarrow \quad (2\cdot \Delta\cdot Q) = \pm\sqrt{[2^{1/N} - 1]}$$

$$\text{But} \quad \Delta = \frac{\omega - \omega_0}{\omega_0}$$

$$\therefore \quad 2\left[\frac{\omega - \omega_0}{\omega_0}\right]Q = \pm\sqrt{[2^{1/N} - 1]}$$

$$2\left[\frac{f - f_0}{f_0}\right]Q = \pm\sqrt{[2^{1/N} - 1]}$$

$$\therefore \quad f_2 - f_0 = \frac{f_0}{2Q} \times \sqrt{[2^{1/N} - 1]},$$

where f_2 is the upper 3-dB frequency

$$\text{and} \quad f_0 - f_1 = -\frac{f_0}{2Q}\sqrt{[2^{1/N} - 1]},$$

where f_1 is the lower 3-dB frequency.

$$\text{Overall bandwidth} \quad [f_2 - f_1] = (BW)_N = (f_2 - f_0) + (f_0 - f_1) = \frac{f_0}{Q}\left[\sqrt{2^{1/N} - 1}\right].$$

The factor $\sqrt{[2^{1/N} - 1]}$ is known as bandwidth shrinkage factor and is denoted by S.

Bandwidth shrinkage factor $S = \sqrt{2^{1/N} - 1}$ of multistage Tuned Amplifiers

Number of stages (N)	2	3	4	5	6	7	8	9
$\frac{f_2^N}{f}$	0.6435	0.5098	0.4349	0.3856	0.3499	0.3226	0.3008	0.2829

Overall bandwidth is 64.35% of the bandwidth of each section of a synchronously Tuned Amplifier with two stages.

When B is the overall bandwidth,

$$\text{Bandwidth of each stage} = \frac{B}{0.6435} = 1.554 \cdot B.$$

Design of synchronously Tuned Amplifier is adopted in IF Amplifier stage of FM radio receiver ($f_0 = 10.7$ MHz).

Synchronously Tuned Amplifier with two tuned circuits at input and output ports in a CASCODE Amplifier using two transistors is shown in Fig. 13.43. Input side tuned circuit has L_1 and C_1 and output side tuned circuit has L_2 and C_2 resonant circuits.

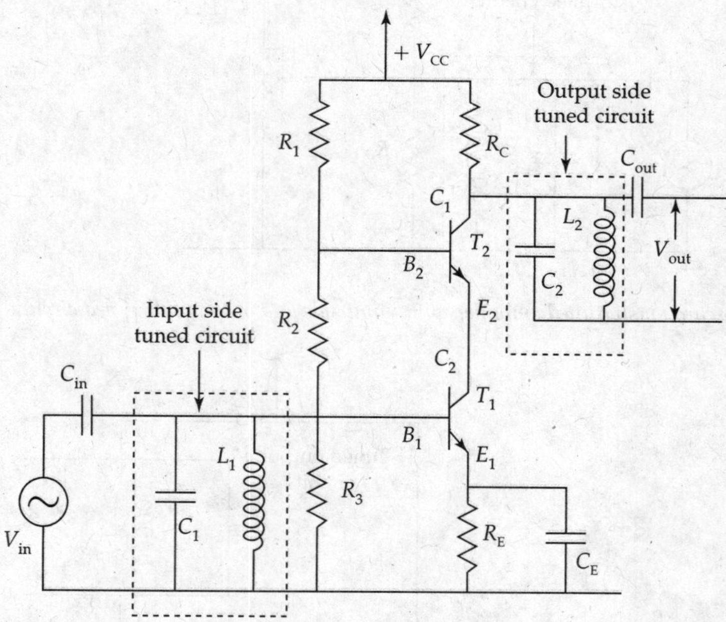

FIG. 13.43 *Synchronously tuned amplifier with input side and output side tuned circuits using BJT*

Both the tuned circuits are tuned to the same central frequency. The central frequency f_0 of the tuned circuits is same.

$$\text{Hence,} \quad \omega_0^2 = \left[\frac{1}{L_1 \cdot C_1}\right] = \left[\frac{1}{L_2 \cdot C_2}\right].$$

13.11.1 MOSFET (Synchronously Tuned) Amplifier (Fig. 13.44)

For Tuned Amplifier, skirt selectivity of 3 dB or less is desirable for most of the communication applications. Additional tuned circuits are connected in cascade to improve skirt selectivity. Complex conjugate matching is adopted in designing Synchronously Tuned amplifier with the combination of CC and CB Transistor Amplifier stages (Fig. 13.44). It is more popular in IC version. It has voltage buffer followed by a gain stage and provides isolation between input- and output-tuned circuits.

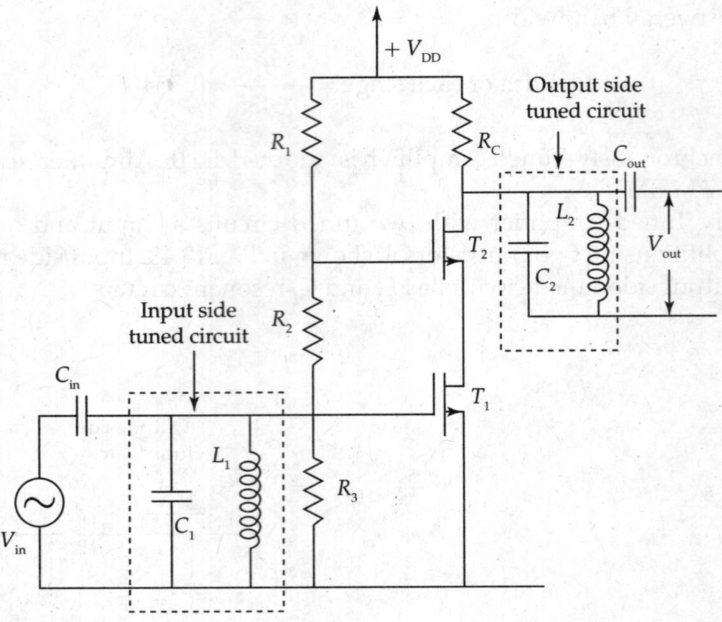

FIG. 13.44 *Synchronously tuned amplifier with input side and output side tuned circuits using MOSFET*

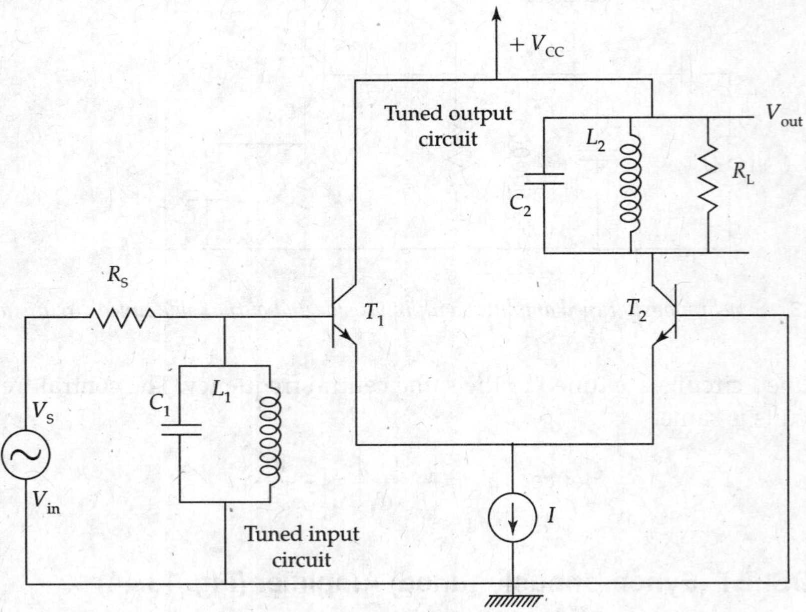

FIG. 13.45 *Common collector and common base configuration amplifiers*

Frequency response of Synchronously Tuned Amplifier (Fig. 13.46)

Frequency response curves show that there is a substantial reduction or rejection of signals at unwanted frequencies around the passband and there is a reduction in bandwidth. The response is sharp, selective and narrow band.

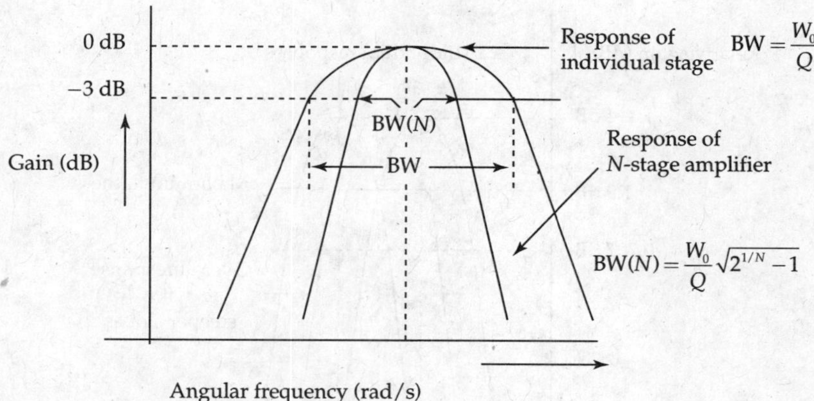

FIG. 13.46 *Frequency response of synchronously tuned amplifier*

EXAMPLE 13.6

IF section of a broadcast band FM radio receiver has an IF of 10.7 MHz and bandwidth of 200 kHz using two tuned circuits connected in CASCODE stage shown in Fig. 13.45. Determine (a) 3-dB bandwidth of each stage. (b) Value of capacitor C_1 for $L_1 = 0.329\ \mu\text{H}$. (c) Value of capacitor C_2 for $L_2 = 2.11\ \mu\text{H}$.

Solution:

(a) Bandwidth of individual stage $= \dfrac{\text{Overall bandwidth}}{\sqrt{2^{1/N}-1}}$

$$= \frac{200\times 10^3}{\sqrt{2^{1/2}-1}} = 311\ \text{kHz}$$

(b) $L_1 = 0.329\ \mu\text{H}$ $\qquad \omega_0^2 = \dfrac{1}{L_1 \cdot C_1}$

$$\therefore\ C_1 = \frac{1}{\omega_0^2 \cdot L_1} = \frac{1}{\left[2\pi(10.7\times 10^6)\right]^2 \times 0.329\times 10^{-6}}$$

$$= \frac{10^{-6}}{\left[4\pi^2 \times 114.49\times 0.329\right]} = 168\ \text{pF}$$

(c) $L_2 = 2.11\ \mu\text{H}$ $\qquad \omega_0^2 = \dfrac{1}{L_2 \cdot C_2}$

$$\therefore\ C_2 = \frac{1}{\omega_0^2 \cdot L_2} = \frac{1}{\left[2\pi(10.7\times 10^6)\right]^2 \times 2.11\times 10^{-6}}$$

$$= \frac{10^{-6}}{\left[4\pi^2 \times 114.49\times 2.11\right]} = 105\ \text{pF.}$$

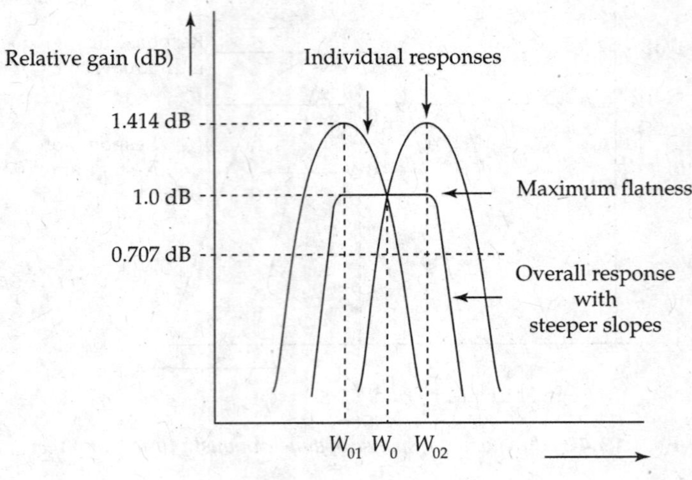

FIG. 13.47 *Angular frequency versus relative gain of stagger tuned amplifier*

13.12 STAGGER-TUNED AMPLIFIER

Figure 13.47 shows the response of Stagger-Tuned Amplifier. Stagger tuning is an improved method over synchronous tuning in multi-Tuned Amplifiers.

1. Shrinkage factor $S_N = \sqrt{2^{1/N} - 1}$ is undesirable.
2. Voltage gain in passband is not flat in synchronously Tuned Amplifiers.

These drawbacks of synchronously Tuned Amplifier are overcome in stagger-Tuned Amplifiers. In stagger tuning, the tuned circuits are slightly staggered (displaced) using centre frequencies slightly different and arranged such that the overall effect is to produce a narrow band with maximally flat response (Butterworth response) around a centre frequency f_0 with steeper fall offs.

Centre frequency of the first stage amplifier

$$\omega_{01} = \omega_0 - \frac{B}{2\sqrt{2}}$$

Centre frequency of the second stage amplifier

$$\omega_{02} = \omega_0 + \frac{B}{2\sqrt{2}},$$

where B is the amplifier bandwidth.

The central frequencies of individual amplifiers are symmetrically staggered by $B/2\sqrt{2}$ around the central frequency ω_0 of stagger-tuned amplifier.

Frequency response curve of 'Stagger-Tuned Amplifier' consisting of two single-stage Amplifiers will be similar to the frequency response of 'single-stage Double-Tuned Amplifier', when each stage of Stagger-Tuned Amplifier is tuned to frequencies synchronously displaced at half the passband from its centre frequency. Response will be maximally flat with steeper slopes by using more tuning circuits with their resonant frequencies as close as possible. Better selectivity (discrimination against signals in the adjacent bands) can be achieved with stagger tuning used in IF stage of TV receivers.

Analysis of Stagger-Tuned Amplifier

The selectivity of a single-tuned direct-coupled circuit

$$\frac{A}{A_{\text{res}}} = \frac{1}{[1 + j \cdot 2 \cdot \Delta \cdot Q]} = \frac{1}{1 + j \cdot x}$$

$$\text{Bandwidth} = 2 \cdot \Delta \cdot f_0 = \frac{f_0}{Q}$$

For a stagger-tuned amplifier with two tuned circuits, the corresponding selectivities are

$$\frac{A}{A_{\text{res1}}} = \frac{1}{[1 + j \cdot (x+1)]}$$

$$\frac{A}{A_{\text{res2}}} = \frac{1}{[1 + j \cdot (x-1)]}.$$

Frequency response of a *Synchronously Tuned Amplifier* showing maximal flatness around a centre frequency f_0 with steeper slopes is shown in Fig. (13.48).

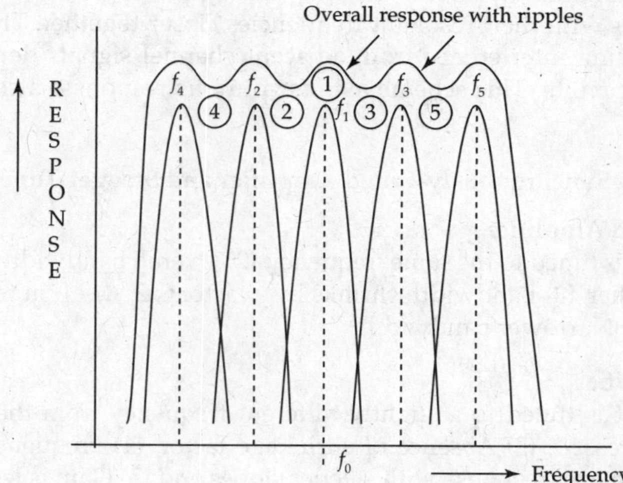

FIG. 13.48 *Frequency response of stagger tuned IF transformers*

Frequency responses of individual Tuned Amplifier stages in IF amplifiers of communication receivers result in several peaks of ripples in the overall response. It can be smoothened by using more tuned circuits with their resonant peaks as close as possible as shown in Fig. 13.48. Use of Stagger Tuning in radio receiver circuits is shown in Figs. 13.47, 13.48 and 13.49.

Application of Stagger-Tuned Amplifier

In a radio receiver circuit (Figs. 13.2 and 13.40), maximum uniform gain over desired passband of 10 kHz with sharp selectivity is achieved by cascaded IF Amplifiers. They use the principle of *Stagger Tuning*.

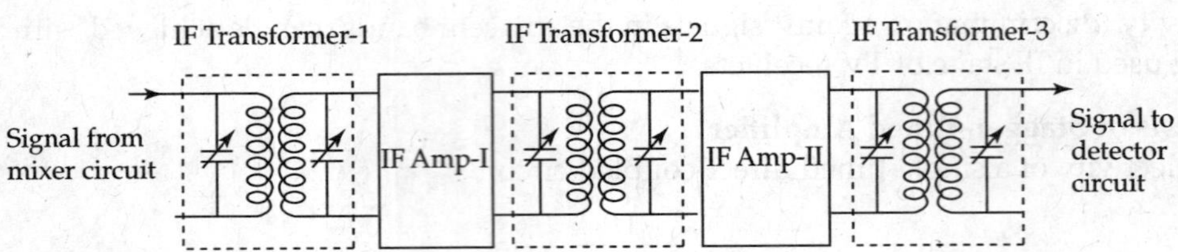

FIG. 13.49 *Three IF double tuned transformers and Two stage IF amplifier*

Principle of stagger tuning of tuned circuits of IF Amplifier

1. Sufficient amplification is provided in IF Amplifiers using Double-Tuned circuits with IF transformers. Radio receivers use an IF Amplifier consisting of two stages with three IF transformers that produces adequate selectivity.
2. An odd number of tuned circuits is used for *stagger-tuning* process. One resonant circuit is tuned to the centre frequency (IF) of 455 kHz. Each successive pair of tuned circuits is tuned with their resonant frequencies staggered at equal intervals from the centre frequency with a passband of 2 kHz.
3. The overall response due to stagger tuning of various tuned circuits contributes to the desired passband accommodating Voice and music (audio) signals.
4. But the response contains ripples at the top. This ripple can be filtered out by adding more tuned circuits with their resonant frequencies closer together. The steepness of the response for avoiding interference from adjacent channel signals depends on the total number of tuned circuits. This scheme avoids spurious responses also to the maximum extent.

Comparison between Synchronously Tuned Amplifier and Stagger-Tuned Amplifier

Synchronously Tuned Amplifier

(1) Each tuned circuit is tuned to the same frequency, (2) Overall bandwidth is less than that of a single-tuned Amplifier, (3) Bandwidth shrinks by a factor $\sqrt{2^{1/N}-1}$ in terms of bandwidth of single stage and (4) Narrower bandwidth.

Stagger-Tuned Amplifier

(1) Each tuned circuit is tuned to a slightly different frequency from the centre frequency, (2) Bandwidth is increased, (3) Absence of shrinkage factor, (4) Frequency response shows maximally flat Butterworth response with steeper slopes and (5) Gain is less when compared to a synchronously Tuned Amplifier.

13.13 STABILISATION TECHNIQUES

Following are the reasons for instability of Tuned Amplifiers due to undesirable oscillations:

- Communication receivers use Tuned Amplifiers with a skirt selectivity of 3-dB or less. To achieve this objective, additional tuned circuits are added in cascaded stages. Each tuned circuit has its own resonant frequency and bandwidth. If interaction is allowed between the stages, it causes instability.
- 'Miller effect' due to capacitance C_μ in Transistor Amplifier circuits causes alignment and tuning problems.

- High gain of Tuned Amplifier, nature of load and HF operation causes positive feedback of a portion of output signal to the input port through the low impedance path of the inter electrode junction capacitance $C_{b'c}$ of the Transistor.

Parasitic oscillations can be minimised by proper shielding, aligning components, using RF chokes, changing circuit parameters by changing Transistors and so on.

Stabilisations of Tuned Amplifier responses

Reduction in Amplifier gain reduces the magnitude of positive feedback. This is achieved by connecting a low-value resistor in parallel with the tuned circuit. This has a side effect of reduction in *Q* factor of the tuned circuit that also further reduces the Amplifier gain. This type of design is not practically suitable.

RF and IF Amplifier stability can be made possible by reducing the positive feedback in the circuit by using one of the following neutralising methods.

Neutralising circuit for stable operation of Tuned Amplifier (Fig. 13.50)

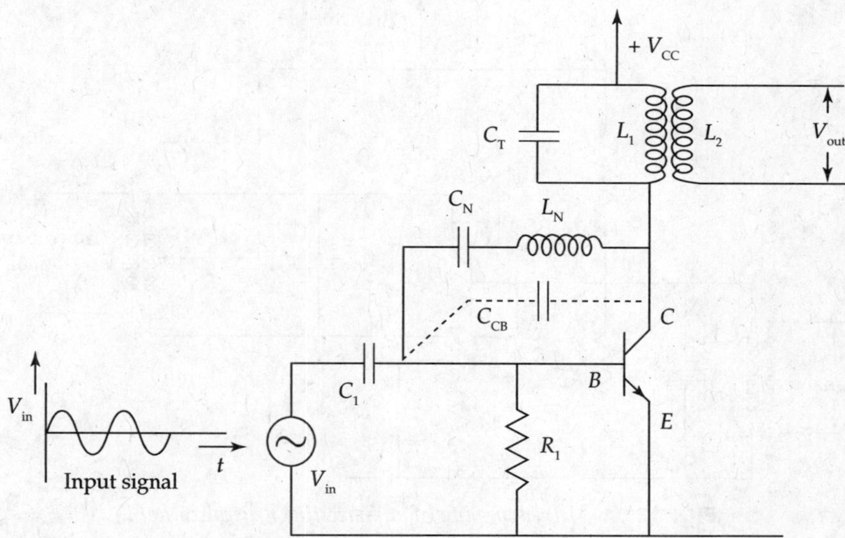

FIG. 13.50 *RF amplifier with stabilisation elements*

- A series combination of one neutralising inductor L_N and a blocking capacitor C_N could be connected between the Collector and the Base of the transistor. (For FET device, the combination of L_N and C_N is connected between the Gate and the Drain.)
- Neutralising inductor L_N forms a parallel resonant circuit with the output junction capacitance C_{CB}. The inductor design is such that it resonates with C_{CB} at the operating frequencies of the Tuned Amplifier. Then the inductor has very high impedance path at radio frequency operation, which stops energy feedback from the output to the input ports of the transistor Amplifier.
- Capacitor C_N does not allow the DC Collector voltage V_C to the input port through the inductor.

This method of stabilisation is also known as coil neutralisation. Coil neutralisation is used in radio transmitters, where neutralisation at single frequency is desired.

Various types of neutralisation circuits

(1) Hazeltine neutralisation (Broadband Neutralisation), (2) Rice Neutralisation (Narrow Band Neutralisation), (3) Cross Neutralisation, (4) Coil Neutralisation and (5) Neutralisation with common feedback.

Broadbanding using Hazeltine neutralisation

Broadband technique of Hazeltine neutralisation is named after its inventor. It is mostly used in Tuned input/Tuned output Amplifier at the front end of RF stages of radio receiver and TV receivers. Feedback effects due to 'Miller feedback capacitance' between Base and Collector (C_μ or $C_{b'c}$) are neutralised by a new capacitance C_N (neutralising capacitance) used in the circuit. Signal current through C_N is equal and opposite to that flowing through inter electrode capacitance $C_{b'c}$ between Collector and Base of the Transistor.

Hazeltine neutralisation circuit for a CE Amplifier (Fig. 13.51)

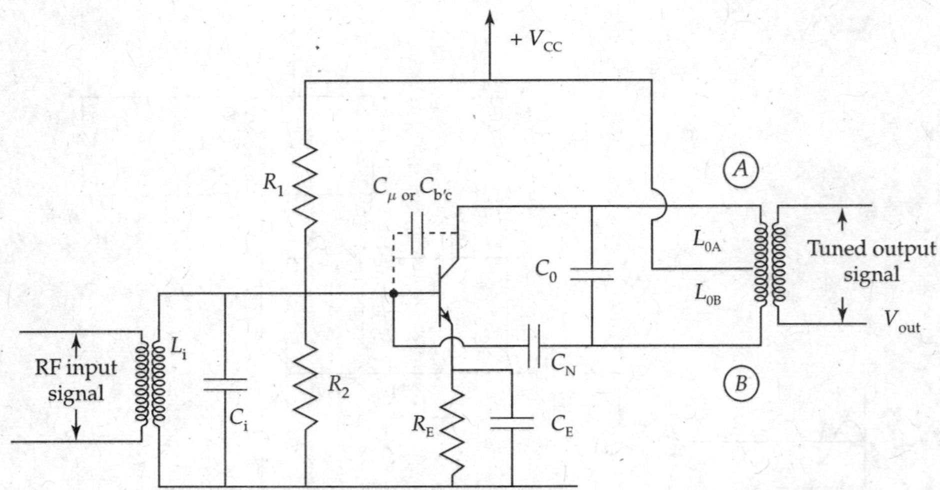

FIG. 13.51 *Tuned amplifier with hazeltine neutralization*

Circuit operation CE Transistor Amplifier has got two tuned circuits, one at input port (L_i in parallel with C_i) and one at output port (Coil L_{AB} in parallel with output-tuning capacitance C_0). Coil L_{AB} of output-tuned circuit is a series combination of L_{0A} and L_{0B}. Inter electrode capacitance C_μ or $C_{b'c}$ is shown in dotted lines. It is also known as *feedback capacitance*, because it provides feedback of energy from output port to input port at high-frequency operation of Amplifier (because of reduction in capacitance reactance of feedback capacitance). Its effect is neutralised by adding a neutralising capacitor C_N connected between the Base of transistor and the bottom end of the coil L_{AB}. Capacitance C_N introduces a signal which is 180° out of phase with the feedback signal through the Collector to Base junction capacitance $C_{b'c}$.

Feedback capacitance C_μ, the neutralising capacitance C_N and the two halves of output coil L_{AB} form bridge circuit with actual output voltage V_{out} of Amplifier and voltage $2 \cdot V_{out}$ across output coil L_{AB} as shown in Fig. 13.52. By slight variation in C_N, bridge is balanced in such a way that feedback effect of Miller capacitance is neutralised and no feedback of output energy to input port of Amplifier occurs.

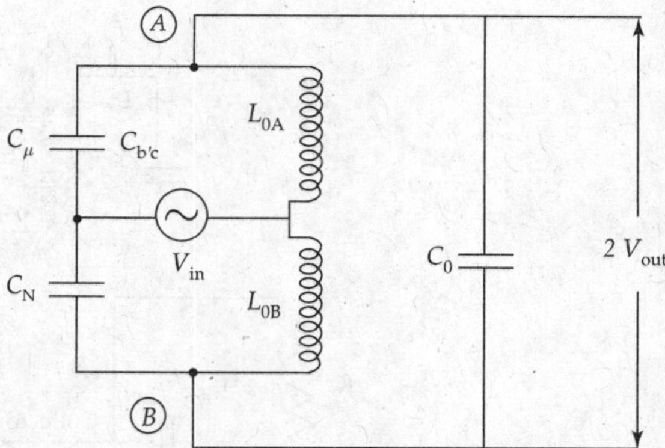

FIG. 13.52 *Bridge circuit showing amplifier output voltage, feedback C_{μ} neutralising capacitor C_N and the two halves of the output coil L_{0A} and L_{0B}*

Under balanced condition

$$C_{\mu} \cdot L_{0A} = C_N \cdot L_{0B}$$

$$\therefore \quad C_N = \frac{C_{\mu} \cdot L_{0A}}{L_{0B}}.$$

Above neutralisation technique is a wideband technique, since neutralization is done independent of frequency of signals through the Amplifier.

EXAMPLE 13.7

Tuned Amplifier has an internal feedback capacitance C_{μ} of 10 pF, which has to be neutralised. It operates at 10 MHz. The output transformer is connected with a tapped primary at $N_A{:}N_B = 1{:}4$. Its primary inductance is 5 μH. If Hazeltine neutralisation is to be used, calculate the size of the neutralising capacitor C_N needed. (JNTU, May/June 2004)

Solution: Data given: $C_{\mu} = 10$ pF, $L_{0A} = 1$ μH, $L_{0B} = 4$ μH

$$C_N = \frac{C_{\mu} \cdot L_{0A}}{L_{0B}} = \frac{10 \times 10^{-12} \times 1 \times 10^{-6}}{4 \times 10^{-6}} = 2.5 \times 10^{-12} = 2.5 \text{ pF}.$$

Narrow band neutralization using a coil: Rice neutralization

Secondary winding of input-tuned circuit L_{AB} consists of equal inductances L_{iA} and L_{iB}. Signal currents through L_{iA} and L_{iB} are equal and opposite in phase so that feedback voltage that may occur across L_{AB} will be zero. A neutralising capacitor C_N is connected between output terminal of the transistor and bottom terminal of input-tuned circuit. It opposes the detrimental effects of Miller feedback capacitance. The balance occurs when $C_N = C_{\mu}$.

Figure 13.53 shows the Rice neutralisation circuit.

Cross neutralisation Fig. (13.54)

Two neutralisation capacitors are connected each between Base of one transistor and Collector of other transistor, to neutralise the effects of feedback signals from Collector to Base capacitances of each transistor.

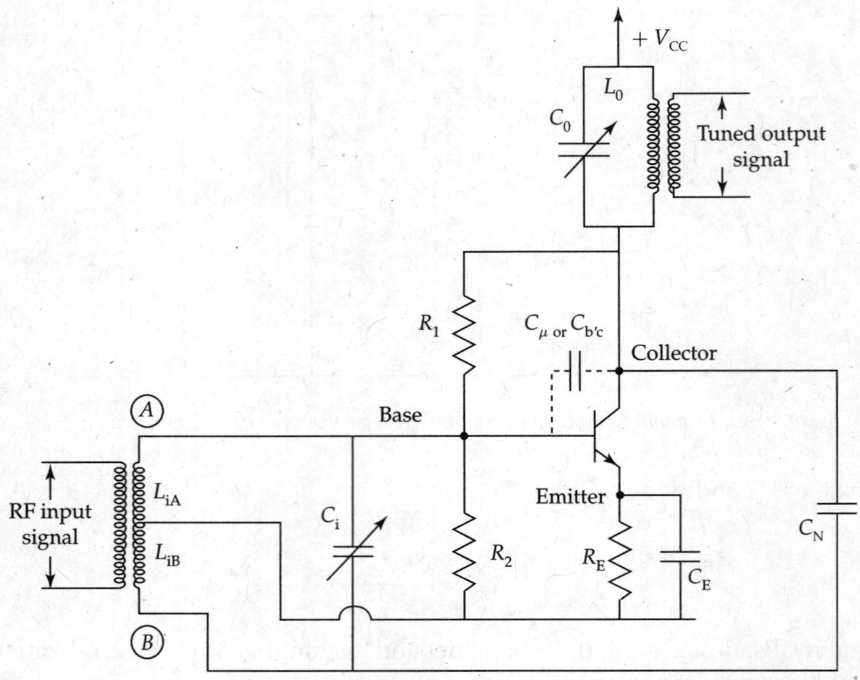

FIG. 13.53 *Tuned amplifier with RICE neutralization*

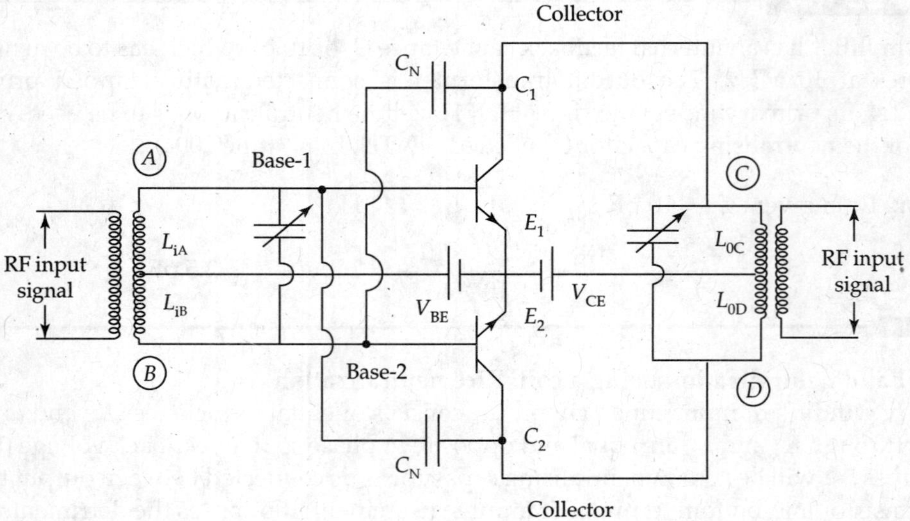

FIG. 13.54 *Transistor tuned amplifier with cross neutralisation*

13.14 RADIO FREQUENCY AMPLIFIERS (TUNED AMPLIFIER)

Class-A Radio Frequency Amplifier (Fig. 13.55)

For Class-A operation, DC-biasing conditions and input signal amplitudes are arranged such that output signal conduction angle is 360°. It has excellent fidelity but very poor

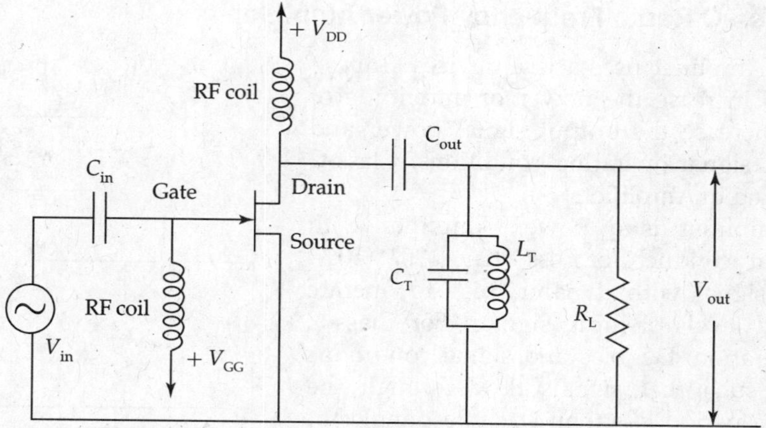

FIG. 13.55 *Single stage Class-A power amplifier using JFET*

efficiency of power conversion. Even in quiescent conditions, power dissipation occurs in device and the circuit components. Power dissipation is more. It is rarely used in RF stages.

13.14.1 Tuned Class-B Amplifier

To increase the output signal power and the maximum signal operating conditions, Class-B operation is used in Amplifiers. DC bias and the signal amplitudes are arranged such that the active devices conduct for 180° or half the cycle of the signal. However, zero cross over distortion has to be eliminated by using an operational amplifier before the complementary symmetry Class-B Amplifier stages.

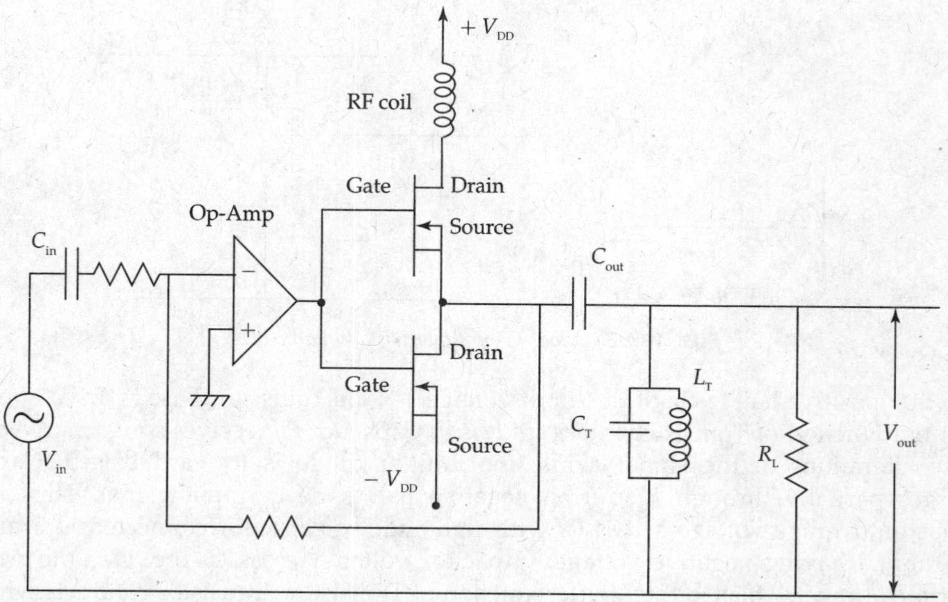

FIG. 13.56 *Complementary symmetry push-pull Class-B power amplifier using MOSFET devices*

13.14.2 Class-C Radio Frequency Power Amplifier

An electronic Amplifier uses active devices (Tubes/BJTs/FETS) to increase the power or amplitude of a signal. To increase the output signal power and the maximum signal operating conditions, Class-C operation is used in Amplifiers.

Class-C Amplifier is a power Amplifier with transistors that conduct for less than 180° (50%) of the input signal with dynamic bias to operate with increased levels of input signals. For Class-C Amplifier operation, DC bias and signal conditions are arranged such that signals flow through the Amplifier for a period less than 180°. Best angles of conductions are between 60° and 120°

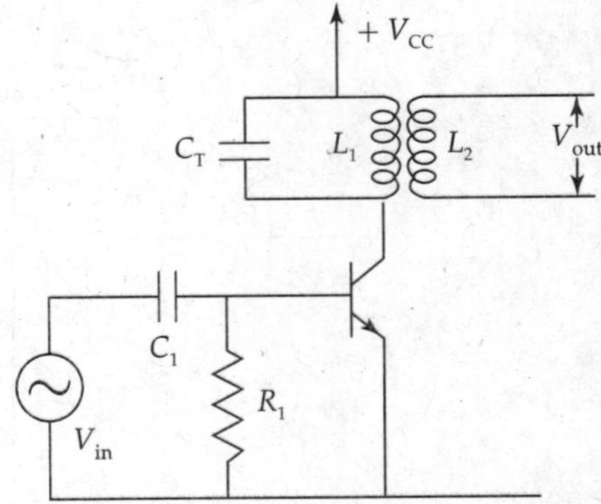

FIG. 13.57 *Class-C RF power amplifier*

As the output signals at Transistor Collector point or Drain point of FET device are pulses, continuous output signals are achieved by using Tuned load circuit in Amplifiers.

For Class-C operation, usual methods of DC bias are not practicable. The DC bias for Class-C operation is provided by R_1–C_1 combination in the input circuit of the Amplifier. Input Base and output Collector currents are pulses conducting for approximately 60°. Using tuned circuit, continuous output signals are obtained.

Principle of 'Dynamic Bias' for Class-C Amplifier operation

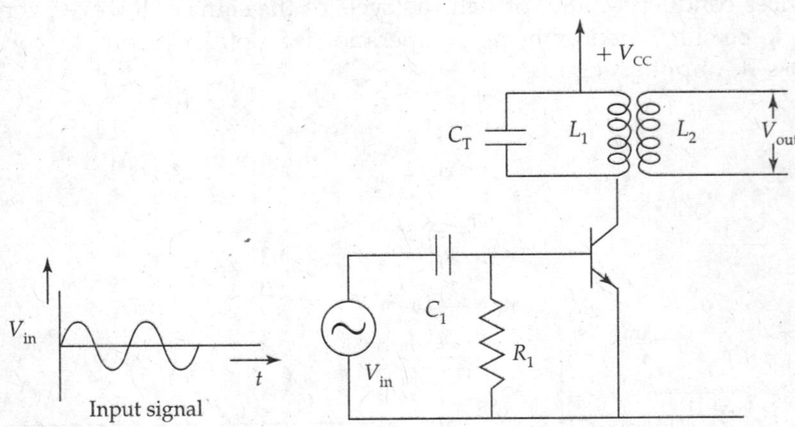

FIG. 13.58 *Class-C amplifier with dynamic bias*

On the first positive half cycle of the input voltage V_{in}, the Transistor Base is driven positive. Then input junction of Transistor is *forward biased*. Capacitor C_1 receives some charge during positive alterations of the signal, while the Emitter diode is forward biased. Capacitor discharges partially through R_1 during negative half cycle of input signal. This process develops an average voltage across C_1 with right side of capacitor connected to Transistor Base terminal having negative voltage. Capacitor voltage across C_1 provides the required magnitude of reverse bias to the Emitter junction of Transistor. Transistor conducts when the positive peaks of input signal voltage overcome the reverse bias voltage across capacitor.

DC Bias provided by R_1–C_1 combination is known as 'dynamic bias'. Input Base current and the resulting Collector current will be pulses existing for small intervals of time. Pulsating output current excites the Tuned circuit. Tuned circuit converts the pulses into continuous sine wave signals, due to charge/discharge cycles of energy between capacitor C_T and inductor L_1 of Tuned circuit. Output voltage of Class-C Amplifier is a continuous sine wave. Use of tuned circuit in Class-C Amplifiers find their application in RF Transmitters.

In cell phones, to conserve battery power, highly efficient power conversions Amplifiers are needed. So, switching voltage regulators use MOSFET devices due to (1) Higher power conversion efficiency; (2) No secondary break-down phenomenon unlike BJTs that suffer from limited operation area due to secondary breakdown; (3) Linear transfer characteristic that supports distortion less device operation.

Advantages

(1) Lower magnitudes of power dissipation in active devices keep the devices cool. (2) Higher values of power conversion efficiency. (3) Find applications in pulsed power Amplifiers for Radar applications. (4) Amplifiers for Wideband CDMA. (5) RFI.D reader circuits. (6) To increase the output power of Radio Transmitters. (7) Reduce distortion due to Tuned loads. (8) Works as a replacement for Travelling wave Tube. (9) Tropo-Scatter Amplifiers.

Disadvantages

(1) Devices conduct for duration of less than 50% of the input signal. So the distortion at the output is high. This means that signal fidelity is worse. (2) Cannot support AM signal operation. (3) Poor dynamic range of operation.

13.15 WIDEBAND AMPLIFIERS

Tuned voltage Amplifier having its frequency response with uniform gain for signals covering a frequency range from a few Hertz to tens of mega Hertz is known as 'Wideband' Amplifier. Wideband Amplifiers were initially used in TV systems to amplify video signals. So, *Wideband Amplifiers* are known as *video Amplifiers* (15 kHz to 5 MHz).

1. Radar Amplifiers require 8 MHz bandwidth. Video Amplifiers require 4 to 6 MHz bandwidth. Wideband frequency response is necessary for the amplification of pulsating video and radar signals.
2. To amplify non-sinusoidal signals such as saw-tooth voltages for horizontal deflection system in CROs. Pulse signal amplifications required in A-type displays and Plan Position Indicator (PPI) indicators in Radar applications use *Wideband Amplifiers*.
3. Wideband IF amplifiers in a Base station for Mobile communications to serve several channels simultaneously and so on.
4. Wideband Amplifier concepts are opposite to narrow band Tuned Amplifiers.

Consider ideal rectangular input pulse with sharp vertical sides and flat top to study the practical aspects of its output signal and design Video Amplifier with better performance.

Variation of reactive impedance in the Amplifier causes certain variations in output signal as shown in Fig. 13.59(b). Output signal response is considered with reference to time taken for amplitude to rise from 10% to 90% of actual input pulse. This time is known as 'rise time'.

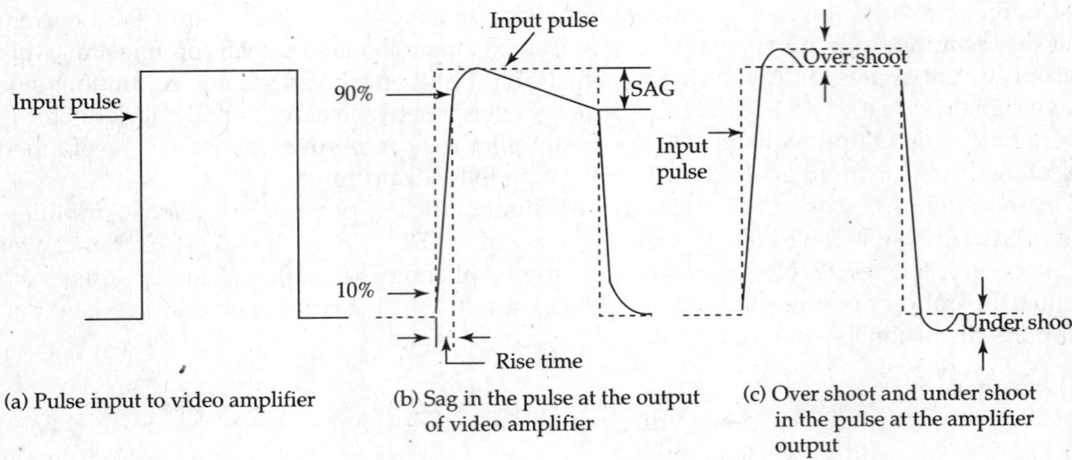

FIG. 13.59 *Video amplifier input and output signal waveforms*

For good reproduction of input signal at the Amplifier output,

$$\text{Rise time } R_T = \frac{0.35}{B} \text{ sec,}$$

where B is the amplifier bandwidth.

1. SAG observed in pulse output wave shape (b) in Fig. 13.59 indicates 'defective low frequency response' of the Amplifier.
2. Output pulse wave shape (c) in Fig. 13.59 showing overshoot and undershoot represents the 'transient response' of the Amplifier. Transient response represents the amount of fastness in output signal response to instantaneous changes in input pulse (simulation to very high-frequency content of video signals).
3. Transient response to pulse inputs is an important criterion for video Amplifiers than its normal frequency response characteristic to sinusoidal signals.
4. Rise times of less than 0.1 μs is ideal for good Television channels. Rise time decides the nature of fidelity of reproduction of signals.
5. Wideband (RF Broadband) Amplifiers of desired power and frequency bands are available in the market for testing TV Transmitters and so on.

Wideband Amplifiers can be designed using one of the three compensating techniques.

1. *High-frequency compensation* to increase the high-frequency range.
2. *Low-frequency compensation* to increase the low-frequency range.
3. *Both low- and high-frequency compensations* to increase overall Bandwidth.

Wideband Amplifier circuits using high-frequency compensation
Small inductance 'L_{SERIES}' is added to load resistance R_L in Amplifier output circuit (Fig. 13.60) to compensate for loss of gain at HF. It is known as *high-frequency compensation.*

Compensating inductance L_{SE} forms a parallel resonant circuit with C_{SH} (Fig. 13.60). Using suitable design value for $L_{SE} = 2\,\omega_2 \cdot R_{L'}$, uniform gain-in during higher frequencies is realised.

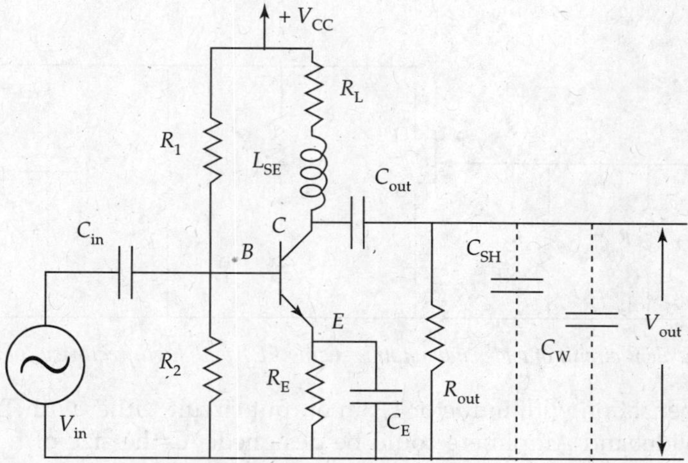

FIG. 13.60 *Wideband amplifier using BJT with high-frequency compensation using L_{SE}*

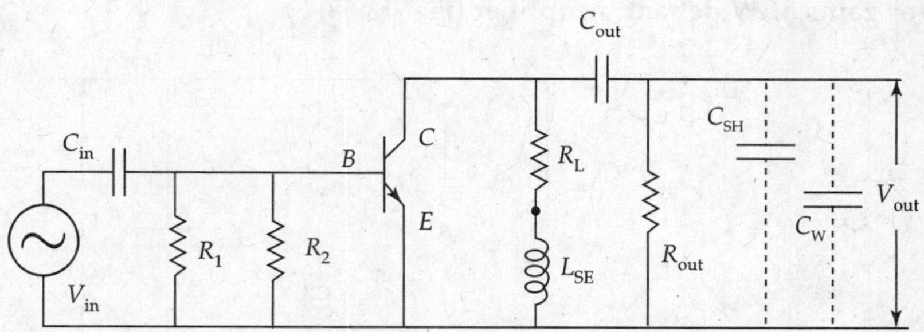

FIG. 13.61 *AC equivalent circuit of wideband amplifier using BJT with high-frequency compensation using L_{SE}*

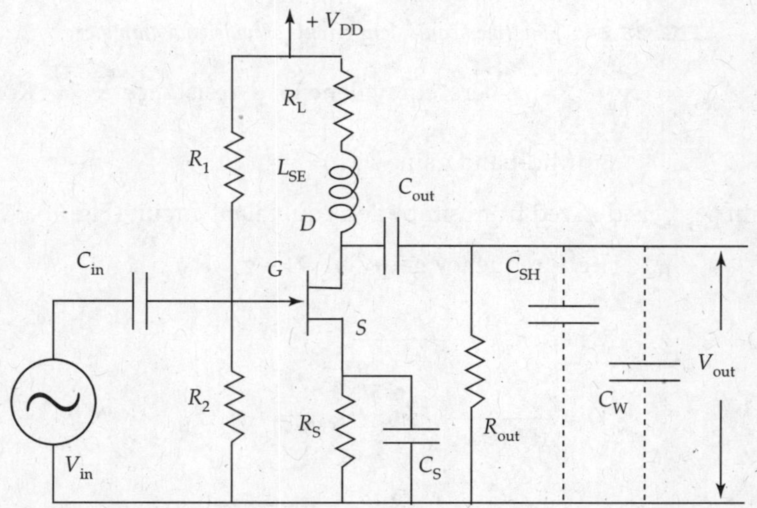

FIG. 13.62 *Wideband amplifier using FET with high-frequency compensation using L_{SE}*

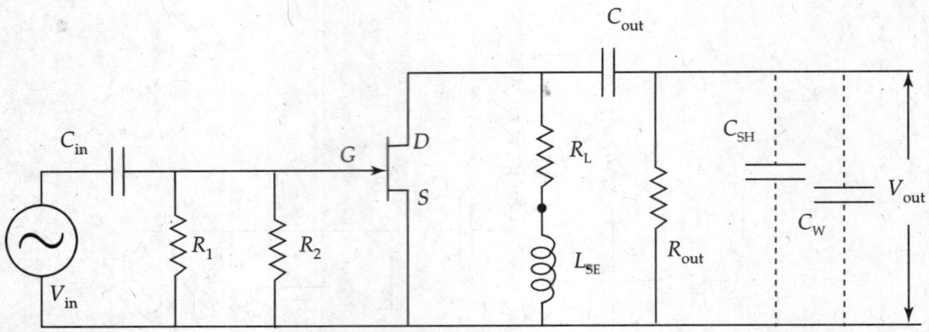

FIG. 13.63 *AC equivalent circuit of wideband amplifier using FET with high-frequency compensation using L_{SE}*

Using HF compensation with inductor L_{SE} in output circuit, either with Transistor or with FET Amplifiers, wideband Amplifiers could be designed. As the size of L_{SE} is increased, Q of the coil also increases. Increase in Load impedance ($R_L + j \cdot \omega \cdot L_{SE}$) at high frequencies increases the Amplifier gain with simultaneous increase in Amplifier bandwidth.

Expressions for gains of Wideband Amplifier (Fig. 13.68)

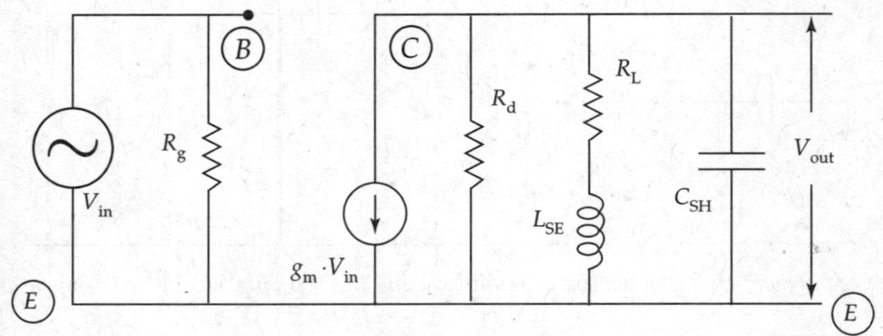

FIG. 13.64 *Simplified equivalent circuit of wideband amplifier*

Mid-band gain $A_{mid} = -g_m \cdot R_{eq}$, where equivalent load resistance $R_{eq} = R_L$ for mid-band frequencies.

$$\text{Mid-band gain} \quad A_M = -g_m \cdot R_L \tag{13.84}$$

Amplifier gain 'A_{high}' is derived from simplified equivalent circuit (Fig. 13.64)

$$\text{High-frequency gain} \quad A_H = -g_m \cdot R_{eq}, \tag{13.85}$$

$$\text{where } R_{eq} = \frac{1}{Y_L + Y_{CSH}}$$

$$Y_L = \frac{1}{R_L + j\omega L} \quad \text{and} \quad Y_{CSH} = j \cdot \omega \cdot C_{SH}$$

$$\therefore \quad A_H = -g_m \cdot R_{eq} = -\frac{-g_m}{Y_L + Y_{CSH}} \tag{13.86}$$

Normalised high-frequency gain

$$\frac{A_{\text{high}}}{A_{\text{mid}}}=\frac{\left(\dfrac{g_{\text{m}}}{Y_{\text{L}}+Y_{\text{CSH}}}\right)}{g_{\text{m}}\cdot R_{\text{L}}}=\frac{1}{R_{\text{L}}(Y_{\text{L}}+Y_{\text{CSH}})} \tag{13.87}$$

$$\frac{A_{\text{H}}}{A_{\text{M}}}=\frac{1}{R_{\text{L}}\left[\dfrac{1}{R_{\text{L}}+j\cdot\omega\cdot L}+j\cdot\omega\cdot C_{\text{SH}}\right]}$$

$$\frac{A_{\text{H}}}{A_{\text{M}}}=\frac{\dfrac{[R_{\text{L}}+j\cdot\omega\cdot L]}{R_{\text{L}}}}{1+j\cdot\omega\cdot C_{\text{SH}}\cdot R_{\text{L}}-(\omega)^2 L\cdot C_{\text{SH}}}$$

$$=\frac{\left[1+j\left(\dfrac{\omega}{\omega_2}\right)Q_2\right]}{\left[1+j\dfrac{\omega}{\omega_2}-\left(\dfrac{\omega}{\omega_2}\right)^2 Q_2\right]}, \tag{13.88}$$

where $\left(\dfrac{\omega}{\omega_2}\right)$ is known as normalised frequency $f_{\text{N}}=\dfrac{f}{f_2}$ and $f_2=\dfrac{1}{2\pi\cdot R_{\text{L}}\cdot C_{\text{SH}}}$.

Normalised high-frequency gain (relative gain) is a function of normalised frequency f_{N}. Flat response to extended bandwidth can be obtained when $R_{\text{L}}=2\cdot\omega_2\cdot L$. Increased half-power frequency f_2 with compensation for Wideband Amplifier will be equal to 1.8 f_2

Low-frequency compensation

- Reduction in voltage gain of an Amplifier in low-frequency region is due to output-coupling capacitor C_{out}. Impedances in the shunt path of Amplifier are very high and do not have any impact on low frequency response of an Amplifier.
- Improvement in voltage gain at low-frequency region of Amplifier could be obtained by using a low-frequency compensation network consisting of a parallel combination of one resistor R_{LF} and a capacitor C_{LF} that is connected between load resistance R_{L} and positive of supply voltage V_{CC} shown in Fig. 13.65.

Circuit operation

- For middle and high-frequency range signals through Amplifier, capacitor C_{LF} offers very low reactance path and virtually provides a short circuit path around R_{LF}. Then the effective load resistance for the Amplifier is R_{L} only.
- For Low-frequency signals, impedance offered by capacitor C_{LF} is so large that it virtually behaves as an open circuit across R_{LF}.
- Then the effective load resistance $R_{\text{L(eq)}}$ to Amplifier is the series combination of two resistors R_{L} and R_{LF}. Hence, equivalent load resistance of the Amplifier increases to $R_{\text{L}}+R_{\text{LF}}$.
- Low frequency voltage gain increases due to increase in load resistance, thus compensating for the loss of gains due to coupling capacitors of the Amplifier.

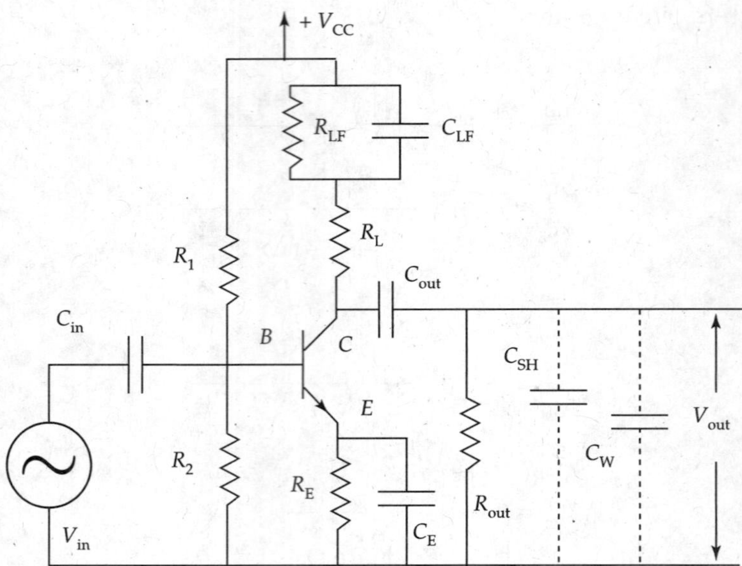

FIG. 13.65 *Wideband amplifier using BJT with low frequency compensation using C_{LF} and R_{LF}*

13.16 APPLICATIONS OF WIDEBAND AMPLIFIERS

1. *Tuned Amplifiers* amplify carrier waves modulated by audio signals in Radio and picture signals in Television transmission systems operate on Wideband signals. They are known as *Wideband Amplifiers.*
2. **Narrow Band Tuned Amplifiers** In communication equipment, speech or music is carried over HF signals using a process of modulation. Information content occupies a narrow band of signals about the centre or carrier waves at RF, VHF and so on. Bandwidth of the information occupies about 10% of carrier frequencies. It is known as narrow band. Such Amplifiers are known as *Narrow Band Tuned Amplifiers.*
3. **Wideband Tuned Amplifiers in Television receivers** For Television picture (video) signal transmission and reception, Amplifiers dealing with TV programs are wideband *or video frequency Amplifiers.* Bandwidth is a large percentage of its centre frequency.
   - Typical 3-dB bandwidth is 5.5 MHz for IF Amplifiers to obtain required frequency response of a TV receiver.
4. **Radar Indicators** The output signal of a Radar receiver will be in the form of visual display using a cathode ray tube in Type A displays. Target detection is made using a visual display obtained by the application of saw-tooth voltage to horizontal plates and radar receiver output for vertical deflection in CRT.

   PPI (Plan Position Indicator) displays use saw-tooth voltages for radial display. Amplification of non-sinusoidal signals (saw-tooth voltages) uses Wideband Amplifiers.
5. Miscellaneous applications Industrial control instrumentation, medical instrumentation, which use non-sinusoidal signals such as square waves, pulses and saw-tooth signals and amplify them during the application require Wideband Amplifiers.

SUMMARY

1. Tuned Amplifiers amplify a narrow band of signals. It functions as Band Pass Filter.
2. Tuned Amplifiers in RF stages (a) Amplify weak signals to increase sensitivity of radio receivers; (b) Additional discrimination is provided against signals in adjacent bands and improves the selectivity and image frequency rejection.
3. Both tuned circuits of Double-Tuned Amplifier are tuned to the same frequency. They provide larger 3-dB bandwidth response with steeper sides and flat top than single-tuned Amplifiers.
4. Double-Tuned Amplifiers provide three types of responses depending upon the degree of coupling between the two tuned circuits of the Amplifier. (a) Critical-coupling response takes place when $KQ = 1$; (b) Over-coupling response occurs when $KQ > 1$; (c) Under-coupling response occurs when $KQ < 1$.
5. For over-coupling situation, the output response of Double-Tuned Amplifiers contains two peaks or double peaks at the top of the response.
6. Tuned Amplifiers with multiple tuned circuits are classified as synchronously Tuned Amplifiers and Stagger-Tuned Amplifiers.
7. Stagger-Tuned Amplifiers are used to achieve maximally flat Butterworth response with flat edges to have good selectivity feature.
8. Voltage gain of a stagger-Tuned Amplifier is less than that of a single-tuned Amplifier. The reason is that increase in bandwidth of stagger-tuned Amplifiers results in reduction in Amplifier gain.
9. Neutralisation is the method to reduce the Miller effect due to the inter electrode capacitance between the Base and the Emitter.

Questions for Practice

1. What is meant by the term Tuned Amplifier and briefly explain the classification of Tuned Amplifiers (JNTU, Nov. 2007).
2. Draw ideal and actual frequency response curves of single-tuned Amplifier (JNTU, Nov. 2007).
3. Draw and explain the significance of gain versus frequency curve of a Tuned Amplifier when they are used in radio Amplifiers (JNTU, Nov. 2007).
4. Draw the circuit diagram and the Small Large-signal AC equivalent circuit of a single-tuned Amplifier using BJT with a tank circuit connected at the input side of the Amplifier (JNTU, Nov. 2006).
5. Draw the equivalent circuit of a capacitance-coupled single-tuned Amplifier and derive the expression for voltage gain (JNTU, May 2005).
6. Why do we use Tuned Amplifiers in the IF and RF stages (JNTU, Feb. 2008)?

7. Draw the high frequency equivalent circuit of a single-tuned capacitance-coupled BJT Amplifier and derive the expression for (a) voltage gain, (b) voltage gain at resonance and (c) 3-dB bandwidth (JNTU, Feb. 2008).
8. Derive the expression for 3-dB bandwidth of a capacitance-coupled single-tuned Amplifier (JNTU, May/June 2005).
9. Draw a simple BJT Tuned Amplifier circuit and its ideal frequency response characteristic (JNTU, May/June 2004).
10. Draw the circuit of single-tuned transformer-coupled JFET Amplifier and analyse its working.
11. (a) Draw and explain the circuit diagram and equivalent circuit using high frequency Hybrid-π model of a single-tuned capacitance-coupled BJT Amplifier.
    (b) Also draw and explain the obtained high frequency equivalent circuit using Miller's theorem (JNTU, Nov. 2006).
12. Draw and explain the circuit diagram of a single-tuned capacitance-coupled Amplifier. Also explain the circuit operation (JNTU, Nov. 2007).
13. (a) Draw the circuit of FET-Tuned voltage Amplifier. Derive the necessary expression to draw the universal resonance curve with all necessary details.
    (b) Design a single-stage FET-Tuned Amplifier for the following specifications: $f_0 = 12$ MHz, bandwidth $B = 10$ kHz and mid-band gain $A_{mid} = -15$. The FET parameters are trans-conductance $g_m = 4$ ms, Drain resistance $r_d = 25\ k\Omega$, Capacitance $C_{GS} = 30$ pF and Capacitance $C_{GD} = C_{DS} = 5$ pF (JNTU, June 2004).
14. Draw the Double-Tuned Transformer-coupled Amplifier circuit. Draw the nature of responses of Amplifier for different values of $KQ = 1$, $KQ > 1$ and $KQ < 1$ (JNTU 2004).
15. Explain how a Stagger-Tuned Amplifier design is superior to synchronously Tuned Amplifier design in the design of multistage Amplifiers? Also draw their circuit diagrams and equivalent circuit diagrams (JNTU, Nov. 2007).
16. Explain the principle of stabilising the Double-Tuned transformer-coupled Amplifier response against the internal feedback (JNTU, May/June 2004).
17. (a) What are the main advantages of Class-C operating mode in RF applications?
    (b) Draw the circuit of Class-C radio frequency Amplifier and explain its operation with necessary waveforms (JNTU, Feb. 2008).
18. (a) Mention the three methods of stabilisation of a single-tuned BJT Amplifier against the feedback capacitance connected between the Base and the Collector.
    (b) Explain in detail various Neutralisation techniques with the help of circuit diagrams (JNTU, Feb. 2008).
19. (a) What is a video Amplifier? Explain the need for video Amplifiers.
    (b) Explain in detail the design considerations of video Amplifiers (JNTU, Feb. 2008).
20. Explain what you mean by Synchronous tuning of Tuned Amplifiers? Draw the frequency response of a synchronously Tuned Amplifier showing the responses of individual stages and overall responses (JNTU, Feb. 2008).

Multiple Choice Questions

1. Tuned Amplifier has the same type of frequency response as that of ______________
   (a) high pass filter (b) low pass filter
   (c) band stop filter (d) band pass filter

2. Tuned Amplifiers are employed as ______________
   (a) audio amplifier (b) HF Amplifier
   (c) RF and IF Amplifiers (d) DC Amplifiers

3. Ideal value of skirt selectivity for good quality Tuned Amplifier is ______________
   (a) 3 or less (b) more than 3 (c) 31.61 (d) 1

4. The bandwidth shrinkage factor for two synchronously Tuned cascaded Amplifiers is ______________
   (a) 0.64 (b) 0.51 (c) 0.43 (d) 0.39

5. Butterworth response can be obtained in ______________
   (a) single-tuned amplifiers (b) double-tuned amplifiers
   (c) synchronously tuned amplifiers (d) staggered-tuned amplifiers

6. Configuration of a Tuned power Amplifier is ______________
   (a) Class-A (b) Class-B (c) Class-C (d) Class-D

7. Frequency response exhibiting flat passband and skirt selectivity is obtained in ______________
   (a) single-tuned Amplifiers (b) double-tuned amplifiers
   (c) synchronously tuned amplifiers (d) staggered-tuned amplifiers

8. Amplifier that do not suffer from Miller effect usually preferred in IC implementation is ______________
   (a) single-tuned (b) double-tuned
   (c) staggered-tuned (d) CC–CB cascade with tuned output stage

Answers to Multiple-Choice Questions

1. (d) 2. (c) 3. (a) 4. (a) 5. (d)

6. (c) 7. (d) 8. (d)

Chapter 14

SWITCHING AND IC VOLTAGE REGULATORS

Learning Objectives

- In Chapter 3, we discussed Voltage Regulators that use discrete components.
- In this Chapter, we describe Voltage Regulators that use integrated components, ICs, considering advancement of technologies and miniaturisation of components.
- We also learn the design aspects of regulator circuits.
- Standby energy sources such as UPS and SMPS are also described, which are quite popular at the times of power failure for commercial and domestic needs.

14.1 INTRODUCTION

14.1.1 IC Voltage Regulators

Electronic circuits need DC power supply. Voltage Regulator circuits provide constant DC output voltages over the designed range of loads and input voltages. IC Voltage Regulators are technologically improved over conventional Voltage Regulators, which use discrete components.

Applications of IC Voltage Regulators in present day technology

- 'Energy Management Systems' and 'energy efficient' embedded systems are being introduced in many systems and will be playing major role in Smart Grid and other areas, all over the world. One example is LP5550. It is a power-wise interface compliant power management unit for reducing electrical power consumption of standalone Baseband processors in mobile phones or other wireless device application processors used in Home Automation (HA) systems.

- GSM/GPRS/EDGE and UMTS cell phones, handheld Radios, PDAs, portable instruments and so on.
- Special systems are being developed for reducing electrical power consumption in many home and industrial applications using wireless techniques.
- Used in Voltage feeder circuits.
- Power source for communication devices.
- Power source for home electronic appliances.

Users can select product type, output voltage and current ratings and package type of IC Voltage Regulators in designing power-supply circuits.

Advantages of IC regulators
(1) Miniaturisation of regulator; (2) Standardization of various building blocks inside the IC; (3) Reduced cost due to mass production; (4) High design flexibility; (5) Good flexibility of connection settings; (6) Improved performance (transient response) characteristics; (7) Functions like current limiting, thermal shutdown; over Voltage protection can easily be embedded in IC regulators. Example: Three-terminal Voltage Regulators have such embedded features.

Linear regulators and switching regulators are available in integrated circuits.

Linear regulator Linear regulator basically contains (1) unregulated DC voltage source as input voltage V_{in}, (2) Voltage-controlled current source, and (3) control circuitry to maintain constant DC output voltage.

Switching regulator Switching regulator contains (1) Unregulated DC voltage source as input voltage V_{in}; (2) Pass transistor to switch the Voltage Regulator circuits ON and OFF. The power dissipation by the Transistor is almost zero. Power efficiency is high; (3) Inverter control circuitry to maintain *constant DC output voltage*; (4) Switching regulators are gaining popularity with latest IC circuit versions and simplicity in design of power-supply circuits. *Switching regulators function as DC-to-DC converters.*

Functions of Switching Regulators

- They are used as DC-to-DC converters to produce varying output voltages. When the output voltage is higher than the input voltage, it is known as *Boost Regulator*. When the output voltage is less than the input voltage, it is known as *Buck Regulator*.
- They generate output DC voltage of opposite polarity to the input voltage. Then such regulator circuits are known as *Invert* or *Buck-Boost Regulator.*
- Due to higher power conversion efficiency, switching regulators are used in Personal Computers, Laptops and Television receivers.

14.2 THREE-TERMINAL IC VOLTAGE REGULATORS

14.2.1 Classification of Linear IC Voltage Regulators

1. **General-purpose regulators** They have fixed output voltages with limited range of DC output voltages and currents.

2. **Precision regulators** They can be operated over a wide range of input and output voltages, with finer granularity.

General-purpose Regulators

They are *Three-Terminal Voltage Regulators* having:

1. Input terminal V_{IN}; (2) Output terminal V_{OUT}; (3) Ground terminal common to both the input and output ports of regulator unit. They provide 5V, 6V, 8V, 12V, 18V etc. up to 50V with current ranges from 0.5 to 3 A.
2. Additional features: (1) Fixed positive or negative output voltages; (2) Current limiting capability; (3) Thermal shutdown.

Limitations

(1) External circuitry is needed to obtain higher Voltages. (2) Complex circuitry is needed for electronic shutdown. (3) Programmability of output voltage is possible, but performance degrades.

Three-pin IC Voltage Regulator circuit (Fig. 14.1)

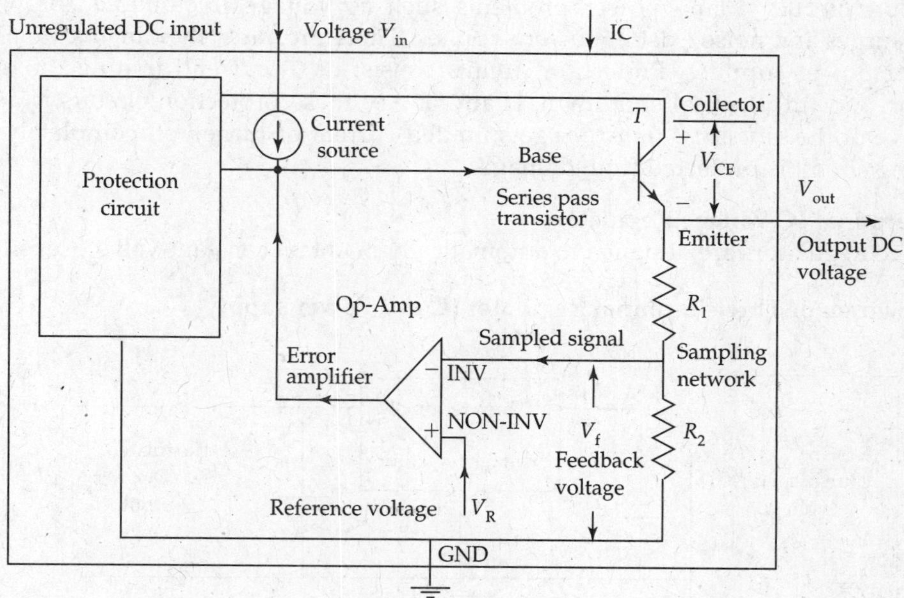

FIG. 14.1 *Various blocks (inside) of three-pin IC voltage regulator*

Various blocks in three-pin IC Voltage Regulator

(1) Sampling Network; (2) Error Amplifier; (3) Series Pass Transistor; (4) Reference Voltage; and (5) Protection Circuit.

Sampling Network (R_1:R_2) *Sampling* (sensing) *network* contains two resistors R_1 and R_2, and senses fractional changes in the output load Voltage V_{out}.

Error Amplifier (Op-Amp) Error Amplifier is an *Operational Amplifier*. Sampled signal is fed to 'inverting terminal' (INV) of *Error Amplifier*. Reference Voltage (V_R) is provided usually from a temperature compensated Zener Diode D_Z. Reference Voltage (V_R) V_Z is fed to 'non-inverting terminal' (NON-INV) of *Operational Amplifier.* It compares a fraction of the sampled output voltage V_f with reference Voltage V_R. Its output is fed to base terminal of *control power transistor T* (Series *Pass Transistor*).

Series Pass Transistor (*T*) Series Pass Transistor is an NPN transistor (*T*) connected between the input and the output ports of power supply. Error Amplifier's output modulates the conduction of *series pass transistor,* according to the sensed variations of output voltage. Output voltage is maintained constant, despite the variations in load conditions. Thus the output voltage is a *regulated DC voltage,* as per the design. Minimum Voltage drop (V_{CE}) is required across the regulator to maintain constant DC output voltage. Minimum Voltage drop V_{CE} is considered as the *low drop out Voltage* (LDO). *Drop out Voltage factor* differentiates between various power-supply circuits.

LDO regulator Voltage Regulator circuits, using a single NPN type pass Transistor, are known as LDO (Low Drop Out) regulator circuits. Minimum drop out Voltage for such regulators is V_{CE}, across the collector and emitter terminals of the series pass transistor.

Protection circuit Many power problems such as Voltage fluctuations (brown outs), Voltage surges and noise exist in power supplies. Therefore, *protection circuits* are necessary for better power supplies. Protection circuits consist of (i) current limiting, (ii) safe area operation and (iii) thermal shutdown. If any one of these protection circuits is activated, base drive of the *series pass transistor* gets limited current or turned off completely and the remaining circuit is protected from damages.

Three-terminal IC Voltage Regulator

IC Voltage Regulators are designed to automatically maintain constant Voltage level.

Block diagram of Three-Terminal Regulator IC in a Power supply

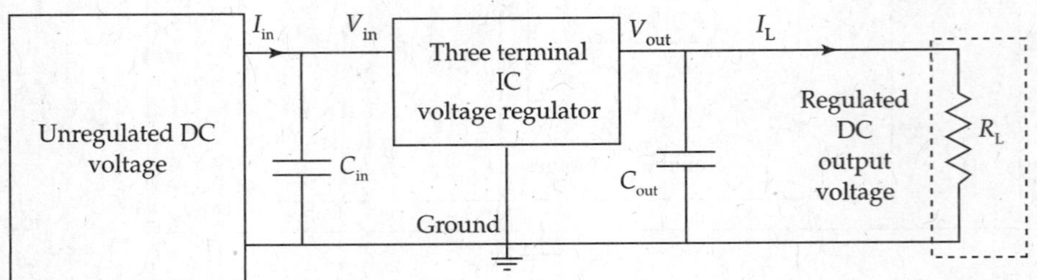

FIG. 14.2 *Pin configuration of three-terminal regulator in a DC power supply circuit*

Basic connections of a three-terminal IC regulator (Fig. 14.2)

- Unregulated DC voltage will be connected to input terminal V_{IN} and ground. Regulated DC output voltage is across V_{OUT} to ground terminals for external use.
- Output voltages can be multiples of input voltage levels.
- Load current Variations cause sudden changes in output voltages. Time taken to correct changes in output voltage is known as *Transient response.* Capacitors C_{in} and C_{out} improve *Transient response* and operational stability of the regulator.

- Power losses in input mains Transformer, Rectifier Diodes and filter circuits are negligibly small. There is lot of power dissipation in the regulator circuit. So, design of regulator circuit with minimum power losses has to be done in practice.
- Power efficiency is an important performance index of Voltage Regulator circuits. (Power efficiency is the ratio of DC output power to AC input power).

Important features of three-terminal IC Voltage Regulators
(1) Require a few external components. (2) Heat Sink for temperature ventilation is very minimal. (3) Available in both plastic and metal packages. (4) They provide either positive or negative output voltages. (5) Output voltage can be fixed or variable.

78XX Series Voltage Regulators (Fig. 14.3)

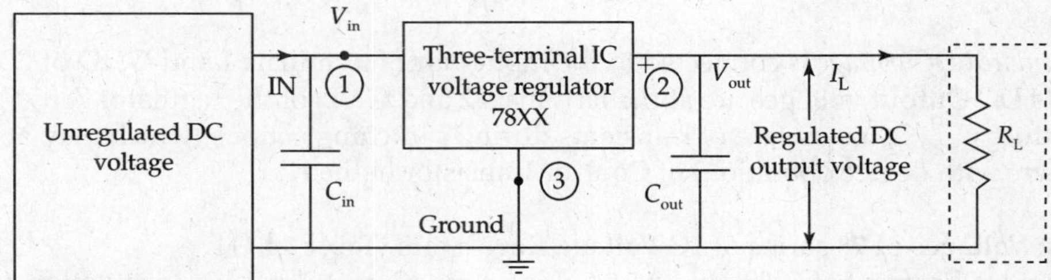

FIG. 14.3 *Pin confjgurations of three-terminal voltage regulator 78XX IC*

- Last two numbers of 78XX series indicate the magnitude of output voltage. As an example, IC 7805 provides +5 V regulated DC output voltage.
- They produce fixed positive Voltages ranging from 5 to 24 V.

Block diagram representation of IC 7805 (Fig. 14.4)

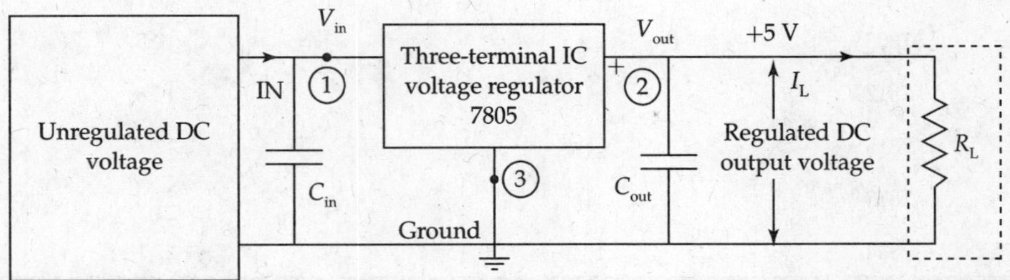

FIG. 14.4 *Pin configurations of three-terminal voltage regulator7805 IC + 5 V DC power supply circuit*

- 7805 IC Voltage Regulator produces a positive output voltage of 5 V.
- Output current could be greater than 0.5 A with proper Heat Sinks to the device.

IC 7805 Voltage Regulator along with unregulated power supply using Bridge Rectifier

- Unregulated DC voltage larger than about 2 V to required DC output voltage of 5 V is produced by a step-down Transformer, Bridge rectifier (containing four Diodes D_1, D_2, D_3 and D_4) as a package unit and shunt capacitor filter C_2 (Fig. 14.5).

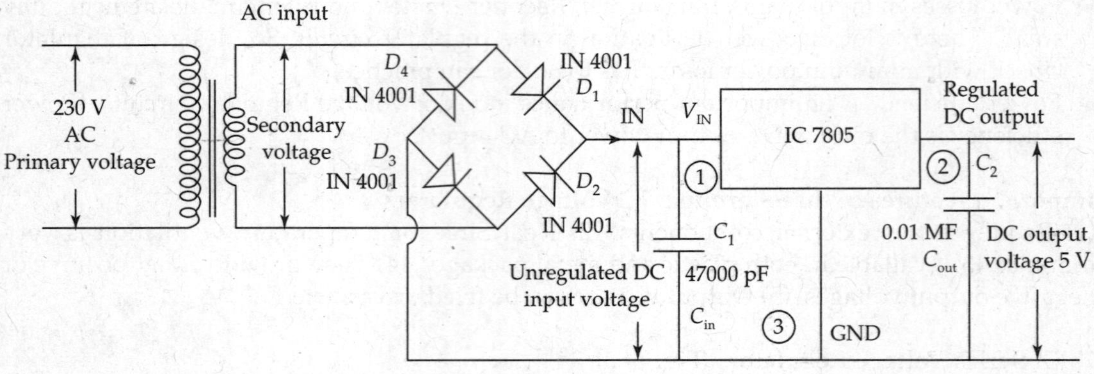

FIG. 14.5 *+5 V power supply using bridge rectifier and voltage regulator IC 7805*

- *Unregulated DC voltage* is connected to the input pair of terminals **1** and **GND** of IC.
- 5 Volts DC Output voltages are at the terminals **2** and **GND** of the regulator 7805.
- Line filter C_{out} (C_2) takes care of transients during switching actions (which may otherwise interfere with device operation) of Control Transistor in the IC.

Working Voltages of 78 series of IC Voltage Regulators (Table 14.1)
Maximum load current of 7800 series is 1 A. Line regulation of 3 mV, load regulation of 15 mV and ripple rejection capability of the order of 80 dB.

Table 14.1 **Working voltages of 78 series ICs (*Positive Voltage Regulators ICs*)**

S. No.	Regulator IC	Normal output voltage (Volts)	Maximum input voltage to IC (Volts)
1	7805	+5	7.3
2	7806	+6	8.3
3	7808	+8	10.5
4	7812	+12	14.6
5	7815	+15	17.7
6	7818	+18	21.0
7	7824	+24	27.1

Circuit diagram of 7815 Voltage Regulator using full-wave Rectifier (Fig. 14.6)
To obtain 15 V DC output voltage, it is provided with unregulated DC input voltage of 18 V (assuming a minimum Voltage drop of 3 V between V_{in} and V_{out}) [$V_{in} = (V_{out} + 3) = (15 + 3) = 18$ V]. Unregulated DC voltage (18 V) is obtained from Full-wave rectifier circuit with AC supply voltage, step-down Transformer (having centre-tapped secondary winding), two rectifier diodes and filter circuit. Unregulated voltage is connected to the input port of 7815 VR IC. Regulated DC output voltage is 15 V. Line filter Capacitor C_2 at output port reduces the high-frequency noise.

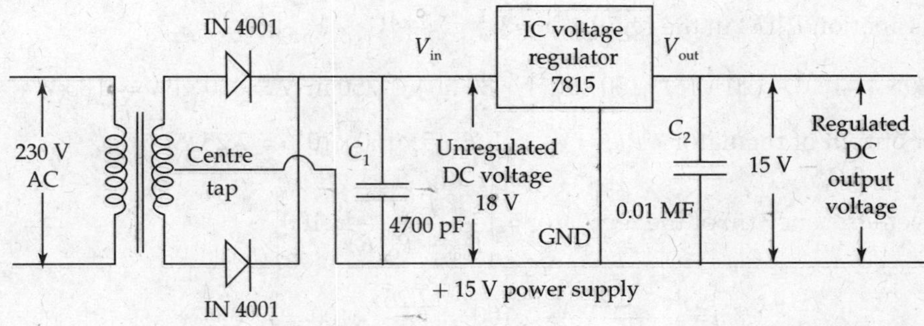

FIG. 14.6 *Using a full-wave rectifier and IC 7815 voltage regulator*

Fixed Negative Voltage Regulators

Voltage Regulator ICs with 79XX series and LM320 provide negative Voltages ranging from –5 to –24 V similar in operation to positive Voltage Regulators.

Table 14.2 Working voltages of 79 series ICs (*Negative Voltage Regulator ICs*)

S. No.	Regulator IC	Normal output voltage (Volts)	Minimum input voltage to IC (Volts)
1	7905	–5	–7.3
2	7906	–6	–8.3
3	7908	–8	–10.5
4	7909	–9	–11.5
5	7910	–10	–12.5
6	7912	–12	–14.6
7	7915	–15	–17.7
8	7918	–18	–21.0
9	7924	–24	–27.1

EXAMPLE 14.1

Unregulated DC voltage of 20 V is applied to IC 7815, which is connected to a load impedance of 100 Ω. Determine (a) Power dissipated in regulator, (b) Regulator efficiency and (c) Find the regulator efficiency, when the input is 24 V.

Solution:

(a) (i) Assume quiescent current $I_Q = 4$ mA

7815 voltage regulator output voltage $V_{out} = 15$ V and $R_L = 100\ \Omega$

$$\text{Load current} \quad I_L = \frac{V_{out}}{R_L} = \frac{15\text{ V}}{100\ \Omega} = 150\text{ mA}$$

Power dissipation $P_D(R)$ in the regulator $= V_{in} \times I_Q + [V_{in} - V_{out}] \times I_L$

$P_D(R) = 20 \times 4 \times 10^{-3} + (20 - 15) \times 150 \times 10^{-3} = 80\text{ mW} + 750\text{ mW} = 830\text{ mW} = 0.83\text{ W}$

(ii) Power output of regulator $P_{out} = V_{out} \times I_L = 15 \times 150 \times 10^{-3} = 2.25\text{ W}.$

(b) % Power efficiency (η) of the regulator $= \left[\dfrac{P_{out}}{P_{out} + P_D(R)} \times 100\right]$

$$\%\eta = \left[\frac{2.25}{2.25 + 0.83}\right] \times 100 = \left[\frac{2.25}{3.08}\right] \times 100 = 73.05\%$$

(c) When $V_{out} = 15$ V, Load current $I_L = \dfrac{V_{out}}{R_L} = \dfrac{15}{100} = 150$ mA

Power dissipated in the regulator $P_D(R)$ when the input $V_{in} = 24$ V

(i) $P_D(R) = I_Q \times V_{in} + [V_{in} - V_{out}] \times I_L = 4 \times 10^{-3} \times 24 + (24 - 15) \times 150 \times 10^{-3}$

$\therefore\ P_D(R) = 96\text{ mW} + 1350\text{ mW} = 1446\text{ mW} = 1.446\text{ W}$

(ii) Output power of the regulator $P_{out} = (V_{out} \times I_L) = 15 \times 150 \times 10^{-3} = 2.25\text{ W}$

$\therefore$ % Regulator efficiency $(\eta) = \left[\dfrac{P_{out}}{P_{out} + P_D(R)}\right] \times 100 = \left[\dfrac{2.25}{2.25 + 1.446}\right] \times 100 = 60.88\%.$

Regulator Efficiency decreases when voltage drop across regulator increases.

EXAMPLE 14.2

Design a constant current source using 7805 Voltage Regulator to deliver 100 mA load current. Given $I_Q = 4.3$ mA and $R_L = 5\ \Omega$ using the circuit in Fig. 14.7.

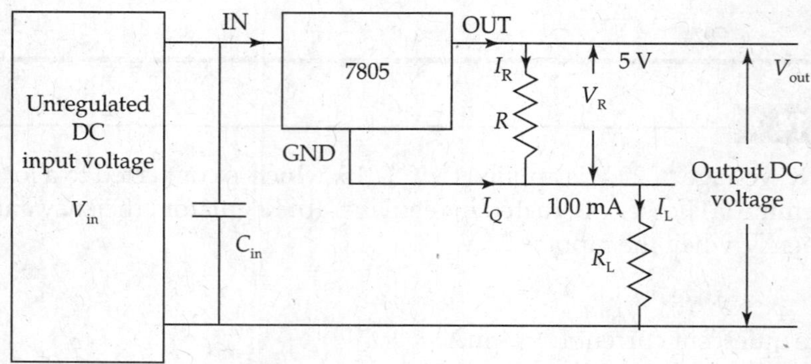

FIG. 14.7 *7805 voltage regulator design*

Solution:

Data: $I_Q = 4.3$ mA, $I_L = 100$ mA and $V_R = 5$ V for 7805 and $R_L = 5\ \Omega$.

From the given voltage regulator circuit,

$$\text{Load current} \quad I_L = \left[\frac{V_R}{R} + I_Q\right]$$

$$\therefore \quad I_L = 100 \text{ mA} = \left[\frac{5}{R} + 4.3 \text{ mA}\right]$$

$$\frac{5}{R} = \left[100 \times 10^{-3} - 4.3 \times 10^{-3}\right] = 95.7 \text{ mA}$$

$$R = \frac{5}{95.7 \times 10^{-3}} = \frac{5}{95.7} \times 10^3 = 52.25\ \Omega$$

$$\text{Output voltage} \quad V_{out} = (V_R + V_L) = (5 + I_L \times R_L)$$

$$\therefore \quad V_{out} = [5 + 100 \times 10^{-3} \times 5]$$

$$= (5 + 0.5) = 5.5 \text{ V}$$

Assuming low voltage drop of 2 V across the regulator,

$$\text{Required magnitude of unregulated voltage} \quad V_{in} = (V_{out} + 2)$$

$$\therefore \quad \text{Unregulated voltage} \quad V_{in} = (5.5 + 2) = 7.5 \text{ V}$$

Adjustable Voltage Regulator LM317 (Fig. 14.8)

LM 317 IC is a positive Voltage Regulator. Unregulated DC voltage (from FW rectifier and filter circuit) is connected to the input port of IC. Fixed resistor R_1 and variable resistor R_2 are connected across the output port of IC. Regulated DC output voltages ranging from 1.2 to 37 V (with load current of 1.5 A) can be obtained.

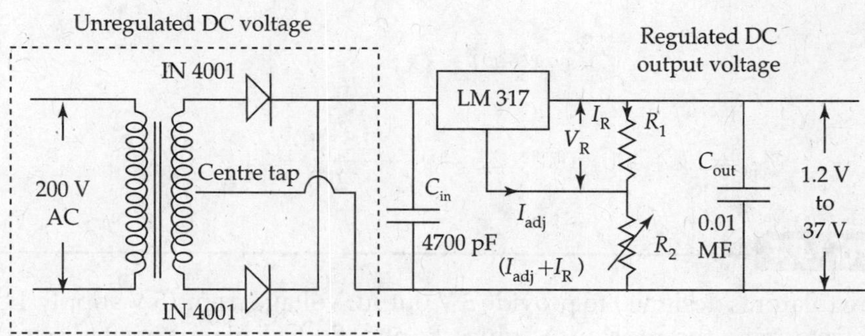

FIG. 14.8 *Adjustable voltage regulator using LM 317 (three-pin regulator)*

A constant Voltage provided by the regulator is connected across R_1. Voltage V_{R1} is used as the reference Voltage V_R. Current through resistor R_1 is the constant reference current I_R. Current through variable resistor R_2 is $(I_{adj} + I_R)$. Variable output voltages can be set by using the external resistors R_1 and R_2 using the following expression:

$$V_{out} = V_R + I_{adj} \times R_2 + I_R \times R_2 \tag{14.1}$$

$$V_{out} = V_R + I_{adj} \cdot R_2 + \frac{R_2}{R_1} \cdot V_R \tag{14.2}$$

Rearranging the terms

$$V_{out} = V_R\left[1+\frac{R_2}{R_1}\right]+I_{adj}\cdot R_2 \quad (14.3)$$

$$V_{out} = [V_{R1} + V_{R2}] = I_R \times R_1 + (I_{adj} + I_R)\, R_2 \quad (14.4)$$

where $I_R = I$.

EXAMPLE 14.3

If $V_R = 1.0$ V, $I_{adj} = 400$ μA, $R_1 = 250\ \Omega$ and $R_2 = 2.5$ kΩ, calculate the magnitude of the output voltage V_0 using the circuit in Fig. 14.9.

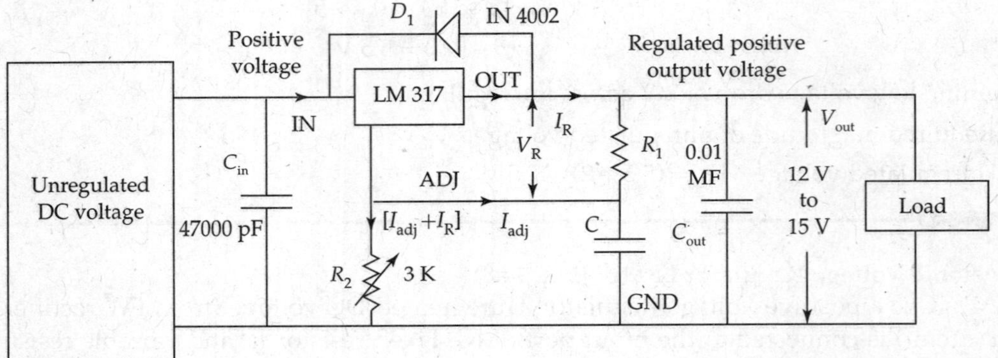

FIG. 14.9 *Adjustable positive voltage regulator using three-pin regulator LM 317*

Solution:

$$V_{out} = \left[V_R + I_{adj}\times R_2 + \frac{R_2}{R_1}\times V_R\right] \quad (1)$$

$$V_{out} = 1.0 + 400\times10^{-6}\times2.5\times10^3 + \frac{2.5\times10^3}{250}\times1.0 = 12.0\text{ V.} \quad (2)$$

EXAMPLE 14.4

An LM317 regulator is designed to provide 5 V output voltage from 15 V supply. Load current is 100 mA. Determine the resistance values R_1 and R_2. Calculate power dissipation in the regulator IC (Fig. 14.9).

Solution: Choosing current I_R through R_1 as 1.25 mA and $V_R = 1.25$ V.

$$R_1 = \frac{V_R}{I_R} = \frac{1.25}{1.25\times10^{-3}} = 1.0\text{ k}\Omega \quad (1)$$

$$R_2 = \frac{(V_{out} - V_R)}{I_R} = \frac{(5-1.25)}{1.25\times10^{-3}} = 3.0\text{ k}\Omega \quad \text{neglecting } I_{adj} \text{ through } R_2. \quad (2)$$

Power dissipation the regulator $= P_D = (V_{in} - V_{out}) \times I_{L(max)} = (15 - 5) \times 100 \times 10^{-3} = 1$ W.

EXAMPLE 14.5

Design an adjustable Voltage Regulator using LM317, for an output voltage $V_0 = 12$ to 15 V and output current of 500 mA, using the circuit in Fig. 14.10. Calculate the voltage as per design specifications. Assume necessary data.

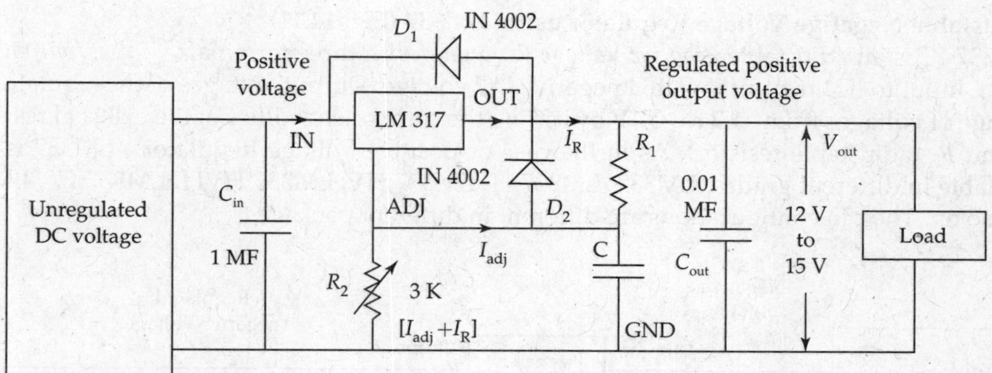

FIG. 14.10 *Adjustable positive voltage regulator using three-pin regulator LM 317*

Solution: For the three-terminal regulators LM317, choosing $I_{adj} = 100\ \mu A$ for LM317 regulator, $V_R = 1.25$ V and $R_1 = 240\ \Omega$

$$\text{Output voltage} \quad V_0 = V_R\left[1+\frac{R_2}{R_1}\right] + I_{adj} \times R_2 \quad \text{refer Eq. (14.2).} \tag{1}$$

$$\text{Case (1) When } V_0 = 12 \text{ V} \tag{2}$$

Substituting in Eq. (1), we get

$$12 = 1.25\left[1+\frac{R_2}{240}\right] + 100\times10^{-6}\times R_2 \tag{3}$$

Solving the above equation for R_2

$$\therefore \quad R_2 = \left[\frac{10.75}{5.3\times10^{-3}}\right] = 2.03 \text{ k}\Omega \tag{4}$$

$$\text{Case (2) When } V_0 = 15 \text{ V} \tag{5}$$

$$15 = 1.25\left[1+\frac{R_2}{240}\right] + 100\times10^{-6}\times R_2 \tag{6}$$

$$R_2 = \frac{13.75}{5.3\times10^{-3}} = 2.59 \text{ k}\Omega \tag{7}$$

To provide adjustable output voltage of 12 to 15 V, resistor R_2 has to be varied from 2.03 to 2.594 kΩ. Standard 3 kΩ potentiometer is used for R_2. To get a load current of 500 mA, we can use T039 packages. Tantalum capacitor C_{in} = 1 MF is used for input bypassing and C_{out} = 1 MF is used to improve transient response. *Protection Diodes are D_1 and D_2.* Assuming low voltage drop of 3 V between V_{in} and V_{out}, the required input voltage is $V_{in} = (V_{out} + 3) = (12 + 3) = 15$ V.

Adjustable Negative Voltage Regulator using LM337 (Fig. 14.11)

LM 337 IC is an adjustable *negative Voltage Regulator* with three terminals – *input, output* and *adjust*. Input to IC is an unregulated negative DC voltage. Output voltages of the regulator are a range of voltages from –1.2 to –37 V by setting the component values of the external resistors R_1 and R_2 (adjustable resistor). As in the case of positive Voltage Regulators, LM337 is also available in different grades: LM337, LM337 H, LM337 HV, LM337 HVH, LM337 LZ, LM337 and so on. The pin configurations are different in different packages.

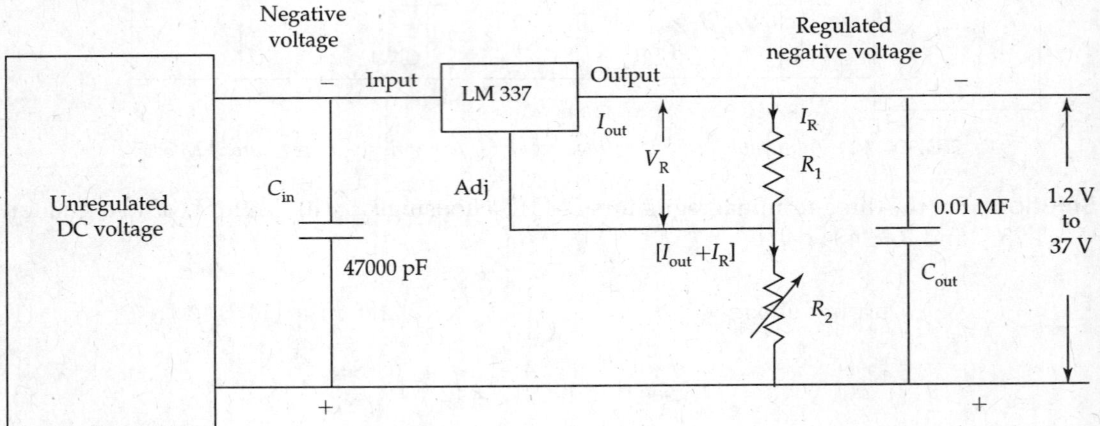

FIG. 14.11 *Adjustable negative voltage regulator using three-pin regulator LM 337*

Limitations of three-terminal voltage regulators

(1) Electronic shut down is external and complex circuit is needed. (2) Output voltage is fixed. (3) External circuitry is needed to obtain higher output voltages.

Precision Voltage Regulators

The above limitations are corrected in *precision Voltage Regulators* such as μA723, LM105 and CA 3085. These precision regulators have more than three terminals. They are extensively used in *series, shunt and switched mode power supplies (SMPS).*

Specifications (performance features) of μA 723 'precision Voltage Regulator'

(1) Output voltage: 2 to 37 V, because the maximum input voltage is 40 V. (2) Maximum output current: 150 mA without external control transistors. (3) Line regulation: 0.03%/V. (4) Load regulation: 0.003%/°C. (5) Ripple rejection: 80 dB. (6) Quiescent current: 2.3 mA. (7) Voltage reference source: Terminal Voltage of Zener Diode. (8) Reference input voltage V_R 7.15 V or 7.5 V (maximum). (9) Maximum operating temperature is 70°C. (10) Power supplies for positive and negative Voltages can be designed.

14.3 IC 723 VOLTAGE REGULATORS

Functional block diagram of μA723 (14 pin DIP) is shown in Fig. 14.12.

- Voltage Regulator IC 723 has Zener Diode D_1 to provide a reference Voltage (V_R) (7.15 V) connected to pin 6.
- Operational Amplifier works as an *Error Amplifier*. Output terminal is internally connected to the base terminal of series pass (control) transistor T_1 and collector terminal of current limiting (sensing) Transistor T_2.
- Supply Voltage V_{CC} is connected to pins 12 and 7.
- Transistor T_1 acts as an Emitter follower. Its regulated output voltage (V_0) is connected to pin 10. It is externally connected to the (INV) inverting input (pin 4) of the Error Amplifier.

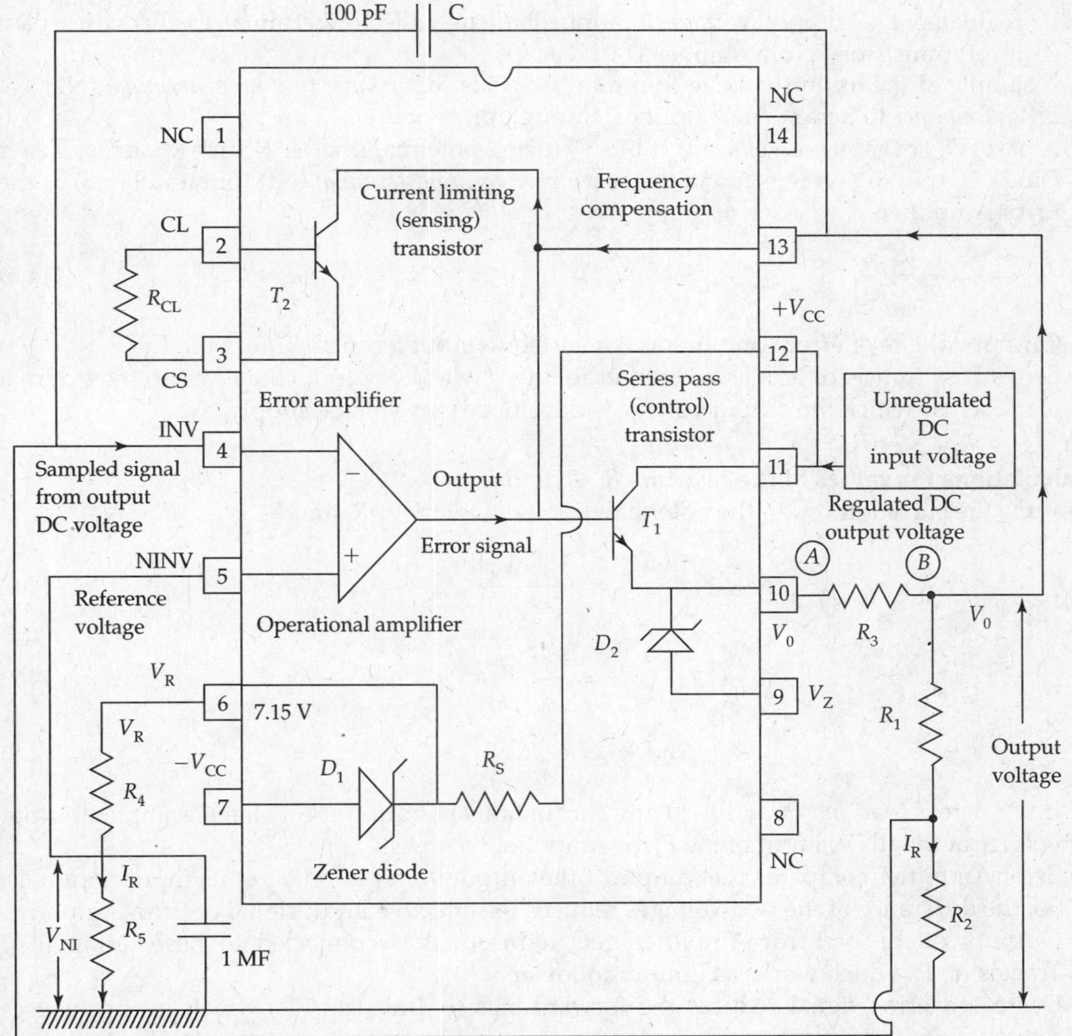

FIG. 14.12 *Pin configuration of IC 723 regulator with internal blocks and external components*

- Error Amplifier compares a sample of the output voltage at the (INV) inverting input terminal and V_R (pin 6) connected to its non-inverting input terminal (NI).
- Error signal from the output of Op-Amp is applied to Base of Transistor T_1.
- Variations in the magnitudes of the error inputs to the error Amplifier in turn cause variations in the (Base current of T_1) conduction of pass transistor T_1.
- Thus, the conduction of the pass transistor T_1 is controlled by the error signal.
- Variations in conduction of the transistor T_1 cause variations in magnitudes of V_{CE} that corrects for the variations in the output voltages of the IC 723 regulator.
- So the output voltage across the load is automatically corrected and the constant output voltage is maintained across the output port of the IC 723 regulator.

14.3.1 Working Principle of Regulator IC 723 (Fig. 14.12)

1. Unregulated DC supply Voltage is applied to the collector terminal of the series pass (control) transistor T_1 (pin number 11 of IC).
2. A sample of the output voltage (pin number 10) is fed to *inverting input terminal* (NIV) of *Error Amplifier* (Operational Amplifier) through pin 4 of the IC.
3. A part (V_{NI}) of *reference Voltage* V_R (7.15 V) (using potential divider R_4 and R_5) across Zener Diode D_1 (pin 6) is connected to the *non-inverting input terminal* (NI) (pin number 5) of the Error Amplifier.
4. $$V_{NI} = V_R \frac{R_5}{[R_4 + R_5]} \tag{14.5}$$
5. Output voltage of error Amplifier is connected to input terminal Transistor T_1.
6. Series Pass transistor acts as an emitter follower with its output connected to the external load (across which the output voltage is developed) for Voltage supply.

Calculations for values of the resistors R_1, R_2 and R_3
Assume the current through the potential divider network of R_1 and R_2 as I_R:

$$R_1 = \frac{[V_0 - V_R]}{I_R} \tag{14.6}$$

$$R_2 = \frac{[V_R]}{I_R} \tag{14.7}$$

$$\text{And } R_3 = R_1 \| R_2 \tag{14.8}$$

- If 'V_0' across load *increases* due to any fluctuations in output DC voltage, sampled output voltage at the (INV) input of the Error Amplifier *increases.*
- Error Amplifier compares the sample of the output voltage and V_R at its input terminals. So, the difference of the two Voltages reduces the effective input signal of Error Amplifier. Output voltage of Error Amplifier gets reduced. It is connected to Base terminal of Transistor T_1, which works as Emitter follower.
- Decreased error signal reduces the forward-bias to Transistor T_1, which in turn causes reduction in its output current, the load current I_L. So, the Voltage across the load reduces, so as to maintain constant output voltage.

Similar explanation could be given when the output voltage 'V_0' across the load decreases. If the output voltage decreases, effective input signal of the error Amplifier increases and causes an increase in the error input signal fed to the pass transistor 'T_1'. So, the forward-bias to the pass Transistor increases. This causes an increase in the load current I_L through the load. Increased load current produces increased output voltage to the designed constant output voltage.

Current limiting Transistor T_2 to protect overload and short circuit conditions

Current limiting function is necessary to take care of short circuit or over load conditions. Current limiting circuit prevents load currents from increasing beyond the maximum permissible designed value I_{max} (over load condition) or short circuit condition.

1. Whenever excessive load current is drawn from a power supply, Transistor T_1 may be damaged. Current limiting Transistor T_2 and external resistor R_{CL} (connected externally between the Base and the Emitter of T_2) provides current limiting functions in the power supply (Fig. 14.12).
2. In normal operation, transistor T_2 is in cut-off state.
3. Current limit terminal CL of the resistor is connected to output terminal V_0 (point A) and current sense terminal CS is connected to the load terminal externally (point B). Load current through R_{CL} produces sufficient forward-bias to switch on current limiting Transistor for designed values of safe load currents. Transistor T_2 gets some of the Base current of T_1 and turns into ON state.
4. Part of the increased load current at the output of Error Amplifier diverts through collector of T_2 and Base current of Transistor T_1 gets reduced. Output current through Transistor (T_1) Emitter decreases. It results in decrease in output load current I_L limiting to its maximum value. This principle of current limiting action is also considered as *current sensing action*.

Calculation of maximum load current $I_{L(max)}$, the regulator IC provides to the load

If the transistor is a silicon Transistor $V_{BE} = 0.7$ V for conduction.

If $R_{CL} = 0.7\ \Omega$, then $I_{L(max)} = \dfrac{V_{BE}}{R_{CL}} = \dfrac{0.7\text{ V}}{0.7\ \Omega} = 1.0$ A.

Power Dissipation P_D by the IC regulator

Regulator Power Dissipation (Inside IC) = $P_D = (V_{in} - V_{out}) \times (I_{L(max)})$,

where V_{in} is the Unregulated DC input voltage and V_{out} is the Regulated DC output voltage.

A graph between load current I_L and load Voltage for a current limited IC 723 regulator is shown in Fig. 14.13. It has built-in current limit of 150 mA. As the load current approaches its maximum permissible value, load Voltage drops rapidly. External Transistors can be added to increase its current handling capability.

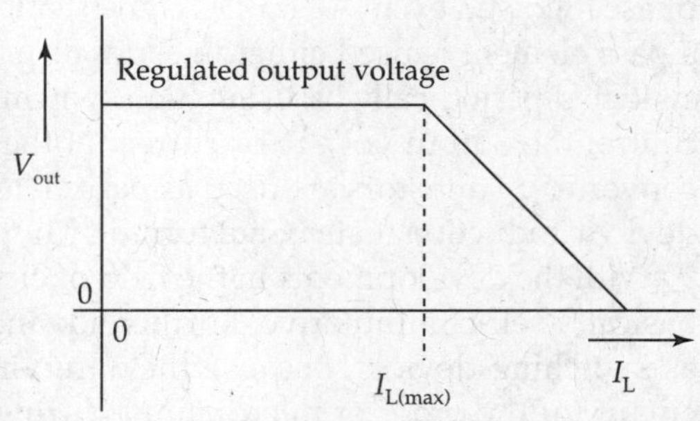

FIG. 14.13 *Load current versus load voltage of current limiting IC 723 voltage regulator*

14.4 DC-TO-DC CONVERTERS

Converters Power conversion process can be classified into four types:
(1) AC-to-DC converter (Rectifier circuits), (2) DC-to-AC converter (Inverter), (3) DC-to-DC converters and (4) AC-to-AC converters (UPS system).

Typical applications of power converters

1. AC-to-DC converters for use in electronic instruments and power supplies.
2. Inverter (AC from DC source) for Home and industries during power failures.
   Inverter Electronic circuit (equipment) that provides AC Power obtained from DC Power source is known as *Inverter*.
   AC supply is not available always in various practical situations. If a DC power-supply source (Dry battery, a storage battery, a sealed maintenance free battery, solar energy converter) is available, DC-to-AC conversion is done utilising an oscillator, switching devices and control circuits to energise the electronic equipments. Such a situation occurs in remote villages, on board spacecraft, satellites and so on. Inverters find their applications in electronic welding circuits also.
3. **DC-to-DC converters** Input voltage to DC-to-DC converter block is a DC voltage and the output voltage is also DC voltage. Output voltage may be higher or lower than input DC voltage. Thus, it is one type of power-converting mechanism used mostly in Computers, Cell phones, Electronic welding circuits, etc.
   DC-to-DC converter circuits are used to supply multiple levels of DC voltages to various blocks in computers. They produce output DC voltages at different magnitudes (from a single input DC source like + 12 V or –12 V) to supply power to motors, +5 V or –5 V DC to operate drive logic circuits, +3.3 V power to low Voltage logic circuits.
   It is similar to a Transformer in AC-to-AC conversion. By stepping up the DC voltage, DC output current is lesser and by stepping down the DC voltage, the DC output current is more when compared to the corresponding values in DC input. This is due to the law of conservation of energy.
4. **AC-to-AC converters** They are used in UPS systems that employ double power conversion process that is from AC mains to DC voltage and DC to AC.

DC-to-DC Converter

DC-to-DC converter employs oscillator or a switching circuit for DC-to-AC conversion in the first phase followed by an AC-to-DC converter to yield required DC output voltage. In this converter, a pass element is used either in series or in shunt path depending upon the configuration and switched periodically by using high switching frequency. By controlling the duty cycle, one can control the output voltage or current at the required level. In the earlier versions of DC-to-DC converters, Thyristors are used as pass elements where in the pulse fed to the Gate controls the device conduction. Later Gate turn-off Thyristors (GTO) as *Choppers* are used.

With the developments in the field of electron devices, BJTs are used as switches, as they are basically self-commutative. Further advances in devices led to the usage of power MOSFETs as switching devices, because they are virtually open during OFF state and virtually short circuit (of the order of micro ohms) during the ON state. Power consumption of MOSFET is negligibly small. Their turn-ON and turn-OFF times are very low when compared to other devices for the reason of their capability of operating at high frequencies. PMOS, NMOS

MOSFETs are used as switching elements. PMOS is preferable when used as a series element, while NMOS FET is used as a shunt element.

Diodes used in the latest converters are axial lead rectifiers employing **Schotkey Barrier** principle or hot carrier metal semiconductor junction type. They are ideally used as freewheeling Diodes (polarity protection Diodes). They have extremely low forward Voltage drop and they are associated with low power loss contributing for higher efficiency of converters. Recent developments are replacing these Diodes by MOSFET switches known as *synchronous rectifiers.*

Various converters use one of the two following types of control strategies to vary the average value of the output. They are (1) Time ratio control and (2) Current limit control.

In time ratio control, the Duty ratio (δ) of the converter circuits is varied in two ways: (1) Constant frequency operation and (2) Variable frequency operation.

- In *constant frequency operation,* turn-ON time T_{ON} is varied while keeping the frequency $f = (1/T)$ constant. This technique is known as *Pulse-width Modulation (PWM).*
- In *variable frequency operation,* time period ($T = T_{ON} + T_{OFF}$) is varied by keeping either T_{ON} constant or T_{OFF} constant. This technique is known as variable frequency modulation. While using variable frequency operation, design of Transformers or inductors gets complicated. As such the PWM technique is more popular.

Classification of DC-to-DC Converters

1. **Fly back converters** Input energy is initially stored in magnetic form. Later, the energy is released to the load.
2. **Forward converters** Input energy goes through the magnetic form and to the load simultaneously.
3. **Push-Pull converters** They are development over forward converters used in push-pull mode.

Another way of classification of DC-to-DC converters

1. **Non-inverting type** Output voltage and input voltage polarities are same.
   - **Buck Converter (step-down regulator)** Output voltage is maintained at a lower level when compared to input voltage level, while maintaining same polarity.
   - **Boost Converter (step-up regulator)** Output voltage is maintained at a higher level when compared to input level, while maintaining the same polarity.
2. **Inverting type** Polarity of the output voltage is opposite to polarity of the input voltage.
3. **Buck-Boost converter (Inverting, step-down/step-up converter)** By adjusting the Duty cycle of the PWM, output voltage can be stepped up or stepped down with different polarities when compared to the input voltages. If the *Duty Cycle* (δ) is less than 0.5, the converter functions as a *Buck converter*. If $\delta > 0.5$, the converter behaves as a *Boost converter*.

Another important classification among the converters

This classification is based on type of isolation between input and output stages.

- Isolating type Fully dielectric between the input and output circuits by using a multiple secondary windings on the Transformer.
- Non-Isolating type No dielectric isolation is used. Voltages can be stepped up or stepped down by a small ratio of less than 4:1.

Further, the CUCK converter and Charge-Pump converters are more popular.

Regulator efficiency of DC-to-DC converters

Efficiency of switching regulator $\eta = \dfrac{\text{Output power to load}}{\text{Input power from source}} = \dfrac{P_{out}}{P_{in}}$

Output load power $P_{out} = \text{Output voltage } (V_{out}) \times \text{Output current } (I_{out})$

$$\therefore P_{LOAD} = P_{OUT} = V_{LOAD} \times I_{LOAD}$$

Input power is the total power drawn from source $= P_{IN}$

Total input power is the product of input DC voltage (V_{in}) and average current drawn at the input port of switching regulator.

$$\therefore \; P_{in} = V_{in} \times I_{in}$$

$\therefore$ DC power conversion efficiency of switching regulator η:

$$\eta = \frac{P_{out}}{P_{in}} = \frac{V_{out} \times I_{out}}{V_{in} \times I_{in}}.$$

Fly Back Converter It is commonly used to obtain high Voltage low output power as well as isolated multiple output voltages. Energy efficiency is inferior when compared to other converter circuits. Fly Back converters can be externally driven or self-oscillating.

Externally driven *fly back converter* (Fig. 14.14) uses MOSFET as a switching device. In practical circuits, output voltage or current feedback is used to control the Duty ratio (ratio of ON time to switching time) that controls the Gate drive of MOSFET.

Primary and Secondary windings of the *fly back Transformer* are tightly coupled. Only one secondary winding is shown and if necessary *multiple secondary windings* can be used to generate isolated *multiple secondary Voltages.*

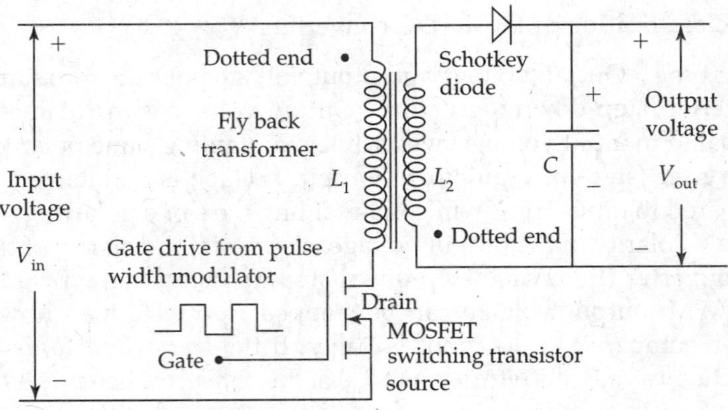

FIG. 14.14 *Externally driven fly back converter*

Operation of the fly back Transformer is different from an ordinary Transformer.

- In a conventional Transformer under load conditions both primary and secondary windings conduct simultaneously.
- A fly back Transformer acts like inductor Transformer wherein the primary and secondary windings are magnetically tightly coupled inductors. At the same time, they do not conduct simultaneously.

- Fly back Transformer is quite compact as switch is driven at very high frequency.
- Transformer secondary Voltage is rectified and filtered to obtain DC voltage V_{out}.

Operation of the fly back converter circuit (Fig. 14.14)

- When the MOSFET switch is ON the current flows through primary winding of the fly back Transformer (dotted end is connected to the positive side of the pulse).
- At the same time, no current flows through secondary winding, as the dotted end of secondary winding becomes high. Diode D gets reverse biased. Filter capacitor provides uninterrupted current to the load, due to previously stored charge.
- When the MOSFET is turned OFF no current flows through the primary winding. Due to the laws of magnetic induction, the polarities of the Voltages across the Transformer windings get reversed. As a result, the Diode in the secondary winding gets forward biased. The rectified Voltage is filtered and fed to the load.
- After several cycles of operation, steady-state operation of the converter is achieved and the output voltage is maintained constant.

Fly back converters are popularly used for relatively low-power applications, such as Cathode Ray Oscilloscopes, High Voltage testers (electronic meggers), Geiger Miller counter and Tubes used in coalbunker level measurement in *Thermal power plants.*

Characteristic features of Fly back Converters

(1) Less number of components, (2) Simple drive circuitry, (3) Lowest efficiency of the order of 60–70% and (4) Maximum ripple content.

Practical fly back circuits use fast recovery Schotkey Diode in series with a combination of a capacitor and a resistor forming a *Snubber* circuit (series RC circuit connected across Diode) connected across the primary winding of fly back Transformer to provide low impedance path for leakage inductance currents and thereby protecting the fly back Transformer against Voltage spikes.

Forward Converter (Fig. 14.15) It offers *lower ripple* than a *fly back converter*. It is useful for power supplies with low output voltages used in TTL ICs, Op-Amps and so on. It is the basic building block over which *push-pull converter* of *high efficiency* and better regulation can be built. Simple circuit of *non-isolated forward converter* (Fig. 14.15).

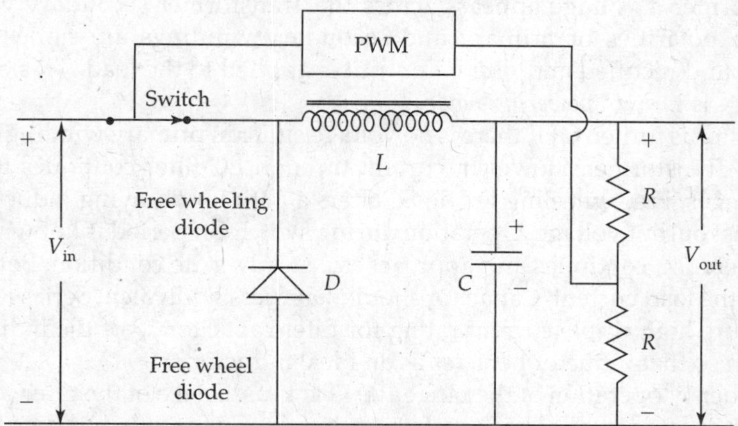

FIG. 14.15 *Non-isolated forward converter*

During the ON time of the switch, energy stored in the inductor is passed into load, as the Diode is reverse biased at the time of T_{ON}. During the OFF period, switch gets opened. Voltage across inductance L gets reversed and the Diode gets forward biased (Diode gets into conduction) permitting the inductor current to continuously circulate for a long time till the switch is turned ON. The Diode is known as *free wheeling Diode or a flywheel Diode*. The cyclic operation continues till the steady state is reached.

Duty cycle δ (ratio of T_{ON} to switching time T) determines the output voltage and input voltage relationships. The above converter is a *Non-Isolated type forward converter*.

Isolated type forward converters (Fig. 14.16) Isolated type forward converter circuits are more popular nowadays for realising low Voltage high current medium power supplies of the order of a few tens of watts (100 W). Basic topology of isolated type Forward converter is shown in Fig. 14.16. Isolation of input and output circuits is achieved by using a Transformer between the two circuits.

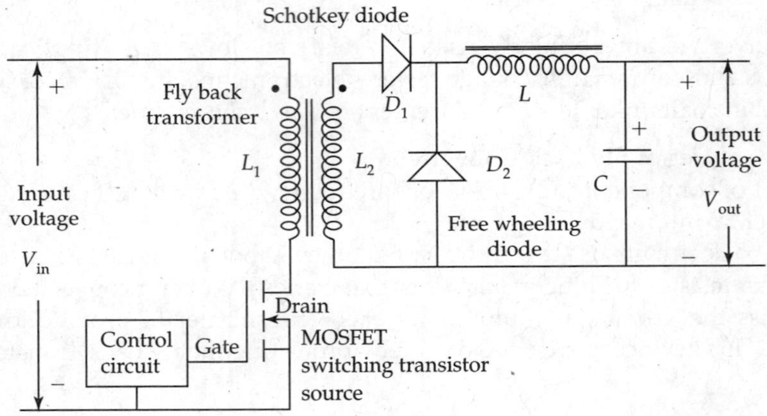

FIG. 14.16 *Basic topology of isolated type forward converter*

When the switch is turned ON due to the control Gate drive, input DC voltage V_{in} is applied to primary winding of the Transformer. Simultaneously, depending upon the turns ratio of Transformer a Voltage appears across the Transformer secondary winding for the reason that the polarities of primary and secondary windings are same. Diode D_1 gets forward biased and rectified and filtered output is applied to the load. This operation when the switch is ON is known *powering mode*.

When the switch is turned OFF, there is no Voltage either in primary winding or in secondary winding of the Transformer. However, current through LC-filter continues to flow without any abrupt changes. Free wheeling Diode D_2 offers a path for decaying inductor current as it is flowing against output voltage. Operation during switch-off period is known as *free wheeling mode*. Cyclic operation continues and approaches a steady-state condition. Both inductor and capacitor share the load current. Capacitor should have less Equivalent Series Resistance (ESR) and ESL to ensure high ripple current rating for filter capacitor. Q of the inductor should be sufficiently high so that inductor behaves as an ideal inductor.

As high-frequency operation of the range 100 kHz is used, size of the filter components and the Transformer become small. The Transformer, inductor, capacitor and the Heat sink for the switching device determine the power-supply volume and the density.

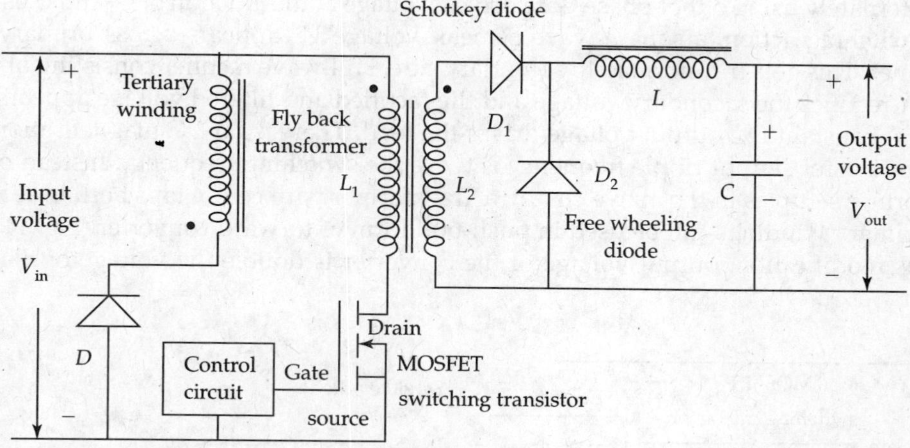

FIG. 14.17 *Topology of isolated type forward converter with tertiary winding*

For practical forward converter, the Transformer should have no air gaps in its flux path. An *extra tertiary winding* is introduced in the Transformer due to the presence of finite magnetising current in a practical Transformer. *Bifilar winding* has to be provided in Transformer primary and *tertiary windings* wound together. They should be capable of withstanding large electrical Voltage stress.

Push-Pull converter It is most widely used converter circuit used for higher power. It is associated with high performance, lower ripple, high efficiency and better regulation. It finds its application in car radios, car CD players and microprocessor-based automatic embedded systems, where the battery Voltage is stepped into a high Voltage used for electronic equipment. They are also used in modern personal computers, laptops and TV to produce multiple Voltages needed for their operation.

Single-ended push-pull converter circuit with external out of phase drives is shown in Fig. 14.18. ON time of both the switching MOSFETs is equal and opposite in phase. They are

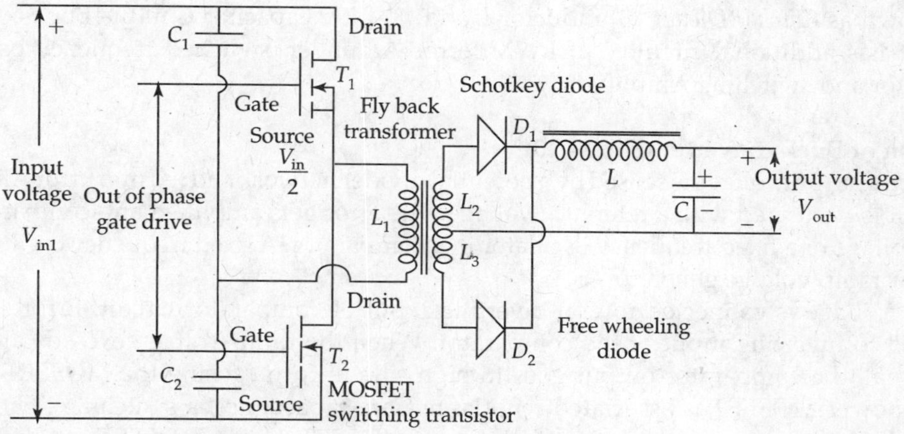

FIG. 14.18 *Single-ended push-pull converter wth external gate drive*

driven alternately using out of phase Gate drives. Voltage at the junction of C_1 and C_2 is 0.5 V_{in}. Due to switching action, a square wave of peak Voltage V_{in} appears across primary of the Transformer. This gets transformed to secondary side. Full-wave rectifier consisting of Diodes D_1 and D_2 rectifies the secondary voltage and the rectified and filtered voltage appears across the load. A constant DC output voltage drives the load. There is no DC magnetisation in the Transformer core. Output ripple frequency is twice the switching frequency. Instead of using a Transformer setup as in the above circuit, a Transformer with centre taps both in secondary and in primary windings can be used in push-pull form of forward converter (Fig. 14.19). By using this modification, output voltage of the converter is double the voltage of the earlier converter.

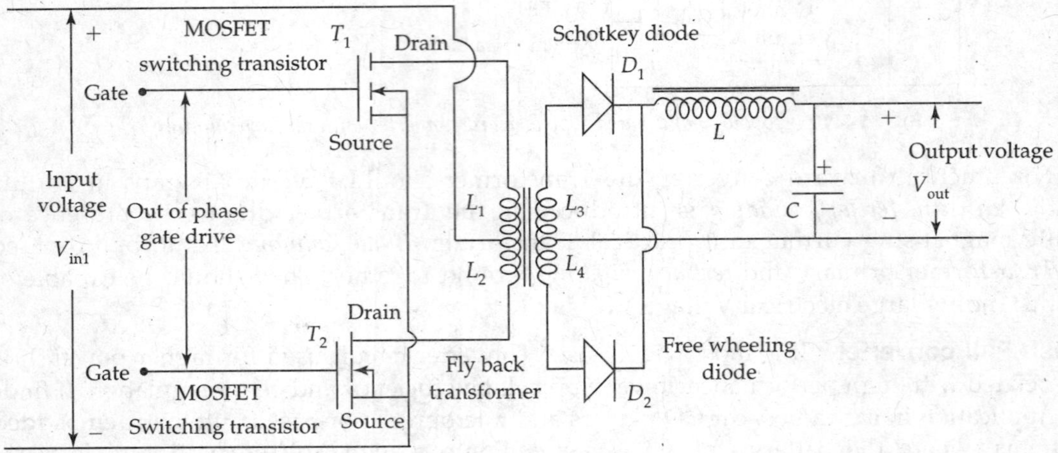

FIG. 14.19 *Push-pull form of forward converter with external gate drive*

Buck Converter (regulator; step-down converter) (Fig. 14.20) Buck switching regulator is used to down convert DC input voltage to a lower output DC voltage of the same polarity. There are four principal components: (1) switching power MOSFET (T), (2) flywheel Diode D (also known as *Catcher Diode*), (3) inductor L and (3)-filter capacitor C in the Buck converter circuit. It has additional circuitry of PWM, error Amplifier, switched frequency oscillator, comparator and switching Amplifier.

Operation of Buck converter (Fig. 14.20)

A fraction of output voltage is sensed by a potential divider network and fed into Error Amplifier, wherein it is compared with a reference Voltage. The error gets amplified and compared with output voltage of a fixed frequency oscillator in a comparator. A fixed frequency oscillator is a triangular ramp voltage generator.

Ramp Voltage is connected to non-inverting input of comparator circuit. Error signal is connected to inverting input of the comparator. When the ramp Voltage exceeds amplified Error signal, the comparator output goes to high side in turn causing the MOSFET to turn OFF. As power MOSFET is associated with large Gate capacitance, a switching Amplifier is interfaced between comparator output and Gate of MOSFET device. MOSFET is switched ON and OFF at the rate of duty cycle (δ) of PWM.

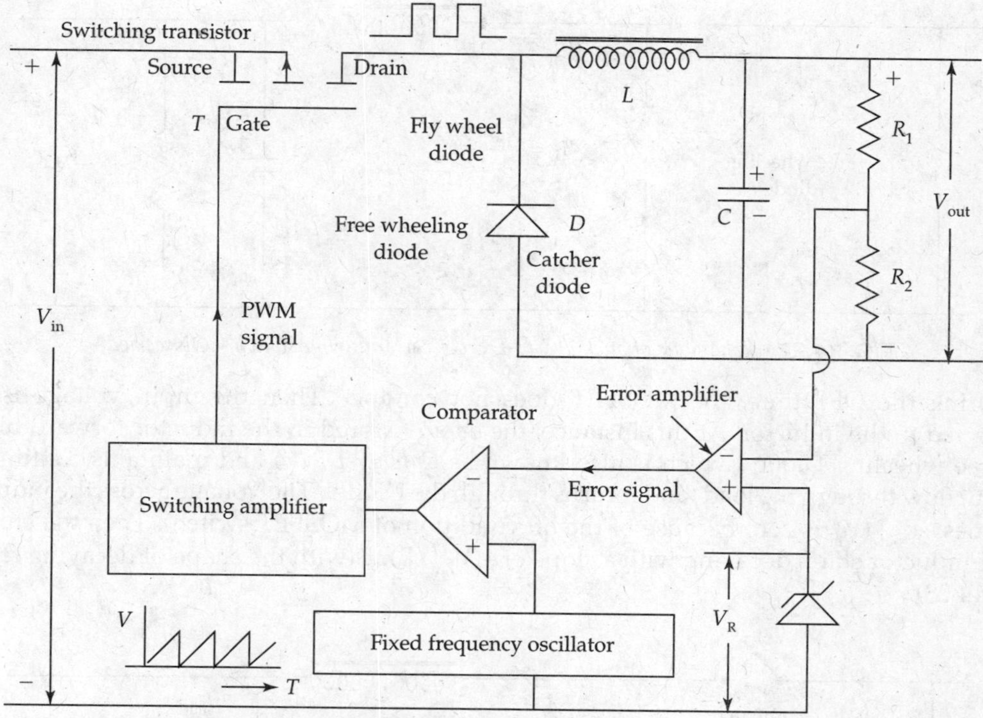

FIG. 14.20 *Buck converter (DC-to-DC converter)*

A simplified diagram of a Buck converter with PWM block is shown in Fig. 14.21. In this Buck converter, according to the duty cycle (δ) of the PWM, the MOSFET device switches ON and OFF so that it connects and disconnects input voltage to the inductor.

MOSFET conducts during the ON time, and then the Diode gets reverse biased (turned off). Then the Voltage across inductor L is ($V_{in} - V_{out}$). Inductor current increases with a slope of $(V_{in} - V_{out})/L$ (Figs. 14.21 and 14.22). (With the shape of increasing ramp) Inductor current flows through the output capacitor and load resistance R_L. The capacitor charges during this on-time period T_{ON} of the device.

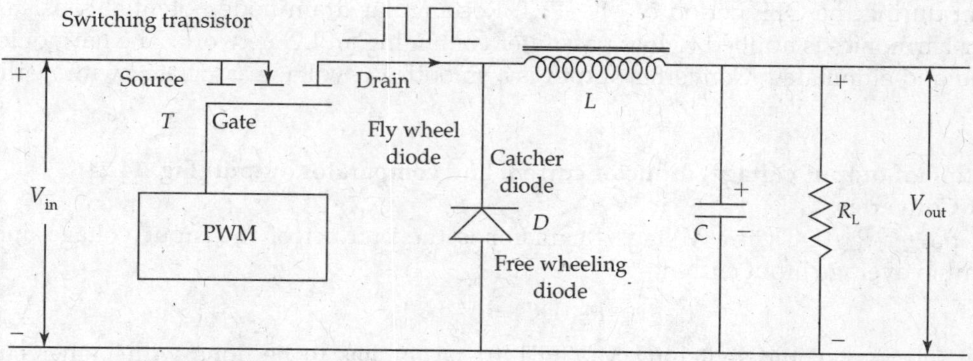

FIG. 14.21 *Basic schematic diagram of Buck (DC-to-DC) converter*

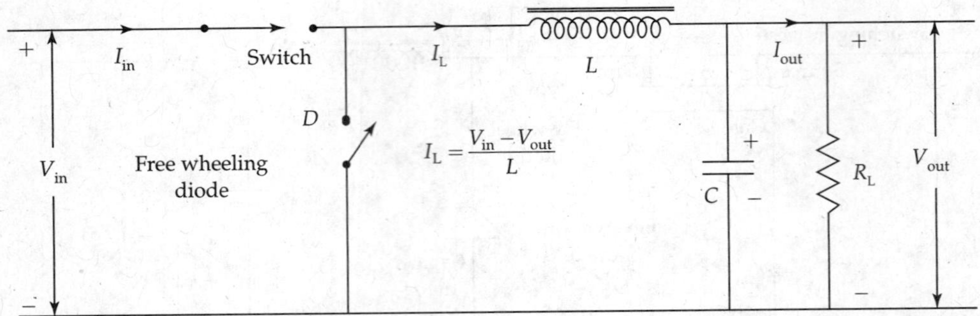

FIG. 14.22 *Behaviour of buck DC-to-DC converter during switch – ON period*

During the OFF time, the MOSFET does not conduct. Then the input voltage is not connected to the inductor. At this instance, the energy stored in the inductor forward biases the free wheeling Diode (which is also known as *Catcher Diode*) and maintains continuous current flow through the load (and returns through the Diode). The Voltage across the inductor becomes $-V_{out}$ (V_{in} is zero because of the off condition of MOSFET switch). Then the current in the inductor starts decaying with a slope of $-(V_{out}/L)$... (with the shape of decaying ramp) (Fig. 14.23).

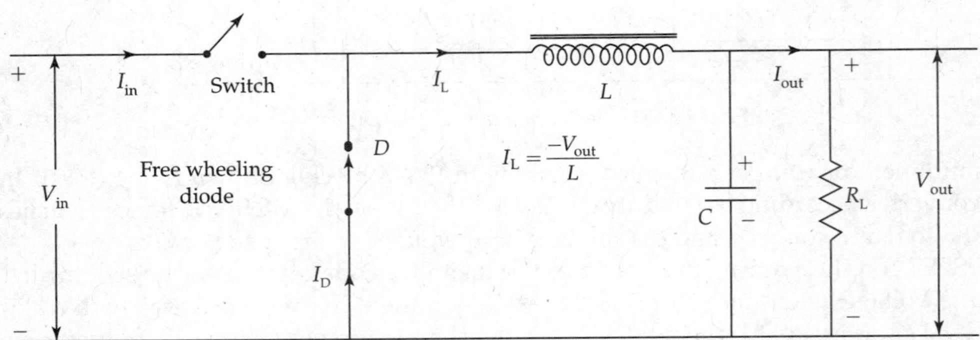

FIG. 14.23 *Behaviour of buck DC-to-DC converter during switch – OFF period*

Energy reservoir formed by inductor *L* and capacitor *C* maintains the load Voltage and current during the OFF period of MOSFET. Rectangular drain node potential consisting of higher harmonics is applied to low pass filter containing an LC network. The harmonics are filtered and eliminated. Converter output is a smooth DC voltage associated with negligible ripple.

Variation of output voltage, inductor current and comparator output (Fig. 14.24) (Buck Converter)

Input power P_{in} (DC) in a Voltage Regulator is the product of the input voltage and the maximum average input current.

$$P_{in}\text{ (DC)} = V_{in} \times I_{in}$$

The selection of the switching MOSFET transistor has to be done with higher current capability than the input or the output current whatever is larger.

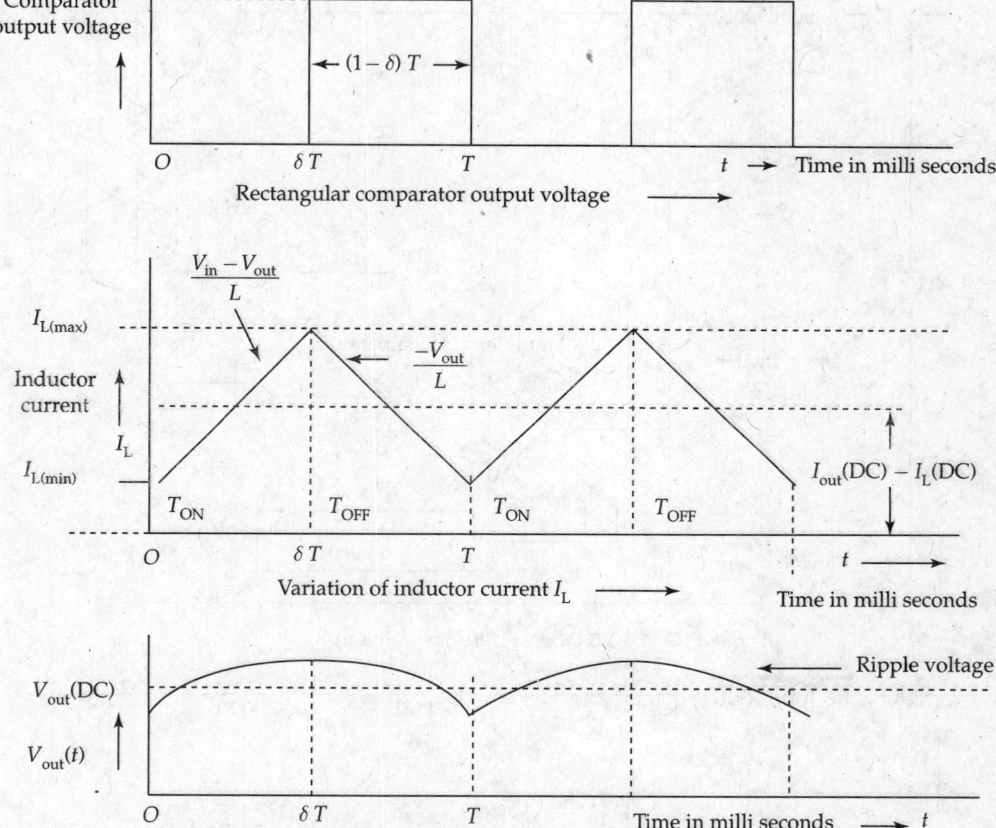

FIG. 14.24 *Various signals in the buck converter*

Output power is the product of the output voltage and the output load current:

$$P_{out}(DC) = V_{out} \times I_{out}.$$

In this case for Buck converter circuit, output DC voltage is less than the input voltage. Naturally, the output current (I_{out}) should be larger than the input current (I_{in}).

Boost Converter (Step-up; Fig. 14.25)

Boost converter is known as step-up converter, since generated output voltage V_{out} is greater than input voltage (V_{in}). N-MOSFET is connected as shunt element in the circuit to function as pass switching element. MOSFET is driven by PWM to turn the device ON and OFF at a rate of switching frequency $f_S = 1/T$.

Generation of PWM signal is similar to that in Fig. 14.20 of Buck converter. Simplified Boost converter circuit with PWM block is shown in Fig. 14.26.

During switch-ON period, current from input source flows through inductor L. Current through inductor increases with a slope of $(-V_{in}/L)$. Diode is reverse biased and the energy is stored in the inductor. This part of working can be understood from Fig. 14.27.

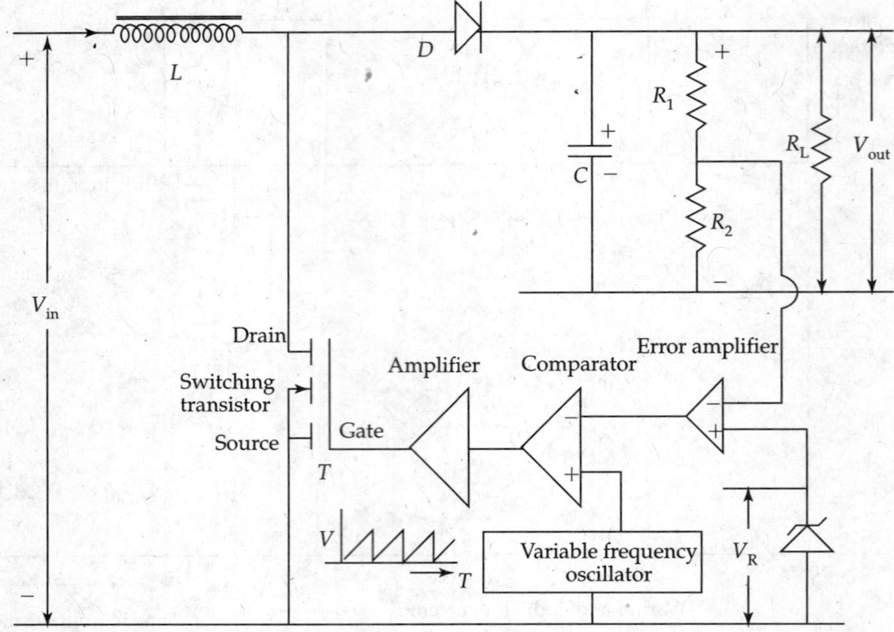

FIG. 14.25 *Boost converter (step-up)*

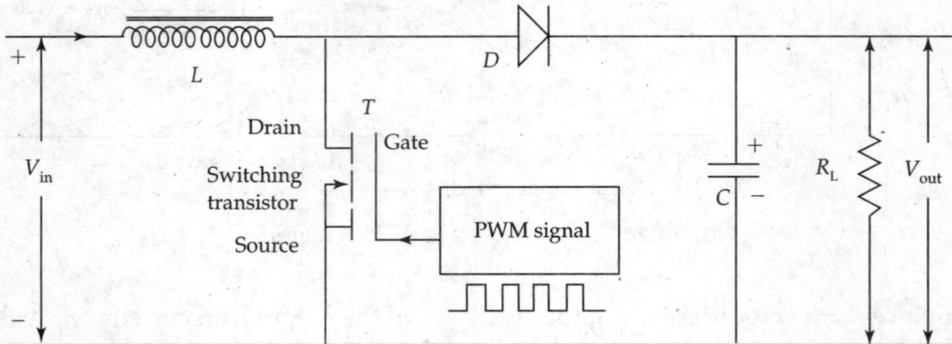

FIG. 14.26 *Simplified boost converter (step-up)*

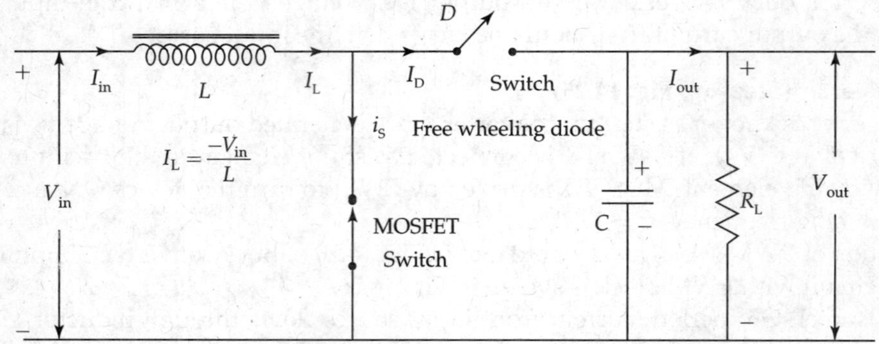

FIG. 14.27 *During switch-on condition of boost converter*

When NMOSFET is turned into OFF-state due to Drain drive Voltage (Variable frequency ramp), the Diode is forward biased. At the instant of turn off of NMOSFET, current I_L through Inductor decreases linearly at a rate of $(V_{in} - V_{out})/L$ through the load and the Diode. Energy reservoir from the inductor L is transferred to the filter capacitor and load R_L. Capacitor charges to a Voltage higher than input DC voltage (Fig. 14.28).

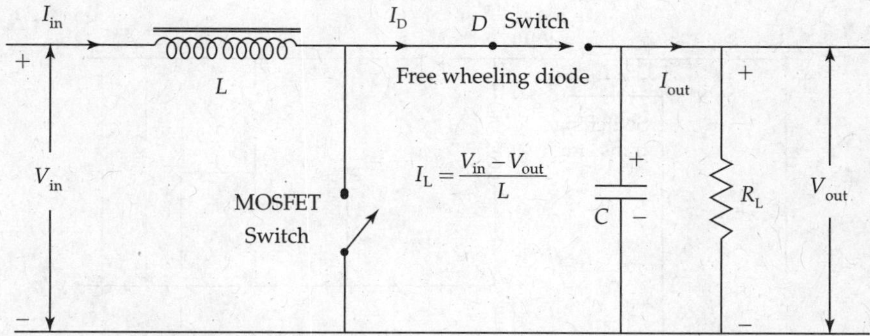

FIG. 14.28 *During switch-off condition of boost converter*

Input power P_{in} (DC) in a Voltage Regulator is the product of the input voltage and the maximum average input current. P_{in} (DC) $= V_{in} \times I_{in}$. Selection of switching MOSFET is done with higher current capability than the input or the output current whatever is larger.

Output power is the product of the output voltage and the output load current:

$$P_{out}\text{ (DC)} = V_{out} \times I_{out}.$$

In this case for Boost converter circuit, output DC voltage is greater than the input voltage. Naturally, the output current I_{out} should be less than the input current I_{in}.

Variation of inductor current in a ***boost converter*** **(Fig. 14.29)**

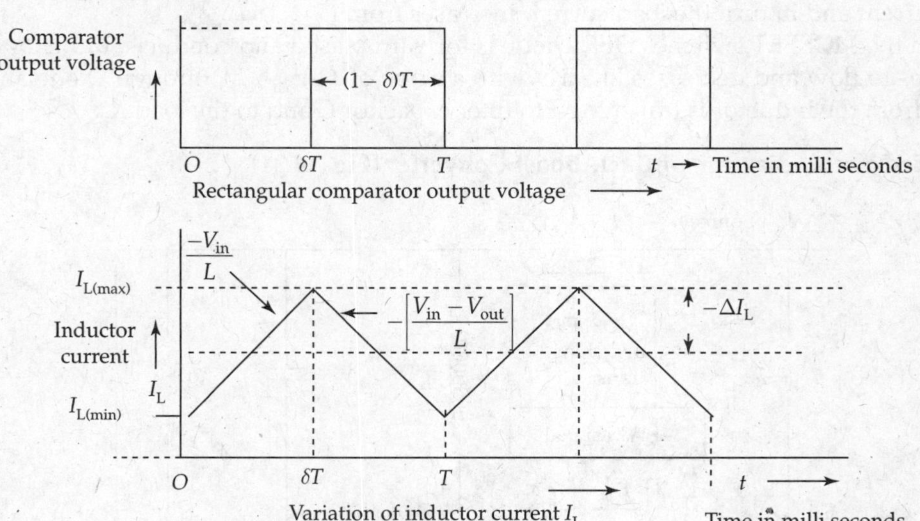

FIG. 14.29 *Inductor current variation in a boost converter*

Buck-Boost converter **[Inverting and step-down/step-up converter] (Fig. 14.30)**

- By adjusting the duty cycle of PWM, output voltage V_{out} can be stepped up (increased) or stepped down (decreased) over the input voltage.
- For $\delta > 0.5$, the converter functions as *Boost Converter.*
- For $\delta < 0.5$, the converter circuit in Fig. 14.29 works as Buck converter.

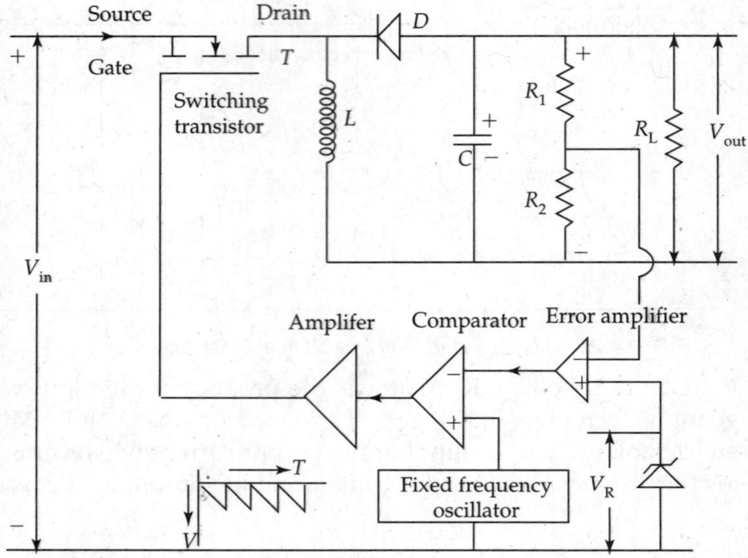

FIG. 14.30 *Simplified block diagram of buck-boost converter*

When the MOSFET is turned ON the Diode D is reverse biased and it is turned OFF (works as an open switch). The Drain current flows through the inductor with a slope of (V_{in} / L). The drain current and in turn the load current increases from $I_{L(min)}$ to $I_{L(max)}$.

When the MOSFET switches OFF, Diode is forward biased and conducts. Inductor current continues to flow and decreases linearly with a slope of $(-V_{out} / L)$ through Diode and load. Energy from the inductor is transferred to filter capacitor C and to the load R_L.

Simplified circuit diagram of Buck-Boost Converter (Fig. 14.31)

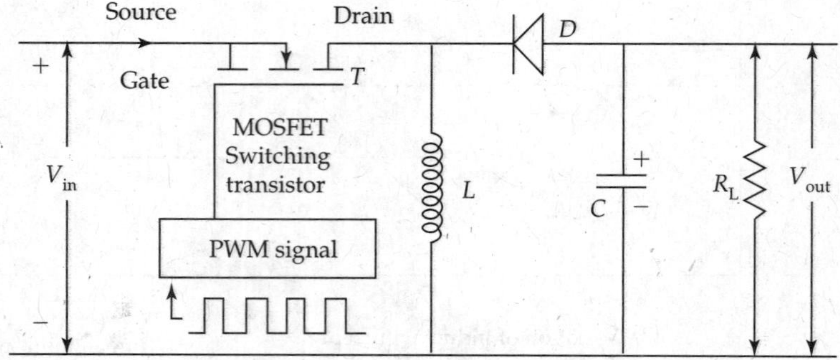

FIG. 14.31 *Simplified block diagram of buck-boost converter*

CUK Converter

CUK converter (pronounced as *chook converter*) is named after its inventor. It is a derived version of the basic *Buck, Boost* and *Buck-Boost converters*. All the advantages of the basic converters can be realised in this CUK converter.

Features

(1) Continuous input current (No input filter is required). (2) Continuous output voltage with minimised ripple. (3) A higher or lesser output voltage compared to the input voltage can be obtained with opposite polarity. (4) The power factor with a well-designed regulator can be improved. (5) CUK converter is a cascaded version of Buck and Boost converters using an additional inductor and a capacitor. This is a distinguishing feature from other converters. Schematic circuit of a CUK converter is shown in Fig. 14.32.

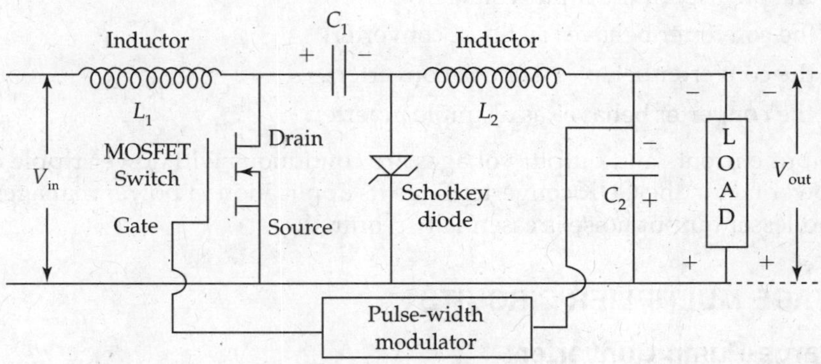

FIG. 14.32 *CUK converter*

When the MOSFET switch is turned ON (as per the variations in duty cycle of PWM) input current flows through inductor L_1 and MOSFET switch. Energy is stored in magnetic field of inductor L_1. Voltage drop across the Inductor opposes the input voltage.

When the MOSFET is turned OFF as per the variations in the duty cycle from the linear PWM, the inductor L_1 opposes the flow of current instantly reversing the EMF across the inductor forcing a continuous input current to flow through capacitor C_1 and into the forward-biased Diode. The capacitor C_1 is the main storage element and it gets charged storing electrical energy.

During the subsequent cycle, when the MOSFET is again switched ON the stored energy in the capacitor C_1 gets transferred (released) by finding a path through L_1 and discharging into the load. Here the inductor L_2 and capacitor C_2 function as an LC-filter minimising the ripple across the output. It is to be observed that during this cycle, the energy is again stored in the inductor L_1. This process gets repeated until a steady-state condition is reached.

The average value of voltage across inductor is given by

$$V_{L1} = \delta V_{in} + (1-\delta)(V_{in} - V_{C1}),$$

where δ is the duty factor of PWM, and V_{C1} is the Voltage across the capacitor C_1. V_{in} = input DC voltage; $V_{L1} = V_{in} - (1 - \delta)\, V_{C1}$.

In the steady state, $V_{L1} = 0$

$$\therefore \quad V_{C1} = \frac{V_{in}}{(1-\delta)}$$

Similarly, the average value of voltage drop across the inductor L_2 is given by

$$V_{L2} = (V_{out} - V_{C1})\delta + (1-\delta)V_{out}$$

$$V_{L2} = V_{out} + V_{C1} \cdot \delta$$

In the steady state, $V_{L2} = 0$

$\therefore V_{out} = -\delta V_{C1}$

Substituting the value of V_{C1} in the above equation, we get

$$V_{out} = -\frac{\delta}{(1-\delta)} V_{in}.$$

The above expression shows that the polarity of the output voltage gets inverted when compared to the polarity of the input voltage.

1. If $\delta > 0.5$, the converter behaves as Boost converter.
2. If $\delta < 0.5$, the converter behaves as Buck converter.
3. If $\delta = 0.5$, the converter behaves as a simple inverter.

As both input currents and output voltages are continuous with lower ripple content and improved power factor, the CUK converter finds its application in power management where low input and lesser output noise is essentially required.

14.5 VOLTAGE MULTIPLIER CIRCUITS

14.5.1 Charge-Pump Converter

The earlier basic switching regulators – Buck, Boost, Buck-Boost, CUK, Flying and Forward converters – operate on the fundamental principle of storing energy in a magnetic field and essentially uses at least one inductor.

In the *Charge-Pump converters,* an entirely different principle for storing electrical energy in a capacitor. The capacitor is known as *flying capacitor* (C_F) or a Bucket capacitor. It is either a dielectric capacitor or electrolytic capacitor having low ESR. Ceramic capacitors possess many advantages in providing fast switching times and efficient filtering capabilities.

During the charging cycle, the capacitor is connected across the unregulated DC input source and gets charged. It cannot charge abruptly (voltage across the capacitor cannot change instantly) and charges exponentially with time. In the discharge mode, stored energy in flying capacitor gets transferred or pumped into another capacitor (reservoir capacitor) C_R and into load. MOSFET devices carry out switching of these capacitors at a high switching frequency. It is the fundamental principle in a *charge-pump converter.*

As no inductors are used and capacitors can easily be integrated in modern IC, charge-pump converters can be made more compact, cheap and more efficient. Efficiency of charge-pump converter is the ratio of output (DC) voltage to input (DC) voltage. Their efficiency is as high as 95%. They are available in present day nano-generation ICs for power management applications, particularly in low-power portable appliances such as cellular telephones, hand-held computers, core supplies for future generation processors, DSP-based power supplies, digital cameras, low Voltage DC bus supplies, USB output ports, powering white LED background lights popularly in the screen of PDAs, in EEPROMs, flash memories, RS232 (Recommended Standard) level shifters.

Voltage Multiplier Circuits using Charge-Pump Converters

They are also used to obtain Voltages of the order ×2 (Voltage double), ×3 (Voltage Tripler), ×(*N*), ... nV_{in} and submultiples of input voltage like ×(1/2), ×(2/3), ×(2/3), ×(4/3) and so on. They are used to obtain output voltages of opposite polarity also. Leading IC manufacturers are using various topologies and switching MOSFET devices for *charge-pump converters*. A few topologies are described below.

Charge-Pump Voltage Doubler (Unregulated Type) (Fig. 14.33)

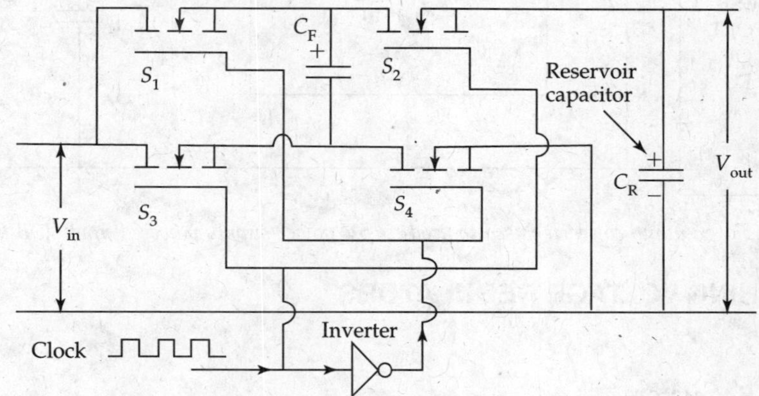

FIG. 14.33 *Charge-pump voltage doubler (unregulated type)*

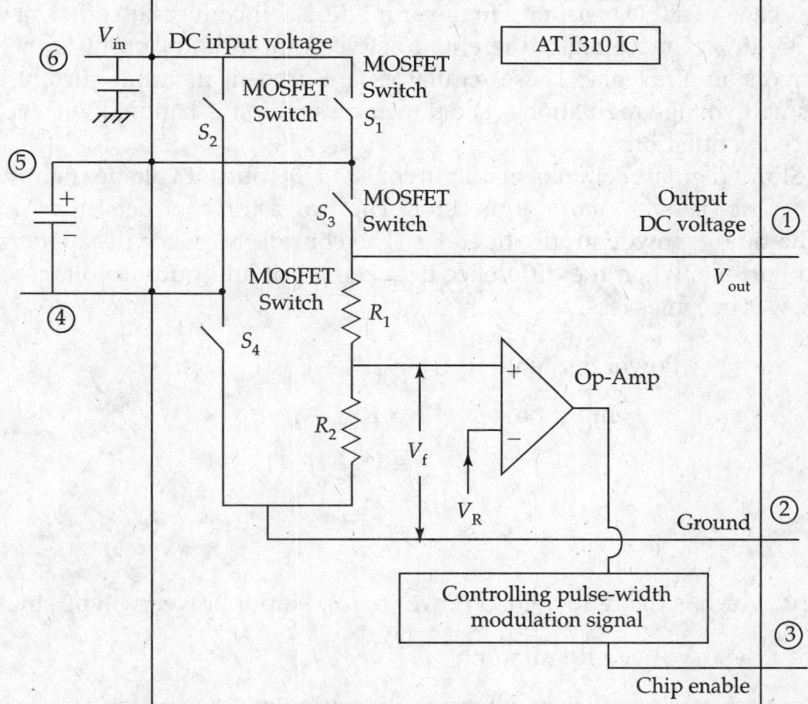

FIG. 14.34 *Typical diagram of charge-pump power converter IC*

Charge-pump converter to produce 5 V power supply (Fig.14.35)

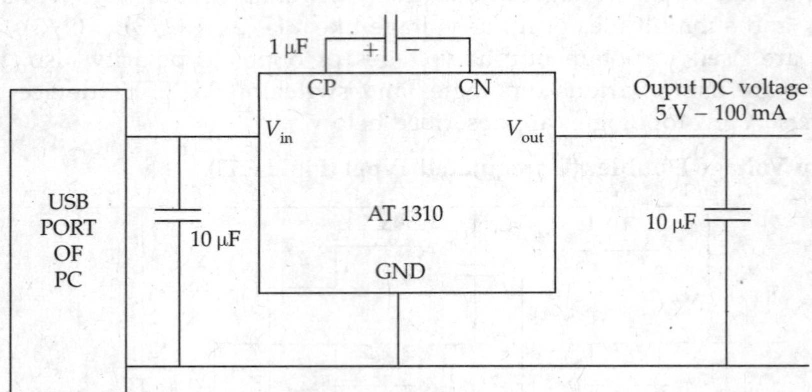

FIG. 14.35 *Charge-pump converter used to produce 5V power supply powered from USB port of a PC*

14.6 SWITCHING VOLTAGE REGULATORS

Introduction

Linear Voltage Regulators

Series and Shunt regulators are known as *dissipative regulators*. The control element conducts continuously in the active region while dissipating more power during the power conversion process. They act as variable resistance in linear mode. Their conversion efficiency is typically of the order 35–45%. The voltage differential between input and output is the main cause for lower conversion efficiency. These regulators are used in medium current applications maintaining small voltage regulations. The power dissipation is handled with heat sinks and arranging forced ventilation.

Series and Shunt regulator circuits maintain constant DC output voltages mainly controlling conduction of *series pass or control* Transistor. The transistor conducts in its active region throughout the time – power supply operates. The control transistor dissipates more power for large load currents when the difference between input and output voltages is more and useful load power becomes less.

$$\text{Power dissipation} \quad P_D = [V_{in} - V_{out}]I_L \text{ watts} \tag{14.9}$$

$$\text{Input power} \quad P_{in} = P_{Load} + P_D \tag{14.10}$$

$$\therefore \quad P_{Load} = [P_{in} - P_D] \tag{14.11}$$

$$\text{Power} \quad \eta = \frac{P_{Load}}{P_{in}} \tag{14.12}$$

Power efficiency, which is the ratio of load power to total input power, will be small.

Limitations of Linear Voltage Regulator

- Lower power conversion efficiency, because of continuous conduction of control transistor to maintain constant DC output voltage from the regulator.

- Fixed output voltages.
- As the operating frequency of AC mains is 50 Hz, transformers and filter circuit elements are large in size.

To eliminate the above major limitations of linear regulator, *switching regulator* in IC form are used in personal computers, where reduced equipment sizes play a major role.

Switching Regulator Switching regulator uses non-dissipative power conversion process. Control element is switched between ON and OFF (in between saturation and cut-off) at very high frequency (known as *switching frequency*). Output voltage is independent of differential between input and output. Thereby the power dissipation becomes very small and power conversion efficiency becomes very high to the order of 95%. Due to higher switching frequency, component sizes will become much smaller. They can be made compact, small in size and weight. There is no need for heat sinks either.

When power efficiency is important, switching Voltage Regulator is used. Conduction intervals of controlling power Transistor are reduced by switching the transistor ON/OFF using PWM, to minimise power dissipation in it. Input voltage (V_{in}) is not permanently connected to regulator circuit. It is connected to load through MOSFET switches that operate ON and OFF, at very high frequencies to limit conduction angles of power Transistors. Higher switching frequencies are used in switching regulator circuit operation. Switching frequencies using MOSFETs are of the order of 200 kHz.

At the same time, Switching frequencies cannot be very high, because it increases power dissipation within control power Transistors that may result in *thermal runaway*.

Advantages of Switching Regulators

(1) Higher power conversion efficiency. (2) Multiple levels of output voltages from single input voltage level. (3) Compactness in size and weight, giving flexibility in design.

Various blocks of a Basic Switching Voltage Regulator

(1) Unregulated DC voltage, (2) Switching device, (3) Sampling network, (4) Reference Voltage, (5) Comparator, (6) Pulse generator, (7) Driver circuit and (8) Filter stage.

Operation of a Switching Voltage Regulator (Fig. 14.36)

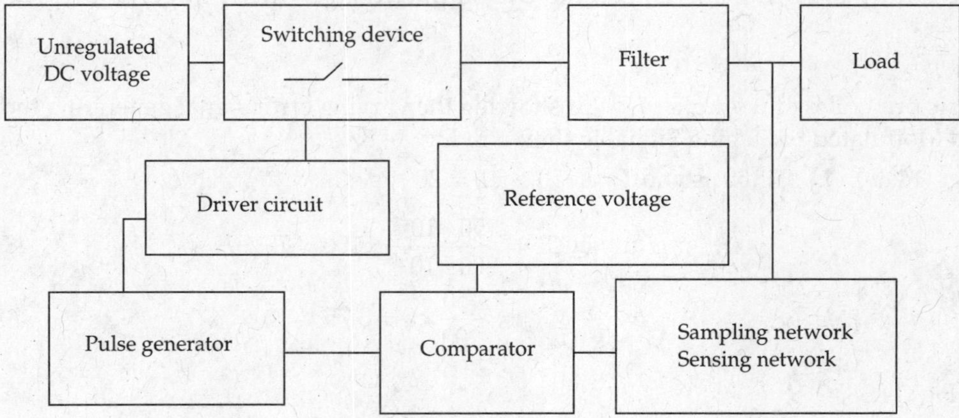

FIG. 14.36 *Biock diagram of switching voltage regulator*

1. **Voltage Source** Switching regulator derives its supply from an unregulated DC source or battery. Source must be capable of supplying required power to supplement losses in inductors, capacitors and switching devices. In the event of power failure, source should be able to supply energy to handle any designed fault-tolerant or recovery operations.
2. **Switching Device** The switching device can be Transistor, DEMOSFET or Thyristor. The switch operates between ON and OFF states, in accordance with the Duty cycle of control signal and thereby regulating average DC voltage at a predetermined value. The switch must be capable of withstanding overloads caused by load faults. It should have *good peak rating*.
3. **Sampling Nework** A feedback mechanism is necessary to control the ON–OFF times of the switching device. Output voltage is sampled using a voltage-divider network. Sampled part of the output voltage is fed to a *comparator*.
4. **Reference Voltage** It uses a temperature compensated Zener Diode.
5. **Comparator** The comparator is fed with two voltages – *sampled* and *reference voltages*. Its function is to produce an error signal. The comparator is usually a *differential Amplifier*.
6. **Pulse Generator** A DC controlled asymmetrical multivibrator is driven by an error signal from the comparator. The resulting pulse train is fed to a Diode to control a switching device. Pulse generator that produces asymmetrical wave *varying in pulse-width* (*pulse-width modulation*, PWM) is used to drive the switching elements.

Pulse-width or *pulse duty factor* of PWM varies based on errors from the comparator. Duty factor of pulse train is typically 10–90%. Maximum pulse width to be handled by a switching transistor is $\leq 0.01f_S$, where f_S is the switching frequency. The *duty factor* or *duty cycle* δ is defined as the ON-period and expressed as a percentage of time. It is the ratio of ON time (t_{ON}) to the time period T of the pulse waveform.

$$\text{Duty cycle} \quad \delta = \frac{\text{ON-time } (t_{ON})}{T \text{ (Time period)}} = t_{ON} \cdot f_S,$$

where f_S is the switching frequency of the pulse train, t_{ON} is the ON time of pulse waveform, T is the Time period of pulse waveform = $t_{ON} + t_{OFF}$, and t_{OFF} is the OFF-period of pulse waveform.

If the duty cycle is made longer, the DC output voltage will be larger and is related by $V_0 = \frac{t_{ON}}{T} \cdot V_{Peak}$.

Basic principles of working of PWM showing the varying error signal generating the Pulse-width Modulated Switching Signal is shown in Fig. 14.37.

In Fig. (14.38) $t_{ON} = 150\ \mu s$ and $T = (150 + 50) = 200\ \mu s$

$$\%\ \text{Duty cycle} = \frac{150 \times 10^{-6}}{200 \times 10^{-6}} \times 100 = 75\%$$

$$\therefore \quad V_0 = \frac{t_{ON}}{T} \cdot V_{Peak} = \frac{150}{200} \times 5 = 3.75 \text{ V}$$

In Fig. 14.38,

$$t_{ON} = 100\ \mu s \ \text{ and } \ T = (100 + 100) = 200\ \mu s$$

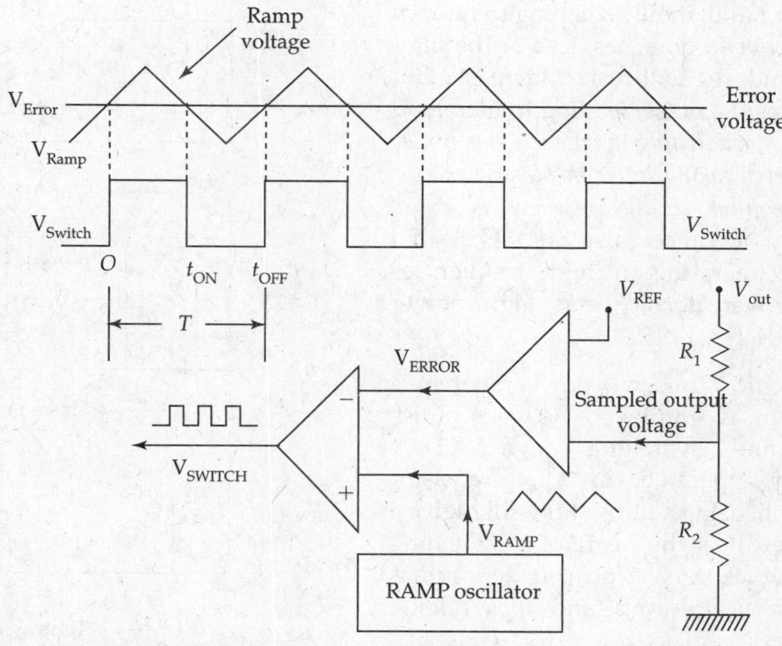

FIG. 14.37 *Varying error signal generates pulse-width modulation signal*

$$\%\text{ Duty cycle} = \frac{100}{200} \times 100 = 50\%$$

$$\therefore \quad V_0 = \frac{100}{200} \times 5 = 2.5 \text{ V}$$

In Fig. 14.38, $t_{ON} = 50\,\mu\text{s}$ and $T = (50 + 150) = 200\,\mu\text{s}$

$$\%\text{ Duty cycle} = \frac{50}{200} \times 100 = 25\%$$

$$\therefore \quad V_0 = \frac{50}{200} \times 5 = 1.25 \text{ V.}$$

From the above discussion, longer the duty cycle, larger is the DC output voltage. Feedback loop corrects changing ON time of PWM and controls switching Transistor to regulate. This fundamental principle is employed in a *switching regulator*.

From equation $\delta = t_{ON} \times f_S$ duty cycle is proportional to the switching frequency. Earlier switching regulators employ 20 kHz as the switching frequency, so as to obtain optimum efficiency and achieve compact size of the regulator. This frequency is not audible to human ear. With the developments in switching devices, present day devices use frequencies as high as 50 kHz, 500 kHz and even 100 MHz. The reason is – inductor size in the filter decreases with increasing frequency, followed by a decrease in ripple. But the frequency increase is not desirable as electrical noise also increases (increased electrical noise radiation reduces the efficiency). With increased frequency, eddy current losses are more than hysteresis losses, and the filter design gets more complicated.

On the other hand, if the switching frequency is chosen at lower frequencies, size of the filter components and regulator size increase. But this also reduces the noise, leading to increased efficiency. *Thus, switching frequency has to be properly considered taking into consideration of the switching device, filter size and noise level.*

Present day advances use MOSFET that is capable of operating at high frequencies to increase power density and high power conversion efficiency (~95%).

7. **Driver Circuit** To drive power switch, a driver circuit is employed between pulse generator and switching device. They are usually operated in CE Transistor configuration, using a transistor with high h_{fe}.
8. **Filter Stage** It is the heart of a switching regulator, hence very critical. It determines *efficiency, transient response* and *ripple voltage,* and *noise considerations.*

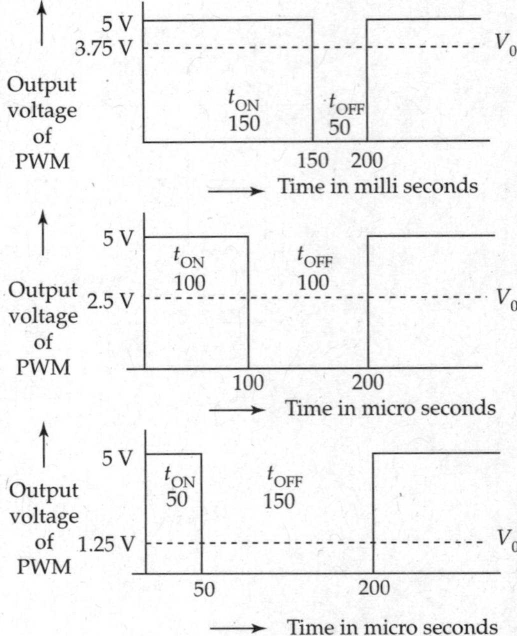

FIG. 14.38 *Variations of output voltages of PWM versus switching tunes*

Three important filter circuit configurations are RC-filter, L-filter and *LC-filter*. While designing filter cores, *torroidal cores* are *preferred over slug-type*. A torroidal core is completely contained within the core and it radiates less noise, whereas for a slug-type core noise radiation is more. Filters are made up of ferrites and *Moly Permalloy (MPP powder)* for torroidal cores, with low core losses and leakage, and high permeability.

Limitations of a Switching Voltage Regulator

1. Switching regulator is not suitable for analog applications. As the device switches continuously between ON and OFF, noise is introduced in neighbouring circuit elements. It also introduces a ripple in the output voltage, thereby degrading the regulator performance.
2. If FET devices are used, their Gate current is zero and power drawn from the source is zero. Another advantage of using FETs in IC regulators is that their ON-resistance is very low. So, higher current IC regulators prefer the use of FET devices. It also further reduces the size of IC regulator package. These advantages demand the use of FETs in IC regulators over BJTs
3. A switching regulator has slow response to load changes. It can be improved by choosing higher switching frequency.
4. Internal heat develops across ESR present in every capacitor, due to power dissipation, when ripple current flows into and out of the capacitor. The capacitor fails, when ripple current is greater than the maximum design value. ESR ranges from 1 to 5 Ω, for stable operation of regulators.
5. Operating currents in switching regulators are low due to the constraint on power dissipation.
6. Ripple voltage across the filter capacitor has a very high frequency > 10 MHz, causing *ringing* in output voltage. Capacitor parasitic effect regulates regulator performance.

7. At the converter input port, a capacitor for high-frequency ripple is used for bypass path. While mounting the filter circuit, shielding is used to reduce interference in the neighbouring circuits.
8. There is a restriction on maximum input voltage rating (~15 V), to take care of power dissipation handling capability in the circuit.

Comparison of performance features of Linear and Switching Regulators

1. **Advantages of Switching Regulators**
   (a) Higher speed of MOSFET power switch causes higher switching efficiency, because of less power dissipation from input to output. (b) Size of devices is small, because of less heat-transfer requirements with low-power switching actions. (c) After DC voltage level conversion, the output voltage can be transferred through a Transformer to another block. This type of transfer mechanism provides electrical isolation from the input circuit.
2. **Disadvantages with Switching Regulators**
   They can be noisy and require energy management in the form of a control loop. Fortunately the solution to these control problems is found integrated in modern switching-mode controller chips.
3. **Advantages of Linear Regulators**
   - Linear regulators provide lower noise and higher bandwidth.
   - Their simplicity can sometimes offer a less expensive solution.
4. **Disadvantages with linear regulators compared to Switching Regulators**
   - Output voltage cannot be greater than the input voltage.
   - Negative DC voltages cannot be developed at the output port.

14.7 UNINTERRUPTIBLE POWER SUPPLY (UPS)

An *Uninterruptible Power Supply* (*UPS*) is a no-break AC power-supply system. It supplies power continuously to the connected load, without interruption as long as the sensitive loads demand it. UPS provides power from the main supply (mains) as long as it is available. In the event of mains failure, the UPS changes over to *internal battery system*. Sensitive loads include microcomputers, semiconductor memories, data storage and processing, telecommunication equipment, airport installations, control and instrumentation of modern power plants, on-line reservations, on-line banking and so on.

One major application of UPS is supplying and protecting present day computers. If AC mains are OFF for some reason, any memory-stored data is lost and running processes will be aborted without saving the context. In order to overcome such situations, computers are recommended to be connected with UPS, so that there will be safe time for the operations to continue to another convenient safe level.

In the absence of UPS, a personal computer may result in keyboard lock-ups, hardware degradations, complete loss of data and burnt motherboards. An outage due to non-availability of mains supply or UPS leads to catastrophic and devastating damages to an application process. Imagine satellite controls, where ground control computers are working dynamically with orbiting satellites, can we imagine connecting to the mains directly without UPS like protection!

Rapid increase of electric loads, almost going up by 3–5 times in residential, and much more in commercial and industrial sectors, power distributors are finding it tough to ensure reliable stream of mains supply. The power gets corrupted with transient surges and sags. UPS solutions, small or large, provide power against black out (no power condition) and brown outs (low voltage condition), ensuring power quality.

14.7.1 Static UPS in Parallel Mode (Fig. 14.39)

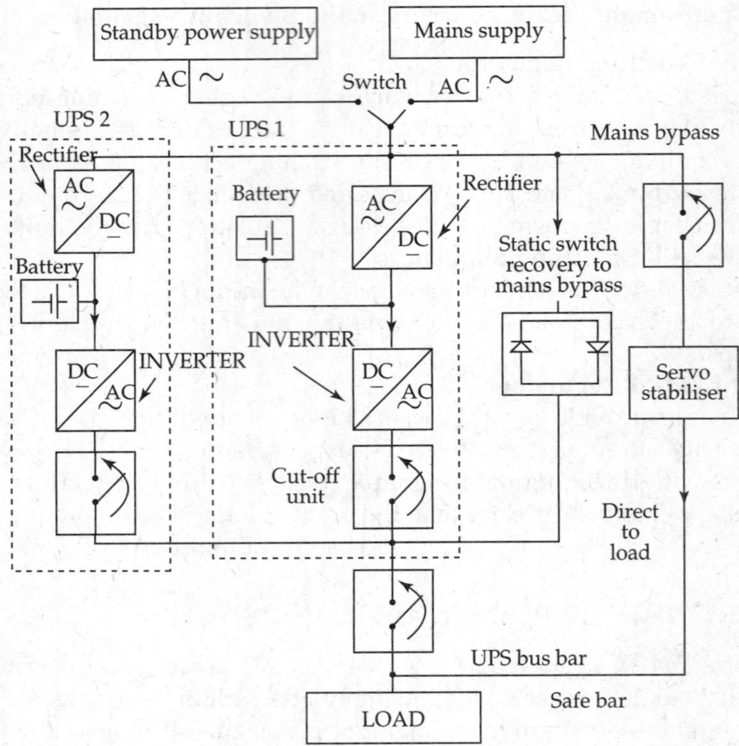

FIG. 14.39 *Static uninterruptible power supply UPS system in parallel*

UPS – Principles of Operation

UPS essentially consists of (i) *rectifier*, (ii) *inverter*, (iii) *static bypass switch*, (iv) *service bypass switch* and (v) *storage battery*. In the event of failure of the main supply due to some disturbance in the grid network, the storage battery is connected in float mode across the inverter input. It supplies energy to the inverter. The inverter converts this DC supply into AC supply to meet the system critical load requirement. When the main supply is restored back, a bump less (smooth) transfer takes place within a fraction of a second restores the UPS operation.

A static bypass switch provides automatic uninterruptible instantaneous transfer of the load from the UPS Bus Bar to the input mains and prevents unpermissible variations of the output caused by faults in UPS installation against sudden load changes.

When large loads are to be fed or improved reliability is required, several UPS modules can be operated in parallel mode to increase passive (circuit wise) and active (operation side) redundancy. Six UPS modules can be paralleled. This operation is known as multi-block parallel system.

The energy stored by the storage battery plays a key role in an UPS operation. The batteries may be automotive, stationary, lead acid, maintenance-free or Nickel Cadmium batteries. The maintenance-free batteries are preferred for large size UPS systems. The batteries should have sufficient AH (ampere–hour) capacity to feed the system loads. They are to be recharged well within a reasonable time to meet the full load under immediate mains outage.

UPS systems can be categorised as the following:

1. ON-Line UPS (True UPS)
2. OFF-Line UPS (Standby mode)
3. Hybrid UPS (Line interactive)

ON-Line UPS It employs double conversion processes, i.e., AC to DC and DC to AC. It provides sine wave output. In the event of failure of main supply, it transfers to battery backup within 3.5 ms. The backup time depends upon the ampere–hour capacity of the batteries. Usually, ON-Line UPS provides 10–12 min of backup. It provides isolation against spikes and surges. It acts as a firewall between mains supply and electronic equipment safeguarding against black outs and brown outs.

OFF-Line UPS A standby backup supply (SBS) solves a minimum number of power quality problems. An OFF-line UPS is known as SBS. These are popularly used with standalone PCs that are used infrequently. Load is powered directly by the input mains power and the battery operation is evoked when mains supply fails. They can provide a backup of 8 h and normally used in large computer centres and hospitals. The capacity of UPS is typically of the order of 5 KVA.

Hybrid UPS It uses *off-line units* of Ferro resonant type. Its output is a square wave or quasi square (trapezoidal). It runs in parallel with mains supply all the time. Large installations like soft ware companies, process control instrumentation use UPS capacities of the order 63.5 KVA and higher.

Drawbacks of UPS

1. Undesirable acoustic noise due to harmonics in the audio range necessitating location of UPS in a separate room.
2. Poor transient response. Sudden changes in load result large transients, under shoot or over shoot lasting up to 10 cycles. It causes havoc in the loads especially in the computer systems.

These drawbacks can be overcome by using higher frequencies such as 15 to 20 kHz employing PWM techniques.

Block diagram of UPS (Fig. 14.40)

- In normal situations DC power to computer hardware or electronic gadgets is provided by DC power-supply units using AC-to-DC converters followed by IC or simple regulator circuits. (For this situation, relay change over contact-1 will be initially closed and the relay contact-2 will be open.)
- When the AC mains are switched OFF during power failures, relays are used to connect another DC power source in the associated UPS to the electronic instruments. Such

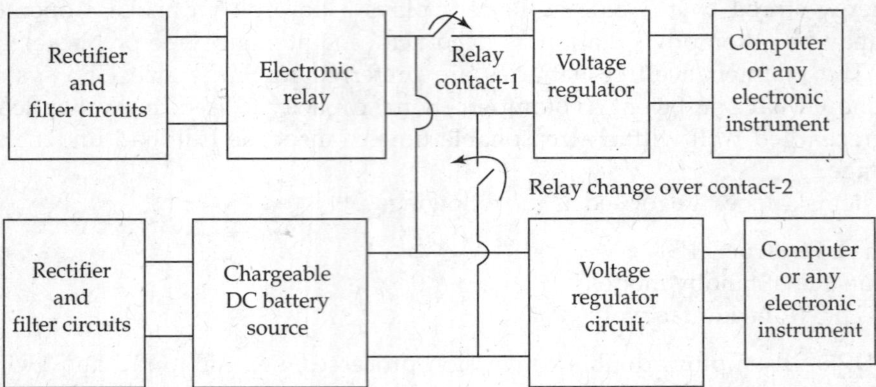

FIG. 14.40 *Block diagram of uninterruptible power supply*

situations are common in IT industry, banks using office automation and so on. (For this situation, relay change over contact-1 will be open and the relay contact-2 will be closed.)

Backup time of UPS during power failure will be from a few minutes to a few hours that depend upon the capability rating of the chargeable DC power source and its associated circuitry. Naturally affordable cost of UPS increases with backup time performance.

14.8 SWITCH MODE POWER SUPPLY (SMPS)

Switch Mode Power Supply (SMPS) is an electronic power-processing unit that has a built-in *switching regulator*. It converts the unregulated AC input from mains (or DC input voltage) to a regulated DC output voltage. It finds application in laptops, scanners, mobiles, zip drives, hubs, printers, TV receivers, computers, VCRs and so on.

SMPS has several features of switching regulators: (1) High conversion efficiency, (2) light in weight and compactness, (3) Electrical isolation of load from source, (4) lower ripple, (5) provides several isolated outputs from a single Voltage input.

SMPS categories

1. AC (in)–DC (out) (off-line rectifier)
2. DC (in)–DC (out) – (DC–DC converters, Voltage Regulators or current regulators)
3. AC (in)–AC (out) – (frequency changers) DC (in)–AC (out)-(Inverters)

14.8.1 Operation of SMPS (Fig. 14.41)

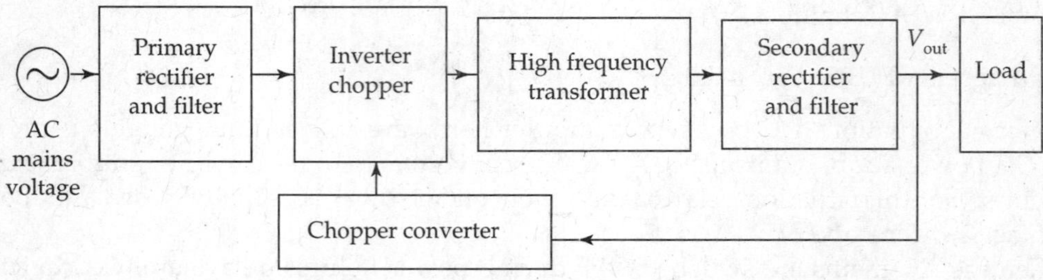

FIG. 14.41 *Block diagram of switched mode power supply*

Various blocks of SMPS

(1) Primary rectifier and primary filter (Input rectifier and filter), (2) Inverter Chopper, (3) High-frequency Transformer, (4) Secondary rectifier and secondary filter (Output rectifier and output filter), (5) Chopper converter and (6) Load.

(1) Primary rectifier and primary filter

AC line Voltage inputs to SMPS are either (1) 120 Volts 60 Hz in USA or (2) 220 Volts 50 Hz in India from the power distribution networks. AC is not transformed before the input of the primary rectifier. AC input is rectified by using silicon rectifiers. If the mains supply is 120 Volts 60 Hz, a Voltage Doubler is used in general. If the incoming supply is 220 Volts AC at 50 Hz, a Full-wave rectifier is used. Rectified Voltage is fed to an LC cascaded filter to obtain a smoother DC voltage with lesser ripple content.

If the SMPS uses incoming DC source (DC voltage equal to $220 \times 1.413 = 330$ V), DC voltage has to be directly fed to the Inverter by automatic switching arrangement. In such cases, the primary rectifier and filter stages are not necessary.

(2) Inverter Chopper (DC-to-AC Converter)

Inverter Chopper converts DC into AC at frequencies above 25 kHz. Switching Transistor acts as a *Chopper*. It is continuously switched between ON and OFF as per the duty cycle of the PWM. Thus, the duty cycle of the chopped DC will influence the AC Voltage generated across the secondary of the *High-frequency Transformer*. To achieve high gain, a high switching frequency is employed to transfer electric power via energy storage components. High gain multi-stage Amplifiers are used because of their lower ON-resistance and higher current carrying capacity.

(3) High-frequency Transformer

Output voltage of the Inverter chopper is a pulsating AC. It is directly connected to the primary winding of High-frequency Transformer. Input to HF Transformer is not a pure sinusoid. Transformer is more compact, because for a given core it is capable of transferring more power without reaching into saturation. It requires fewer turns. Transformer is small and compact as it operates at high frequency. It has one drawback of increase in Skin Effect of the conductor with increase in frequency. They use *Ferrite Torroidal cores* to reduce size and weight. Winding costs for torroidal cores are more. Bindings are made of branded Litz wire (A type of cable wire specially designed to minimise skin effect at higher operating frequencies.). HF Transformer is the most critical stage in SMPS and distinguishes from other regulators. Waveforms in SMPS are high-speed PWM signals. HF Transformer windings must be capable of supporting higher harmonics due to skin effect. They cause a major power loss.

(4) Secondary Rectifier and Secondary Filter

The Voltage obtained from the secondary of the HF Transformer is connected to a secondary rectifier for another cycle of rectification and filtering to obtain specified DC voltage. Normally, Silicon Diode is used for Voltages above 10 V with sufficient PIV (*Peak Inverse Voltage*).

For lower Voltages, Schotkey Diodes are used as they possess faster recovery time and lower Voltage drop while conducting. Secondary filter is really LC-filter because of its advantages over RC- and L-filters. Smaller inductors and capacitors are needed.

(5) Chopper Converter

Chopper Technology is the simplest form of high-frequency conversion. It consists of a power Transistor switch and a Catcher Diode (Fig. 14.42). A sensing circuit in SMPS detects the variations in output voltage. Detected variations are connected to *Chopper converter*. Those variations are compared with a reference Voltage in the Chopper converter. Its output is a PWM signal that is applied to *Inverter Chopper*. Its duty cycle varies in accordance with the error that causes increase or decrease in the conduction of *Inverter chopper*.

When there are small changes in the load Voltage, the chopper converter tends to keep the output voltages to the desired values. When an increased Voltage is sensed, the chopper converter reduces the duty cycle of PWM signal. It in turn causes a decrease in Voltage in the secondary of the high-frequency Transformer so that the output voltage drops back to predetermined original value.

Simple Block diagram of SMPS

SMPS provides controlled output voltage with negligible ripple and acts as good DC source.

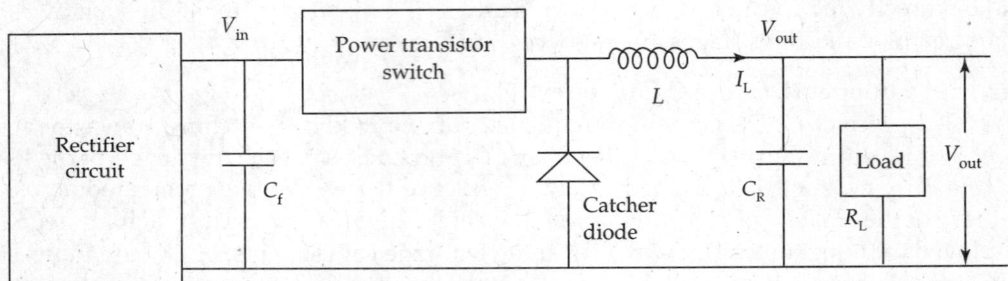

FIG. 14.42 *Simple block diagram of switched mode power supply*

Unregulated DC voltage V_{in} (larger than the required DC output voltage V_{out}) is applied to SMPS. Filter capacitor C_f does filtering to V_{in}. Power control MOSFET switches connect the input voltage to the load through the Inductor at a very high frequency (switching frequencies of the order of 50 kHz). High-frequency Switching reduces the power dissipation in the power Transistor.

Continuous load currents to maintain constant DC output voltage could be obtained by the *Control Power Transistor* with the help of Inductor, Capacitor and Catcher Diodes in action. Inductor is situated in the output path to smooth the variations in load current. Inductor *L* and Capacitor C_R work as *low pass filter* to smooth the ripple content in V_{in} and provide constant DC output voltage.

Other salient features of SMPS

(1) Input current of SMPS has higher harmonic content and a relatively low-power factor. Power factor conversion circuits are necessary. Using elaborate filter banks filters the harmonic content. (2) Specialised control techniques are adopted such as sensing of the output by an Opto-coupler. Certain SMPS are used in TV receivers such as plasma display devices.

Drawbacks of SMPS

1. The design of SMPS is quite complicated when compared to a switching regulator.
2. Presence of harmonic distortion and high-frequency ripple. The harmonics cause additional heating in the wiring and neighbouring circuits.
3. Low pass filter must block electromagnetic interference of high frequency.

4. SMPS tend to act as a dipole in radio transmitters. The emanated high frequencies cause disturbances in radio, TV, PC and peripherals connected on the same phase. SMPS has to be shielded in metal casing.
5. Failure rate of filter capacitors in SMPS is more than in any power supply, because of the high-frequency oscillations.
6. SMPS is prone for electrical shocks. To prevent this, ground plane earth is to be provided.

Special features for Power Supplies

Computers, Control and various Communication systems of today need specialised power supplies. In particular, the systems used in Telecommunication Towers, ground communication, Telecommunication Satellites – Hybrid power supplies and power supplies for IT industry have different considerations.

About half a century back, with the developments of linear discrete circuitry, power supplies using Vacuum Tubes, Mercury Arc Rectifiers, Silicon Control Rectifiers and Semiconductor Diodes went into the 'Black Hole'. Thyratron-controlled Rectifiers employing closed loop control were the major workhorses.

Later on the development of IGFETs and MOS devices using VLSI, the SMPS with higher efficiencies are the major sources of power supplies. Using Integrated Circuits, miniaturised power supplies are in use since last three decades. Specialised power supplies – the analogue cousin of SMPS-Digital Power Supplies – are entering into the field of modern power supplies.

Power supplies for radio link apparatus in telecommunication towers

Radio stations, repeater stations, base stations of mobile communication systems have to be built-in places, where mains supply from power distribution network does not exist. They require Mains independent power supplies. Mains independent power supplies are constructed in special shelters such as power rooms. Hybrid systems like a diesel generator or solar generator or a wind generator and batteries are provided depending upon the field situation. Diesel generators use diesel oil as fuel and emit Sox and Nox emissions in the neighbouring environment.

Energy stores are equipped with lead acid batteries or stationary alkaline batteries or special Nickel–Cadmium batteries with capacities as high as 1250 AH. Conventional batteries are now a days replaced by maintenance-free batteries. These batteries need immediate standby system, control and distribution panel and DC distribution board.

In some places low-power steam turbines, thermo-electric generators or solar batteries supplement diesel generators. The solar batteries must be provided to cater for the hours of darkness in the absence of sunlight for sun-less days. At the mountain ridges, open plains, coastlines in funnel-shaped villages, deserts and islands, hybrid power supplies use wind generators and batteries. In certain hybrid power supplies for example in ground communication systems, inverters are used in half load parallel arrangement with a static switch with an arrangement to revert-to-mains. They are also provided with redundancy.

The radio links or base stations of mobile stations use high towers for mounting Antenna systems. They are prone to lightning. Special measures are adapted to safe guard against lightning protection. They need special type of Earthing arrangements or Earth grid arrangement where a number of earth electrodes forming Earthing-mat to keep the earth resistance well within 1 Ω. In rocky soils counter pose Earthing has to be provided to safe guard against lightning strokes and overload and so on. Surge suppressor Diodes, Zener Diodes provide limitation of over Voltages.

SUMMARY

1. IC regulators are popular as they are associated with reduced cost of manufacturing ICs, high reliability and flexible design.
2. Advanced communications through IPAD, Cell phones with 3G technology are available because of IC power supplies on-card on a PCB to feed few circuits.
3. Flexibilities like thermal shutdown, over Voltage protection, current limiting, etc. can be standardised as building blocks in IC regulators.
4. Thermal shutdown and current limiting facilities are internally provided and electronic shutdown is externally provided for the IC voltage regulators.
5. DC-to-DC converters are used to change DC voltages from one level into another level using Buck, Boost, Buck-Boost, CUK and Charge-pump converters.
6. Buck converter steps down the input voltage.
7. Boost converter steps up the input voltage.
8. In Buck-Boost and CUK converter circuits, the input voltage can be either stepped up or down with inverted polarity at the output.
9. A charge-pump converter is based on the principle of stored energy in the form of electric charge in a capacitor.
10. Charge-pump converter produces small Voltages of the order of $\times\frac{1}{2}$, $\times\frac{3}{2}$, $\times\frac{4}{3}$, $\times\frac{2}{3}$ of input voltage x, depending upon the circuit configuration.

Questions for Practice

1. Using a diagram, explain the working of three-terminal IC Voltage Regulator.
2. (a) List out the important features of three-terminal Voltage Regulators.
   (b) Draw the circuit diagram of a three-terminal Voltage Regulator as a current source and explain its working.
   (c) Draw the circuit diagram of a Voltage Doubler circuit and explain its operation. Also sketch the input and output waveforms. (JNTU, Mar. 2006)
3. (a) What are the limitations of three-terminal Voltage Regulators?
   (b) Draw the circuit of IC 7812 Voltage Regulator circuit along with unregulated Bridge circuit and derive the expression for load current. (JNTU, Mar. 2006)
4. (a) Draw the circuit of IC 7805 Voltage Regulator and explain its working.
   (b) Using 7805 IC Voltage Regulator, design a current source to deliver 0.25 A current to 48 Ω 10 W load. (JNTU, Nov. 2005)
5. (a) Draw the internal block schematic and pin configurations of IC 723 Voltage Regulator.
   (b) Draw the circuit diagram of IC 7812 Regulator along with current boosting circuit and explain its operation. Derive the expression for load current. (Mar. 2006)

6. (a) Draw the circuit of a half-wave Voltage doubler and explain its operation. Sketch the input and output waveforms. What is its output voltage under no load Conditions?
   (b) Draw the circuit of IC 7815 Voltage Regulator along with unregulated circuit. Derive the expression for load current. (Nov. 2005)
7. (a) Draw the diagram of IC 723 Voltage Regulator and explain its principle.
   (b) With a neat circuit explain the working of Voltage multiplier. What are its applications? (JNTU, Nov. 2008)
8. Write short notes on (a) Constant current limiting, (b) Fold back current limiting, (c) Crowbar protection and (d) Thermal shutdown.
9. With a neat circuit diagram explaining the operation of Buck-Boost converter.
10. Distinguish the various performance features of linear regulator and SMPS.
11. Why power MOSFETs and IGBTs are preferred in UPS systems?
12. What is UPS? Explain how it differs from regulated power supply? (JNTU, Feb. 2008)
13. Using three-pin Voltage Regulator, design current source that will deliver 0.25 A current to 48 Ω 10 W load. Data: $I_Q = 4.2$ mA and $V_R = 5$ V. (JNTU, Feb. 2008)
14. (a) What is catcher Diode and explain the necessity of catcher Diode in Switch Regulator with the help of circuit diagram? (JNTU, Nov. 2006)
    (b) List the operating ratings and electrical characteristics of IC 723.
15. (a) Explain the significance of Low Pass Filter in Switching Regulator.
    (b) What are the limitations of switching regulators?
    (c) Why switching frequencies are limited in Switching Regulator and also explain how to overcome this? (JNTU, Nov. 2006)

Multiple Choice Questions

1. An inverter converts ______________.
   (a) AC to DC (b) AC-to-AC (c) DC to DC (d) DC to AC
2. A converter transforms ______________.
   (a) AC/DC (b) AC/AC (c) DC/DC (d) DC/DC
3. A rectifier converts ______________.
   (a) AC/DC (b) AC/AC (c) DC/DC (d) DC/AC
4. UPS system converts ______________.
   (a) AC to AC (b) DC to AC (c) AC to DC (d) DC to DC
5. A Zener diode is used in power-supply circuits for ______________.
   (a) voltage regulation (b) protection
   (c) RF suppression (d) current limiting
6. The pass element used in modern switching regulators ______________.
   (a) PNP transistors (b) NPN transistors (c) Thyristor (d) MOSFET

7. The switching devices are located in shunt path for ____________.
   (a) buck converter
   (b) boost converter
   (c) buck-boost converter
   (d) forward converter

8. The converter operating by storing energy as electric charge in a capacitor is ____________.
   (a) buck converter
   (b) harge-pump converter
   (c) fly back converter
   (d) forward converter

9. The converter using a transformer to store the energy is ____________.
   (a) fly back converter
   (b) buck converter
   (c) boost converter
   (d) buck-boost converter

10. The DC–DC converter most popularly used in modern multi-voltage switch mode power supplies as in PCs and TVs is ____________.
    (a) fly back converter
    (b) forward converter
    (c) charge-pump converter
    (d) buck-boost converter

11. The greatest number of power quality problems is eliminated in ____________.
    (a) SMPS
    (b) switching regulators
    (c) UPS
    (d) linear power supplies

12. Highest level of power protection for the serious home, office user is by ____________.
    (a) off-line SBS (standby power supply)
    (b) line interactive SBS
    (c) on-line UPS
    (d) switched mode power supply

13. The type of modulation used in DC–DC converter is ____________.
    (a) amplitude modulation
    (b) pulse-width modulation
    (c) pulse-position modulation
    (d) frequency modulation

14. The type of converter a development from doubling and Voltage multiplying rectifier circuits is ____________.
    (a) charge-pump converter
    (b) fly back converter
    (c) buck-boost converter
    (d) forward converter

Answers to Multiple-Choice Questions

1. (d) 2. (a & c) 3. (a) 4. (a & b) 5. (d)

6. (d) 7. (b) 8. (b) 9. (a) 10. (b)

11. (c) 12. (c) 13. (b) 14. (a)

Chapter 15

SPECIAL PURPOSE ELECTRONIC DEVICES

Learning Objectives

To get familiarity of structural details and working principles of special devices

- Tunnel Diode
- Photo Diode
- Varactor Diode
- Schottky Barrier Diode
- Light Emitting Diode
- Silicon Control Rectifier

15.1 TUNNEL DIODE

15.1.1 Introduction

Tunnel Diode (Fig. 15.1) is also known as Esaki Diode, named after Leo Esaki for the discovery of 'electron tunnelling phenomenon' in Tunnel Diodes.

Tunnel Diodes are two terminal devices. Fundamental difference between ordinary P–N Diode and Tunnel Diode is doping of P- and N-type semiconductor materials. In P–N Diode, either P-side or N-side is heavily doped, whereas Tunnel Diode is doped heavily on both sides. Because of heavy doping, Tunnel Diodes have P–N junctions with very narrow depletion region widths about 100 Å (1 Angstrom unit = 10 nm), in Silicon, Germanium and

Gallium Arsenide materials provided with high concentrations of doping of impurity elements (one impurity atom for every 10^3 atoms of intrinsic semiconductor material).

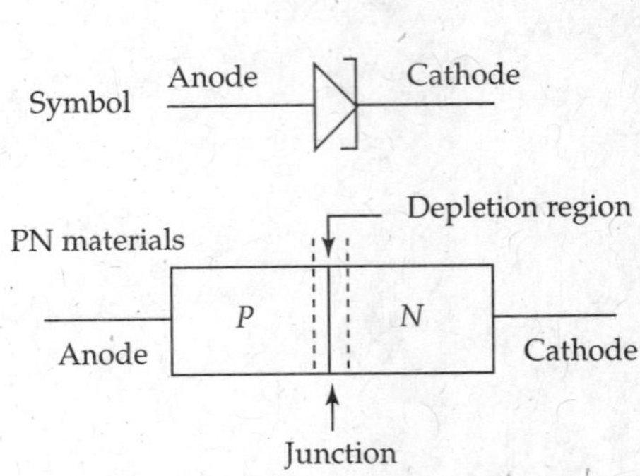

FIG. 15.1 *Tunnel diode*

Typical Features of Tunnel Diode

- Concept of conduction is different from the normal Diode conduction.
- Tunnel Diode is a special purpose high-frequency device working on the principle of 'quantum mechanical tunnelling', when an electron crosses through the narrow junction (1 millionth of an inch) with speed of light, which is quite different from the charge carrier movement in the other semiconductor devices. It caused revolution in electronics industry.
- Depletion region width is very small compared to normal Diodes, resulting in reduced transit times suitable to work with HF signals up to a few Giga Hertz.
- Tunnel Diode has negative resistance characteristic, allowing it to work as an amplifier, oscillator and a high frequency trigger switch with applications in various communication equipment, computers, TV, etc.
- Tunnel Diodes used with reverse bias as fast switching rectifiers are known as 'Back Diodes'.
- Doping levels of P-and N-type semiconductors of Tunnel Diode are increased to such magnitudes that its reverse breakdown voltage is zero. Further, the Diode conducts in reverse direction also without breakdown (Fig. 15.3).

15.1.2 Principle of Operation of Tunnel Diode

If the doping in P- and N-materials increases, *the Fermi level* E_{FN} for N-material moves up towards the conduction band and if the doping is heavy enough the Fermi energy level E_{FN} can enter the conduction band and up to E_{FN} all energy states will be filled with electrons. Similarly for a P-material the Fermi level E_{FP} enters the valence band allowing empty states in the valence band as shown in Fig. 15.2.

When a sandwich is made of such heavily doped P- and N-materials, empty energy levels exist in the valance band on the P-material side and filled energy levels appear in the conduction band on the N-material side.

A possibility for the empty energy levels on one side of the P–N junction to face the filled energy levels on the other side can occur and conduction is possible through tunnelling of charge carriers through the depletion region. This phenomenon is called 'quantum mechanical tunnelling'. Expressed differently the probability for a charge carrier having lesser energy than the barrier height may be very low but is non-zero.

This non-zero probability may be very low value but if concentrations are high enough, there can be enough number of carriers on the other side of the depletion region barrier with energies less than the height of built-in voltage. Charge carriers may not be able to scale the height but they can bore a tunnel through the space charge region and appear on the other side of the transition region. This is very much possible in reality and the process is called

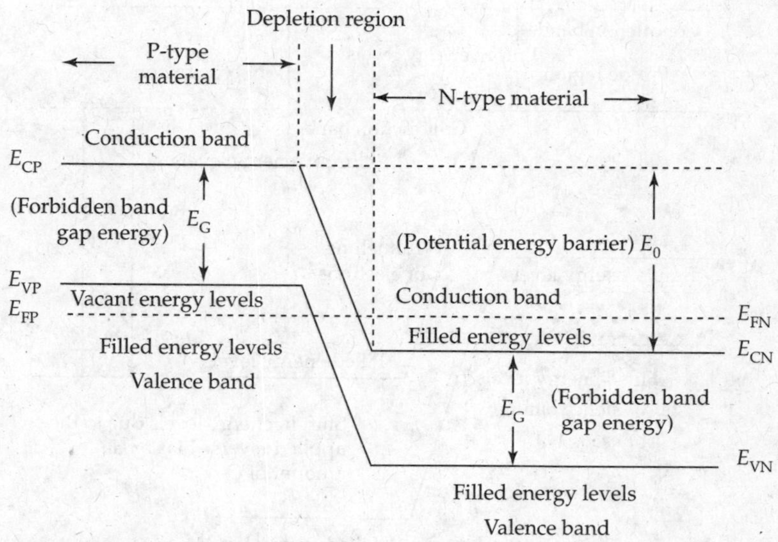

FIG. 15.2 *Energy band diagram for unbiased P–N junction of tunnel diode*

'quantum mechanical tunnelling'. P–N junction Diodes using this property are called 'Tunnel Diodes'.

For normal Diode conduction, it was assumed that conduction is possible whenever electrons enter the conduction band and Holes move into the valence band. *This means that filled energy levels are created nearer to the conduction band in the forbidden gap and vacant energy levels are created in the forbidden gap just above the valence band, thus making conduction possible with very small voltages under forward bias conditions.* This need not necessarily be the only method by which conduction is possible. It is sufficient if vacant energy levels and filled energy levels encounter each other.

Figure 15.2 shows the Energy Band Diagram (EBD) of an unbiased Tunnel Diode. It is seen that not only the conduction band in the P-material, but also the valence band contains empty or vacant energy levels. Similarly in the N-material the conduction band contains filled energy levels up to the Fermi level. *But still, there can be no conduction under unbiased conditions since on either side of junction* unfilled energy levels face unfilled energy levels and filled energy levels face filled energy levels.

15.1.3 Energy Band Diagram for Reverse-Biased Tunnel Diode (Fig. 15.3)

When a Tunnel Diode is reverse biased, its built-in potential is increased by reverse bias V_R volts. Increase in energy due to V_R causes a shift in energy level $(E_{FP} - E_{FN}) = Q \cdot V_R$.

Filled energy levels in P-type region encounters empty energy levels on N-type region above E_{FN}, thus allowing the electrons to Tunnel from P- to N-region of the Diode through the depletion region. As the reverse bias increases, E_{FN} goes down further and further making the region of filled energy levels on the P-region larger and larger. The tunnelling current (reverse current) increases with reverse bias voltage V_R and is limited only by the manufacturer's specifications. This part of Tunnel Diode characteristic is shown in the third quadrant of Fig. 15.8.

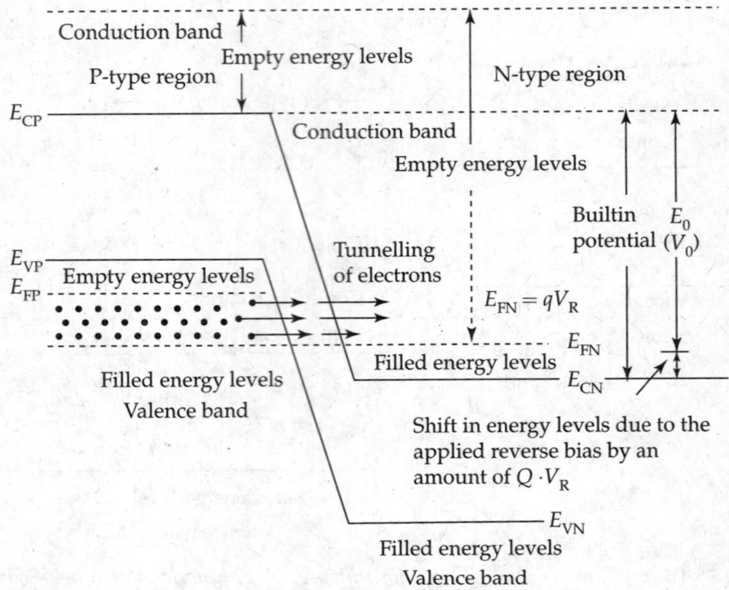

FIG. 15.3 *Energy band diagram for reverse biased P-N junction of tunnel diode*

15.1.4 Energy Band Diagrams for Forward-Biased Tunnel Diode

When a Tunnel Diode is forward biased (FB), potential barrier height is reduced to the extent of FB (V_f) and Fermi Energy level E_{FN} is pulled upwards relative to E_{FP} so that $(E_{FN} - E_{FP}) = QV_f$.

Then filled energy levels on N-region move up and encounter vacant energy levels in P-region above E_{FP}. Tunnelling of electrons occur from N- to P-region. With increase in forward bias more and more filled energy levels on N-type region face more and more unfilled energy levels favouring an increase in tunnelling current (Figs. 15.3 and 15.4).

There is a maximum limit for tunnelling current of value I_P (I_P = Peak forward tunnelling current) as shown in Fig. 15.8. This occurs when the maximum possible filled energy levels face maximum possible unfilled energy levels on the other side as shown in Fig. 15.4.

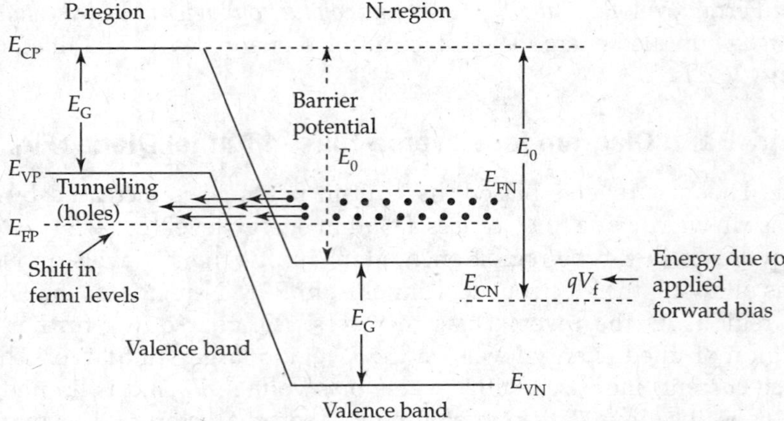

FIG. 15.4 *Energy band diagram (EBD) for forward bias (forward tunneling) (more current in forward direction)*

A further increase in forward bias pulls E_{FN} further up. Then, the number of the filled energy levels encountering vacant energy levels on the other side starts decreasing until the filled energy levels on N-type region gape at forbidden gap on P-type region and Tunnelling stops and current reaches I_{min} as shown in Figs. 15.5 and 15.6.

Further increase in forward bias beyond V_V on the characteristic in Fig. 15.8 starts normal Diode conduction and the current increases with voltage. Between I_{max} (I_P) and I_{min} (I_V) the current decreases with increase in forward bias voltages and exhibits a negative resistance. I_P:I_V may be of the order of 20:1. From this ratio, between the maximum or peak tunnelling current I_P and the valley or minimum current I_V, range of negative resistance can be predicted. So, this ratio is useful as a figure of merit for the Tunnel Diode. From $V_{Forward} = 0$ and up to V_{Peak}, the characteristic exhibits a positive resistance region, and between V_{Peak} and V_{Valley} the *Diode has a negative resistance* and from V_V and further up the Tunnel Diode has again a *positive resistance*. V_{Peak} to V_{Valley} voltage range predicts the voltage spread between the two positive resistance regions of the Tunnel Diode.

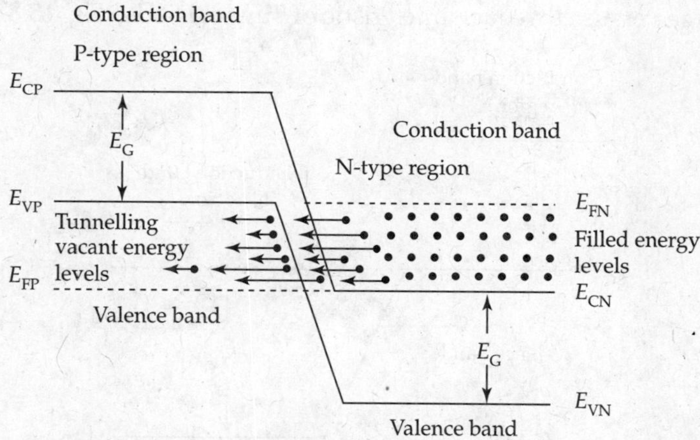

FIG. 15.5 *Maximum current for forward Bias*

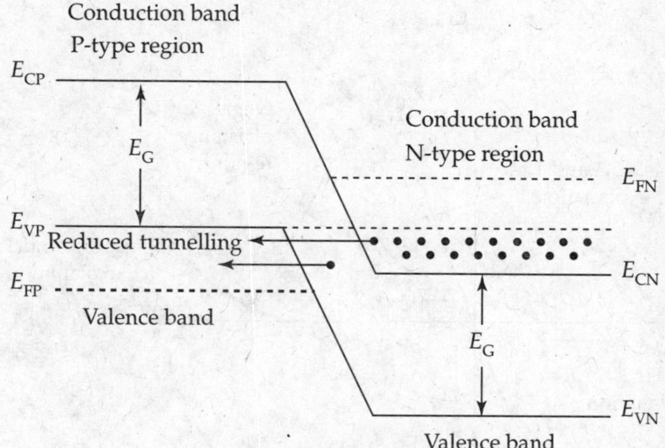

FIG. 15.6 *Further more forward bias but reduced tunnelling current*

15.1.5 Voltage–Current Characteristic of Tunnel Diode

When forward bias is applied to the Tunnel Diode,

- At the beginning stages of increasing forward bias, electric field at junction will be very large (due to very narrow depletion region width) and conditions become favourable for electrons in N-material for Tunnelling through space charge region (energy barrier) with speed of light and spontaneously appear on P-material (Figs. 15.3, 15.4 and 15.7).
- When the forward bias is increased further, filled energy levels in N-type region and unfilled energy levels in P-type region misalign. So, Diode current slowly drops for increasing forward bias voltages up to a certain point. During this forward bias operation, as the current decreases for increasing voltages, Diode exhibits 'negative resistance'. Characteristic region, when the Diode has negative resistance is known as 'negative resistance region' (Figs. 15.6 and 15.8).
- When the forward bias is further increased, the Tunnelling effect ends (Fig. 15.7) and the 'Tunnel Diode' begins to work as a normal Diode (Fig. 15.8) after the voltage V_V on the characteristic.
- Complete voltage versus current characteristic of Tunnel Diode (Fig. 15.8).

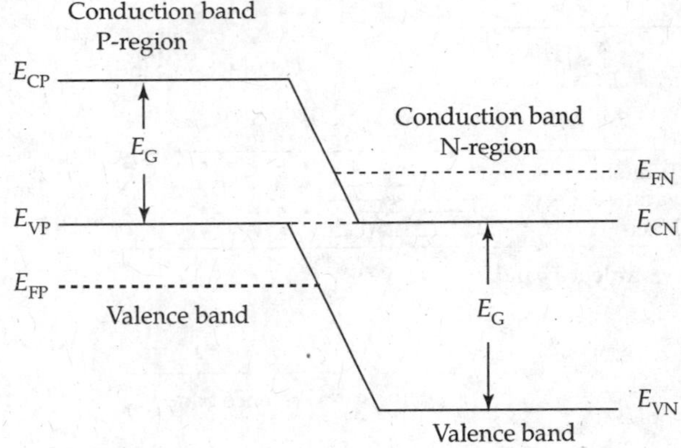

FIG.15.7 *Forward tunnelling ends*

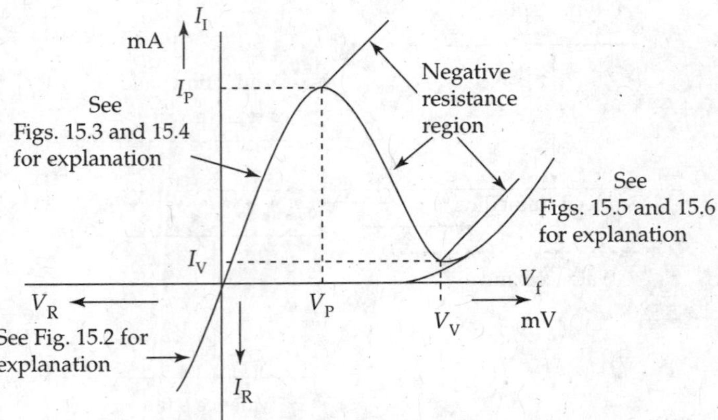

FIG.15.8 *Voltage-current characteristic of tunnel diode*

15.1.6 Schematic Symbol of Tunnel Diode and its Equivalent Circuit

The symbol for Tunnel Diode and its equivalent circuit with typical parameters are shown in Fig. 15.9. The parameters of the Tunnel Diode are given below:

- L_S represents the lead inductance;
- R_S is the equivalent series resistance;
- C_j is the junction capacitance;
- R is the slope of the negative resistance region of the Tunnel Diode.

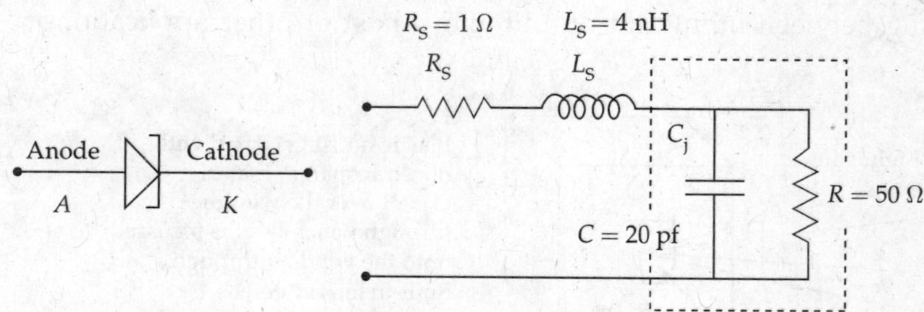

FIG. 15.9 *Tunnel diode and its equivalent circuit*

The frequency of operation of a Tunnel Diode far exceeds the frequency that can be calculated on the basis of depletion region width. Such behaviour can be explained as following. The electron hitting one edge of the barrier is supposed to emit a photon, which travels at the speed of light and on arriving at the other edge of the barrier emit another electron. Thus it is as though the photon is the particle that travels the transition region instead of the electron, which is bulkier, compared to the photon.

Applications of Tunnel Diode:

1. Oscillators for UHF range of the order of tens of Giga hertz,
2. For low access time memories,
3. Pressure to current Transducers,
4. As mixers, detectors and converters for UHF receivers,
5. As a high-speed switch of the order of Nanoseconds switching time,
6. Microwave frequency amplifiers, and
7. For space applications as Tunnel Diode operation is not effected by radiations.

Advantages

1. Low noise Figure,
2. High frequency response and
3. Lower dissipation.

Disadvantages

1. Low voltage and current levels of operation of the device
2. Being a two-terminal device, no isolation between input and outputs.

15.2 SEMICONDUCTOR PHOTO DIODE

15.2.1 Construction Details of Semiconductor Photo Diode (Fig. 15.10)

A Semiconductor Photo Diode is a P–N junction which is operated under reverse bias and exposed to light energy. Photoconductors provide a change in conductivity proportional to exposed optical energy. Photo Diode is an Opto-electronic device that produces an electronic output (current) for light energy input. Its applications span a wide range including sound recording on films, street light control, and Bar code detection on consumer products, CD reading in computer applications, fibre optic communications and as optic-isolators, which reduce coupling between input and output and a host of other applications in electronics industry.

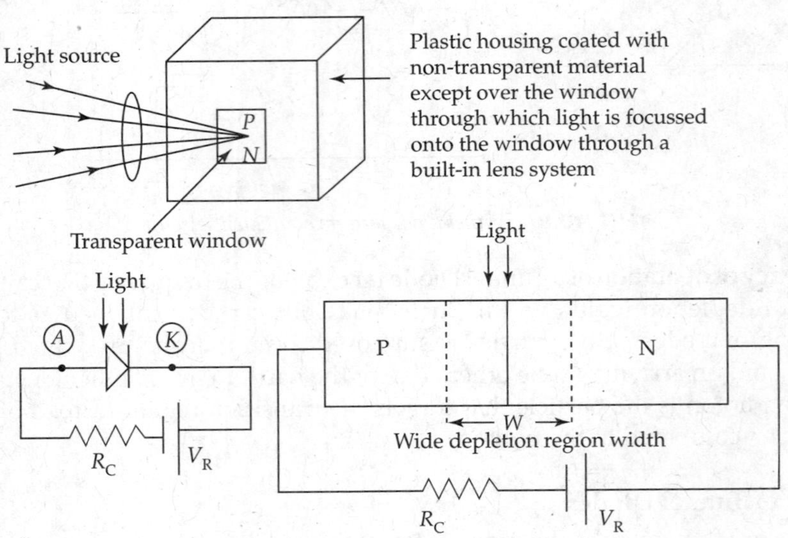

FIG.15.10 *Construction details of photo-diode*

It is an established fact that the reverse saturation current of a conventional Diode gets doubled for every 10°C rise in temperature. But temperature rise is caused by heat, i.e., thermal radiation, which is electromagnetic in nature. *The difference between heat and light is only of the range of frequencies of radiation.*

So, it is natural to expect the reverse saturation current to change when a Diode junction is exposed to visible or invisible radiation. For instance, infrared energy is in the lower invisible frequency domain and the ultra violet radiation is in the upper invisible region of the Electromagnetic wave spectrum for photoelectric devices.

A P–N junction is housed in a clear plastic housing with inbuilt lens across an open window through which light can be focused onto the centre of the junction. But for the window the other areas are coated with non-transparent paint.

15.2.2 Principles of Working of Photo Diode

P–N Diode has depletion region width *W* depending on doping on N- and P-regions. Depletion region width increases with increase in reverse bias and a normal reverse saturation current I_0 flows. If the junction is irradiated uniformly or illuminated by photons with h_f greater than E_G, new covalent bonds break forming Hole–electron pairs that increase reverse saturation current by a magnitude I_L. Larger the quantum of radiation, larger changes in reverse saturation current occur. Change in reverse saturation current is almost directly proportional to incident radiation. The current that flows in the circuit in the absence of radiation is called the **Dark current**. Usual dark current is of the order of 10 μA. Smaller the dark current better is the device performance. Signal current will be the difference between Dark current and current under incidence of radiation I_L. Smaller the dark current, larger is the signal current from the device.

15.2.3 Photo Diode Characteristics

Photo Diode characteristics comprise variation of current with reverse-biased voltage for different illumination (light) strengths, i.e., *lumens* as shown in Fig. 15.11.

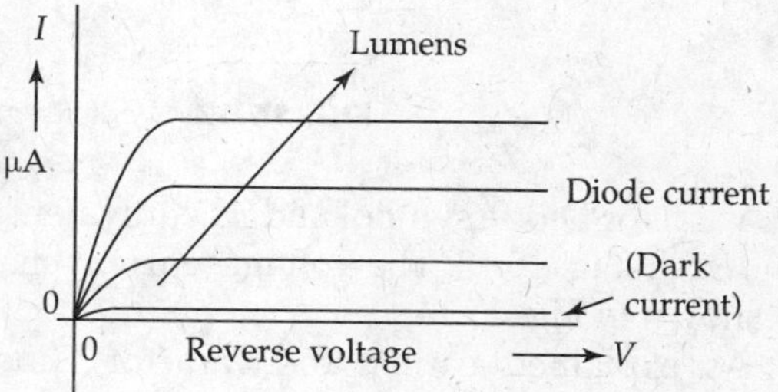

FIG. 15.11 *Photo diode characteristics*

The expression for the current in a Photo Diode can be represented as follows:

$$I_T = I_L + I_0\left[1 - e^{V_B/\eta V_T}\right],$$

where I_T represents the total current, I_L is the current due to incident illumination (optical generation of Hole–electron pairs), I_0 is the reverse saturation current of the Diode, and V_B is the reverse Bias Voltage. In the absence of illumination, I_R can be seen to be just the reverse saturation current I_0 for large values of reverse bias V_B or simply the dark current which is of the order of 10 μA. The incident illumination produces excess electron–Hole pairs proportional to illumination as explained earlier.

In the absence of radiation (illumination is cut-off or Zero illumination), voltage across R_L is almost negligible if the Dark current is small (Dark current is of about 10 μA). When light falls on the window of photo Diode, current increases in proportion to incident light energy (when h_f is greater than E_G) and a voltage is developed across R_L, which constitutes optic signal (Fig. 15.12). Optical signals can generate a proportionate voltage output and can serve as an opto-electronic converter. For instance, 0s and 1s (in the form of no light and light) can produce lower and higher voltages corresponding to binary digits.

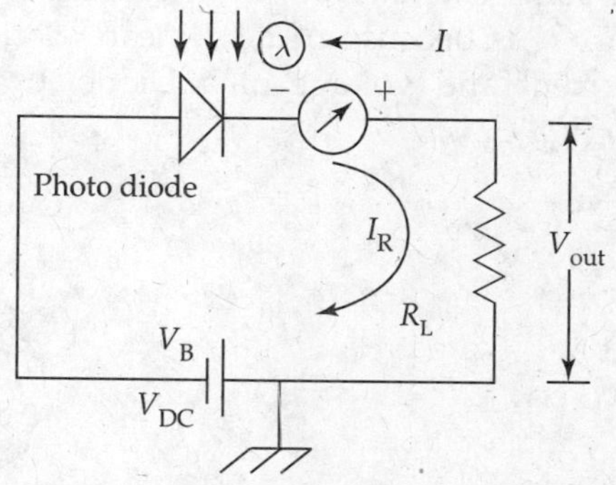

FIG.15.12 *Photo diode application circuit*

15.3 VARACTOR DIODE

15.3.1 Principle of Working of Varactor Diode and its Equivalent Circuit

Semiconductor device name 'Varactor' for the Varactor Diode is a shortened form of 'variable reactor'. Varactor is also known as varicap (variable capacitance), since the capacitance of the semiconductor Diode can be changed using a voltage. As the reverse bias to the Diode increases, depletion region width *W* increases and the capacitance C_T decreases and vice versa. Doping profile near the junction also has some effect on the value of capacitance when reverse biased. Hence, a reverse-biased P–N junction so designed to act as a voltage variable capacitor is called a *Varactor.*

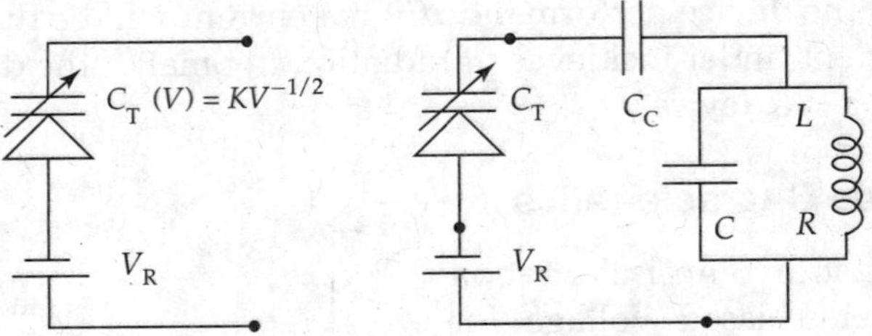

FIG. 15.13 *Varactor diode with its equivalent circuit*

Varactor Diode symbol and its equivalent circuit are shown in Fig. 15.13. The Voltage–current characteristic is shown in Fig. 15.14.

As explained earlier, the varactor Diode can be used for reactance control by voltages. Figure 15.15 shows the profile of capacitance variation with voltage for forward and reverse-biased conditions. From the characteristic, it can be observed that forward bias has to be avoided, because it causes excessive current, which is undesirable for any capacitor operation at high currents. It is not linear and it is a disadvantage to some extent. Non-linearity with respect to voltage for C_D is because of non-linear relationship between current and voltage of a Diode. C_T is proportional to $(V_R)^{1/T}$.

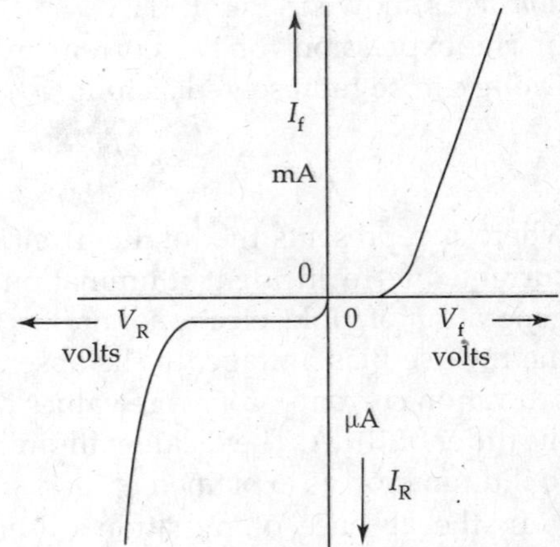

FIG. 15.14 *Reverse bias characteristic of varactor diode*

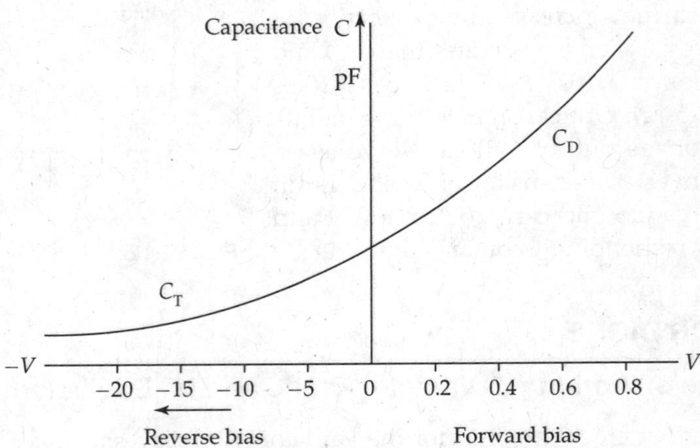

FIG. 15.15 *Capacitance variation profile under forward and reverse bias conditions*

Applications of Varactor Diode

1. Voltage-controlled tuning. As shown in Fig. 15.15 by varying the reverse bias V_R to the Diode, the capacitance shunting the tuned circuit or the tank circuit can be tuned to a range Switching of frequencies. C_C prevents the DC voltage entering the tank circuit.
2. A varactor or a set of varactor Diodes find their use as substitute for variable capacitor in channel the tuning stages of TV and radio receivers.
3. Automatic frequency control circuits in radio receivers.
4. Frequency modulation.
5. Parametric amplifiers as low noise microwave amplifiers.
6. Microwave frequency multiplications.
7. Ultra-high-speed operations.

15.4 SCHOTTKY BARRIER DIODE

15.4.1 Schottky Barrier Diode Characteristics

Schottky Diode (Schottky Barrier Diode (SBD)) or surface barrier or hot Diode is formed from a metal and semiconductor. Depending upon the metal and the type of semiconductor used, when the metal work function is smaller than that of the semiconductor, $\phi_m \langle \phi_S$ the semiconductor contacts with the metal are ohmic when the work function of the metal ϕ_m is greater than the work function of the semiconductor ϕ_S a depletion region is formed in the N-type semiconductor near the junction. This depletion region has positive charges due to the uncompensated ions and there will be equal number of negative charge in the metal at the junction. This forms a barrier or contact potential at the junction. It is a rectifying contact.

This rectifying property allowing easy current flow in one direction makes the possibility of Schottky Barrier Diode. Current conducts from metal anode to the semiconductor in SBD. Therefore, cathode symbol is shaped in the form of *S*, denoting Schottky Barrier Diode. Figure 15.16 shows that voltages drop of Schottky Barrier Diode is 0.3 to 0.5 V compared to 0.7 V in Silicon P–N junction Diodes. Gallium Arsenide SBD exhibits a forward voltage drop of about 0.7 V and they are used in GaAs circuits.

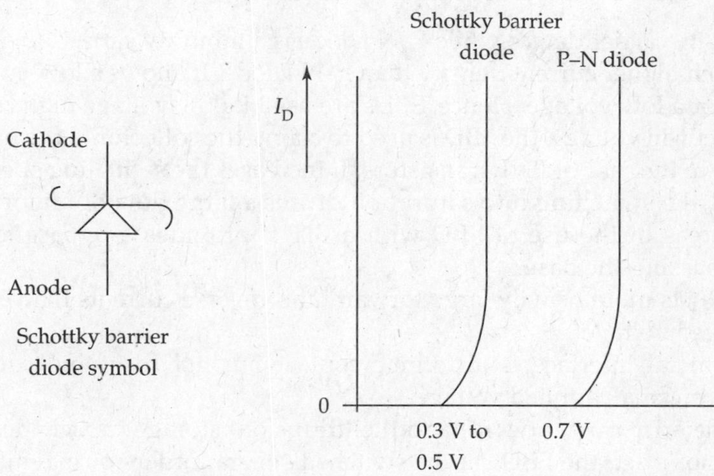

FIG. 15.16 *Schottky barrier diode symbol and characteristics*

15.4.2 Principle of Operation of Schottky Barrier Diode

Structural Details of Schottky Barrier Diode

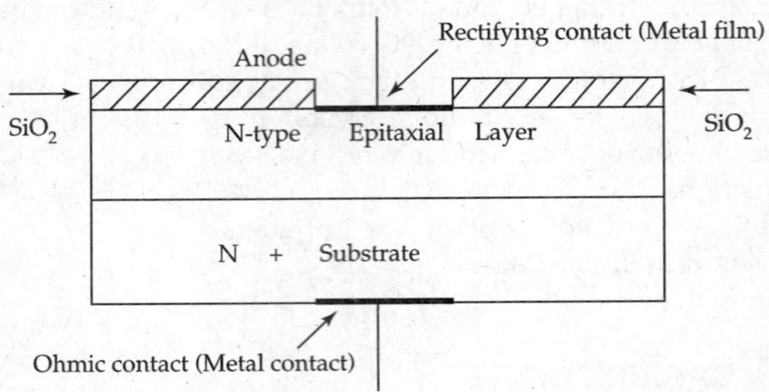

FIG. 15.17 *Structural details of Schottky barrier diode*

Working Principle The metals used in the construction of Schottky barrier Diode (hot-carrier Diode) are molybdenum, platinum, chrome and the semiconducting material is normally N-type silicon. Metal is deposited on the heavily doped semiconductor region forming a terminal known as ohmic terminal. As described previously, the hot majority carriers of N-type semiconductors, i.e., electrons, flow into the metal and simultaneously form a surface barrier potential or contact potential between the two materials. On application of a forward voltage, the contact potential (surface barrier voltage) will be reduced and heavy flow of electron current is established through the device. The metal to semiconductor junctions behaves like a Diode and is known as surface barrier Diode (hot-carrier Diode).

SiO_2 layer is an insulating layer to help to reduce the surface electric field and to improve reverse characteristic.

Schottky diode current equation is similar to that of normal semiconductor P–N diode:

$$I = I_0[e^{eV/KT} - 1]. \quad (15.1)$$

Comparison Between Schottky Diode and P–N Diode

1. SBD is a majority carrier device while P–N Diode is a minority carrier device.
2. SBD has a much higher current density than P–N Diode. It allows a low-cut-in voltage and high current for a low voltage. Hence, SBDs are used in low voltage high current rectifiers.
3. Due to lower cut-in voltage, the SBD is used to clamp the collector to the base of transistor at about 0.4 V in the case of PNP transistors. It increases the switching speed of transistor. In case of BJT, the switching into saturation causes a large flow of minority carriers into the base whereas in the use of SBD with a BJT it provides a bypass for the minority carrier by bypassing the base.
4. The SBD conducts more heavily large forward and reverse currents than a P–N Diode for the same applied voltage.
5. The cathode–metal interface is an ohmic contact but not rectifying contact under the influence of even small applied voltage.
6. As SBD is a majority carrier device and the (diffusion) storage capacitance with minority carriers does not exist, the SBD can be switched several orders of magnitude faster than P–N Diode. This is the major advantage of SBD.

7. The drawback of an SBD is the flow of large leakage current in the reverse direction when compared with the P–N junction Diode.
8. The forward voltage drop is 0.3 V in an SBD and it is 0.7 V in P–N junction Diode.

Applications of Schottky Diodes

(1) Mainly used in high-speed and high-frequency applications associated with low noise figure, (2) Low voltage, high current rectification, (3) Mainly used in ICs – Schottky TTL, (4) Power monitoring of low level radio frequencies, (5) High-frequency detectors, (6) Doppler Radar mixers, (7) A/D converters, (8) AC to DC converters and (9) Switching power supplies.

15.5 LIGHT EMITTING DIODE

15.5.1 Principle of Operation of LED

LED is an acronym for Light Emitting Diode. In ordinary P–N Diodes, the semiconductor materials used are indirect band gap semiconductors such as Silicon and Germanium. During the process of current flow, the recombination of charges releases heat to the lattice. LED uses semiconductor materials, which provide light source. Its function is to convert a forward current into light. For fabricating the LEDs, P–N junctions using a semiconductor of the type known as direct band gap materials such as Gallium Arsenide (GaAs) and InP are used. In a forward-biased LED, minority carriers are injected across their junction diffusing into P- and N-regions. Recombination takes place between these diffused minority carriers and majority carriers. Such recombination result in the emission of photons and radiation occurs.

This spontaneous emission gives rise to light emission. The light emitted is proportional to the forward current of the device due to the number of recombination taking place in LED. The LED is an electro-luminescent device. The commercial LEDs will emit different colours – Red, Orange, Yellow and Green satisfying the eye. GaAs is used for infrared LEDs and it is also used for visible LEDs. GaAs exhibits a very high probability for direct radiative transition. GaP and GaAsP are used for visible LEDs.

15.5.2 Identification and Symbol of LED (Fig. 15.18)

Structural Details (Construction) of LED

An N-type layer is grown on a substrate. The process of diffusion deposits a P-type layer on this N-type layer. At the outer edges of P-type layer, metal film contacts are made. Light is

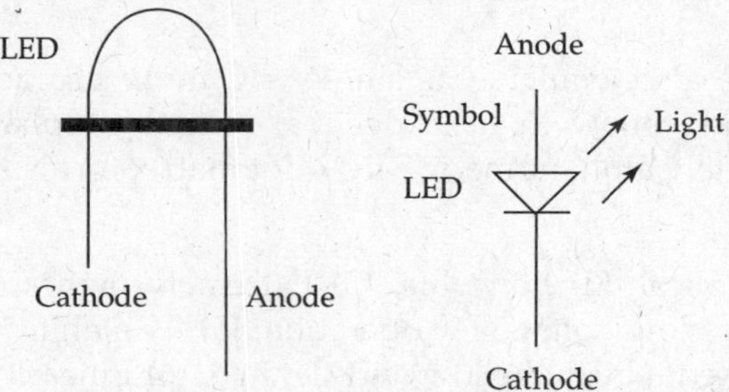

FIG. 15.18 *LED identification and symbol*

emitted from the freer upper surface. A metal film preferably gold coating is deposited at the bottom of N-type layer as shown in Fig. 15.19.

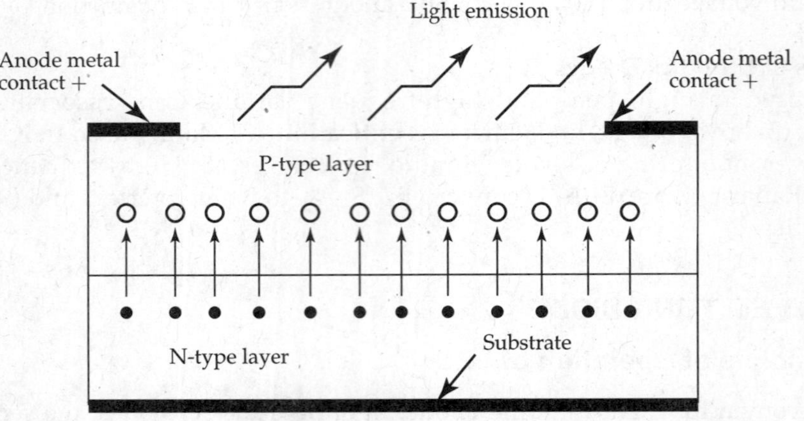

FIG. 15.19 *Construction details of light emitting diode (LED)*

Seven-Segment Displays

Because of the long life, reliability, low cost, Low driver requirements, Wide-operating ranges, high speed and appearance of LEDs, the visible LED displays are available as seven-segment displays.

They use light pipe fabrication technique, a reflecting light cavity is placed over the LED chip. The emitted light is reflected from the reflecting cavity. The block diagram of a typical seven-segment display is shown in Fig. 15.20.

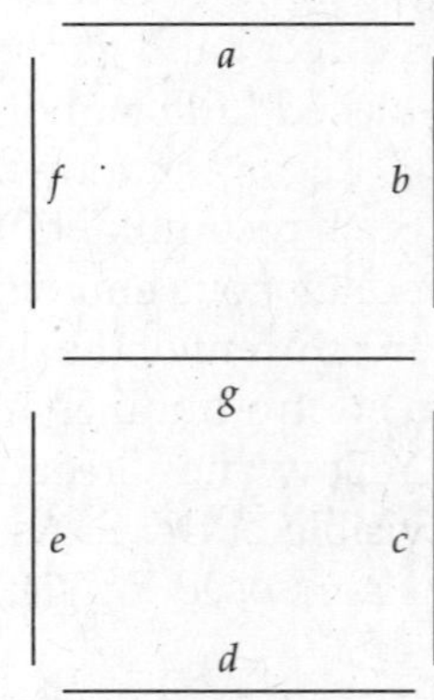

FIG. 15.20 *Seven segment LED display*

A decoder driver is required for the operation of the seven-segment LED display. The individual LED has either a common anode or a common cathode type, or individual driver switches, which sinks the LED in forward condition. The seven segments a, b, c, d, e, f and g are arranged in clockwise direction as shown in the seven-segment display in Fig. 15.21. The decoder/driver provides the blanking inputs to operate. According to the BCD input applied to the decoder/driver, the display will be from 0 through 9.

The LED displays are also available as single and multi character displays in various colours – red, green and yellow in hexa decimal partially alphanumeric or completely alphanumeric or completely alphanumeric 5 by 7 dot matrix forms.

Advantages of LEDs

(1) Low cost and economical, (2) Long life, (3) Extremely high speed of the order a few nanoseconds, (4) Requires low voltage for operation, (5) Availability of a variety of spectral output colours such as Red, Green, Yellow and Orange, (6) Linearity of power output with forward current and (7) Compatible with integrated circuits.

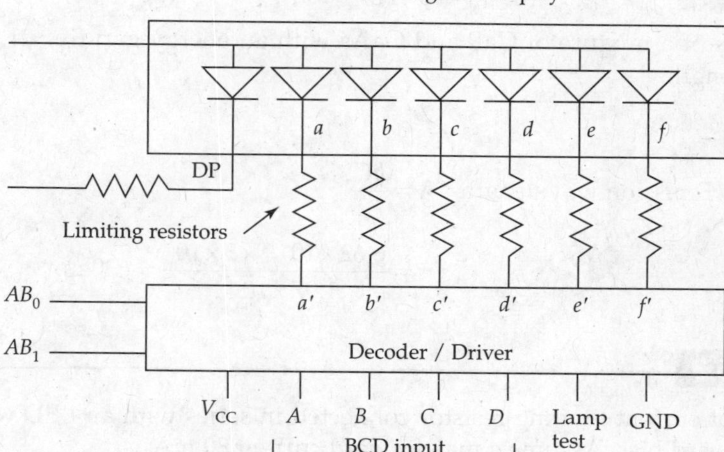

FIG. 15.21 *Block diagram of an LED display with decoder/driver*

Disadvantages of LEDs

(1) The radiant output power and wavelength are sensitive to temperature variations. (2) Over voltage or over current damages the LEDs. (3) Wide optical bandwidth. (4) Theoretical overall efficiency can be achieved only in special cooled and pulsed conditions.

Applications of LED

(1) Design of numerous types of displays such as in digital voltmeters and laboratory instruments. (2) Availability as opto-isolators (package of LED and photo Diode) popularly used in medical instrumentation for diagnosis and surgery to reduce the risk of electrical shock and in the design of digital system. (3) As a source of light in optical communication system and fibre optic communication. (4) In recording and reading of compact discs. (5) In industrial instrumentation. (6) In seven-segment displays and alphanumeric displays. (7) As a coherent source in aviation hazard for tall buildings, antenna tower, powerhouse cooling towers and chimneys. This can be done by bunching a number of LEDs and placing them in plastic having a high refractive index. (8) In harbours, docks and airports for navigation and guidance. (9) Embedded devices like TV remotes, garage doors opening, etc., intruder alarms, burglar alarms, fire alarms. (10) Digital control systems. (11) Barcode readers. (12) Commercial displays like arrival and departure of flight timings and its advertising. (13) As a natural choice of multiplexing. (14) A high speed light source of known wavelength. (15) Cameras, calculators, automobile instrument panels and electrical appliances.

EXAMPLE 15.1

Find the emission wavelength of an LED with GaAs.

Solution: Data: $h = 6.62 \times 10^{-34}$; Velocity of light $C = 3 \times 10^8$ m/s; $E_G = 1.45$ eV

$$\text{Emission wavelength} \quad \lambda = \frac{hC}{E_G} = \frac{6.62\times10^{-34}\times3\times10^{8}}{1.45\times1.6\times10^{-19}} = 8500\ \text{Å}.$$

EXAMPLE 15.2

An LED is made of a mixture of GaP and GaAs with an energy gap $E_G = 1.97$ eV. Find the emission wavelength.

Solution:

$$\text{Emission wavelength} \quad \lambda = \frac{hC}{E_G}$$

$$= \frac{6.62\times10^{-34}\times3\times10^{8}}{1.97\times10^{-19}} = 630 \text{ nm}.$$

EXAMPLE 15.3

Find the value of current limiting resistor connected in series with an LED where 5 V DC is applied as a forward bias. Assume a max forward current 80 mA.

Solution: Let R_S be the current limit resistance. Assume that a current of 10 mA is flowing through LED and a voltage rating of 1.22 V for the LED.
Voltage across $R_S = 5 - 1.22 = 3.78$ V

$$\text{Current limiting resistance} \quad R_S = \frac{3.78 \text{ V}}{80\times10^{-3}} = 47.5\ \Omega.$$

EXAMPLE 15.4

An LED is driven by a 18 V DC source. LED current is 16 mA. Find the value of the limiting resistor?

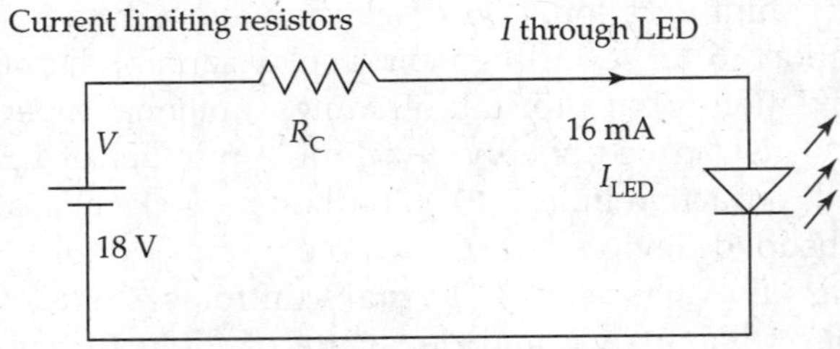

FIG. 15.22

Solution: Assume a forward voltage drop $V_f = 2$ V across the Diode.

$$\text{Current limiting resistor} \quad R_C = \frac{(V = V_f)}{I_{LED}} = \frac{(18-2)}{16\times10^{-3}} = 1\,\text{k}\Omega.$$

EXAMPLE 15.5

LED is irradiating at a divergence angle θ on to photo detector at a distance D (Fig. 15.23). Calculate (a) Irradiated area and (b) Incident radiance H (flux at the detector).

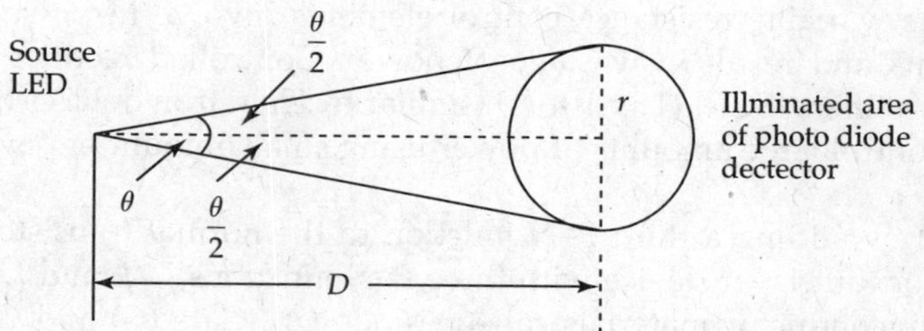

D = Separating distance between LED source and Photo diode detector
θ = angle of divergence of light from LED expressed in radians
r = radius of circular area illuminated
A = Radiated area

FIG. 15.23 *Point source geometry of an LED irradiating at a divergence angle over an area*

Solution: Let D = LED-to-photo detector separation
θ = angle of divergence of light from LED expressed in Radians
r = radius of circular area illuminated and
A = radiated area

From Fig. 15.22, radius $r = D \cdot \tan\dfrac{\theta}{2} \cong \dfrac{D\theta}{2}$.

Angle θ is in radians and very small

$$A = \pi r^2 = \frac{\pi D^2 \theta^2}{4}.$$

Let P_0 is the Power output of LED.
Assuming the LED as a point source

$$\text{Incident radiance} \quad H = \frac{P_0}{A} = \frac{4P_0}{\pi D^2 \theta^2}.$$

EXAMPLE 15.6

A photo detector is separated from an LED at a distance of 2″. An LED with a power output of 1 mW is illuminated at 30° divergence angle. Assuming the LED as a point source, find the irradiance H.

Solution: $D = 2'' = 2 \times 2.54 = 5.08$ cm; $\theta = 30° = 0.524$ radians; $P_0 = 1$ mW

$$H = \frac{4P_0}{\pi D^2 \theta^2} = \frac{4 \times 1 \times 10^{-3}}{\pi (5.08)^2 (0.524)^2} = 0.024 \text{ mW/cm}^2.$$

15.6 SILICON CONTROL RECTIFIER

15.6.1 Principle of Working of Silicon Control Rectifier (Thyristor)

In industrial power control applications, such as powers delivered to electric motors or induction heating elements or relay controls, large output powers have to be controlled. In high-power circuits, variable resistance control elements involve in power dissipations in controlling elements and result in wastage of power. Controlled rectifier devices such as silicon-controlled rectifiers (SCR) (Thyristors) (similar to Thyratron devices of Vacuum tubes) are developed to control large amounts of power using small amount of powers in the 'Gate' circuit.

SCR is obtained by adding a third P–N junction to the normal transistor structure. SCR is a silicon PNPN or four-layer device with three P–N junctions J_1, J_2 and J_3 with three leads attached to the semiconductor materials called Anode (*A*), Gate (*G*) and Cathode (*K*). SCR structure and circuit symbol are shown in Fig. 15.24. SCR symbol has gate terminal added to the cathode of conventional Diode symbol. Diode arrow points in the possible easy-flow direction for current through SCR. Gate voltages control the conduction of SCR. Term Anode implies that terminal so named is normally connected to a voltage source in such a manner as to make it positive relative to the cathode for the device functioning.

SCR with four PNPN layers of semiconductors can be considered as three Diodes D_1, D_2 and D_3 as shown in Fig. 15.25. These Diodes with Junctions J_1, J_2 and J_3 are considered with biasing voltages that forward or reverse bias the junctions for understanding the working of SCR device depending on polarity of Anode voltage V_{AA} and Gate voltages V_{GG}. Conduction or non-conduction states of SCR are predicted from applied operating voltages.

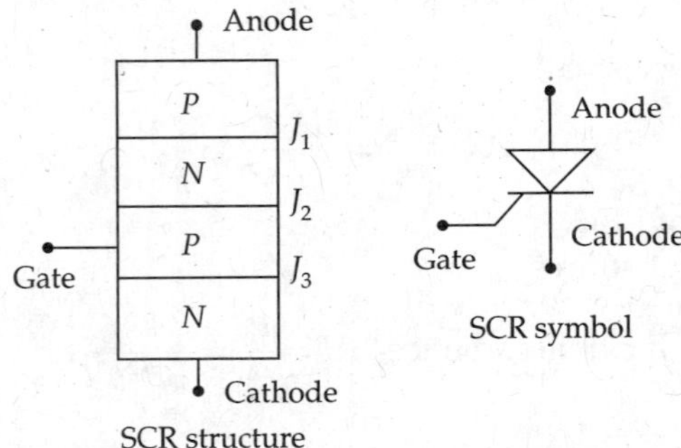

FIG. 15.24 *Silicon controlled rectifier structure and symbol*

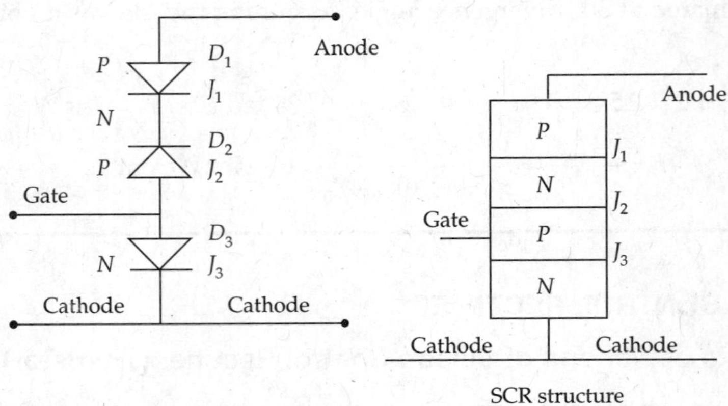

FIG. 15.25 *SCR considered as three diodes with three junctions*

15.6.2 Basic Circuit Diagram to Obtain SCR Characteristics

For negative anode-to-cathode voltage V_{AK}, central junction J_2 in Fig. 15.26 is prone to be forward biased, but outside junctions J_1 and J_3 are reverse biased. As a result, this is a 'reverse blocking' Diode and characteristic for negative voltage V_{AK} appears as shown in the third quadrant of Fig. 15.27. OFF state resembles that of a reverse-biased Diode. No reverse current flows unless the *Avalanche-Breakdown* voltage is exceeded.

For positive V_{AK}, two junctions J_1 and J_3 are forward biased and central Junction J_2 is reverse biased. For positive V_{AK}, current at low voltages is limited by reverse-biased behaviour of junction J_2 (Central junction J_2 acts as a Dam). Only a small forward leakage current I_{FX} flows through SCR until Anode voltage is made very much positive relative to Cathode.

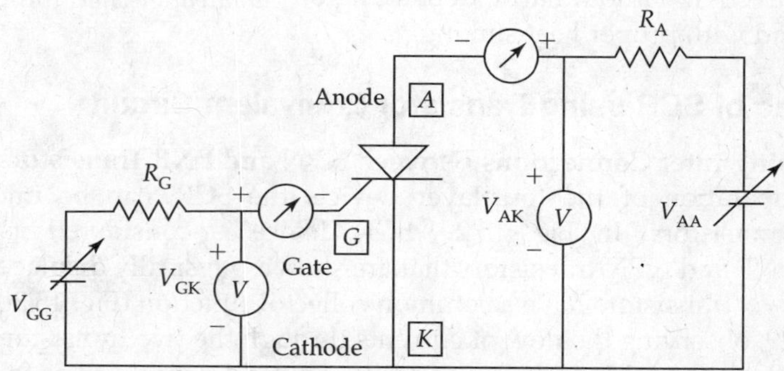

FIG. 15.26 *Basic circuit to obtain SCR characteristics*

As the applied voltage increases, the current increases slowly until the break over voltage V_{FBO} (for the central junction) is reached. At this point, it is interesting that 'once the Dam is broken', so to speak, current floods easily through the device (the current rises abruptly) with only a small voltage drop of the order of 1 V (the voltage across SCR drops sharply) across the SCR from anode to cathode. Then the Diode has switched to the 'ON-state', i.e., the SCR has changed from an open-switch mode to a mode much like forward-biased Diode or a closed switch. Before forward break over the central junction J_2 is in the blocking state (due to reverse bias to junction J_2) and at the V_{FBO} the central junction J_2 breaks under reverse breakdown.

However, when a gate current is supplied to SCR, the forward break over voltage (but not the reverse breakdown voltage) is reduced in proportion to the gate current. If a gate current greater than some gate–trigger current I_{GT} is supplied to the gate, the device goes into the break over or conduction state for less positive anode–cathode voltages. The gate must be positive by only about 1 V relative to cathode so that gate makes trigger current I_{GT} to flow. This voltage forward biases the bottom P–N Junction J_3. Low-power SCRs operate with $I_{GT} < 0.1$ mA and high-power SCRs operate with $I_{GT} > 15$ mA, whereas some minimum current is required in some cases.

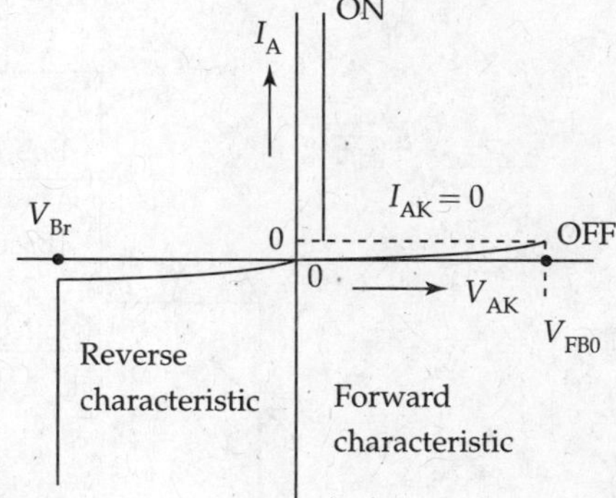

FIG. 15.27 *SCR forward and reverse characteristics*

The voltage for reverse breakdown is about the same as the SCR's forward break over voltage (for $I_G = 0$). As was true for normal Diodes, the reverse breakdown situation can ruin by overheating the Diode, if the current is allowed to become large.

SCR is not ruined in the forward direction break over case since V_{AK}, the anode–cathode voltage, quickly drops to about 1 V and the device is designed to carry large currents ($V_{AK} \cdot I_A$ product is then very small), as shown in the SCR characteristics in Fig. 15.28. Typical SCRs can pass a current of 1 to 15 A in the forward direction without harm, depending on the SCR rating and with proper heat sinking.

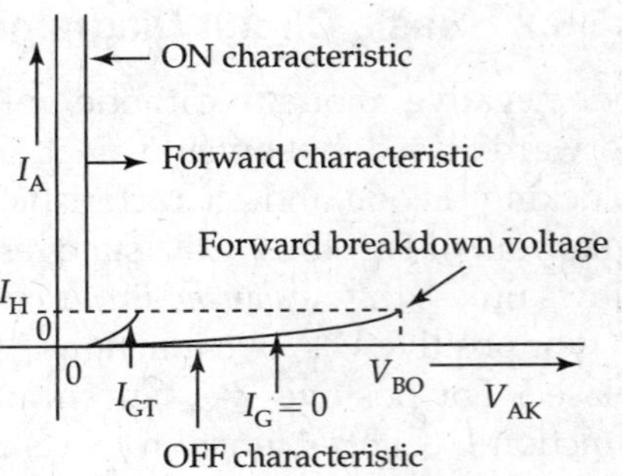

FIG. 15.28 *Firing charactenstrs of silicon controlled rectifier (SCR)*

15.6.3 Working of SCR using Transistor Equivalent Circuit

SCR Concept Using Inter Connections Between NPN and PNP Transistors

The switching operation of the four-layer device, the SCR, can be understood as two interconnected transistors. In Fig. 15.29, SCR device is considered as a back-to-back combination of PNP and NPN transistors that are shown physically displaced but electrically connected. The two Transistors have a common collector Junction (Fig. 15.30).

From Fig. 15.29, observing the flow of currents through the two transistors T_1 and T_2, it can be seen that $I_A = I_{E1} = I_{E2}$ and $I_A = (I_{C1} + I_{C2})$, where I_{C1} and I_{C2} are given as follows:

$$I_{C1} = \alpha_1 I_{E1} + I_{C01} = \alpha_1 I_A + I_{C01} \tag{15.2}$$

$$I_{C2} = \alpha_2 I_{E2} + I_{C02} = \alpha_2 I_A + I_{C02} \tag{15.3}$$

From Fig. 15.29, the collector currents will be obtained by summing the currents into the Transistor T_1. Substituting the values of I_{C1} and I_{C2} in the equation

$$I_A - \alpha_1 I_A - I_{C01} - \alpha_2 I_A - I_{C02} = 0 \quad \text{using } I_A = I_{C1} + I_{C2}.$$

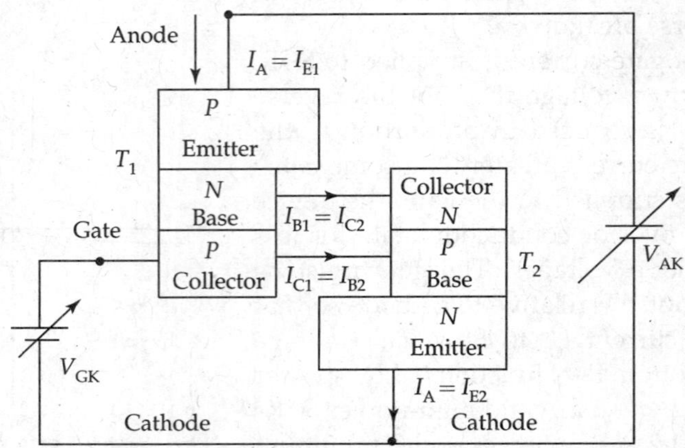

FIG. 15.29 *SCR – two back to back connected transistors concept*

$$\therefore \quad I_A\,[1-(\alpha_1+\alpha_2)] = I_{C0},$$

where $I_{C0} = I_{C01} + I_{C02}$. (15.4)

$$\therefore \text{ Anode Current } \quad I_A = \frac{I_{C01}+I_{C02}}{1-(\alpha_1+\alpha_2)} = \frac{I_{C0}}{1-(\alpha_1+\alpha_2)}. \tag{15.5}$$

At low currents, α is small. As I_A increases with an increase in V_{AK}, α_1 and α_2 increase. As the quantity $(\alpha_1 + \alpha_2)$ approaches unity, current tends to increase without limit and break over occurs. After switching ON the Diode voltage is smaller than 1 V and only the resistance in the external circuit limits the current.

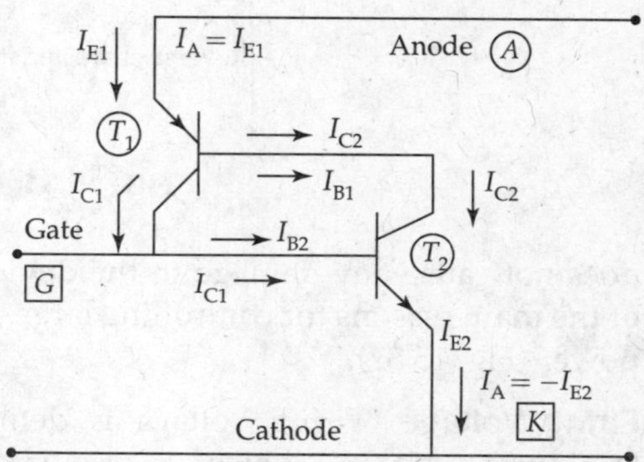

FIG. 15.30 *SCR shown as two interconnected transistors*

Once in conduction state, SCR remains so even after gate current/voltage is removed. SCR remains in the ON-state until the anode voltage is reduced so that the anode current falls below a small holding current I_H (typically a few milliamperes). SCR will be certainly turned off if the anode is actually made negative relative to the cathode. However, a negative voltage to SCR gate will not turn-off the SCR. Turn-on time for SCR is of the order of 1 μs, while turn-off time after reverse bias is about 20 μs.

Introduction of Gate current of the order of a few microamperes switches the SCR (Thyristor) to the ON-state at voltages lower than V_{BO} for the corresponding Diode. Such an SCR provides flexible switching and since the Gate current is required only momentarily very efficient control of large currents in the anode circuit can be achieved with virtually a few microamperes of Gate current that too momentarily. Therefore, firing voltage is a function of Gate current I_G. In comparison with power Transistors, SCRs (Thyristors) are designed with thick Base layers for high-voltage capability and necessary low α characteristic. Currents may be larger because the total junction areas are available for conduction.

15.6.4 Silicon-controlled Rectifier Characteristics (Fig. 15.31)

From the forward characteristic of SCR device, during positive voltages of V_{AK} from zero to forward break over voltage V_{BO}, the device functions as an *open switch* as the central P–N junction J_2 is under reverse bias and not allowing the current flow as a Dam, even if the outer junctions J_1 and J_2 are under forward-biased conditions.

During this region, the anode current is of the order of a few microamperes and hence SCR behaves as an open switch with a very high resistance of the order of mega ohms, the SCR being in the off state or no conduction state. Once the anode voltage reaches the forward break over voltage V_{BO}, the Dam effect of central junction J_2 is broken resulting in flooding of anode current through the device. Now the device is in the ON-state or conducting state and acts as a *closed switch*. At the same time, the voltage across SCR drops down to saturation voltage of about 1 V, with a consequent decrease of power dissipation across the device, (product of anode current and saturation voltage across SCR about 1 V). This

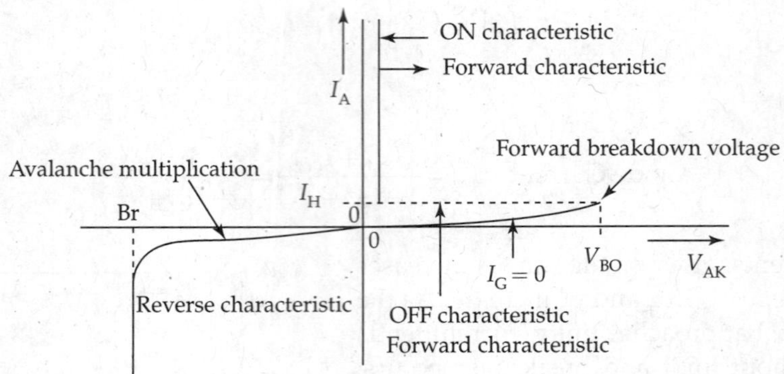

FIG. 15.31 *SCR firing characteristics*

does not cause any damage to the device, i.e., one of the main reasons for controlling large amounts of power (Fig. 15.32).

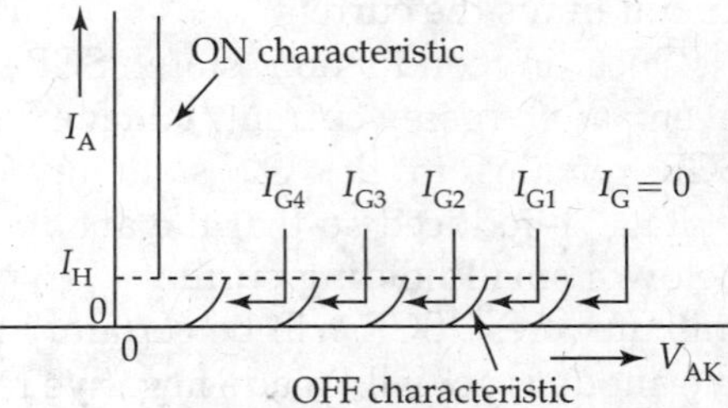

FIG. 15.32 *SCR characteristics for different gate currents*

Firing Voltage Firing Voltage is defined as the minimum voltage, which is required between Anode and Cathode of Thyristors (SCR) to trigger in to conduction. Firing voltage is slightly more than ON-state voltage.

Latching Current (I_L) Latching current is the minimum current required to latch the device from OFF-state to ON-state. It can also be said to be the minimum current required to trigger the device. Typical values of latching current may range from 8 to 10 mA.

Holding Current (I_H) Holding current is defined as the minimum current required for holding the device in conduction. Below this value of current the device cannot conduct and the device returns to the OFF state. The typical values of holding current of a device may range from 3 to 5 mA.

Gate Current (I_G) It is the current, which flows into Gate terminal to control the device.

Turn-ON Time Though Thyristors are very fast switching devices, they do not switch instantly, rather they takes finite time to reach its full conduction from the time the trigger signal is applied. In other words, it is called *response time*. The typical values of T_{ON} range from 150 to 200 ns.

Turn-OFF Time A reverse voltage has to be applied across the device to turn-off from conduction state of the device. It takes a finite time after application of reverse voltage to switch the device OFF (commuted). This time is called the turn-off time of the device. The typical value of turn-off time is 200 ns.

Ratings Peak Inverse Voltage (V_{PIV}) is defined as the maximum voltage, which the device can safely withstand in OFF condition. V_{PIV} depends on temperature and ON-state voltage. Voltage drop across Thyristors when conducting is very low. This voltage is called the ON-state voltage of Thyristors. The typical voltage lies between 1 and 4 V.

SUMMARY

1. Because of the special property of negative resistance on Tunnel Diode Characteristics, it finds applications in high-frequency oscillators, amplifiers and switches.
2. Schottky Diode is used as a low voltage high current rectifier.
3. Schottky Barrier Diode is much faster in switching than the P–N Diode for the reason that the storage capacitance associated with minority carrier current does not exist.
4. As the Schottky Barrier Diode has the properties of speed and low voltage, it finds applications in logic families, high-frequency mixers, rectifiers, modulators, detectors, waveform generators and fast pulse processing systems.
5. The varactor Diode structure is a P–N Diode with a graded junction to produce desired capacitance versus reverse-biased voltage variations.
6. Varactor Diodes are often used at RF (Radio frequencies) to tuned LC circuits eliminating unreliable mechanically tuned tuners in TV and FM receivers.
7. LEDs are coherent light sources. They are available in a variety of spectral colours.
8. LEDs have long life, low cost, extremely high-speed electronic device.
9. LEDs are mainly used in visual displays as indicators and as light sources in optical communication systems. Visible LED displays are popularly used as seven-segment Displays.
10. SCRs find most of its applications in power electronics.

Questions for Practice

1. Explain Tunnelling phenomenon in Tunnel Diodes.
2. Draw the energy band diagrams of Tunnel Diode and explain them.
3. Draw the Tunnel Diode characteristic with salient features on it and explain them.
4. Explain the significance of the ratio of peak current to valley currents on the Tunnel Diode Curve and its relation to the Figure of merit of Tunnel Diodes. Explain its significance as the device to operate as a high-speed switch.
5. State the advantages and disadvantages of Tunnel Diode.
6. Mention the Principle of working of Varactor Diode with its applications.
7. Draw the Photo Diode characteristics and mark the dark current region and the current region due to light energy. Mention the applications of the device.
8. Define the situation of Dark current of a Photo Diode.
9. Explain the operation of SBD and its use in faster switching applications.
10. Briefly explain the operation of LED. What are its advantages and disadvantages?

11. Sketch the seven elements of seven-segment display. Show which are ON and OFF for each number from 0 through 9.
12. State and explain typical applications of LEDs.
13. How does a voltage variable capacitance (Varactor) Diode operate? Write the advantages of using electronic variable capacitance over mechanically variable capacitor in tuner circuits in TV and FM receivers.
14. Explain the operation of Silicon-controlled rectifier with reference to its equivalent Circuit. Draw SCR characteristics and mention the inferences from Characteristics.

Multiple Choice Questions

1. Schottky Diode is ______________.
   (a) low voltage, high current rectifier
   (b) low voltage, low current rectifier
   (c) high voltage low current rectifier
   (d) none of these
2. The main disadvantage of Schottky Diode is ______________.
   (a) the property that permit low cut-in voltage or high currents at low voltage
   (b) large currents in the reverse direction
   (c) larger forward current than in a P–N Diode
   (d) device current is due to flow of majority carriers
3. The typical application of a Schottky barrier Diode is ______________.
   (a) RF oscillator, amplifier, switch
   (b) efficient harmonic generator
   (c) HF mixers, rectifiers, modulators, detectors, waveform generators
   (d) matched RF attenuator
4. The structure of a Schottky barrier Diode is ______________.
   (a) P–N junction
   (b) metal to semiconductor junction
   (c) P–N junction with very heavy doping
   (d) same as a Tunnel Diode with a low peak current
5. The following statements are made with reference to Schottky Diode. Identify the statement which is not relevant.
   (a) majority carrier device
   (b) minority carrier device
   (c) hot-carrier Diode
   (d) turns ON and OFF very fast
6. The structure of a Tunnel Diode is ______________.
   (a) metal to semiconductor junction
   (b) P- and N-regions separated by intrinsic region
   (c) P–N junction with very high doping
   (d) P–N Diode with graded junction

7. A Tunnel Diode is a P–N Diode with ______________.
   (a) high impurity concentration in both P-region and N-region
   (b) turns on for a very low forward voltage
   (c) can also be formed from a metal and semiconductor
   (d) depletion layer capacitance
8. Tunnel Diode characteristic differs from ordinary P–N Diode w.r.t the following.
   (a) characteristics are non-linear
   (b) exhibits negative resistance region
   (c) characteristics are linear
   (d) none of these
9. The specific property of a Tunnel Diode is ______________.
   (a) zero voltage breakdown
   (b) negative resistance region
   (c) long storage time
   (d) large capacity charge with reverse-biased voltage
10. The typical application of a Tunnel Diode is ______________.
   (a) used under reverse bias as a voltage rectifier
   (b) high-frequency oscillator, amplifier, switch
   (c) matched EF attenuator
   (d) voltage reference source
11. The specific property of a varactor Diode is ______________.
   (a) negative resistance region
   (b) long storage time
   (c) zero reverse voltage breakdown
   (d) large capacity charge with reverse breakdown voltage
12. Typical application of varactor Diode is ______________.
   (a) matched RF attenuator
   (b) efficient harmonic generator
   (c) HF oscillator, amplifier, switch
   (d) replaces a mechanically tuned capacitor and automatic frequency control in FM receivers
13. The material used for visible LED is ______________.
   (a) Gallium Phosphide (GaP)
   (b) GaAsP
   (c) GaA
   (d) none of these
14. Diode mostly used in optical communication is ______________.
   (a) LED
   (b) P–N diode
   (c) tunnel diode
   (d) Schottky diode
15. Thyristors consist of ______________.
   (a) three semiconductor layers
   (b) four alternate layers of P–N junctions with three terminals
   (c) four terminals
   (d) none of these

16. Isolate the statement not relevant to Photo Diode.
    (a) P–N junction can be exposed to light
    (b) utilising LED the photo Diode can be used in fibre optic data communication Systems
    (c) associated with extremely high dark resistance
    (d) regenerative process
17. The industrial workhorse in power electronics is ______________.
    (a) thyratron (b) Schottky barrier diode
    (c) tunnel diode (d) UJT
18. The following statements are made in reference to LEDs. Identify the statements, which are true.
    (a) forward voltage of LED is 2 to 3 V.
    (b) for adequate brightness LED requires 10 to 20 mA
    (c) the relative luminous intensity is more than 1
    (d) none of these

Answers to Multiple-Choice Questions

1. (a)	2. (a & b)	3. (c)	4. (b)	5. (b)
6. (c)	7. (a)	8. (b)	9. (b)	10. (b)
11. (d)	12. (d)	13. (a & b)	14. (a)	15. (b)
16. (b)	17. (a)	18. (a & b)		

Index

D

N

O

P

Q

T

U

V